Manual of Geographic Information Systems

Edited by
Marguerite Madden

The American Society for
Photogrammetry and Remote Sensing
Bethesda, MD

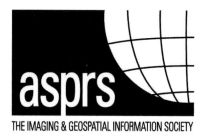

THE IMAGING & GEOSPATIAL INFORMATION SOCIETY

Published by
American Society for Photogrammetry and Remote Sensing
5410 Grosvenor Lane, Suite 210
Bethesda, Maryland 20814

ISBN 1-57083-086-X

www.asprs.org

Printed in the United States of America

TABLE OF CONTENTS

Section 5 Analysis and Modeling

Section 6   Blending Technologies: Remote Sensing, GPS and Visualization

Section 7   GIS and the World Wide Web

Section 8   GIS Reaches Out: Applications

FOREWORD

Jack Dangermond

I can recall the time, not long ago, when there were no manuals of GIS, no textbooks of GIS and no journals devoted to GIS; a time when the existing literature consisted mostly of papers presented at conferences and final reports of the few projects that had used what has come to be called GIS technology.

Now, after only a generation, the ASPRS *Manual of Geographic Information Systems* captures, within the scope of one large volume, the explosive growth of the GIS field, surveying what has been accomplished in the last 50 years, what is being done today and what is on the horizon for tomorrow.

GIS is interdisciplinary; geospatial data are brought together through the cooperation and collaboration of workers in many scientific and technical fields, from government, industry, and academia. This *ASPRS GIS Manual* facilitates photogrammetry and remote sensing working more closely with GIS; that will inevitably produce important new scientific and technological developments.

GIS is about using the geographic approach: the science and technology of capturing, storing, managing, analyzing, modeling, integrating and then applying—particularly to decision making—geographic or geospatial information of many kinds, including, as primary data sources, remotely-sensed data and imagery.

The challenging times we live in have played an important role in fostering the growth of GIS.

We need to better understand and better manage our planet. Growing human population, unsustainable use of natural resources, loss of biodiversity, human hunger and poverty, climate change, rapid urbanization, wars, terrorism, energy use, natural disasters, food production, human health—and many other issues—are problems we must address.

GIS plays a role in dealing with each of these.

In dealing with these problems large resources have been devoted to creating new science and technology, including both the science and technology of remote sensing and of geospatial information.

The development of these new capabilities has also been driven by rapid advances in technology, including more powerful computer processors on ever smaller microchips, high capacity storage devices, high speed wide-band communications, improved display devices, faster graphics processors, new visualization techniques, and the remarkable evolution and rapid growth of the Internet, including wireless network access.

These new capabilities have strongly influenced photogrammetry and remote sensing.

Remotely-sensed data are increasingly available and affordable; they come from a widening array of sources and sensing devices. Satellites can now collect staggering amounts of data and transmit high definition television; autonomous and unmanned vehicles provide platforms for remote sensing on land, sea and in the air, and at ever higher spatial resolution. We are on the threshold of creating an instrumented universe in which pervasive remote sensing will be complemented by vast numbers of networked sensors and measuring instruments, located throughout the human and natural environments.

These developments provide many challenges. Terabyte, petabyte and even larger sized image databases need to be managed and used effectively and securely. National and global spatial data infrastructures need to be created. Users want access to authoritative data in near real time, in 3D, and in high definition. They need searchable metadata and efficient geospatial browsers to locate these data.

These and other challenges provide opportunities for GIS science and technology to complement photogrammetry and remote sensing. The rapid growth of GIS applications, of GIS use and of the GIS industry all show that GIS is up to the task.

Professionals from many disciplines, including photogrammetry and remote sensing, have critical roles to play in these developments. They must provide technical mastery: developing improved technology and methods, devising QA/QC procedures, providing authoritative content, setting standards, assuring interoperability, increasing efficiency and managing costs. They are needed in policy making, management and to advise decision makers. They must also do the fundamental research and educate and train the next generation of professionals.

Also, in recent years, because of the capabilities and ease of use of free intuitive viewers on Internet websites like Google Earth and Microsoft Virtual Earth, the use of imagery and maps by non-professional and casual users has greatly increased. These "neogeographers" also create content and mashups of content. Professionals have an important role to play in responding to these developments.

I hope this *Manual* will bring all geospatial professionals closer together. We need their cooperation and collaboration in problem solving; we need them to lead our efforts to meet the challenges we face.

As we confront these great challenges with such capable science and technology I look forward to an exciting future.

Jack Dangermond
President
Environmental Systems Research Institute, Inc. (ESRI)

PREFACE

Marguerite Madden

Volumes of ASPRS manuals line my bookshelf from the 1968 first edition of the *Manual of Color Aerial Photography* I inherited from Dr. C.P. Lo's library to the 2006 *Remote Sensing of the Marine Environment,* the sixth volume of the third edition of the *Manual of Remote Sensing.* Throughout my career I have turned to these manuals for definitions, explanations, history, leading researchers, references, case studies and future directions in photogrammetry, remote sensing and digital characterization of the terrain. Although the subject of geographic information systems (GIS) is mentioned here and there within these renowned volumes, until now there has never been an ASPRS manual devoted solely to GIS. I was thrilled when the ASPRS Publications Committee, Kim Tilley, ASPRS Communications Director and Associate Executive Director, and Jim Plasker, ASPRS Executive Director, accepted my proposal to edit the ASPRS *Manual of Geographic Information Systems.* Through my involvement in ASPRS as the Director of the GIS Division and in activities of the International Society for Photogrammetry and Remote Sensing (ISPRS) Commission IV, Geodatabases and Digital Mapping, I already knew the experts I would invite to write the chapters I envisioned would make up such a manual. I outlined the topics and contacted the authors. One-by-one they accepted my invitation to add yet another writing obligation to their busy schedules, and the ASPRS *Manual of Geographic Information Systems* was born.

The 62 chapters have been organized in the following 8 major sections.

- Introduction, Background and Overview
- Data Models, Metadata and Ontology
- GIS Data Quality, Uncertainty, Accuracy and Standards
- Spatio-Temporal Aspects of GIS
- Analysis and Modeling
- Blending Technologies: Remote Sensing, GPS and Visualization
- GIS and the World Wide Web
- GIS Reaches Out: Applications

In this manual, top researchers in the field of GIS, along with emerging scholars who will see GIS advance with technologies we have yet to imagine, have told the story of a discipline that truly integrates the geospatial world. GIS is the glue that binds geospatial techniques, data, algorithms, models, maps and people together. It is the master plan that lays a foundation for all that is spatial in character. GIS provides the framework for data connected to location, the language needed for spatial conversation and the analysis tools for discovery of geographic place, proximity, dimensions, trends and correlations.

As many before me have more eloquently stated, the lines that formerly delineated the boundaries of traditional photogrammetry, remote sensing, land surveying, geodesy, cartography and computer science are now blurred, largely due to GIS. Students no longer have the luxury of choosing one of these fields to study, at the exclusion of the others. Today's students must be versed in all aspects of data acquisition by satellite sensors, digital cameras, personal cell phones and geo-sensor networks. They must have a solid understanding of the concepts and theories of geographic information science, an appreciation of spatial uncertainty and error, and practical experience manipulating and displaying geographic data. They should know of the pioneers who have gone before them. Given the relatively short history of GIS, many of these pioneers are still active in the industries, universities, agencies and professional organizations that gave birth to GIS. These leaders continue to shape the future of GIS. I encourage students to seek out the authors of the chapters in this *Manual of Geographic Information Systems* and engage them in discussions that will inevitably lead to invaluable connections and creative thinking. ASPRS is an organization that welcomes practitioners, researchers and developers to join in the central theme of the advancement of GIS. I hope that this volume will take a step towards true integration of geospatial technologies and give GIS the recognition it deserves as the discipline that brings us all to the same table.

Many thanks go out to my colleagues at the Center for Remote Sensing and Mapping Science (CRMS) and the Department of Geography at the University of Georgia (UGA), especially Dr. Thomas Jordan, CRMS Associate Director, Janna Masour, our GIS Research Professional, Donna Johnson, CRMS Accountant, and all of the graduate/undergraduate students and international visiting scientists, without whom we could not conduct our research. Dr. Roy Welch, Director Emeritus of CRMS and Past President of ASPRS, introduced me to GIS in 1987 through Dr. Joseph Berry's workshop on pMAP and ESRI Arc/INFO training on a Prime minicomputer. For the past 26 years we have conducted countless projects that integrate GIS with remote sensing, photogrammetry, GPS, spatial analysis, modeling and geovisualization for landscape analysis and assessment of human impacts on the environment. Special thanks go out to Dr. Jordan and Hunter Allen, UGA Geography graduate student, who provided the beautiful images for the cover from our National Park Service-funded research in Great Smoky Mountains National Park.

This book would not exist without the efforts of the ASPRS team headed by Kim Tilley. Rae Kelly, ASPRS Assistant Director of Publications, kept us all on track with spread sheets and regular conference calls. Matthew Austin, ASPRS Publication Production Assistant, performed an awesome job of graphic design and layout. Dr. Tina Cary, ASPRS Technical Editor, President of Cary and Associates and ASPRS Past President, and her associate, Linda Duffy, performed amazing technical editing of all 62 chapters. Dr. Qingmin Meng, Associate Director of Technology at the University of North Carolina at Charlotte's Center for Applied Geographic Information Science, did a great job on compiling, editing and organizing the digital content for the accompanying DVD. Without this team, you would not be holding a copy of the ASPRS *Manual of Geographic Information Systems* in your hands today.

Marguerite Madden
Professor and Director
Center for Remote Sensing and Mapping Science (CRMS)
Department of Geography
University of Georgia
Athens, Georgia

AUTHOR AFFILIATIONS

Section 1

Chapter 1
Barbara J. Ryan, *World Meteorological Organization Space Programme, Switzerland, formerly US Geological Survey*
Jon C. Campbell, *US Geological Survey*

Chapter 2
Rachael A. McDonnell, *University of Oxford, UK*
Peter A. Burrough , *University of Oxford, UK*

Chapter 3
Paul A. Longley, *University College London, UK*
Michael F. Goodchild, *University of California, Santa Barbara*
David J. Maguire, *ESRI, Inc.*
David W. Rhind, *City University London, UK*

Chapter 4
Michael N. DeMers, *New Mexico State University*

Chapter 5
Charles Olson, *Michigan Technological Research Institute*

Section 2

Chapter 6
Martien Molenaar, *International Institute for Geo-Information Science and Earth Observation (ITC), The Netherlands*
Peter van Oosterom, *Delft University of Technology, The Netherlands*

Chapter 7
Xiaobai Yao, *University of Georgia*

Chapter 8
E. Lynn Usery, *US Geological Survey*
Michael P. Finn, *US Geological Survey*
Clifford J. Mugnier, *Louisiana State University*

Chapter 9
Chandra Giri, *US Geological Survey*
David Hastings, *United Nations*
Bradley Reed, *US Geological Survey*
Ryutaro Tateishi, *Chiba University, Japan*

Chapter 10
Michael Moeller, *Westview Middle School, South Carolina, formerly with National Oceanic and Atmospheric Administration (NOAA) Coastal Services Center*

Chapter 11
Frederico Fonseca, *University of Maine*
Gilberto Câmara, *National Institute for Space Research (INPE), Brazil*

Chapter 12
David J. Cowen, *University of South Carolina*
Nancy von Meyer, *Fairview Industries*
Bob Ader, *US Bureau of Land Management*

Section 3

Chapter 13
Wenzhong (John) Shi, *Hong Kong Polytechnic University*

Chapter 14
Russell G. Congalton, *University of New Hampshire*

Chapter 15
Peter Fisher, *University of Leicester, UK*

Chapter 16
Thomas W. Owens, *US Geological Survey*

Chapter 17
George Percivall, *Open Geospatial Consortium*
Arliss Whiteside, *BAE Systems*

Chapter 18
S. Thomas Purucker, *US Environmental Protection Agency*
Heather E. Golden, *US Environmental Protection Agency*
Gerard F. Laniak, *US Environmental Protection Agency*
L. Shawn Matott, *University of Waterloo, Ontario, Canada, formerly with US Environmental Protection Agency*
Daniel J. McGarvey, *US Environmental Protection Agency*
Kurt L. Wolfe, *US Environmental Protection Agency*

Section 4

Chapter 19
Yanfen Le, *Northwest Missouri State University*
E. Lynn Usery, *US Geological Survey*

Chapter 20
Paul R. T. Newby, *The Photogrammetric Record, UK, formerly with the Ordnance Survey, UK*

Chapter 21
Keith Murray, *Ordnance Survey, Great Britain, UK*
Duncan Shiell, *Ordnance Survey, Great Britain, UK*

Chapter 22
Matthew Dunbar, *University of Kansas*

Chapter 23
Costas Armenakis, *York University, Toronto, Canada*
Eva Siekierska, *Natural Resources Canada, Ottawa, Canada*

Chapter 24
Bing Xu, *University of Utah*
Peng Gong, *University of California, Berkeley*

Chapter 25
Mei-Po Kwan, *The Ohio State University*

Section 5

Chapter 26
Kevin M. Johnston, *ESRI, Inc.*

Chapter 27
Giorgos Mountrakis, *State University of New York, College of Environmental Science and Forestry*

Chapter 28
Peter M. Atkinson, *University of Southhampton, UK*
Christopher D. Lloyd, *Queen's University, UK*

Chapter 29
Joseph K. Berry, *Berry & Associates // Spatial Information Systems and the University of Denver*

Chapter 30
Paul A. Longley, *University College London, UK*

Chapter 31
Anthony Gar-On Yeh, *University of Hong Kong*
Xia Li, *University of Hong Kong*

Chapter 32
C.P. Lo (deceased), *University of Georgia*
Xiaojun Yang, *Florida State University*

Chapter 33
Xiaojun Yang, *Florida State University*

Chapter 34
Raymond D. Watts, *US Geological Survey*
Giorgos Mountrakis, *State University of New York, College of Environmental Science and Forestry*

Chapter 35
James E. Vogelmann, *US Geological Survey, formerly with Science Applications International Corporation (SAIC)*
Donald O. Ohlen, *US Geological Survey, formerly with Science Applications International Corporation (SAIC)*
Zhi-liang Zhu, *US Geological Survey*
Stephen M. Howard, *US Geological Survey, formerly with Science Applications International Corporation (SAIC)*
Matthew G. Rollins, *US Geological Survey, formerly with US Department of Agriculture Forest Service*

Section 6

Chapter 36
Marguerite Madden, *University of Georgia*
Thomas R. Jordan, *University of Georgia*
Minho Kim, *University of Georgia*
Hunter B. Allen, *University of Georgia*
Bo Xu, *California State University, San Bernadino*

Chapter 37
Manfred Ehlers, *University of Osnabrueck, Germany*
Karsten Jacobsen, *University of Hannover, Germany*
Jochen Schiewe, *University of Bonn, Germany*

Chapter 38
Michael F. Goodchild, *University of California, Santa Barbara*

Chapter 39
Barend Köbben, *International Institute for Geo-Information Science and Earth Observation (ITC), The Netherlands*

Chapter 40
William Cartwright, *Royal Melbourne Institute of Technology (RMIT) University, Australia*

Chapter 41
Katy Appleton, *University of East Anglia, UK*
Andrew Lovett, *University of East Anglia, UK*

Chapter 42
Robert Edsall, *Arizona State University*
Gennady Andrienko, *Autonomous Intelligent Systems Institute, Germany*
Natalia Andrienko, *Autonomous Intelligent Systems Institute, Germany*
Barbara Buttenfield, *University of Colorado, Boulder*

Chapter 43
John J. Kosovich, *US Geological Survey*
Jill J. Cress, *US Geological Survey*
Drew T. Probst, *Formerly with Science Applications International Corporation (SAIC)*
Thomas P. DiNardo, *US Geological Survey*

Chapter 44
Brian N. Davis, *US Geological Survey*
Brian G. Maddox, *US Geological Survey*

Chapter 45
Norbert Haala, *University of Stuttgart, Germany*
Martin Kada, *University of Stuttgart, Germany*

Chapter 46
Yongwei Sheng, *University of California, Los Angeles*

Section 7

Chapter 47
Mark L. DeMulder, *US Geological Survey*
Gail A. Wendt, *US Geological Survey*

Chapter 48
Ming-Hsiang Tsou, *San Diego State University*

Chapter 49
Maged N. Kamel Boulos, *University of Plymouth, UK*

Chapter 50
Shunfu Hu, *Southern Illinois University*

Chapter 51
Vincent Tao, *Microsoft Virtual Earth, USA/York University, Toronto, Canada*
Steve H.L. Liang, *University of Calgary, Calgary, Canada*

Section 8

Chapter 52
Tonny J. Oyana, *Southern Illinois University*
Jessica Gaffney Clark, *Southern Illinois University*

Chapter 53
Jean-Claude Thill, *University of North Carolina at Charlotte*

Chapter 54
Liz Kramer, *University of Georgia*
Alexa J. McKerrow, *US Geological Survey and North Carolina State University*
Leonard G. Pearlstine, *National Park Service, formerly with University of Florida*
Frank J. Mazzotti, *University of Florida*
David M. Stoms, *University of California Santa Barbara*
Jill Maxwell, *University of Idaho*

Chapter 55
Shailesh Nayak, *Ministry of Earth Sciences, India*

Chapter 56
Carol A. Johnston, *South Dakota State University*
Terry N. Brown, *University of Minnesota-Duluth*
Tom Hollenhorst, *US Environmental Protection Agency*
Peter T. Wolter, *University of Minnesota-Madison*
Nicholas P. Danz, *University of Wisconsin-Superior*
Gerald J. Niemi, *University of Minnesota-Duluth*

Chapter 57
Steven D. Fleming, *US Military Academy, West Point*
Michael D. Hendricks, *US Military Academy, West Point*
John A. Brockhaus, *US Military Academy, West Point*

Chapter 58

Eileen B. Allen, *State University of New York, Plattsburgh*
Raymond P. Curran, *Independent Consultant, Adirondack Information Group*
Sunita S. Halasz, *Adirondack Research Consortium*
Stacy McNulty, *State University of New York, College of Environmental Science and Forestry*
John W. Barge, *New York State Adirondack Park Agency*
Andy Keal, *Formerly with the Wildlife Conservation Society*
Michale J. Glennon, *Wildlife Conservation Society*

Chapter 59

E. Lynn Usery, *US Geological Survey*
David D. Bosch, *US Department of Agriculture*
Michael P. Finn, *US Geological Survey*
Tasha Wells, *University of Georgia*
Stuart Pocknee, *University of Georgia*
Craig Kvien, *University of Georgia*

Chapter 60

Trent M. Hare, *US Geological Survey*
Randy L. Kirk, *US Geological Survey*
James A. Skinner Jr., *US Geological Survey*
Kenneth L. Tanaka, *US Geological Survey*

Chapter 61

Minna A. Lönnqvist, *University of Helsinki, Finland*
Emmanuel Stefanakis, *Harokopio University, Greece*

Chapter 62

David DiBiase, *Pennsylvania State University*
Michael DeMers, *New Mexico State University*
Ann Johnson, *ESRI, Inc.*
Karen Kemp, *Independent Scholar, formerly with University of Redlands*
Ann Taylor Luck, *Pennsylvania State University*
Brandon Plewe, *Brigham Young University*
Elizabeth Wentz, *Arizona State University*

DVD

Qingmin Meng, *University of North Carolina at Charlotte*

SECTION 1
Introduction, Background & Overview

CHAPTER 1
The Future Geography of GIS

Barbara J. Ryan and *Jon C. Campbell*

Applications of geographic information systems—GISs as they are widely known—have multiplied exponentially over the last several decades. Indeed, today there are few professional practitioners of geography, cartography, or any field of endeavor with spatial aspects, that have not used GIS. By facilitating analysis of disparate data sets and enabling comparisons to be made of seemingly unrelated classes of information, these systems have opened new vistas for investigating the world around us. Moreover, GIS has, by accelerating the payoff of good planning in government and business, by advancing efficiencies in transportation, communications, and other industries, and by contributing to many consumer applications of geographic technology, significantly stimulated economic growth in this country and worldwide. Other chapters in this manual characterize the multi-faceted applications and benefits of GIS. In this chapter we would like to underscore the importance of fostering geographic knowledge—the first word element in the acronym GIS—as a national goal.

Geography, more than any other science, is an integrative science. In her keynote address at the Association of American Geographers Centennial Meeting in 2004, Rita Colwell, former Director of the National Science Foundation, spoke of geography as "the original multi-disciplinary discipline, the ultimate field of confluence." Colwell also cited the ancient mathematician, geographer, and astronomer Ptolemy's description of the role of geography in mapping places as an attempt to construct "a view of the whole." What higher purpose can we possibly imagine for a modern GIS than this classical vision—to construct a "view of the whole?" To reach this ideal, however, we must begin to think more broadly of GIS and its role in society.

First, we must demand still more research and development at the interface between our traditional, or as a National Research Council report recently called them "industrial-strength" information technologies and the ultimate benefactors of geographic analysis—the policy makers, lawmakers, resource managers, and the public at large. As a community, we should not allow ourselves to become part of a professional tendency to focus on the computer technology aspects, the "tools," of GIS. These tools, to be sure, are absolutely essential for geographic analysis. Professionals, however, often talk about these tools as an end unto themselves, rather that a means to an end. Our specialized terminology distances us from the very people we wish to serve. Increased research in the human dimension of information technology and further development of "thin" clients, "middleware," or other tools that function as translators for a lay audience are sorely needed. These research and development efforts should be shouldered by a broad array of both public and private organizations in order to realize the full range of societal benefits that GIS can offer.

Secondly, we must strive for the widest public access to geographic data—the fuel for continued development and implementation of GIS. Recent developments by companies like Google and Microsoft[1] to display views of the Earth at multiple scales are increasing, by orders of magnitude, the numbers of people able to experience geographic data. Just as the development of in-car navigation systems was stimulated by public (in this case mostly federal) investment in geographic data, additional value-added services and applications will

---

1. The use of trade names is for descriptive purposes only, and does not imply specific endorsement by the federal government or ASPRS.

be stimulated by these online Earth views. Much, if not most, of the satellite and airborne imaging displayed in these Earth views has been publicly financed, and therefore, highlights the importance of public/private partnerships in demonstrating the untapped potential of geographic data in public electronic discourse.

Third, and most importantly, we, as GIS professionals and as global citizens, should support national commitments in education to foster spatial literacy. Spatial thinking is integral to everyday life. People, natural objects, and our built environment exist somewhere in space. The interactions of people with their environment can more easily be understood in terms of locations, distances, directions, shapes, and patterns. Spatial thinking, and its cultural extension, geographic literacy, is essential in preparing our children to be citizens of a global community. Improved geographic education in public schools will not only equip students for work and life in the 21st century at a personal level, but it will lead to a heightened use of GIS by the broadest possible array of practitioners—professionals and generalists alike—who, because they are well-prepared to think spatially about social, financial, and environmental issues from local to global scales, can make important contributions for the public good.

Thinking spatially in any field of inquiry, especially when accelerated by the power of geographic information systems, can bring Ptolemy's vision to life— "to construct a view of the whole." This is the highest goal of geography. With such a view we can see even further to new horizons.

Other Sources

Auch, R., J. Taylor and W. Acevedo. 2004. Urban Growth in American Cities – Glimpses of U.S. Urbanization. US Geological Survey Circular 1252, 52 pp.

Bernhardsen, T. 1999. *Geographic Information Systems: An Introduction.* New York: John Wiley & Sons, Inc., 448 pp.

Brunn, S. D., S. L. Cutter and J. W. Harrington, Jr., eds. 2004. *Geography and Technology.* Dordrecht, The Netherlands: Kluwer Academic Publishers, 613 pp.

Craig, W. J. 2005. White knights of spatial data infrastructure: the role and motivation of key individuals. *URISA Report* 16 (2):5–13.

Halsing, D., K. Theissen and R. Bernknopf. 2004. A Cost Benefit Analysis of *The National Map.* US Geological Survey Circular 1271, 40 pp.

Heywood, I., S. Cornelius and S. Carver. 2002. *An Introduction to Geographical Information Systems,* 2nd edition. Upper Saddle River, New Jersey: Prentice Hall, 296 pp.

Interagency Working Group on Earth Observations. 2005. *The U.S. Integrated Earth Observation System, Strategic Plan for the U.S.* The National Science and Technology Council, Committee on Environment and Natural Resources, 149 pp.

Longley, P. A., M. F. Goodchild, D. J. Maguire and D. W. Rhind. 2005. *Geographic Information Systems and Science,* 2nd edition. Chichester, UK: Wiley, 536 pp.

Loveland, T. R., editor. 2004. Ecoregions for environmental management. *Environmental Management* vol. 34, supplement 1. New York: Springer-Verlag, 147 pp.

McMahon, G., S. P. Benjamin, K. Clarke, J. E. Findley, R. N. Fisher, W. L. Graff, L. C. Gunderson, J. W. Jones, T. R. Loveland, K. S. Roth, L. E. Usery and N. J. Wood. 2005. Geography for a Changing World – A Science Strategy for the Geographic Research of the U.S. Geological Survey, 2005-2015. US Geological Survey Circular 1281, 54 pp.

National Research Council. 2005. Learning to Think Spatially: GIS as a Support System in the K-12 Curriculum. Washington, DC: National Academies Press, 248 pp.

Ogrosky, C. E., editor. 2003. Special Issue: *The National Map. Photogrammetric Engineering and Remote Sensing* 69 (10):1077–1200.

Richardson, D. and R. Colwell. 2004. The new landscape of science: a geographic portal. *Annals of the Association of American Geographers* 94 (4):703–708.

Ryan, B. J. 2002. One year after [the events of September 11th] – a USGS perspective. *Geospatial Solutions* 12 (19):56–57. <http://www.geospatial-solutions.com/geospatialsolutions/Public+Policy/One-Year-After-A-USGS-Perspective/ArticleStandard/Article/detail/29506> Accessed 21 April 2008.

Ryan, B. J. and J. Campbell. 2004. *The National Map* of the USA – developments past to present. *GIM International* 18 (9):14–16.

Turner, B.L. II. 2005. Geography's profile in public debate "inside the Beltway" and the national academies. *The Professional Geographer* 57 (3):462–467.

CHAPTER 2
Principles of Geographic Information Systems

Rachael A. McDonnell and *Peter A. Burrough*

2.1 Introduction

During the last 25 years, information technology and, in particular, its application in the practices, research and education that fall under the banner of "Geographic, or Spatial, Information Systems" (GISs/SISs), has undergone great changes. Close examination, however, shows that many of the principles applied in present-day GISs are still similar to those developed in earlier decades, although during the interim there has been an exponential rise in the worldwide adoption of the technology and a vast expansion of the user community. From a technological point of view, both the computer software and the hardware have improved immeasurably, and the necessary six or more volumes of command-line operation manuals have been replaced by sophisticated graphical user interfaces supporting intuitive learning, and by online help facilities and tutorials. Yet much of the functionality and many spatial operations of these systems remain the same: map overlay, the search for combinations of data; counts of cells in nominal classes; measurements of length, area and distance; creation of digital elevation models and their derivatives, and so on. In more academic terms, the old debates on vector and raster spatial data models have been replaced by discussions on objects and fields or ontological definitions. While the rapidity of changes in GISs and their role in spatial information systems has been seen by some as breathtaking, several of the old problems remain unsolved. Arguably the biggest difficulty, in both theoretical and technical terms, is the development of a data model to support multivariate and multi-scale space-time representations and temporal analysis.

This chapter gives an overview of the principles and the development of GISs over the last few decades. It highlights the basic approaches to representing and using digital spatial data in a computational environment.

2.2 Definitions of the Principles of GISs and SISs

To start with, it is necessary to think about what is actually meant by the terms "Geographic Information System" and/or "Spatial Information System." Much debate has followed a "tool versus science" line of argument (Wright et al. 1997) and arguably the most-used definitions are tool-based ones, such as:

> *GIS is a powerful set of tools for collecting, storing, retrieving at will, transforming and displaying spatial data from the real world for a particular set of purposes. These tools are used to manipulate, and operate on standard geographical primitives, such as points, lines and areas, and/or continuously varying surfaces, known as fields.* (Burrough and McDonnell 1998, page 11)

This citation captures the notion that there is a group of software modules (tools), with associated hardware, data and users, which is able to manage and support the analysis of standardized geographic data, thereby gaining greater understanding of uniformly described spatial phenomena and processes. This notion was derived from and built on conventional approaches to mapping, photogrammetry and computer-aided design.

By implication, the term "Geographic Data" means "data for which the location of the spatial objects under examination is related to a national or international topographic grid." The term "SIS" is often used in a wider context than GIS to denote many types of spatial analysis for which the data are spatially registered to a local or worldwide grid, such as may be carried out by using multivariate statistical analysis or numerical models of spatial or temporal change.

In the early 1990s this somewhat technocratic approach to the principles of spatial information was found to be limited and did not embrace the full extent and complexity of the research, education and technological thinking that was taking place (Goodchild 1992). Many developments in these areas have led to interesting debates on a more complete definition of GISs. It had been appreciated that GISs were more than a database of digitized thematic maps and orthophotographs that have been linked in a relational database to give them the potential for providing the user with a spatially dynamic view of areas from a local to a global world. Therefore, in the acronym GIS should the *"S"* component of principles be *Systems, Science* or even *Services* (Longley et al. 2005)?

Since the major academic journal changed its name to the *International Journal of Geographical Information "Science,"* this term has been increasingly used in the context of the more fundamental and theoretical aspects of geographic information such as how it is conceptualized, modeled, represented and analyzed. This approach has brought in and enriched academic GISs with thinking and methods from a disparate set of disciplines such as mathematics, philosophy, geostatistics, social and computer sciences, geography and geodesy, not to mention geophysics and ecology.

The motivations and motivators for this debate were many and varied. During the 1990s, there has been a maturing in the academic research output in GIS studies, with a move away from exploration of database storage models, analytical methods and technology applications into more theoretical and complex frameworks for spatial data. There was also somewhat of a backlash from others working within geography and related disciplines to the growing dominance and particularly demands on resources from those teaching and researching using GISs. Some have also argued that the move was linked to a need to gain intellectual credibility (Pickles 1999).

While we are considering definitions it is useful at this stage to mention other related terms to GIS that are found in the literature such as geomatics and geocomputation. The former is very much concerned with the science and mathematics of collecting, interpreting and analyzing data related to the Earth's surface. Geocomputation describes work that is undertaken to develop an understanding of spatial and temporally dynamic (and frequently inexact) phenomena and processes using a range of computational tools—such as statistics, cellular automata, agent-based modeling, exploratory spatial data tools—that are not part of a GIS per se, but they use components of GISs, such as the data management and storage and visualization techniques of spatial data.

2.3 Principles and Developments in GIScience

If GISs are distinctive because they handle spatial data, it is first important that we understand what and how these data are defined. Geographic phenomena require two descriptors to represent the real world; *what* is present (attribute data), and *where* it is (locational data). For many phenomena they also often require a time stamp, the *when*. To link these basic principles we also need to understand what it is beside, what includes, or what links to each other etc. Also, are they disjoint or continuous, connected, or linked? Are the objects of interest spatially exact or crisp or inexact, fuzzy or stochastic? What spatial operations must or may be carried out on these phenomena?

Transforming this complexity of geographic data into the bits and bytes of computer form lies at the heart of difficulties in defining position and associated spatial and attribute descriptors. GISs therefore need to provide database capabilities that can store the "what and where" information as well as the relationship data between individual phenomena. For example, if a stream is represented as a series of lines then it is important to state that these individual lines join together to form a whole entity of a river.

2.3.1 From Vectors and Rasters Through Continuous Fields and Objects to Scale Fuzzy Boundaries: The Ontological Debates

The conceptualizations of geographic data over the last 25 years of GIS technology have taken over many ideas from traditional topographic mapping and remote sensing. The former describes space in terms of primitive objects such as points, lines and areas (otherwise known as polygons) and this, known as the vector model, has been adopted in many GISs. Attributes are linked to these geographic primitives to carry data on the kinds of points, lines or areas represented or linked. The all-important spatial relations of connectivity, etc., are explicitly defined within the database and are known as the topology.

In other systems, ideas from remote sensing have been used in which the geographic space may be thought of as a checker board in which each square (or other tessellated shape) is classified in terms of presence or absence, or degree, of a particular phenomenon. In this raster model the spatial links of the entities with each other are implicit and based on the occurrence of a same-classified cell next door.

During the 1980s there was much debate on the relative benefits and disadvantages of the vector and raster models and comparisons were made in terms of their accuracy of representation, data storage efficiency, speed of access to the data, analytical functionality and aesthetic value of the output amongst others (Burrough and McDonnell 1998). Little attention was paid to spatial analysis as such—that had to wait. Today there is still a distinction and most systems support either one or other of the data models.

Following on from this somewhat technocratic approach to spatial data, the middle 1990s were dominated by the "ontological debates" that cover a period of scientific development in which GIScience researchers started to develop rules for describing spatial information in a more sophisticated way than these simple primitives. In the GIS/Spatial Information literature of today you will encounter many articles on ontology, a field of study with a long history in philosophy that in more recent times has developed understandings emanating from information engineering (Winter 2001; Agarwal 2005). Ontological thinking focuses on what things exist before language or knowledge, how we define things in space, and how these things interact with each other.

Early discussions on ontology in GIScience focused on different concepts of space with the plenum and atomic ontologies (Couclelis 1992 gives a very clear explanation) with the former conceptualizing the world as a space-and-time framework within which "things" exist (which is the object model) and the latter conceiving of groups of space/time attributes being constructed as phenomena (the field model); there are obvious links with the earlier vector and raster models.

In recent years there has been a focus in research on defining what is in a space and how and where we place boundaries around the various types of geographic entities. These deepening philosophical debates have been stimulated by a number of factors such as the increasing access to data collected by others, the increase in inter-disciplinary studies that involve using data collected by those working in different intellectual frameworks, as well as the developing field of exploratory spatial data analysis in which phenomena are defined from the data (Winter 2001). Given that we cannot measure/observe "everything" in an area, our disciplinary background will determine what we measure and how we conceptualize the environment we see. The soil scientist will measure different attributes (mineralogical content, soil structure, etc.) than a hydrologist (hydraulic conductivity, nitrate content) when considering the same phenomenon, in this case soil. An interesting debate is whether it is possible to derive a set of universally held definitions for phenomena in an area and have them be the primitives on which further work can be based.

Linked to these ideas is the debate on scale that has always been with us but is receiving increasing attention these days. This is concerned with the effects of scale on how we observe and conceptualize phenomena and processes. Scale may be perceived as the cartographic ratio of distance on a map to that on the Earth's surface, but it also relates to concepts such as support, resolution, extent and coverage, i.e., how many measurements were taken, over what area, and at what spacing to define a unit of what size? (Lam and Quattrochi 1992; Cao and Lam 1997; McMaster and Sheppard 2004). The size of the smallest distinguishable unit (space or time) of a data set has an important impact on what may be explored and understood, whether a census district, a particular grid cell size, or some typical minimum representative unit area such as those frequently used in habitat and hydrological studies. This minimum unit determines the limits of our ability to observe small or large forms of patterns or processes. It also influences how an entity should be represented in a database, whether as a point or line at one scale or a polygon object at another. Similarly it may be defined by none, one or more classified cells in the field model.

If we consider boundaries in a more general sense, it becomes clear that while some are obvious and fixed, such as that of a house (these are known as *bona fide* objects), others are less easy to define, such as political or habitat boundaries, as they are either human constructs (these are known as *fiat* objects) or the delimits are gradual and mixed over space. While the field-data model allows us to be somewhat imprecise in defining anything like a boundary, a major area of development in GIScience has been the use of ideas of uncertainty and fuzziness in our definitions of objects.

2.3.2 Fuzzy Geographic Objects

Conventional geographic objects such as houses, roads and soil units have been almost exclusively represented by points, lines, polygons and pixels. At high resolution, almost everything can be represented by arrays of cells, as digital photography has demonstrated. Nevertheless, many geographic objects are themselves indistinct in attributes, boundaries and shape or form (Burrough and Frank 1996)—a truly representational GIS will also carry information about inexactness and the role it has to play in real life. Fuzziness is distinct from statistical variation, in which the uncertainty follows a well-described probability distribution, and is probably commoner than we think. It occurs in many areas and provides a means of describing the possibility of variation, rather than chance.

2.3.3 The Third and Fourth Dimensions

The discussion so far has focused essentially on two-dimensional entities. The use of data which have *x, y* and *z* defined positions has been important in many applications, especially in the built environment and in geological surveys. Developments in data-gathering technology such as digital photogrammetry and geophysical survey techniques have ensured more truly three-dimensional data are now available.

The spatial data models used to define this data are essentially extensions of the vector and raster concepts used in two dimensions. The equivalent to the pixel of the raster model is known as voxel; voxels are regular tessellated 3D shapes usually cubes or cuboids which are classified. Developing the equivalent in the vector model is more complicated; solid objects such as buildings are generated using the basic *x, y* representations of the footprint and then developing a solid from a limited number of height samples (Raper 2000).

Turning to defining data with a temporal dimension, this subject is perhaps one of the most thorny problems still to be tackled in GIScience. The widely used conceptual object and field models demand that the temporal dimension of spatial data is included almost indirectly with each time shot recorded as a different layer of data. There is no connection between the individual spatial primitives between the different time slices and any change analysis needs to be performed through operations such as overlay (Peuquet 1999 gives a good review). While advancements in technology are now supporting more sophisticated visualizations of an area over time (this is discussed in more detail in Section 2.4.4) the theoretical framework needed for spatio-temporal data modeling has yet to be sufficiently developed for incorporation into working GISs.

2.4 Principles and Developments in the GIS Toolbox

While the developments in GIScience have been marked by a deepening and increasingly philosophical approach, the changes in GISs have been rapid and hugely transforming. These have been fuelled by developments in computer capabilities, in particular processing power and graphical user interfaces (GUIs). One of the most important trends in the 1990s has been the change from a command-driven to a more user-friendly graphical interface.

While these changes have liberated GIS technicians the world over, surely it is the inter-connectivity of computers across the world that has led to the most exciting developments in GISs in the last 25 years. In the early days of GISs each computer system used a distinctly different coding system (and computer) for the data, so linking these systems into a common network was difficult or nearly impossible. However, by the end of the 20th century inter-active computer science had progressed so far that large data sets in a GIS could be easily downloaded and integrated in popular browser-based mapping tools. Today we can sit in an internet café in Accra, Ghana, or a wireless-enabled coffee shop in Bahrain and download mapped data gathered in a census somewhere in another hemisphere, or view images from a satellite orbiting this or another planet. This has been more than just a technical achievement. The GIS theorists needed to progress from linkages of simple objects to sophisticated data exchange that included information (metadata) about the kinds of data being exchanged.

2.4.1 Developments in Data Collection and Availability

Building a geographic database has always been tedious. The georeferencing framework has to be first established and then the various data sets must be converted to digital form and brought into the database management system before any analysis can take place (Burrough and McDonnell 1998 and other textbooks give an overview of the data input process). Now in the early 21st century, tedium has been relieved by so much geographic data being readily available and exchangeable. In many parts of the world the old map-based records of geographic data have been transformed by digitization or scanning into seamless layers of digital spatial information that may be readily input into a GIS.

The collection of new field survey data has also been transformed with the development of lasers (for accurate distance measuring), and the Global Positioning System (GPS), which give us an x, y and z position nearly anywhere on the Earth's surface. Innovations in field-based technology have meant that the collection of attribute values and the positioning of geographic data have never been so easy, or so accurate. Developments in remote sensing devices, which collect values for reflected and/or emitted electromagnetic radiation on a regular basis, have also had a considerable impact on the availability of geographic data. The increasing spatial resolution, temporal frequency and spectral resolution of the various remote sensing devices give a wider range of applications.

Satellite technology has been important not only for collecting data but also for transferring it. Automatic gauging sites, whether measuring atmospheric gases, river levels or seismic waves, are situated in remote locations with the data sent via telemetry through satellite communication systems to receiving centers, which may be thousands of miles away.

With the availability of such an array of data it has obviously been important to develop common standards so that whichever GIS has been chosen, the data may be readily imported. Previously, various tabular proprietary database systems were used that limited the use of a data set to a particular software system, but now open transfer standards have been adopted by both the data providers and the vendors of commercial GISs. These transfer standards permit many different datasets to be accessed and "data warehouses" have been established across the internet where they may be obtained for free or purchased. For example, many government departments allow other departments to access "their" data resources to support informed decision making.

The development of common standards along with the relative ease of transferring digital files across network technology has also led to the development of Spatial Data Infrastructures (SDIs), which are initiatives to promote consistent means of exchanging data across regions, nations and the globe (Masser 2005).The aims of such initiatives are to not only support better decision making through the provision of more data, but also to bring about significant savings in the costs of data collection, as it would help eliminate replication of effort. These developments in data collection and provision have ensured that during the last decade, there has been a steady increase in the usage of data that has been collected by one organization for the benefit of others. It is obviously important that information on the methods, timing, georeferencing system, and subsequent processing of the survey data are also exchanged. Primary data collectors are therefore increasingly providing metadata, i.e., information about the data, along with the database. This information obviously makes the data more understandable, and impacts how and by whom the data are used.

2.4.2 Developments in GIS Technology

While GISs as a technology can be traced back in time to the mid 1970s, it was really in the 1980s that commercial systems became readily available to those with sufficient monetary resources. These systems developed from the computer-aided mapping and design systems of the period. The early systems had their functionality built on proprietary software. These were beasts of systems that required dedicated learning to begin to master in any way. They were command-driven with specific syntax unique to each one. Data storage was based on individually developed data management systems. The result was that everyone's GIS worked differently. Thankfully there were actions taken in the 1990s towards openness and general standards (as just described). Huge amounts of spatial data are currently stored in relational or object-oriented database management systems using widely adopted and commercially available software.

There have been similarly liberating trends in the actual computing hardware with workstations (often having UNIX-based operating systems) being replaced by personal computers (PCs) (usually having Windows operating systems). In many medium to large organizations the GIS technology used is dominated by large data-holding computers, which are linked across networks to smaller PC-based processing units. These allow the users of the data access without having to manage and store the information on their own computers. This is obviously efficient and opens up the user community.

The development of the functionality in GISs might be described as being in a mature stage, with more consolidation rather than new ideas occurring at this time. The functionality supports querying, inventorying, basic modeling and trend analysis. Queries may be based on simple location or attribute information, such as how much of a particular phenomenon or the number of occurrences. In more sophisticated analysis the data may be integrated across the layers and used in Boolean operations.

This functionality has always been deterministic but developments in software are supporting a more inductive approach to analysis. In the inductive approach, the data are explored to determine if any objects, patterns, or structures exist that were not known before. In other words the starting point for knowledge development is the data. Various data mining methods have been developed, such as functions to identify spatial autocorrelation or clustering (O'Sullivan and Unwin 2002).

In many respects, some of the innovation in GISs has been in broadening the functionality by linking to other types of geographic data software. For example, many GISs now offer specific functionality to undertake spatial data generation through geostatistical interpolations, while others support direct linkages to outside specialist software. These packages, such as Gstat (Pebesma 2006) and R (O'Sullivan and Unwin 2002), allow data to be generated for a whole area from a set of point samples and include operations based on local and globally defined relationships, such as inverse distance weighting and kriging.

2.4.3 GISs and Dynamic Modeling

In many applications there is a need to represent the processes of change over both space and time. One might need to represent the impact of a change in the central bank base rate on industry in an area, the variations in water quality along a river, or the spread of a disease. In all these cases the nature of the phenomena—both form and values—will vary over time. This is difficult to model given the database structures available in current GISs, and there has been a certain degree of ingenuity involved in developing such systems.

The approaches used depend initially on the spatial data model used. For some applications, often environmental, a field-based model has been adopted and the layers of information are combined using various mathematical functions to represent the processes. Certain commercial GISs provide these capabilities as complete integrated software packages, but modelers are also able to use specially developed systems that provide defined blocks of functions, which make building environmental models easier. In the more specific modeling environments the ideas of cellular automata are also bringing new ways to represent processes that are partly controlled by interactions with neighboring grid cells (Utrecht University 2006; Karssenberg 2002a, 2002b). The raster approach inherent in these techniques uses the GIS as a data management and visualization machine. In addition, the GIS in a linked programming environment can be used for developing modeling capabilities in which the processes are defined as modules. This approach has been particularly popular for object-based or multi-scaled data (see Maguire et al. 2005).

2.4.4 Output Visualization Developments

Visualizing spatial data is one of the important steps in knowledge development. Today GISs offer many different options in output, with the user firmly in control of choices such as viewpoints, scale, and color options. Traditional approaches such as choropleth mapping are easily supported, as are tables and graphic representations such as pie charts, either with or without a spatial component. In more recent years computer technology has supported rapid improvements in methods that give more user-controlled interaction and submersion, such as through animation and virtual reality. This gives the user a different experience of an area than gained from a two-dimensional map and, importantly, allows the visualization of dynamics over both space and time. Familiar graphical imaging has been used, for example in computer games, to visualize spatial data to gain a more in-depth understanding of an area.

2.5 The User Community

Humans are an important component of GISs and over the last 25 years there has been an incredible expansion in both depth and breadth of their use of GISs. For example, the in-car navigation used by many drivers all over the world demonstrates the very simple application of spatial analysis of digital vector data. Alternatively if one is taking public transport, such as the bus, it is possible, at a particular stop, to determine where the next bus or train is and when it is likely to arrive. These two technologies bring together GPS with a database of street maps and basic GIS processing to deliver up-to-the-minute spatially based information for a particular transport activity.

There are several aspects to the human dimension in GIScience and Systems. The development of the GIS database is based on many decisions, i.e., what data are included, which georeference framework is used, how the database should be structured, etc. These answers depend upon the reason for establishing the system and the areas of interest it is likely to address. In many organizations, the users involved with the development aspect are different from those who apply the data and functionality in problem solving. These users may have little direct connection in developing the system, but in their own work regularly need the information provided through analysis or modeling of the integrated geographic data. The third component of the user community may be identified as those whose lives are affected by the use of a GIS. These people are not directly connected to systems but may have aspects of their work affected by the techno-scientific nature of the handling of geographic data. It is interesting to look at the changes in the user community over time.

The early users of GISs were often the programmers of the system to whom people would come with a problem; the programmers would undertake the work needed to create a solution using the software. As the systems became commercial a growing body of computer-literate users learnt the various command-lines needed to undertake the data development, management and analysis.

In recent years many of the commercial GISs have been developed using a modular approach, with some parts, such as the map-making bits, very intuitive and user friendly, requiring little training to produce an output. In the more complex analysis modules, a greater understanding of the principles of GISs is needed for effective and accurate usage. There are thus many types of GIS users today, ranging from applied users to those involved with structuring and developing networked databases or programming in the supported development languages.

It could be argued that the main users of spatial data will continue to be those working in government and the developments in networking capabilities have ensured that more spatially referenced information is now available to support planning, policy and decision making than ever before. Understanding what is where and how it has changed or will change is important to such organizations. GISs offer the abilities to store, manage, integrate and analyze vast quantities of data to this end. This role may be in day-to-day operations, strategic development or disaster relief.

In commercial organizations GISs are used increasingly in both the planning and operation of activities. As more data have become available and the user-friendliness of the systems improved, then the range of applications and the background of the users have become ever wider. Developing an understanding of the population in an area and the market opportunities associated with that population underlies many of the applications. Geographic data might also be used by organizations moving within an area so that these movements may be efficient in terms of time and money.

From these broad examples it becomes obvious that there are many reasons for the proliferation of GIS usage, including availability of spatial data, relative decreasing costs of software and hardware, rapidly increasing numbers of trained personnel, and a much greater awareness of the usefulness of GISs for analysis and decision support.

2.5.1 Avenues for Communication and Education

The spaces in which people talk and read about GISs have changed dramatically over the last few decades. During the late 1980s and early 1990s the stage was very much dominated by large GIS conferences where software vendors and users presented the results of their latest work (e.g., GIS/LIS and EUROGIS conferences). As the business has matured into a multimillion-dollar industry, these conferences have been largely replaced by specialized meetings, with many either being a particular software user conference or a themed academic gathering.

To support an understanding of the underlying concepts and principles of GISs, textbooks in the 1980s and early 1990s (Burrough 1986; Peuquet and Marble 1990; Tomlin 1990; Antennuci et al. 1991; Laurini and Thompson 1992) have been augmented by over fifty general textbooks and thousands of specialized books that cover a vast range of theoretical and applied subjects. The very graphic nature of the subject has meant that these books are both attractive and informative.

For more in-depth treatment of the subject, academic and trade journals have been important, such as the *International Journal of Geographical Information Science* (originally *System*) and *Mapping Awareness.* A series of such publications has now been developed that covers both the general and subject-specific areas. The outlets for academic research, such as *Transactions in GIS,* reflect the increasingly high intellectual quality of articles, which increase the impact factor ratings. The language of many of these journals has been English, but in recent years with the wide expansion in user communities there have been a number of publications in other languages, particularly in developing countries to support the growing skilled workforce. The web, however, has brought the most changes to disseminating information on GISs and the often high-graphical content of many postings is well-supported by this medium

In terms of learning, there are two main routes. The most common is through training offered by specific software vendors in which the basic principles of geographic data and GISs are introduced, followed by more in-depth coverage of the use and application of specific functions of database development and analysis. Today this training is offered through the internet, or more traditional courses are offered through training companies or colleges/universities. This has ensured a rapidly expanding body of users who have a working knowledge of the various different software systems.

The second route of learning is through higher education in which there is more development of both the academic principles and the technical skills of both GIScience and Systems. These are offered at both undergraduate and postgraduate level, with a small but increasing input through secondary school education.

2.5.2 Understanding the Social Context and Impact of GIS Use

The use of GISs and associated digital spatial data, in the ever-increasing areas of application, does not take place without having an effect on people. Anecdotal evidence as well as more formal research has shown that this impact, which may be both negative and positive, is different across the various strands of a particular society. For example, in recent discussions a non-governmental organization (NGO) worker from a hill-farming area of India spoke of the impact on women in a community, created by pinning simple GIS maps of plot ownership on the village school posts. Over the next few years the women became more involved with the governance in their village with an implied "empowerment" resulting from the tangible evidence of their land ownership. Contrary impacts have also been found, particularly in situations where the techno-scientific nature of the GIS, access and training issues, and the difficulties in defining computationally important information (often social or cultural) have led to marginalization of particular groups of people. It is therefore important in any GIS development to consider who will be supported and who will be marginalized or disenfranchised.

Manual of Geographic Information Systems

2.6 The Journey and the Way Ahead

The value GISs/SISs bring to understanding things and processes that are spatial has been shown across the globe and even in interplanetary work. The ability to bring together geographic data and combine, interrogate and model allows ever more sophisticated problems to be addressed. GISs not only have a role in linking social and natural science data, but also different kinds of information over time, thereby helping to resolve the big framework problems. Given the wide adoption of the technology by governments and industry it is possible to imagine that every second of every day a decision is being made based on GIS outputs. This has been aided by sufficient successful applications, increasing accessibility of data and growing affordability of hardware and software, so that now the business justification for adoption is much easier to make.

The 19th century socialist, Havelock Ellis (1859–1939) stated, "What we call 'Progress' is the exchange of one nuisance for another…" The development of GISs is still at a maturing stage with a number of riddles still to be solved on the technical front, in particular the ability to represent dynamic space-time phenomena. But perhaps the biggest "nuisance" that has emerged in the last decade and has still to be faced is the understanding of the cultural and social impacts of introducing these systems into areas that previously used only limited techno-scientific approaches to decision making. It is important that GISs are used to the benefit of all, and not just a few.

References

Agarwal, P. 2005. Ontological considerations in GIScience. *International Journal of Geographical Information Science* 19:501–536.

Antennuci, J., K. Brown, M. J. Croswell, M. J. Kevany and H. Archer. 1991. *Geographic Information Systems: A Guide to the Technology.* New York: Van Rheinold.

Burrough, P. A. 1986. *Principles of Geographical Information Systems.* Oxford: Oxford University Press.

Burrough, P. A. and A. U. Frank. 1996. *Geographic Objects with Indeterminate Boundaries.* London: Taylor and Francis.

Burrough, P. A. and R. A. McDonnell. 1998. *Principles of Geographical Information Systems.* Oxford: Oxford University Press.

Cao, C. and N. S-N. Lam. 1997. Understanding the scale and resolution effects in remote sensing and GIS. In *Scale in Remote Sensing and GIS,* edited by D. Quattrochi and M. F. Goodchild, 57–72. Baton Rouge: CRC Lewis Publishers.

Couclelis, H. 1992. People manipulate objects (but cultivate fields): Beyond the raster-vector debate in GIS. In *Theories and Methods of Spatio-temporal Reasoning in Geographic Space,* edited by A. U. Frank, I. Campari and U. Formentini, 65–77. Pisa: Springer-Verlag.

Goodchild, M. F. 1992. Geographical information science. *International Journal of Geographical Information Systems* 6:31–45.

Karssenberg, D. 2002a. Building dynamic spatial environmental models. *Nederland Geografische Studies 305.* Utrecht: KNAG/Faculteit Ruimtelijke Wetenschappen Universiteit Utrecht.

———. 2002b. The value of environmental modelling languages for building distributed hydrological models. *Hydrological Processes* 16:2751–2766.

Lam, N. S-N. and D. A. Quattrochi. 1992. On the issues of scale, resolution and fractal analysis. *Professional Geographer* 44: 88–98.

Laurini, R. and D. Thompson. 1992. *Fundamentals of Spatial Information Systems.* London: Academic Press.

Longley, P., M. F. Goodchild, D. Maguire and D. Rhind. 2005. *Geographic Information Systems and Science,* 2d ed. Chichester, UK: John Wiley & Sons.

Maguire, D., M. Batty and M. F. Goodchild, eds. 2005. *GIS, Spatial Analysis and Modeling.* Oakland, Calif.: ESRI Press.

Masser, I. 2005. *GIS Worlds: Creating Spatial Data Infrastructures.* Oakland, Calif.: ESRI Press.

McMaster, R. B. and E. Sheppard. 2004. Introduction: Scale and geographic inquiry. *Scale and Geographic Inquiry: Nature, Society and Method,* edited by E. Sheppard and R. B. McMaster, 1–22. Oxford: Blackwell.

O'Sullivan, D. and D. Unwin. 2002. *Geographic Information Analysis.* Hoboken, New Jersey: John Wiley & Sons.

Pebesma, E. J. 2006. Gstat Home Page. <http://www.gstat.org/> Accessed 23 January 2008.

Peuquet, D. 1999. Time in GIS and geographical databases. *Geographical Information Systems,* Vol. 1, *Principles and Technical Issues,* edited by P. A. Longley, M. F. Goodchild, D. J. Maguire and D. W. Rhind, 91–103. Chichester, UK: John Wiley & Sons.

Peuquet, D. and D. Marble, eds. 1990. *Introductory Readings in Geographic Information Systems.* London: Taylor and Francis.

Pickles, J. 1999. Arguments, debates and dialogues: The GIS-social theory debate and the concern for alternatives. *Geographical Information Systems,* Vol. 1, *Principles and Technical Issues,* edited by P. A. Longley, M. F. Goodchild, D. J. Maguire and D. W. Rhind, 49–60. Chichester, UK: John Wiley & Sons.

Raper, J. 2000. *Multidimensional Geographic Information Science.* London: Taylor & Francis.

Tomlin, D. 1990. *Geographic Information Systems and Cartographic Modelling.* Englewood Cliffs, New Jersey: Prentice-Hall.

Utrecht University. 2005. PCRaster Website: Home. <http://pcraster.geo.uu.nl/> Accessed 23 January 2008.

Winter, S. 2001. Ontology: Buzzword or paradigm shift in GI Science? *International Journal of Geographical Information Science* 15:587–590.

Wright, D. J., M. F. Goodchild and J. D. Proctor. 1997. Demystifying the persistent ambiguity of GIS as "tool" versus "science." *Annals of the Association of American Geographers* 87:346–362.

CHAPTER 3
Geographic Information Science

Paul A. Longley, Michael F. Goodchild,
David J. Maguire and *David W. Rhind*

3.1 Geographic Information Systems and Science

Science is concerned with the discovery and organization of knowledge. By employing scientific methods, scientists create empirically testable theories and models of observable events. It is common to subdivide science into natural sciences—including biological, environmental and physical—and social sciences—economics, psychology, politics, etc. Science can be practiced in both a pure and applied way. Geographic information science (henceforth GIScience) deals with the science of information that can be used to describe, explain and hence predict the forms and processes that characterize the Earth surface and near surface. In practice, it is often impossible to study events and occurrences without considering both natural and human-induced processes (e.g., the impacts of deforestation on subsequent environmental disasters), and thus GIScience spans natural-social and pure-applied scientific divides. GIScience unifies the ways in which we study the surface of the Earth and the impact of those living upon it, using geography as an organizing framework to structure the information that is available to us.

GIScientists use models, or simplifications, to represent aspects of the seemingly infinite complexity of the Earth in digital information systems (see also Longley, Chapter 30, this volume). Some of these models are predominantly descriptive—data or information models—and are primarily used in studies concerned with surface morphology or form. These are used to store vast amounts of information about the Earth. Other models are mainly analytical and are used to understand the processes that operate upon it. These processes may be static or dynamic, and process models may be used both inductively and deductively to search for, and test, theories and hypotheses about geographic events. Because of the complexity of environmental and social systems there are only a few geographic laws of nature. This can frustrate our attempts to model the complexities of the real world, to explain how things work, and to make predictions about future outcomes.

Core to the remit of Geographic Information Systems (GIS) is our ability to create and share representations of the world that often lie beyond our direct experience. But how do we know that a shared representation of the world is fit for the purpose for which it has been designed or applied? This question may in turn invite questioning of the provenance of the data that have been used to create the representation, the ways in which diverse data sets have been structured and assembled together, or the decision-making context in which an application is developed. Other contributions to this manual have addressed the rapid development of geographic information (GI) technologies, developments in software and organizational work flow, changes to the data economy of GIS, and broadening of the range of settings in which basic GIS skills may be acquired (see McDonnell and Burrough, Chapter 2, this volume; see DeMers, Chapter 4, this volume; Longley et al. 2005b). Implicit in these various discussions is the notion that all but the most basic concepts in GIS—such as distance measurement, topology, proximity and connectivity—are fast changing, and hence transitory. Yet this is not the full story, for the apparent rapid evolution of GI software conceals rather a lot that is enduring about the application of GIS, in terms of core organizing principles and techniques.

This is, of course, widely recognized, and many of the enduring characteristics of GIS receive attention in this manual, alongside discussion of what is new. Our own objective here is to introduce the remit of GIScience as weaving many of these different threads into a distinctive area of research and applications activity. The context for this is disarmingly simple: if the real world is of seemingly infinite complexity and the process of representation requires us selectively to discard nonessentials, then the field of GIScience is necessary to guide us in this task, and to suggest ways in which the quality of the resulting partial representations may be evaluated. This requires us to move beyond basic principles and routine applications, to "the science behind GIS" (Goodchild 1992). The developing field of GIScience provides much of the direction, growth and impact of GIS and related technologies. If information science studies the fundamental issues arising from the creation, handling, storage, and use of information, then the remit of GIScience may be defined as study of the fundamental issues arising from geographic information, as a well-defined class of information in general. Other terms have much the same meaning: *geomatics* and *geoinformatics*, *spatial information science*, or *geoinformation engineering*. Each of these terms suggests a scientific approach to the fundamental issues raised by the use of GIS and related technologies, though they all have different disciplinary roots and emphasize different ways of thinking about problems.

Invocation of the term "GIScience" implicitly differentiates between straightforward spatial query operations (such as those involving distance measurement, overlay analysis, buffering, optimal routing and neighborhood analysis) and the much broader and generalized range of transformations, manipulations and techniques that form the bedrock to the field that is often described as *spatial analysis* (see Longley, Chapter 30, this volume). The term is also used to describe the use of GIS in support of scientific research, in the social or environmental sciences, in accordance with the norms and practices of science. Furthermore, discussion of GIScience also requires that we consider the scientific activities associated with the field, its developing organizational infrastructure, and the types of uses to which it may be put.

Insofar as we are concerned with the successful application of GIScience, the social, organizational/management and even political contexts in which decisions are made also need to be considered. This is often apparent when considering the big challenges facing humankind—such as climate change, natural hazards like tsunamis, major migrations, or global terrorism, to name but a few. For example, the precise location of the December 2005 tsunami off Indonesia was predicted in a published map in 1979, yet no warning systems were installed or other preparations made.

3.2 Core Organizing Principles and Remit of GIS and GIScience

GIS helps us to manage what we know about the world—to hold it in forms that allow us to organize and store, access and retrieve, manipulate and synthesize, visualize and communicate spatial data. Many of the routine operations that are core to GIS implicitly invoke hypotheses with respect to the data and information that are available—in the spirit of deductive reasoning. Induction also plays an (increasingly) important part in GIS-based analysis, whereby data "mining" is used to identify what to leave in (and, hence, what to take out of) a representation, and what weight to assign that which is left in. Yet these complementary procedures of induction and deduction can often raise questions that are at the same time frustrating and profound. For example, how does a GIS user know that the results obtained are accurate? How might the quality of the input data be ascertained, with respect to other validatory sources that might be available to us? How can we be sure that the visual medium of a GIS does not obscure the underlying messages of a representation? What principles might help a GIS user to design better maps? How can GIS be fine-tuned to assimilate the limits of human perception, cognition and decision making? Some of these are questions of GIS design, and others are about GIS data, methods and system design. GIScience takes this a stage further, towards developing models that improve our understanding of the effects of underlying processes upon unique places. They all arise from practical use of GIS, but also relate to its core underlying principle and techniques.

A good starting point to develop an empirical understanding of the remit of GIScience is the "First Law of Geography," often attributed to geographer Waldo Tobler. This can be succinctly stated as "everything is related to everything else, but near things are more related than distant things." This statement of geographical regularity is key to understanding how events and occurrences are structured over space, and presents an empirical agenda for GIScience. It can be formally measured as the property of *spatial autocorrelation* (Longley et al. 2005a) which, along with the property of *temporal autocorrelation* ("the past is the key to the present") established as the uniformitarianism principle by Hutton over 200 years ago, suggests a fundamental geographic premise: the geographical context of past events and occurrences can be used to predict the future. However, the observable regularity in spatial and temporal structure that allows us to generalize and predict—on the basis of observed distributions of events and occurrences—falls far short of the degree of regularity worthy of being described as laws of GIScience. To take a building metaphor, in attempting to explain evolving geographic phenomena, no representation is ever founded upon rock; but rather, the structures of our representations are erected upon piles that have been sunk sufficiently deeply into the mud to sustain them. Thus a representation of the local market for a grocery store may reduce the complexity of human decision making to distance minimization in ways that appear crass, simplistic or primitive, but the point is that such simplification may be both necessary and sufficient to devise usable forecasts of store turnover. From this perspective, "good GIS" is about retaining the most significant spatial and temporal events, and discarding detail that may be considered to be irrelevant for practical purposes. What we know already about the problem (our substantive expertise, such as how far people are prepared to travel to buy groceries, or how grocery shopping fits into daily activity patterns) may be used deductively to frame the problem, and principles from GIScience help guide us towards identifying the precise attenuating effect of distance (upon store patronage, for example).

3.3 GIScience Agendas

GIScience has evolved significantly in recent years—some idea of the range of current interests in the field can be gained from the program of the GIScience biannual conference series (Longley 2005a). Efforts to enumerate the constituent issues of GIScience began with the US National Center for Geographic Information and Analysis in the early 1990s and the twenty systematic research initiatives that it spawned. Since then the University Consortium for Geographic Information Science has assumed responsibility for developing an extensive research agenda (McMaster and Usery 2005), which has been modified over time to keep up with a changing and expanding set of issues. There are currently (October 2006) thirteen long-term issues:

- Spatial Data Acquisition and Integration
- Cognition of Geographic Information
- Scale
- Extensions to Geographic Representations
- Spatial Analysis and Modeling in a GIS Environment
- Uncertainty in Geographic Data and GIS-Based Analysis
- Future of the Spatial Information Infrastructure
- Distributed and Mobile Computing

- GIS and Society: Interrelation, Integration, and Transformation
- Geographic Visualization
- Ontological Foundations for Geographic Information Science
- Remotely Acquired Data and Information in GIScience
- Geospatial Data Mining and Knowledge Discovery.

A different, and disarmingly simple, way of viewing the remit of GIScience is provided by the Varenius project; for a summary see Longley et al. 2005a, Chapter 1. Here, GIScience is viewed as anchored by three concepts— the individual, the computer, and society. These form the vertices of a triangle, and GIScience lies at its core. The various terms that are used to describe GIScience activity (such as those enumerated in the previous paragraph) can be thought of as populating this triangle. Thus research about the individual is dominated by cognitive science, with its concern for understanding of spatial concepts, learning and reasoning about geographic data, and interaction with the computer. Research about the computer is dominated by issues of representation, the adaptation of new technologies, computation, and visualization. And finally, research about society addresses issues of impacts and societal context.

It is possible to identify a number of themes that link the different taxonomies and characterizations of the field. First, it follows that if we select what to leave in and what to take out of a representation, then any representation is necessarily incomplete and hence presents an inherently *uncertain* view of the world. Ongoing research in GIScience is attempting to understand the outcomes of choice, convention and chance upon what a representation does *not* reveal about the world. Formal theories have been developed based in the frameworks of geostatistics and spatial statistics, implementing many ideas of geometric probability. Techniques have been devised for simulating uncertainty in data, and for propagating uncertainty through GIS operations in order to provide confidence limits on results.

Second, research in GIScience is having profound implications for "low order" as well as "high order" geographic concepts. GIScientists have attempted to write a formal theory of geographic information, replacing the somewhat intuitive and informal world of raster and vector data and topological relationships that existed prior to the 1990s. Formal theories of topological relationships between geographic objects have been developed and GIScientists have formalized the fundamental distinction between object-based and field-based conceptualizations of geographic reality. The narrow and pragmatic view of GIS has hitherto been of software dominated by either raster or vector data structures, in turn reflecting the analytic roots to particular suites of software (McDonnell and Burrough, Chapter 2, this volume). Yet these new ideas are now becoming embedded in the standards and specifications promulgated by the International Organization for Standardization (ISO) and the Open Geospatial Consortium (OGC).

Third, GIScientists have investigated the social contexts in which GIS are applied, and the ways in which the technology both empowers and marginalizes. This work was stimulated in the early 1990s by a series of critiques of GIS from social theorists, following which it became clear that the broader social impacts of the technology were an important subject of investigation. Critics have worked alongside GIScientists to develop insider views of the ways in which GIS may be acquired and manipulated by the powerful, sometimes at the expense of the powerless. Active research communities in GIS and Society and Public-Participation GIS attest to the compelling nature of these arguments. Such research, especially of both closely and loosely coupled interactions between human activities and natural events, is of course also being carried out by many others. In the extreme case, terrorist actions designed to undermine financial systems would have a downstream effect on the capacity of some nations to sustain and protect their citizenry—the first duty of the state. Perhaps surprisingly, GIScience can be embedded successfully in modeling to assess probabilities and amelioration options.

From a broader perspective, developments in GIScience can be defined through their relationship to other, larger disciplines. Information science studies the nature and use of information, and in this context GIScience represents the study of a particular type of information. In principle all geographic information links location on the Earth's surface to one or more properties, and as such the field of GIScience is particularly well-defined. For this reason, many have argued that geographic information provides a particularly suitable testbed for many broader issues in information science. For example, the development of spatial data infrastructure in many countries has advanced to the point where its arrangements can serve as a model for other types of data infrastructure. Metadata standards, geoportal technology, and other mechanisms for facilitating the sharing of geographic data are sophisticated, when compared to similar arrangements in other domains.

3.4 Conclusion

At its core, GIS is concerned with the development and transparent application of the explicitly spatial organizing principles and techniques of GIScience, in the context of appropriate management practices. GIS is a practical problem-solving tool that is judged by the "success" of its applications. The spatial dimension to problem solving is special because it poses a number of unique, complex and difficult challenges that are investigated and researched through GIScience. Together, these provide a conduit for committed applications specialists to pursue their interests through vocation, in academic, industrial, and public service settings alike. All sciences have their tools and systems: astronomers use telescopes to view stars and information systems to record their characteristics; biologists use electron microscopes to visualize the structure of cells and supercomputers to simulate ecological systems; and computer scientists develop new computer architectures using design software. GIScientists also have their tools—geographic information systems—which are a fundamental and integral part of pursuing GIScience. These tools are applied across the whole arena of the natural and social sciences to enhance our understanding and explore better ways of making the world a more sustainable place.

References

DeMers, M. N. 2008. A hitchhiker's guide to geographic information systems: A multidisciplinary introduction. In *Manual of Geographic Information Systems* , edited by M. Madden, Chapter 4 in this volume. Bethesda, Md.: ASPRS.

Goodchild, M. F. 1992. Geographical information science. *International Journal of Geographical Information Systems* 6:31–45.

Longley, P. A. 2008. Spatial analysis and modeling. In *Manual of Geographic Information Systems* , edited by M. Madden, Chapter 30 in this volume. Bethesda, Md.: ASPRS.

Longley, P. A., M. F. Goodchild, D. J. Maguire, and D. W. Rhind. 2005a. *Geographic Information Systems and Science*. 2d ed. Chichester: John Wiley & Sons.

———. 2005b. *Geographical Information Systems: Principles, Techniques, Management and Applications*. 2d ed, abridged. Hoboken, New Jersey: John Wiley & Sons.

McDonnell, R. and P. Burrough. 2008. Principles of geographic information systems. In *Manual of Geographic Information Systems* , edited by M. Madden, Chapter 2 in this volume. Bethesda, Md.: ASPRS.

McMaster, R. B. and E. L. Usery, eds. 2005. A *Research Agenda for Geographic Information Science*. New York: Taylor & Francis.

CHAPTER 4

A Hitchhiker's Guide to Geographic Information Systems: A Multi-disciplinary Introduction

Michael N. DeMers

4.1 Introduction

Geographic Information Systems (GISs) have been around in one form or another since the early 1960s. As with many applied technologies, their origins are not always easily defined and they possess an inherent ability to adapt to new conditions, new uses, and disciplinary hybridization. Although geographers often point out the fundamental geographic nature of GIS (DeMers 2008; Dobson 1991), without computer science GIS would not exist (Marble 1999), and in the absence of mathematics, cartography—itself a keystone component of GIS—would still be in its infancy (Resnikoff and Wells 1973). Evolving in tandem with other geospatial technologies such as remote sensing, cartography, spatial statistics, and spatial analysis, the hybrid nature of GIS exhibits itself through often nearly identical algorithms—so much so that the borders between the contributing disciplines have been permanently blurred (Fisher and Lindenberg 1989). More recently, practitioners of other disciplines attempting to apply the technology to their own problem sets have had a substantial, often profound impact on the analytical capabilities of GIS. Such disciplines include, but are not limited to, landscape ecology (McGarigal and Marks 1995), criminal justice and crime mapping (National Law Enforcement and Corrections Technology Center 2005, 2007), epidemiology (Graham et al. 2004), geology (Ray 2004) and many more. This list is quite large and is growing very rapidly.

This hybridization has resulted in the evolution of Geographic Information Systems from a predominantly application-based orientation to a more disciplinary focus in its own right—now commonly referred to as Geographic Information Science. In a sense this means that we have all become hitchhikers on the GIS highways and byways—hence my choice of title for this chapter. In my twenty-plus years of teaching GIS I have found that increasingly those entering the university classroom to learn GIS lack many of the fundamental skills and much of the disciplinary background necessary to quickly succeed in GIS. This problem is becoming more severe as more disciplinary specialists already in the workforce are becoming aware of the utility of this empowering technology and are expected to become instant experts in GIS. In fact there are currently so many paths into GIS it is becoming nearly impossible to decide how or where to begin. Duane Marble recognized the need for a multi-path entry into GIS training and education (Marble 1998) where students at different levels and with widely varying disciplinary backgrounds could enter the GIS academic arena with as little delay as possible. His efforts culminated in *The Strawman Report* (UCGIS 2003) that later resulted in the final definition of the *GIS&T Body of Knowledge* (DiBiase et al. 2006). It has not, as yet, yielded a multi-path strategy for GIS education, but the foundation for such a strategy has been laid. This chapter serves as a preliminary survival guide for those attempting this arduous task prior to the formal definition of the discipline or subsequent implementation of these educational strategies.

4.2 What Is a GIS and What Does It Do?

A GIS is a software package, its database, people, institutions, social and legal contexts and related suppliers and vendors all designed to provide an environment for solving geographical problems. While this definition seems unnecessarily broad, encompassing many aspects not always thought relevant to the immediate problems at hand, such a wide-ranging definition is becoming increasingly relevant as the technological components of GIS permeate society. Beyond the software system that drives the algorithms, GIS often requires massive amounts of data, much of which must be converted from analog to digital form, people to operate the software and perform analysis on the data, an institutional setting that promotes and organizes its use, social and legal contexts that determine its appropriateness and utility, and suppliers and vendors who provide software, hardware, data, user support, and graphical user interfaces to keep the systems functioning. It consists then of both internal and external components (Figure 4-1).

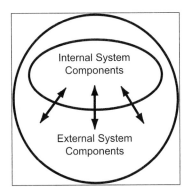

Figure 4-1 Internal versus external GIS system components. Notice that the flows between internal and external components are in both directions.

While the non-analytical operations of the GIS are important, most users obtain GIS software with the explicit intention of using its analytical power to help them with the archiving, retrieval, analysis, visualization, and output of data related to their own disciplines. Yet despite this desire to employ the power of GIS many neophytes are unaware of the capabilities of the software. Below is a list of the general categories of data manipulation and analytical tasks for which GIS software is designed—and can be used once the data are input (Table 4-1). For a general description of when the primary analytical components of the software are most appropriately applied, Andy Mitchell (1999, 2005) provides a nice, recipe-style approach to analysis in his two books.

It is commonplace that initial attempts at analysis often leave the GIS practitioner with a sense of something substantial missing from one's educational background. This feeling is often not so much a limited knowledge of the GIS software as it is a lack of experience in thinking about problems spatially. Industry leaders continue to tell me that they can teach people how to use the software. They view this as a necessary investment. What they are unwilling to do is to educate employees in the recognition of spatial problems, identification of spatial components of problems, and application of the technology to the solution of the problems.

Table 4-1 Typical data manipulation and analysis techniques available in most high end GIS software, with short descriptions.

TASK	SHORT DESCRIPTION
Registration and rectification	Creation of geodetic framework. Assures that all map locations are reasonably accurate, are co-registered, and are registered to real Earth coordinates
Query and retrieval	Boolean, mathematical and logical searches for maps based on map types or internal map attributes
Measurement	Simple and complex measures of absolute and relative length, size, area and buffers
Pattern analysis	Descriptions of the absolute and relative locations of objects in geographic space
Classification and reclassification	Nominal, range-graded, and numerical assignments and reassignments based on attributes
Network analysis and route finding	Shortest, fastest, and least cost paths along networks. Includes the concepts of impedance and gravity models.
Location and allocation	Finding best location for market share and allocating portions of networks based on population and other linked attributes
Overlay	Logical, mathematical and selective map overlay operations
Neighborhood analysis	Description of local sites based on relative position, condition, and other measures of nearby geographic space
Shape analysis	Euclidean and fractal geometries, integrity measures, texture, and sinuosity
Geostatistical analysis	Spatial statistics, including nearest neighbor analysis, mean vector
Surface analysis	Interpolation, hydrological and volumetric analysis, view-shed, sound-shed, travel-shed, texture, and other analyses of statistical surfaces (especially physical surfaces)

Beyond this ability to recognize and solve spatial problems are the multifaceted aspects of GIS that often go unnoticed. Many who are new to the business take the analytical power of the GIS software package for granted, without considering its complexity, the need for basic spatial cognition and computational skills, the multiple data structures and models necessary for analysis, the technological impact on organizations, legal aspects of the output from analysis, and ethical considerations revolving around the collection and use of geospatial data. These topics make up the larger body of knowledge (Table 4-2) currently being defined by a University Consortium for Geographic Information Science (UCGIS) Curriculum Taskforce.

Table 4-2 The University Consortium for Geographic Information and Science (UCGIS) *GIS&T Body of Knowledge* topical list (DiBiase et al. 2006). Although a working document subject to change, this does outline the overall topics covered by the emerging science of geographic information.

CS. Conceptualization of space
CS1 Characteristics of space
CS2 Spatial thinking
CS3 Field–based vs. object–based views of geographic space
CS4 Spatial relationships

FS. Formalizing spatial conceptions
FS1 Effects of scale
FS2 Data modeling
FS3 Representation of inexact information

SM. Spatial data models and data structures
SM1 Basic storage and retrieval structures
SM2 DBMS & the relational model
SM3 Tessellation data models
SM4 Vector data models
SM5 Multiple scale representation/models
SM6 Object–based models
SM7 Temporal representation/models
SM8 Query operations & query languages
SM9 Metadata
SM10 Data exchange & interoperability

DE. Design aspects of GI S&T
DE1 Scientific modeling in a spatial context
DE2 GI S&T applications: I – Conceptual system design
DE3 GI S&T applications: II – System implementation design

DA. Spatial data acquisition, sources and standards
DA1 Remote sensing
DA2 Field data collection
DA3 Sample design
DA4 Data quality
DA5 Surveying
DA6 Photogrammetry

DM. Spatial data manipulation
DM1 Data format conversions
DM2 Generalization and aggregation
DM3 Transaction management of spatio–temporal data

EA. Exploratory spatial data analysis
EA1 GIS analytic functionality
EA2 Descriptive spatial statistics
EA3 Scientific visualization
EA4 Data mining

CA. Confirmatory spatial data analysis
CA1 Spatial statistics
CA2 Geostatistics
CA3 Spatial econometrics
CA4 Analysis of surfaces
CA5 Transportation modeling and operations research
CA6 Simulation and dynamic spatial modeling

CG. Computational geography (geocomputation)
CG1 Uncertainty
CG2 Computational aspects of dynamic spatial modeling and neurocomputing
CG3 Fuzzy sets
CG4 Genetic algorithms and agent–based models

CV. Cartography and visualization
CV1 Conceptualizing spatial visualizations and presentations
CV2 Building spatial visualizations and presentations
CV3 Evaluating spatial visualizations and presentations

OI. Organizational and institutional aspects of GI S&T
OI1 Managing GIS operations and infrastructure
OI2 Organizational structures and procedures
OI3 GI S&T workforce themes
OI4 Institutional aspects

PS. Professional, social, and legal aspects of GI S&T
PS1 Aspects of information and law
PS2 Public policy aspects of geospatial information
PS3 Economic aspects of geographic information science & technology
PS4 Legal and ethical responsibility for the generation and use of information
PS5 Control of information – Information as property
PS6 Control of information – Dissemination of information

4.3 Educational Opportunities and Approaches

No single source of GIS education or training will suffice for such a large and diverse discipline. Moreover, the choices for education and/or training are largely dependant on the following:

- Background in GIS and related disciplines
- Career path
- Desired depth and breadth of knowledge
- Stage or point of entry into the career path

Students who take an introductory level GIS course at a community college or university may find themselves surprised by both the depth and breadth of material they are expected to absorb. This is often as true of those planning to teach community college GIS technology courses as it is among first time students in community colleges. Beyond the classroom experience, they are often overwhelmed when they go to their first GIS conference and observe the wide topical diversity of paper presentations, workshops, and seminars. Fortunately this frequently suggests a need for continuing education and training not just in GIS, but in relevant disciplinary specialties and in the foundational sciences of GIS as well.

This illustrates the importance of educational background to success of the GIS professional at the early career stage. Among the more important subjects seem to be basic cartographic principles (e.g. map projections, scale, generalization, and cartographic design), basic computer science, and mathematical essentials. For those new to GIS a background in cartography seems reasonable—even obvious. It is less obvious that there is a need for a computer science and mathematics background. This stems from the inherent complexity and power of the software that suggests its analytical comprehensiveness. As GIS software continues to become more modular, more algorithmically and computationally robust, and less integrated, the ability to write new computer code, especially macros and graphical user interfaces, becomes more crucial. This requires the GIS user to become familiar with the mathematics of spatial analysis and the computer techniques and data structures within which the code must be written.

Before writing code it is even necessary that the student be capable of recognizing the analytical limitations of the software, and possess a subsequent understanding of the need for such computational activities. Once disciplinary specialists become comfortable with recognizing and acting on spatial problems, they often formulate questions that stretch the capabilities of GIS. Take for example the development of algorithms designed to analyze directional spacing of discrete events such as crime along a strip mall or the orientation of crime hot-spots (Figure 4-2). While the general nearest-neighbor statistic describes multi-direction spatial clustering, the underlying spatial structure of the strip mall forces events to occur in a linear directionally restricted pattern. Such concepts are often forthcoming from disciplinary specialists. These hitchhikers often benefit from examining volumes of GIS applications in their own knowledge domain. In turn, they contribute by forcing the GIS community to think about certain spatial aspects of problems they don't always encounter in their own application areas. In this case, a new set of algorithms has resulted (Levine 2007). Many new algorithms based on these originals are also finding their way into new versions of existing commercial GIS software.

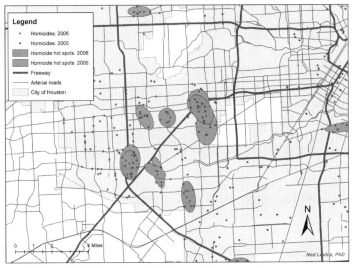

Murder in Southwest Houston: 2005 and 2006
Location of Homicides and Homicide Hot Spots

Figure 4-2 Hot spot analysis showing the orientation of point distributions. Dr. Ned Levine provided this map of homicides for Southwestern Houston, Texas based on crime data provided by the Houston Police Department. This map shows how point distributions can have orientations, but also that they can change through time. Mark Gahegan produced this map as part of a GIS design class. See included DVD for color version.

Two additional groups of applications specialists that have contributed substantially to the conceptual framework of GIS are the landscape ecologists and wildlife researchers. The landscape ecologists have, almost from the development of their field as a science, begun to envision the environment as a matrix filled with patches, edges, lines, and a host of other environmental shapes. Much of their interest is in the description and analysis of these patterns, the processes that created them, and the impact such patterns have on ecological function. Such spatial thinking has resulted in a new set of spatial statistical techniques encapsulated in the set of public domain algorithms known as Fragstats (McGarigal and Marks 1995). These algorithms introduced metrics for analysis of patch density, patch size and variability, edge quantification, shape evaluation, core area determination, numerous patch-related nearest-neighbor statistics, diversity, contagion and interspersion. None of these was originally envisioned in most commercial GIS products. Today these algorithms are available as non-commercial software running on UNIX-based machines as well as PCs. A simple version of Fragstats that provides a simple approach to such analysis is called Patch Analyst. This package is a free download, includes many of the key spatial metrics, and interacts with commercial GIS software as well (Figure 4-3).

Wildlife scientists had a different set of spatial problems to deal with, many of which involved the analysis of non-static point data. Telemetry devices attached to wild animals enabled researchers to track the animals, but few algorithms existed to analyze the spatial and temporal nature of point objects that both occupy areas and move in geographic space. Researchers in the US Geological Survey – BRD, Alaska Biological Science Center developed a robust set of statistics and algorithms to analyze the ranges (areas) occupied by wildlife and the paths that they use in their everyday movements (USGS 2002; Hooge and Eichenlaub 2000). These algorithms were subsequently incorporated as extensions into existing professional GIS software.

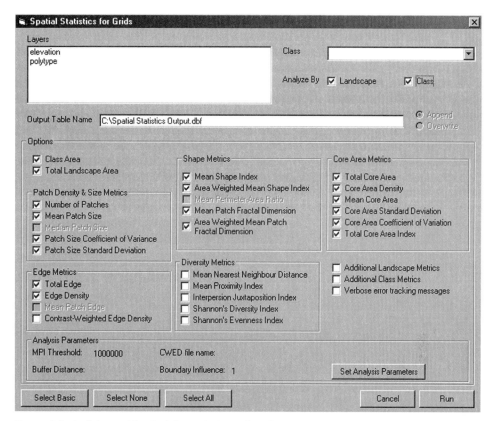

Figure 4-3 A picture of the Patch Analyst user interface showing the selection of spatial statistics metrics available for calculation. Notice how one can select the particular metrics that might be desired by checking the appropriate boxes.

The previous examples are merely a sampling of the contributions made to GIS by researchers outside the traditional GIS community. More importantly, they demonstrate that many GIS hitchhikers already posses many of the skills and tools they need to understand what GIS is about and how it can be applied to their own knowledge domain. Disciplines such as epidemiology, environmental toxicology, geology, anthropology and archaeology, history, military studies and a host of others have all contributed substantially to an increasingly robust set of tools available to the GIS hitchhiker.

Another source of insight—beyond additions to the GIS toolkit from other disciplines—comes from those who go into the environment being modeled to reconcile the available algorithms with how the real world actually functions. Among the classic examples of this approach is the seminal work of Peter Fisher who evaluated various algorithms for determining view-sheds. While showing that different software produced different results, not any of which was totally accurate (Fisher 1991, 1994) he began to question whether any of these algorithms accurately depicted how we actually see our landscape from a particular vantage point (Fisher 1993). The important lesson for the GIS hitchhiker is that there is often no "correct" solution to a given problem, but rather reasonable or defensible versus unreasonable or indefensible solutions.

Whatever path such innovations take, they all lead to an ever increasing complexity of the software. Moreover, they also contribute to the growth of the theoretical underpinnings of the discipline. As this continues, the pressures on GIS professionals to obtain both education and training will also increase. This is just as true for seasoned GIS veterans as for beginners.

Because there are many levels of GIS professional and many different needs, there is an attendant need for multiple, flexible paths of learning.

4.4 Paths to Learning

4.4.1 University

Over the last four decades the university has been the traditional source of education and training for those entering the GIS workforce. Small four year colleges and larger universities with graduate programs continue to offer classes in the theory and the technology of GIS as well as more advanced courses in design, geographic information science, and computational geography. Sources of information on GIS programs at universities around the world can be found at Petrie (Undated) and on the sites of some professional organizations and GIS software vendors.

Once the premier source for such education, universities and colleges are now taking a somewhat less prominent role, especially for the increasing number of busy professionals who do not have the luxury of pursuing a four-year degree. Some colleges and universities are responding to the changing market by offering certificate programs and other short-term, intense programs to accommodate non-traditional students. The number of such programs will certainly increase in response to demand.

4.4.2 Community Colleges

The community colleges have also begun filling in the educational void for those needing to get into the workforce quickly. Geographic information system technology and training programs are appearing wherever there is a market. In some cases these programs are being supported by industry personnel who share their technical experience with the students. Alliances have been established between nearby university programs where the community colleges sometimes act as feeder schools for the university programs—while, alternatively, the university faculty members share their expertise by training the community college faculty in GIS. Some of these programs are generic—teaching the basics of the software and hardware to a common student body. Others are more focused, offering GIS courses aimed at particular topical areas like criminal justice or environmental studies. The number of community colleges offering Geographic Information Science and Technology (GIS&T) is growing rapidly and more articulation agreements are being proposed, all providing increasing opportunities for access to education and training for the non-traditional student. Such programs are often quite helpful for non-GIS professionals needing to add the toolkit to their existing knowledge.

4.4.3 K-12

There are opportunities for younger students in some regions where vendors and elementary and secondary teachers are joining forces to teach many of the basic principles of GIS analysis in both analog and digital environments (DeMers 1999). The primary limitations for these programs seem to be lack of educators knowledgeable about the concepts of spatial analysis and GIS, and technological limitations, especially a lack of computers for upper level students who actually use the software. As computers continue to decrease in price, this suggests a need to develop teaching institutes for elementary and secondary students to learn the fundamentals of GIS. Basic lists of K-12 teacher resources can be found online (see for example ESRI undated; Berrien County Intermediate School District undated). GIS software vendors often have specific K-12 workshops, paper sessions, and discussions, all aimed at disseminating the technology into the educational systems. Many also have free datasets, software, and sample exercises for educators. Eventually there will likely be a large portion of the GIS practitioner workforce whose exposure to the technology and its applications has occurred well before they are of college age.

4.4.4 Vendor Software Training

Efforts to guarantee that users continue to purchase their software and employ their services include software training provided by the GIS software vendors. These training opportunities are not generally geared toward geographic information science, theory, or design, but some are. Most focus either on how to use the software or on the appropriate and efficient applications for particular industries. Such training is made available at vendor headquarters, at regional centers, or even at the client's place of work. The advantage of these sessions is that they can be focused on specific techniques or software systems. Their major shortcoming is that they are not always sequenced for optimum learning for all participants, many of whom have wildly different levels of knowledge and experience. This is particularly true when attendees take courses out of sequence due to timing difficulties.

4.4.5 Workshops and Seminars (Including Webinars)

Workshops and seminars at professional meetings include training opportunities provided both by industry trainers and subject-matter application experts. Professional, academic, and vendor meetings frequently provide both formal and informal opportunities for learning. Additionally, vendors frequently provide opportunities for specific problem solving on a one-on-one basis. While not technically part of the educational infrastructure, the vendors are the most knowledgeable about their own software, and are also able to provide reference and reading material that the student can take home after the training.

When the technology permits, web-based seminars (webinars) provide opportunities for synchronous distance learning of software applications and conceptual material. Audio questions and answers, video images, shared software, and text-based interactions are all possible with a variety of web-enabling software such as CENTRA, WebEX, GoToWebinar, Windows Messenger, Netbriefing, Yahoo, and a host of others. Each package carries its own advantages and disadvantages beyond the scope of this discussion. In general, the following are factors that should be considered when selecting the appropriate software for webinars:

- Can the participants share their desktops?
- Can the software integrate e-mail and instant messaging software?
- Does the provider guarantee 24/7 availability?
- Is this the most cost-effective method?
- Does the software support audio and video?
- Is the software easy to use?
- Is the software easy to install?

4.4.6 e-Learning

While webinars are effective for occasional short-term educational opportunities, they may not be applicable to long-term learning venues. The software used for webinars, however, is readily applied to an approach to education called e-learning. Vendors, community colleges, four year colleges and universities, as well as education consultants are beginning to embrace the use of the internet for both synchronous (webinar-like) and asynchronous education (e.g. ESRI's virtual university). Generally this form of education is far cheaper in human resources for educational institutions, readily available for those who cannot travel to training sites, and often results in highly structured and well-thought-out courses (Painho et. al. 2002). GIS vendors have been involved in e-learning for years, providing on-line classes, completion certificates, searchable library resources and more. Even traditional universities are investing substantial resources in this form of education through distance education programs (Pennsylvania State University 2008; Regents of the University of California 2008). Some of these offer actual degree programs—others provide certificates (Barnhart 1997). Many are approved by the Veterans Administration and other agencies that offer grants and loan opportunities for retraining.

4.4.7 Internships and On-the-Job Training

A more commonplace method of learning both software and GIS concepts is through student internships and on-the-job training. Within the US government there are numerous internship programs that allow students to learn GIS-related technologies and, upon completion of their degrees, obtain employment with the agency. These programs are often supported by the universities, but are initiated by the agency. Some universities have found it useful to formalize and integrate the internship within the existing university programs. An excellent example of the potential success of such programs comes from Arizona State University (Wentz and Trapido-Lurie 2001). The internship has the advantage of providing immediate, as needed, on demand skills for the employer at reduced or no cost. The result is a trained employee that not only has the necessary skills, but also knows how those skills are applied to the organization's programs. Some GIS businesses themselves offer internship programs that are not necessarily linked to universities, but can be. As with government agencies, the result is trained employees.

The University of Redlands has taken a different route to practical training. Taking advantage of its proximity to a major GIS vendor, many of its educators are themselves GIS professionals. This essentially takes the internship directly to the student because these instructors know exactly what is necessary.

4.4.8 Self Learning

Not the easiest approach to learning GIS is that of obtaining the software, texts, and supporting materials and teaching oneself. This approach is most often taken by busy professionals who need to learn GIS while employed. It has the merits of being self-paced and focused on the apparent educational needs. Where this approach often fails, however, is that it lacks any formal structure to the learning process, often providing the learning in a haphazard and asynchronous manner. Still, for those who are aware of the skills needed for their own industry, and have the drive to pursue the necessary training, it can be effective. Those who have proven successful taking this approach often reach out to the GIS industry through some of the other educational opportunities already discussed. This is especially true of the use of on-line courses and training.

4.5 Modeling: General Considerations

Joseph Berry and his Ph.D. student, C. Dana Tomlin (1983) coined the term cartographic model to describe the process of using the GIS to systematically solve geospatial problems. Tomlin refined and finally defined the term as "A set of ordered map operations that act on raw data (plus derived and intermediate map data) to simulate a spatial decision-making process" (Tomlin 1990). As this suggests, the cartographic modeling process is cyclical. That means that one moves through the four GIS subsystems—input, storage and retrieval, analysis, output —iteratively until a final model solution is achieved. There are so many commands in modern GIS packages that deciding which of these to use and when to apply them is often very confusing. We must first recognize that all models, including cartographic models, can be broken down into more elemental sub-models, each of which can subsequently be linked in an appropriate sequence and through correct analytical processes to simulate environmental or socio-economic conditions. Thus, depending on the conditions of our model and its relevant circumstances we appropriately eliminate commands that do not contribute to our solution.

4.5.1 Modeling: The Importance of Domain Knowledge

Among the more important prerequisites to effective GIS modeling is a thorough knowledge of the subject matter to which the software's algorithms are to be applied. No matter what you are trying to model, or how you are trying to make decisions, it is not enough to know the particulars of your GIS software. Take for example, the landscape ecologist who is attempting to model the flow of plant propagules through a landscape composed of several interactive landscape structures, such as grassy areas interspersed with tree-lines (with holes), on the edge of a ranch, with row crops, etc. She must know the types of plant propagules, how they move, the affects of entrainment, wind speed and turbulence effects, etc. While the modeling of flows has some similarities among modeling domains such as hydrology, geomorphology, toxicology, and even fire modeling, knowing the specifics of your own modeling environment allows you to break the model down into its constituent parts far more easily than the broadly trained GIS modeler.

Decomposing your GIS model into smaller components does more than simplify the task of cartographic modeling: it also forces you to think about the components that make up the system you are modeling as well. By doing so you can also begin to weigh the relative importance of components and individual elements based on your knowledge of how the system works in the real world. It also forces all of us to reconsider preconceptions of our model environment and to expose portions of our system that we know little about. In this way the model helps us understand our system and the system teaches us how to build the model. Another important aspect of decomposing our models is that, as each component is identified, we also must determine what types and levels of data we need to build our models. This is a preferred method as opposed to identifying data first, then trying to build our model around that (DeMers 1999).

4.5.2 Modeling: The Importance of Flowcharting

As we begin the process of model building and identifying its relevant components and data needs, it is often difficult to keep track of what we are doing. Most modern GIS software contains some form of modeling language that allows the operator to sequence the operations and map layers contained within the GIS to simulate some form of spatial decision-making process. The command line approach, still available for those who choose to use it, has, for the most part, been replaced by more graphical approaches. Today, the more advanced GIS software uses a graphical user interface (GUI) that resembles most others except that the map portions of the GUI allows the user to populate these objects with real map data and to run a complete model with the touch of a single icon. This also allows for rapid changes, incremental model development and iterative model implementation as well.

The primary problem with the modeling languages, even those with strong flowcharting-based GUIs, is that the flowcharts are often far better designed for model implementation that for model formulation. A formulation flowchart can be as simple or as complex as you require, can be broken down into a series of easy-to-understand submodels, and need only be reversed using software-specific flowcharting programs to implement. It is best to do a formulation flowchart first, working with the assumption that all the data elements (individual maps) that you need will be readily available. Then, once you have completed the formulation flowchart you can identify all the necessary data elements, then quickly determine which of these data elements are not available and will need to be obtained, ignored, or have surrogates put in their place.

4.6 Careers and Job Opportunities

The opportunities for GIS positions in all sectors—government, private industry, military, non-government agencies, universities, etc. —are increasing as rapidly as the discipline itself. The types of positions are also changing as GIS moves from integrated applications to distributed, object-oriented toolkits with more specialized applications where only the necessary tools are employed for a particular problem. A brief anecdotal sample of on-line jobs databases further illustrates that, at least in the United States, the distribution of these jobs seems to be relatively uniform when compared to population. This is not necessarily the case worldwide, but as educational opportunities spread across the globe, the opportunities for international employment increase. Unfortunately, like many other technical jobs US geospatial positions are being shipped overseas (Leighton 2004). This is partly due to a lack of trained professionals. The community colleges are undoubtedly going to play a major role in filling this void, providing professionals with needed skills.

4.7 Credentials: Certificate versus Degree

Among the more contentious issues surrounding the Geographic Information Science and Technology disciplines are those revolving around the awarding of credentials, certificates, documents of completion, course credit, degrees, minors, and the myriad other forms of acknowledgement of a level of achievement. Two major professional certification programs are available for the GIS professional. The American Society for Photogrammetry and Remote Sensing (ASPRS) maintains certification processes for the following:

- Certified Photogrammetrist
- Certified Mapping Scientist – Remote Sensing
- Certified Mapping Scientist – GIS/LIS
- Certified Photogrammetric Technologist
- Certified Remote Sensing Technologist
- Certified GIS/LIS Technologist

Each of these has specific requirements; all rely on a combination of demonstrated professional competence, years of experience, and education (ASPRS 2008). The Urban and Regional Information Systems Association (URISA) is a founding member of the newly developed GIS Certification Institute. Those having the necessary credentials—again a combination of competence, experience, and education—receive a certification as a GIS Professional (GISP). The surveying professionals have their own certification process, as do geodetic scientists and engineers, and many others. There are, no doubt, additional efforts going on throughout the United States and around the world, and this writing is not meant to recommend one over another.

In addition to the professional certification process, community colleges and universities are providing a growing number of certificates. To date there is no control over what skills and knowledge these certificates entail or regarding the credentials of the instructors or programs themselves. This latter issue is part of the difficulty that employers face as they attempt to hire GIS professionals with very specific skills.

4.8 Note to Employers

Over the past two decades I have had many occasions to interact with potential employers of my own students through telephone interviews, face-to-face meetings, job advertisements, letters of reference and the like. I have seen employers request skills more akin to traditional computer scientists and engineers, GIS jobs mistakenly advertised as GPS surveying positions, qualifications requiring six years of experience with software that was only three years old, and salary offers tens of thousands of dollars below those offered for far less complex skills. I've seen industries who would not hire geographers because they "didn't know computer science" and computer scientists not receive offers because they "didn't know geography." The position titles themselves represent a cacophony of misunderstanding including such titles as GIS hardware network specialist, GIS analyst, GIS program specialist, GIS expert, and many more. Huxhold (2006) has grouped these into six general categories: managers, coordinators, programmers, modelers, specialists, technicians. For years the discipline progressed without any formal method of legitimizing its own professionals. The *GIS&T Body of Knowledge* (DiBiase et al. 2006) is the first attempt to quantify the content of the discipline with over 1600 specific learning objectives. While this document has limitations and is meant to be a living, changing document, it represents a major step forward in defining what skills and competencies GIS professionals should have with a four year university degree specializing in GIS. Moreover, this will allow a formal linkage between needed job skills and university level GIS graduates.

4.9 Software Vendors

In the 1970s, literally hundreds of GIS software packages were available. Many of these were highly experimental—many of these failed. Today, the field of commercial software vendors is far more limited. Yet there is still a selection of software that specializes in data structures (e.g., raster versus vector), focus (e.g. remote sensing, transportation, GIS analysis, etc.). To provide a list would probably reveal my own prejudices and experience. Instead I encourage the reader to use internet search engines and directories to learn about available commercial software. Choosing a GIS software package today is not as complex a process as in the past when packages were limited by data structures, incompatible with existing forms of digital spatial data, tied to specific hardware, or priced beyond potential client's abilities to pay. Today the choice is more often based on preference, functional requirements, or knowledge and training of current employees. Some will choose packages that focus on the manipulation of imagery; others need software that focuses on linear data and networks, while others will choose their software based on compatibility with other data types.

4.10 Data Clearinghouses

No GIS operates without spatial data. Among the most important skills of a GIS data analyst is an ability to locate and identify timely, cost-effective, relevant data for a particular project and study area. For specialized projects, especially those that are small in scope or highly focused in modeling needs, the data must be obtained *in situ* or from existing cartographic documents. There are, however, substantial sources of free data such as those available at the geocommunity website (http://data.geocomm.com/), and subject-specific sources such as the Conservation GIS Portal (http://www.conservationmaps.org/index.jsp). Many US states have substantial spatial datasets that are available for little or nothing through the Federal Geographic Data Clearinghouse search engine http://clearinghouse1.fgdc.gov/. Other sources of inexpensive or free digital geographic data include the geography network (http://www.geographynetwork.com/data/index.html), federal agencies throughout the world, universities, and more. [These four web sites were all accessed 15 May 2008.]

Beyond the free or at-cost spatial data vendors there is a growing industry of commercial data vendors including the GIS software vendors themselves. It is not always easy to locate data even from commercial vendors who wish to sell them. This has led to the creation of a related industry of data detectives whose primary service is to find appropriate digital data sets for GIS projects. These professionals exist because there is no consistent comprehensive index for spatial data, making the identification of data increasingly elusive. Even librarians, who excel at locating elusive information, are now becoming GIS hitchhikers (Reid et al. 2004).

4.11 Conclusions

We are all hitchhikers on the GIS train. While this creates some difficulties deciding how to obtain the necessary training and education, it also promotes cross pollination that has proven very successful at moving both the technology and the theoretical roots of GIS forward. The availability of digital data is a blessing in that it saves time and energy, but its proliferation has in itself produced a new industry of GIS data detectives specializing in finding relevant, cost-effective solutions for GIS analysis. There are fewer software vendors than in the past, but modern GIS software packages are now making it considerably easier to model complex systems because they employ flowcharting software that includes the ability to populate the flowchart components with spatial data. This not only systematizes the problems being modeled, it forces the modeler to compartmentalize the problem. This in turn assists in the identification of potential missing data sets, and permits the recreation of models when new or improved data sets are found. Because we are all hitchhikers on the GIS bandwagon, the future of employment is dependant on some form of rational identification of employee qualifications. This will prevent the hiring of unqualified employees, and will also prevent employers from asking for qualifications that do not exist, or in some cases cannot be attained. Above all, the discipline is becoming far more hybridized by the input and interactions of all of us hitchhikers. This trend is not only highly probable, but welcome. With it, the future of GIS, both as a set of techniques and technologies, and also as an integrative discipline, is assured.

References

ASPRS (American Society for Photogrammetry and Remote Sensing). 2008. ASPRS Certification Program. <http://www.asprs.org/membership/certification/> Accessed 30 June 2008.

Barnhart, P. A. 1997. *The Guide to National Professional Certification Programs,* 2nd edition. Amherst, Mass.: HRD Press.

Berrien County Intermediate School District. Undated. <http://www.remc11.k12.mi.us/bcisd/classres/gis.htm> Accessed 15 May 2008.

DeMers, M. N. 1999. Integrating GIS into the K-12 curriculum. Education West 3 (2):24–25.

———. 2008. *Fundamentals of Geographic Information Systems,* 4th edition. New York: Wiley.

DiBiase, D., M. DeMers, A. Johnson, K. Kemp, A. Taylor Luck, B. Plewe and E. Wentz, editors. 2006. *Geographic Information Science and Technology Body of Knowledge.* Washington, DC: Association of American Geographers. <http://www.aag.org/bok/> Accessed 16 May 2008.

Dobson, J. E. 1991. The G in GIS: Geography is to GIS what physics is to engineering. *GIS World* 4 (1):80–81.

ESRI. Undated. GIS for Schools. <http://www.esri.com/industries/k-12/index.html> Accessed 16 May 2008.

Fisher, P. F. 1991. First experiments in viewshed uncertainty: the accuracy of the viewshed area. *Photogrammetric Engineering and Remote Sensing* 57 (10):1321–1327.

———. 1993. Algorithm and implementation uncertainty in viewshed analysis. *International Journal of Geographical Information Systems* 4:331–347.

————. 1994. Probable and fuzzy models of the viewshed operation. *Innovations in GIS* 1:161–175.

Fisher, P. F. and R. E. Lindenberg. 1989. On distinctions among cartography, remote sensing and geographic information systems. *Photogrammetric Engineering and Remote Sensing* 55 (10):1431–1434.

Graham, A. J., P. M. Atkinson and F. M. Danson. 2004. Spatial analysis for epidemiology. *Acta Tropica* 91 (3):219–225.

Hooge, P. N. and B. Eichenlaub. 2000. Animal movement extension to Arcview. Ver. 2.0. Alaska Science Center - Biological Science Office, U.S. Geological Survey, Anchorage, Alaska, USA.

Huxhold, W., editor. 2006. *Model Job Descriptions for GIS Professionals.* Chicago: Urban and Regional Information Systems Association.

Leighton, J.F.G. 2004. Jobs in GIS: here today, offshore tomorrow? *GIS Educator* Spring 2004:3–4.

Levine, N. 2007. CrimeStat: A Spatial Statistics Program for the Analysis of Crime Incident Locations (v 3.1). Ned Levine & Associates, Houston, TX, and the National Institute of Justice, Washington, DC. March. <http://www.icpsr.umich.edu/NACJD/crimestat.html/about.html> Accessed 23 May 2008.

Marble, D. F. 1998. Urgent need for GIS technical education: rebuilding the top of the pyramid. *ArcNews.* 20 (1):1, 28–29.

————. 1999. Developing a model, multipath curriculum for GIScience. *ArcNews* 21 (2):1, 31.

McGarigal, K. and B. J. Marks. 1995. FRAGSTATS: spatial pattern analysis program for quantifying landscape structure. USDA Forest Service Gen. Tech. Rep. PNW-351.

Mitchell, A. 1999. *The ESRI Guide to GIS Analysis. Vol. 1.* Redlands, Calif.: ESRI Press.

————. 2005. *The ESRI Guide to GIS Analysis. Vol. 2.* Redlands, Calif.: ESRI Press.

National Law Enforcement and Corrections Technology Center. 2005. Guide to using and understanding CMAP CASE. <http://www.crimeanalysts.net/CASE%20Analysts%20Guide.pdf> Accessed 20 May, 2008.

————. 2007. Crime Analysis Spatial Extension (CASE). <http://www.crimeanalysts.net/case.htm>. Accessed 20 May 2008.

Painho, M., P. Cabral, M. Piexato and P. Pires. 2002. E-teaching and GIS: ISEGI-UNL learning experience. Pages 118–128 in *Proceedings, Third European GIS Education Seminar (EUGISES),* held in Gerona, Spain 12-15 September 2002. <http://www.knmi.nl/samenw/cost719/documents/eteach.pdf> Accessed 26 June 2008.

Pennsylvania State University. 2008. Penn State | Online. <http://www.worldcampus.psu.edu/> Accessed 23 May 2008.

Petrie, G. Undated. University & College Depts. & Institutes. In *Web Links Database – Geoinformatics (Covering Surveying, Geodesy, Imaging, Photogrammetry, Remote Sensing, Mapping, Cartography & GIS).* <http://www.weblinks.spakka.net/db/680> Accessed 30 June 2008.

Ray, P.K.C. 2004. GIS in Geosciences: The Recent Trends. <http://www.gisdevelopment.net/application/geology/mineral/geom0012.htm> Accessed 15 May 2008.

Regents of the University of California. 2008. Online Courses in Geographic Information Systems. <http://www.extension.ucr.edu/sciences/geo/online.html> Accessed 23 May 2008.

Reid, J. S., C. Higgins, D. Medychkyj-Scott and A. Robson. 2004. Spatial Data Infrastructures and Digital Libraries: Paths to Convergence. *D-Lib Magazine* 10 (5), May 2004. ISSN 1082-9873. <http://www.dlib.org/dlib/may04/reid/05reid.html> Accessed 15 May 2008.

Resnikoff, H. L. and R. O. Wells, Jr. 1973. *Mathematics in Civilization.* New York: Rinehart and Winston.

Tomlin, C. D. 1983. *A Map Algebra.* In *Proceedings, Harvard Computer Graphics Conference,* held Cambridge, Massachusetts, 31 July – 4 August 1983. Cambridge, Mass.: Harvard.

Tomlin, C. D. 1990. *Cartographic Modeling and GIS.* Upper Saddle River, New Jersey: Prentice Hall.

UCGIS (Task force for the development of model undergraduate curricula). 2003. *Development of Model Undergraduate Curricula for Geographic Information Science & Technology: The Strawman Report.* <http://www.ucgis.org/priorities/education/priorities/final%20strawman%20text.pdf> Accessed 16 May 2008.

USGS. 2002. Alaska Science Center-Biological Science Office: GIS Tools. <http://www.absc.usgs.gov/glba/gistools/#ANIMAL%20MOVEMENT> Accessed 23 May 2008.

Wentz, E. A., and B. Trapido-Lurie, 2001. Structured internships in geographic information science education. Journal of Geography 100 (4):140–144.

CHAPTER 5

The Early Days of GIS
Charles Olson's Interview
with Maury Nyquist

Denver, Colorado, 26 May 2004

In this chapter, ASPRS Fellow Chuck Olson interviews Maury Nyquist, who was ASPRS President in 1994 and is also an ASPRS Fellow.

Olson: How/where/when did you first become involved in Remote Sensing/GIS?

Nyquist: How, where and when did I first become interested in remote sensing and GIS is easy. I started working for the National Park Service (NPS) in 1974. We were in a group called the Branch of Science and Consultative Services. This was when NEPA[1] had just been enacted and the NPS had a "what me" moment and said we couldn't possibly need to do anything with NEPA compliance because "we're the good guys." In fact, they needed to do a lot. The Service at the time was basically being run by architects, engineers, landscape architects and rangers. So there were a lot of designs and plans that probably were not the best for the resource and the mission of the Park Service. To make a long story short, we were doing compliance documents after the planning and designs had already been completed. Finally the "planner and designer types" thought maybe it would be a good idea if they'd include this bunch of natural scientists more fully in the planning and design process at the beginning, instead of at the end.

So, we started working on plans and designs, as well as compliance, and the first thing that hit me square in the eye was there were very limited data. What data that did exist were seldom comprehensive and/or current. There might be a thousand different disparate little studies on whatever charismatic mega-fauna lived in that park, but little about other resources. There might be some observations of plants and animals, check lists of fauna and flora in the parks, but nothing mapped, by and large. And what was mapped was done by USGS[2] for their purposes, or some other organization for theirs. It tended to be crude. It was not organized. It was the days of paper, folks. This was back before the computer revolution. So, about two years into this, I was really frustrated and I started looking around to see what vegetation mapping had been done in the parks. I found that there were only a small percentage of parks that had any type of vegetation map, no matter how crude, incomplete or out-of-date. So I started working with NASA[3]-Earth Resources Lab (ERL) which, at that time, was in Slidell, Louisiana. It is now located at the Stennis Space Center in Mississippi. At that time, the goal for me was first to get mapped information on vegetation for the parks because vegetation was, to me, the critical element. You combine vegetation with topographic variables and one could do numerous types of application. Basically, I was thinking like a GIS, but GIS wasn't "developed" yet, so it was just what you did in your head. I said, wouldn't it be wonderful if we could use the new Landsat satellite which had been launched in 1972. The NASA people were looking for warm bodies to help figure out

---

1. NEPA: The National Environmental Policy Act, 42 U.S.C. 4321 *et seq.*
2. US Geological Survey, part of the Department of Interior.
3. National Aeronautics and Space Administration.

what it was good for. A lot of people had a good idea what it could potentially be good for, but there weren't many "success stories" at that time. So I started working with them and it was a really neat partnership, because I knew virtually nothing about computers and remote sensing at that time. I did understand biology and how things portrayed out on the land, and how these patterns showed up in the data. You also have to remember this was the time of the first display monitors versus line printer outputs and NASA actually had color monitors, because they were designing and writing their own software. The work was being done on large minicomputers, while some of it was running on mainframes.

Anyway, this looked like a real opportunity to do some good for the Park Service. The NASA people were looking to broaden their scope in the use of this information. So, I started working with them, and within about a half a year I had cobbled together something that was called an ASVT[4] project. NASA had already worked a couple of times with the Park Service, but they were kind of failures. It was just the wrong mix of people working, I think, more than anything. It was more the engineers and computer programmers with natural scientists, ecologists like myself, brought in peripherally. So we started working on a new NPS area called Big Thicket. It was probably the worst place to work on this type of experiment, because of the diversity of vegetation. There were four major biomes coming together in a very short geographic proximity—everything from swamps to almost semi-desert grasslands, but most of it was sort of swampy terrain. I got good support from the Superintendent, Tom Lubbert. He had almost no information, and he said, "If the maps are only half right, it's better than the nothing I have now." So we worked on Big Thicket for a couple of years with people like Armond Joyce, Bill Cibula, and many others. Wayne Mooneyhan, who was ERL's Director then, gave us good support. Wayne and I became good friends, as I did with many other people there. Anyway the project turned out to be a success. NASA liked it, the Park Service liked it, so we extended the ASVT because I said, "That could be just pure luck and we've got to try it in different environments."

At that point we broadened the ASVT to work in Death Valley,[5] Shenandoah[6] and Olympic.[7] One of the things that became very apparent at that time, and this was probably now about 1978, was that the need to be able to add other data to the multispectral data, to be able to do modeling was an absolute necessity. I remember talking to Ronnie Pearson, one of NASA's top-notch programmers, about if we could just do modeling, we could really make an improvement in the classifications. So he came up with a program, PCAL (programmable calculator) and we were the first ones to try it out. It really worked phenomenally well on Olympic. We were coming up with a pretty good land-cover map of basic SAF[8] Forest Cover Classes (i.e. the major dominant species) and accuracies were in the upper 80 percents, which was quite good. Shenandoah was not quite as good, because the topographic variables didn't have quite as dramatic an effect on the vegetation as they did in the Pacific Northwest. Death Valley was basically a superficial geology map because there wasn't enough vegetation there to detect. Nevertheless the geologic information was very useful. Anyway, a long story short, the Olympic project was really outstanding at that time. It was one of the first projects that they had been able to achieve such good results. Bill Cibula and I co-published it in *PE&RS*, and it got good reviews and people were quite impressed with it.

So, the whole rationale behind this was not just to be able to use remote sensing but was really to be able to use integrated geospatial technologies as we know them today. At that time the technology was not quite there yet and the mentality wasn't quite there yet either. People were still arguing about raster and vector. The vector people couldn't understand how

---

4. Applications Systems Verification and Transfer.
5. Death Valley National Park is in California and Nevada.
6. Shenandoah National Park is in Virginia.
7. Olympic National Park is in the state of Washington.

you could live with all those coarse cells; and the raster people said well how can you deal with just the outlines of things without all of the variability in between? So there were lots of geospatial technology wars, especially within the Department of the Interior.

About this time, I realized we needed to institutionalize [GIS] in the Park Service. I was finally allowed to hire staff, but there was no way in the world the NPS was going to cough up the million dollars that was needed for the hardware and other requirements, at that time. So we started a very good manual PI[9] group, one of Chuck Olson's students – Susan Stitt; Nancy Thorwardson from Oklahoma State; another one from Michigan, Steve Bracken I think was also one of Olson's students; LouAnn Jacobson out of the University of New Mexico; a couple of Park Service people that wanted to work in this area, Ralph Root and Gary Waggoner – started working for me. So we had a very well-schooled group of people, some with a lot of training. Ralph, for example, had a masters and Ph.D. in geospatial technologies, although most of it was remote sensing because little of the "GIS stuff" was invented yet. Susan, Steve, LouAnn and Nancy were all working on masters degrees and finished up. The bottom-line was that these students were on Graduate Co-op Appointments and I was able to hire them non-competitively after successful completion of their degrees. I hired them, and we had a real outstanding cadre that complemented each other because we were multi-disciplinary and each person added their own strengths to the mix. We worked on a lot of different things in our shop. We also worked with the EROS[10] Data Center, we worked with the EPA[11] lab down in Las Vegas, and …. Tom Mace's shop in Vegas and Don Lauer's group up at EROS Data Center – on a number of things in order to get the processing capability so we could continue on in the digital processing domain. So that was the early vestiges of a Remote Sensing/GIS Program within the Park Service.

I don't remember the exact dates, but part of the deal with the ASVT with NASA was that at the end of it they were going to try to enable, as much as they could, the Park Service to have stand-on-its-own capabilities with equipment and software. We were really imbued with the idea, me especially, of using public-domain software, because software is very expensive and, because of all the operating system differences, you always had to rewrite things. The goal was that they were going to give us a minicomputer, one that they were going to surplus. I made this bet with the manager of the Denver Service Center (DSC) who later became Deputy-Director of the Park Service and was also Acting-Director of the Park Service. So I told the DSC manager that what we needed from the Park Service was to build a facility – because you had to have the special air-conditioned computer rooms with handicapped access. He bet that we would never get it because of the cost and other things. And I said, "I'll bet you a six-pack on it." Well, to make a long story short, we got the computer system, they built the room, and during the annual Christmas Party he made me come into his office and grudgingly gave me the six-pack. While I was there he said, "If I had known of all of the high-level politics going on behind the scenes, I would have never bet." And I said, "Just remember this, Denny, I never bet beer unless it's a sure thing."

So, we got our minicomputer system. I hired more staff, which came and went over the years, but people like Susan and Ralph worked for/with me for over twenty-five years. We had good careers and great times together. At the time of this interview I should mention that I'm retiring within the next week. All of my staff have been taken care of in other groups and will be carrying on their tradition of outstanding work.

_____

8. Society of American Foresters.
9. Photo interpretation.
10. At that time, EROS stood for Earth Resources Observation System, now it is Earth Resources Observation and Science.
11. Environmental Protection Agency.

Manual of Geographic Information Systems

At that time I was working very closely with a colleague named Harvey Fleet. He was very much into the digital cartography—the vector side of things. So we had both the raster side and the vector side because the technologies really weren't integrated yet. We were working together and worked mightily to become a base-funded, Washington office. At this time, everything had to be done on a project-by-project basis. There was no continuity. At the end of the day, the data would go to the planning teams or the Parks, and it would disappear or get lost in a black hole someplace. We wanted to implement what is now know as the life-cycle management approach to handling data and related things. So we worked hard and within a year – this was '84 or '85 – we eventually became a base-funded, Washington headquarters office, the GIS Division, which had the integration of raster and vector GIS, image processing, and other remote sensing and geospatial technologies known at the time.

Because of our experience with public-domain software and different computer systems, we were always strong advocates of open systems, but we also were realists. At that time, UNIX came on the scene. I don't remember the exact dates, but I think it was sometime in the late-'80s, when Sun had just developed the precursor of Sun-1. However, I don't remember what they called their micro-computers before the Sun-1. These micro-computers gave you tremendous increases in power, great decreases in cost, little need for special operating conditions, because they were basically like what we know as desktops today, and essentially a public-domain operating system (UNIX) that eventually moved to any number of different platforms. That was Sun's goal; that was also our goal. So in essence our visions converged and that was the path we embarked upon. It turned out to be a very beneficial path, because, sure enough, that sort of vision started to play out very shortly and we were able to go from -- I think it was less than three or four computer systems at places that were doing geospatial technologies within the Park Service – to about a hundred and twenty-five in about a five-year period. We were doing it all by the ones and twos, installations in parks or other offices, with people on my staff and on Harvey's staff doing hardware/software installations and teaching people the operating system requirements, as well as doing all of the data base development and related things.

Olson: How did the GIS Division get started in ASPRS?

Nyquist: When I came on as a Division Director of the Remote Sensing Applications Division in 1988, somebody nominated me for the Executive Committee right out of the bag. I hadn't even served a day on the Board, yet I got elected to the ExCom. I was really in it up to my neck, but it was both intriguing and exciting, and I could see it was a way to make things happen within the organization. I took to it, and I enjoyed doing it. Somewhere in that period, I remember the Division Directors had a mid-night revolt because we were bound and determined to create a new division called the GIS Division, because there was that whole realm of geospatial technologies that wasn't remote sensing, wasn't photogram-metry, wasn't primary data acquisition, and it wasn't professional practice. It was new and it was different. Sure, people were using database approaches and modeling in remote sensing applications, but there were a lot of people that weren't doing any remote sensing and were just taking the results of remote sensing or other mapping and were just doing pure GIS. A lot of them didn't want to know about all that other "stuff" that went on. So we actually went around at Virginia Beach and we grabbed everybody we could out of every bar and every place where people were hanging around and got them to sign a petition to the Board to establish the GIS Division. However, that didn't go over well with ACSM who was a partner at the time, and some of the ASPRS Board members didn't like it much either. They thought such an action was premature, because "GIS might just be a fad." So, there was about a year of it sitting there percolating, but I think within a year the ASPRS GIS Division was formed. The rest is history, as they say.

SECTION 2
Data Models, Metadata and Ontology

CHAPTER 6
Conceptual Tools for Specifying Spatial Object Representations

Martien Molenaar and *Peter van Oosterom*

6.1 Introduction

The Earth's surface can be considered as a spatio-temporal continuum in which processes of different kinds take place, such as the development of vegetation covers, geologic processes, demographic processes, the development of land use, etc. Each process is the result of interacting forces caused and affected by internal and external factors and will lead to spatial patterns of terrain characteristics which will change with time. Two major classes of such processes can be distinguished:

1. The first class refers to the processes with a field characteristic, i.e., the strength of the interacting forces depends on their positions within the field and the resulting pattern also can be expressed in terms of position-dependent field values;
2. The second class is based on the behavior of (spatially interacting) objects, the resulting pattern of such a process can be expressed by the spatial distribution and the state of these objects.

This chapter will explain some principle characteristics of data models for the representation of object-structured terrain situations in a database environment. The concepts presented here have, to a large extent, been based on Molenaar (1998) and a large part of the text of this chapter is an upgrade and further elaboration of Molenaar (2000). Each represented terrain situation can be considered as a state of a type 2 process at some specified moment and it can be considered a time slice of the spatio-temporal continuum in which such a process has been defined. Time will be dealt with only marginally in this chapter. (See Section 6.4 for further discussion of spatio-temporal GIS.) The emphasis, instead, will be on the representation of spatial objects with their geometric and non-geometric (thematic) aspects. The main aim of this chapter is to make the reader aware of some important decisions or choices that have to be made to formulate models for such spatial object representations.

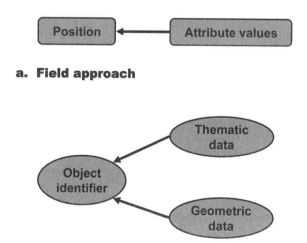

Figure 6-1 Two types of spatial representations (modified from Molenaar 1998).

For most applications, the thematic aspects of a terrain description are of prime importance. This means that data querying and processing will be seen primarily in a thematic perspective. The analysis of the geometric aspects of the data will be secondary and it will depend on the thematic problem formulation.

Two principal structures are used for linking thematic and geometric data in a GIS environment (Peuquet 1990; Laurini and Thompson 1993; Chrisman 1997; Burrough and McDonnell 1998):

1. For the field approach, thematic terrain characteristics are represented by attributes with position-dependent values. The thematic attributes are therefore directly linked to position attributes. This structure has been represented in Figure 6-1a.

2. For the object-structured processes, terrain features or objects are defined, each having a geometric position and shape and several non-geometric characteristics. These objects are represented in an information system by means of an identifier to which the thematic data and the geometric data are linked, as in Figure 6-1b.

   Note: the reader should take care not to confuse the terminology used here with the terminology used for modern techniques in computer science; there are of course relationships between the concepts used in these different fields but there are also important differences in the way the terminology is used.

The representation of a field in a geodatabase requires that the spatial continuum be discretized. This is generally done in the form of points or finite cells often in a regular grid or raster format where the thematic attribute values are evaluated for each point or cell. This is the raster approach. The alternative approach is that the linear structure of terrain features is represented geometrically in the form of chains of points or vertices linked by edges. This is the vector approach.

One should be careful not to equate the field approach with the raster approach or the object approach with vector approach. The concepts of fields and objects refer to the semantics of the represented phenomena, whereas raster and vector data refer to the representation of their geometric aspects (Molenaar 1998). It is very well possible to represent the geometry of field phenomena in a vector structure and to represent the geometry of objects in a raster format.

The remainder of this chapter will deal with the representation of geospatial objects in spatial information systems and is largely based on the presentation of the subject matter in Molenaar (1998). Section 6.2 will elaborate some general characteristics of spatial objects and some of their geometric aspects. Section 6.3 explains several semantic aspects of object modeling:

- Section 6.3.1 discusses the role of thematic object classes and classification hierarchies for organizing the thematic object descriptions.
- Section 6.3.2 describes how aggregation hierarchies can be used to link objects at different levels of complexity.
- Section 6.3.3 explains similarities and differences between object aggregations and object associations.
- Section 6.3.4 focuses on the semantic aspects of three-dimensional object models
- Section 6.3.5 adds the temporal dimension to the spatial objects, resulting in so-called spatio-temporal models
- Section 6.3.6 explains the important role of constraints for refining the semantics in models

Section 6.4 discusses spatial object models in the context of the model phases and involved layers of models. It also gives a short overview of modeling tools such as the Unified Modeling Language (UML) and the Extensible Markup Language (XML), along with standards from the International Organization for Standardization (ISO) and the Open Geospatial Consortium (OGC). Finally, Section 6.5 explains how object definitions and object descriptions are related to the context in which a spatial database will be used.

6.2 Spatial Objects

In the object-structured approach, the link between the thematic data and the geometric data is made through an object identifier as in Figure 6-1b. The definition of the objects will primarily be made within a thematic or application context: the objects within a cadastral environment will obviously be quite different from the objects in a topographic survey or in a land use analysis at a continental scale. The relevant thematic aspects will be specified within a specific application context. The thematic descriptions of the objects will then be organized in attribute structures which are different for different types of objects or object classes, Figure 6-2.

If geometric information is required, then several decisions have to be made. For each class of objects a decision should be made whether the objects will be represented as point-, line- or area objects, see Figure 6-2 (Molenaar 1998):

- For point objects only the position is stored.
- For line objects the position and shape are stored, "length" is the only size measure given.
- For area objects position and shape will be given, the length of the perimeter and the area are the size measures.

This choice should depend primarily on the role that the objects play in the terrain description and analysis rather then on the actual appearance of the objects or the scale or resolution of the terrain description. A river might be considered as an area object by the authority responsible for the maintenance, whereas that same river might be considered as a line object when it is seen as a part of an hydrological network for flow analysis. A city might be represented as a point object when it is considered as a node in air traffic networks, but the same city will represented as an area object in the case of urban land use mapping.

A decision should be made about the required accuracy of the geometric description of the terrain objects. This concerns the location accuracy, the precision of the shape description and the required geometric detail. Furthermore there is the choice of the geometric format, i.e. whether geometry will be handled in a raster or a vector format (Figure 6-3). This is a

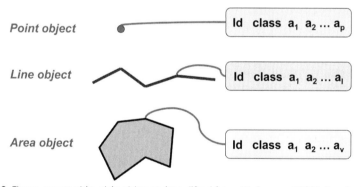

Figure 6-2 Three geometric object types (modified from Molenaar 1998). See included DVD for color version.

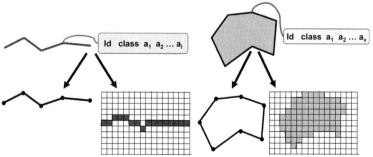

Figure 6-3 Raster and vector representations of spatial objects. See included DVD for color version.

choice for the most convenient format (i.e., it is not a matter of principle) and will depend on the available software and/or on the type of geometric queries and operations that should be performed. Some will be easy to handle in a raster format (like overlaying) whereas others might be easier to handle in a vector format (like many topology-related queries).

Spatial terrain objects are generally part of spatial complexes consisting of many topologically related objects, i.e., consisting of a contiguous set of mutually adjacent area objects intersected by line objects and containing point objects as in Figure 6-4. The representation of the geometric aspects of these objects should therefore be based on a joint common set of geometric elements that contain the topologic information to support the querying and analysis of these spatial relationships (Egenhofer and Herring 1992; De Floriani et al. 1993). The vector structure is very suitable for the implementation of such topologic structures (Molenaar 1998).

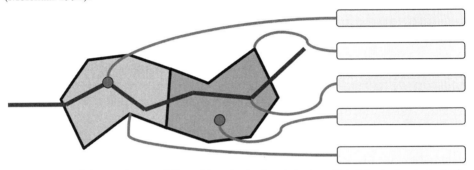

Figure 6-4 Spatial complex consisting of topologically related spatial objects. See included DVD for color version. See included DVD for color version.

Two constraints can be formulated for the specification of spatial objects, especially area objects:
1. A spatial data set, or map, should have no spatial ambiguity. The objects of a particular mapping domain, or thematic field, should be specified so that the area objects do not overlap and they should be spatially disjoint. Each point of the mapped area belongs to only one area object.
2. The spatial data set, or map, should be complete. The objects of a particular mapping domain, or thematic field, should be specified so that the area objects form a complete coverage of the mapped area. There should be no holes in the map and all points of the mapped area belong to an area object.

The combination of these two constraints implies that the area objects of a spatial data set, or map, form a spatial partition of the mapped area.

6.3 Semantic Object Modeling

The previous section already mentioned the fact that terrain objects can be organized in classes according to their thematic aspects. Objects classes define characteristics which their members have in common. This principle has been exploited in computer science in techniques implemented in object oriented programming languages and object oriented database management techniques (Smith and Smith 1977; Cox 1987; Hughes 1991). In computer science, classes define description structures for their member objects and operations that can be applied to them (see also Brodie 1984; Brodie and Ridjanovic 1984). The concepts in computer science have been interpreted and modified by several authors for spatial database applications (Egenhofer and Frank 1989; Kemp 1990; Nyerges 1991; Molenaar 1993, 1998; Fritsch and Anders 1996). Many of them emphasize the adaptation of database techniques to the requirements for operations on spatial data. Some authors adapt the concepts to define semantic data models for describing spatial phenomena. Their emphasis is not on the definition of database operations but rather on the development of conceptual tools for describing the real world and their activities are considered to be in the domain of geo-information theory rather then in computer science. The discussions in this section also will be along this line (Molenaar 1998).

6.3.1 Terrain Object Classes and Generalization Hierarchies

Suppose that a farm consists of several lots, some of which are used as arable land and others as pasture land. The fact that we make a distinction between these two sorts of fields clearly indicates that they are different because the farmer manages these two sorts of land differently. The farmer needs different types of information for their management and this means dealing with two different object classes. Let the objects be identified by an identification number (id). Let the thematic description for arable land be given by the attributes crop type, sowing date, herbicide, fertilizer and crop type last year. Let the description of pasture land be given by grass type, (monthly) biomass production and fertilizer.

A table can be generated for each class. In Figure 6-5, the columns W1, W2 and W3 represent the thematic attributes of the class of pasture land and the columns A1 through A5 are the thematic attributes of arable land; attribute names are given separately. Each class has its own attribute structure and for every object of a class a value will be assigned to each attribute. These values must fall within the range of the attribute domain, which must be defined prior to the actual assignment of attribute values.

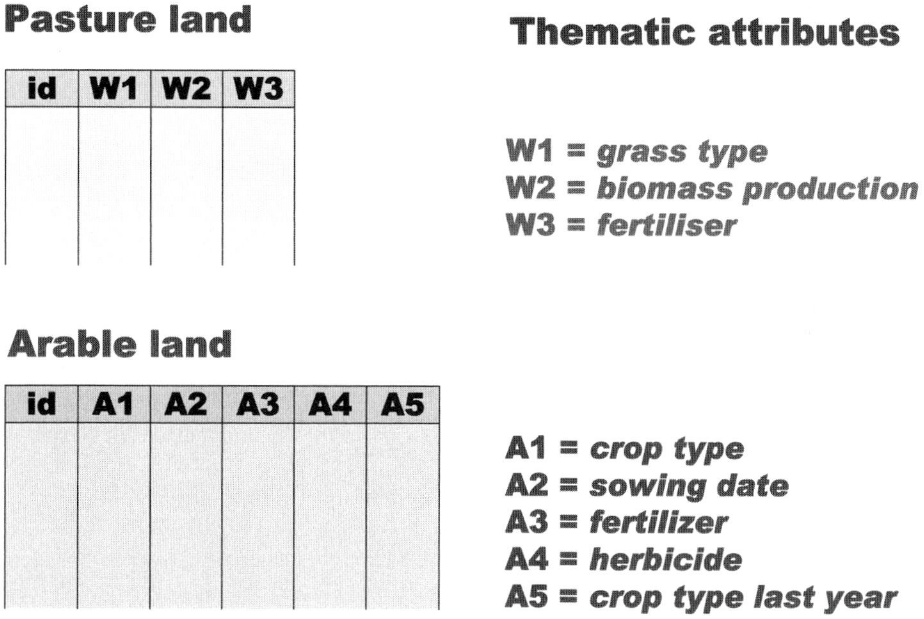

Pasture land

id	W1	W2	W3

Thematic attributes

W1 = *grass type*
W2 = *biomass production*
W3 = *fertiliser*

Arable land

id	A1	A2	A3	A4	A5

A1 = *crop type*
A2 = *sowing date*
A3 = *fertilizer*
A4 = *herbicide*
A5 = *crop type last year*

Figure 6-5 Tables for two classes of agricultural land use (modified from Molenaar 1998). See included DVD for color version.

This relationship between objects, classes and attributes has been represented in the diagram in Figure 6-6. Each class has its own unique list of attributes (attribute structure), each object in a class has a list consisting of one value for every attribute. We will assume that each object belongs to only one class, so that the attribute structure of an object is completely determined by the class to which it belongs, i.e., an object inherits the attribute structure of its class. Different classes have different attribute structures, but that is not to say that all attributes are different. We will extend the list of attributes of both classes in our example with area, soil water level and soil type.

Now we have two possibilities for adjusting the table structure of Figure 6-5. We can extend both existing tables with the new attributes (Figure 6-7a) or we can create a new table with these new attributes (Figure 6-7b). The latter implies that all objects that appeared in the two original tables must also appear in the new table. This new table, called farm lot, is a more generalized description of the objects. The distinction between arable land and pasture land is then in fact a further thematic specification of the objects. This is apparent in the fact that per class a more detailed specification of attributes is added to the less specific class of farm lot, as can be seen clearly in the extended lists of attributes for the tables in Figure 6-7a. We speak of farm lot as a generalized class or super-class above the classes of arable land and pasture land. An object that belongs to the class pasture land does not only have the attributes of this class, but also those of the super-class farm lot.

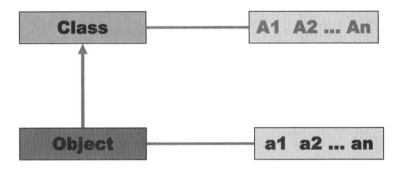

a_i is attribute value of A_i
attributes evaluated per object

Figure 6-6 Diagram representing the relation between objects, classes and attributes (modified from Molenaar 1998). See included DVD for color version.

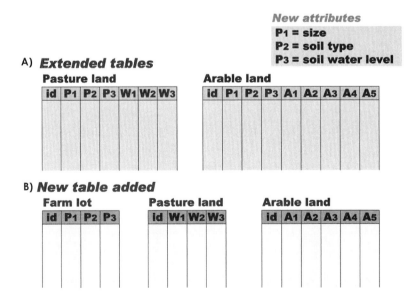

Figure 6-7 Two ways to deal with common attributes for different object classes: A) tables extended for common attributes; B) a separate table for the common attributes (modified from Molenaar 1998). See included DVD for color version.

Figure 6-8 represents a hierarchy of classes created in this way. All classes in a hierarchy can be distinguished by their own unique attribute structure. Within a hierarchic line, these structures are handed down, i.e., objects that belong to a specific class not only have the attributes of that class but also all those of its super-class(es). In a strict hierarchy the relation between a given level and the level above it is always n:1 (many to one), thus a super-class can have many lower-level classes but a class on a lower level belongs only to one super-class on the next higher level. Thus when descending through a hierarchy, we see that at each level an increasingly detailed part of the object's attribute structure is defined. We speak in this case of specialization. At the object level there is no further extension of the attribute list, but the values are assigned to the attributes. When we ascend in a hierarchy, the description of objects becomes less specific and we speak of generalization.

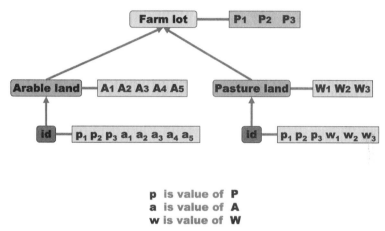

Figure 6-8 Class hierarchy for agricultural objects (modified from Molenaar 1998). See included DVD for color version.

Let U be the collection of all spatial objects represented in a spatial database; we call U the universe of the database. A classification system should be set up so that it is complete and exclusive, i.e., in such a way that all objects of U belong to exactly one class at the lowest level of the system. In that case, the classes at this level form a thematic partition of U (see P1 in Figure 6-9). This implies that within a hierarchy each object also belongs to exactly one class at each higher level of the system so that each level forms a thematic partition of U (see P2 and P3 in Figure 6-9). Then each object of U receives its attribute structure via one and only one inheritance line.

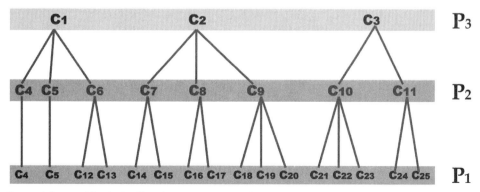

Figure 6-9 An object classification system consisting of several class hierarchies (modified from Molenaar 1998). See included DVD for color version.

Different hierarchical lines, or inheritance lines, can co-exist in one classification system and form one or more trees, as in Figure 6-9. The upwards relation in a classification hierarchy is called an 'ISA' relation. Thus capital ISA city ISA built up area and Amsterdam is an object instance of the object class Capital. The example shows that an ISA relation assigns objects to a class and its super-classes. Classes and super-classes are typified by their attribute structures. A classification system structured in this manner is a manifestation of the concept 'thematic field'. It is represented by a classification system, i.e., a collection of classification hierarchies, with their classes, super-classes and their hierarchical relations. Furthermore, it includes the attribute structure of classes and super-classes, and the attribute domains. A thematic terrain description is the complete set of objects with a list of attribute values per object. In an information system, such a thematic field is always defined in a specific user's context. This context is determined by many factors, including—but not limited to—the mapping discipline, the point in time or era of the mapping, and the scale or aggregation level.

Terrain objects occur at the lowest level in a classification hierarchy and can, therefore, be seen as the elementary objects within the thematic field represented by a given classification system. This implies that the decision whether or not to consider certain objects as elementary must be made within the context of such a thematic field. In other words, this is a context-dependent decision. Objects that are elementary within one thematic field are not necessarily elementary within another.

6.3.2 Aggregation Hierarchies

The fact that the previous subsection dealt with elementary objects implies that there can also be composite objects, or aggregates. These can be defined within the framework of aggregation hierarchies. An aggregation hierarchy describes the way in which composite objects are built up from elementary objects and how these composite objects, in turn, can be combined to form even more complex objects. Figure 6-10 shows an aggregation hierarchy. In the first step from level 1 to level 2, farm fields are combined to form lots. Next these lots are combined with a farmyard to form a farm. In the third step, a number of farms are combined to form an agricultural district.

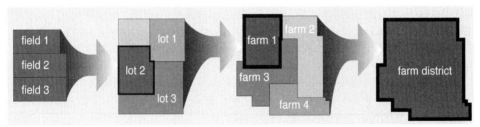

Figure 6-10 The aggregation of agricultural objects (modified from Molenaar 1998). See included DVD for color version.

An aggregation hierarchy has a bottom-up character, in the sense that elementary objects from the lowest level are combined to compose increasingly complex objects as one ascends in the hierarchy. The compound objects inherit the thematic data from their constituent objects and there are generally two types of rules for constructing composite objects. First, there are rules indicating the object classes of which a given compound object can be composed. Second, there are rules indicating which lower level objects can be included in a composite object on the next level. In a GIS, these latter rules are often based on topological relations between objects. The agricultural district in Figure 6-10, for example, is formed from farms that are mutually adjacent and fall within a communal boundary.

This means that aggregation types can be determined by their construction rules. If elementary objects are combined to form a composite object, their attribute values are often aggregated as well. Farm yield is the sum of field yields, and district yield is the sum of farm yields, as in Figure 6-11. We speak of upwards inheritance in aggregation hierarchies as 'PART OF' relations. For example, 'St. James Park is PART OF Westminster is PART OF London.' The PART OF relations connect groups of objects with a certain aggregate and possibly on a higher level with another even more complex aggregate, and so on.

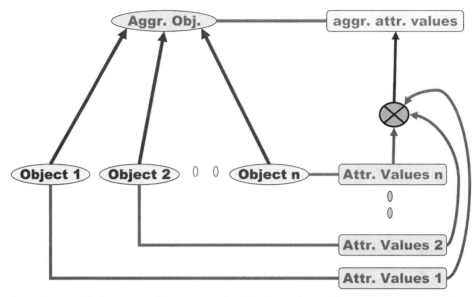

Figure 6-11 Attribute values of the composite object are derived from attribute values of elementary objects (modified from Molenaar 1998). See included DVD for color version.

Another characteristic which distinguishes an aggregation hierarchy from a classification hierarchy is that within a thematic field an elementary object can belong to one and only one class, and thus has only one line of inheritance within a classification structure. It may be part, however, of several different aggregates. A set of aggregate types thus need not be either exclusive or complete. This implies also that not all elementary objects are necessarily a part of an aggregate. For example, hydrological systems and shipping routes are non-exclusive aggregates of waterways. A river can be part of a hydrological system existing of rivers lakes and streams. This same river can also belong to the shipping routes made up of rivers, lakes and canals. However, it does make sense to define aggregates within a hierarchy in such a way that the aggregates of one type are mutually exclusive, i.e., that elementary objects belong to one aggregate of a given type. In this restricted sense the relationships between objects and aggregates are many-to-one, m:1. Thus, a house can only be part of one neighborhood, which can only be part of one municipality. In the same manner, a river can only belong to one hydrological system.

6.3.3 Object Associations

The two types of hierarchies discussed in the previous subsections have clear descriptions. A classification hierarchy has a top down, step-wise introduction of attribute structures for terrain objects. Classes are collections of objects with the same attribute structure. An aggregation hierarchy is defined by the construction rules that describe how the objects on a given level are composed of objects from the next lower level. From the bottom up, the levels have an increasing complexity.

A third way of organizing terrain objects is by more-loosely-defined object associations. The formation of object associations is not bound by a strict set of rules, but objects are grouped on the basis of some common aspects. The relationships that are formed are not necessarily m:1 (many to one) relations, but also may consist of m:n (many to many) relations. This implies that the associations of one type need not be exclusive. The following example should clarify this principle.

The neighborhoods in Figure 6-12 form associations. The neighborhood of plot 3 is formed by all the other plots that are adjacent to it. In this respect, the composition of an association resembles the composition of an aggregate. The difference, however, is these plots also belong to other neighborhoods, e.g., plot 4 has a neighborhood which overlaps the neighborhood of 3.

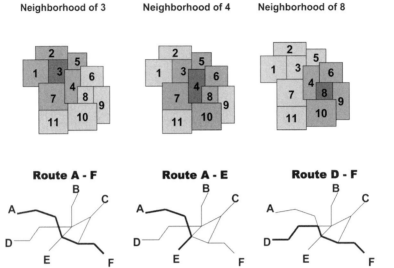

Figure 6-12 Examples of object associations (modified from Molenaar 1998). See included DVD for color version.

The road network in Figure 6-12 also can be regarded as an aggregate; the routes in this network however are another example of associations. The route from A to F consists of different roads or roads segments. The roads (and segments) that are a part of the route from A to F also can be a part of other routes such as from A to E or from D to F. Thus the routes are not mutually exclusive.

These examples show that associations consist of m:n relations. That means that a given object can be a part of several associations of the same type and they do not exclude each other as in the case of aggregations. The relation between an object and an association is called a 'MEMBER OF' relation. For example, 'plot 5 is a MEMBER OF the neighborhood of plot 4'.

6.3.4 Three-dimensional Object Models

Interest in three-dimensional (3D) modeling is growing, both in applications and in science. Typical examples include 3D cadastres (Stoter et al. 2002), telecommunications (Kofler 1998) and urban planning (Cambray 1993). At the same time, geographic information systems (GISs) are changing into integrated architecture in which administrative and spatial data are maintained in one environment in line with the presented modeling theory in Section 6.3.1. It is for this reason that mainstream GIS and database management systems support spatial data types according to the 'Simple Feature Specifications for SQL' described by the OGC. However, these specifications are 2D, as indeed are nearly all the current systems.

The development of 3D applications is mainly due to the growing need for multifunctional use of space by, for example, buildings above roads and railways and bridges and tunnels. As in the two-dimensional (2D) situation, two modeling approaches can be followed (and both are needed) based on:

1. Topological structure where lower level topological primitives (nodes, edges and faces) are used to represent volume, surface, line or point objects (Molenaar 1990; Pilouk 1996).

2. Three dimensional geometric primitives such as polyhedron, sphere, tetrahedron and voxel (Arens et al. 2005).

The absence of real 3D (topological and geometric) primitives in modeling creates major problems. The systems do not recognize 3D spatial objects, because they do not have a 3D primitive to model them. This results in functions that do not work properly; for example, there is no validation for the 3D object (Figure 6-13).

When 3D objects are stored as a set of polygons, no relationship exists between the different polygons that define the object. Besides the fact that validation is impossible and that any set of polygons can be inserted, another disadvantage is that the same coordinates are listed several times (causing inconsistency risks) and there is no information about the outer or inner boundaries (shells) of the polyhedron.

a   **b**   **c**   **d**

Figure 6-13 Examples of valid (a and b) and invalid (c and d) 3D objects.

6.3.5 Spatio-temporal Models

In the introduction we referred to the Earth's surface as a spatio-temporal continuum in which processes of different kinds take place. This implies that each represented terrain situation should be considered as a state of such a process at some specified moment, i.e., it can be considered a time slice of the spatio-temporal continuum. Clearly the world is not static and changes in objects do occur. Of course these changes should be reflected in the information systems and the models on which these systems are based. Good overviews of handling spatio-temporal data can be found in Langran (1992), Al-Taha et al. (1994), CHOROCHRONOS (1996–2000): A Research Network for Spatiotemporal Database Systems, and Abraham and Roddick (1999). Some papers emphasizing modeling include Worboys (1994), Egenhofer and Golledge (1998), Tryfona and Jensen (1999), van Oosterom et al. (2000) and Peuquet (2002).

When dealing with temporal change of objects, questions such as the following have to be answered (Langran 1992):
- Where and when did a change occur?
- What type of changes occurred?
- What is the rate of change (trend)?
- What is the periodicity of change (if any)?
- Where was this object two years ago?
- How has this area changed over last five years?
- What processes underlie a change?

These questions deal with temporal aspects of spatio-thematic phenomena as represented in the triangle of Figure 6-14a. In order to deal with spatio-temporal data, a system should provide functions such as the following (Langran 1992):
- inventory—complete description at a certain moment in time;
- analysis—explain, exploit, forecast spatial developments and processes;
- updates— supersede outdated spatial and thematic information with new versions;
- quality control—monitor and evaluate new data and check if these are consistent with old data;
- scheduling—identify threshold states which trigger predefined actions; and
- display—generate maps or tables of a temporal process.

A number of basic temporal modeling concepts that pertain to objects will now be introduced. The smallest time unit is called chronon; in a sense this may be compared to the resolution in the spatial domain, such as pixel size. A moment in time is represented with a point on the time line, which runs from left (history) to right (future). There is one very special moment in time and that is now, which is always on the move on the time line. The time line also may branch off to represent different plans/scenarios/predictions (Figure 6-14b). A time interval is the segment on the time line between two moments in time or epochs. A time interval may be used to represent the time a version of an object is valid. The term frequency is used to represent how often certain patterns (of reoccurring events) do happen and may be expressed, for example, in Hertz (cycles per second, hz or 1/sec). Similar to the spatial domain, temporal topology is used in modeling. For moments in time (points on the time line) this is very simple: before, equal or after. The relationships between two intervals of time are more interesting since they can have such topology relationships as disjoint, touch, overlap, include and equal. Even more complex possibilities can be imagined, such as "overlap and end at the same time."

Space-Time-Theme triangle   **More time lines**

Figure 6-14 Basic spatio-temporal concepts: space-time-theme triangle (a) and timeline (b).

Data granularity related to time is an important aspect when modeling a spatio-temporal dataset as it can range from 'coarse, more redundancy' to 'fine, less redundancy' as the following four examples of data granularity illustrate:

1. Complete universe level or whole data set, e.g., based on aerial photography acquired every 6 years, a new topographic map is produced and the whole data set has the same time stamp,
2. Object class level (that is thematic), e.g., in future topographic maps, certain object classes may be updated more frequently than other object classes such as the road objects updated every 2 years,
3. Object instance level, e.g., individual parcels are updated in the cadastral map on a daily basis such as a complete parcel with all its attributes is renewed, while other objects in the database may remain the same, and
4. Object attribute level; e.g., when measuring every hour the ground water level at a specific point location the other attributes of the ground water measurement object remain the same, but the ground water level attribute is updated very frequently.

For each event there are several moments at which time could be registered:
- when it happened in the real world (real world or user time),
- when it was observed in reality (date photo),
- when it was included in the database (system transaction time),
- when it was last checked in reality,
- the date (time) of the signature/registration/postmark time,
- when an error (in historic data set) was discovered and corrected (two moments),
- when it was last displayed (to user on screen/map).

It is possible to include multiple times in one model and an often-used approached is the bi-temporal model, which includes both user and system time, as discussed below. The changes in a spatial object database (such as in the example of the cadastre, the third example above) are of a discrete type in contrast to more continuous changes in natural phenomena, often represented via the field model (Cheng and Molenaar 1998). There are two types of models for representing this:

1. State orientation—every object is extended with some time attributes (tmin and tmax);

2. Event orientation—store/document the changes (which attribute did change, why, when).

Both state- and event-oriented models are suitable for querying the changes in the past such as 'give changes in map between t1-t2' (Figure 6-15). However, it is not easy to retrieve the situation at any given point in time ('give map at moment t') in the event-oriented approach. Therefore, the state-based approach is used more often and is sometimes mixed with the event approach to document the changes. More information on event-based modeling is given in Chen and Jiang (1998).

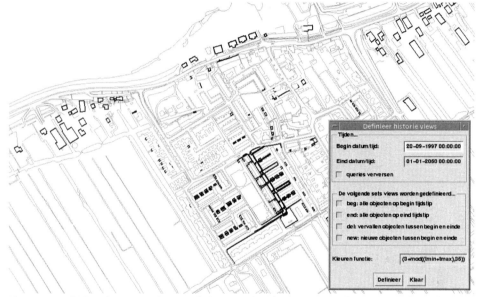

Figure 6-15 Object instance changes in interval. See included DVD for color version.

In state-based temporal modeling (with object instance granularity) a minimal approach is to extend every object with two additional attributes: tmin and tmax as in the Postgres model (Stonebraker and Rowe 1986). The objects are valid from and including tmin and remain valid until and excluding tmax. Current objects get a special tmax value of max_time, indicating they are valid now. These are system times. Furthermore, a model can be extended with the user time attributes valid_tmin and valid_tmax, which would make it a bi-temporal model (Figure 6-16).

When a new object is inserted, the current time is set as the value for tmin, and tmax receives the special value of max_time. When an attribute of an existing object changes, this attribute is not updated, but the complete record, including the object identifier (oid), is copied with the new attribute value. Current time is set as tmax in the old record and as tmin in the new record. This is necessary to be able to reconstruct the correct situation at any given point in history. The unique identifier (key) is the pair (oid, tmin) for every object version in space and time.

In the bi-temporal model, the valid time (user or real world time) and the system time (transaction time) are both maintained. If both time intervals (user and system) are used together, then this results in a time rectangle. Assuming that the system time is after the user time, the lower left (and upper right) is normally below the diagonal of the diagram, which depicts system and user time; see Figure 6-16.

Manual of Geographic Information Systems

Bi-temporal model

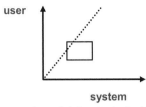

Figure 6-16 Time interval in bi-temporal model. See included DVD for color version.

It should be noted that despite the fact that the state-based, object instance level, temporal model was explained in greater detail, no single temporal model is the best for every situation. The same is true for spatial models. Sometimes a raster model is the best, and sometimes a vector model is the best as with the difference in natural vs. man-made objects. The model that was explained in greater detail was emphasized because it functions relatively well in many situations.

6.3.6 The Role of Constraints in Models

Constraints are important in every GIS modeling process, but until now, constraints have received only ad hoc treatment depending on the application domain and the tools used. In a dynamic context, with constantly changing geo-information, constraints are very relevant. Indeed, any changes arising should adhere to specified constraints, otherwise inconsistencies (data quality errors) will occur. In GIS, constraints are conditions which must always be valid for the model of interest. This chapter argues that constraints should be part of the object class definition, just as with other aspects of that definition, including attributes, methods and relationships. Furthermore, the implementation of constraints (whether at the front-end, database level or communication level) should be driven automatically by these constraint specifications within the model. Figure 6-17, for example, illustrates an edit front-end that includes constraints.

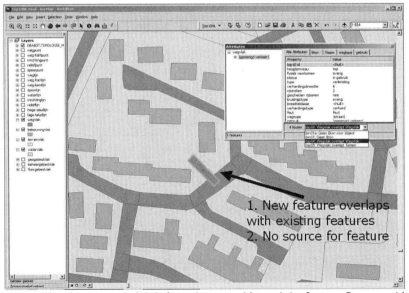

Figure 6-17 Violating a constraint during a topographic update. Source: Topographic Service Kadaster, the Netherlands, from van Oosterom (2006). Copyright © 2006. Dynamic & Mobile GIS: Investigating Changes in Space and Time. ISBN 0-8493-9092-3. Reproduced by permission of Taylor & Francis, a division of Informa plc. See included DVD for color version.

In certain applications some functions—linear programming in spatial decision support systems, survey least squares adjustment, cartographic generalization, editing topologically structured data, etcetera—partially support constraints. However, the constraints are not an integral part of the system and the constraint specification and implementation are often one and the same, and deep in the application's source code. The result is that the constraints are hidden in some subsystems (with other subsystems perhaps unaware of these constraints) and it may be very difficult to maintain the constraints in the event that changes are required. This is true for information systems in general, including GISs, but is especially true for dynamic environments with changing objects, where the support of constraints is required but presents a challenge.

Cockcroft (2004) advocated an integrated approach to handling integrity based on a repository which contains the model together with the constraints. Constraints should be part of the object class definition, similar to other aspects of the definition. The repository is used both by the database and the application as a consistent source of integrity constraints. In order to better understand constraints and their use, it is important to classify them, including their spatial or dimensional aspects. Classification of the different types of (spatial) constraints reveals a complex taxonomy. Cockcroft (1997) presents a 2D taxonomy of (spatial) constraints. The first axis is the static versus transitional (dynamic) distinction and the second axis is the classification into topological, semantic and user constraints. While the transitional aspect of integrity constraints is relevant, in this chapter this is considered to be the 'other side of the same coin.' Van Oosterom (2006), based on four case studies, refines the second axis of Cockcroft's taxonomy by recognizing five sub-axes (or five different criteria) for the classification of integrity constraints:

1. the number of involved objects/classes/instances;
2. the type of properties of objects and relationships between objects involved: topologic (neighborhood or containment), metric (distance or angle between objects), temporal, thematic or mixed;
3. the dimension (related to the previous axis): 2D, 3D or mixed time and space, that is, 4D;
4. the manner of expression: 'never may' (bush never may stand in water) or 'always must' (tree always must be planted in open soil);
5. the nature of the constraint can be 'physically impossible' (tree cannot float in the air) or 'design objective' (bush should be south of tree).

With respect to the first sub-axis of the constraint taxonomy, 'the number of involved objects,' the following cases can be identified:

1. one instance (restrictions on attribute values of a single instance),
2. two instances from the same class (binary relationship),
3. multiple instances of the same class (aggregate),
4. two instances from two different classes (binary relationship) or
5. multiple instances from different classes (aggregate).

Further, the fourth sub-axis, 'the manner of expression,' only has practical value for communicating the constraints between the users. Once the objects and the constraints are formally defined, the expressions 'never may' and 'always must' can be represented by one constraint that is more efficient from an implementation point of view. For example, the constraint 'a tree never may stand in water, or street or house' is equivalent to the constraint 'a tree always must be in a garden or park' under the assumption that there are only five possible ground objects of water, street, house, garden and park.

6.4 Model Phases, Layers and Available Tools

After the above discussion of several aspects of conceptual modeling of spatial objects, other aspects could be added, such as fuzzy or uncertain objects, topology relationships or topology structures (e.g., planar partitions and linear networks) because it is beneficial to realize the bigger picture. Two 'dimensions' of modeling will be discussed below, model phases and model layers. Next, example modeling tools (or standards) will be discussed: UML (including OCL) and XML.

6.4.1 Model Phases

Conceptual modeling is often considered the first level of modeling, giving a model description of the important aspects from the user point of view (semantic aspects). If the involved persons do agree on the conceptual model, then the next step is to translate this into a logical model (Figure 6-18). Before this translation takes place, however, one first has to decide on the type of database platform. Currently the most popular choices are the relational model and the object-relational model. The latter one is an hybrid database form based on the relational model and extended with features of the object-oriented model (such as the ability to add new data types). Other alternatives could be the pure object-oriented database model, network and hierarchical models. The latter two could be considered old-fashioned and do not occur very often anymore. Though one could argue with the advent of XML databases that the hierarchical model is making a comeback. When the conceptual model has been described in a formal manner and a specific logical model is chosen (e.g., object-relational) it is possible to automate the translation to a large extent (e.g., table definitions could be generated automatically). The last step in modeling is then translating the logical model into a physical model. In this step the logical model is used as input and refined with storage and access related measures: exact data types (maximum length of fields), physical clustering, indexing, and perhaps also partitioning, distribution, replication and definition of materialized views.

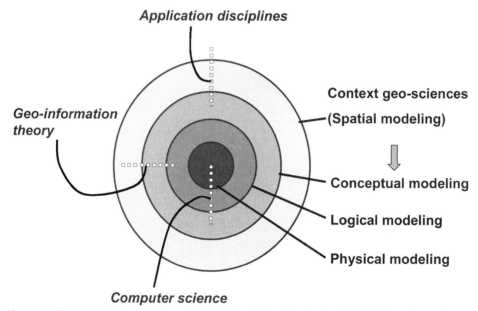

Figure 6-18 Layers (left) and phases (right) of modeling. See included DVD for color version.

6.4.2 Model Layers

Models are not usually built from scratch, but they are built on top of other models. This process can be repeated a number of times and every time the next layer can use the elements defined at the previous layers. Figure 6-18 above shows three such layers: 1) computer science—at this level basic data types and structures are defined including strings, numbers and dates; 2) geo-information theory—how to model spatial-temporal concepts for 2D and 3D geometry and topology primitives; and 3) application disciplines—many different domains such as hydrography, topography, cadastre, soil, geology, transportation, agriculture and environment. Agreeing on these (levels) of models has several advantages: First of all, one does not have to develop everything single-handedly. Second, agreeing on concepts also will enable meaningful exchange of data based on these models. Besides the three mentioned layers, there is at least one other layer that is the final model of a specific application; e.g., the exact model of the Netherlands Topographic Data Set.

6.4.3 Modeling Tools and Standards

When presenting or trying to describe a model, one always faces the question of how to describe this model for domain experts—non-technical end-users or managers who are not modeling experts. This question reappears in every context where models are developed and also applies to the previous sections of this chapter. Textual descriptions alone are difficult to understand as the model structure may not be visible. For the purpose of conveying model structure, all kinds of diagrams have been developed with 'boxes and arrows.' However, the 'boxes and arrows' may have different meanings in different diagrams, making general understanding—even by modeling specialists—difficult. Therefore, the Object Management Group (OMG)) standardized the main types of diagrams and the meaning of 'boxes and arrows' (Booch et al. 2005). The result was the Unified Modeling Language (UML), which is nowadays the state-of-the-art approach for object-oriented modeling (OMG 2002).

UML is a graphic language which gives a wide range of possibilities for representing objects and their relationships. In general, the language can be used for modeling business processes, classes, objects and components, as well as for distribution and deployment modeling. UML consists of diagram elements (e.g., icons, symbols, paths and strings) which can be used in nine different types of diagrams. The most relevant diagram for this discussion of data modeling is the UML class diagram. It provides formalism for describing the objects/classes, with their attributes and behavior, and relationships between these objects such as association, generalization and aggregation. A UML class diagram describes the types of objects and the various kinds of structural relationships that exist among them such as associations (composites, part-whole) and subtypes (specialization-generalization). Furthermore, the UML class diagrams show the attributes and operations of a class and the constraints that apply to the way objects are connected (Booch et al. 2005). UML class diagrams are reasonably well-suited to describe a formal and structured set of concepts, that is, an 'Ontology' (Gruber 1995). Experiences in several different application domains show that it is still not easy to read these diagrams. A good solution for this is the use of 'Literate Modeling,' in which UML diagrams are embedded in text explaining the models. More details and discussion on Literate Modeling, with examples from British Airways, can be found in Arlow et al. (1999).

Having shown the importance of constraints in different applications and having presented a refined taxonomy (in Subsection 6.3.6), the question remains how to specify the constraints. First of all, the specification of the constraints has to be intuitive for the user and the constraints have to be included in the object model. This model should be as formal as possible to enable users to derive constraint implementations within the different subsystems (e.g., edit, store, exchange). Formal modeling is an essential part of every large project; it is also helpful in small and medium-sized projects. Using a formal model facilitates communicating ideas with other professionals as well as describing clear, unambiguous views on implementation strategies. Despite their potential for formalizing objects and processes, UML class diagrams are typically not sufficiently refined to provide all the relevant aspects of constraints. Constraints are often initially described in natural language and practice has shown that this results in ambiguities. In order to write unambiguous constraints, a non-graphic language is provided within UML for the further modeling of semantics or knowledge frameworks, namely, the Object Constraint Language (OCL) (OMG 2002). When an OCL expression is evaluated, it simply returns a binary value. The state of the system will not change when the evaluation of an OCL expression returns false. The advantage of using OCL is that—as with UML class diagrams—generic tools are available to support OCL and it is not GIS-specific. OCL has been used successfully in the context of GIS, an example being the IntesaGIS project with the GeoUML model specifying the 'core' geographic database for Italy (Belussi et al. 2004). The context of an invariant is specified by the relevant class; e.g., the object class 'parcel' is the context of the constraint 'the area of a parcel is at least 5 m2.' It also is possible within a constraint to use the association between two classes. For example, the constraint that every instance of the object class 'parcel' must have at least one owner could be depicted as an association with the class 'person.' OCL enables one to formally describe expressions and constraints in object-oriented models and other object modeling artifacts. Below are two examples in UML/OCL syntax for the above mentioned constraints (keywords in bold print):

context Parcel **inv** minimalArea:
self.area > 5

context Parcel **inv** hasOwner:
self.Owner -> notEmpty()

Figure 6-19 shows the UML class diagram with the objects and the constraints (depicted as associations) used for SALIX-2 (Louwsma 2004). In principle, there is no difference between a 'data model' relationship (association, aggregation, specialization) and a 'data model' integrity relationship constraint. Both are depicted as lines in the UML class diagram. From a high level conceptual (or philosophical) point of view, the difference may be very small. However, normal associations are often indented, in subsequent implementations, to be explicitly stored in one or both directions. The relationship constraints, on the other hand, should not result in such an explicit storage, but in a consistency rule in the implementation environment. In order to distinguish between the two, normal relationships are depicted in black, while integrity relationship constraints are depicted in color. In the diagram notes can be used to explain the constraints on relationships and/or properties. These notes can contain either UML/OCL or natural language text.

Going back to the three different layers of modeling and the relationship to standardized tools, one could state the basic primitives in UML class diagram to form the first (computer science) layer. The second layer is formed by the spatial schema defined by OGC (and ISO 2003). The third layer can be formed by any standardized application domain; for example, the Core Cadastral Domain Model by the FIG (Lemmen et al. 2005). All three are at the conceptual level and use UML for describing the models. At the logical and physical level similar equivalents can be detected within the database and data exchange worlds. For example, the XML schema defines the basic primitives at the first level (XSD 2004). At the second level there is Geography Markup Language (GML) (ISO 2004) and at the third level domain specific XML schemas can be found. Similarly, in the database world these three levels can be found:

1. basic SQL (Date and Darwen 1997);
2. geo-information additions according to the OpenGeospatial 'Simple Feature Specification for SQL' (OGC 2005); and
3. specific database template models for various domains.

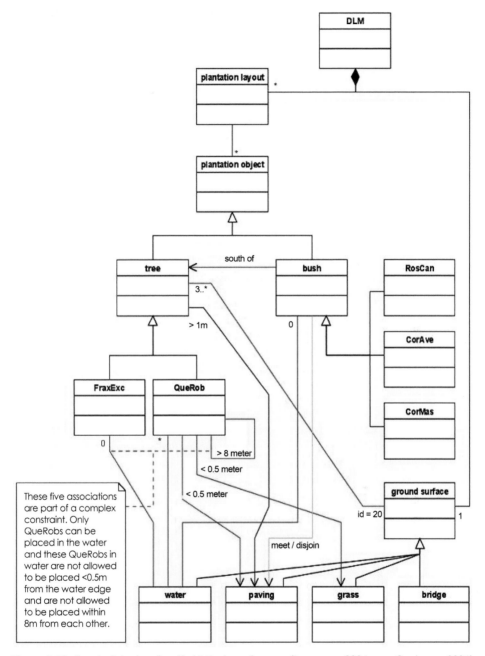

Figure 6-19 Constraints visualized in UML class diagram (Louwsma 2004; van Oosterom 2006). Copyright © 2006 Dynamic & Mobile GIS: Investigating Changes in Space and Time. ISBN 0-8493-9092-3. Reproduced by permission of Taylor & Francis, a division of Informa plc. See included DVD for color version.

6.5 Conclusion: Object Definitions and Context

In Sections 6.2 and 6.3 we described, in general, how a terrain object is represented in an information system via an identifier with associated geometric and thematic data. Section 6.3.1 explained that such objects are meaningful within a certain classification system. Therefore, before a database can be built, the classification structure must be chosen. This choice must be made within a use context with the following characteristics.

The first aspect considers the discipline(s) of the users. One can be dealing, for example, with a soil map or a demographic study or a cadastral system. Each discipline has its own definition of terrain objects, with classes and attributes. These definitions depend not only on the mapping discipline but also on the scale or aggregation level which is used. It makes quite a difference whether certain phenomena (e.g., land use or vegetation) are considered at a local, a regional, a national or a global scale. At each level, different elementary objects are relevant. Furthermore, elementary objects at one level may be aggregates of elementary objects at another level. For example GIS at a municipal level can contain data referring to houses, streets and parks, while a GIS at a national level contains municipalities or built-up areas.

Another aspect of the use context concerns the type of use that is to be made of the data. It makes quite a difference if data are to be used for management purposes or for the analysis of a terrain situation or for processes such as planning and design activities. All of these activities have their own standards for data and terrain descriptions, but that is not to say that there is no overlap.

A final aspect is the point in time in which the terrain description is made. In many cases the value or relevancy of information depends on the era. We can see from an agricultural point of view that the need for information about soils has changed in the course of time. Whereas the major interest used to be in the suitability of soils for certain crops, nowadays there is more interest in the capacity to bind certain chemical elements, with an eye on environmental effects. In cadastres the original task was to collect and store data for raising land tax and/or for the protection of owners' titles, but presently there is an increasing request for economic data such as the dynamics of real estate prices and the number of transactions and mortgages. An operational definition of 'use context' has been given in Bishr (1997); this definition has been based on a formal data schema of Molenaar (1998), specifying that geometric object descriptions should be related to the hierarchical classification models, as in Subsection 6.3.1. The definition of context by Bishr (1997) is, in fact, a meta model describing the semantics of the spatial data model which have been specified for a particular application. The relevance of data always depends on the context in which the data will be used.

If the data are modeled with the concepts presented in this chapter, then a context will be expressed through the semantic definitions of the objects and their actual descriptive structures. This includes, in such a context, the elementary objects with their classes at the different hierarchical levels. Several class hierarchies also may co-exist, i.e., the collection of classes may form one or more trees. If the classes of these hierarchies are defined so that at each level each object of the database is a member of exactly one class, then they form a thematic partition per level. This implies that each object inherits its attribute structure through exactly one inheritance line of such a system. A classification system structured in this manner is a representation of a 'thematic field.' It is characterized by a hierarchical classification system of classes and super-classes with their hierarchical relations. Further characteristics include the attribute structure of classes and super-classes, and the attribute domains. A thematic terrain description is the complete set of objects with a list of attribute values per object. In an information system, such a thematic field is always defined in a specific user's context.

Furthermore, the choice of geometric type must be made for these objects. This again depends on the role the objects are to play in a terrain description. A river can be regarded as a line object in a hydrological database, while the same river can be handled as an area object in the database of an organization that manages waterways. In the same way, a city can be seen as an area object in a database for demographic studies, whereas the same city appears as a point object, or a node, in a database for continental transport lines. Thus the decision which geometric aspects of a given class of terrain objects are relevant (i.e., the choice of treating objects as points, lines or areas) always depends on the user's context. This implies that the choice of which objects should be regarded as elementary with their relevant thematic and geometric characteristics also depends on the user's context.

Generally the definition and identification of elementary objects follows their thematic specification as expressed through the classification system. The specifications of a spatial data set are unambiguous if two constraints are fulfilled:

1. The classification system is designed so that the classes at each hierarchic level form a thematic partition of the universe of the spatial data base; and
2. The objects have been specified so that they collectively form a spatial partition of the mapped area (according to Section 6.2).

In that case the thematic object classes generate a spatial partition so that there are no spatial overlaps of thematic classes, and therefore there is, in principle, no ambiguity in the thematic classification of areas.

Within such a context, decisions also must be made as to which object aggregates and associations are relevant. Although not necessarily explicitly stored in a database, the decisions may be implied in the form of generic models—rules and procedures to generate these aggregates and associations. Together these choices and decisions define spatial object representations that characterize the spatial patterns of the Earth's surface that, in turn, reflect the interaction of internal and external processes.

References

Abraham, T. and J. F. Roddick. 1999. Survey of spatio-temporal databases. *GeoInformatica* 3: 61–99.

Al-Taha, K. K., R. T. Snodgrass and M. D. Soo. 1994. Bibliography on spatiotemporal databases. *International Journal of Geographical Information Systems* 8:95–103.

Arens, C., J. Stoter and P. van Oosterom. 2005. Modelling 3D spatial objects in a geo-DBMS using a 3D primitive. *Computers & Geosciences* 31 (2):165–177.

Arlow, J., W. Emmerich and J. Quinn. 1999. Literate modelling - capturing business knowledge with the 'UML.' Pages 189–199 in *The Unified Modeling Language: <<UML>>'98: Beyond the Notation*. Edited by Jean Bézivin and Pierre-Alain Muller. First International Workshop, Mulhouse, France, June 3-4, 1998, Selected Papers. Lecture Notes in Computer Science vol. 1618. Berlin: Springer.

Belussi, A., M. Negri and G. Pelagatti. 2004. GeoUML: A geographic conceptual model defined through specialization of ISO TC211 standards. In *Proceedings 10ᵗʰ EC GI & GIS Workshop—ESDI: State of the Art*, Warsaw, Poland, 23-25 June 2004. Ispra, Italy: Institute for Environment and Sustainability, JRC, European Commission <http://www.ec-gis.org/Workshops/10ec-gis/papers/> Accessed 31 January 2007.

Bishr, Y. 1997. *Semantic Aspects of Interoperable GIS*. ITC Publication Series, No. 56. Enschede, The Netherlands. 154 pp.

Booch, G., J. Rumbaugh and I. Jacobson. 2005. *The Unified Modeling Language User Guide,* 2ⁿᵈ edition. Boston: Addison-Wesley Technology Series.

Brodie, M. L. 1984. On the development of data models. In *On Conceptual Modeling*, edited by M. L. Brodie, J. Mylopoulos and J. W. Schmidt, 19–47. Berlin: Springer Verlag.

Brodie, M. L. and D. Ridjanovic. 1984. On the design and specification data base transactions. In *On Conceptual Modeling*, edited by M. L. Brodie, J. Mylopoulos and J. W. Schmidt, 277–306. Berlin: Springer Verlag.

Burrough, P. A. and R. A. McDonnell. 1998. *Principles of Geographical Information Systems*. Oxford: Oxford University Press.

Cambray, B. 1993. Three-dimensional modeling in a geographical database. Pages 338–347 in *Proceedings of AUTO CARTO 11: Eleventh International Symposium on Computer-Assisted Cartography*, held in Minneapolis, Minn. October 1993. Bethesda, Md.: ASPRS.

Chen, J. and J. Jiang. 1998. An event-based approach to spatio-temporal data modelling in land subdivision systems. *GeoInformatica* 2:387–402.

Cheng, T., and M. Molenaar. 1998. A process-oriented spatio-temporal data model to support physical environmental modelling. Pages 418–430 in *Proceedings of the 8th International Symposium on Spatial Data Handling*, held in Vancouver, British Columbia. Edited by T. K. Poiker and N. R. Chrisman. Burnaby, British Columbia: International Geographical Union.

CHOROCHRONOS. 1996–2000. A Research Network for Spatiotemporal Database Systems <http://www.dbnet.ece.ntua.gr/~choros> Accessed 10 January 2007.

Chrisman, N. R. 1997. *Exploring Geographic Information Systems*. New York: John Wiley & Sons.

Cockcroft, S. 1997. A taxonomy of spatial data integrity constraints. *GeoInformatica* 1 (4):327–343.

———. 2004. The design and implementation of a repository for the management of spatial data integrity constraints. *GeoInformatica* 8 (1):49–69.

Cox, B. J. 1987. *Object Oriented Programming*. Reading, Mass.: Addison-Wesley.

Date, C. J. and H. Darwen. 1997. *A Guide to the SQL Standard*, 4th edition. Reading, Mass.: Addison-Wesley.

De Floriani, L., P. Marzano and E. Puppo. 1993. Spatial queries and data models. In *Spatial Information Theory, a Theoretical Basis for GIS*, edited by A. U. Frank and I. Campari, 113-138. Berlin: Springer-Verlag.

Egenhofer, M. J. and A. U. Frank. 1989. Object oriented modeling in GIS: Inheritance and propagation. Pages 588–598 in *AUTO CARTO 9 Proceedings. Ninth International Symposium on Computer-Assisted Cartography*, Baltimore, Md., 2-7 April 1989. Falls Church, Va.: ASPRS.

Egenhofer, M. J. and R. G. Golledge, editors. 1998. *Spatial and Temporal Reasoning in Geographic Information Systems*. Oxford: Oxford University Press. 276 pp.

Egenhofer, M. J. and J. R. Herring. 1992. *Categorizing Binary Topological Relationships Between Regions, Lines, and Points in Geographic Databases*. Technical report, Department of Surveying Engineering. Orono: University of Maine.

Fritsch, D. and K. H. Anders. 1996. Objectorientierte Konzepte in Geo-Informationssystemen. *Geo-Informations-Systeme* 9 (2):2–14.

Gruber, T. R. 1995. Toward principles for the design of ontologies used for knowledge sharing. *International Journal of Human-Computer Studies* 43 (5):907–928.

Hughes, J. G. 1991. Object-oriented Databases. New York: Prentice Hall.

ISO. 2003. ISO/TC 211, ISO 19107:2003. *Geographic Information—Spatial Schema*. Geneva, Switzerland: ISO.

———. 2004. ISO/TC211, ISO 19136 *Geographic Information—Geography Markup Language*. Geneva, Switzerland: ISO.

Kemp, Z. 1990. An object-oriented model for spatial data. Pages 659–668 in *Proceedings of the 4th International Symposium on Spatial Data Handling, Zurich, July 1990*. Edited by K. Brassel and H. Kishimoto. Zürich: University of Zürich.

Kofler, M. 1998. *R-trees for the Visualizing and Organizing Large 3D GIS Databases*. Ph.D. Dissertation. Graz, Austria: Technical University. 131 pp.

Langran, G. 1992. *Time in Geographic Information Systems*. London: Taylor & Francis.

Laurini, R. and D. Thompson. 1993. *Fundamentals of Spatial Information Systems*. London: Academic Press.

Lemmen, C., P. van Oosterom, J. Zevenbergen, W. Quak and P. van der Molen. 2005. Further progress in the development of the core cadastral domain model. Proceedings of

'*From Pharaohs to Geoinformatics*', FIG Working Week 2005 and GSDI-8, Cairo, Egypt, April 16-21, 2005. Frederiksberg, Denmark: International Federation of Surveyors (FIG) <http://www.fig.net/pub/cairo/papers/ts_11/ts11_01_lemmen_etal.pdf> Accessed 31 January 2007.

Louwsma, J. H. 2004. *Constraints in geo-information models; Applied to geo-VR in landscape architecture.* MSc Thesis. Geodetic Engineering, Delft University of Technology, The Netherlands. 104 pp.

Molenaar, M. 1990. A formal data structure for 3D vector maps. Pages 770–781 in *Proceedings EGIS*, Amsterdam, April 10-13, 1990, vol. 2. Edited by J. Harts, H.F.L. Ottens and H. J. Scholten. Utrecht, The Netherlands: EGIS Foundation.

———. 1993. Object hierarchies and uncertainty in GIS or Why is standardisation so difficult. *Geo-Informations-Systeme.* 6:22–28.

———. 1998. *An Introduction to the Theory of Spatial Object Modelling for GIS.* London: Taylor & Francis. 229 pp.

———. 2000. Conceptual tools for specifying geospatial descriptions. In *Geospatial Data Infrastructure – Concepts, Cases and Good Practice*, edited by R. Groot and J. Mclaughlin, 151–173. Oxford: Oxford University Press.

Nyerges, T. L. 1991. Representing geographical meaning. In *Map Generalisation: Making Rules for Knowledge Representation*, edited by B. P. Buttenfield and R. B. McMaster, 59–85. London: Longman.

OGC. 2005. OpenGeospatial Consortium, Inc. *OpenGIS® Implementation Specification for Geographic Information - Simple Feature Access - Part 2: SQL Option*, Version: 1.1.0. OpenGIS Project Document OGC 05-134, 22 November 2005.

OMG. 2002. Object Management Group. *Unified Modeling Language Specification* (Action Semantics), UML 1.4 with action semantics. January 2002.

Peuquet, D. J. 1990. A conceptual framework and comparison of spatial data models. In *Introductory Readings in GIS*, edited by D. J. Peuquet and D. F. Marble, 250–285. London: Taylor & Francis.

———. 2002. *Representations of Space and Time.* New York: Guilford Press. 380 pp.

Pilouk, M. 1996. *Integrated Modelling for 3D GIS.* Ph.D. Dissertation, ITC, The Netherlands. 200 pp.

Smith, J. M. and D.C.P. Smith. 1977. Database abstractions: Aggregation and generalization. *ACM Transactions on Database Systems* 2:105-133.

Stonebraker, M. and L. A. Rowe. 1986. The design of POSTGRES. Pages 340–355 in *Proceedings of the 1986 ACM SIGMOD International Conference on Management of Data*, May 28-30, 1986, Washington, DC. Edited by Carlo Zaniolo. New York: ACM Press.

Stoter, J. E., M. A. Salzmann, P.J.M. Van Oosterom and P. Van der Molen. 2002. Towards a 3D cadastre. *Proceedings FIG ACSM/ASPRS*, Washington DC, USA. Frederiksberg, Denmark and Bethesda, Md.: FIG/ACSM/ASPRS. Digital CD. 12pp.

Tryfona, N. and C. S. Jensen. 1999. Conceptual data modelling for spatiotemporal applications. *GeoInformatica* 3:245–268.

van Oosterom, P.J.M., B. Maessen and C. W. Quak. 2000. Generic query tool for spatio-temporal data. *International Journal of Geographical Information Science* 16 (8):713–748.

van Oosterom, P.J.M. 2006. Constraints in spatial data models, in a dynamic context. Chapter 4 in *Dynamic & Mobile GIS: Investigating Changes in Space and Time*, edited by J. Drummond, R. Billen, D. Forrest and E. João, 104–137. London: Taylor & Francis.

Worboys, M. F. 1994. Unifying the spatial and temporal components of geographical information. Pages 505–517 in *Advances in GIS: Proceedings of the 6ᵗʰ Symposium on Spatial Data Handling*, Edinburgh, Scotland, September 1994. Edited by T. Waugh and R. Healey. London: Taylor & Francis.

XSD. 2004. *XML Schema.* <http://www.w3.org/XML/Schema> Accessed 15 January 2007.

CHAPTER 7
Qualitative Spatial Reasoning and GIS

Xiaobai Yao

7.1 Introduction

Qualitative spatial relations are fundamental concepts in human spatial cognition and spatial reasoning. Qualitative spatial reasoning (QSR) studies ways to represent qualitative spatial relations, to analyze them, and to draw conclusions with them. QSR is increasingly important to geographic information systems (GIS), as well as other geospatial technologies, on several accounts. First of all, human beings often have qualitative abstractions rather than complete a priori quantitative knowledge about space (Cohn and Hazarika 2001). It is a logical expectation that modern geospatial technologies have the functionality to make use of fundamental spatial concepts, such as qualitative spatial relations. Secondly, human beings have the striking ability to make sensible judgments even when they only know qualitative information about a situation (Kuipers 2004). It is desirable to incorporate or take advantage of such abilities in GISs. Finally, knowledge discovery of vast geographical data has received enormous research efforts in the past few years due to the still dramatically increasing volumes of data which are often mixes of qualitative and quantitative data. Therefore, representing and "mining" the qualitative information along with the metric data is an emergent research challenge (Santos and Amaral 2004).

Current GISs are built on a quantitative foundation with spatial representation and spatial analysis primarily designed in metric systems. Quantifiable spatial relations are readily embedded in current GISs. Typically, various metric distances, such as Euclidean distance and network distances, are among the most popular metric measures of spatial relations in GIS. While current GISs have been inherently tuned for metric representations, the capabilities for the representation and analysis of qualitative spatial relations are very limited. Research challenges are particularly seen for theoretical and technological development for qualitative spatial reasoning in GIS.

This chapter will focus on modeling approaches to the representation and reasoning of qualitative spatial relations in GIS. However, it is worthwhile to note that QSR research efforts have actually been made from various perspectives for a wide range of applications. Researchers from a variety of disciplines have contributed to the understanding and modeling of the ways in which human beings learn and reason about space. Reported investigations include those by behavioral and cognitive scientists on how people perceive and reason about space (e.g., Lynch 1960; Golledge and Rushton 1976; Montello and Frank 1996; Montello 1997; Lloyd 1997; Golledge 1998), and those by linguists and geographers on the relationship between natural language and perceptual representation of space (e.g., Lakoff 1987; Talmy 1988; Mark and Frank 1992; Kemmerer 1999). More recently, there come lines of research that lead to computational models of spatial cognitive maps (e.g., Kuipers 1983; Kuipers and Levitt 1988; Yeap 1988; Gopal et al. 1989; Touretzky et al. 1993; Ghiselli-Crippa et al. 1996) and formal models about the representation and reasoning of qualitative spatial relations, which is the focus of this chapter.

This chapter starts with a review and critique of qualitative spatial reasoning models that are either currently used or could potentially be implemented in GIS. This is followed by a recent experiment of bringing some QSR models into GIS. The chapter concludes with a vision of next-generation GIS having QSR capabilities and a discussion of some theoretical and technical barriers for further research.

7.2 Qualitative Spatial Reasoning

Qualitative spatial relations express in natural language the relations between spatial features. Three fundamental spatial relations and formal models of QSR on them are examined in this section. They are: 1) topological relations that describe neighborhood and incidence; 2) direction relations (e.g., north, south) and orientation relations (e.g., left and right) that describe angular order; and 3) distance relations (proximity) such as "near" and "far."

7.2.1 Topological Relations

Topology is a branch of mathematics concerning the properties of geometric figures or solids that are not changed by homeomorphisms (bicontinuous one-to-one transformations) such as stretching or bending. In this paper, we are concerned with topological relations between spatial figures as represented in GIS. A naïve view of topology can be seen as geometry on a rubber sheet, as topological relations are preserved even if the two involved spatial figures are translated, rotated, or scaled (Egenhofer 1991; Chen et al. 2001). The studies and associated theories of topological spaces show up in several branches of mathematics including general topology, graph theory and more. Most influential are general topology (or point-set topology) and algebraic topology. General topology defines and studies useful properties (connectivity, containment, continuity) of spaces and maps (Csaszar 1963), while algebraic topology provides a powerful tool to study topological spaces (Spanier 1966).

In geographic information science (GIScience), research efforts have focused on the formalization of topological relations between geographical features. Two families of topological models, the family of intersection models and the family of region connection calculus (RCC) models, have been proposed to represent and reason about topological relations in metric geographical systems.

Initially developed by Egenhofer and colleagues (Egenhofer and Franzosa 1991; Egenhofer and Herring 1991), the family of intersection models became most influential in the GIScience community. The traditional intersection framework provides a formal presentation of topological relations in GIS for subsequent expansions of the family (Shariff et al. 1998; Papadias et al. 1999; Chen et al. 2001). There are 4- and 9-intersection models in the framework, depending on the choices of primitives. In the 9-intersection model, each spatial feature is associated with three primitive sets: interior (denoted by $A°$), exterior (denoted by A^-), and boundary (denoted by ∂A). Figure 7-1 illustrates the primitives and clearly shows that the union of the interior, boundary, and exterior is the universe of discourse. The 4-intersection works very similarly while keeping two primitives only, namely the interior and boundary.

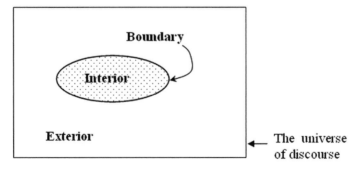

Figure 7-1 The topological primitives of interior, boundary, and exterior in the 9-intersection model.

In the 9-intersection model, the topological relation of any two spatial features boils down to the values of the nine intersections of primitive sets (interior, boundary, and exterior) of the two features. For two entities A and B, the nine intersections are shown in Equation 1 (adapted from Egenhofer 1991):

$$R_9(A,B) = \begin{bmatrix} A^\circ \cap B^\circ & A^\circ \cap \partial B & A^\circ \cap B^- \\ \partial A \cap B^\circ & \partial A \cap \partial B & \partial A \cap B^- \\ A^- \cap B^\circ & A^- \cap \partial B & A^- \cap B^- \end{bmatrix} \qquad (7\text{-}1)$$

The model considers values (i.e., empty or non-empty) of the nine intersections. Combinations of the nine intersections concisely describe topological properties.

Although the traditional intersection framework is theoretically sound and it works perfectly well for textbook situations, it still faces challenges of more complicated situations when it comes to the issue of handling topology-relation spatial terms in natural languages. One type of such complication is caused by the richness and sophistication of spatial terms in natural languages. Many topology-related spatial terms also have implications on metrics. For example, "the road bypasses a city" and "the road ends just outside of the city" have the same topological configuration but clear metric differences. The presence of both topology and metric properties in the same spatial terms suggests that interpreting these topology-and-metric spatial terms in GIS inevitably involves the use of metric principles. Shariff et al. (1998) studied this problem and added a computational extension of the intersection model to account for both topological and metric properties. The extension uses data from a human-subject survey involving a set of approximately 60 English spatial terms that contain both topology and metric implications. Although the above study concerns spatial relations between linear and areal objects, the same problem exists for the point-polygon or polygon-polygon spatial relations. For instance, both "in" and "at the center of" imply the same topological relation (containment), whereas the two linguistic expressions provide different levels of detail in metrics.

The imprecision in a person's cognition and interpretation of spatial relations is attributed to another type of complication. This is particularly relevant because inexact or inaccurate location information is commonly seen in spatial databases and in human spatial reasoning and queries. Figure 7-2 illustrates two topological relationships that are difficult to distinguish. The relationships sketched in this diagram look very similar. The only difference between them is the size of the overlapping area in the "overlap" relation.

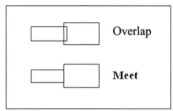

Figure 7-2 Examples of similar topological relations.

If the overlapping area keeps shrinking until it becomes hardly distinguishable from a line, the "overlap" will be so similar to the "meet" that people cannot discriminate between the two diagrams. One may perceive the topology of any of the two cases either as "meet" or as "overlap." Several solutions have been proposed in response to this issue. Egenhofer and Mark (1995a) developed two models, the snapshot model and the smooth-transition model, to measure the similarity between topological line-region relations. This model constructs the so-called conceptual neighborhood by setting a pair of topological relations as neighbors according to certain criteria. For example, the smooth-transition model defines the criteria of neighbor as that in which the two topological relations can be transformed to each other by smooth deformations. Based on the conceptual neighborhood construction, Papadias et al. (1999) further developed a quantifiable topological similarity measure. With this measure, one can examine the similarity between a topological relation and each of the other topological relations in a system. All other relations that are sufficiently similar (meaning within a certain threshold) to the tested relation will be included for consideration as a possible alternative to this relation. These constructions extend the capability of the traditional intersection model by accommodating the uncertainty associated with the inherent fuzziness in natural language or with imprecision in human cognition.

The other family of topological models based on Region Connection Calculus (RCC) (Randell et al. 1992; Cohn et al. 1997) and so is called the RCC model. The RCC approach takes regions of space as the only primitive. It defines a set of Jointly Exhaustive and Pairwise Disjoint (JEPD) dyadic relations between pairs of regions. The JEPD relations are then used as primitive elements for more sophisticated constructions. A set of relations is said to be JEPD when the following holds true: for any two spatial features, one and only one of the relations holds. Therefore, the set provides an exhaustive qualitative classification of possible dyadic relations. Two RCC schemes exist in the literature, known as RCC5 and RCC8, each named with the number of basic JEPD relations defined in the scheme. Building on the mathematical foundation of predicate calculus, the RCC model supports a set of theorems and functions with the primitive dyadic relation $C(x,y)$ and the JEPD relations as variables. These theorems and functions become quite an expressive sub-language of RCC (Cohn et al. 1997), which makes the model flexible and powerful. With compositions of these basic relations, a wide range of topological relations and spatial properties can be modeled.

By and large, the intersection model and RCC model work similarly in the representation and modeling of topological relations. Both the intersection family and the RCC family have been extended in a very similar way to represent and reason about regions with indeterminate boundaries. Cohn and Gotts (1996) presented the egg-yolk theory based on the RCC model. In this theory, as shown in Figure 7-3, the egg is the maximal extent of an uncertain region and the yolk is the minimal extent of the uncertain part. The egg white is the indeterminacy. The outline of the region can extend anywhere within the gray area (egg white). The intersection model is extended similarly by Clementini and Di Felice (1996) to model the topological relations between objects with undetermined boundaries. The approach introduces the concept of "region with broad boundaries" which consists of an inner boundary and an outer boundary. With this approach, the represented region is like a standard homogeneous region with a hole in it. The inner and outer boundaries represent the minimum and the maximum extent of the region itself, which expresses the indeterminacy of the region boundaries.

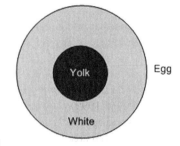

Figure 7-3 The egg/yolk interpretation (adapted from Cohn and Gotts 1996).

7.2.2 Cardinal Directions and Orientation Relations

Cardinal directions refer to the primary directions such as north, south, east, west, and the derivative directions northeast, northwest, southeast, and southwest. Spatial orientation here refers to spatial direction information with respect to a specific spatial feature or a specific reference system. Examples include left, right, front, back, same, opposite, etc. Both orientation and cardinal direction information concern angular relations expressed in natural language. Because of the close relevance to quantitative angular measure, they are also believed semi-qualitative. Despite the similarity between direction and orientation relations, the major difference between them lies in the underlying frame of reference (FofR) that is implied by a direction or orientation relation. Cardinal directions are situated in the three-dimensional geographical coordinate systems. Yet the orientation relations do not require a fixed reference system (Freksa 1992). A spatial term of orientation could have its own frame of reference, which may range from a global geographic reference system to a subject-oriented, feature-specific reference system. In fact, many a spatial term of orientation (e.g., front, left) applies a certain local reference system which is usually defined with a local object (spatial feature or human subject) of interest. In this regard, we can say cardinal direction is a specific type of spatial orientation. Thus, they are reviewed here together.

Previous studies investigated the properties of cardinal directions and orientations. Peuquet and Zhan (1987) observed the "triangular" property – the area that is covered for any given direction increases with distance so that the shape of the coverage of a direction is triangular. Frank (1992, 1996) describes some properties of directions with a set of formal algebraic axioms. Freksa (1992) discussed the "periodicity" property of this type of spatial relation.

This property refers to the fact that the directions are ordered in a circular pattern. The periodicity property works in concert with the proposed idea of "conceptual neighborhood" of direction/orientation relations. Given a specific qualitative resolution of direction/ orientation relations (e.g., left, front, right, back), conceptual neighbors are the direction/ orientation relations in a direct transition from a given relation. In the above example, both "right" and "left" are conceptual neighbors of the relation "front."

Let us first look at QSR models for cardinal directions, and then discuss orientation reasoning with special focus on the topic of frame of reference.

In modeling cardinal directions, some studies consider four primary directions (north, south, east, west) while others include eight directions (adding northeast, northwest, southeast, southwest). As the choice of the number of directions does not affect the modeling methods, eight directions will be used for illustration purpose. Three approaches have been proposed to qualitative reasoning with cardinal directions. Figure 7-4 illustrates the centroid-based approach. In this scheme, the direction between two objects is determined by the angle between the two object-to-centroid lines (Frank 1992, 1996; Papadias et al. 1999).

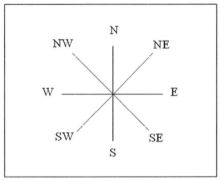

Figure 7-4 Cardinal directions defined by half-planes and cones (adapted from Frank 1996).

The second approach, projection-based-directions with neutral zone (Frank 1992, 1996), treats the central area around the origin as a neutral zone within which cardinal directions are not decided the same way as other zones. As illustrated in Figure 7-5a, this approach divides the entire space into nine regions: the central neutral area (around the reference point where directions will be measured), four regions where only one primary direction applies, and four regions where two primary directions apply. The approach first defines the four directions (north, south, east, west) as pair-wise opposites and each pair divides the plane into two half-planes, as shown in Figure 7-5b. The other four directions (northwest, northeast, southwest, southeast) further divide the half-planes into four quarters, as shown in Figure 7-5c. By adding the neutral zone, this approach alleviates the problem of the centroid-based approach that when two points are close to each other, the determination of direction may be very sensitive to measurement accuracy. Although the idea of "neural zone" is proposed for the projection-based-directions scheme, there is no reason why the same idea cannot be extended to the centroid-based approach as well.

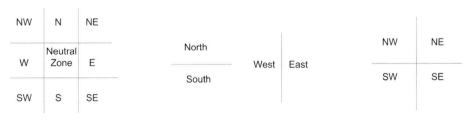

(a) directions with neutral zone   (b). Half-planes of (N,S) and (E,W)   (c) More directions defined by half-planes

Figure 7-5 Cardinal directions defined with neutral zone (adapted from Frank 1996).

The third approach extends the above-mentioned two approaches with fuzzy logic since each of the above two schemes has its advantages and disadvantages. One common weakness of both schemes is that the clear-cut boundary of adjacent directions makes both schemes intolerant to data inaccuracy. This problem is dealt with in the third approach. Because it is not cognitively plausible to make clear-cut spatial boundaries for the territory of each cardinal direction, Papadias et al. (1999) modified the previous two approaches by expanding the crisp boundary to a fuzzy boundary. A fuzzy boundary itself contains a range of angular directions, each with a defined membership grade to describe its degree of suitability for the specific cardinal direction. Based on fuzzy subset theory that was defined by Zadeh (1965) and Kaufmann (1975), the membership function was defined in the following format:

$$A = \int_U \mu_A(x)/x \qquad (7\text{-}2)$$

where A is a fuzzy subset in the universe of discourse U. In this equation, $x \in U$ and $\mu_A(x)$ indicate the grade of membership of x in A. $\mu_A(x)/x$ is an expression of the ordered pair $(x, \mu(x))$. In this case, x is any metric angular direction measure. Papadias et al. (1999) applied the trapezoidal function of x to capture membership grade $\mu_A(x)$. In this approach, a metric angular direction may be assigned to multiple cardinal directions with different membership grades. Hence, the territories of the cardinal directions are not exclusive, but overlapping. This approach increases the system's tolerance to errors and inaccuracy of the spatial database.

Spatial orientation shares many modeling issues with that of the cardinal directions except for the frame of reference (FofR), which is fundamental to the concept of orientation. Frame of reference is embedded in language as a means of representing the locations of entities in space (Klatzky 1998) by implying some details of underlying coordinate systems and geometry. The notion of "frame of reference" is crucial to the study of spatial cognition across disciplines and modalities (Levison 1996). The classifications and terminologies are different across disciplines or even within the same discipline. In response to the somewhat confusing and inconsistent terminologies with respect to classifications of reference frames, Frank (1998) proposes to classify reference frames by three parameters: that of the reference object (the ground), the orientation of the reference frame, and the handedness of the reference frame (right or left-handed).

Figure 7-6 illustrates some examples of FofR for a pair of primary objects (POs) and reference objects (ROs). Let us examine two sentences "the tree is in front of the house" and "the tree is to the left of the house." As illustrated in Figure 7-6a, the first sentence suggests an intrinsic FofR with the orientation term "front." The FofR is said to be intrinsic because it makes use of the inherent properties (front/back) of the referenced object. The second sentence brings up a relative FofR because the orientation ("left") is relative to the observer, as shown in Figure 7-6b. Figure 7-6c displays an example of absolute FofR. In this example, no external information is necessary except for the locations of the primary and reference objects.

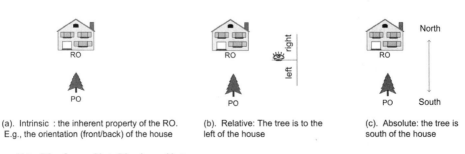

(a). Intrinsic : the inherent property of the RO. E.g., the orientation (front/back) of the house

(b). Relative: The tree is to the left of the house

(c). Absolute: the tree is south of the house

Note : RO: reference object,; PO: primary object

Figure 7-6 Three examples of frames of reference. See included DVD for color version.

Freksa (1992) developed an intriguing approach to representing and reasoning about qualitative orientation information. In this approach, a one-dimensional feature is represented by an oriented line with an ordered pair of locations a and b. Then the space around the vector ab is partitioned into a finite number of qualitative orientation measures. Based on this representation and also borrowing the idea of Allen (1983)'s interval-based temporal reasoning, Freksa (1992) further developed an inference scheme for orientation-based spatial reasoning. The scheme infers orientation relations between two vectors from known orientation relations associated with the vectors.

7.2.3 Proximity Relations

Proximity is the spatial relation about distance expressed in natural languages. It is also referred to as qualitative distance or linguistic distance. Two distinct characteristics of proximity, the inherent fuzziness and context-contingency, are emphasized in the literature. One or both characteristics are also the key factors in modeling proximity in relation to metric distances. Most previous studies use fuzzy logic to capture the inherent fuzziness of proximity spatial relations (e.g., Robinson 1990; Gahegan 1995; Worboys 2001; Guesgen and Albrecht 2000; Guesgen 2002a; Yao and Thill 2005). Robinson (1990, 2000) used a computerized machine learning method to conceptualize nearness between a city and adjacent cities by a question-answering scheme. Worboys (2001) also introduces several approaches to analyzing proximity data. The approaches are based on premises of crisp set, fuzzy set, and rough set, respectively. The parameters are estimated from survey data. Guesgen and Albrecht (2000) and Guesgen (2002a, 2002b) present an approach that explicitly represents a proximity relation by a fuzzy set. However, no attempt is made to estimate the membership functions. Alternatively, they merely take an inverse quadratic function.

Another stream of research looks at the context-contingency of the proximity relations. A significant body of literature endorses the concept that human beings reason about proximity while considering the context of the proximity perception (e.g., Gahegan 1995; Hernandez et al.1995; Worboys 1996, 2001; Yao and Thill 2005). In the light of observations drawn from psychometric testing of perceived proximity, Gahegan (1995) suggests the scale of the area in consideration, the attractivness of the perceived object, and the network reachability as context factors that influence human perception of proximity. Starting from the earlier investigations, Yao and Thill (2005) classify context factors into two categories, namely subjective factors and objective factors. Subjective factors are those variables serving as surrogates for personal characteristics of a distance perceiver. Examples include the perceiver's levels of familiarity with the area and/or the route, the perceiver's time and financial budget, demographic characteristics, and socio-economic characteristics. Objective factors are those context variables that are independent of the distance perceiver. Scale of the study area, type of activity, intervening opportunities, transportation mode, and traffic conditions are examples of objective context factors. The figure-ground relationship, an important concept in spatial cognition and related fields of studies such as cartography, also contributes to the context of a proximity distance perception. Here figure refers to the more prominent feature, while ground is the less important feature in a proximity perception. The figure-ground relationship in a proximity relation has close ties to the asymmetry property of proximity relations, which was first observed by Sadalla and colleagues (1980). The asymmetry property says that there is the unequal human perception of nearness between two locations, with more significant reference points or landmarks being more often perceived to be near to other points than vice versa (Duckham and Worboys 2001). The property was also confirmed by more recent empirical findings. For instance, in a case study of distance perception, Worboys (2001) identified many cases where place x is perceived as near place y, while place y is not perceived as near place x. The asymmetry property reinforces the importance of context in proximity modeling. What-is-the-figure-and-what-is-the-ground is part of the context to be considered. The other possible context factor is the frame of reference which is in place when a proximity relation involves orientation type of information. Similar to the previous discussion of the role of FofR in spatial orientation, researchers (e.g., Hernandez et al. 1995; Clementini et al. 1997) examined how FofR serves as a means to describe orientation type of context in proximity modeling.

7.3 Analyzing Qualitative Spatial Relations in GIS

7.3.1 Challenges of Having QSR Capabilities in GIS

Revealed in the above review, analyzing qualitative spatial relations in GIS is a rather thorny task. There are theoretical barriers as well as methodological hurdles. Two theoretical challenges are particularly discussed here. The first is the inherent fuzzy nature of natural language and human perception, as it is not consistent with current GISs that are fundamentally quantitative systems. The second is the context-contingency with the interpretation of most qualitative measures. Fuzziness has been discussed in modeling efforts for all three types of qualitative spatial relations. Context-contingency is another significant research challenge common for the qualitative spatial relations in regard to the translation between qualitative and quantitative measures. With human perception of spatial relations expressed in natural language, the interpretation of a qualitative measure heavily depends on the physical environment in which the qualitative measure is perceived and the human environment (personal, cultural, etc.) by which the qualitative measure is taken. Discussions of context issues are emphasized in many studies of qualitative spatial relations, explicitly or implicitly (e.g., Gahegan 1995; Worboys 2001; Yao and Thill 2005). Yet context-contingent computational models with real-world data have been very limited. In addition to the theoretical issues, there are also technical barriers for the implementation of QSR models in GIS. One significant technical barrier concerns the prevalent GIS data models and data structures. Current data models, data structures, and query algorithms have been designed for metric-based location data. Yet many QSR models have different requirements on data representation and retrieval, particularly due to the fuzziness nature in qualitative spatial information. Other technical issues include user interface and computational load, especially in comparison with the increasingly higher demand of real-time responses. The need is pressing for studies to remove theoretical and technological barriers that hold back the realization of qualitative spatial reasoning power in GIS.

Given the theoretical and technical challenges, current GISs have very limited, if any, QSR capabilities. In fact, most of the reasoning power that is enabled by the existing QSR models has not been brought to life in current GIS packages and services. Table 7-1 shows the result of a brief survey about the qualitative spatial representation and reasoning capabilities in some most popular GIS packages and online services. Proximity and cardinal directions/ orientation are not handled at all in any of these GIS packages. For topological relations, all three GIS packages in the survey can identify the topological relations as in the traditional 9-intersection model. However, it should be noted that the GIS packages are based on graph theory and geometric principles, rather than the Intersection Model or the RCC Model. As a result, what is missing in these packages is the extensibility of the Intersection and RCC Models. The rich set of extended Intersection Models and RCC Models and the associated extra power in representation and reasoning, particularly concerning inherent uncertainties, are currently unavailable and also will not be easy to implement in GISs. The two popular online GIS services shown in Table 7-1, Google Local and MapQuest, can handle a few topological relations such as connectivity between two linear features (e.g., street intersections) and limited proximity relations such as "near" or "nearby." They are limited in the sense that the "near" relation is not handled in a context-sensitive manner, but instead seems to be directly translated into a buffer area of a certain distance that is predefined by the package. In sum, the survey tells us that qualitative spatial relation representation and reasoning capability is extremely limited in current GIS packages and applications.

Table 7-1 A brief survey of qualitative spatial relation representation and reasoning capabilities in some major GIS packages and online GIS services.

GIS package or GIS services	Topology		Proximity	Cardinal Directions & Orientation
	Basic topological relations (e.g., as those defined in the Intersection model)	Expanded topological representational and reasoning power for natural language input		
ArcGIS (ESRI)	Yes	No	No	No
TransCAD (Caliper)	Yes	No	No	No
ERDAS Imagine (Leica Geosystems)	Yes	No	No	No
Google Local (maps.google.com)	Some	No	Very Limited	No
MapQuest (www.mapquest.com)	Some	No	Very Limited	No

7.3.2 A Case-study of Interpreting Proximity Relations in GIS

A study of constructing proximity models based on case-specific data and then implemented in GIS will be discussed. The purpose is two-fold. First, it shows recent research advancement in response to the two theoretical barriers of QSR. Second, it demonstrates the feasibility of implementing context-contingent QSR in GIS.

Because current GISs operate in quantitative environments, a natural solution to representing and analyzing proximity relation in GIS is to establish a mapping mechanism for the translation between qualitative and quantitative distance measures. For such a mapping mechanism, it is essential to account for the inherent characteristics of proximity relations, the fuzziness and the context-contingency. Nonetheless, most existing computation models account for either the fuzzy nature or the inherent context-contingency of proximity relations, but not both. Furthermore, in previous studies modeling for the fuzziness, the form of the membership function is typically pre-defined without proving the validity. This case study involves new strategies to cope with fuzziness and context-contingency. It introduces two modeling approaches that take case-specific inputs in the constructed proximity models. In addition, the second approach captures the fuzziness without presuming the form and parameters of the fuzzy relationship in the model. Finally, the case study further demonstrates the feasibility of implementing the proposed QSR into GIS.

This study starts with a questionnaire survey among undergraduate students at the University at Buffalo, the State University of New York, while the modeling methodologies are developed to be generally applicable. Collected data include distances (both linguistic and corresponding metric distances) and context information for hypothetical trips perceived by 95 randomly selected undergraduate students at the university. Each student answered a questionnaire which contains questions about 19 hypothetical home-based trips in the metropolitan area. For each hypothetical trip, a trip scenario is predefined in the survey design or chosen by each survey participant. Participants are asked to indicate their perception of each trip length given the trip scenario. The perception can be expressed by one of the five linguistic distance measures: very near, near, normal (not so near, not so far), far, and very far. The actual network distance calculated along the shortest path is a surrogate of metric trip distance. The context factors collected in the survey consist of personal characteristics of the perceiver (subjective context variables) and circumstances of the metric distance that is perceived (objective context variables). To be specific, these context variables include scale, type of activity, topological closeness, intervening opportunities, transportation mode, the familiarity level of the environment, financial and time budget, as well as personal and demographic characteristics of the perceiver.

Based on the survey data, two approaches were proposed to model the relationship between the qualitative measures and metric distance measures mediated by the context variables. The first is an ordered logit model composed of the following fitted regression curves based on the survey data (Yao and Thill 2005):

$$
\begin{aligned}
\mathbf{logit}(p_1) &= 0.916 - y \\
\mathbf{logit}(p_1)+(p_2) &= 2.467 - y \\
\mathbf{logit}(p_1)+(p_2)+(p_3) &= 4.513 - y \\
\mathbf{logit}(p_1)+(p_2)+(p_3)+(p_4) &= 7.283 - y \\
p_5 &= p_1 - p_2 - p_3 - p_4
\end{aligned}
\tag{7-3}
$$

where p_1 is the probability that the linguistic distance be "very near," p_2 is the probability that it be "near," and so on, each corresponds to one of the five proximity measures in the questionnaire survey. Here y stands for the summation of contributions from all significant context variables, weighted by the corresponding regression coefficients:

$$
\begin{aligned}
y = {}& 0.207\ DISTANCE - 0.633\ ACTIVITY1 + 0.638\ ACTIVITY2 + 0.114\ ACTIVITY3 \\
& - 0.613\ ACTIVITY4 - 0.038\ ACTIVITY5 + 2.287\ MODE1 + 1.024\ MODE2 + 1.882\ MODE3 \\
& + 2.398\ FAMILIARITY1 + 1.893\ FAMILIARITY2 + 1.766\ FAMILIARITY3 + 0.925\ FAMILIARITY4 \\
& + 1.294\ ENVIRONMT1 + 0.303\ ENVIRONMT2 + 0.430\ ENVIRONMT3 - 1.68\ CAR + 0.525\ MALE \\
& + 0.419\ ETHNICITY1 + 1.168\ ETHNICITY2 - 0.397\ ETHNICITY3 + 0.862\ ETHNICITY4
\end{aligned}
\tag{7-4}
$$

The coefficients also reveal the degree of impact each context factor has on the probability of the proximity measures.

A second approach, the neuro-fuzzy approach (Yao and Thill 2007), aims to develop a systematic way to explore the fuzzy relations between qualitative proximity measures and the corresponding metric measures without making unproven assumptions of the fuzzy membership functions. The starting point of the approach is that the membership function of the fuzzy relation is a function of the two distance measures and of the context variables of the proximity measures.

$$
\mu_{ij} = f(L_i, M_j, C_1, \ldots, C_n)
\tag{7-5}
$$

This approach then uses neuro-fuzzy inference to derive the fuzzy membership functions that best fit the sample data. A neuro-fuzzy system is a hybrid system that integrates fuzzy logic and neural networks. Particularly in this practice, the fuzzy relationship as defined in Equation 7-5 can be seen as fuzzy functions comprised of IF-THEN fuzzy rules. The variables and parameters of such fuzzy rules are specified through training processes that seek to maximize the model's fitness to existing sample cases (known as training cases). The choices of variables are selected through a stepwise strategy. Combinations of any number of candidate variables are tested and trained starting from the least number of input choices. The approach adds more input context variables until no improvement of model prediction accuracy can be achieved. Figure 7-7 illustrates an example of a 2-input-variable neuro-fuzzy system structure.

Each of the above two approaches yields a satisfactory model that translates between proximity measures and metric distance measures, while the neuro-fuzzy approach gives slightly better performance. Readers are referred to the original articles for details. Although the

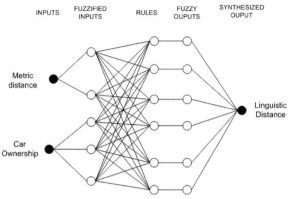

Figure 7-7 An example of a neuro-fuzzy system configuration for proximity modeling (Yao and Thill 2007).

models in the prototype are designed for a certain group of people in a given city, the modeling approaches are generally applicable to different groups of people in any geographical region.

The models are subsequently implemented in a prototype GIS for case-specific proximity reasoning capabilities. The interface of a query example illustrated in Figure 7-8a shows the user interface allowing users to query with proximity measure. The query in this example is to "find nearby parks" from the reference point as shown in Figure 7-8b. The result contains multiple parks around the reference location and each has a membership grade to the concept of "near." The multiple locations are displayed with proportional symbols in Figure 7-8b. Because the context variables and associated parameters are based on the survey data from undergraduate students at the University at Buffalo, the query results will be most plausible to these subjects. However, it is possible to apply the same modeling procedures for data collected from individual or group users. Thus the same GIS prototype can use individual-based or group-based QSR models.

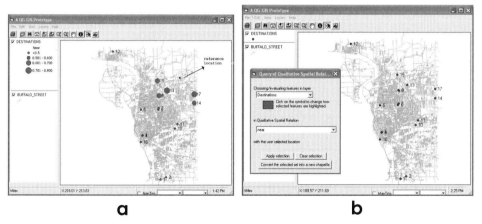

Figure 7-8 A prototype GIS with some proximity spatial reasoning capabilities (Yao and Thill 2006). See DVD for color version. See included DVD for color version.

7.4 Conclusion

Envisioning the significant role of qualitative information in next-generation GIS applications for the general public, I argue that research on QSR becomes increasingly more important as it is driven by the rapid development of new enabling information and communication technologies. Ubiquitous computing with location-aware devices is offering the general public a profusion of information on the spatial environment around them.

Easy access of the general public to vast spatial information brings along opportunities and research challenges of providing qualitative-data-handling capabilities in these GIS services. A major contribution can be made by incorporating qualitative spatial reasoning capabilities in GIS as endorsed by the notion of Naïve Geography (Egenhofer and Mark 1995b), a seminal discourse that addresses the important issue of accommodating common-sense geographic queries. Kuipers (2004) also emphasized the importance of current information technology to use common sense reasoning and to make useful conclusions out of it. For similar considerations, Egenhofer and Kuhn (1998) suggest retooling the old paradigms of desktop GISs for new application needs. The time is ripe to instill new theories and tools into GIS to represent and analyze qualitative spatial information.

Retooling GIS to make it ready for quantitative data handling is by no means a trivial endeavor. The fundamental barriers are rooted in the mismatch between the inherent fuzzy nature in qualitative measures and the pursuit of high precision in metric systems. Qualitative spatial relations (e.g., north, near) generally have substantially coarser granularity than corresponding metric measures in a GIS. This brings about uncertainty in translating between qualitative spatial relation and the metric measures. Furthermore, human subjectivity is inevitably introduced in qualitative measures in the human perception process. The human factors together with the relevant objective environmental conditions which have been collectively called context factors, have significant impacts on a person's perception of some qualitative spatial relations (e.g., proximity measures). These innate characteristics of qualitative spatial relations suggest that the translation (into metric forms) of a qualitative spatial relation faces uncertainty, and may ideally need additional case-specific contextual information. The characteristics also make it clear that representing and reasoning about qualitative spatial relations in a metric environment call for knowledge and efforts from a variety of perspectives including linguistics, geography, cognitive science and information technologies. Moreover, there also are some technical hurdles for the implementation of QSR models in GIS. Future research efforts in all these areas are expected towards the realization of qualitative spatial reasoning capabilities in future GIS.

References

Allen, J. F. 1983. Maintaining knowledge about temporal intervals. *Communications of the ACM 26*, 11:832–843.

Chen, J., C. Li, Z. Li and C. Gold. 2001. A Voronoi-based 9-intersection model for spatial relations. *International Journal of Geographical Information Science* 15, 3:201–220.

Clementini, E. and P. Di Felice. 1996. An algebraic model for spatial objects with indeterminate boundaries. Pages 155–170 in *Geographic Objects with Indeterminate Boundaries*, edited by P. A. Burrough and A. U. Frank. Bristol, Penn.: Taylor & Francis.

Clementini, E., P. Di Felice and D. Hernandez. 1997. Qualitative representation of positional information. *Artificial Intelligence* 95. 2:317–356.

Cohn, A. G. and N. M. Gotts. 1996. The 'egg-yolk' representation of regions with indeterminate boundaries. Pages 171–187 in *Geographic Objects with Indeterminate Boundaries*, edited by P. A. Burrough and A.U. Frank. Bristol, Penn.: Taylor & Francis.

Cohn, A. G., B. Bennet, J. Gooday and N. M. Gotts. 1997. Qualitative spatial representation and reasoning with the region connection calculus. *Geoinformatica* 1, 3:275–316.

Cohn, A. G. and S. M. Hazarika. 2001. Qualitative spatial representation and reasoning: an overview. *Fundamental Informaticae* (The Netherlands: IOS Press) 46, 1–2:1–29.

Csaszar, A. 1963. *Foundations of General Topology*. New York: The Macmillan Company.

Duckham, M. and M. Worboys. 2001. Computational structure in three-valued nearness relations. Pages 76–91 in *Proceedings: Spatial Information Theory, Foundations of Geographic Information Science*. International Conference, COSIT 2001. *Lecture Notes in Computer Science 2205*, edited by D.R. Montello. Berlin: Springer.

Egenhofer, M. J. 1991. Reasoning about binary topological relations. Pages 143–160 in *Proceedings of the 2nd Symposium on Large Spatial Databases*, Lecture Notes in Computer Science, 523. New York: Springer-Verlag.

Egenhofer, M. J. and R. Franzosa. 1991. Point-set topological relations. *International Journal of Geographical Information Systems* 5, 2:161–174.

Egenhofer, M. J. and J. Herring. 1991. *Categorizing Binary Topological Relations between Regions, Lines, and Points in Geographic Databases*. Technical Report, Department of Surveying Engineering, University of Maine.

Egenhofer, M. J. and D. M. Mark. 1995a. Modeling conceptual neighborhoods of topological line–region relations. *International Journal of Geographical Information Systems* 9, 5:555–565.

Egenhofer, M. J. and D. M. Mark. 1995b. Naive geography. Pages 1–15 in *Spatial Information Theory: A Theoretical Basis for GIS*, edited by A. U. Frank and W. Kuhn. Berlin: Springer-Verlag.

Egenhofer, M. J. and W. Kuhn. 1998. Beyond desktop GIS, paper presented at *GIS PLAN-ET*, Lisbon, Portugal.

Frank, A. 1992. Qualitative spatial reasoning about distances and directions in geographic space. *Journal of Visual Languages and Computing* 3:343–371.

———. 1996. Qualitative spatial reasoning: Cardinal directions as an example. *International Journal of Geographical information Systems* 10(3):269–290.

———. 1998. Formal models for cognition – Taxonomy of spatial location description and frames of reference. Pages 293–312 in *Spatial Cognition: An Interdisciplinary Approach to Representing and Processing Spatial Knowledge*, edited by C. Freksa, C. Habel, and K. F. Wender. Berlin: Springer.

Freksa, C. 1992. Using orientation information for qualitative spatial reasoning. Pages 162–178 in *Theories and Methods of Spatio-Temporal Reasoning in Geographic Space*, edited by A. U. Frank, I. Campari and U. Formentini. Berlin: Springer-Verlag.

Gahegan, M. 1995. Proximity operators for qualitative spatial reasoning. Pages 31–44 in *COSIT `95 Proceedings: Spatial Information Theory - A Theoretical Basis for GIS, Lecture Notes in Computer Science* 988, edited by A.U. Frank and W. Kuhn. Berlin: Springer.

Ghiselli-Crippa, T., S. C. Hirtle and P. W. Munro. 1996. Connectionist models in spatial cognition. In *The Construction of Cognitive Maps*, edited by J. Portigali, 87–104. Dordrecht: Kluwer Academic Publishers.

Golledge, R. G. 1998. *Wayfinding Behavior: Cognitive Mapping and Other Spatial Processes*. Baltimore: Johns Hopkins University Press.

Golledge, R. G. and G. Rushton (eds). 1976. *Spatial Choice and Spatial Behavior*. Columbus: Ohio State University Press.

Gopal, S., R. Klatzky and T. Smith. 1989. NAVIGATOR: A psychologically based model of environmental learning through navigation. *Journal of Environmental Psychology* 9:309–331.

Guesgen, H. and J. Albrecht. 2000. Imprecise reasoning in geographic information systems. *International Journal for Fuzzy Sets and Systems* 113, 1:121–131.

Guesgen, H. W. 2002a. Reasoning about distance based on fuzzy sets. *Applied Intelligence* 17:265–270.

———. 2002b. Fuzzyfying spatial relations. Pages 1–16 in *Applying Soft Computing in Defining Spatial Relations*, edited by P Matsakis and LM Sztandera. Berlin: Springer-Verlag.

Hernandez, D., E. Clementini and P. Di Felice. 1995. Qualitative distances. Pages 45–57 in *COSIT `95 Proceedings: Spatial Information Theory - A Theoretical Basis for GIS, Lecture Notes in Computer Science 988*, edited by A.U. Frank and W. Kuhn. Berlin: Springer-Verlag.

Kaufmann, A. 1975. *Introduction to the Theory of Fuzzy Subsets*. New York: Academic Press.

Kemmerer, D. 1999. "Near" and "far" in language and perception. *Cognition* 73(1):35–63.

Klatzky, R. L. 1998. Allocentric and egocentric spatial representations: Definitions, distinctions, and interconnections. Pages 293–312 in *Spatial Cognition: An Interdisciplinary Approach to Representing and Processing Spatial Knowledge*, edited by C. Freksa, C. Habel, and K. F. Wender. Berlin: Springer.

Kuipers, B. 1983. Modeling human knowledge of routes: Partial knowledge and individual variation. Pages 1–4 in *Proceedings of AAAI 1983 Conference, the Third National Conference on Artificial Intelligence*, Washington, DC, August 22-26, 1983. Menlo Park, Calif.: AAAI Press.

———. 2004 Making sense of common sense knowledge. *Ubiquity* 4(45):2.

Kuipers, B. and T. S. Levitt. 1988. Navigation and mapping in large-scale space. *AI Magazine* 9:25–43.

Lakoff, G. 1987. *Women, Fire, and Dangerous Things: What Categories Reveal about the Mind*. Chicago: University of Chicago Press.

Levison, S. C. 1996. Frames of reference and Molyneux's question: Crosslinguistic evidence. Pages 109–170 in *Language and Space*, edited by P. Bloom, M. A. Peterson, L. Nadel, and M. F. Garrett. Cambridge, Mass.: MIT Press.

Lloyd, R. 1997. *Spatial Cognition: Geographic Environments*. Boston: Kluwer Academic.

Lynch, K. 1960. *The Image of the City*. Cambridge: MIT Press.

Mark, D. M. and A. U. Frank. 1992. *NCGIA Research Initiative 2: Language of Spatial Relations*. Closing Report. NCGIA Technical Report Series.

Montello, D. R. and A. U. Frank. 1996. Modeling directional knowledge and reasoning in environmental space: Testing qualitative metrics. Pages 321–344 in *The Construction of Cognitive Maps*, edited by J. Portugali. Dordrecht: Kluwer Academic.

Montello, D. R. 1997. The perception and cognition of environmental distance: Direct sources of information. Pages 297–311 in *Spatial Information Theory: A Theoretical Basis for GIS*, edited by S.C. Hirtle and A.U. Frank. Berlin: Springer-Verlag.

Papadias, D., N. Karacapilidis and D. Arkoumanis. 1999. Processing fuzzy spatial queries: A configurations similarity approach. *International Journal of Geographical Information Science* 13, 2:93–118.

Peuquet, D. J. and C.-X. Zhan. 1987. An algorithm to determine the directional relationship between arbitrarily-shaped polygons in the plane. *Pattern Recognition* 20:65–74.

Randell, D. A., Z. Cui and A. G. Cohn. 1992. A spatial logic based on regions and connection. Pages 165–176 in Proc. *3rd International Conference on Knowledge Representation and Reasoning*. San Mateo, California, 1992. San Francisco, Calif.: Morgan Kaufmann

Robinson, V. B. 1990. Interactive machine acquisition of a fuzzy spatial relation. *Computers and Geosciences* 16, 6:857–872.

Robinson, V. B. 2000. Individual and multipersonal fuzzy spatial relations acquired using human-machine interaction. *Fuzzy Sets and Systems* 113:133–145.

Sadalla, E. K., W. J. Burroughs and L. J. Staplin. 1980. Reference points in spatial cognition. *Journal of Experimental Psychology: Human Learning and Memory*. 6(5):516–528.

Santos, M. Y. and L. A. Amaral. 2004. Mining geo-referenced data with qualitative spatial reasoning strategies. *Computers & Graphics* 28:371–379.

Shariff, A. R., M. J. Egenhofer and D. M. Mark. 1998. Natural-language spatial relations between linear and areal objects: The topology and metric of English-language terms. *International Journal of Geographical Information Science* 12, 3:215–246.

Spanier, E. 1966. *Algebraic Topology*. New York: McGraw-Hill Book Company.

Talmy, L. 1988. How language structures space. In *Cognitive and Linguistic Aspects of Geographic Space*, edited by D. Mark, B-1–B-11. NCGIA Technical Report 88-3.

Touretzky, D. S., A. D. Redish and H. S. Wan. 1993. Neural representation of space using sinusoidal arrays. *Neural Computation* 5(6):869–884.

Worboys, M. F. 1996. Metrics and topologies for geographic space. Pages 365–376 in *Advances in Geographic Information Systems Research II: Proceedings of the International Symposium on Spatial Data Handling*, Delft, The Netherlands. Edited by M. J. Kraak. and M. Molenaar. London: Taylor & Francis.

Worboys, M. F. 2001. Nearness relations in environmental space. *International Journal of Geographical Information Science* 15, 7:633–651.

Yao, X. and J. C. Thill. 2005. How far is too far? – A Statistical approach to context-contingent proximity modeling. *Transactions in GIS* 9, 2:157–178.

Yao, X. and J. C. Thill. 2006. Spatial Queries with Qualitative Locations in Spatial Information Systems. *Computers, Environment and Urban Systems*. 30, 4:485–502.

Yao, X. and J. C. Thill. 2007. Neuro-fuzzy modeling of context-contingent proximity relations. *Geographical Analysis*. 39:forthcoming.

Yeap, W. K. 1988. Towards a computational theory of cognitive maps. *AI* 34(3):297–360.

Zadeh, L. 1965. Fuzzy sets. *Information and Control*. 8(3):339–353.

CHAPTER 8
Coordinate Systems and Map Projections

E. Lynn Usery, Michael P. Finn and *Clifford J. Mugnier*

8.1 Introduction

Transformation of geographic data is necessary to support the development of a common coordinate framework from which geographic information system (GIS) operations, such as overlay, spatial buffering, and other analyses, can be performed. According to Keates (1982), we recognize three different types of transformations, the first two of which are mathematical and, therefore, reversible, and the third is non-mathematical and irreversible. The first of these transformations is from the spherical or ellipsoidal Earth to a plane coordinate system and is referred to as map projection. The second is transformation from the three-dimensional Earth form to a two-dimensional form. The final transformation is generalization from the real world to a representation and includes selection, simplification, symbolization and induction (Robinson et al. 1995). In this chapter, we will focus on the first and second types of transformations and the mathematical procedures that allow coordinate transformation to provide common reference frameworks for GIS. We also will briefly examine geometric correction of map and image data, which uses transformation of data from one plane coordinate system to another, but also is essential for GIS.

8.2 Geodesy

Transformation of the spherical or ellipsoidal surface of the Earth to a two-dimensional form falls under the fields of geodesy and map projections; areas of study with well-developed theory and implementation. In this chapter, only some basic aspects of this theory will be discussed to achieve the necessary basis for providing common frameworks for GIS. The study of spherical or ellipsoidal transformations from the Earth's surface to a two-dimensional representation requires the use of four interrelated concepts: ellipsoid, datum, map projection, and coordinate system. Each of these is discussed below.

8.2.1 Ellipsoids

The coordinate frame of reference for a geographic dataset is defined by a reference ellipsoid, a representation of the Earth in which the semi-major and semi-minor axes are of defined length (Figure 8-1). The term spheroid often is used synonymously with ellipsoid (Snyder 1987; Iliffe 2000); however, geodesists often use the terms separately reserving spheroid for association with a global datum on the ellipsoid. In this discussion we will use

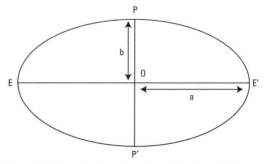

Figure 8-1 Terminology for ellipsoids of revolution: EE′ is the major axis; PP′ is the minor axis, a is the semi-major axis and b is the semi-minor axis.

the term ellipsoid since this is the more common term for the basic figure of the Earth that is used for map projection. Common ellipsoids and their characteristics are shown in Table 8-1; a description of the correct use of the ellipsoids listed in Table 8-1 is given in Snyder (1987). A more complete listing of world ellipsoids is found in Mugnier (2004).

Table 8-1 Selected official ellipsoids and their characteristics. Adapted from Snyder 1987.

Name	Date	Equatorial Radius (a) in meters	Polar Radius (b) in meters	Flattening (f)	Use
GRS 80	1980	6,378,137	6,356,752.3	1/298.257	Basis of NAD 83
WGS 72	1972	6,378,135	6,356,750.5	1/298.26	NASA; Dept. of Defense; oil companies
WGS 84	1984	6,378,137	6,356,752.3	1/298.257	Basis of GPS
Australian	1965	6,378,160	6,356,774.7	1/298.25	Australia
Krasovsky	1940	6,378,245	6,356,863.0	1/298.3	Soviet Union
International	1924	6 378,388	6,356 911.9	1/297	Remainder of the world
Hayford	1909	6 378,388	6,356 911.9	1/297	Remainder of the world
Clarke	1880	6,378,249.1	6,356,514.9	1/293.46	Most of Africa; France
Clarke	1866	6,378,206.4	6,356,583.8	1/294.98	North America; Philippines
Airy	1830	6,377,563.4	6,356,256.9	1/299.32	Great Britain
Bessel	1841	6,377,397.2	6,356,079.0	1/299.15	Central Europe; Chile; Indonesia
Everest	1830	6,377,276.3	6,356,075.4	1/300.80	India; Burma; Pakistan; Afghan; Thailand; etc.

Since the Earth is properly represented as an oblate ellipsoid (spheroid), the primary parameters defining the geometric representation are the Equatorial radius (*a*) and the Polar radius (*b*) (Table 8-1, Figure 8-1). Using a and b, we define the flattening factor (*f*) as:

$$f = (a\text{-}b)/a \tag{8.1}$$

The flattening factor is a measure of the oblateness of the ellipsoid and since an approximate factor for the Earth is 1/298, it is not visible to the naked eye even in satellite views. Therefore, the oblateness shown in Figure 8-1 is exaggerated to allow the visual interpretation of the ellipsoid shape. We define the first eccentricity (*e*), another fundamental measure of the shape (characteristic) of an ellipsoid of revolution, from the flattening factor as:

$$e = (2f\text{-}f^2)^{\frac{1}{2}} \tag{8.2}$$

We can then define the geodetic coordinates — latitude (φ), longitude (λ), and height above the ellipsoid (*h*) — as shown in Figure 8-2. Geodetic coordinates may be transformed to Earth-centered Cartesian coordinates, X, Y, and Z using the following equations:

$$X = (v + h) \cos \varphi \cos \lambda \tag{8.3}$$

$$Y = (v + h) \cos \varphi \sin \lambda \tag{8.4}$$

$$Z = \{1\text{-}e^2) v + h\}\sin \varphi \tag{8.5}$$

where
$$v = a/(1\text{-}e^2\sin^2 \varphi)^{\frac{1}{2}} \tag{8.6}$$

For complete mathematical development of geodetic coordinates and transformations with Cartesian coordinates, see Mugnier (2004).

The most common ellipsoids currently (2007) used with geographic data are Clarke 1866, World Geodetic System 1984 (WGS 84), and Geodetic Reference System 1980 (GRS 80). The Clarke 1866 ellipsoid is the basis for most maps created in the US before the

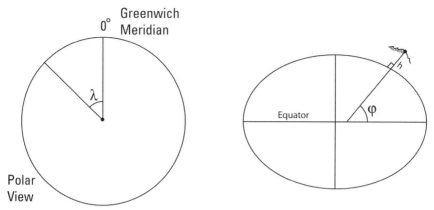

Figure 8-2 Representation of geodetic coordinates, latitude (φ), longitude (λ), and height above the ellipsoid (h).

1980s, primarily because it was designed to fit North America. The Clarke 1866 ellipsoid is referenced to the Earth's surface with geodetic measurements and is the basis of the North American Datum of 1927 (NAD 27). The WGS 84 and GRS 80 ellipsoids were established by satellite positioning techniques, are referenced to the center of mass of the Earth, i.e., geocentric, and provide a reasonable fit to the entire Earth. The WGS 84 datum provides the basis of coordinates collected from the Global Positioning System (GPS), although modern receivers transform the coordinates into almost any user selected reference datum.

8.2.2 Datums

A datum is the basis of a coordinate system and defines an initial point. A datum can be local or global depending on the initial point and whether or not the datum is referenced to an ellipsoidal representation of the Earth. A horizontal datum allows specification of latitude and longitude or x, y Cartesian coordinate locations relative to the initial point. A vertical datum allows specification of height above or below the initial point. For a global datum, the initial point is a point on the surface of the Earth, as with NAD 27, which uses the triangulation station at Meads Ranch, Kansas, as an initial point. Such a datum is referenced as a geodetic datum, and requires another point to establish a reference angle to align the coordinate system. For NAD 27, the reference point is the nearby triangulation station, Waldo in Kansas. For a geocentric datum, established by satellite positioning, the initial point is the center of the Earth and no reference angle is required (Snyder 1987; Iliffe 2000). For detailed treatment of datum concepts, including complete mathematical development, see Mugnier (2004).

8.3 Coordinate Systems

We can define an ellipsoidal coordinate system called the geographical reference system (Figure 8-3) of latitude (φ) and longitude (λ) once we have defined the datum. Note that each ellipsoidal system is different based on the choice of datum; thus, a specification of latitude and longitude location is not sufficient without knowing the datum. Differences in projected plane coordinates can be hundreds of meters as in the United States where the difference between Universal Transverse Mercator (UTM) coordinates on NAD 27 and NAD 83 may be as much as 200 m (Welch and Homsey 1997).

With a datum and projection defined, we can then define plane coordinate systems. A coordinate system uses the initial point of the datum in the projection chosen, and establishes X and Y coordinates based on a grid system of the selected units of measure. Common plane coordinate systems are based on a set of Cartesian axes usually referenced as X and Y

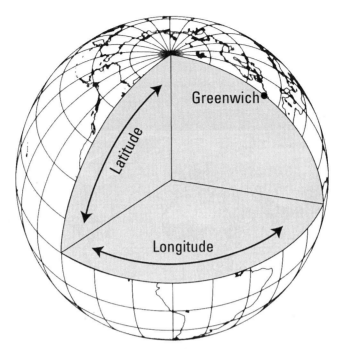

Figure 8-3 Geographic reference system with coordinates of latitude and longitude.

or Eastings (E) and Northings (N) with units measured in feet or meters. A right-handed Cartesian coordinate system defines the origin at 0, 0 and X increases to the right and Y increases to the top. Two common US systems for large-scale (high-resolution) applications are the UTM, a worldwide system, and the State Plane Coordinate systems of the United States, its territories and possessions.

In map projection terminology, we define the scale factor at the origin, m_o, as the maximum scale distortion in the projection. It is the ratio of the scale along a meridian or parallel at a given point to the scale at a standard point or along a standard line that is made true to scale (Snyder 1987). The UTM system is based on projections of six-degree zones of longitude, 80° S to 84° N latitude and the scale factor is specified for the central meridian of the zone.[1] The scale factor for each UTM zone along the central meridian of the projection is 0.9996, yielding a maximum error of 1 part in 2,500. In the northern hemisphere, the X coordinate of the central meridian is offset to have a value of 500,000 meters instead of 0, normally termed the "False Easting." The Y coordinate has 0 set at the Equator. In the southern hemisphere the False Easting is also 500,000 meters with a Y offset of the Equator or False Northing equal to 10,000,000 meters. These offsets force all coordinates in the system to be positive.

The State Plane Coordinate system, available only in the United States and its territories and possessions, also uses Eastings and Northings as the coordinate axes. It also is projected in zones to preserve accuracy with State Plane Coordinate zones designated by states. The maximum width of a zone (the part of the Earth surface projected with its own unique central meridian) is 158 miles wide, which allows a higher accuracy of transformation from the ellipsoid to the plane than the UTM system. The zone width allows maintenance of a m_o

---

1. In the Universal Military Grid System, the polar areas, north of 84° N and south of 80° S, are projected to the Universal Polar Stereographic Grid with the pole as the center of projection and a $m_o = 0.9994$. They are termed "North Zone" and "South Zone."

of approximately 0.9999, or an accuracy of 1 in 10,000. The projection from the ellipsoid also is dependent on the shape of the state. States with an east-west long axis, Tennessee, for example, use the Lambert Conformal Conic projection for each zone. States with a north-south long axis, Illinois, for example, use the Transverse Mercator projection for each zone. Three states, New York, Florida, and Alaska, use both projections since these states have parts extending both E–W and N–S. Zone 1 of Alaska uses the Hotine Rectified Skew Ortho-morphic (RSO) Oblique Mercator projection. Coordinate measurement units of State Plane Coordinates depend on the datum. For NAD 27, the measurement units are US Survey Feet (as opposed to the International Foot defined as 610 nm smaller than the US Survey Foot); while newer systems cast on NAD 83 have an official unit of the meter. Often NAD 83 coordinates also are expressed in feet, but depending on the state, some now use US Survey Feet, and others the International Foot. Some states, such as Wisconsin and Minnesota, have established plane coordinate systems for each county specifically for use with GIS applica-tions. The traditional m_0 at the origin, normally associated with the State Plane Coordinate system zones, is modified for height above the ellipsoid so that field survey measurements will correspond closely with the county GIS scale factor and thus reduce hand computations necessary for conversion by the GIS analyst.

A final plane coordinate system of relevance to geographic data synthesis and modeling, particularly for satellite images and photographs, is an image coordinate system. A digital image system is not a right-handed Cartesian coordinate system since usually the initial point (0,0) is assigned to the upper left corner of an image. The X coordinate, often called sample, increases to the right, but the Y coordinate, called the line, increases down. Units commonly are expressed in picture elements or pixels. A pixel is a discrete unit of the Earth's surface, usually square with a defined size, often expressed in meters. Photogrammetric applications, however, transform the origin (0,0) of each image to correspond to the center of perspective, or the intersection of the optical axis of the lens with the image plane. For frame imagery, that is the center of the image. For pushbroom imaging sensors, different geometry is used and modeled, and commonly is associated with rational functions.

8.4 Map Projections

Since the Earth is spherical or, more correctly, ellipsoidal, and usually we work with plane coordinate representations, geographic data must be projected from ellipsoidal coordinates to plane coordinates. This transformation is referred to as map projection, which is defined as a systematic transformation of ellipsoidal coordinates of latitude and longitude to a plane coordinate representation and mathematically,

$$x = f_1\,(\varphi, \lambda) \qquad\qquad (8\text{-}7)$$

$$y = f_2\,(\varphi, \lambda) \qquad\qquad (8\text{-}8)$$

The transformation is implemented from a "generating globe," which is a reduced scale model of the Earth as either a sphere, an ellipsoid, or an "aposphere." The projection trans-formation always results in error, with only a single point, circle, or one or two lines where the scale relation to the generating globe is true. While error always is a result of the trans-formation, specific properties can be preserved, e.g., angular relations in small areas, polygon areas, such as continents, or specific directions, such as straight lines away from the North or South Poles. Angles and area, however, cannot be preserved simultaneously in the same projection since they are mutually exclusive transformations. Maps with angles preserved are called conformal projections. Maps with areas preserved are called equal-area or equivalent projections. Equal area projections also are called authalic, meaning that at any point the scales in two orthogonal directions are inversely proportional, which forces equal areas.

One can understand map projection by examining the transformation of spherical coordinates to the geometric figures: cylinder, cone, and plane. A graphical illustration of these transformations is shown in Figure 8-4. Note that for a cylinder only a single line of contact exists in a tangent projection (two lines for a secant projection). It is only along this line that the true scale of the generating globe, and thus along the same imaginary line on the Earth's surface, is retained. This line is said to have a scale factor of 1. Any other line is projected away from the sphere and possesses a scale factor larger than 1. Note that the greater the distance away from the line of contact, the Equator in a normal aspect cylindrical projection, the larger the scale factor and thus the greater the distortion. A similar line for a conic projection is shown in Figure 8-4, but for an azimuthal (zenithal) projection, only a single point on the tangent plane retains a scale factor of 1. Complete documentation of map projection theory is available in Snyder (1987), Pearson II (1990), Bugayevskiy and Snyder (1995), Yang et al. (2000) and Canters (2002). A history of map projection development is available in Snyder (1993) and a documentation of characteristics of various projections is provided in Snyder and Voxland (1989).

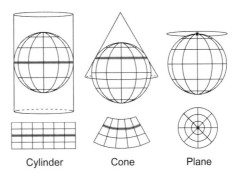

Cylinder   Cone   Plane

Figure 8-4 Geometric figures used for map projection from sphere or ellipsoid to the plane. All are shown in the normal aspect with lines or points where the scale factor = 1 is highlighted.

8.4.1 Classification

Projections are classified by a variety of methods including geometry, shape, special properties, projection parameters and nomenclature (Canters 2002). The geometric classification is based on the patterns of the graticule, the grid of parallels of latitude and meridians of longitude, that result from a perspective projection of the sphere on a cylinder, cone or plane as shown in Figure 8-4. These projections are referred to as cylindrical, conical, azimuthal (also occasionally called zenithal) and aphylactic (meaning none of the former). Although commonly referred to as developable because of the apparent ability to develop these projections from a perspective projection of the sphere, the spacing of the parallels in the patterns is derived from differential calculus. This process allows for the preservation of specific characteristics and minimizes distortion, such as angular relations (shape) or area. The geometric classification into cylindrical, conical and azimuthal is not complete since many projections fit none of these classes. Thus, the classification commonly is expanded to include pseudocylindrical, projections with straight parallels as with the cylindricals, but curved meridians; pseudoconical with parallels as curved arcs, as with the conicals, but with parallel length adjusted so meridians are not straight arcs; and polyconical, with non-concentric circular arcs for the parallels (Canters 2002). Projections that do not fit these six classes are referred to as non-conicals. A complete description of these geometric patterns and their associated names can be found in Lee (1944).

Another common projection classification system is based on the shape of the graticule. Maurer (1935), cited in Canters (2002), developed a hierarchical system including five levels primarily based on the appearance of the meridians and parallels. For a description of the system, see Maurer (1935) or Canters (2002). Starostin et al. (1981) also presented a classification system based on the shape of the graticule as described in Bugayeskiy and Snyder (1995). This system is similar to Maurer's with classes based on the shape of the parallels and symmetry of the graticule.

Subdivisions in Lee's (1944) classification are based on special properties. Goussinsky (1951) also produced a system based on special projection properties. Within his system, the five classes—nature, coincidence, position, properties and generation—are not mutually exclusive, but within each class, the types of properties preserved are exclusive. For example, class 3, position, includes direct, transverse and oblique. Maling (1968) proposed 11 special properties to use in map projection.

Tobler (1962) used a general approach based on parametric classification in four groups. The groups are based on whether or not the plane coordinates x and y are based on a time, formation of latitude, longitude, or both. Maling (1992) included geometric classes that relate to Tobler's parametric classes and the traditional geometric approach. A complete description of Tobler, Maling and other classification systems is available in Canters (2002). As is obvious from this brief discussion, map projections can be classified in many ways. The International Cartographic Association (ICA) Commission on Map Projections has a current (2006) project to establish a standard classification and naming system (ICA 2006).

8.4.2 Suggested Projections

The selection of an appropriate map projection for a given application depends on a variety of factors, including the purpose the map, the type of data to be projected, the area of the world to be projected and scale of the final map. Advice on selection is available from a variety of print and web sources, including Finn et al. (2004) and USGS (2006). In the discussion below that provides a description of specific projections, we will distinguish between large-scale (small areal extent) and small-scale (large areal extent) applications. In GIS, large-scale data sets commonly are projected with a conformal projection to preserve angles and shape. For such applications, area distortion is so small over the geographic extent that it is negligible and an area preserving projection is not needed. Whereas there is no sharp boundary to determine large-scale from small-scale applications, we will use an area of 150,000 square kilometers, roughly the size of some US states, and a scale of 1:500,000 as a convenient breakpoint. Commonly, large-scale data files are used in GIS applications of limited geographic extent, e.g., a watershed, a county or a state. The two most commonly used projections for these scales are the Lambert Conformal Conic and the Transverse Mercator, which are the basis of the UTM and most of the State Plane coordinate systems discussed earlier in this chapter. Later in this chapter, we will describe these projections and several others used for small-scale applications for areal extents of states, regions, countries, continents or the entire globe. The descriptions are adapted and dates of presentation and authors are taken from Snyder and Voxland (1989) unless otherwise noted.

8.4.3 Description of Specific Projections

The following section details characteristics of a selected set of projections. The name and creator, or inventor, is detailed for each projection. We outline specific characteristics including properties preserved, shapes of parallels and meridians, the lines or points of true scale, the extent of the Earth that can be shown and—for specific cases—particular charac-teristics that make the projection unique. A graphic representation of each projection is included with world data and a set of distortion circles plotted on the graphic, usually for one quarter of the Earth coverage, since the distortion circles repeat the same patterns in other quadrants. The distortion is plotted as a Tissot Indicatrix (Tissot 1881; Canters 2002). The intersection of any two lines on the Earth is represented on a map with an intersection at the same or different angles (Figure 8-5). At almost every point, there is a right angle intersection of two lines in some direction, which also is shown as a right angle on the map. All other line intersections at that point will not be at right angles, unless the map is conformal at that point. The greatest deviation from the correct angle is the maximum angular deformation (ω). For a conformal map, the value of ω is zero. We use plots of small circles on the maps to indicate distortion; if the circles all remain circles, but change in size, the map is conformal and does not preserve area. If the circles change shapes to ellipses of the same size, the map preserves area, but does not maintain angular relations. On maps with changes in both the shape and size of the circles, neither area nor angular relations are preserved. The reader should note that in the projection figures the circles and ellipses may not appear to be the same area when in fact they are. It is well-known that the human eye is poor at estimating the relative size of geometric symbols, such as circles and ellipses, and methods to psychologically scale such symbols have been developed and used in cartography (Flannery 1971). However, in the figures in this chapter, exact areas are used without psychological scaling, and thus some circles and ellipses will appear to be of different size when in fact they are equal.

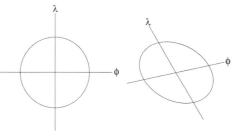

Figure 8-5 A graphic illustration of Tissot's Indicatrix, right. An infinitely small circle on the Earth (left) appears as an ellipse on many maps. Only on conformal maps will the figure remain a circle on the map.

8.4.3.1 Cylindrical

8.4.3.1.1 Mercator

The Mercator projection is a cylindrical conformal projection developed by Gerardus Mercator in 1569. It was developed to show loxodromes or rhumb lines, which are lines of constant bearing, as straight lines. The Mercator projection made it possible to navigate a constant course based on drawing a rhumb line on the chart. The projection has meridians as equally spaced parallel lines, while parallels are shown as unequally spaced straight parallel lines, closest near the Equator and perpendicular to the meridians. The North and South Poles cannot be shown. Scale is true along the Equator (tangent case) or along two parallels equidistant from the Equator (secant case). Significant size distortion occurs in the higher latitudes as shown by the circle sizes in Figure 8-6. The Mercator projection was defined for navigational charts and is best used for navigation purposes. It is a standard for marine charts.

8.4.3.1.2 Transverse Mercator

The transverse aspect of the Mercator projection is a projection where the line of constant scale is along a meridian rather than the Equator. The central meridian, each meridian 90° from the central meridian, and the Equator are straight lines. Other meridians and parallels are complex curves, concave toward the central meridian and nearest pole, respectively. The Poles are points along the central meridian. The projection has true scale along the central meridian or along two meridians equidistant from and parallel to the central meridian in the secant case. Conceptually, it is created by projecting onto a cylinder wrapped around the globe tangent to the central meridian or secant along two small circles equidistant from the central meridian. It commonly is used for large-scale, small area, presentations; many of the world's topographic maps from 1:24,000 scale to 1:250,000 scale use this projection. It is the basis of the UTM coordinate system and many of the State Plane Coordinate systems for states with an elongated north-south axis. The Transverse Mercator projection using the zero degree longitude at Greenwich as the central meridian is shown in Figure 8-7.

8.4.3.1.3 Lambert Cylindrical Equal Area

The Cylindrical Equal Area projection, first presented by Johann Heinrich Lambert in 1772, became the basis for many other similar equal area projections including the Gall Orthographic, Behrmann, and Trystan-Edwards. From Lambert's original projection with the line

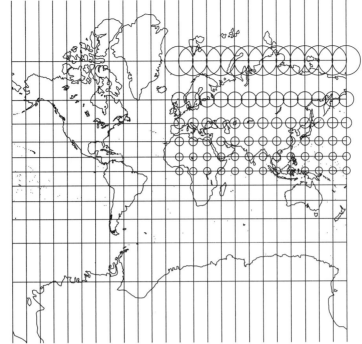

Figure 8-6 Mercator projection with distortion circles (Tissot's Indicatrix) in the upper right corner, illustrating that size distortion is greater at higher latitudes.

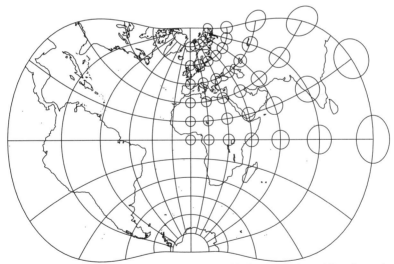

Figure 8-7 Transverse Mercator projection with the central meridian through Greenwich; Tissot's Indicatrices in the upper right illustrate that scale is constant along the prime meridian, and how distortion in this projection is greater for locations farther from the Equator and central meridian.

of constant scale along the Equator, one simply makes the projection secant at two small circles (parallels). Each of the above projections uses different parallels as the lines of constant scale. Lambert's Cylindrical Equal Area projection has meridians that are equally-spaced straight parallel lines 0.32 times as long as the Equator. Lines of latitude are unequally spaced parallel lines furthest apart near the Equator, and are perpendicular to the meridians. The projection maintains equal areas by changing the spacing of the parallels. Significant shape

Manual of Geographic Information Systems

distortion, however, results from maintaining equal areas with the distortion greater in high latitudes near the poles as shown by the ellipses in Figure 8-8. While this projection is not often used, it is a standard to describe map projection principles in textbooks. It has also served as a prototype for other projections, as described earlier in this chapter.

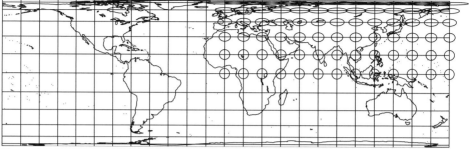

Figure 8-8 Lambert Cylindrical Equal Area projection.

8.4.3.2 Conical

8.4.3.2.1 Lambert Conformal Conic

The Lambert Conformal Conic (LCC) projection, presented in 1772, shows meridians as equally spaced straight lines converging at a common point, which is one of the poles. Angles between the meridians on the projection are smaller than the corresponding angles on the globe. Parallels are unequally spaced concentric circular arcs centered on the pole of convergence of the meridians, and spacing of the parallels increases away from the pole. The pole nearest the standard parallel is a point; the other pole cannot be shown. Scale is true along the standard parallel or along two standard parallels in the secant case. Scale also is constant, although not true, along any given parallel. The projection is free of distortion only along one or two standard parallels. Shapes are maintained at the expense of area as shown by the perfect circles of different sizes in Figure 8-9. The LCC projection is extensively used for large-scale mapping of regions with an elongated axis in the east-west directions and in mid-latitude regions. It is the projection for the State Plane Coordinate system for US states with an east-west axis, such as Tennessee. It also is a standard of the US Geological Survey (USGS) for State Base Maps at 1:500,000 scale, and for maps of the 48 US contiguous states.

8.4.3.2.2 Albers Equal Area

The Albers Equal Area projection, presented by Heinrich Christian Albers in 1805, has meridians as equally spaced straight lines converging at a common point, which normally is beyond the pole. Angles between the meridians are less than the true angles and parallels are unequally spaced concentric circular arcs centered on the point of convergence of the meridians. Spacing between the parallels decreases away from the point of convergence, the poles being circular arcs. Scale is true along one or two standard parallels. The scale factor at any given point along a meridian is the reciprocal of the scale factor along the parallel, thus preserving area at the expense of shape. This is shown in Figure 8-10 where circles maintain size but change in shape to ellipses away from the standard parallel. The projection is free of angular and scale distortion only along the one (tangent case) or two (secant case) standard parallels. The Albers Equal Area projection is used to show areas of east-west extent in applications where preservation of area is important. It commonly is used for equal-area maps for the 48 contiguous US states and is the projection upon which the *National Atlas of the United States* (www.nationalatlas.gov) is based.

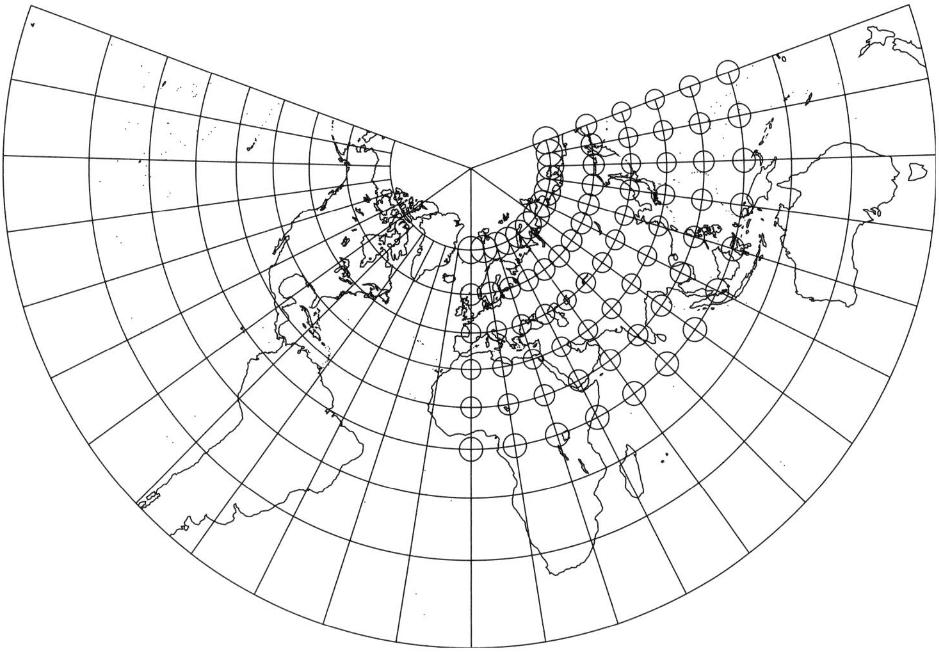

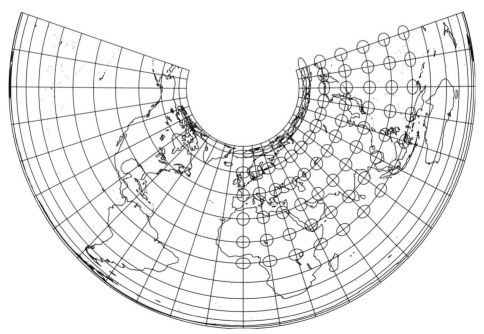

Figure 8-9 Lambert Conformal Conic projection preserves shape, as shown by the fact that Tissot's Indicatrix is everywhere a circle. Variation in the size of circles shows that area is not preserved.

Figure 8-10 Albers Conical Equal Area projection.

8.4.3.2.3 *Polyconic*

The polyconic projection was originated by Ferdinand Rudolph Hassler of the US Coast and Geodetic Survey for plane table and alidade coastal mapping, and was easy to construct from simple tables while in the field. The projection uses many cones for the projection, one along each parallel, hence the name "poly" conic. The central meridian is a straight line with all others appearing as complex curves. The Equator is the only parallel that is a straight line, with others as non-concentric circular arcs spaced at true distances along the central meridian. Scale is true along the central meridian and along each parallel. The projection is free of distortion only along the central meridian and results in significant distortion if the range is extended far to the east and west. While the projection preserves neither area nor shape (termed aphylactic), it was the only projection used by the USGS for topographic maps until the 1950s. One reason for this usage was the ease of construction of the projection for quadrangle maps from tables of rectangular coordinates. These tables may be used from any polyconic projection on the same ellipsoid by applying the proper scale and central meridian. Therefore, for each quadrangle map the same tables could be used. These quadrangle maps for the same ellipsoid and for the same central meridian at the same scale will fit exactly from north to south. They also fit exactly east to west, but cannot be mosaicked in both directions simultaneously unless only one central meridian is held for an entire map series. Such variations of aphylactic projections are called quadrillages. The Polyconic projection also was used for the Progressive Military Grid for the military mapping of the United States in the 15-minute format. This grid later was incorporated into the World Polyconic Grid that was referenced to the Clarke 1866 ellipsoid, measured in yards, and used for artillery fire control mapping during World War II. A graphic illustration of the projection is shown in Figure 8-11. This projection is not recommended for regional or global maps since better projections are available.

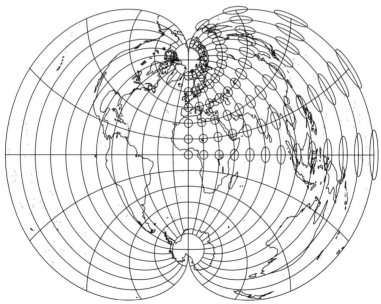

Figure 8-11 The Polyconic projection applied to a world data set.

8.4.3.3 Azimuthal

8.4.3.3.1 Orthographic

The Orthographic projection, developed by the Egyptians and Greeks by the 2nd century B.C., is a perspective azimuthal (planar or zenithal) projection that is neither conformal nor equal-area. It is used in polar, Equatorial and oblique aspects, and results in a view of an entire hemisphere of the Earth. In the polar aspect, shown in Figure 8-12, meridians are equally spaced straight lines intersecting the central pole. Angles between meridians are true. Parallels are unequally spaced circles centered on the pole, which is a point. Spacing of the parallels decreases away from the pole. Other aspects are described in Snyder and Voxland (1989). Scale is true at the center and along the circumference of any circle with its center at the projection center. Such circles are parallels in the polar aspect of the orthographic projection. Scale decreases radially with distance from the center. Distortion circles are shown in Figure 8-12, which also shows the globe-like look of the projection. The orthographic projection is essentially a perspective projection of the globe onto a tangent plane from an infinite distance (orthogonally). It is commonly used for pictorial views of the Earth as if seen from space.

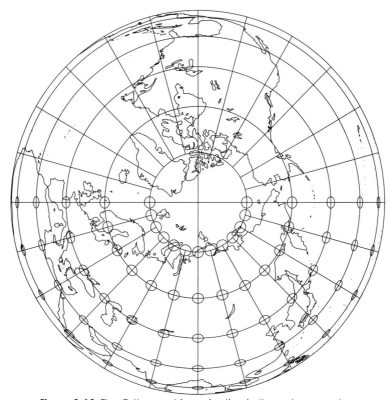

Figure 8-12 The Orthographic projection in the polar aspect.

8.4.3.3.2 Stereographic

The Stereographic projection, also developed by the Egyptians and Greeks by the 2nd century B.C., is a perspective azimuthal projection that preserves angles, i.e., is conformal. As with the Orthographic projection, the polar, Equatorial and oblique aspects result in different appearances of the graticule. The polar aspect is achieved by projecting from one pole to a plane tangent at the other pole (Figure 8-13). In this aspect, meridians

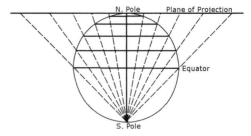

Figure 8-13 Projection from the South Pole onto a plane tangent at the North Pole creates the Stereographic projection.

are equally-spaced straight lines intersecting at the pole with true angles between them. Parallels are unequally spaced circles centered on the pole represented as a point. Spacing of the parallels increases away from the pole. The projection commonly is used only for a hemisphere. It can be used to show most of the other hemisphere (Figure 8-14) at an accelerating scale. Scale is true only where the central latitude crosses the central meridian or along a circle concentric about the projection center, and scale is constant along any circle with the same center as the projection. The Stereographic projection is used in the polar aspect for topographic maps of the polar regions. The Universal Polar Stereographic is the sister projection to the UTM for military mapping. This projection is in current (2007) use in oblique ellipsoidal form in a number of nations throughout the world, including Canada, Romania, Poland and The Netherlands (Thompson et al. 1977). This projection generally is chosen for regions that are roughly circular in shape, and it normally is used only in the secant case where the scale factor is less than 1.0. Different countries have different mathematical developments that include the Stereographic Double, the Roussilhe Stereographic, and various truncations of the Hristow Stereographic. East and West hemisphere maps commonly use the Equatorial aspect of the Stereographic projection.

8.4.3.3.3 Gnomonic

The Gnomonic projection is a perspective azimuthal projection that is neither conformal nor equal area. The Greek, Thales, possibly developed it around 580 B.C. The name derives from the point of projection being at the center of the earth where the mythical "gnomes" live. It has the unique feature that all great circles, including all meridians and the Equator, are shown as straight lines. As with other azimuthals, the graticule appearance changes with the aspect. In the polar aspect, meridians are equally spaced straight lines intersecting at the pole with true angles between them. Parallels are unequally spaced circles centered on the pole as a point. Spacing of the parallels increases from the pole. The Equator and opposite hemisphere cannot be shown. The projection, which can be viewed conceptually as projected from the center of the globe on a plane tangent at a pole or another point, only can show less than a hemisphere. Scale is true only where the central parallel crosses the central meridian and increases rapidly with distance from the center. Distortion circles (Figure 8-15) show that the projection is neither conformal nor equal area. Its usage results from the special feature of great circles as straight lines, and thus assists navigators and aviators in determining the shortest and most appropriate courses.

8.4.3.3.4 Lambert Azimuthal Equal Area

The Lambert Azimuthal Equal Area projection, developed by Johann Heinrich Lambert in 1772, is a non-perspective azimuthal equal area projection. In the polar aspect, meridians are equally spaced straight lines intersecting at the central pole with true angles between them. Parallels are unequally spaced circles centered at the pole as a point. Parallel spacing decreases away from the pole. The projection can be used for the entire Earth with the opposite pole appearing as a bounding circle with a radius 1.41 times that of the Equator. Scale is true at the center in all directions and decreases rapidly with distance from the center along radii and

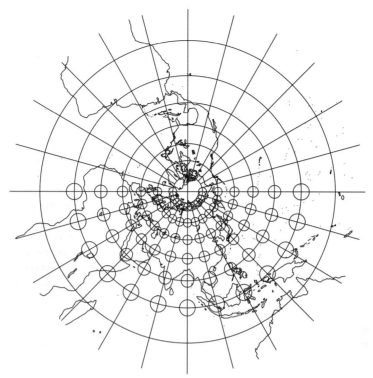

Figure 8-14 Stereographic projection showing more than one hemisphere of data.

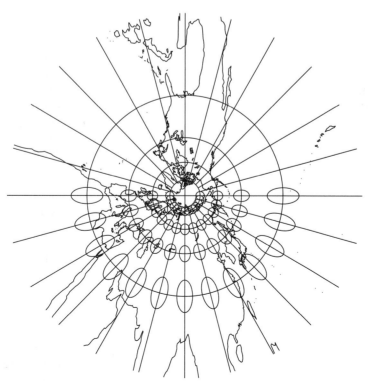

Figure 8-15 The Gnomonic projection in the polar aspect. The variation is both size and shape of the Tissot's Indicatrices show that it does not preserve either.

Manual of Geographic Information Systems

increases with distance in a direction perpendicular to radii. A projected northern hemisphere with distortion circles showing the equal area preservation, but shape distortion into ellipses is shown in Figure 8-16. The Lambert Azimuthal Equal Area projection often is used in the polar aspect for atlases of the polar regions. The Equatorial aspect is used for the East and West hemisphere maps, and is best used for equal-area maps of regions of approximately circular extent with the projection centered on the center point of the region.

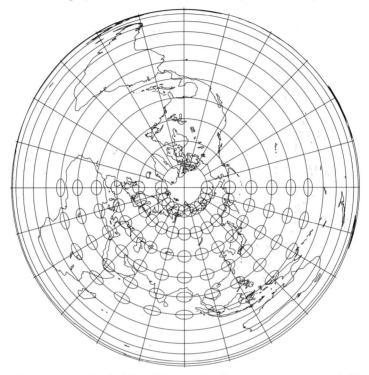

Figure 8-16 The Lambert Azimuthal Equal Area projection preserves area and distorts shape.

8.4.3.3.5 Azimuthal Equidistant

The (ellipsoidal) Hatt Azimuthal Equidistant projection has origins similar to that of the Roussilhe Oblique Stereographic. Both Philippe Eugene Hatt and Henri Roussilhe were Chief Hydrographers of the French Navy, and both men devised projections for use in the hydrographic surveys of near-shore waters and harbors. Because of the prestige associated with the papers published by both men in *Annals Hydrographique* in the 19th and early 20th centuries, a number of countries adopted one or the other projection for their own grids (Takos 1978). The Hatt Azimuthal is merely based on a polar coordinate origin point from which clockwise azimuths from north are measured to points. To define the distance to the points specific series expansions for the geodesic are used (Figure 8-17). The USGS used a modified Azimuthal Equidistant for geological mapping of Yemen, and in later years, John P. Snyder used Clarke's Long-Line Geodesic for the Azimuthal Equidistant USGS series of Micronesia.

8.4.3.4 Pseudocylindrical

8.4.3.4.1 Mollweide

The Mollweide is a pseudocylindrical equal area projection developed by Carl Mollweide in 1805. The central meridian is a straight line one-half as long as the Equator, thus forming an elliptical area of projection for the entire globe. Meridians 90° East and West of the central meridian form a circle. Other meridians are equally spaced semi ellipses intersecting at the

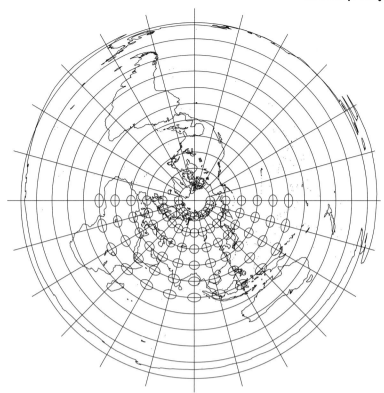

Figure 8-17 The Azimuthal Equidistant projection in a polar aspect.

poles and concave toward the central meridian. Parallels are unequally spaced straight parallel lines perpendicular to the central meridian, farthest apart near the Equator with spacing changing gradually. The poles are shown as points. Scale is true along latitudes 40°44' North and South and constant along any given latitude. The entire globe projected and centered on the Greenwich meridian is shown in Figure 8-18. The distortion circles indicate preservation of area since all are the same size, and distortion of shape since they become ellipses toward the poles. It occasionally has been used for world maps, particularly thematic maps where preservation of area is important. Goode (1925) combined it with the sinusoidal projection to create the Homolosine. Different aspects of the Mollweide have been used for educational purposes, and the projection was used in *The Times Atlas* in England.

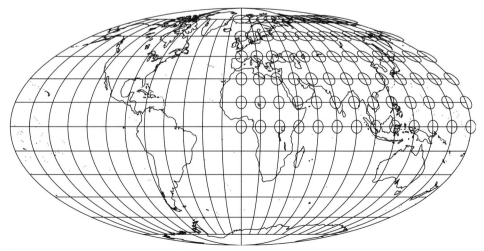

Figure 8-18 The Mollweide projection.

Manual of Geographic Information Systems

8.4.3.4.2 *Sinusoidal*

The Sinusoidal is an equal-area, pseudocylindrical projection developed in the 16th century and used by various cartographers in atlases. It also is known as the Sanson-Flamsteed projection for later users and is the oldest of the pseudocylindrical projections. The central meridian is a straight line one-half as long as the Equator. Other meridians are equally spaced sinusoidal curves intersecting at the North and South Poles and concave toward the central meridian. The parallels are equally spaced straight lines perpendicular to the central meridian. The Poles are shown as points. The scale is true along the central meridian and along every parallel. The sinusoidal projection preserves area, but distorts shapes (Figure 8-19), with the greatest distortion occurring near outer meridians and in high latitudes. The Equator is free of distortion. It has been used for maps of South America and Africa, and sometimes for world maps. It was combined with the Mollweide by Goode (1925) to create the Homolosine projection.

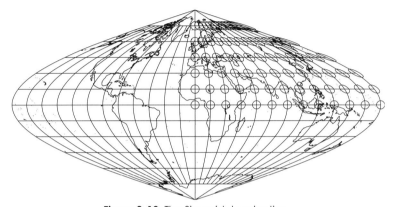

Figure 8-19 The Sinusoidal projection.

8.4.3.4.3 *Robinson*

Presented by Arthur H. Robinson in 1963 at the request of Rand McNally and Company, the Robinson projection is a pseudocylindrical projection that is neither conformal nor equal area. It uses a set of tabular coordinates rather than mathematical formulas to project coordinates. Robinson created it to improve the world view. The central meridian is a straight line 0.51 the length of the Equator. Other meridians are equally spaced, and resemble elliptical arcs concave toward the central meridian. Parallels are equally spaced straight parallel lines between 38° North and South, with space decreasing beyond these latitudes. The poles are shown as lines 0.53 times the length of the Equator. Scale is true along the 38° latitudes North and South, and is constant along any given latitude. There is no point completely free of distortion, and both size and shape change as shown by the circles in Figure 8-20. The Robinson projection is used for world maps by Rand McNally in their *Goode's World Atlas* (Veregin 2006).. The National Geographic Society adopted it for world maps for a time during the 1990s.

8.4.3.5 Other Projections

Regions that are not predominately elongated along the cardinal directions, not circular in shape, and too large for local projections present a conundrum to the cartographer and the geodesist. The Transverse cylindrical of Professor Rosenmund for Switzerland (Mugnier 2001) and the Oblique Mercator of French General Jean Laborde for Madagascar (Mugnier 2000) were based on double projections that utilized an equivalent sphere. Laborde's development is based on the Gauss-Schreiber Transverse Mercator projection. Hotine (1946, 1947) first introduced the development of the oblique Mercator on the ellipsoid through the "aposphere," a surface of constant curvature and thence to the plane. Hotine's original implementation was for the peninsula of Malaya and the island of Borneo, but this projection also has been used as a grid in numerous areas elsewhere, including Alaska Zone 1 of the State Plane Coordinate Systems on both NAD 27 and NAD 83.

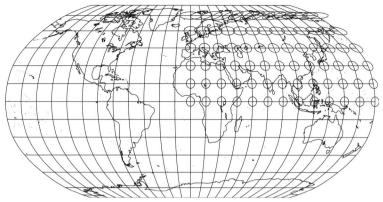

Figure 8-20 The Robinson projection.

8.4.3.5.1 Van der Grinten

The Van der Grinten projection (also called Van der Grinten 1), presented by Alphons J. van der Grinten of Chicago in 1898, is a polyconic projection that is neither conformal nor equal area. The projection has a straight central meridian whereas other meridians are circular and equally spaced along the Equator, concave toward the central meridian. Parallels are circular arcs, concave toward the nearest pole, with the Equator as a straight line exception. The Poles are points. Scale is true along the Equator and increases rapidly with distance from the Equator. The projection has significant distortion near the poles (Figure 8-21). The projection encloses the entire world in a circle. The US Department of Agriculture, the USGS, and the National Geographic Society are a few of the organizations that have used it for world maps.

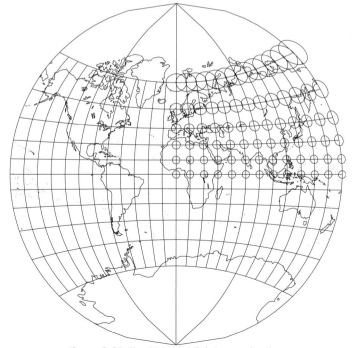

Figure 8-21 The Van der Grinten projection.

Manual of Geographic Information Systems

8.4.3.5.2 Cassini-Soldner

The Cassini-Soldner projection is a relic of the 19th century mapping efforts of the Europeans and their colonies (Clark 1973; Iliffe 2000). Largely replaced by the Transverse Mercator Projection, the Cassini-Soldner occasionally is still found in former British colonies that describe cadastral records and/or hydrocarbon exploration/production concessions with this grid. The Cassini-Soldner is an aphylactic projection also in that it is neither conformal nor equivalent. Survey computations on the developed surface are especially problematic, particularly with respect to the conversion between geodetic distances measured on the ground, and grid distances measured on the developed surface. The Cassini-Soldner projection centered over the zero degree latitude and longitude is shown in Figure 8-22.

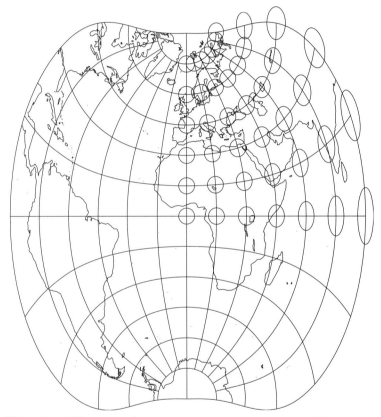

Figure 8-22 The Cassini-Soldner projection centered on zero degrees latitude and longitude. Note that only a part of the Earth appears in the projection.

8.4.3.5.3 Space Oblique Mercator

The Space Oblique Mercator projection, conceived by Alden P. Colvocoresses in 1973 and developed mathematically by John P. Snyder in 1977, was designed to map the ground track of a satellite and maintain conformality. Meridians and parallels are complex curves at slightly varying intervals to account for the motion in time of the satellite (Figure 8-23). The Poles are points. The scale is true along the ground track, but varies about 0.01 percent within the normal sensing range of the satellite. There is no distortion along the ground track and distortion is constant along lines of constant distance parallel to the ground track. The projection is conformal to within a few parts per million for the sensing range. The projection is used for satellite images including Landsat and others.

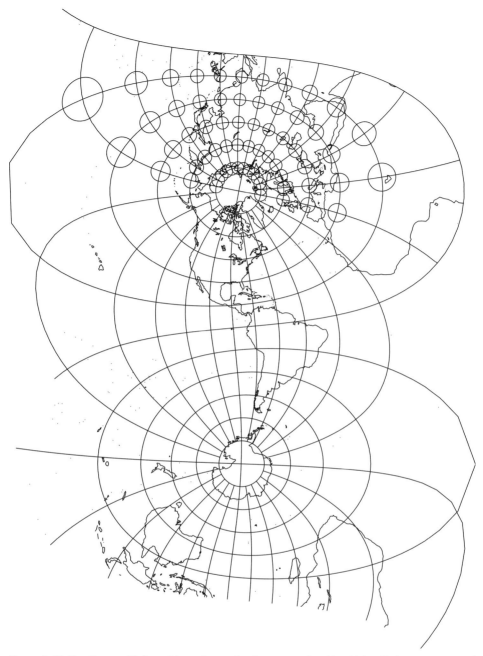

Figure 8-23 The Space Oblique Mercator protection conceived by Alden Colvocoresses and derived mathematically by John P. Snyder.

8.5 Plane Coordinate Transformation

As GIS users are well aware, geographic data often exist in a plane coordinate system, but not in the particular plane system we wish to use. If the datum, projection, and associated parameters are known, the data can be reprojected to the desired projection and coordinate system. In this situation, the data are first inversely projected to the geographic reference system of latitude and longitude, and then forward projected to the desired system. This reprojection operation allows exact control of the transformation, and the accuracy and errors involved.

The data are frequently in plane coordinates, but the datum or projection is not known. An example is a photograph or image that exists in an image coordinate system that we want to transform to UTM to match other data. Other examples are digitized or scanned maps and images. For these data, an approximate transformation to a set of known coordinates can be performed. A common approach is to locate ground control points (GCPs) in the image and a reference system and develop a polynomial transformation between the two. Coordinates in the image system can be located visually and measured directly on a computer screen. Coordinates of the same points in a reference system can be located from a map or in the field with a GPS receiver. The mapping of points between the two systems and an explanation of the process for establishing the equations to be solved for the transformation is illustrated in Figure 8-24 and 8-25, respectively. The mapping in Figure 8-25 shows the process to transform an image scanned at 1,024 by 1,024 pixels to UTM coordinates on NAD 83. The root-mean-square error of the transformation is ±0.69 m. Note that in the example a simple first order polynomial is used, which is sufficient for most current (2007) geographic data sets. Higher order polynomials can be used to eliminate higher distortion, but require larger numbers of control points. The minimum and recommended numbers of GCPs for polynomials of degrees 1-5 are shown in Table 8-2.

The described transformation between plane coordinate systems can be used with point (vector) data or image (raster) data representations. For vector data, the transformation is complete since attributes are associated with the transformed points, lines, or polygons. For raster data, since we are transforming discrete cells, we must determine how the new cell values will be assigned. For these data, we must resample the original digital numbers (DNs) or raster cell values to match the new geometry of the transformed image. A graphic example of the concept of this resampling is shown in Figure 8-26. It is shown as an inverse resampling, from the output coordinate space to the input coordinate space, since this is the common implementation. The exact raster organization desired, i.e., the number of rows, the number of columns, and the pixel size is assumed, and the image is mapped from this assumed space back to the original coordinate space. Once the exact location in the input space is known, the appropriate DN or raster cell value is placed in the output coordinate (pixel) position. Common resampling approaches are nearest neighbor (assume the value of the closest pixel), bilinear interpolation (a distance-weighted average of the four nearest values), or cubic convolution (a distance-weighted average of the 16 nearest pixels). Details of these resampling methods are available in a variety of sources (Jensen 2005). Steinwand et al. (2005) have developed resampling techniques for thematic (categorical) data that allow users to select minimum, maximum, modal and other statistical or user-specified values from those available for the sample area in the input raster dataset. The techniques provide significantly better output results than traditional nearest neighbor methods.

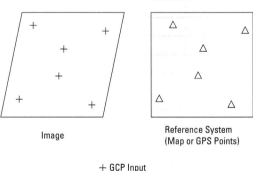

+ GCP Input

△ GCP Reference

Figure 8-24 Mapping of points between two raster GIS or image and map systems using a series of ground control points (GCPs) (Welch and Usery 1984).

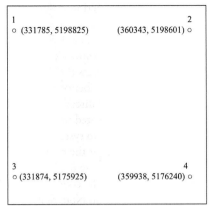
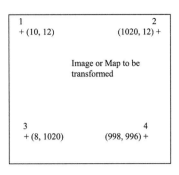

$$X^t = a_0 + a_1 x + a_2 y$$
$$Y^t = b_0 + b_1 x + b_2 y$$

From the first point:

$$331785 = a_0 + a_1(10) + a_2(12)$$
$$5198825 = b_0 + b_1(10) + b_2(12)$$

From the second point:

$$360343 = a_0 + a_1(1020) + a_2(12)$$
$$5198601 = b_0 + b_1(1020) + b_2(12)$$

From the third point:

$$331874 = a_0 + a_1(8) + a_2(1020)$$
$$5175925 = b_0 + b_1(8) + b_2(1020)$$

From the fourth point:

$$359938 = a_0 + a_1(998) + a_2(996)$$
$$5176240 = b_0 + b_1(998) + b_2(996)$$

Solve for the unknowns a_0, a_1, a_2, b_0, b_1, b_2 by simultaneous solution with least squares adjustment. Apply coefficients to all other points in the input image to create complete transformed geometry of the image. For raster images, resample to generate new gray level or color values.

Figure 8-25 Implementation method for polynomial transformation for image or map data.

Table 8-2 Number of GCPs for plane coordinate transformation

Polynomial Order	Minimum Number of GCPs	Recommended Number for Effective Least Squares
1	3	6
2	6	10
3	10	15
4	15	21
5	21	30

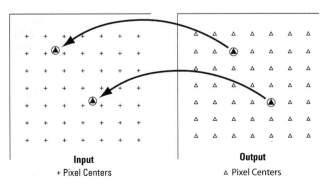

Input
+ Pixel Centers

Output
△ Pixel Centers

Figure 8-26 Resampling example using nearest neighbor concept from the output image mapped into the input image.

Manual of Geographic Information Systems

8.4.1 Three-Dimensional to Two-Dimensional Transformations

Three-dimensional (3D) and two-dimensional (2D) transformations occur with geographic data since the surface of the Earth is not a perfect sphere or ellipsoid. Thus, data acquired over a 3D surface must be transformed to a plane representation. The transformation process is shown in Figure 8-27. As with map projection, this transformation is completely mathematical and model error sources can be determined exactly. Figure 8-28 provides an image example. A photograph of the Tenth Legion, Virginia, area is shown in Figure 8-28a and the same photograph after it has been orthorectified to remove distortion resulting from tilt and relief is shown in Figure 8-28b. Note that roads on the photo appeared curved because of terrain relief, but in the orthophotograph the roads appear straight. The procedures for these transformations are the subject of photogrammetry and are explained in complete detail in the ASPRS *Manual of Photogrammetry* (McGlone 2004).

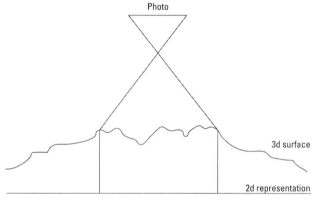

Figure 8-27 Graphic example of the 3D to 2D transformation process.

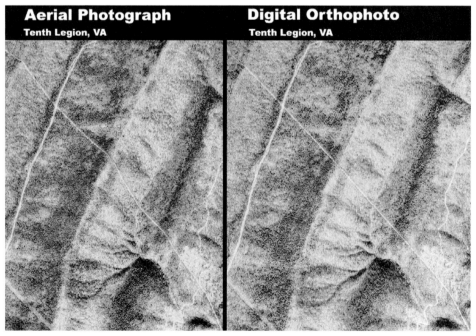

Figure 8-28 Photo example of the 3D to 2D transformation. Note the road appears curved in the (left) uncorrected image and straight in the (right) terrain-corrected image.

8.6 Conclusions

Coordinate transformations are the basis of achieving a common frame of reference for geographic information analysis in GIS. The requirement of a common ellipsoid, datum, map projection, and finally plane coordinate system make it possible to use plane geometry for all types of spatial overlay and analysis. The methods of transformation are many and varied, and can be accomplished as rigorous mathematical transformations or as simple approximations. The accuracy of the resulting analysis, however, can only be as good as the accuracy of the data. Geographic data projection from the ellipsoidal Earth to a plane coordinate system always results in error in area, shape, and other properties. With appropriate selection of a projection, the user can preserve desired characteristics at the expense of others. In this chapter, basic concepts of coordinate systems and map projections were examined. For a more in-depth treatment, the reader is referred to the texts and sources referenced in this section and listed below.

References

Bugayevskiy, L. M. and J. P. Snyder. 1995. *Map Projections: A Reference Manual*. London: Taylor and Francis, 248 pp.

Canters, F. 2002. *Small-Scale Map Projection Design*. London: Taylor and Francis, 336 pp.

Clark, D. 1973. *Plane and Geodetic Surveying - Volume Two - Higher Surveying*. London: Constable, 292 pp.

Finn, M. P., E. L. Usery, S. T. Posch and J. C. Seong. 2004. A decision support system for map projections of small scale data. US Geological Survey Scientific Investigation Report, 2004-5297.

Flannery, J. J. 1971. The relative effectiveness of some common graduated point symbols in the presentation of quantitative data. *The Canadian Cartographer* 8:96–109.

Goode, J. P. 1925. The Homolosine projection: A new device for portraying the Earth's entire surface. *Annals of the Association of American Geographers* 15:119–125.

Goussinsky, B. 1951. On the classification of map projections. *Empire Survey Review* 11 (80):75–79.

Hotine, B. M. 1946. The orthomorphic projection of the spheroid. *Empire Survey Review* 8 (62):300–311.

Hotine, B. M. 1947. The orthomorphic projection of the spheroid. *Empire Survey Review*, 9 (63):3–35, (64):52–70, (65):112–123, (66):157–166.

ICA. 2006. Commission on Map Projections, International Cartographic Association. <http://www.csiss.org/map-projections/> Accessed 24 January 2007.

Iliffe, J. C. 2000. *Datums and Map Projections*. Caithness, Scotland: Whittles Publishing, 150 pp.

Jensen, J. R. 2005. *Introductory Digital Image Processing: A Remote Sensing Perspective*, 3rd Ed. Upper Saddle River, New Jersey: Prentice Hall, 526 pp.

Keates, J. S. 1982. *Understanding Maps*. New York: John Wiley, 122 pp.

Lee, L. P. 1944. The nomenclature and classification of map projections. *Empire Survey Review* 7 (51):190ff.

Maling, D. H. 1968. The terminology of map projections. *International Yearbook of Cartography* 8:11–65.

Maling, D. H. 1992. *Coordinate Systems and Map Projections*, 2nd edition. Oxford: Pergamon Press.

Maurer, H. 1935. *Ebene Kugelbilder: Ein Linn⊠sches System der Kartenentwürfe*. Petermanns Mitteilungen, Ergänzungsheft no. 221.

McGlone, J. C., editor. 2004. *Manual of Photogrammetry*, 5th edition. Bethesda, Md.: American Society for Photogrammetry and Remote Sensing, 1168 pp.

Mugnier, C. J. 2000. Grids and datums: The Republic of Madagascar. *Photogrammetric Engineering and Remote Sensing*, 66 (2):142–144.

Mugnier, C. J. 2001. Grids and datums: The Swiss Confederation. *Photogrammetric Engineering and Remote Sensing* 67 (8):897–898.

Mugnier, C. J. 2004. Object space coordinate systems. In *Manual of Photogrammetry*, 5th edition, edited by J. C. McGlone, 181–210. Bethesda, Md.: American Society for Photogrammetry and Remote Sensing.

Pearson II, F. 1990. *Map Projections: Theory and Applications*. Boca Raton, Florida: CRC Press, 384 pp.

Robinson, A. H., J. L. Morrison, P. C. Muehrcke, A. J. Kimmerling and S. C. Guptill. 1995. *Elements of Cartography*, 6th edition. New York: John Wiley and Sons, 674 pp.

Rosenmund, Max, 1903, Die Änderung des Projektionssystems der schweizerischen Landesvermessung: Bern, Switz., Militärdepartements, Abteilung für Landestopographie, Verlag, 147 p. [Oblique Mercator projection of ellipsoid as used for large-scale mapping of Switzerland.]

Snyder, J. P. 1987. *Map Projection: A Working Manual*. US Geological Survey Professional Paper 1395. Washington, D.C.: US Government Printing Office, 383 pp.

Snyder, J. P. and P. M. Voxland. 1989. *An Album of Map Projections*. US Geological Survey Professional Paper 1453. Washington, D.C.: US Government Printing Office, 249 pp.

Snyder, J. P. 1993. *Flattening the Earth: Two Thousand Years of Map Projections*. Chicago: University of Chicago Press, 365 pp.

Starostin, F. A., L. A. Vakhrameyeva and L. M. Bugayevskiy. 1981. Obobshchennaya Klassifikatsiya Kartograficheskikh Proyektsiy Po Vidu Izobrazheniya Meridianov I Paralleley, Izvestiya Vysshikh Uchebnykh Zavedeniy. *Geodeziya I Aerofotos'emka* 6:111–116.

Steinwand, D. R., M. P. Finn, J. R. Trent, E. L. Usery and R. A. Buehler. 2005. Re-projecting raster data of global extent. *Proceedings, Auto-Carto 2005: A Research Symposium*, Las Vegas, Nevada. Gaithersburg, Md.: Cartography and Geographic Information Society.

Takos, I. K. 1978. *The Azimuthal Equidistant Projection of Hatt* (in Greek). Greece: Hellenic Military Geographical Service, 21–52.

Thompson, D. B., M. P. Mephan and R. R. Steeves. 1977. The Stereographic Double Projection. *University of New Brunswick Technical Report No. 46*, 47 pp.

Tissot, N. A. 1881. *Mémoire sur la Représentation des Surfaces et les Projections des Cartes Géographiques*. Paris: Gauthier-Villars.

Tobler, W. R. 1962. A classification of map projections. *Annals of the Association of American Geographers* 52:167–175.

USGS. 2006. Cartographic Research, <http://carto-research.er.usgs.gov/>, US Geological Survey, Rolla, Missouri. Accessed 23 January 2007.

Veregin, H., ed. 2006. *Rand McNally Goode's World Atlas,* 21st edition. Chicago: Rand McNally & Company. 371 pp.

Welch, R. and A. Homsey. 1997. Datum shifts for UTM coordinates. *Photogrammetric Engineering and Remote Sensing* 63 (4):371–375.

Welch, R. and E. L. Usery. 1984. Cartographic accuracy of Landsat-4 MSS and TM image data. *IEEE Transactions on Geoscience and Remote Sensing*, GE-22 (3):281–288.

Yang, Y., J. P. Snyder and W. R. Tobler. 2000. *Map Projection Transformation Principles and Applications*. London: Taylor and Francis, 367 pp.

CHAPTER 9
Status and Future of Global Databases

Chandra Giri, David Hastings, Bradley Reed and *Ryutaro Tateishi*

9.1 Introduction

Global environmental and social problems have recently attracted public attention. It is now widely recognized and accepted that accurate and up-to-date information of the world is essential to help solve these problems (Htun 1997; Rhind 1997; MA 2005). In response, increasing amounts of global data are becoming available, largely driven by four main factors: 1) strong interest and commitment from national governments and international organizations; 2) a desire by the scientific community to answer complex environmental and socioeconomic problems; 3) requirements for military applications; and 4) advancements in computer hardware/software and information technologies.

Recent advancements in remote sensing, geographic information systems (GIS), global positioning systems (GPS) and information technology (IT) have stimulated the creation of numerous global databases, which are important to ecosystem and societal studies. Several national governments and other governmental groups such as the United States, Japan, India, and the European Union have been providing large volumes of remote sensing data for global change studies. Similarly, international organizations such as the United Nations Environment Programme (UNEP) and the Food and Agriculture Organization (FAO) of the United Nations (UN) have been generating large volumes of global data needed for environmental assessment and sustainable development. Scientific communities eager to answer key research questions such as how the Earth's environment has changed over time, and whether its rate of environmental change is accelerating as a result of human activity, also have generated new data sets (Giri 2005). Similarly, global data sets needed and prepared for military applications are becoming available for civilian applications.

Consequently, the amount, extent, availability and utility of global data continue to expand rapidly. The user community, however, remains uncertain about what data are available, whether data even exist, how data were produced and how to effectively use global data. This chapter aims to address these issues focusing on conceptual issues, past efforts, the present situation, and methodological considerations.

9.2 What Is Global Database?

9.2.1 The Concept of "Global"

The Merriam-Webster Online dictionary (www.m-w.com) defines "global" as "of, relating to, or involving the entire world." Thus, the term 'global' embraces the "entire Earth" including both land and water surfaces extending from 900N to 900S and 1800W to 1800E. However, in practice, the term is sometimes used loosely to represent "near-global" or "quasi-global" data sets. As the term "World Series" is used for an annual competition between two baseball leagues constrained to two countries (i.e., USA and Canada), "global" has been used to describe data that are highly uneven in character, or entirely lacking in many countries, rural areas or otherwise "off-the-map" areas in a particular theme. Conversely, a global tropical forest database covering tropical regions of the world, point coverage of major cities around the world, well-core drilling data consisting of only 100 drilling sites but scattered around the world and species distribution data of mammals, amphibians, reptiles, etc. also are classified as global data. So, what is the meaning of "global" in global database? The answer can be explained in terms of geographic coverage, thematic coverage, global integration and global completeness.

Figure 9-1 A conceptual diagram of the concept of global database. See included DVD for color version.

9.2.1.1 Geographic Coverage

Simply defined, a "global" database covers the entire world. Many data sets which omit Antarctica, or are collected by a satellite mission lacking global coverage, might thus be called quasi-global. Examples include data from the National Oceanic and Atmospheric Administration (NOAA) Global Vegetation Index products and the Shuttle Radar Topography Mission (SRTM) imagery or elevation data, as the latter do not extend to the poles. Land use classifications from satellite images marred by cloud cover and detailed data sets emphasizing land areas but omitting tidal zones are also quasi-global.

9.2.1.2 Thematic coverage

A global data set treats a theme in a globally appropriate manner, making it fully and globally useful. A soils database covering the world, but with legend characterizations developed mostly by experts in temperate zones, risks inadequate consideration for equatorial, tropical, boreal and polar issues. Global considerations include: 1) what issues/problems the data for those areas must be able to address; and 2) how the theme should be represented to be globally useful. Many themes, such as topography or soils, are inconsistently characterized among various areas. Such inconsistency may result from differing legend schemes, observational approach or detail. Variations among countries or environments and different cultural concepts of a theme impact the compilation and use of global data sets. Consistently appropriate global thematic coverage is arguably the greatest challenge for many global database compilations.

9.2.1.3 Global Integration

Some people have argued that the global collection of imagery from a satellite mission, such as Landsat, constitutes a global database. However, many such collections lack a means of operationally geo-locating, mosaicking, and integrating such data. Such missions may have global or near-global coverage, but arguably are not global data sets. The Advanced Very High Resolution Radiometer (AVHRR) was the first system aimed at operational geo-location and mosaicking. Yet AVHRR's onboard clock is allowed to drift by as much as one second around the actual time—resulting in geo-locational errors of almost 7 km by this factor alone. Procedures are now available for adjusting for this error. However, this shows the evolutionary nature of the community's effort to produce fully global databases.

9.2.1.4 Global Completeness

Some themes of a global database may cover only parts of the world. For example, a global database of habitats for snake species or a global database of tropical forests of the world would only cover certain parts of the world. Nevertheless, the database could arguably be called global if it were globally complete.

In summary, the concept of a fully global data set has evolved considerably in recent years, thanks to several international meetings, working groups, and global database compilation efforts and evaluations (Singh 1994; Tateishi and Hastings 2000, 2002). This increased understanding of the subject should greatly assist future global database efforts.

9.2.2 The Concept of "Database"

A *database* is commonly considered to be a relatively thorough, systematically organized compilation of data or information related by some common characteristic. A data set is often considered to be a more simply organized collection of data, such as one with less temporal or thematic comprehensiveness, than a database. For example, a systematically organized collection of digital national boundaries might be considered a data set, whereas such a collection of national and subnational boundaries (perhaps to several lower hierarchical levels of administration) might constitute a database of administrative boundaries. Similarly, a systematically organized digital representation of roads might constitute a data set (whether or not the attributes might help differentiate national, provincial, or local, divided or unpaved, year of construction and last maintenance of features like bridges or signaling), while such a collection representing road, rail and air transport (including infrastructure such as fueling stations, rest areas, capacity restrictions and traffic levels) might constitute a transportation database. A georeferenced, multispectral satellite image might constitute a data set, while daily or monthly time series developed in a systematic fashion, which includes systematically derived, stored and retrievable parameters (such as sea- and land-surface characteristics) might constitute a database.

9.2.3 The Concept of "Global Environmental Database"

Global environmental databases are databases of environmental parameters covering the globe. Environmental parameters consist of not only physical parameters such as climatic and soil data, but also socioeconomic parameters such as human population and energy consumption by humans. All environmental parameters are characterized for geographic locations or regions. That is, environmental parameters are geospatial data. 'Global' is a concept concerning horizontal area and although there is no explicit definition about the vertical range of 'global environmental databases,' it is generally similar to that of the biosphere.

Some global environmental parameters are produced and published as one data set, while others are published as a database that includes multiple data sets. Examples of environmental parameters for global environmental databases (e.g., NASA 2007) are listed in Table 9-1.

Table 9-1 Environmental parameters for global environmental databases.

Physical Parameters			Socioeconomic Parameters
Land	**Ocean**	**Atmosphere**	
topography land cover biophysical (e.g., leaf area index, biomass) soils hydrology biodiversity	ocean depth salinity/density sea surface temperature ocean waves sea ice marine sediments	temperature precipitation drought severity aerosols humidity indices	governmental boundaries population distribution land use oil and gas production livestock reclamation/restoration

9.3 Brief History

Although data sets of the known world have a history of at least two millennia—including Al-Idrisi's maps and treatises of 1154 and 1161 (Parry 2004)—organized attempts at compiling global spatial databases are a much more recent phenomenon. Albrecht Penck (1858-1945), a German geographer, proposed to prepare a worldwide system of maps on the Millionth Scale to the 5th International Geographical Congress in Berne in 1891. This initiative, called the International Map of the World (IMW) or "The Millionth Map," aimed to prepare paper maps at the scale of 1:1,000,000 showing both human and physical features. Deliberations continued, and in 1913 detailed specifications were prepared and agreed upon. Individual countries were responsible for the compilation and publication of maps in their territory. Under the umbrella of the International Geographical Union (IGU), with the Secretariat at the Ordinance Survey in London, the project was adopted by the UN in 1953.

The effort by numerous mapping agencies of developing standards and applying those standards, as they separately developed maps in this series, was daunting. Nevertheless, such disparate organizations as mapping authorities in Australia, Brazil, France, the Soviet Union, the United Kingdom, the US Army Map Service and the American Geographical Society, contributed maps to the series. The UN produced periodic reports on the progress of the effort, including a listing of maps contributed to the series by many countries (United Nations 1966, 1975).

By the mid 1970s, only half of the Earth's surface was mapped because many countries did not have enough resources or expertise to complete the project. This initiative, however, fostered cooperation by many countries (Collins and Rhind 1997; Anderson and Kline 2000). The greatest success was the Map of Hispanic America completed by the American Geographical Society from 1920 to 1945. The project was officially ended in 1986 as dwindling interest and activity by then had limited progress (Thrower 1996). Even after the IMW, the UN continued to work at preparing coarse-scale global data sets. In 1994, the UNDP and the UNEP organized an International Symposium on Core Data Needs for Environmental Assessment and Sustainable Development Strategies (Estes et al. 1994). They identified ten high priority core data sets: land use/land cover, demographics, hydrology, infrastructure, climatology, topography, economy, soils, air quality and water quality (Singh 1994).

Anderson and Kline (2000) defined the concept of a reference framework:

> *A reference framework serves to orient and locate data so that it is in the correct context to better enable understanding. A critical aspect of the framework is the coordinate system chosen and the geodetic control used to relate it to the surface of the earth… The scientific understanding of earth and atmospheric processes requires the ability to precisely locate observations from different sensors and different times so that the interaction (interconnection) between different variables can be understood. This is the role of the reference framework.* (p. 26)

A reference framework—usually consisting of seashore lines, rivers/lakes, elevation, political boundaries and transportations—can be considered as basic data in a global environmental database. The attempt to produce a global reference framework started well before the time of growing concern for the global environment. Initiatives are discussed further below.

According to Anderson and Kline (2000, p. 2), "The International Map of the World (IMW), also known as the Carte Internationale du Monde au Millionème, was the first truly global collaborative mapping effort." Proposed by Albrecht Penck at the 5th International Geographical Congress in 1891, the specifications for the IMW went through several permutations due to technological innovation and interruptions by two World Wars. The prime meridian was chosen to be Greenwich, rather than Paris. Information to be included in the

IMW was population, bathymetry, elevation, roads and railways, and political boundaries. The IMW was a large undertaking for a world not quite prepared for the necessary effort and, unfortunately, it was only partially successful.

During World War II and thereafter, both US and former Soviet Union military and intelligence services prepared a number of maps of the world. The Soviet mapping project called "Karta mira" produced topographic maps covering both terrestrial and marine areas at scales of 1:2,500,000 and 1:1,000,000. For several cities, topographic maps at the scale of 1:10,000 also were prepared. These maps were not available to the general public until the break-up of the Soviet Union.

In the 1960s, US military agencies produced a number of global data sets—including the World Data Bank (WDB) I and II, Tactical Pilotage Chart (TPC), Operational Navigation Chart (ONC), Jet Navigation Chart (JNC) and Global Navigation & Planning Chart (GNC). The nominal scale of the WDB data sets is 1:2,000,000. The scales of TPC, ONC, JNC and GNC are 1:500,000, 1:1,000,000; 1:2,000,000 and 1:5,000,000, respectively. The World Data Bank consists of data layers representing land boundaries, rivers and political boundaries, while TPC, ONC, JNC and GNC consist of topographic maps with aeronautical information overlain. A thinned version of WDB II, called Micro World Data Bank, was derived and distributed on floppy diskette in 1988. This was arguably the first global database for personal computers (Dinkins 1988; Pospeschil 1992). NIMA also produced ARC Digitized Raster Graphics (ADRG) digital data by rasterizing paper maps with worldwide coverage of GNC, JNC, ONC and TPC products.

TERDAT was perhaps the pioneering global multi-thematic raster database (Cumming and Hawkins 1981). This database was derived by manually estimating and recording parameter values from ONCs on a 10-minute latitude-longitude grid. Its themes included elevation (minimum, maximum and modal value per 10-minute cell), primary and secondary land cover classes, percentage of water cover, percentage of urban development and number and direction of topographic ridges. The data tended to be rudimentary, with large flat areas where elevation values were absent in the source maps. TERDAT was used extensively in generating ETOPO5, ELEVBATHY and, to a lesser degree, in TerrainBase.

A world map showing ETOPO5 was immensely popular and the database was incorporated in a large number of data packages. ELEVBATHY was a parallel effort to ETOPO5, which was the source of a high-resolution (for the day) computer-derived color stereo-pair world map (Hastings 1986; Hastings and Cordell 1988). TerrainBase (Row and Hastings 1994) was an advancement in the collection, management, documentation and access to global data. Over 20 elevation and bathymetric models were collected, evaluated, mosaicked and documented with the most extensive monograph on global Digital Elevation Models (DEMs) to date (Row et al. 1995). TerrainBase's development was managed in the Geographic Resources Analysis Support System—Geographic Information Science (GRASS-GIS), a GIS that pioneered easy handling of diverse data sets with varying area coverage, grid spacing and registrations.

The 1980s saw several global data sets on vegetation and land use data including Matthews (1983), Olson et al. (1983) and Wilson and Henderson-Sellers (1985). Since the 1980s, new global data sets have been prepared by a number of organizations. This was made possible, in part, due to the increased availability of remote sensing data and advancements in GIS technology. Selected data sets produced during this period are discussed in the following subsections.

9.3.1 Digital Chart of the World (DCW)

The Digital Chart of the World (DCW) created by the US Defense Mapping Agency is a digital database of nearly 200 feature types populated from the 1:1,000,000 scale Operational Navigation Chart (ONC) series (Danko 1992). The DCW Version 1.0, published in July 1992, included roads, railroads, rivers, lakes, streams, major utility networks, cross-country pipelines, communication lines, airports, elevation, coastlines, international boundaries, populated places and geographic names. (Anderson and Kline 2000). Detailed information about DCW can be found at http://www.lib.ncsu.edu/gis/dcw.html.

9.3.2 World Vector Shoreline (WVS)

The World Vector Shoreline (WVS) is a global digital data file containing shorelines, international boundaries and country names at a nominal scale of 1:250,000. For many years, this was a standard global database for many NIMA applications. The main source material for the WVS was the Digital Landmass Blanking (DLMB) data derived primarily from the Joint Operations Graphics and NIMA coastal nautical charts (Anderson and Kline 2000). Further information can be found at http://www.csc.noaa.gov/shoreline/world_vec.html.

9.3.3 Global Environmental Databases

Attempts to produce global data of various environmental parameters were accelerated in the 1980s by growing concern for the environment and the launches of many Earth-observation satellites. Different types of data were developed by scientific groups interested in topics such as oceanography, land cover, land use, biodiversity, soil, hydrology, climate and so forth. The description of the data production of each environmental parameter is beyond the scope of this manual; Tateishi and Hastings (2000, 2002) describe the development of each type of environmental data set in detail.

As various global environmental data were produced, databases of each field also were produced; i.e., global oceanographic database, global soil database, etc. Furthermore environmental data of different fields were collected. The UNEP established the Global Resource Information Database (GRID) in 1985 for the development of a global environmental database. A similar attempt by a different initiative was made by the International Geographical Union (IGU). It started the Global Databases planning project to examine the provision of databases to the International Geosphere-Biosphere Programme (IGBP) in 1988 (Mounsey and Tomlinson 1988). However, this attempt by IGU was not successful in distributing the planned databases.

Other global environmental data such as land cover have been produced by the International Geosphere-Biosphere Programme Data and Information System – Cover (IGBP DISCover) (Loveland et al. 2000); University of Maryland Global Land Cover Classification (Hansen et al. 2000 and 2002), MODerate Resolution Imaging Spectroradiometer (MODIS) Land Cover (Friedl et al. 2002) and Global Land Cover-2000 (Bartholome and Belward 2005). Additional MODIS data biophysical products include Leaf Area Index and Net Primary Production (Justice et al. 2002), while global population data such as Gridded Population of the World (GPW) (CIESIN 2000) and LandScan (Dobson et al. 2000) also resulted from the availability of global remote sensing data.

These represent only a small portion of the data sets prepared since 1980. A detailed list and description of other global data sets can be found in Tateishi and Hastings (2000, 2002).

9.4 Importance of Global Databases

The Earth's environment has been changing since time immemorial from natural forces and more recently by anthropogenic impacts. Scientific evidence suggests that such changes are accelerating and are occurring across multiple scales, ranging from local to global. Improved scientific analysis using global databases is necessary to better understand global environmental changes and their implications. Global data sets can be used to measure, map, monitor, model and predict such changes and their impact to ecosystems and society.

Global data sets are needed for a wide variety of uses including investigations of climate change, biodiversity, environmental assessment and modeling. For example, a human population distribution map can be used to identify populations affected by natural disasters such as tsunamis and cyclones. Similarly, global land cover data can be used to identify the remaining habitat of wildlife species, and global elevation data can be used in delineating watersheds, estimating runoff and other applications in hydrology.

Several past and on-going global programs and projects including the IGBP, the Intergovernmental Panel on Climate Change (IPCC), Millennium Ecosystem Assessment (MA), the Land Degradation Assessment in Drylands project (LADA), and the Global Biodiversity Assessment (GBA) have either helped develop, use, plan or emphasized the need for such data sets (Heywood 1995; LCLUC 2002; IPCC 2002; MA 2005; LADA 2002). Similarly, commitments from leading research groups and organizations—such as the National Aeronautics and Space Administration (NASA), US Geological Survey (USGS), FAO, UNEP and Joint Research Center of the European Commission (JRC)—to produce such data sets are encouraging. The user bases of global data also are broadening.

International treaties and agreements—such as Agenda 21, the Framework Convention on Climate Change, the Kyoto Protocol, the United Nations Convention to Combat Desertification and the Biodiversity Convention—also have emphasized the importance of the availability of time series global land data sets (CIESIN 2000; Cihlar 2000). The United Nations Development Programme (UNDP) and UNEP, in 1994, identified 10 high-priority core data sets that are essential for environmental assessments and sustainable development: land use/land cover, demographics, hydrology, infrastructure, climatology, topography, economy, soils, air quality and water quality (UNDP 1994; UNEP 1994).

Several US global change research programs also have reiterated the importance of global data. The Land Cover Land Use Change (LCLUC) program, an interdisciplinary scientific theme within NASA's Earth Science Enterprise (ESE), aims "to identify the current distribution of land cover types and track their conversion to other types" (LCLUC 2002) particularly in relation to the impact of land cover land use change on biogeochemical cycles (e.g., carbon and nitrogen) and the hydrological cycles. In 2001, the National Research Council (NRC) in its publication "Grand Challenges in Environmental Sciences" described the need to: 1) develop long-term, regional databases of land use, land cover, and related social information; 2) formulate spatially explicit and multi-temporal land-change theory; 3) link land-change theory to space-based imagery; and 4) develop innovative applications of spatial simulation techniques for assessing environmental dynamics and land use and land cover change.

In addition to these major global and US global change research programs, several scientists and researchers also have highlighted the need for global data products (Sellers et al. 1997; Nemani et al. 1996; Reed 1997; Brown et al. 1999; Ramankutty and Foley 1999; Zhu et al. 1999; Hansen et al. 2000; Rosenqvist et al. 2000; Achard et al. 2002).

9.5 Current Situation

Global data sets available as maps and charts are being replaced by digital data sets. Recent advancement is GIS has facilitated converting, capturing, storing, and analyzing vast amounts of global data consisting not only of geographic and cartographic data, but also of socio-economic and political data. The availability, accessibility, and more importantly, the application of global data have improved significantly in the last 20–25 years. These data sets were primarily prepared by government agencies, international organizations and Non-Governmental Organizations (NGO).

The increased availability is mainly due to the recent developments in GIS technology and availability of remotely sensed data. GIS has provided us with the ability to scan/digitize, store, manipulate and retrieve information in ever-increasing quantities. GIS also helped integrate biophysical and socio-economic data. For example, researchers have produced historical global land cover data by analyzing historical statistical inventories such as census data, tax records and land surveys data. In the recent past, we have witnessed an exponential increase in the quantity of data and the range of data products being produced by earth observation satellites. These remotely sensed data are available with improved spatial, spectral and temporal resolutions that are being used routinely in preparing global data sets.

Global data are being used for a number of applications including climate change, biodiversity conservation, environmental assessment, modeling, disaster management, homeland security, and human health. Although the number and breadth of application areas have expanded, users of global data are not distributed evenly throughout the world (Figure 2). For example, in response to a voluntary request from the IGBP DISCover project; a number of users have registered online, providing information such as name, affiliation, address, geographic area of interest and potential application of the data set. As of September 2005, approximately 1,300 users from 89 countries (Figure 2) have registered and downloaded data with a volume of ~500 gigabytes from the National Center for Earth Resources Observation and Science (EROS) website (http://edcdaac.usgs.gov/glcc/glcc.html). Selected global or quasi-global mapping initiatives are described below.

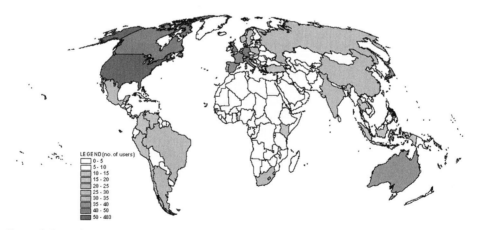

Figure 9-2 Registered users of Global Land Cover Characterization database worldwide. See included DVD for color version.

9.5.1 Global Mapping

In response to Agenda 21 adopted during the United Nations Conference on Environment and Development held in Brazil in 1992, the Ministry of Land, Infrastructure and Transport (formerly Ministry of Construction) of Japan initiated the Global Mapping project in 1992. The main goal of the project is to develop 1:1 million scale or 1-km resolution global data in cooperation with national and international organizations. The project is coordinated by the International Steering Committee for Global Mapping (ISCGM), which consists of heads of National Mapping Organizations with representatives of regional and international organizations, and academic institutions. Since its inception in 1996, a number of national mapping agencies have joined and contributed to the project.

According to the ISCGM, the Global Mapping project is expected to facilitate the implementation of global environmental agreements and conventions, help monitor environmental phenomena and encourage economic growth within the context of sustainable development. The data are being prepared with standardized specifications and consist of 8 layers: political boundaries, drainage, transportation, population centers, elevation, land cover, land use and vegetation. The ISCGM aims to develop global maps as a core database by 2007 and update the database every five years.

9.5.2 United Nations Geographic Information Working Group (UNGIWG)

The United Nations Geographic Information Working Group (UNGIWG) has been working to generate validated information and maps regarding international and administrative boundaries for all the UN member countries. Their primary goal is to provide data to the UN community. The working group collects and compiles boundary information and maps from national mapping agencies (and other sources) at the scale of 1:1 million. The Second Administrative Level Boundaries (SALB) project was launched in 2001; its goal is to provide validated data and information by the national mapping agencies including the historic changes observed since 1990 at the 1st sub-national level and since 2000 at the 2nd sub-national level. The project aims to produce the following data products: 1) a list of administrative units as observed in January 2000; 2) maps of historic changes dated from 1990 to 2006 at the 1st administrative level and from 2000 to 2006 at the 2nd administrative level; and 3) administrative boundaries maps with one map for each period of representatively observed between 2000 and 2006. Further information about UNGIWG can be found at http://www.ungiwg.org/.

9.5.3 Shuttle Radar Topography Mission (SRTM)

The Shuttle Radar Topography Mission (SRTM) project spearheaded by NGA and NASA is an international project with a goal to collect topographic data covering approximately 80% of Earth's land surfaces. The project created a quasi-global data set that is the most detailed digital elevation data available at 90 m spatial resolution. Data were collected during an 11-day mission on-board the space shuttle *Endeavour* in February 2000. An Interferometric Synthetic Aperture Radar instrument acquired the data with 16-m absolute vertical height accuracy at 30-m postings. Detailed information about SRTM can be found at http://www2.jpl.nasa.gov/srtm/ and http://srtm.usgs.gov/.

9.5.4 Global Population Distribution Data

Two global population distribution data sets are being prepared on a regular basis: Gridded Population of the World (GPW) and LandScan. GPW provides information on residential populations whereas LandScan data provides information about ambient population on an annual basis. GPW Version 3 data are available for both population counts (raw counts) and population densities (per square km) (Figure 9-3). The LandScan data consist of global data prepared using census data available at sub-national level and modeling based on proximity to roads, slope, land cover, nighttime lights and other information (Dobson et al. 2000). GPW is prepared by the Center for International Earth Science Information Network (CIESIN), Columbia University, while LandScan is prepared by the Oak Ridge National Laboratory (ORNL). Further information about GPW and LandScan can be found at http:// beta.sedac.ciesin.columbia.edu/gpw/ and http://www.ornl.gov/sci/landscan/, respectively.

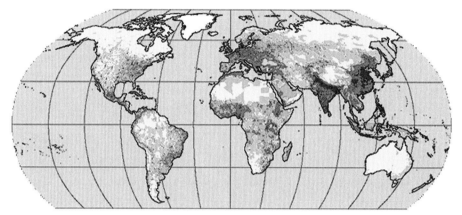

Figure 9-3. Gridded Population of the World (GPW) Version 3 showing population density in 2000. Source: http://beta.sedac.ciesin.columbia.edu/gpw/global.jsp. See included DVD for color version.

9.5.5 GLOBCOVER

The GLOBCOVER project is funded by European Space Agency (ESA) and is being implemented by a consortium led by Medias France in association with FAO, UNEP, JRC, IGBP, and Global Observation of Forest Cover and Land Cover Dynamics (GOFC-GOLD). The main objective of the project was to prepare a global land cover database of the world for the year 2005, using recently available Medium Resolution Imaging Spectrometer (MERIS) 300-m data. The global map data were released in early 2007 (ESA 2007). The project produced two versions of GLOBCOVER maps: Version 1 used automatic image processing methodologies and Version 2 was produced by involving national and regional partner institutions. The latter was implemented by JRC and contains global as well as regional data sets. More information about GLOBCOVER can be found at http://www.esa.int/.

9.5.6 International Vector Data (VMAP)

The International Vector Data (VMAP1) seamlessly covers the entire world and is an updated and improved version of the NIMA DCW mentioned above. The primary source for the database is the 1:1,000,000-scale ONC series co-produced by the military mapping authorities of Australia, Canada, United Kingdom, and the US. VMap Level 0 also is available and includes major road and rail networks, hydrologic drainage systems, utility networks, major airports, elevation and depth contours, coastlines, international boundaries, vegetation and populated places (NGA 2004).

9.6 Recent Trends

Based on the above discussion, six trends in recent global data generation are identified:
    1) to produce higher resolution global data such as VMAP1 while serving the community who need coarse resolution data;
    2) to produce digital data sets in standard formats that can be shared through Internet and digital media;

3) to produce and distribute data in collaboration with national and international agencies;

4) to fulfill the requirements of a broad range of applications;

5) to recognize the need for validated products; and

6) to support the user community.

As an example, the development of global land cover data sets and their trends in recent years are listed in Table 9-2. The first generation of digital global land cover data sets were generated using published maps, field data and aerial photographs. For example, the Matthews vegetation data set was produced by compiling more than 100 existing map sources supplemented by a large collection of aerial photographs (Matthews 1983). The Olson vegetation data representing the world's major ecosystem complexes was compiled from patterns of pre-agricultural vegetation, modern aerial surveys, and intensive biomass data from research sites (Olson et al. 1983). Similarly, global vegetation and soil data were prepared by collating existing natural vegetation, forestry, agriculture, land use and soil maps (Wilson and Henderson-Sellers 1985).

All three of these data sets are of coarse scale (0.5° to 1° resolution) and prepared primarily to serve the global climate change and modeling communities. Matthews' vegetation data were produced to provide important input to climatic models in various studies conducted at the Goddard Institute for Space Studies (GISS), Columbia University. Olson vegetation was developed for use as a reference base to interpret the role of vegetation in global CO2 cycling; a basis for improved estimates of carbon content of soil and vegetation, and as a means of correcting estimates of carbon released into the atmosphere in recent times due to continuing changes in vegetation patterns (Olson et al. 1983). The Wilson and Henderson-Sellers data were produced for use in general circulation climate models.

Table 9-2 Availability of global land cover data.

Data	Year	Resolution	Available at this URL 12 June 2007
Matthews Global Vegetation and Land Use	1983	108 × 108 km	http://www.giss.nasa.gov/
Olson Land Cover and Vegetation	1983	54 × 54 km	http://www.grid.unep.ch/
Wilson and Henderson–Sellers Global Land Cover	1985	108 × 108 km	http://www.ngdc.noaa.gov/
Tateishi and Kajiwara Global Land Cover	1991	16 × 16 km	http://www.ngdc.noaa.gov/
DeFries and Townshend-Global Land Cover	1995	108 × 108 km	http://glcf.umiacs.umd.edu/
GLCC (IGBP DISCover)	1997	1 km × 1 km	http://edc2.usgs.gov/glcc/
UMD Land Cover	2000	1 km × 1 km	http://www.geog.umd.edu/landcover/1km-map.html
MODIS Land Cover	2003	1 km × 1 km	http://edcdaac.usgs.gov/
Vegetation Continuous Fields	2003	1 km × 1 km	Unavailable due to errors in the algorithm. When fixed, data will be available at http://glcf.umiacs.umd.edu
GLC2000	2004	1 km × 1 km	http://www-tem.jrc.it/glc2000/
Vegetation Continuous Fields	2003	500m × 500m	http://glcf.umiacs.umd.edu/
GLOBCOVER (forthcoming)	200?	300m × 300m	http://www.esa.int/
GeoCover-LC™	2003	30m × 30m	http://www.mdafederal.com/geo-cover

The second generation of global land cover data sets began with the production of the NOAA AVHRR-based global land cover products, e.g., the work of Tateishi and Kajiwara (1991), DeFries and Townshend (1994), Loveland and Belward (1997) and DeFries et al. (1998). Tateishi et al. (1991) used a clustering approach to classify nominal 16-km monthly Global Vegetation Index from AVHRR data of 1986 to 1989, generating 13 land cover types. DeFries and Townshend (1994) used a maximum-likelihood classification approach to classify monthly Normalized Difference Vegetation Index (NDVI) data of 1987 by selecting training samples from previous land cover maps. Eleven land cover types were generated to fulfill the requirements of climate models. Loveland et al. (2000) performed unsupervised classifications on monthly AVHRR NDVI composites acquired between April 1992 and March 1993 followed by iterative labeling to produce the IGBP DISCover. Similarly, DeFries et al. (1999) used the same data sources as IGBP DISCover to produce the University of Maryland (UMD) Land Cover using a decision tree classifier.

The AVHRR-based land cover analyses provided substantial improvements over previous global land cover data sets in terms of spatial resolution, classification system, accuracy and their applications. The IGBP/DISCover, for example, created the highest resolution global land cover data ever produced and introduced a flexible classification system; the data were validated quantitatively. Through the participatory approach, the IGBP land cover classification system was developed. Alternative land cover classes representing global ecosystems, the USGS land use/land cover system; a simple biosphere model, a simple biosphere 2 model, a biosphere-atmosphere transfer scheme and a vegetation life form classification system also were presented. The spatial resolution improved from several kilometers to one kilometer.

Currently, Boston University (BU) has been working on the production of the global land cover data set, which is based on MODIS resampled 1 km data. A preliminary version of this data set is now available at http://edcdaac.usgs.gov/modis/mod12q1.html. At the same time, Vegetation Continuous Field (VCF) data produced by Hansen et al. (2002) is also available. The VCF uses an innovative approach to land cover characterization explained in DeFries et al. (1999) and Hansen et al. (2002). Both BU and the University of Maryland (UMD) have developed automated land cover mapping and characterization methodologies. Similarly, JRC in association with a number of partner organizations around the world prepared the Global Land Cover 2000 (GLC-2000) data with the main objective of preparing a harmonized land cover data set of the world for the year 2000. Also ISCGM is producing Global Land Cover by National Mapping Organizations (GLCNMO) using MODIS data of 2003 by cooperation with a number of national mapping organizations.

The third generation of global land cover data began with the production of MODIS 500-m global VCF and GLOBCOVER data. The first releases of percent tree cover data are now available at http://glcf.umiacs.umd.edu/. Besides VCF data, areal estimate of life form (proportion of woody vegetation, herbaceous vegetation, or bare ground), leaf type (proportion of woody vegetation that is needle leaf or broadleaf), and leaf longevity (proportion of woody vegetation that is evergreen or deciduous) also will be included in the future data release. The percent tree cover data set is expected to be particularly useful for climate change modeling and forest cover assessment. The classification is based on regression tree classifier using training data derived from high-resolution imagery (Hansen et al. 2002).

Efforts to produce a fourth generation of global land cover data are already underway. The GeoCover-LC™ was implemented by EarthSat (Earth Satellite Corporation now MDA Federal, Inc.) under a NASA/NGA contract (MDA 2007). The main purpose of the project was to prepare a high resolution land cover data set of the world using 30-meter spatial resolution Landsat data. The project includes Landsat TM (Thematic Mapper) and ETM+ (Enhanced Thematic Mapper Plus) data acquired in 1990 and 2000 for global scale change analyses.

9.7 Methodological Considerations

9.7.1 Data Source and Data Capture

As discussed above in the presentation of global mapping efforts through recent 4th generation products, global data are produced using a wide variety of data sources including published maps, GIS data, ground surveys, socioeconomic data and remotely sensed data. Many researchers in the early 1980s collected published maps from countries and used those maps as a primary source for preparing global maps. They also used aerial photographs and ground survey to fill gaps in certain areas where published maps were not available. One of the major drawbacks to this approach is that data collected from numerous sources may vary greatly in scale and quality. Other problems often encountered include the age and variability of source maps, map projection systems and classification system/class definitions, along with missing metadata information. Availability of comprehensive remotely sensed data made it possible to prepare global data in a consistent manner by overcoming some of these limitations. Recently, a number of global databases were produced using field inventory (e.g., global biodiversity data) and socioeconomic (e.g., LandScan) data.

The Digital Bathymetric DataBase 5-minute (DBDB5) global raster data, for example, was developed in the early 1980s by the US Naval Oceanographic Office, by interpolating digitized contours from 1:5,000,000-scale bathymetric charts for assisting in the development of bathymetric profiles along naval cruises. It is the primary source of bathymetric data for ETOPO5, ELEVBATHY and TerrainBase. ETOPO5 was a mosaic of several regional DEMs at 5- and 10-minute grid spacing, with TERDAT used where better data were unavailable. Similarly, the Digital Terrain Elevation Data (DTED) began inauspiciously with an experimental conversion of 1:250,000 scale topographic maps to grids by the US Army Map Service (AMS). The goal of this effort was to make industrial moulds for plastic relief maps of the existing paper 1:250,000 map series. The maps were intended to help military personnel visualize terrain more effectively than was possible with traditional paper contour maps. Initial prototypes included Puerto Rico, for which a plastic relief map was made in this manner in the late 1950s. Based on this work, a digitized map of the US was created by the Army Map Service in the 1960s. These plastic relief maps were popular until AMS discontinued the product.

The process began with the development of a methodology for physically tracing the contour values etched into copper printing plates for those maps, editing the resultant data, then using contour-to-grid interpolation software to produce grids spaced at every three arc-seconds of latitude and longitude. The compromises in accuracy resulting from the original objectives, and budget for the project, impacted the scientific use of the national data set for decades. However, the effort was the beginning of the global DTED goal to make the best practicable digital elevation data sets available for the US Department of Defense. Various data sources are used, depending on the best available source at the time of production including aerial photographs, satellite imagery and hardcopy maps.

Ground data play an important role in preparing and updating global data. Operational Navigation Charts were globally and even locally inconsistent in their representation. For example, a very few smokestacks were listed. Conceptually, this has raised a tremendous potential for networks of volunteers contributing to populate such data layers with more examples—eventually leading to a substantial number of data sets representing each theme—such as a global database of smokestacks (each with its own standard attribute sets, developed by a team of experts on the GIS representation of smokestacks). Lately, uploading ground data using the internet is rapidly growing.

9.7.2 Scale

Global data are generally prepared at the scale of 1:1,000,000 or smaller, compromising broad-area coverage for high detail of features. Recently, global data are being prepared at large spatial scales as well. For example, VMAP1 data are now available at 1:250,000 scale. GLOBCOVER data are being prepared using 300-m resolution satellite data, and some MODIS data products are available at 250- and 500-m resolutions. Considering the range of needs by the user community, global data at multiple scales are actually preferred. For some applications such as climate change modeling, coarser resolution data at the scale of several kilometers is sufficient. For other applications, 1:250,000 or better resolution data are desirable. To meet the requirements of diverse user communities, data such as GPW are available in multiple scales.

It is important to note that even within a single data set, spatial resolution may vary from one location to another. For example, the horizontal grid spacing of GTOPO30 is 30 arc-seconds which is approximately 1 km at the equator. The resolution, however, decreases in the east/west (longitudinal) direction as latitude increases. At its extreme, the ground distance for 30 arc-seconds of longitude at the South Pole in Antarctica is zero.

9.7.3 Data Format

Global data are available in raster and/or vector format. Examples of global vector data include World Data Bank II, Digital Chart of the World (DCW) and VMAP, while raster data include DTED, DBDB5, ETOPO5, GTOPO30 and GLOBE. The majority of global data sets are available in raster format. The ISCGM aims to produce the global data in raster or vector format as appropriate for the type of ground feature being represented. For example, data sets of continuous information such as elevation, land cover, land use and vegetation will be produced in raster format and linear or relatively discrete features such as drainage system, transportation, political boundaries and populated places will be produced in vector format. In some cases (e.g., GeoCover), data are available in both raster and vector format. The utility of the global data increases when the data are available in both raster and vector formats primarily because users will have more options to choose from.

9.7.4 Files and Tiles

Global data are available in single or multiple files. If the data volume is too large, global data are often produced and distributed in quadrangles or tiles. The tiles can be divided in a number of ways depending on the data volume and requirements of the project. DCW consisted of many millions of arcs/nodes and megabytes for major layers such as drainage, coastlines and national boundaries. Indeed, the numbers of features, and the size of the files, made fully global compilation of all drainage features beyond the scope of virtually every GIS software package and host computer system available at the time. Thus the developers of DCW parsed the data into 5 x 5-degree latitude-longitude tiles (DMA 1992). Similarly, VMAP0, VMAP1, GTOPO30 and GLOBE divided the world into 8, 234, 33 and 16 tiles, respectively. MODIS land products are being distributed by the LPDAAC on a 1200 × 1200 km tiling scheme. In some cases, it is also desirable to develop a two-tiered tiling scheme. For example, one edition of VMap0 had a two-tired tiling scheme: 15 × 15-degree tiles for continental landmasses and 30 × 30-degree tiles for ocean areas. A latter edition of VMAP0 used the tiling scheme as shown in Table 9-3 (DMA 1995). Alternatively, data also are produced and distributed by continents.

Table 9-3 Tiling scheme for VMAP0.

Latitude	Tile Size (Degrees Latitude by Degrees Longitude)	Origin (Latitude north and south, Longitude)
0 – 0	5° × 5°	0°,0°
40 – 50	5° × 6°	40°,0°
50 – 60	5° × 8°	50°,0°
60 – 65	5° × 10°	60°,0°
65 – 70	5° × 12°	65°,0°
70 – 75	5° × 15°	70°,0°
75 – 80	5° × 20°	75°,0°
80 – 90	5° × 90°	80°,0°

9.7.5 Boundary Issues

One of the crucial issues in developing global data is how to deal with national and international boundaries. There are many disputed boundaries around the world and to confound this further, the boundaries continue to change. At the subnational level, re-districting and alterations of subnational divisions are constantly underway. In addition, political changes create new countries. Thus, generating up-to-date national and international boundary data is a daunting task. In preparing the GPW, boundary data from DCW was used primarily to ensure consistency at international borders. Although DCW data are far from perfect, the data are being widely used for global GIS analysis. The UNGIWG has prepared national and international boundary coverage for the use by UN agencies only. These data are not released to other non-UN users because of the sensitive issues associated with disputed international boundaries. The ISCGM plans to work with the boundaries as supplied by the national mapping agencies. According to ISCGM, if two boundaries are submitted, both data sets will be included in the final product.

9.7.6 Projection System

Global data can be produced in a wide variety of projection systems. The IMW suggested using the modified polyconic projection. Later, the suggestion was made to use the Lambert conformal conic projection to ensure conformity with the world aeronautical chart program. The Karta mira produced topographic maps in azimuthal projection for high latitudes and conical for areas between the Equator and 60 degrees. Similarly, topographic maps prepared by the USSR during the 1970s and 1980s used the Krassovsku ellipsoid, the Gauss conformal projection and the Kronstadt datum. The ISCGM aims to produce the global data in the International Terrestrial Reference Frame-94 (ITRF94) coordinate system with GRS80 ellipsoid as the reference coordinate system. SRTM data are available in the native geographic coordinate system and also in the Universal Transverse Mercator (UTM) coordinate system. MODIS data products—provisional and validated Terra and Aqua satellite data sets—are available in Integerized Sinusoidal and Sinusoidal Projections.

The main criterion to select one projection over another is the ability to routinely reproject the data with minimal error involved in the transformation. Principally, there are two major sources of errors: 1) errors associated with transformation of spherical three-dimensional objects to two-dimensional objects; and 2) errors associated with the reprojection and transformation of digital data. A spatial data set is repeatedly altered when it is projected several times and this is particularly so in the case of raster data. Commonly encountered errors in raster data conversion are changes in pixel values, elimination of pixels or duplication of pixels. Errors introduced in global data compared to local data are significant because of the large size of the area involved. Distortion may not be that high for local data compared to other sources of errors and inaccuracies.

Equal area projections are generally considered better for global raster database because they preserve area characteristics, and pixel areas are more accurate and equivalent (Usery et al. 2002). After a quantitative analysis of traditional map projection distortions, Steinwand et al. (1995) provided a set of equal area map projections specifically suitable for continental, regional and global data sets. The projections recommended for global data sets include: the Wagner IV projection, the Wagner VII projection, the Interrupted Mollweide projection, and the Goode's Homolosine projection (Figure 9-4). However, even the Goode's Homolosine projection system was found to replicate some pixel values, distort some areas and exhibit gaps at the line of 180 degrees E (W) longitude (Yang et al. 1996).

While vector data are not subject to the pixel-related errors listed above for raster data, equal area projections are generally better for vector data also, because only such projections enable the user to compare areas occupied by categories of interest.

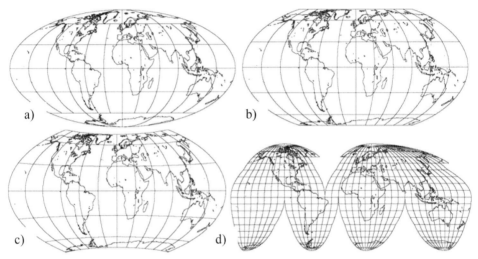

Figure 9-4 Global Projections: a) Wagner IV projection; b) Wagner VII projection; c) Interrupted Mollweide projection; and d) Goodes Homolosine projection.

A loss of information and distortion occurring during image warping and reprojection of raster data depends on the spatial resolution of the raster data (Steinwand et al. 1995). For coarse spatial resolution data, distortion is higher compared to high resolution data.

Seong (1999), Usery et al. (2001) and Usery and Seong (2001) investigated specific effects of raster cell size and latitudinal position on the accuracy of thematic attributes. According to their findings, all of the equal-area projections yield adequate accuracies with pixels of 8 km or smaller. With 8- to 16-km pixels, the Mollweide projection yields greater accuracy, and with 50-km pixels, the Lambert Cylindrical Equal-Area projection yields greater accuracy. Seong (1999) also examined the effects of latitude on the accuracy of projecting raster data and found that the Mollweide projection retained accuracy better at all latitudes, although with various discrepancies. Because of these limitations, overlaying of two or more global data sets becomes extremely difficult (see Figure 9-4).

9.7.7 Naming Conventions, Format and Minimum Mapping Unit

Countries have full formal names and short names and different names in different languages. Naming conventions for countries in global databases often follow the ISO (International Organization for Standardization) 3166 codes. The latest version (1997) provides a unique two-letter code and three-letter code for each country, detailed description of which can be found at www.iso.org. For specific uses, a two- or three-digit numeric code is given primarily to save bits in computer storage (instead of using the full name of the country). Another issue is what do we do with disputed territories? Is Taiwan a country or a part of China? Furthermore, countries may change their names over time (e.g., Burma, Yugoslavia) and may change political status by gaining independence. Multiple names, disputes, name changes and independence create problems in preparing standardized naming conventions.

Data can be distributed in a wide variety for formats such as ARC/INFO Interchange (.e00), Band Interleaved by Line (BIL), Band Sequential (BSQ), Hierarchical Data Format (HDF), shapefile and GeoTIFF. DCW data are available in their original data format called Vector Product Format or VPF. VPF was designed with a limited number of stakeholders in mind; however, this format was not received well by the GIS community at large. The result was considerable opportunities for entrepreneurs to help customers ingest DCW into their particular operating GIS. DCW evolved into VMAP, which is now available in a variety of formats. Since VMAP is unrestricted in use and redistribution, it has become a de facto standard for most applications needing such data. It also has spawned a plethora of value-added commercial efforts, supplementing data themes such as roads, utilities, and other data deemed to have commercial potential. MODIS data products are available in Hierarchical Data Format-Earth Observing System (HDF-EOS) format. HDF-EOS is the standard data format for all NASA EOS data production developed at the National Center for Supercomputing Applications (NCSA) at the University of Illinois.

Raster data are usually produced in 4-, 8-, 16-, or 32-bit[1] data in signed or unsigned integer format. For example, ISCGM aims to produce vegetation, land cover and land use layers with 8-bit unsigned data and the elevation data with 16-bit unsigned data. The elevation data are in big-endian byte order, meaning the most significant byte is stored first.

For global data, the minimum mapping unit may range from a few square kilometers to several hundred square kilometers. For example, the DCW does not map the islands in the ocean with less than 1 square kilometer area. Metadata are usually provided with the global data sets. Besides metadata, a header file may accompany the raster file.

9.7.8 Errors and Accuracy

Errors in global data may come from a number of sources including source data, data conversion, projections and the algorithm used. If the global data are prepared from a number of data sources collected from national and regional institutions, data may not be consistent across the globe. This is because data coming from a number of cartographic sources may vary widely in reliability, detail, completeness, precision, scale and methodology. These problems also were encountered in constructing the FAO Soil Map of the world which was prepared by digitizing approximately 11,000 paper maps at the scale of 1:5,000,000 (Zobler 1986). Grid values were assigned by selecting the dominant soil unit in 1 degree x 1 degree grids. Grid cell values assume that the soil type is homogenous within a cell, which may not be the case in reality. In this case, errors might have been introduced during digitizing, as well as during averaging.

---

1. One byte is 8 bits, so 16-bit and 32-bit data are multi-byte data. Some computer systems are designed to have the most significant byte stored first (analogous to reading from left to right); these are big-endian systems. Computer systems designed to have the most significant byte stored last are little-endian systems.

Similarly, DTED Level-1 data contained input from various cartographic sources, as well as from various types of imagery. If individual map sources had errors in recorded horizontal datum, or had misplaced contour lines (as could happen during compilation or in printing), the resultant DEM would be misplaced horizontally in an undocumented fashion. If individual sources of imagery had errors in image navigation, the resultant DEMs could be mislocated (if these errors were not corrected during development of the respective DEMs).

Under these circumstances, it is extremely helpful to provide some measure of reliability of the map across the globe as was provided for the FAO soil map and GPW. The published FAO Soil Map of the World contains inset maps showing three categories of reliability for the source data used to prepare the map. This information helps users understand where the reliability is higher and where it is lower.

Data producers should provide information on the quality of their data to all users. The inclusion of such information has been problematic due to the lack of standards. For some data sets, accuracy is provided. For example, the World Vector shoreline data's specification for positional accuracy is that 90% of all identifiable shoreline features should be located within 500 m of their true geographic position with respect to the WGS-84 datum. Absolute horizontal accuracy of the DCW hypsography is reported to be 2040 m rounded to nearest 5 m at 90% circular error. Vertical error is considered to be 610 m for contours and 30 m for spot elevations (DMA 1992). However, the errors can vary locally. In the DCW data, errors such as miscoded polygons, mismatched border segments and bad polygon topology for some areas of the world were reported. Population censuses are taken in different years in different countries. Thus, it is non-trivial to prepare a global population database for a base year such as 2000 or 2005. Under such circumstances, providing temporal reliability will be useful. The temporal reliability is inversely proportional with the reference years (in case of GPW, the reference year was 1990 and 1995). The index of reliability for GPW was calculated separately for two base years using the following equation (CIESIN 2000).

$$Rb = MIN\left[2, \frac{1}{MAX\,(0.5,\,|b - Y_1|)} \; + \; \frac{1}{MAX\,(0.5,\,|b - Y_2|)}\right]$$

where Rb is reliability in a given base year, b is a base year, Y_1 is the year of the first estimate and Y_2 is the year of the second estimate

9.7.9 Validation

Validation of global data sets can be done at multiple levels. To start with, qualitative validation can be performed to identify and correct gross classification errors. One method is to divide the data into regularly spaced grids and visually check each grid with secondary data. This method is extremely useful for identifying and correcting gross errors arising from misclassification. This can be followed by quantitative validation using a sampling approach. For example, accuracy assessment of GLCC is based on the validation of 25 random samples stratified by 15 of the 17 land cover classes (water and snow and ice classes were not validated). Three regional expert image interpreters analyzed the high resolution satellite data covering each sample to determine the true land cover types. A majority decision rule was used to determine sample accuracy (Scepan 1999). Possible registration error in high resolution satellite data, averaging high resolution pixels to represent a coarse resolution pixel, knowledge of the expert image interpreter and the approach of major decision rules might have introduced errors during the validation.

9.7.10 Standards and Protocols

To maintain specific standards and protocols at the global scale, it is important to follow standard use of administrative boundary data for all countries, naming conventions for all countries, an editing protocol, a coding scheme, a metadata profile, accuracy and reliability assessment and a validation process. Standard international borders are necessary to ensure that boundaries of neighboring countries are compatible to one another and the editing protocol ensures quality and consistency. Naming conventions and a coding scheme are needed to identify each unit mapped through time and space. Accuracy, reliability and validation are needed to ensure data quality and utility in different parts of the world, while metadata provide information regarding the creation and use of the data.

9.8 Challenges

While significant progress has been achieved in developing global-scale geographic databases, a number of challenges remain; these can be categorized as administrative/political, practical and technical.

A major administrative/political challenge constantly faced by global database producers is that of political will in creating, maintaining and improving global databases. In times of tight federal budgets, the focus of spending funds for mapping and database creation becomes more and more narrow until it is limited to specific national interests (i.e., one's own country/union) or to regions where the broader national interest is paramount (i.e., strategic locations). It is difficult to obtain the resources necessary for building and maintaining databases for applications that have long-term applications, such as change and/or trend analysis over regions where the short-term applications are not obvious. Even if a long-term program is established, it is a challenge to maintain long-term consistency with international cooperators. For example, the points of contact may change, resulting in costly start-up time to get the replacement up to speed on critical issues.

Maintaining consistent data standards presents another administrative/political challenge. It is difficult to harmonize multiple data sets that have been gathered from countries with their own approaches toward data creation, measurement conventions, thematic content and metadata. Extensive and intensive negotiations and compromise are necessary to resolve such challenges and this can be costly in terms of both time and money. The ISCGM Global Map project discussed earlier serves as a prime example of the commitment that is necessary to achieve such harmonization.

Data access policies of many data providers vary widely but, encouragingly, are starting to become more liberal. For example, the US Government policy calls for cost recovery, with data prices based on the cost of filling user requests rather than on the value of the data. In many areas there is a distinction between "essential" data, such as ownership, administrative boundaries and roads, and "non-essential" value-added data. Essential data are often accessible free of charge, while non-essential data may have a charge. In reality, it is difficult to sort through all of the various data access issues, and this, unfortunately, may discourage data usage. With rapid developments in computing and the Internet, many mapping agencies are re-examining data policies. Accessibility of global data has increased significantly in the last 10-15 years because many of the global data are freely available over the Internet. Although methods to improve data exchange such as improved networks and open data policies have helped, Internet accessibility in many developing countries is still limited.

Practical challenges include the difficulty of obtaining supporting information of consistent global database quality from one country to another. Where the global data sets are a composite of nationally or regionally collected information, the quality of the data will naturally vary. However, seldom is the quality (or accuracy) explicitly stated and it is therefore difficult to make a broad statement about the quality of the particular database.

Because many global databases are constructed from existing country- or regional-level information, there are often issues in data consistency across borders. This leads to an appearance of a "data fault-line" at administrative boundaries. While it may not be clear which side of the fault-line has more reliable data, the credibility of the database as a whole is affected by the disagreeable appearance of such features. Often these cross-border discrepancies are the result of differing personal or cultural styles of local or regional mappers. In some projects, decisions to reduce the number or diversity of compilers may facilitate completion of the project, but may intellectually weaken the resultant global data set.

Despite the magnitude and rate of change in computing power, working with large, global-scale data sets is still a technical challenge. Database size and resolution play an important role in the usefulness of the database—if the resolution is too coarse, it is not useful for many applications; if the resolution is fine, the database may be too large to use effectively. Easily accessible data archive and retrieval also present challenges for effectively using global databases. When there is a large amount of data (e.g., moderately high resolution data, such as MODIS) collected and stored on a frequent time-step (daily, 8-day, 16-day, etc.) for multiple products and multiple years, the challenges for the NASA Distributed Active Archive Centers (DAACs) to archive and distribute the data becomes enormous. Many of the end users of such data seek an easier method to gain access to the data over their study area of interest. Seamless map servers, that permit users to define exactly the study area and data sets of interest, are becoming more popular for distributing large-area data sets.

Validating global databases presents a challenge on many levels. First, the vast areas that must be covered make it impossible to do extensive validation efforts. The time and cost commitment for validation may exceed the costs associated with creating the database itself and it is often difficult to obtain sufficient funding for validation efforts. Since extensive validation is often impossible, creative sampling approaches must be undertaken. For example, the MODIS Land Science Team has taken an approach that contributes to and leverages several ongoing international field validation activities. The hope is to use the validation data to attach uncertainty information to each of the MODIS Land products (Morisette et al. 2002). The main approach is to collect aircraft and ground data to compare with the satellite derived products. The scientists working on validation have established over 30 "Core Sites" around the world where detailed validation information for a variety of satellite-derived products is collected.

Figure 9-5 shows a concept of integrated global environmental databases. It is an ideal database system which is not yet realized, comprising various global environmental databases that use consistent administrative boundaries, naming conventions, editing protocols and coding schemes. They have been assessed for reliability, validated and contain proper metadata. Individual global environmental databases such as an oceanographic database or soil database, would then be linked to a core database. The Global Mapping project by ISCGM is trying to develop such a core database by including reference framework data such as land cover and land use.

A mechanism to connect distributed global environmental databases into one virtual database is realized by developing data infrastructure including data standardization, interoperability and user interfaces. The project "Digital Earth," which aims at a virtual representation of our planet, or the association, Global Spatial Data Infrastructure (GSDI), are potential communities to attain this mechanism to realize an ideal integrated global environmental database.

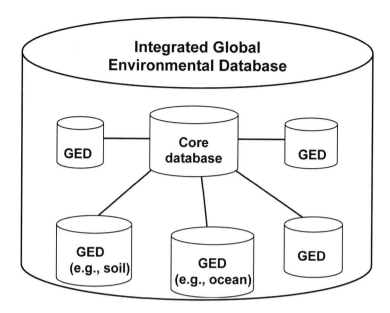

Figure 9-5 Concept of the Integrated Global Environmental Database.

9.9 Conclusions

Global databases are needed to help solve global environmental and social problems of our time. Such data are becoming increasingly available due to four main reasons: 1) strong interest and commitment from national governments and international organizations; 2) a desire to answer complex environmental and socioeconomic problems by the scientific community; 3) increasing requirements for military and humanitarian applications; and 4) recent advancements in computer hardware/software and information technology.

Currently the world is far from the development of an ideal integrated global environmental database because of administrative/political, practical and technical challenges. The International Society for Photogrammetry and Remote Sensing (ISPRS) Working Group IV/6 (1996-2000) and IV/8 (2000-2004) noticed that each database development initiative has common problems requiring common solutions. The Working Group surveyed the present situations and future directions and published their results as two books by Tateishi and Hastings (2000, 2002).

Although data sets of the known world have a history of at least two millennia, organized attempts at compiling and maintaining global spatial databases are a much more recent phenomenon. Since the 1980s, a number of new global data sets have been prepared, mainly due to the availability of remotely sensed data and advancements in GIS technology. A number of global data sets available in maps and charts are being replaced by digital data sets. New data sets also are being prepared using socioeconomic and field inventory data. As a result the number and breadth of global data and their applications have widened. While significant progress has been achieved in developing and using global GIS data, a number of challenges remain. In the quest to produce and use better quality global data, both data producers and users are expected to overcome these challenges and make significant progress in the future.

References

Achard, F., H. D. Eva, H. J. Stibig, P. Mayaux, J. Gallego, T. Richards and J. P. Malingreau. 2002. Determination of deforestation rates of the world's humid tropical forests. *Science* 297:999–1002.

Anderson, K. E. and K. D. Kline. 2000. A reference framework for global environmental data. In *Global Environmental Databases – Present Situation; Future Directions*, edited by R. Tateishi and D. A. Hastings, 21–40. Hong Kong: GeoCarto International Centre.

Bartholome, E. and A.S. Belward. 2005. GLC2000: A new approach to global land cover mapping from Earth observation data. *International Journal of Remote Sensing* 26 (9): 1959–1977.

Brown, J. F., T. R. Loveland, D. O. Ohlen and Z. Zhu. 1999. The global land-cover characteristics database: The users' perspective. P*hotogrammetric Engineering and Remote Sensing* 65 (9):1069–1074.

CIESIN (Center for International Earth Science Information Network). 2000. *Gridded Population of the World (GPW)*, Version 2. Columbia University; International Food Policy Research Institute (IFPRI); and World Resources Institute (WRI). Palisades, NY: CIESIN, Columbia University. <http://sedac.ciesin.columbia.edu/plue/gpw> Accessed 13 June 2007.

Cihlar, J. 2000. Land cover mapping of large areas from satellites: Status and research priorities. *International Journal of Remote Sensing* 21 (6–7):1093–1114.

Collins, M. and J. Rhind. 1997. Developing global environmental databases: Lessons learned about framework information. In *Framework for the World*, edited by D. Rhind, 120–129. Cambridge: GeoInformation International.

Cumming, M. J. and B. A. Hawkins. 1981. TERDAT: the FNOC system for terrain data extraction and processing. *Technical Report MII Project M-254* (second edition). Monterey, California: US Navy Fleet Numerical Oceanographic Office.

Danko, D. M. 1992. Global data: The Digital Chart of the World. *Geo Info Systems 2* (1):29-36.

DeFries, R. S. and J.R.G. Townshend. 1994. NDVI-derived land cover classification at global scales. *International Journal of Remote Sensing* 15:3567–3586.

DeFries, R. S., M. Hansen, J.R.G. Townshend and R. Sohlberg. 1998. Global land cover classifications at 8km spatial resolution: the use of training data derived from Landsat imagery in decision tree classifiers. *International Journal of Remote Sensing* 19 (16):3141–3168.

DeFries, R. S., C. B. Field, I. Fung, G. J. Collatz and L. Bounoua. 1999. Combining satellite data and biogeochemical models to estimate global effects of human-induced land cover change on carbon emissions and primary productivity. *Global Biogeochemical Cycles* 13:803–815.

Dinkins, R.E. 1988. Map retro on a shoestring. *Geography and Map Division Bulletin* 153 (September 1988):3–10. Special Libraries Association, Geography and Map Division (SLA G&M).

DMA (Defense Mapping Agency). 1992. *Digital Chart of the World (DCW)*. Fairfax, Virginia: Defense Mapping Agency.

DMA. 1995. *Vector Smart Map Level-0 (VMap0)*. Fairfax, Virginia: Defense Mapping Agency.

Dobson, J. E., E. A. Bright, P. R. Coleman, R. C. Durfee and B. A. Worley. 2000. LandScan: A global population database for estimating populations at risk. *Photogrammetric Engineering & Remote Sensing* 66 (7):849–857

ESA (European Space Agency). 2007. European Space Agency Ionia GlobCover Portal <http://ionia.terradue.com/index.asp> Accessed 13 June 2007.

Estes, J. E., J. Lawless and D. W. Mooneyhan, eds. 1994. *Report of the International Symposium on Core Data Needs for Environmental Assessment and Sustainable Development Strategies*, held in Bangkok, Thailand, November 15-18, 1994. Sponsored by UNDP, UNEP, NASA, USGS, EPA and USRA. New York, New York: UNDP/UNEP. Vol. 1, 59 pp. Vol. II, 130 pp.

Friedl, M.A., D. K. McIver, J.C.F. Hodges, X. Y. Zhang, D. Muchoney, A. Strahler, H.C.E. Woodcock, S. Gopal, A. Schneider. and A. Cooper. 2002. Global land cover mapping from MODIS: Algorithms and early results *Remote Sensing of Environment* 83 (1-2):287–302.

Giri, C. 2005. Global land cover mapping and characterization: Present situation and future research priorities. *Geocarto International* 20 (1):35–42.

Hansen, M. C., R. S. Defries, J.R.G. Townshend and R. Sohlberg. 2000. Global land cover classification at 1km spatial resolution using a classification tree approach. *International Journal of Remote Sensing* 21 (6-7):1331–1364.

Hansen, M. C., R. DeFries, J. Townshend, R. Sohlberg, C. Dimiceli and M. Carroll. 2002. Towards an operational MODIS continuous field of percent tree cover algorithm: Examples using AVHRR and MODIS data. *Remote Sensing of Environment* 83:303–319.

Hastings, D. A. 1986. *Stereo Pair World Map*. Boulder, Colorado: National Oceanic and Atmospheric Administration, National Geophysical Data Center.

Hastings, D. A., and L. Cordell. 1988. Earth's surface in digital, shaded relief color stereo: An initial geological assessment. *Geological Society of America Abstracts with Programs*, 18 (6):630.

Heywood, V., ed. 1995. *Global Biodiversity Assessment*. United Nations Environment Programme. Cambridge: Cambridge University Press. 1140 pp.

Htun, N. 1997. The need for basic map information in support of environmental assessment and sustainable development strategies. In *Framework for the World* edited by D. Rhind, 111–119. Cambridge: GeoInformation International.

IPCC (Intergovernmental Panel on Climate Change). 2002. <http://www.ipcc.ch/> Accessed 13 June 2007.

Justice, C. O., J.R.G. Townshend, E. Vermote, R. Wolfe, N. El Saleous, and D. Roy. 2002. Status of MODIS, its data processing and products for terrestrial science applications. *Remote Sensing of Environment* 83:3–15.

LADA (Land Degradation Assessment in Drylands). 2002. <http://www.fao.org/ag/agl/agll/lada/default.stm> Accessed 13 June 2007.

LCLUC (Land Cover Land Use Change). 2002. *NASA's Land Cover Land Use Change Program*. <http://lcluc.umd.edu/> Accessed 13 June 2007.

Loveland, T. R. and A. S. Belward. 1997. The International Geosphere-Biosphere Programme (IGBP) data and information system global land cover data set (DISCover). *Acta Astronautica* 41 (4-10):681–689.

Loveland, T. R., B. C. Reed, J. F. Brown, D. O. Ohlen, Z. Zhu, L. Yang and J. W. Merchant. 2000. Development of a global land cover characteristics database and IGBP DISCover from 1 km AVHRR data. *International Journal of Remote Sensing* 21 (6-7):1303 –1330.

MA (Millennium Ecosystem Assessment). 2005. About the Millennium Ecosystem Assessment. <http://www.millenniumassessment.org/en/index.aspx> Accessed 13 June 2007.

Matthews, E. 1983. Global vegetation and land use: New high-resolution data bases for climate studies. *Journal of Climate and Applied Meteorology* 22:474–487.

MDA. 2007. GeoCover-LC: Landsat imagery land cover classification. <http://www.mdafederal.com/geocover/geocoverlc/> Accessed 13 June 2007.

Morisette, J. T., J. L. Privette and C. O. Justice. 2002. A framework for the validation of MODIS land products. *Remote Sensing of Environment* 83 (1-2):77–96.

Mounsey, H. and R. F. Tomlinson, eds. 1988. Building Databases for Global Science: The Proceedings of the First Meeting of the International Geographical Union Global Database Planning Project. Meeting held at Tylney Hall, Hampshire, UK, 9-13 May 1988. London: Taylor & Francis. 434 pp.

NASA (National Aeronautics and Space Administration). 2007. Global Change Master Directory. <http://gcmd.gsfc.nasa.gov/index.html> Accessed 15 June 2007.

Nemani R. R., S. W. Running, R. A. Pielke and T. N. Chase. 1996. Global vegetation cover changes from coarse resolution satellite data. *J. Geophys. Res.* 101:7157–7162.

NGA (National Geospatial-Intelligence Agency). 2004. VMap0 and VMap1 Boundaries. <http://geoengine.nga.mil/geospatial/SW_TOOLS/NIMAMUSE/webinter/product_legends.html> Accessed 13 June 2007.

NRC (National Research Council). 2001. Grand Challenges in Environmental Sciences. Washington, DC: National Academy Press. <http://www.nap.edu/catalog/9975.html> Accessed 14 June 2007.

Olson, J. S., J. A. Watts and L. J. Allison. 1983. *Carbon in Live Vegetation of Major World Ecosystems*. Report ORNL-5862. Oak Ridge, Tennessee: Oak Ridge National Laboratory.

Parry, J. V. 2004. Mapping Arabia. *Saudi Aramco World* 55 (1): 20–37. <http://www.saudiaramcoworld.com/issue/200401/mapping.arabia.htm> Accessed 13 June 2007.

Pospeschil, F. 1992. *Micro World Databank II (MWDBII): Coastlines, Country Boundaries, Islands, Lakes, and Rivers*. Digital vector data at 1-minute resolution. In *Global Ecosystems Database Version 2.0*. Boulder, Colorado: NOAA National Geophysical Data Center.

Ramankutty, N. and J. A. Foley. 1999. Estimating historical changes in global land cover: croplands from 1700 to 1992. *Global Biogeochemical Cycles* 13 (4):997–1027.

Reed, B. C. 1997. Applications of the U.S. Geological Survey's global land cover product. *Acta Astronautica* 41 (4-10):671–680.

Rhind, D., ed. 1997. *Framework for the World*. Cambridge: GeoInformation International.

Rosenqvist, A., M. Shimada, B. Chapman, A. Freeman, G. De Grandi, S. Saatchi and Y. Rauste. 2000. The global rain forest mapping project - a review. *International Journal of Remote Sensing*. 21 (6-7):1375–1387.

Row, L. W. and D. A. Hastings. 1994. *TerrainBase Worldwide Digital Terrain Data on CD-ROM*, Release 1.0. Boulder, Colorado: NOAA National Geophysical Data Center.

Row, L. W., D. A. Hastings and P. K. Dunbar. 1995. *TerrainBase Worldwide Digital Terrain Data – Documentation Manual*, CD-ROM Release 1.0. Boulder, Colorado: NOAA National Geophysical Data Center.

Scepan, J. 1999. Thematic Validation of High-Resolution Global Land-Cover Data Sets. *Photogrammetric Engineering and Remote Sensing* 65 (9):1051–1060.

SEDAC (Socioeconomic Data and Applications Center). 2007. Gridded Population of the World and the Global Rural - Urban Mapping Project. <http://beta.sedac.ciesin.columbia.edu/gpw/global.jsp>. Accessed 14 June 2007.

Sellers, P. J., R. E. Dickinson, D. A. Randall, A. K. Betts, F. G. Hall, J. A. Berry, G. J. Collatz, A. S. Denning, H. A. Mooney, C. A. Nobre, N. Sato, C. B. Field and A. Henderson-Sellers. 1997. Modeling the exchange of energy, water and carbon between continents and the atmosphere. *Science* 275:502–509.

Seong, J. C. 1999. *Multi-temporal, integrated global GIS database and land cover dynamics, Asia, 1982-1994*. Athens: University of Georgia, Ph.D. Dissertation.

Singh, A. 1994. Pages 302–305 in *United Nations Environment Programme (UNEP) and International Union of Forestry Research Organization (IUFRO) International Workshop on Developing Large Environmental Databasess for Sustainable Development*, Nairobi, Kenya, July 1993. GRID Information Series 22. Sioux Falls, South Dakota, January 1994.

Steinwand, D. R., J. A. Hutchinson and J. P. Snyder. 1995. Map projections for global and continental data sets and an analysis of pixel distortion caused by reprojection. *Photogrammetric Engineering and Remote Sensing* 61 (12):1487–1499.

Tateishi, R. and D. A. Hastings, eds. 2000. *Global Environmental Databases – Present Situation; Future Directions*. Volume 1. International Society of Photogrammetry and Remote Sensing (ISPRS). Hong Kong: Geocarto International Centre. 233 pp. <http://www.geocarto.com/features.html> Accessed 13 June 2007.

Tateishi, R. and D. A. Hastings, eds. 2002. *Global Environmental Databases – Present Situation; Future Directions*. Volume 2. Hong Kong: Geocarto International Centre. 154 pp.

Tateishi, R. and K. Kajiwara. 1991. Land cover monitoring in Asia by NOAA GVI data. *Geocarto International* 6 (4):53–64.

Tateishi, R., K. Kajiwara and T. Odajima.1991. Global land-cover classification by phenological methods using NOAA GVI data. *Asian-Pacific Remote Sensing Journal* 4 (1):41–50. <http://www.ngdc.noaa.gov/seg/cdroms/ged_iib/datasets/b04/reprints/tk2.htm> Accessed 13 June 2007.

Thrower, N. 1996. *Maps and Civilization*. Chicago: University of Chicago Press.

UNDP (United Nations Development Programme). 1994. Report of the Secretary-General. <http://www.un.org/Docs/SG/SG-Rpt/ch3c-1.htm> Accessed 14 June 2007.

UNEP (United Nations Environmental Programme). 1994. Report of the North American UNEP/GRID Users' Meeting, May 12-13, 1994, Sioux Falls, South Dakota, USA. <http://www.na.unep.net/publications/report.html> Accessed 14 June 2007.

United Nations. 1966. *International Map of the World on the Millionth Scale, Report for 1965*. New York: United Nations.

United Nations. 1975. *International Map of the World on the Millionth Scale, Report for 1973*. Publication ST/ESA/SER.D/16. New York: United Nations.

Usery, E. L., M. Finn and D. Scheidt. 2002. Projecting global raster databases. *Proceedings of Symposium on Geospatial Theory, Processing and Applications*, Ottawa, Canada, 9-12 July 2002. <http://carto research.er.usgs.gov/projection/pdf/procedings_498.pdf> Accessed 14 June 2007.

Usery, E.L., J. C. Seong, D. R. Steinwand and M. P. Finn. 2001. Methods to achieve accurate projection of regional and global raster databases. *U.S. Geological Survey Open-File Report 01-383*.

Usery, E.L. and J.C. Seong. 2001. All equal area map projections are created equal, but some are more equal than others. *Cartography and Geographic Information Science* 28 (3):183–193.

Wilson, M. F. and A. Henderson-Sellers. 1985. A global archive of land cover and soils data for use in general circulation climate models. *Journal of Climatology* 5: 119–143.

Yang, L., Z. Zhu, J. Izaurralde and J. Merchant. 1996. Evaluation of North and South America AVHRR 1-Km data for global environmental modelling. In *Proceedings of the Third International Conference/Workshop on Integrating GIS and Environmental Modeling*. Santa Fe, New Mexico, USA, 21-25 January 1996. <http://www.ncgia.ucsb.edu/conf/SANTA_FE_CD-ROM/sf_papers/yang_limin/my_paper.html> Accessed 14 June 2007.

Zhu, Z., E. Waller, R. Davis and M. Lorenzini. 1999. Forest cover mapping for the Forest Resources Assessment 2000 of the Food and Agriculture Organization. Pages 520–525 in *Proceedings of the ASPRS Annual Conference*, Portland, Oregon, May, 1999 (CD-ROM). Bethesda, Maryland: ASPRS.

Zobler, L. 1986. A World Soil File for Global Climate Modelling. NASA Technical Memorandum 87802. New York, New York: NASA Goddard Institute for Space Studies.

CHAPTER 10
Metadata as a Component of Data

Michael Moeller

10.1 Introduction

For the purposes of this chapter, the reader is assumed to have little to no knowledge of metadata and its role in data management. The term geospatial as used in this chapter also refers to any data that are measured or collected at or regard a specific site or region on the Earth. This includes data collected at depths within the ocean and at various altitudes in the atmosphere.

10.2 So, What is Metadata?

A common definition of metadata often heard is metadata is "data about data." That definition more times than not produces cocked heads and curious, confused looks. A more appropriate description of metadata might be "information about data." A plain and simple example can be seen in a photograph. Do you have a box of pictures at home with nothing written down about them? Why should you document them? You took the picture; you know what the picture shows. But imagine your grandchildren inheriting that box of pictures. Do you think they will know anything about the pictures? Providing information about the "who," "what," "where," "when" and "how" of the picture will provide valuable information for any of your family members or others who may have possession of those photos in the future.

Another example of metadata is the information contained in the library card catalog system. Imagine going to the library and trying to find a book without this system. The card catalog entry for a book contains information about that book, such as title, author, publication date, number of pages, etc. It also includes keywords that can be used by the search system to help you locate it. Geospatial metadata is very similar to the card catalog information, but if written properly it is a much more robust document, as it includes additional information such as quality, accuracy and scale.

10.2.1 Metadata as a Component of Data

Properly documenting your data provides vital information to interested parties. These interested parties may be internal or external to your organization, but either way, well-written, fully developed metadata affords them the opportunity to discover details about the data that will allow them to make a decision as to whether they want the data, and if so, how they can access, transfer and use the data. The example in Table 10-1 illustrates this point.

Table 10-1 Example of a portion of an environmental sensitivity index data set.

RARNUM	CONC	SEASON_ID	ELEMENT
1	20	1	Bird
3	HIGH	2	Fish
9	MED	1	Fish
2	LOW	1	Reptile

This is a portion of an environmental sensitivity index (ESI) data set. If you are a scientist who routinely works with ESI data, then you might understand what the columns in the data set represent. But if you have acquired this data set to include in a project and have never seen the data before, then you will need information that will help you understand the data. More importantly, you need to understand how best to incorporate these data into your project. Metadata serves that function. In this example, you would discover that "RARNUM" is the unique combination of species, their concentration and seasonality. You would find that "CONC" refers to the density of the species at a given location and that "SEASON_ ID" references a code list related to the particular seasons. Finally, you would learn that "ELEMENT" describes the particular biologic group to which each belongs. Without this information contained in the metadata, these data may well be worthless to you.

As evidenced by this example, metadata is the component that describes the data. It documents characteristics of the data such as content, condition, and quality. New federal regulations require federal agencies to document the quality of their data under the Data Quality Act - Public Law 106-554, Section 515. Metadata can help agencies meet the requirements of this new regulation by documenting their data quality within their metadata.

Metadata is most commonly associated with digital spatial data; however, it can be used to describe any data, such as remotely sensed imagery, GPS data, biological data or in situ data. Think for a moment about a data set that does not have a tie to a location on the Earth's surface. Can you think of one? When considering metadata for your data, whatever they may be, why start from scratch? Using a current standard, such as the Federal Geographic Data Committee's (FGDC) Content Standard for Digital Geospatial Metadata (CSDGM) can save time and money, and the CSDGM is flexible enough to allow for changes to fit the needs of any specific data community.

As you consider metadata for a project, a common question that often arises is "At what level of granularity should the metadata be written?" In other words, can one project level record suffice, or does each individual data element need to be documented? The answer to that question is that it depends. It may be possible to capture all the appropriate information in one project level record, but more often than not additional records may have to be written to describe various components of the data set. One course of action is to write a project level record at the beginning of the project, and use that to build templates for other more specific metadata within the project as it progresses. Remember, there is no right or wrong answer to the question of granularity. It is organization- and project-specific, but at whatever level you choose to produce metadata, make sure it is fully developed and well-written.

Metadata is a critical component of a complete data set. As such, it should never be viewed as a separate entity. It is and should remain an integral component of the data. No data set should be considered complete without a fully-developed, well-written metadata record or records.

10.2.2 The Value of Metadata – Doing Business the Old Way

With an understanding of what metadata is, the next questions that logically arise are "Why write metadata?" "What is the value of writing metadata?" "What is metadata used for?" When the FGDC put together the CSDGM, they stated four uses of metadata written using that standard—data discovery, assessment, access and use. This traditional focus tended to be on the benefits to external users of the metadata. A user could access the FGDC clearing-house (no date) and search the metadata for data sets of interest. Once a record was found that appeared to be of interest, the information in the metadata could be used to find out if the data was fit for the planned use, how to access and copy the data, and then perhaps how to best use the data. The idea behind this approach was that as more data set providers documented their data and made that information available, more people would be able to share data. The system relies on maximum participation to derive the greatest benefits.

From an internal perspective, the traditional role of metadata was one of 'inheritance,' which simply refers to metadata's ability to help preserve a data set's usefulness through time. For example, imagine an individual who has worked with a particular type of data for years and then suddenly leaves that organization. If that individual did not adequately document the data, then it is possible that some, if not all, of that data's usefulness will be lost.

10.2.3 A New View of Metadata

There has been a shift in recent years away from the traditional external and internal roles of metadata as a mechanism of data discovery and a means of documenting inheritance. This shift is focusing on the use of metadata as an aid to data management (Wayne 2001). As this shift occurs, the value of metadata, both external and internal, will shift.

As an organization's data holdings increase, data management becomes critical. Metadata can assist in building an efficient data management program. External values of metadata now become internal values. Organizations can use metadata to document their data holdings, and then use this metadata internally for data discovery, assessment, access and use. This has several benefits, including a reduction in labor and/or data duplication, and these types of benefits translate to more economic efficiency.

The ability to maintain a record of a data set's currency will help ensure that project source data are not out-of-date. Metadata can aid in tracking editing and update frequencies, as well as the usage for a data set's source files, along with where and how often a data set is being distributed.

Cost is always an issue, and data managers are constantly looking for ways to better manage their resources. Metadata also can be used as an aid to monitoring the data development process. Over time, data managers can use metadata to make more accurate cost predictions for future projects based on past experiences. To realize the full potential of metadata under this new concept, the creation of metadata must become integral to the data development process. The question is "How?"

Integrating metadata into the data development process may require a bit of a sales pitch to the appropriate members within an organization. If faced with making such a sales pitch, highlight these key functions of metadata:

- **Data archive**
  Data are the most expensive components of a GIS. In fact, some estimates place the cost of creating the database as high as 75% of the total cost of operating a GIS (Demers 2004). Metadata is a means of preserving the value of data investments. This is of particular significance to organizations or government agencies that may experience high staff turnover rates.

- **Data assessment**
  GIS data development has shifted from data producers to data consumers. From a consumer perspective, metadata is the "truth in labeling" required to assess available data products. From the producer's perspective, metadata is a means of declaring data limitations and serves as a form of liability insurance.

- **Data management**
  Metadata enables organizations to retrieve in-house data resources by specific criteria for global edits and annual updates, as well as maintaining some oversight on the data development process.

- **Data discovery**
  Metadata is the primary means of locating available geospatial data resources via the Internet. It is also considered a primary public information resource as it provides a non-technical means of presenting technical information.

- **Data transfer**
  Metadata is increasingly used by software systems as a means of properly ingesting data and by analysts for properly displaying data.
- **Data distribution**
  By building metadata in compliance with national standards, you can promote your organization or agency by participating in the Global Spatial Data Clearinghouse. Participation promotes your agency and frees staff from answering data inquiries.

Incorporating metadata into the data development process begins with a fundamental shift in the way an organization looks at life. Given the reality that government agencies operate differently than private industry, the operational paradigm shift that must occur within these agencies is often difficult. To facilitate this shift, the focus should be directed towards building administrative, technical and organizational support.

10.3 Building Support within Your Organization

10.3.1 Administrative Support

To generate support from the administration of your organization, highlight these ideas:
- Metadata preserves data investment. The cost of data acquisition and processing is a large portion of a project's budget. It would be a shame to lose that investment because the data were not properly documented.
- Metadata limits liability. One of the most-often-overlooked features of a well-written metadata record is its use as a legal document. The various constraints of a data set should be clearly defined within a metadata record. This does not mean it will keep an organization out of court, but metadata does add one more layer of documentation to support the organization's case.
- Metadata helps manage data resources. Metadata can help with data management issues such as data currency, data utility, monitoring of the development process and estimating data development costs.
- Metadata aids in external data acquisition. Metadata can be searched from existing external and/or internal clearinghouse systems to find data of interest.
- Metadata facilitates data access and transfer. Once a data set is found, metadata should contain information on access and transferring that data.
- Metadata provides for efficient data distribution. Writing metadata in accordance with current federal standards allows organizations to participate in the National and Global Spatial Data Clearinghouse.

10.3.2 Technical Support

The primary responsibility for metadata creation will most likely fall to your technical staff. As such, it is vital to stress the individual benefits of metadata creation along with the institutional benefits. These benefits include:
- **Workload reduction over the long term.** Metadata contains information that can aid in quickly locating and retrieving data resources. The metadata does this through the use of specific criteria such as keywords, bounding coordinates, time period, data type, entities and attributes.
- **Field fewer data inquiries.** Well-written metadata contains within it the information most people will need to evaluate, access, and use your data. Providing this information will help cut down on staff response time to data inquiries.
- **Document individual contributions.** The initial metadata record produced for a given data set establishes the core content that will persist, with updates, for the life of the data

set. This provides data developers with an opportunity to document their efforts and contributions. The metadata record also serves as a tangible performance indicator that may be incorporated into organizational and individual evaluations.

It is important to include your technical staff in the decision-making process pertaining to metadata program design and feasibility analysis. They may bring unique insight into this process. Such inclusion also will foster a spirit of ownership in your metadata program, thereby increasing the chances the program will succeed.

There are a couple of other points to consider when it comes to building strong staff support, including:

1. Incorporate metadata expectations into job descriptions and performance standards. By establishing the ground rules up front, you will avoid ambiguities later on. This works well for new hires, but existing staff will have to be approached a bit differently; and
2. Provide staff development opportunities by providing the three 'T's – training, tools and time. In most situations, your staff will be carrying a full workload. To ensure that they will be able to become contributors to the metadata program, it is important to make sure they have the time to get trained in metadata creation, as well as time to become familiar with the metadata tools that are available.

Much of the angst regarding metadata is associated with related standards including the FGDC CSDGM and the International Organization for Standardization (ISO) Geographic Information – Metadata standard. The standards are extensive and somewhat overwhelming because they are written to address a wide range of geospatial data types (e.g., imagery, GIS files, GPS data, geocoded databases) developed by a wide range of organizations. Individual organizations can address this problem by building custom metadata templates. Templates are built by extracting those metadata fields pertinent to the organization and the specific data types and geographies of the organization. In addition, libraries can be built to provide information about contacts, sources and methodologies common to the organization.

A suggested method for building such templates includes:

1. Adopt all mandatory fields specified by the national metadata standard.
2. Adopt all 'mandatory if applicable fields' pertinent to the data type or organization.
3. Identify 'optional fields' of interest to the organization.
4. Create a pilot record from the draft template.
5. Have the pilot reviewed and revised by administrators, analysts, technicians, and contributing scientists.
6. Identify those fields that tend to remain consistent. This may include:
   • access and use constraint statements
   • data distribution methods and contacts
   • contact information
   • north, south, west, east bounding coordinates
   • coordinate system and datum
   • place keywords
   • native data set environment
   • source citations

After the templates have been designed, go ahead and map those fields included in the template to the individuals or groups within your organization that are responsible for that information. This mapping has the effect of destroying the perception that metadata is a huge, onerous task, and turns it into a series of manageable steps within the data development process. These smaller steps also are easier to incorporate into the process as it proceeds, writing as you go if you will, and eliminates the need for a much larger and more involved effort at the end of the project. Here is an example of how this mapping might occur.

Data Development Stage	Metadata Information (by section and element)

Data Planning	Identification Information
	    Title, Originator, Abstract, Purpose, Keywords,
	    Time Period
	Data Organization
	    Point, Raster, Vector
	Spatial Referencing
	    Coordinate System and Datum
	Entity and Attributes (planned)

Data Processing	Data Quality
	    Completeness, Positional Accuracy,
	    Geoprocessing Steps

Data Analysis	Data Quality
	    Attribute Accuracy, Analysis Steps
	    Entity and Attributes (Results)
	    Metadata Reference

10.4 Implementing Your Metadata Program

To implement your metadata program effectively, you will need to use a tool or tools to collect the metadata. In its simplest form, metadata can be collected using a pencil and pad of paper. In fact, some of your metadata may be sitting on your desk in just such format. However, it is much more likely that you will employ something a bit higher on the technology ladder. Let us take a look at some options.

- **Form Documents.** Whether hardcopy or digital, these are perhaps the most versatile means of capturing information at various stages of the data development process. These forms can be customized to facilitate an efficient collection effort, and may be formatted to limit the exposure of most personnel to the metadata standard. Customized forms give you control over content, and allow you to guide the contributor through the information collection process.
- **Database / Spreadsheets.** These tools allow you to store metadata information in data tables. Column headings relate to specific metadata fields and you can develop records for each data set. Within the relational database realm, you can build in the relationship between the workflow discussed earlier to specific data tables. One big advantage of working with a database is the ability to integrate production rules such as populating mandatory fields, using 'pick lists' for sources and contacts, and automating the QA/QC routines.
- **Shareware Metadata Products.** In order to comply with Executive Order 12906, government agencies began developing in-house tools for metadata creation. Most of these are available to the public. Some were developed for very specific data, while others were broader in their approach. For a description of several that are available, see the Appendix at the end of this chapter for a list of links to online sources.
- **Commercial Metadata Products.** Commercial metadata products fall into two categories: 1) stand-alone; and 2) GIS-internal. Stand-alone products generally allow the user to 'harvest' some metadata information directly from the geospatial data set and provide a user-interface for additional data entry. These programs are typically robust production tools that facilitate the building of templates and libraries and enable interaction with a range of data types. As such, they are particularly useful to organizations that produce and manage data using multiple data development software packages. GIS-internal products also provide a data entry interface, but due to their proprietary nature, are able to harvest more information directly from the data set.

When considering the purchase of metadata production software, some things to consider include the following:
- Is the software easy to understand and use?
- Is it built to existing and applicable metadata standards?
- Does it automatically capture and update metadata from the data set?
- Does it allow for the use of digital forms for customized data entry?
- Does it support global updates and edits?
- Does it provide a means for optional and/or custom viewing formats?
- Does it package the metadata with the data?
- Does it support required metadata import and export formats?

10.4.1 Targeting Success

Success should be the ultimate goal of your metadata program. To help reach that goal, consider putting in place procedures and policies that will help guide participation, streamline operations, and encourage compliance. Develop these policies and procedures with the following in mind:

Assign responsibilities. Once you have your workflow mapped and you have decided on a toolset, begin assigning responsibilities for the collection and management of the metadata.
- Managers can be responsible for documenting metadata information mapped to the data planning stage, coordinating overall collection efforts, and enforcing policies.
- Technicians can be responsible for documenting metadata information mapped to the data processing stage. They can build data source citations and contact information libraries.
- Scientists and field staff can be responsible for reviewing and revising metadata information pertinent to data collection methods and findings.
- Analysts can be responsible for documenting metadata information mapped to the data analysis stage and assisting technicians in metadata documentation.
- Information technology / system managers can be responsible for developing and maintaining the metadata collection 'tool,' managing and updating metadata records, and providing support within and external to the organization for clearinghouses.

Assign priorities. Trying to decide which data set to document first? If possible, start with current products, and then work back through historical data sets. Develop a plan that considers:
- What is the core, or framework, value of the data set to the organization?
- What is the utility of the data set within the organization?
- How many external requests do you receive for the data set?
- What historical significance is the data set to your organization?

Establish administrative guidelines. Guidelines assist employees to understand and implement the program objectives. The following are examples of administrative guidelines:
- Organizational compliance is defined by the use standards and templates.
- Standardized language should be established for metadata distribution liability and access/use constraint statements. (Check with your legal staff on this one.)
- When using contractors, develop boilerplate language with respect to metadata creation as a criterion for acceptance of externally developed contract deliverables of data.
- Develop job descriptions for remote sensing/GIS positions that include metadata skills.
- Require units within your organization to publish their metadata holdings.
- Publish a metadata 'Standard Operating Procedure' (SOP) document that outlines and specifies the policies and procedures your organization adopts.

Implement and Advocate. Advocating participation will help integrate metadata into the data development process. Here are some suggestions to help increase participation:

- Provide your staff with appropriate metadata training and time to learn how to use the metadata tools you have acquired.
- Encourage your staff to publish their efforts in professional and research journals.
- Provide incentives such as prizes and awards.
- Utilize project 'punch lists' that indicate that the work is not complete until the metadata is done.
- Present metadata as a management priority during staff and management meetings, memos, and presentations.

10.5 Discovering Data with Metadata

As mentioned earlier, one of the primary internal and external values of metadata is its use for data discovery. In the external world of metadata, the FGDC has established a clearinghouse network to help users search for and find data of interest, much the way a library catalog helps you find a book you want. On its Web site <www.fgdc.gov/dataandservices/>, the FGDC provides a description of the clearinghouse network. It states:

> *The Clearinghouse Network is a community of distributed data providers who publish collections of metadata that describe their map and data resources within their areas of responsibility, documenting data quality, characteristics, and accessibility. Each metadata collection, known as a Clearinghouse Node, is hosted by an organization to publicize the availability of data within the NSDI. The metadata in these nodes is searched by the geodata.gov portal to provide quick assessment of the extent and properties of available geographic resources.* Accessed 2 August 2007.

The Clearinghouse has six gateways that allow a user to access the entire collection of internet servers, or nodes. Imagine a room with six doors and inside this room are more than 425 file cabinets, each containing a unique collection of metadata. Whatever door you choose to enter through, you have access to all of these file cabinets. This is exactly how the gateways function. Each gateway gives the user access to the same collection of clearinghouse nodes.

The FGDC also participates in a larger activity known as the Geospatial One-Stop, or GOS, Initiative. The GOS Initiative provides access to a variety of digital geospatial information and services through the geodata.gov portal. The FGDC Web site provides a description of the portal:

> *The geodata.gov portal is operated in support of the Geospatial One-Stop Initiative to provide "one-stop" access to all registered geographic information and related online access services within the United States. Geographic data, imagery, applications, documents, web sites and other resources have been catalogued for discovery in this portal. Registered map services allow casual users to build online maps using data from many sources. Registered data access and download services also exist for use by those interested in downloading and analyzing the data using GIS or viewer software.* Accessed 2 August 2007.

10.6 Standardized Metadata

When you stop and think about it, we interact with standards on a daily basis. Think about putting gas in your car, measuring ingredients to bake bread or plugging a lamp into a wall socket. Each of these actions is controlled or influenced by standards. Metadata should be no different. Standardized metadata ensures consistency. It serves as a uniform summary description of the data set.

The US agency responsible for developing the metadata standard at the federal level is the FGDC, which is a 19-member interagency committee composed of representatives from the Executive Office of the President, along with Cabinet-level and independent agencies. The FGDC is responsible for developing the National Spatial Data Infrastructure (NSDI) in cooperation with organizations from state, local and tribal governments, the academic community and the private sector. The NSDI encompasses policies, standards and procedures for organizations to cooperatively produce and share geographic data.

The FGDC created the CSDGM (Content Standard for Digital Geospatial Metadata) to address the requirements found in Executive Order 12906. This Presidential mandate states "…each agency shall document all new geospatial data it collects or produces, either directly or indirectly, using the standard under development by the FGDC, and make that standardized documentation electronically accessible to the Clearinghouse network." To accomplish this, the CSDGM utilizes common terms and definitions within a common language and structure. It establishes the names and definitions of the various elements in the standard. The primary objectives of the standard are to help users: 1) determine if a particular data set is available; 2) assess the data for their intended use; 3) access the data; 4) transfer the data; and 5) understand how the data are used.

Metadata, written in the CSDGM format, can answer questions such as these:

Who?
> Who collected the data?
> Who processed the data?
> Who wrote the metadata?
> Who to contact for questions?
> Who to contact to order?
> Who owns the data?

What?
> What are the data about?
> What project were they collected under?
> What are the constraints on their use?
> What is the quality?
> What are appropriate uses?
> What parameters were measured?
> What format are the data in?

Where?
> Where were the data collected?
> Where were the data processed?
> Where are the data located?

When?
> When were the data collected?
> When were the data processed?

Why?
> Why were the data collected?

How?
> How were the data collected?
> How were the data processed?
> How do I access the data?

How do I order the data?
How much do the data cost?
How was the quality assessed?

Take a moment to consider the questions above. Can you answer all of them for your data? Do they all apply to your data? What other questions might be asked about your data? Answers to the questions above will produce a fairly robust metadata record.

10.6.1 Using the CSDGM

The FGDC's CSDGM is a collection of 334 data elements (logically primitive data items) and compound elements (groups of data elements representing higher-level concepts). It is arranged into numbered chapters called "sections." Each section is organized into a series of elements that define the information content for metadata to document a data set. Each section provides:

- Section name and definition
- Compound element names and definitions
- Data element names and definitions
- Production rules, or conditionality, for each element

The FGDC has also produced a workbook version of the standard which presents the production rules of the standard in a graphical format. These graphics illustrate:

- The structure of the standard
- Element groupings
- Element conditionality
- Element repeatability

As mentioned, the standard is organized using numbered chapters called "sections." Each section includes:

- Section name and definition
- Compound element names and definitions
- Data element names, definitions and the valid values possible for each data element
- Conditionality of each element (i.e., "Mandatory," "Mandatory if Applicable," or "Optional")

A compound element is a group of data elements and other compound elements. All compound elements are described by data elements, either directly or through intermediate compound elements. Compound elements represent higher-level concepts that cannot be represented by an individual data element. A data element is a logically primitive item of data. The form for the definition of a data element is:

Data element name -- definition.
Type: (choice of "integer" / "real" / "text" / "date" / "time")
Domain: (describes valid values that can be assigned)

The information about the values for that data element includes a description of the type of value and a description of the domain of values. The data element type describes the kind of value required. The choices include "integer" for integer numbers, "real" for real numbers, "text" for ASCII characters, "date" for day of the year and "time" for time of day. The domain describes valid values that can be assigned to the data element. The domain may specify a list of valid values or restrictions on the range of values that can be assigned to a data element.

An example of the definition of the data element "Abstract" is:

Abstract -- *a brief narrative summary of the data set.*
Type: text
Domain: free text

10.6.2 The Graphical Representation of the CSDGM Production Rules

As mentioned, the workbook version of the CSDGM has converted the production rules into graphics (Figure 10-1). These graphics provide a much easier and faster way of navigating through the various sections of the standard.

As you can see in Figure 10-1, there are seven main sections and three supporting sections. The supporting sections are never used by themselves, but rather plug in at various places within the other sections.

Sections are depicted using a three-dimensional "button" icon (⬚), compound elements are represented as two-dimensional boxes (▢), and data elements are shown as three-dimensional boxes (⬓).

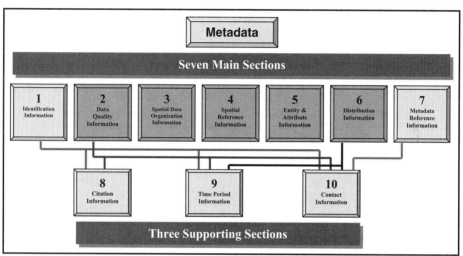

Figure 10-1 The seven main sections and three supporting sections of the CSDGM, in graphical form. See included DVD for color version.

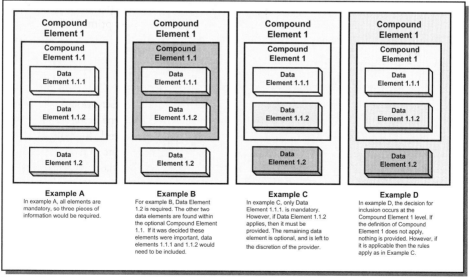

Figure 10-2 Examples of how color is used to denote conditionality. See included DVD for color version.

Manual of Geographic Information Systems

The production rules of the standard are colored to depict the three levels of conditionality (Figure 10-2).

- **Mandatory** elements are colored yellow. Information for these elements must be provided. If the information is not known for a mandatory data element, the entry "Unknown" or a similar statement should be given.
- **Mandatory if applicable** elements are colored green. Information for these elements must be provided if the data set exhibits the characteristic defined by the element.
- **Optional** elements are colored blue. Information for these elements is provided at the discretion of the data set producer.

The FGDC CSDGM, as mentioned earlier, is a flexible standard. The CSDGM can be extended to address the specific needs of a data community. To extend the CSDGM means that the conditionality of an element or elements is tightened (never loosened) or new elements are added. Once the standard has been extended to address specific data needs, the next process that may be addressed is having the extension approved by the FGDC. Once an extension is reviewed and approved by the FGDC, it becomes an official "profile" of the CSDGM. Several different profiles have already been developed, including the Shoreline Profile and Biological Data Profile. For more information on creating extensions and profiles, refer to Appendix D of the CSDGM.

As previously discussed, federal agencies and those organizations receiving federal monies are required to use the FGDC CSDGM. Other metadata standards are in use by other countries and organizations, including the ISO 19115 standard and the Dublin Core standard. Some organizations have even created their own standards specific to their operations. But to do that requires a good deal of time and patience, and it will only work well within that organization. If you are just starting out and are looking for a way to capture metadata, consider existing standards such as the FGDC CSDGM, which offers an existing structure as well as the option to modify that structure should the need arise.

If you decide to use the FGDC CSDGM and need some training or guidance, be sure to check the FGDC Web site. It includes a list of available metadata trainers along with a calendar of classes being taught around the country.

10.7 Things to Remember

Hopefully the information provided in this chapter will get you started with your adventures in metadata, or have helped answer some questions you may have had. There are a few things to remember when considering metadata:

- Metadata is a document that describes your data's content, condition and quality.
- Metadata is an integral component of your data and should never be viewed as a separate entity.
- Metadata's traditional external value includes data discovery, assessment, access and use.
- Metadata also is a powerful tool that an organization can utilize to build a strong internal data management structure.
- In order to be successful, this internal management structure requires building support at the administrative, technical and organizational levels.
- Standardized metadata ensures consistency in documentation and aids in the efficient discovery, access, transfer and use of data.

For additional information about metadata, check out the compendium of links in the following Appendix.

References

DeMers, M. N. 2004. Fundamentals of Geographic Information Systems, 3rd edition. New York: John Wiley & Sons.

Federal Geographic Data Committee. 1998. Appendix D: Guidelines for creating extended elements in the content standard for digital geospatial metadata. Pages 74–76 in *Content Standard for Digital Geospatial Metadata* (revised June 1998). FGDC-STD-001-1998. Washington, DC: Federal Geographic Data Committee. <http://www.fgdc.gov/standards/projects/FGDC-standards-projects/metadata/base-metadata/v2_0698.pdf> Accessed 12 April 2007.

FGDC. No date. FGDC clearinghouse. <http://clearinghouse3.fgdc.gov/> Accessed 9 March 2007.

Wayne, L. 2001. Metadata in action: Expanding the utility of geospatial metadata. <http://www.fgdc.gov/metadata/documents/MetadataInAction.doc> Accessed 12 March 2007.

Appendix: Links to More Information about Metadata

All links in this appendix were accessed 12 March 2007.

Metadata Information

Formal Metadata – Information and Tools

<http://geology.usgs.gov/tools/metadata/>

Start with this site. It has a wealth of good information to get you going, including links to download several metadata tools.

Federal Geographic Data Committee (FGDC)

<www.fgdc.gov>

These folks wrote the standard. You can view information on various FGDC metadata standards that have been approved or are in development, or you can connect to the FGDC Clearinghouse to search hundreds of nodes to discover geospatial data.

Content Standard for Digital Geospatial Metadata Workbook

<http://www.fgdc.gov/metadata/documents/workbook_0501_bmk.pdf>

The workbook complies with the latest version of the FGDC Content Standard for Digital Geospatial Metadata (CSDGM), FGDC-STD-001-1998, and contains both textual and color graphical information about the standard.

Biological Data Profile Workbook

Information on the Biological Data Profile to the FGDC metadata standard and other metadata activities at the National Biological Information Infrastructure (NBII) can be found at <http://159.189.176.5/portal/community/Communities/Toolkit/Metadata/>

The workbook itself is at <http://www.nbii.gov/images/uploaded/151871_1162508205217_BDP-workbook.doc>

The National Biological Information Infrastructure (NBII) offers the Biological Data Profile Workbook. Patterned after the FGDC-CSDGM workbook, this document contains all FGDC elements plus biological elements and helpful tables and charts.

Metadata Quick Guide

<http://www.fgdc.gov/metadata/documents/MetadataQuickGuide.pdf/view?>

This is a quick reference guide for writing high-quality metadata. It also includes information on the required use of International Organization for Standardization (ISO) keywords in FGDC metadata.

Metadata Tool Review

Metadata Tools for Geospatial Data

<http://sco.wisc.edu/wisclinc/metatool/>

Before you decide on a tool, you might want to check out these great tool reviews by Hugh Phillips.

Metadata Creation Tools

ArcView® Metadata Collector v2.0 Extension

<www.csc.noaa.gov/metadata/download.html>

This extension is a nice metadata entry tool for those who are still using ArcView 3.x.

Tkme: Another editor for formal metadata

<http://geology.usgs.gov/tools/metadata/tools/doc/tkme.html>

Developed by Peter Schweitzer of the U.S. Geological Survey (USGS), this editor was designed to simplify the process of creating metadata that conform to the FGDC's metadata standard.

MetaScribe: A New Tool for Metadata Generation

<http://www.csc.noaa.gov/metadata/metascribe/>

If you have a lot of metadata to write that is fairly redundant, this tool from NOAA Coastal Services Center can help you simplify the process through the use of metadata templates.

NNDC Metadata Home

<http://www.ngdc.noaa.gov/metadata/>

Introduction to the NOAA Metadata Manager and Repository (NMMR), a powerful system for archiving, editing, and publishing FGDC-compliant metadata records. This system enables users to create and manage FGDC-compliant metadata using a series of html interfaces. Records can be uploaded or manually entered into the system and validated on-line using Metadata Parser.

Metavist 2005

To read about the software, see "Creating FGDC and NBII metadata with Metavist 2005." <http://ncrs.fs.fed.us/pubs/2737>. To order the software, see <http://ncrs.fs.fed.us/pubs/products/metavist>.

Metavist 2005, a software tool for the metadata archivist, is used to create metadata compliant with two of the Federal Geographic Data Committee (FGDC) metadata standards—"FGDC Content Standard for Digital Geospatial Metadata" (FGDC 1998) and "FGDC Biological Data Profile of the Content Standard for Digital Geospatial Metadata" (FGDC 1999).

Metadata Enterprise Resource Management Aid

<http://www.ncddc.noaa.gov/Metadata/Tools/>

The National Coastal Data Development Center (NCDDC) provides this tool to develop, validate, manage and publish metadata records via secure Internet access. Validation includes enhanced features to check dates and ensure valid URLs.

NPS Metadata Tools for ArcGIS

"NPS Metadata Tools and Editor: Version 1.1"

<http://science.nature.nps.gov/nrdata/tools/>

This National Park Service extension streamlines importing existing and legacy metadata into ArcCatalog and offers the user the ability to parse metadata with Metadata Parser (see below), as well as spell-check records.

GeoMedia Catalog Editor Utility
<http://www.intergraph.com/geomediacatalog/GeoMedia_Catalog_Editor_Utility.asp>
Earlier versions were called Spatial Metadata Management System (SMMS). This free utility allows users of GeoMedia to create, edit, view and publish standardized spatial metadata. This is a commercial product.

Metadata Formatting and Reviewing Tools

CNS: A pre-parser for formal metadata
<http://geology.usgs.gov/tools/metadata/tools/doc/cns.html>
If you are not using a metadata-specific tool to create your metadata, you can use this pre-parser (cns comes from Chew and Spit) to create a metadata file that is properly formatted so that it can be parsed.

MP: A compiler for formal metadata
<http://geology.usgs.gov/tools/metadata/tools/doc/mp.html>
After running your metadata through Chew and Spit (CNS) (see above), use this tool to check for technical errors. Once all errors are taken care of, you may use this tool to publish your metadata in a number of different formats such as text, HTML, XML or SGML.

Enumerated Domain Helper
<http://geology.usgs.gov/tools/metadata/tools/doc/ctc/edom.shtml>
This tool converts a textual table into enumerated domain metadata elements, which can then be copied and pasted into a metadata record.

MP Batch Processor
<http://support.intergraph.com/Geospatial/Downloads/Tools.asp?ID=48&SORT=Title>
This program provides a windows interface for running Metadata Parser (MP) or Chew and Spit (CNS) (see above). Multiple files can be processed at one time utilizing this tool.

Geospatial Metadata Validation Service
<http://geo-nsdi.er.usgs.gov/validate.php>
Once again, Peter Schweitzer comes to our aid, incorporating the Metadata Perser (MP) (see above) tool he developed into an on-line validation service. This tool will recognize elements from the 1998 FGDC standard, biological data profile, shoreline profile and remote sensing profile.

Crimson Editor
<http://www.crimsoneditor.com>
This is a basic text editor, the category of tool you should always use if you are not using a metadata creation tool. Never, ever use a word processing program (e.g., Word, WordPerfect, Writely, etc.). Such programs apply formatting to the document that will then choke the validation tools.

Keyword Thesauri and Attribute Label Definition Sources

National Land Cover Data (NLCD) Land Cover Class Definitions
<http://landcover.usgs.gov/classes.php>

Global Change Master Directory
<http://gcmd.gsfc.nasa.gov/Resources/valids/>

Geographic Names Information System (GNIS)
<http://geonames.usgs.gov/domestic/index.html>

Cowardin U.S. Fish and Wildlife Service Wetlands Classification System
<http://www.water.ncsu.edu/watershedss/info/wetlands/class.html>

Anderson Land Cover Classification System
<http://landcover.usgs.gov/pdf/anderson.pdf>

NBII Systematics
<http://nbii.gov/portal/server.pt?open=512&objID=249&parentname=CommunityPage& parentid=2&mode=2&in_hi_userid=2&cached=true>

Glossary of Geologic Terms
<http://www.geotech.org/survey/geotech/dictiona.html>

The Natural Resources Conservation Service Soil Taxonomy Database
<http://soils.usda.gov/technical/classification/taxonomy/>

Spatial Data Standard for Facilities, Infrastructure, and Environment (SDSFIE)
<http://www.sdsfie.org/SDSFIEHome/tabid/36/Default.aspx>

Canada Core Subject Thesaurus (CST)
<http://en.thesaurus.gc.ca/>

Aquatic Sciences and Fisheries Abstracts (ASFA)
<http://www4.fao.org/asfa/asfa.htm>
This is an abstracting and indexing service covering the world's literature on the science, technology, management, and conservation of marine, brackish water, and freshwater resources and environments, including their socioeconomic and legal aspects.

Metadata Examples
Perhaps the most difficult part of getting started with metadata creation can be trying to figure out what is supposed to go in the data element fields. Examples of well-written metadata can be extremely useful in the writing process. This list provides only a few of the many metadata repositories that are available.

Federal Geographic Data Committee (FGDC) Clearinghouse
<http://clearinghouse1.fgdc.gov/>
The FGDC Clearinghouse is a distributed network of over 250 metadata nodes that can be searched using a variety of criteria, including spatial, temporal and keywords.

USGS Geoscience Data Catalog
<http://geo-nsdi.er.usgs.gov>
This is the catalog of earth science data produced by U.S. Geological Survey (USGS) that allows searches by geographic region, subject area or full text.

NOS Data Explorer
<http://www.nos.noaa.gov/dataexplorer/welcome.html>
Data Explorer offers interactive mapping tools that allow users to locate National Ocean Service (NOS) products in any area in the United States and its territories through a metadata catalog.

Coastal Services Center Coastal Information Directory (CID)
<http://www.csc.noaa.gov/CID/>
This is a search engine that allows users to search for data sets containing information relevant to coastal issues. This interface provides metadata in its original form, or in a more reader-friendly format.

NBII Metadata Clearinghouse
<http://mercury.ornl.gov/nbii/>
The National Biological Information Infrastructure (NBII) Clearinghouse is an initiative to help you locate, evaluate, and access biological data and information from a distributed network of cooperating data and information sources.

Environmental Information Management System
<http://www.epa.gov/eims/index.html>
The U.S. Environmental Protection Agency (EPA) has developed a system that stores, manages and delivers metadata for data sets, databases, documents, models, multimedia, projects and spatial information.

Geodata.gov
<http://gos2.geodata.gov/wps/portal/gos>
Geodata.gov is part of the Geospatial One-Stop E-Gov initiative, providing access to geospatial data and information. Metadata on this site follows either FGDC or International Organization for Standardization (ISO) standards.

Metadata Publication and Discovery
Metadata plays two major roles—data discovery and data compatibility. When written in compliance with the FGDC's Content Standard for Digital Geospatial Metadata, metadata records are used to fuel the FGDC's Clearinghouse system, which allows users to find data sets of interest.

Publishing Your Metadata
Once you have created your metadata, publish it by either submitting it to an existing FGDC Clearinghouse node <http://clearinghouse1.fgdc.gov> or set up your own FGDC Clearinghouse node. The FGDC provides technical references <http://www.fgdc.gov/dataandservices/implementation> including an on-line tutorial <http://www.fgdc.gov/dataandservices/isite_tutorial> to help those interested in establishing a node presence on the National Geospatial Data Clearinghouse system.

Additional Links
Compliments of Bruce Westcott. All of the following links were accessed 12 March 2007.

Introduction to Metadata
<www.getty.edu/research/conducting_research/standards/intrometadata/index.html>

Metadata for the Rest of Us
<http://www.sdvc.uwyo.edu/metadata/rockies.html>

Metadata Themes: The Basics
<http://www.tdan.com/i005fe01.htm>

The Metadata Discussion List
<http://lists.geocomm.com/mailman/listinfo/metadata>

Saving Florida's Fish and Wildlife: Research Institute Embraces Metadata Management
<http://tinyurl.com/39lxbc>

Socioeconomic Data and Applications Center
<http://sedac.ciesin.columbia.edu/metadata/links.html>

CHAPTER 11
Geo-Ontologies

Frederico Fonseca and *Gilberto Câmara*

11.1 Introduction

In order to understand how people see the world and how, ultimately, mental conceptualizations of the comprehended geographic features are represented in a computer system, we must develop abstraction paradigms. The result of the abstraction process is a general view of the process from the real object to its computer representation. Different levels of abstraction allow the development of specific tools for the different types of problems at each level. Creating a solid conceptual model is at the foundation of system design practice. Lately, ontologies were brought into the discussion on modeling. For instance, Guarino (1998, p. 10) says that "every (symbolic) information system (IS) has its own ontology, since it ascribes meaning to the symbols used according to a particular view of the world." Wand and Weber (2004, p. v) argue that since theories of ontology are tools that help us describe a specific world (i.e., the target of an IS), "our information systems will only be as good as our ontologies."

The subject of ontology also is an important field of research in geographic information science (Mark 1993; Frank 1997; Smith and Mark 1998; Bittner and Winter 1999; Rodríguez et al. 1999; Bishr and Kuhn 2000; Câmara et al. 2000; Frank 2001; Kuhn 2001; Kavouras et al. 2005). Ontologies have been used as a means of knowledge-sharing among different user communities, thus improving interoperability among different geographic databases. Information can be integrated based primarily on its meaning by integrating ontologies that are linked to sources of geographic information. The use of an ontology, translated into an active, information-system component, leads to Ontology-Driven Information Systems (ODIS) (Guarino 1998) and, in the specific case of Geographic Information Systems (GIS), it leads to what is called Ontology-Driven Geographic Information Systems (ODGIS) (Fonseca and Egenhofer 1999). In GIS, the use of ontologies is diverse. They can be used to deal with aerial images (Câmara et al. 2001) or with urban systems (Fonseca et al. 2000), for instance. Ontologies are theories that use a specific vocabulary to describe entities, classes, properties and functions related to a certain view of the world. They can be a simple taxonomy, a lexicon or a thesaurus, or even a fully axiomatized theory (Gruber 1995; Guarino and Giaretta 1995).

11.2 Ontology and ontologies

The two uses of the term that we need to contrast are: 1) the way the word is used in Philosophy; and 2) the most current use of the term in Artificial Intelligence, Computer Science, and Information Systems. In Philosophy, Ontology is the basic description of entities in the world, the description of what would be the truth, and the term is used with an upper-case O. Guarino (1998) considers the philosophical meaning of ontology to be a particular system of categories that reflects a specific view of the world. Smith (1998) notes that since ontology for a philosopher is the science of being, of what is, it is inappropriate to talk of a plurality of ontologies as software engineers do. To solve this problem, Smith suggests a terminological distinction between referent or reality-based ontology (R-ontology) and elicited or epistemological ontology (E-ontology). R-ontology is a theory about how the whole universe is organized and corresponds to the philosopher's point of view. An E-ontology fits the purposes of software engineers and information scientists, and is defined as a theory about how a given individual, group, language or science conceptualizes a given domain.

Researchers that use the philosophical meaning of ontology are resorting to the theory of Ontology, the ontology methods, the tools and theories developed within the philosophical discipline of Ontology to find the basic constructs of information systems. They are investigating what information systems are as a concept. From their findings they are able to draw the primitives that conceptual models should use if we are to build better information systems. On the other hand, when Guarino is talking about ontology-driven information systems, the ontologies he is referring to are computational ontologies or ontologies of the second kind we discussed above, ontologies in the original Artificial Intelligence (AI) sense. These are real artifacts that explain a domain. Guarino calls them "engineering artifacts."

Indeed, even the more restricted term computational ontology has been used with more than one meaning in the literature. In this section we review some of the meanings and argue that computational ontologies are theories that explain a domain. In this we agree with Guarino and Giaretta (1995), Smith (2003), and Wand and Weber (2004).

Guarino and Giaretta (1995) recommend that we restrain ourselves to the meaning of the term ontology which aims towards a theory instead of the simple specification of particular epistemic states. Analyzing ontology as a theory, they say (p. 30) that "an ontological theory differs from an arbitrary logical theory (or knowledge base) by its semantics, since all its axioms must be true in every possible world of the underlying conceptualization." Here they are trying to clarify a common use of the term ontology. Gruber (1995) gave a definition of ontology as a "specification of a conceptualization" based on Genesereth and Nilsson's (1987) work. One of the interpretations of Gruber's definition that Guarino wants to avoid is that a conceptualization would define a state of affairs. Guarino (1998) uses an example of the relations among a set of blocks on a table. In Gruber's definition, an ontology would specify for instance that block A is over block B and block C is on the side of block A. Guarino says that the problem with this notion of conceptualization is that it refers to common relations in the blocks' world, i.e., extensional relations. These relations depict a specific state of affairs. In this case they are reflecting a specific arrangement of blocks on the table. Guarino thinks we need to address the meaning of these relations instead of the current situation on the table. He says that an ontology should describe intensional relations such as the meaning of above for instance. Guarino summarizes with the definition 'C = <D, W, R>' in which C is a conceptualization, D is a domain, W is a set of relevant state-of-affairs or possible worlds, and R is a set of conceptual relations on the domain space <D, W>. After clarifying what a conceptualization is, he gives a new definition of an ontology:

> ...an ontology is a logical theory accounting for the intended meaning of a formal vocabulary, i.e. its ontological commitment to a particular conceptualization of the world. The intended models of a logical language using such a vocabulary are constrained by its ontological commitment. An ontology indirectly reflects this commitment (and the underlying conceptualization) by approximating these intended models. (p. 7)

Smith (2003) says that in the current context of research on information sharing an ontology is seen as a dictionary of terms expressed in a canonical syntax. In this use it is implied that ontology is a common vocabulary shared by different IS communities. Smith then gives a definition of an IS (or computational) ontology: "an ontology is a formal theory within which not only definitions but also a supporting framework of axioms is included (perhaps the axioms themselves provide implicit definitions of the terms involved)" (Smith 2003, p. 158).

Wand and Weber (2004) say that although many ontologies restrict themselves to be more a taxonomy than a theory, they still have predictive and explanatory tones. They say "if phenomena are classified correctly according to the theory, humans will be better able to understand and predict the phenomena and thus work more effectively and efficiently with the phenomena" (p. iv). Nevertheless, ontologies as theories of Ontology or as computational ontologies are important models of the world and the activity of ontology engineering has been compared to modeling.

11.3 What's Special about Spatial

"What is special about spatial?" (Anselin 1989; Egenhofer 1993) or what is special about geo-ontologies? A geo-ontology has to provide a description of geographic entities, which can be conceptualized in two different views of the world (Couclelis 1992; Goodchild 1992). The field view considers spatial data to be a set of continuous distributions while the object view conceives the world as occupied by discrete, identifiable entities. Representing geographic entities–either constructed features or natural variation on the surface of the earth–is a complex task. These entities are not merely located in space, they are tied intrinsically to space. They take from space some of its structural characteristics (Smith and Mark 1998). A geo-ontology is different from other ontologies because topology and part-whole relations play a major role in the geographic domain. Geographic objects can be connected or contiguous, scattered or separated, closed or open. They are typically complex and have constituent parts (Smith and Mark 1998). The topological and containment relations between objects have led to the introduction of mereology (Husserl 1970), which describes the relation between parts and wholes. For a review of mereology see Simons (1987) and Casati and Varzi (1999). Smith (1995) introduced mereotopology, which extends the theory of mereology with topological methods.

The development of ontologies of the geographic world (i.e., geo-ontologies) is important to allow the sharing of geographic data among different communities of users. Nevertheless, before we share digital data it is necessary to collect and organize it. Conceptual schemas are built in order to abstract specific parts of the real world and to represent schematically what data should be collected and how they must be organized. In the next sections we review the most recent work on geo-ontologies and geographic data models, in order to gain insight on how the distance between ontologies and conceptual schemas can be shortened.

Spatial databases intend to be a representation of geographic space. But what exactly constitutes geographic space? The most widely accepted conceptual model for geographic information science considers that geographic reality is represented as either fully definable entities (objects) or smooth, continuous spatial variation (fields). The object model represents the world as a surface occupied by discrete, identifiable entities, with a geometrical representation and descriptive attributes. The field model views the geographic reality as a set of spatial distributions over the geographic space. As some authors have already pointed out (Couclelis 1992), the field and object models have an underlying common notion, which is the implicit reliance on Cartesian (or absolute) space as an a priori frame of reference for locating spatial phenomena. In this view, Cartesian space is simply a neutral container within which all physical processes occur. The primitive notion on a Cartesian space is the idea of georeferenced location. Each entity of space is associated to one or more locations on Earth, and spatial relations are derived from the location. The alternative to absolute space is to consider a relative notion of space (Couclelis 1997), constituted through the spatial relations arising among geographic entities. In the framework of relative space, the primitive notion is that of the spatial relation between entities. Spatial interaction models and location-allocation models used in transportation are examples of applications that use the relative notion of space.

Current GIS technology embodies an absolute view of space, since the most common geometric representations available in GIS – such as grids, triangulated irregular networks (TINs) and planar vector maps – are all based on the notion of a georeferenced location. It is therefore not surprising that the notions of objects and fields, as defined in the current GIS literature, can be generalized into a single formal definition.

Nunes (1991) pointed out that the first step in building a next-generation GIS would be the creation of a systematic collection and specification of geographic entities, their properties, and relations. Ontology playing a software specification role was suggested by Gruber (1991). Wiederhold (1994) suggested the use of ontologies as the common point among diverse user communities. Ontology plays an essential role in the construction of GIS, since it allows the establishment of correspondences and interrelations among the different domains of spatial entities and relations (Smith and Mark 1998). Frank (1997) believes that the use of ontologies will contribute to better information systems by avoiding problems such as inconsistencies between ontologies implicitly embedded in GIS, conflicts between the ontological concepts and the implementation, and conflicts between the common-sense ontology of the user and the mathematical concepts in the software. Harvey (1999) warns that bringing fundamental semantic concerns early into the design process is important. Bittner and Winter (1999) say that the usual role of ontologies in modeling spatial uncertainty is to support object extraction processes. Kuhn (1993) asks for spatial information theories that look toward GIS users instead of focusing on implementation issues. Another semantic approach to integrate geographic information is GeoCosm (Ram et al. 2001), a web-based prototype to integrate autonomous distributed heterogeneous geospatial data. They employ a canonical model that integrates diverse conceptual schemas. An ontology is used to help in solving conflicts among information sources.

Fonseca (2001) proposed a framework that uses ontologies as the foundation for the integration of geographic information. By integrating ontologies that are linked to sources of geographic information, Fonseca created a mechanism that allows geographic information to be integrated based primarily on its meaning. Since the integration may occur across different levels, he also created the basic mechanisms for changing the level of detail. The use of an ontology, translated into an information system component, is the basis of Ontology-Driven Geographic Information Systems (ODGIS).

11.4 Geo-Ontologies

A geo-ontology describes entities, semantic relations, and spatial relations:

1. entities can be assigned to locations on the surface of the Earth;
2. semantic relations between these entities include, e.g., hypernymy—relation of class to subclass, hyponymy—relation of subclass to class, mereonomy—part of a whole, and synonymy—same as; and
3. spatial relations between entities (e.g., adjacency, spatial containment, proximity and connectedness).

A geo-ontology also has two basic types of concepts: concepts that correspond to physical phenomena in the real world and concepts that correspond to features of the world that we create to represent social and institutional constructs. We call the first type physical concepts and the second type social concepts. It is important to note that both result from human conventions. As discussed in the literature, although the description of physical features may vary according to cultural and social conventions, they represent variations on the surface of the Earth rather than social conventions per se (Frank and Mark 1991; Mark and Egenhofer 1994; Mark et al. 1999; Smith and Mark 2003).

The physical concepts can be further subdivided into:
• Concepts that are associated with individual geographic objects, each of which has a clearly defined boundary such as qualitative differentiations or spatial discontinuities in the physical world. These are equivalent to the notion of bona fide objects (Smith and Mark 1998). Examples: lake and mountain.
• Concepts that are assumed to be continuous in space (fields). Examples: temperature, slope, pollution and population density.

The social and institutional concepts can be further subdivided into:
• Concepts describing individual objects created by institutional and legal conventions. These are equivalent to the notion of fiat objects or non-naturally demarcated geographical entities of Smith and Mark (1998). Examples: parcel and borough.

- Concepts that are assumed to be continuous over space and represent socially agreed conventions. Examples: social exclusion, infant mortality, homicide rate and human development.

A concept in a geo-ontology is defined by a name, a definition and a set of attributes. A geo-ontology is a set of terms and a set of semantic and spatial relations between terms. The set of semantic relations is created by the semantic components present in the definitions of the terms of a geo-ontology. For instance, the definition of a stream being a flow of water in a channel or bed as a brook, rivulet, or small river and the definition of a creek being a small, often shallow or intermittent tributary to a river would lead to the consequent semantic relation of hyponymy that a creek is a stream.

Another example is that the definition of a basin being a region drained by a single river system and the definition of a valley being an extensive area of land drained or irrigated by a river system would lead to the consequent semantic relation of similarity in which a basin is similar to a valley.

The set of spatial relations is created by the spatial components present in the definitions of the terms of a geo-ontology. The spatial nature of the terms generates spatial relations between terms in a geo-ontology. For instance, the definition of affluent as a stream or river that flows into a larger one leads to the consequent spatial relation that an affluent is connected to streams.

A second example of spatial relations is in the definition of a valley as an elongated lowland between ranges of mountains, hills or other uplands, often having a river or stream running along the bottom that leads to the consequent spatial relation of mountains being adjacent to valleys.

As the above examples show, a geo-ontology has to take into consideration not only semantic relations such as synonymy, similarity, mereonomy and hyponymy, but also spatial relations such as adjacency, spatial containment and connectedness. Given that both semantic and spatial relations are conceptual components of geo-ontologies, defining these relationships plays a critical role in the integration of geographic information.

References

Anselin, L. 1989. What Is Special About Spatial Data? Alternative Perspectives on Spatial Data Analysis. Santa Barbara, Calif.: NCGIA Technical Report.

Bishr, Y. A. and W. Kuhn. 2000. Ontology-based modelling of geospatial information. Pages 24–27 in *Proceedings of the 3rd AGILE Conference on Geographic Information Science*, Helsinki, Finland, May 2000. Edited by A. Ostman, M. Gould and T. Sarjakoski.

Bittner, T. and S. Winter. 1999. On ontology in image analysis in integrated spatial data-bases. In *Integrated Spatial Databases: Digital Images and GIS - Lecture Notes in Computer Science 1737*, edited by P. Agouris and A. Stefanidis, 168–191. Berlin: Springer-Verlag.

Câmara, G., M. Egenhofer, F. Fonseca and A.M.V. Monteiro. 2001. What's in an image? Pages 474–488 in *Spatial Information Theory. Foundations of Geographic Information Science*. Edited by D. R. Montello. Proceedings, International Conference COSIT 2001, Morro Bay, Calif., September 19-23, 2001. Lecture Notes in Computer Science Volume 2205/2001. Berlin: Springer.

Câmara, G., A. Monteiro, J. Paiva and R. Souza. 2000. Action-driven ontologies of the geographical space: Beyond the field-object debate. Pages 52–54 in GIScience 2000—Program of the First International Conference on Geographic Information Science, Savannah, Ga., October 28–31, 2000, chaired by M. Egenhofer and D. Mark.

Casati, R. and A. Varzi. 1999. *Parts and Places*. Cambridge, Mass.: MIT Press.

Couclelis, H. 1992. People manipulate objects (but cultivate fields): Beyond the raster-vector debate in GIS. In *Theories and Methods of Spatio-Temporal Reasoning in Geographic*

Space, edited by A. U. Frank, I. Campari and U. Formentini, 65–77. Lecture Notes in Computer Science 639, New York: Springer-Verlag.

Couclelis, H. 1997. From cellular automata to urban models: New principles for model development and implementation. *Environment and Planning B: Planning and Design* 24:165–174.

Egenhofer, M. 1993. What's special about spatial?-Database requirements for vehicle navigation in geographic space. *SIGMOD RECORD* 22(2):398–402.

Fonseca, F. 2001. *Ontology-Driven Geographic Information Systems*. Ph.D. Thesis. Orono: University of Maine.

Fonseca, F. and M. Egenhofer. 1999. Ontology-driven geographic information systems. Pages 14–19 in *Proceedings, 7th ACM Symposium on Advances in Geographic Information Systems*, held in Kansas City, Mo., November 1999. Edited by C. B. Medeiros. <http://www.spatial.maine.edu/~fred/fonseca_acmgis.pdf> Accessed 12 February 2007.

Fonseca, F., M. Egenhofer, C. Davis and K. Borges. 2000. Ontologies and knowledge sharing in urban GIS. *Computer, Environment and Urban Systems* 24(3):232–251.

Frank, A. 1997. Spatial ontology. In *Spatial and Temporal Reasoning*, edited by O. Stock, 135–153. Dordrecht, The Netherlands: Kluwer Academic.

Frank, A. 2001. Tiers of ontology and consistency constraints in geographical information systems. *International Journal of Geographical Information Science* 15(7):667–678.

Frank, A. and D. Mark. 1991. Language issues for GIS. In *Geographical Information Systems, Volume 1: Principles*, edited by D. Maguire, M. Goodchild and D. Rhind, 147–163. London: Longman.

Genesereth, M. R. and N. J. Nilsson. 1987. *Logical Foundations of Artificial Intelligence*. Los Altos, Calif.: Morgan Kaufmann.

Goodchild, M. 1992. Geographical data modeling. *Computers and Geosciences* 18(4):401–408.

Gruber, T. 1991. The role of common ontology in achieving sharable, reusable knowledge bases. Pages 601–602 in *Proceedings of the 2nd International Conference on Principles of Knowledge Representation and Reasoning (KR'91)*, Cambridge, Mass., April 22-25, 1991. Edited by J. F. Allen, R. Fikes and E. Sandewall. San Francisco: Morgan Kaufmann Publishers.

Gruber, T. R. 1995. Toward principles for the design of ontologies used for knowledge sharing. *International Journal of Human Computer Studies* 43(5/6):907–928.

Guarino, N. 1998. Formal ontology and information systems. In *Formal Ontology in Information Systems*, edited by N. Guarino, 3–15. Amsterdam, The Netherlands: IOS Press.

Guarino, N. and P. Giaretta. 1995. Ontologies and knowledge bases: Towards a terminological clarification. In *Towards Very Large Knowledge Bases: Knowledge Building & Knowledge Sharing*, edited by N.J.I. Mars, 25–32. Amsterdam, The Netherlands: IOS Press.

Harvey, F. 1999. Designing for interoperability: Overcoming semantic differences. In *Interoperating Geographic Information Systems*, edited by M. Goodchild, M. Egenhofer, R. Fegeas and C. Kottman, 85–98. Norwell, Mass.: Kluwer Academic.

Husserl, E. 1970. *Logical Investigations*. Translated by J. N. Findlay from the second German edition of Logische Untersuchungen. London: Routledge and Kegan Paul - Humanities Press.

Kavouras, M., M. Kokla and E. Tomai. 2005. Comparing categories among geographic ontologies. *Computers and Geosciences* 31(2):145–154.

Kuhn, W. 1993. Metaphors create theories for users. In *Spatial Information Theory*, edited by A. Frank and I. Campari, 366–376. Lectures Notes in Computer Science 716. Berlin: Springer-Verlag.

Kuhn, W. 2001. Ontologies in support of activities in geographical space. *International Journal of Geographical Information Science* 15(7):613–631.

Mark, D. 1993. Toward a theoretical framework for geographic entity types. In *Spatial Information Theory*, edited by A. Frank and I. Campari, 270–283. Lectures Notes in Computer Science 716. Berlin: Springer-Verlag.

Mark, D. and M. Egenhofer. 1994. Calibrating the meanings of spatial predicates from natural language: Line-region relations. Pages 538–553 in *Sixth International Symposium on Spatial Data Handling (SDH '94)*, Edinburgh, Scotland, September 1994. Edited by T. Waugh and R. Healey. <http://www.spatial.maine.edu/~max/Calibration.pdf> Accessed 12 February 2007.

Mark, D., B. Smith and B. Tversky. 1999. Ontology and geographic objects: An empirical study of cognitive category. Pages 283–298 in *Spatial Information Theory-Cognitive and Computational Foundations of Geographic Information Science*. Edited by C. Freksa and D. Mark. Proceedings, International Conference COSIT 1999, Stade, Germany. Lecture Notes in Computer Science Volume 1661. Berlin: Springer-Verlag.

Nunes, J. 1991. Geographic space as a set of concrete geographical entities. In *Cognitive and Linguistic Aspects of Geographic Space*, edited by D. Mark and A. Frank, 9–33. Norwell, Mass.: Kluwer Academic.

Ram, S., V. Khatri, L. Zhang and D. D. Zeng. 2002. GeoCosm: A semantics-based approach for information integration of geospatial data. Pages 152–165 in *Revised Papers from the HUMACS, DASWIS, ECOMO, and DAMA on ER 2001 Workshops*. Edited by H. Arisawa, Y. Kambayashi, V. Kumar, H. C. Mayr and I. Hunt. 21st International Conference on Conceptual Modeling, Yokohama, Japan, November 27-30, 2001. Lecture Notes in Computer Science; Volume 2465. London: Springer-Verlag.

Rodríguez, A., M. Egenhofer and R. Rugg. 1999. Assessing semantic similarity among geospatial feature class definitions. Pages 1–16 in *Interoperating Geographic Information Systems*. Edited by A. Vckovski, K. Brassel and H.-J. Schek. Second International Conference, INTEROP'99, Zurich, Switzerland, March 10-12, 1999. Lecture Notes in Computer Science 1580. Berlin: Springer.

Simons, P. 1987. *Parts: An Essay in Ontology*. Oxford: Clarendon Press.

Smith, B. 1995. On drawing lines on a map. Pages 475–484 in *Spatial Information Theory—a Theoretical Basis for GIS*. Edited by A. Frank and W. Kuhn. Proceedings, International Conference Cosit'95, Semmering, Austria, September 21-23, 1995. Lecture Notes in Computer Science 988. Berlin: Springer Verlag.

Smith, B. 1998. An introduction to ontology. In *The Ontology of Fields*, edited by D. Peuquet, B. Smith and B. Brogaard, 10–14. Santa Barbara, Calif.: National Center for Geographic Information and Analysis.

Smith, B. 2003. Ontology. In *The Blackwell Guide to the Philosophy of Computing and Information*, edited by L. Floridi, 155–166. Malden, Mass.: Blackwell.

Smith, B. and D. Mark. 1998. Ontology and geographic kinds. Pages 308–320 in *Proceedings of the Eighth International Symposium on Spatial Data Handling*, held in Vancouver, British Columbia, Canada. Edited by T. K. Poiker and N. Chrisman. Burnaby, British Columbia: International Geographical Union.

———. 2003. Do mountains exist? Towards an ontology of landforms. *Environment and Planning B: Planning and Design* 30(3):411–427.

Wand, Y. and R. Weber. 2004. Reflection: Ontology in information systems. *Journal of Database Management* 15(2):iii–vi.

Wiederhold, G. 1994. Interoperation, mediation and ontologies. Pages 33–48 in *International Symposium on Fifth Generation Computer Systems (FGCS94)*; Workshop on Heterogeneous Cooperative Knowledge-Bases, Tokyo, Japan, December 1994. Edited by K. Yokota. Tokyo: ICOT.

CHAPTER 12

Parcel Data

David J. Cowen, Nancy von Meyer, and *Bob Ader*

12.1 Introduction

The purpose of this chapter is to examine the status of land parcel data in the context of geographic information systems (GIS). The chapter highlights the importance of parcel data to support decision making and the unique aspects of parcel data in terms of creation, maintenance and distribution. It also provides a survey of the current status of parcel data in the US and describes a possible strategy for improving the coordination of parcel data within the nation. The chapter relies heavily on work of the Federal Geographic Data Committee Subcommittee for Cadastral Data <www.nationalcad.org>.

GIS technology and institutions have evolved over the past quarter century in response to the demands for handling the difficult problems associated with building and maintaining a set of high resolution land parcels that represent the ownership, use and value of property. Of particular note have been advances in the integration of high-resolution imagery with vector based features that can support the detail required in an urban setting. With these new tools it is possible to geographically register, create and modify parcel features with a high level of precision. It is also possible to fully integrate the precise field measurements made by surveyors and represented on deeds to compile accurate parcel databases. Improved procedures for incorporating computer-aided design (CAD) drawings prepared by architects and engineers directly into GIS greatly facilitates the exchange of information between the developers and the parcel data producers.

A number of recent advancements in geospatial techniques have positively influenced parcel database development, manipulation and application. For example, advanced editing features enable parcel data producers to build topologically integrated polygons using exact distance and angle measurements. Relational database management environments now enable large numbers of parcels to be created and maintained in a true enterprise model with multiple editors simultaneously working on different versions. Major advancements in surveying procedures based on Global Positioning Systems (GPS) and other methods have dramatically improved the quality of ground measurements. Greatly improved mathematical models have effectively solved the problems associated with conversion between coordinate systems and integration of GIS data themes in different map projections. Improved cartographic functions and web-based mapping environments have greatly facilitated the options for displaying, locating and retrieving parcel information. As a result, it is possible to create and maintain a parcel-based GIS that can support a long term service-oriented set of government functions.

12.2 Definitions

12.2.1 Land Parcel

When discussing land parcels it is important to be clear about the related terminology. The most appropriate definition for a parcel was developed for *Multipurpose Land Information Systems: The Guidebook* (Brown and Moyer 1990–1996):

> *A parcel is an unambiguously defined unit of land within which a bundle of rights and interests are legally recognized in a community. A parcel encloses a contiguous area of land for which location and boundaries are known, described and maintained, and for which there is a history of defined, legally recognized interests.* (Epstein and Moyer 1993, pp. 3-2)

With modern geographic information systems the requirement for parcels to be considered as contiguous areas is not as essential as it was in 1993 and in fact non-contiguousness is a relatively common situation. The boundaries of a parcel are legally defined in a deed or on a survey plat and can be located on the ground by a surveyor. These parcels are the primary units for managing information about land rights and interests (Figure 12-1).

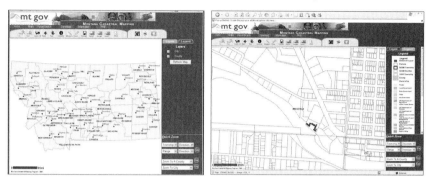

Figure 12-1 Internet access to Montana statewide parcel data system <www.mt.gov>.

In most local government land information systems, the real estate tax parcel is the primary unit of data management. These parcels support the property tax system that is a major source of revenue and they are linked to all decisions relating to land use. The aggregate set of parcel information—representing the distribution and ownership of the taxable assets of a community—forms the basis for land use and zoning decisions. The parcels also represent the location of residences, businesses and public lands. In other words, almost every aspect of government and business can be associated with a land parcel. While there are a few notable exceptions relating to infrastructure and human services, even these can typically be added or incorporated into the parcel system without compromising the tax parcel management.

12.2.2 Cadastre

Parcels are closely related to the term cadastre which was defined by the National Research Council (NRC) in 1980 as:

> …a record of interests in land, encompassing both the nature and extent of these interests. An interest in land (or property right) may be narrowly construed as a legal right capable of ownership or more broadly interpreted to include any uniquely recognized relationship among people with regard to the acquisition and management of land. (National Research Council 1980, p. 5)

Using this definition, there can actually be several different cadastres. In a sense each field in a parcel database constitutes a theme which may be associated with a separate cadastre. For example, the parcel ownership information may be maintained by a recorder of deeds and a parcel mapper who is interested in the juridical or legal cadastre. On the other hand, the tax assessor who is required to know the attributes and qualities that affect the value associated with each parcel maintains a fiscal cadastre. There also may be cadastres associated with land use or land quality. The importance of the cadastre is well stated by Enemark (2002) in the case of a national cadastre in Denmark:

> The modern cadastral system is primarily concerned with detailed information at the individual land parcel level. As such it should service the needs both of the individual and of the community at large. Benefits arise through its application to e.g.: guarantee of ownership and security of tenure and credit; facilitate

efficient land transfers and land markets; management of assets; town and country planning; development and environmental control. In short, benefits arise through cadastral applications for land management in general. (Enemark 2002, p. 2)

While we might assume that the management of land parcels and related information is a modern phenomenon, this is far from the case. (Refer to The Need for a Multipurpose Cadastre by the National Research Council (1980) for an excellent history of the evolution of cadastre.) Once it was recognized that land was a scarce commodity with value related to location, society developed ways to delineate parcels and collect property taxes. According to the French etymologist Blondheim, the term cadastre is probably derived from the Greek word katastichon meaning notebook. (Von Simmerding 1969). In Latin, the term gradually evolved to captastrum or register of territorial taxation units into which Roman provinces were divided.

There is evidence that Egyptian Pharaohs and priests principally generated revenues through taxes on the land. In fact clay tablets from the ruins of Sumerian villages noted "charges against the land, maps of towns and tracts of land, area computations, and most notably court trials adjudicating ownership and boundary disputes." (NRC 1980, p. 5, based on Richeson 1966) The Greeks and Romans also established elaborate land-record systems in support of land taxation policies (Richeson 1966). The Domesday Book of Norman England that was created in 1085-1086 also was a cadastral project. It covered most of England with the exception of the four northern counties and the cities of London and Winchester. It is also interesting to note that Napoleon appointed the mathematician Delambre to chair a Commission given the task:

> *To survey . . . more than 100 million parcels, to classify these parcels by the fertility of the soil, and to evaluate the productive capacity of each one; to bring together under the name of each owner a list of the separate parcels which he owns to determine, on the basis of their total productive capacity, their total revenue and to make of this assessment a record which should thereafter serve as the basis of future assessment.* (Simpson 1976 quoted in NRC 1980, p. 6)

This historical evidence demonstrates that cadastral systems have been an integral part of governing and government decision making for centuries. The concepts in this chapter are based on the presumption that cadastral systems are needed to support modern governments just as they were essential in the past. The questions facing the "modern" multi-purpose cadastre include: 1) how are the landownership records integrated in the information system; 2) by what standards are attributes and geometry integrated into the system; and 3) who will be responsible for the maintenance and distribution of the data?

12.3 Parcel Support for Decision Making

As GIS tools, computing environments and data resources have matured there has been increased use of parcels to support everyday decision making. Although there were some exemplary parcel-based systems in existence in 1980, they pale in comparison to the robust systems of today. There is increased awareness that parcels are the optimal unit of analysis for questions relating to ownership and use of land, real estate and insurance, value and taxation, public administration (accessibility, equity, logistics), business and marketing, impacts from events (hurricanes, tornadoes, fires, floods, landslides, etc.) and issues relating to tribal lands. In fact scores of everyday decisions in the public and private sector rely on current and accurate parcel level information. As examples:

- Who owns the land?
- What is the current land use?
- What is the permitted land use?
- What were historical land uses that might affect current land use?
- Where are historically important buildings and artifacts located?
- What are the current assessed value and tax rate?
- What is the market value for comparable properties?
- What factors on and surrounding the property affect or influence value?
- What are the lending practices on a parcel and are they fair?
- What is the tax base to support public education or other tax-base-funded activities?
- Are the assessed values in line with market values?
- Where is agricultural use occurring?
- Is the land use in agreement with the zoning?
- Is the property in a floodplain or a wetland? Near endangered species?
- What property is within 100 m (or other distance) of a proposed zoning change?
- Where are new schools (libraries, police stations, fire stations) needed?
- What and who would be impacted by a new highway?
- Where should voting districts be delineated?
- Where should sewage treatment (landfills, jails, communication towers, etc.) be located?
- How do I get access to land for timber harvesting or oil extraction?
- What properties are qualified for noise reduction measures?
- Who is potentially affected by a gas plume or oil spill?
- Where are the assets at high risk for wildland fires?
- What are the exact boundaries for incorporated areas?
- Who owns the mineral, gas or oil rights?
- Where is the demand for pool chemicals, new roofs, riding lawn mowers, chain saws, etc.?
- What property is affected by an event (hurricane, tornado, fire, flood, landslide, etc.)?
- Where are sacred tribal grounds located?
- Who owns the tribal rights and interests?

12.3.1 Public Technology Inc.

In a 2003 study of the use of GIS in local governments by Public Technology Inc. (PTI, now Public Technology Institute), a survey of 1,156 local governments documented that GIS is an integral part of their work environment (PTI 2003). A few important trends were identified in the survey.

> *On the horizon, GIS technology will become a key component of every govern-ment applications system. In addition to the visual analysis of data, a key driver for enterprise GIS applications is that location is the connection point for the interoperability of desperate (sic) systems.*
> - *77% of respondents use GIS technology to view aerial photography.*
> - *70% use GIS technology to support property record management and taxation services.*
> - *57% of respondents use GIS technology to provide public access information.*
> - *41% use GIS technology to support capital planning, design, and construction.*
> - *38% use GIS technology to support permitting services.*
> - *38% use GIS to support emergency preparedness and response activities.*
> - *33% use GIS to support computer aided response activities.* (PTI 2003, p. 8)

With 70% of the local governments using GIS to support property record management and taxation services, the importance of cadastral information in any GIS built to support decision making is apparent. Not only is it important to help those without automated systems get started, it is equally important to help those 70% with installed systems move ahead as easily as possible.

12.3.2 Cadastral Data Business Cases

The FGDC Subcommittee for Cadastral Data has developed a series of business cases identifying some specific business processes that rely on parcel or cadastral information. These business cases were developed in conjunction with specialists from many fields such as the emergency management and response communities that deal with wildland fire and hurricanes. The goal was to identify the specific cadastral information needed for each business operation and to define the relevant data content. The resulting list of fields and attributes forms the basis of the committee's core data standard. The subcommittee identified the following generic business functions relating to parcel data:

Navigation and Discovery – Identifying and locating parcels.

Regional Integration – The process of assembling parcel data from varied sources across jurisdictional boundaries.

Inventory and Analysis – Identify where parcel data are missing or areas with potential data integration issues.

Emergency Event Planning – Preparing for the response to emergency events.

Emergency Event Preparation – The actual preparation and provisioning in response to a developing event.

Emergency Event Response – Specific reactions to an event.

Emergency Event Recovery – Actions after the response and in the process of recuperation.

Emergency Event Mitigation – Long term, ongoing effort to lessen the impact of disasters.

These business functions and their relationship to parcel data are summarized in Table 12-1.

12.3.3 Real Estate and Insurance

For the average citizen, land parcels are most closely related to real estate transactions and property taxes with the tax bill calculated from value, mil rates and tax districts. The trend is to develop a sophisticated system of computer assisted mass appraisal (CAMA) that utilizes the geographic location of the parcels and house/property characteristics to estimate the market value based on real estate transactions. This is probably the single most important application for parcel data. Many states require local jurisdictions to keep these assessments in line with current market values. This is often calculated as the Equalization Rate (ER) which is a ratio between the Sum of Roll's Assessed Values for all Taxable Property and the Total Full Market Value for the same properties. The importance of this function has been the prime motivation for states such as Tennessee (Stage and von Meyer 2006) to create their own state level parcel databases. In order to support informed decision making in the real estate sector many communities have built special applications to support discovery of existing market conditions. See, for example, a listing of comparable properties provided by Richland County, South Carolina (Figure 12-2)

Figure 12-2 Richland County, South Carolina, parcel-based comparable sales system.

Table 12-1 Cadastral Business Cases.
Source: Federal Geographic Data Committee Cadastral Subcommittee 2006.

	Navigation and Discovery	Econ. Devel. and Regional	Emergency Event Planning	Emergency Event	Emergency Event Response [2]	Emergency Event Recovery	Emergency Event Mitigation	Energy Management
Cadastral NSDI – Parcels								
Metadata	✓	✓	✓	✓	✓	✓	✓	✓
Parcel Outline (Polygon)[1]	✓	✓	✓	✓	✓	✓	✓	✓
Parcel Centroid (Point)[1]	✓	✓	✓	✓	✓	✓	✓	✓
Parcel ID	✓	✓	✓	✓	✓	✓	✓	✓
Source Reference		✓				✓	✓	✓
Source Reference Date		✓	✓			✓	✓	✓
Surface Owner Type	✓	✓	✓	✓	✓	✓	✓	✓
Improved	✓	✓	✓	✓	✓	✓	✓	✓
Owner Name (Surface Management Agency)				✓		✓	✓	✓
Assessment / Value for Land Information			✓			✓	✓	✓
Assessment / Value for Improvements Information			✓			✓	✓	✓
Assessment / Value Total			✓			✓	✓	✓
Primary Assessment / Value Classification			✓			✓	✓	✓
Secondary Assessment / Value Classification			✓			✓	✓	✓
Tax Bill Mailing Address						✓	✓	✓
Parcel Street Address		✓	✓	✓		✓	✓	✓
Parcel Area			✓			✓	✓	✓
Parcel Zoning			✓					✓
Public Parcel Name		✓	✓	✓		✓	✓	✓
Subsurface Owner Type								✓
Subsurface Management Agency or Owner Name								✓
Cadastral NSDI – Cadastral Reference								
States/Counties	✓	✓	✓	✓	✓	✓	✓	✓
Municipalities	✓	✓	✓	✓	✓	✓	✓	✓
PLSS Townships		✓	✓					✓
PLSS Section (Township Division)		✓	✓					✓
Survey System Area (subdivisions)		✓	✓		✓	✓		✓
Survey Named Area (Ohio)		✓	✓					
Secondary Survey Named Area (Ohio)		✓	✓					
Grid/Cell Reference		✓	✓			✓	✓	✓

NOTES
1. The standard intends that either a centroid point or a polygon is available but both are not needed
2. Emergency Response uses maps and data developed during event planning.

In the private sector, Zillow represents a company that has capitalized on the interest in parcel-level information and has successfully demonstrated that it is possible to efficiently handle 67,000,000 parcels (Figure 12-3). When it was first launched in February 2006, the site quickly became one of the most popular sites in the US and continues to perk interest from average citizens and businesses alike (Wall Street Journal 2006).

Figure 12-3 Zillow parcel-based real estate value system. <www.zillow.com>

Closely related to the real estate market are issues relating to property insurance. The entire process of definition of risk, determination of proper premiums, property appraisal, risk mitigation, damage assessment and recovery are reliant on parcel level information. The importance came into focus as a result of the 2005 hurricane season. Digital records for property ownership along the Gulf Coast following Hurricanes Katrina and Wilma were largely nonexistent and public agencies were left scrambling to assemble some form of property information that could identify the location, value, ownership and extent of damage to thousands of pieces of property and structures. The Government Accountability Office recently reported that the absence of critical documented property records resulted in millions of dollars in fraudulent claims (Government Accountability Office 2006).

Many of the issues that have delayed recovery along the Gulf Coast relate to the issue of flood insurance. As part of national policy, the Federal Emergency Management Agency (FEMA) has established Special Flood Hazard Areas (SFHA) which are:

> *The land area covered by the floodwaters of the base flood is the Special Flood Hazard Area (SFHA) on [National Flood Insurance Program] NFIP maps. The SFHA is the area where the NFIP's floodplain management regulations must be enforced and the area where the mandatory purchase of flood insurance applies.* (FEMA 2006)

The determination of whether a piece of property requires flood insurance requires GIS to overlay the SFHA polygons with a set of parcels. Many of the recovery problems along the Gulf Coast relate directly to this relationship. Parcels outside the SFHA were not required to be insured against damage from floods and the owners have discovered that their insurance policies do not cover losses from such flooding. Several companies such as First American Flood Data Services offer services that integrate "disaster data from FEMA with our own proprietary property data to help you understand the location of all your properties in relation to the declared FEMA Disaster Areas, and to assess the potential damage of each" (First American Flood Data Services 2006).

Manual of Geographic Information Systems

Throughout history, land parcels have been defined on stone tablets, stakes in the ground, legal documents, paper maps and now features in a digital database. Increasingly, they form the core of service-oriented government that focuses on sound decision making. In a modern environment they also are the center of attraction for a curious public that enjoys using a web browser to access information about real estate through sites such as Zillow.com.

12.4 Evolution of Parcels Within GIS

12.4.1 Multipurpose Cadastre

Over the past thirty years, various terms have been linked to the role that land parcels play in the general GIS environment. The term "multipurpose cadastre" was in vogue in 1980 when the National Research Council published the classic study, Need for a Multipurpose Cadastre (National Research Council 1980). In that report,

The multipurpose cadastre system is designed to overcome the difficulties associated with these more limited approaches by: (1) providing in a continuous fashion a comprehensive record of land-related information and (2) presenting this information at the parcel level. The multipurpose cadastre is further conceptualized as a public operationally and adminis-tratively integrated land-information system, which supports continuous, readily available, and comprehensive land-related information at the parcel level. (National Research Council 1980, p. 13)

The Multipurpose Cadastre consisted of the following components:
1. A reference frame, consisting of a geodetic network;
2. A series of current, accurate large-scale maps;
3. A cadastral overlay delineating all cadastral parcels;
4. A unique identifying number assigned to each parcel; and
5. A series of registers, or land data files, each including a parcel index for purposes of information retrieval and linking with information in other data files (McLaughlin 1975).

The 1980 National Research Council report depicted the multipurpose cadastre in relationship with other themes (Figure 12-4). This diagram is important because it illustrates how parcels were viewed in relationship to addresses, zoning, floodplains and even utilities.

The term "multipurpose cadastre" is not widely used in the US today; however, the term is widely used by the international surveying community. In fact, the term appears promi-nently in the International Federation of Surveyors (FIG) report, *Cadastre 2014,* which discusses the aspects of a multipurpose cadastre that integrate parcels to support "facilities management, base mapping, value assessment, land use planning, and environmental impact assessment" (FIG 1998).

12.4.2 Land Information Systems

Within the US the term multipurpose cadastre lost favor with the maturation of Geographic Information Systems. By 1990, the term Land Information System (LIS) was in common use. This term was defined by Epstein and Brown (1990, p. 1-5) as follows:

> *The data, products, services, the operating procedures, equipment, software , the people – the sum of all the elements that systematically make information about land available to users.*

While there was no inventory of the status of land parcel systems during that period, Dane County, Wisconsin was clearly one of the major testing grounds for building a multipurpose LIS. An active group of academic researchers and public employees embraced the concept and helped implement some of the most sophisticated examples. Many of these individuals

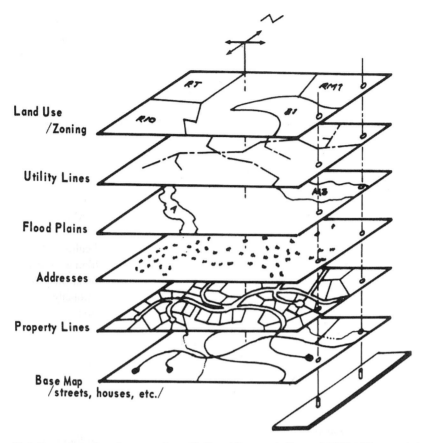

Figure 12-4 Parcel overlay diagram. From National Research Council 1980. With permission of the National Academy of Sciences.

helped prepare the massive 25-chapter *Multipurpose Land Information Systems: The Guidebook*. A prominent component of this guidebook was the stack of layers for the town of Westport in Dane County, illustrated in Figure 12-5. This diagram is still widely regarded as the classic example of a multipurpose LIS for that period. It is important to note that this model is a true intergovernmental one. In fact, Moyer (1990, p. 7-15) made the following comments about what the diagram represents:

> *Building an LIS that is complete, comprehensive, and responsive requires the cooperation of all organizations that are organized vertically, to ensure the horizontal benefits of LIS are fully realized. The importance of the cooperation of governmental units at all levels of government can be seen....*

In the multipurpose LIS, the parcel layer sits on top of a stack of other GIS layers and is maintained by county surveyors. In terms of technology, the layers were actually digital representations of themes which were often at different scales. Although they were maintained independently, it was possible to perform digital overlays to extract information about parcel information and other themes such as soil information that impacted agricultural productivity Therefore, by the early 1990s it was possible to create, manage and integrate multiple themes within a local government Land Information System. By employing polygon overlay procedures it was also possible to intersect the layers in the stack and generate new information such as the proportion of a parcel that fell in a floodplain or the quality of soils within the parcel. In other words, by 1990 a true GIS-based urban LIS existed with parcels being a key component.

Manual of Geographic Information Systems

Over the past decade in North America the term "LIS" has given way to the more generic term, GIS. At the same time many users continue to refer to LIS for any large scale (1:1200) GIS database application that includes land parcels. Clearly the series of GIS/LIS meetings that ended 1998 capitalized on the term GIS/LIS to attract the widest possible attendance. It should also be noted that the Minnesota GIS/LIS Consortium still exists to provide, as its Mission Statement says, "a forum for communicating information to, and improving cooperation among, those interested in Geographic Information Systems (GIS) and Land Information Systems (LIS) in the State of Minnesota" <http://www.mngislis.org/>.

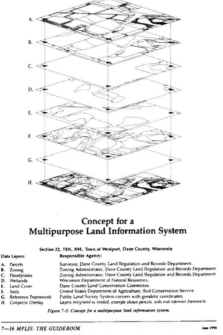

Figure 12-5 caption:

Figure 12-5 Illustration of Multipurpose Land Information System in Westport, Dane County, Wisconsin. From Brown and Moyer 1993.

12.4.3 Modern Parcel GIS Data Model

In the current setting, there is wide-spread appreciation that GIS applications have moved to a much higher level of granularity or resolution than was possible in 1990 or even 2000. This strongly demonstrates that decision making has moved to a level that must include information about ownership, use and value of land parcels. This represents a major shift in emphasis that has had a major impact on the relationship between federal mapping organizations and local government. Instead of finding ways to ingest federal data or to digitize existing maps, the current emphasis has shifted to developing sophisticated tools to handle the complex problems associated with creating and maintaining very high-resolution information derived from surveys, legal documents, imagery, geocoding and a wide range of field-based operations. These requirements have resulted in rigorous specifications for a parcel-based GIS data model. In a modern GIS environment, parcels become spatial entities that exist and must behave within a general enterprise that supports a wide range of applications and business functions. In the US the development of this data model has been fostered by the Federal Geographic Data Committee Subcommittee for Cadastral Data that has developed an approved data standard complete with entity relationship diagrams (Figure 12-6). In the current model, each parcel is a discrete object that is part of a larger database that integrates with related information. Parcels are directly integrated with other themes and do not just float as an independent layer. As the parcels are created, they can be forced to align with certain features such as transportation and utility rights of way, political boundaries and rivers. The parcel features are directly impacted by changes in other information. For example, the geographic position of all parcels may change on the basis of improved survey measurements. Parcel boundaries are automatically adjusted when a new road impacts the right of way or other easements. The parcels also must conform to a set of topological rules that impact how new parcels can be added and how utility lines can be connected to structures. At the same time, other themes that are dependent on parcel representation such as land use or zoning can be synchronized with parcel data. Another important and often complicating aspect of the Data Content Standard is the modeling of the individual rights and interests as separate entities. This means that rights and interest in the chain of title differ from limitation on the use of land such as zoning imposed by a government entity. In many county operations the parcels in the GIS are the real estate tax parcels and therefore the subtle modeling required to handle tracing specific rights and interests is not needed.

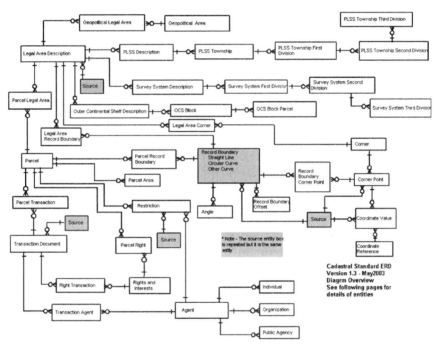

Figure 12-6 Overview of Entity Relationship Diagram – a four-part diagram for entity and attribute definitions. From FGDC Subcommittee on Cadastral Data 2003, p. 15.

12.5 Components of a Modern Parcel Data Model

Several software vendors and consultants have worked to develop software tools and related policies and procedures that would be used to implement the modern parcel model. For example, ESRI has a well-defined parcel data model (Figure 12-7) that is described in depth in Designing Geodatabases: Case Studies in GIS Data Modeling (Arctur and Zeiler 2004). The model also is described in GIS and Land Records: The ArcGIS Parcel Data Model by von Meyer (2004, p. 27) who identifies seven components of the model (1) Parcel Framework, (2) Corners, (3) Boundaries, (4)Tax Parcels, (5) Rights and Interests, (6) Related Use features and (7) Administrative Areas. The US Department of Interior Bureau of Land Management (BLM) and US Department of Agriculture (USDA) Forest Service are developing a version of the model to develop the National Integrated Land System (NILS). Many local governments such as Oakland County, Michigan are now utilizing a version of this data model and related software tools. The current model differs from earlier ones in several ways. From a technology viewpoint there is an emphasis on high-resolution orthophotography to provide a highly accurate and detailed background for the system. These images, with resolutions of six or even four inches, allow direct assignment of coordinates to features on the ground. In combination with accurate survey measurements, these images also provide an excellent basis for quality control. Most modern systems will utilize GIS coordinate geometry (COGO) software tools to create parcels directly from the legal descriptions of the property. These coordinate geometry tools create parcel polygon representations that depict the geographic location and dimensions of parcels with high precision.

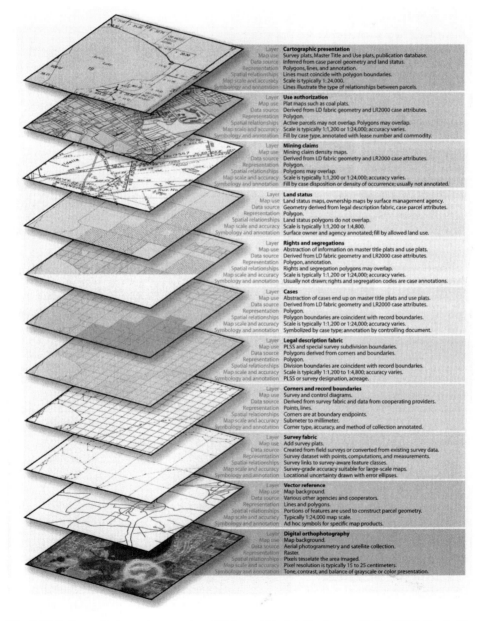

Figure 12-7 Thematic layers in a modern GIS Parcel Model. (Arctur and Zeiler 2004) Used with permission of ESRI. See included DVD for color version.

12.5.1 Legal Descriptions

Parcels are legally described on a deed or plat. As such, they may be considered the first component of the parcel data model. Consequently, it can be argued that the modern parcel data model brings together the diverse communities of photogrammetry, remote sensing, surveying, database management and GIS. Since parcels must be surveyed they provide a basis for the professional land surveyor in the general workflow of parcel creation and maintenance. The importance of this function is well defined by the International Federation of Surveyors (FIG). Stig Enemark, 2007-2010 President of FIG, provides an interesting discussion of the relationship between measurement science and land management within the context of general spatial information management. He states that:

> *Traditional education of surveyors has focused on geometry and technology more than on land use and land administration. Taking a land administration approach to surveying education, there is a need to change the focus from being seen very much as an engineering discipline. There is a need for a more managerial and interdisciplinary focus as a basis for developing and running adequate systems of land administration. A future educational profile for land administrators should be composed by the areas of Measurement Science and Land Management and supported by and embedding in a broad interdisciplinary paradigm of Spatial Information Management.* (Enemark 2004, p. 15)

He suggests there are at least four steps in the creation of parcel data:
1. Property definition – the legal/economic/physical concept;
2. Property determination – process of determination; general/fixed boundaries;
3. Property formation – process, institutions and actors; the role of the surveyors;
4. Property transfer – process, institutions and actors; legal consequences. (Enemark 2004, p. 5)

The importance of professional surveyors in the creation and maintenance of parcel data is clearly spelled out in FIG's projections about the future of cadastre systems, Chapter 3 in its report *Cadastre 2014* (Table 12-2).

Table 12-2 Projections of the Status of Cadastre Systems in 2014. From FIG 1998.

Statement 1	Cadastre 2014 will show the complete legal situation on land, including public rights and restrictions.
Statement 2	The separation between maps and registers will be abolished. (The division of responsibilities between surveyor and solicitor in the domain of cadastre will be seriously changed.)
Statement 3	The cadastre mapping will be dead. Long live modeling. (In 2014 there will be no draftsmen and cartographers in the domain of the cadastre.)
Statement 4	Paper-and-pencil cadastre will be gone! (The modern cadastre has to provide the basic data model. Surveyors all over the world must be able to think in terms of models and to apply modern technology to handle such models.)
Statement 5	Cadastre 2014 will be highly privatized! Public and private sector are working closely together! (Public systems tend to be less flexible and customer-oriented than those of private organizations. The private sector will gain in importance. The public sector will concentrate on supervision and control.)
Statement 6	Cadastre 2014 will be cost recovering! Cost/ benefits analysis will be a very important aspect of cadastre reform and implementation. Surveyors will have to deal more with economic questions in future.

Many methods or systems are used for legal descriptions. The two most common are metes and bounds (which are a sequence of distances, directions and monument calls) and area descriptions which are descriptions related to an already defined area of land (such as reference to a lot in a plat or an area of the Public Land Survey System (PLSS)). There are many variations and other types of legal descriptions, but these two are the most common. The American Congress on Surveying and Mapping (ACSM) defines a legal description as a description recognized by law which definitely locates property (ACSM 1978). In the US there are many forms for legal descriptions and the exactness of any legal description may be subjective. Some of the types of legal descriptions are as follows (von Meyer 2004):

- *Area Descriptions* define an extent by calling out an existing area or an area that can be derived from the hierarchical division of an existing area. An example of one of the most commonly recognized area descriptions is "Lot 3, Block 6 of Green Acre Subdivision in Happy Valley Township, Raccoon County." Another common area description in the US uses the PLSS and its hierarchical nesting to define an area of land. These legal descriptions call out or describe an area of land that has been previously described in another document and does not create a remainder or otherwise divide the existing area.
- *Perimeter Descriptions* describe an extent by defining the boundaries around an area. A single perimeter description encloses a single area. As with the area description, perimeter descriptions in and of themselves do not create remainder areas. One of the most commonly occurring perimeter descriptions is a series of bearings and distances around an area. Within the classification of perimeter descriptions there are three categories: bounds, metes, and metes and bounds.
- *Strip Descriptions* are included as a separate category because they have some special concerns in terms of data modeling and information content. A strip description is a description of a linear feature with an offset to one side or both sides. Strip descriptions are common in utility, roadway and pipeline descriptions. Sometimes these descriptions are in support of easements. Strip descriptions also include a point with a radius or diameter or some other shape around it, such as the legal descriptions that might limit the use of land around a municipal well in well-head protection zoning.

There are many other types of legal descriptions that combine elements of measurements, bounds and areas. These descriptions typically create a remainder area. This means they also have the characteristic of creating junior and senior rights. For example, a parent parcel will be senior and some portion of the parcel that is divided from it will be junior—later in time than the parent.

Finally, because the parcels are legally defined, their boundaries are not directly visible on aerial imagery. In some cases, physical features such as fences, roads, tree lines or water boundaries may coincide with a parcel boundary and may be visible, but the final definition of the legal area is dependent on constructing the boundaries from information in a legal description.

12.5.2 Cadastral Reference

The parcel data model is based on cadastral reference information which is the set of information that allows parcel-level information to be registered to other data themes and to be tied to features on the ground. Cadastral reference information is composed of the spatial reference data (geodetic control and orthophotography) and survey frameworks such as the PLSS, parcel map grids, subdivision boundaries or municipal boundaries. Other reference information may come from roads, hydrography or other features. Parcels are nested into and tied to the cadastral reference. It is the cadastral reference elements that are needed to support query, mapping and navigation. These elements are part of legal descriptions and include information about survey systems, such as subdivisions, geopolitical areas, land grants and the PLSS (Figure 12-8).

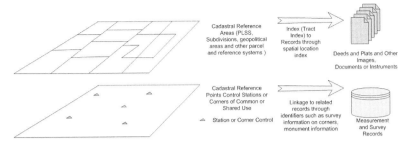

Figure 12-8 Cadastral reference components.

12.5.3 Automation and Conversion of Parcel Features

A major decision in creating a parcel database in a GIS environment relates to the geometric representation of the features. While a parcel represents a closed area on the Earth's surface, it may not always be possible or even desirable to represent parcels as polygons. Furthermore, the GIS representation is only a graphic and not a substitute for the legal survey. Therefore, in many jurisdictions collecting a point to represent a parcel feature is a relatively quick and easy way to get started. By placing a point for each tax parcel record, many of the basic spatial analysis and display activities can be accomplished without the higher cost of a complex editing. In fact, parcel points can be placed on top of orthoimagery or a road network and can be linked to the attribute information in the real estate system through a point identifier such as the tax parcel identification number. Smaller jurisdictions often start with the parcel points and then add the polygons as staff become more experienced with GIS operations. As a system matures, the parcel points can be used to tie attributes to polygons.

Ultimately, parcel polygons will provide the most benefit to parcel data producers. The polygon representations of legal descriptions can identify problems in legal descriptions, assure that the county has border-to-border coverage and can be used to create the traditional tax map polygons. At the same time, the parcel-based GIS can assist a jurisdiction in providing a series of spatial and non-spatial error checks. For example, every entry in the assessor's tax roll should have a one-to-one match with a parcel in the GIS environment. In fact, geographic reconciliation of these databases often provides the necessary return on investment for a jurisdiction to support the system. New parcels are typically added to the tax roll and properties that do not have a private owner are discovered. The polygon representation provides a valuable check to determine whether utility connections are linked to the proper property and provide an important way to monitor whether land use is in agreement with zoning. The downside of collecting parcel polygons is the initial cost to construct the parcel boundaries, the relatively long time it takes to finish the collection, and the higher level skill sets and GIS software needed to maintain the data over time.

The decision on how to create a polygon-based parcel database is closely related to the financial and technical resources of the producer. According to Arctur and Zeiler (2004) the creation of a digital version of parcels can be accomplished by one of four methods with the differences based on the source materials and the procedures. The methods are: 1) vector conversion of existing tax maps; 2) heads-up digitizing of raster versions of scanned tax maps using orthorectified imagery for reference; 3) creating polygons from the legal survey descriptions; or 4) using direct field-based cadastre measurements. The tradeoffs among these four methods are associated with positional accuracy, compilation cost, maintenance cost and the ability to integrate the data with other themes. The simplest and fastest method involves the scanning and vectorizing of existing tax maps. It was once common practice to mount tax maps on a digitizing table and to manually digitize the line work and then try to spatially adjust the lines. Since the original tax maps were often sketches of parcels and other features the conversion to GIS did nothing to improve the positional accuracy or validity of the polygons. They were never intended to become part of a geographically accurate digital database that would be used in conjunction with other geographically registered features. Therefore, they generally represent poor source materials without any control points and often without any coordinate reference system. If the actual maps are on paper they also suffer from the normal problems associated with unstable media. Nevertheless, if there is a good control network it is possible to spatially adjust (i.e., rubber sheet) these vector representations into a reasonable mosaic of GIS themes with real world coordinates. These mosaics, however, may not align very well with a highly accurate set of street centerlines. With all their inherent accuracy problems, they also can be very difficult to maintain, especially when new parcels are inserted from an accurate surveyed set of metes and bounds. Today, the preferred method is to scan and georeference existing maps to known control points and interactively align them with imagery on a high-resolution display. Using heads-up digitizing and employing topological rules, a set of parcels derived from existing tax maps can be created that provides an acceptable parcel base without gaps and overlaps.

Ideally, a parcel database would be created directly from the legal description using COGO tools and adjusted with field measurements. Entering numerical data from a survey plan or plat into a GIS provides the greatest consistency and least distortion, resulting in greater integrity for individual parcel geometry. This is very expensive, with costs of four to twenty times more expensive than manual digitizing (Hohl 1998). Nevertheless, some counties such as Dupage County, Illinois have built 100% of the parcel data using COGO software procedures. Basically, COGO software transforms field measurements into accurate geographic positions and spatial relationships using mathematical calculations (Parrish 2002). The COGO software tools are compatible with standard surveying techniques. They can include attributes of traverse data such as direction, distance, radius, delta, tangent arc length and arc direction.

It also should be noted that using COGO tools does not ensure that a topologically consistent set of parcels will be created. In fact, most parcels were surveyed as independent features; therefore, it is highly unlikely that two adjacent parcel polygons will align perfectly and some manual editing will be required to reconcile neighboring parcels. Hopefully, the small gaps and overlaps can be reconciled using a set of spatial rules regarding acceptable sizes. The process of entering each parcel from its legal description may reveal varying quality of survey measurement surveys and may even require legal proceedings to resolve conflicts. COGO techniques are used to directly input some features (e.g., parcels, roads, right-of-way) that can be used to anchor the entire system or fabric. Once the adjustments are made, the completed parcels are usually left in-tact and adjustments are made to the entire system. Topological rules can be enforced to handle future parcel splits. Furthermore, many communities require developers to submit subdivision plans in digital format such as CAD drawings and even real world coordinates.

12.5.4 Parcel Attribute Information

Associated with each parcel is a set of attributes and a unique identification number. In the US these attributes constitute public information about land ownership and provide a means for individuals, private businesses and public employees to locate, discover and monitor conditions, activities and events relating to the land. Parcel information describes society's relationship to the land and generally includes information about the value, ownership, land use and zoning, address and legal description. According to von Meyer (2004) the information consists of:

> *all the documents, maps and information that depict rights and interests in land. It includes things like the deed that describes the owner (the grantee) and who sold the land (the grantor), the legal description of the extent of the land*

and any included or excluded rights and uses such as restrictive covenants in a
subdivision that describe the size and setback of homes or other structures. It also
includes information on the value of the land and improvement, the address and
the property tax amount.(von Meyer 2004, p. 5)

The goal of compiling parcels into a GIS database is not to create a legally binding document, but a set of mutually exclusive and exhaustive polygons that can be used to represent and query a wide range of information. The selection of parcel attribute data is a local government decision and is closely related to the existing organizational structure. For example, the Metro GIS that serves the Minneapolis/St. Paul area maintains 65 different attributes for each parcel.

The Federal Geographic Data Committee (FGDC) Cadastral Subcommittee has defined a minimum set of attributes about land parcels that is used for publication and distribution of cadastral information by cadastral data producers for use by applications and business processes (Table 12-3).

Table 12-3 Recommended Minimum Set of Parcel Attributes.
Source: FGDC Subcommittee Cadastral NSDI Reference Document, 2006

Metadata
Parcel Outline (Polygon) CAD
Parcel Centroid
Parcel ID
Source Reference
Source Reference Date
Owner Type
Improved
Owner Name
Assessment/Value for Land Information
Assessment/Value for Improvements Information
Assessment Value Total
Primary Assessment/Value Classification
Secondary Assessment/Value Classification
Tax Bill Mailing Address
Parcel Street Address (may be many)
Parcel Area
Parcel Zoning (may be many)
Public Parcel Name

In a general sense, parcel information or attributes describe the human relationship to the land. Ownership information specifies the interests that are owned, the shape and extent of ownership, and the value and assessment of the land and improvements. For leases, parcel information describes the period of time of the lease and specifies the permitted uses of the property during the lease period. On public lands, parcel information identifies individuals or organizations that have been granted rights to use that same land and any specified limitations on use. The parcel GIS database can generate an infinite variety of reference and thematic maps. For example, the typical tax map may only display parcels, rights-of-way and parcel numbers, but can include dimensions and street addresses (Figure 12-9).

Figure 12-9 Detailed view of parcels for Waukesha County, Wisconsin. <http://www.waukeshacounty.gov/cm/Business Units/Parks and Land Use/Land Information System/>

12.5.5 Linkage of Geometry and Attributes

In a typical database environment, the GIS representation of a parcel would only require the unique parcel identification number. Parcel information is then linked to the parcel geometry through the unique parcel ID. In most land information systems the guiding principle for attribute management with the parcel data is to maintain only the minimum number of attributes needed to use the parcel data with the associated geometry. This approach maximizes the use of existing systems and provides the best integration of the GIS data in most cases. A good practice is to have the graphical data and keys created and maintained by the digital mapping office. Information about ownership, use, value and characteristics can be maintained by separate offices. In practice, the core information typically handled by the assessor would consist of information required for generating a property tax bill including acreage, current owner, characteristics about the structures (number of rooms, size and date of improvements), market value and assessed value.

In reality, the maintenance of the attribute information pertaining to parcels is a complex process that requires collaboration and sound workflow management. Information must be synchronized in space and time, as well as ownership and rights. In Denmark, this level of synchronization is performed at the national level (Engemark 2002). Parcel numbers are associated with files maintained by the building and dwelling register, the municipal register of property, the central population register, the register of plans, etc. In order to sustain integrated and correctly linked attributes, the parcel maintenance workflow must consider all departments that tie data to the parcels and the attributes they use to relate to the parcels. The levels of GIS integration with the parcel attribute data may be as follows.

GIS Centric – In this approach, the land records transaction begins with a map and is housed and driven through a graphic interface to that map. This interface includes the ability to query by parcel number, site address, subdivision name, or any other location-based attribute. Map editing also takes place within this interface as a component of the workflow. Once a parcel or set of parcels is selected, the user would be able to select the attribute operations to be performed, such as ownership change, new parcel geometry, assessment attributes, re-valuation analysis or other operations. Screen interfaces can be used to enter information into the database and changes can be seen immediately in the map interface.

GIS Embedded – In this approach, an application such as Computer Assisted Mass Appraisal (CAMA) operates as a stand-alone attribute management system, much as it does today in many jurisdictions. The parcel mapping and maintenance take place in a GIS setting outside of the attribute management. The map display is live and attributes on the map change as the production system updates attributes.

GIS Linked – In this approach, both the map maintenance and attribute management are independent operations. The GIS information and the attribute information are linked at the time of publication, which can occur on a regular basis through a system-launched routine.

Regardless of the level of integration of the GIS with the attribute information, the minimum attributes for the geographic feature in the GIS should be based on the following factors.

- Primary key to track unique records
- Foreign key to link to other features or related attributes or images
- Cartographic presentation
- Subtyping (separate domains of values or default values for subtypes)
- Topology rules
- Feature level metadata
- Maintenance and editing notes or codes
- Versioning or workflow management

12.5.6 Production Versus Publication

In any parcel GIS production mode, there should be separate environments for production (operation and maintenance) and publication (distribution and access) (von Meyer et al. 2000). The basic taxonomy consists of two highly interrelated but separate environments. With advances in GIS database management, these two environments may be the same database, but may reflect different states or conditions of the database. The production environment is the maintenance of databases and maps, and production of information products. The publication environment is the distribution and publication of maintained GIS data to users outside the data production or data maintenance community. In terms of work flow, dynamic data should be in a production environment, while data being distributed or published for others to use are considered to be in a publication environment. In fact, often one department's published information can become another's production information. For example, land surveyors may produce highly accurate adjusted coordinate values that are published or made available to parcel mappers who incorporate this published information into their parcel map production environment. Other differences in these two environments include the permissions for editing or changing data, the linkages between related tables, and the connections to image files such as digital pictures of parcels. It is also good practice to create metadata describing maintenance steps and to save notes on geometry construction.

12.6 Unique Aspects of Parcels

Within the context of GIS land parcels differ from other types of information in several important ways.

12.6.1 Parcels Are Normally the Smallest Administrative Unit in Society

The land parcels play an important role in society. While there are some sub parcel exceptions for structures, land uses (e.g., agriculture), soil types or elevation zones, each individual parcel represents a unique administrative unit. It is the unit for which legal, real estate and financial decisions are made. At the same time, the parcels can be aggregated or dissolved using GIS to create other meaningful polygons. For example, census blocks and all related census geographical units including incorporated areas can normally be created from parcels. By dissolving the parcels in such a manner, it is possible to generate all non-private lands—including transportation rights-of-way. Using this approach, local governments can accurately track jurisdictional boundaries and annexations.

12.6.2 Parcels Are the Best GIS Container for Other Features and Linkages

While parcels represent the appropriate administrative polygon for numerous operations and decisions they also represent a container or feature for linking and spatially associating many other features. Because the parcel represents the association between a landowner, legally defined boundaries and the ground, it provides the link to a series of rights and interests. In fact, it is the key component for spatial and attribute searches that address accountability of government. For example, buildings, businesses, utility hookups, facilities, hazardous chemicals, archeological sites, mineral resources and grazing rights should be associated with a parcel and its owner. In many cases the linkage can be made through database associations via parcel ID, street address or owner. The spatial representation, however, is tied directly to the parcel.

12.6.3 Parcels Are Dynamic

Private property is constantly being bought and sold and tracts of land are often subdivided. Therefore, the legal and financial representation of a community's parcel fabric can change on an hourly basis. Parcel attributes associated with ownership, value and use are impacted by every property purchase and sale, every revaluation and re-financing, as well as every change in land use and every permit. It is challenging to capture and represent the current state of the parcels in a GIS database. For this reason, many jurisdictions have established a GIS-based maintenance program that manages the set of parcels on a transaction basis. In a perfect world this type of data would be handled by a rigorous series of interrelated information systems that monitor transactions. More so than with other GIS themes, it is important to capture and retain the time of events. While this was once quite a challenge, this type of temporal information is incorporated into modern GIS software. These functions allow one to retrieve and display the database as it existed at any previous time. This is particularly valuable for mapping and monitoring land use. The importance of maintaining parcels in real time was highlighted in a recent article about the Chicago Metropolitan Agency for Planning (CMAP) (Sanders 2006). The Web Projects manager describes the role of parcels in the regional information system based on enterprise architecture and second generation web services.

> *City, county, regional, state and federal agencies can make data available via web services that can be called by other servers at any time. This means that all partners can incorporate the most current data available into their own data systems. The data can be fetched as needed from the most authoritative source, then displayed on a web form, pulled into a predictive model or used to calculate aggregate statistics. Users of our data system conduct property surveys for a variety of zoning, economic development and other purposes. They walk up to the property with a smart phone, open a browser, select a property address from the list and click "Go". Our web server issues consecutive data requests to various county and city services, renders the resulting data into HTML and sends the whole batch down to the user's browser. All within less than a second. The latest data, straight from the source. Since the Assessor's teams continuously sweep through various parts of the county updating assessments, this is an important feature.*
>
> *Web services are not just for data exchange with external partners; they also work well for cross-departmental sharing. A city government wishing to provide a one-page parcel profile for use by city employees might create a web form that pulls together property ownership, permits, physical characteristics, business licenses, court records, crime data, any public financing or subsidies, historic value of the structure, building condition and many other attributes. But typically these bits of information are housed and owned by various data stewards across several departments. A series of light-weight web services could be deployed as interfaces between departments.* (Sanders 2006, pp. 1 and 10)

12.6.4 Linkage to Street Addresses

One of the most important attributes for parcels is the site address or situs. It provides the legal basis for location and navigation to a piece of property or a structure. Addresses are critical for the delivery of mail, the routing of emergency vehicles and conducting the Decennial Census. This seemingly simple attribute is, in fact, a very complex one. The Urban and Regional Information Systems Association (URISA) has accepted the task of developing a new FGDC standard for street addresses (URISA 2006). It must be noted that the system of assigning street addresses is not always a simple one-to-one relationship. Since addresses are attached to buildings or structures, the assignment of a site address to a parcel can become complicated. Indeed, any apartment complex will require that several addresses be linked to a single parcel polygon. Within a local government environment, the assignment of addresses to parcels may be handled by the engineering office that ensures that street names are unique and addresses conform to local regulations regarding spacing and format. They may also be assigned by the planning department through permitting processes or through a municipal government within the county. Ideally when this transaction is completed the new information about new parcels, streets and addresses is communicated to the E911 system, utility companies and ultimately the US Postal Service. Building site addresses in a standard database with a domain of values can reduce many of the commonly occurring errors or problems with site addresses. Variations of the address such as listing the direction as North, N or NO, for example, can be standardized in the database. Likewise it is sometimes problematic when street names are directions (East Street). Having a standardized list of road names, road types and directions reduces errors found in many legacy database systems. It is interesting to note that Australia currently supports a nationwide web-based (Google Map) address matching system based on land parcels (Figure 12-10). In the US, several private companies such as Proxix Solutions and Navteq are now offering geocoding based on parcel databases for parts of the country.

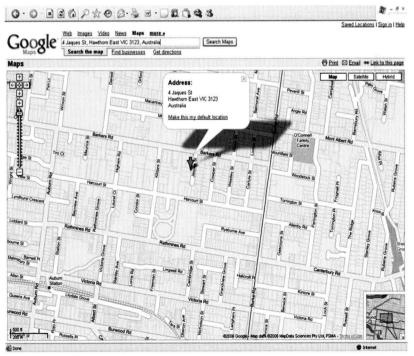

Figure 12-10 Australian parcel data as part of nationwide parcel address matching system for Google Maps. <www.maps.google.com.au>

12.6.5 Lack of Enforced Standards

Since parcel maintenance functions are relegated to local government in the US they are not governed by any systematic set of standards. Furthermore, since they have evolved over the past thirty years there are many unique systems that have their own idiosyncrasies reflecting individual expertise and capabilities of a community. In general, the only standards that are established are at the state level for tax map production and associated calculation of property values and taxes. Therefore, although the Federal Geographic Data Committee has developed a standard, there is no mandate for enforcement of the standard. It should also be noted that digital parcel databases do not even exist for about two-thirds of the counties.

12.6.6 Inconsistent Practices for Complex Parcels

Parcel databases also must deal with property ownership that can exist in three dimensions. The obvious example is a multi-floor condominium where parcels can be assigned to one or more floors and ownership of the surrounding land may be held in group ownership. Some jurisdictions capture individual floor plans and link these to the tax parcel polygons. Other jurisdictions have a series of "stacked" polygons representing the varying interests in a single extent. How to design for and manage parcels with multiple owners and three-dimensional geometry requires more consideration than can be handled in an overview.

12.6.7 Reconciling Data at County and State Boundaries

While most counties are only interested in maintaining parcels to meet internal business needs there are times when the data are critical for regional planning or emergency response purposes. A good example of regional coordination of parcel data exists in the Minneapolis/ St. Paul area where several local governments are represented by the Metropolitan Council of the Twin Cities. In the mid-1990s, this regional government was given the responsibility for operating transit and sewer services for the region. In order to meet these needs, it established a consortium to deal with reconciling the geometry and 65 attributes for a data set containing nearly 1 million parcels across seven counties. As they state:

> *Many government organizations straddle county boundaries and had been un-able to get easy access to a standardized parcel dataset. The Mosquito Control District, for example, uses parcel data to keep track of homeowners' requests about spraying their areas. It has also been used to guide disease response and reduce risk in neighborhoods following identification of encephalitis in one resident. School districts use the data for bus routing and long-range planning. And of course the Council uses it for land-use, transportation, and sewer planning.* (Minnesota GIS/LIS Consortium 2006)

12.6.8 Significant Distribution Issues

Local governments typically maintain parcel data as part of their normal business practices. Increasingly, sophisticated public administrators have adopted parcel data systems as the optimal manner to be accountable to elected officials, local businesses and citizens. While the parcel data system has been established to support internal operations, it also is one of the most valuable data assets of a community. Consequently local governments are faced with important decisions regarding how to distribute the parcel data to external users.

With today's technology, distribution can take many forms. Average users directly access parcels and the related attribute information through a web mapping service. Since information pertaining to a parcel is in the public domain, many communities will support web-based searches based on owner, parcel ID or address. There are no standards to regulate this system and wide variations in the systems and level of detail exist. For some applications, a user may require a complete version of the parcel database. Policies and procedures relating to distribution of parcel data vary considerably. At one extreme, communities allow full access to owner information, tax payments and images of documents. For example, Delaware County, Ohio, continuously maintains its cadastre based on transactions and updates its website every week. At the other extreme, jurisdictions only distribute their parcel data under strict licensing agreements and hefty fees.

Conflicting policies regarding the distribution of parcel data have recently resulted in a fairly heated debate. Over the past couple of years there have been some significant legal decisions and opinions over the distribution of parcel data. In 2005 the Connecticut Supreme Court made an important decision that attracted national attention. It decided that the City of Greenwich had to release GIS data to a private entrepreneur. In reaching this decision the court rejected the City's claim that trade secret exemption could apply to the electronic GIS maps. It determined that all information contained in the maps is available from town departments, therefore they are not secret. The court also rejected the claims that release of the information could pose a risk to public safety. The case is significant on the national level because of the interest of journalists. In fact, the Reporters Committee for Freedom of the Press Society of Environmental Journalists Investigative Reporters and Editors, Inc. filed an important legal brief in the case as an Amicus Curiae. The brief draws the following argument:

> *Publicly funded computerized Geographic Information System (GIS) records, and the maps generated from GIS systems data, have become a basic tool for government study and decision-making in fields such as environmental policy, public safety, and health. The public also requires access to GIS records and maps relied upon by government officials in order to conduct its own study and to monitor, criticize, and, as warranted, challenge decisions based upon that data. Journalists represented by Amici play a key watchdog role in this process. They must be able to access original computerized GIS data and maps used by official decision-makers and disseminate them to the public. Thus, amici have a vital interest in ensuring that the government places no improper restrictions on the public's right to obtain those records.*
>
> *GIS data and maps are, without question, public records. The Connecticut Freedom of Information (FOI) law guarantees all requesters access to nonexempt public records in whatever format they choose. In the current proceeding, the Town of Greenwich ("Greenwich") has claimed that computerized GIS records should be exempt from this clear mandate of Connecticut FOI law. Greenwich alleges–without specific or convincing evidence–that public access to these records would hurt town security, trade secrets, and information technology systems security.*
>
> *Amici agree with the Connecticut Freedom of Information Commission (FOIC) that none of these exemptions are applicable. Moreover, Greenwich's circular argumentation fails to support its refusal to permit access to GIS records in electronic form. And it is irrelevant that some requesters may intend commercial gain through access to GIS records.* (Reporters Committee for Freedom of the Press 2004, pp.1-2)

Because of the size of the city and its long history, a 2006 decision in Los Angeles, California also attracted considerable national attention. In this case, the Los Angeles County assessor made a major change in policy when he reacted to the California State Attorney General opinion that

> *parcel boundary maps maintained in electronic format by a governmental entity are subject to public inspections and copying under provisions of the Public*

> *Records Act and, therefore, must be provided for the cost of duplication in accordance with the parameters set forth in the California Public Records Act.* (Auerbach and Wolfe 2006)

As a result the county discontinued its program of marketing and selling the data to the private sector; it now distributes the entire parcel database to everyone who places an order, with no distinction based on the purchaser or purpose. A modest fee (less than $10) covers the cost of the DVD and shipping.

12.7 Current Status of Parcel Data in the US

Through the efforts of the FGDC Subcommittee Cadastral Data, we have a fairly accurate inventory of the current status of parcel data programs in the US. Stage and von Meyer (2003 and 2005) conducted surveys that provide an interesting benchmark for such systems. They found that all states, with the exception of Alaska, distribute the responsibility of collecting parcel data to local governments with varying degrees of oversight and support provided by a state agency. Twelve states indicated that they centrally manage parcel data and eight states indicated that the geometry is centrally managed. Many states require local governments to submit all or a portion of their real estate tax information to a state auditing agency (typically the State Department of Revenue) that is responsible for ensuring equity of assessments across jurisdictions.

12.7.1 State and Local Government

In most states within the US, the responsibility for maintaining tax parcels rests at the county level with a total of 2,925 counties acting as the primary responsible entity for collecting and managing parcel data. There are a few states that have taken on the responsibility of building the parcel geometry, Montana being a prime example where the state is managing the parcel boundaries for 48 of 56 counties. Stage and von Meyer (2005) estimate that the total number of privately owned parcels in all 50 states and the District of Columbia is approximately 144.3 million with the average number of persons per parcel being about 2. This density ranges from 0.3 persons per parcel in Wyoming to 3.5 in New York. Another perspective on density can be acquired by looking at the number of parcels per square mile, the average for all fifty states being 80. New Jersey and Rhode Island reported the highest parcel density with each having 373 parcels per square mile, and Alaska the least at 1.7 parcels per square mile followed closely by South Dakota at 4 parcels per square mile.

By examining this information over the past few years Stage and von Meyer (2003 and 2005) have been able to identify some trends in the development of parcel databases. They have estimated that the number of parcels increased by 2% from 141.3 to 144.2 million. Of these, the percentage that have been converted into digital data has increased from 61% to 68%. The number of states with a large-scale orthoimagery program increased from 8 to 16 while the small-scale orthoimagery programs decreased from 30 to 22. Eighteen states indicated that they had some type of parcel management program to assist local governments. There appears to be an increased emphasis by states to support the efforts of local governments by acquiring large-scale imagery and by the creation of programs to assist in their modernization efforts. The National States Geographic Information Council (NSGIC) Digital Imagery for the Nation initiative demonstrates the widespread need for this orthoimagery.

A major conclusion of the FGDC Subcommittee for Cadastral Data is that states that have parcel management programs have been able to exceed the national average of conversion. These programs support the use of standards, cooperative ventures and land records modernization in communities that do not have the resources to implement a conversion program. Seventeen states and the District of Columbia support such a statewide program and eleven states indicated their programs were substantial efforts targeted at achieving complete statewide conversion of parcel data to digital maps. Another finding is that seven states with well established parcel conversion programs have an average of 86% of the parcels in digital forms. This is 18% percentage points above the national average of 68%.

The FGDC Subcommittee for Cadastral Data also has observed that the conversion of parcel data into a format that can be used in a GIS continues to grow. Although the total number of parcels converted is approaching 70%, it seems likely that most of the conversion to GIS is taking place in the more urban areas. On the other hand, a major digital divide exists in terms of the development of parcel data. It is estimated that 76% (2,389) of the counties in the US are not likely to have the expertise and resources to spatially enable their parcel data. In many counties, there are simply not sufficient financial, human or technical resources to create and maintain a parcel database. The states will necessarily play a key role in this effort through the development of standards, providing training, technical support skills and in some instances taking on the responsibility of developing and maintaining parcel boundary files as is happening in Montana, Tennessee and Alabama. At the same time small counties such as Roseau County, Minnesota, with a population of 16,338 recently invested $200,000 in a three year effort to develop a parcel database. (Minnesota GIS/LIS Consortium 2006, p. 12).

12.7.2 Federal Geographic Data Committee Interests in Parcel data

12.7.2.1 Cadastre Framework Layer

In 1994, Executive Order 13286 established the National Spatial Data Infrastructure (NSDI) as the technology, policies, standards and human resources necessary to acquire, process, store, distribute and improve utilization of geospatial data (Federal Register 1994). As a part of the implementation of that Executive Order, the Office of Management and Budget—under Circular No. A-16 Revised (Office of Management and Budget 2002)—established the FGDC and gave it the responsibility to lead and support the NSDI strategy. One of the first tasks of the FGDC was to define a national system of framework data layers including geodetic control, orthoimagery, elevation and bathymetry, transportation, hydrography, cadastral and governmental units. The FGDC established subcommittees to oversee the development of standards for each of these framework layers and appointed a lead agency (Figure 12-11). The BLM was given responsibility for the coordination and use of cadastral information.

Although parcel data are clearly considered part of a cadastral framework layer, a strong case could be made that parcel-level information is critical to the accurate delineation of several other NSDI themes.

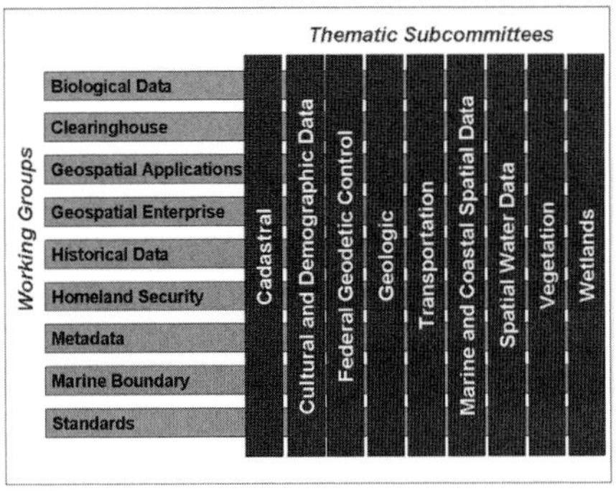

Figure 12-11 Federal Geographic Data Committee Working Groups and Thematic Subcommittees.

Manual of Geographic Information Systems

Buildings and Facilities (General Services Administration): The facility theme includes federal sites or entities with a geospatial location deliberately established for designated activities. A facility database might describe a factory, military base, college, hospital, power plant, fishery, national park, office building, space command center or prison. Facility data are submitted from several agencies, since there is no one party responsible for all the facilities in the Nation and facilities encompass a broad spectrum of activities. The FGDC promotes standardization on database structures and schemas to the extent practical. (Buildings and facilities must rest on a parcel of land which has an owner, value and use.)

Cadastral (Department of the Interior (DOI), BLM): Cadastral data describe the geographic extent of past, current, and future right, title and interest in real property as well as the framework to support the description of that geographic extent. The geographic extent includes survey and description frameworks such as the Public Land Survey System, as well as parcel-by-parcel surveys and descriptions. (The land parcel is the fundamental entity in cadastral data.)

Cultural and Demographic Statistics (Department of Commerce (DOC), US Census Bureau (USCB)): These geospatially referenced data describe the characteristics of people, the nature of the structures in which they live and work, the economic and other activities they pursue, the facilities they use to support their health, recreational and other needs, the environmental consequences of their presence, and the boundaries, names and numeric codes of geographic entities used to report the information collected. (Structures and activities are associated with land parcels; ownership of those parcels impacts use.)

Cultural Resources (DOI, National Park Service): The cultural resources theme includes historic places such as districts, sites, buildings and structures of significance in history, architecture, engineering, or culture. Cultural resources also encompass prehistoric features as well as historic landscapes. (Cultural features are located on land parcels; in fact many sites are defined by the parcel boundary.)

Governmental Units (DOC, USCB): These data describe, by a consistent set of rules and semantic definitions, the official boundary of federal, state, local and tribal governments as reported/certified to the USCB by responsible officials of each government for purposes of reporting the Nation's official statistics. (Incorporated areas are defined by parcel boundaries; in fact many local governments track annexations on the basis of parcels and define the jurisdictional boundaries by dissolving parcels.)

Housing (Department of Housing and Urban Development (HUD)): HUD's database maintains geographic data on homeownership rates, including many attributes such as HUD revitalization zones, location of various forms of housing assistance, first-time homebuyers, underserved areas and race. Data standards have not yet been formalized. (HUD has recognized that parcels are critical to tracking information about housing units)

Federal Land Ownership Status (DOI, BLM): Federal land ownership status includes the establishment and maintenance of a system for the storage and dissemination of information describing all title, estate or interest of the federal government in a parcel of real and mineral property. The ownership status system is the portrayal of title for all such federal estates or interests in land. (The parcel is the only appropriate unit to track federal land ownership and interests on the land.)

12.7.2.2 Federal Geographic Data Committee (FGDC) Subcommittee for Cadastral Data

The BLM, in conjunction with the FGDC, formed the national Subcommittee for Cadastral Data. This subcommittee serves as the focal point for federal activities relating to parcel or cadastre information. Funding for the activities of the subcommittee has been provided by the BLM, the designated federal custodian for cadastre information. This subcommittee is charged with organizing the coordination of interest in cadastre information from stakeholders at all levels of government. The subcommittee members meet regularly to review the objectives and priorities and it works closely with agencies and interested parties. In addition to the development of the content standard, the committee has completed two surveys of parcel activities at the state and local levels, analyzed the role of parcel data to assist with emergency response activities, documented best practices and developed business model templates with appropriate metrics (FGDC 2006).

Some of the highlights of the most current activities include developing data element standards for energy, hurricane response, wildland fire response, homeland security and data discovery for GeoSpatial One Stop based on extracts from the Cadastral Data Content Standard. The Cadastral NSDI implementation includes defining data stewardship relationships necessary to collect, publish and maintain cadastral information. The Subcommittee has adopted a focus on states for this stewardship. A business plan template that includes a framework for inventorying the current status and needs of all cadastral producers in a state has been developed with state cadastral coordinators. For example, Arkansas has developed the State of Arkansas Cadastral Spatial Data Infrastructure Business Plan. "The publication format, attributes and publishing cycle for parcel data in Arkansas is based upon the FGDC Cadastral Data Core Content Standard." (Arkansas Assessment Coordination Department and The Arkansas Geographic Information Office 2006, p.7).

12.7.3 Federal Parcel Programs

12.7.3.1 National Integrated Lands System (NILS)

The National Integrated Land System (NILS) is a joint project between the BLM and the USDA Forest Service. NILS is billed as

> *the first step toward providing a common solution for the sharing of land record information within the government and the private sector…implies the development of a common data model and a set of GIS tools that unify the worlds of surveying and GIS* (Cone 2003, p. 227).

The goal of NILS is to "improve the accuracy and quality of data so as to create standard land descriptions and cadastral data that can be used by anyone." From a technological and organizational perspective, NILS is being created using several items that did not exist in 1980 or even 1990. These include: 1) off-the-shelf GIS software to handle the complexities of parcel data including the input of measurements from surveyors; 2) the FGDC Cadastral Data Content Standard and the metadata standard; and 3) object-oriented software (Figures 12-12 and 12-13)

The NILS, therefore, is a specific implementation of an integrated parcel data maintenance model at the federal government level. It is designed to provide a process to collect, maintain, and store parcel-based land and survey information that meets the common, shared business needs of land title and land resource management. The NILS project is being developed in four modules:

1) Survey Management (SM) is a set of applications that provides surveyors with the ability to manage survey data collected in the field;

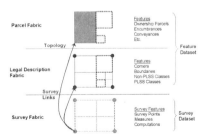

Figure 12-12 National Integrated Land System (NILS) Conceptual Data Model. <http://www.blm.gov/style/medialib/blm/wo/MINERALS__REALTY__AND_RESOURCE_PROTECTION_/nils.Par.23626.File.dat/NILS-implem-v1-10-02workflow.pdf>

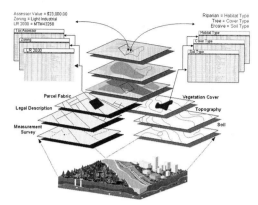

Figure 12-13 Interrelationship of survey, legal and parcel fabric in the National Integrated Land System (NILS). <http://www.blm.gov/wo/st/en/prog/more/nils.html>

2) Measurement Management (MM) is a desktop GIS application that allows surveyors to analyze and adjust surveyed data from the field;

3) Parcel Management (PM) is a desktop GIS application that provides tools for land managers to create and manage parcel features and their legal area descriptions; and

4) GeoCommunicator (GC) is an Internet website for cadastral survey and land management information and data from the NILS and the BLM's Land and Mineral Use Authorization system.

The objectives of these modules is to provide the user with tools to manage land records and cadastral data in a "Field-to-Fabric" manner. The user can use field survey measurement data directly from the survey measuring equipment, manipulate these data into lines and points, and create legal land and parcel descriptions to be used in mapping and land record maintenance. NILS intends to unify the worlds of surveying and GIS. This unification process is fundamental for land records managers of cadastral mapping databases to improve the accuracy and quality of the data, to create standard land descriptions and cadastral data that can be used by anyone. (BLM 2006a)

12.7.3.2 Geographic Coordinate Data Base (GCDB)

The importance of accurate ground references has also been recognized by the federal government. The BLM has created the Geographic Coordinate Data Base (GCDB) which is a collection of geographic information representing the PLSS.

The GCDB is computed from BLM survey records (official plats and field notes), local survey records and geodetic control information on a township basis. The survey boundaries are delineated by computing the geographic positions of township, section, aliquot part, government lot and special survey corners. Next, official land descriptions are assigned to each land unit in the grid. The records are then reformatted so GIS software can be used to view the PLSS information spatially. (BLM 2006b)

12.7.3.3 Department of Agriculture Common Land Units

One of the most interesting federal parcel activities from the viewpoint of photogrammetry and remote sensing is the Department of Agriculture's Farm Services Agency (FSA) Common Land Units. Under this program the USDA is partnering with Rural Development and the Natural Resource Conservation Service (NRCS) to capture and maintain a layer of more than 16 million farm and field boundaries. This is an important program with respect to rural parcel data on a national basis and the effort to share parcel level geospatial information. As the USDA supporting documentation states:

Specifically, the Deputy Under-Secretary, FFAS, and the RMA and FSA Administrators established a cross-functional team to implement a common information system that will eliminate the need for producers to report the same information to FSA and to reinsured companies; create efficiencies for producers, the agencies, and reinsured companies; and reduce the need for data reconciliation. The common information system (CIS) will enable the sharing of customer land use related information by utilizing USDA's e-Gov initiative and the Office of Management and Budget's (OMB) Geospatial One-Stop Initiative. The system is based on the common land unit (CLU), which identifies all farm fields, range land, and pasture land in the United States. USDA customers report and receive services related to land location, such as insurance, commodity payments, loans, conservation plans, and program contracts. (Department of Agriculture 2003)

12.8 The Need for a Cadastral NSDI Model

The uncoordinated approach to coordinating parcel data in the US was the subject of the classic 1980 National Research Council study The Need for a Multipurpose Cadastre and is the focus of a recently released NRC study National Land Parcel Data: A Vision for the Future (National Research Council 2007). A review of the changing status of land parcel data and what happened to the recommendations from the 1980 study are covered in an article "A Retrospective Look at the Need for a Multipurpose Cadastre" (Cowen and Craig 2003) The 1980 report made several important recommendations to improve the coordination and stewardship of parcel data. For example, the 1980 report (National Research Council 1980) recommended:

> *… that federal legislation be prepared to authorize and fund a program to support the creation of a multipurpose cadastre in all parts of the Nation* (p. 3)

> *… that the Office of Management and Budget designate a lead agency for the multipurpose cadastre.* (p. 3)

It is very interesting to note the 1980 committee's optimism about creating a national parcel database when they concluded: "Current technology is adequate in most cases for the surveying, mapping, data collecting, filing and dissemination of information" (National Research Council 1980, pp. 101-102).

The 2007 study provides an assessment of the current situation and offers a series of nine recommendations for implementation of a Nationally Integrated Land Parcel Data System. It also provides a vision that would allow internet access to separately-maintained sets of land parcels. This system would include the following attributes:

- A National Land Parcel Coordinator, working with coordinators for federal lands, Indian lands, and each state, would oversee the development and integration of consistent land parcel data across the nation.
- National land parcel data would be in the public domain, but no information would be provided about private ownership, use or value, in order to protect privacy and confidentiality.
- Built on already existing state and local parcel data systems, the envisioned system would link a series of servers maintained by local and state governments. The system would ideally be able to seamlessly assemble accurate parcel information for any part of the nation at a given point in time.
- Each parcel would be treated as a unique entity whose information would be maintained by local government officials. (National Academy of Sciences 2007)

The common thread to the two NRC studies is the optimism of the committee members. At the same time the two studies provide a realistic assessment of the institutional barriers that would have to be crossed to form the type of coordinated and federated arrangements that would make such an approach become a reality and bring the United States in line with many developed nations.

12.8.1 Cadastral NSDI Model

The FGDC Cadastral Subcommittee has served as the focal point for parcel data coordination in the US. The efforts of this committee have been highlighted in the new National Research Council report (National Research Council, 2007). Collectively these activities provide a blueprint for a rational approach for the coordination of parcel data in the US. In the largest sense, this blueprint calls for a Cadastral NSDI Model. Given the complexities of cadastral information and the diverse and distributed nature of the collection, maintenance and publication of cadastral information they believe that this NSDI model should be built on some important guiding principles. These include developing and maintaining it through partnerships and collaboration with federal, tribal, state and local governments and the private sector. The Cadastral NSDI will leverage the work that is being done on a daily basis by thousands of organizations and individuals and will assist in making that data available to a broad range of customers for purposes even beyond those which the data were originally intended to support. It should also utilize existing standards that will allow for the full integration of cadastral information with other data themes and to allow cadastral information collected by different agencies to be seamlessly stitched together to form a national fabric. The Cadastral NSDI also should be maintained by stewards at all levels of government who are responsible for property rights and the management of information about property rights. Accordingly, the Cadastral NSDI would not disrupt or impinge on these local operations; instead it recognizes the importance of local governance and supports the maintenance of cadastral data by the appropriately identified data steward. It is important that the Cadastral NSDI be utilized for decision making and the success of the program will be judged on how well it supports business applications across the nation. This is particularly important in times of emergency, when a host of decisions depend upon the ownership, value and use of the land. The Cadastral NSDI should also leverage existing efforts. This principle acknowledges that all of the data exist in either hard copy or automated format. It is used by the data stewards for on-going daily operational systems. Therefore, the Cadastral NSDI does not require new data. Instead existing data should be enhanced to be in a form that meets the needs of business applications and consumers.

The objectives and strategies for the Cadastral NSDI are built on these guiding principles and should be revisited on a regular basis to assure that the Subcommittee is meeting the needs of the nation and is responsive to the ever-changing technology environment. The strategy to implement the Cadastral NSDI is to build on the partnerships and collaboration at all levels of government. These should help identify the recognized sources of cadastral data. For example, in western states the BLM may develop and maintain the PLSS cadastral reference and rights and interest on public lands, while local governments develop and maintain privately held rights and interests to support real estate tax systems. By identifying and recognizing the shared responsibilities of data stewardship, data stewards can work collaboratively with each other. They also should develop and maintain points of contact that can increase the efficiency of communication and reduce duplication of effort. The partners should develop and maintain a data publication and distribution architecture with the goal to assemble pieces of information from diverse sources into a single cohesive data theme that is available through state-based portals. The partners also should identify resources and funding alternatives for collection, maintenance and publication of cadastral data. It is important to document best practices for cadastral programs and to use business applications and decision support requirements to guide the priorities for the development. This will require specification of data requirements, data profiles, publication standards and standard queries or views of the data for business functions. The final strategy encourages the use of the latest technology to facilitate the maximum collaboration and sharing of information through portals and other internet-based approaches.

12.9 Conclusion

In summary, this chapter has reviewed the development of parcel-based GIS processes and institutional framework over the past quarter of a century. Clearly, a robust GIS model now exists that can serve the unique and demanding requirements of the parcel community. It also is clear that parcels are increasingly recognized as the critical component in local government information systems and such systems are becoming essential even in small communities. At the same time the federal government has identified parcels as a framework layer in the NSDI and the FGDC Cadastral Subcommittee has been actively working on standards, identification of best practices and business models. It also has initiated programs such as NILS and GCDB to address some of the technical issues that currently impede the integration of parcel data. Nevertheless, the US does not have a unified approach to the coordination of parcel data across all levels of government. The cadastral NSDI is offered as a set of guiding principles and strategies to rectify this situation.

References

ACSM (American Congress on Surveying and Mapping). 1978. *Definition of Surveying Terms*. Gaithersburg, Maryland: ACSM.

Arctur, D. and M. Zeiler. 2004. *Designing Geodatabases: Case Studies in GIS Data Modeling*. Redlands, Calif.: ESRI Press, 393 pp.

Arkansas Assessment Coordination Department and The Arkansas Geographic Information Office. 2006. *Cadastral Spatial Data Infrastructure Business Plan*. Little Rock, Ark.: State of Arkansas.

Auerbach, R., and D. L. Wolfe. 2006. Letter to The Honorable Board of Supervisors, County of Los Angeles, dated 7 March 2006. <http://www.opendataconsortium.org/documents/LAC_Change_Parcel_Policy.pdf> Accessed 18 April 2007.

BLM (Bureau of Land Management). 2006a. National Integrated Land System (NILS). <http://www.blm.gov/nils/> Accessed 18 April 2007.

BLM. 2006b. Geographic Coordinate Data Base. <http://www.blm.gov/gcdb/> Accessed 18 April 2007.

Brown, P. M. and D. D. Moyer, editors. 1990-1996. *Multipurpose Land Information Systems: The Guidebook* (2 volumes). Rockville, Md.: US Department of Commerce, National Oceanic and Atmospheric Administration, National Geodetic Survey Division, Federal Geodetic Control Committee.

Cone, L. 2003. The National Integrated Land System. *Surveying and Land Information Science* 63(4):227–234.

Cowen, D. J. and W. J. Craig. 2003. A retrospective look at the need for a multipurpose cadastre. *Surveying and Land Information Science* 63(4):205–214.

Department of Agriculture. 2003. *Audit Report: USDA Implementation of the Agricultural Risk Protection Act of 2000*. Report No.50099-12-KC. Office of Inspector General, Great Plains Region. <http://www.usda.gov/oig/webdocs/50099-12-KC.pdf> Accessed 18 April 2007.

Enemark, S. 2002. Land Administration in Denmark. Copenhagen, Denmark: The Danish Association of Chartered Surveyors <http://www.ddl.org/thedanishway/LandAdm_01.pdf> Accessed 18 April 2007

Enemark, S. 2004. Building land information policies. United Nations, FIG and PC IDEA Inter-Regional Special Forum: Building of Land Information Policies in the Americas, Aguascalientes, Mexico, 26-27 October 2004. 20 pp. <http://www.fig.net/pub/mexico/papers_eng/ts2_enemark_eng.pdf> Accessed 18 April 2007.

Epstein, E. and P. Brown. 1989. Introduction to multipurpose land information systems. In *Multipurpose Land Information Systems: The Guidebook* (2 volumes), edited by P. M. Brown and D. D. Moyer. Rockville, Md.: US Department of Commerce, National Oceanic and Atmospheric Administration, National Geodetic Survey Division, Federal Geodetic Control Committee.

Epstein, E. F. and P. M. Brown. 1990. Land interests. In *Multipurpose Land Information Systems: The Guidebook* (2 volumes), edited by P. M. Brown and D. D. Moyer, 4.1–4.15. Rockville, Md.: US Department of Commerce, National Oceanic and Atmospheric Administration, National Geodetic Survey Division, Federal Geodetic Control Committee.

Epstein, E. and D. Moyer. 1993. The parcel map. In *Multipurpose Land Information Systems: The Guidebook* (2 volumes), edited by P. M. Brown and D. D. Moyer, 13.1–13.23. Rockville, Md.: US Department of Commerce, National Oceanic and Atmospheric Administration, National Geodetic Survey Division, Federal Geodetic Control Committee.

FEMA (Federal Emergency Management Agency). 2006. <http://www.fema.gov/NFIPKeywords/description.jsp?varKeywordID=85> Accessed 18 April 2007.

FGDC (Federal Geographic Data Committee). 2006. <http://www.fgdc.gov/participation/working-groups-subcommittees> Accessed 18 April 2007.

FGDC Subcommittee on Cadastral Data. 2003. Cadastral Data Content Standard for the National Spatial Data Infrastructure, Version 1.3 - Third Revision. <http://www.nationalcad.org/data/documents/CADSTAND.v.1.3.pdf> Accessed 23 May 2007.

Federal Geographic Data Committee Cadastral Subcommittee. 2006. Cadastral NSDI Reference Document. <http://www.nationalcad.org/data/documents/Cadastral%20NSDI%20Reference%20Document%20v10.pdf> Accessed 21 May 2007.

Federal Register. 1994. Executive Order 13286, published in the March 5, 2003, Federal Register, Volume 68, Number 43, pp. 10619–10633. <http://a257.g.akamaitech.net/7/257/2422/14mar20010800/edocket.access.gpo.gov/2003/pdf/03-5343.pdf> Accessed 21 May 2007.

FIG (International Federation of Surveyors). 1998. *Cadastre 2014* <http://www2.swisstopo.ch/fig-wg71/cad2014.htm> Accessed 18 April 2007.

First American Flood Data Services. 2006. <http://fafds.floodcert.com/news/index.asp?ID=78> Accessed 21 May 2007.

Government Accountability Office. 2006. Hurricanes Katrina and Rita Disaster Relief: Continued Findings of Fraud, Waste, and Abuse, GAO-07-252T. <http://www.gao.gov/new.items/d07252t.pdf> Accessed 18 April 2007.

Hohl, P. 1998. *GIS Data Conversion: Strategies, Techniques, and Management*. Clifton Park, New York: OnWord Press, an imprint of Thomson Delmar Learning.

McLaughlin, J. D. 1975. *The Nature, Design and Development of Multi-Purpose Cadastres*, Ph.D. Thesis, University of Wisconsin, Madison, Wisconsin.

Minnesota GIS/LIS Consortium, 2006. GIS/LIS News: Newsletter of the Minnesota GIS/LIS Consortium Spring, Issue 44, p.12.

Moyer, D. 1990. Why implement a multipurpose land information system? In *Multipurpose Land Information Systems: The Guidebook* (2 volumes), edited by P. M. Brown and D. D. Moyer, 7.1–7.19. Rockville, Md.: US Department of Commerce, National Oceanic and Atmospheric Administration, National Geodetic Survey Division, Federal Geodetic Control Committee.

National Academy of Sciences. 2007. *National Land Parcel Data: A Vision for the Future.* (Report in Brief) Washington, D.C.: National Academy Press. 4 pp.

National Research Council. 1980. *The Need for a Multipurpose Cadastre.* Washington, D.C.: National Academy Press. 112 pp. <http://books.nap.edu/openbook.php?isbn=NI000560> Accessed 27 April 2007.

National Research Council. 2007. *National Land Parcel Data: A Vision for the Future.* Washington, D.C.: National Academy Press. 172 pp.

Office of Management and Budget. 2002.Circular A-16, 19 August 2002. *Coordination of Geographic Information and Related Spatial Data Activities.* <http://www.whitehouse.gov/omb/circulars/a016/a016_rev.html> Accessed 21 May 2007.

Parrish, S. 2002. Double corners and multiple monuments. *ACSM Bulletin.* July–August: 21.

Public Technology, Inc. 2003. *National GIS Survey Results: 2003 Survey on the Use of GIS Technology in Local Governments.* October 2003. Washington, D.C.: Public Technology, Inc. (now Public Technology Institute). Send email to info@pti.org to request report.

Reporters Committee for Freedom of the Press. 2004. Brief and Appendix of Amici Curiae Supreme Court State of Connecticut S.C. 17262. <http://www.rcfp.org/news/documents/20041111-greenwich.pdf> Accessed 21 May 2007.

Richeson, A. W. 1966. *English Land Measuring to 1800.* Cambridge, Mass.: MIT Press.

Sanders, G., 2006. Get ready for real time up-to-the-minute data exchange. URISA NEWS Issue 215 September / October. pp. 1–10.

Simpson, S. 1976. *Land Law and Registration.* Cambridge: Cambridge University Press.

Stage, D. and N. von Meyer. 2003. *An Assessment of Parcel Data in the United States.* Federal Geographic Data Committee's Subcommittee on Cadastral Data.

Stage, D. and N. von Meyer. 2005. National Parcel Data Inventory - Introduction <http://www.nationalcad.org/2005survey/intro.asp> Accessed 21 May 2007.

Stage, D. and N. von Meyer. 2006. An Assessment of Best Practices in Seven State Parcel Management Programs. Prepared for the FGDC Cadastral Data Subcommittee. <http://www.nationalcad.org/data/documents/3StateParcelMgtProgFinal.pdf> Accessed 29 May 2007.

URISA (Urban and Regional Information Systems Association). 2006. Draft Street Address Data Standard. <http://urisa.org/about/initiatives/addressstandard> Accessed 21 May 2007.

von Meyer, N. 2004. *GIS and Land Records: The ArcGIS Parcel Data Model.* Redlands, Calif.: ESRI Press. 169 pp.

von Meyer, N., S. Oppmann, N. Bushor and C. Lucas. 2000. Production and publication: a concept for geographic information environments. *Wisconsin Land Records Quarterly* pp. 6–10. For updates, see Fairview Industries web site. <www.fairview-industries.com> Accessed 2 August 2007.

von Simmerding, F. 1969. Verwendung und Herkunft des wortes Kataser. Z. *Vermessungswesen* 94:333–341. Stuttgart, Germany: Verlag Konrad Wittwer.

Wall Street Journal. 2006. A new web site for real-estate voyeurs. 8 Feb. 2006: D1.

Zillow.com. 2006. <http://www.zillow.com> Accessed 19 April 2007.

SECTION 3

GIS Data Quality, Uncertainty and Standards

CHAPTER 13
Spatial Data Quality and Uncertainty

Wenzhong (John) Shi

13.1 Introduction

In the field of geographic information science (GIScience), spatial data quality—including uncertainty—is identified as one of the fundamental properties, together with space and time, in GIScience and spatial analysis. This chapter addresses the issue of spatial data quality and uncertainty by introducing the concepts of uncertainty and error, dimensions of spatial data, the elements of spatial data quality, sources of uncertainty, modeling uncertainties in spatial data, and modeling uncertainties in spatial analysis.

13.2 Uncertainty

13.2.1 The Concept of Uncertainty

Uncertainty is a word widely used in many fields, such as physics, statistics, economics, metrology, psychology and philosophy. The connotations of uncertainty can be different from one field to another. The term uncertainty used in GIScience is mainly related to the uncertainty used for the fields of metrology and statistics.

In GIScience, the uncertainty of a measurement can be described by giving a range of values such that the range possibly includes the true value of the measured object. The measurement can be, for instance, the height of a building, the length of a river, the area of a forest, etc. For example, the height of Tai Mao Mountain in Hong Kong is measured to be 957 meters with an accuracy of ±0.5 meter. This means the true value of the mountain height is possibly a value within the range 956.5-957.5 meters. The uncertainty of the mountain height measurement value is given by the range.

Uncertainty can be either imprecision, ambiguity or vagueness. Imprecision refers to the level of variation associated with a set of measurements. Ambiguity is associated with one-to-many relations, i.e., difficulty of making the decision to which class an object should belong. Vagueness is associated with difficulty of making sharp or precise distinction in the real world. Detailed descriptions on ambiguity and vagueness are given by Fisher (1999).

Uncertainties are described differently in different mathematical theories. Probability and statistics theories are employed for handling the uncertainty of imprecision. The fuzzy set theory and the fuzzy measure theory are used to assess two different uncertainty categories. Recent research shows that fuzzy measure can be used to describe vagueness while ambiguity can be treated with the discord measure, confusion measure and non-specificity measure. For describing ambiguity in a crisp set, Hartley (1928) introduced a measure of information now known as Hartley's measure. Shannon (1948) introduced a general uncertainty measure (Shannon's entropy) based on probability theory. Fuzzy topology theory is applied for modeling uncertain topologic relations between spatial objects (Shi and Liu 2004). Mathematical theories for assessment of uncertainties in spatial and aspatial data are summarized in Figure 13-1.

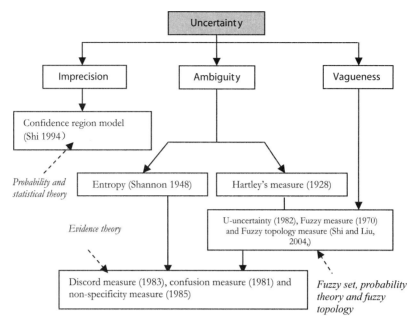

Figure 13-1 Uncertainties measured based on various mathematical theories (Adapted and further extended based on Klir and Folger 1988).

13.2.2 Errors

In many cases, error is used as a synonym of uncertainty. In fact, in terms of connotation, error is narrower than uncertainty. Furthermore uncertainty is a natural term compared with error. Uncertainty may not necessarily be caused by mistakes, it may be due to incomplete information, for example.

Error refers to the statistical deviation or mistake. The observation conditions and the instability of the observer, the measurement instrument and the real world are the major sources of error in the spatial data capture process. Errors can be classified into the following three categories: (a) systematic errors, (b) random errors and (c) gross errors.

Systematic Errors

After performing a series of measurements under the same circumstances, one may notice that the magnitude and sign of errors follow a regular pattern. This type of error is called systematic error. The systematic error arises from the performance of measurement equipment. It can be partially eliminated by improving the performance of measurement equipment.

Random Errors

When measured values obtained from a series of measurements under the same conditions contain irregular errors of magnitude and sign, these irregular errors are called random errors. The random errors are due to occasional or uncertain factors such as a random environmental change, the discrimination by the human eye or others. It is difficult to estimate the influence degree for each occasional factor on the measured value as the occasional factor causes an irregular change of one measured value. Although the occasional error makes the measured value move to left or right by chance, in a large number of measurements random errors 'cancel each other out,' meaning they have no effect on the mean.

Gross Errors

Gross errors are mistakes that occur in the process of identifying, hearing or memorizing the object to be measured by an observer, or having a human mistake introduced in a computation process. The gross errors are generally significantly larger than the systematic errors or the random errors. They seriously affect the reliability of measured results. Therefore, an appropriate precaution or measure should be taken to reduce or eliminate the influence of gross errors. On the one hand, the observer should be trained so that he/she will compute and manipulate data cautiously. On the other hand, whether the surveying technology selected is applicable and scientifically reasonable should be carefully considered.

13.3 Spatial Data Quality

13.3.1 Dimensions of Spatial Data

An entity in the real world is normally described by spatial data in various dimensions. Before we can describe uncertainties of spatial data, we should first understand dimensions of spatial data. Here, spatial data is used as the synonym of geographic data, or geomatics data. A spatial object in a geographic information system (GIS) can be uniquely determined by a set of spatial data in the following five dimensions:

- space
- time
- scale
- attribute
- relation

As illustrated in Figure 13-2, objects A and B are uniquely determined by their "coordinates" in each of the five dimensions.

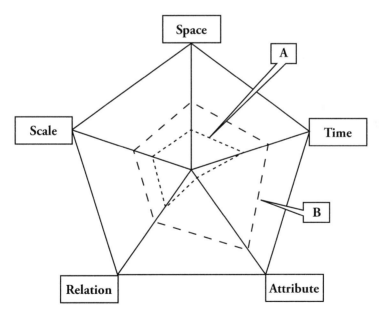

Figure 13-2 The objects A and B are determined in the five dimensions.

The meaning of the five-dimensional graph is explained through the following example of land parcels in GIS (see Figure 13-3). Here positional information of a land parcel, for example Land 1002, is given by the coordinates of its boundary point stream in a geographic *space*—here it is in a map projection "Hong Kong Grid 80." Information about *time* of the Land 1002 is given by its Start_Time of 1945 and End_Time of 1998. The map *scale* is

1:1,000. The *attributes* of the Land 1002 include, for example, the Owner_Name of John Li and his Address of "BCD." The *relation* of Land 1002, is described by its adjacent neighbors: Land 1001, Land 1003, Land 1004 and Land 1005.

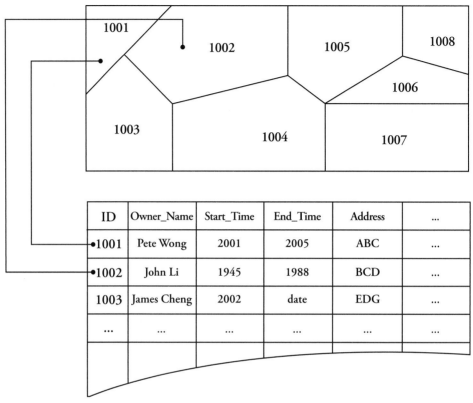

ID	Owner_Name	Start_Time	End_Time	Address	...
•1001	Pete Wong	2001	2005	ABC	...
•1002	John Li	1945	1988	BCD	...
1003	James Cheng	2002	date	EDG	...
...	...	...	...	...	...

Figure 13-3 An example of spatial data on land parcel in the dimensions of space, time, attribute, scale and relation.

13.3.2 Spatial Data Quality

13.3.2.1 Quality Components

Accuracy
Accuracy is the extent to which an estimated value approaches the true value or the value considered to be true, i.e. the degree to which it is free from bias. According to ISO 3534-1 (1993), accuracy is defined as the closeness of agreement between a test result and the accepted reference value. Here, a test result can be either observations or measurements.

Precision
In GIS, we have two streams of models for spatial objects, (a) object-based and (b) field-based. Precision is normally used for the first category of models, and resolution is used for the second category.

Precision refers to the ability of a measurement to be consistently reproduced. In statistical terminology, precision is a measure of the dispersion (usually referred to as standard derivation) of observations about a mean. Generally, it is difficult (if not impossible) to obtain a true value for an object, such as point coordinates. Thus, instead of using point accuracies, point precision is often used as a quality measure for point coordinates. Theoretically, precision is different from accuracy. However, in many cases, a precision measure is used as an accuracy measure.

The precision of stochastic variables is expressed by their variance-covariance matrix. The precision of a vector **y** of point coordinates depends on the precision of the measurement vector **x** and the functional relationship between the coordinates and measurements. If the coordinates are computed from the measurements with a functional relation g, then **y** = g(**x**).

After linearization of the g function, the variance-covariance matrix of the derived coordinates can be calculated based on the law of variance propagation $\Sigma_{yy} = J_{yx}\Sigma_{xx}J_{yx}^{T}$, where J_{yx} is a Jacobian matrix of the g function with respect to the measurements.

On the other hand, precision also refers to the number of significant digits to which a value has been reliably measured. For example, for a point P we have two sets of measurement of its coordinates by using two different survey instruments – Equipment A and Equipment B: (a) P [385.6637, 22.0787] and (b) P′ [385.66, 22.08] respectively. We can thus state that the measurements from Equipment A have a higher precision than that of B, since the measurement of Equipment A has more significant digits than that of B.

Resolution

In computer science or remote sensing, resolution is used to describe how finely-detailed an image is. It can be indicated by pixels per square inch, dots per square inch, lines per millimeter, etc. on an image or a computer-generated display. Resolution of an image is very important in remote sensing and photogrammetry, because it affects the quality of feature extracted from the images by computer processing or visual interpretation.

Completeness

The National Committee for Digital Cartographic Data Standards (NCDCDS) standard defines completeness as an attribute describing the relationship between the objects represented in a data set and the abstract universe of all objects (Morrison 1988).

The definition of completeness was further extended by Brassel et al. (1995) and they give a more comprehensive definition: the degree of completeness describes to what extent the entity objects within a data set represent all entity instances of the abstract universe.

Logical Consistency

Logical consistency refers to the degree of adherence to logical rules of data structure, attribute and relationships. Here, data structure can be conceptual, logical or physical (ISO19113 2005).

Reliability

Statistically, reliability refers to the closeness of the initial estimated value(s) to the subsequent estimated values. In this statistical sense, reliability is different from logical consistency. We have relatively less research on reliability of spatial data quality in GIS so far.

The assessment of quality information may include a measure of the reliability of the quality information. This type of information is recorded in a quality evaluation report based on ISO19114 (2005).

13.3.2.2 Elements of Spatial Data Quality

According to the dimensions of spatial data which are space, time, scale, attribute and relation, and the quality components of spatial data which are accuracy, precision, resolution, completeness, logical consistency and reliability, we can now form a full matrix on elements of spatial data quality. There are a total 30 spatial data quality elements in this matrix. However, we can only clearly define some of them at this stage of the concept development for this Chapter, and the rest of them (which are marked as "--" in the Table 13-1) need to be further investigated in the future of research in GIScience.

In the remainder of this section, we introduce those elements of spatial data quality that can be well-defined at this stage.

Table 13-1 The matrix of quality elements of spatial data.

	Space	Time	Attribute	Scale	Relation
Accuracy	Positional Accuracy	Temporal Accuracy	Attribute Accuracy	--	--
Precision	--	--	--	--	--
Resolution	Spatial Resolution	Temporal Resolution	Thematic Resolution	--	--
Completeness	--	--	--	Completeness	--
Logical Consistency	Locational Consistency	Temporal Consistency	Domain Consistency	--	Topologic Consistency
Reliability	--	--	--	--	--

Positional Accuracy

Positional accuracy refers to the accuracy of the position of features in GIS. The positional accuracy of a spatial feature is largely dependent on the data type under consideration. Topographic features—e.g., a house corner, a center line of a street, etc.—are normally surveyed with high positional accuracy. Another type of feature, such as the boundary of a woodland or soil parcel, may have very low positional accuracy. Two error sources are involved in the latter types of line measurement: (a) line recognition error and (b) measuring error, where the former is more significant. Positional error of line in this case can be very significant.

Positional accuracy can be further defined as (a) absolute accuracy which refers to closeness of reported coordinate values to the reference values or the values accepted as true; and (b) relative accuracy which refers to the closeness of the relative positions of the measured features to their respective relative positions accepted as true. The coordinates of spatial data can also be classified as horizontal and vertical coordinates, we can thus define horizontal accuracy and vertical accuracy respectively.

Attribute Accuracy

Attribute accuracy is determined by the closeness of the attribute values to their true value or the value considered to be true. There are two types of attribute data: quantitative data, for example, the pH value assigned to a land parcel, and qualitative data, for example a land cover type assigned to a land parcel. Correspondingly, there are two types of attribute accuracy: quantitative accuracy and qualitative accuracy.

Classification of a remote sensing image is a commonly used method to obtain attribute data for a GIS. Classification accuracy refers to a comparison of the classes assigned to features to a universe of discourse, such as a reference dataset or that considered to be true data.

Attribute uncertainty of a GIS object is related to the data capturing process. For example, the procedure for understanding an object in the real world is shown in Figure 13-4. If an object in a GIS is obtained from a classified image, one of the major error sources is remote sensing classification error. In some cases, human interpretation is involved. The uncertainty in this case is dependent on human skill and experience of cognition, and varies from one person to another.

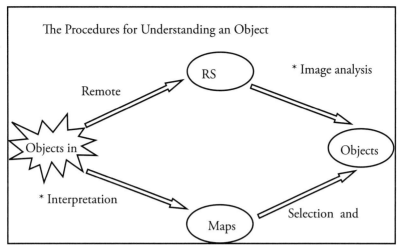

Figure 13-4 The procedure for understanding an object from real world to GIS (from Shi 1994).

In these procedures, the sources which introduce attribute uncertainty are as follows. Uncertainty of nature itself may cause difficulties in the definition of classes. Uncertainty may be introduced through cognition or automated classification. Uncertainty may also be introduced through posterior data manipulation.

Temporal Accuracy
Temporal accuracy refers to the accuracy of the temporal attributes and temporal relationships of features (ISO 19113 2005).

Logical Consistency
According to ISO 19113, logical consistency refers to the degree of adherence to logical rules of data structure, attribution and relationships. Here data structure can be conceptual, logical or physical data structures. Logical consistency can be further classified as
  • Conceptual consistency: adherence to rules of the conceptual schema,
  • Domain consistency: adherence of values to the value domains,
  • Format consistency: degree to which data is stored in accordance with the physical structure of the dataset, and
  • Topological consistency: correctness of the explicitly encoded topological characteristics of a dataset.

Completeness
Completeness refers to presence and absence of features, their attributes and the relationships of spatial data in comparing what is defined in the data model or what is in the real world. Completeness can be indicated by the following two error indicators: (a) error of commission —data present in a dataset that is not present in the data model or the real world; and (b) error of omission—data that is present in the data model or the real world is absent in the dataset.

13.3.3 An Example of Using Spatial Data Quality Elements

The following is an example of describing the quality of building objects in 1:1000 GIS data based on the elements of spatial data quality, and even their sub-elements (further detailed quality elements).

Table 13-2 An Example of Data Quality Elements and Sub-elements for Buildings.

Quality elements	Quality sub-elements	Description by examples
Completeness	Commission error	Buildings with area less than 4 m² are presented in Building Polygon layer of 1:1000 data set.
	Omission error	Buildings with area equal to or larger than 4 m² are absent from the Building Polygon layer.
Positional accuracy	Horizontal accuracy	RMSE of a building polygon based on a comparison of the horizontal coordinates of all the nodes of its footprints of a building in GIS with the corresponding reference values.
	Vertical accuracy	RMSE of a building polygon based on a comparison of the vertical coordinates of all the nodes of its footprints of a building in GIS with the corresponding reference values.
Attribute accuracy	Classification correctness	Correctness that a building or related features is correctly classified as one (or more) building-related features.
	Non-quantitative attribute correctness	The Name of a building polygon may be correct or wrong in a GIS.
	Quantitative attribute correctness	The value of the field "Building Top Level" of a Building Polygon may be correct or wrong.
Logical consistency	Conceptual consistency	A tower is described to be under its podium.
	Domain consistency	The classification of feature code for a building polygon is beyond any of the following given classes: BP, BAP, BUP, IBP, OSP, PWP, TSP.
	Format consistency	Building names in title case—Hong Kong Airport—are consistent, while a name such as "HONG KONG Airport" is not consistent in format.
	Topological consistency	When the outline of a building polygon is closed, the topology is consistent; when the outline is not closed, the topology is not consistent.

13.4 Sources of Uncertainty

The fundamental source of uncertainties of spatial data in GIS is the difference between the complex and continuous real world and its simplified and discrete representation in a computer environment. Furthermore, uncertainties can also be generated from error of measurements, and uncertainties can be propagated or even amplified in spatial analysis or data processing in GIS.

In this section, we classify the uncertainty sources of spatial data into the following four categories: (a) inherent uncertainties in the real world, (b) limitations and uncertainties in human cognition of the real world, (c) measurement errors and (d) uncertainties arising through spatial analyses and spatial data processing. Spatial data may be affected by more than one source out of these four. The uncertainty sources and their impacts are illustrated in Figure 13-5.

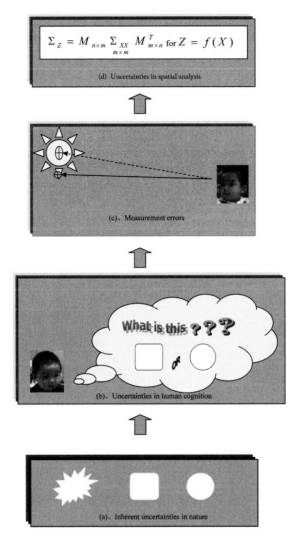

Figure 13-5 Main sources of uncertainty in spatial data (adapted from Shi 2005a).

13.4.1 Inherent Uncertainties in the Real World

The real world is complex, huge and nonlinear, and spatial data in GIS is used to describe the real world in a computer environment which is limited and discrete. Numerous spatial entities in the real world are uncertain themselves, in spatial, scale and other domains. In the spatial domain, for example, many spatial entities do not have deterministic boundaries or their boundaries cannot be identified easily. These boundaries are not sharp ones rather a zone with gradual class change (e.g. the soil boundary, or the forest or grassland boundary). Even if a spatial entity is recognized to be a class over its domain, its interior part may still be uncertain - irregular or comprised of heterogeneous classes.

In the scale domain, the cognition's depth is variant under scaling. The measured results for an object with the identical measurement technology of different resolutions may be different. The world observed with the micron-based or nano-based measurement technology is completely different from the macroscopic world. For example, the length of Hong Kong's coast line measured in meters can be very different from the length measured in millimeters.

13.4.2 Uncertainty Due to Human Cognition

Geospatial cognition is an essential step for spatial data capture and spatial analyses in GIS. Compared to the complexity of the real world, human cognition is very limited. Due to the limitations of our knowledge and methodology in comprehending the real world, we know very little about it.

Due to the complexity of spatial entities in the real world, only those important spatial features are measured and represented in our GIS. Attribute data of these spatial features are also measured and captured based on certain theories, technologies and methods. These theories, technologies and methods implicitly or explicitly include a certain level of abstraction and approximation to the real world attribute. Practically, this approximation is limited by human cognition; and what in spatial database is an abstracted description of the real world. Therefore, the spatial entity's computer representation loses some details of the real world object. This can be uncertainty in the spatial entity due to the spatial data sampling or classification. For instance, consider a grass and forest class obtained from a classification of a remotely sensed image: the boundary between grass and forest is unclear – a type of uncertainty from our cognition.

13.4.3 Uncertainty from Measurement

13.4.3.1 Spatial Data Capture Methods

To recognize the real world, a human being needs to capture data about the real world first. With the development of equipments, our capabilities of capturing data about real world have been improved. Historically, the invention of the telescope motivated the development of astronomy; the microscope led to advances in biology and medicine; the aeronautical engine contributed to the first industrial revolution; the electrometer led to the second industrial revolution mainly in Germany during the late 19th to early 20th century. Nowadays, advances in computing technology are leading to an information revolution in the human community. Scientific inventions directly cause a leap in human cognition of the real world and accordingly result in a new uncertainty problem.

The evolution of spatial data capture technologies has been experienced from ground survey through aerial photogrammetry to satellite mapping. The accuracy of spatial data capture is getting higher and higher.

Specifically, the spatial data capture methods for GIS include the following: data conversion, interpolation methods, map digitization (including vector digitization and raster scanning), photogrammetry (including analog, analytical and digital photogrammetry); field surveying methods (including utilizing total station, distance measuring equipment, and global positioning systems (GPSs)), laser scanners (including terrestrial and air-borne laser scanners), remote sensing technologies (including using multispectral images, high spatial resolution images and radar images).

The spatial data capture technologies are classified as direct methods and indirect methods (Shi 1994). The direct method involves fewer in-between processes and conversions; thus, the chance of introducing errors is smaller. In contrast, the indirect method requires many in-between processes and conversions and so the chance of introducing errors is relatively higher.

13.4.3.2 Analysis of Spatial Data Capture Methods

Accuracy of the spatial data is subject to the spatial data capture method applied. Generally speaking, a spatial data capture method includes the survey equipment applied, the survey scheme adopted, the post data processing method, and operation of the equipment by an operator. As a result, spatial data capture technology is still under restriction: the level of technology development for metrology is not such as to let us reach any desired level of accuracy.

Therefore, uncertainty is one feature of spatial data captured from any data collection method.

In the following, we briefly analyzes the main sources of error introduced in spatial data capture by means of map digitization, photogrammetry, laser scanning, total station, GPS and remote sensing.

Map Digitization
The accuracy of digitization is associated with many factors: the quality of the raw paper map (including inherent errors in the source map and errors due to paper deformation subject to a humid climate); the density and complexity of spatial features on the map; the skill level and operation method of the digitization operator; the instrument quality (such as the error of digitization instrument or scanner); the digitization software's ability in error checking; and the accuracy of control points used for coordinate transformation.

Aerial Photogrammetry
Accuracy of the photogrammetry product is affected by the following factors: accuracy and scale of the aerial photographs, accuracy of ground control points, accuracy and reliability of the algorithms used, such as image matching, image registration, automatic aerial triangulation, and internal and external orientation accuracy. Although digital photogrammetry involves many automatic processes, human operators are still involved in practical photogrammetric map production. Therefore, the reliability of the operators is also one factor affecting the accuracy of photogrammetry products. Presently, photogrammetry has reached the geometric accuracy of decimeter level; the horizontal accuracy is higher than the vertical accuracy. Some photogrammetry technologies can provide a vertical accuracy up to 0.3 m or better.

Laser Scanner
Accuracy of airborne laser scanning data is associated with several factors: accuracy of GPS and IMU (inertial measurement unit), and the power and accuracy of the laser emitter. In many cases, the point set obtained from the laser scanner cannot be used directly. It needs further processing to extract desired information. The performance of the extraction algorithms greatly affects quality of the information obtained.

The point cloud scanned with a ground-based laser scanner has a geometric accuracy ranging between millimeter and centimeter. The airborne laser scanner provides the point accuracy of decimeter in general, depending on flight height and quality of the laser scanner.

Total Station
A total station measures angles and distances optically and electronically. The accuracy of a total station's observations is then affected by integrated factors of the optical and electronic technologies. The integrated factors include the instrument function, the operator's skill level, the measurement technology and plan, the instrument and weather condition during measurement. Among them, the weather condition has a great impact on the accuracy of total station's observations: variations in humidity and temperature alter the index of refraction. Moreover, air pressure and the temperature influences the accuracy of distance measurement. Some weather conditions such as rain or snow can affect the efficiency of the electronic distance measurement.

At present, a high-performance total station has an angle-measuring accuracy of 0.5 second and a distance measuring accuracy of at least 1 mm + 1 ppm (part per million).

GPS
Many factors may affect GPS accuracy: the atmosphere (where the ionosphere and troposphere cause GPS signal delay); the multipath effect (a result of signals from a satellite reaching the antenna over more than one path due to different surface reflections); the number of GPS satellites in view; the geometric distribution of the satellites; the GPS and receiver clock errors; the GPS orbital errors; environmental disturbance (such as a high voltage power transformer station, or vegetation with elevation angle greater than 15 degrees

from the horizontal); and human factors such as the SA (selective availability) policy.

Positioning accuracy of GPS is closely related to positioning mode and measurement type used. With a standalone GPS positioning mode using pseudorange measurement, GPS positioning accuracy is around 10-20 m. Differential GPS (DGPS) mode utilizes a reference station to reduce the errors in GPS measurement; positioning accuracy on the order of a meter can be achieved. As the precision of carrier phase measurement can reach millimeter level, GPS positioning accuracy with carrier phase can reach centimeter level or higher. The GPS Real-Time Kinematic (RTK) technique is a special technique of GPS positioning using carrier phase, which can be used in real-time with positioning accuracy of centimeter level.

Remote Sensing Technology

Accuracy of spatial data captured from remote sensing is subject to the spatial resolutions of the satellite images used. The atmospheric conditions and the instability of remote sensors and platforms directly cause errors of remotely sensed images, and these errors are sometimes difficult to control. Although some errors can be eliminated by an appropriate method for the spectrum and geometric correction, errors still exist. Many factors affect the quality of remote sensing images. The quality mainly depends on raw image resolutions (including spatial resolution, spectral resolution, radiometric resolution and temporal resolution). As remote sensing technology developed, the spatial resolution has been continuously improved: from 80 m for Landsat MultiSpectral Scanner (MSS), 30 m for Landsat Thematic Mappter™, 10 m for SPOT, 5 m for SPOT-5, 1 m for Ikonos, 0.67 m for QuickBird and up to 0.50 m for WorldView-1 imagery. The spectral resolution has progressed from one band (called panchromatic) through the multispectral of several bands to the hyperspectral of over 100 bands. Similarly, radiometric resolution has improved from 6-bit to 12-bit, and temporal resolution has improved as well.

In addition, the quality of final mapping products from remote sensing images is affected by other factors, such as the accuracy of ground control points, the reliability of classification method, the quality of the mathematical model for geometric correction, terrain conditions and the resolution of the remote sensing image, etc.

13.4.4 Uncertainties in Spatial Analyses

Spatial analyses refer to the techniques for analyzing, simulating, forecasting and controlling geographical events and spatial processes based on spatial distribution. Spatial analysis or data processing includes, for example, spatial and non-spatial data updating, conversion, analysis, statistical processing, etc. Each of these spatial analysis or data processing steps may introduce, propagate or even amplify uncertainties in the spatial data.

Spatial analysis is a commonly used technique in GIS, however it is not error-free. The error generated from the spatial operation will significantly affects the quality of the resulting data set(s). This leads the research on error propagation of spatial analysis. The researchers aim to develop methods to quantify the errors propagated through the spatial operation or analysis.

Error propagation in a spatial analysis is defined as a process where error is propagated from the original data set(s) to the resulting data set(s) that is (are) generated from a spatial operation. The concept of error propagation in spatial analysis is illustrated in Figure 13-6.

Error propagation of a spatial analysis is dependant on a) the source data and error in source data, b) the spatial operation used, and c) the presentation method of the resulting data from the spatial operation.

There are two approaches for modeling error propagation: analytical approach (for example based on error propagation law in statistics) and simulation approach (such as Monte Carlo simulation), for either a raster-based or vector-based spatial analysis environment.

In general, the analytical method is adequate in assessing the error propagation when the analytical function of the output of a GIS operation and the input variables can be explicitly

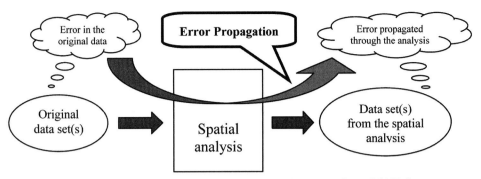

Figure 13-6 Error propagation through a spatial analysis (Adapted from Shi 2007).

defined, while the simulation method is more suitable for the cases where GIS operation is complex and difficult to be defined by a precise analytical function. The analytical and the simulation models are complementary to each other.

13.5 Modeling Uncertainties in Spatial Data and Analysis

In this section, we introduce the mathematical methods for quantifying uncertainties in spatial data and analysis. A framework for modeling uncertainties in spatial data and analysis is illustrated in Figure 13-7. Uncertainties include: (a) uncertainty modeling for spatial *data,* which covers error modeling for positional uncertainties in spatial data, modeling thematic uncertainties, the accuracy of digital elevation models, and techniques and methods to improve the quality of satellite images, and (b) uncertainty modeling for *spatial analyses,* quality control for spatial data, and dissemination of uncertainties. In the following, we will elaborate the main modeling methods covered by the framework.

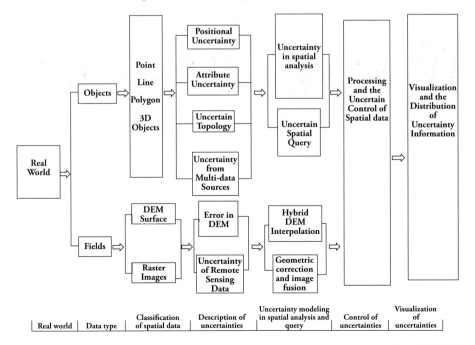

Figure 13-7 A framework for modeling uncertainties in spatial data and analysis (from Shi 2005b).

Manual of Geographic Information Systems

13.5.1 Modeling Positional Uncertainty

13.5.1.1 Error Modeling for Points

The *point* is one of the fundamental representations of geographical objects in GIS. It may represent a bus stop on a road map, a land parcel on a small-scale map, and others. The point feature is also a composite of a line or a polygon in GIS. Positional error of the point feature has thus an impact on positional error in other spatial features in GIS. Uncertainty of a point can be indicated by its variance-covariance matrix, and can be further described by either an error ellipse or a circular normal model.

Standard error ellipses
If the errors in the coordinates of a point are normally distributed, then their two-dimensional probability density function has a shape similar to a bell, and we can represent the precision of point coordinates by the standard ellipses (Mikhail and Ackermann 1976).

The error ellipse is a common way to describe the error at the point. Let Q_{21} denote a point with the x-directional variance of σ_x^2, the y-directional variance of σ_y, and the covariance of . Then, the correlation of this random point in the x- and y-directions is denoted by ρ_{xy}.

If the error at point Q_{21} follows a normal distribution, its probability density function will be given as

$$f(x,y) = \exp\{-[x-\mu_x)^2/\sigma_y^2 + (y-\mu_y)^2/\sigma_y^2 - 2\rho_{xy}(x-\mu_x)(y-\mu_y)/(\sigma_x\sigma_y)] \tag{13-1}$$
$$/2(1-\rho_{xy}^2)\}/\left(2\pi_x\sigma_y\sqrt{1-\rho_{xy}^2}\right)$$

The major and minor semi-axes (*E* and *F*) as well as the orientation θ of the corresponding error ellipse are

$$E = \sigma_x^2 + \sigma_y^2 + \left[(\sigma_x^2 + \sigma_y^2)^2 + 4\sigma_{xy}^2\right]$$

$$F = \frac{\sigma_x^2 + \sigma_y^2 - \left[(\sigma_x^2 - \sigma_y^2)^2 + 4\sigma_{xy}^2\right]}{2} \tag{13-2}$$

$$\tan(2\theta) = \frac{2\sigma_{xy}}{\sigma_x^2 - \sigma_y^2}$$

Then, the probability Pr that the point is inside the error ellipse is the volume of the two-dimensional error curved surface over the error ellipse (as shown in Figure 13-8):

$$Pr(x,y \subset \Omega) = \iint_\Omega f(x,y)\,dxdy = 4\int_0^r \int_0^{\frac{\pi}{2}} f(r,\theta)\,drd\theta = 0.393. \tag{13-3}$$

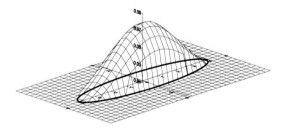

Figure 13-8 Two-dimensional error curved surface over the error ellipse.

The circular normal model

If the accuracies in the x and y directions are identical and independent, the ellipse becomes a circle, again centered on the point. We can also use this modified model if the precision in x and y directions does not differ much. The result is a circular normal model of point positional error. The positional error of a point is seen in terms of a bell-shaped surface centered over the point. The probability that any point is the true location is measured by the height of the surface of the bell over the point (Goodchild 1991). Based on the normal distribution model and the circular normal model, a number of statistical terms can be defined which can serve as error indicators.

Error indicators for one- and two-dimensional points

When the error is considered to be random, the frequency distribution of the errors is mathematically modeled by a normal distribution. The error of a one-dimensional coordinate, such as vertical coordinate (h), is evaluated to be a one-dimensional normal distribution. Two-dimensional positional data are expressed by a bivariate normal distribution. This distribution can be normalized to express equivalent circular error probabilities. Thus, horizontal positional accuracy can be expressed by a circular distribution.

Based on the linear distribution the following error indicators are defined for one-dimensional points (CCOSAM 1982): standard error, probable error, linear map accuracy, near certainty error and root-mean-square error.

The circular error distribution is a simplification of the elliptical error distribution for the convenience of probability calculation. For a two-dimensional point, the circular distribution is used to define the following error indicators: (CCOSAM 1982): circular standard error, circular probable error, circular map accuracy, standard circular near certainty error and root-mean-square error.

13.5.1.2 Error Modeling for Lines

According to the nature of uncertainties included in a line in GIS, Shi (1994) classified the geographic lines into two types: Type I line and Type II line. The fundamental difference between these two types of lines is this: there are "real points" in the real world which construct a Type I line, while there is no real world point for a Type II line, and thus we have to determine or interpret these points for a Type II line by ourselves. Therefore, compared with the Type I line, a Type II line has an extra error source: determining the points for the line in the real world.

Table 13-3 A Comparison of Type I and Type II Lines.

Type I	Type II
There are specific points in the real world to define the line feature.	There is NO specific point in the real world to define the line feature. People have to recognize and determine the points.
Uncertainty sources of the line in GIS: - measurement error - manipulation error	Uncertainty sources of the line in GIS: - line interpretation error - measurement error - manipulation error
Examples: - cadastral boundaries - building boundaries	Examples: - soil boundaries - forest boundaries

The uncertainty of a Type I line is mainly from measurement and manipulation errors. Thus, the uncertainty of Type I lines can be relatively easy to quantify. For example, we may assume as statistical properties that coordinate errors are independent and measurements follow a normal distribution. In addition to the measurement and manipulation errors that a type I line possesses, the dominant uncertainty source of a Type II line is line interpretation error. Because this error source involves human interpretation, it is relatively difficult to quantify. In the following, we focus our discussion on the error modeling for Type I lines.

Epsilon band error model of line

For modeling the error of a line feature, the epsilon band model was proposed by Perkal (1956, 1966); the "epsilon band" is a buffer of constant width (epsilon) on either side of a line. Suppose there exists some abstract, true version of a line; then the model proposes that the true line will lie within a band width of epsilon on either side of the measured line.

The epsilon band model is defined on certain assumptions: (1) each error effect relevant to a particular digital line in a GIS can be treated as a random variable, perturbing the true line to obtain the observed line, and (2) the processes of generating a digital line in a GIS can be treated as being independent.

Given a cartographic line as a straight line approximation, it might be supposed that the true line lies within a constant tolerance, epsilon, of the measured line. For a straight line segment, this locus is simple, consisting of the union of a rectangle parallel to the segment, twice the width of epsilon, with circles of radius epsilon centered at each end point. By the union of this simple figure, more complex lines can be handled (Chrisman 1982). The epsilon model is shown in Figure 13-9, where the dashed line represents the true location of the line to be measured; the central solid line is the measured line; the two parallel periphery lines describe the region of an epsilon band; and the true line lies within the epsilon band region of the digitized line. The epsilon band can also be described as the area occupied by rolling a ball along the line.

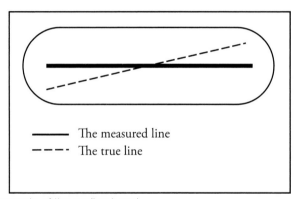

Figure 13-9 An example of the epsilon band.

N. R. Chrisman (1982) further developed the theoretical epsilon band model for practical application. He applied observed measurements for the epsilon band to model positional error in a GIRAS (Geographic Information Retrieval and Analysis System) digital file of the US Geological Surveys Land Use/ Land Cover data. The bandwidth was determined from a statistical function of those positional errors on the line accumulated from the first stage to the final stage of data capture. The error-band model can be applied during the execution of many spatial operations easily. However, it is sensitive to any outlier of a line because in this model, the true location of a line is supposed to be definitely located inside the error band.

An alternative approach to determine the bandwidth was suggested by Goodchild and Hunter (1997). The bandwidth was estimated by calculating the proportion of a measured location of the line lying within the epsilon band, such that the proportion must be equal to or larger than the predefined tolerance (say, 0.95). This modified epsilon band is a more appropriate error-band model than the previous version in terms of the outlier sensitivity.

Zhang and Tulip (1990) and Caspary and Scheuring (1993) further studied the error-band model for lines based on error propagation law. Dutton (1992) simulated the probability distribution of a line segment by using Monte Carlo simulation methods.

The confidence region error model of lines

With the assumption of uncorrelated positional errors for the endpoints of a line, Shi (1994) proposed the confidence region model for line. The model was first developed to assess the positional error for a line in two-dimensional spatial and then extended for n-dimensional lines (Shi 1998).

The confidence region J of a line segment is such that all points ζ_r with $r \in [0, 1]$ are contained in J with a probability that is larger than a prescribed confidence level y, i.e., $P(\zeta_r \in J_r$ for all $r \in [0, 1]) > y$. Here, J is the union of sets J_r for all $r \in [0, 1]$. One region J_r is a set of points $(x, y)^T$ with x satisfying $X_r - c \leq x \leq X_r + c$ and y satisfying $Y_r - d \leq y \leq Y_r + d$

where $c = k^{1/2}[((1-r)^2 + r^2)\sigma_{11}]^{1/2}$ and $d = k^{1/2}[((1-r)^2 + r^2)\sigma_{22}]^{1/2}$

The parameter k is dependent on the selected confidence level γ and can be looked up in a chi-square distribution table, $k = \varkappa_{22};(1+\gamma)/2$, e.g., if $y = 0.90$, $(1 + y)/2 = 0.95$, $k = 5.99$.

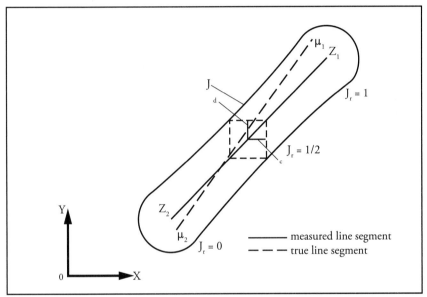

Figure 13-10 The confidence region of a line segment.

The shape of the confidence region is different from the epsilon model that assumes the identical positional error at each line's point. In the confidence region model, positional errors for all points on the line can be different.

Furthermore, the error model for a line segment with the conditions that the two endpoints have correlated positional error was developed (Shi and Liu 2000). It gives the analytical relationships between the band shape and size with the error at the two endpoints and relationships between them.

13.5.1.3 Error Models for Curves

A method to describe positional error on a curve feature is given by Shi (2009). The error in the curve feature is proposed to be assessed with two error models: the ε_o error model and the ε_m error model. The ε_o error model is used to measure the error in any point on the curve in the direction of the normal to the curve at that point. The ε_m error model can describe the maximum error from the point to the curve. The third-order basic spline curve, for example, is taken to illustrate how the proposed approach could be implemented in assessing positional error on this spline curve.

For modeling positional uncertainty of an irregular curve, the model Tepdem (theory of the equivalent probability density error model) was applied (Shi 2009). The geometric shape of the Tepdem model is determined by the scale coefficient for the standard error ellipse at any point on the fitting curve: $\lambda_{(i,k)}$ where natural number $i \leq n$ and $k = 0, 1, \ldots, m(i)$. The size factor $\lambda_r \in [0, +\infty)$ is determined by the scale coefficient for the standard error ellipse at the feature point $P(x_r, y_r)$, $0 \leq r \leq n$ with the least amplitude of vibration of the 2D probability density distribution function.

13.5.1.4 Error Modeling for Polygon Objects

Positional error for a polygon (also known as an areal feature in GIS) is caused by positional error in its boundary points. The positional error for the polygon can be modeled based on the following two approaches: the first one is based on the error propagation from the component points of the polygon and the second one is based on the error band models for the polygon. Accuracy of a polygon can be indicated by its area, perimeter, gravity, etc. The following is an example of polygon accuracy indicated by its area accuracy.

Suppose a polygon A is of n vertices $P_i = (x_i, y_i)$ and n edges $P_1P_2, P_2P_3, \ldots, P_{n-1}P_n$, and P_nP_1. The covariance matrices of the individual vertices are expressed as

$$\Sigma_{P(i)P(i)} = \begin{bmatrix} \sigma^2_{x(i)} & \sigma_{x(i)y(i)} \\ \sigma_{x(i)y(i)} & \sigma^2_{y(i)} \end{bmatrix}. \tag{13-4}$$

The area S of the polygon is computed from

$$S = \frac{1}{2} \sum_{i=1}^{n} [x_i(y_{i+1} - y_{i-1})] = \frac{1}{2} \sum_{i=1}^{n} [x_i \Delta y_{i-1,\,i+1}]. \tag{13-5}$$

The differential of the area is given as

$$dS = \frac{1}{2} \sum_{i=1}^{n} [\Delta y_{i-1,\,i+1} dx_i + \Delta x_{i-1,\,i+1} dy_i]. \tag{13-6}$$

This differential is then used to compute the variance of the area of the polygon:

$$\sigma^2_{S1} = \frac{1}{4} \sum_{i=1}^{n} [\Delta y^2_{i-1,\,i+2} \sigma^2_{x(i)} + \Delta x^2_{i-1,\,i+1} \sigma^2_{y(i)} + 2\Delta x_{i-1,\,i+1} \Delta y_{i-1,\,i+1} \sigma_{x(i)y(i)}]. \tag{13-7}$$

When $\sigma^2_{x(i)} = \sigma^2_{y(i)} = \sigma^2_0$ and $\sigma_{x(i)y(i)} = 0$, the variance is simplified to

$$\sigma^2_{S1} = \frac{1}{4} \sum_{i=1}^{n} [\Delta y^2_{i-1,\,i+2} + \Delta x^2_{i-1,\,i+1}] \sigma^2_0 = \frac{1}{4} \sum_{i=1}^{n} l^2_{i-1,\,i+1} \sigma^2_0 \tag{13-8}$$

where $l_{i-1,\,i+1}$ is the distance between vertices P_{i-1} and P_{i+1}. For the n-sided regular polygon, $l_{i-1,\,i+1}$ is identical for all $i = 1, 2, \ldots, n$. Its area has the standard deviation of

$$\sigma_{S1} = \sqrt{n} \sin\left[\frac{\pi}{2} - \frac{\pi}{n}\right] \cdot l \cdot \sigma_0. \tag{13-9}$$

13.5.2 Modeling Attribute Uncertainty

In this section, we will discuss error modeling for two types of attribute data: categorical (or nominal) attribute data and continuous attribute data.

13.5.2.1 Error Modeling for Categorical Attribute Data

One of the methods to obtain an attribute value for GIS is through classification. This is particularly true for remote-sensing-image-based classification. A classification is a procedure of deciding whether an object (pixel) o belongs to class *A*. Therefore the uncertainty of a classification may include three aspects: (a) defining a class, (b) measuring an object, and (c) making the decision that object o belongs to class *A*.

Uncertainty may exist in class definition. Class *A* may not be well-defined and one may not be able to identify the class clearly as *A* or another class, *B*. Soil classifications for example are typically fuzzy. Measuring an object *o* is the second source of uncertainty, discussed extensively earlier in this chapter.

In making the decision $o \in A$, our confidence may vary within the area of a polygon. For example, at the center of a polygon, we can have high confidence that the class is *A*, but at the boundary of the polygon we may not be so sure. A probability vector model was defined for describing the uncertainty of data from a maximum likelihood classification. The concept of a probability vector can give us a description of the spatial distribution of uncertainty on a per-pixel basis. The uncertainty can, for example, be derived during a maximum likelihood classification. Based on the probability vector model, four parameters (Shi 1994) can be further defined, and used as indicators to describe the spatial distribution of attribute uncertainties in a classified image.

Sampling-based test method

Basically there are three statistical measures reported in the literature dealing with categorical attribute uncertainty which are based on an experimental approach: (a) percentage correctly measured for the whole set (Rosenfield 1986); (b) percentage range correctly measured per class or for the whole set at a given confidence level (Aronoff 1985; Hord and Brooner 1976); and (c) percentage correctly measured per class (Greenland et al. 1985; Rosenfield and Fitzpatrick-Lins 1986). The attribute data can be an area-class map, a classified remotely sensed image, etc.

These methods are suitable for various kinds of discontinuous attribute data assessment. In remote sensing classification assessment, the discontinuous attribute data accuracy can be assessed by evaluating a sampling of the classification result. The classes determined by classification are defined as reference data. By comparing the measured data with the reference data, we can form an error matrix. This matrix describes the quality of the classification as a whole, and that of individual classes. Based on this matrix, detailed descriptions such as error of omission, error of commission and Kappa Coefficient of classification accuracy can be derived. For detailed description on these, please refer to Congalton and Green (1999) and Congalton (2008).

13.5.2.2 Error Modeling for Continuous Attribute Data

Due to the fact that both positional data and continuous attribute data relate to continuous random variables, the methods of modeling error propagation for continuous attribute data are similar to the methods used for positional data. That is, error propagation law or the Monte Carlo simulation can also be applied to the analysis of error propagation of continuous attribute data. The only difference is that we need to reformulate the problem: the random variables need to be changed from coordinate measurements to attribute measurement variables within a spatial location. The transformation is as follows.

$$y_k(s) = g_k(x_1(s), x_2(s), \ldots, x_n(s)) \qquad (13\text{-}10)$$

where $x_1(s)$, $x_2(s)$, …,$x_n(s)$ are random variables of attribute values with known stochastic characteristics at location s. $y_k(s)$ is a functional variable at location s. $y_k(s)$ is related to $x_1(s)$, $x_2(s)$, …,$x_n(s)$ through function g_k. The problem is now to derive the stochastic characteristics of $y_k(s)$ based on the known statistical characteristics of $x_1(s)$, $x_2(s)$, …, $x_n(s)$. Then we can derive the behavior of the random variable $y_k(s)$. Based on this transformation, we can apply the error propagation law for the error modeling.

13.5.3 Modeling Uncertainty in Spatial Analysis

13.5.3.1 Modeling Uncertainty in Overlay Analysis

In GIS-based spatial analysis, there are two types overlay analyses: (a) raster-based overlay and (b) vector-based overlay analysis. Heuvelink (1998) conducted an intensive study on error propagation in environmental modeling with GIS, which was based on raster GIS. For vector-based overlay spatial analysis, Shi et al. (2004) introduced an analytical error model derived based on the error propagation law, and a simulation error model. For each of these two error models, the positional error in the original or derived polygons is proposed to be assessed by three error measures: (a) the variance-covariance matrices of the polygon vertices, (b) the radial error interval for all vertices of the original or derived polygons, and (c) the variance of the perimeter and that of area of the original or derived polygons.

13.5.3.2 Modeling Uncertainties in Buffer Spatial Analysis

Zhang et al. (1998) derived an uncertainty propagation model of a buffer in a vector environment based on the epsilon band (ε-band) and the E-band models. For each error band, an absolute uncertainty and a relative uncertainty for the buffer around a spatial feature are derived based on positional uncertainty at all vertices of the spatial feature.

A further study on modeling error propagation in a buffer spatial analysis for a vector-based GIS was given by Shi et al. (2003). The buffer analysis error is defined as the difference between the expected and measured locations of a buffer. Four error indicators and their corresponding mathematical models, in multiple integrals, were proposed for describing the propagated error in a buffer spatial analysis. These included the error of commission, error of omission, discrepant area and normalized discrepant area.

13.5.3.3 Modeling Uncertainties in Line Simplification

Positional uncertainty of a simplified line is caused by (a) positional uncertainty in an initial line and (b) uncertainty in deviation of the simplified line from the initial line. The initial line may contain measurement uncertainties, model uncertainties, and other uncertainties (Burrough and McDonnell 1998; Cheung and Shi 2004). Accuracy of two nodes of the straight line is determined by accuracy of the field surveying (between a few centimeters and several decimeters) and that of the digitizing process (varying between 0.1 and 0.3 mm) (Caspary and Scheuring 1993).

Tveite and Langaas (1999) proposed to use the buffer-based uncertainty model to assess the uncertainty in line simplification. One of the quantitative measures of line similarity is areal offset defined as the area of offset between the initial line and the corresponding simplified line, also known as distortion polygon (White 1985; McMaster 1987; Buffenfield 1991). Dividing the areal offset by the length of the simplified line is a uniform distance distortion (Veregin 2000). Both the areal offset and the uniform distance distortion provide a single quantitative index of the deviation of the simplified line from the initial line. The displacement vector is another measure of line similarity, which is drawn from the simplified line to the initial line in the perpendicular direction to the simplified line (McMaster 1986).

White (1985) proposed common-point comparison, in which the number of points common to the initial line and simplified one is computed as a measure of the deviation of the simplified line from the initial line. Veregin (2000) also proposed a similar measure – critical distance. The critical distance was defined as the proportion of points on the initial line distant not greater than a specified weed threshold from the simplified line.

Cheung and Shi (2004) modeled the overall uncertainty in the simplified line by the overall processing uncertainty that integrates both the propagated and the modeling uncertainties in the line simplification process. Three uncertainty indices and corresponding mathematical solutions were proposed for each of the uncertainty types by measuring its mean, median and maximum values. For modeling uncertainty the overall processing uncertainty, the mean deviation, the median deviation and the maximum deviation were proposed.

13.6 Concluding Remarks

This chapter has addressed the issue of spatial data quality and uncertainty, where uncertainty in the spatial data and spatial analysis is the leading quality problem.

The concept of uncertainty and errors, particularly associated with spatial data, were first given. The meaning of uncertainty is wide and includes imprecision, ambiguity and vagueness. Each of these aspects of uncertainty can be modeled based on the corresponding mathematical theories. To fully understand spatial data quality, we started with defining the dimensions of spatial data (in Section 13.3.1), which include space, time, attribute, scale and relation where the latter two dimensions are normally not mentioned in the literature. Quality of a spatial data set is specifically determined by the quality of the spatial data set in each of these five dimensions. The quality components were also defined and include accuracy, precision, resolution, completeness, logical consistency and reliability. With the five dimensions and the six quality components, we can form a total of 30 elements of spatial data quality, as indicated in Section 13.3.2.2. However, we can only clearly define a small portion of the spatial data quality elements at this stage, for example 6 spatial data quality elements are defined in the ISO 19113 standard: positional accuracy, thematic (attribute) accuracy, temporal accuracy, completeness and logical consistency. There is room to further explore new elements for spatial data quality, such as on reliability and in the scale dimension of spatial data.

Uncertainty in spatial data may possibly be generated from the following four sources: (a) inherent uncertainties in the real world, (b) uncertainty in the process of human cognition, (c) error generated during measurement and (d) uncertainty propagated or amplified in spatial analyses and data processing in GIS, as described in Section 13.4. The uncertainty source (a) is out of our control, and our understanding in this regard is still very limited. Our philosophy of dealing with this type of uncertainty perhaps should have two parts: firstly, try to improve our understanding of the uncertainty in the real world gradually and, secondly, to learn to live with uncertainty.

Compared to uncertainty in the real world (a), we have made a relatively better progress in understanding uncertainty in human cognition (b). However, we are still far from a full understanding of this uncertainty. Much progress is expected in this area in the future, and GIS specialists with geography background and experts with domain knowledge can potentially make significant contributions to this area.

Among the four sources of uncertainties, the best progress has been made in modeling error from measurement – uncertainty type (c). This is basically due to the fact that in many cases, the uncertainty type (c) can be described mathematically. Furthermore the knowledge in surveying error processing is relatively mature, and can be adapted for dealing with uncertainty type (c). Sections 13.5.1 and 13.5.2 give a very brief summary indicating the knowledge we have gained so far in this area.

The research on modeling uncertainty propagation or amplification in spatial analyses and spatial data processing in GIS – dealing with type (d) uncertainty – needs to be further investigated. Section 13.5.3 highlighted the progress made in this area. It is one of the major future development directions for research of spatial data quality and uncertainty – to control quality of spatial data for GIS products and to model uncertainty propagation in spatial analyses and data processing.

References

Aronoff, S. 1985. The minimum accuracy value as an index of classification accuracy. *Photogrammetric Engineering and Remote Sensing* 51 (1):1687–1694.

Brassel, K., F. Bucher, E.-M. Stephan and A. Vckovski. 1995. Completeness. In *Elements of Spatial Data Quality,* edited by S. C. Guptill and J. L. Morrison, 81–108. Pergamon Press.

Burrough, P. A. and R. A. McDonnell. 1998. *Principles of Geographical Information Systems for Land Resources Assessment.* Oxford, UK: Clarendon Press.

Buttenfield, B. P. 1991. A rule for describing line feature geometry. In *Map Generalization: Making Rules for Knowledge Representation,* edited by R. B. McMaster and B. P. Butterfield, 150-171. Harlow, UK: Longman.

Caspary, W. and R. Scheuring. 1993. Positional accuracy in spatial databases. *Computer, Environment and Urban Systems* 17:103–10.

CCOSAM (Canadian Council on Surveying and Mapping). 1982. National Standards for the Exchange of Digital Topographic Data, II - Standards for the Quality Evaluation of Digital Topographic Data.

Cheung, C.K and W. Z. Shi. 2004. Estimation of the positional uncertainty in line simplification in GIS. *The Cartography Journal* 41 (1):37–45.

Chrisman, N. R. 1982. A theory of cartographic error and its measurement in digital data base. Pages 159-168 in *Proceedings of International Symposium on Computer-Assisted Cartography (Auto-Carto 5),* Crystal City Virginia, August 22-28, 1982.

Congalton, R. and K. Green. 1999. *Assessing the Accuracy of Remotely Sensed Data: Principles and Practices.* Boca Raton, Fla.: CRC/Lewis Press. 137 pages.

Congalton, R. 2008. Accuracy assessment of spatial data sets. In *Manual of Geographic Information Systems,* edited by M. Madden, Chapter 14 in this volume. Bethesda, Md.: ASPRS.

Dutton, G. 1992. Handling Positional Uncertainty in Spatial Databases. Pages 460–469 in *Proceedings of 5th International Symposium on Spatial Data Handling,* Charleston, 1992.

Fisher, P. F. 1999. Models of uncertainty in spatial data. In *Geographic Information Systems, Vol.1,* edited by P. A. Longley, M.F. Goodchild, D.J. Maguire and D.W. Rhind, 191–205. New York: John Wiley & Sons, Inc.

Goodchild, M. F. 1991. Issues of quality and uncertainty. In *Advances In Cartography,* edited by J. C. Muller, 113–139. New York: Elsevier.

Goodchild, M. F. and G. J. Hunter. 1997. A simple positional accuracy measure for linear features. *International Journal of Geographical Information Science.* 11 (3):299–306.

Greenland, A., R. M. Socher and M. R. Thompson. 1985. Statistical Evaluation of Accuracy for Digital Cartographic Database. Proceedings of International Symposium on Computer-Assisted Cartography (Auto-Carto 7), ASP-ACSM, Washington DC, 1985.

Hartley, R.V.L. 1928. Transmission of information. *The Bell Systems Technical J.* 7:535–563. <http://www.dotrose.com/etext/90_Miscellaneous/transmission_of_information_1928b.pdf> Accessed 4 October 2008.

Heuvelink, G. B. M. 1998. *Error Propagation in Environmental Modeling with GIS.* New York: Taylor & Francis. 127 pages.

Hord, R. M. and W. Brooner. 1976. Land use map accuracy criteria. *Photogrammetric Engineering and Remote Sensing* 42 (5):671–677

ISO 19113. 2005. Geographic Information – Quality Principles, BS EN ISO 19113:2005.

ISO 19114. 2005. Geographic Information – Quality Evaluation Procedures, BS EN ISO 19114:2005.

ISO 3534-1. 1993. Statistics - Vocabulary and Symbols - Part 1: Probability and general statistical terms.

Klir, G. J. and T. A. Folger. 1988. *Fuzzy Sets, Uncertainty, and Information.* New York: Prentice-Hall. 355 pages.

McMaster, R. B. 1986. A statistical analysis of mathematical measures of liner simplification. *The American Cartographer* 13 (2):103–116.

McMaster, R. B. 1987. Automated line generalization. *Cartographica* 27 (2):74–111.

Mikhail, E. M. and F. Ackermann. 1976. *Observations and Least Squares.* New York: IEP–Dun-Donnelley. 497 pages.

Morrison, J. 1988. The proposed standard for digital cartographic data. *The American Cartographer* 15 (1):129–135.

Perkal, J. 1956. On epsilon length. *Bulletin de l'Academie Polonaise des Sciences.* 4:399–403.

Perkal, J. 1966. On the Length of Empirical Curves: Discussion Paper 10, Ann Arbor MI, Michigan Inter-University Community of Mathematical Cartographers.

Rosenfield, G. H. and K. FitzPatrick-Lins. 1986. A coefficient of agreement as a measure of thematic classification accuracy. *Photogrammetric Engineering and Remote Sensing* 52 (2):223–227.

Rosenfield, G. H. 1986. Analysis of thematic map classification error matrices. *Photogrammetric Engineering and Remote Sensing* 52 (5):681–686.

Shannon, C. E. 1948. The mathematical theory of communication. *The Bell System Technical Journal*, 27:379–423,623–656.

Shi, W. Z. 1994. *Modeling Positional and Thematic Uncertainties in Integration of Remote Sensing and Geographic Information Systems.* Enschede, The Netherlands: ITC Publication No.22. ISBN 90 6164 099 7. 147 pages.

———. 1998. A generic statistical approach for modeling errors of geometric features in GIS. *International Journal of Geographical Information Science* 12 (2):131–143.

———. 2005a. *Principle of Modeling Uncertainties in Spatial Data and Analysis.* Beijing: Science Press. ISBN: 7-03-015602-1. 408 pages (in Chinese).

———. 2005b. Towards uncertainty-based geographic information science (part A) - modeling uncertainty in spatial data. Pages 14–26 in *Proceedings of 4th International Symposium on Spatial Data Quality*, 25th-26th August 2006, Beijing, China. The Hong Kong Polytechnic University Press.

———. 2007. Error propagation. In *Encyclopedia of Geographic Information Science,* edited by K. K. Kemp. Thousand Oaks, Calif.: Sage Publications, Inc. 584 pp.

———. 2009. *Principle of Modeling Uncertainty in Spatial Data and Analysis.* London: CRC Press.

Shi, W. Z. and W. B. Liu. 2000. A stochastic process-based model for positional error of line segments in GIS. *International Journal of Geographical Information Science* 14 (1):51–66.

Shi, W. Z. and K. F. Liu. 2004. Modeling fuzzy topological relations between uncertain objects in GIS. *Photogrammetric Engineering and Remote Sensing* 70(8):921–929.

Shi, W. Z., C. K. Cheung and C. Q. Zhu. 2003. Modeling error propagation in vector-based buffer analysis. *International Journal of Geographic Information Science* 17 (3):251–271.

Shi, W. Z., C. K. Cheung and X. H. Tong. 2004. Modeling error propagation in vector-based overlay spatial analysis. ISPRS *Journal of Photogrammetry and Remote Sensing* 59 (1-2):47–59.

Tveite, H. and S. Langaas. 1999. An accuracy assessment method for geographical line data sets based on buffering. *International Journal of Geographical Information Science* 13 (1):27–47.

Veregin, H. 2000. Quantifying positional error induced by line simplification. *International Journal of Geographical Information Science* 14 (2):113–130.

White, E. R. 1985. Assessment of line-generalization algorithms using characteristics points. *The American Cartographer* 12 (1):17–27.

Zhang, G. Y. and J. Tulip. 1990. An Algorithm for the Avoidance of Sliver Polygons and Clusters of Points in Spatial Overlay. *Proceedings of 4th International Symposium on Spatial Data Handling*, Zurich, Switzerland. pp. 141–150.

Zhang, B., L. Zhu and G. Zhu. 1998. The uncertainty propagation model of vector data on buffer operation in GIS. *ACTA Geodaetica et Cartographic Sinica* 27:259–266 (in Chinese).

CHAPTER 14
Accuracy Assessment of Spatial Data Sets
Russell G. Congalton

14.1 Introduction

The ability to validate or assess the accuracy of spatial data is vital to the further development and acceptance of spatial data analysis by an ever-growing community of users. There are a number of reasons why this assessment is so important, including:

- The need to know how well you are doing and to learn from your mistakes (e.g., what is the accuracy of that land cover map and are there systematic errors in the map that can be corrected?)
- The ability to quantitatively compare methods or techniques (e.g., is the accuracy of the digital elevation model (DEM) generated from technique A significantly better than the DEM generated from technique B?)
- The ability to use the information resulting from your spatial data analysis in some decision-making process (e.g., if I use this zoning map in my decision to develop a new part of the city, but the map is of low accuracy, what impact will that have on the decisions that I make?)

There are many examples in the literature as well as an overwhelming selection of anecdotal evidence to demonstrate the need for assessing the accuracy of spatial data. For instance, many different groups have mapped or quantified the amount of tropical deforestation in the Amazon Basin of South America (e.g., Skole and Tucker 1993). Estimates have differed by almost an order of magnitude. Which estimate is correct? Without a valid accuracy assessment it is impossible to know. Several federal, state, and local agencies have created wetland maps for Wicomico County on the Eastern Shore of Maryland. Techniques used to make these maps included satellite imagery, aerial photography (at various scales and film types), and ground sampling, all with varying classification schemes and wetlands definitions. Comparing the various maps yielded very little agreement about where wetlands actually existed. Without a valid accuracy assessment it is impossible to know which of these maps to use for any decision-making or policy analysis process.

It is no longer always sufficient to conduct a spatial data analysis and then simply print out the final map. Instead, in many situations, it is necessary to take some steps towards assessing the accuracy or validity of that spatial data analysis or that map. There are a number of ways to assess the accuracy of the spatial data. These methods are as follows:

- Qualitative assessment
  - Metadata review
  - Logical consistency check
  - Completeness verification
  - Visual inspection/review
- Quantitative assessment
  - Positional accuracy
  - Thematic accuracy

It is the goal of this chapter to review these methods of assessing the accuracy of spatial data and to encourage their widespread use.

14.2 Accuracy Assessment Methods

14.2.1 Qualitative Assessment

A qualitative assessment of your spatial data analysis or map is the first step in the assessment process. There are a number of valuable assessment procedures including: (1) reviewing the metadata (i.e., evaluating the lineage), (2) checking if the results make sense (i.e., logical consistency), (3) verifying completeness, and finally (4) performing a visual inspection/review. All of these methods can provide some insight into the accuracy of the spatial data/map.

14.2.1.1 Reviewing the Metadata

Most GIS projects do not collect all the spatial data used in the project. Instead, many layers are obtained from a variety of sources including federal, state, and local government agencies, non-governmental organizations (NGOs), and private groups. Metadata or data about the spatial data are included with the spatial data to allow the user to know the lineage of how that data layer was created. Chapter 10 in this manual is devoted to the discussion of metadata (see Moeller, Chapter 10, this volume).

Ideally, metadata should be reviewed when deciding to add that data layer to a GIS project. However, metadata can also be reviewed upon completion of the spatial analysis/map to qualitatively evaluate the potential error contributed by each data layer in the analysis. Upon careful consideration and review of the metadata and the specific analysis techniques used in a GIS project, it may be possible to compile some type of error budget for the project. An error budget is a method for identifying those parts of the spatial analysis project most prone to error, which can be most easily fixed or controlled, and which are most difficult or costly (Lunetta et al. 1991).

Table 14-1 presents the results of performing an example error budget analysis for a GIS project. The table is generated one column at a time beginning with a listing of the possible sources of error for the project. In order to make effective use of any GIS, it is important to understand the errors associated with the spatial information (Goodchild and Gopal 1989). According to Burrough and McDonnell (1998), errors associated with spatial information can be divided into three groups as follows: (1) user error, (2) measurement/data error, and (3) processing error. User errors are those errors that are probably most obvious and are more directly in the control of the user. Measurement/data errors deal with the variability in the spatial information and the corresponding accuracy with which it was acquired. Finally, processing errors involve errors inherent in the techniques used to input, access and manipulate the spatial information.

First, the various components that comprise the total error are listed in Table 14-1. Second, each component is qualitatively assessed to determine its relative contribution to the overall error. Third, our ability to deal with this error is evaluated. It should be noted that some errors may be very large but are easy to correct while others may be rather small and quite costly to correct. In this example, an error index is created directly by multiplying the error contribution potential by the error control difficulty (see Table 14-1). Combining these two factors allows one to establish priorities in dealing with error and understand the accuracy of the final spatial data analysis/map.

14.2.1.2 Checking for Logical Consistency

Logical consistency is really a check to see if the spatial data/map makes sense. Typically, there will be certain locations within the spatial data of which you as the analyst have intimate knowledge. Therefore, you will have certain expectations for the results of the analysis in those areas or perhaps an educated guess at what the map should look like. Do the results of the analysis make sense in those areas that you know? For example, are there

Table 14-1 Example of an error budget analysis to evaluate and control error.

Source of Error	Error Contribution Potential	Error Control Difficulty	Error Index	Error Priority
User Error				
Age of Data	3	2	6	7
Scale of Data	3	1	3	3
Coverage/Extent/Footprint	1	2	2	1
Surrogate/Derived Layer	2	4	8	8
Measurement/data Error				
Instrument Error	2	5	10	11
Field/Collection Error	1	3	3	3
Natural Variation	1	5	5	6
Processing Error				
Precision	1	3	3	3
Interpolation	2	4	8	8
Generalization	3	3	9	10
Data Conversion	3	4	12	12
Digitization	1	2	2	1

Error Contribution Potential: relative potential for this source as contributing factor to the total error (1 = low, 2 = medium, and 3 = high).
Error Control Difficulty: given the current knowledge about this source, how difficult is controlling the error contribution (1 = not very difficult to 5 = very difficult).
Error Index: an index that represents the combination of error potential and error difficulty.
Error Priority: order in which methods should be implemented to understand, control, reduce, and/or report the error due to this source based on the error index.

buildings located in the middle of a lake? Have you selected an area for timber harvesting in a densely wooded residential neighborhood? Is a wetland area labeled as suitable for future development?

Testing for logical consistency usually requires comparing the results to reasonable expectations. If, during this checking, the results of the spatial analysis or map fail to make sense, then it is better to go back and begin the analysis over instead of continuing with more quantitative accuracy assessment.

14.2.1.3 Verifying Completeness

Completeness is similar to logical consistency in that it involves some review and comparison of the spatial data or map to determine. The spatial data or map can be said to be complete if it contains the information it was supposed to represent. If none of the information that the analysis was supposed to represent appears, then the problem is likely more one of logical consistency. However, if some of the desired features are present, but not all, then the issue is completeness. For example, a spatial data layer or map may be incomplete due to some generalization imposed by the analysis. If a minimum mapping unit is imposed on some spatial data as a result of the scale of the data collection, this layer may be missing some key features that the analyst expected to see in the results. In this case, the analysis may not be flawed, but the data are left incomplete because of external limitations.

Manual of Geographic Information Systems

14.2.1.4 Visual Inspection/Review

Actually looking at the spatial data/map is a key component for assessing the accuracy. This visual inspection might include all or some of the methods discussed above, including reviewing the metadata, checking the logical consistency, and verifying the completeness. Visual inspection is a necessary first step, but it is not sufficient. In other words, it is very important to perform a visual assessment of your map and to be convinced that it looks right. After all, it would not make sense to further assess a map that does not even look right. However, it is not appropriate to conclude your assessment with only a visual inspection. It is simply not sufficient. Many maps that "looked good" were later found to have serious errors as a result of further quantitative accuracy assessment. In the example in Figure 14-1, it is important that the water in the image be labeled water in the land cover map. It is also possible to compare other map classes to determine if the visual inspection makes sense. If the map fails the visual inspection then the analysis should be redone before any further accuracy assessment is undertaken.

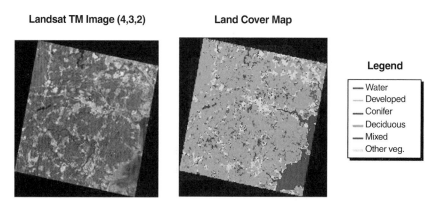

Figure 14-1 Example of visual inspection of a map as an important, but not the only, step in accuracy assessment/validation. See included DVD for color version.

14.2.2 Quantitative Assessment

It is a well-known but often forgotten fact that the accuracy of spatial data is a function of both positional accuracy and thematic accuracy. For example, it is possible to be in the correct location and mislabel (incorrectly measure or classify) the attribute. It is also possible to correctly label the attribute but be in the wrong location. In either case, error is introduced into our spatial data or map. These two factors are not independent of each other and great care needs to be taken to not only assess each of the factors, but also to control them to minimize the errors.

14.2.2.1 Positional Accuracy

Traditionally, when most of us think about the accuracy of a map or the accuracy of spatial data, we are thinking about positional accuracy. Positional accuracy is a measure of how far a certain feature on a map is from the true ground location (Bolstad 2002). Figure 14-2 shows an example of positional accuracy. The road layer is displayed over the top of an orthorectified digital image. The intersections can clearly be seen on the image and yet it is obvious that the road layer does not exactly align with the imagery (i.e., there is positional error).

Positional accuracy is most often computed when registering a spatial data layer to a map projection/coordinate system. In this situation, a series of corresponding points are chosen on the spatial data layer to be registered and compared to a reference data layer, map, or image. Once the registration is completed, a measure of the fit of the spatial data layer to the

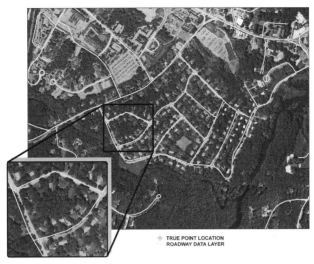

Figure 14-2 Example showing the roads layer not accurately positioned over the orthorectified digital imagery. See included DVD for color version.

reference layer is computed. This value is called the root mean square error (RMSE). The root mean square error is the standard value for reporting positional error and is a good indicator of how well the two data layers register to each other.

However, there is a very important issue here. The calculation of RMSE during the registration process is not an independent assessment of positional accuracy for that spatial data layer or map. Instead, it is simply a measure of how well the chosen points used in the registration fit to the reference layer. In other words, this RMSE represents the best-case scenario of positional accuracy for this layer. What is required is to perform an independent assessment based on the collection of a separate set of points not used in the registration process to obtain a valid positional accuracy.

Therefore, a standard method for measuring positional accuracy incorporating RMSE has been developed by the Federal Geographic Data Committee (FGDC). This standard is called the National Standard for Spatial Data Accuracy (NSSDA) and provides guidelines for the number and distribution of sample points along with an explanation of the computation of RMSE and associated values for the proper determination of positional accuracy (FGDC 2005). These procedures outlined by the FGDC should be followed any time a positional accuracy assessment is conducted.

14.2.2.2 Thematic Accuracy

The key element of thematic accuracy assessment is the creation of an error matrix (Congalton and Green 1999). An error matrix is a square array of numbers organized in rows and columns that express the number of sample units assigned to a particular category relative to the actual category as indicated by the reference data (Table 14-2). The columns usually represent the reference data while the rows represent the map or spatial data layer. Reference data are assumed correct and can be collected from a variety of sources appropriate for that spatial data type. Some possible sources of reference data include but are not limited to photo interpretation, videography, ground observation, and ground measurement. An error matrix is a very effective way to represent accuracy in that the accuracies of each thematic category are plainly described along with both the errors of inclusion (commission errors) and errors of exclusion (omission errors) present in the map or spatial data layer.

Manual of Geographic Information Systems

Table 14-2 Example error matrix for assessing thematic accuracy.

		Reference Data			row total	Land Cover (Thematic) Categories
		V	W	U		
	V	44	8	6	58	V = Vegetation
Map or Spatial Data Layer	W	3	33	5	41	W = Water
	U	2	4	36	42	U = Urban
column total		49	45	47	141	OVERALL ACCURACY = 113/141 = 80%

PRODUCER'S ACCURACY			USER'S ACCURACY		
V = 44/49 =	90%		V = 44/58 =	76%
W = 33/45 =	73%		W = 33/41 =	80%
U = 36/47 =	77%		U = 36/42 =	86%

The error matrix can then be used as a starting point for a series of descriptive and analytical statistical techniques. Perhaps the simplest descriptive statistic is overall accuracy, which is computed by dividing the total correct (i.e., the sum of the major diagonal) by the total number of sample units in the error matrix. In addition, individual category accuracies can be computed in a similar manner. However, this case is a little more complex in that one has a choice of dividing the number of correct sample units in that category by either the total number of sample units in the corresponding row or the corresponding column. Traditionally, the total number of correct sample units in a category is divided by the total number of sample units of that category as derived from the reference data (i.e., the column total). This accuracy measure indicates the probability of a reference sample unit being correctly labeled and is really a measure of omission error. This accuracy measure is often called "producer's accuracy" because the producer of the map or spatial data layer is interested in how well a certain area can be labeled. On the other hand, if the total number of correct sample units in a category is divided by the total number of sample units that were classified in that category, then this result is a measure of commission error. This measure, called "user's accuracy" or reliability, is indicative of the probability that a sample unit classified on the map or spatial data layer actually represents that category on the ground (Story and Congalton 1986).

In addition to these descriptive techniques, an error matrix is an appropriate beginning for many analytical statistical techniques. This is especially true of the discrete multivariate techniques. Starting with Congalton et al. (1983), discrete multivariate techniques have been used for performing statistical tests on the classification accuracy of digital remotely sensed data. Since that time many others have adopted these techniques as the standard accuracy assessment tools (e.g., Rosenfield and Fitzpatrick-Lins 1986; Hudson and Ramm 1987; Campbell 1987; and Lillesand et al. 2008). While very appropriate and commonly used to assess the accuracy of remotely sensed data, these techniques are equally valid for any spatial data layer or map that contains thematic information.

One analytical step to perform once the error matrix has been built is to "normalize" or standardize the matrix using a technique known as "MARGFIT" (Congalton et al. 1983). This technique uses an iterative proportional fitting procedure that forces each row and column in the matrix to sum to one. The rows and column totals are called marginals, hence the technique name MARGFIT. In this way, differences in sample sizes used to generate the matrices are eliminated and therefore individual cell values within the matrix are directly comparable. Also, because the iterative process totals the rows and columns, the resulting normalized matrix is more indicative of the off-diagonal cell values (i.e., the errors of omission and commission) than is the original matrix. The major diagonal of the normalized matrix can be summed and divided by the total of the entire matrix to compute a normalized overall accuracy.

Another discrete multivariate technique of use in thematic map accuracy assessment is called KAPPA (Cohen 1960). The result of performing a KAPPA analysis is a KHAT statistic (an estimate of KAPPA) that is another measure of agreement or accuracy. The values can range from +1 to -1. However, since there should be a positive correlation between the thematic map or spatial data layer and the reference data, positive KHAT values are expected. Landis and Koch (1977) characterized the possible ranges for KHAT into 3 groupings: a value greater than 0.80 (i.e., 80%) represents strong agreement; a value between 0.40 and 0.80 (i.e., 40–80%) represents moderate agreement; and a value below 0.40 (i.e., 40%) represents poor agreement.

The power of the KAPPA analysis is that it also provides two statistical tests of significance. Using this technique, it is possible to test if an individual thematic map or spatial data layer is significantly better than if the map had been generated by randomly assigning labels to areas. The second test allows for the comparison of any two matrices to see if they are statistically significantly different. In this way, it is possible to determine that one algorithm is different than another one, and based on a chosen accuracy measure (e.g., overall accuracy) to conclude which is better. In other words, one could test if algorithm A used to generate a digital elevation model is statistically better than the results produced by algorithm B.

The above descriptive and analytical techniques are based on the error matrix. An assumption made here is that the matrix was properly generated and is therefore indicative of the map or spatial data layer it represents. If the matrix was not properly created, it is useless or at best anecdotal evidence. Certain statistical considerations are required in order to assure that this assumption is valid. Developing a statistically rigorous accuracy assessment requires choosing an appropriate sampling scheme, sample size, sampling unit, maintaining independence between the training and reference data, and considering the effects of spatial autocorrelation (Congalton 1991). The following section provides a discussion of the various components that must be considered when planning a thematic accuracy assessment. In most cases, the exact procedure chosen is a compromise between what is statistically valid and what is practically achievable.

Sampling Scheme: There are numerous possible sampling schemes used in collecting accuracy assessment data including: simple random sampling, systematic sampling, stratified random sampling, cluster sampling, and stratified systematic unaligned sampling. Each scheme has its own advantages and disadvantages. Randomness provides powerful statistical properties that are important for further analysis of the results. Systematic and cluster sampling can provide practical advantages. It is important to understand each scheme and apply the one most appropriate for the situation. The analysis undertaken must then match the sampling scheme chosen. In many cases, stratified random sampling provides the best combination of statistical validity while still being practical and efficient.

Independence: It is critical that the data collected for developing the map or spatial data layer be independent (separate) from the data used in the accuracy assessment. This factor is important for positional accuracy assessment as well as thematic accuracy assessment. After the ground reference data are collected, a stratified random sample of the data should be selected for accuracy assessment and put aside and not looked at until after the map or spatial data layer has been generated. The remaining data can then be used for training/creating the map or spatial data layer. It is important to stratify the data by map class to insure that sufficient training and accuracy samples exist for each map class.

Sample Size: Sample size is dictated by the need to express accuracy in an error matrix. The sample size must be large enough for the error matrix estimates to have adequate precision. An error matrix does not fall into the right/wrong binomial scenario but rather a multinomial situation in which there is one correct for each class and n-1 wrongs (where n is the number of map classes). Therefore, experience as well as the multinomial equation shows that approximately 50 samples (30 as an absolute minimum) per map class are required to adequately populate an error matrix (Story and Congalton 1986).

Sampling Unit: The key factor in determining the sampling unit is positional accuracy. It is very important that the sampling unit be of sufficient size to be accurately located on both the map/spatial data layer and on the reference data. If the sample unit is too small, then errors due to positional inaccuracies will result in a thematic error. Sample unit should be selected to minimize positional error. For example, there are three common sampling units used in assessing the accuracy of remotely sensed data. They are (1) the pixel, (2) a grouping of pixels, or (3) a polygon. It should be noted that the pixel should not be used as the sampling unit because of our inability to accurately locate it on the ground (even using GPS) and on the imagery. When single pixels are used much of the thematic error indicated by the error matrix is actually a result of positional error that was uncontrolled during the assessment process. Either a grouping of pixels, such as a 3x3 block, or a polygon should be selected as the sample unit depending on the specific needs of the project.

Spatial Autocorrelation: Spatial autocorrelation is a measure of the influence, positive or negative, that some characteristic at a certain location has on its surrounding neighbors. Spatial autocorrelation is an important consideration when deciding which sampling scheme to employ. If there is positive correlation between samples then it is important for precision of the accuracy estimates to space the samples far enough apart to minimize this correlation. This issue is particularly important for certain schemes such as cluster sampling and systematic sampling.

14.3 Conclusions

This chapter presents a variety of techniques that can be used to validate or assess the accuracy of spatial data layers/maps. Although it is important to perform a visual examination of the spatial layer/map, it is not sufficient. Other techniques, both qualitative and quantitative, are necessary to fully understand and evaluate the accuracy. Error budgeting is a very useful exercise in helping to realize error and consider ways to minimize it. Quantitative accuracy assessment provides a very powerful mechanism for both descriptive and analytical evaluation of the spatial data. It is critical to understand the relationship between positional accuracy and thematic accuracy for any spatial data layer or map. Failure to consider these accuracies simultaneously will lead to erroneous conclusions about the true accuracy of the spatial data layer or map.

As our use of spatial data continues to grow, so must our use of these tools for evaluation. If you are a novice spatial data user, please consider the techniques proposed here and implement as many as you can. If you are an advanced spatial data user, there is no excuse for not employing these techniques to better evaluate your analysis. It is not acceptable to languish in a mode of "it looks good," but rather let us struggle forward to advance the use of spatial data in all aspects of our work.

References

Bolstad, P. 2002. *GIS Fundamentals*. White Bear Lake, Minn.: Eider Press.

Burrough, P. and R. McDonnell. 1998. *Principles of Geographical Information Systems*. New York: Oxford University Press.

Campbell, J. 1987. *Introduction to Remote Sensing*. New York: Guilford Press.

Cohen, J. 1960. A coefficient of agreement for nominal scales. *Educational and Psychological Measurement* 20:37–46.

Congalton, R. 1991. A review of assessing the accuracy of classifications of remotely sensed data. *Remote Sensing of Environment* 37:35–46.

Congalton, R. and K. Green. 1999. *Assessing the Accuracy of Remotely Sensed Data: Principles and Practices*. Boca Raton, Fla.: CRC/Lewis Press.

Congalton, R. G., R. G. Oderwald and R. A. Mead. 1983. Assessing Landsat classification accuracy using discrete multivariate statistical techniques. *Photogrammetric Engineering and Remote Sensing* 49:1671–1678.

FGDC (Federal Geographic Data Committee). 2005. Geospatial positioning accuracy standards, parts 1-5. <http://www.fgdc.gov/standards/projects/FGDC-standards-projects/accuracy> Accessed 25 March 2008.

Goodchild, M. and S. Gopal, eds. 1989. *The Accuracy of Spatial Databases*. New York: Taylor & Francis.

Hudson, W. and C. Ramm. 1987. Correct formulation of the kappa coefficient of agreement. *Photogrammetric Engineering and Remote Sensing* 53:421–422.

Landis, J. and G. Koch. 1977. The measurement of observer agreement for categorical data. *Biometrics* 33:159–174.

Lillesand, T., R. Kiefer and J.W. Chipman. 2008. *Remote Sensing and Image Interpretation*, 6th ed. New York: John Wiley & Sons.

Lunetta, R., R. Congalton, L. Fenstermaker, J. Jensen, K. McGwire and L. Tinney. 1991. Remote sensing and geographic information system data integration: Error sources and research issues. *Photogrammetric Engineering and Remote Sensing* 57:677–687.

Moeller, M. 2008. Metadata as a component of data. In *Manual of Geographic Information Systems*, edited by M. Madden, Chapter 10 in this volume. Bethesda, Md.: ASPRS.

Rosenfield, G. and K. Fitzpatrick-Lins. 1986. A coefficient of agreement as a measure of thematic classification accuracy. *Photogrammetric Engineering and Remote Sensing* 52:223–227.

Skole, D. and C. Tucker. 1993. Tropical deforestation, fragmented habitat, and adversely affected habitat in the Brazilian Amazon: 1978–1988. *Science* 260:1905–1910.

Story, M. and R. Congalton. 1986. Accuracy assessment: A user's perspective. *Photogrammetric Engineering and Remote Sensing* 52:397–399.

CHAPTER 15
The Representation of Uncertain Geographic Information

Peter Fisher

15.1 Introduction

Geographic information is made up of terabytes of information, and every day that information is being supplemented by satellite sensors at a vast rate. The information can be described as being made up of two components:

- Measurements or data, and
- Interpretations (strict use of the word information)

It is therefore not surprising that geographic information should be ingrained with uncertainty. The uncertainty starts with the question: are we sure that what we are measuring exists? Followed by: is it measurable? Which might be rephrased as: does what we are measuring relate to what we want to record? And so on. Views on all these topics vary. This chapter will attempt to reflect these variations, and explore how we can accommodate the problems.

It should also not be surprising that the areas of uncertainty in their different forms have attracted a huge amount of research and publication. Some key books in the area are: Goodchild and Gopal (1989), Guptill and Morrison (1995), Heuvelink (1998), Shi et al. (2002), Zhang and Goodchild (2002) and Petry et al. (2005). There have also been some application-specific discussions of uncertainty in spatial information (Hunsaker et al. 2001). There are a number of journal special issues related to uncertainty in one way or another (e.g., Robinson et al. 2003), and almost any issue of the key journals in the field (especially the *International Journal of Geographical Information Science* and *Transactions in GIS*) will be found to contain at least one paper related to the topic.

Fundamental to understanding uncertainty in spatial information, it is necessary to start with the philosophical stance of Geographic Information Scientists. To typify these individuals, they can perhaps be described as belonging to the Realist school of philosophy, many with strong training and leanings towards a positivist stance, and some more inclined to critical realism. That is to say, they are agreed that the world is real and can be experienced by people (realists; Figure 15-1). It is possible to make objective measurements of that world, and it is possible to make repeated measurements coming up with the same answer. Indeed, the extreme Positivists would say that everything is objectively measurable, and if it is not then it is not worth measuring. Furthermore, they would say that the information classes are meaningful and reflect exactly what they intend them to reflect, and other users will understand the exact meaning. On the other hand, while the Realists broadly agree with this proposition for some phenomena, they would argue that it is not always the case. The more extreme Critical Realists say that all information is relative to what it is intended to mean and what any user may read into it. Note that this is not the same as saying that all opinions are correct. In other words, given the same set of data, any two users might generate different but valid and correct interpretations or measurements (Figure 15-1). This leads to a "semantic heterogeneity" (Figure 15-1) in the understandings of the meanings of the information, whether that is in the meaning of the theme of the information and/or in the classes within a particular theme.

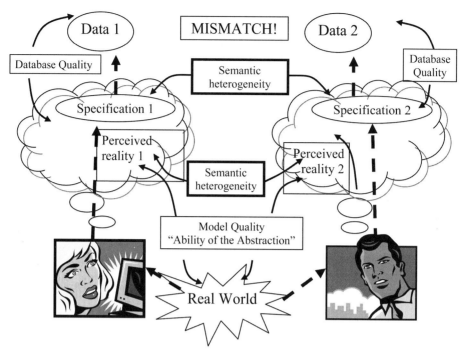

Figure 15-1 The basic problem of uncertainty in data arises because two people using the same information have different perceptions of the information, for whatever reason. Persons 1 and 2 could be original collectors of the data, providers of the data, or users of the data.

Reconciling these different philosophical positions is not easy, but the last 25 years have seen a strong movement towards the critical realist view, which believes that two different people can hold different but correct conceptual models of the same thing. This means that the emphasis in uncertainty research has shifted from measurement error to semantic uncertainty. Therefore the conceptualization of the information is at the heart of uncertainty research currently, but error remains very important.

This Chapter is structured as follows. Section 15.2 sets the scene by discussing the types of information, and Section 15.3 presents metadata as it relates to uncertainty. In Section 15.4 some causes of uncertainty are outlined, and in Section 15.5 types of uncertainty are discussed. Section 15.6 looks at data quality as the official record of uncertainty and points out the lack of equivalence of the concepts expressed and the status of uncertainty research. Section 15.7 examines the idea of an audit of uncertainty, and, finally, in Section 15.8 some ways of working with uncertainty are addressed.

15.2 Information Types

Geographic information can be described as being made up of two components: measurements and interpretations. These are also sometimes known as data and information, but what characterizes geographic information, whether it is measurement or interpretation, is that it is a record of the situation at a particular time at the surface of the Earth. It is information that would, in the past, have been presented and, indeed, stored on paper maps—maps of topographic and thematic information.

Geographic information (as with many other types of information) comes in two different flavors:

Primary information: the information is collected in the course of a scientific (natural or social) inquiry. It is analyzed within that inquiry and the results of the analysis presented. The information is then disposed of, or archived by the investigator and never again accessed.

Secondary information:

- Inquiry information: An individual investigator and an inquiry team collect data for a specific study with a particular question or questions in mind. They report on the analysis of that data and then place the information with a publicly accessible archive.

- Inventory Information: An agency of some sort (the Ordnance Survey, the US Geological Survey, a Soil Survey organization, the Census Agency, etc.) collects a data set either for a particular primary purpose (the Census Agency, the Geological Survey) or because the information is deemed to be nationally important (the Ordnance Survey). The information is intended for use by a general group of users (either those who are granted access, or those who can purchase access).

The importance of recognizing the separation of primary and secondary information and of inventory and inquiry information is simply this. Information collection of any sort involves the application of myriad assumptions and hypotheses. In primary and inquiry information there is some hypothesis that is being investigated that, given the experience and the knowledge of the investigator or investigators, determines the actual classes of information to be collected, the properties to be measured, the questions to be asked, and the meaning in the question. As long as those data are unique to the investigator, it is the responsibility of that investigator to communicate to a readership in policy documents or peer-reviewed literature the meanings of the variables and the questions. If the data are to be passed on to others as secondary information then the information itself becomes what is known as a Boundary Object (Harvey and Chrisman 1998); it is created to be archived in the first place and then to be used by anyone who comes along and thinks it may be relevant to the issue in which they are interested, whether that is siting a nuclear power plant, or planning the route to visit their Granny. Increasingly when agencies offer funding for research they require investigators to make any information they collect available to future users (ESRC Data Archive, http://www.data-archive.ac.uk/; NERC Environmental Information Centre, http://www.nerc.ac.uk/research/sites/data/terrestrial.asp; Archaeological Data Service, http://ads.ahds.ac.uk/). This is seen as an efficient use of the research funds—preventing the need for re-collection of information. Unfortunately it also means that information, which might have been collected for one purpose, may be used for many others. It requires the documentation of detailed reasons for selection of variables in the information and of the investigator's meaning and understanding of the variables. There is a long-lasting assumption that the users of such information will have experience in the subject of the information, but this assumption is increasingly unsustainable. One of the biggest problems in all information processing is communicating the assumptions underlying the information to the users. This should enable the collector and preparer of the information to override the preconceptions of subsequent users as to the meanings embedded in the information. This is the role of metadata, which comes in a number of forms, including handbooks that are independent of the data and ***structural metadata*** that are supposedly embedded with the data.

15.3 Metadata

15.3.1 Metadata

Metadata are the information about a particular piece of information. Metadata are usually divided into three different types:

- **Descriptive** is that which identifies the existence and ownership of the information;
- **Administrative** is that which locates the information within the organization, together with information on personnel who may be responsible for it or expert in its use; and
- **Structural** is information which explains the information (Library of Congress 2008).

Of principal interest in a discussion of uncertainty and error are the structural metadata, which include data quality and semantic descriptions of the information.

15.3.2 Handbooks

Handbooks are in the published literature available through libraries and bookshops provided by either commercial publishers (Handbook of the Census; for the 1981 UK Census, Rhind 1983; for the 1991 UK Census, Openshaw 1995; for the 2001 UK Census, Rees et al. 2002) or by specialist (e.g., government agencies; Soil Survey Staff 1993). These documents are rich in domain-specific knowledge, but they tend to be written for the specialist using domain-specific vocabularies. Although they may set out to explain one restricted vocabulary they do so using another. There is therefore a large investment for a novice user in facilitating the comprehension of the content. Typically handbooks describe general characteristics of a whole class of information in a specific country such as the Census in England and Wales or the US Soil Survey (Soil Survey Staff 1993).

15.3.3 Survey Reports

Another source of metadata is the traditional (through most of the twentieth century) production of a report or memoir to accompany natural resource surveys (soils, geology, land use and vegetation). Thus geological surveys are always of a particular area, inventorying the rocks that occur in that area and showing on a map the relative arrangement of those rocks, and in particular showing the lithostratigraphy of the rocks according to the contemporary understanding of the sequence. For any area a report accompanies the map. The report documents the rock types within the area giving much background information, interpretations of the rocks and aspects of possible uses. Similar reports are usually produced for soil surveys and sometimes for vegetation surveys. A selection of US Soil Survey county reports can be consulted at http://soils.usda.gov/survey/online_surveys/. Regretfully, the tradition of producing reports is waning so many land cover mappings are being done without any report being prepared, and the mass distribution of natural resource mappings has dissociated existing reports from the maps.

Unfortunately neither handbooks, memoirs nor metadata are reliable methods for communicating the surveyor's semantic understanding of the information and his or her expertise in the data because they are all equally easily ignored.

15.4 Causes of Uncertainty

15.4.1 Technical Issues that Affect Uncertainty

A number of technical concerns about how data are constructed significantly contribute to the uncertainty in the information, although they are not necessarily measured or reported as part of the result.

15.4.1.1 Measurement Errors

The most widely recognized reason for uncertainty in information is that measurements of a property do not correspond to some idea of the true value that should have been recorded. Included in the idea of such errors are things like transcription error due to human or automated recording of information and the imprecise measurement of a quantitative value.

15.4.1.2 Resolution and Minimum Mapping Unit

A primary spatial issue of uncertainty is the scale at which information is recorded. On a paper map, scale is a fundamental property of the information portrayed. The scale is expressed as a ratio of the length of a line on a map to the length of the same line on the ground (1:50,000, for example). This concept has been inherited by many digital data sets but it has no direct meaning in the data set because the data can be represented at any scale on paper or on the screen. (A digital data set with a nominal scale of 1:50,000 can be shown at 1:100,000, 1:10,000, 1:1,000, or any other scale, for example). The scale in the product or database name is supposed to convey:

- a normal (or safe) map reproduction scale for the data; and
- a set of metadata which will explain the map units and what is included in the data set and, by inference, what is not, from the user's familiarity with the paper map products at that scale—in other words it is a quick and easy way of conveying structural metadata.

A more meaningful concept than scale of digital databases is the concept of resolution. The resolution of a database does, however, have various meanings. For a vector database it is usually the precision at which the vector lines represent the same lines on the ground; if you like, the accuracy of the line or the chance of finding the line on the ground in the position recorded in the database. Thus if a data set of lines has a resolution of 2 m, then you would expect that the line on the ground occurs within 2 m either side of the same line in the database. The same argument could be made for a point data set.

A concept related to resolution for a polygonal data set is the Minimum Mapping Unit (MMU). The line work forming the outlines of the polygons would have a resolution just as the lines referred to above have, but there would also be a minimum size of polygon that would be included in the data set. If a polygon existed that was smaller than the MMU then it would not be recorded—usually the ground would be allocated to the adjoining polygon type or types. MMU is a fundamental property of any polygonal data set, but you will rarely find it quoted.

15.4.1.3 Pixelized Data

Raster data are a special case of the MMU problem, which is to say the information comes with a defined MMU that is at least the size of the pixel in the imagery.

Earth surface information is collected by satellite systems that provide data in raster format. The ground area measured by such systems depends on the sensor itself, the altitude of the satellite, and the look angle. The pixel size for optical sensors on satellites in 2008 ranges from 40 cm for GeoEye-1 to 1 km for AVHRR (Advanced Very High Resolution Radiometer) data. One of the most common applications for this imagery is mapping of land cover classes using automated classification algorithms. In these the spectral information collected for each pixel in the image is assigned to a class (Figure 15-2). In the process of interpreting imagery to identify land cover classes, a number of simplifying assumptions that are not at all realistic are made about the arrangement of pixels with respect to land cover types. Such mappings have been completed for many countries around the world (see Section 15.5.2.3 Semantics).

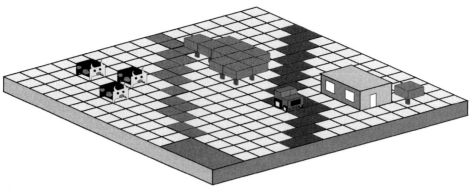

Figure 15-2 The classification of satellite imagery implicitly makes the assumption that all objects of interest are the size of the pixels in the imagery pixelization (after Fisher 1997). See included DVD for color version.

15.4.1.4 Semantic Heterogeneity

Another cause of semantic heterogeneity within data sets is that the traditional paper maps portray different mapping unit types as the same theme (Soils, for example).

In soil mapping, there is a regular arrangement of types of mapping unit to scale that has no specific hierarchical arrangement; a map unit at one scale may be present within many map units at another scale, and locations in the same mapping unit at one scale may be in any number of mapping units at another scale. The same happens in terms of spatial arrangement as we move through the levels of Land Use and Cover classification (Anderson et al. 1976). On the one hand, land cover concepts do not uniquely belong to land use concepts—grass or trees can be in many land use types (Figure 15-3); on the other hand, residential land at one scale of analysis does not all have to be in urban or built-up areas (Table 15-1).

On the other hand, the spatial-hierarchical arrangement of enumeration areas for the census data is much more ordered: Output Areas (2001; Rees et al. 2002) or Enumeration Districts (1991 and earlier; Rhind 1983; Openshaw 1995) are grouped into Wards, which are grouped into Districts, which are grouped into Council areas (Counties and Unitary authorities) and into Regions. There is a many-to-one relationship between each level in the hierarchy, and all locations in one enumeration region in the first level of the hierarchy will necessarily be in the same region at the top level of the hierarchy. But this is a human-political imposed organized hierarchy, defined by so-called *fiat* boundaries. It is not explicitly for mapping but for political resource allocation and representation.

15.4.2 Policy Issues that Affect Uncertainty

One of the main reasons for the creation of uncertainty in spatial information is the cross-organization, cross-discipline, cross-border and multi-temporal incompatibility of semantic definitions of information. Many of the demographic definitions used in one country are different from those in another country. Within the European Union, Eurostat has the responsibility to standardize socioeconomic data for EU member states, e.g., What is the criteria for an individual to be included in the unemployment count in each country? What are the bankruptcy criteria in each country; is it meaningful to compare bankruptcies in EU countries? OECD and UN have the same task for wider international groups of countries.

Table 15-1 USGS Land use and Land cover classification (after Anderson et al. 1976).

Level 1			Level 2		
1	Urban or built-up land	(Use)	11	Residential	(Use)
			12	Commercial and service	(Use)
			13	Industrial	(Use)
			14	Transportation, communication and utilities	(Use)
			15	Industrial and commercial complexes	(Use)
			16	Mixed urban or built-up land	(Use)
			17	Other urban or built-up land	(Use)
2	Agricultural land	(Use)	21	Cropland and pasture	(Use)
			22	Orchards, groves, vineyards, nurseries, and ornamental horticultural areas	(Use)
			23	Confined feeding areas	(Use)
			24	Other agricultural land	(Use)
3	Rangeland	(Cover)	31	Herbaceous rangeland	(Cover)
			32	Scrub and brush rangeland	(Cover)
			33	Mixed rangeland	(Cover)
4	Forest land	(Cover)	41	Deciduous forest land	(Cover)
			42	Evergreen forest land	(Cover)
			43	Mixed forest land	(Cover)
5	Water	(Cover)	51	Streams and canals	(Cover)
			52	Lakes	(Cover)
			53	Reservoirs	(Use)
			54	Bays and estuaries	(Cover)
6	Wetland	(Cover)	61	Forested wetland	(Cover)
			62	Nonforested wetland	(Cover)
7	Barren land	(Use)	71	Dry salt flats	(Cover)
			72	Beaches	(Cover)
			73	Sandy areas other than beaches	(Cover)
			74	Bare exposed rock	(Cover)
			75	Strip mines, quarries and gravel pits	(Use)
8	Tundra	(Cover)	81	Shrub and brush tundra	(Cover)
			82	Herbaceous tundra	(Cover)
			83	Bare ground tundra	(Cover)
			84	Wet tundra	(Cover)
			85	Mixed tundra	(Cover)
9	Perennial snow or ice	(Cover)	91	Perennial snowfields	(Cover)
			92	Glaciers	(Cover)

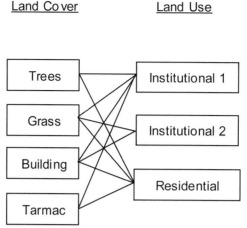

Figure 15-3 Many-to-many relations of land cover and land use. Many land covers contribute to any one land use and more than one land use can be composed of the same suite of land covers. Furthermore, not all instances of the same land use type will necessarily have the same land covers (from Fisher et al. 2005).

A notorious cross-border problem is between Belgium and the Netherlands, which use different datums for their national elevation measurements, meaning that there is approximately a 1 m virtual cliff in elevation between the two countries (Kwadijk and Sprokkereef 1998, p. 51).

There are well-known political reasons for a country to change the definitions of statistical counts such as unemployment. For example between the 2003 and 2004 reports of terrorist activity by the US Department of State, the criteria for judging an incident as being a terrorist incident were changed:

> *A Republican congressional aide, who also spoke on condition of anonymity, said it would be unfair of Democrats to claim terrorism was getting worse under the Bush administration, stressing that the 2004 and 2003 numbers were not counted in the same way and hence were not comparable.* (White 2005)

Of course, these sorts of changes mean that it is impossible to compare counts over time, and they are almost always done under the spurious excuse of improved accuracy.

In the British Isles the Land Cover mapping programs in 1990 and 2000 were done for very different reasons on each occasion, as is documented by Comber et al. (2003). As a result mappings at the two different dates have incompatible classification schemes. Indeed there are many-to-one, one-to-many and many-to-many relationships between the two dates.

Similarly, the US Department of Agriculture produced in 1975 one of the most influential systems of soil classification (Soil Survey Staff 1975). They included 10 high-level Soil Orders. In the 1999 2nd edition there are 12 orders (Soil Survey Staff 1999). Have the soils changed? No. Human understanding has changed, or rather the human view of how solid should be classified has changed. A new generation of surveyors has had experience working with the 10-class scheme and found it does not well describe the soils they encounter. Actually this is not as disturbing as it may appear, because the new orders are largely a reassignment of types lower in the classification scheme, and at the lowest level of classification it results in no change at all. The complexity of the variety of national soil classification schemes can be identified from a detailed examination of the ITC web page (http://www.itc.nl/~rossiter/research/rsrch_ss_class.html).

15.5 Types of Uncertainty (based on Fisher et al. 2005)

There are a number of models of uncertainty relevant to spatial information (Figure 15-4) (Fisher 1999; Fisher et al. 2005).

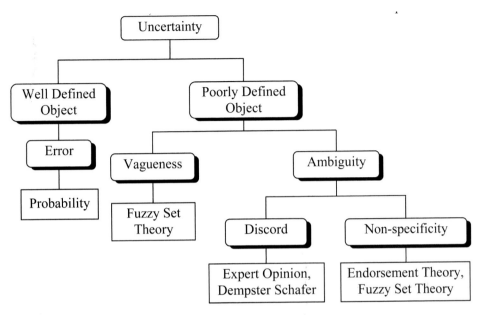

Figure 15-4 A conceptual model of uncertainty in spatial data (after Fisher 1999, adapted from Klir and Yuan 1995: 268, with revisions).

If information is clearly defined, in as much as threshold values can be identified in both measurable attributes and in space that clearly delineates concepts, then the class of object is definable. A building is a clear example of this, as are humanly defined areas such as legal land ownership parcels and census enumeration areas—areas defined by fiat (humanly made) boundaries. For well-defined objects, any uncertainty in measurements should be considered a matter of error.

Not all phenomena stored in geographic databases are well defined. They exhibit the "*Problem of Definition,*" which is to say that it is not possible to clearly and unequivocally state where the boundary in either attribute or in spatial dimensions lies. It could be here or it could equally be there. In these cases it is more typically possible to identify locations that epitomize the classes either side of the boundary, but the position of the boundary itself is a "*Matter of Degree.*" At one location it is more like one class and at another it is more like another, but to say that here is the switch over is not possible.

The treatment of such cases has been argued over for many years, but much work has been done on the use of fuzzy sets to address this problem, and the so-called Sorites paradox (the paradox of the heap that asks when a collection of sand grains is a heap of sand) can be used as a diagnostic test of when this problem occurs (see Fisher 2000 for an extended discussion of the Sorites Paradox).

Other examples of the Problem of Definition occur when there is ambiguity over the definition of a phenomenon. One person or process can define an object as belonging to one class while another person or process defines it as another. If one operator uses one threshold or one set of criteria to define the phenomenon while the other operator uses different thresholds or criteria, then both may be exactly correct definitions, but they may result in different classes or extents of the object. The definition may be considered unequivocal by each individual or process, but either they or some independent observer cannot decide between them. In this case the attribution of the object is considered ambiguous. The most extreme form of ambiguity leads to literal discord and violence—most clearly in the political arena when countries disagree over the sovereignty of territory.

15.5.1 Measurement Uncertainty

Measurement uncertainty means that we have conceived of something as measurable; it can effectively be represented by a value, measured on a continuous scale. The value we determine for it may be the correct value. If we make the measurement a second time (if nothing has changed) we will get a second reading that should be the same as the first (this is a problem for dynamic systems such as meteorological or fluvial systems). Any reading will be made to the precision at which the observer is capable of capturing the information with the instrumentation they have. Any difference between the measurements will be a matter of error in the readings. We have to agree that there is a measurable phenomenon, and that the measurement we are going to make is representative of the phenomenon of interest.

There are a number of terms associated with errors, each of which is in common use but they have specific meanings, too (and are summarized in Figure 15-5):

- *Accuracy* – describes the degree to which the measurement matches the intended value.
- *Precision* – has two principal meanings: it can mean that multiple measurements of the same thing are closely similar, whether they are accurate or not; and it can also mean the precision with which a measurement is calculated with a particular device—thus computers calculate to 32 or 64 bit precision, or a programmer may allocate double precision to a particular task yielding 64 or 128 precision, or a DEM may be stored in meters as integer values—and higher precision (e.g., centimeters) should not be interpreted from the values.
- *Bias* - is the systematic variation in values from the intended values.

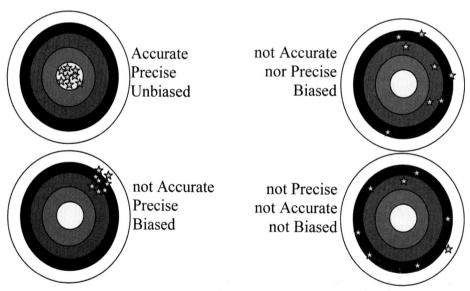

Accurate
Precise
Unbiased

not Accurate
nor Precise
Biased

not Accurate
Precise
Biased

not Precise
not Accurate
not Biased

Figure 15-5 Three aspects of error are shown: Accuracy, Precision, and Bias. See included DVD for color version.

Probability theory has been designed for this purpose and is well established in the scientific method. It is taught in most introductory science courses. That does not mean that everyone is comfortable with the mathematics, but it is well known.

Most surveying is conceived in this way. A surveyor wishes to make a measurement of a point—they have conceived it as a point with some significance and some reason for measurement. The assertion is that if they return to exactly that point they will record the same location with the same coordinates and differences in the coordinates are described as error. Thus the concept of ***Root-Mean-Square Error (RMSE)*** is common among surveyors, as is the ***Circular Map Accuracy Standard (CMA)***.

The RMSE of a point observation is determined by taking a number of measurements of the coordinates of a point, and the actual location of the point. The distances between the actual location and the measurements are determined, and the square root of the sum of the squares of those deviations is found.

CMA is simply an application of the RMSE in the production of map information, allowing quality assurance in data production. It states that the RMSE will be less than the CMA value for all points measured in a data set.

The concept of the point RMSE has been widely applied to modeling the accuracy of vector GIS (Shi 1998). This work has been extended to a theory of measurement-based GIS (Leung et al. 2004)

The concept of the surface of the Earth is also widely agreed; it is agreed to be measurable, and height above a datum is an appropriate way to do it. Therefore the RMSE is also used for reporting the accuracy of Digital Elevation Models (DEM). The DEM RMSE is given by:

$$\text{RMSE} = \sqrt{\frac{\Sigma(z - w)^2}{n}} \tag{15-1}$$

where z is the elevation recorded in the DEM;
w is the elevation measured at the higher precision; and
n is the number of locations tested.

If we make the assumption that the mean error is 0 (as is stated in the documentation of some accuracy specifications for DEMs), then this formula is the same as the formula for the standard deviation of the population, s:

$$s = \sqrt{\frac{\Sigma(\bar{z} - (z - w))^2}{n}} \tag{15-2}$$

where $\bar{z}$ is the mean error.

Many researchers have recommended the adoption of the mean error and the standard deviation as the basis of reporting error, superseding the RMSE. This equivalence has significance in the discussion in Section 15.8.1.

One of the fundamental problems of the RMSE as a measure of spatial accuracy is that it has no spatial component. It is stationary over the study area (it has no record of variation over that area), which can be as large as a whole country. In the UK the Ordnance Survey used to report the accuracy of the Panorama DEM as having a RMSE (2–3 m) for the whole country (Fisher 1998, p. 217)!

This measure has no spatial component, but if the error is in a measurement across a field of values (such as the ground surface), then there is every reason to suppose that the error will be spatially dependent, i.e., autocorrelated. Figure 15-6A shows the spatial distribution of errors in a DEM model when the values in the DEM are compared with surveyed spot heights. Notice the tendency for points in proximity to each other to have similar values. Also illustrated are the statistical distribution of the errors (the histogram), the relationship to slope (which is good for negative errors but poor for positive; Figure 15-6C) and variograms fitted and observed for the spatial distribution of errors. The experimental variogram (Figure 15-6D) shows the uncertainty increasing with lag distance from a nugget variance of about

30 (note that variance is the square of the standard deviation, so this is approximately equal to the RMSE = 5.7; Figure 15-6B). It rises to a maximum of about 52 before falling to a sill of 48 (Figure 15-6D). From this more complete description of the error it is possible to do far more than with the RMSE (see Section 15.8.1).

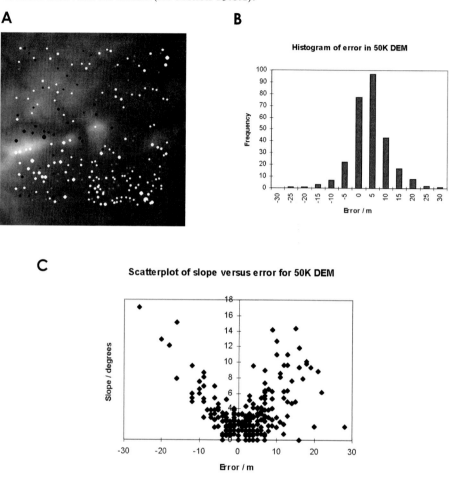

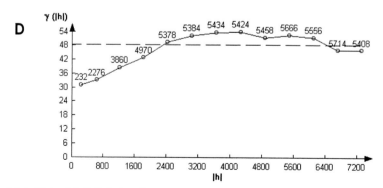

Figure 15-6 A) An OS digital elevation model is shown with superimposed graduated circles showing errors between the DEM and spot heights; white dots are positive and black dots are negative. The diameters are proportional to the magnitude of the error; B) the histogram shows a small bias (mean is greater than zero) in the errors; C) poor relation between slope and error (but better for negative errors); and D) the experimental variogram (after Fisher 1998).

15.5.2 Categorical Uncertainty

Many spatial data sets show categories of objects. As with other measurements, categories are subject to uncertainty of a number of types.

If the categories are defined as Boolean classes (they are present and they definitely occur—there can be no doubt), then the principal problem is that they will suffer from attribution error. The object class may be misidentified. When this is the case, it is typical to compare a sample of locations that are assessed independently (sometimes called ground truth) with those that are classified in the database. The cross tabulation of the two yields the confusion matrix, which has been the subject of considerable research (Congalton and Mead 1983; Congalton and Green 1999) and is currently being widely reevaluated, and reinterpreted (Power et al. 2001; Hagen 2003; Hagen-Zanker et al. 2005). McGwire and Fisher (2001) have shown how this may be extended to a spatially explicit model of error using a geostatistical method known as indicator kriging.

15.5.2.1 Vagueness

Many geographic phenomena are poorly defined; they suffer from the ***Problem of Definition***. If so then they can be diagnosed by the ***Sorites Paradox***, by asking the simple question: whether increasing the threshold variable for inclusion in the set by one value will make a crucial difference. Thus if woodland is defined as being any area where trees currently give more than 50% canopy cover, then is the area (ecologically, aesthetically, or in any other terms you wish to define) significantly different when trees form 51% of the cover (woodland) and when they form 49% of the cover. Notice that to define woodland as having 50% canopy cover, it is necessary to define an area over which the measurement is made.

If in a particular circumstance there is no realistic difference between woodland with 51% and 49% canopy, then the definition of woodland may be specific (in terms of the 50% canopy) but it is meaningless. That is to say that it is possible to measure canopy cover and to precisely delineate those locations where trees form more than 50% of the cover, but it is not meaningful to do so. From this we can recognize that woodland is a ***vague concept***, and define a degree to which any location is woodland—so we might say that less than 40% canopy is definitely not woodland while more than 65% cover is definitely woodland. We then define a transition zone that can be expressed as a ***Fuzzy Set Membership*** (Zadeh 1965).

At the heart of ***fuzzy set theory*** is the concept of a fuzzy membership. A Boolean set is defined by the binary coding {0,1}, whereby an object that is in the set has code 1, and an object that is not has code 0. A fuzzy set membership, on the other hand, is defined by any real number in the interval [0,1]. If the value is closer to 0, then the object is less like the concept being described, if closer to 1 then it is more like. The fuzzy membership is commonly described in formulae by μ, and, particularly, an object x (such as a particular location) has fuzzy membership of the set A, $\mu(x)_A$.

There are two well-known methods for populating fuzzy sets (Robinson 2003). The most common way of populating a fuzzy set is known as a ***Semantic Import Model***. Here some form of expert knowledge is used to assign a membership on the basis of the measurement of some property, d. There are two main ways this can be done.

- The first method involves defining a set of paired values called fuzzy numbers that specify the critical points of the membership function defining the limits of the transition from memberships of $\mu=0$ to $\mu = 1$. In the case of woodland discussed above, this might read ({40%,**0**}, {50%,**0.5**}, {65%,**1**}) (Figure 15-7A).
- The second method is to specify a formula that relates the changes in the value of d to continuous variation in μ. The continuous membership function is defined through a formula such as that specified in Equation 15-3, which actually defines the function illustrated in Figure 15-7B that moves from a value near to $\mu = 0$ up to a value of

$\mu = 1$. It can be reformulated in various ways. This formulaic nature of the membership function has been used in many different forms in the geographic information science literature.

$$\mu(d) = \begin{cases} 1 & \text{for } d \geq b_1 \\ \dfrac{1}{\left(1 + \left(\dfrac{d - b_1}{b_2}\right)^2\right)} & \text{for } d < b_1 \end{cases} \qquad (15\text{-}3)$$

where $\mu(d)$ is the fuzzy membership of an object in a set;

d is the value of the property used to define the membership function;

b_1 is a threshold value of the property at the limit of any object definitely being a member of the set (65); and

b_2 is a value that describes the decay with decreasing d; in the curve in Figure 15-7, $b_2 = 0.008$, which puts the value 45 close to the membership value 0.5, sometimes called the cross-over point.

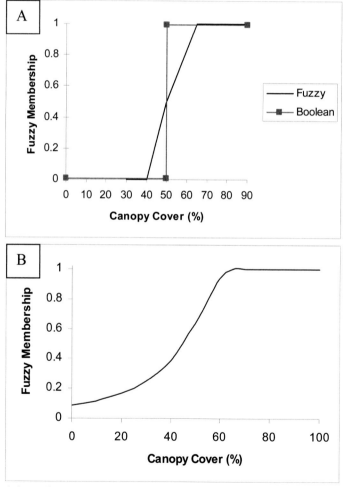

Figure 15-7 A) shows the general form of a fuzzy set (as discussed in the text) specified by fuzzy numbers superimposed on a possible Boolean version of the same set; B) shows the form of a fuzzy set that might be equivalent to that shown in A) but specified from a formula such as that given in Equation 15-3.

The other popular method of defining a fuzzy set membership is through automated measurement of class similarity, known as the ***Similarity Relation Model***. While the ***Semantic Import Model*** of fuzzy memberships requires specification of the membership function, the Similarity Relation Model uses automated classification procedures to search the data for membership values. A variety of fuzzy classification strategies have been suggested (Bezdek 1981), although two are best known—Fuzzy c Means (Bezdek et al. 1984) and, more recently, fuzzy neural networks (Wilkinson et al. 1995). Both methods take as input a multivariate data set of p variables by q cases. According to their own algorithms, they optimize the identification of a predetermined number of groups, c, and for each of the q cases the degree to which any case resembles the properties of each of the c groups is reported as a fuzzy membership value in the range [0,1]. Fuzzy c means have been applied to a large number of different types of geographic information, including soils information (Odeh et al. 1990; Powell et al. 1991; McBratney and De Gruijter 1992) and land cover mapping (Fisher and Pathirana 1990; Foody 1995, 1996; Lees 1996; Wilkinson et al. 1995).

Note that by defining the fuzzy set on the basis of a variable that varies over space (tree cover) the fuzzy membership itself varies over space. Fuzzy memberships can also be applied to the properties of polygonal information where the assignment of the polygon to a set is vague, but the spatial extent is not. Fisher (2000) provides a general review of fuzzy models of geographic information.

15.5.2.2 Ambiguity

Ambiguity occurs when two processes or people assign the same object to different classes or assign two different extents to the same object. This can arise for a number of different reasons. First we can have a particular algorithm for the identification of the process, but we are not specific about how that algorithm is applied. This can be as simple as one person applying one threshold to a particular parameter and another person applying a different threshold (land over 15° slope is steep, or land over 20° slope is steep—both are correct to a particular person but what angle constitutes a steep slope?). This clearly relates back to the idea of vagueness—a steep slope is a vague class, and can be handled by fuzzy set theory.

At the same time, there may be a similar threshold for extraction that is not so readily interpreted as a fuzzy set. In Landserf (www.landserf.org), the geomorphometric extraction of a peak is based on curvature and slope parameter; if the slope is less than a threshold, the location can be classed as a peak. But there are no hard and fast guidelines as to what spatial extent the analysis should be conducted over. A location that might be classified as a peak at one resolution might be classified as a ridge or a planar slope at another. However, a peak is conceptually a vague object and might be expected to have a vague extent. One way of treating this is shown by Fisher et al. (2004). They used the variation of the assignment of locations to peaks at different resolutions to explore the extent of peakness for mountains in the English Lake District. The approach is not totally successful for the latter purpose, but for some mountains it proved relatively successful (Figure 15-8). In other cases, generally the major mountains, the ridges surrounding caused considerable interference (Figure 15-9).

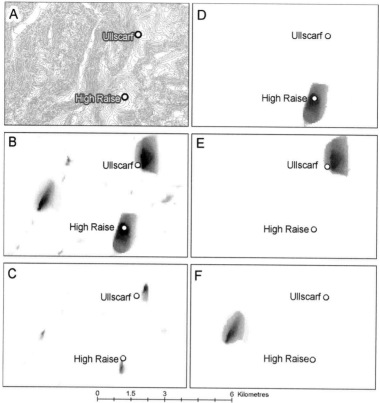

Figure 15-8 Some successfully mapped but minor peaks. The peakness of gray is overlaid on the contour map, and in the maps the individual peaks have been selected as separate raster objects. The place names are only approximate locations (after Fisher et al. 2004).

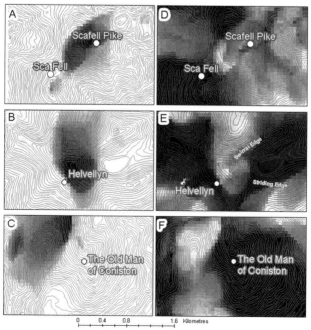

Figure 15-9 In the cases of some large mountains like Sca Fell, and especially the Old Man, the peakness (left hand figures, A, B and C), are interfered with by the ridgeness of the ridges on which they lie (D, E and F). Scafell Pike shows up very well with respect to its location and the peak. Helvellyn is something between the two (after Fisher et al. 2004).

They also suggested that by using the map of peakness for Helvellyn it was possible to answer completely new questions about the perception of landscape, viz. by determining the reverse viewshed (the area that can see a location rather than the area that can be seen from the location) of each degree of peakness of Helvellyn it is possible to show the degree to which it is possible to see Helvellyn as a peak (Figure 15-10A). Of course, once a geographic object is stored as a fuzzy set, it is possible to execute fuzzy logic. The intersection of two fuzzy sets is the minimum fuzzy membership of the two:

$$\mu_{P \cap Q} = \min (\mu_P, \mu_Q) \tag{15-4}$$

where μ_p is the possibility of the location being a peak and
μ_q is the possibility of the location being able to see Helvellyn as a peak.

This interrogation is shown in Figure 15-10B. In spite of the problem of associating the area of high peakness with Helvellyn on the map (Figure 15-10B), the area indicated is still the zone of highest peakness, and so when viewing Helvellyn it is still the degree to which the mountain will be viewable (seen) as a peak. Much more work could be done on this form of ambiguity, caused by non-specificity, especially in algorithms. Indeed, many people implicitly do this type of analysis while struggling to find the best value for some parameter, when they have no clear *a priori* idea of what would qualify as a best value or how they would judge the result to show it is best.

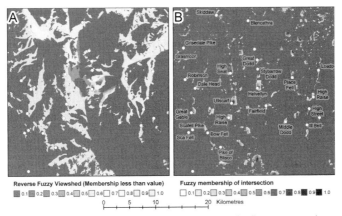

Figure 15-10 A) Locations from which it is possible to see Helvellyn as a peak, and the degree to which it will appear to be a peak, and B) the possibility of seeing Helvellyn as a peak while standing on a peak (after Fisher et al. 2004). See included DVD for color version.

15.5.2.3 Discord and Semantics

The other form of ambiguity is known as discord. Discord is the assignment of a single target object to one class by one individual (or process) and to another class by another individual (or process). Discord is usually realized in the semantics of the naming of phenomena. Two phenomena can be called the same thing but their properties or description may be different. Equally, they may be called different things when a fuller description would show them to be similar. Discord can be common in the instance of people classifying land or soils, but is epitomized by political discord over territory that can have terrible consequences such as war and Diaspora.

There are many examples of discordant uncertainty. A striking example is the mapping of land cover in Britain. The classifications schemes used in the Land Cover Map of Great Britain 1990 and Land Cover Map 2000 are shown in Figure 15-11. Relationships between the two schemes can be seen as one-to-one, many-to-one, one-to-many, and, even, many-to-many. In the area of the SK tile (Figure 15-12) of the British Ordnance Survey stretching from Leicester in the south to Sheffield in the North, there were 12 pixels (less than 1 ha) of bog in 1990 when bog was defined by permanent water and waterlogging, and 120,728 pixels (75 km²) in 2000 when it was based on the depth of peat being greater than 0.5 m. It is hard to conceive of a more confused situation (Comber et al., 2005, p. 49).

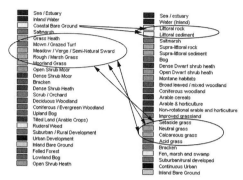

Figure 15-11 The land cover classes in the 1990 (on the left) and 2000 (on the right) classifications of land cover of Britain. Colors are the closest correspondence between the classification schemes. See included DVD for color version.

Figure 15-12 Maps of land cover in 1990 (on the left) and 2000 (on the right) for part of the 100 × 100 km SK tile (west of Sheffield). Colors are the closest correspondence between the classification schemes. Copyright: Natural Environment Research Council. Acknowledgement: Centre of Ecology and Hydrology, Monks Wood. See included DVD for color version.

Working with this discordant classification is very problematic. Land cover mapping is widely used in change analysis and several studies have examined methods for affecting such a comparison. Ahlqvist et al. (2003) examined the fusion of two discordant land cover mappings using rough and fuzzy sets. Comber et al. (2004, 2005) explored the discordance in the British land cover mappings in 1990 and 2000, and articulated a method of analysis based on statistical analysis and semantic look-up tables. They showed complete success in locating inconsistent information, but only limited success in mapping change. Ahlqvist (2005) looked at a similar problem: remapping the US land cover map according to the European CORINE classification scheme.

15.5.3 Spatial Uncertainty

One of the original views of uncertainty was that of positional versus attribute error (Guptill and Morrison 1995). This is preserved in the current framework in the form of positional measurement error, on the one hand, and in terms of semantic uncertainty, spatial fuzziness, etc., on the other. The most important concept introduced by this work was that of boundary error. When you have two (or more) databases showing different subjects and they were digitized (or collected) independently, it is an issue when the same feature has a different position. For a point data set, the point representing the same location appears at different geographic coordinates in the two databases. In a line data set, the two representations of the same line (a road for example) do not coincide (Figure 15-13A). The same problem is introduced for polygons, but the failure of the lines to intersect leads to sliver polygons (Figure 15-13B) when the two data sets are overlaid (which may actually happen). Objects can also be clearly identified, but the ground area distinguishable of a house, for example, can be very different in a ground survey as compared to an aerial survey (Figure 15-13C).

Recognizing situations in which boundaries are supposed to be coincident and objects are actually the same is a major research activity. It relies primarily on semantic descriptions of the objects (river or road names or the street addresses of buildings, for example), but as discussed in Section 15.5.2 this is not necessarily easy.

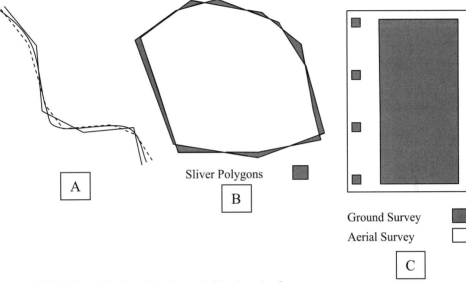

Sliver Polygons

A

B

Ground Survey
Aerial Survey

C

Figure 15-13 Uncertainties arising from digitized vector lines.

15.6 Data Quality: The Record of Uncertainty

In its present form data quality has been created by the producers of spatial data (FGDC 1998). It mimics the concerns of institutional producers of cartographic products, but is informed by the need to transfer spatial information. The principal organizations specifying transfer standards are either industrial or institutional, and the process is dominated by two sectors: the National Mapping Agencies (NMAs) and the software companies. The process is now furthered by the Open GIS Consortium (Buehler and McKee 1998), and has been additionally developed in the areas of ISO 9000 (Dassonville et al. 2002), and more recently in ISO 19113 (ISO, 2002). There are six principal areas of data quality listed in Table 15-2. Four are widely recognized (FGDC 1998), and two others are suggested in chapters in the book edited by Guptill and Morrison (1995). These are briefly defined below. Unfortunately few of these terms are comparable with understandings of uncertainty. Indeed, only the two error terms that refer to measurement error and to category error have direct equivalents (Table 15-3).

Table 15-2 Some recognized aspects of Data Quality (after Fisher 2003).

Data Quality	Lineage	
	Accuracy	Positional
		Attribute
	Completeness	
	Logical Consistency	
	Semantic Accuracy	
	Currency	

Table 15-3 Similarity relations among Uncertainty and Data Quality (after Fisher 2003).

Uncertainty	Data Quality	
Error	Accuracy	Positional
		Attribute
	Completeness	
Vagueness, Discord and Ambiguity ?	Semantic accuracy	
Error, Discord, Vagueness and Ambiguity?	Currency	
Discord	Logical Consistency	
?	Lineage	

15.6.1 Lineage

Lineage is documentation of the history of the information recording its creation and processing, people involved in those processes, and dates. This is most correctly viewed now as metadata. It is very difficult to do anything with this; indeed only the GeoLinius program (Lanter and Veregin 1992) has been developed on the basis of it. Quality reporting through lineage is not central to the issues; rather the propagation of uncertainty through a number of operations is the objective.

15.6.2 Accuracy

The accuracy of measurements both in plan and in attribute is one of the fundamental fields. Accuracy is a report of the degree to which the information *included* in the database is in a correct representation of its position on the ground.

15.6.3 Completeness

Completeness in the context of data quality refers to whether the data set is a complete representation of what it purports to be. Are all the roads included in the data set of roads?

15.6.4 Logical Consistency

Logical consistency reports on whether a database is self-consistent. Thus, is a network of roads consistent so that it can be used in network analysis? Does a data set of polygons exhaustively cover the area of study without holes, which are not meant to be there, and without unintended overlaps? It could also be used to report on whether the soils mapped in a region are consistent with the model of soil occurrence as it relates to the climate and hypsography of a region.

15.6.5 Semantic Accuracy

Late in the day, Salgé (1995) introduced the idea that actually a database may not be semantically correct. At its simplest, this means that, for example, woodlands defined in the database as oak woodlands may not actually match the definition of oak woodlands on which the database is based. This concept of quality has still had very little impact upon the standards process and data quality specifications (FGDC 1998; Buehler and McKee 1998).

15.6.6 Currency

Currency is clearly important in the use of information. While the date of creation of a database is a fundamental metadata field (FGDC 1998), currency is neither estimated nor are the consequences of a data set out of currency even considered.

15.7 Lineage or an Audit of Uncertainty

With respect to any single spatial database there is usually a long history of development. This should be express through the *lineage* of the information, but concerns with whom did what and when in the lineage may make it hard to reconstruct the decisions. Therefore the idea of an ***Audit of Uncertainty*** has been suggested (Fisher and Tate 2006).

In an audit all possible causes of uncertainty in the creation of a database will be documented and shown as a flow chart, from conceptual uncertainty to measurement error (Figure 15-14). Each is introduced at a particular stage and each contributes to the overall uncertainty. The audit may demonstrate that it is not actually possible to trace the consequences of all possible forms of uncertainty since they are lost in the history of the data (which may go back a long way). In other cases, it may indeed be possible to track every step in the construction of the data. The audit trail is also important when verifying the accuracy or uncertainty of data. It can be used to specify exactly where the accuracy comparison is being made and, therefore, how the accuracy is being tested.

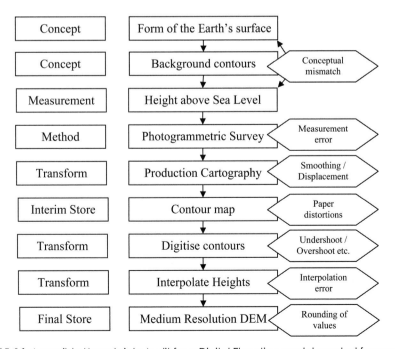

Figure 15-14 A possible Uncertainty Audit for a Digital Elevation model created from contours.

15.8 Working with Uncertainty

Ultimately the description of uncertainty in the form of data quality is supposed to enable the user of a set of information to be able to judge the fitness of that data for their use. This is not necessarily the case, however. To transform any measure of uncertainty into an evaluation requires considerable computing and subtlety. Here we will discuss two methods, one based on probability and one based on fuzziness. Both are methods of working with uncertainty models. The first allows the judgment of suitability to be made, while the second enables the user to model some of the uncertainty in the decision process.

15.8.1 Probability Models

As an example, it is common to assess the visual impact of proposed wind farms as part of a wider environmental impact assessment. It is straightforward to find the binary viewshed that answers this question (Figure 16). As part of the assessment it should be normal practice to establish the impact of DEM error on the area from which the wind farm can be seen, given that the visible area is determined from a line of sight through a flawed DEM. There is a risk that lines of sight identified as having a clear view of the wind farm will not, and equally those that are reported as outside the viewshed may actually be able to see it. We can achieve an estimate of the risk of this by **_Monte Carlo Simulation_** (Figure 15-15), which uses a probability-based estimate of error in the DEM (see Section 15.5.1). This statistical model is used to simulate random fields that each might resemble the error in the DEM. We can evaluate the effect of that error on the visibility by averaging the viewsheds to yield the probable viewshed.

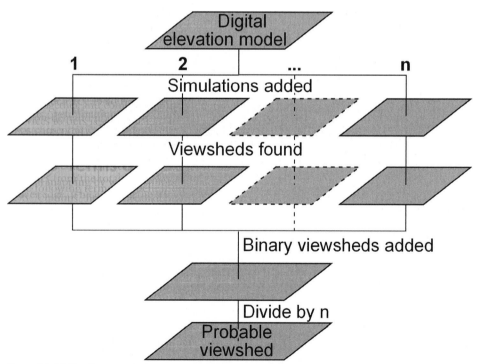

Figure 15-15 The flow model for determining the Probability Viewshed by Monte Carlo simulation.

We noted in Section 15.5.1 that the RMSE = s when the mean error is zero; therefore, the simplest error model we can use is to generate random numbers drawn from a normal distribution to populate a white-noise field error model (assuming there is no spatial autocorrelation to conform with the reporting of data quality). The resulting probable viewshed is shown in Figure 15-16B, and should be compared to the binary version in Figure 15-16A. It is possible to simulate very large positive spatial autocorrelations in the error field. This yields the error field shown in Figure 15-16C. There is a huge difference between Figures 15-16B and 15-16C—the first shows a pattern with very few high-probability areas, while the second has almost no low-probability areas. The one is as unlikely as the other. Using the variogram (Figure 15-6D) in the simulation of the error fields, and conditioning the simulation with the actual measured errors at the spot heights (Figure 15-6A), we can derive the probable viewshed in Figure 15-16D. This has a much more complex distribution of error that is vastly more believable. A much greater amount of area than in Figure 15-16B has a high probability of being in view, and a much larger area has some level of probability. It also makes use of all the information on error; none is omitted from consideration. This is the model that should be presented to an inquiry, a manager or householders. From this last probable viewshed it is possible to identify the risk of anywhere being able to see the wind farm. Holmes et al. (2000) use geostatistics to examine the effect of DEM error on GIS-derived hydrological parameters.

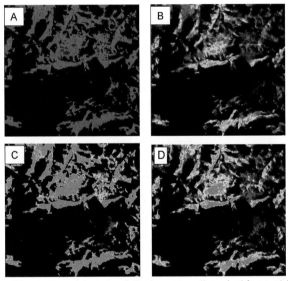

Figure 15-16 A) The binary area (viewshed) that can see the wind farm at the green crosses; B) the equivalent probable viewshed using the RMSE only for error simulation; C) the equivalent probable viewshed using the RMSE and simulating very high spatial autocorrelation; D) the equivalent probable viewshed conditioned to the variogram and the observed errors at spot heights (based on Fisher 1998). See included DVD for color version.

This type of analysis is widespread in the modeling community as a way of evaluating the sensitivity of any model to the uncertainty in model parameters. Aerts et al. (2003) present an interesting example that, like the one discussed here, looks at the sensitivity of the outcomes to a single input spatial variable. Very few people look at the influence of spatial patterns of uncertainty in more complex situations (where for example multiple spatial data types are uncertain and included in the analysis), although it is accommodated in the modeling structure of PCRaster (Karssenberg and de Jong 2005). One example where multiple spatial data sets are included in a single analysis is given by Davis and Keller 1997. Random or systematic sampling of the distributions of the parameters reveals critical values of the parameters in the model (Crosetto and Tarantola 2001) and most especially in the GLUE (Generalized Linear Uncertainty Estimation) model (Beven and Binley 1992).

15.8.2 Fuzzy Models

Apart from fuzzy models of space extending our understandings of geographic phenomena such as mountains or land covers, fuzzy set theory can also be used to refine our analysis of space. For example, we can take a simple model of site location (sometimes known as sieve mapping). This is based on the work of Langford (1991) known as *Getting Started with GIS*. It is a site location analysis for a sports ground in a location near Loughborough, UK. The ground originally needed to meet the following criteria:

- Within 500 m of the urban area
- Within 450 m of the Motorway, A roads or B roads
- Slopes less than 2.5%
- Grade III land (pasture and scrubland) only
- The parcel must be more than 2.5 ha.

The first four are based on mapped information while the last can be calculated as part of the GIS processing. The whole can be constructed as a cartographic model summarized in Figure 15-17, and in the final step five land plots are determined to be suitable based on these criteria.

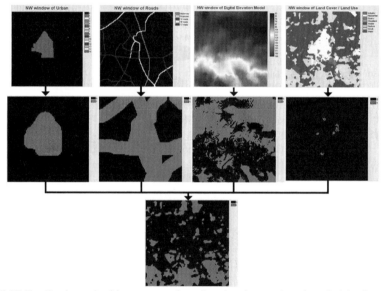

Figure 15-17 The Boolean decision process from four land mappings to suitable sites greater than 2.5 ha. See included DVD for color version.

This is what is called a Boolean Analysis; any location is either suitable or unsuitable. However, we can ask why is a location 501 m away from the urban area unsuitable, when a location 500 m away is suitable? Why is a location with 2.3% slope suitable when a location with 3% slope is unsuitable (especially when construction machinery can easily level the land)?

In short, the problem can be reformulated as a fuzzy decision problem. It is therefore possible to identify values for all mapped criteria so that the ideal locations can be identified (which may be less than the thresholds above) and the maximum possible values can also be identified. These are break points in a fuzzy membership function. Table 15-4 shows the threshold values and Figure 15-18 illustrates the comparison between the fuzzy and Boolean model of distance from the roads.

Table 15-4 The thresholds of ideal and possible criteria for fuzzy membership functions of suitable land.

Criteria	Maximum Ideal	Maximum Possible
Urban proximity	100	1000
Road proximity	50	600
Slope	1%	5%
Land	III (pasture and scrubland	IV (woodland too)

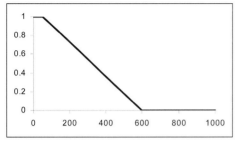 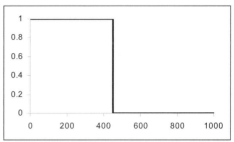

Figure 15-18 Fuzzy and Boolean membership functions for the road criterion in the land selection process.

The fuzzification of the criteria can be fit into the cartographic model (Figure 15-19) for the decision process. The separate criteria (now known as factors because they are not binary variables) can be merged using a minimize operation (actually the same operation as in the Boolean analysis; Figure 15-19A). From these, the plots that meet the area criteria can be isolated (Figure 15-19B). A map of the fuzzy memberships in those land plots are shown in Figure 15-19C. The outcomes are more varied and require more explanation than in the Boolean analysis.

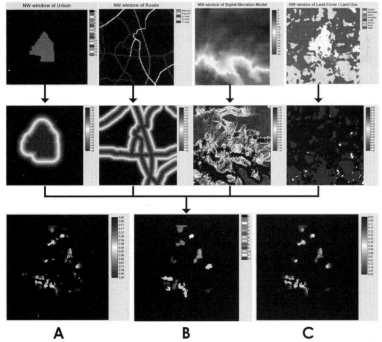

A   B   C

Figure 15-19 The Fuzzy decision process from four land mappings to three alternative mappings of degrees of suitability. See included DVD for color version.

Manual of Geographic Information Systems

Twelve plots are greater than 2.5 ha with some degree of suitability for the purpose. From the cumulative area graph of fuzzy memberships (Figure 15-20) we can see that one land plot, Plot 4, is the largest at every fuzzy membership level except 0.1, and usually it is much larger than any other plot. This plot is co-located with the largest single plot in the Boolean analysis, which demonstrates conformity between the analyses. However, the second most suitable plot from the fuzzy analysis, Plot 7 (the largest plot with 0.7 membership) is not even identified in the Boolean analysis. The fuzzy analysis would appear to provide a much more complex decision space, which means that not only is the decision itself more complex but decision making can be more subtle. Clearly tracking back to why plots have been identified in either the Boolean or the fuzzy analysis is also possible, but it is suggested that the fuzzy approach is a clear improvement over the Boolean.

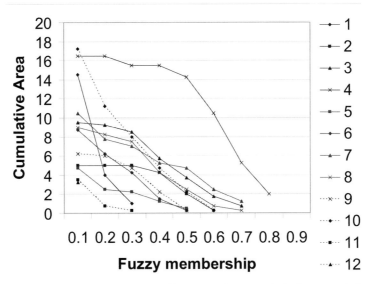

Figure 15-20 The areas of suitable plots identified in the fuzzy analysis. The cumulative area is shown according to fuzzy suitability. See included DVD for color version.

Fuzzy decision-making within GIS operations is increasingly used. Banai (1993), for example, documents how aspects of fuzzy set theory can be used in locating a landfill site. Fuzzy models are one of the methods suggested by Bonham-Carter (1995) for locating exploitable mineral resources. Some GIS software (Idrisi, for example) integrate functions to facilitate this approach, although the suitability analysis done here could also be done in a number of other GIS packages.

15.9 Conclusion

A large number of issues are related to uncertainty that can occur in geographic information. As we have seen uncertainty has many realizations and many aspects. The situation of spatial information is far from unique, and what is stated here could be said of many different types of information. However, it is unusual to find the conceptualization and categorization problems together with measurement problems and the positional issues all in a single context, and possibly entering a single analytical situation.

Perhaps the most important part of understanding and modeling uncertainty is conceptualizing the uncertainty itself. This starts with defining the exact source of possible uncertainty in a particular situation. That done, it is possible to recognize where errors or misunderstandings can occur and why, and it may be possible to use methods of fuzzy set theory or probability combinations to examine different possible outcomes in the analysis of the uncertain information.

The issues of uncertainty extend well beyond those discussed here. At one level semantic heterogeneity, semantic uncertainty or spatial uncertainty, as discussed, are all problems for interoperability of two or more data sets. On first examination, uncertainty can make the analysis of information problematic at best and invalid at worst, but if modeled and accommodated the analysis can actually derive more useful and more believable results than when done without that consideration.

One important aspect of uncertainty in spatial information that has been omitted from discussion here is that of the uncertainty that arises from processing information. Not all Geographic Information Systems produce the same results; different areas are defined by what appear to be the same operation in different systems; different numerical values may be defined for the same spatial object when the same analysis is conducted in more than one GIS. The results of analysis may actually be ambiguous. This issue has been documented by a number of pieces of research, but a general approach to the problem has not been developed.

Uncertainty and errors associated with geographic information and geographic information processing have made headlines on a number of occasions, such as when drivers following the instructions of their in-car navigation systems have driven into a river or onto a railroad track. A 1998 episode in Germany, in which a driver drove down a ferry ramp and into a river, clearly states the consequences of semantic uncertainty. In the German topographic databases, river crossings are not distinguished and so bridges are not separated from ferry crossings.

References

Aerts, J.C.J.H., M. F. Goodchild and G.B.M. Heuvelink. 2003. Accounting for spatial uncertainty in optimization with spatial decision support systems. *Transactions in GIS* 7 (2):211–230.

Ahlqvist, O. 2005. Using uncertain conceptual spaces to translate between land cover categories. *International Journal of Geographical Information Science* 19 (7):831–857.

Anderson, J. R., E. E. Hardy, J. T. Roach and R. E. Witmer. 1976. *A Land Use and Land Cover Classification System for Use with Remote Sensor Data.* US Geological Survey, Professional Paper 964, Reston, Va. <http://landcover.usgs.gov/pdf/anderson.pdf> Accessed 3 November 2008.

Banai, R. 1993. Fuzziness in geographical information systems: Contribution from the analytic hierarchy process. *International Journal of Geographical Information Systems* 7 (4):315–329.

Beven, K. J. and A. Binley. 1992. The future of distributed models: Model calibration and uncertainty prediction. *Hydrological Processes* 6:279–298.

Bezdek, J. C. 1981. *Pattern Recognition with Fuzzy Objective Function Algorithms.* New York: Plenum Press.

Bezdek, J. C., R. Ehrlich and W. Full. 1984. FCM: The fuzzy c-means clustering algorithm. *Computers & Geosciences* 10:191–203.

Bonham-Carter, G. 1995. *Geographic Information Systems for Geoscientists: Modeling with GIS.* Oxford: Pergamon.

Buehler, K. and L. McKee, eds. 1998. *The Open GIS Guide: Introduction to Interoperable Geoprocessing and the OpenGIS Specification.* 3rd ed. Draft. <http://gis.geo.hm.edu/klauer/gi/OpenGISGuide980629.pdf> Accessed 3 November 2008.

Comber, A. J., P. F. Fisher and R. A.Wadsworth. 2003. Actor network theory: A suitable framework to understand how land cover mapping projects develop? *Land Use Policy* 20:299–309.

———. 2004. Integrating land cover data with different ontologies: Identifying change from inconsistency. *International Journal of Geographical Information Science* 18 (7):691–708.

———. 2005 Comparing statistical and semantic approaches for identifying change from land cover datasets. *Journal of Environmental Management* 77, 47–55.

Congalton , R. G. and K. Green. 1999. *Assessing the Accuracy of Remotely Sensed Data; Principles and Practices*. Boca Raton, Fla.: Lewis.

Congalton, R. G. and R. A. Mead. 1983. A quantitative method to test for consistency and correctness in photo-interpretation. *Photogrammetric Engineering and Remote Sensing* 49:69–74.

Crosetto, M. and S.Tarantola. 2001. Uncertainty and sensitivity analysis: Tools for GIS-based model implementation. *International Journal of Geographical Information Science* 15 (5):415–437.

Dassonville, L., F. Vauglin, A. Jakobsson and C. Luzet. 2002. Quality management, data quality and users, metadata for geographical information. Pages 202–215 in *Spatial Data Quality*. Edited by W. Shi, P. F. Fisher and M. F. Goodchild. London: Taylor & Francis.

Davis, T. J. and C. P. Keller. 1997. Modeling uncertainty in natural resource analysis using fuzzy sets and Monte Carlo simulation: Slope stability prediction. *International Journal of Geographical Information Science* 11 (5):409–434.

FGDC (Federal Geographic Data Committee). 1998. *Content Standard for Digital Geospatial Metadata*. FGDC-STD-001-1998, National Technical Information Service, Computer Products Office, Springfield, Virginia, USA.

Fisher, P. F. 1997. The pixel: A snare and a delusion. *International Journal of Remote Sensing* 18 (3):679–685.

———. 1998. Improved modeling of elevation error with geostatistics. *GeoInformatica* 2 (3):215–233

———. 1999. Models of uncertainty in spatial data. Pages 191–205 in *Geographical Information Systems: Principles, Techniques, Management and Applications*. Vol 1. Edited by P. Longley, M. Goodchild, D. Maguire and D. Rhind. New York: Wiley and Sons.

———. 2000. Sorites paradox and vague geographies. *Fuzzy Sets and Systems* 113 (1):7–18.

———. 2003. Data quality and uncertainty: Ships passing in the night! Pages 17–22 in *Proceedings of the 2nd International Symposium on Spatial Data Quality '03*. Hong Kong Polytechnic University.

Fisher, P. F. and S. Pathirana. 1990. The evaluation of fuzzy membership of land cover classes in the suburban zone. *Remote Sensing of Environment* 34:121–132.

Fisher, P. F. and N. J. Tate. 2006. Causes and consequences of error in digital elevation models. *Progress in Physical Geography* 30:467–489.

Fisher P. F., T. Cheng and J. Wood. 2004. Where is Helvellyn? Multiscale morphometry and the mountains of the English Lake District. *Transactions of the Institute of British Geographers* 29:106–128.

Fisher P. F., A. J.Comber and R. Wadsworth. 2005. Nature de l'incertitude pour les données spatiales (Approaches to uncertainty in spatial data). Pages 47–63 in *Qualité de l'Information Géographique*. Edited by R. Devillers and R. Jeansoulin. Paris: Hermes Press.

Foody, G. M. 1995. Land cover classification by an artificial neural network with ancillary information. *International Journal of Geographical Information Systems* 9:527–542.

———. 1996. Approaches to the production and evaluation of fuzzy land cover classification from remotely-sensed data. *International Journal of Remote Sensing* 17:1317–1340.

Goodchild, M. and S. Gopal, eds. 1989. *Accuracy of Spatial Databases*. London: Taylor & Francis.

Guptill, S. C. and J. L. Morrison, eds. 1995. *Elements of Spatial Data Quality*. Oxford: Elsevier.

Hagen, A. 2003. Fuzzy set approach to assessing similarity of categorical maps. *International Journal of Geographical Information Science* 17 (3):235–249.

Hagen-Zanker, A., B. Straatman and I. Uljee. 2005. Further developments of a fuzzy set map comparison approach. *International Journal of Geographical Information Science* 19 (7):769–785.

Harvey, F. and N. R. Chrisman. 1998. Boundary objects and the social construction of GIS technology. *Environment and Planning A* 30 (9):1683–1694.

Heuvelink, G.B.M. 1998. *Error Propagation in Environmental Modeling with GIS.* Research Monographs in GIS Series. London: Taylor and Francis.

Holmes, K. W., O. A.Chadwick and P. C. Kyriakidis. 2000. Error in a USGS 30m DEM and its impact on terrain modeling. *Journal of Hydrology* 233:154–173.

Hunsaker, C., M. Goodchild, M. Friedl and P. Case, eds. 2001. *Spatial Uncertainty in Ecology.* New York: Springer.

ISO. 2002. Geographic information: quality principles, ISO 191113:2002, International Organization for Standardization, Zurich.

Karssenberg, D. and K. de Jong. 2005. Dynamic environmental modeling in GIS: 2. Modelling error propagation. *International Journal of Geographical Information Science* 19: 623–637.

Kwadijk, J. and E. Sprokkereef. 1998. Development of a GIS for Hydrological modelling of the River Rhine. Pages 47–55 in *European Geographic Information Infrastructures: Opportunities and Pitfall.* Edited by P. Burrough and I. Masser. London: Taylor & Francis.

Langford, M. 1991. *Getting Started in GIS.* Booklet and disks. Leicester, UK: Midlands Regional Research Laboratory.

Lanter, D. P. and H. Veregin. 1992. A research paradigm for propagating error in layer-based GIS. *Photogrammetric Engineering and Remote Sensing* 58 (7):825–833

Lees, B. 1996. Improving the spatial extension of point data by changing the data model. In *Proceeding of the Third International Conference/Workshop on Integrating GIS and Environmental Modeling.* Santa Barbara, Calif., National Center for Geographic Information and Analysis, 1996. CD-ROM <http://www.ncgia.ucsb.edu/conf/SANTA_FE_CD-ROM/sf_papers/lees_brian/santafe2.html> Accessed 3 November 2008.

Leung, Y., J.-H. Ma and M. F. Goodchild. 2004. A general framework for error analysis in measurement-based GIS Part 1: The basic measurement-error model and related concepts. *Journal of Geographical Systems* 6 (4):325–354.

Library of Congress. 2008. Metadata Object Description Schema. <http://www.loc.gov/standards/mods/> Accessed 6 November 2008.

McBratney, A. B. and J. J. De Gruijter. 1992. A continuum approach to soil classification by modified fuzzy k-means with extragrades. *Journal of Soil Science* 43:159–175.

McGwire, K. and P. F. Fisher. 2001. Spatially variable thematic accuracy: Beyond the confusion matrix. Pages 308–329 in *Spatial Uncertainty in Ecology.* Edited by C. Hunsaker, M. Goodchild, M. Friedl and P. Case. New York: Springer.

Odeh, I.O.A., A. B. McBratney and D. J.Chittleborough. 1990. Design and optimal sample spacings for mapping soil using fuzzy k-means and regionalized variable theory. *Geoderma* 47:93–122.

Openshaw, S. 1995. *Census User's Handbook.* Chichester: Wiley & Sons.

Petry, F., V. Robinson and M. Cobb. 2005. *Fuzzy Modeling with Spatial Information for Geographic Problems.* New York: Springer.

Powell, B., A. B. McBratney and D. A. MacLeod. 1991. The application of ordination and fuzzy classification techniques to field pedology and soil stratigraphy in the Lockyer Valley, Queensland. *Catena* 18:409–420.

Power, C., A. Simms and R. White. 2001. Hierarchical fuzzy pattern matching for the regional comparison of land use maps. *International Journal of Geographical Information Science* 15 (1):77–100.

Rees, P., D. Martin and P. Williamson. 2002. *The Census Data System*. Chichester: Wiley & Sons.

Rhind, D. 1983. *The Census Users Handbook*. London: Routledge.

Robinson, V. 2003. A perspective on the fundamentals of fuzzy sets and their use in geographic information systems. *Transactions in GIS* 7:3–30.

Robinson, V., F. Petry and M. Cobb. 2003. Fuzzy sets in geographic information systems. *Transactions in GIS* 7 (1):1.

Salgé, F. 1995. Semantic accuracy. Pages 139–151 in *Elements of Spatial Data Quality*. Edited by S. C. Guptill and J. L. Morrison. Oxford: Elsevier.

Shi, W. 1998. A generic statistical approach for modeling error of geometric features in GIS. *International Journal of Geographical Information Science* 12 (2):131–144.

Shi, W., M. F. Goodchild and P. F. Fisher, eds. 2002. *Spatial Data Quality*. London: Taylor & Francis.

Soil Survey Staff. 1975. Soil Taxonomy: A Basic System of Soil Classification for Making and Interpreting Soil Surveys. US Government Printing Office.

———. 1999. Soil Taxonomy: A Basic System of Soil Classification for Making and Interpreting Soil Surveys. 2nd ed. US Government Printing House, Pittsburgh, Penn. <http://soils.usda.gov/technical/classification/taxonomy/> Accessed 3 November 2008.

———. 1993. Soil survey manual. Soil Conservation Service. US Department of Agriculture Handbook 18. <http://soils.usda.gov/technical/manual/> Accessed 3 November 2008.

White, Paul. 2005. Major terror attacks triple in '04 by U.S. count. <http://www.alipac.us/ftopicp-11341.html> Accessed 6 November 2008.

Wilkinson, G. G., F. Fierens and I. Kanellopoulos. 1995. Integration of neural and statistical approaches in spatial data classification. *Geographical Systems* 2:1–20.

Zadeh, L. A. 1965. Fuzzy sets. *Information and Control* 8:338–353.

Zhang, J. and M. F. Goodchild. 2002. *Uncertainty in Geographical Information*. London: Taylor & Francis.

CHAPTER 16
Elliptical Method for Estimating Positional Accuracy

Thomas W. Owens

16.1 Introduction

A valid positional accuracy estimation method, which calculates an estimate of the difference between test coordinates and reference coordinates, should be reproducible, statistically valid, sufficient, and easy to calculate to ensure it is practically applicable. Reproducible means that similar results will be obtained over several tests and accuracy statements can be compared for different sets of data. Statistically valid means that the method provides a probability estimate about the accuracy statement. Sufficient means that all relevant parameters are described. Finally, a positional accuracy estimation method must be relatively easy to calculate. In this era of computers and spreadsheets, it is possible to calculate complex formulae with relative ease.

This chapter presents a positional accuracy estimation method that is an alternative to current standard methods. This alternative method, referred to here as the *elliptical method,* meets the four criteria listed above, while standard estimation methods do not fully meet these criteria.

16.2 Current Positional Accuracy Estimation Methods

The United States Federal Geographic Data Committee (FGDC) maintains the National Standard for Spatial Data Accuracy (NSSDA) reporting standard for the United States Federal Government. The NSSDA uses root-mean-square error (RMSE) to estimate positional accuracy. RMSE is the square root of the average of the set of squared differences between test coordinate values and reference coordinate values for identical points. Accuracy is reported in ground distances at the 95% confidence level (which is determined by adjusting the calculated RMSE by a conversion factor), and means that 95% of the positions in the data set will have an error with respect to true ground position that is equal to or smaller than the reported accuracy value. This is a reporting standard and not a positional accuracy standard. The FGDC encourages organizations to establish positional accuracy standards for their spatial data (see FGDC 1998 for details on calculating the RMSE).

The NSSDA is the primary accuracy reporting standard found in geospatial literature. It supersedes the 1947 National Map Accuracy Standard (US Bureau of the Budget 1947) and the 1990 ASPRS Accuracy Standards for Large-Scale Maps (ASPRS 1990). The NSSDA is used by organizations such as the US Geological Survey, the US Forest Service, and the State of Minnesota Planning organization.

The NSSDA meets three of the four criteria listed above for a valid accuracy standard: it is reproducible, statistically valid, and easily calculated. However, the NSSDA is not sufficient; it does not describe all parameters of the error function. Using a circle assumes that the X and Y variance in error are identical. Variance in the X direction and variance in the Y direction are related, but not necessarily identical, so this cannot be assumed. The NSSDA also assumes that systematic error (bias in measurement due to equipment error or environmental conditions) has been eliminated, which also cannot be assumed. It is for these reasons that the author proposes an alternative method for estimating positional accuracy that meets all four criteria.

16.3 Elliptical Method for Estimating Positional Accuracy

When a sample of test coordinates is compared to control locations (which are assumed to be more accurate), the quantity of interest is the distance of test coordinates from the control coordinates in the X and Y directions. The X and Y distances from the test coordinate and the control coordinates is a bivariate quantity, and the joint probability distribution function can be described by a probability ellipse with the center at $\overline{X}, \overline{Y}$. An ellipse is an appropriate shape for a joint probability distribution function because it has two dimensions; it is not rectangular because the joint probability of points occurring in the corners is small, and it is generally not circular, because X and Y are not necessarily the same (Batschelet 1981).

If the individual dimensions $(X$ and $Y)$ are linear and normally distributed, they can be described by two statistics: (1) the mean, and (2) the standard deviation. The ellipse can be described with five statistics: the sample means of (1) X, and (2) Y, and the sample standard deviations of (3) X, (4) Y, and (5) the correlation coefficient between X and Y. The quantities X and Y are jointly distributed, where X and Y depend on each other, but different pairs are independent of each other in a sample. From these statistics, ellipses can be constructed.

Elliptical measures of positional accuracy assume that the parent populations are normally distributed. This assumption can be tested by the Chi-squared goodness of fit for normality (for methods to use the Chi-squared test see Zar 2006).

There are three ellipses that need to be calculated: the standard ellipse, the confidence ellipse (Batschelet 1981), and the tolerance ellipse (Chew 1966). The standard ellipse is a descriptive tool used to visualize the shape of the ellipse and its orientation and cannot be used for statistical inference. The confidence ellipse estimates accuracy—how close an estimate is to the true value. If the confidence ellipse covers the origin, there is no systematic error in the sample. If it does not cover the origin, there is systematic error in the sample. The tolerance ellipse estimates precision by developing a confidence interval with a percentage of the population sampled enclosed by the ellipse.

Five statistics are needed to construct the ellipses: $\overline{X}, \overline{Y}, S_x^2, S_y^2$, and the correlation coefficient r where $r = Cov(X, Y)/S_x S_y$ and $Cov(X, Y) = 1/(n-1)\Sigma(X_i - \overline{X})(Y_i - \overline{Y})$. The center of the ellipse is $\overline{X}, \overline{Y}$.

1) **Standard ellipse**—the function is described by:

$$S_y^2(X-\overline{X})^2 - 2rS_xS_y(X-\overline{X})(Y-\overline{Y}) + S_x^2(Y-\overline{Y}) = (1-r^2)S_x^2S_y^2 \qquad (16\text{-}1)$$

The standard ellipse is a descriptive tool used to visualize the distribution and serves the same purpose as $\overline{x} \pm s$ does for univariate functions.

An ellipse is defined by a major semi-axis (a), a minor semi-axis (b), and an angle θ of translation, which is the angle that the major axis is offset from the X axis (and the angle the minor axis is offset from the Y axis). The standard ellipse parameters are (for ease of calculation, some coefficients are shortened):

$$\text{Major semi-axis } a = [2D/(A + C - R)]^{1/2}, \qquad (16\text{-}2)$$

$$\text{Minor semi-axis } b = [2D/(A + C + R)]^{1/2}, \text{ and} \qquad (16\text{-}3)$$

$$\text{Angle of translation } \theta = \arctan[2B/(A - C - R)]. \qquad (16\text{-}4)$$

where $A = s_y^2$, $B = -rs_xs_y$, $C = s_x^2$, $D_s = (1 - r^2)s_x^2s_y^2$ and $R = [(A - C)^2 + 4B^2]^{1/2}$

2) **Confidence ellipse**—the function is described by:

$$S_y^2(X-\overline{X})^2 - 2rS_xS_y(X-\overline{X})(Y-\overline{Y}) + S_x^2(Y-\overline{Y}) = (1-r^2)S_x^2S_y^2n^{-1}T^2 \qquad (16\text{-}5)$$

The confidence ellipse serves the same purpose as $\bar{x} \pm tsn^{1/2}$ does for univariate functions.

Note that the right side of Equation 16-5 has an added term, T^2, which is based on the familiar F value for the univariate solution where

$$T^2 = 2[(n-1)/(n-2)]F_{2,n-2}. \qquad (16-6)$$

$F_{\acute{a}/1,\,n-2}$ denotes the critical value from the F one-tailed distribution with $n-2$ degrees of freedom and a significance level of $\acute{a}$.

The shortened coefficients are as follows:

$$\text{Major semi-axis } a = [2D/(A + C - R)]^{1/2}, \qquad (16-7)$$

$$\text{Minor semi-axis } b = [2D/(A + C + R)]^{1/2}, \text{ and} \qquad (16-8)$$

$$\text{Angle of translation } \theta = \arctan[2B/(A - C - R)] \qquad (16-9)$$

where $A = S_y^2$, $B = -rS_xS_y$, $C = S_x^2$, $D_s = (1 - r^2)S_x^2S_y^2 n^{-1}T^2$ and $R = [(A - C)^2 + 4B^2]^{1/2}$.

It is important to note whether or not the confidence ellipse includes the origin of the graph (0, 0). This graphically shows whether or not a systematic difference exists between the sample and the expected result of no difference. This can be statistically tested using Hotelling's one-sample test (Batschelet 1981).

3) **Tolerance ellipse**–the function is described by (from Owens and McConnville 1996 and Chew 1966):

$$s_y^2(X - \overline{X})^2 - 2rs_xs_y(X - \overline{X})(Y - \overline{Y}) + s_x^2(Y - \overline{Y}) = (1 - r^2)s_x^2s_y^2 n^{-1}H \qquad (16-10)$$

Where H is based on the non-central χ^2 distribution and is approximated by Table 2 in Chew (1966). This function uses the same calculation coefficients as above except for D_s:

$$D_s = (1 - r^2)s_x^2s_y^2H \qquad (16-11)$$

The remaining coefficients a, b, and θ are calculated using the same equations as for the normal and confidence ellipses.

Consider the case shown below in Figure 16-1, which shows the distance in meters of test locations from their control locations on the X and Y axes. The data in this figure are in Table 16-1, a list of 40 test coordinates in the Twin Cities derived from digital terrain model data sets created using photogrammetric techniques. The reference (control) points were collected from Global Positioning System (GPS) receivers with a reported horizontal accuracy of 10–15 mm. The Minnesota Department of Transportation's photogrammetric unit produced these data, which are used to plan and design roadways and roadway improvements (Minnesota Planning 1999, page 11).

The NSSDA standard, as stated in the report, is 0.181 meters, which means that 95% of the positions in the test data set will have an error with respect to the control positions that is equal to or smaller than 0.181 meters. Figure 16-2 shows how the NSSDA standard appears on the graphed data.

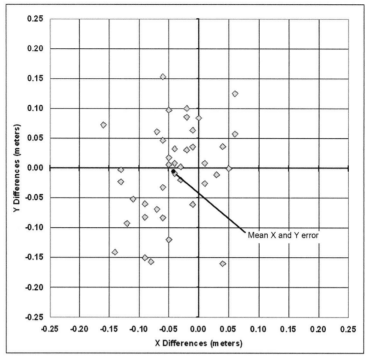

Figure 16-1 Graph of *X* and *Y* differences between test locations and control locations in meters. The black round point is the average *X* and *Y* difference. Data are from Table 16-1.

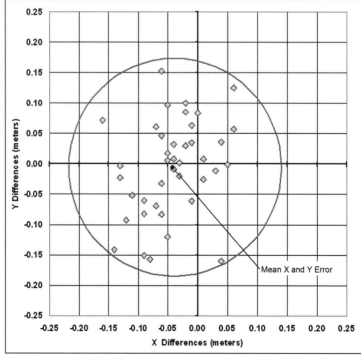

Figure 16-2 NSSDA data standard for reporting positional error from Table 16-1. 95% of the points will fall within the circle, which has a radius of 0.181 meters and is centered on the average *X* and *Y* error (the black round point).

Table 16–1 Differences in meters between the coordinates from measured (test) points and control points in spatial data derived from 1:3000 scale aerial photography. Courtesy: Minnesota Department of Transportation.

Point Number	X Control	Y Control	X Test	Y Test	X Difference	Y Difference
1	178247.28	48326.075	178247.37	48326.135	−0.09	−0.06
2	178249.23	48287.228	178249.17	48287.171	0.06	0.06
3	178456.79	48337.408	178456.73	48337.283	0.06	0.13
4	178715.82	48542.511	178715.88	48542.543	−0.06	−0.03
5	179047.54	48657.388	179047.65	48657.44	−0.11	−0.05
6	179227.78	48336.177	179227.8	48336.147	−0.02	0.03
7	179238.56	48671.457	179238.69	48671.48	−0.13	−0.02
9	180257.36	48337.972	180257.39	48337.97	−0.03	0.00
10	180426.36	48445.001	180426.36	48444.917	0.00	0.08
11	180568.35	48523.693	180568.48	48523.696	−0.13	0.00
12	180680.73	48275.075	180680.78	48274.978	−0.05	0.10
13	180676.31	48413.085	180676.38	48413.154	−0.07	−0.07
14	180654.46	47955.055	180654.47	47954.992	−0.01	0.06
15	180843.48	48505.391	180843.56	48505.548	−0.08	−0.16
17	181338.97	48313.103	181339.11	48313.244	−0.14	−0.14
18	181283.2	48174.063	181283.25	48174.057	−0.05	0.01
19	181075.07	48171.737	181075.09	48171.637	−0.02	0.10
20	181495.79	48043.414	181495.85	48043.497	−0.06	−0.08
21	181679.58	48242.779	181679.59	48242.744	−0.01	0.04
22	181673.86	48579.533	181673.82	48579.693	0.04	−0.16
24	181937.26	48136.264	181937.3	48136.256	−0.04	0.01
26	182085.95	48127.717	182085.96	48127.778	−0.01	−0.06
27	182243.61	48032.915	182243.57	48032.879	0.04	0.04
28	182289.49	48729.272	182289.56	48729.211	−0.07	0.06
29	182259.51	48630.614	182259.63	48630.707	−0.12	−0.09
30	182277.52	48410.278	182277.57	48410.398	−0.05	−0.12
32	182590.79	48437.482	182590.88	48437.633	−0.09	−0.15
33	182494.13	48422.78	182494.22	48422.862	−0.09	−0.08
34	182410.21	48672.544	182410.24	48672.564	−0.03	−0.02
35	182740.18	48307.436	182740.15	48307.447	0.03	−0.01
36	182771.78	47967.3	182771.77	47967.292	0.01	0.01
37	183067.28	48044.513	183067.27	48044.539	0.01	−0.03
38	183242.23	47952.797	183242.18	47952.798	0.05	0.00
39	183458.2	47885.194	183458.24	47885.162	−0.04	0.03
40	183778.2	48230.799	183778.26	48230.753	−0.06	0.05
41	183886.38	47924.349	183886.4	47924.264	−0.02	0.08
42	184394.5	48083.648	184394.54	48083.657	−0.04	−0.01
43	184644.38	48068.904	184644.44	48068.751	−0.06	0.15
44	184804.19	48192.963	184804.35	48192.891	−0.16	0.07
45	185120.62	48201.523	185120.67	48201.505	−0.05	0.02
				Average	−0.042	−0.006

Manual of Geographic Information Systems

Calculating the ellipses

Using the data from Table 16-1, the coefficients for the standard ellipse are as follows:

$$A = s_y^2 = 0.006, \ B = -\text{Cov}(X,Y) = -0.001,$$

$$C = s_x^2 = 0.003, \ D_s = (1 - r^2)s_x^2 s_y^2 = 0.00002,$$

$$R = [(A - C)^2 + 4B^2]^{1/2} = 0.004,$$

$$\text{Major semi-axis } a = [2D/(A + C - R)]^{1/2} = 0.082,$$

$$\text{Minor semi-axis } b = [2D/(A + C + R)]^{1/2} = 0.051,$$

$$\text{Angle of translation } \theta = \arctan[2B/(A - C - R)] = 1.217,$$

and when converted from radians to degrees = 69.72°.

The ellipse is centered on –0.042 m on the X-axis and –0.006 m on the Y-axis, the major semi-axis is 0.082 m, the minor semi-axis is 0.051 m, and the angle of translation is 69.72°.

Next we turn to calculating the coefficients for the confidence ellipse. In our example, we have two variables, 40 samples, and are using $\acute{\alpha} = 0.05$ (95% confidence level), so $F_{.05/1, 38} = 3.24$. (F values calculated from statistical functions tables, see Rodriguez 1998). Plugging in our values to the equations $T^2 = 2[(n - 1)/(n - 2)]F_{2,n-2}$, $T^2 = 2[(40–1)/40–2)]3.24 = 6.66$.

$$A = s_y^2 = \ 0.006, \ B = -\text{Cov}(X,Y) = -0.001,$$

$$C = s_x^2 = 0.003, \ D_s = (1 - r^2)s_x^2 s_y^2 n^{-1} T^2 = 0.000003,$$

$$R = [(A - C)^2 + 4B^2]^{1/2} = 0.004,$$

$$\text{Major semi-axis } a = [2D/(A + C - R)]^{1/2} = 0.035,$$

$$\text{Minor semi-axis } b = [2D/(A + C + R)]^{1/2} = 0.022,$$

$$\text{Angle of translation } \theta = \arctan[2B/(A - C - R)] = 1.217,$$

and when converted from radians to degrees = 69.72°.

The confidence ellipse is centered on –0.042 m on the Y-axis and -0.006 m on the X-axis, the major semi-axis is 0.035 m, the minor semi-axis is 0.022 m, and the angle of translation is 69.72°.

The coefficients for the tolerance ellipse are:

$$H_{\acute{\alpha}=0.05, \ n = 38} = 8.17 \text{ (from Table 2 in Chew 1966)}$$

$$A = s_y^2 = .006, \ B = -r s_x s_y = -.001$$

$$C = s_x^2 = 0.003, \text{ and } D_s = (1 - r^2)s_x^2 s_y^2 H = 0.001, \text{ and}$$

$$R = [(A - C)^2 + 4B^2]^{1/2} = 0.004$$

Again, the ellipse parameters are as follows:

$$\text{Major axis } a = [2D/(A + C - R)]^{1/2} = 0.233,$$

$$\text{Minor axis } b = [2D/(A + C + R)]^{1/2} = 0.146, \text{ and}$$

$$\text{Angle of translation } \theta = \arctan[2B/(A - C - R)] = 1.217,$$

and when converted from radians to degrees = 69.72°.

The tolerance ellipse is centered on –0.042 m on the *Y*-axis and –0.006 m on the *X*-axis, the major semi-axis is 0.233 m, the minor semi-axis is 0.146 m, and the angle of translation is 69.72°.

Figure 16-3 presents the ellipses that were calculated above.

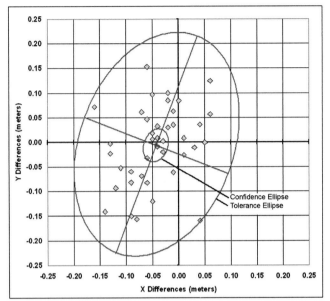

Figure 16-3 Positional errors from Table 16-1, with calculated ellipses.

The confidence ellipse, which is the smaller ellipse, does not encompass the origin (0, 0), which graphically shows that the sample mean is significantly different than zero. It is centered on (0.042, –0.006), has a major axis of .070 meters, a minor axis of 0.44 meters and an angle of 69.7°. This graphically shows that a systematic difference exists between the sample and the expected result of no difference. This can be statistically tested using Hotelling's one-sample test. The larger ellipse, named the tolerance ellipse, contains 95% of the population sampled (at the 0.05 level), is centered on the same point as the confidence ellipse, has a major axis of 0.466 meters, and a minor axis of 0.295 meters with an angle of 69.7°. Thus, if an additional point were taken, we are 95% confident that it would fall within this tolerance ellipse.

This method of analyzing the positional accuracy meets all four of the criteria discussed in the introduction of this section; it is repeatable, statistically valid, sufficient, easily calculated, and it has the added merit of being graphically intuitive. It shows the differences in the test sample from the control data, and how *X* and *Y* vary. It graphically shows if there is a systematic (as opposed to uncontrollable random) error, which is a crucial piece of information the NSSDA does not show.

Here is a second example from Minnesota Planning's Positional Accuracy Handbook (1999). The positional accuracy of Washington County's parcel base data set was tested. Covering 425 square miles with approximately 82,000 parcels of land ownership, the data set has features typically found in half-section maps, including plat boundaries, lot lines, right-of-way lines, road centerlines, easements, lakes, rivers, ponds and other requirements of county land record management. Readily available GPS equipment capable of producing submeter results prompted use of field-measured locations for control. Such a level of accuracy would meet the NSSDA stipulation of using an independent source of data of the highest accuracy feasible and practical to evaluate the accuracy of the test data set. The resultant data are shown in Table 16-2:

Table 16–2 Differences in meters between the coordinates from test points and control points in spatial data derived from parcel base data. Courtesy: Minnesota Department of Transportation.

Point Number	X Control	Y Control	X Test	Y Test	X Difference	Y Difference
2	512838.5	265305.5	512832.1	265304.6	6.42	0.88
3	513804.6	265289.0	513779.1	265292.1	25.45	–3.09
4	506995.3	259036.3	506987.0	259039.1	8.37	–2.86
5	505890.0	267608.1	505900.1	267586.4	–10.10	21.67
6	499522.9	268070.5	499517.0	268057.6	5.89	12.93
7	500886.3	277084.7	500889.9	277076.8	–3.56	7.89
9	506832.2	284524.3	506833.0	284524.2	–0.76	0.06
15	512469.9	300556.5	512494.5	300550.2	–24.59	6.32
16	499541.7	295469.8	499542.9	295470.4	–1.19	–0.59
19	495674.4	295158.3	495672.6	295155.5	1.81	2.85
20	493348.4	283897.9	493356.4	283893.7	–8.03	4.25
21	486511.1	275873.7	486512.7	275878.2	–1.62	–4.56
22	483617.1	275899.3	483617.9	275902.5	–0.74	–3.28
23	479455.9	291709.0	479472.7	291683.7	–16.76	25.30
24	469037.3	298365.9	469025.2	298366.2	12.13	–0.30
25	456160.1	300964.3	456172.4	300971.0	–12.28	–6.68
26	453048.4	300995.9	453051.8	301016.3	–3.48	–20.41
31	471995.9	289610.8	472008.0	289606.5	–12.03	4.29
32	471828.5	289748.1	471845.7	289734.9	–17.14	13.18
33	473084.2	283256.8	473083.9	283250.2	0.33	6.62
34	459897.8	254995.3	459900.2	254990.2	–2.40	5.14
35	475603.3	244363.6	475603.0	244371.5	0.37	–7.89
36	489350.1	256106.4	489350.2	256110.2	–0.07	–3.80
36	467667.6	272602.0	467674.1	272610.6	–6.47	–8.54
37	483572.5	269361.1	483572.6	269357.2	–0.04	3.94
37	452112.0	277311.0	452108.6	277322.3	3.47	–11.30
38	494171.3	238673.6	494160.1	238666.1	11.19	7.56
38	451973.5	269628.3	451977.1	269625.6	–3.63	2.63
40	505295.9	223453.4	505293.2	223446.2	2.76	7.19
41	497444.7	218442.2	497461.9	218479.5	–17.23	–37.27
41	473066.7	264154.8	473069.3	264153.8	–2.58	1.05
42	481800.1	213775.5	481797.1	213762.8	2.96	12.72
43	475144.9	233082.6	475146.2	233082.4	–1.39	0.22
44	466236.2	211022.2	466238.0	211022.3	–1.75	–0.08
45	475253.3	189933.6	475247.4	189931.4	5.92	2.14
46	472999.9	175394.1	473000.8	175391.2	–0.95	2.91
47	461164.5	163210.1	461162.2	163207.4	2.32	2.74
49	460948.6	140008.2	460948.0	140006.3	0.58	1.93
50	496582.7	147523.5	496567.2	147537.0	15.46	–13.52
51	474434.7	126212.9	474434.6	126207.5	0.15	5.39
52	460963.8	118776.2	460964.1	118775.2	–0.32	1.02
53	493944.3	106859.1	493949.9	106859.2	–5.63	–0.14
54	500142.5	206064.8	500140.4	206063.0	2.09	1.79
55	513038.7	203149.3	513036.5	203144.5	2.21	4.79
56	498843.6	189987.1	498848.6	189984.9	–4.99	2.22
57	516059.2	180143.2	516059.1	180136.5	0.19	6.68
58	492428.1	173572.2	492427.6	173557.3	0.45	14.91
59	500207.8	173314.2	500207.1	173312.6	0.66	1.63
60	512300.0	162002.5	512306.3	162006.0	–6.30	–3.46
62	513787.8	128008.1	513805.5	128012.0	–17.72	–3.89
				Average	–1.45	1.26

Manual of Geographic Information Systems

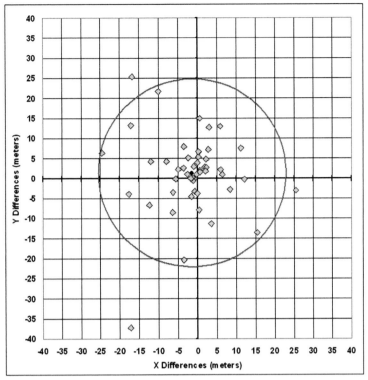

Figure 16-4 NSSDA data standard for reporting positional error from Table 16-2. 95% of the points will fall within the red circle, which has a radius of 23 meters and is centered on the average error of (–1.45, 1.26) as shown by the black round point.

Calculating the ellipses

Using the data from Table 16-2, the coefficients for the standard ellipse are as follows:

$$A = s_y^2 = 0.0094.39, \ B = -\text{Cov}(X, Y) = 4.28,$$

$$C = s_x^2 = 76.21, \ D_s = (1 - r^2)s_x^2 s_y^2 = 7174.55,$$

$$R = [(A - C)^2 + 4B^2]^{1/2} = 20.09,$$

$$\text{Major semi-axis } a = [2D/(A + C - R)]^{1/2} = 9.76,$$

$$\text{Minor semi-axis } b = [2D/(A + C + R)]^{1/2} = 8.67,$$

$$\text{Angle of translation } \theta = \arctan[2B/(A - C - R)] = -1.35,$$

$$\text{and when converted from radians to degrees} = -77.38°.$$

The standard ellipse is centered on –1.45 m on the X-axis and 1.26 m on the Y-axis, the major semi-axis is 9.76 m, the minor semi-axis is 8.67 m, and the angle of translation is –77.38° (Figure 16-4).

Next we turn to calculating the coefficients for the confidence ellipse. In our example, we have two variables, 51 samples, and are using $\acute{\alpha} = 0.05$ (95% confidence level), so $F_{.05/1, 49} = 3.179$. (F values calculated from statistical functions tables, see Rodriquez 1998). Plugging in our values to the equations $T^2 = 2[(n-1)/(n-2)]F_{2,n-2}$, $T^2 = 2[(51-1)/51-2)]3.24 = 6.49$.

$$A = s_y^2 = 94.39, B = -\text{Cov}(X,Y) = 4.28,$$

$$C = s_x^2 = 76.21, D_s = (1 - r^2)s_x^2 s_y^2 n^{-1} T^2 = 1292.96,$$

$$R = [(A - C)^2 + 4B^2]^{\frac{1}{2}} = 20.09,$$

$$\text{Major semi-axis } a = [2D/(A + C - R)]^{\frac{1}{2}} = 4.15,$$

$$\text{Minor semi-axis } b = [2D/(A + C + R)]^{\frac{1}{2}} = 3.68,$$

$$\text{Angle of translation } \theta = \arctan[2B/(A - C - R)] = -1.35,$$

and when converted from radians to degrees $= -77.38°$.

The confidence ellipse is centered on -1.45 m on the Y-axis and 1.26 m on the X-axis, the major semi-axis is 4.15 m, the minor semi-axis is 3.68 m, and the angle of translation is $-77.38°$. The coefficients for the tolerance ellipse are:

$$H_{\dot{\alpha}=0.05, n=49} = 7.86 \text{ (from Table 2 in Chew 1966)}$$

$$A = s_y^2 = 94.39, B = -rs_x s_y = 4.28$$

$$C = s_x^2 = 76.21, \text{ and } D_s = (1 - r^2)s_x^2 s_y^2 H = 56391.96, \text{ and}$$

$$R = [(A - C)^2 + 4B^2]^{\frac{1}{2}} = 20.29$$

The tolerance ellipse parameters are as follows:

$$\text{Major semi-axis } a = [2D/(A + C - R)]^{\frac{1}{2}} = 27.37,$$

$$\text{Minor semi-axis } b = [2D/(A + C + R)]^{\frac{1}{2}} = 24.32, \text{ and}$$

$$\text{Angle of translation } \theta = \arctan[2B/(A - C - R)] = -1.35,$$

and when converted from radians to degrees $= -77.38°$.

The tolerance ellipse is centered on -1.45 m on the Y-axis and 1.26 m on the X-axis, the major semi-axis is 27.37 m, the minor semi-axis is 24.32 m, and the angle of translation is $-77.38°$.

The confidence and tolerance ellipses are shown in Figure 16-5.

The confidence ellipse encompasses the origin, which means there is no systematic error in this sample set. The major and minor axes in the tolerance ellipse are 54.7 and 48.6 meters, respectively, which means they are very similar, but not identical. The angle of the ellipse is $-77.38°$, which means that the X and Y errors are arrayed along this axis.

It is evident that the elliptical analysis produces rich, informative results and provides sufficient, critical statistical information on systematic error. A proposed elliptical reporting standard would include the average X and Y distances from the reference, the major and minor axes of the confidence and tolerance ellipses and their offset angles, along with a graph showing the distribution of errors and the ellipses. An accuracy standard should include, at a minimum, the requirement that the data contain no systematic error.

Acknowledgement

Data provided courtesy of the Minnesota Land Management Information Center.

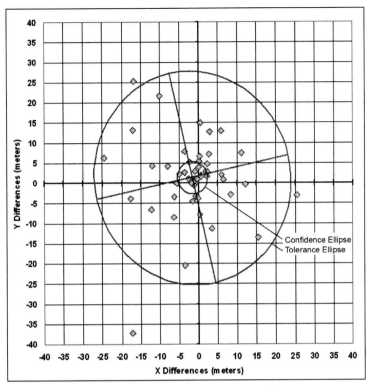

Figure 16-5 Positional errors from Table 16-2, with calculated ellipses.

References

ASPRS (American Society for Photogrammetry and Remote Sensing) Specifications and Standards Committee. 1990. ASPRS accuracy standards for large-scale maps. *Photogrammetric Engineering and Remote Sensing* 56 (7): 1068–1070.

Batschelet, E. 1981. *Circular Statistics in Biology.* New York: Academic Press.

Chew, V. 1966. Confidence, predictive, and tolerance regions for the multivariate normal distribution. *American Statistical Association Journal* 61:605–617.

FGDC (Federal Geographic Data Committee). 1998. FGDC-STD-007.3-1998 *Geospatial Positioning Accuracy Standards Part 3: National Standard for Spatial Data Accuracy.* Washington, D.C.: Federal Geographic Data Committee. <http://www.fgdc. gov/standards/projects/FGDC-standards-projects/accuracy/part3/chapter3> Accessed 25 March 2008.

Minnesota Planning. 1999. *Positional Accuracy Handbook.* St. Paul, Minn.: Minnesota Planning. <http://www.mnplan.state.mn.us/pdf/1999/lmic/nssda_o.pdf> Accessed 25 March 2008.

Owens, T. and D. McConville. 1996. *Geospatial Application: Estimating the Spatial Accuracy of Coordinates Collected Using the Global Positioning System.* National Biological Service, Environmental Management Technical Center, Onalaska, Wisconsin, April 1996. LTRMP 96-T002. <http://www.umesc.usgs.gov/documents/reports/1996/96t002.pdf> Accessed 25 March 2008.

Rodriguez, C. 1998. Online statistical tables. University of Albany. <http://omega.albany. edu:8008/mat108dir/Problems/Tables.html> Accessed 22 March 2008.

US Bureau of the Budget. 1947. *United States National Map Accuracy Standards.* Washington, D.C.: US Bureau of the Budget.

Zar, J. 2006. *Biostatistical Analysis,* 5th ed. Englewood Cliffs, NJ: Prentice Hall.

CHAPTER 17
Standards for Geospatial Interoperability

George Percivall and *Arliss Whiteside*

17.1 Introduction

The widespread application of computers and use of geographic information systems (GIS) have led to the increased analysis of geographic data within multiple disciplines. Based on advances in information technology, society's reliance on such data is growing. Geographic data sets are increasingly being shared, exchanged, and used for purposes other than their producers' intended ones. GIS, remote sensing, automated mapping and facilities management (AM/FM), traffic analysis, geopositioning systems, and other technologies for Geographic Information (GI) are entering a period of radical integration.

Standards for geospatial interoperability provide a framework for developers to create software that enables users to access and process geographic data from a variety of sources across a generic computing interface within an open information technology environment.

- "a framework for developers" means that the International Standards are based on a comprehensive, common (i.e., formed by consensus for general use) plan for interoperable geoprocessing.
- "access and process" means that geodata users can query remote databases and control remote processing resources, and also take advantage of other distributed computing technologies such as software delivered to the user's local environment from a remote environment for temporary use.
- "from a variety of sources" means that users will have access to data acquired in a variety of ways and stored in a wide variety of relational and non-relational databases.
- "across a generic computing interface" means that standard interfaces provide reliable communication between otherwise disparate software resources that are equipped to use these interfaces.
- "within an open information technology environment" means that the standards enable geoprocessing to take place outside of the closed environment of monolithic GIS, remote sensing, and AM/FM systems that control and restrict database, user interface, network, and data manipulation functions.

This chapter summarizes the most significant aspects of the OGC (Open Geospatial Consortium) web services (OWS) architecture. This architecture is a service-oriented architecture, with all components providing one or more services to other services or to clients. Because that architecture is not yet completed, some aspects are not described here, and other aspects may change in the future.

The definition of services includes a variety of applications with different levels of functionality to access and use geographic information. While specialized services will appropriately remain an area for proprietary products, standardization of the interfaces to those services allows interoperability between proprietary products. Developers of geographic information systems and software will use these standards to provide general and specialized services that can be used for all geographic information. The approach of this Chapter and the referenced standards is integrated with the approaches being developed within the more general world of information technology.

17.2 OWS Architecture Overview[1]

The OGC web services (OWS) architecture, which the OGC is currently developing, includes five significant groups of properties:

a) Service components are organized into multiple tiers.

    1) All components provide services, to clients and/or other components, and each component is usually called a service (with multiple implementations) or a server (each implementation).

    2) Services (or components) are loosely arranged in four tiers, from Clients to Application Services to Processing Services to Information Management Services, but un-needed tiers can be bypassed.

    3) Services can use other services within the same tier, and this is common in the Processing Services tier.

    4) Each tier of services has a general purpose, which is independent of geographic data and services.

    5) Each tier of services includes multiple specific types of services, many of which are tailored to geographic data and services.

    6) Servers can operate on (tightly bound) data stored in that server and/or on (loosely bound) data retrieved from another server.

b) Services use is usually chained.

    1) Services can be chained with other services and often are chained, either transparently (defined and controlled by the client), translucently (predefined but visible to the client), and opaquely (predefined and not visible to the client), see Subclause 7.3.5 of ISO 19119.

    2) Services are support to facilitate defining and executing chains of services.

    3) Some service interfaces support server storage of operation results until requested by the next service in a chain.

c) Services communication uses open Internet standards.

    1) Communication between components uses standard World Wide Web (WWW) protocols, namely HTTP GET, HTTP POST, and SOAP.

    2) Specific server operations are addressed using Uniform Resource Locators (URLs).

    3) Multipurpose Internet Mail Extensions (MIME) types are used to identify data transfer formats.

    4) Data transferred is often encoded using the Extensible Markup Language (XML), with the contents and format specified using XML Schemas.

d) Service interfaces use open standards and are relatively simple.

    1) All services support open standard interfaces from their clients, often OGC-specified service interfaces.

    2) OGC web service interfaces are coarse-grained, providing only a few static operations per service.

    3) Service operations are normally stateless, not requiring servers to retain interface state between operations.

    4) One server can implement multiple service interfaces whenever useful.

    5) Service interfaces share common parts whenever practical.

    6) Service interfaces can have multiple specified levels of functional compliance, and multiple specialized subset and/or superset profiles.

    7) Standard XML-based data encoding languages are specified for use in data transfers.

    8) Geographic data and service concepts are closely based on the ISO 191XX series of standards.

---

1. A list of abbreviations and acronyms is provided at the end of the chapter.

9) Standard specifications are used for defining and referencing well-known coordinate reference systems (CRSs).

e) Server and client implementations are not constrained.

1) Services are implemented by software executing on general purpose computers connected to the Internet. The architecture is hardware and software vendor neutral.

2) The same and cooperating services can be implemented by servers that are owned and operated by independent organizations.

3) Many services are implemented by standards-based Commercial Off The Shelf (COTS) software.

4) All services are self-describing, supporting dynamic (just-in-time) connection binding of services supporting publish-find-bind.

17.3 Services Tiers

17.3.1 Overview

Except for clients, all OWS architecture components provide services, to clients and/or to other components. Each such component is usually called a service when multiple implementations are expected, and each implementation is called a server (or service instance). These components are thus usually called services or servers in this chapter.

Clients are software packages that provide access to a human user, or operate as agents on behalf of other software. Software that provides access to a human user can be thin (e.g., a web browser), thick (a large application), or "chubby" (in between).

All services (or components) are loosely organized in four tiers, as shown in Figure 17-1. This organization is loose in that clients and services can bypass un-needed tiers, as indicated by some arrows. Services can use other services within the same tier, and this is common especially in the Processing Services tier. Also, some services perform functions of more than one tier, when those functions are often used together and combined implementation is more efficient. Assignment of such combined services to tiers is somewhat arbitrary.

NOTE: Complete separation of services into tiers is not required, especially when separation would be inefficient.

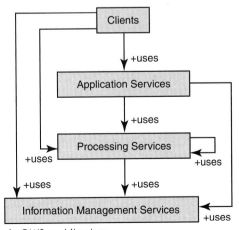

Figure 17-1 Service tiers in OWS architecture.

This OWS architecture is designed for use where data is important and often voluminous. Servers can operate on (tightly bound) data stored in that server and/or on (loosely bound) data retrieved from another server. Most data is stored by the servers in the Information Management Services tier, but some data (can be and often) is stored in other services and servers.

Manual of Geographic Information Systems

Each tier of services has a general purpose, as indicated by the names in Figure 17-1. That tier name is independent of geographic data and services, since some tier services are not specific to geographic data or services. Each tier of services includes multiple specific types of services, many of which are tailored to geographic data and services. Some of the specific services included in each tier are discussed in the following subsections.

17.3.2 Application Services Tier

The Application Services tier contains services designed to support Clients, especially thin client software such as web browsers. That is, these Application Services are designed for use by clients instead of each client directly performing these often-needed support functions. The services in the Application Services tier are used by Clients, and can use other services in the Application Services, Processing Services, and Information Management Services tiers. The specific services included in this tier include (but are not limited to) the services listed in Table 17-1.

Table 17-1 Some Application Services.

Service Name	Service Description
Web portal services	Services that allow a user to interact with multiple application services for different data types and purposes
WMS application services	Services that allow a user to interact with a Web Map Service (WMS) to find, style, and get data of interest
Gazetteer application services	Services that allow a user to interact with a Gazetteer service
Geographic data discovery services	Services that allow a user to locate and browse metadata about geographic data, interacting with a catalog
Geographic data extraction services	Services that allow a user to extract and edit feature data, interacting with images and feature data
Geographic data management services	Services that allow a user to manage geospatial data input and deletion, interacting with Information Management Services
Map style management services	Services that allow a user to create, edit, and manage map styles, interacting with Information Management Services
Map symbol management services	Services that allow a user to create, edit, and manage map symbols, interacting with Information Management Services
Data structure management services	Services that allow a user to create, edit, and manage data structures, interacting with Information Management Services
Feature generalization application services	Services that allow a user to interact with Processing services for feature generalization
Coverage generalization services	Services that allow a user to interact with Processing services for coverage generalization
Other application services	Services that allow a user to interact with other Processing and Information Management services
Chain definition services	Services to define a service chain and enable it to be executed by the workflow enactment service; may also provide a chain validation service
Workflow enactment services	Services to interpret chain definitions and control instantiation of servers and sequencing of activities, maintaining internal state information associated with various services being executed
Access control services	Services that control access to other servers, for privacy, intellectual property, and other reasons
Usage accounting services	Services that keep track of the usage of other servers, for billing and other purposes

17.3.3 Processing Services Tier

The Processing Services tier contains services designed to process data, sometimes both feature and image (coverage) data. The services in the Processing Services tier are used by clients and by services in the Application Services tier. These services can use other services in the Processing Services and Information Management Services tiers. The specific services included in this tier include (but are not limited to) the services listed in Table 17-2.

Table 17-2 Some specific Processing Services.

Service Name[a]	Service Description
SLD (Styled Layer Descriptor) Web Map Service (WMS) [b]	Dynamically produces spatially referenced maps from geographic feature and/or coverage data, returning client-specified pictorial renderings of maps in an image format (not actual feature data or coverage data)
Web Terrain Service (WTS)[b]	Dynamically produces client-specified perspective views from geographic feature and/or coverage data, returning client-specified pictorial renderings of data in an image or graphics format
Web 3D Service (W3DS)	Dynamically produces client-specified perspective views from geographic feature data, returning perspective of feature data in a graphical format
Web Coordinate Transformation Service (WCTS)[b]	Transforms the coordinates of feature or coverage data from one coordinate reference system (CRS) to another, including "transformations," "conversions," rectification, and orthorectification
Web Image Classification Service (WICS)	Performs classification of digital images, using client-selected supervised or unsupervised image classification method
Feature Portrayal Service (FPS)	Dynamically produces client-specified pictorial renderings in an image or graphics format of features and feature collections usually dynamically retrieved from a Web Feature Server (WFS)
Coverage Portrayal Service (CPS)	Dynamically produces client-specified pictorial renderings in an image or graphics format of a coverage subset dynamically retrieved from a Web Coverage Service (WCS)
Geoparser Service	Service to scan text documents for location-based references, such as a place names, addresses, postal codes, etc., for passage to a geocoding service.
Geocoder Service	Service to augment location-based text references with position coordinates
Geolinking Service (GLS)[b]	Service that links geospatial data
Geolinked Data Access Service (GDAS)	Service that uses linked geospatial data
Geographic data extraction services	Services supporting extraction of feature and terrain information from images
Dimension measurement services	Services that compute dimensions of objects visible in an image or other geospatial data
Route determination services	Determine optimal path between two specified points based on input parameters and properties contained in a Feature Collection; may also determine distance between points and/or time to follow path
Proximity analysis services	Given a position or geographic feature, finds all objects with a specified set of properties that are located within a user-specified distance of the position or feature
Change detection services	Services to find differences between two data sets that represent the same geographical area at different times
Data alignment services	Service that adjusts sensor geometry models to improve the match of a coverage (or image) with other coverages and/or known ground positions
Feature generalization services	Service that reduces spatial variation in a feature collection to counteract the undesirable effects of scale reduction
Coverage generalization services	Service that reduces spatial variation in a coverage to counteract the undesirable effects of scale reduction
Format conversion services	Service that converts data from one format to another, including data compression and decompression
Semantic translation services[a]	Service that converts data from one set of semantics to another
[a]Names ending in "Service" are currently specified specific services. Names ending in "services" are types of services that are not yet specified. [b]Can process both feature and image (coverage) data.	

17.3.4 Information Management Services Tier

The Information Management Services tier contains services designed to store and provide access to data, normally handling multiple separate data sets. In addition, metadata describing multiple data sets can be stored and searched. Access is usually to retrieve a client-specified subset of a stored dataset, or to retrieve selected metadata for all data sets whose metadata meets client-specified query constraints.

The services in the Information Management Services tier are used by clients and by services in the Application Services and Processing Services tiers. These services can use other services in the Information Management Services tier. The specific services included in this tier include (but are not limited to) the services listed in Table 17-3.

Table 17-3 Some Specific Information Management Services.

Service Name[a]	Service Description
Web Map Service (WMS)[b]	Dynamically produces spatially referenced maps of client-specified ground rectangle from one or more client-selected geographic data sets, returning pre-defined pictorial renderings of maps in an image or graphics format
Web Feature Service (WFS)	Retrieves stored features and feature collections that meet client-specified selection criteria
Web Coverage Service (WCS)	Retrieves client-specified subset of client-specified coverage (or image) dataset
Catalog Service for the WEB (CSW)[c]	Retrieves object metadata stored that meets client-specified query criteria
Gazetteer Service	Retrieves location geometries for client-specified geographic names
Universal Description, Discovery and Integration (UDDI) Service	Allows a client to find a web-based service
Standing order services	Allows a user to request data over a geographic area be disseminated when it becomes available, including reformat, compress, decompress, prioritize, and transmit information requested through standing queries or profiles
Order handling services	Allows clients to order products from a provider, including: selection of geographic processing options, obtaining quotes on orders, submission of order, statusing of orders, billing, and accounting

[a]Names ending in "Service" are currently specified specific services. Names ending in "services" are types of services that are not yet specified.
[b]Can store and access both feature and image (coverage) data.
[c]Many specific profiles of the CSW are expected to be specified and implemented, for metadata for many different types of data sets, and also for storing and accessing small whole data sets.

A Catalog Service for the WEB (CSW) server can store metadata (and perhaps also data sets) for one or more types of data sets, including (but not limited to) the types listed in Table 17-4.

Table 17-4 Some Specific Types of Data Sets.

Type Name	Type Description
Service	Definition of a service or server
Feature	Geographic features and feature collections (including composite features)
Coverage (including image)	Geographic coverage or image, can be a gridded coverage or another type of coverage, can be georectified or not
Styled Layer Descriptor (SLD) document	Specifies client-controlled styling for map portrayal of features and images (coverages)
Map Symbol	Defines map display symbols
Web Map Context document	Specifies a composite, symbolized map view that can be saved, restored, and transmitted to other viewers, using WMS only
OWS Map Context document	Specifies a composite, symbolized map view that can be saved, restored, and transmitted to other viewers, using WFS, WCS, WMS, etc.
Query template	Template for OGC Common Catalog Query Language queries
Filter Encoding (FE) template	Encodes queries for features or other data meeting specified constraints
URN Definitions XML document	Encodes in XML definitions of OGC-defined URNs
GML Application Schema	Geography Markup Language (GML) application schema and/or profile for specific application
OWS XML Schema	XML Schema used by an OWS or other service
General XML Schema	Any XML Schema
UML model	Any UML (Unified Modeling Language) model
Web Service Description Language (WSDL) document	Specifies web service interface
Business Process Execution Language (BPEL) document	Specifies process sequences for a specific purpose
Accounting record	Records usage of servers and other resources, for billing and other purposes

17.4 Services Chaining

In many cases, multiple services must be used together to perform a useful function. The OWS architecture thus supports "chaining" together of multiple servers, and such chaining is frequently used. This chaining is not limited to a linear chain; a network of services can also be "chained." Within such a chain, most servers input the data that is output from the previous server in the chain. Services can be chained transparently (defined and controlled by the client), translucently (predefined but visible to the client), and opaquely (predefined and not visible to client), see Subclause 7.3.5 of OGC 02-006, ISO 19119.

To facilitate service chaining, some services are defined to support defining and executing chains of services. Also, some Processing Service interfaces are designed to support retrieving the data to be processed from another service, which can be an Information Management Service or another Processing Service.

To allow more efficient execution of server chains, some service interfaces support server storage of operation results until requested by the next service in a chain.

17.5 Services Communication

17.5.1 Overview

Communication between clients and services, and between services, uses only open non-proprietary Internet standards. That is, the OWS architecture uses the Internet or equivalent as its distributed computing platform (DCP). More specifically, communication between components uses standard World Wide Web (WWW) protocols, namely HTTP GET, HTTP POST, and Simple Object Adaptor Protocol (SOAP). Specific operations of specific servers are addressed using Uniform Resource Locators (URLs). Multipurpose Internet Mail Extensions (MIME) types are used to identify data transfer formats. The data transferred is often encoded using the Extensible Markup Language (XML), with the contents and format carefully specified using XML Schemas.

17.5.2 HTTP GET Operation Requests

In many cases, a request to perform an operation by a service is transferred as a Hypertext Transfer Protocol (HTTP) GET message. That GET message is addressed to an HTTP Uniform Resource Locator (URL), where that URL locates a specific operation of a specific server. A URL for an HTTP GET request is in fact only a URL prefix, to which additional parameters are appended to construct a valid operation request. The prefix defines the network address to which operation request messages are sent, and may also identify a configuration of that server.

A query is appended to the URL prefix to form a complete request message. Each OWS operation request has mandatory and/or optional request parameters. Each parameter has a defined name, and has multiple allowed values. To formulate the query part of the URL, the mandatory request parameters, and any desired optional parameters, are appended as name/value pairs in the form "name=value&" (parameter name, equals sign, parameter value, ampersand). In the OGC, this parameter encoding is often referred to as keyword value pair (KVP) encoding.

17.5.3 HTTP POST Operation Requests

Less frequently, a request to perform an operation by a service is transferred as a Hypertext Transfer Protocol (HTTP) POST message. That POST message is addressed to a (possibly different) HTTP Uniform Resource Locator (URL), where that URL locates a specific operation of a specific server. A URL for an HTTP POST request is a complete URL (not merely a prefix as in the HTTP GET case).

Clients transmit request parameters to the URL in the body of the HTTP POST message. An OWS does not require additional parameters to be appended to the URL in order to construct a valid target for the operation request. When HTTP POST is used, the operation request message is normally encoded as an XML document, formatted as specified by one or more XML Schemas. The operation request message can alternately be KVP encoded, in the body of the HTTP POST message.

17.5.4 HTTP Operation Responses

After receiving an operation request, a server replies with a response message corresponding exactly to the request, or sends an exception report if unable to respond correctly. Responses to operation requests are the same whether the request is transferred by HTTP GET or POST. In most cases, the operation response is encoded in XML, using XML Schemas to specify the correct response contents and format. This is true for both normal and exception operation responses, which are separately specified for each OWS service.

All XML Schemas contain documentation of the meaning of each specified element, attribute, and type. All of these documentation elements are considered normative, unless labeled "informative." Almost all of the concrete XML elements defined in these OWS Schemas can be used without separate XML Schemas, whenever no content extensions or restrictions are needed. An additional XML Schema is used whenever element contents extension is required, and should be used in some other cases to specify needed restrictions.

A server may send an HTTP Redirect message (using HTTP response codes as defined in IETF RFC 2616 (Fielding et al. 1999)) to an absolute URL that is different from the valid request URL that was sent by the client. HTTP Redirect causes the client to issue a new HTTP request for the new URL. Several redirects could in theory occur. Practically speaking, the redirect sequence ends when the server responds with an operation response. The final response shall be an OWS operation response that corresponds exactly to the original operation request, or an exception report.

17.5.5 MIME Types Use

Response messages are accompanied by the appropriate Multipurpose Internet Mail Extensions (MIME) type for that message. A list of MIME types in common use on the internet is maintained by the Internet Assigned Numbers Authority (IANA). A server can support parameterized MIME types, and this is common to more completely identify the specific format. In addition to parameterized MIME types, servers usually offer the basic un-parameterized version of the format, for clients that do not understand the parameterized MIME type.

Response messages are accompanied by other HTTP entity headers as appropriate. In particular, the Expires and Last-Modified headers provide important information for data caching. Content-Length may be used by clients to know when data transmission is complete and to efficiently allocate space for results, and Content-Encoding or Content-Transfer-Encoding may be necessary for proper interpretation of the results. When returning a large XML document, some form of data compression should be supported; since client-server communication transfer speeds will be considerably faster if the document is compressed.

17.5.6 SOAP Operation Requests and Responses

In some cases, a request to perform a specific operation by a specific server can be transferred in a Simple Object Adaptor Protocol (SOAP) operation request message. That SOAP request message is addressed to an HTTP Uniform Resource Locator (URL), where that URL locates a specific operation of a specific server. In this case, the operation request parameters are encoded in XML just like they can be for HTTP POST. When SOAP is used, the response from that operation is transferred in a SOAP operation response message, and is XML encoded just like responses to HTTP POST requests.

17.6 Service Interfaces

OGC web service interfaces use open standards and are relatively simple. In addition to being well-specified and tested for interoperability, the OGC-specified service interfaces are coarse-grained, providing only a few static operations per service. For many services, only three service operations are specified. One server can implement multiple service interfaces whenever useful.

NOTE 1: OGC web service interfaces are not fine-grained interfaces. Fine-grained, object-oriented interfaces typically provide tens of operations per service to be implemented and exercised, with some interface objects being dynamically created and destroyed.

NOTE 2: One service interface is not required to support all the abilities of one server. When useful, one server can implement more than one service interface. For example, some

WCS, WMS, and WFS data servers are expected to also implement CSW service interfaces cataloging the data sets available from that server.

The OGC web service interfaces are usually stateless, so session information is not passed between a client and server. Clients retain any needed interface state between operations.

NOTE 3: Client-service sessions are not used, meaning servers are not required to retain interface state between operations. This also simplifies use in a dynamic network, where a server can stop operation or fail.

The OGC web service interfaces share common parts whenever practical, allowing those parts to be specified and implemented only once. For example, all OWSs have a mandatory GetCapabilities operation to retrieve server metadata. That server metadata includes four required sections, with the contents and format of three sections common to all services, and part of the fourth section common to most services. In addition, many service interfaces have multiple specified levels of functional compliance, or multiple specialized subset and/or superset profiles.

NOTE 4: The interface parts are NOT separately and independently specified and developed. In addition to the GetCapabilities operation, all OWSs have one mandatory operation to get a data subset, and most have one optional operation to get the description of a dataset or object. Interface compliance levels and profiles inherently share interface parts wherever practical.

Standard XML-based data encoding formats and languages are used in many server-to-client and client-to-server data transfers. The formats and languages specified include (but are not limited to) those listed in Table 17-5. In these formats and languages and elsewhere, the geographic data and service concepts are closely based on the ISO 191XX series of standards.

NOTE 5: The ISO 191XX data and service concepts were developed by international groups of experts, so are carefully formulated and are being widely used. The concepts or semantics used by different services are not independently developed and specified.

Table 17-5 Some standardized encoding formats and languages.

Specification name	Description
Filter Encoding (FE)	Encodes WFS queries for features or other data meeting specified constraints
OGC Common Catalog Query Language	Encodes catalog queries for objects meeting specified constraints
Styled Layer Descriptor (SLD)	Encodes client-controlled styling for map portrayal of features and coverages (images)
Geography Markup Language (GML)	Language defined using XML Schemas based on the ISO 191XX series of standards, to be used to specify application-specific XML Schemas
Coordinate Reference Systems (part of GML)	Encodes definitions of coordinate reference systems, coordinate systems, datums, and coordinate transformations (and conversions)
Web Map Context	Encodes the context of a user application including multiple WMS service references
OWS Context	Encodes the context of a user application including multiple OWS service references
URNs using OGC URN namespace	Standardized Universal Resource Identifiers (URNs) referencing most well-known coordinate reference systems (CRSs) and grid CRSs
Sensor Model Language (SensorML)	Encodes descriptions of remote and *in-situ* sensors, including imaging and environmental sensors
Web Service Description Language (WSDL)	Encodes web service interfaces
Business Process Execution Language (BPEL)	Encodes process sequences for specific purposes

17.7 Server Implementation

Servers and client implementations are not constrained except for supporting the specified service interfaces. Each can be implemented by software executing on any general purpose computer connected to the Internet or equivalent. The architecture is hardware and software vendor neutral. The same and cooperating services can be implemented by servers that are owned and operated by independent organizations.

NOTE: Cooperating servers and clients need not be owned and operated by one organization or by formally cooperating organizations.

All OWS services and clients are implemented by available standards-based Commercial Off-the-Shelf (COTS) software. This commercial software can sometimes be used without requiring major software development, or can be adapted to specific needs with limited software development.

17.8 Service Trading (Publish – Find – Bind)

All OGC architecture services are self-describing, supporting dynamic (just-in-time) connection binding of servers using service trading. Service trading addresses discovery of available service instances. Trading facilitates the offering and the discovery of interfaces which provide services of particular types. A trader implementation records service offers and matches requests for advertised services. Publishing a capability or offering a service is called "export." Matching a service request against published offers or discovering services is called "import." This can also be depicted in an equivalent manner as the "Publish – Find – Bind" (PFB) pattern of service interaction. The fundamental roles are:

a) Trader (Registry) - registers service offers from exporter objects and returns service offers to importer objects upon request according to some criteria.

b) Exporter (Service) - registers service offers with the trader object

c) Importer (Client) - obtains service offers, satisfying some criteria, from the trader object.

NOTE: In the OWS architecture, a Registry is implemented using the Catalog Service for the WEB (CSW) service interface.

The Trading function is elaborated in document ISO/IEC 13235-1. Most importantly, a trader supports dynamic (i.e. run-time) binding between service providers and requesters, since sites and applications are frequently changing in large distributed systems. The fundamental roles and interactions are depicted in Figure 17-2[2]. The equivalent PFB terminology is shown as well (the six items in parentheses). A trader registers service offers from exporter objects and returns service offers upon request to importer objects according to some criteria.

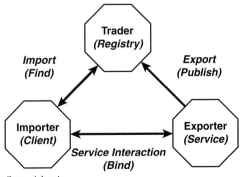

Figure 17-2 Service trading objects.

---

2. Many readers will recognize Figure 17-2: most of the recent web services white papers include similar diagrams that map onto it directly. In many cases 'find,' 'bind,' and 'publish' substitute for 'import,' 'service interaction,' and 'export,' respectively.

Manual of Geographic Information Systems

a) To publish a service offer, an Exporter gives a Trader a description of a server, including a description of the interface at which that service instance is available.

b) To find suitable server offers, an importer asks a trader for a server having certain characteristics. The trader checks the previously registered descriptions of servers, and responds to the importer with the information required to bind with a server. Preferences may be applied to the set of offers matched according to service type, constraint expressions, and various policies. Use of preferences can determine the order used to return matched offers to the importer.

c) To bind a service, an Importer applies information received from the Trader to bind to a server. The Client then proceeds to use that server.

17.9 Architecture Tiers Relationship to ISO 19119

The four tiers of the OWS architecture are loosely related to the geographic services categories specified in Subclause 8.3 of ISO 19119 Geographic information—Services (copied in OGC Abstract Specification Topic 12). This relationship is:

a) The Application Services and the Clients together provide the "human interaction services" described in Subclause 8.3.2.

b) The Processing Services provide the "processing services" described in Subclause 8.3.5. It would be possible to divide these services into the "spatial," "thematic," "temporal" and "metadata" categories described in Subclause 8.3.5.

c) The Information Management Services provide the "model/information management services" described in Subclause 8.3.3.

In this description of the OWS architecture, the storage of data sets is not separated from catalogs or registries that store and search metadata, because they are not separated in ISO 19119. Also, these services are not separated because that distinction is frequently fuzzy. Furthermore, one server can implement a dataset storage and retrieval interface plus a catalog interface for storing and searching the metadata for those data sets.

Some services perform functions of more than one tier, when those functions are usually used together and combined implementation is more efficient. Assignment of such combined services to tiers is somewhat arbitrary.

17.10 Terms and Definitions

The following terms and definitions apply to this chapter.

aggregate service
chained services appear as a single service which handles all coordination of individual services behind the aggregate service [paraphrased from ISO 19119]
NOTE: The user has no awareness that there is a set of services behind the aggregate service.

client
software component that can invoke an **operation** from a **server**

coordinate reference system
coordinate system which is related to the real world by a datum [ISO 19111]

coverage
feature that acts as a **function** to return values from its **range** for any **direct position** within its spatial, or **spatio-temporal domain**
EXAMPLE: Examples include a **raster** image, polygon overlay, or digital elevation matrix.

feature
abstraction of real world phenomena [ISO 19101]

function
rule that associates each element from a **domain** (source, or domain of the function) to a unique element in another domain (target, co-domain, or **range**) [ISO 19107]

geographic information
information concerning phenomena implicitly or explicitly associated with a location relative to the Earth [ISO 19128]

interface
named set of operations that characterize the behavior of an entity [ISO 19119]

opaque chaining
see aggregate service

operation
specification of a transformation or query that an object may be called to execute [ISO 19119]

parameter
variable whose name and value are included in an operation **request** or **response**

request
invocation of an **operation** by a **client**

response
result of an **operation**, returned from a **server** to a **client**

server
a particular instance of a **service** [ISO 19119 edited]

service
  i) distinct part of the functionality that is provided by an entity through interfaces [ISO 19119]
  ii) capability which a service provider entity makes available to a service user entity at the interface between those entities [ISO 19104 terms repository]

service chain
sequence of **services** where, for each adjacent pair of **services**, occurrence of the first action is necessary for the occurrence of the second action [ISO 19119]

service instance
see **server**

service metadata
metadata describing the **operations** and **geographic information** available at a **server** [ISO 19128]
NOTE: Most of this service metadata is specific to one server implementing a service type.

translucent chaining
execution of the chain is managed by a workflow service (or multiple workflow services) [paraphrased from ISO 19119]
NOTE: The user's involvement in the steps of the chain is mostly watching the chain execute the individual services that are visible to the user. The defined chain exists prior to the user executing the pattern.

transparent chaining
user defines and controls the order of execution of the individual services [paraphrased from ISO 19119]
NOTE: Details of the services are not hidden from the user.

user defined chaining
see transparent chaining

workflow-managed chaining
see translucent chaining

17.11 Conventions

17.11.1 Abbreviated Terms

BPEL   Business Process Execution Language
COTS   Commercial Off-the-Shelf
CRS   Coordinate Reference System
GML   Geography Markup Language
HTTP   Hypertext Transfer Protocol
ISO   International Organization for Standardization
KVP   Keyword Value Pair
MIME   Multipurpose Internet Mail Extensions
OGC   Open Geospatial Consortium
OWS   OGC Web Service, or Open Web Service
SOAP   Simple Object Adaptor Protocol
SQL   Structured Query Language
UDDI   Universal Description, Discovery and Integration Service
UML   Unified Modeling Language
URI   Universal Resource Identifier
URL   Uniform Resource Locator
URN   Universal Resource Name
WCS   Web Coverage Service
WCTS   Web Coordinate Transformation Service
WFS   Web Feature Service
WICS   Web Image Classification Service
WMS   Web Map Service
WSDL   Web Services Description Language
WTS   Web Terrain Service
XML   Extensible Markup Language

Other Sources

Berners-Lee, T., N. Fielding and L. Masinter, eds. 1998. Uniform Resource Identifiers (URI): Generic Syntax. IETF RFC 2396, August 1998. <http://www.ietf.org/rfc/rfc2396.txt> Accessed 17 February 2008.

Biron, P. V. and A. Malhotra. 2001. XML Schema Part 2: Datatypes. W3C Recommendation 2 May 2001. <http://www.w3.org/TR/2001/REC-xmlschema-2-20010502/> Accessed 18 February 2008.

Bray, T., J. Paoli, C. M. Sperberg-McQueen, E. Maler and F. Yergeau, eds. 2006. Extensible Markup Language (XML) 1.0 (Fourth edition) <http://www.w3.org/TR/REC-xml> Accessed 12 March 2008.

Doyle, A. and C. Reed, eds. 2001. Introduction to OGC Web Services. <http://portal.opengeospatial.org/files/?artifact_id=14973> Accessed 17 February 2008.

Fallside, D. C. 2001. XML Schema Part 0: Primer. W3C Recommendation 2 May 2001. <http://www.w3.org/TR/2001/REC-xmlschema-0-20010502/> Accessed 18 February 2008.

Fielding, R., J. Gettys, J. Mogul, H. Frystyk, L. Masinter, P. Leach and T. Berners-Lee, eds. 1999. Hypertext Transfer Protocol – HTTP/1.1. IETF RFC 2616, June 1999. <http://www.ietf.org/rfc/rfc2616.txt> Accessed 17 February 2008.

Freed, N. and N. Borenstein, eds. 1996. Multipurpose Internet Mail Extensions (MIME) Part One: Format of Internet Message Bodies. IETF RFC 2045, November 1996. <http://www.ietf.org/rfc/rfc2045.txt> Accessed 17 February 2008.

Horton, R. M. 2005. Spatial Interoperability Demonstration Project: Notional Architecture Executive Summary. <http://portal.opengeospatial.org/files/?artifact_id=10687> Accessed 18 February 2008.

IANA (Internet Assigned Numbers Authority). 2007. MIME Media Types. <http://www.iana.org/assignments/media-types/> Accessed 17 February 2008.

Lieberman, J., ed. 2003. OpenGIS Web Services Architecture. OGC 03-025. <http://portal.opengeospatial.org/files/?artifact_id=1320> Accessed 17 February 2008.

Moats, R. 1997. URN Syntax. IETF RFC 2141, May 1997. <http://www.ietf.org/rfc/rfc2141.txt> Accessed 17 February 2008.

Percivall, G., ed. 2002. The OpenGIS Abstract Specification Topic 12: OpenGIS Service Architecture, Version 4.3. OGC 02-006. <http://portal.opengeospatial.org/files/?artifact_id=1221> Accessed 17 February 2008.

Thompson, H. S., D. Beech, M. Maloney and N. Mendelsohn, eds. 2001. XML Schema Part 1: Structures. W3C Recommendation 2 May 2001. <http://www.w3.org/TR/2001/REC-xmlschema-1-20010502/> Accessed 18 February 2008.

Whiteside, A. 2000. Guidelines for Successful OGC Interface Specifications. OGC 00-014r1. <http://portal.opengeospatial.org/files/?artifact_id=929> Accessed 12 March 2008.

Whiteside, A., ed. 2005. OpenGIS® Web Services Common Specification, Version 1.0.0. OGC 05-008. <http://portal.opengeospatial.org/files/?artifact_id=8798> Accessed 18 February 2008.

Whiteside, A., ed. 2005. URNs of Definitions in OGC Namespace. OGC 05-010. <http://portal.opengeospatial.org/files/?artifact_id=8814> Accessed 18 February 2008.

CHAPTER 18
Free and Open Source GIS Tools: Role and Relevance in the Environmental Assessment Community

S. Thomas Purucker, Heather E. Golden, Gerard F. Laniak,
L. Shawn Matott, Daniel J. McGarvey and *Kurt L. Wolfe*

The presence of an explicit geographical context in most environmental decisions can complicate assessment and selection of management options. These decisions typically involve numerous data sources, complex environmental and ecological processes and their associated models, risk assessment and cost-benefit considerations, as well as competing stakeholder interests that reflect different decision objectives (Linkov et al. 2004). When evaluating alternatives, decision makers must balance these different technical criteria to make an objective decision. Geographic information system (GIS) output is often used to provide this level of decision support, and the mapping of environmental stressor sources, potential exposures, and risk is an essential part of the process. Maps of spatial distributions of stressors—whether derived from empirical sources or estimated using models—assist in collating information sources so they can be interpreted by decision makers. Given the wealth of data inputs and decision options, it is no surprise that numerous spatial support systems have been designed that assist decision making in a GIS environment (Ascough et al. 2002). These systems can display combinations of stressors, receptors, and locations of concern for environmental assessment graphically; they function as decision support toolkits for efficient environmental assessment and can be updated quickly to meet different decision-making objectives.

An environmental decision support software toolkit consists of integrated modules that assist environmental assessors and decision makers in various arenas. These include analyzing and documenting environmental data and models; identifying environmental problems and related uncertainties; promoting interaction among participants in decision making; facilitating simple database queries; and presenting model results through clear, visual outputs. The primary purpose of an environmental toolkit is to run the incorporated statistical and environmental science models. However, interface features that aid data entry, model execution, data interpretation, consensus building among participants, and intelligent display of results are important factors in the acceptance of a toolkit by assessors and decision makers. Also important is a track record of successful application and adequate quality assurance review, testing, and documentation.

Two main areas of application for these systems are data analysis and environmental modeling. Traditionally, the statistical and process modeling relationships commonly used in site assessments have ignored or downplayed relationships between data points and spatially explicit processes. However, environmental models of physical and biological processes increasingly incorporate spatially auto-correlated patterns of stressor processes and distributions. Adequately characterizing these spatial representations is central to communication, outreach, planning, assessment, and resource management. Predictive models of stressor dynamics and effects are the foundation of forecasting change and of assessing proactively how ecosystem functions and services likely will respond to natural and human stressors. Efforts to quantify these patterns and incorporate them into a conceptual site model that uses this information can facilitate protective, yet cost-effective, decisions (Crumbling et al. 2004). Thus, environmental decision support toolkits should include models and algorithms that can generate or recognize spatial correlation in environmental data, incorporate spatial effects into a conceptual site model, and manage uncertainty associated with spatial stressor distributions. This typically involves dynamic coupling of traditional site assessment tools with a GIS—beyond simple creation of data sets that can be post-processed independently by a GIS or building a subset of additional functions *de novo* within an available GIS.

This trend has driven a steady increase in available GIS-based environmental decision support software toolkits. Free and open source software (FOSS) is a significant portion of these capabilities. This chapter surveys common environmental assessment tools used with GIS. It does not aim to be comprehensive, given the number of available tools, but to go into detail on a few selected tools. Although some of the tools can be linked to a proprietary GIS environment, the focus here is on FOSS approaches and available freeware (proprietary software distributed freely). Surveyed topics include tools for accessing environmental data sets, performing aquatic and terrestrial assessments, modeling atmospheric deposition processes, and handling uncertainty. This chapter concludes with a discussion of the advantages and limitations of FOSS GIS approaches for environmental assessments.

18.1 Data Management and Frameworks

Integrated data management that passes information between models without extensive manipulation by the user is important to the success of computerized systems that support complex environmental decisions. Modern environmental modeling systems are designed to integrate or "link" multiple models, each characterizing a component of the environmental system of interest (e.g., linking an atmospheric model with a watershed model and a surface water model). To manage the interdependent execution of the linked models, they are typically contained within a framework or infrastructure. Currently, many support systems employ a "null integration strategy" that assigns responsibility for data transfer between models to the user (Denzer 2005). Although this may be acceptable to users with the necessary technical skills, it unnecessarily limits access for other participants. Also, automation may be required for highly parameterized models, making the application of complex models logistically difficult unless automated data transfer capabilities are available. Therefore, a prerequisite for integrated environmental modeling is the ability to automate data access, retrieval, processing, display, and formatting of disparate data sets.

Several software frameworks for applying complex modeling systems exist. The USEPA's Data for Environmental Modeling (D4EM) project is a set of open source software tools that allows environmental model developers to populate input files with data available from disparate sources like those listed above. FRAMES (Framework for Risk Analysis in Multimedia Environmental Systems) is an infrastructure that contains the USEPA's 3MRA (Multi-media, Multi-receptor, Multi-pathway Risk Assessment) (USEPA 2003) modeling system. The USEPA's BASINS (Better Assessment Science Integrating point and Nonpoint Sources) (USEPA 2001a) illustrates the diversity of GIS-enabled model evaluation tools. BASINS is a freely available GIS-based watershed modeling environment that has been implemented on both proprietary (i.e., ESRI ArcView) and open-source (i.e., MapWindow) GIS platforms. The USDA's Object Modeling System (OMS) (David et al. 2002) and the Department of Defense's Advanced Risk Assessment Modeling System (ARAMS) (Dortch 2001; Dortch and Gerald 2002) also serve as frameworks for integrated modeling. Each software infrastructure is designed to facilitate an application of multiple science models, despite sometimes containing components not initially designed for interoperability. Assessments conducted with these modeling systems range from simulating the movement of non-point source contaminant through watersheds and the subsequent impact on surface waters, to estimating the national scale impacts of industrial waste disposal on human and ecological receptors.

Within integrated modeling systems, data must be transferred from one model to another. As an example, the 3MRA modeling system includes 17 different environmental simulation components. Consequently, model execution requires not only a significant amount of environmental data, but also that data be expressed in a consistent manner throughout the system. Geographic data is particularly challenging because it is difficult to maintain spatial attributes that describe the extent and interconnectedness of various physical features within the modeling domain (i.e., air, water, watersheds, ecological habitats, etc.). Data must be collected and organized not only from a systems perspective, considering all model components used in the application, but also from the individual model perspective. Consider a watershed model that produces sediment fluxes and runoff from individual sub-basins combined with a surface water model that receives these fluxes and simulates the resulting impacts in each segment of its associated network. To couple these two models, a consistent watershed sub-basin identification method is needed—one that associates each contributing area with the appropriate surface water segment. This connectivity among physical features of a model significantly increases the burden of data preparation.

D4EM is designed for desktop modeling applications (e.g., 3MRA) accessing internet sources of data. The system architecture is component-based, allowing for different data sources and approaches to be "plugged" into the framework, and contains libraries of routines, each with a documented application programming interface (API). The software is model/modeling system independent, i.e., it is accessible to any modeling system that implements a thin interface (passing data with minimal processing) between D4EM and a modeling system user interface. D4EM contains no proprietary software and is being developed as an open source collaboration for use and enhancement by the larger environmental science community. To ensure that data sources and their processing steps are preserved, the design collects and catalogs metadata. Metadata contains descriptive information about the execution of each step of data access, retrieval, and processing. Metadata accompanies the data packet throughout D4EM, an essential quality assurance feature that provides transparency and reproducibility.

The software stack of D4EM includes a user interface, components of data methods and tools, and a database with results of data collection. The components include Use Case libraries; the Data Manager library; the Data Manager Extension libraries; and the GIS library. Each library has an API, allowing direct program access to any D4EM functionality. MapWindow (2008) is used as D4EM's open source GIS. MapWindow is lightweight, extensible, and freely distributable, while providing the necessary functionality for basic GIS needs of environmental modeling. It includes a GIS processing engine (MapWinGIS ActiveX control), a user interface/map display application, and a series of "plug-ins" that add specific functionalities (e.g., geo-processing, additional user interface features, etc.). The MapWinGIS includes a GIS API for vector and raster geoprocessing. MapWindow is an integral part of functionality for utilities and methods included in other D4EM libraries.

Data Manager Extensions (DME) contain utilities for accessing and retrieving data from a range of sources. The DME API provides a common interface for requesting data. On the data source side, DME has software specific to individual data sources, each containing its own protocol for accessing and retrieving data. Increasingly, standard-based web services are used to make DME connectivity and data querying relatively straightforward. The Data Manager library contains utilities for processing raw data into formats that match the input needs of an environmental model(s). The principal categories of data management are statistical characterization and GIS manipulation.

In D4EM, the Use Case library contains high level algorithms that process and organize logical units of data needed by models. For example, watershed models require sub-basins to be delineated on the basis of digital elevation data. The watershed delineation use case is an algorithm that manages acquiring necessary elevation data and processing it into a form that provides geographic boundaries of each resulting drainage sub-basin. Use-case algorithms request data and related processing from the Data Manager (DM). The DM, in turn, requests data from DME, which retrieves the data and returns it to the DM for any necessary processing, after which finished data is returned to the use case. Use cases are intended to be reusable and extensible, that is, intended to produce well-defined groupings of data needed by classes of models, not just an individual model. Thus, a use case may be designed to collect census data, such as the specific distribution and demographics of human populations within an area of interest. When well designed, the use case can be employed by multiple models requiring the same data. When completed, a use case places resulting data and associated metadata into a D4EM database for transfer to specific model input files. Variable names within the D4EM database conform to a standard nomenclature. To transfer data from D4EM to a model or modeling system, the modeler must write a use case whose function is to map and transfer data from the D4EM database into specific model input files and formats (Wolfe et al. 2007).

18.2 Aquatic Assessment Tools

FOSS GIS tools for aquatic assessment range from stand-alone add-ins to web-hosted databases with graphical querying interfaces to fully integrated, dynamic modeling toolsets. A variety of task-specific add-ins/plug-ins can be downloaded from websites such as Jenness Enterprises (Jenness and Engelman 2008), the ESRI Support Center (ESRI 2007), and the MapWindow portal (MapWindow 2008). For example, the "Basin1" add-in (Petras 2003) will delineate a catchment boundary, relative to any user-defined point location along a stream/river channel within a landscape (represented by a digital elevation model [DEM]), then generate the stream network within that catchment. Hornby (2006) developed add-ins that determine "Strahler and Shreve stream order" for every segment within a network. Jones (2007) can be used to interpolate inundated surface areas from a triangulated irregular network (TIN), river network, and user-specified flows. However, the majority of aquatic add-ins are currently distributed in the Arc Macro Language or as ArcView version extensions (written in Avenue script), and must be converted and recompiled in Python (ESRIC 2002) or Visual Basic (Tonias and Tonias 2002) before being implemented in other programs, such as ArcMap or MapWindow.

One impressive example of a bundled, web-based GIS is the New Zealand Ministry of Fisheries' (2007) "National Aquatic Biodiversity Information System" (NABIS). NABIS allows users to query and visualize a wide array of environmental and biological data, including land use, bathymetry, climate, and distributions of approximately 150 species of marine fish, invertebrates, mammals, and birds. It also displays management boundaries and depicts commercial fishery results, stratified by year, species, and catch method. Most notably, hyper-linked metadata is appended to the entire data set. Queries are submitted by specifying object values (e.g., "all orange roughy, *Hoplostethus atlanticus*, captured within 5 km of the coast, in 2006"), or through the graphical interface (e.g., digitizing polygons). Data can be viewed and exported in tabular format, and new data can be uploaded and visualized using formatted text files; new data cannot be implemented in dynamic calculations, however, nor can users edit existing data. NABIS is an interactive web browser, not a platform for spatial analysis or dynamic modeling. That said, it is remarkably well designed and maintained, with powerful visualization capabilities. Little or no prior GIS experience is necessary to utilize NABIS, making it immediately accessible to natural resource managers and the general public.

Integrated sets of aquatic assessment tools are also available to perform more complex tasks. For example, The Nature Conservancy's "GIS Tools for Stream and Lake Classification and Watershed Analysis" (TNC 2006) draws from a suite of physical habitat data (e.g., stream networks, DEMs, geology, and land use) to delineate water bodies (i.e., lakes, rivers, streams, and wetlands) with unique ecosystem properties. To facilitate this process, TNC (2006) utilizes Arc Macro Language and Visual Basic functions. First, stream network data (preferably taken from the previously discussed NHD data set; USGS 2007) are preprocessed to ensure that connectivity, flow routing, and directionality are correctly assigned throughout the drainage network; automating this process is a major accomplishment in itself. Next, a host of physical parameters (e.g., slope, elevation, and stream order) are calculated for every stream segment, lake, and wetland. These data can then be superimposed on thematic maps (e.g., water chemistry and flow stability) to determine "macrohabitat" types, at multiple nested spatial scales. Finally, results are exported as text files that can be used in multivariate analyses to identify watersheds with common or unique ecosystem characteristics.

The Nature Conservancy toolset was designed to solve a specific problem: identifying areas of high conservation priority, such as aquatic systems with exceptional biodiversity, when biological data are scarce or nonexistent (Groves et al. 2002). Given the *ad hoc* nature of this task, it is not surprising that the tools have a distinctly piecemeal quality. To utilize the full toolset, users must call each function in a particular sequence. Moreover, they must distribute the workload among four program interfaces: ArcInfo (Arc Macro Language routines can also be implemented in ArcMap with a script tool, or by recompiling in Python), ArcView, Microsoft Access, and a suitable statistics package (any program that performs multivariate ordinations and cluster analyses will work). Therefore, substantial GIS experience is necessary to use the toolset. Fortunately, an excellent tutorial is distributed with the toolset that allows users to implement the tools quickly and efficiently. By modifying and expanding the existing tools, it may also encourage users to develop customized functionality.

18.3 Terrestrial Assessment

Basic GIS capabilities found in a number of available packages can be sufficient to conduct terrestrial assessments without additional functionality. FOSS GIS in this category include Diva-GIS, a system often used for mapping and analyzing biodiversity data, such as species' distributions. Quantum GIS is a cross-platform open source GIS system providing mainstream support for vector, raster, and geodatabase formats. uDig is an internet-aware spatial data viewer/editor and provides a common Java platform for building spatial applications with open source components. SAGA (System for Automated Geoscientific Analyses) is a hybrid GIS with bundled geoscientific libraries capable of supporting terrestrial analyses, including terrain analysis, geostatistics, and dynamic environmental processes. GRASS (Geographic Resources Analysis Support System), the most well-known FOSS GIS, is capable of geospatial data management and analysis, image processing, graphics/maps production, spatial modeling, and visualization.

There are also a number of available spatial data analysis libraries, many of which implement sophisticated geostatistical methods. These can extend available GIS packages to perform terrestrial data analyses and assessments with minimal modification. Examples include GSLIB (Geostatistical Software LIBrary), a collection of geostatistical programs, which has been incorporated into many software packages. Gstat, an open source computer code for geostatistical modeling, prediction and simulation, has also been interfaced to various GIS. The popular R language for spatial data analysis contains contributed libraries accessible from a number of GIS, including GRASS, SAGA, and ArcGIS. GeoBUGS (Geospatial extensions to Bayesian inference Using Gibbs Sampling) employs and displays spatial models within WinBUGS, allowing for Bayesian analyses within a spatial setting. STARS (Space-Time Analysis of Regional Systems) is a Python-based open source package for the analysis of areal data over time. The open source versions of these libraries can be recompiled and linked to available (open source or proprietary) GIS.

An example of a stand-alone freeware GIS designed for environmental assessments is Spatial Analysis and Decision Assistance (SADA) (Stewart and Purucker 2006; Purucker et al. 2008). The software is typically applied for decision-making support at soil remediation sites driven by risk assessment-based cleanup criteria. These assessments are amenable to spatial decision support approaches (Nyerges et al. 1997; Thayer et al. 2003), but with limitations (Woodbury 2003). They contain an explicit spatial context for the distribution of stressors and environmental exposures. Risk assessment and management of such sites is further complicated by the fact that they contain multiple contaminants in several media, with each combination differing in its spatial distribution. Also, depending on cleanup context and the exposure/risk to receptors, a number of regulatory decision criteria may need to be met. Thus, many sites have multiple contaminants at or above regulation threshold levels. GISs like SADA, when integrated with environmental assessment algorithms, are capable of spatial analyses at sites with multiple contaminants where remediation progress could be impeded by inadequate characterization (Preston 2002). GIS tools with embedded spatial analysis and risk assessment features allow data to be efficiently organized and can facilitate prioritizing areas and toxicants for remediation. Continuing development of these tools improves data collection and analysis methods to conduct ecological (Chow et al. 2005; Gaines et al. 2005) and human health (Cech and Montera 2000; Bién et al. 2004; Hooker and Nathanail 2006) risk assessments.

SADA integrates a stand-alone GIS system, statistical analysis techniques, and risk assessment methods. It is used to address site-specific concerns when characterizing a contaminated site, assessing risk, specifying the location of future samples, and designating areas of concern. While decision support tools that incorporate environmental process models into risk assessment methodologies do exist (e.g., Babendreier and Castleton 2005; Dortsch 2001; USDOE 2004), SADA is a publicly available GIS system that integrates data analysis, emphasizing spatial analysis capabilities, with risk assessment methods aimed at typical environmental assessors. SADA's decision-making support features are sufficient for it to be considered an environmental software toolkit (Bartell 2003; Holland et al. 2003). The software design process ensures that the user interface and incorporated algorithms are easy to implement, that assumptions behind the models are transparent and easily exportable, and that site information is collected in a single database file that is readily transferable between computers.

Once geographic and analytical data are imported, they can be displayed in the program's native GIS, an approach that provides a number of advantages in conducting environmental assessments. Individual sites can be defined using polygon and layer tools, while sampling locations and analytical results can be displayed with GIS layers imported from ESRI shape files or AutoCAD DXF files. Data management capabilities within the GIS are available, including subsetting data by date ranges and discrete sampling event queries, implementing statistical proxy methods for non-detects imported at the analytical detection limit, and resolving duplicate data using several methods. Many common environmental assessment statistical methods for evaluating contaminated sites are available, including univariate measures of relative standing, central tendency, and dispersion (USEPA 2006). Graphical displays for posting plots, histograms, and ranked data plots, and non-parametric hypothesis tests versus decision criteria or reference concentrations are also available. Spatial interpolation procedures can also be used to delineate cleanup areas at a site, while achieving the twin goals of minimizing cleanup volume and meeting target remediation levels. Implementation requires inputs of contaminant analytical data, a model of the spatial distribution of contamination, and human health or ecological exposure information and models, along with decision-maker cleanup criteria and spatial scale definition. The resulting remedial design is spatially-explicit and optimal for the inputs and decision objectives (Purucker et al. 2008).

18.4 Modeling Atmospheric Inputs of Contaminants

Using GIS for estimating deposition rates of atmospheric contaminants to watersheds, water bodies, and terrestrial ecosystems is a rapidly developing area of applied research. Depending upon the detail required for an application (e.g., watershed nutrient input-output budgets, watershed fate and transport modeling, ecosystem exposure modeling) and the importance of understanding transport and transformation dynamics of atmospheric constituents, two broad mathematical approaches exist for estimating rates of atmospheric deposition across extensive spatial scales. The first approach is statistical, developed from a given set of data (e.g., measured precipitation concentrations); the second is process-based, which aims to capture the underlying physical processes of a system in mathematical terms. Furthermore, models assessing atmospheric deposition are typically based upon emission sources or measured precipitation chemistry and deposition (USEPA 2001b). Several user-friendly GIS-based tools have evolved from these approaches. The examples given here will be limited to nutrient deposition (e.g., nitrogen and sulfur compounds) due to the number of tools that focus on other chemical compounds.

The ClimCalc model is one statistical tool that models atmospheric nutrient deposition across heterogeneous landscapes within a GIS framework (Ollinger et al. 2001). ClimCalc provides grid-based outputs of climate (monthly time step) and annual wet and dry atmospheric deposition of nitrogen and sulfur compounds throughout the northeastern US using simple regression (Aber et al. 1995; Ollinger et al. 1993; Ollinger et al. 1995). Precipitation and temperature data are collected from 30 years of weather station data, and precipitation chemistry data for the models are gathered from measured data from the National Atmospheric Deposition Program (NADP). Predictor variables are mostly based on geographic position, such as latitude and longitude. The model is applied to 1-km resolution GIS grid cells and focuses on regional variability of atmospheric deposition. Although it is a good predictor of broad spatial patterns of atmospheric nutrient inputs to ecosystems, local effects on deposition from factors such as elevation, aspect, and distance from large water bodies are not included. Outputs from the model can be defined at a specific site by providing local coordinates or by generating a 1-km resolution GIS coverage (see Ollinger et al. 1993; Ollinger et al. 1995).

The NADP provides the most comprehensive national network of precipitation chemistry and atmospheric deposition monitoring. Precipitation chemistry at each NADP site is collected weekly and deposition estimates are calculated using these data; after quality assurance/quality control measures are implemented, the data are publicly posted (NADP 2007). As part of this database, NADP also generates and posts national isopleth maps of annual precipitation chemistry and deposition of multiple atmospheric compounds and elements as ArcView grid files. These annual gridded maps are developed from network monitoring data using geostatistical spatial interpolation techniques and provide spatially explicit estimates of deposition and concentration across the diverse national landscape. However, the technique for extrapolating data beyond measured points should be considered when choosing an appropriate spatially explicit GIS deposition database. For example, under contemporary conditions (2002–2004) in New York State, interpolation methods were found to be slightly less accurate for estimating spatial variability of wet nitrogen deposition rates, compared to a regression-based model that incorporates factors such as geographic position and elevation (Golden and Boyer 2008).

In addition to statistical approaches, several process-based models capture dynamic interactions of sources, transformation processes, and deposition of atmospheric contaminants and provide output that can be utilized with the functionality of GIS software. The Community Modeling Air Quality (CMAQ) model is a physically-based model of deposition that elucidates sources and controls on deposition (Byun and Ching 1999). The current CMAQ version, 4.6, models the deposition of several air quality pollutants at multiple scales (currently 144 km² and 1,296 km², with 16 km² being refined), based on the National Emissions Inventory and on atmospheric and land surface processes that regulate transport, transformation, and deposition of pollutants. CMAQ is a component of the Models-3 system framework, and its main purpose is to support air quality monitoring applications, from scientific inquiries to policy applications. GIS-based gridded output from CMAQ can be accessed by running the model (NOAA/USEPA 2007a) or implementing a user-friendly Watershed Deposition Tool (WDT) (NOAA/USEPA 2007b). The WDT provides ArcGIS shape file output for deposition of dry and wet, reduced and oxidized nitrogen and sulfur, in addition to total dry and wet mercury, for 2001 and for projected deposition in 2010, including implementation of the Clean Air Mercury Rule provisions. WDT's ability to provide only annual deposition from 2001 and 2010 limits opportunities to assess trends in deposition over time. However, for many applications, WDT's simplicity of access outweighs the skill level and learning time required for implementing CMAQ.

The Regional Modeling Systems for Aerosols and Deposition (REMSAD) is similar to CMAQ, but includes different atmospheric transport algorithms (ICF-International 2006). The model was originally intended as a coarse (national) scale model for generating information on the distribution of atmospheric particulate matter, the deposition of pollutants to land and water bodies, and changes resulting from implementing air quality policies (ICF-International 2005). REMSAD, based on the variable-grid Urban Airshed Model (UAM-V), has evolved extensively. It currently functions as a three-dimensional grid model, simulating the physical and chemical processes of atmospheric pollutants and consequent wet and dry deposition of each chemical species, and provides a gridded output of each on a selected time scale. The improvements to REMSAD make it technologically rigorous with a potentially steep learning curve for those unfamiliar with modeling techniques; a simplified output similar to WDT for CMAQ is not yet available. The ability of CMAQ versus REMSAD for calculating one year (2001) of nitrate concentrations and aerosol sulfur is assessed by Gégo et al. (2006). Both models simulate similar nitrate concentrations, but CMAQ appears to simulate aerosol nitrate slightly better. Both skill of the GIS modeler and purpose of the study should guide the user when choosing between REMSAD and CMAQ for modeling deposition of pollutant species.

18.5 GIS Tools for Uncertainty Evaluation

Every environmental model contains a degree of subjectivity that ensues from selection of a specific quantitative model, and the parameter constants and distributions used to represent the system of interest. This subjects models to criticism, particularly in a decision-making context. Technical validation and verification of a closed mathematical model of an open environmental system under study is difficult, if not impossible (Oreskes et al. 1994). Under decision-making conditions involving many principals who are representing different interests, it is difficult to select objectively a unique approach for the modeled system. Consequently, model selection and evaluation criteria for decision making can differ from standard, scientific approaches. Acceptance of a quantitative decision support tool that implements environmental models becomes a negotiation with different or additional criteria used to evaluate the model (Haag and Kaupenjohann 2001). In particular, acceptance of a model can be based as much on its comprehensibility and relevance to decision makers as on traditional measures of science. Regardless, for GIS and decision tools to be useful in a policy-relevant context, elucidating data and model uncertainty is challenging but necessary. Sophisticated geostatistical packages and GIS extensions (discussed earlier) are often integral to addressing spatial variability, model uncertainty, and data transformation errors.

Output associated with a given model is affected by numerous upstream sources of uncertainty, including model input (e.g., forcing functions, such as a time-series of rainfall data), model parameters (i.e., various unobserved or unobservable constants and conditions), and model structure (e.g., the degree to which the model equations and initial and boundary conditions represent the real world). Model evaluation methods include uncertainty analysis, sensitivity analysis, and parameter estimation. Specific model assessment tools and algorithms often represent a combination of these methods. Formal uncertainty analysis (UA) methods propagate sources of uncertainty through a given model to generate statistical moments or complete probability distributions for various model outputs. Classical Monte Carlo UA schemes wrap a stochastic shell around a possibly calibrated model. More recent approaches (e.g., Hierarchical Bayes) explicitly condition model output on available data and efficiently assimilate new data as it becomes available. Less formal, more qualitative UA methods include scenario analysis and alternative futures. Sensitivity analysis (SA) studies the degree to which model output is influenced by changes in input. By measuring input importance, SA methods identify critical areas where knowledge or data are lacking. In this way, SA assists decision makers in prioritizing future research and data collection. Parameter estimation (e.g., calibration) methods attempt to infer appropriate values or distributions of uncertain model parameters using available data and knowledge. These methods typically involve an iterative sampling or search algorithm that repeatedly evaluates a model and calculates an objective or likelihood function based on correspondence between model output and available data.

In a GIS environment, data uncertainty is associated with many spatial issues and complications. It may be regarded as pre-existing (i.e., raw, or introduced prior to the use of a given data set by a given user) or may be introduced through a series of one or more transformations performed within the GIS. Sources of raw data uncertainty include spatial, temporal, and population variability, and errors in data classification, projection, location, and value. Raw uncertainty is sometimes explicitly recorded in a given geodatabase (e.g., via entries for the mean and standard deviation), but is more commonly not reported or reported via the metadata.

As a GIS manipulates a given data set, raw data uncertainty can be augmented significantly. For example, uncertainty accumulates as a result of data aggregation (i.e., spatial or temporal averaging, or feature simplification) and disaggregation (i.e., spatial refinement or discretization). Such operations are often necessary to align the spatial scales and structures of various data sets and models. Other GIS manipulations that can increase uncertainty include data re-projection, data re-classification, and manual adjustment of data values and locations.

Another common use of a GIS is development of derived data sets, representing a mathematical combination or analysis of one or more existing data sets. The uncertainty of a derived data set is possibly a non-linear combination of uncertainties in the parent data sets. Examples of derived data sets include watersheds delineated via analysis of digital elevation models, raster coverages interpolated or extrapolated from a sparse set of point measurements (e.g., via geostatistical kriging), and buffer zones centered on possibly uncertain point or polygon coordinates.

Unlike model uncertainty, whose analysis requires complex tools, treatment of data uncertainty is generally straightforward. During data manipulation, low-order uncertainty accumulation can be managed by careful bookkeeping. For example, keeping track of both the mean and the variance of a spatially aggregated data set may be sufficient. Preserving higher order statistical properties, such as spatial correlation, is much more difficult and requires specialized techniques. When developing derived data sets, uncertainty propagation can usually be handled via simple analytical equations rather than the sampling schemes that typify model-based uncertainty propagation. One area where complications arise is the aggregation and disaggregation of classification data. For such "set-membership" data sets, conventional uncertainty measures are inadequate and the GIS community has embraced fuzzy set theory as a promising alternative.

Various "GIS-enabled" model evaluation tools are freely available. Most of these do not distinguish between spatial and non-spatial information. Instead they are GIS-enabled in that they have been linked to a given GIS via software "wrappers" (i.e., code that links a given GIS with an external model evaluation tool). GIS-enabling wrappers are commonly distributed as GIS extensions, plug-ins, or DLLs (dynamically linked libraries).

For the BASINS model, several tools are available for parameter estimation and uncertainty and scenario analysis (van Griensven and Meixner 2006). These include PEST (Model-independent Parameter Estimation) (Doherty 2004, 2007), ParaSol (Parameter Solutions) (van Griensven and Meixner 2007), SUNGLASSES (Sources of Uncertainty Global Assessment using Split SamplES) (van Griensven and Meixner 2004), LH-OAT (Latin Hybercube Sampling – One factor At a Time) (van Griensven et al. 2006), CANOPI (Confidence Analysis of Physical Inputs) (van Griensven and Meixner 2006), GenScn (Generation and analysis of model simulation Scenarios) (Kittle et al. 1998), and UNCSIM (UNCertainty SIMulator) (Reichert 2005).

Additional popular or emerging model evaluation tools have been or could be GIS-enabled easily. Some of these include UCODE (Universal Code for Inverse Modeling) (Poeter and Hill 1999; Poeter et al. 2005), DAKOTA (Design Analysis Kit for Optimization and Terascale Applications) (Eldred et al. 2006), GLUE (Generalized Likelihood Uncertainty Engine) (Beven and Binley 1992), OSTRICH (Optimization Software Tool for Research In Computational Heuristics) (Matott 2005), SimLab (Simulation Laboratory for Uncertainty and Sensitivity Analysis) (Saltelli et al. 2004), and BATEA (Bayesian Total Error Analysis) (Kavetski et al. 2002; Kavetski et al. 2006a, b). An EPA website has been developed to encourage comprehensive treatment of uncertainty in environmental modeling (USEPA 2008). The site contains numerous code, executable and documentation links for a variety of model evaluation tools.

A tool discussed earlier that assists with tracking data uncertainty is D4EM, a GIS-enabled tool used in conjunction with the MapWindow open source GIS project. D4EM exercises various use cases to locate, download, and transform environmental data sets into formats that serve as inputs for a given environmental model. During each step of a given use case, D4EM tracks and records data set alterations to a log file. Review of the log file helps to rigorously assess data transformation errors. Another example, the Data Uncertainty Engine (DUE) (Brown and Heuvelink 2007), is designed to analyze data and generate data realizations (i.e., random samples that respect an underlying "best-fit" probability model of the data). DUE handles a variety of file formats and accommodates spatial, temporal, and attribute uncertainties.

18.6 Discussion and Conclusion

GIS systems are important for communicating environmental spatial information to decision makers. Environmental assessments require a high degree of integration within and among modeling components, prompting a significant data interchange that must be handled by the (often GIS-based) software architecture. While the global environmental assessment software market is significant enough to draw many proprietary GIS applications, there is not always a significant commercial market for more technical scientific environmental applications. GIS computing technologies (web-enabled, desktop, and embedded) and their use in environmental applications respond less to the dichotomy between FOSS and proprietary systems than to the solutions that ultimately meet end-user and interoperability requirements. While proprietary software and FOSS solutions can respond to "market" needs, FOSS can also serve as test beds for early applications that eventually will go mainstream and for those that will never have a significant commercial market. For these reasons, FOSS GIS applications have established themselves over the past decade as an important part of the environmental computing landscape, and their influence is likely to grow as assessment techniques become more technical and specialized.

End-user requirements often influence relevant interoperability requirements. Effective decision support software is accessible to all study participants and provides full submodel documentation to maximize model transparency. Various methods, from intensive interaction between modelers and decision makers to web-enabled applications, allow accessibility. Programs that facilitate accessibility via easy-to-use graphical user interface, by free distribution on operating systems used by a majority of environmental decision makers (e.g., Microsoft Windows), and by encompassing all site information in a manner easily distributed among different computers and users, have advantages in environmental applications even if they impose technical limitations. Programs requiring significant financial investment, multiple input files, or limited to implementation on less common operating systems impede portability and use by decision makers, which limits acceptance of a decision support tool for environmental applications. In this arena, well-documented and tested proprietary software (free or pay) and FOSS applications are on equal footing to provide necessary information that supports environmental decisions. However, complex, implementation-specific workflows require significant flexibility and component interoperability, which, in turn, require access to source code; therefore, FOSS systems provide modification and quality assurance capabilities not available in proprietary software.

Acknowledgments

Comments from Craig Barber and Fran Rauschenberg improved the manuscript. This paper has been reviewed in accordance with the US Environmental Protection Agency's peer and administrative review policies and approved for publication. Mention of trade names or commercial products does not constitute endorsement or recommendation for use.

References

Aber, J. D., S. Ollinger, C. Federer, P. Reich, M. Goulden, D. Kickligher, J. Melillo and R. Lathrop Jr. 1995. Predicting the effects of climate change on water yield and forest production in the northeastern United States. *Climate Research* 5:207–222.

Ascough II, J. C., H. D. Rector, D. L. Hoag, G. S. McMaster, B. C. Vandenberg, M. J. Shaffer, M. A. Weltz and L. R. Ahuja. 2002. Multicriteria spatial decision support systems: Overview, applications, and future research directions. In *Integrated Assessment and Decision Support, Proceedings of the First Biennial Meeting of the International Environmental Modelling and Software Society.* Edited by A. E. Rizzoli and A. J. Jakeman. Manno, Switzerland: iEMSs.

Babendreier, J. E. and K. J. Castleton. 2005. Investigating uncertainty and sensitivity in integrated, multimedia environmental models: Tools for FRAMES-3MRA. *Environmental Modelling & Software* 20:1043–1055.

Bartell, S. 2003. Effective use of ecological modeling in management: The toolkit concept. In *Ecological Modeling for Resource Management.* Edited by V. H. Dale. Berlin: Springer.

Beven, K. and A. Binley. 1992. The future of distributed models: Model calibration and uncertainty prediction. *Hydrological Processes* 6 (3):279–298.

Bién, J. D., J. ter Meer, W. H. Rulkens and H.H.M. Rijnaarts. 2004. A GIS-based approach for the long-term prediction of human health risks at contaminated sites. *Environmental Modeling and Assessment* 9:221–226.

Brown, J. D. and G.B.M. Heuvelink. 2007. The Data Uncertainty Engine (DUE): A software tool for assessing and simulating uncertain environmental variables. *Computers and Geosciences* 33 (2):172–190.

Byun, D. W. and J.K.S. Ching, eds. 1999. *Science Algorithms of the EPA Models-3 Community Multiscale Air Quality (CMAQ) Modeling System.* US Environmental Protection Agency, Office of Research and Development, Washington, D.C., EPA/600/R-99/030.

Cech, I. and J. Montera. 2000. Spatial variations in total aluminum concentrations in drinking water supplies studied by geographic information systems (GIS) methods. *Water Research* 34:2703–2712.

Chow, T. E., K. F. Gaines, M. E. Hodgson and M. D. Wilson. 2005. Habitat and exposure modeling for ecological risk assessment: A case study for the raccoon on the Savannah River site. *Ecological Modelling* 189:151–167.

Crumbling, D., J. Hayworth, R. Johnson and M. Moore. 2004. The triad approach: A catalyst for maturing remediation practice. *Remediation: The Journal of Environmental Cleanup Costs, Technologies and Techniques* 15 (1):3–19.

David, O., S. L. Markstrom, K. W. Rojas, L. R. Ahuja and I. W. Schneider. 2002. The object modeling system. Pages 317–331 in *Agricultural System Models in Field Research and Technology Transfer.* Edited by L. Ahuja, L. Ma and T. A. Howell. Boca Raton, Fla.: Lewis Publishers, CRC Press.

Denzer, R. 2005. Generic integration of environmental decision support systems–state-of-the-art. *Environmental Modelling & Software* 20:1217–1223.

Doherty, J. 2004. *PEST: Model-independent Parameter Estimation, User Manual,* 5th ed. Brisbane, Australia: Watermark Numerical Computing.

———. 2007. *Addendum to the PEST Manual.* Brisbane, Australia: Watermark Numerical Computing.

Dortch, M. S. 2001. Army Risk Assessment Modeling System (ARAMS). In *Assessment and Management of Environmental Risks: Cost-efficient Methods and Applications.* Edited by I. Linkov and J. Palma-Olivera. Boston: Kluwer Boston, Inc.

Dortch, M. S. and J. S. Gerald. 2002. Army risk assessment modeling system for evaluating health impacts associated with exposure to chemical. In *Brownfield Sites: Assessment, Rehabilitation and Development.* Edited by C. A. Brebbia, D. Almorza and H. Klapperich. Southampton, UK: WIT Press.

Eldred, M. S., A. A. Giunta, S. L. Brown, B. M. Adams, D. M. Dunlavy, J. P Eddy, D. M. Gay, J. D. Griffin, W. E. Hart, P. D. Hough, T. G. Kolda, M. L. Martinez-Canales, L. P. Swiler, J. Watson and P. J. Williams. 2006. *DAKOTA, A Multilevel Parallel Object-Oriented Framework for Design Optimization, Parameter Estimation, Uncertainty Quantification, and Sensitivity Analysis: Version 4.0 Users Manual.* Report No.: SAND2006-6337. Albuquerque, NM and Livermore, Calif.: Sandia National Laboratories.

ESRI (Environmental Systems Research Institute). 2007. ESRI Support Center. <http://arcscripts.esri.com/> Accessed 7 April 2008.

ESRIC (Environmental Systems Research Institute Canada). 2002. AVPython: Python language support for ArcView GIS. ESRI Canada Limited. <http://avpython.sourceforge.net/> Accessed 7 April 2008.

Gaines, K. F., D. E. Porter, T. Punshon and I. L. Brisbin Jr. 2005. A spatially explicit model of the wild hog for ecological risk assessment activities at the Department of Energy's Savannah River site. *Human and Ecological Risk Assessment* 11:567–589.

Gégo, E., P. S. Porter, C. Hogrefe and J. S. Irwin. 2006. An objective comparison of CMAQ and REMSAD performances. *Atmospheric Environment* 40:4920–4934.

Golden, H. E. and E. W. Boyer. 2008. Contemporary estimates of atmospheric nitrogen deposition to the watersheds of New York state. Manuscript submitted for review.

Groves, C. G., D. B. Jensen, L. L. Valutis, K. R. Redford, M. L. Shaffer, J. M. Scott, J.V. Baumgartner, J. V. Higgins, M. W. Beck and M. G. Anderson. 2002. Planning for biodiversity conservation: Putting conservation science into practice. *BioScience* 52:499–512.

Haag, D. and M. Kaupenjohann. 2001. Parameters, prediction, post-normal science and the precautionary principle—a roadmap for modelling for decision-making. *Ecological Modelling* 144:45–60.

Holland, J., S. M. Bartell, T. G. Hallam, T. Purucker and C.J.E. Welsh. 2003. Role of computational toolkits in environmental management. In *Ecological Modeling for Resource Management.* Edited by V. H. Dale. Berlin: Springer.

Hooker, P. J. and C. P. Nathanail. 2006. Risk-based characterization of lead in urban soils. *Chemical Geology* 226:340–351.

Hornby, D. 2006. Create Strahler or Shreve stream order for braided vector river networks. <http://arcscripts.esri.com/details.asp?dbid=14708> Accessed 7 April 2008.

ICF-International. 2005. User's guide to the Regional Modeling System for Aerosols and Deposition (REMSAD), Version 8. <http://www.remsad.com/documents/remsad_users_guide_v8.00_112305.pdf> Accessed 7 April 2008.

———. 2006. REMSAD Regional Modeling System, ICF International and UAM-V. <http://www.remsad.com/> Accessed 7 April 2008.

Jenness, J. and L. Engelman. 2008. Jenness Enterprises. <http://www.jennessent.com/> Accessed 7 April 2008.

Jones, S. 2007. Hydrological flood analysis and surface modeling. Eagle Technology Group, Auckland, New Zealand. <http://arcscripts.esri.com/details.asp?dbid=14960> Accessed 7 April 2008.

Kavetski, D., S. W. Franks and G. Kuczera. 2002. Confronting input uncertainty in environmental modeling. Pages 49–68 in *Calibration of Watershed Models.* Edited by Q. Duan, H. V. Gupta, S. Sorooshian, A. N. Rousseau and R. Turcotte. Washington, DC: AGU.

Kavetski, D., G. Kuczera and S. W. Franks. 2006a. Bayesian analysis of input uncertainty in hydrological modeling: 1. Theory. *Water Resources Research* 42 (3):W03407.

———. 2006b. Bayesian analysis of input uncertainty in hydrological modeling: 2. Application. *Water Resources Research* 42 (3):W03408.

Kittle, J. L., A. M. Lumb, P. R. Hummel, P. B. Duda and M. H. Gray. 1998. A tool for the generation and analysis of model simulation scenarios for watersheds (GenScn). *Water Resources Investigation Series.* WRI 98-4134. Reston, Va.: US Geological Survey.

Linkov, I., A. Varghese, S. Jamil, T. P. Seager, G. Kiker and T. Bridges. 2004. Multi-criteria decision analysis: A framework for structuring remedial decisions at contaminated sites. In *Comparative Risk Assessment and Environmental Decision Making.* Proceedings of the NATO Advanced Research Workshop, Rome, Italy. Edited by I. Linkov and A. Ramadan.

MapWindow. 2008. MapWindow GIS: Open Source Programmable Geographic Information System Tools. <http://www.mapwindow.org/index.php> Accessed 7 April 2008.

Matott, L. S. 2005. *Ostrich: An Optimization Software Tool, Documentation and User's Guide,* Version 1.6. University at Buffalo, Department of Civil, Structural, and Environmental Engineering, Buffalo, NY.

National Atmospheric Deposition Program (NADP). 2007. National Atmospheric Deposition Program (NRSP-3). NADP Program Office, Illinois State Water Survey. <http://nadp.sws.uiuc.edu/> Accessed 22 April 2008.

New Zealand Ministry of Fisheries. 2007. National Aquatic Biodiversity Information System. New Zealand Ministry of Fisheries. <https://www.nabis.govt.nz/nabis_prd/index.jsp> Accessed 22 April 2008.

NOAA/USEPA (National Oceanographic and Atmospheric Administration and US Environmental Protection Agency Partnership). 2007a. Community Multiscale Air Quality (CMAQ), Atmospheric Sciences Division. <http://www.epa.gov/asmdnerl/CMAQ/> Accessed 7 April 2008.

————. 2007b. Watershed Deposition Tool, Atmospheric Sciences Division. <http://www.epa.gov/asmdnerl/Multimedia/depositionMapping.html> Accessed 7 April 2008.

Nyerges, T., M. Robkin and T. J. Moore. 1997. Geographic information systems for risk evaluation: Perspectives on applications to environmental health. *Cartography and Geographic Information Systems* 24 (3):123–144.

Ollinger, S., J. Aber, S. Millham, R. Lathrop and J. Ellis. 1993. A spatial model of atmospheric deposition for the northeastern U.S. *Ecological Applications* 3 (3):459–472.

Ollinger, S., J. Aber, C. A. Federer, G. M. Lovett and J. Ellis. 1995. *Modeling Physical and Chemical Climate of the Northeastern United States for a Geographic Information System.* Radnor, Pa.: US Department of Agriculture, Forest Service.

Ollinger, S., J. Aber, G. M. Lovett and C. Federer. 2001. ClimCalc, Complex Systems Research Center. <http://www.pnet.sr.unh.edu/climcalc/> Accessed 7 April 2008.

Oreskes, N., K. Shrader-Frechette and K. Belitz. 1994. Verification, validation, and confirmation of numerical models in the earth sciences. *Science* 263:641–646.

Petras, I. 2003. Basin1. Department of Water Affairs & Forestry, Pretoria, South Africa. <http://arcscripts.esri.com/details.asp?dbid=10668> Accessed 7 April 2008.

Poeter, E. P. and M. C. Hill. 1999. UCODE, a computer code for universal inverse modeling. *Computers & Geosciences* 25 (4):457–462.

Poeter, E. P., M. C. Hill, E. R. Banta, S. Mehl and S. Christensen. 2005. UCODE_2005 and Six Other Computer Codes for Universal Sensitivity Analysis, Calibration, and Uncertainty Evaluation. TMWRI 6-A11. Denver, Col.: US Geological Survey.

Preston, B. L. 2002. Hazard prioritization in ecological risk assessment through spatial analysis of toxicological gradients. *Environmental Pollution* 117:431–445.

Purucker, S. T., R. N. Stewart and C.J.E. Welsh. 2008. SADA: Ecological risk based decision support system for selective remediation. In *Decision Support Systems for Risk Based Management of Contaminated Sites.* Edited by A. Marcomini, G. W. Suter and A. Critto. In press: Springer-Verlag.

Reichert, P. 2005. UNCSIM - A computer programme for statistical inference and sensitivity, identifiability, and uncertainty analysis. Pages 51–55 in *Proceedings of the 2005 European Simulation and Modelling Conference (ESM 2005).* Edited by J.M.F. Teixeira and A. E. Carvalho-Brito. Porto, Portugal: EUROSIS-ETI.

Saltelli, A., S. Tarantola, F. Campolongo and M. Ratto. 2004. *Sensitivity Analysis in Practice: A Guide to Assessing Scientific Models.* New York: John Wiley & Sons.

Stewart, R. N. and S. T. Purucker. 2006. SADA: A freeware decision support tool integrating GIS, sample design, spatial modeling, and risk assessment. In *Proceedings of the Third Biennial Meeting of the International Environmental Modelling and Software Society,* Burlington, Vt.

Thayer, W. C., D. A. Griffith, P. E. Goodrum, G. L. Diamond and J. M. Hassett. 2003. Applications of geostatistics to risk assessment. *Risk Analysis* 23 (5):945–960.

TNC (The Nature Conservancy). 2006. GIS tools for stream and lake classification and watershed analysis. The Nature Conservancy, Freshwater Initiative. <http://conserveonline.org/workspaces/gistoolsfreshwater> Accessed 7 April 2008.

Tonias, C. N. and E. C. Tonias. 2002. *Avenue Wraps: A Guide for Converting Avenue Scripts into VB/BVA Code.* Rochester, New York: The CEDRA Press.

USDOE (US Department of Energy). 2004. RESRAD-BIOTA: A Tool for Implementing a Graded Approach to Biota Dose Evaluation, User's Guide, Version 1. Interagency Steering Committee on Radiation Standards, DOE/EH-0676.

USEPA (US Environmental Protection Agency). 2001a. Better Assessment Science Integrating Point and Non-Point Source BASINS Version 3.0. Office of Water. Washington, D.C. EPA-823-H-01-001.

————. 2001b. Frequently asked questions about atmospheric deposition: a handbook for watershed managers, EPA-453/R-01-009, US Environmental Protection Agency, Office of Wetlands, Oceans, and Watersheds and Office of Air Quality Planning and Standards, Washington, D.C.

————. 2003. Multimedia, Multipathway, and Multireceptor Risk Assessment (3MRA) Modeling System. Volume I. Modeling System and Science Office of Research and Development - National Exposure Research Laboratory and Office of Solid Waste. Washington, D.C., EPA-530-D-03-001a.

————. 2006. Data Quality Assessment: Statistical Methods for Practitioners. Office of Research and Development, Washington, D.C., EPA/QA/G-9S.

————. 2008. Model Evaluation Website. <http://www.epa.gov/athens/research/modeling/modelevaluation/index.html> Accessed 12 May 2008.

USGS (US Geological Survey). 2007. The National Hydrography Dataset: concepts and contents. U.S. Geological Survey, Reston, Va. <http://nhd.usgs.gov/data.html> Accessed 7 April 2008.

van Griensven, A. and T. Meixner. 2004. Dealing with unidentifiable sources of uncertainty within environmental models. In *iEMSs 2004 International Congress on Complexity and Integrated Resources Management.* Edited by C. Pahl, S. Schmidt and T. Jakeman. Osnabrück, Germany.

————. 2006. Methods to quantify and identify the sources of uncertainty for river basin water quality models. *Water Sci. Tech.* 53 (1): 51–59.

————. 2007. A global and efficient multi-objective auto-calibration and uncertainty estimation method for water quality catchment models. *Journal of Hydroinformatics* 9 (4): 277–291.

van Griensven, A., T. Meixner, S. Grunwald, T. Bishop, M. Diluzio and R. Srinivasan. 2006. A global sensitivity analysis tool for the parameters of multi-variable catchment models. *Journal of Hydrology* 324 (1-4):10–23.

Wolfe, K. L., R. S. Parmar, G. F. Laniak, A. B. Parks, L. Wilson, J. E. Brandmeyer, D. P. Ames and M. H. Gray. 2007. Data for Environmental Modeling (D4EM): Background and example applications of data automation. *Proceedings of International Symposium on Environmental Software Systems,* Prague, Czech Republic.

Woodbury, P. B. 2003. Do's and don'ts of spatially explicit ecological risk assessments. *Environmental Toxicology and Chemistry* 22:977–982.

SECTION 4
Spatio-Temporal Aspects of GIS

CHAPTER 19
Adding Time to GIS

Yanfen Le and *E. Lynn Usery*

19.1 Introduction

Geographical data are multidimensional, including dimensions of space, time, and attributes. To map multidimensional geographic data to a two-dimensional (2D) space, dimensions have to be reduced. For example, the three-dimensional (3D) Earth surface is projected to a 2D map space. There are some good examples of mapping the three dimensions to two, such as Minard's well-known map of Napolean's March on Moscow, which presents two dimensions of space and time with several attributes, including the temperature, size of the army by numbers, and the direction of the march (Kraak 2003). Commonly, time is fixed in cartography and the traditional geographic information system (GIS) rooted in cartography is static in nature (Usery 1996; Galton 2001).

All geographic phenomena are changing. Some change slowly and are relatively stable, such as elevation and administrative boundaries, whereas others evolve rapidly, such as storms and human activity. These two types of phenomena can be divided as entities and processes, respectively (Usery 2000). For example, Mount Everest was measured at 8,850 m recently and is growing slowly at about one cm a year (Getis et al. 2006). A GIS database, which models real-world phenomena and processes in a computer information system, also should evolve with the changing world. After almost 40 years of development, space and attributes are well studied in GIS, but time is not. It is the purpose of this section to present a review of representation of time in GIS, implementation issues for databases and GIS, and temporal GIS concepts including analysis and visualization.

19.2 Background and History

The pioneering work on time in geography began in the 1970s (Hägerstrand 1970; Thrift 1977); thereafter, little was done until the publication of several PhD theses in the last decade (Hazelton 1991; Kelmelis 1991; Al-Taha 1992). Langran's (1992) *Time in Geographic Information Systems* is regarded as a landmark in temporal GIS (Peuquet 2002). Among all spatio-temporal representations, the traditional layer-based framework is the only one supported by traditional GIS. However, previous research indicates that layer-based approaches are insufficient for representing temporal information because only snapshots or changes are represented in this schema (Langran 1992; Peuquet and Duan 1995). Time, as well as space and attributes, is required to be added to GIS.

Among the various spatio-temporal approaches proposed heretofore, early studies focus on entities. For example, many early spatio-temporal models were designed for cadastre (Hunter and Williamson 1990; Al-Taha 1992; Chen and Le 1996). This primarily occurred because the spatio-temporal changes of land information are relatively simple to represent, and necessary for cadastral databases. Temporal GIS also is gaining more attention in transportation agencies when the focus is moving from construction phases to facilities management (Koncz and Adams 2002). Recently, there are growing efforts on modeling continuous processes, including storm events (McIntosh and Yuan 2005a), personal daily journeys (Kwan et al. 2003), and environmental health (Mark et al. 2003).

Along with research in GIS, time also is studied in computer science, especially in the areas of databases and cognitive science. The database community is interested in spatio-temporal models, as well as access methods and implementations (Koubarakis and Sellis

2003). In cognitive science, there is a growing body of literature working on spatio-temporal information (Galton 2001; Frank 2003; Grenon and Smith 2004). Galton (2003, 2004) identified key desiderata for a spatio-temporal ontology, and discussed the distinction between field- and object-based approaches to space, time, and space-time. Grenon and Smith (2004) argued dynamic spatial ontology should combine a purely spatial ontology and a purely spatio-temporal ontology. Research on cognitive science provides ontological support to GIS researchers because "a representational theory that closely reflects human cognition is highly efficient and minimally complex from a computing standpoint" (Yuan et al. 2004, p. 148).

19.3 Time and Spatio-temporal Change

There exist different concepts of time. These concepts pose requirements for the representation of time in GIS. How people think about time in the real world is important to the way that time is modeled in a computer information system.

19.3.1 Time

Absolute vs. Relative Time

In Newton's absolute view, time is composed of instants and exists independently of the frame of time. In Leibniz's relative view, objects are located relative to each other instead of in a single dimensional time axis. The absolute view predominated until the adoption of Einstein's relativity theory at the beginning of this century (Kern 1983). Leibniz's relative view is not contradictory, but complementary to Newton's absolute view. Peuquet (1994) argued that absolute time is objective, and relative time is subjective.

Linear vs. Cyclical Time

Gould (1987, p.10) discussed the dichotomy: "linear and cyclical time, or time's arrow and time's cycle." Linear time, the primary metaphor of history, views history as an unalterable sequence of unrepeatable events, and views time as a straight directional line stretched from past through present and to future. Cyclical time means there is no direction of time, and events are components of repeating cycles. Gould (1987, p. 13) recognizes that "the dichotomy is oversimplified and both of them are needed." Therefore, "it is possible to define four metaphors of time: linear (past, present, and future), cyclical (season, day/night, etc.), branching (inter-entity time dependencies), and multidimensional time (world time, survey time, valid time, display time)" (Claramunt and Thériault 1995, p. 25). Similarly, Worboys (1998) identified three types of temporal topology: linear, branching, and periodical (Figure 19-1). Cyclical is one example of relative time, whereas linear time is absolute in nature.

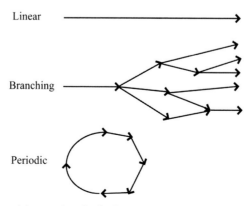

Figure 19-1 Linear, branching and periodic times.

Discrete vs. Continuous Time

Time, as an independent variable, can be measured continuously or discretely. Theoretically, it is possible to measure continuous time at an arbitrary level of precision (Frank 1998; Worboys 1998). Discrete time is measured at selected instants or intervals, whereas continuous time is measured always. Continuous time is appropriate for continuously changing phenomena that have an underlying model, assuming interpolation of any time point from measurement at sampled time points (Peuquet 1994). For example, the Weather Channel uses a continuous time variable to show weather prediction based on weather conditions at certain previous times. Discrete time is appropriate for phenomena that evolve only at a particular time instead of all the time. For example, land ownership in a cadastral database uses discrete time. Peuquet (1994) argued that time is continuous in nature and is divided into discrete units for the purpose of measurements. These discrete units are also called temporal granularity.

World Time vs. Database Time

A fourth metaphor of time, multidimensional time, also has been investigated (Jensen et al. 1995; Date et al. 2003). World and database time are considered as two dimensions (Langran 1992; Jensen et al. 1995; Worboys 1998). Other dimensions include display time, survey time, etc. World time means a fact is true in the real world, whereas database time indicates a specific phenomenon is recorded in the database. In temporal database management systems, world and database time also are called valid and transaction time, respectively (Date et al. 2003). Jensen and Snodgrass (1996) argued that world time is needed if changes to the past are important, whereas database time is required if rollback to a previous database state is necessary. A data model that supports both world and database time is called bitemporal (Jensen and Snodgrass 1996; Worboys 1998).

Point Time vs. Interval Time

Point time is an instant or a single point in the time dimension, whereas interval time is a period of time with a starting and an ending point in the time dimension. Point time also is termed as instant (Snodgrass 2000). The concepts of point and interval time are similar to those of a point and an interval in space. Actually, many terms used in time, such as temporal resolution and temporal interpolation, are comparable to those in space (Peuquet 1994).

Four Modes of Temporal Explanation

Harvey (1969) analyzed four modes of temporal explanation. To Harvey, narrative mode is the weakest, but also is the best when there are few historical data. Explanation by reference to time treats time as an independent variable rather than a parameter. Explanation by reference to hypothesized process, which assumes a specific mechanism and an artificial time scale, encounters some conceptual difficulties. Explanation by reference to an actual process is a scientific approach, although it is inferior in human geography.

Temporal Granularity

Temporal granularity is the unit of measurement in the time dimension. For example, the temporal granularity of birthday is a day, whereas that of a flight schedule is minutes. Dyreson et al. (1995) stated that the mixture of different temporal granularities in a single database is common but causes problems. In temporal database management systems, a chronon is defined as the shortest duration of time, just as a cell is the smallest and non-decomposable section of space (Jensen et al. 1995). Temporal granularity is related with temporal resolution and temporal scale.

19.3.2 Spatio-temporal Changes

Spatio-temporal change occurs in space with time. Claramunt and Thériault (1995, p. 25) studied the evolution of entities with space and time, and identified "three main types of spatio-temporal processes: change of a single feature, functional relationships between entities, and evolution involving several entities." Wang and Cheng (2001) distinguished three types of spatio-temporal change: discrete, stepwise, and continuous (Figure 19-2).

Peuquet (1999) argued that temporal duration, frequency, and pattern are important temporal characteristics. She identified four types of spatio-temporal changes: continuous, majorative, sporadic, and unique, and defined four temporal distribution patterns: steady, oscillating, random, and chaotic. A continuous change takes place throughout the time interval; a majorative change occurs during most of the time; a sporadic change happens only some of the time; and a unique change occurs only once during the time interval. In Peuquet's (1994) view, the four temporal distribution patterns (steady-state, oscillating, random, and chaotic) are parallel to the four spatial distribution patterns (regular, clustered, random, and chaotic).

Usery (1996) identified temporal relationships including rate of change, speed, and direction of change. McIntosh and Yuan (2005b) described geographic events and processes using six indices: elongation, orientation, distribution, percent growth, granularity of change, and relative movement. With these indices, they assessed similarity of complex geographic processes such as storms.

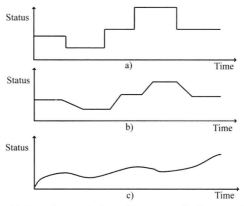

Figure 19-2 a) Discrete, b) stepwise and c) continuous spatio-temporal changes.

19.4 Representing Time in GIS

The concept of representation can be explained at three levels: data model, formalization, and visualization (Yuan et al. 2004). A data model conceptualizes the real world in a computer information system (Goodchild 1987), and constrains what can be formalized and visualized (Yuan et al. 2004). Therefore, spatio-temporal representation is the key to adding time to GIS. Among the large amount of literature in temporal GIS, the majority is on spatio-temporal data models. The closer a spatio-temporal data model is to our concepts of space and time, the easier spatio-temporal knowledge will be represented, and the faster it will be accessed.

Time has been studied in GIS since 1970. Hägerstrand (1970) developed the concept of time geography and proposed to represent time using space-time paths. Thrift (1977) proposed time as an additional dimension to space. Langran (1992) described four temporal GIS models: the space-time cube, sequential snapshots, a base state with amendments, and the space-time composite. Later research proposed various approaches to modeling time in GIS, including event-based (Claramunt and Thériault 1995; Peuquet and Duan 1995; Chen and Jiang 1998; Worboys 2005), process-based (Yuan 1997; McIntosh and Yuan 2005a), feature-based (Usery 1996; Le 2005a), time-based (Peuquet 1999), activity-based (Wang and Cheng 2001; Frihida et al. 2002), object-oriented (Raper and Livingstone 1995; Wachowicz 1999), identity-based (Hornsby and Egenhofer 2000), and integrated (Peuquet 1994; Koncz and Adams 2002) approaches. Some of these methods are similar in concept.

19.4.1 Space-time Cube and Space-time Paths

A space-time cube is a 3D cube representing 2D space and one-dimensional time. The 2D space progresses along the time dimension. This view is the same as Thrift's (1977) idea, which treats time as an additional dimension to space. Time is linear and continuous in the space-time cube. The space-time cube is suitable for fields with continuous spatio-temporal change (Ohsawa and Kim 1999). The space-time cube is easy to conceive from the perspective of a static GIS data model. However, Langran (1992) argued the space-time cube is difficult to implement because of philosophical and conceptual problems and the status of computer technologies.

Langran (1992) defined the space-time cube using Hägerstrand's (1970) space-time paths. Each object has a 3D trajectory or a space-time path in the space-time cube. The history of an object can be accessed by tracing its trajectory. In our opinion, a space-time path is an identifiable entity or process in the space-time cube or hyperspace, as an object is in the space; therefore, space-time paths are only one representation of a space-time cube. In addition to objects, there are fields in spatial representation. Since similarities exist between space and space-time (Galton 2003), another view should exist that represents the space-time cube as a continuous field. For example, temporal information of the atmosphere or temperature is continuous in the space-time cubes, and the field representation fits better than space-time paths.

With advancement in computer technology, including object-oriented programming, the space-time path has evolved into other approaches such as object-orientation (Wachowiz 1999) and activity-based representation (Wang and Cheng 2001). Recently, space-time paths have been studied to model a person's journey (Kwan et al. 2003; Yu 2006). A person's journey is relatively simple to represent using space-time paths for two primary reasons: 1) the spatial dimension of a person's location is a point, which is easy to represent compared to a line or polygon; and 2) there is no split, merge/combination, addition, or deletion of object identity, so a personal journey can be modeled with a polyline in the space-time cube. In the space-time cube, a point only moves or changes location and there is no change in shape. For a line or polygon, it cannot only move together, but also change point-by-point. In other words, a line or polygon can change shape in the space-time cube because each point of the line or polygon can move in a different direction at a different speed. It is difficult to represent a complex process, such as a storm, with a space-time path (McIntosh and Yuan 2005a). Moreover, lines and polygons cannot only change shape, but also add, delete, split or merge/combine; therefore, personal location is a simple case of space-time paths.

19.4.2 Sequential Snapshots

The sequential snapshots approach, introduced by Armstrong (1988), is the only spatio-temporal representation currently supported by most commercial GIS (Peuquet 1999). This approach uses a temporal series of spatially registered snapshots $\{S_i\}$ to represent the spatial-temporal progress (Figure 19-3). Each snapshot models one status of the real world at that specific time. Discrete and point time are employed in this approach. The temporal distances between any two consecutive times, t_i and t_{i+1}, may be the same or different. Snapshots also are viewed as a series of time-slices from the space-time cube (Langran 1992). This approach can be employed for field and object data. Since all snapshots are spatially registered and history is tracked by location, this method also is called the location-based approach.

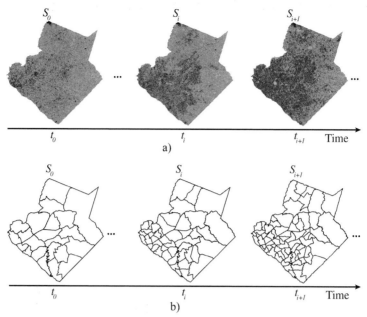

Figure 19-3 Sequential snapshots for a) fields and b) objects.

The sequential snapshots approach is conceptually straightforward (Peuquet 1999). For each time t_i, it is easy to retrieve the status of the real world at any spatial location. However, we cannot tell for sure the status of the real world between t_i and t_{i+1}. Furthermore, the accumulated changes between t_i and t_{i+1} are not explicitly modeled. Langran (1992) argued that snapshots are the temporal equivalence of spaghetti in space. This approach can only model accumulated changes between any consecutive times, t_i and t_{i+1}, by comparison between these two snapshots based on location. The comparison procedure is time consuming if the computer is not efficient. Also, data volume quickly increases since each snapshot records the whole area whenever change takes place at that location. This brings about redundancy in a database, especially for geographic phenomena that are not changing frequently.

19.4.3 Base State with Amendments

Base state with amendments (Figure 19-4) is an alternative to the sequential snapshots approach. It starts with the original snapshot at the time of beginning, t_0, as base status, and records only changes of events under observation between any two consecutive times. This change-only, time- or event-based temporal approach models change instead of world status (Langran 1992). The same types of time concepts, point and discrete time, are used in this approach as they are in the snapshots approach. This approach can be used for field and object data.

The base state with amendments is better than the sequential snapshots approach because it models change instead of world status and has minimal redundancy (Langran 1992). It is an important improvement if the computer is slow and the storage space is limited. Accumulated changes between two point times are easy to access. World status can be retrieved by amending changes to the base status. This approach is better for location-based queries than for object-based ones (Peuquet and Duan 1995).

Ohsawa and Kim (1999) and Le (2005a) argued the current data are the most frequently used and need to be accessed efficiently, so it should be more meaningful to take the snapshot at the current or latest time (t_n) rather than the beginning time (t_0) as the base state.

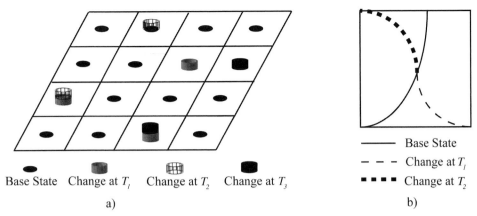

Figure 19-4 Base state with amendments for a) fields (after Langran 1992) and b) objects.

19.4.4 Space-time Composite

The space-time composite was developed by Langran and Chrisman (1988). This approach begins with a base map, and decomposes the base map using the overlay method when the real world evolves. The space-time composite is fragmented little by little with time.

Langran (1992) argued that this approach used an atemporal method to deal with a temporal problem. Changes are represented implicitly in this approach. The space-time composite is conceptually straightforward. To access world status, all fragments valid at that point in time are retrieved, and then common boundaries between these fragments are resolved by rebuilding spatial topology. The problem with this approach is the fragmentation. The updating process is relatively more time consuming than other methods; therefore, space-time composite representation has not been further studied.

19.4.5 Event- and Time-based Approaches

Events are things, conditions, processes, or objects that exist (Claramunt and Thériault 1995). Several event-based approaches have been proposed by Claramunt and Thériault (1995), Chen and Le (1996), Peuquet and Duan (1995), and Chen and Jiang (1998); some are different in essence, whereas others are similar. The common idea behind these four approaches is explicitly presenting the temporal successive relationships using backward and forward pointers in the database.

Claramunt and Thériault (1995) presented an event-oriented approach for vector data using extended versioning. To them, ordered events are more useful than precisely dated events, and versioning is a mechanism for ordering history. Extended versioning is good at describing the history involving several entities. Temporal successive relationships are explicitly recorded using extended versioning in this approach. Using a different versioning mechanism, Chen and Le (1996) proposed another approach for land information systems. Conceptually, these two approaches are similar. With these versioning mechanisms, the history of features can be tracked by following the backward or forward pointers, and the world status at a given time can be accessed by retrieving valid features at that time.

Peuquet and Duan (1995) developed an event-based spatio-temporal data model for raster data (Figure 19-5); they called their approach a time-based data model. The sequential snapshots and base state with amendments approach are all based on time. Conceptually, this model is an implementation of base state with amendments. The original framework of base state with amendments is implemented by explicitly representing the temporal successive relationships using backward and forward pointers, and the amendments are realized by change components. This agrees with the opinion that explicitly ordered events are better than implicitly dated versions (Claramunt and Thériault 1995; Wachowicz 1999). This framework is implemented in a temporal GIS data management system called TEMPEST, which is applied in the Apoala project and is claimed to have fast access to data (GeoVISTA 2003). This model was an improvement in spatio-temporal representation when computer storage space was an issue in the mid-1990s.

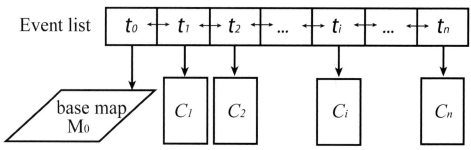

Figure 19-5 Event-based spatio-temporal data model (after Peuquet and Duan 1995).

Chen and Jiang (1998) proposed another event-based spatio-temporal data model for vector cadastral data (Figure 19-6). Several particular events relating to cadastral application are identified and modeled in this approach. There are temporal succession relationships between these predefined events. According to Harvey (1969), this approach is scientific because it refers to the actual process in cadastre.

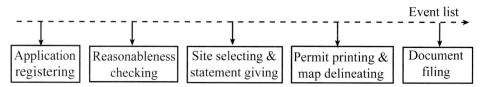

Figure 19-6 Event flow (after Chen and Jiang 1998).

19.4.6 Object-, Feature-, Entity-, Identity-, or Activity-based Approaches

A feature is an entity in the real world and an object in the information system (Tang et al. 1996). Object-oriented (Wachowicz 1999), feature-based (Usery 1996), entity-based (Peuquet 1999), identity-based (Hornsby and Egenhofer 2000), and activity-based (Wang and Cheng 2001) approaches are all conceptually similar. All of them treat an identifiable object, feature, entity, process, or activity as a base for spatio-temporal modeling. The object-oriented approach has been explored by many researchers. This type of approach, which originated in computer engineering, allows for a cohesive representation of time. The Spatial Architecture and Interchange Format (SAIF), an object-oriented spatio-temporal model, has been adopted as a Canadian national standard (SAIF 2006).

Wachowicz (1999) proposed an object-oriented approach for modeling the evolution of public boundaries. She studied Hägerstrand's (1970) space-time paths in the space-time cube. By identifying discrete spatio-temporal change in the spatio-temporal path for a public boundary, she developed a spatio-temporal data model based on space-time paths and object-orientation. A version mechanism is used to facilitate history tracking. Wachowicz's space-time path of public boundaries (Figure 19-7) is similar to the event list in Chen and Jiang's (1998) event flow.

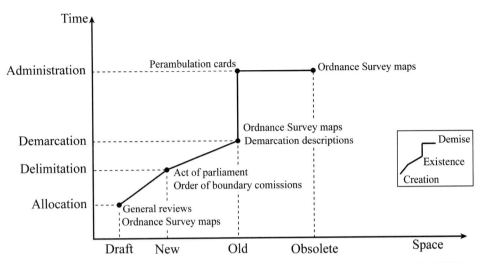

Figure 19-7 An example of space-time path of a public boundary (after Wachowicz 1999).

Wang and Cheng (2001) developed a spatio-temporal approach for transport activities of a single person. They identified three types of spatio-temporal behavior (continuous, discrete, and stepwise), and two types of activities that interact with locations ("stay at" and "travel between"). These activities are treated like an object in the spatio-temporal approach. An extended-entity-relationship diagram is employed to represent the conceptual framework of the hierarchical relationships in the activity-based transport data model. Wang and Cheng (2001) state this specific approach could answer time-, person-, activity-, and location-based questions. The transport activity in this study is the same as the human activity in Kwan et al. (2003) and Yu (2006), but the latter provides better visualization within the space-time cube.

Usery (1996, 2000) proposed a feature-based data model with three basic dimensions of space, theme, and time (Table 19-1). Yuan (1996) proposed a similar model referred to as the three-domain model. In Usery's (1996) model, each dimension has attributes and relationships. Temporal relationships, such as "was_a" and "will_be," are employed to represent the temporal succession relationships between features. Attributes related with changes, such as change rate and erosion speed, also are considered temporal relationships. This approach needs to be further explored in the temporal aspect. For example, it does not address any explicit version mechanism so that temporal relationships between features are difficult to track. Also, attribute-level timestamps put forward in this model are excellent in concept, but will slow down the operation of accessing the entire record when implemented in a traditional database system. Usery (2004) studied the implementation of the feature-based data model in a feature library in which temporal information is organized based on change: {time$_i$, {space/theme$_j$, value}}, and the timestamp is shared by any change in space and/or themes taking place at the same time. This data structure is efficient in change tracking, but is slow in recovering the status of a feature. This approach requires a fully developed object-oriented database management system to facilitate data storage, accessing, indexing, and management. The feature-based data model has advantages over others because it captures all three dimensions of a feature. For example, Hägerstrand's (1970) space-time path only emphasizes the space and time dimension, but does not include the themes explicitly.

Table 19-1 Feature-based GIS conceptual model (after Usery 1996).

	Space	**Theme**	**Time**
Attributes	ϕ, λ, Z Point, line, area, surface, volume, pixel, voxel, …	Color, size, shape, pH, …	Date, duration, period, …
Relationships	Topology, direction, distance, …	Topology, is_a, kind_of, part_of, …	Topology, is_a, was_a, will_be, …

Le (2005a) furthered Usery's feature-based model. She argued that time should not be treated as equally as the space and theme. In Le's feature-based temporal model, a feature is a temporal feature that can have different temporal spaces and themes during its lifespan/duration (Figure 19-8). The feature, the space, and the themes are all functions of time. Although discrete functions are employed in the case study, theoretically, these functions can be discrete or continuous. The explicit versioning mechanism is enabled using the "Was/Became" relationships between temporal features, temporal spaces, and temporal themes. The many-one relationships are implemented in an object-relational database using nested tables. In Le's (2005a) feature table, one record represents a temporal feature containing its history, so it is straightforward to retrieve the entire history of a feature in one step.

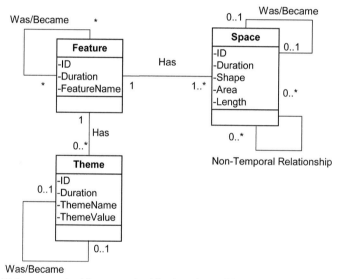

Figure 19-8 UML conceptual framework of feature-based temporal model (after Le 2005a).

Hornsby and Egenhofer (2000) developed a spatio-temporal model based on the object identity, which is distinct from an object's properties. Change, rather than world status, is modeled using a change description language. In fact, the idea of object identity is implicitly applied in other spatio-temporal approaches. For example, each feature in Usery's (1996) feature model has a unique feature identifier.

19.4.7 Combined/Integrated Approaches

Most of the approaches discussed above are location- or feature-based, and are efficient in answering only location-based or feature-based spatio-temporal questions. A location-based framework is good at tracking histories at one location or sets of locations, whereas feature-based schemata allow for histories related with each feature. In reality, a spatio-temporal data model is required to provide temporal information from multiple perspectives (Peuquet 1994, 1999; Yuan 1999). For example, the activity-based approach proposed by Wang and Cheng (2001) can provide time-, person-, activity-, and location-based views.

Peuquet (1988) proposed a Dual framework with location-based and object-based representations. Based on the Dual framework, Peuquet (1994) developed the Triad, a framework with three representations of object-, location-, and time-based views. These three representations are important to efficiently solve "what," "where," and "when" questions, which correspond to "themes," "space," and "time," respectively; but geographic information has to be duplicated in this three-view representation. Galton (2001, p. 184) argued "the Triad framework of Peuquet is more purely schematic and cannot avoid unacceptable duplication of information in implementation."

Koncz and Adams (2002) presented a multi-dimensional location referencing system (MDLRS) data model for management of transportation facilities. The MDLRS data model provides four representations for "what," "where," "when," and "how" questions, which are important questions in geography studies.

19.4.8 Event- and Process-oriented Approach

Worboys (2005) identified four stages: static, snapshots, object change, and event and action, in the development of spatio-temporal modeling. He drew the "things-happenings" distinction in space-time and argued that "happenings" are as important as "things." In his event-oriented approaches, "happenings" are represented as events. Although the event-oriented approach sounds like event-based models, it is closer to process-based approaches in nature. Worboys' (2005, p. 19) approach is good at spatio-temporal reasoning and appropriate for clearly-defined processes or activities, such as a vehicle's motion—"brake the vehicle, then it slows down and stops." However, it is difficult to determine specific events in geographic reality. For example, a land parcel changes from agriculture to urban. Why does this happen and how does one define the events? There are so many potential reasons that it is impossible to define a simple deterministic process like the vehicle's motion. In another example, from two maps of the same location in 1980 and 2000, one reads: 1) there is a county, 2) its boundary is bigger in 2000, and 3) its name has changed. What map readers can learn from the maps is about a county, an entity in reality, which has different space and different attributes at different times. Moreover, even in a simple activity like braking, the vehicle does not necessarily slow down if the brake fails or the vehicle drives down-hill. According to Harvey (1969), explanation by reference to an actual process is inferior in human geography because it is complex and difficult to identify; therefore, care must be taken when this approach is applied to human studies. Moreover, this approach provides only a conceptual model and there is no detail in implementation, such as visualization and how to manage the data in a database.

In reality, some applications cannot be represented using a single approach, and many applications use fields and objects. Therefore, multiple approaches may be employed for a single application or system based on spatial data, characteristics of the time, types of spatio-temporal changes, and the problem being posed on the system. For example, Yuan (1997) employed four data models to represent the lifecycle of wildfire. Among these four conceptual data models, three are supported by current GIS. They are location-based snapshots for raster data, entity-based snapshots for vector data, and fire mosaics representing the temporal sequence of burns for vector data. The fourth conceptual model, the layer-based model for wildfire, is inadequately supported. A fire is treated as a feature in that model. With these models, the lifecycle of wildfire is modeled by the information cycle. Since these four data models are different in spatial and temporal resolutions, data of one model must be spatially and temporally aggregated and disaggregated before they are used in another model. Although Yuan employed four models for raster and vector data, those models do not work together but independently in the wildfire lifecycle. Another difference between Yuan's and others' studies is that a wildfire is a continuously moving entity in a space-time hyperspace, and is much more difficult to represent.

McIntosh and Yuan (2005a) introduced a hierarchical framework with event, process, sequence, and zone for complex dynamic geographic phenomena (Figure 19-9). In this framework, an event contains one or more processes, a process includes sequences, and a sequence is made up of zones. This framework represents a complex dynamic process with a temporal series of snapshots, so it is not truly continuous in nature. This agrees with Peuquet's (1994) argument that time is continuous in nature, but divided into discrete times for measurements. Representation of dynamic process is recognized as one of the research objectives by University Consortium for Geographic Information Science (UCGIS) because of its complexity (Yuan et al. 2004).

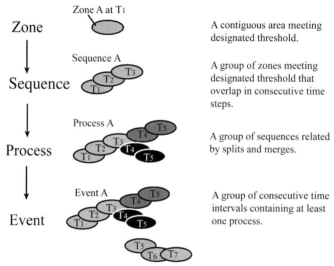

Figure 19-9 Temporal framework (after McIntosh and Yuan 2005a).

19.4.9 Other Approaches

Dragicevic and Marceau (2000) proposed to model time using a fuzzy set approach. Temporal interpolation is performed using fuzzy membership, which is derived between two consecutive snapshots. This approach is good for areas undergoing continuous and dynamic changes, including atmosphere, vegetation, and the rural/urban environment.

Spatio-temporal data models also are studied by the database community. Medak (1999) described a feature's life using its lifestyles. Using spatio-temporal constraints, Tryfona et al. (2003) extended the entity-relationship (E/R) model and the unified modeling language (UML). An E/R model is an abstract representation of the structure of a database using entity sets, attributes, and relationships. The UML is a standard notation for describing object-oriented models (Gomma 2005). Güting et al. (2003) and Grumbach et al. (2003) proposed modeling spatio-temporal data based on data types and constraints, respectively. These data models are at an abstract level (Güting et al. 2003) and, therefore, do not capture the appropriate ontological commitment for time in GIS. Erwig et al. (1998) proposed to model temporal objects with an abstract data type (ADT), and argued that the ADT approach is more versatile and more controllable than the traditional flat view. An ADT contains two parts: one or more domains, and mathematical operations defined on elements from the domains (Lewis and Denenberg 1991). An ADT is neither a data structure nor a data type. When it is implemented in a particular language such as Oracle, the domains can be data types.

19.4.10 Summary

The spatio-temporal models described above are listed in Table 19-2. The names of models are given by the developer, and there is no naming convention. Most approaches are based on a single data model and for spatial data in either field- or object-form. Although some approaches can be used for fields and objects, they are not the best choice for both, and usually are implemented differently in the two cases. For example, base state with amendments (Figure 19-4) has distinct implementations for fields and objects, but is better for fields.

Table 19-2 Spatio-temporal models.

Model		Applicable to			
Name	Developer(s)	Fields	Objects	Particular applications	Continuous change
space-time cube	Thrift (1977)	✓	✓		✓
sequential snapshots	Armstrong (1988)	✓	✓		
space-time paths	Hägerstand (1970)	✓	✓		✓
space-time paths	Kwan et al. (2003); Yu (2006)		✓	human activity	✓
base state with amendments	Langran (1992)	✓	✓		
space-time composite	Langran and Chrisman (1988)	✓	✓		
event-based	Chen and Jiang (1998)		✓	cadastre	
event-based	Peuquet and Duan (1995)	✓			
event-oriented	Claramunt and Thériault (1995)		✓	cadastre	
event-oriented	Worboys (2005)		✓		✓
process-based	Yuan (1997); McIntosh and Yuan (2005a)	✓	✓	wildfire, storm	✓
object-oriented	Wachowicz (1999); SAIF (2006)		✓	boundary	
feature-based	Usery (1996)	✓	✓		
feature-based	Le (2005a)		✓		
identity-based	Hornsby and Egenhofer (2000)		✓		✓
activity-based	Wang and Cheng (2001)		✓	human activity	✓
the Dual framework	Peuquet (1988)	✓	✓		
the Triad framework	Peuquet (1994)	✓	✓		
multidimensional transportation	Koncz and Adams (2002)		✓	transportation	
the fuzzy set approach	Dragicevic and Marceau (2000)	✓			✓

19.5 Integrating Time into a Geographic Database

Geographic data are one type of spatial data that deals with space at geographic scales. A database is a critical issue in managing geographic data. It provides advantages over simple files including: standard query language, index for faster searching, storage management, security, data sharing, controlled update, and backup and recovery (Longley et al. 2005). If time is to be added to GIS, it should be integrated into the geographic database.

Three primary types of database management systems (DBMS) have been used in GIS: relational (RDBMS), object-relational (ORDBMS), and object-oriented (OODBMS) (Longley et al. 2005). An RDBMS contains simple tables. It works well with simple data, but cannot handle complex data types such as geographic data. The OODBMS initially was designed to provide object-oriented (OO) features in a database. It is excellent in managing complex objects and support behavior/functions on objects, but it is not commercially successful for two reasons. First, OODBMS is weak at backwards compatibility to RDBMS, which includes more than 90 percent of the existing database applications. Second, many commercial RDBMS are extended to ORDBMS by providing OO features in RDBMS. The ORDBMS is supported by most DBMS vendors including Oracle, IBM DB2, IBM Informix, and Microsoft SQL Server. For example, a spatial data type is available in these ORDBMS.

19.5.1 Adding Time to Non-spatial Databases

In non-spatial temporal databases, there are two ways to incorporate time into traditional static databases. One is using an ORDBMS or OODBMS, and the other is extending an RDBMS. Temporal information is added as timestamps in RDBMS. In an extended RDBMS, timestamps can be added at the table-, record-, or attribute-level. Conceptually, a timestamp at the table-level is straightforward, but has much redundancy. A timestamp at the attribute-level has minimal redundancy, but it requires time to retrieve all attributes for one feature.

Currently, there are three ways, called dimensions, to manage slowly changing attributes in a commercial database (Kimball Group 2000; Lee 2005). In the type 1 dimension, a record simply is updated and the original data are lost. In the type 2 dimension, a new record is added to the table, and the original and the new data are present. In the type 3 dimension, two more columns are added, one for the new data and another for the change time. However, the type 3 dimension cannot hold all historical information. If the attributes change more than once, only the last two versions of the attribute values, the current and the one previous, are kept in the table. For example, Table 19-3 tracks a person who lived in Georgia before 1995, then moved to California, and finally settled in Washington in 2003.

To facilitate the query of time in the database, temporal query has been studied in the past decade. There are studies focused on temporal databases, including temporal topology, temporal query language, and indexing (Sellis 1999; Jensen and Snodgrass 1996; Date et al. 2003). Allen (1984) identified seven types of temporal topological relationships between two interval times: before, equal, meet, overlap, during, start, and end. Jensen et al. (1995) extended the structured query language 92 (SQL-92) to the temporal structured query language 2 (TSQL2).

Table 19-3 An example of type 1, 2, and 3 slowly changing dimensions.

a) Original table, 1990

ID	Name	State
100	Jason	Georgia

b) In type 1 dimension, 2003

ID	Name	State
100	Jason	Washington

c) In type 2 dimension, 2003

ID	Name	State
100	Jason	Georgia
101	Jason	California
102	Jason	Washington

d) In type 3 dimension, 2003

ID	Name	Original state	Current state	Change time
100	Jason	California	Washington	2003

19.5.2 Integrating Time into a Geographic Database

The GIS community has utilized research results from computer science to incorporate time into geographic databases. For example, a type 2 dimension with record-level timestamps traditionally is employed in a temporal geographic database. So far, most spatio-temporal representations are implemented in RDBMS using simple data types. Data structures defined by McIntosh and Yuan (2005b) are shown in Figure 19-10. In the event, process, sequence, and zone tables, timestamp is added at record-level with "StartTime" and "Duration." Solid lines associate records between tables, whereas dashed lines relate records within tables. Since an event may contain many processes, and a process may contain many other processes, these associations are many-one relationships. These data structures are simple, and the tables can be linked together using RELATE or JOIN operations. However, it is inefficient to retrieve an event from the tables because RELATE and JOIN are the least efficient operations among the query languages.

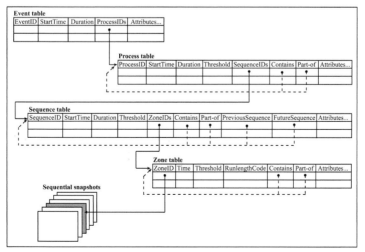

Figure 19-10 Data structures for events, processes, sequences and zones (after McIntosh and Yuan 2005b).

Manual of Geographic Information Systems

The OODBMS and ORDBMS are better than RDBMS for integrating time into geographic databases. Wachowicz (1999) used OODBMS to store and manage temporal boundaries data. Le (2005a) designed an object-relational feature table for temporal entities (Table 19-4). In Table 19-4, one record represents the history of one temporal feature. An ADT, including nested tables, lists, and references, is defined for temporal space and themes. Explicit versioning between features is realized using nested tables, "Predecessor" and "Successor." Temporal queries are facilitated with functions defined on FeatureTime, SpaceTime, and ThemeTime.

Table 19-4 An example of a temporal feature (after Le 2005b).

Feature					Space							Themes				
Feature ID	Feature Type	Feature Time	Predecessor	Successor	Space ID	Space Time	Space Info	Area	Length	Pre	Suc	Theme i (i=0, 1, ... , n)				
												Theme ID	Theme Time	Theme Value	Pre	Suc
25	Highway	[1993, now]	Pres / Pre Time: 12, 16 / 1995 ; 13 / 2000	Sucs / Suc Time: 28, 30 / 2000	1	[1993, 1995)	...	...	...	null	7	1	[1993, 1995)	...	null	5
					7	[1995, 2000)	...	...	...	1	18	5	[1995, now)	...	1	null
					18	[2000, now]	...	...	...	7	null		...			

19.6 Temporal GIS

Temporal GIS are GIS capable of working with temporal geographic data. Among the various spatio-temporal representations proposed heretofore, only the layer-based models, including the sequential snapshots and the change-based approaches, are supported by existing commercial GIS software (Langran 1992; Peuquet and Duan 1995). For example, the national historical geographic information system (NHGIS) represents spatio-temporal census boundaries with sequential snapshots and can be accessed using existing systems (NHGIS 2005).

For the majority of temporal GIS unsupported by commercial GIS, the researchers who introduce a spatio-temporal model have to develop their own temporal GIS. Because of the diversities in spatio-temporal frameworks and the specific purpose of a single application, most temporal GIS introduced so far have been designed for one particular application, such as cadastre (Al-Taha 1992; Chen and Le 1996), wildfire (Yuan 1997), atmosphere (Peuquet 1994, 2002), public boundaries (Wachowicz 1999), human activities (Yu 2006), and storms (McIntosh and Yuan 2005a). For example, Peuquet and Duan (1995) developed TEMPEST for their event-based model. As a result, although some GIS software with temporal capabilities has been introduced for specific data models, they can only work with data represented in that particular model. Le (2005b) argued there should be a temporal GIS capable of manipulating temporal geographic data represented in different spatio-temporal models because there is a need to work with data modeled in different spatio-temporal frameworks. For example, temporal data of transportation, population, and land use are required for post-analysis of the planning of a highway, and these data may be represented in different data models.

A temporal GIS can be developed from scratch, or extended from an existing system. It is easy to extend an existing system because time and effort can be saved, but the new system will be constrained by the existing system. For example, existing GIS software is limited in using ADT (ESRI 2004). Peuquet (2006) argued it was "new models in old bottles" by implementing a new system with existing GIS. Le (2005b) developed a new prototype GIS that supports ADT, including nested tables.

19.6.1 Spatio-temporal Analysis

Spatio-temporal data models determine the way and efficiency with which the data are accessed and manipulated. For approaches that model world status, such as snapshots, it is straightforward to retrieve the world status, but difficult to query changes. For approaches that model changes, such as base state with amendments, it is easier to retrieve changes than world status. For feature-based approaches, the history of single feature is much easier to access than the history on one location. For location-based frameworks, it is straightforward to track the history on the same location. For combined or integrated approaches, such as the Triad framework, it is efficient to retrieve histories based on location, feature, and time; therefore, the combined or integrated approaches are more flexible than others in data access. Of course, integrated approaches are more complex and bring about more redundancy compared to others.

Various studies have been completed on spatio-temporal query (Langran 1992; Worboys 1994). Temporal relationships and spatio-temporal topology (Peuquet 1994, 1999; Muller 2002) have been examined even more than the spatio-temporal query language. Peuquet (1994) argued there are three methods for spatio-temporal analyses: qualitative, quantitative, and visual, and the retrieval of spatio-temporal data is a fundamental form of spatio-temporal qualitative analysis. To Yuan (1997), temporal query and analysis included change rate or frequency, life expectancy, and temporal relationships. Based on a unified spatio-bi-temporal model, Worboys (1994) classified possible spatio-temporal operations including equality, subset, spatio-temporal projection, spatial selection, and temporal selections. Oosterom et al. (2002) developed a generic query tool for spatio-temporal data, and employed four database views: thematic, temporal, spatial, and aggregate for easy querying.

So far, most research on spatio-temporal analysis focuses on spatio-temporal query. McIntosh and Yuan (2005b) studied the characteristics of events and process in a storm and compared similarities of complex events using six indices. This kind of work is important in spatio-temporal data analysis and knowledge discovery.

To speed up spatio-temporal query, indexing, partitioning, and other mechanisms are required (Langran 1992; Claramunt and Thériault 1995). There is a need for interdisciplinary efforts so that research results from the database community, such as TSQL2, can be applied to temporal GIS (Peuquet 2006).

19.6.2 Visualization of Spatio-temporal Data

Spatio-temporal query, visualization, and analysis are closely related. Visualization and analysis use the results from query as input data. Geographic visualization (GeoVis) provides a three-dimensional and temporal view for knowledge communication and exploration unlike static graphics or maps (MacEachren 1995).

In GeoVis, spatio-temporal resolution and scale are important, and visualization with different spatio-temporal resolution or granularity can be quite different. For example, Wang and Cheng (2001) assumed that an individual had a daily activity pattern. At the temporal scale of hours, the person is at different locations at certain times, but at the temporal scale of days, we may say the person is always at home every night. If the temporal scale is 100 years, maybe the person will never exist, or appear just once or twice in the database. The person does not change in the real world. What changes is the temporal scale that we use to represent and view the real world in the information system. Peuquet (1994) argued that the ideal temporal scale and resolution depends upon the specific phenomenon under observation, and the problem posed about it.

Advanced visualization techniques, such as time animation (Kraak and Klomp 1995) or time as a third dimension, can facilitate spatio-temporal analysis (Oosterom et al. 2002). By animating change of specific modeling parameters, significant spatio-temporal change patterns can be explored, and human-environment-relationships, such as urban growth, can be discovered. For example, Kwan et al. (2003) uses a space-time prism to visualize the space-time accessibility of a person in a space-time cube. Any position in the prism is a possible spatio-temporal location that the person can reach. The space-time paths of a person and its prism can be used for modeling transport demands based on individual activity. Le (2005b) designed a prototype temporal GIS for visualization of spatio-temporal data in snapshot and animation format. Different from regular animation, each frame in the animation is not pre-defined, but retrieved from the spatio-temporal database on the fly.

The human-computer interface is a significant issue in visualization of spatio-temporal data. In the Apoala project, linear and cyclical interaction tools are combined in the graphic interface to facilitate locating and understanding the time dimension (Brewer et al. 2000; GeoVISTA 2003). Study by Edsall et al. (1997) indicated there were no significant differences in performance and efficiency between three legend types: text, linear, and cyclical. Le (2005b) integrated sound in her time animation to show the time distance explicitly. When the audio button is selected and the audio device is open, the system will generate one beep to represent one time interval, which is four years in this example.

19.7 Conclusion

This chapter discusses issues in adding time to GIS by providing an extensive review of spatio-temporal representation, database and GIS implementation issues, and temporal GIS concepts including analysis and visualization. Spatio-temporal representation is the key to temporal GIS. The current trend is representing continuous spatio-temporal changes, but it is relatively difficult, and there has been no true continuous representation yet (Peuquet 2006). The implementation of spatio-temporal representations should not be restricted to RDBMS and existing commercial GIS software because of their limit in using ADT. Analysis and visualization of spatio-temporal data need to facilitate knowledge communication and exploration.

References

Allen, J. F. 1984. Towards a general theory of action and time. *Artificial Intelligence* 23:123–154.

Al-Taha, K. K. 1992. Temporal reasoning in cadastral systems. Ph.D. thesis, Department of Surveying Engineering, University of Maine.

Armstrong, M. P. 1988. Temporality in spatial databases. Pages 880–889 in *Proceedings of GIS/LIS'88,* 2. Bethesda, Md.: American Congress of Surveying and Mapping.

Brewer, I., A. M. MacEachren, H. Abdo, J. Gundrum and G. Otto. 2000. Collaborative geographic visualization: Enabling shared understanding of environmental processes. Pages 137–144 in *IEEE Information Visualization Symposium 2000,* held in Salt Lake City, Utah.

Chen, J. and J. Jiang. 1998. Event-based spatio-temporal database design. Pages 105–109 *International Archives of ISPRS,* Vol. 32, Part 4.

Chen, J. and Y. Le. 1996. Defining and representing temporal objects for describing the spatio-temporal process of land subdivision. Pages 48–56 in *International Archives of ISPRS, XXXI-B2,* held in Vienna, Austria.

Claramunt, C. and M. Thériault. 1995. Managing time in GIS: An event-oriented approach. Pages 23–42 in *Recent Advances in Temporal Databases: Proceedings of the International Workshop on Temporal Databases.* Edited by J. Clifford and A. Tuzhilin. Zurich, Switzerland: Springer-Verlag.

Date, C. J., H. Darwen and N. A. Lorentzos. 2003. *Temporal Data and the Relational Model.* New York: Morgan Kaufmann Publishers.

Dragicevic, S. and D. J. Marceau. 2000. A fuzzy set approach for modeling time in GIS. *International Journal Geographical Information Science* 14:225–245.

Dyreson, C. E., R. T. Snodgrass and M. D. Soo. 1995. Temporal granularity. Pages 345–382 in *The TSQL2 Temporal Query Language.* Edited by R. T. Snodgrass. Boston: Kluwer Publisher.

Edsall, R. M., M. J. Kraak, A. MacEachren and D. J. Peuquet. 1997. Assessing the effectiveness of temporal legends in environmental visualization. Pages 677–685 in *Proceedings: GIS/LIS '97,* held in Cincinnati, Ohio.

Erwig, M., R. H. Güting, M. Schneider and M. Vazirgiannis. 1998. Abstract and discrete modeling of spatio-temporal data types. Pages 131–136 in *Proceedings, ACM GIS '98,* held in Washington, D.C.

ESRI. 2004. FAQ: Does ArcSDE support Oracle Tables with ADT (abstract data type) columns? <http://support.esri.com/index.cfm?fa=knowledgebase.techarticles.articleShow&d=23143> Accessed 3 October 2006.

Frank, A. U. 1998. Different types of "times" in GIS. Pages 40–62 in *Spatial and Temporal Reasoning in Geographic Information Systems.* Edited by M. J. Egenhofer and R. G. Golledge. Oxford: Oxford University Press.

———. 2003. Ontology for spatio-temporal database. Pages 9–77 in *Spatio-Temporal Database,* number 2520 in Lecture Notes in Computer Science. Edited by M. Koubarakis, T. Sellis, A. U. Frank, S. Grumbach, R. H. Güting, C. S. Jensen, N. Lorentzos, Y. Manolopoulos, E. Nardelli, B. Pernici, H. J. Schek, M. Scholl, B. Theodoulidis and N. Tryfona. Berlin: Springer.

Frihida, A., D. J. Marceau and M. Thériault. 2002. Spatio-temporal object-oriented data model for disaggregate travel behavior. *Transactions in GIS* 6 (3):277–294.

Galton, A. 2001. Space, time, and the representation of geographical reality. *Topoi* 20:173–187.

———. 2003. Desiderata for a spatial-temporal geo-ontology. Pages 1–12 in *Spatial Information Theory: Foundations of Geographic Information Science (COSIT 2003),* number 2825 in Lecture Notes in Computer Science. Edited by W. Kuhn, M. F. Worboys and S. Timf. Berlin: Springer.

———. 2004. Fields and objects in space, time, and space-time. *Spatial Cognition and Computation* 4 (1):39–67.

GeoVISTA. 2003. The Apoala Project. GeoVISTA Center of Pennsylvania State University. <http://www.geovista.psu.edu/grants/apoala/tgis.htm> Accessed 3 October 2006.

Getis, A., J. Getis and J. Fellmann. 2006. *Introduction to Geography.* 10th ed. New York: McGraw-Hill, New York.

Gomma, H. 2005. *Designing Software Product Lines with UML: From Use Cases to Pattern-based Software Architectures.* Boston: Addison-Wesley.

Goodchild, M. F. 1987. A spatial analytical perspective on geographic information systems. *International Journal of Geographical Information Systems* 1:327–334.

Gould, S. J. 1987. *Time's Arrow and Time's Cycle: Myth and Metaphor in the Discovery of Geological Time.* Cambridge, Mass.: Harvard University Press.

Grenon, P. and B. Smith. 2004. SNAP and SPAN: Towards dynamic spatial ontology. *Spatial Cognition and Computation* 4 (1):69–104.

Grumbach, R. H., M. H. Bohlen, M. Erwig, C. S. Jensen, N. Lorentzos, E. Nardelli, M. Schneider and J.R.R. Viqueira. 2003. Spatio-temporal data models and languages: An approach based on data types. Pages 117–176 in *Spatio-Temporal Database,* number 2520 in Lecture Notes in Computer Science. Edited by M. Koubarakis, T. Sellis, A. U. Frank, S. Grumbach, R. H. Güting, C. S. Jensen, N. Lorentzos, Y. Manolopoulos, E. Nardelli, B. Pernici, H. J. Schek, M. Scholl, B. Theodoulidis and N. Tryfona. Berlin: Springer.

Güting, S., M. Koubarakis, P. Rigaux, M. Scholl and S. Skiadopoulos. 2003. Spatio-temporal models and languages: An approach based on constraints. Pages 177–201 in *Spatio-Temporal Database,* number 2520 in Lecture Notes in Computer Science. Edited by M. Koubarakis, T. Sellis, A. U. Frank, S. Grumbach, R. H. Güting, C. S. Jensen, N. Lorentzos, Y. Manolopoulos, E. Nardelli, B. Pernici, H. J. Schek, M. Scholl, B. Theodoulidis and N. Tryfona. Berlin: Springer.

Hägerstrand, T. 1970. What about people in regional science? *Paper of the Regional Science Association* 14: 7–21.

Harvey, D. 1969. *Explanation in Geography.* New York: St Martin's Press.

Hazelton, N.W. J. 1991. Integrating time, dynamic modeling and geographical information systems: Development of four-dimensional GIS. Ph.D. thesis, The University of Melbourne.

Hornsby, K. and M. J. Egenhofer. 2000. Identity-based change: A foundation for spatio-temporal knowledge representation. *International Journal of Geographical Information Science* 14:207–224.

Hunter, G. and I. P. Williamson. 1990. The development of a historical digital cadastral database. *International Journal of Geographical Information Systems* 4: 169–179.

Jensen, C. S. and R. Snodgrass. 1996. Semantics of time-varying information. *Information Systems* 21:311–352.

Jensen, C. S., R. Snodgrass and M. D. Soo. 1995. The TSQL2 data model. *The TSQL2 Temporal Query Language.* Edited by R. T. Snodgrass. Boston: Kluwer.

Kelmelis, J. A. 1991. Time and space in geographic information: Toward a four-dimensional spatio-temporal data model. Ph.D. thesis, Pennsylvania State University.

Kern, S. 1983. *The Culture of Time and Space.* Cambridge, Mass.: Harvard University Press.

Kimball Group. 2000. Kimball Design Tip #8: Perfectly Partitioning History with the Type 2 Slowly Changing Dimension. <http://ralph12.securesites.net/html/designtips.html> Accessed 3 October 2006.

Koncz, N. A. and T. M. Adams. 2002. A data model for multi-dimensional transportation applications. *International Journal Geographical Information Science* 16:551–569.

Koubarakis, M. and T. Sellis. 2003. Introduction. Pages 1–8 in *Spatio-Temporal Database,* number 2520 in Lecture Notes in Computer Science. Edited by M. Koubarakis, T. Sellis, A. U. Frank, S. Grumbach, R. H. Güting, C. S. Jensen, N. Lorentzos, Y. Manolopoulos, E. Nardelli, B. Pernici, H. J. Schek, M. Scholl, B. Theodoulidis and N. Tryfona. Berlin: Springer.

Kraak, M. J. 2003. Geovisualization illustrated. *ISPRS Journal of Photogrammetry & Remote Sensing* 57:390–399.

Kraak, M. J. and A. Klomp. 1995. A Classification of Cartographic Animations: Towards a Tool for the Design of Dynamic Maps in a GIS Environment. <http://cartography.geog.uu.nl/ica/Madrid/kraak.html> Accessed 3 October 2006.

Kwan, M. P., D. G. Janelle and M. F. Goodchild. 2003. Accessibility in space and time: A theme in spatially integrated social science. *Journal of Geographical Systems* 5 (1):1–3.

Langran, G. 1992. *Time in Geographic Information Systems.* London: Taylor & Francis.

Langran, G. and N. Chrisman. 1988. A framework for temporal geographic information. *Cartographica* 25:1–14.

Le, Y. 2005a. Representing time in base geographic data. Ph.D. thesis, University of Georgia, Athens, Ga.

———. 2005b. A prototype temporal GIS for multiple spatio-temporal representations. *Cartography and Geographic Information Science* 32 (4):315–329.

Lee, C. H. 2005. Slowly Changing Dimensions. <http://www.1keydata.com/datawarehousing/scd.html> Accessed 3 October 2006.

Lewis, H. R. and L. Denenberg. 1991. *Data Structures & Their Algorithms.* New York: HarperCollins Publishers.

Longley, P. A., M. F. Goodchild, D. J. Maguire and D. W. Rhind. 2005. *Geographic Information Systems and Science.* 2nd ed. Chichester, England: John Wiley & Sons.

MacEachren, A. M. 1995. *How Maps Work: Representation, Visualization, and Design.* New York: Guilford Press.

McIntosh, J. and M. Yuan. 2005a. A framework to enhance semantic flexibility for analysis of distributed phenomena. *International Journal of Geographical Information Science* 19 (10):999–1018.

———. 2005b. Assessing similarity of geographic processes and events. *Transactions in GIS.* 9 (2):223–245.

Mark, D., M. J. Egenhofer, L. Bian, P. Rogerson and J. Vena. 2003. Spatio-temporal GIS Analysis for Environmental Health. <http://www.geog.buffalo.edu/~dmark/research.html> Accessed 3 May 2005.

Medak, D. 1999. Lifestyles–an algebraic approach to change in identity. Pages 19–38 in *Spatio-temporal Database Management.* Edited by M. H. Böhlen, C. S. Jensen, and M. O. Scholl. Berlin: Springer.

Muller, P. 2002. Topological spatio-temporal reasoning and representation. *Computational Intelligence* 18: 420–450.

NHGIS. 2005. National Historical Geographic Information System. <http://www.nhgis.org> Accessed 3 October 2006.

Ohsawa, Y. and K. Kim. 1999. A spatiotemporal data management method using inverse differential script. Pages 542–553 in *Advances in Database Technologies.* Edited by Y. Kambayashi, D. L. Lee, E. P. Lim, M. K. Mohania and Y. Masunga. Singapore: Springer.

Oosterom, P. V., B. Maessen and W. Quak. 2002. Generic query tool for spatio-temporal data. *International Journal Geographical Information Science* 16:713–748.

Peuquet, D. J. 1988. Representations of geographic space: Toward a conceptual synthesis. *Annals of the Association of American Geographers* 78:375–394.

———. 1994. It's about time: A conceptual framework for the representation of temporal dynamics in geographic information systems. *Annals of the Association of American Geographers* 84:441–461.

———. 1999. Time in GIS and geographical databases. Pages 91–103 in *Geographical Information Systems.* 2nd ed. Edited by P. A. Longley, M. F. Goodchild, D. J. Maguire and D. W. Rhind. New York: John Wiley & Sons.

———. 2002. *Representations of Space and Time.* New York: Guilford Press.

———. 2006. Extensions to Representation. UCGIS 2006 Summer Assembly. <http://www.ucgis.org/priorities/research/2006research/chapter_5_update.pdf> Accessed 3 October 2006.

Peuquet, D. J. and N. Duan. 1995. An event-based spatiotemporal data model (ESTDM) for temporal analysis of geographical data. *International Journal Geographical Information Systems* 9:7–24.

Raper, J. and D. Livingstone. 1995. Development of a geomorphological spatial model using object-orientated design. *International Journal of Geographical Information Systems* 9:359–384.

SAIF. 2006. Spatial Archive and Interchange Format. <http://ilmbwww.gov.bc.ca/bmgs/pba/saif/> Accessed 6 June 2008.

Sellis, T. 1999. Research issues in spatio-temporal database systems. Pages 5–11 in *Proceedings: Advances in Spatial Databases: 6th International Symposium, SSD '99.* Hong Kong, China: Springer.

Snodgrass, R. T. 2000. *Developing Time-oriented Database Applications in SQL.* San Francisco: Morgan Kaufmann Publishers.

Tang, A. Y., T. M. Adams and E. L. Usery. 1996. A spatial data model design for feature-based geographical information systems. *International Journal of Geographical Information Systems* 10 (5):643–659.

Thrift, N. 1977. *An Introduction to Time Geography.* London: Geo-Abstracts.

Tryfona, N., R. Price and C. S. Jensen. 2003. Conceptual models for spatio-temporal applications. Pages 79–116 in *Spatio-Temporal Database,* number 2520 in Lecture Notes in Computer Science. Edited by M. Koubarakis, T. Sellis, A. U. Frank, S. Grumbach, R. H. Güting, C. S. Jensen, N. Lorentzos, Y. Manolopoulos, E. Nardelli, B. Pernici, H. J. Schek, M. Scholl, B. Theodoulidis and N. Tryfona. Berlin: Springer.

Usery, E. L. 1996. A feature-based geographic information system model. *Photogrammetric Engineering & Remote Sensing* 62: 833–838.

———. 2000. Multidimensional representation of geographic features. Pages 240–247 in *International Archives of Photogrammetry and Remote Sensing,* Vol. XXXI, Part B4, Commission 4.

———. 2004. Multidimensional Data Modeling for Feature Extraction and Mapping <http://carto-research.er.usgs.gov/feature_extraction/ppt/acsm2004.ppt> Accessed 25 February 2007.

Wachowicz, M. 1999. *Object-oriented design for temporal GIS.* London: Taylor & Francis.

Wang, D. and T. Cheng. 2001. A spatio-temporal data model for activity-based transport demand modeling. *International Journal of Geographical Information Science* 15:561–585.

Worboys, M. F. 1994. A unified model for spatial and temporal information. *The Computer Journal* 37:26–34.

———. 1998. A generic model for spatio-bitemporal geographic information. Pages 25–39 in *Spatial and Temporal Reasoning in Geographic Information Systems.* Edited by M. J. Egenhofer and R. G. Golledge. Oxford: Oxford University Press.

———. 2005. Event-oriented approaches to geographic phenomena. *International Journal of Geographical Information Science* 19 (1):1–28.

Yu, H. 2006. Spatial-temporal GIS design for exploring interactions of human activities. *Cartography and Geographic Information Science* 33 (1):3–19.

Yuan, M. 1996. Modeling semantical, temporal, and spatial information in geographic information systems. Pages 334–347 in *Geographic Information Research: Bridging the Atlantic.* Edited by M. Craglia and H. Couclelis. London: Taylor & Francis.

———. 1997. Use of knowledge acquisition to build wildfire representation in Geographical Information Systems. *International Journal of Geographical Information Science* 11:723–745.

———. 1999. Use of a three-domain representation to enhance GIS support for complex spatiotemporal queries. *Transactions in GIS* 3 (2):137–159.

Yuan, M., D. M. Mark, M. J. Egenhofer and D. J. Peuquet. 2004. Extensions to geographic representations. Pages 130–156 in *A Research Agenda for Geographic Information Science.* Edited by R. B. McMaster and E. L. Usery. Boca Raton, Fla.: CRC Press.

CHAPTER 20

Database Revision: Experiences of the Ordnance Survey of Great Britain

Paul R. T. Newby

20.1 Introduction and Overview

In March 1995, the Ordnance Survey of Great Britain (OS), based in Southampton, England, became the first national mapping organization in the world to complete its digital database, at the very large map scales of 1:1250 (in urban areas), 1:2500 (in populated rural areas) and 1:10,000 (in mountains and moorland), throughout its area of operation that covers the whole land area of England, Wales and Scotland, a total of some 230,000 km^2 (somewhat smaller than Oregon or Wyoming and larger than Utah or Idaho). Since the inception of the digital production program in the early 1970s, every completed digital map sheet had immediately entered the standard OS procedure of continuous revision. Thus what eventually became known as database update was embedded in the digital cartographic system from the very beginning, in contrast to the vast majority of digital mapping applications in which the initial population of a databank appeared to be the prime objective, with its revision left as an issue for the distant future. Nevertheless it should not be imagined that Britain's digital mapping processes are lost in some 1970s time warp: the business of digital mapping has been characterized not only by continuous revision of its content but also continual change and evolution of practice and product, always aimed very specifically at supporting both the national interest and customer needs, as well as the (not always compatible) objective of keeping costs to a minimum. The development and maintenance of digital mapping have never been easy.

OS data today (early 2006) exists in two quite distinct forms: one, still known as Land-Line, recognizable in principle to the earliest customers for digital mapping in the 1970s, and the other, OS MasterMap, in the form of a topologically structured, object-oriented, seamless database suitable for any GIS application. The data is delivered both by OS itself and by a wide range of private sector "value-added resellers" and system suppliers, now collectively known as "licensed partners," to customers who are in turn licensed to make use of the data and receive updates. Updates may be in the form of resupply of the full data set for an area of interest or of "change-only," to be applied by the customer's software to the data already held. It is the customer's choice whether to receive updates at intervals based on elapsed time or on amount of surveyed change, or simply to call for the latest version of the data whenever it is required. In addition, an "annual snapshot" of the entire database ensures that periodic records of the state of the nation's topography are permanently archived for historical purposes.

Meanwhile the process of re-engineering to meet the next generation of user needs continues unabated. Over all of this period since the first experiments in the 1960s, practice has run well ahead of both geographic information theory and hardware and software systems available in the marketplace. It has thus provided a large-scale real-world example, and on occasions a testbed for theoreticians worldwide, as well as for other practical data producers who have followed in the OS's wake.

In this contribution the author attempts to deliver a personal view of these developments, with a chronology of their practical realization over the past 35 years, with particular reference to the data revision process, from the perspective of both the provider and the customer. The period from 1985 to 1994, when the author was directly involved in the developments, represents a convenient middle phase, midway between the original concept and its early execution, and the consolidation of the complete database through to the present day. In that middle period, the initial difficulties of creating a digital map for the whole country had been overcome, the product itself was beginning to gain broad acceptance, the desirable future direction for the database was becoming clear, many of the mature systems that are in place today had already been conceived, and the problems of their development and introduction were being investigated, even though the initial population of the database was not yet quite complete. During this period also, the world outside Great Britain began to recognize the need for digital mapping systems to incorporate the update processes that would eventually be required (Newby 1996a).

To complete this introduction it should be explained that at the end of the nineteenth century Great Britain possessed a superb series of reasonably up-to-date large-scale maps but that this national asset was allowed to wither away due to inadequate resources being made available for map revision during the first third of the twentieth century. The Report of the Davidson Committee (Davidson 1938) brought the decline to a halt with a series of far-sighted recommendations that were swiftly accepted and then implemented after the Second World War. This led to a massive program of resurvey from 1946 until the early 1980s ("The 1980 Plan"). For various technical and economic reasons the postwar 1:2500 scale mapping of rural areas was carried out as an "overhaul" of existing maps, recast on Davidson's new National Grid coordinate system but falling short of a resurvey to modern standards (see Ordnance Survey 1972; Matthews 1976; Leonard 1982; Newby 1990a). The new maps, once published, were immediately placed under the system of continuous revision also recommended by Davidson, which was so well established as to be taken for granted by the time of the first digital mapping experiments in the 1960s. Although many reviews and many changes of terminology have come and gone, the principles of continuous revision (CR) of important features and of the careful measurement of real-world change exemplified by the "house unit" (HU), now generally called "unit of change," remain in place to this day.

Finally it should be noted that Northern Ireland (part of the United Kingdom but distinct from Great Britain) has its own separate mapping organization, the Ordnance Survey of Northern Ireland (OSNI), based in Belfast. Digital mapping and database developments there have to some extent paralleled those in Great Britain, with the same original basic scales of mapping, a later start on digitizing but earlier introduction of some limited database concepts (Brand 1986), a complex present-day database of vector data with some topological structure and attributes, still evolving but providing full coverage of the whole (much smaller) territory, all updated through systems of continuous and periodic revision that closely resemble those of the Ordnance Survey of Great Britain. The Northern Ireland case and its ongoing development will not be examined in this contribution, but more information can be found, for example, in http://www.osni.gov.uk.

20.2 Earliest Days—Automated Cartography Leads to Digital Mapping, 1966–72

By the late 1960s it was recognized that the techniques of computer-assisted drafting, which were being introduced in engineering and architecture, might also have an application in map production (Sowton 1991). In the UK, David Bickmore, an atlas editor at Oxford University Press, was able to secure funding for an Experimental Cartography Unit (ECU) at the Royal College of Art (initially in Oxford, subsequently in London). One of the ECU's early projects, in 1969–70, was a joint study with the OS that included the production of experimental maps of the Bideford area in southwest England. The objectives of the study were to convert 100 existing OS large-scale (1:2500) maps into computer-compatible form on magnetic tape and to derive smaller scale maps from the data (Bickmore 1971). It is clear even in this short article that Bickmore himself was already thinking in terms of several of the crucial aspects of digital cartography for future developments, including the relationships between areas that would lead naturally into the geographic information systems of the future, the addition of information about the provenance of the data itself (what would later be termed *metadata*), the inclusion of the third dimension and the challenge of revising the cartographic data to keep it up to date.

The OS recognized the Bideford trial as a broadly successful demonstration of possibilities, and used it to justify its continued activity in the digital cartographic field. However, the process of generalization to smaller scale mapping was judged to be ahead of its time and neither practical nor cost-effective in the context of OS requirements. For example, to revise one sheet of one-inch-to-one-mile mapping (then soon to be replaced by mapping at 1:50,000 scale) using the basic 1:1250 and 1:2500 scale maps, it could be necessary to bring as many as 7,200 separate sheets of urban 1:1250 up to date simultaneously; this would never happen under the OS continuous revision process as it operated at that time. Perhaps in part due to the clashes of personality described by Rhind (1988), and in part to the practical difficulties encountered in the ECU experiment, OS then proceeded rapidly to develop its own digital mapping system with the prime aims of automating, improving and reducing the cost of large-scale mapping flowlines, including revision flowlines, and secondary aims of creating a new data product for a new kind of customer, and, only in the long term, perhaps also deriving the full range of map scales from the large-scale data.

Adequate credit seems rarely to have been given to the bold and far-sighted decisions by the then Director General (Maj.-Gen. Brian St G. Irwin) and to the determination and skill of his Deputy Director of Planning and Development (Col. Robin C. Gardiner-Hill), although Sowton (1991) and Rhind (1988) each relate some of the story. Doubtless with the aid of a small team of juniors including both military officers and experienced OS cartographic draftsmen, Gardiner-Hill wrote and implemented the software—in machine code—that formed the basis of the OS digital mapping system for many years to come. One of those junior officers, Capt. Michael St G. Irwin (son of the Director General) gave a lecture on the OS's perspective on the Bideford project to a meeting of the Royal Institution of Chartered Surveyors (RICS) on 12 February 1970. The publication that followed (Irwin 1970) makes it clear that, far from initial data capture being the sole objective, ability to update the data was seen from the outset as the fundamental requirement of a digital cartographic system for the OS. The paper goes into some detail on future possibilities for interactive editing, as well as the practical minutiae of locating the part of the magnetic tape requiring revision. Irwin also envisaged raster digitizing as an alternative to point-by-point capture of vector data, although he did not use either of those terms at that time.

Gardiner-Hill's own presentation at the Commonwealth Survey Officers' Conference (1971) and the OS Professional Paper based on it (1972) describe some of the practical details of the system but also emphasize its goals and advantages. The latter included easy update (with an early hint of the "change-only update" concept, which only came to fruition almost thirty years later); user-defined outputs including scale and detail to be shown; and the seamless structure in which sheet lines would cease to have any meaning. He also gave details of two early forays into very small-scale digitizing, with a clear intention of publication and annual revision of the 1:625,000 scale Route Planning Map. This digital task only came back into view some 12 years later, whereas the large-scale work (at 1:1250 and 1:2500 scales) led directly into a decision to implement a pilot production project, initially for 1:2500 scale maps. Success in this project would lead, in process of time, to the long-term objective of all topographic data being banked in digital form, continually updated and readily accessible for presentation in any required form (Gardiner-Hill 1972).

20.3 Routine Production and Continuing Development, 1972–85

It is curious that published sources differ on the start date of the digital cartographic production flowline. Newby (1992) and the OS MasterMap User Guide (Ordnance Survey 2003, 2005a) both indicate 1971, while 1973 has been often quoted in many contexts (see for example Sowton 1991 and Rhind 1997). McMaster et al. (1986) state that the first sheets from the flowline were published in 1974. However, Thompson (1978) and Proctor (1986) are both precise and consistent in stating that the pilot production flowline was implemented in September 1972. Successive OS Annual Reports, which should be authoritative, indicate development activity from 1966/67 onwards, through to beginning to digitize a pilot area in 1971/72: by the end of this reporting (fiscal) year the development system "with the technical problems ... largely resolved" had been handed over to the Cartography Division for evaluation in the pilot production environment. The key to moving into routine production was the newly available Ferranti Master Plotter, with a light spot projector capable of matching the accuracy and line quality of manual cartography (Gardiner-Hill 1972; Atkey and Gibson 1975; Sowton 1991). The first digital map to emerge from the pilot production flowline was printed as a 2 km^2 paper sheet at 1:2500 scale in March 1973, at which time 53 sheets (each covering 0.25 km^2) at 1:1250 scale and 395 km^2 at 1:2500 scale were at various stages of production. Whatever the exact chronology (and clearly the production process from first input of documents to databanking, printing and publication was not going to be instantaneous, indeed Thompson (1978) discloses that the elapsed time could reach 12 to 14 months), it seems indisputable that this was the world's first mass production flowline for topographic map digitizing. The resulting maps accumulated, slowly at first, to form the databank; they were immediately subjected to the OS practice of continuous revision; they have been revised digitally ever since in accordance with the rules and processes in force at any given time; and they remain in the National Topographic Database (NTD) to this day. The remainder of this contribution addresses these revision processes as they have evolved over more than 30 years up to the present time.

While Rhind (1988) felt able to assert that no concept of areas nor of any relationship between cartographic entities was included in the original OS digital mapping system, it is certain that such matters were being thought about in the earliest days, but were recognized as an obstacle to progress in achieving the immediate objectives of cartographic automation to produce a map that was capable of revision. In a final contribution before his retirement, Gardiner-Hill (1973) discussed the possible alternative data structure in which lines would have no significance in their own right but would represent the boundaries between areas. He stated that it was inevitable that OS would code lines according to the linear feature they represented, but suggested a conceptual mechanism for changing the structure where areas are also required. He also remarked that a topographic database for Great Britain would be far easier to look after and keep up to date than its current graphic counterpart, a store containing 220,000 map-sheet-sized glass negatives!

External consultants were soon employed, from 1974 onwards, on a project to devise means of restructuring the data into "real world objects" (see, for example, Atkey and Gibson 1975; Thompson 1978; McMaster et al. 1986; Sowton 1989; Rhind 1997). Although the project proved to be much more difficult than originally envisaged, and eventually ran years behind schedule, some successes were achieved. These included valuable collaboration with one particular local authority (Dudley Metropolitan Borough [MB]) and the creation of the first of a proposed set of "user languages," User Language (Land), to enable the formation of land parcels from the data and queries about objects within it. However, the complications, slow progress, excessive cost and lack of demand from other customers led to the development project being signed off without any immediate lasting result some five years later in 1979, although user evaluation in collaboration with Dudley MB continued for several years thereafter, and Sowton (1991) noted that the land parcel element of the project was undoubtedly the precursor of GIS in Great Britain. The considerable interval before the next foray into structured data, with the Topographic Database Project of 1985–87, was almost certainly beneficial for the establishment, acceptance and gradual evolution of revision processes for the simple unstructured "spaghetti" data generated during the years from 1972.

Other landmark events during the first 12 years of digital mapping, culled from successive OS Annual Reports, included, in 1973/74, a new procedure for the supply of unpublished survey information (SUSI) (initially only to local government authorities but eventually to any customer) by copying the current up-to-date field survey revision document after an agreed amount of change had been added to it. With a threshold far lower than the criterion for production of a new printed edition (which was generally set at 300 HU of surveyed change), this SUSI process provided the foundation for the future supply of current data on a variety of graphic media including paper, film and microfilm, followed some years later by digital data. The following year it was stated that up-to-date copies of field documents would always be available for purchase at short notice (an advance on the specific change threshold introduced in the previous year); and by 1975/76 the service was being provided at OS field survey offices rather than documents having to be sent to headquarters for the purpose. The policy of continuous revision (CR) in "fast change" areas and periodic revision (PR) in "slow change" areas was also explicitly articulated in the 1974/75 Annual Report and implemented in 1975.

In General Irwin's final Annual Report of 1976/77, at the end of one of the most significant periods of twentieth-century OS history, he reminded readers of the specter of the inadequate revision of the early twentieth century, and he stated that even in the present time of reduced resources, continuous and periodic revision would be maintained; if necessary this would even be at the expense of slowing down the new surveys of the 1980 Plan. This Report also disclosed that, following investigations into the technical and logistical problems of maintaining the digital databank, a new update system had been introduced early in 1977, based on fixed levels of physical change on the ground, linked with the threshold of 50 HU, which was by then in force for the production of survey information on microfilm (SIM). Investigations into the problems and costs of update at even lower levels of change were also in hand in conjunction with a major local authority (LA). Updating at the 50 HU level continued to be noted for several years, through to 1979/80, while the emphasis of initial digitizing shifted from piecemeal small areas across the country to an attempt to cover large contiguous areas for the benefit of those LAs expressing most serious interest in becoming digital data users; experiments in interactive editing also began. The stewardship of Walter Smith, who followed Irwin, as the first external appointee to the post of Director General, was marked by consolidation of the digital mapping process, the long-awaited completion of the post-war remapping program, a heavy emphasis on the resulting complete set of modern large-scale maps as an essential component of the infrastructure of the nation, and a considerable amount of forward thinking on how that national asset should be maintained and developed in future.

With the impending completion of the post-war remapping, the start of the 1980s saw a major investigation, the "study of revision." This was the most complete analysis yet of users' requirements for up-to-date survey information and of the OS's current and possible future processes for providing it. The study was undertaken just before digital update at the survey stage became a serious prospect (as opposed to subsequent digitization of the surveyor's graphic product, which was already in force), although the day was envisaged when that situation might change.

The eventual report of the study of revision (Ordnance Survey 1981) included a vast mass of conclusions and recommendations that mainly covered administrative aspects of the updating process. However, it did represent a major step in the evolution of the categorization of types of change (active; fast and slow; major and minor; etc.) and of processes of map update (intelligence collection; continuous, periodic or cyclic revision; "sweep;" etc.). All of these categories generally reflect the link between population, economic activity and consequential change to map detail. The study also reflected the gradual introduction of the concept of the relative importance of different kinds of change to customers; it recognized that customers are in fact a disparate body of users, and that the OS response to them had in the past necessarily been generalized, tailored only broadly to meet the wide spectrum of their needs, and had certainly been inflexible in the years of intensive remapping after the Second World War. It called for the reintroduction of an age/change criterion that would lead to a periodic, systematic sweep of whole maps even in areas of slow or minor change, in addition to the piecemeal continuous revision of important but localized change as soon as "economic survey" became practical. It confirmed the view that the well-established field techniques of graphic survey (largely without modern instrumentation) remained most appropriate and cost-effective for small areas of change, but envisaged greater future use of aerial photography involving a range of photogrammetric methods at varying levels of rigor, and of field instrumentation including automated data capture and recording. The study also envisaged the future development of interactive editing systems that would allow the field surveyor to digitize his work swiftly after completing his revision surveys, and the consequent possibility of "digital SUSI" instead of the current purely graphic processes for supplying copies of unpublished survey information.

Suggested ways of achieving the necessary revision surveys economically in future, on reasonable time scales, included the hitherto almost unthinkable ideas of employing external contractors and of accepting building developers' "as-built" surveys into the archive. The former became commonplace for some, but not all, classes of OS work fairly soon afterwards, while the latter, continually investigated over the ensuing 20 years, is only now in the twenty-first century becoming a routine reality. The report also called for future examination of the assumption that current methods were rigorous enough to ensure that there would be no long-term degradation of the archive as a result of revision; the existing routine programs of accuracy testing continued to be developed to cover this concern.

Finally, the study of revision also reached two interesting negative conclusions. It dismissed suggestions for legislation to make all topographic change compulsorily reportable to OS by the developer responsible; although the idea was occasionally revived over the years (including once by this author) it never again seems to have received serious high-level consideration. It also concluded that action to improve the accuracy of 1:2500 scale "overhaul" maps could not be justified in terms of user needs, except in some expanding urban and peri-urban areas. Despite this firmly stated conclusion, the matter inevitably continued to be of great concern to some customers and also to OS managers faced with the high cost of revising the substandard "overhaul" maps, compared with the relatively straightforward revision of modern rigorous surveys. Over the years, remedial surveys of various kinds were undertaken in many areas, without however addressing the overall problem (Newby and Proctor 1990; Farrow 1992). It was not until pressure from GPS users became irresistible that the positional accuracy improvement program (PAI) was launched in 2001, with a view to solving the problem once and for all (Holland and Allan 2001). This process, mainly using modern photogrammetric techniques, will be complete during 2006; customers who have linked their own data to substandard coordinates are being helped to remedy their own compatibility problems, and the "overhaul" will at last be transformed from ongoing issue to history.

Subsequent Annual Reports show the slow but inexorable uptake of ideas aired in the study of revision. In 1981/82 (the fiscal year in which the study itself reported) decisions were expected later in 1982. This Annual Report also noted that the evaluation of the digital data restructuring software (resulting from the long-drawn-out 1970s project) was completed and that an interactive graphics editing trial had been successful; meanwhile a review of digital mapping development strategy by an external consultant had recommended more effort to develop the digital update system. 1982/83 saw the introduction of a new revision policy based on CR for "primary" and PR for "secondary" change, respectively, while investigations continued on the acceptance of "as-built" external data. While it was noted that addition of surveyed change to digital data at the SIM update level was in force, a strong requirement was seen for improvements in the provision of an effective and cost-efficient on-demand digital data revision service equivalent to the graphic SUSI service. The "completion report" on the implementation of the 1970s restructuring project noted that the system was proving costly but that it was in fact in use and being maintained in Dudley MB. The postwar basic scales remapping program, so long styled "the 1980 Plan," was finally completed during 1983 despite the delays due to scarce resources mentioned earlier, but the achievement passed almost without remark in Annual Reports—the organization naturally moved on, to the revision process that had already long been its prime preoccupation.

In 1983/84 it was announced that the annual new edition of the "Routeplanner" (1:625,000 scale) map would soon be produced from a topologically structured digital network database and that this would be updated in successive years in future, thus reviving the intention first expressed in the early 1970s. The landmark was achieved for the 1986 edition, published in late 1985. The availability of the 1:625,000 scale topological road network very soon led to the creation, by NextBase Ltd., of the outstandingly successful route-planning software originally known as Autoroute, which eventually gained worldwide acceptance under a variety of names after the company was bought by Microsoft in 1995. The 1982 large-scales revision policy continued in force, resulting in a gradual shift of survey manpower from CR to PR. User-led amendments to digital map specifications were blamed for a slowing of digital production output during the year, a doubtless unintended consequence of greater consultation with customers, but discussions began on the possibility of a greatly accelerated digitizing program with the involvement of the private sector. Studies took place on the problems of sheet edge matching (originally envisaged by Gardiner-Hill [1971, 1972] as inconsequential under his system, but inevitably increasing as more contiguous blocks of work of varying age and survey history were built up); on interactive edit stations; and on "measuring and modeling change." Although the consultants' report on the latter study (Elston et al. undated) did not include clear conclusions, recommendations or indeed any working predictive model, it did embody a statistical estimation of "change since last full revision" and of annual rates of change in Great Britain. Moreover, some of the concepts developed, including the new subdivision of primary and secondary change into "surveyed," "known but not surveyed" and "unknown," proved useful and informed subsequent internal OS activity in the analysis and management of the revision problem for many years to come.

Following on from that study, the Annual Report for 1984/85 noted the finding that annual change was then running at about 1.25 million units (HU), of which 600,000 HU were "primary," while some 700,000 units were actually surveyed during the reporting year. The databank now contained almost 25,000 digital maps, of which just over 2,000 had been produced during the year, progress again being slowed by user-led enhancements to the data format, which now included dual feature coding of detail such as building fronts, which also form a road edge. This snapshot conveniently closes the opening dozen years of digital mapping: this Annual Report also noted the beginnings of far-reaching developments in database design, transfer standards and digital activity in field and photogrammetric survey that would ultimately lead to the systems in force today.

20.4 Modern Revision and Modern Database Concepts, 1985–94

The year 1985 was marked by the beginning of development work that would have the most far-reaching implications for the OS's digital mapping, its data structure and its update processes. Most important of all, because of its impact on the future prospects for GIS in Great Britain, was the "Topographic Database" project that began in August of that year (McMaster et al. 1986; Haywood 1989; Sowton 1991). The project took several years to complete, and not all of its recommendations were adopted immediately or simultaneously, in part because of the technical difficulties of updating a topologically structured database in that period. However, it led first, in the early 1990s, to a conversion of the existing unstructured cartographic data to the clean, seamless and potentially topological model whose data, known originally as Land-Line.93 but now simply as Land-Line, is still in use in 2006 by the majority of customers, and, ultimately, to the object-oriented and fully topological structure of OS MasterMap that was launched in 2001, and which will eventually be adopted, explicitly or implicitly, by all users. At the same time, a national effort to develop transfer standards for digital mapping was coordinated by OS in conjunction with local authorities, utilities and other major user groups; these efforts led over the years to a succession of national and international standards that have certainly assisted both OS and the user community, but which do not have any explicit bearing on the update experience. Meanwhile, in the short term, the existing overall map specification and format were frozen, pending completion of the database project and to enable substantial acceleration of the initial digitization program.

Falling in behind the database project, although nominally started two months earlier, in June 1985, the "Photogrammetric Digital Data Capture" project was aimed at developing systems for the integration of photogrammetric resurvey and revision data into the digital archive. It soon became clear that the revision aspects were considerably more complicated, so this became the subject of a separate project. This was the stage at which this author became directly involved in these matters, both as an OS manager and eventually, from 1988 to 1996, as Chairman of the ISPRS Working Group dealing with map and database revision. A series of readily accessible publications, mainly in the *International Archives of Photogrammetry and Remote Sensing* and in *The Photogrammetric Record,* describe the ongoing developments in this field at frequent intervals over the succeeding ten years (Newby and Walker 1986; Proctor 1986; Proctor and Newby 1988; Newby and Proctor 1990; Newby 1990a, b, c, d, 1991; Farrow 1992; Newby 1992, 1994, 1996a, b), so only a brief overview is given later in this section.

The third link in the OS production chain, the field surveyor, was also being exposed to digital activity as never before. Interactive graphics editors were at last being introduced into headquarters digital cartographic flowlines; at about the same time the successful trial of a separate editing system known as the Digital Field Update System (DFUS) led directly to routine production use and its further development and phased introduction to some 700 staff based at about 140 field offices over a period of several years. At the outset the data traffic between headquarters and field offices was on cassette tapes; later this changed to off-peak overnight use of the public telephone network until finally a dedicated satellite-based communications system was adopted. The initial thrust of field developments was simply to ensure that new survey detail drawn on the old-style field document, the Master Survey Drawing (MSD), was digitized and databanked as soon as it was available. In an echo of the advent of the Master Plotter that enabled the launch of the original digital cartographic flowline in 1972, once sufficiently fast and accurate plotters were found to be available in 1986 it became possible to conceive a totally digital local field office. Here a new temporary revision document for field use would be produced from the existing digital data, and the new revision (including metadata indicating the provenance of the new information) would be databanked via DFUS immediately after survey, instead of maintaining the MSD as the archive (Coote 1988, 1989a, b; Sowton 1989, 1991).

Trials of the DFUS principle in 1987 led to the high-profile "Project 88" in the new city of Milton Keynes, launched by a government minister who ceremonially visited the office there, to shred an original MSD, in early 1988. This project was so immediately successful that within the year MSDs were being formally discarded from other field offices equipped with DFUS and suitable plotters. Thus the task of long-term maintenance of MSDs was eliminated. This labor-intensive process had included the "penning" and accumulation of new surveyed detail (as well as tidy deletion of superseded features) over what could be a period of many years until the "300 HU of change" criterion for a new edition of the map was reached. Although it took several years for this change to be implemented nationwide, it was certainly more radical than appears at first sight, because it implied that the OS topographic archive, hitherto enshrined in the surveyor's ink-on-plastic MSD, would henceforth consist only of the digital data comprising the latest edition of the map and all subsequent update. At least in principle, the archive would then also become truly seamless (no longer being reliant on physical marks on individual documents). In practice the digital data was still structured in "tiles" based on the original map sheet layout, and edge match problems continued to accumulate until a massive automated sheet-edge cleaning program was eventually implemented in conjunction with the line-junction cleaning process required for the Land-Line.93 product.

By the time of the 1987/88 Annual Report, Director General Peter McMaster was able to introduce the outcome of the topographic database project as the way forward for OS immediately (with the extension of the prototype database to a pilot area) and for all data users in the longer term, with the clear objective of suitability for GIS applications. The Milton Keynes "fully digital office" experiment had advanced far enough for him to be able to boast of the users' choice of cartographic colors and scales of plot that could be bought from the local office within two days of revision surveys on the ground. Meanwhile close liaison with representatives of the utilities and other users had led to major simplification of the hitherto always extremely long list of feature codes. Some 200 different feature codes at inception (Gardiner-Hill 1973) had swollen over the years to more than 300 for various reasons, while user opinions over a long period had favored a radical reduction to totals ranging from about 15, or even fewer, up to about 40 (Sowton 1991). The compromise adopted (Ordnance Survey 1988) allowed 35 different feature codes, which greatly aided both users and revisers. Subsequent developments did however lead to a gradual reversion to a longer list, and today's OS MasterMap User Guide (Ordnance Survey 2003, 2005a) shows almost as many feature codes as were in the original 1970s list; the simplified coding from 1988 seems now to have been subsumed in the "descriptive group feature classification attribute" in the new object-oriented data structure.

All of these themes continued to feature in Annual Reports through subsequent years. Data conversion for the pilot areas of the structured topographic database enabled supply of the GIS-ready digital data to collaborating local authorities and utilities, as well as the development of the new family of graphic products known as "Superplan." The need for more advanced editing software for the proper handling of topologically structured data was also being addressed (Sowton 1989). Removal of MSDs from field offices continued, as did developments in digital field and photogrammetric survey processes. In his final Annual Report as Director General, 1990/91, Peter McMaster could boast of a radical acceleration in digitization such that the databank now contained over 100,000 maps (a two-fold increase in less than two years), thanks to very rapid expansion both of the employment of contractors by OS and the acceptance into the databank of maps digitized by contractors directly for the utilities. In digital update, both the capability and the service to customers had been expanded, an increasing number of users had signed up to maintenance agreements for their holdings of digital map data, and over a million house units of change were surveyed during the year. Development of the National Transfer Format (NTF) was complete and it had been adopted for the generality of data supply, with the popular DXF format, common in CAD/CAM applications, also available as a "value-added" alternative. The entire contents of the databank that had already been accumulating for some 18 years were subjected to a validation process by software that examined the data for consistency and corrected it, with the aid of human intervention only where necessary. Spin-off products from structured large-scale data, such as road centerline networks (OSCAR), began to be created and offered to customers in areas where they were available. On the basis of experience of handling the fully structured data of the pilot topographic database, systems were being developed for a program of semi-automatic data cleaning of the whole databank. This process would see the end of the old "spaghetti" data model and lead to seamless data with unique line-junction coordinates, an essential half-way house on the road to the topological destination. This "clean" data would become the OS's prime large-scale product, under the Land-Line.93 banner.

The photogrammetric developments during this period, aimed at incorporating the results of photogrammetric data capture directly into the digital archive, have already been mentioned. Early publications by Proctor (1986) and Newby and Walker (1986) empha-sized the difficulty of introducing an apparently straightforward technical development into a very well-established production environment with a heavy burden of history. Proctor in particular gave details of some of the minutiae of the processes of the photogrammetry itself, in conjunction with feature coding, field completion and the editing and merging of digital data using the variety of edit stations and software that had by now been introduced. An immense team effort by workers in all the production areas involved led eventually to the conclusion that all of the various combinations of equipment already available within the organization, ranging from computer-assisted analog stereo plotters to the newest analytical plotters, could be used cost-effectively for new surveys (Newby and Proctor 1990; Newby 1990a, d).

Meanwhile the much more complicated problems of data integration involved in photogrammetric revision of digital maps were also being addressed. Here, the two most cost-effective processes were found to be the most primitive (pure analog plotting with subse-quent digitization and merger of detail) and the most sophisticated (analytical stereo plotters with online interactive editing at the data capture stage, with or without optical superimpo-sition of the existing digital map). With optical superimposition, the latter process employed the most expensive single item of equipment in the OS inventory at the time. The interme-diate technological steps of computer-assisted analog or analytical stereo plotters without an online interactive edit capability were found to be less suitable for revision than for *ab initio* surveys (of which some still continued even after the completion of the postwar remapping program, when, for example, urban development led to the upgrading of 1:2500 basic scale mapping to 1:1250). Much more detail of these results, together with the background of current administrative and revision policy, including a cost model for revision at various thresholds of change on the ground, is given by Newby (1994). A by-product of the optical superimposition process, even given the OS preference for normal-angle (305 mm or 12 inch focal length) photography for large-scale mapping, was of course that the existing digital data required the addition of information on the third dimension, in order to ensure that the two-dimensional image of the existing map was correctly projected in the stereo plotter optics. This requirement led to a whole new flowline for digitizing the graphic contours, which were already available for the whole country at 1:10,000 scale, and soon to the new spin-off digital contour and terrain model (DTM) products that remain on sale today under the name Land-Form PROFILE. At the same time, developments had begun in fully digital photogrammetry (initially using scanned diapositives) that led in due course to an automated aerial triangulation process, as well as a digital monoplotting flowline for map revision (Newby 1990a, b, c, 1991, 1996b; Vincent and Logan 1996; Murray 1997). The monoplotting process used a digital orthophoto based on the DTMs mentioned above, viewed in conjunction with an interactive editor; it operated successfully in routine production from 1995 to 2000.

In the field context, some minor use was made of "instrumental detail survey" using electronic total stations to provide digital data that was also integrated with other digital revision data during this period (Proctor 1986). However, the vast majority of urban revision was still carried out most cost-effectively using low-technology graphical survey methods based on alignments and short-distance taped measurements in conjunction with the MSD mounted on a portable drawing board (known as a sketching case). In 1986, portable computers with graphic screens were becoming available. This author was privileged to take part in an OS research and development meeting early that year when the device was conceived that would enable the future surveyor to revise the digital map directly in the field; and the author lays personal claim to having devised the name, Portable Interactive Edit Station (PIES), by which it was known throughout the initial development process (Sowton 1991; Farrow 1992). Although a working prototype was produced considerably more quickly than first expected, it took much longer to bring the technology to a form in which it could be introduced generally in the field. The author was not alone in being chagrined to find, some time after he moved from a development to a production management role, that the appellation PIES had been replaced by PRISM, acronym for the bulky and clumsy name, Portable Revisioning and Integrated Survey Module, now given to what eventually became an outstandingly successful, compact and lightweight interactive editor (Greenway 1994; Murray 1997). With the addition of a GPS capability, PRISM remains the foundation of OS database update in the field today. The one development envisaged during this early period that has not yet reached the current version is the inclusion of a digital orthophoto background; the elegance of this handheld digital photogrammetric revision concept remains attractive but will only enter practice when the data volumes involved are manageable within the confines of a cost-effective tool.

It has been shown that the gradual acceptance of digital data captured by photogrammetry or field survey processes—rather than only digitizing graphic detail from a survey document—was driven mainly by very careful measures of cost. The commonsense approach of retaining digits already captured—logical on grounds of retention of accuracy as well as intuitively more cost-effective—did not always prevail in practice at first, in the face of existing smooth, effective flowlines that had taken many years to develop (Newby and Walker 1986). However, it has also been seen that the team effort by researchers and diverse production workers led ultimately to workable and cost-effective exploitation of emerging technologies, to the medium- and long-term benefit both of customers and of the organization itself.

This period also saw a revolution in quality management at OS. This was variously driven by external pressures from the British government for the introduction of quality systems in all industries; by modern management theory that emphasized the importance of customer–supplier relationships within, as well as beyond, the organization; by the essential need to codify relationships with the new breed of data gatherers, the private sector contractors being increasingly employed by OS and by the utilities; and finally by the need for rigorous quality control procedures in the subsequent management of the new product, digital map data. OS could justly claim that it had been well served by the fourfold quality management traditions embedded in its 200-year history: eighteenth-century scientific discipline; nineteenth-century professional ethics; self-checking methods in land surveying, which sprang from both those roots; and extremely close supervision and repetitive checking. These last were employed for craft-based activities such as cartography and printing, as well as for detail survey that could never be made self-checking in the same way as the framework surveys on which the detail was based. However, in the space of a few years, management measures including "Total Quality Management" (TQM) and British and international standards for quality systems (ISO 9000) were introduced, together with new standardized statistical systems for inspection of batches (ISO 2859), the latter in conjunction with the National Joint Utilities Group (NJUG) (NJUG 1988). All were overlaid on existing systems such as routine positional accuracy testing, although at the same time many repetitive checks on the work of responsible individuals were now judged uneconomic and were discontinued. The author was directly involved in many of these initiatives, which are described in greater detail in Newby (1992). In the specific context of the maintenance of a digital mapping system, the major lesson drawn from the first 20 years of experience of the discipline was that data integrity and data security are the fundamental issues. Digital mapping had proved to be both very complicated and very unforgiving of sloppy procedures; if inconsistencies in data are not to overwhelm a database and render it unusable, very rigorous control procedures are necessary, and expensive remedial action may still be required from time to time. Most of the principles and some of the explicit practice worked out during this period remain in force to this day.

The retirement of Peter McMaster in late 1991 and the appointment of David Rhind to the office of Director General from January 1992 truly marked the end of the old era and the start of a new. As an external appointee from a senior academic position in the GIS field, Rhind felt able to discard much of the burden of history, to take an entirely new look at the nation's needs in the last decade of the twentieth century, and to argue the case with government for OS to change radically while greatly accelerating the costly activities that were already in hand. His opening Annual Report (1991/92) already included, firstly, a firm commitment to completing the digital databank of all basic scale map data by the end of 1994/95 (requiring a considerable further acceleration to build on what had already been achieved in the past two years), and, secondly, the introduction as a firm policy of the concept of "low threshold revision" under which significant change on the ground would in future be detected and surveyed as it occurs (Farrow 1992; Newby 1994). The revision categories and thresholds drawn up during this transitional period remain recognizable as the basis of database update policy and practice to this day (Ordnance Survey 2003, 2005a). This Annual Report also records the tests of the PRISM device described earlier and the dependence of field updating on the ability to transfer map data between field and headquarters using public telephone lines overnight, an average of 800 large-scale maps being transferred nightly during the year. New digital spin-off products, "Boundary Line" showing all administrative boundaries in Great Britain, and "ED-Line," a snapshot of boundaries of enumeration districts (EDs) for census purposes, had been produced in collaboration with relevant parts of central government and would self-evidently require updating in future whenever statutory boundaries are required.

Although a somewhat disturbing trend began of announcing, as totally new, activities that had been either in hand or at least under development for some considerable time (Newby 2001)—numerous examples can be found including the "launch" of Superplan and the progressive removal of MSDs—Rhind himself was normally punctilious in acknowledging that the achievements of his era were built on the foundations laid by his predecessors. Genuinely new in 1992/93 were the negotiation of a "service level agreement" (SLA) with the entire local authority community that would assist the take-up of digital data and simplify royalty payments, as well as a recasting of the sales agent network and appointment of new, fully computer-based, "National Agents." At the same time the field offices gave up their role as sales outlets, releasing their staff to concentrate on the updating of the mapping itself and transferring the job of selling OS large-scale products entirely to the private sector. The name Land-Line was introduced to cover large-scale digital map data products. The contents of the existing databank—the "picture" of the graphic map resulting from the original 1972 specification although subsequently modified in detail, and whose feature coding had been simplified in 1988 (Ordnance Survey 1988)—became known as Land-Line.88; while the "clean-data" version, now being produced and ready for implementation of full topological structure, would be known as Land-Line.93. After a very long gestation period due to problems of update and of consistency in the underlying Royal Mail data, a new geographic-referenced postal address product known as ADDRESS-POINT was also announced; this would be treated in future as a component of the National Topographic Database (NTD) together with the Land-Line topographic data and OSCAR (road centerlines). Thus would eventually end the distinction—maintained for over twenty years—between databank (what already existed) and database (a future topologically structured version that would enable GIS-style queries and analysis).

1993/94 showed good progress towards fulfilling the technical commitments made in 1992: national cover of OSCAR was already complete and therefore also subject to a frequent revision cycle; ADDRESS-POINT was now available; some 1,200 digital maps per day were being delivered to customers; and in terms of topographic survey update, around 3,200 maps per day were passing between 80 field offices, headquarters and 15 sales agents. In a muted echo of the very successful and high-profile Project 88 in the field office environment, and arising from recommendations of the 1980s topographic database project, a trial known as "Project 93" went one step further than the "clean" Land-Line.93 specification by taking truly object-based data to a pilot field office for maintenance in accordance with normal revision policy for several months during the calendar year 1993. However, it would not be until 2001 that these efforts towards object orientation would come to fruition in the OS MasterMap product. Meanwhile the main preoccupation of management over the preceding two years had been with a radical reorganization and staff reduction (mainly affecting middle management and lower staff levels), which took formal effect on 1 January 1994 (Rhind 1997).

This period was also marked by the publication in the UK and USA of a major two-volume work on GIS (Maguire et al. 1991) in which the issue of revision of geographic data sets somehow went largely unrecognized. Neither "revision" nor "update" featured in their own right in the index; the single page citation of "database update" covered only "temporal characteristics of remotely sensed data" (Davis and Simonett 1991). Despite this, the second volume did in fact contain the full chapter on activities at OS (Sowton 1991) that has already been repeatedly cited in this memoir, in which he naturally discussed at length the revision processes embedded in the OS approach to digital map data. The UK institutional setting in which OS operates was also discussed at considerable length in the first volume by both Coppock and Rhind (1991) and Chorley and Buxton (1991), but the requirement for database update was at best implicit. Intriguingly, the monumental second edition of the work (Longley et al. 1999) does index two highly theoretical contributions on change and update (Peuquet 1999; Egenhoffer and Kuhn 1999), which both essentially concluded that these problems were still rather difficult at that time—the antithesis of the practical OS approach. Sowton's chapter was omitted from this second edition; a new OS contribution (Smith and Rhind 1999) made some reference to current rapid updating policies but again escaped indexing under that head.

While this review is largely based on the purely practical developments in national mapping by the OS, it is fair also to mention two external publications that certainly influenced subsequent developments in Great Britain. The OEEPE research project of 1992–94 on updating complex digital topographic databases (Gray 1995) was led by the Ordnance Survey of Northern Ireland. OSGB did not participate directly, but useful concepts such as the "versioned object" (Winstanley 1995) covering successive states of real-world objects emerged from it and were echoed by Woodsford (1996) of Laser-Scan Ltd., supplier of OSGB's headquarters editing system of choice over a period of many years. Although the

GIS then available did not handle time elegantly, OS metadata already allowed storage of some relevant information about features that had been modified. The versioned object would provide a practical escape from theoretical arguments that time should be treated as a third or fourth dimension in geographic databases, and it can be recognized in the OS data structures that emerge in the following section.

20.5 Complete Digital Coverage and the Introduction of OS MasterMap, 1995–2006

Fiscal year 1994/95 duly produced the long-awaited completion of total large-scale digital coverage, when in March 1995 Great Britain became unquestionably the first administration in the world to achieve this. By now the total number of maps affected had increased to almost 230,000, mainly through the inexorable process of urbanization and consequent upgrade of 1:2500 scale sheets (1 km²) to 1:1250 scale (each sheet occupying 0.25 km²); the final thrust of the initial digitizing program had, however, been to complete the data set for the 1:10,000 scale mapping of the least populated and least developed mountain and moorland areas. The program of data cleaning for seamless cover and unique junction coordinates to the Land-Line.93 specification would be completed later in 1995. At this point it seemed scarcely necessary to mention that maintenance of the digital data had become the *raison d'être* of OS, for this had long been the prime task of the majority of its workforce. However, it was duly noted that some 934,000 HUs of revision had been surveyed during the year. It was also a good moment to put on record that, in addition to the update task, periodic re-engineering of the database was still envisaged, in order to meet the developing needs of GIS.

Subsequent Annual Reports through the late 1990s presented new revision policies of rolling five- and ten-year update cycles even for moorland areas; the introduction of the digital photogrammetric monoplotting and field "PRISM" update systems described in the previous section; and the encouragement of private sector marketing of digital geographic data through the recognition of value-added resellers and licensed system suppliers. After many years of simply updating the database (obviously with elaborate backup systems but, strangely, no explicit public archive of historic data nor any formal remit to maintain such an historical record) a decision was made in 1995/96 to take an annual "snapshot" of the National Topographic Database (NTD) and to publish it—and reinforce the assertion of copyright—through the mechanism of placing it in the UK Legal Deposit Libraries. Fleet (1999) noted that this annual snapshot now provided the preferred archive for historical purposes; he estimated the total annual volume at 38 GB (or 57 CD-ROMs) of which files that change in any year (and would alone be provided in future) might amount to 23 GB, the overall average rate of change being only 4 HU per "tile" per annum. He compared this process favorably with libraries' earlier need to refer to individual published map sheets for historic purposes: under the former new edition criterion of 300 HU of change, even local snapshots could be at extremely long intervals and neighboring maps would not normally be contemporaneous.

A more detailed technical view of developments in revision at the beginning of this period is given by Vincent and Logan (1996). Against the background of the update task and current policy for it, they described the range of procedures available and the monoplotting system then being introduced. They also discussed the revolution in progress in the field, with the introduction of direct recording of new survey detail using the PRISM pen computer (here incorrectly styled "Portable Ruggedized Integrated Survey Module," a further example of the frequent variations in terminology that could sometimes be baffling to OS insiders and outsiders alike). Be that as it may, the Annual Report for 1997/98 was able to boast of the completion of the PRISM program, thereby making Great Britain "the first country in the world to computerize the whole mapmaking and updating process," following the decision made as long ago as 1991 to attempt to combine surveying and digitizing into one process, and some twelve years after the PIES device was conceived. After a phased introduction of the equipment and techniques, a massive training program ensured that every field surveyor (now down to just over 450 in number) was ready by February 1998 to download the required section of the National Topographic Database on to his (or her) pen computer running PRISM software, to update that section on-site and transmit the results back to HQ, where the revised version would be made available within 48 hours to any customer who required it. Thus the situation pioneered for local customers in the Milton Keynes office in 1988 was formally extended nationwide, ten years later, as the culmination of "the biggest change in surveying in 50 years."

In a clear, comprehensive and forward-looking review of current research and future prospects, with particular reference to geospatial data applications, which he presented to the annual conference of the Association for Geographic Information, Murray (1997) emphasized the needs of customers, highlighting their requirements for the integration of diverse data sets and for improvements in the process and delivery of update, while remarking that none of these ideas were actually new. He further emphasized that funding considerations would influence the fine balance between research aimed at meeting short-, medium- and long-term needs. He included useful tabulations giving details of the various components of the NTD, data volumes and other statistics for each, and the possible increases in volume that could result from future developments in database structure. For example, the topography and other components of the current base data could swell from around 80 GB to some four times that volume. The long-term objective was now clearly identified as an object-oriented database with associated "object server," which could eventually integrate the diverse products of that period and thus come close to realizing the visions of the pioneers of the early 1970s. However, he pointed out that this scenario would have to be developed while simultaneously continuing to deliver and maintain the existing 80 GB portfolio; non-contiguous portions amounting to some 10% of the whole database were undergoing update at any one time, currently leading to annual supply of more than 200 GB of data to customers.

During these years the idea gradually emerged that the database itself was scale-free; users could employ it at any scale of representation, while having due regard to the original scale and method of survey (and ongoing revision) that was by now incorporated in metadata, which itself formed part of the database (McMaster et al. 1986; Sowton 1989). The concept of a "National Geospatial Data Framework" (NGDF)—initially mooted as early as 1995 and characterized by Murray (1997) as a somewhat utopian commitment—eventually took root in 1999 as the rather different Digital National Framework (DNF), based on five principles concerned with interoperability within the geographic information community. While there has been ambiguity on whether the DNF is a concept, a model, a standard or the underlying topographic framework data itself, or merely a description of best practice (Brayshaw 2005), the principles have been clear enough since its inception (Holland and Allan 2001; Ordnance Survey 2004, 2005b; Murray et al. 2005; DNF 2005). A number of other terms, including "national spatial data infrastructure" and many variants, have been used during this period to cover broadly the same range of concepts, but the DNF is now well established as the key to overcoming the major obstacles to the realization of the potential of GIS that were inherent in questions of interoperability and consistency of data. The term National Topographic Database (NTD) was already well fixed during this period, but as the database continues to evolve and to acquire additional non-topographic components there may be a temptation in future for further changes, with the possibility that "topographic" may finally give way to "geospatial" or "geographical" or even "geographic."

Several changes of chief executive in rapid succession in the period 1998 to 2000, which certainly involved significant discontinuities both in staffing at senior management levels and in relationships with national and local government, did not seriously interrupt technical progress. Indeed, shortly before he resigned after barely a year as Director General of OS, Geoff Robinson issued an upbeat press release outlining his personal assessment of the organization's relations with its customers and the direction in which he had already started to lead it (Robinson 1999); his intentions expressed in that document largely survived his departure and some of the actions noted below could be traced to decisions taken during his period in office. Annual Reports tell of the start of delivery of small parcels of digital data by e-mail and progress towards the long-standing goal of an object-oriented and fully topologically structured database. Meanwhile the NTD, as it already stood, consisted of over 2 million features on 230,000 maps occupying 36 GB of storage, with some 1,000 tiles being updated every day. The NTD lay behind not only the flagship Land-Line product but also formed the foundation for the ADDRESS-POINT, OSCAR and Boundary-Line data sets as well as the OS's full range of paper maps. Simplified licensing arrangements and reduced prices introduced in response to consultations with users led to encouragingly increased take-up of data. The positional accuracy improvement (PAI) project for the 1:2500 overhaul was launched in 1999/2000, and efforts were in hand to ease customers' possible difficulties with handling change in the data, which did not represent any change on the ground. The introduction of "Collection of Data from External Sources" (CODES) allowed the speedy incorporation of developers' plans into the database after many years of discussion and experiment.

Inextricably linked with the relentless drive towards achieving full cost recovery (in place of the annual government subsidy that had already been progressively reducing ever since the Second World War), the long-drawn-out negotiation of a "national interest mapping service agreement" (NIMSA) (Ordnance Survey and Department of the Environment 1996; Rhind 1997) was also finally completed and implemented in 1999/2000. NIMSA would in future fund OS activities in the national interest, which could not be justified on purely commercial grounds, and would thus be of fundamental importance to the maintenance of large-scale mapping in the less populated areas of the country. NIMSA ensures that data sets that would normally be bought by relatively few customers will always in fact be available both for routine sale and to deal with emergency situations whenever and wherever they occur. This funding immediately enabled a reduction in the period of rural cyclic revision to only five years and in 1999/2000 alone 32,000 individual maps with relatively minor change were brought up to date using aerial photography. The benefit of NIMSA is chiefly in the provision of updated maps and aerial photography, but it also extends to various other uneconomic tasks such as meeting OS obligations under the Welsh Language Act. Inevitably there is no permanent guarantee of NIMSA funding, which has to be renegotiated at intervals with the government of the day, but meanwhile the principle at least has been clearly established and the system of payment has proved its worth in practice, covering some 12% of OS turnover in 2004/05 and a rather higher proportion in some earlier years (NIMSA Review Group 2005).

Also germane to its long-term relationship with government, in 1999 OS commissioned external consultants to investigate the value of the economic infrastructure built on its data. In addition to substantial but unquantifiable qualitative benefits to national life, it was estimated that between £79 and 136 billion (worth US$128–220 billion at that time), or some 12 to 20%, of Britain's business and public services rely to a significant extent on OS data (as measured by "gross value added," the main component of gross domestic product) (OXERA 1999). Although it was admitted that even the upper limit of £136 billion might be an underestimate, the complex and fascinating report and its main quantitative conclusion have been translated into a simple and often repeated one-line formula, that "OS data underpins around £100 billion of economic activity in Great Britain each year" (then equivalent to around US$162 billion; now at the time of writing US$174 billion). It goes without saying that the ongoing update of this data is, and will remain, of fundamental national importance.

Arriving from a successful commercial geographic background in 2000, Vanessa Lawrence (the first female Director General in the 210-year history of the OS) immediately made plain her commitment to the twenty-first century themes of "e-business" and of geographic information systems in support of location-based services. Her first Annual Report (2000/01) also noted progress in the re-engineering of the National Topographic Database, which would lead to the launch in late 2001 of OS MasterMap, the fully structured, object-oriented version of the OS database, within the principles of the DNF mentioned earlier. Among many benefits foreseen was the provision of the change-only update facility already envisaged by Gardiner-Hill at the very start of digital mapping in 1971 and discussed as a development goal throughout the intervening thirty years. Meanwhile the field update process was being enhanced by the use of real-time kinematic (RTK) GPS, not only to provide the control framework as hitherto, but also for detail survey; this would in due course be introduced to the PRISM field tool, which had itself been significantly upgraded during the year in preparation for the new demands of object-based referencing.

Apart from the justifiably triumphant launch of OS MasterMap in November 2001, the Annual Report for 2001/02 also noted the launch of a new data product to be known as "Pre-Build." This was the logical culmination of the earlier CODES activity. Once the principle of receiving developers' data for map updating purposes had finally been accepted, it was just one extra step to make it immediately available to customers such as the gas, electricity, water and telecommunications utilities, which most especially need the data prior to completion of major building developments. Those within OS who had objected, perhaps over many years, that builders' plans invariably represented what they might build, rather than what would actually appear on the ground and require to be surveyed later for inclusion on the map, must have been permitted a wry smile at having the argument so creatively turned against them. It is certain that OS will have introduced adequate quality management procedures to ensure that whatever data is supplied by developers does in fact meet OS standards for the portrayal of completed development, before its eventual inclusion in the definitive topographic database. Over the following years OS MasterMap was expanded from the underlying topography layer to include additional components styled the "Address Layer" (the former ADDRESS-POINT), the "Imagery Layer" (orthorectified aerial photography) and the "Integrated Transport Network (ITN) Layer," initially a roads network data set embodying the various levels of the long-standing OSCAR product, but intended ultimately to cover all modes of transportation. By 2003/04, Lawrence could boast not only annual update totaling around 1.5 million changes to the geographic database (or some 5,000 changes every day), but also, from the users' perspective, that 23 terabytes of data had been delivered to customers.

The most extensive information on OS MasterMap, now described as "the definitive digital map of Great Britain," including its update regime, is available on the Internet in the form of a two-volume User Guide (Ordnance Survey 2005a); an earlier single-volume version was produced in paper form (Ordnance Survey 2003). No attempt will be made to abstract details here, although references to some of the update-related content will have been discerned in earlier paragraphs and this discussion will continue below. A clear outline of OS MasterMap and the DNF in their institutional setting is given by Murray and Shiell (2003) and again more recently by the same authors (see Murray and Shiell, Chapter 21, this volume). Papers presented to meetings of the International Federation of Surveyors (FIG) have also explained the place of OS MasterMap and the DNF within OS's e-business strategy and have emphasized both the obligation to maintain the base data and the cost of doing so (Murray 2002; Murray et al. 2005).

Very useful previews of OS MasterMap from the technical point of view, with closely related content, had been provided by Murray (2000) and Holland and Allan (2001), the latter with particular reference to database update, although as these articles were both published before its official launch the new product was not explicitly named. Holland and Allan (2001) began by introducing the DNF as the overriding framework for the development of OS MasterMap, together with the important new concept of a unique topographic identifier (TOID). Every feature will possess a TOID that will remain unchanged throughout its lifetime, to provide an unambiguous and permanent reference to it, and thus act as a hook to which users may in future attach their own data concerning the feature. While the TOID will remain invariant, features that are subjected to change in the real world will be revised in the database and their version number will be incremented, within a complex set of life cycle rules that go somewhat beyond the "versioned object" concept mentioned in the previous section. Holland and Allan also explained that the data would be supplied in GML, the OpenGIS Geography Markup Language of the Open Geospatial Consortium; the old tile (map sheet) structure would give way to a feature-based structure representing real-world objects, and customers would thus be able to request data for any desired area of interest, or any individual (object-oriented) theme. The National Grid, based on the OSGB36 datum with a single transverse Mercator (TM) projection, would remain unchanged as the sole coordinate reference system, but transformations to and from GPS-based systems would be available at various levels of precision and would become increasingly transparent and routine.

At this point, just when the database itself was finally becoming truly seamless and virtually scale-free, it is appropriate to reproduce Holland and Allan's (2001) figures for the numbers of tiles held at each of the historic basic scales of survey. As indicated earlier in this memoir, the areas involved have been continually changing, and will continue to change over the years. This is due mainly to ongoing urbanization and consequent upgrading of map scale (normally through a resurvey), although changes in the perceived boundary between populated rural and unpopulated mountain and moorland areas have also at times led to fluctuations in the opposite direction. It should also be noted that many tiles of basic 1:10,000 (mountain and moorland) mapping contain "rural" tiles surveyed and mapped at 1:2500 scale. Moreover many tiles inevitably contain areas of water. It is therefore purely coincidental that the total number of tiles (map sheets of size varying with original scale of survey as indicated below) is almost identical with the reputed total land area of Great Britain in square kilometers, 230,000 km², noted in the opening paragraph of this memoir. After delivering these warnings, it is fair to quote the exact numbers given by Holland and Allan (2001) as follows:

1:1250	Urban	$66,953 \times 0.25$ km²;
1:2500	Rural	$157,872 \times 1$ km²;
1:10,000	Mountain and moorland	$4,049 \times 25$ km²;
Total number of basic scale tiles		228,874.

Holland and Allan (2001) went on to explain that while the bulk of urban update (1:1250 scale) continues to be by field survey using the PRISM pen computer, most rural revision employs photogrammetric methods. The monoplotting process was phased out at the end of 2000 in favor of new digital photogrammetric workstations operating in a stereoscopic viewing, interactive editing environment. The very long-serving Laser-Scan edit system (favored over many years, originally in part because of its ability to handle attributes at the point level) had been upgraded so as to be able to operate in the topologically structured, object-oriented environment of OS MasterMap and the DNF. The same paper also gives some technical details of the PAI program (which incidentally is said to present less difficulty to the user in an OS MasterMap environment, thanks to the TOID, than in the former data structure) and forecasts the imminent availability of orthorectified imagery as an additional layer of the new product, as well as the possible future inclusion of aspects of the third dimension.

What Holland and Allan (2001) and other OS authors have tended to gloss over is the importance, and technical difficulty, of the changes to the update process demanded by a topologically structured, object-oriented, seamless database. This must be the primary reason why successive attempts to introduce such a database, in the 1970s, 1980s and 1990s (Sowton 1989, 1991), have only finally come to fruition in the twenty-first century. As mentioned very much earlier in this contribution, to update the unstructured "spaghetti" data of the 1970s, or even the clean, seamless, "unique- junction coordinate" data of Land-Line.93, is a relatively undemanding process, equivalent to editing for errors in the originally captured data. With no relationships between objects, revision has no effect beyond the point or line in question; if the old and the new work are both accurate, the revision will fit; the only issue of any seriousness arises where original errors are found, as was frequently the case with the maps that resulted from the 1:2500 overhaul; and the revisers became adept at dealing with that problem within the accepted limitations of those maps. With a topologically structured and object-oriented database, every feature touched in the revision process affects the topology and relationships of the whole object of which it forms part. Thus, the addition, modification or deletion of a point or line spreads ripples through the database in a way that unstructured data never knew, and the preservation of the integrity and consistency of the database as a whole becomes a major challenge. In designing systems to handle this problem, there are in principle two choices: either to capture and edit survey data in some relatively simple form, and then update the database in bulk at intervals; or alternatively and more elegantly, but certainly with greater difficulty, to revise the topology as well as the geometry "on the fly." The practical resolution of this dilemma, when much of the update is performed on a compact field device, PRISM, carrying a necessarily small portion of the entire database, continues to exercise OS, and a two-stage edit process has been required up to now.

The author has been informed (Murray 2005, 2006a) that a "corporate editor," dating back to the developments of the mid- and late- 1990s, which led to OS MasterMap, has hitherto been interfaced with the various different in-house systems such as PRISM and Laser-Scan's LAMPS2, while ongoing development and procurement processes have also endeavored to incorporate other proprietary editors and data models. However, massive development work is now well advanced, to replace these systems and bring OS's field, photogrammetric, job management and central database systems together with a unified suite of editors for the future. Even the labor-intensive but pragmatically successful house unit count might eventually give way to an automated means of measuring and managing change within the unified system.

The present difficult steps recall the long history of successive generations of hardware and software for digital mapping. Just one of many examples over the years was noted by Sowton (1989), but in this contribution the author has attempted as far as possible to avoid reference either to procurement difficulties or to individual products selected at a given time. Changes driven by the increased capabilities, and complexity, of digital mapping and geographic information technology have led to successive procurements in which a difficult balancing act—between the desires for standardization, or diversification, or continuity, or the personal preferences of the main participants, or rigid adherence to international and national rules on tendering procedures—has not always produced an easy outcome nor a direct route to improvements in processes. As has always been the case at OS, practice has continued to run ahead of geographic information theory and of the offerings of hardware and software in the marketplace. Thus, as ever, there is no clear roadmap, nor obvious vehicle, for the route to GIS nirvana. Indeed, the metaphor of nirvana is especially appropriate because there is no final real-world destination. It is certain that development of both the processes and the products of OS will continue, as they always have, in response to the needs of customers, the national interest and the financial imperatives of the time.

In the last Annual Report available at the time of writing, for 2004/05 (Ordnance Survey 2005b), the rate of change appeared to be stable at 5,000 units per day, but data deliveries had risen to exceed 40 terabytes during the year (this major increase may have been assisted by the completion of the Imagery Layer for England—though Wales and Scotland would require further seasons of better weather, even with the newly acquired digital aerial camera). The seemingly impossible target of "99.6% of significant real-world features being represented in the database within six months of completion" was substantially exceeded, with an actual performance of 99.85%, by a field revision workforce then reduced to some 330. The national GPS infrastructure has been further developed so as to provide cm-level positioning to those surveyors (now only 290 in number [Murray 2006b])—and to any other customer of the GPS network. The TOID is singled out for special mention (or rather, the TOIDs, representing almost half a billion topographic and other features); OS permits their royalty-free use in order to encourage the widest possible reuse of information once captured and to minimize duplication of effort or of data. Progress is clearly being made in weaning already satisfied customers to OS MasterMap from the proven Land-Line.93 format (now known simply as Land-Line), although OS is committed to maintaining the latter until at least 2008 while carrying out extensive consultations about its eventual elimination.

Virtually all OS mapping, including the smaller scales paper products (the full national range covers 1:10,000, 1:25,000, 1:50,000, 1:250,000 and 1:625,000 while some tourist maps employ intermediate scales), is now created, updated and available in digital form, and the smaller scale data sets are increasingly capable of automatic derivation from the large-scale database, although technical obstacles still remain on the road to full automatic map generalization. Through a range of agreements with central and local government, including a "pan-government agreement" with 210 signatories as well as the NIMSA discussed earlier, it is clear that uniform, consistent and up-to-date geographic data can be readily accessed by those who need it, in a way unthinkable a few years ago. The legion of "value-added resellers" and system suppliers are now united under the banner of "Licensed Partners," some 180 in number. There has been a move towards offering both customers and partners access to on-demand web services as well as periodic supply of update via the web or on "traditional" media such as CD-ROMs and DVDs. Licensed customers have the right to both automatic and on-demand updates; major customers who hold full national coverage can enjoy the "Managed GB Sets" service both on demand and automatically at six-weekly or three-monthly intervals; a choice of full resupply or change-only update is available. Meanwhile, at the other end of the scale, small customers are being encouraged to take up the possibilities of digital mapping by way of simplified licensing and online access to the database through the OS "Options" network. Collaboration with local authorities continues as ever, with Dudley Metropolitan Borough Council (pioneers of structured data in the 1970s and 80s) still among the leaders with their GIS–MO project ("Getting Information Simply–Mapping Online"), which gives every Council employee direct access to the OS database on his or her desktop.

Meanwhile, both in-house and collaborative research and development continue apace, with the still incompletely solved small-scales generalization problem in hand with European partners, and novel data capture techniques, automated change detection and the addition of the third dimension to OS MasterMap all under investigation internally. Development of the "seamless data collection, maintenance and management system" is the current major item of capital investment, with ambitions that clearly go far beyond the development of a unified editor for topologically structured geographic data discussed in earlier parts of this memoir. Underpinning all of this confident annual review were the assumptions that the data used by all customers is actually up to date—the subject of this contribution to *Manual of Geographic Information Systems*, that it is suitable for onward analysis in geographic information systems, and that the OS is also now financially self-supporting. It is very satisfying that in Great Britain in 2006 these assumptions hardly need to be explicitly stated.

20.6 Acknowledgements

The author was employed at Ordnance Survey Headquarters in Southampton, UK, from 1973 to 1977 and 1985 to 1994. This memoir could not have been written in 2005/06 without the assistance, support and encouragement of the Ordnance Survey of today, in the persons of Keith Murray, Head of Geographic Information Strategy, and Cathy Layton, Librarian. The author warmly thanks them both. The article includes many references to trademarks and registered trademarks of Ordnance Survey and of other organizations, and these are all hereby acknowledged.

20.7 Abbreviations

AGI	Association for Geographic Information
CAD/CAM	Computer-Aided Design/Computer-Aided Manufacturing
CODES	Collection of Data from External Sources
CR	Continuous Revision (never "Cyclic"—see also PR)
DFUS	Digital Field Update System
DNF	Digital National Framework
DXF	Drawing Interchange Format; Data Exchange File
ECU	Experimental Cartography Unit (of the Royal College of Art)
ED	Enumeration District (for census purposes)
FIG	Fédération Internationale des Géomètres, International Federation of Surveyors
GB	Great Britain; gigabyte
GIS	Geographic Information System(s)
GML	OpenGIS Geography Markup Language
GPS	Global Positioning System
HU	House Unit
ISO	International Organization for Standardization
ISPRS	International Society for Photogrammetry and Remote Sensing
LA	Local Authority (any UK local government entity)
MB	Metropolitan Borough (one particular form of UK urban local authority)
MSD	Master Survey Drawing
NG	National Grid (Cartesian coordinate system embodied in OSGB36)
NGDF	National Geospatial Data Framework
NIMSA	National Interest Mapping Service(s) Agreement
NJUG	National Joint Utilities Group
NTD	National Topographic Database
NTF	National Transfer Format
OEEPE	European Organization for Experimental Photogrammetric Research (EuroSDR)
OS	Ordnance Survey of Great Britain
OSCAR	OS road network data products (Ordnance Survey Centre Alignment of Roads)
OSGB	Ordnance Survey of Great Britain
OSGB36	Datum and projection used by OSGB for all national mapping purposes
OSNI	Ordnance Survey of Northern Ireland
PAI	Positional Accuracy Improvement
PR	Periodic Revision (sometimes described as cyclic revision)
PRISM	Portable Revision Integrated Survey Module (several variants used over the years)
RICS	Royal Institution of Chartered Surveyors
RTK	Real-Time Kinematic (GPS)
SIM	Survey Information on Microfilm
SLA	Service Level Agreement
SUSI	Supply of Unpublished Survey Information

TM   Transverse Mercator (projection)
TOID   Topographic Identifier
TQM   Total Quality Management

References

Atkey, R. G. and R. J. Gibson. 1975. Progress in automated cartography. *Proceedings of the Conference of Commonwealth Survey Officers,* Paper J3, August 1975, Cambridge, UK.

Bickmore, D. P. 1971. Experimental maps of the Bideford area. Pages 217–223 in *Proceedings of the Conference of Commonwealth Survey Officers,* Part 1, Paper E1, August 1971, Cambridge, UK.

Brand, M.J.D. 1986. The foundation of a geographical information system for Northern Ireland. Pages 4–9 in *Proceedings of AutoCarto London,* Volume 2, *Digital Mapping and Spatial Information Systems.* Edited by M. Blakemore. 14–19 September 1986, London, UK.

Brayshaw, J. 2005. OSNet GPS consultation workshop. *Presentation at Ordnance Survey OSNet™ Utilities Sector Consultation Workshop,* held 5 October 2005, Solihull, UK. <http://www.ordnancesurvey.co.uk/oswebsite/business/sectors/utilities/docs/JamesBrayshaw.zip> (Accessed 19 June 2008).

Chorley, R. and R. Buxton. 1991. The government setting of GIS in the United Kingdom. Pages 67–79 in *Geographical Information Systems.* Vol. 1, *Principles.* Edited by D. J. Maguire, M. F. Goodchild and D. W. Rhind. Harlow, UK: Longman and New York: Wiley.

Coote, A. M. 1988. Current developments in field-based digital mapping systems at Ordnance Survey. *Mapping Awareness Conference,* January 1988, Oxford, UK.

———. 1989a. Current developments in field-based digital mapping systems at Ordnance Survey. Pages 113–119 in *The Association for Geographic Information Yearbook 1989.* Edited by P. J. Shand and R. V. Moore. London: Taylor & Francis and Oxford, UK: Miles Arnold.

———. 1989b. Managing a large spatial archive. *GIS–A Corporate Resource–Proceedings of AGI 1989,* Paper B.3, 11–12 October 1989, Birmingham, UK.

Coppock, J. T. and D. W. Rhind. 1991. The history of GIS. Pages 21–43 in *Geographical Information Systems.* Vol. 1, *Principles.* Edited by D. J. Maguire, M. F. Goodchild and D. W. Rhind. Harlow, UK: Longman and New York: Wiley.

Davidson, Rt. Hon. The Viscount (Chairman). 1938. *Final Report of the Departmental Committee on the Ordnance Survey.* London: HMSO.

Davis, F. W. and D. S. Simonett. 1991. GIS and remote sensing. Pages 191–213 in *Geographical Information Systems.* Vol. 1, *Principles.* Edited by D. J. Maguire, M. F. Goodchild and D. W. Rhind. Harlow, UK: Longman and New York: Wiley.

DNF. 2005. Digital National Framework. <http://www.dnf.org/Pages/technical guidance/> Accessed 19 June 2008.

Egenhoffer, M. J. and W. Kuhn. 1999. Interacting with GIS. Pages 401–412 in *Geographical Information Systems.* 2nd ed. Vol. 1, *Principles and Technical Issues.* Edited by P. A. Longley, M. F. Goodchild, D. J. Maguire and D. W. Rhind. Chichester, UK and New York: Wiley.

Elston, M., D. Rhind, S. Finch and B. Hedges. Undated, about 1985. Measuring and modelling change. Unpublished report. Social and Community Planning Research and Birkbeck College, London, UK.

Farrow, J. E. 1992. Ordnance Survey revision problems and solutions. Pages 73–79 in *Proceedings of the ISPRS and OEEPE Joint Workshop on Updating Digital Data by Photogrammetric Methods.* Edited by P.R.T. Newby and C. N. Thompson. 15–17 September 1991, Oxford, UK. OEEPE Official Publication No. 27, Frankfurt am Main, Germany.

Fleet, C. 1999. Ordnance Survey digital data in UK legal deposit libraries. *Liber Quarterly* 9:235–243.

Gardiner-Hill, R. C. 1971. Automated cartography in the Ordnance Survey. Pages 235–241 in *Proceedings of the Conference of Commonwealth Survey Officers,* Part 1, Paper E3, August 1971, Cambridge, UK.

———. 1972. *The Development of Digital Maps.* Ordnance Survey Professional Paper New Series No. 23, Ordnance Survey, Southampton, UK.

———. 1973. Data structure for digital mapping used by the Ordnance Survey. *Proceedings of International Cartographic Association Commission III,* August 1973, Budapest, Hungary.

Gray, S. 1995. *Updating of Complex Digital Topographic Databases.* Edited by S. Gray. OEEPE Official Publication No 30, Frankfurt am Main, Germany.

Greenway, I. 1994. The use of pen computers for revision of large scale mapping. *Proceedings of 20th FIG Congress,* Commission 5, Paper 505.4., Melbourne, Australia.

Haywood, P. E. 1989. Structured topographic data–the key to GIS. *GIS–A Corporate Resource–Proceedings of AGI 1989,* Paper B.1. 11–12 October 1989, Birmingham, UK.

Holland, D. A. and L. E. Allan. 2001. The Digital National Framework and digital photogrammetry at Ordnance Survey. *Photogrammetric Record* 17 (98):291–301.

Irwin, M. St G. 1970. The application of automated cartography in the Ordnance Survey. *Chartered Surveyor* 102 (10):467–473. Reprinted as Ordnance Survey Technical Paper No 21.

Leonard, J. P. 1982. Revision of Ordnance Survey 1:2500 scale maps. *Photogrammetric Record* 10 (60):681–685.

Longley, P. A., M. F. Goodchild, D. J. Maguire and D. W. Rhind. 1999. *Geographical Information Systems.* 2nd ed. Chichester, UK and New York: Wiley.

Maguire D. J., M. F. Goodchild and D. W. Rhind. 1991. *Geographical Information Systems.* Harlow, UK: Longman and New York: Wiley.

Matthews, A.E.H. 1976. Revision of 1:2500 scale topographic maps. *Photogrammetric Record* 8 (48):794–805.

McMaster, P., P. E. Haywood and M. Sowton. 1986. Digital mapping at Ordnance Survey. Pages 13–23 in *Proceedings of AutoCarto London,* Vol. 1. 14–19 September 1986, London, UK.

Murray, K. 1997. Anticipating trends in geospatial data applications–research developments at Ordnance Survey. *Geographic Information–Exploiting the Benefits. AGI '97 Conference,* Paper 8.4. 7–9 October 1997, Birmingham, UK.

———. 2000. Building a new geospatial infrastructure: The Digital National Framework. *AGI Conference at GIS 2000,* Paper T1.1. 26–28 September 2000, London, UK.

———. 2002. A new geo-information framework for Great Britain. *XXII FIG International Congress and ACSM/ASPRS Annual Conference,* Session TS 3.3. 19–26 April 2002, Washington, D.C.

———. 2005. Meeting with author at Ordnance Survey, Southampton, 13 May 2005.

———. 2006a. Email and telephone conversation with the author, 12 April 2006.

———. 2006b. Email to author, 27 March 2006.

Murray, K. J. and D. Shiell. 2003. A new geographic information framework for Great Britain. *Photogrammetric Engineering and Remote Sensing* 69 (10):1175–1182.

———. 2008. Evolving the geographic information infrastructure in Great Britain. In *Manual of Geographic Information Systems* , edited by M. Madden, Chapter 21 in this volume. Bethesda, Md.: ASPRS.

Murray, K., B. Munday and I. Bush. 2005. Enabling information integrity within spatial data infrastructures–the Digital National Framework concept. *From Pharaohs to Geoinformatics: FIG Working Week 2005 and GSDI-8*, TS49.1, SDI Data Issues. 16–21 April 2005, Cairo, Egypt.

Newby, P.R.T. 1990a. Photogrammetric developments in the Ordnance Survey in 1990. *Photogrammetric Record* 13 (76): 561–576.

————. 1990b. Digital terrain modelling as a means to digital photogrammetric revision. *International Archives of Photogrammetry and Remote Sensing* 28 (3/2):629–638.

————. 1990c. New initiatives in image digitising at the Ordnance Survey. *International Archives of Photogrammetry and Remote Sensing* 28 (3/2):639–656.

————. 1990d. Digital map revision at the Ordnance Survey. *International Archives of Photogrammetry and Remote Sensing* 28 (4):146–154.

————. 1991. Digital photogrammetry at the Ordnance Survey. Pages 234–243 in *Digital Photogrammetric Systems, Proceedings of ISPRS Inter-Commission Working Group II/III Conference*, Munich 1991. Karlsruhe, Germany: Wichmann.

————. 1992. Quality management for surveying, photogrammetry and digital mapping at the Ordnance Survey. *Photogrammetric Record* 14 (79):45–58.

————. 1994. Revision policy and practice for Great Britain's topographic database. *International Archives of Photogrammetry and Remote Sensing* 30 (4):260–267.

————. 1996a. Working Group IV/3 report and review of progress in map and database revision. Invited Paper, ISPRS Congress, Vienna. *International Archives of Photogrammetry and Remote Sensing* 31 (B4):598–603.

————. 1996b. Digital images in the map revision process. *ISPRS Journal of Photogrammetry and Remote Sensing* 51:188–195.

————. 2001. Editorial. *Photogrammetric Record* 17 (98):221–223.

Newby, P.R.T. and D. W. Proctor. 1990. Revision of large-scale maps at the Ordnance Survey. *ISPRS Journal of Photogrammetry and Remote Sensing* 45:137–151.

Newby, P.R.T. and A. S. Walker. 1986. The use of photogrammetry for direct digital data capture at Ordnance Survey. *International Archives of Photogrammetry and Remote Sensing* 26 (4):228–238.

NIMSA Review Group. 2005. *The National Interest Mapping Services Agreement Annual Report 2004–05.* Office of the Deputy Prime Minister and Ordnance Survey, Southampton, UK. <http://www.ordnancesurvey.co.uk/oswebsite/aboutus/reports/nimsa/> Accessed 3 February 2006.

NJUG (National Joint Utilities Group). 1988. *NJUG Publication No. 13: Quality Control Procedure for Large Scale Ordnance Survey Maps Digitised to OS 1988 Version 1.* London: NJUG.

Ordnance Survey. 1972. *The Overhaul of the 1:2500 County Series Maps.* Professional Papers New Series No. 25, Ordnance Survey, Southampton, UK.

————. 1981. Report of the study of revision. Unpublished internal report. October 1981, Southampton, UK; Summary report of the study of revision. Unpublished internal report. August 1981, Southampton, UK.

————. 1988. *OS 1988–Contractors Specifications for Digital Mapping.* Ordnance Survey, Southampton, UK.

————. 2003. *OS MasterMap User Guide v3.0–3/2003.* Ordnance Survey, Southampton, UK.

————. 2004. *The Digital National Framework–A White Paper.* Ordnance Survey, Southampton, UK.

————. 2005a. *OS MasterMap User Guide.* Part 1 Product Specification v6.0.1.–10/2005; *OS MasterMap User Guide.* Part 2 Reference Section v6.0–03/2005. <http://www.ordnancesurvey.co.uk/oswebsite/products/osmastermap/guides/userguide.html> Accessed 15 February 2006. At the time of writing this chapter, the 2005 guide in two parts was the current version of the document; by 2008 it had evolved into the six parts that can be accessed (19 June 2008) at <http://www.ordnancesurvey.co.uk/oswebsite/products/osmastermap/userguides/docs/OSMMTopoLayerUserGuide.pdf>.

————. 2005b. *Annual Report and Accounts 2004–05.* London: The Stationery Office.

Ordnance Survey and Department of the Environment. 1996. *Results of the Consultation Exercise on the National Interest in Mapping.* Southampton, UK: Ordnance Survey.

OXERA. 1999. *The Economic Contribution of Ordnance Survey GB.* Public version of final report. Oxford Economic Research Associates, Oxford, UK. <http://www.ordnancesurvey.co.uk/oswebsite/aboutus/reports/oxera/oxera.pdf> Accessed 3 February 2006.

Peuquet, D. J. 1999. Time in GIS and geographical databases. Pages 91–103 in *Geographical Information Systems.* 2nd ed. Vol. 1, *Principles and Technical Issues.* Edited by P. A. Longley, M. F. Goodchild, D. J. Maguire and D. W. Rhind. Chichester, UK and New York: Wiley.

Proctor, D. W. 1986. The capture of survey data. Pages 227–236 in *Proceedings of AutoCarto London,* Vol. 1. 14–19 September 1986, London, UK.

Proctor, D. W. and P.R.T. Newby. 1988. Revision of large scale maps at the Ordnance Survey. Invited paper. ISPRS Congress, Kyoto. *International Archives of Photogrammetry and Remote Sensing* 27(B4):298–307.

Rhind, D. 1988. Personality as a factor in the development of a discipline: The example of computer-assisted cartography. *American Cartographer* 15 (3):277–289.

———. 1997. Facing the challenges: Redesigning and rebuilding Ordnance Survey. Pages 275–304 in *Framework for the World.* Edited by D. Rhind. Cambridge, UK: GeoInformation International.

Robinson, G. 1999. Mapping out a radical new business strategy for the new millennium. Ordnance Survey press release. Tuesday 28 September 1999, Southampton, UK.

Smith, N. S. and D. W. Rhind. 1999. Characteristics and sources of framework data. Pages 655–666 in *Geographical Information Systems.* 2nd ed. Vol. 2, *Management Issues and Applications.* Edited by P. A. Longley, M. F. Goodchild, D. J. Maguire and D. W. Rhind. Chichester, UK and New York: Wiley.

Sowton, M. 1989. Digital data: The future for Ordnance Survey. Pages 493–504 in *AutoCarto 9, Proceedings of Ninth International Symposium on Computer Assisted Cartography.* 2–7 April 1989, Baltimore, Maryland. Falls Church, Virginia: ASPRS & ACSM.

———. 1991. Development of GIS-related activities at the Ordnance Survey. Pages 23–38 in *Geographical Information Systems.* Vol. 2, *Applications.* Edited by D. J. Maguire, M. F. Goodchild and D. W. Rhind. Harlow, UK: Longman and New York: Wiley.

Thompson, C. N. 1978. Digital mapping in the Ordnance Survey 1968–1978. *ISPRS Commission IV symposium.* 2–6 October 1978, Ottawa, Ontario. *International Archives of Photogrammetry* 22-4.

Vincent, G. N. and I. T. Logan. 1996. Ordnance Survey policy and practical implications for the revision of rural mapping. *Photogrammetric Record* 15 (88):503–517.

Winstanley, A. C. 1995. Appendix VIII: Updating complex digital topographic databases, an object oriented solution. Pages 63–82 in *Updating of Complex Digital Topographic Databases.* Edited by S. Gray. OEEPE Official Publication No 30, Frankfurt am Main, Germany.

Woodsford, P. A. 1996. Spatial database update–a key to effective automation. *International Archives of Photogrammetry and Remote Sensing* 31(B4): 955–961.

In addition to the above, reference has been made to *Ordnance Survey Annual Reports* covering the entire period from 1966/67 to the present. It should be noted that these *Reports* cover fiscal years (ending on 31 March) rather than calendar years. The latest available at the time of writing, 2004/05, is also explicitly cited above.

CHAPTER 21

Evolving the Geographic Information Infrastructure in Great Britain

Keith Murray and *Duncan Shiell*

21.1 National Context for Geographic Information (GI)

The need to understand one's location has been a necessity for mankind as far as history can recall. Even today, one of the first tasks in inter-planetary science is to map the object of attention, to locate features, name them and to create a terrain model to determine the relative height of objects. On this base other information can then be attached, such as landing sites, the potential sources of minerals, water and even the potential for life. However, when we map the Moon (Masursky et al. 2004), Mars or other objects we are not bounded by political frontiers. Although the level of detail is relatively sparse compared to Earth, this condition permits a consistent approach to be made to planetary mapping and in locating the objects and features of those planets and moons.

We have never had that luxury here on Earth. Since the mathematics and instruments required to accurately and economically model and map the Earth have only been available over the last 2-300 years, each nation has developed its own way of creating, developing and using information about location. This has resulted in a very wide variety of implementations, each highly influenced by the following parameters:

- Historical influence
- Political dimension
- Economic position
- Applications and users
- Information infrastructure

Within the global geographic information (GI) industry we all recognize that the power of GI can play a major role in a wide range of functions within society and industry (security, navigation, asset management, tourism, agriculture, land management, etc.). Many also feel that although GI is now better recognized as a useful information tool, we still have some way to go before we reach the full potential in its use and exploitation by society at large.

All this comes at a time when the globalization of all we do, from activities such as tourism and travel to international trade and attempting to manage climate change, is driving a growing need to create a coherent information infrastructure to support these needs consistently at a global level.

The following sections will describe the position in Great Britain using the above five parameters to set the context and background to describe what is done, why it is done that way and how it is now evolving to support and integrate GI into mainstream business processes in Great Britain.

21.2 Historical Influence

21.2.1 Founding

Ordnance Survey was established in the British Isles in 1791 as a military organization and was housed in the Tower of London. It was created to meet a pressing need at that time. This was in the defense of the nation against the invading forces of Napoleon. Mapping of Kent in the southeast part of the country at 1 inch to 1 mile scale (1:63,360) was undertaken and this contributed to Napoleon remaining firmly on his side of the English Channel (Figure 21-1). Thus the first application area (military) was established and so too was Ordnance Survey. From 1824 6-inch (1:10,560) surveys were commenced in Ireland, and a new geodetic survey known as the Principal Triangulation was completed. New more-detailed mapping commenced at 25 inch (i.e., 1:2500) from 1858 onwards. In urban areas, with populations of over 4,000 people, surveys at 1:500 scale were also incorporated. Here the level of detail was exacting and even included interior features of some buildings. Extensive research and debate (lasting 20 years!) was undertaken before the scales above were selected. Military personnel staffed the organization and hence labor was to some extent readily available. Consequently the entire British Isles was mapped and contoured in great detail by the 1890s. A major revision of the mapping was well advanced when the First World War broke out and much of the work was then put on hold as resources were required elsewhere.

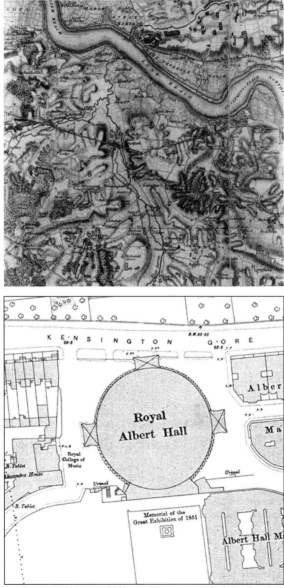

Figure 21-1 "One-inch" map of Kent (1801) and Town map (originally at 1:500 scale). See included DVD for color version.

21.2.2 Mapping and Land Information

The 1920s witnessed great post-war changes; in particular several new Acts of Parliament introduced legislation to establish a better infrastructure to support land registration, land valuation, local government and town planning. However during this period of intensified need for accurate mapping, required to underpin the new legislation, it was unfortunate that Ordnance Survey was in decline due to post-war economic cutbacks resulting in staff numbers being reduced to 1,000.

An accident of history also arose from the successful defeat of Napoleon in the 1790s. Some years later in the early 1800s Napoleon introduced a new system in France to raise taxes and ensure a stable flow of funds to the French state; this was known as the "cadastral system." This also introduced a way of referencing rights to land and land value, it underpinned taxation, and the principles were later extended to support the transfer of property in other countries. The model mutated and today there are several forms of cadastre where their functions vary (Burmantje 2005). The cadastral system concept spread quickly across the Empire but clearly it did not reach the British Isles. Here a different form of land registration developed to support property ownership and guaranteed tenure to land based on the "general boundaries" rule. Land and property valuation likewise developed its own path in Britain and Ireland. Consequently today different organizations are responsible for various aspects of land information and they have developed very effective and efficient mechanisms to execute and deliver their roles and responsibilities to society at all levels.

The lack of foresight within government in the 1920s became apparent several years later, in the early 1930s. The new legislation was placing a demand on mapping, which through lack of investment was unable to meet expectations and requirements. In 1935 a fundamental review of national mapping in Great Britain was commissioned which made its recommendations in 1938 (Ordnance Survey 1938). Still today, some of the recommendations appear to be farsighted. While the move to a single map projection may have been an obvious step (this is known as the "National Grid" today), the grid was to be based on the metric system and this was a major decision that framed a new approach to map referencing. A retriangulation (Harley 1975) was followed by new mapping of cities and towns (1:1250), adjusted and revised rural mapping (1:2500), and new mapping of uncultivated areas (1:10,560, though this was an anachronism that was changed a couple of decades later to 1:10,000). The decision to map rural areas at 1:2500 proved to be a false economy and these areas have recently been resurveyed to better meet positional accuracy needs and support consistency with GPS and digital orthoimagery. This review established the parameters for the basic topographic and geodetic base that we are still evolving today.

21.3 Political Dimension

The 1920s saw the independence of the Republic of Ireland from 1922, with six counties remaining within the United Kingdom, forming the province of Northern Ireland. This led to the formation of two fully independent sister organizations in addition to Ordnance Survey (which concentrated on Great Britain, i.e., England, Wales and Scotland). These are "Ordnance Survey Ireland" and "Ordnance Survey of Northern Ireland." While independent, the three separate organizations still maintain a close working relationship (Murray et al. 2002) and work together under a Memorandum of Understanding.

This political event also resulted in the splitting of associated land information functions across the different regions and today in addition to the three mapping agencies there are four land registry bodies (land law is also different in Scotland than in England and Wales) and land valuation bodies. Within the last decade devolution in Wales and Scotland has led to the establishment of the Welsh Assembly and in Scotland a Parliament which is responsible for enacting its own legislative program. However, some legislation and policy remains applicable to the whole of the United Kingdom. Ordnance Survey is responsible for the survey and mapping of England, Wales and Scotland.

Concurrently co-operation within Europe has been getting stronger and the growing number of European directives requires a pan-European GI infrastructure. Legislation originating within the European Parliament, once adopted, requires transposition into legislation within member states within a given period, for example the Water Framework Directive (WFD) and the Public Sector Information Directive. The WFD seeks to improve a wide range of environmental factors across Europe and is a major piece of legislation in its own right. The plans to build a European Spatial Data Infrastructure (ESDI) have been developing for several years. At the time of writing (2005), this is encapsulated within the INSPIRE legislation and is going through the European Parliamentary procedure to provide a framework (INSPIRE undated) whereby relevant GI can be integrated and presented in a consistent way to support directive reporting, e.g., river quality at sample points as part of the WFD.

21.3.1 From Local to Global

The requirements to improve democracy at the local level, e.g., through the devolution of power to Wales, Scotland, and Northern Ireland, introduces smaller national units of operation. At the same time there is also the parallel consolidation of member states into the European Union. This is forcing the need for greater data sharing between local councils through to the European Commission. This places greater demands on re-use and information sharing across all these communities. It is within this context that the now wholly civilian Ordnance Survey operates and responds to the ever-changing needs and demands.

At the national level, government is now much more alive to the benefits and power of GI. This applies across all parts of government from local authorities who are big users and creators of information to central government bodies such as the Office for National Statistics (ONS), the Department for Environment, Food and Rural Affairs (DEFRA), the Land Registry and Registers of Scotland (RoS), the Department for Transport (DfT), Office of Deputy Prime Minister (ODPM), the Environment Agency (EA), Scottish Environmental Protection Agency (SEPA), Welsh Assembly and many more. However while GI is important to all these organizations, realistically it is a means to an end, a way of getting the job done more effectively in delivering their mainstream services. This can place a major constraint on what can be done and how fast changes in methods to support better interoperability can be made.

Although there is a very strong and active market for GI in Great Britain there is no overall body responsible for the coordination of geographic information at this level or across the United Kingdom. The establishment in 2005 of a Geographic Information Panel to advise government on geography issues (GIP 2008) may be a step in that direction, but more can also be achieved at the organizational level if there is an agreement to work towards common goal(s).

21.4 Economic Position

From the deliberation of the scales of mapping in the 19th century, to the financial cutbacks in the 1920s and to the present day, debate has persisted about the ongoing costs of maintaining a robust national reference base. It is not necessarily the cost of creating the survey, which may have been undertaken several decades ago, but of its maintenance. This presents the main cost challenge since the economic and practical factors involved in locating and updating small distributed fragments of change and ensuring that the new information fits with the original survey, without degrading it, impose very different working parameters from an ab-initio survey. As a rule of thumb annual maintenance, *for the revision of just the most important changes (i.e., not all),* will cost something in the order of 40-50% of the original survey costs, ***annually***.

To avoid the stop/start problems associated with government budgets (e.g., the 1920s problem) and in recognizing the value of the information being collected, the sale of map-based products was established over one hundred years ago to support the income of Ordnance Survey. An example of early innovation in this field was an experimental aerial image mosaic of Salisbury in 1921 (Figure 21-2). At the time this image map cost 21 shillings (or about £1.10/€1.50/$1.90 in 2005 currencies). Publications attract their own production costs and margins are always tight. The costs of maintaining the detailed mapping and the database have always been expensive compared with all other operational costs across the organization. Today it is also recognized that a dedicated sales and marketing workforce is needed to promote and help users understand what the information is capable of and how they can get the most out of it.

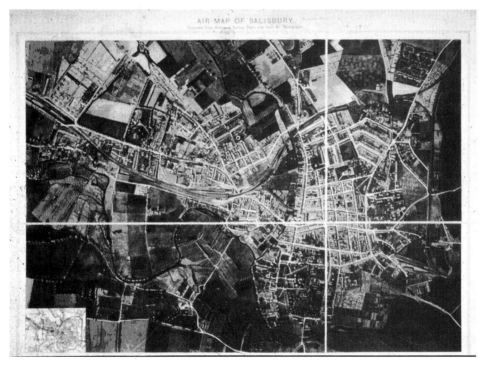

Figure 21-2 Air Map of Salisbury, Wiltshire (1921). See included DVD for color version.

Over the last 20-30 years the trend across government in the United Kingdom, almost irrespective of the party in power, has been to reduce the burden on the taxpayer. No one gets elected on a policy of raising taxes—or improving GI for that matter! Unlike many government departments Ordnance Survey has been able to market material that has value across the public and private sectors and the many application areas that transcend this economic divide such as transport, planning, land and property, etc. Many of the new initiatives within the commercial sector are often driven by legislation of one form or another such as the Traffic Management Act, Home Information Packs, and the Land Registry Act which enabled the first steps towards e-conveyancing.

The emergence of government Trading Funds in the United Kingdom over the last decade has supported the drive to reduce costs by making Trading Fund organizations wholly dependent on the income they earn and contracts they sign. They are also obliged to pay a percentage return on investment to HM Treasury.

As a consequence of this, organizations that operate under a Trading Fund model now have to better understand the needs of users. If they fail here, their users will and can seek alternatives elsewhere. Increasing competition through the use of lower-cost technology and labor sourced from the global marketplace is forcing government trading organizations to focus very clearly on meeting customer needs.

While this imposes massive pressure on the Trading Fund organization to respond in short timescales, to cut costs and to modernize, it does have a silver lining. This is the ability and to a certain extent the freedom to invest within given parameters, in developing the national dataset to meet modern needs. For Ordnance Survey it is unlikely that the funds to develop OS MasterMap would have been made available in the timescales recently achieved if Ordnance Survey had to compete with other government priorities on transport, health and education for example, even though the investment costs are trivial in relative terms (Figure 21-3). There is a variable contract to support a defined set of tasks that would not necessarily be undertaken if these were assessed purely in full cost recovery terms, e.g., update of mapping in rural areas, scientific work, national metadata service, etc. This operated under a separate contract known as the National Interest Mapping Services Agreement. (NIMSA).

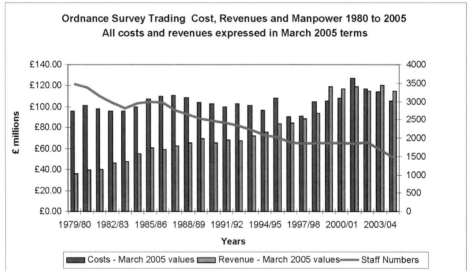

Figure 21-3 Ordnance Survey economic development 1980-2005 (adjusted to 2005). See included DVD for color version.

The Trading Fund model is not universally appreciated, though this is at least in part due to a lack of understanding. Within the model it is necessary to ensure a fair and balanced market for the information in which Trading Fund organizations operate. Checks and safeguards have been established by government, and voluntary schemes to support this have also been established, for example the "Information Fair Trader Scheme" (IFTS) operated by the Office of Public Sector Information (OPSI 2008). In addition, several new legislative instruments have been transposed into member-state statutes in recent years from overarching European legislation and these establish baselines across all sectors, not just for GI but for activities across all industries. The legislation includes European-wide Competition law, Access to Environmental Information, Copyright directive and more recently the Public Sector Information directive, operative from the 1st July 2005.

21.4.1 Formal Agreements

During the 1990s it was recognized that the task of negotiating and signing contracts with lots of individual organizations was a major task and the concept of a service level agreement was developed. The initial group to take this up was the local authority community. Consisting of over 500 different organizations, an overarching package was agreed which gave them better value for money in terms of maintained datasets provided. It also gave the organizations a stable agreement for three to four years thus minimizing the amount of effort required to put the agreement in place, and a welcome period of stability. This model was also adopted by the majority of utility companies and then by central government in the form of the "Pan-government Agreement" This accommodates over 550 different central government departments and organizations and has encouraged significant take up of GI by many of these organizations since 2002.

More recently at all levels of government, procurement has now moved on to a new level whereby services for the provision of national reference information are advertised by invitation to tender. This is obligatory for large contracts under European legislation. The renewal of the local authority agreement in 2004 initiated this approach in Great Britain, followed by the renewal of the Pan-government agreement for 2006 and more recently the European Commission has advertised for selected pan European datasets.

Funding and procurement methods are a rapidly evolving aspect in the provision of the national framework for GI. As our recent history has demonstrated several times in the past, it is critical to be able to provide fit-for-purpose information at the right time. Users of reference information have realized their buying power and their ability to use the market to obtain the information they need. This will inevitably further stimulate the information industry and while it may drive down costs, the potential downside is the possible lack of interoperability introduced by competing datasets and data providers at a time when this has been identified as critical to our information-hungry society.

It is also apparent that insufficient funding affects many other countries, making the national vision for a modern GI infrastructure often unaffordable. Hence it is vital that a healthy and equitable balance is maintained to ensure that national frameworks are maintained to support the economic prosperity of the nation on a fair and equitable basis. Costs and funding models are likely to become a larger issue in the future as the requirements to meet tomorrow's infrastructure needs in an information-dependent society will require more investment. This will be essential in meeting the need for a maintained level of industrial strength and reliability on a par with other information industries.

21.5 Applications and Users

21.5.1 Transition from the World of Paper Mapping

In the paper map world the professional users had two choices. Either they drew their information on the paper map or they made a copy of the map and added their information to it. With the first option there was a limit to what they could add before the map become a mass of lines and ultimately illegible, while the second option offered greater flexibility, for example the user could change scale at this stage or deselect some detail. This was of course a very labor-intensive process. When digital mapping was introduced from the 1970s users tended to replicate their paper processes, but in a digital way. Everyone knew there was a bigger and brighter vision for GI (Ordnance Survey 1974) but the technology and costs of achieving it, as well as the level of skill required, made it all largely prohibitive. That is until recently.

Labor costs have gone up in the developed world while in the wider global marketplace skills have risen, but labor rates there have not yet caught up. At the same time technology costs have gone down everywhere. All of this has changed the equation significantly; and after several decades of gradual development we are now much closer to truly realizing the potential of GI, probably for the first time. This itself will take several years and will continue to evolve. The move from paper mapping to digital mapping took almost 30 years in Great Britain for the vast majority of organizations to migrate and longer in some other countries. The transition from digital (backdrop) mapping to GI should take less time but it would be unrealistic if we expected it to take less than a decade. The length of the period will be determined by many factors but ultimately it will be driven by the market and by user needs. It will not necessarily be driven by datasets or new products from systems vendors; if these organizations have the vision that matches users' needs in the future they will have the potential to be successful as enablers.

Likewise, users and data providers will adopt new methods and practices when it is appropriate for them to do so (Figure 21-4). Costs, skills and investments need to be matched against organizational business objectives in both the public and private sectors. If an opportunity to invest in a new process or new data is missed, the chance may not arise again for another five to seven years. Fortunately there are always some organizations that have the vision to successfully pioneer by exploiting the benefits and drivers to:

- Save current operating costs
- Offer a better service
- Offer a new capability

These are the leaders who push back the boundaries, and others follow.

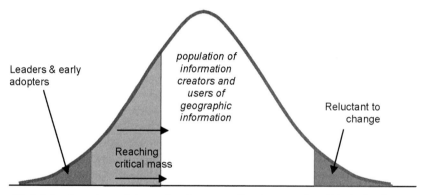

Figure 21-4 Adoption of new information & new methods across the GI industry. See included DVD for color version.

21.5.2 Paradigm Shift (to Geographic Information Flows)

The ability to link information about an object, and to understand its characteristics no matter which organization maintains and holds the data, is a paradigm shift driven by the need to maintain accurate up-to-date information as well as reduce overall costs (and wasteful duplication). In the past the hardcopy map base was used as a fixed "publication" for a period of months or years. It provided a reference for user information such as their geographic "view"— for example, a cadastral parcel, planning zone, postal district. This would have been overlaid on to map base. The information world is driven more by information processes and events along the lifecycle of a process (e.g., salary deposited into bank account, mortgage paid from bank account). As an example of a process we can take the development of a new property where someone has an idea to build a house on a piece of land (Table 21-1). This demonstrates at a high level who is involved in maintaining records that are of geographical relevance.

Table 21-1 Cross-organizational information flows in planning, developing and occupying a property in Great Britain.

Lifecycle stage	GI action of some kind by:			
	Local Authority	Central Government	Utility	NMA
seek and obtain planning permission,	Outline planning			
buy the land,	Search	Register		Survey
design the building and grounds,	Detailed planning		Plan capacity and assets	
construct the building(s)	Building	Temporary address		
request utility services are connected up			Plan & Connect services	
someone moves in and occupies the property		Value property		
the property is registered in the land registry and/or cadastre	Assign address in consultation with other bodies	Register the property		Survey the property & link the address
receive post and deliveries immediately	Use the official address including other deliveries			
pay tax on the property and/or land	Recover tax			
navigate to/from the property	Use the address and/or other information to navigate			
the owner/occupier might sell and move out – this will involve marketing the property, searches, surveys and then sales and alteration of the land register/cadastre and someone else will move in (address and property extents may change e.g. conversion to apartments etc.) and so on until the property is demolished and the land is reused	Amend records	Amend records	Amend records	Maintain records

Processes flow through organizations in logical sequences of events involving geography at regular intervals. This can be seen in the case of a new development (which land does the planning application refer to? what kind of development is permitted in this area? which neighbors are affected? are there utility services nearby and do they contain sufficient capacity? Then there is a need to issue a plot number for the property, change this to a postal address when it is built, deliver services, e.g., refuse collection when the property is occupied, deliver post, etc.). Much of this was done in a paper map world but the steps were slow and there was massive duplication. This in itself was not a major problem when labor costs were low, but today duplication is unaffordable and worse still it is a major obstacle to intelligent data sharing and accuracy.

The concept of a reference base remains in the new paradigm – *indeed in a world where many players are exchanging information about the same object, this is more important than ever it was.* For operational and maintenance purposes there is a need to conceptually separate the "reference base information" (such as a building or address) from the "user's/application" information (such as who lives here? what is the water temperature? what was the traffic flow at 10 A.M. ? etc.). Another element of the paradigm shift is the change from direct referencing by coordinates to one of referencing and attaching user information to objects recognized easily and universally by users, such as farms, property parcels or even individual features such as a building or segments of highway. The process flow also affects the way that the reference base is maintained, as we will see later. This introduces another (temporal) dimension in the update of GI information since the stages of the lifecycle (design, as-built,

Manual of Geographic Information Systems

modification, demolition, etc.) of the entities being updated will now need to be taken into account.

Everything happens somewhere and generally some people may know about it, and if it's important many more people need to know about it. *How we share information in future and make a step change in the use of GI, while keeping the costs within what is affordable, is the most interesting and challenging development in the industry at the current time—indeed it lies at the heart of the use of GI.*

Users expect high quality and low cost, and while these expectations are generally mutually exclusive, this need not be the case. With the correct balance of funding and user rights management it is possible for the citizen to access detailed information, e.g., in support of local community services such as planning. Work within the Open Geospatial Consortium (OGC) is currently investigating a framework for Digital Rights Management in GI (Vowles 2005) to support this in the future where transactions are undertaken in greater number using e-business channels.

21.6 The Information Infrastructure

21.6.1 Roles of Players in the Use of GI

As we have seen, in addition to the national mapping agency in Great Britain, many other organizations create or use GI to a greater or lesser extent. Generally these fall into one of three groups as shown in Figure 21-5. Some organizations may perform more than one role.

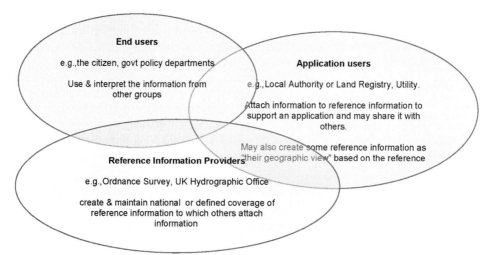

Figure 21-5 The roles of participants in the GI industry in Great Britain. See included DVD for color version.

21.6.2 Roles of Providers and Users in GI in Great Britain

In Figure 21-5, it is clear that there is some overlap between organizations. For example the Land Registry (LR) will use the reference information maintained by Ordnance Survey. However the LR has a specific "geographic view" it requires for its business, i.e., the property extent. This is referenced to the map base, but can and does (within LR) provide another level of reference information to which ownership and mortgage information can be attached. This is also common in local authorities who develop land information systems (land terriers), in environmental departments who manage rivers, for example, and many others. While some agencies may make their information available for others to use (e.g., statistical Output Areas) caution must always be exercised since what may be fit for purpose for one (internal) application may invariably not be suitable for someone else, who could use it for a different purpose altogether (Figure 21-6).

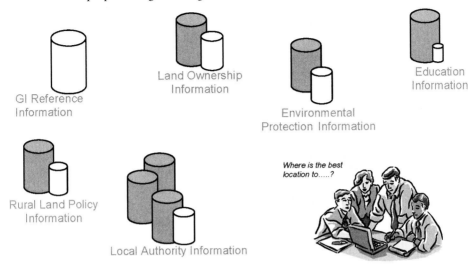

GI Reference
Information

Land Ownership
Information

Environmental
Protection Information

Education
Information

Rural Land Policy
Information

Local Authority Information

Where is the best
location to.....?

Figure 21-6 Distributed geographic information.

Note that the percentage of GI to business information will vary but it is generally a means to an end, rather than the end goal for a business in its own right.

21.6.3 The Digital National Framework

In the late 1990s after several technical research trials and challenging business cases, Ordnance Survey took a step back to visualize where the GI market in Great Britain was heading and what its own role should be. This led to a refocusing back to the core business and a withdrawal from "value add" projects such as co-publications and new information services (for example, the National Land Information Service) which could, and sometimes did, compete with commercial sectors players. In discussion with industry leaders the concept of the Digital National Framework (DNF) emerged. The OS MasterMap Topographic layer was the first tangible manifestation of the development of DNF and one upon which other components of the framework could be built and referenced.

The aim of DNF is to develop "an industry standard for integrating and sharing business and geographic information from multiple sources." The objective is to minimize duplication and promote reuse (definitive datasets) by using open architectures and a structured approach through common methods of object referencing. This is in response to users who need to derive reliable results from reliable information based on attainable cost-effective solutions, while ensuring flexibility to meet the needs of different communities and/or applications over time.

DNF works by tracking and encapsulating best practice within a coherent framework. The methods include referencing/linking information supported by a range of tools and services (e.g., feature cataloging) so as to accelerate the use of GI fully integrated with the growing information industry. The release of OS MasterMap did not immediately realize a host of new applications, but eventually applications started to emerge, based on business re-engineering (DNF 2005 (see case studies); Murray et al. 2005) and not simply an extension of the digital mapping paradigm. This in turn has led to further development of the DNF model based on proven practice and industry experience to deal with the inevitable issues that arise and this is expected to continue into the future.

DNF is guided by a small number of basic principles (DNF 2005) promoting reuse, recording information at the highest resolution possible and using generalization to simplify data for lower resolution publication. DNF is also supported by a growing documented architecture based around the view that each user "has their own view of the world" and a common geographic reference base enables these views to be joined up. This model (Figure 21-7) ensures that a level of data integrity is attained which is essential for reliability and automated industrial level applications. The user's "view" of the world is represented as a geographic object which can in turn be referenced or cross-referenced to existing objects. The documentation is being developed "by users for users" and will incorporate proven formal standards as appropriate such as those of the International Organization for Standardization (ISO), the European standards body CEN, the Open Geospatial Consortium, and others.

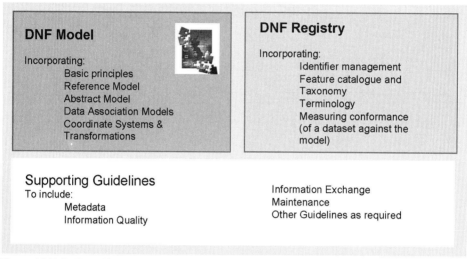

Figure 21-7 DNF Architecture under development (October 2005).

At the heart of the DNF model is the method of data association. This is performed using unique identifiers and simple cross-referencing to enforce referential integrity. In OS MasterMap the identifiers are known as TOIDs, but it's possible for users to create their own "view" using collections of the base topographic objects and assign that object an identifier as well (or they may already hold an identifier in their text-based database). Of course not all user views can be accommodated by aggregations of the objects in the national topographic database and additional geometry is often required. This is treated in the same way as any topographic object and included to form the geographic view that meets the user's needs. The techniques to support these methods are currently being developed by the DNF Expert Group. This is a cross-industry group whose members are all experienced in geographic referencing and applying DNF methods.

In Figure 21-8, Organization A references a property via several real world objects in the reference base. These references are passed to Organization B and the extent referenced in the same way ensuring perfect data integrity. Organization B then records occupation details against the property and passes this back to Organization A (or any other participating organization). This minimizes the traffic in geographic objects and promotes reuse of definitive data and integration with mainstream information services.

Downstream, it may be possible for systems integrators and developers to use the unique identifiers to access several objects from different organizations and to execute processes based on their guaranteed fit with a common base. In the meantime a major benefit for the user is maintenance of their own view with the underlying topographic base, the improvements in data integrity and the ability to share information with a level of confidence. This is not to say that all problems have been solved but with the pragmatism brought by industry

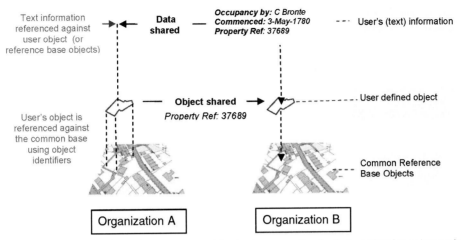

Figure 21-8 Data sharing using common object references. See included DVD for color version.

practitioners there is a level of optimism that already the current solutions will meet a high percentage of current and immediate needs.

21.6.4 Ordnance Survey and OS MasterMap

Ordnance Survey has committed itself to establishing the foundation layers of the DNF model; the 2005 status of OS MasterMap is summarized in Table 21-2. Each of the detailed datasets is being migrated to form a set of interoperable layers (Figure 21-9) of reference information using data association where applicable.

Table 21-2 Status of OS MasterMap in 2005.

OS MasterMap dataset	Availability	Status and Linkages
Topographic Objects	From 2001	430,000,000 objects under seamless maintenance
Integrated Transport Network – Highways	From 2002	6.5 million features, 3.5 million road links and 2.9 million nodes. The features are cross referenced with the appropriate highway objects in the topographic layer.
Address	From 2002, under further development	26,000,000 postal addresses cross referenced to respective topographic building objects
Imagery	From 2003, full cover of Great Britain by 2006	High resolution color orthoimage geometrically interoperable with other layers.
Digital Terrain Model	To be migrated post - 2007	Variable accuracy model based on a new photogrammetric survey (replacing existing data) and LIDAR data available from 2005. It is geometrically interoperable with other layers.
Boundaries (administrative and electoral)	To be scheduled	

Figure 21-9 Example of OS MasterMap, illustrating the geometric interoperability of the layers—Topography, Addresses, Integrated Transport Network-Roads and Orthoimagery. See included DVD for color version.

At the time of writing the internal database within Ordnance Survey is also being redeveloped to support an integrated maintenance and storage model along with a job management system to help better support update and other programmed tasks. For this reason no further layers will be added to OS MasterMap until the completion of this major investment, scheduled for 2010.

Further research is also ongoing in developing procedures to derive smaller scale publications by automated means. For the foreseeable future it is not expected that this will be fully automated and intermediate databases will be necessary for several years. This is an interesting illustration regarding the vision of the potential of GI established in the 1970s, where it was expected that automated derived mapping was to develop quickly, but has taken rather longer!

21.6.5 National Initiatives

Several new national initiatives are at early stages of development currently. This includes the National Spatial Address Infrastructure, which aims to bring together the different address views maintained by several key organizations (Royal Mail, Ordnance Survey, Valuation Office and local authorities). The National Underground Asset Group (NUAG undated; Ordnance Survey 2005) has been formed to develop and support a common infrastructure to join up underground plant and assets. Work has continued on developing the integrated coastal zone initiative (SeaZone Solutions undated) and is now looking at an integrated hydrological model to meet the needs of the next 10-20 years. Collaboration across government is essential in all this work. Each of these initiatives seeks to employ the DNF model and ultimately link the information to the OS MasterMap Topography layer as a common reference base.

21.6.6 Collaboration and Applications

There are several examples of users identifying ways of solving immediate business problems by employing the new infrastructure offered by DNF and OS MasterMap.

Highways – in 2003-04 Oxford County Council re-engineered its highway referencing from an isolated model to the new framework by cross referencing its highway attribute information and linking this to the identifiers in OS MasterMap Integrated Transport Network (ITN). This allows the transfer of attributes to other organizations across government and beyond, using the same base, regarding highway information.

Land and Property – Dudley Metropolitan District Council has re-engineered its twenty-year-old property terrier to reuse the OS MasterMap topographic objects and has developed a method of dynamic building of the property extents from references when required. (Roberts et al. 2005). Dudley's GI is now incorporated in a corporate information service and reaches almost 3,000 users across the council. Pilot work with 100,000 property parcels with the Land Registry has demonstrated that it is possible to automatically re-engineer and incorporate user-defined geometry to improve data integrity and data sharing capabilities.

Environmental – The largest program to date, using the DNF area referencing model, was undertaken by the UK office of Black & Veatch for the Countryside Agency in referencing and registering 1.3 million hectares of land now open to the public for access under the Countryside and Rights of Way Act 2000. Open access land was given a collection identifier and referenced those topographic objects (TOIDs) contained within its boundary. To this, a comprehensive audit trail was attached comprising all records from the public consultation process.

21.6.7 Europe and International

The European scene has also been developing rapidly since the turn of the century. The INSPIRE initiative is now European law and this will be transposed into member states by May 2009. A supporting work program to develop the European Spatial Data Infrastructure (ESDI) is commencing concurrently and is expected to develop in stages. This will start with the development of Implementation Rules but we can expect it to take several years to establish the many components envisaged in the work program.

To meet a key component of this need the mapping agencies across Europe have developed plans to establish a pan-European reference base known as EuroSpec. (EuroGeographics 2008). This—like INSPIRE—is a major, if not massive challenge, but early work has shown that such collaboration is possible through the (partly manual) creation of the Seamless Administrative Boundaries of Europe (SABE) dataset and more recently the first steps towards pan-European datasets at 1:1 million (EuroGlobalMap) and 1:25,000 (EuroRegionalMap)

The European mapping agencies are also supported by a jointly funded and supported spatial data research body EuroSDR (www.EuroSDR.net). A broad spectrum of research is supported by a rolling research program and is undertaken in the shape of workshops and projects. The aim is to share knowledge and advance the state of the art across members on all kinds of GI development from capture to dissemination.

21.7 Summary

It is clear that mapping in Great Britain has developed significantly from the days when it created its first map series to defend the nation against invasion, to today where we are working towards harmonization of our information with other European countries.

Contrary to popular opinion, spatial data/information infrastructures are already established in most countries, though these will vary significantly from one country to another. The origins will initially have been driven by historical events but are today more influenced by the political, and economic positions that the organizations find themselves in and the room for maneuver that senior management enjoy in responding to or even influencing market trends.

The increased level of awareness and globalization is exerting pressure for nations to achieve more and doing this for less. It is also true to say the transition from "offering what is available" to "determining what the market needs" is also accelerating within mapping agencies.

Following on from this, the information society demands solid, industry strength services underpinned by GI. This is the major challenge of the next decade, and is now being recognized as the paradigm shift we all have to make.

In Great Britain the development of the Digital National Framework in which several key players are participating, underpinned by OS MasterMap, creates a family of users working together for the common goal of reuse and easily integrated information. The DNF seeks to exploit the paradigm shift from mapping to mainstream information and information processes.

Adopters demand short- to medium-term pay back, as well as assurance that it will better position their businesses in providing definitive information and data sharing for the future. There is evidence that this is happening, but the transition will take time as investment cycles run in periods of 5 to 7 years or more. External developments also stimulate change and this can include anything from legislation to better managed resources (e.g., water across Europe) to responding to disasters such as the tsunami in 2004 and hurricane events in 2005.

The knowledge of "what is where" and how it is connected to other information is vital in the cycle of planning and prevention, in responding to emergencies, in site restoration and in setting future policy. Here for the future there are many new opportunities and possibilities for national databases, whoever manages them and however they are distributed.

And finally, in the paradigm shift for geography towards a mainstream information world many things will change, not least that people will need joined up information, including geography – this does not necessarily mean mapping as we know it today.

References

Burmantje D.A.J. 2005. Spatial data infrastructures and land administration in Europe. In *Proceedings of the GSDI/FIG Working Week* held in Cairo, Egypt, 16-21 April 2005. Available on CD. <http://www.fig.net/pub/cairo/papers/ps_02/ps02_03_burmantje.pdf> Accessed 6 October 2008.

Digital National Framework (DNF). 2005. Welcome to the Digital National Framework. <http://www.dnf.org> Accessed 6 October 2008. See Case studies from Dudley Metropolitan Borough (Land & Property) and Oxfordshire County Council (Transport).

EuroSDR. Undated. EuroSDR – Spatial Data Research. <http://www.eurosdr.net> Accessed 21 October 2008.

EuroGeographics. 2008. Welcome to the EuroGeographics Website. <http://www.eurogeographics.org> Accessed 6 October 2008.

GIP (Geographic Information Panel). 2008. Welcome to GI Panel. <www.gipanel.org.uk> Accessed 21 October 2008.

Harley, J. B. 1975. Ordnance Survey Maps, a descriptive manual.. Southampton, UK: Ordnance Survey.

INSPIRE. Undated. INSPIRE Directive. <http://www.ec-gis.org/inspire/> Accessed 6 October 2008.

Masursky, H., G. W. Colton and F. El-Baz. 2004. Scheme for preparation of lunar maps. Appendix C in *Apollo over the moon, a view from orbit.* <http://www.hq.nasa.gov/office/pao/History/SP-362/app.c.htm> Accessed 5 October 2008.

Murray, K. J., C. Bray and T. Steenson. 2002. Better connected – the three Ordnance Surveys improve georeferencing links. *Proceedings of the 2002 Association for Geographic Information Conference,* held in London 17-19 September 2002. London: The Association for Geographic Information.

Murray K. J., B. Munday and I. Bush. 2005. Enabling information integrity within SDI's – the digital national framework concept. In *Proceedings of the GSDI/FIG Working Week* held in Cairo, Egypt, 16-21 April 2005. Available on CD. <http://www.fig.net/pub/cairo/papers/ts_21/ts21_01_murray_etal.pdf> Accessed 6 October 2008.

NUAG (National Underground Assets Group). Undated. Welcome. <http://www.nuag.co.uk/> Accessed 21 October 2008.

Office of Public Sector Information (OPSI). 2008. The Information Fair Trader Scheme. <http://www.opsi.gov.uk/ifts/> Accessed 6 October 2008.

Ordnance Survey. 1938. Final Report of the Departmental Committee on the Ordnance Survey. London: HMSO. 39 pages.

Ordnance Survey. 1974. Report on The Feasibility Study into the Restructuring of Ordnance Survey Digital Map Data by PMA Consultants Ltd. 160 pages.

Ordnance Survey. 2005. Building a framework for buried services. <http://www.ordnancesurvey.co.uk/oswebsite/media/news/2005/sept/utilities.html > Accessed 21 October 2008.

Roberts, P., L. Ratcliffe and B. Higgs. 2005. Creating a change management driven approach to ensure data integrity in a dynamic corporate management environment. *Proceedings of the 2005 Association for Geographic Information Conference,* held in London 8-10 November 2005. London: The Association for Geographic Information.

SeaZone Solutions. Undated. Marine Geographic Information Solutions from Instrument to Desktop. <http://www.seazone.com/index.php> Accessed 21 October 2008.

Vowles, G. 2005. Geodigital rights management - the conceptual model, and Ordnance Survey Pathfinder and Magnesium Projects – Progress and Results. NMCA's and the Internet II, Joint Workshop of EuroGeographics and EuroSDR, Electronic Delivery and Feature Serving, held Frankfurt, Germany, 23-25 February 2005.

Other Sources

Ordnance Survey. 2004. Ordnance Survey Framework Document. 32 pages. <http://www.ordnancesurvey.co.uk/oswebsite/aboutus/reports/> Accessed 6 October 2008.

Ordnance Survey. 2005. Annual Report and Accounts 2004-05. 72 pages. <http://www.ordnancesurvey.co.uk/oswebsite/aboutus/reports/> Accessed 6 October 2008.

CHAPTER 22

Analyzing and Visualizing 60 Years of Forest-cover Change in Northeast Kansas[1]

Matthew Dunbar

22.1 Introduction

One common form of spatio-temporal GIS (geographic information system) analysis involves investigating landcover change. In its simplest form, landcover change analysis can compare two dates of information, providing a "before and after" look at one or many landcover types. As the temporal and spatial resolution of the landcover data layers increase, all aspects of the research simultaneously increase in complexity. Throughout the stages of data collection, analysis and finally communicating results, multi-temporal high-resolution landcover change analysis has many issues to consider. This section presents one research study to illustrate several of the problems and potential solutions involved in such a spatio-temporal GIS investigation.

22.1.1 Research Summary

Large regions of prairie historically covered the American Midwest, but this landcover has been significantly altered (Whitney 1994). Prior to European settlement, habitats within the prairie's eastern ecotone were a mixture of tallgrass prairie and eastern deciduous forest. A combination of environmental factors, including soil type, topography, fire and climate, largely dictated the historical dominance of one cover type over another (Anderson 1990). Human interactions with the landscape have since affected the balance of these controlling variables and also the stability of plant communities in the region (Fitch and McGregor 1956). We now understand that, throughout the prairie-forest ecotone, woody species can invade grassland habitats that are not burned, grazed, cultivated or mowed (Holt et al. 1995), and Abrams et al. (1986) have suggested that forest expansion into the grasslands of this region has occurred within the last 100 years. This environment was selected for study because of an observed need for further detailed landscape level analysis documenting the structural patterns of change. For this study, aerial photography collected approximately every 10 years from 1940 through 2000 was used to map, analyze and visualize the changes in forest cover within two study areas located in the prairie-forest ecotone of northeastern Kansas.

22.1.2 Study Area

The USGS 7.5 minute Midland topographic quadrangle (quad), located in northeastern Kansas, falls within the tallgrass prairie-eastern deciduous forest ecotone (Küchler 1974). The quad is located at the convergence of Douglas, Jefferson and Leavenworth counties. The first study area focused on the 24 complete 1-mile square Public Land Survey System (PLSS) sections that make up the southern half of the quad, covering approximately 62 km² (Figure 22-1). Land use in this area is highly varied. While it is primarily a rural landscape with a great deal of pasture and cropland, the area also includes an ecological reserve, home sites, businesses, an interstate highway and a municipal airport.

---

1. This chapter is a modified version of the author's Master's Thesis: Dunbar, M. 2005. Mapping, Analyzing, and Visualizing 60 Years of Forest-cover change in Northeast Kansas. M.A. Thesis, University of Kansas, Lawrence, Kansas, 136 pp.

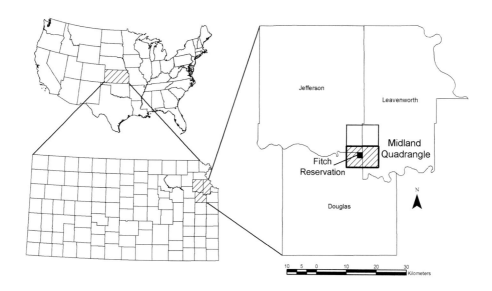

Figure 22-1 Study areas: Southern half of the USGS 7.5 Minute Midland Quadrangle (62 km²) and Fitch Natural History Reservation (2.6 km²).

For over fifty years, the Fitch Natural History Reservation has been maintained as a long-term ecological succession site for studying natural history, ecology, and natural succession patterns (Fitch and McGregor 1956). The presence of the Fitch Reservation at the center of the Midland quad study area provides a unique opportunity to compare landscape level forest-cover changes to that found within a local environment where human impacts have been excluded. To make this comparison, a second study area was defined as a single PLSS section (Section 4, Township 12S, Range 20E), 2.58 km² in size and composed almost entirely of the reservation.

22.1.3 Outline

The purpose of this research was to provide a thorough description of 60 years of forest-cover change within the ecotone or transition zone between the tallgrass prairie and eastern deciduous forest. The main research has been divided into three phases, covered in Sections 22.2 through 22.4. For each phase, relevant background information is reviewed, the specific goals and objectives of the study are outlined, and the study methodology and results are presented. Section 22.2 describes how aerial photography, collected approximately every 10 years from 1940 through 2000, was used to map changes in forest cover. Section 22.3 discusses how landscape metrics and comparisons between slope and forest occurrence were used to create a quantitative description of the changing spatial patterns of forest cover over the past 60 years. Section 22.4 demonstrates how advanced realistic visualization techniques were used to render a variety of images and animations depicting forest-cover change. To conclude this chapter, Section 22.5 summarizes the research and its relevance to spatio-temporal GIS studies.

22.2 Creating Data

In order to draw the most complete picture of landcover change, a data set should be chosen that extends back in time as far as possible and provides enough detail to adequately address the research goals. Historical aerial photographs have long been recognized as an important source of information for studies of vegetation dynamics due to their high spatial resolution and long temporal coverage, stretching back to the 1930s (Dunn et al. 1991). While mapping vegetation change with aerial photography is quite common (Iverson 1988; Simpson et al. 1994; Loehle et al. 1996; Boettcher and Johnson 1997), several studies have specifically looked at aerial photography to aid in the study of landcover fluctuations in ecotonal vegetation communities (Baker et al. 1995; Mast et al. 1997).

Traditional methods for classifying historical air photos have merit for certain applications, but they are limited in several ways. With high-resolution panchromatic data, manual classifications of landcover boundaries are accurate but time-consuming for all but small areas. In contrast, pixel-based digital classification techniques are fast but often less accurate than manual methods. Examining these digital classification techniques shows that many still rely on concepts developed in the early 1970s, and they do not take into account the spatial context of the pixels (Blaschke et al. 2000). When looking at high-resolution imagery, pixels near one another are more likely to belong to the same landcover class. This concept leads to the conclusion that investigating images as groups of similar pixels, or objects, would be more meaningful than examining images one pixel at a time (Blaschke and Strobl 2001). The idea of segmenting images into homogenous regions for the development of meaningful image objects is not new (Haralick et al. 1973), but due to technological advances object-oriented classification is now gaining acceptance in the field of remote sensing as a more efficient method (Benz et al. 2004; Hall et al. 2004; Walter 2004).

With any remote-sensing-based study of landcover change, the accuracy assessment of classification results is an integral analysis step (Congalton and Green 1999). The need for a highly accurate classification is even greater when the intended application is the analysis of landscape pattern using spatial metrics (Shao and Wu 2004). When a classification is derived from a recently collected remotely sensed data set, ground sampling can be used to acquire reference data to validate the accuracy of the classification (Jensen et al. 1997). However, with a classification derived from historical aerial photography, recently collected field data would not provide a meaningful reference source for validating the classification. If pre-existing classifications for each study year are not available, another valuable approach involves using other skilled air photo interpreters to verify the classification results (Skirvin et al. 2004).

The goal for this phase of the research was to use the historical landcover record captured by aerial photography over the past 60 years to accurately map forest-cover change in a 62 km^2 area in the prairie-forest ecotone of northeastern Kansas (southern Midland quad). While manual classification of aerial photography is an accepted standard for creating such a data set, this project focused on developing a more efficient but equally accurate classification procedure. Object-oriented classification was selected as the most appropriate method to match the spectral and spatial qualities of this aerial imagery and to meet the needs of the landscape pattern analysis and visualization phases of the research. To achieve the main goal, four specific objectives were outlined:

- Locate historical aerial photography and convert the prints to a digital format;
- Process the imagery data to create a completely mosaicked and georeferenced digital data set appropriate for classification;
- Develop and implement an object-oriented classification scheme for extracting forest cover from the historical aerial photography;
- Carry out an accuracy assessment to determine how well the classification process performed.

Manual of Geographic Information Systems

22.2.1　Locating Photography

Imagery completely covering the southern half of the Midland quad was acquired from 1941, 1954, 1966, 1976, 1991 and 2002. All aerial photography used was taken in the summer with leaf-on conditions. Black and white digital orthophoto quarter quads (DOQQs) from 1991 and color infrared imagery from 2002, both at 1-meter resolution, were available in a pre-processed digital format from the Kansas Applied Remote Sensing (KARS) program's archives. The historical data from 1941, 1954, 1966, and 1976 were located as hardcopy black and white photographs archived at the Douglas, Jefferson and Leavenworth County Planning and/or Farm Service Agency offices. These photographs were scanned at a resolution sufficient to match the 1991 and 2002 imagery.

22.2.2　Processing Imagery

The 1991 and 2002 imagery were acquired in a rectified format, requiring no additional processing. The more recent historical photographs, from 1976 and 1966, were geo-rectified using the 1991 DOQQ as the reference image. The older historical photographs, from 1954 and 1941, used the newly geo-rectified 1966 imagery for reference. This shift in the reference imagery for the rectification of the older photography was done because it became increasingly difficult to identify ground features in the photos, such as roads and buildings, in common with the 1991 DOQQ. Once rectified, the yearly data sets were still made up of multiple individual images. Before classification, the images were mosaicked together into a complete data set to reduce mismatch along the edges of images during classification. The mosaicking process resulted in six georeferenced aerial photography data sets at 1-meter resolution from 1941 to 2002 with an approximate temporal interval of 10 years (Figure 22-2).

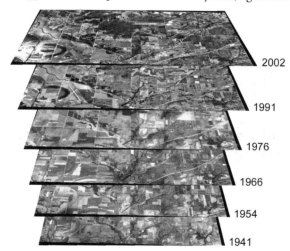

Figure 22-2 Georectified aerial photography of the southern half of the USGS Midland Quadrangle.

22.2.3　Forest Classification

The classification approach used for this study was a simple binary forest/non-forest scheme to focus on the research goal of describing changing forest extent. All tree species were grouped together under the general landcover class of forest. All remaining landcover was classified as non-forest. In reality four forest or woodland alliances, of the 40 vegetation alliances of Kansas (Lauver et al. 1999), encompass nearly all of the woody habitat in this area. By far the most dominant forest-cover alliance is the Oak-Hickory Forest. The other common forest-cover types include the Mixed Oak Ravine Woodland, Mixed Oak Flood-plain Forest, and Ash-Elm-Hackberry Floodplain Forest.

This research used commercial object-oriented classification software to extract a multi-temporal forest-cover/non-forest-cover data set from the six dates of aerial imagery. To begin, each image was segmented into appropriately sized image objects, designed to capture a range of woody habitat on the landscape from large forest-covered areas down to individual trees (Figure 22-3). Each segmented image was then classified with a clustering approach that groups objects together based on the similarity of their spectral mean and standard deviation. Manually interpreted training objects were used to direct the clustering, resulting in the classification of all image objects as forest or non-forest cover. In order to create a more accurate record of forest cover, the last stage in the classification process involved a manual cleanup of the classified image.

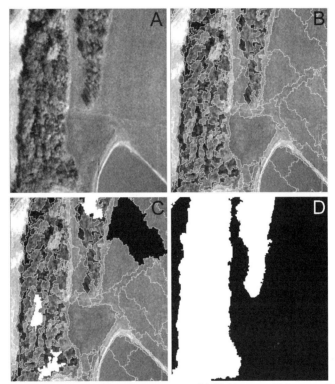

Figure 22-3 Object oriented classification process: A) Initial image, B) Segmented image, C) Training samples, and D) Classified image (Figures C and D, forest samples and classification are white and non-forest are black).

22.2.4 Accuracy Assessment

To perform an accuracy assessment of the classified aerial photography, two custom software applications were written: 1) An application for creating stratified random point samples and 2) An application that can be given to skilled air photo interpreters to classify these random point samples. Five photo interpreters were each given 100 random points, for every study date year (500 sample points/year). The classification results for all five interpreters were combined and statistics, such as overall accuracy, user's / producer's accuracy, and K_{hat}, were generated in the form of error matrices (Table 22-1). The overall accuracy is relatively high for an automated classification, with no year falling below 94% and a combined accuracy for all years at nearly 96%.

Table 22-1 Accuracy assessment results for the classification of six years of historical aerial photography. Overall accuracy is shown in bottom right of each table, producer's accuracy along the last row, user's accuracy in the last column, and K hat below each table.

1941		Interpreters		
		F	NF	User
Classified	F	70	10	87.5%
Data	NF	17	403	96.0%
	Producer	80.5%	97.6%	94.6%

K_{hat} = 80.6%

1954		Interpreters		
		F	NF	User
Classified	F	92	3	96.8%
Data	NF	8	397	98.0%
	Producer	92.0%	99.3%	97.8%

K_{hat} = 93.0%

1966		Interpreters		
		F	NF	User
Classified	F	82	8	91.1%
Data	NF	19	391	95.4%
	Producer	81.2%	98.0%	94.6%

K_{hat} = 82.5%

1976		Interpreters		
		F	NF	User
Classified	F	89	6	93.7%
Data	NF	13	392	96.8%
	Producer	87.3%	98.5%	96.2%

K_{hat} = 88.0%

1991		Interpreters		
		F	NF	User
Classified	F	98	12	89.1%
Data	NF	6	384	98.5%
	Producer	94.2%	97.0%	96.4%

K_{hat} = 89.3%

2002		Interpreters		
		F	NF	User
Classified	F	107	3	97.3%
Data	NF	19	371	95.1%
	Producer	84.9%	99.2%	95.6%

K_{hat} = 87.8%

Overall Accuracy for all 6 Dates:	95.9%

22.3 Analyzing Data

One approach to the quantitative analysis of changing forest-cover patterns is the use of spatial indices or metrics as suggested by the field of landscape ecology (Franklin and Forman 1987; LaGro 1991; Loehle et al. 1996; Petit and Lambin 2002). This discipline seeks to understand the interrelationships between spatial patterns in landcover and ecological processes (Turner et al. 2001). To perform a quantitative analysis of spatial patterns on the landscape, landcover must be broken down into its most basic component: the patch. On classified raster images, a patch is a grouping of contiguous pixels of the same cover type (Forman and Godron 1981). Landscape ecology suggests that landscape metrics, such as the number, size, and shape of patches, can indicate more about the functionality of a landcover type than the total area of cover alone (Forman 1995).

Multi-temporal landscape classifications can also be used to study the relationships between landcover change and attributes of the underlying landscape. One of the most common landscape characteristics for this type of analysis is elevation data (Fu et al. 1994; Mast et al. 1997). Specifically, comparisons made between slope and landcover have been used to show that forest expansion rates often vary with topographic position (Bragg and Hulbert 1976; Knight et al. 1994). The results of these comparisons can be effectively combined with the measures describing patch structural characteristics to provide a more detailed description of the patterns of landcover change.

The goal of this phase of the research was to examine the dynamics of the spatial extent of forest cover within the prairie-forest ecotone of northeast Kansas from 1941 to 2002. Forest-cover change was analyzed at both a human-modified landscape level, the southern half of the Midland quadrangle, and a local level within a controlled setting, the Fitch Natural History Reservation. The maps of forest cover generated from high-resolution (1-meter) historical aerial photography served as the data source for forest change analysis, giving a spatially rich look at the landscape over 60 years. In order to meet the primary goal, landscape metric analysis and slope/forest-cover comparisons were used to answer three specific questions:

- Has forest cover expanded, as suggested, in the past 60 years and at what rate?
- What are the spatial patterns of forest-cover change?
- What effect does topography play on the change in forest cover?

22.3.1 Forest Expansion

Of the various categories of landscape metrics, area is the most elemental and arguably the single most important piece of information describing a landscape, because it provides an overall measure of the extent of landcover change. For this reason, area metrics were given the most attention in this study. The first area-based metric calculated was the total forest area, which is the sum of the areas of all forest patches. Several other values were derived from the total forest area to aid in comparison between years of forest-cover data. The percent of the study area occupied by forest was calculated by dividing the total forest area for each year by the total land area within each study site. The forest area for each study date was also compared to the area from previous study dates, to calculate both the percent change in forest cover per year and total percent change in forest area since the first date (1941) of aerial photography.

The rate of forest-cover change within the larger southern Midland quad study area suggests the general trend of woody vegetation change within the highly human-impacted regions of the prairie-forest ecotone. Overall, forest cover has increased by 40.7% from 1941 to 2002, shifting from 15.7% of the total landscape to 22.1% (Table 22-2). Rates of increase were quite variable across the study interval, with major increases occurring from 1941 to 1954 and 1976 to 1991, and little to no change occurring between other study dates.

Table 22-2 Basic measurements of forest area for the southern Midland quadrangle and Fitch Reservation study areas.

Southern Midland Quadrangle

Data Year	Forest Area (ha)	Percent of Total Study Area	Change per Year	Change from 1941
1941	973.6	15.7%	---	---
1954	1170.0	18.9%	1.6%	20.2%
1966	1136.3	18.3%	-0.2%	16.7%
1976	1172.1	18.9%	0.3%	20.4%
1991	1365.0	22.0%	1.1%	40.2%
2002	1370.0	22.1%	0.0%	40.7%

Fitch Natural History Reservation Section

Data Year	Forest Area (ha)	Percent of Total Study Area	Change per Year	Change from 1941
1941	155.9	60.0%	---	---
1954	169.4	65.3%	0.7%	8.7%
1966	199.0	76.7%	1.5%	27.7%
1976	218.6	84.2%	1.0%	40.3%
1991	240.6	92.7%	0.7%	54.4%
2002	238.8	92.0%	-0.1%	53.2%

In contrast to the larger Midland study area, the controlled environment of the Fitch Natural History Reservation offers a glimpse of landcover change in the absence of such management practices as burning, grazing, and farming. Throughout the dates captured with the aerial photography, forest cover increased a total of 53.2% and growth in forest-cover area was observed for all study periods with the exception of a slight decline between 1991 and 2002. Compared to the Midland study area, the Fitch Reservation showed both a larger percent of total forest cover at the beginning of the study period in 1941, and a larger overall increase throughout the study's duration.

22.3.2 Spatial Patterns of Change

Along with raw forest area, two additional area-based landscape metrics were calculated: the number of patches and the area-weighted mean patch size. The number of patches is the total count of all of the forest patches within the study area. This measure is most valuable as an indicator of the degree of landscape fragmentation. For example, when forest area holds constant and the number of patches increases, the landscape can be considered more fragmented. The other area-based metric calculated was the area-weighted mean patch size. While a normal mean patch size is calculated as the sum of the area of all forest patches divided by the total number of patches, area-weighted mean patch size is weighted by area such that larger patches have more influence. This decision to use an area-weighted mean was made after observing that the traditional mean was so impacted by the overwhelmingly large number of small patches in the forest-cover data set that it held no analytical value. The area-weighted mean patch size is calculated as the sum of all patch areas multiplied by the proportional abundance of that patch (patch area divided by the sum of all patch areas). The values calculated for the area-weighted mean patch size provide a synthesis of the information contained in the metrics of total forest area and number of patches.

In addition to the area-based metrics, one metric was selected from each of three other categories: core area, edge and shape. Core area measures the interior habitat (non-edge) of a landcover class by removing a user-defined buffer distance from the edge of each patch. As well as providing a measure of core habitat, core area may also be used to measure the effects of eliminating small patches from the overall cover area. For this study, a 5-m buffer was chosen for calculating core area, because a 10-m diameter object is the average size for a single adult tree crown in this area. Rather than presenting total core area, the percentage of the total forest found within the core area was used to show the effect of removing tree-sized patches from the total forest cover. The edge metric chosen for this study was total forest-cover edge, which equals the sum of the lengths of all edge segments of each forest-cover patch. Examining the total forest edge or perimeter over time indicates the changing degree of boundary effects between forest cover and other landcover types. The final metric considered for this study was the shape index. This measurement provides a relative indication of the compactness of patches on the landscape. Specifically, the shape index equals the patch perimeter or edge divided by the minimum perimeter possible for a maximally compact patch (a square when analyzing raster data) of the corresponding patch area. The shape index is equal to 1.0 when the patch is maximally compact, and increases without limit as the patch shape becomes more irregular (Forman and Godron 1986). An area-weighted version of the shape index was used in this analysis. The calculation of, and rationale for, using an area-weighted mean is the same as for the area-weighted mean patch size.

Landscape metrics calculated for both study areas provide a detailed form of quantitative analysis of forest-cover change that highlight specific aspects of this transitional environment. For the southern Midland quad, total forest area has already been discussed; it is presented with the other metrics for comparison (Figure 22-4). The second metric calculated was the number of patches. This metric appears relatively stable until 2002, but this is due to the gross scale exaggeration required to fit the 2002 value of nearly 9000 patches on the graph. Comparing each year's forest-cover map to the aerial photography for that year revealed that the classification technique used for this study captured higher levels of detail, in the form of more individual trees, within the color infrared imagery used in 2002 than the panchromatic black and white imagery from all other years. Because of this, all metrics generated for 2002 will be analyzed relative to this bias. Looking again at the total number of patches for all years other than 2002, the Midland quad was most fragmented in 1941. After an abrupt drop in 1954, the landscape has returned to nearly the same number of patches in 1991 as were present at the beginning of the study. The final area-based metric, the area-weighted mean patch size, shows a constant increase from 1941 until 1991, with a slight decrease in 2002 (caused by the large number of patches identified in the color infrared imagery for that year). As a synthesis of total area and number of patches, the area-weighted mean patch size results show that the landscape has seen an overall increase in the cohesion of forest covered area.

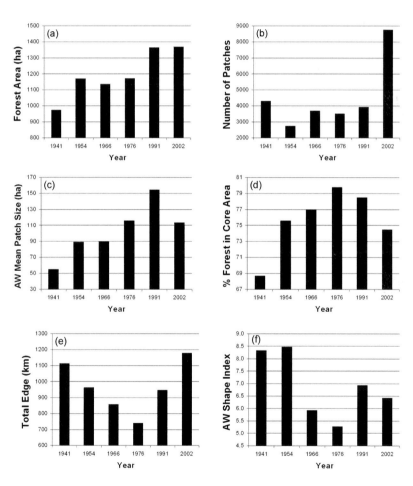

Figure 22-4 Landscape metrics calculated for the southern Midland quad study area: (a) forest area, (b) number of patches, (c) area-weighted mean patch size, (d) percent of total forest contained in core area, (e) total edge, (f) area-weighted mean shape index

The percent of forest in core area was the next metric calculated. As core area was calculated by eliminating a 5-m buffer from the edge of all patches, and given that a 10-m diameter object is the average size for a single adult tree crown in this area, this measure indicates the relative proportion of forest area comprised of objects larger than an individual tree. The lowest percent of forest in core areas was found in 1941, suggesting that this year had the largest relative number of individual tree sized patches of all study years. The percent-in-core-area value increased to a peak in 1976, dropped slightly in 1991, then fell even further in 2002 under the impacts of increased numbers of small patches mapped that year. These increasing values of percent forest in core area further validate the observed grouping of forest area into larger unified stands. The results for the core area metric were found to be inversely correlated with the next metric, total edge. Total edge decreased in the southern Midland quad from 1941 to 1976, but increased through 1991 and 2002. The final metric used in this analysis was the area-weighted mean shape index, which measures the shape deviation from a square, with high values indicating a more complex shape. The results indicate that the overall forested area was most irregular in shape during the first two decades of the study, became the most simplified during the next two decades, and increased slightly in complexity for the last two study dates.

The landscape metrics calculated for the smaller Fitch Natural History Reservation study area produced a more coherent description of changing forest-cover patterns and were less impacted by the different data type used for 2002 (Figure 22-5). As discussed earlier, forest cover area increased constantly through the study period on the Fitch Reservation. The number of forest patches decreased from 1941 to 1954, suggesting the clumping of growing forest patches. The number of patches grew substantially between 1954 and 1966, when a large number of new trees appeared, and remained constant through 1976. Then, between 1976 and 1991, the number of patches dropped dramatically as most of the forest area became joined into one large forest patch. The area-weighted mean patch size grew slowly but remained relatively small for the first half of the study period, then grew by over 100% between 1966 and 1976, and remained relatively large for the latter half of the study. The slow increase at first was caused by individual patches growing in size; the large growth event occurred as the forest began coalescing into a few large patches; then, as patches grew together into the bigger stands, mean patch size remained large. The percent-cover-in-forest-area metric is also indicative of this tendency of forest patches to grow together.

The percent forest in core area grew throughout the study area, until falling slightly in 2002 due to an increase in number of patches (resulting from the use of a different type of data for mapping). The total edge metric fell throughout the study until 2002, quantifying the tendency of patches to group together and thus decrease the overall patch perimeter. Finally, the area-weighted shape index indicates that forest cover on the Fitch Reservation grew to a maximum complexity between 1941 and 1954, became less complex throughout the next 30 years as the forest patches coalesced into stands covering over 90% of the area, then appeared to grow more complex again in 2002 (again due to the increase in small patches caused by the color infrared data source).

22.3.3 Topographic Impact

Digital Elevation Model (DEM) data were employed to examine the relationship between topography and changing forest cover. A 10-m DEM for the entire study area was acquired from the US Geological Survey's National Elevation Dataset (NED). This elevation data represented the finest resolution available for the entire area at the time. The DEM was used to calculate percent slope for the study area, which was simplified into a data set containing slope classes of 5% intervals from 0–5% to 25–30%, and a final slope class for all values greater than 30%.

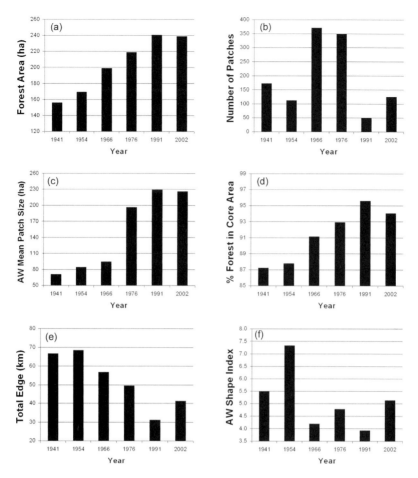

Figure 22-5 Landscape metrics calculated for the Fitch Reservation: (a) forest area, (b) number of patches, (c) area-weighted mean patch size, (d) percent of total forest contained in core area, (e) total edge, (f) area-weighted mean shape index.

Two different methods were used to provide a comprehensive picture of how slope and forest cover are related for both the southern Midland quad and inset Fitch reservation data sets. The first calculation examined the percentage of the total forested area for each year that was found on each slope class. This result can be misleading, however, because the total area of each slope class decreases in this region as the slope interval increases in steepness. To compensate for this, the second calculation determined the percent of each slope interval that was forested for each study year.

The first comparison, performed in the southern Midland quad, demonstrates that the majority of the total forest area was found on land with a slope less than 15% for the duration of the study (Figure 22-6). Taking into account that there is less total land area present on steeper slopes, it is also useful to examine the percent of land within each slope interval that was forest cover (Figure 22-7). This comparison draws a slightly different picture, where a greater proportion of the total land area is occupied by forest as the slope of the land increases. These data also demonstrate that the largest percent increase in forest cover, depicted as growth within a slope class through the 60 years of this study, occurred on the steepest slope.

Manual of Geographic Information Systems

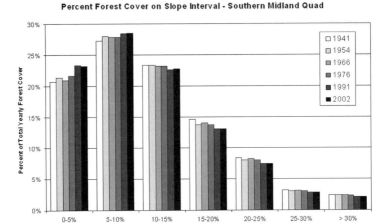

Figure 22-6 Percent of forest cover on each slope interval for the southern Midland quad, subdivided within each slope interval by study year.

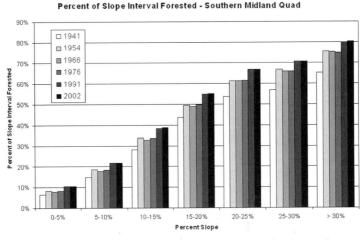

Figure 22-7 Percent of each slope interval that was forest cover for the southern Midland quad, subdivided within each slope interval by study year.

An identical set of comparisons was made between the forested area within the Fitch Reservation and the slope of the landscape. Examining the percent of total forest cover on each slope interval shows a more-pronounced shift towards more forest area on steeper slope intervals on the Reservation (Figure 22-8) than on the larger Midland quad area (Figure 22-6). The percent of each slope interval that was forested demonstrates that there was a faster rate of forest increase throughout the study period on the flatter slopes in the Fitch Reservation than on the steeper slopes (Figure 22-9). At the beginning of the study in 1941 on the Fitch Reservation, more than 80% of the land with a slope greater than 30% was forested while less than 30% of the land between 0-5% slope was forested. Throughout the duration of the study, the highly sloped landscape increased slightly in forest cover to 95%, while the flattest areas increased to over 80% forest cover. These results suggest that on the Fitch Natural History Reservation in 1941 forest cover had already occupied a great deal of the land at the highest slope intervals. Throughout the study, more and more of the less steep landscape became forested at a rapid rate. The difference between these findings and those on the larger southern Midland quad are certainly the result of different land management practices.

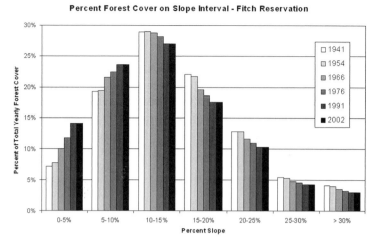

Figure 22-8 Percent of forest cover on each slope interval for the Fitch Reservation, subdivided within each slope interval by study year.

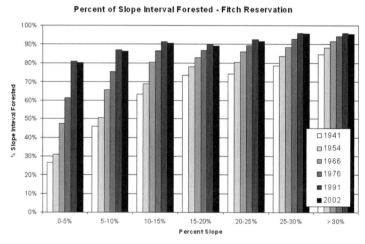

Figure 22-9 Percent of each slope interval that was forest cover for the Fitch Reservation, subdivided within each slope interval by study year.

22.4 Visualizing Data

The visualization of natural landscapes has been increasingly used to deliver the results of environmental change studies and management plans, especially concerning forested environments (Tang and Bishop 2002). Geographic visualizations have also been used to form hypotheses and explore data more effectively than traditional graphic representations (Hearnshaw and Unwin 1994). Until recently, forest visualization efforts have focused primarily upon illustrating static concepts or the possible outcomes of management actions (Bishop and Karadaglis 1997; McGaughey 1997; Buckley 1998). This form of time series visualization compares two or more individual images, generated to represent specific points in time. By animating these static visualizations to move the viewpoint through a 3D landscape, the visualizations can more clearly communicate spatial relationships based on the fact that human vision is "hardwired" with special sensors to detect motion (Gregory 1997). Along with animation through space, visualizations can also use animation to move

Manual of Geographic Information Systems

the viewer through time to provide a dynamic representation of changing landcover. With recent increases in computer speed and software availability, forest visualization techniques are beginning to include the communication of change analysis studies using animation (Stoltman et al. 2002).

The goal of this final phase of the research was to demonstrate the potential of computer visualization as a tool for analyzing and communicating the results of landscape change studies. The classified aerial photographs of the prairie-forest ecotone of northeast Kansas were used to generate photo-realistic static images and animations depicting this landscape through time.

22.4.1 Static Visualizations

The first visualizations developed from the Kansas forest data set were static images displaying the distribution of forest cover at specific dates corresponding to the maps of forest cover generated from historical aerial imagery. These stills are based on the idea that the spatial extent of forest cover is more easily understood when it is displayed as a collection of 3D tree objects. Using commercial visualization software, appropriate ground cover and vegetation image objects were combined to mimic a generic forest and grassland vegetation cover for this region. This style of geographic visualization provides a more familiar and interpretable way of comparing multiple dates of landcover than a traditional GIS-viewer. Static visualizations of a 2.5 km² subset of the study area were rendered from three different vantage points and compiled together to produce a mass-multiple display comparison (Figure 22-10). This non-animated style of visualization can effectively complement quantitative patch structural measurements describing the same area.

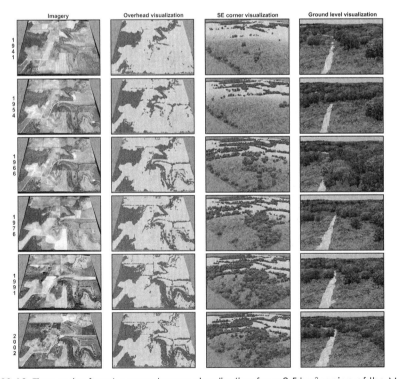

Figure 22-10 Time-series forest-cover change visualization for a 2.5 km² region of the Midland quad study area. See included DVD for color version.

22.4.2 Animated Visualizations

Visualizations using animation provide a greater sense of the process of landcover change recorded by multi-temporal imagery. By representing real-world time in years with animation time in seconds, this display method represents change in the same way that we are familiar with viewing it in the real world. The animated visualizations were created using a different approach than the static renderings. This was done because the resulting product needs to describe the change between years, rather than the before and after snapshots produced by the still frames. Within a GIS, classification comparisons were produced between each consecutive date pair of forest-cover data. This process reduced the six dates of forest/non-forest classified imagery to five change comparison images with classes defined as unchanged non-forest, forest addition, forest removal or unchanged forest cover. In the visualization software, instead of using static landcover representations, as with the still visualizations, animated landcover representations were used to match the classification comparison results. The animated landcover types represented the unchanged classes with grassland or tree cover, the forest addition cover class with growing trees and the forest removal class with trees shrinking and disappearing from the landscape.

The animated visualization used the same classified data sets as the static visualizations, but recreated the changing forest cover as a dynamic process rather than individual moments in time. By animating the appearance or removal of trees in much the same fashion that forest cover would change in the real world, viewers of these animations can experience the events of the landscape history of this area. Used in connection with quantitative patch-structure measurements, the animations provide a visual reference in a qualitative format. To demonstrate this concept, the final Kansas animation was amended to include graphs of several landscape metrics (area, number of patches, average patch size and total edge), which are displayed in an animated fashion that progresses in time with the visualization

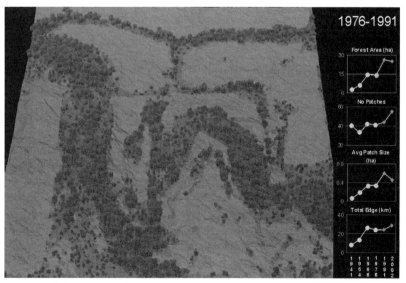

Figure 22-11 One frame of an animated visualization showing forest cover in a visual and quantitative form. See included DVD for color version.

22.5 Conclusions

This study produced several valuable results pertaining to forest-cover change in the prairie-forest ecotone as well as the methods used to investigate it. The object-oriented forest-cover classification showed an overall classification accuracy across all photo years of 96% when compared with manual interpreters. The analysis of this classification indicated that forest cover increased over the past 60 years both on the larger southern Midland quad area (40.7% increase) and within the Fitch Natural History Reservation (53.2% increase). Landscape-metric analysis suggested that forest cover is becoming more cohesive: the average forest patch size is increasing in both study areas. In the southern Midland quad, forest cover is more prevalent and increasing at a faster rate on areas of greater slope. On the Fitch Reservation, the forest has been expanding more rapidly on less steep land, as areas of higher slope have already been forested and there are no limiting human factors to restrict growth on the flatter land. Finally, computer visualizations were shown to be an effective tool for communicating spatio-temporal forest-cover patterns in a clear, concise and effective manner, complementing the numerical results generated from quantitative analysis.

More generally, the three phases of this study illustrated a number of important concepts, pertinent to any spatio-temporal GIS analysis. The first phase demonstrated one method for producing a multi-temporal GIS data set: digitizing and classifying historical aerial photography. With any historical data set, the acquisition and processing of data can be quite time consuming. Automated processing, such as object-oriented classification, can be helpful when dealing with large volumes of historical data. When compiling historical data, accuracy should always be assessed as errors will be compounded when comparing multiple dates of information. One additional issue that was illustrated regards data continuity. While the color infrared imagery from 2002 may have provided a superior classification, the results from this imagery complicated analysis as any multi-temporal study is constrained by the most general data set (black-and-white imagery in this case).

The second phase of this research explored the changing extent of forest cover using several spatio-temporal analysis techniques, including landscape metrics and comparison between landcover and slope. As demonstrated, the variety of methods available is wide, ranging from simple analysis (area of landcover) to complex analysis (landscape metrics) and even comparison with other spatial variables (slope). For any study, data qualities and research goals dictate the most appropriate specific analysis technique(s). For example, an area-weighted mean patch size was chosen for this study due to the high spatial resolution of the aerial photography and the study objective of producing a general description of the spatial extent of forest cover. Combining multiple techniques can be essential in drawing the most complete conclusions from a multi-temporal data set.

Computer visualizations were used in the third phase of this research to render a variety of images and animations depicting the spatio-temporal forest-cover change. Visualizations were shown to provide a more familiar view of the landscape than the statistical results of quantitative analytical methods commonly used in spatio-temporal research. Animations demonstrate the key benefit to merging visualization and multi-temporal GIS: GIS permits the archiving and classification of the landscape at various points in time, while visualization can tie each of these snapshots together by representing the detected change as a process using visually recognizable objects.

References

Abrams, M. D., A. K. Knapp and L. C. Hulbert. 1986. A ten-year record of aboveground biomass in a Kansas tallgrass prairie: Effects of fire and topographic position. *American Journal of Botany* 73 (10):1509–1515.

Anderson, R. C. 1990. The historic role of fire in the North American Grassland. In *Fire in North American Tallgrass Prairies,* edited by S. L. Collins and L. L. Wallace, 8–18. Norman, Okla.: University of Oklahoma Press.

Baker, W. L., J. J. Honaker and P. J. Weisberg. 1995. Using aerial photography and GIS to map the forest-tundra ecotone in Rocky Mountain National Park, Colorado, for global change research. *Photogrammetric Engineering and Remote Sensing* 61 (3):313–320.

Benz, U. C., P. Hofmann, G. Willhauck, I. Lingenfelder and M. Heynen. 2004. Multi-resolution, object-oriented fuzzy analysis of remote sensing data for GIS-ready information. *ISPRS Journal of Photogrammetry and Remote Sensing* 58:239–258.

Bishop, I. D. and K. Karadaglis. 1997. Linking modeling and visualization for natural resources management. *Environment and Planning and Design* 24:245–358.

Blaschke, T., S. Lang, E. Lorup, J. Strobl and P. Zeil. 2000. Object-oriented image processing in an integrated GIS/remote sensing environment and perspectives for environmental applications. In *Environmental Information for Planning, Politics and the Public (vol. 2),* edited by A. Cremers and K. Greve, 555–570. Marburg, Germany: Metropolis-Verlag.

Blaschke, T. and J. Strobl. 2001. What's wrong with pixels? Some recent developments interfacing remote sensing and GIS. *GIS - Zeitschrift für Geoinformationssysteme* 6:12–17.

Boettcher, S. E. and W. C. Johnson. 1997. Restoring the pre-settlement landscape in Stanley County, South Dakota. *Great Plains Research* 7:27–40.

Bragg, T. R. and L. C. Hulbert. 1976. Woody plant invasion of unburned Kansas bluestem prairie. *Journal of Range Management* 29 (1):19–24.

Buckley, D. J. 1998. The virtual forest: Advanced 3-D visualization techniques for forest management and research. *Proceedings of ESRI 1998 User Conference* July 27-31, San Diego, California.

Congalton, R. G. and K. Green. 1999. *Assessing the Accuracy of Remotely Sensed Data: Principles and Practices.* Boca Raton, Fla: Lewis Publishers, 137 pp.

Dunn, C. P., D. M. Sharpe, G. R. Guntenspergen, F. Stearns and Z. Yang. 1991. Methods for analyzing temporal changes in landscape pattern. In *Quantitative Methods in Landscape Ecology,* edited by M. G. Turner and R. H. Gardner, 173–198. New York: Springer-Verlag.

Fitch, H. S. and R. L. McGregor. 1956. The forest habitat of the University of Kansas Natural History Reservation. *University of Kansas Museum of Natural History Publications* 10 (3):77–127.

Forman, R.T.T. 1995. Some general principles of landscape and regional ecology, *Landscape Ecology* 10 (3):133–142.

Forman, R.T.T. and M. Godron. 1981. Patches and structural components for a landscape ecology. *BioScience* 31 (10):733–740.

———. 1986. *Landscape Ecology.* New York: Wiley, 640 pp.

Franklin, J. F. and R.T.T. Forman. 1987. Creating landscape patterns by forest cutting: Ecological consequences and principles. *Landscape Ecology* 1 (1):5–18.

Fu, B., H. Gulinck and M. Z. Masum. 1994. Loess erosion in relation to land-use changes in the Ganspoel Catchment, Central Belgium. *Land Degradation and Rehabilitation* 5 (4):261–270.

Gregory, R. L. 1997. *Eye and Brain: The Psychology of Seeing,* 5th edition. Oxford: Oxford University Press, 297 pp.

Hall, O., G. J. Hay, A. Bouchard and D. J. Marceau. 2004. Detecting dominant landscape objects through multiple scales: An integration of object-specific methods and watershed segmentation. *Landscape Ecology* 19 (1):59–76.

Haralick, R., K. Shanmugan and I. Dinstein. 1973. Textural features for image classification. *IEEE Transactions on Systems, Man and Cybernetics* 3 (1):610–621.

Hearnshaw, H. M. and D. J. Unwin. 1994. *Visualization in Geographical Information Systems.* New York: John Wiley & Sons, 260 pp.

Holt, R. D., G. R. Robinson and M. S. Gaines. 1995. Vegetation dynamics in an experimentally fragmented landscape. *Ecology* 76 (5):1610–1624.

Iverson, L. R. 1988. Land-use changes in Illinois, USA: The influence of landscape attributes on current and historic land use. *Landscape Ecology* 2 (1):45–61.

Jensen, J. R., X. Huang and H. E. Mackey, Jr. 1997. Remote sensing of successional changes in wetland vegetation as monitored during a four-year drawdown of a former cooling lake. *Applied Geographical Studies* 1 (1):31–44.

Knight, C. L., J. M. Briggs and M. D. Nellis. 1994. Expansion of gallery forest on Konza Prairie Research Natural Area, Kansas, USA. *Landscape Ecology* 9 (2):117–125.

Küchler, A. W. 1974. A new vegetation map of Kansas. *Ecology* 55 (3):586–604.

Lauver, C. L., K. Kindscher, D. Faber-Langendoen and R. Schneider. 1999. A classification of the natural vegetation of Kansas. *The Southwestern Naturalist* 44 (4):421–443.

LaGro, J., Jr. 1991. Assessing patch shape in landscape mosaics. *Photogrammetric Engineering and Remote Sensing* 57 (3):285–293.

Loehle, C., B. Li and R. C. Sundell. 1996. Forest spread and phase transitions at forest-prairie ecotones in Kansas, U.S.A. *Landscape Ecology* 11 (4):225–235.

Mast, J. N., T. T. Veblen and M. E. Hodgson. 1997. Tree invasion within a pine/grassland ecotone: An approach with historic aerial photography and GIS modeling. *Forest Ecology and Management* 93:181–194.

McGaughey, R. J. 1997. Techniques for visualizing the appearance of timber harvest options. *Forest Operations for Sustainable Forest Health and Economies, 20th Annual Meeting of the Council on Forest Engineering,* July 28-31, Rapid City, South Dakota.

Petit, C. C. and E. F. Lambin. 2002. Impact of data integration technique on historical land-use/land-cover change: Comparing historical maps with remote sensing data in the Belgian Ardennes. *Landscape Ecology* 17:117–132.

Shao, G. and W. Wu. 2004. The effects of classification accuracy on landscape indices. In *Remote Sensing and GIS Accuracy Assessment,* edited by R. S. Lunetta and J. G. Lyon, 209–220. Boca Raton, Fla.: CRC Press.

Simpson, J. W., R.E.J. Boerner, M. N. DeMers, L. A. Berns, F. J. Artigas and A. Silva. 1994. Forty-eight years of landscape change on two contiguous Ohio landscapes. *Landscape Ecology* 9 (4):261–270.

Skirvin, S. M., W. G. Kepner, S. E. Marsh, S. E. Drake, J. K. Maingi, C. M. Edmonds, C. J. Watts and D. R. Williams. 2004. Assessing the accuracy of satellite-derived land-cover classification using historical aerial photography, digital orthophoto quadrangles, and airborne video data. In *Remote Sensing and GIS Accuracy Assessment,* edited by R. S. Lunetta and J. G. Lyon, 115–131. Boca Raton, Fla.: CRC Press.

Stoltman, A. M., V. C. Radeloff, D. J. Mladenoff and B. Song. 2002. Computer visualization of pre-settlement forest landscapes in Wisconsin. *Proceedings of 17th Annual Symposium of the International Association for Landscape Ecology,* April 23-27, Lincoln, Nebraska.

Tang, H. and I. D. Bishop. 2002. Integration methodologies for interactive forest modeling and visualization systems. *The Cartographic Journal* 39 (1):27–35.

Turner, M. G., R. H. Gardner and R. V. O'Neil. 2001. *Landscape Ecology in Theory and Practice.* New York: Springer-Verlag, 401 pp.

Walter, V. 2004. Object-based classification of remote sensing data for change detection. *ISPRS Journal of Photogrammetry and Remote Sensing* 58:225–238.

Whitney, G. C. 1994. *From Coastal Wilderness to Fruited Plain.* New York: Cambridge University Press, 485 pp.

CHAPTER 23
Visualization of Spatio-temporal Change

Costas Armenakis and *Eva Siekierska*

23.1 Introduction

Spatio-temporal change can be defined as the differences in the characteristics of geographic features between different times t_i and t_j ($j>i$). The change can be either in the spatial properties, such as in position, size, shape, orientation, and/or in the attribute properties of a feature. Change can be determined by comparison methods between two or more spatio-temporal datasets. The change can be discrete or continuous in both space and time.

Understanding and interpreting the differential and temporal nature of change can be achieved through the visualization process using appropriate representation schemas that provide computer-generated virtual images of the actual changes, in which images emulate the brain's synthetic process (McCormick et al. 1987). Thus, visualization permits the representation of spatio-temporal change through a visual passive or interactive reconstruction process, where the actual change is mentally reconstructed through the brain's synthetic ability by relying on cognitive pattern recognition.

23.2 Representation of Spatio-temporal Change

Change occurs when a geographic feature is transformed from state S_i to state S_j over a time period $\Delta t=t_j-t_i$. The representation of the temporal evolution of a feature is therefore equivalent to the visualization of a dynamic process. Dynamic processes or events can be visualized as a series of static time-stamped representations or via visualization techniques that create the perception of continuous sequential representation. Visualization of change should also be an exploratory tool for understanding the change (Andrienko et al. 2002; Andrienko and Andrienko 2007). The visualization of spatio-temporal changes should provide answers to analytical questions, such as: "what is the current situation", "what was the situation at a past time t", "what are the differences between the geographic state at t_i and the geographic state at t_j", "what caused the changes", "how have the changes occurred", and "what will be the situation in the future time t_f under certain conditions". With respect to the change itself, the magnitude, direction, rate, duration, and trend of the change need to be represented in answer to analysis queries. Therefore, the appropriate visualization of the spatio-temporal changes not only enhances the inference process and allows the users to generate, display, view and manipulate data relying on cognitive pattern recognition, but also enhances data exploration and the understanding process.

Systems for the visualization of spatial changes should allow the user to control the visualization process, view, analyze and understand the evolution of temporal patterns, control the motion, study the evolving images at the user's own pace, interrupt the sequential progression of images, visually and numerically compare multi-temporal images, and consult and integrate information from various sources. It should be performed in an active cartographic environment (Armenakis 1996), with the following main characteristics:

- visualization of the evolution of a feature over time in a continuous reconstruction animated mode;
- visualization of the feature's state at various specific and discrete time points (time snapshots);
- visual comparison of at least two and preferably three consecutive temporal states of the feature's characteristics by juxtaposition or overlay;

- visualization of the differences – the changes themselves – in the feature's characteristics between consecutive time intervals;
- visualization of the elements of spatial changes, such as magnitude, direction, duration, speed, acceleration and trends;
- ability to "drive" the visualization process through interactive functionality.

The following definitions are related to the change of geospatial elements (coverage area or feature objects) and their visualization (Langran and Chrisman 1988). State: the condition (status) of a geospatial element at a given point or interval in time; event: the instant in time at which an occurrence causes changes to the state of the geospatial element; version: the form (configuration) of the state during a time period; mutation: the period in time at which a geospatial element undergoes change; duration: the time interval during which a version or a state lasts. The visualization of change can be performed either using time-stamped states (state mode) or time-stamped state differences (differential mode) or a feature-based versioning approach (Armenakis et al. 2006).

23.2.1 State Mode Change Representation

In the state mode, the status of the geographic domain is visualized as discrete time snapshots in consecutive time-ordered sequence. The data are organized, stored and displayed in time-series snapshots:

$$S_0, S_1, S_2, \ldots, S_i$$

where, S_i is the individual time-snapshot
and $i = 0, 1, 2, \ldots, t$, with 0 indicating the initial state and t the current state.

It is therefore a time series of concatenated time snapshots of the geographical domain. All versions of the elements comprising each state carry the same time-stamp. The state frame mode allows the direct visualization of a geographical state at discrete and specific instants in time. This affects the visualization of sequentially displayed static states (succession of frames) as they might not be perceived by the viewers as a continuous form of presentation, unless interpolated states of some form are created during the display. For continuous temporal changes, the sampling time points are taken at frequent time intervals, to represent the continuity of the spatio-temporal paths of the geographical data accurately and completely.

The comparison between states, to visualize changes which occur within a given time interval, is determined as:

$$\Delta S_{ij} = S_j - S_i \qquad (23\text{-}1)$$

where, ΔS_{ij} is the change between states S_j and S_i, and $j > i$.

The storage of all temporal time snapshots is of concern. No matter the degree of changes, a great portion of unchanged data will have to be stored every time an updated state is produced. This creates an unnecessary storage redundancy and results in longer processing times.

23.2.2 Differential Mode Change Representation

23.2.2.1 Forward

In the forward differential mode, the initial time snapshot frame (state) and the subsequent differential elements representing the changes from one state to the next are used for the visualization of changes. The stored and displayed data for the forward differential mode are

$$S_0, \Delta S_1, \Delta S_2, \ldots, \Delta S_i$$

where, S_0 is the initial state,
ΔS_i are the recorded changes over time, and
$i= 1, 2,\ldots, t$ indicates that changes occur at times $1,\ldots, t$ between states i and i-1.

For visualization each state is determined as:

$$S_t = S_o \cup \left(\bigcup_{i=1}^{t} \Delta S_i \right) \tag{23-2}$$

The changes between t_i and t_j for the visualization, for example the changes between t_1 and t_3, can be determined as:

$$\Delta S_{i,j} = \bigcup_{i=i+1}^{j} \Delta S_i \tag{23-3}$$

For the determination of the changes, the required processing is less than that required in the case of absolute states (static mode). In the latter, the entire state files need to be merged and compared to produce the differences, while in the differential mode only the differential files need to be retrieved.

The involved differential elements should have integral topological structure, that is, they must not depend on the preceding state. This is necessary for the visualization of the changes within the 'delta' files. The storage requirements for the differential mode are minimal. This approach is particularly useful where only certain elements change over time, while most state elements remained unchanged.

23.2.2.2 Backward

Backward differential mode is also an option, where the current time snapshot and the preceding differential elements are used. This approach is based on a continually updated state. That is, whenever a change occurs it is immediately incorporated into the current state, and is stored as a time-indexed differential element. The process is expressed as:

$$S_{t-1} = S_t \cup \Delta S_{t-1,t} \tag{23-4}$$

The elements for the visualization of change are:
  i) the current state file: S_t
  ii) the difference files: $\Delta S_{t-1,t}$, $\Delta S_{t-2,t-1}$, ..., $\Delta S_{1,2}$, $\Delta S_{0,1}$
The approach is quite attractive as the present state is readily available when needed and there is no redundant storage of data. In addition, it is easy to extract the changes which occurred between time periods, and it is possible to indirectly reconstruct any previous geographical situation.

23.2.3 Feature Versioning Representation

This representation is based on the spatio-temporal evolution of the status of the feature components of the geospatial space. In this object-based approach the spatial trajectory and the attributes of a changing feature are represented over time. Unchanged features maintain their initial version over time. The representation of the state at time t is reconstructed by integrating all the concurrent objects' versions at this instance of time. Obviously, this approach is more flexible as it is possible to represent both the entire area and selected feature elements (feature time series). It also reduces storage requirements since it stores only the feature versions, therefore unchanged versions over time remain as they are. The visualization of changing features can be performed by displaying the trajectories of changing features over the fixed background of unchanged features. Changes in attributes can be displayed by changing the various display parameters for the features. A schematic space-time cube example of visualizing feature versions is given in Figure 23-1, where feature F_1 has changed location by Δs, shape (square to ellipse) and attribute (black to gray) between times t_o and t_i, while feature F_2 has remained unchanged from t_o to t_i.

Manual of Geographic Information Systems

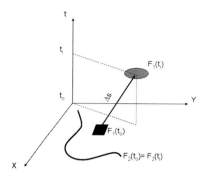

Figure 23-1 Schematic visualization of feature versioning.

23.3 Cartographic Visualization of Spatial Changes

The visualization of spatial change in mapping will benefit from the dynamic cartographic representations. These will enable the user to reconstruct, visualize, the temporal changes of geographical phenomena and events by creating an illusion of the spatio-temporal transformation and evolution.

The different approaches to visualize change depend on the process which generates the time-dependent data. Various computer techniques can be applied to create the perception of continuous motion. Most electronic maps replicate the cartographic design principles of conventional maps, and thus portray a static abstract picture of reality using graphic forms. Although they can be used to find answers to questions dealing with location and theme, they leave the analysis procedures to the user, and they treat time as a static variable. The effective portrayal of the spatial changes on maps requires special attention to represent effectively the unique properties of the time parameter. The current methodologies for change visualization need to utilize the opportunities provided by modern computer technologies but at the same time should incorporate the legacy of conventional cartography (Andrienko et al. 2003).

The involvement of the users in the visualization process must also be considered. Interactive systems based on predefined processes provide quick response to user queries. Proactive approaches allow users to decide their own exploratory paths, and to access, retrieve, and process various sources of information during the analysis. One approach to build an active cartographic environment is to integrate motion – animated maps – and various multiple types of data – images, text, sound, numerical data – with conventional and innovative cartographic representations. The concept of hypermedia can provide the tool for exploration, while the visualization of spatio-temporal data will facilitate dynamic representations (Siekierska and Armenakis 1999).

The human brain perceives change as either movement or change in shape, texture, color, sound or feel. Thus, historically maps were very limited to show change through time because of their static nature and being two dimensional space representations. Maps were used with a single slice of time to show change phenomena. Also challenging was to show interactions or flows between places (Tufte 1990). Most commonly, representation of change is demonstrated by several states at once through two of more maps, displayed side by side. For example, change in urban growth would have been represented by one map showing spatial extent of the city in one specific time period and a second map showing its current boundaries. In this type of change visualization users are to make visual comparison and interpretation of presented data. Another type of change visualization in static maps is to indicate amount or rate of change using a single choropleth map, where change is represented as a ratio and thus assigned color or texture. For example census subdivisions with the darkest color would symbolize the greatest change in population numbers, while the light areas would signify the least change.

With advances in computer technology, multimedia are used to represent a multitude of changes in a dynamic environment. Cartographic animation has become a very effective visualization technique to intuitively represent dynamic geographical phenomena (Buziek 1999). Through cartographic animation one can show interrelations amongst geospatial data components, location, attribute and time, and 2-, 3- and 4-dimensional representations. Spatio-temporal changes of dynamic phenomena are best visualized through dynamic maps, enabled by computer technology (Peterson 1995; Oberholzer and Hurni 2000). In dynamic maps the real time is compressed or scaled into changing display. Change may be represented by non-moving occurrences where events are added and deleted at places through time or by moving objects where movement is animated on the screen.

23.3.1 Graphical Variables for Change Representation

Cartographically, change can be visualized through various graphical variables in the form of animation (Blok 2005). These graphical variables, first studied in the 1960s by Jacques Bertin, are location, value, hue, size, shape, spacing, and orientation (Bertin 1967; Slocum et al. 2005). Location emphasizes where the symbol is and thus is determined primarily by the space. It is also a primary means of showing spatial relations. The human brain computes relations such as "is within" or "crosses" instantly from the eye's perceived image of the map. Hue, value and saturation are three dimensions of color. Hue, frequently referred to as color, is important aesthetically, and usually represents qualitative differences. When using hue in change visualization, it is important to be aware of color association and use appropriate hues, for example green for positive change and red for negative change. Value is associated with the lightness or darkness of the color. In visualization of change, value is very important as the eye tends to be led by light and dark patterns. Value is usually used to represent quantitative differences and traditionally darker symbols are associated with "more." Saturation, the third dimension of color, is a measure of the vividness of a color. There are also other terms describing saturation with slightly varied definitions: chroma, colorfulness, purity, and intensity (Brewer 1999).

Size is another visual variable and the size of the symbol conveys quantitative differences. The human brain has difficulty inferring quantity accurately from the size of a symbol. For example, if proportional circles are used to portray city population, doubling the radius of a circle (quadrupling its area) is perceived as indicating more than twice the population, but not four times. Thus, the brain infers population from the approximate size of symbol and not from the precise values of radius or area. Shape is a geometric form of the symbol used to differentiate between object classes. Shape is also used to convey nature of the attribute, for example population indicated by images of people, or urban density by house symbols. Spacing, arrangement, and density of symbols in a pattern are used to show quantitative differences. A classic example is a dot density to show population, while the orientation of a pattern is used to show qualitative differences.

When visualizing change through maps one must take into consideration graphical limits (Vasiliev 1996). In terms of spatial limits, symbols or phenomena that are changing must be in close proximity for a human eye to notice. If it is shape, the change must be observable. If for some reason it is not, color or change in scale may enhance the ability to perceive change. However, display pixels have a set size, and thus a finite number of spatial locations.

Visualizing change through color requires particular attention and careful planning. Because computer monitors are able to display millions of possible colors, it is important that the selection of color be limited in order to be comprehensible. Also the changes in color need to be sufficiently significant for a human eye to distinguish (DiBiase et al. 1992). Depending on the purpose of visualization, certain changes may need to be highlighted or emphasized through expanding or limiting the range of luminance and contrast and using brighter and more vibrant colors. Important to consider is color association in order to prevent faulty analysis, especially in the case of complex and long animations. Particular attention needs to be paid to non-interactive animation where viewers do not have the tools to control sequence or control the speed of animation (Peterson 1995).

Change is directly linked to time (Peuquet 1994). Time is typically treated as an attribute to be mapped. However, according to MacEachren (1994), treating time as an attribute limits the potential of dynamic maps displays. Kraak and MacEachren (1994) advocated treating time as a cartographic variable to be manipulated, the same as size, hue and spacing. Several authors (DiBiase et al. 1991; Szegö 1987; MacEachren 1994) have identified four fundamental dynamic variables: duration, rate of change, order and phase. The duration is controlled by user interaction (viewing time) of a given image (e.g., short or long). The rate of change is depicted as duration of different animation frames (e.g., slow-constant, fast-constant or steadily increasing). Matching animation frame order with the temporal order and nature of the depicted phenomena is the most natural way of ordering dynamic variables. However, with dynamic maps one can use time order to represent in a symbolic way any order of interest based on selected attributes. Phase has been defined as a rhythmic repetition of certain events (Szegö 1987). According to MacEachren (1994), the addition of new variables to the classical set of cartographic variables as defined by Bertin (1967), will likely make the most substantial impact on maps as a visualization tool.

23.3.2 Active Cartographic Environment for Change Representation

To view and understand the spatio-temporal patterns, the user needs to control the animated motions, study the evolving images at the user's own pace, interrupt the rolling of images, visually compare multi-temporal images, seek various alternative solutions, and consult and integrate various sources of information. Thus, an active cartographic environment equipped with "steering" capabilities and interactive control over the computations by modifying various parameters will support the user's exploratory visualization of changes (Gimblett and Ball 1991). For example, the following functionality should be available:

- interactive operations;
- animated displays capable of being started and viewed in forward and backward play mode;
- start, restart, pause and resume animation from current position; the animation be paused at any frame followed by a restart or return to the beginning;
- frame by frame forward and backward motion;
- retrieval and display of temporal states either by selected times (e.g., use of a time slide bar) or by frame number;
- kinetic control for displaying the animation at variable user-selected speeds; this can be achieved by expressing the key-frame number as a function of time;
- dynamic legend to explain the variations in the temporal displays;
- the animation be paused to permit hypermedia links to other information sources, including bookmarking of certain nodes/anchors for direct access and return;
- standardized user interface, with icons and metaphors for intuitive operations, with on-line help and navigation schemas for guidance;
- good overall performance (e.g., quality, speed, consistency).

Figure 23-2 shows the *Territorial Evolution of Canada* (Armenakis 1996; Siekierska and Armenakis 1999), an example of integrating both cartographic animation and hypermedia to depict the changes of the provincial and territorial boundaries. Initially it was developed using HyperCard and SuperCard (Armenakis 1993), and then it was fully developed using MacroMind Director and Lingo language.

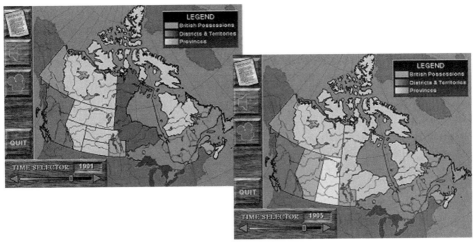

Figure 23-2 Visualization of the provincial and territorial boundary evolution of Canada from 1901 (upper left) to 1905 (lower right). See included DVD for color version.

23.3.3 Electronic Map Design for Change Representation

The design of electronic maps effectively displaying change and providing access to multiple types of information is not an easy task (Kraak and Klomp 1995; Kraak et al. 1997; Harrower 2003). Electronic mapping provides various intriguing possibilities for representing spatio-temporal events (Buttenfield and Ganter 1990). The changes in phenomena or events over space and time can be visualized: a) in a kinematic sense, where displayed map elements are in motion, and/or b) in a metamorphosis sense (morphing), where displayed map elements change their properties (e.g., shape, size, orientation, color). The illustration of motion and changes on time-maps can be performed using a variety of point-, line-, and symbol-types, labels and temporal glyphs, flow-linkage arrows, blinking symbols, distortions of shape and color, shading and fading, changes of hue or texture, changes in transparency and opacity, progressive zooms, rotations, scrolls and pans, and successive perspective views. Various combinations of these display techniques may be required and are desirable to produce the necessary visual effects. With increasing capability of electronic displays new techniques to represent change are becoming available. For example, the experimental designs of Figures 23-3 and 23-4 illustrate how opacity can be used to show the process of changes and how color-coding is more suitable to represent a discrete change.

Figure 23-3 Spatiotemporal evolution of Iqaluit, Nunavut, Canada using image opacity (interface design by Ken Francis). See included DVD for color version.

Manual of Geographic Information Systems

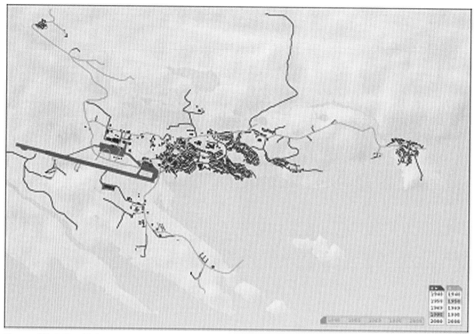

Figure 23-4 Spatiotemporal evolution of Iqaluit, Nunavut, Canada using color–coded changes (design by Ken Francis). See included DVD for color version.

23.4 Display Methods for Spatio-temporal Changes

23.4.1 Static Representation of Change

In conventional cartography, time, and thus change, is treated basically as fixed variable. Visual display of temporal changes can be performed in various forms (Szegö 1987; Monmonier 1990; Vasiliev 1997) such as:

a) *aspatial time-series graphs* (charts or plots) of the type $y = f(t)$, which relate time t with the attribute y. Because these graphs relate time and attribute domains, usually labels, legends or symbols are used to provide a link to spatial data (e.g., plot showing the temporal fluctuations of clear-cut areas);

b) *static time-point maps,* where time is the fixed frame on which all geographic information is attached. This is what is called "snapshot," where the map displays all the current information for a particular time-point. Here the term time-point refers to both the mathematical definition and the human perception of time (e.g., road maps of 2005 cover an annual period, however 2005 is a point on the temporal reference axis);

c) *static time-interval maps,* where time is the variable to be depicted through the changes of the geographical phenomena and events. Usually, this representation is applicable for a period of time during which information about magnitude, direction, and rate of change of locations and themes can be extracted (e.g., expansion of drought areas over period of time).

In addition to single static time-maps, *multiple static time-maps* can also be used. Their use supports the simultaneous visual comparison between two or more juxtaposed maps. The examination may involve either similar or different types of maps (e.g., comparison between two positional maps, or comparison between a positional map and an aspatial time graph).

23.4.2 Dynamic Cartography

Dynamic cartography has the capability of displaying motion and changes over time. This offers possibilities for cartographic representations of temporal changes, which are difficult or even impossible to apply using static paper or electronic maps. Interactive computer graphics systems and video technology make possible the idea of *maps-in-motion*. Maps-in-motion significantly extend the capabilities for the visual display of temporal change, a variable of inherent dynamic nature.

These maps allow the continuous consecutive dynamic display of temporal static maps (snapshots) in sequential mode, thus adding realism to the display process. The incorporation of maps-in-motion for displaying geographical variables can effectively show their temporal evolution, using, for example, the spatial motion of symbols, the playback of frames, or color changes.

Motion and change in map-in-motion cartographic applications can be expressed using computer animation techniques. There are considerable differences between static and kinematic maps (Ormeling 1995). The latter involve high display speeds, larger amount of display information and continuous changes of map contents, which greatly affect the user as they can create perception and understanding problems (Kousoulakou and Kraak 1992). For example, the user's viewing attention will concentrate on particular processes or areas of the map or symbols that change, while missing the global impact of changes and even reducing the ability for general orientation.

While the forms of static time-maps are applicable to both paper and electronic mapping, the production of maps-in-motion comes mainly from electronic mapping applications. The technological developments have made the idea of *animated cartography* feasible (Campbell and Egbert 1990; Armenakis 1992; DiBiase et al. 1992; Asche and Herrmann 1993; Peterson 1995; Ogao and Kraak 2002; Harrower 2002, 2003; Kraak 2007).

The creation of virtual reality images for the display of time-dependent geographical information enables us to perform dynamic presentations applicable to mapping. Animated maps significantly extend the mapping capabilities by including the visual display of time, a variable of an inherently dynamic nature. These maps allow the continuous display of time-dependent data in chronological order, thus adding realism to the display process. Because spatio-temporal data are characterized by motion and change, animated maps are considered part of the core of an active cartographic environment. Animated mapping is possible through the computer-generated motion of displayed information (Muller et al. 1988). Computer animation is well suited to represent dynamic processes. Animation supports dynamic data exploration and visualization. Dynamic representations of the evolution of phenomena extend our knowledge and understanding of our environment, due to a much better perception of both the nature of the changes, and the dynamic behavior of the real world. Computer generated animation significantly contributes to the mapping of time-dependent geographical information.

23.5 Computer Animation

The potential of cartographic computer animation has been recognized for some time (Moellering 1980; Siekierska 1984). Recent advances in computer processing power and graphics make it attractive and feasible. Computer animation is a development with great cartographic capabilities that is most suited to present dynamic processes. Computer animation can be defined as computer-generated motion that imitates the key-frame techniques of traditional animation, by using key positions of graphic data in space, instead of key-drawings. In key-framing the user creates a series of positions and the times at which they occur. In order to produce the animation, the computer can be also used to generate the in-between positions through interpolation. Linear interpolation creates evenly spaced frames. Piecewise polynomials, and especially spline techniques, usually provide smoother transitions between frames (continuous first and second derivatives) (Steketee and Badler 1985).

Key-frame animation provides the user with an overall picture of the evolving events at distinct time intervals. However, the creation of the in-between frames at frequent time intervals may cause computational and rendering problems, since real-time animation requires the creation and display of 30 frames/sec or more. Certain trade-offs between spatial resolution and time-resolution are inevitable. We may have to choose between displaying a few frames at high resolution, or many frames at a lower resolution. For single moving features, the path of motion of the object through time is generated from given point positions. Although key-framing may be used, it is more efficient to superimpose the locations of the points of the trajectory on the existing electronic map. These point positions can be computed in real-time (e.g., from dynamic models, real-time vehicle tracking) or can be generated from the database (e.g., routes of explorers, population movements, transportation networks). Motion kinetic control provides the ability to change the speed and acceleration of the motion at different time intervals. This can be achieved by expressing the key-frame number as a function of time. At specific time intervals, within a given time period, the corresponding in-between framing positions are generated, based on motion parameters. If we wish to see these frames displayed faster or slower, we have only to modify the function that relates key-frames and time. Otherwise, we would have to generate new position frames at the required times to satisfy acceleration or deceleration requests (Steketee and Badler 1985). It should be pointed out here that computer animation may reconstruct discrete temporal real-world phenomena as being continuous; the reverse process is also possible.

Besides key-frame animations, which require interpolators, the following animation techniques can be also applied to dynamic mapping of changes:

- *frame by frame playback* animation, where all frames (time snapshots) are created in advance and stored (Figure 23-5). Usually co-registered temporal frames are used, which are generated from either time-stamped data or multi-temporal observations, such as time-series remotely sensed images (Harrower 2002). At a later stage, all frames are displayed in sequential order giving the illusion of continuity of changes from frame to frame. It serves to display evolution of specific historical data (e.g., chronological development of cities);

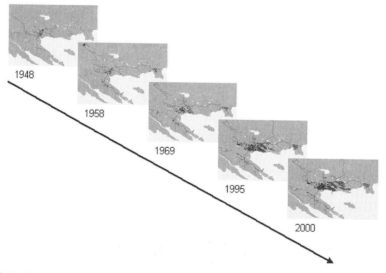

Figure 23-5 Principle of the frame playback animation. See included DVD for color version.

- *color-table animation,* where the colors in the look-up table are loaded at each step of the animation (Figure 23-6). The values of the pixels on the graphics screen change as the look-up table is reloaded at the end of each raster display cycle. The appropriate images are loaded into the refreshed buffer (Foley et al. 1990). It can support the reconstruction of changes in land-cover mapping, such as the encroachment of pollution or the depletion of natural resources; see, for example, color fading to display depletion of oil fields on the Electronic Atlas system (Siekierska 1983);

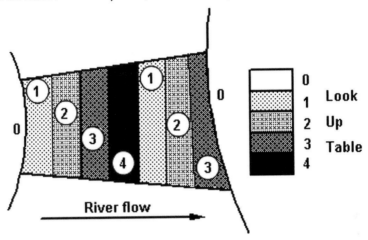

Figure 23-6 Color-table animation.

- *graphic scripted animation,* where a storyboard is written by the user, describing the stages of motion (features, position, duration, color, scale, etc.), and the control system interprets and executes the commands to produce animation (Figure 23-7). This ordering of the sequence of events over time provides interaction between objects and parameters and enables synchronization (Monmonier 1989; Wilhelms 1987; Gilbert and Richmond 1981). Graphic scripted animation can facilitate cartographic designs for the production of dynamic electronic maps. An example of display ordered by duration attributes is the electronic map of the territorial evolution of Canada (Siekierska 1990);

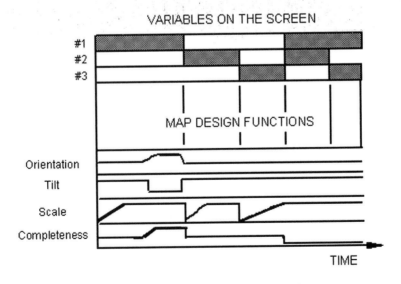

Figure 23-7 Graphic-scripted animation (after Monmonier 1989).

Manual of Geographic Information Systems

- *double-buffering animation,* which is based on the flipping of the display between the primary (visible) and the secondary (hidden) image buffers (Brinkman 1990; Campbell and Egbert 1990; Foley et al. 1990). While an image is being drawn on the screen using the main screen memory, the next image-frame is being created and stored using the non-displayable portion of the screen memory. The switch between visible and virtual displays results in an animated display of sequential frames. The double-buffering technique requires additional screen memory;

- *exclusive or (xor) animation,* which is based on logical (bitwise) operations between a pair of pixels in the generation and display regions. These operations between pixels, called RasterOp or Write Mode, write their results in the display (destination) region (Foley et al. 1990). The "xor" operation, on bi-level displays, inverts the value of the original pixels. Thus, it is possible to display (draw) objects moving across the screen without destroying the original background pixels, which are restored (redrawn) during the next drawing cycle (Campbell and Egbert 1990);

- *sprite animation,* where simple-to-draw objects move along restricted set of paths on the screen (Gersmehl 1990). A sprite is a small rectangular region of memory, whose location is specified in registers in the frame buffer (Foley et al. 1990). The sprite moves around on the top of the background image, when the locational values in the registers change. Sprites can be used to animate persistent or repetitive motion patterns.

- *fly-bys over 4D space,* where the corresponding temporal versions of spatial objects (*x, y, z, t*) are displayed as a function of time as time *t* changes (Siekierska et al. 2001).

Change maps of geographical phenomena can be viewed on a graphics screen in temporal sequence. For example, the evolution of a phenomenon or event can be viewed using symbols moving across the electronic map. These types of electronic maps are appropriate for the visualization of moving patterns (e.g., sea ice motion, image maps displaying the changes of the normalized difference vegetation index (NDVI) across a continent over a period of time). To display the elements of change, such as variations in the rate of change, variable symbology patterns can be used (Andrienko et al. 2002).

Computer animation covers all changes that have visual effects. Therefore, it can also exist without the actual motion as defined in the kinematic sense. Besides time-considered animation, where the computer calculates the in-between positional frames and moves an object along a trajectory including rotations and translations, we may have time independent animation as well. In cartographic applications, timeless computer animation systems can be used: a) to apply virtual camera operations like scroll, pan, zoom or tilt, b) to change colors and light intensity, c) as graphic editors, to interactively create, paint, store, retrieve and modify drawings, d) in metamorphosis, where an object is transformed into another, and, e) to synchronize motion with sound (Magnetat-Thalman and Thalman 1985).

23.5.1 Tools for Computer Animation

Various tools are available for generating cartographic animation for change visualization. Their applicability depends on the type of animation used and the degree of interactivity. For example, for a passive view of temporal snapshots playback a simple PowerPoint-based slide show may be sufficient. However, for exploratory animated visualization of spatio-temporal changes, authoring and geodata systems are employed. Most systems can also generate applications accessible over the Internet.

The authoring tools are very useful for the development of the interfaces and controls of the animated change images. The images of change are generated elsewhere using photogrammetric, image analysis and GIS systems. Common authoring systems are MacroMedia Director, Virtual Reality Modeling Language (VRML) and GeoVRML, and its successor Extensible 3 Dimensions (X3D). Examples of spatio-temporal applications using these authoring tools are the interactive multimedia *Atlas of Switzerland* (Hurni et al. 1999), the

VoxelViewer prototype visualization system for exploring remotely sensed data (Harrower 2002), the web-based Dynamic Visualization System (DVS) (Williams et al. 2006) and the dynamic terrain visualization and historic evolution of Iqaluit, Nunavut, Canada (Natural Resources Canada 2005).

Geodata systems use georeferenced geodatabases. One of the emerging geodata-driven vector graphics standards for interactive mapping and change visualization functionality is the so-called Scaleable Vector Graphics (SVG). SVG is an open, vendor neutral, XML-based (Extensible Markup Language) vector graphics standard for the depiction of resolution- and device-independent two-dimensional vector and raster data. This 2D graphics format generates output for many media types, including printing and wireless mobile applications, but it is optimized for Web-based applications. SVG allows fully interactive web-based mapping applications. SVG allows for three types of graphic objects: vector graphic shapes (e.g., circles, rectangles, and paths consisting of straight lines and curves), images and text. The advantage of SVG-generated data over OGC-based WFS (Web Feature Service, a standard of the Open Geospatial Consortium) is that the repetitive querying to a web map server using web feature service requests is eliminated when using SVG, since data has been downloaded to the client initially. SVG is used to mark-up graphical data in the same way that XML and HTML are used to mark-up text.

SVG supports animation though start time, duration, and attribute values. It implements animation by supporting the modification of color values, coordinate values, and transformation values of the graphic objects. It also permits motion of an object along a path.

The animation is controlled in SVG with the following animation parameters:
- *Begin:* specifies the time or event for activating the animation and includes the play/stop buttons.
- *Duration:* specifies the interval between start and end of the animation.
- *Repeat/count:* controls animation cycle.
- *Interpolation:* specifies the transition frames between start and end. The options are: discrete with no interpolation steps in-between, just one from start to end position; linear with linearly interpolated steps between start and end positions; and paced with even steps along the distance between start and end points.
- *Freeze:* stops animation at the end of its completion cycle.
- *Start/end values:* parameters for the individual animations.

The animation types supported by SVG are:
- *Opacity:* animates the transparency of the objects. Examples are blinking (to draw attention) and fade-in/out (to make objects progressively appear/disappear).
- *Color:* changes color of object during the duration of the animation.
- *Scale:* changes the size of objects during the animation.
- *Rotation:* rotates the objects from a starting angle to an ending angle to show object direction changes.
- *Motion:* shows map symbols moving along the path of linear features.
- *Progressive drawing:* draws gradually the course of linear or areal objects.
- *Morphing:* shows progressive changes in the shape of the object.

To represent a spatio-temporal change usually the animation types are combined so the user can better visualize the phenomenon. Shape change and motion can be used to represent for example the evolution of flood over time. An example of using SVG animation is in the visualization of the dynamic glacier processes (Isakowski 2003), where colored circles (increased/decreased values) of variable radius (magnitude of elevation change) display the elevation changes of the ice thickness. Vectors show the surface velocities, with vector length representing the velocity as the vectors move in the 2D space. The glacier extent is visualized by successively displaying the temporal boundary lines one after the other, while hiding the previous ones.

23.6 Visualization of Attribute Changes

The most common method for visualization changes of the attribute data is performed with the choropleth map methods, where the attribute differences—similarly to the actual attribute values—are encoded to the corresponding spatial features as shades of color. Trends of the attribute data can be visualized using time-attribute graphs. Preferably, both spatial and aspatial temporal attribute representations are shown and explored in an integrated visualization mode (Andrienko and Andrienko 2007).

Figure 23-8 shows the relative changes in the gross domestic product in the selected European countries between 1992 and 1993 (Andrienko et al. 2002). The changes are shown as differences in blue and brown shades, where the degree of darkness shows the amount of difference between the represented value and the reference value. A time plot shows the time graph of the displayed attribute. The visualization of the attribute differences has been combined with animation, where the differences are re-computed and shown on the map.

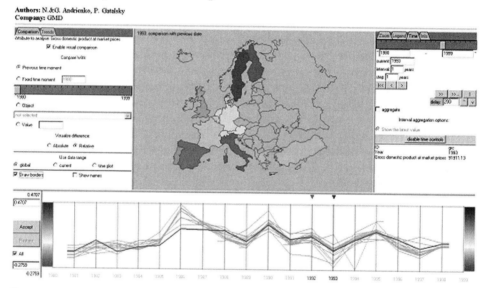

Figure 23-8 Visualization of change in attribute data (GDP per country)(Andrienko et al. 2002). See included DVD for color version.

23.7 Closing Remarks

Monitoring spatio-temporal changes and change analysis are important tasks towards quantifying and understanding the continuously changing environmental conditions. Geographic Information Systems increasingly focus on change to improve future projections and modeling in practically all aspects of Earth sciences, including physical geography, which addresses, amongst others, climate change issues, and economic geography, which focuses on the studies of societal change.

The visualization process contributes to the understanding and interpretation of spatio-temporal variations using appropriate representation schemas and computer-generated virtual images of the changes that support the mental reconstruction of the actual changes. Dynamic cartography and computer animation have been successful in the visualization of time-dependent events. Many of the applications are of passive type, where the user just observes the evolution of the event on the screen trying to understand and interpret general patterns of change, such as overall magnitude, direction, and duration. This serves very well for example cases of multi-temporal Earth observation images, where changes in vegetation cover, polar ice cover or flood extents can be evident.

However, the visual perception of change using only dynamic maps and images will be enhanced with the integration of analytical and interactive functionality that allows data exploration and quantification of the change parameters based on access, query and manipulation of the database information of the changing features/objects accompanied by user interactive participation. Improvements are also welcome in the design of the graphical variables of change representation and the systems interfaces, which are the means for the eye-brain process to perceive the generated image and understand the various spatial or thematic relationships.

In this chapter various theoretical and practical aspects for the visualization of time-dependent geospatial information have been examined. Change representation models have been presented, with the feature-based time-series being the approach to further explore. The cartographic environment and graphical variables for the visualization of changes have been discussed together with various methods and techniques for the static and dynamic display of spatio-temporal changes. Because of the dynamic nature of spatial change, emphasis was placed on the computer-generated cartographic animation methods. The visualization of attribute changes was also addressed.

The ever-evolving digital technologies facilitate the handling of large amount of data. Therefore the visualization of change has become technically more feasible and more effective.

References

Andrienko, N., G. Andrienko and P. Gatalsky. 2002. Data and task characteristics in design of spatio-temporal data visualization tools. *Proceedings of the ISPRS Commission IV-SDH-CIG Joint International Symposium on Geospatial Theory, Processing and Applications,* available on CD-ROM. Held in Ottawa, Canada, 9-12 July 2002. <http://www.isprs.org/commission4/proceedings02/pdfpapers/040.pdf> Accessed 21 August 2008.

———. 2003. Exploratory spatio-temporal visualization: an analytical review. *Journal of Visual Languages and Computing* 14 (6):503–541.

Andrienko, G. and N. Andrienko. 2007. Multimodal analytical visualization of spatio-temporal data. In *Multimedia Cartography,* 2nd edition, edited by W. Cartwright, M. P. Peterson and G. Gartner, 327–346. Berlin: Springer-Verlag.

Armenakis, C. 1992. Electronic mapping of time-dependent data. Pages 445–454 in *Proceedings of the ASPRS/ACSM/RT'92 Annual Convention, Vol. 3.* Held in Washington, DC, August 1992. Bethesda, Md.: ASPRS.

———. 1993. Map animation and hypermedia: tools for understanding changes in spatio-temporal data. Pages 859–868 in *Proceedings of the Canadian Conference on GIS 1993,* CIG. Held in Ottawa, Canada, March 1993.

———. 1996. Mapping of spatio-temporal data in an active cartographic environment. *Geomatica* 50 (4):401–413.

Armenakis C., A. Müller, E. Siekierska and P. Williams. 2006. Visualization of spatial change. In *Geographic Hypermedia: Concepts and Systems,* edited by E. Stefanakis, M. Peterson, C. Armenakis and V. Delis, 347–367. Berlin: Springer-Verlag.

Asche H. and C. M. Herrmann. 1993. Electronic mapping systems – a multimedia approach to spatial data use. Pages 1101–1108 in *Proceedings of the 16th International Cartographic Conference, Vol. 2.* Held in Cologne, Germany, 3-9 May 1993.

Bertin, J. 1967. *Sémiologie Graphique: les diagrammes, les réseaux, les cartes.* Paris: Mouton.

Blok, C. A. 2005. Dynamic visualization variables in animation to support monitoring. *Proceedings of the 22nd International Cartographic Conference,* available on CD-ROM. Held in A Coruña, Spain, 9-16 July 2005. International Cartographic Association.

Brewer, C. A. 1999. Color use guidelines for data representation. Pages 55–60 in *Proceedings of the Section on Statistical Graphics,* held in Baltimore, Maryland, 1999. Alexandria, Va.: American Statistical Association.

Brinkman, R. M. 1990. 3-D graphics from Alpha to Z-buffer. *BYTE* July: 271–278.

Buttenfield, B. P. and J. H. Ganter. 1990. Visualization and GIS: What should we see? What might we miss? Pages 307–317 in *Proceedings of the 4th Int. Symposium on Spatial Data Handling, Vol. 1.* Held in Zürich, Switzerland, 23-27 July 1990.

Buziek, G. 1999. Dynamic elements of multimedia cartography. In *Multimedia Cartography,* edited by W. Cartwright, M. P. Peterson and G. Gartner, 231–244. Berlin: Springer.

Campbell, C. S. and S. L. Egbert. 1990. Animated cartography: thirty years of scratching the surface. *Cartographica* 27 (2):24–46.

DiBiase, D., A. M. MacEachren, C. Reeves and A. Brenner. 1991. Animated cartographic visualization in Earth Systems Science. Pages 223–232 in *Proceeding of the 15th International Cartographic Association Conference "Mapping the nations"* held in Bournemouth, UK, 23 Sept-1 October 1991. International Cartographic Association.

DiBiase D., A. M. MacEachren, J. B. Krygier and C. Reeves. 1992. Animation and the role of map design in scientific visualization. *Cartography and Geographic Information Systems* 19 (4):201–214.

Foley, J. D., A. van Dam, S. K. Feiner and J. F. Hughes. 1990. *Computer Graphics: Principles and Practice,* 2nd edition. Reading, Mass.: Addison-Wesley Publishing Company, 1174 pp.

Gersmehl, P. J. 1990. Choosing tools: nine metaphors of four-dimensional cartography. *Cartographic Perspectives* (5):3–17.

Gimblett, H. R. and G. L. Ball. 1991. Adaptation in natural systems: an intelligent action model which incorporates artificial intelligence (AI) within a spatial dynamic modeling framework. Pages 996–1008 in *Proceedings of the Canadian Conference on GIS,* held in Ottawa, Canada, March 1991.

Gilbert J. C. and J. Richmond. 1981. Combining computer animation and television presentation: a case study - The open university mathematics course. In *Computers for Imagemaking,* edited by D. R. Clark, 105–130. Oxford, UK: Pergamon Press.

Harrower, M. 2002. Visualizing change: using cartographic animation to explore remotely-sensed data. *Cartographic Perspectives* (39):30–42.

———. 2003. Tips for designing effective animated maps. *Cartographic Perspectives* (44):63-65.

Hurni, L., H-R Bär and R. Sieber. 1999. The Atlas of Switzerland as an interactive multimedia atlas information system. In *Multimedia Cartography,* edited by W. Cartwright, M. P. Peterson and G. Gartner, 99–112. Berlin: Springer-Verlag.

Isakowski, Y. 2003. Visualisation of dynamic glacier processes with SVG animation. *Proceedings of the 2nd Annual Conference SVG Open 2003,* held Vancouver, Canada, 13-18 July 2003. <http://www.svgopen.org/2003/papers/DynamicGlacierProcesses/index.html> Accessed 21 August 2008.

Kousoulakou, A. and M. J. Kraak. 1992. Spatio-temporal maps and cartographic communication. *Cartographic Journal* 29 (2):101–108.

Kraak M. J. and A. M. MacEachren. 1994. Visualization of the temporal component of spatial data. Pages 391–409 in *Proceedings of the 6th International Symposium on Spatial Data Handling (SDH 94), Vol. 1,* held in Edinburgh, Scotland, 1994.

Kraak, M. J. and A. Klomp. 1995. A classification of cartographic animations: towards a tool for the design of dynamic maps in GIS environment. Pages 29–35 in *Proceedings of the seminar on teaching animated cartography,* held in Madrid, Spain, 30 August – 1 September, 1995. Edited by F. Ormeling, B. Köbben and R. Perez Gomez. Utrecht: International Cartographic Association.

Kraak, M. J., R. Edsall and A. M. MacEachren. 1997. Cartographic animation and legends for temporal maps: exploration and/or interaction. Pages 253-260 in *Proceedings of the 18th ICA Conference ICC 1997,* Vol. I. Held in Sweden, 23-27 June, 1997.

Kraak, M. J. 2007. Cartography and the use of animation. In *Multimedia Cartography,* 2nd edition, edited by W. Cartwright, M. P. Peterson and G. Gartner, 317–325. Berlin: Springer-Verlag.

Langran, G. N. and N. Chrisman. 1988. A framework for spatio-temporal information. *Cartographica* 25 (3):1-14.

MacEachren, A. M. 1994. Time as cartographic variable. In *Visualization in Geographical Information Systems,* edited by H. M. Hearnshaw and D. J. Unwin, 115–130. New York: John Wiley & Sons Ltd.

Magnetat-Thalman, N. and D. Thalman. 1985. *Computer Animation-Theory and Practice.* Tokyo: Springer-Verlag. 240 pp.

McCormick, B. H., T. A. DeFanti and M. D. Brown. 1987. Visualization in scientific computing. *Computer Graphics* 21 (6):6–26.

Moellering, H. 1980. The real-time animation of three-dimensional maps. *The American Cartographer* 7 (1):67–75.

Monmonier, M. 1989. Graphic scripts for the sequenced visualization of geographic data. Pages 381–389 in *Proceedings of the GIS/LIS'89,* held in Orlando, Florida, 26-30 November 1989. Bethesda, Md.: ASPRS.

———. 1990. Strategies for the visualization of geographic time-series data. *Cartographica* 27 (1):30–45.

Muller, J.-P., T. Day, J. Kolbusz, M. Dalton, S. Richards and J. C. Pearson. 1988. Visualization of topographic data using video animation. Pages 602–615 in *Proceedings of the XVIth Congress of ISPRS, Vol. B4, Com. III/IV,* held in Kyoto, Japan, July 1988.

Natural Resources Canada. 2005. Dynamic terrain visualization: cartographic visualization on the Internet. <http://maps.nrcan.gc.ca/visualization/results/terrain_visual.html> Accessed 4 October 2008.

Oberholzen, C. and L. Hurni. 2000. Visualization of change in the interactive multimedia Atlas of Switzerland. *Computers & Geosciences* 26 (1):37–43.

Ogao, P. J. and M. J. Kraak. 2002. Defining visualization operations for temporal cartographic animation design. *International Journal of Applied Earth Observation and Geoinformation* 4 (1):23–31.

Ormeling, F. 1995. Teaching animated cartography. *Proceedings of the Seminar on Teaching Animated Cartography,* held in Madrid, Spain 30 August - 1 September 1995. Edited by F. J. Ormeling, B. J. Köbben and R. Perez Gomez. Madrid/Utrecht: International Cartographic Association. <http://cartography.geog.uu.nl/ica/Madrid/ormeling.html> Accessed 4 October 2008.

Peterson, M. P. 1995. *Interactive and Animated Cartography.* Englewood Cliffs, N.J.: Prentice Hall, 257 pp.

Peuquet, D. J. 1994. It's about time: a conceptual framework for the representation of temporal dynamics in geographic information systems. *Annals of the Association of American Geographers* 84 (3):441–461.

Siekierska, E. M. 1983. Towards an electronic atlas. Pages 464–474 in *Proceedings of the Auto-Carto VI Conference* held in Ottawa, Canada, 16-21 October 1983.

———. 1984. Towards an electronic atlas. Cartographica 21 (2,3):110–120. (Reprint of the paper from the Auto-Carto VI Conference.)

———. 1990. Electronic atlas of Canada and electronic mapping projects. Pages 45–51 in *Proceedings of the ICA National Atlases Commission Meeting,* Beijing, China, publication of the China Cartographic Publishing House.

Siekierska, E. and C. Armenakis. 1999. Territorial evolution of Canada: an interactive multimedia cartographic presentation. In *Multimedia Cartography,* edited by W. Cartwright, M. P. Peterson and G. Gartner, 131–140. Berlin: Springer-Verlag.

Siekierska E., K. Francis, J-L. Moisan, D. Mouafo, A. Muller and J. Shang. 2001. Cartographic solutions for visualization of the northern city of Iqaluit, Nunavut, Canada. Pages 84–94 in *Proceedings of the Workshop on Maps and the Internet,* Guangzhou, China, publication of the Publishing House of Journal of South China Normal University.

Slocum, T. A., R. B. McMaster, F. C. Kessler and H. H. Howard. 2005. *Thematic Cartography and Geographic Visualization,* 2nd edition. Upper Saddle River, N. J.: Pearson Education.

Steketee, S. N. and N. I. Badler. 1985. Parametric key-frame interpolation incorporating kinetic adjustment and phasing control. *Computer Graphics (SIGGRAPH '85)* 19 (3):255–262.

Szegö, J. 1987. *Human Cartography: Mapping the World of Man.* Stockholm: Swedish Council for Building Research. 237 pp.

Tufte, E. R. 1990. *Envisioning Information.* Cheshire, Conn.: Graphics Press.

Vasiliev, I. 1996. Design issues to be considered when mapping time. In *Cartographic Design – Theoretical and Practical Perspectives,* edited by C. Wood and C.P. Keller, 137–147. New York: Wiley.

———. 1997. Mapping time. Monograph 49. *Cartographica* 34 (2):1–51.

Wilhelms, J. 1987. Toward automatic motion control. *IEEE Computer Graphics & Applications* 7 (4):11–22.

Williams P., E. Siekierska, C. Armenakis, F. Savopol, C. Siegel and J. Webster. 2006. Visualization and hypermedia for decision making. In *Geographic Hypermedia: Concepts and Systems,* edited by E. Stefanakis, M. Peterson, C. Armenakis and V. Delis, 309–328. Berlin: Springer-Verlag.

CHAPTER 24

Spatial Temporal Modeling of Endemic Diseases: Schistosomiasis Transmission and Control as an Example

Bing Xu and *Peng Gong*

24.1 Infectious Disease Transmission, Globalization and Environmental Change

The emergence of new infectious diseases and re-emergence of previously controlled infectious diseases have attracted a significant amount of attention from scientists, professionals, politicians and the general public. The relationship among infectious disease, globalization, and environmental change, however, is a very complicated one that presents a difficult challenge to scientists from many disciplines covering biological and medical, social and environmental sciences that study pathogens, human environment and the impact of environmental change. Infectious disease dispersion is becoming more rapid and more extensive due to economic globalization. The impact of infectious diseases is often related to the population movement of the entire world. Severe Acute Respiratory Syndrome (SARS) rapidly spread over 30 countries and regions during a period of less than half a year from the beginning of 2003, leading to over 8,000 infected people and over 700 deaths (CDC undated). The West Nile virus, originating from Uganda, was found in New York in 1999, and had spread to over 44 states by 2002; in 2003 and 2004, the West Nile virus had infected over 12,000 people, killing 350 (CDC 2008). On average, interpandemic influenza took 5.2 weeks to spread across the lower United States during 1972 to 2002 (Viboud et al. 2006). After battling schistosomiasis for many years along the Yangtze River Basin, many counties in China had the disease under control for some time. However, there have been recent resurgences in many counties. In 2004 alone, seven counties that used to have schistosomiasis under control had resurgences of the disease (Liang et al. 2006).

Human activities in combination with natural forces (solar radiation and geologic changes) are causing global environmental change at various space and time scales. Environmental changes join force with human activities causing the speedup of exotic-species invasion and spread of infectious diseases (Figure 24-1). Transmission of infectious disease in this context is clearly a complex problem involving both natural and social factors.

However, factors that dominate the spreading mechanism of a particular infectious disease are mostly unknown. For example, though the direct reason of SARS dispersal is due to human air travel, we have little understanding of its origin, transmission channel, and media. On the other hand, many other diseases are being carried around spatially by trade and tourism, but do not spread in their new environment. Therefore, we would like to ask which diseases will spread around the globe or a certain region successfully via globalization. What are the social and physical causes? What are their origins, destinations, and spreading channels? What is the likelihood of survival and endemics of a pathogen under new environment?

In order to answer those questions, we need to develop models that can predict the transmission of infectious diseases. We need to understand the history and current endemic region of an infectious disease. We are certain that the increase of human spatial connectivity through socioeconomic interactions among humans from different locations, and environmental change resulting from globalization, are two dominant reasons for the intensification of the spread of infectious diseases. An intensification of infectious diseases can be identified

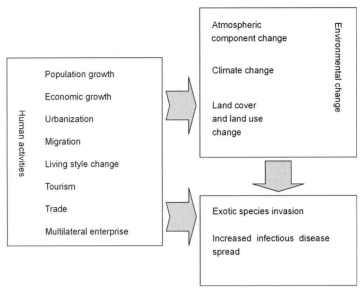

Figure 24-1 Spread of infectious disease by the driving force of human activities and environmental change.

in association with environmental change. Geographic information systems (GIS) and remote sensing are effective tools for the study of spatial connectivity and environmental change, respectively.

24.2 Use of GIS, Remote Sensing and Related Technology in Infectious Disease Studies

There are three stages in predicting the transmission of infectious diseases: (1) identification of the pathogen, its animal host, and its pathway of transmission among the hosts; (2) determining the spatial transmission pattern of each infectious disease, particularly the relationship among the distribution of the disease, its environment and host behaviors; (3) understanding the dynamic process of the transmission of the disease, using models calibrated with field survey data (Rogers and Randolph 2003). The epidemiological model thus established will have the capability to predict the dispersal of the virus and its likelihood of transmission in a new environment. However, each of these stages is difficult to complete. The work at stage one is a diagnosis and initial exploration of the disease. The problem is a biological one.

The second stage involves the survey and quantitative description of the spatial and temporal pattern of an infectious disease, followed by an analysis of the relationship of the disease with its physical and social environment. Vector-borne diseases are strongly influenced by environmental factors. Changes in environmental conditions will alter the distribution of vectors of infectious diseases. Geographic information systems (GIS), remote sensing, global positioning systems (GPS), and statistical methods are most suited to dealing with problems at this stage (Schroder 2006). Researchers are advocating the use of a landscape ecology approach to solving epidemiological problems with these technologies (Kitron 1998).

Remotely sensed data provide us with information about the condition and dynamics of land cover, human land use, surface temperature, soil moisture, and vegetation growth, among other environmental parameters. Such information is very useful in indirectly predicting the abundance of virus transmission vectors, including mosquitoes, ticks, mice and snails. GPS provides us with a convenient way of locating field survey data and geo-referencing. GIS provides a spatial database containing environmental, social and epidemiological data. Such data provide a basis for statistical analysis. For example, with GIS one can explore the statistical relationship between infectious disease data and environmental data, and then map the risk level in the area of interest (Kolivras 2006). Infection risk mapping is usually based on environmental suitability analysis. This is done by searching for the environment whose conditions meet the requirements of a particular infectious disease (Ron 2005). There are four cases that can result from the comparison between suitability for and actual distribution of a particular disease: first, the actual distribution is located in the suitable area; second, the disease is found in unsuitable areas; third, the suitable area does not have the actual disease; fourth, no disease in unsuitable areas. Though the first and second cases are the most reasonable ones, the third should not be considered incorrect as it is totally possible for a disease that has not reached its maximum spatial extent to leave some suitable areas with no disease. Only when a suitable area is classified as an unsuitable area, is the case incorrect as this will cause ignorance of resource allocation in disease prevention and control. More can be done with GIS (Kistemann et al. 2002). For example, when examining the infection risk distribution of schistosomiasis, one can evaluate the spatial autocorrelation among different residential groups; such information is helpful in the spatial control of the disease transmission (Zhou et al. 1996; Spear et al. 1998).

The third stage is based on the two previous stages. The goal is to establish a quantitative process-based biological model. Such models can be calibrated with field measurements and surveys. In general, these models are limited to modeling the biological reproduction cycle of a particular disease, and relevant hosts (e.g., Anderson and May 1991). However, there are also studies that directly model the dynamics between the environment and the transmission vector. For example, the relationship between the climate and mosquito populations can be modeled. Rogers and Randolph (2003) found that the surface temperature was nonlinearly related to the mortality of Gambia tsetse fly with a one-month time lag.

It is worth mentioning that the three stages of study are usually independently undertaken. It is particularly true for the second and third stages. This has largely limited the understanding of the impact of the environment on the transmission of infectious diseases. Such a limitation further hampers disease control and prevention.

In this chapter, we use schistosomiasis as an example to illustrate the important roles that GIS and remote sensing can play in modeling the interaction between the environment and disease transmission. We assimilate parameters derived from remote sensing and GIS into a dynamic process model to construct an innovative spatial temporal model for schistosomiasis transmission.

24.3 A Conceptual Mathematical Model of the Life Cycle of Schistosomiasis

According to the World Health Organization, schistosomiasis disease is endemic in 74 countries with approximately 120 million people infected and over 600 million people at risk. In China, there are approximately 800,000 people infected and over 60 million at risk. In recent years, the patterns of endemic occurrences and control of schistosomiasis have been changing due to changes in socioeconomic and natural factors. On the one hand, some endemic areas have intensified the risk of schistosomiasis infection. On the other hand, some areas that were not endemic areas in the past, for example, in the mountainous areas of Sichuan Province, where snails are present but have had no historically reported schistosomiasis infection, have become endemic areas. How and where new snail habitats will emerge according to recent environmental changes have increasingly attracted research attention.

Our model is obtained by adding a spatial connectivity component, and simplifying the temporal dynamics from a detailed dynamics model in Liang et al. (2002). According to the life cycle of schistosomiasis in Figure 24-2, we can build such a conceptual model. We first describe the number of adult worms in the final host as a function of time

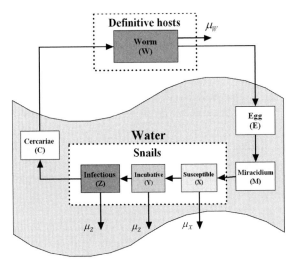

Figure 24-2 The life cycle of schistosomiasis. Eggs, E, are excreted with the stool of humans and other animal hosts (i.e., water buffalo and other cattle). They then hatch into miracidia, M, in water with suitable temperature. Miracidia must penetrate into snails within 48 hours in the water, causing the susceptible snails, X, to be infected into Y. The incubation of the miracidia will last 30-60 days to asexually reproduce into cercaria, C. The number of infected snails that survive to this stage are noted as Z. Cercaria must penetrate the skin of their final host in the water within 48 hours and find their final destination near the liver organ of the host and grow into the adult schistosome, W, where they pair to reproduce eggs and complete the life cycle. Humans and cattle get infected by schistosomiasis through contact with water contaminated by cercaria (e.g., work in the field, grazing, washing, swimming, etc.). See included DVD for color version.

$$\frac{dW_i}{dt} = \beta_i c_i - \mu_w w_i - \pi_i w_i \tag{24-1}$$

where W_i is the worm load in village i (or country, or group of humans, or cattle) at time t; β_i is the infection rate through water contamination with cercaria (density c_i); μ_w is the natural death rate of schistosomes in the host; π_i is the death rate of schistosomes by medical treatment. In the life cycle, egg production by unit time is modeled by

$$e_i = \frac{1}{2} hgn_i W_i \phi \tag{24-2}$$

where e_i is the average egg production of n_i infected people in group i. Half of the adult worms produce eggs of quantity h, produced by each worm. The number of eggs contained in each gram of stool is g. Because the life cycle of the adult worm is longer than the other forms, we will not use a differential equation to model the number of eggs. The successfully hatched eggs into miracidia is determined by

$$m_i = \sum_{j=1}^{n} \frac{\alpha Z_j a_j S_{ij}}{b_i} \tag{24-3}$$

where m_i is the density of miracidia in group i, which includes the import and export of miracidia from and to other groups; the hatching rate is Σ and the area of water surface is b_i; the redistribution coefficient among neighboring groups is S_{ij}; n represents the number of groups. The key to the spatial dispersion of schistosomes is to determine the spatial distribution coefficients as a function of the spatial interaction at different levels. This is determined by the spatial interaction processes at different scales. In general, a hierarchical scheme is needed to construct the spatial distribution coefficient. Because we lack the spatial interaction data at multiple scales, we only use a distance (e.g., 1.5 km) to determine the interconnection among neighboring groups through the ditch networks. This is done with GIS (see the next section). The infection of snails by miracidia in the water is described by

$$\frac{dZ_i}{dt} = \rho m_i x_i - \mu_z Z_i \tag{24-4}$$

where Z_i is the number of infected snails that are shedding cercaria; this is determined by the total snail density, x_i, and the density of miracidia, m_i, the infection rate of snails, ρ, and the death rate of snails, μ. Snail density, x_i is obtained with remote sensing methods (see Section 24.4). Finally, the cercaria production is determined by

$$c_i = \sum_{j=1}^{n} \frac{\sigma Z_j a_j S_{ij}}{b_i} \tag{24-5}$$

where σ is the daily cercaria production rate of each infected snail; a_j is the water area of snail habitat. The above model omitted the infected latent snail, Y, as shown in Figure 24-2. To obtain Z from Y, one only needs to multiply with a death rate of infected snails. The above five equations provide us with a spatio-temporal model for schistosomiasis transmission.

The model described above is based on the following assumptions: (1) the accessible and associated immunization among host groups during the infection process is ignored; (2) there is no density dependence in schistosomiasis infection, this assumption is reasonable when the modeling period is not long (~5 years); (3) there is no relationship between the number of miracidia that infect snails and the number of cercaria shed by infected snails; (4) the aggregation distribution parameter, k, in host groups is constant; (5) the host population during the modeling process is constant; (6) annual climate does not change (although the model can accommodate climate change with available data). Values for each parameter in this model are listed in Table 24-1.

Table 24-1 Parameter ranges in the schistosomiasis transmission model.

Parameters	Interpretation and unit	Ranges	References
τ_w	Development time of worms in human hosts (day)	20 – 40	(Anderson and May 1991)
μ_w	Worm natural mortality (/day)	0.000183 – 0.0014	(Anderson and May 1991)
b	Eggs excreted (/worm pair /gram feces)	0.768 – 2.72	(Hubbard et al. 2002)
μ_s	Snail mortality rate (/day)	0.0023-0.007	(Zhao et al. 1995)
μ_z	Patent and latent snail death rate (/day)	0.0063 – 0.033	(Xie et al. 1990)
σ	Cercarial production (/sporocyst/day)	20 – 50	(Pesigan et al. 1958; Qian et al. 1997)
p_i	Efficacy of praziquantel	0.8 – 0.95	(Stelma et al. 1995; Liang et al. 2000)
β	Schistosome infection rate (/cercaria/ m^2 contact)	0.0001 – 0.5	Model calibration
ρ	Snail infection (/miracidium/m^2 surface water)	0.000001 – 0.0005	Model calibration
w_{0i}	Initial worm burden in the ith group	Data estimate	Local data
z_0	Initial density of infected snails	Data estimate	Local data and satellite image
x_0	Initial density of susceptible snails	Data estimate	Local data and satellite image
κ_{0i}	Initial worm aggregation parameter	Data estimate	Local data

Distributions for all parameters are uniform except for α and ρ, which have log-uniform distributions.

24.4 Spatial Interaction Determination Between Neighboring Villages Through GIS

We selected an endemic area of schistosomiasis in Xichang surrounding Qionghai Lake for our study area. The area has 227 natural villages covered by one scene of Ikonos imagery (11km × 11km) (27°47'-27°50' N, 102°14'-102°18' E) (Figure 24-3), and is located in the western mountainous area in Sichuan Province at an elevation between 1500m and 2700m. We constructed a digital elevation model (DEM) with a grid size of 15 m, based on a stereo pair of ASTER images acquired in August of 2002 (Figure 24-4). The precision of this DEM is assessed by taking GPS measurements of 29 points in the study area, resulting in an average error of less than 6 m. This level of accuracy is sufficient for our analysis.

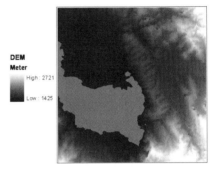

DEM
Meter

High : 2721

Low : 1425

Figure 24-3 The Ikonos image of Xichang acquired in December of 2000. The false color image is made from a combination of green, red and near infrared bands displayed with a color gun of blue, green and red, respectively. The yellow areas are the ditches where field snail surveys were conducted in 19 villages.

Figure 24-4 The digital elevation model of the study area derived from a stereo pair of ASTER images. The red area is the Qionghai Lake extracted from image analysis of the image in Figure 24-3.

See included DVD for color versions of Figures 24-3 and 24-4.

We digitized the boundaries of 227 villages, and input them into a GIS database. In the GIS, we calculated the geometric center of each natural village, and then calculated the inter-village distance between the neighboring villages. We also calculated the number of neighboring villages for each village. With the boundaries of the natural villages overlaid on top of the DEM, we calculated the average elevation for each village. The slope between the center points and the direction of ditch water flow between the neighboring villages was determined by comparing the average elevation between neighboring villages. The amount of miracidia and cercaria exchange and retention was determined by water flow direction, slope between neighboring villages, and the area of each individual village. Miracidia and cercaria flowed in the direction of the water flow. Therefore, only villages at lower elevations received inputs from villages at higher elevations.

We constructed a village-village spatial connection matrix S. The ith row and jth column, S_{ij}, represents the number of miracidia and cercaria transport from the ith village to the jth village. In this study we did not consider the migration of snails and eggs, because the active movement of snails is rather limited in space, and the eggs stored in stool are mostly accumulated by individual farmers and applied to their own fields near their house. The interaction of eggs between neighboring villages is limited. Passive movement of snails is usually caused by the transport of agricultural products, but such numbers are usually low. Therefore, we ignored this. The diagonal elements in S represent the retention rates of the villages. The retention rate, S_{ii}, is related to the area and slope of the village. The calculation is done by setting an upper and lower bound (0.3-0.9), and by building a linear model of slope and area. The retention rate is then calculated as the average between the outcomes of the slope function and area function. The Sichuan Institute of Parasitic Diseases conducted some observations of the viability of cercaria and miracidia with respect to the hydrological condition in the study area, finding that their distance of movement during half of a lifetime in water is 400m. However, the average diameter of natural villages is 500m. Therefore, we did not consider indirectly connected villages. Clearly, S is not symmetric. If $S_{ij}>0$ then $S_{ji}=0$, because miracidia and cercaria can only flow from higher places to lower places. S_{ij} is estimated by

$$S_{ij} = (1 - S_{ii})\omega_j$$
$$\omega_j = \beta_{ij}/\sum_{k=1}^{ne} \beta_{ik} \qquad\qquad (24\text{-}6)$$

where ne represents the effective neighbors of the ith village, that is, those villages having a higher elevation than village i; β is the slope; and ω is the normalization factor.

Applying the above method, we get the spatial interaction matrix as displayed in Figure 24-5. Substituting S_{ij} into equations 24-3 and 24-5 allows us to redistribute the miracidia and cercaria numbers among the 227 villages. Xu et al. (2006) introduced a simpler version of this spatial interaction matrix construction method.

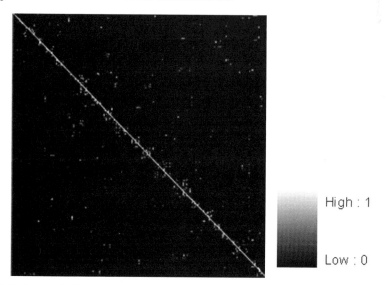

High : 1

Low : 0

Figure 24-5 The spatial interaction (connectivity) matrix constructed in GIS with data from a DEM and village boundaries. This matrix has 227 rows and 227 columns.

24.5 Snail Density Estimation with Remote Sensing

Previous research involving remote sensing for schistosomiasis control primarily concentrated on the mapping of potential snail habitats. Xu et al. (2004) attempted to construct a statistical relationship between field survey snail densities and land cover information derived from remote sensing data, producing a snail density for the entire study area. They used 4-m resolution Ikonos multispectral imagery and a DEM derived from ASTER images, spatially densified into 4-m grids in a land cover classification. As a result, 16 land cover types were obtained (Figure 24-6). Field validation indicates that the average accuracy of this map is 89%. Classification accuracies of the major cover types such as residential areas, floodplains, crop areas, riverbeds, lowland terraces, highland terraces, and forest areas are all greater than 87%. The land cover types are intentionally schemed so that they are not sensitive to season. In this manner, it is possible to establish statistical relations with imagery obtained from different seasons for snail density estimation. In order to find out if detailed land cover information with such a classification scheme can be derived from remote sensing data at lower resolutions, the land cover data derived from the 4-m data were converted to fractional cover data based on an aggregation of 7×7 pixels. They then applied a linear model to estimate snail density,

$$SA = a_1 f_1 + a_2 f_2 + \dots a_i f_i \dots a_k f_k \tag{24-7}$$

where f is the fraction of area for a particular land cover type calculated from the 7×7 pixel window; SA is the estimated snail density; and a's are the coefficients of the model. The snail density for each village was calculated. Using over 10,000 snail sampling points collected in the ditches from 19 villages, they found that the R2 value between survey data and the estimation data could be as high as 0.87.

Because it is very time- and labor-intensive to do snail surveys in the field, we only did field surveys in the summer of 2001. We surveyed every ditch for 19 villages at 10-m intervals. At each sampling site, we placed a quadrant frame (0.11 m²) to survey the snail density. A shortcoming of this experiment was that we did not have independent samples to validate our statistical model, since we used all the sampling data when building the multivariate statistical model in Equation 24-7. Additionally, snail density changes with season. Another shortcoming of this research was that we did not use satellite data from multiple times to estimate snail density variation in time. However, since the purpose of this study is mainly to test the feasibility of the conceptual model for spatial temporal dynamics simulation, this is sufficient. With Equation 24-7, we can calculate the snail density for each village. By then using the village area, the snail number in each village can be obtained (Figure 24-7). Snail number can then be applied in equation 24-4.

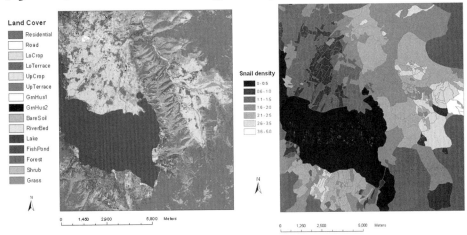

Figure 24-6 Land cover map of the Xichang area based on Ikonos and DEM data. See included DVD for color version.

Figure 24-7 Snail density map of the study area estimated with land cover fraction data. See included DVD for color version.

24.6 Simulation Results

Based on model parameter adjustments, we used Equations 24-1 to 24-5 to simulate the schistosomiasis transmission dynamics. With spatial connectivity, our temporal dynamics model became a spatial temporal model. We tested the model from 15 June 2000 and ran it on a daily basis for 5 years. The worm load in year 1 and year 5 is shown in Figure 24-8. Clearly, if there is no schistosomiasis control the worm load increases annually. The areas showing no change are forested areas with no human settlement. The total number of simulated worm load after 5 years is 789,467.

(a)   **(b)**

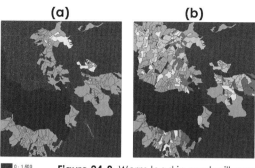

▉ 0 - 1,500	
▓ 1,501 - 4,000	
░ 4,001 - 5,000	
░ 5,001 - 7,000	
▓ 7,001 - 14,000	
▉ Lake	

Figure 24-8 Worm load in each village as simulated with the spatial temporal model. (a) model simulation results for the first year and (b) model simulation results for the fifth year. See included DVD for color version.

The effect of schistosomiasis control is easy to examine with the spatial temporal model. If only the patients in 5 villages can be treated for 1 week, our model can help us to answer which villages should be selected to maximize the control effect. For example, we can select the villages with the greatest worm load for treatment. Or, we can select the villages with the strongest spatial connectivity. We can also consider both, that is, those villages with high worm load and also high connectivity with a large number of effective neighbors. Figure 24-9 compares two treatment plans: treating the 5 villages with the greatest worm load and treating the 5 villages with high worm load and high spatial connectivity. Treating the 5 villages with the greatest worm load caused a reduction of 13,211 worms, while treating the 5 villages with both high worm loading and high spatial connectivity led to a worm load reduction of 17,505. From Figure 24-9, the influence of the later treatment plan can reach many more villages. Therefore, it is necessary to compare different control strategies by considering spatial connectivity. The advantage of this spatio-temporal model for schistosomiasis transmission and control is that it allows us to develop and compare various control plans, in order to select the optimal ones to support control decision making.

(a)   **(b)**

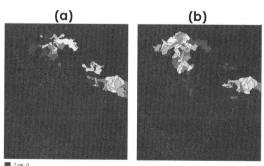

Low: 0

High: 3256

Figure 24-9 Simulation results from the spatio-temporal model treating patients in 5 villages for one week each year. Two different village selection plans were compared: worm load reduction resulting from (a) treating 5 villages with the greatest worm load, and from (b) treating 5 villages with high worm load and high spatial connectivity. See included DVD for color version.

Figure 24-10 compares the simulation results without schistosomiasis control to results with the patient treatment from the second village selection plan. Only the cercaria number and worm load are shown in the figure. Each curve represents results for one village in the 5 year simulation period. There is a clear distinction between the two simulations for cercaria production and worm loading in each village. Because treating 5 villages primarily kills the worm in hosts, some of the worm load drops in the curves can be clearly seen in Figure 24-10d.

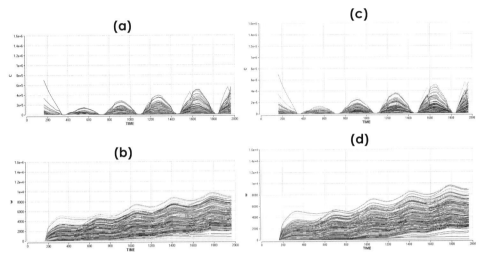

Figure 24-10 The simulation results for (a) cercaria production and (b) worm loading, based on the spatio-temporal model without any control.

The simulation results for (c) cercaria production and (d) worm loading, based on the spatial temporal model with a control plan of treating patients in 5 villages that have high worm load and high spatial connectivity. See included DVD for color version.

24.7 Summary and Discussions

The above results demonstrated three things:

1. A conceptual model for the spatio-temporal schistosomiasis transmission dynamics can be realized.
2. Remote sensing and GIS are indispensable components in such spatio-temporal models.
3. It is possible to use a spatio-temporal schistosomiasis transmission model in supporting spatial decisions, to improve the effectiveness of schistosomiasis control.

In comparison with most existing GIS applications, our conceptual framework allows the assimilation of GIS data and remote sensing data from multiple sources and times. Previous efforts involving GIS have mostly been focused on the function of GIS as a database or some preliminary analysis tools. For example, within a GIS, empirical models between environmental factors and occurrence risk of infectious diseases are built through statistical regression. Others use spatial statistical techniques in health data interpolation, hot spot detection and spatial cluster identification (Munch et al. 2003). GIS analyses are hardly dynamic although being able to deal with time in a GIS is an important feature—recognized some time ago—for solving health related problems and for the establishment of early warning systems (Kistemann et al, 2002). Our conceptual model provides a natural framework to involve time in the modeling. This is one of the important features in the work reported here. In addition, the spatio-temporal modeling framework can be expanded to include other spatial connectivity such as population migration and material transport. More detailed transmission at the individual level can also be accommodated by embedding in the host infection part in Figure 24-2 a different model such as a cellular automata model or an agent-based model. Such models have been used alone in health studies (Patlolla et al. 2006; Venkatachalam and Mikler 2006)

A lot of work needs to be done in order to make the model practical. Firstly, we need to further investigate snail density estimation methods based on remotely sensed data from multiple sources. Secondly, spatial connectivity exists at multiple scales and among different environmental factors, and more work needs to be done in this aspect. For example, the spatial unit in this study is natural villages; but, populations can be divided at even finer units, such as at the occupational level, family level, or even at the individual level. On the other hand, scaling up from the village unit to the township and county level still remains to be resolved. Thirdly, more field data need to be collected to validate the models developed.

The transmission of schistosomiasis only represents one type of transmission process for an infectious disease interacting with vectors, intermediate host and various environmental factors. Each infectious disease has its original endemic area. Its spatial transmission mainly relies on the natural forces, such as climate variation, vegetation succession, and atmospheric and ocean circulation. Human activity promotes the transmission of infectious disease to new environments, by facilitating species invasion. Thus, human activity in the infectious disease transmission system acts as a positive feedback. We must have a better understanding of this positive feedback system, establish a better prediction model, and improve our prevention capacity. The model proposed in this study sheds lights on the spatio-temporal modeling of other infectious diseases. When the biological and environmental processes between the origin of the disease, its vector and host are relatively clear, we can adopt such models to predict the spatial and temporal dynamics of an infectious disease. For example, the plague, hemorrhagic fever with renal syndrome, and lyme disease can be modeled using the conceptual framework proposed here. Remote sensing and GIS can play important roles in supporting the spatial decisions in controlling infectious disease transmission.

Acknowledgements

This research is partially funded by grants from National Science Foundation of China (30590372), the 10th 5 year key project (2004BA718B06) and NIH (RO1-AI-43961). We are grateful to the research group led by Professors Xueguang Gu and Dongchuan Qiu of the Sichuan Institute of Parasitic Diseases of China, and to Professor Bob Spear, Professor Song Liang and Dr. Edmund Seto from the School of Public Health at University of California Berkeley for their help, discussions and data provision.

References

Anderson R. M. and R. M. May. 1991. *Infectious Diseases of Humans: Dynamics and Control.* London: Oxford University Press.

CDC (Centers for Disease Control and Prevention). Undated. Severe Acute Respiratory Syndrome (SARS). <http://www.cdc.gov/ncidod/sars/> Accessed 17 October 2008.

CDC (Centers for Disease Control and Prevention). 2008. West Nile Virus. <http://www.cdc.gov/ncidod/dvbid/westnile/> Accessed 17 October 2008.

Hay, S. I., C. Guerra, A. J. Tatem, P. M. Atkinson and R. W. Snow. 2005. Urbanization, malaria transmission and disease burden in Africa. *Nature Reviews: Microbiology* 3 (1):81–90.

Hubbard, A., S. Liang, D. Maszle, D. Qiu, X. Gu and R. C. Spear. 2002. Estimating the distribution of worm burden and egg excretion of Schistosoma japonicum by risk group in Sichuan Province, China. *Parasitology* 125 (3):221–231.

Kistemann, T., F. Dangendorf and J. Schweikart. 2002. New perspectives on the use of Geographical Information Systems (GIS) in environmental health sciences. International *Journal of Hygiene and Environmental Health* 205 (3):169–181.

Kitron, U. 1998. Landscape ecology and epidemiology of vector-borne diseases: Tools for spatial analysis. *Journal of Medical Entomology* 35 (4):435–445.

Kolivras, K. N. 2006. Mosquito habitat and dengue risk potential in Hawaii: a conceptual framework and GIS application. *Professional Geographer* 58 (2):139–154.

Liang, S., D. Mazsle, R. Spear. 2002. A quantitative framework for a multi-group model of Schistosomiasis japonicum transmission dynamics and control in Sichuan, China. *Acta Tropica* 82:263–277.

Liang, S., C. Yang, B. Zhong and D. Qiu. 2006. Re-emerging schistosomiasis in hilly and mountainous areas in Sichuan, China. *Bulletins of World Health Organization* 84:139–144.

Liang, Y. S., G. C. Coles and M. J. Doenhoff. 2000. Short communication: Detection of praziquantel resistance in schistosomes. *Tropical Medicine & International Health.* 5 (1):72.

Munch, Z., S.W.P. Van Lill, C. N. Booysen, H. L. Zietsman, D. A. Enarson and N. Beyers. 2003. Tuberculosis transmission patterns in a high-incidence area: a spatial analysis. *International Journal of Tuberculosis and Lung Disease* 7 (3):271–277.

Patlolla P., V. Gunupudi, A. R. Mikler and R. T. Jacob. 2006. Agent-based simulation tools in computational epidemiology. In *Innovative Internet Community Systems,* edited by T. Böhme, V. M. Larios Rosillo, H. Unger and H. Unger, 212–223. 4th International Workshop IICS 2004, Guadalajara, Mexico, 21-23 June 2004, revised papers. Lecture Notes in Computer Science 3473. New York: Springer.

Pesigan, T. P., N. G. Aristón, J. J. Jáuregui, E. G. Garcia, A. T. Santos, B. C. Santos and A. A. Besa. 1958. Studies on Schistosoma japonicum infection in the Philippines. 2. The molluscan host. *Bulletin of World Health Organization.* 18 (4):481–578.

Qian, B.-Z., J. Qian, D.-M. Xu and M. V. Johansen. 1997. The population dynamics of cercariae of Schistosoma japonicum in Oncomelania hupensis. *Southeast Asian Journal of Tropical Medicine and Public Health* 28 (2):296–302.

Rogers, D. J. and S. E. Randolph. 2003. Studying the global distribution of infectious diseases using GIS and RS. *Nature Reviews: Microbiology* 1:231–237.

Ron, S.R. 2005. Prediction of the distribution of amphibian pathogen *Batrachochytrium dendrobatidis* in the new world. Biotropica 37 (2):209–221.

Schroder, W. 2006. GIS, geostatistics, metadata banking, and tree-based models for data analysis and mapping in environmental monitoring and epidemiology. *International Journal of Medical Microbiology* 296: Suppl.40:23–26.

Spear, R., P. Gong, E. Seto, Y. Zhou, B. Xu, D. Maszle, S. Liang, G. Davis and X. Gu. 1998. Remote sensing and GIS for schistosomiasis control in mountainous areas in Sichuan, China. *Geographic Information Sciences* 4:14–22.

Stelma, F. F., I. Talla, S. Sow, A. Kongs, M. Niang, K. Polman, A. M. DeElder and B. Gryseels. 1995. Efficacy and side effects of praziquantel in an epidemic focus of Schistosoma mansoni. *American Journal of Tropical Medicine and Hygiene* 53:167–170.

Venkatachalam S. and A. R. Mikler. 2006. An infectious disease outbreak simulator based on the cellular automata paradigm. In *Innovative Internet Community Systems,* edited by T. Böhme, V. M. Larios Rosillo, H. Unger and H. Unger, 198–211. 4th International Workshop IICS 2004, Guadalajara, Mexico, 21-23 June 2004, revised papers. Lecture Notes in Computer Science 3473. New York: Springer.

Viboud, C., O. N. Bjørnstad, D. L. Smith, L Simonsen, M. A. Miller and B. T. Grenfell. 2006. Synchrony, waves, and spatial hierarchies in the spread of influenza. *Science* 312 (5772):447–451.

Xie, F. X., G. L. Yin, J. Z. Wu, Y. Duan, X. Zhang, J. Yang, K. Qian, H. Tan, J. Zheng and R. Zhang, 1990. Life span and cercaria shedding of schistosome-infected snails in mountainous region of Yunnan. *Chinese Journal of Parasitology and Parasitic Diseases* 8 (1):4–7.

Xu, B., P. Gong, G. Biging, S. Liang, E. Seto and R. Spear. 2004. Snail density prediction for schistosomiasis control using Ikonos and ASTER images. *Photogrammetric Engineering and Remote Sensing* 70(11):1285–1294.

Xu, B., P. Gong, E. Seto, S. Liang, C. Yang, S. Wen, D. Qiu, X. Gu and R. Spear. 2006. A spatial-temporal model for assessing the effects of intervillage connectivity in schistosomiasis transmission. *Annals of the Association of American Geographers* 96 (1):31–46.

Zhao, W.X., X. G. Gu, F. S. Xu, Y. X. Li, L. G. Zhao, H. Z. Yun, X. J. Li and X. F. Zhou. 1995. An ecological observation of Oncomelania hupensis robertsoni in Xichang, Daliang Mountains, Sichuan. *Sichuan Journal of Zoology* 14 (3):119–121.

Zhou Y., D. Maszle, P. Gong, R. C. Spear and X. Gu. 1996. GIS based spatial network models of schistosomiasis infection. *Geographic Information Sciences* 2:51–57.

CHAPTER 25
Space-Time Paths
Mei-Po Kwan

25.1 Movement of People in Space and Time

Analysis of the movement of objects or people in space and time has become an important theme in Geographic Information Science (GIScience) in the past decade or so (e.g., Hornsby and Egenhofer 2002). Recent studies not only explore new methods for analyzing human movements (e.g., Shoval and Isaacson 2007); some studies also analyze people's "travel" in cyberspace (Ren and Kwan 2007). An effective means for analyzing human movements in space and time is the construction and interactive visualization of the paths of these movements in 3D space, where the X-Y axes represent geographic space and the Z axis represents the progression of time. This chapter on space-time paths focuses mainly on the analysis and visualization of the movements of people in space and time. This area of research covers a wide range of applications such as migration, residential mobility, shopping, travel, and commuting behavior (Kwan 2004; Kwan and Ding 2008).

All conventional quantitative analysis of people's movement faces a major difficulty: individual movement in space-time is a complex trajectory with many interacting dimensions (Kwan 2000b). These include the location, timing, duration, sequencing and type of activities and/or trips. This characteristic of human movements has made the simultaneous analysis of its many dimensions difficult. Two different approaches were adopted to resolve this problem in past research. On one hand, some studies focused on a few component dimensions of these movements at a time (e.g., Golob and McNally 1997; Goulias 1999). On the other hand, there are studies that view the various dimensions of space-time paths as a multidimensional whole and use multivariate methods to derive generalized behavioral patterns from a large number of variables (e.g., Bhat and Singh 2000; Golob 1985; Ma and Goulias 1997a, b; Recker et al. 1987).

The development and application of these quantitative methods have enhanced our understanding of people's movements in significant ways. For instance, through the use of multivariate group identification methods, such as clustering or pattern recognition algorithms, complex patterns in the original data set can be represented by some general characteristics and organized into a relatively small number of homogeneous classes. Further, once human movement patterns are represented in terms of a limited number of categories, they can be related to a large number of attributes of the individuals or households that generate them and used as a response variable in statistical models. While these quantitative or statistical methods are useful for modeling purposes and for discovering the complex interrelations among variables, they also have their limitations.

Since many statistical methods used in past studies (e.g., log-linear models) are designed to deal with categorical data, organizing the original data in terms of discrete units of space and time has been a necessary step in most analyses of human movement. Discretization of temporal variables, such as the start time or duration of activities, involves dividing the relevant span of time into several units and assigning each activity or trip into the appropriate class (e.g., dividing a day into eight or 12 temporal divisions into which activities or trips are grouped). Discretization of spatial variables, such as distance from home, involves dividing the relevant distance range into several "rings." Since both the spatial and temporal dimensions are continuous, results of any analysis that are based upon these discretized variables may be affected by the particular schema of spatial and/or temporal divisions used. The problem may be serious when dealing with the interaction between spatial and temporal variables, since two discretized variables are involved.

Further, few of these methods were designed to handle real geographic locations of human movement in the context of a study area. Often, the spatial dimension is represented by some measures derived from real geographic locations (e.g., distance or direction from a reference point, such as home or workplace of an individual). Further, locational information of activities or trips was often aggregated with respect to a zonal division of the study area (e.g., traffic analysis zones or census tracts). Using such zone-based data, measurement of location and/or distance involves using zone centroids where information about specific activity locations in geographic space and their spatial relations with other urban opportunities is lost (Kwan and Hong 1998). Lastly, as detailed data about people's daily movements and activities have become more readily available in recent years, effective methods for exploring these data are also urgently needed (McCormack 1999). Without them, the researcher may need to model human movement without a preliminary understanding of the behavioral characteristics or uniqueness of the individuals in the sample at hand. This can be costly in later stages of a study if the model's specifications fail to take into account the behavioral anomalies involved.

To overcome these three difficulties in the analysis of human movement, effective conceptual framework and analytical methods are needed. In this chapter, I present the time-geographic perspective as a framework that provides a sound conceptual basis for a range of analytical methods that overcome some of these difficulties. I then discuss some of the GIS-based time-geographic methods that were developed in recent years. These include three-dimensional (3D) geovisualization and geocomputational methods. Usefulness of these methods is illustrated through examples drawn from recent studies by myself and other researchers. I also attempt to show that GIS provides an effective environment for implementing time-geographic constructs and for the future development of operational methods in time-geographic research. I suggest that GIS-based time-geographic methods are effective means for the study of human activities and movements in space and time in the urban context.

25.2 Space-Time Paths and Time Geography

An effective conceptual framework is needed for mitigating the difficulties in the analysis of human movement in space and time. First, we need a conceptual framework for developing methods that avoids the discretization of the spatial and temporal variables. Second, the framework should be attentive to and could provide a basis for taking real geographic locations of human movements and activities into account. Third, these methods should facilitate exploratory spatial data analysis (ESDA), which helps to make modeling efforts more focused and fruitful in later stages of a study. Time geography, whose central construct is the space-time path, offers such a conceptual framework for developing methods for analyzing human movement.

Time geography was developed by a group of Swedish geographers at Lund University in the 1950s and 1960s, including Torsten Hägerstrand, Tommy Carlstein, Bo Lenntorp, and Don Parkes (Kwan 2004). Important constructs in time geography, such as stations, projects, space-time paths and prism constraints, are well articulated in Carlstein et al. (1978), Hägerstrand (1970), Parkes and Thrift (1975), Thrift (1977) and Lenntorp (1978). The first in-depth analytical treatment and operationalization of time-geographic constructs was provided by Lenntorp (1976).

In time geography, an individual's activities and trips in a 24-hour day are conceived as a continuous temporal sequence in geographic space. The trajectory that traces this activity sequence is referred to as a space-time path, while the graphical representation of the three-dimensional space in which this path unfolds is referred to as the space-time aquarium. The number and location of everyday activities that can be performed by a person are limited by the amount of time available and the space-time constraints associated with various obligatory activities (e.g., work) and joint activities with others. These constraints largely arise from the spatial or temporal rigidity associated with certain types of activities people undertake in their daily lives. These activities are called fixed activities (for example, work or visiting a doctor) because it is difficult to change the place or time to perform them, and as a result they also tend to restrict a person's freedom to undertake other spatially and temporally flexible activities. This time-geographic conception is valuable for understanding human activities and movements in space-time because it integrates the temporal and spatial dimensions of human activity patterns into a single analytical framework. Although time, in addition to space, is a significant element in structuring individual activity patterns, past approaches mainly focus on either their spatial or temporal dimension. The significance of the interaction between the spatial and temporal dimensions in structuring individual daily space-time trajectories is often ignored.

Time geography not only highlights the importance of space for understanding the geographies of everyday life. It also allows the researcher to examine the complex interaction between space and time and their joint effect on the structure of human movement in particular places (Cullen et al. 1972). It can be applied in a wide range of fields and research areas. Since the early 1990s, the perspective has been particularly useful for understanding women's activity-travel behavior, because it helps to identify the restrictive effect of space-time constraints on their activity choice, job location, travel, as well as occupational and employment status (Kwan 1999a, b, 2000a; Laws 1997; Tivers 1985). Time geography has also been used as a framework for the study of migration and mobility behavior (Odland 1998), exposure to health risk, and the everyday life of the elderly, children, and homeless people (e.g., Mårtensson 1977; Rollinson 1998). Many transportation researchers have also found the time-geographic perspective useful for modeling human activity-travel behavior.

Despite the usefulness of time geography in many areas of social science research, there are very few studies that actually implemented its constructs as analytical methods up to the mid-1990s—with the notable exception of Bo Lenntorp's Program Evaluating the Set of Alternative Sample Path (PESASP) simulation model. The limited development of time-geographic methods was largely due to the lack of detailed geographic and individual-level data, as well as analytical tools that could realistically represent the complexities of an urban environment (e.g., the transportation network and spatial distribution of urban opportunities). For example, a study in the late 1970s that dealt with 286 urban opportunities spent about three months in the manual construction of a digital street network that only had 939 nodes and 2,395 arcs. So the time needed for constructing the geographic data for time-geographic studies was considerable. Another difficulty was that the algorithms used to implement time-geographic methods were computationally intensive.

However, with increasing availability of digital geographic databases of urban areas and georeferenced individual-level data, as well as improvement in the representational and geocomputational capabilities of Geographic Information Systems (GISs), it is now more feasible than ever before to operationalize and implement time-geographic constructs. Further, the use of GIS also allows the incorporation of large amounts of geographic data that are essential for any meaningful analysis of human movement in space-time. Because of these changes, time-geographic methods are undergoing a new phase of development as several recent studies indicate (Kwan 1998, 2000b; Miller 1999; Ohmori et al. 1999; Takeda 1998; Yu 2006; Weber and Kwan 2002). Although the primary focus of these studies is on individual accessibility, there are many areas in social science research where time-geography can be fruitfully applied.

25.3 Analyzing Human Movement with Geocomputational Methods

The time-geographic framework provides a solid foundation for the analysis and visualization of human movement in space and time. Time-geographic methods have been used in the analysis of human movement in two particular forms: geocomputational methods and GIS-based three-dimensional (3D) geovisualization (Kwan 2004). Geovisualization is the use of concrete visual representations and human visual abilities to make spatial contexts and problems visible. Through involving the geographic dimension in the visualization process, it greatly facilitates the identification and interpretation of spatial patterns and relationships in complex data in the geographic context of a particular study area. Geocomputation refers to a wide range of methods involving the use of new computational tools and methods to depict geographic variations of phenomena across scales (Longley 1998). It encompasses various computational techniques, including expert systems, fuzzy sets, genetic algorithms, cellular automata, neural networks, fractal modeling, visualization, and data mining. Many of these methods are derived from the field of artificial intelligence and the more recently defined area of computational intelligence (Couclelis 1998). The availability of affordable high-speed computing and the development of GIS technologies in recent years have greatly facilitated the application of geocomputation in time-geographic research.

Using geocomputation for the analysis of human movement and space-time paths is most prominent in recent studies on individual accessibility (Kim and Kwan 2003; Kwan 1998, 1999b; Miller 1999; O'Sullivan et al. 2000; Weber 2003; Weber and Kwan 2002, 2003). It involves the development and application of dedicated algorithms for computing certain characteristics of space-time paths (such as space-time accessibility measures) within a GIS environment. Space-time accessibility measures are largely based on the analytical framework formulated by Lenntorp (1976) and Burns (1979). They are based on the time-geographic construct of a potential path area, which is the geographic area that can be reached within the space-time constraints established by an individual's fixed activities. It is the area that an individual can physically reach after one fixed activity ends, while still arriving in time for the next fixed activity. All space-time accessibility measures are derived from certain measurable attributes of this area (e.g., number of opportunities it includes).

Because of the need to represent real-world complexities and to deal with the large amount of geographic data, a GIS provides an effective environment for implementing geocomputational algorithms for space-time accessibility measures. With modern GIS technologies and increasingly available disaggregated data, highly refined space-time measures of individual accessibility can be operationalized. Several studies in recent years have developed and implemented geocomputational algorithms based on the time-geographic perspective. Drawing upon my recent research, several examples are discussed below to illustrate the application of geocomputation in the analysis of human movement.

25.3.1 Early Network-based Geocomputational Algorithm

Kwan (1998) represents a recent attempt in the geocomputation of space-time accessibility measures. It examined individual access to urban opportunities for a sample of 39 men and 48 women in Columbus (Ohio, USA). Data for the study came from three main sources. The first source is an activity-travel diary data set collected by the author through a mail survey in 1995. In addition to questions about the activity-travel characteristics of the respondent, data of the street addresses of all activity locations and the subjective spatial and temporal fixity ratings of all out of home activities were collected (Kwan 2000a). The second source of data is a digital geographic database of the study area that provides detailed information about all land parcels, their attributes, and other geographic features of the study area. Among the 34,442 non residential parcels in the database, 10,727 parcels belonging to seven landuse categories were selected as the urban opportunities in the study. The third data source is a detailed digital street network of the study area. The network database contains 47,194 arcs and 36,343 nodes of Columbus streets and comes with comprehensive address ranges for geocoding locations.

Using these data, 20 conventional measures of the gravity and cumulative opportunity variants were evaluated using the home locations of the 87 individuals as origins and 10,727 property parcels as destinations. Distances were computed using point-to-point travel times through a digital street network. Three space-time measures were also computed for each individual using a geocomputational algorithm written in ARC Macro Language (AML) and implemented in ArcInfo GIS. These three measures evaluate the size of the space that can be reached, the number of opportunities that can be reached, and the size or attractiveness of those opportunities. The algorithm used in the study was based on the one developed in Kwan and Hong (1998). Although it only provides an approximate solution of the exhaustive set of reachable opportunities, it is computationally more tractable. It uses the intersection of a series of paired arc-allocations to generate individual network-based potential path areas (PPAs), each of which is defined by the space-time coordinates of two fixed activities (see Figure 25-1 for a schematic representation of the algorithm).

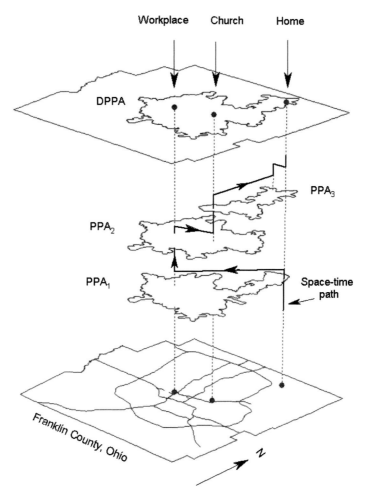

Figure 25-1 A schematic representation of Kwan's algorithm. See included DVD for color version.

The results of the study reveal the contrast between conventional and space-time measures. While the values produced by most gravity and cumulative opportunity measures were highly correlated and produced similar spatial patterns, space-time measures were very different. Gravity measures tended to replicate the geographic patterns of urban opportunities in the study area by favoring areas near major freeway interchanges and commercial developments, while cumulative opportunity measures emphasized centrality within the city by showing the downtown area to be the most accessible place. In contrast, space-time measures produced different spatial patterns, and the patterns for men resembled the spatial distribution of opportunities in the study area while the women's patterns were considerably different. The study shows that space-time measures are capable of revealing individual differences that are invisible when using conventional accessibility measures.

25.3.2 Extension to Incorporate Variable Travel Speeds over Transport Networks

To take into account the effect of the spatial and temporal variations in travel speeds and facility opening hours on individual accessibility, Weber and Kwan (2002, 2003) developed a second-generation algorithm for computing space-time accessibility measures. The study used a new geographic database with the enhanced geocomputational algorithm. The activity-travel diary data set used was collected through the Activity and Travel Survey in Portland Metropolitan Area in Oregon (USA) in 1994 and 1995. The data set logged a total of 129,188 activities and 71,808 trips undertaken by 10,084 respondents. Among the respondents, 101 men and 99 women were selected for the study. In addition, a digital street network with estimates of free flow and congested travel times (with 130,141 arcs and 104,048 nodes) and a comprehensive geographic database of the study area were used. A digital geographic database containing 27,749 commercial and industrial land parcels was used to represent potential activity opportunities in the study area.

The analytical procedures involved creating a realistic representation of the temporal attributes of the transport network and urban opportunities in the study area, as well as developing a geocomputational algorithm for implementing space-time accessibility measures within a GIS environment. The algorithm was developed and implemented using Avenue, the object-oriented scripting language in the ArcView 3.x GIS environment. Five space-time accessibility measures were computed. The first is the length of the road segments contained within the daily potential path area (DPPA). The second is the number of opportunities within the DPPA. The total area and total weighted area of the land parcels within the DPPA is the third and fourth space-time accessibility measures computed. Finally, to incorporate the effect of business hours on accessibility measures, opportunity parcels were assumed to be available (and could therefore be accessible to an individual) only from 9:00 A.M. to 6:00 P.M. This creates the fifth accessibility measure.

The results show that link-specific travel times produce very uneven accessibility patterns, with access to services and employment varying considerably within the study area. The time of day activities were carried out has also been shown to have an effect on accessibility, as evening congestion sharply reduced individual's access throughout the city. The effect of this congestion on mobility is highly spatially uneven. Further, the use of business hours to limit access to opportunities at certain times of the day shows that non-temporally restricted accessibility measures produce inflated values by treating these opportunities as being available at all times of the day. It is not just that incorporating time reduces accessibility, but that it also produces a very different, and perhaps unexpected, geography of accessibility (Weber and Kwan 2002). This geography depends much on individual behavior and so cannot be discerned from the location of opportunities or congestion alone. The study observed that the role of distance in predicting accessibility variations within cities is quite limited (Kwan and Weber 2003).

25.3.3 Extension to Take into Account Activity Duration and Network Topology

In an attempt to render earlier geocomputational algorithms more realistic, several enhancements were conceived and implemented by the third-generation algorithm developed by Kim and Kwan (2003). First, space-time accessibility is extended as a measure of not only the number of accessible opportunities, but also the duration for which these facilities can be enjoyed given the space-time constraint of an individual and facility opening hours. Second, more realistic travel times are incorporated through better representation of the transportation network, such as one-way streets in downtown areas and turn prohibition—besides incorporating the effect of congestion and location—and segment-specific travel speeds. Third, ways are developed to better incorporate other factors such as facility opening hours, minimum activity participation time, maximum travel time threshold, and delay times. The study seeks to enhance space-time accessibility measures with more rigorous representation of the temporal and spatial characteristics of opportunities and human activity-travel behavior.

A new GIS-based geocomputational algorithm was developed to implement these enhancements. The key idea of the algorithm is to efficiently identify all of the feasible opportunities within the space-time prism using several spatial search operations in ArcView GIS, while limiting the spatial search boundary with information about the travel and activity participation time available between two fixed activities. This algorithm was developed based upon numerous tests of the computational efficiency of different methods and a series of experiments using a large activity-travel diary data set and a digital street network. The GIS algorithm for deriving the potential path area (PPA) and for calculating space-time accessibility was implemented using Avenue in ArcView GIS. The study shows that space-time accessibility measures that do not consider the effect of facility opening hours and activity duration threshold will tend to over-estimate individual accessibility.

25.4 Visualizing Space-Time Paths

Using GIS-based 3D geovisualization in time-geographic research is a relatively recent phenomenon. In earlier studies, 2D maps and graphical methods were used to portray human movement in space and time (e.g., Chapin 1974; Tivers 1985). Individual daily space-time paths were represented as lines connecting various destinations. Using such kind of 2D graphical methods, information about the timing, duration and sequence of activities and trips was lost. There is, however, noticeable change in recent years. As more georeferenced activity-travel diary data become available, and as more GIS software incorporates 3D capabilities, GIS-based 3D geovisualization has become a more feasible approach for analyzing space-time paths.

Forer (1998), for instance, implemented space-time paths and prism on a 3D raster data structure for visualization and computational purposes. Their method is useful for aggregating individuals with similar socioeconomic characteristics and for identifying behavioral patterns. Kwan (2000b) and Kwan and Lee (2004) implemented 3D geovisualization of space-time paths and aquariums using vector GIS methods and activity-travel diary data. These studies indicate that GIS-based geovisualization can be a fruitful method for time-geographic research. Further, implementing 3D visualization of human movements in the form of space-time paths can be an important first step in the development of GIS-based geocomputational procedures that are applicable in many areas of social science research. The following subsections provide an overview of some of these methods from my recent research.

25.4.1 Space-time Aquarium and Space-time Paths

The space-time aquarium, first conceived by Hägerstrand (1970), is the earliest 3D method for the visualization of human space-time paths. In a schematic representation of the aquarium, the vertical axis is the time of day and the boundary of the horizontal plane represents the spatial scope of the study area. Individual space-time paths are portrayed as trajectories in this 3D aquarium. Although the schematic representation of the space-time aquarium was developed long ago, it has never been implemented using real activity-travel diary data. The main difficulties include the need to convert the activity data into "3Dable" formats that can be used by existing visualization software, and the lack of comprehensive geographic data for representing complex geographic objects of the urban environment.

The first study that implemented the space-time aquarium and space-time paths in a 3D GIS environment using individual-level activity-travel diary data is Kwan (1999a). Based on a subsample of 72 European Americans from the data set she collected in Columbus (Ohio, USA), the study examines the effect of gender on the space-time patterns of out-of-home non-employment activities. Visualization of the space-time paths of three groups of research participants reveals their distinctive activity patterns, and this insight guided the structural equation modeling in the later phase of the study. The results of the study show that the structure of one's daily activity patterns and day-time fixity constraint depends more on one's gender than on some conventional variables of household responsibilities, such as the presence or number of children in the household.

Kwan (1999a) did not use a large geographic database of the study area due to the limited 3D visualization capabilities of GIS available when the study was conducted (although the study did incorporate a transportation network). The first study that incorporated a comprehensive geographic database for the 3D geovisualization of space-time paths was that of Kwan (2000b). This study used the Portland activity-travel diary data set mentioned earlier and a comprehensive geographic database of the Portland metropolitan region (Oregon, USA). The GIS database provides comprehensive data on many aspects of the urban environment and transportation system of the study area. It has data for about 400,000 land parcels in the study area. The digital street network used, with 130,141 arcs and 104,048 nodes, covers the four counties of the study area (i.e., Clark, Clackamas, Multnomah and Washington).

With these contextual data incorporated into the GIS, the activity-travel data can be related to the geographic environment of the region during visualization. To implement 3D geovisualization of the space-time aquarium, four contextual geographic data layers are first converted from 2D map layers to 3D format and added to a 3D scene. These include the metropolitan boundary, freeways, major arterials, and rivers. For better close-up visualization and for improving the realism of the scene, outlines of commercial and industrial parcels in the study area are converted to 3D polygons and vertically extruded in the scene. Finally, the 3D space-time paths of the African- and Asian-Americans in the sample are generated and added to the 3D scene. These procedures finally created the scene shown in Figure 25-2.

As shown in Figure 25-2, the overall pattern of the space-time paths for these two groups reveals heavy concentration of day-time activities in and around downtown Portland. Using the interactive visualization capabilities of the 3D GIS, it was observed that many individuals of these two ethnic groups work in downtown Portland and undertake a considerable amount of their non-employment activities in areas within and east of the area. Space-time paths for individuals who undertook several non-employment activities in a sequence within a single day tend to be more fragmented than those who have long work hours during the day. Further, ethnic differences in the spatial distribution of workplace are observed using the interactive capabilities provided by the geovisualization environment. The space-time paths of Asian Americans are more spatially scattered throughout the area than those of the African Americans, whose work and non-employment activities are largely concentrated in the east side of the study area.

A close-up view from the west of the 3D scene is given in Figure 25-3, which shows some of the details of downtown Portland in areas within and around the "loop" and along the Willamette River. Portions of some space-time paths can also be seen in the figure as well. With the 3D parcels and other contextual layers in view, the figure gives the researcher a strong sense about the geographic context through a virtual reality-like view of the downtown area.

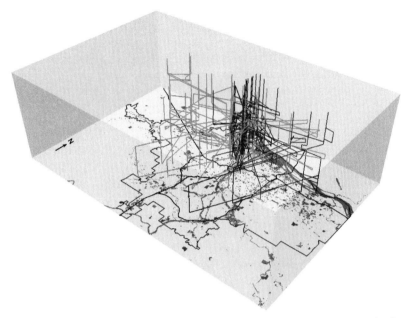

Figure 25-2 Space-time aquarium showing the space-time paths of African- and Asian-Americans in the sample. See included DVD for color version.

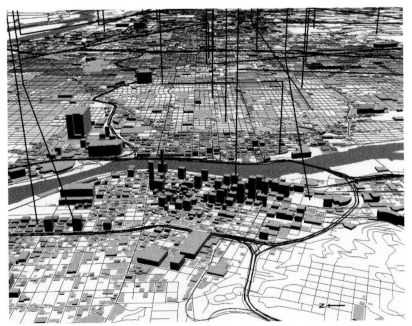

Figure 25-3 A close-up view of downtown Portland, including portions of some space-time paths. See included DVD for color version.

Manual of Geographic Information Systems

25.4.2 Space-time Paths Based on GPS Data

Although the 3D space-time paths shown in Figures 25-2 and 25-3 are helpful for understanding the activity patterns of different population subgroups, these paths are not entirely realistic since they only connect trip ends with straight lines and do not trace the travel routes of an individual. This limitation is due to the lack of route data in the Portland data set. When georeferenced activity-travel data collected by GPS are available and used in the geovisualization environment, the researcher can examine the detailed characteristics of an individual's space-time behavior as actual travel routes can be revealed by this kind of data (Kwan 2000c; Kwan and Lee 2004). Figure 25-4 illustrates this possibility using the GPS data collected in the Lexington Area Travel Data Collection Test conducted in 1997 (Battelle 1997). The original data set contains information of 216 licensed drivers (100 male, 116 female) from 100 households with an average age of 42.5. In total, data of 2,758 GPS-recorded trips and 794,861 data points of latitude-longitude pairs and time were collected for a 6-day period for each survey participant.

To prepare for 3D geovisualization, three contextual geographic data layers of the Lexington metropolitan area are first converted from 2D map layers to 3D format and added to a 3D scene. These include the boundary of the Lexington metropolitan region, highways and major arterials. As an illustration, the 3D space-time paths of the women without children under 16 years of age in the sample are generated and added to the 3D scene. These procedures finally created the scene shown in Figure 25-4. The overall pattern of the space-time paths for these women indicates that their trips were undertaken using largely highways and major arterials. There is some regularity as indicated by the daily repetition of trips in more or less the same time throughout the 6-day survey period. This suggests that distinctive patterns of space-time behavior can be revealed by 3D geovisualization.

There are difficulties in the analysis of these GPS data due to the computational intensity of processing and visualizing large space-time data sets (Kwan 2001a; Lowe 2003). For instance, the original GPS data file for the 100 households contains 794,861 data points of latitude-longitude pairs and time (Kwan 2000c; Murakami and Wagner 1999). It takes up about 230 megabytes of disk space in the format provided on the data CD. Manipu-

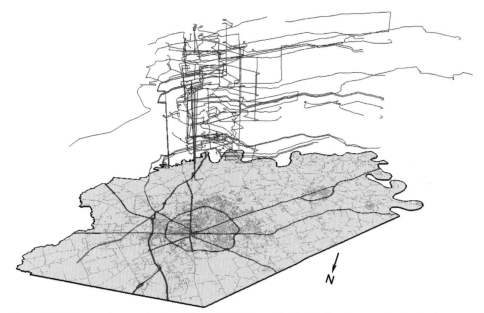

Figure 25-4 Space-time paths based on GPS data collected in Lexington, Kentucky. See included DVD for color version.

lating files of this size can be taxing for the computer hardware normally available to social scientists. Although improvement in computing power in the near future will reduce this problem, much research is still needed to develop more efficient algorithms and data manipulation methods for handling large GPS data sets.

25.4.3 Cyberspatial Activities and Space-time Paths

Three-dimensional geovisualization has also been applied to visualize human activities in both the physical world and cyberspace based on the notion of human extensibility (Kwan 2000d). The concept of the individual as an extensible agent was first formulated by Janelle (1973), where extensibility represents the ability of a person to overcome the friction of distance through using space-adjusting technologies, such as transportation and communication. Human extensibility not only expands a person's scope of sensory access and knowledge acquisition, it also enables a person to engage in distantiated social actions whose effect may extend across disparate geographic regions or historical episodes (Kwan 2001b). To depict human extensibility that include activities in both the physical world and cyberspace, Adams (1995) developed the extensibility diagram using the cartographic medium. The diagram, based on Hägerstrand's space-time aquarium, portrays a person's daily activities and interactions with others as multiple and branching space-time paths in three dimensions, where simultaneity and temporal disjuncture of different activities are revealed. This method can be used to represent a diverse range of human activities in both the physical and virtual worlds, including telephoning, driving, emailing, reading, remembering, meeting face-to-face, and television viewing.

Although the extensibility diagram is largely a cartographic device, most of its elements are amenable to GIS implementation. Kwan (2000d) developed a method for implementing the extensibility diagram using 3D GIS. The study used real data about a person's physical activities and cyberspatial activities (e.g., email messages and web browsing sessions). The focus is on incorporating the multiple spatial scales and temporal complexities (e.g., simultaneity and disjuncture) involved in individual hybrid-accessibility. The following example, derived from the case examined in Kwan (2000d), illustrates the procedures for constructing the multi-scale 3D extensibility diagram and its use in a GIS-based 3D geovisualization environment. The first step is to determine the most appropriate spatial scales and extract the relevant base maps from various digital sources.

Consider a person who lives and works in Franklin County (Ohio, USA), and engages in cyberspatial activities (e.g., sending and receiving email) involving cities in the northeastern region of the US and other countries (e.g., South Africa and Japan). To prepare the GIS base maps at these three spatial scales (local, regional and global), a map of Franklin County and a regional map of 15 US states in the northeastern part of the country were first extracted from a commercial geographic database. Franklin County is the home county of the person in question, whereas the US region extracted will be used to locate the three American cities involved in her cyber-transactions. These cities are Chicago in Illinois, Maywood in New Jersey and Charlotte in North Carolina. At the global scale, the world map layer was derived from the digital map data that came with ArcGIS.

After performing map-scale transformations to register these three map layers to the person's home location in Franklin County, these 2D map layers were converted to 3D shape files and added to an ArcGIS 3D Analyst scene as 3D themes. After preparing these map layers, 3D shape files for the person's space-time paths were generated using Avenue scripts and added to the 3D scene. These procedures finally created the multi-scale extensibility diagram shown in Figure 25-5. It shows how various types of transactions at different spatial scales can be represented in a 3D GIS environment.

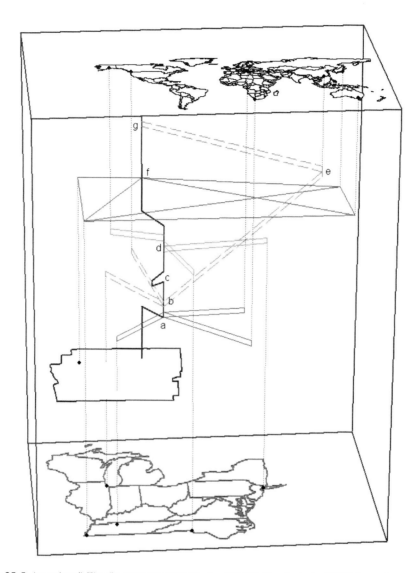

Figure 25-5 An extensibility diagram rendered using 3D GIS. See included DVD for color version.

Figure 25-5 shows five types of activities undertaken on a particular day. On this day, the person in question worked from 8:30 A.M. to 5:30 P.M. and had a one-hour lunch break at a nearby restaurant (*c* on the diagram). She subscribes to a web-casting service where news is continuously forwarded to her web browser. On this day, she read some news about Yugoslavia, South Africa and Nashville, Tennessee (*a* on the diagram), before she started work. An hour later she sent an email message to three friends located in Hong Kong, Chicago and Vancouver (*b*). The friend in Chicago read the email two hours later and the friend in Vancouver read the email five hours later. The friend in Hong Kong read the email 13 hours later and replied immediately (*e*). The reply message from this friend, however, was read at 2:00 A.M. at the person's home (*g*). In the afternoon, she browsed web pages hosted in New York, Charlotte and Anchorage, Alaska (*d*). She was off from work at 5:30 P.M. and spent the evening at home. At 9:00 P.M. she started an ICQ (real-time chat) session with friends in Tokyo; Melbourne; Memphis, Tennessee; and Dublin, Ohio (*f* on the diagram).

As Figure 25-5 indicates, very complex interaction patterns in cyberspace can be represented using multiple and branching space-time paths. These include temporally coincidental (real-time chat) and temporally non-coincidental (emailing) interactions; one-way radial (web browsing), two-way dyadic or radial (emailing), and multi-way (chat) interactions; incoming (web casting) and out-going (emailing) transactions. The method is thus capable of capturing the spatial, temporal and morphological complexities of a person's extensibility in cyberspace.

In addition, the visualization functions available in ArcGIS 3D Analyst also enable one to interactively explore the 3D scene in a very flexible manner (e.g., the scene is visible in real-time while zooming in and out, or rotating). This allows for the selection of the best viewing angle and is a very helpful feature, especially when visualizing very complex space-time paths. To focus only on one type of transaction or activity at a particular spatial scale, one can select the relevant themes for display while keeping the other themes turned off. Further, when the three sets of paths and base maps are displayed at the same time, they can be color coded to facilitate the visualization. In the original color 3D scene, each segment of the space-time paths are represented using the same color as the relevant base map (e.g., blue for Franklin County and local activities), conveying a rather clear picture of the spatiality and temporal rhythm characterizing the person's activities on that day. But in the black-and-white version presented in Figure 25-5, spike lines are used to identify the location involved in each transaction.

25.5 Conclusion

As shown by the examples in this chapter, GIS-based time-geographic methods are powerful means for analyzing human space-time paths. These include two broad classes of methods: three-dimensional (3D) geovisualization and geocomputational methods. Usefulness of these methods is illustrated through examples drawn from recent studies by myself and other researchers. Besides these two types of methods, there are also many methodological innovations in the study of human space-time paths in recent years. There are studies that use an object-oriented approach for extracting and visualizing space-time paths. There are studies that attempt to reconstruct space-time paths from large geographic databases using neural networks and self-organizing maps. Researchers have also used agent-based modeling and pattern aggregation techniques like sequence alignment to extract meaningful patterns of space-time paths (e.g., Shoval and Isaacson 2007). These are promising areas for future research in GIS-based time-geographic methods.

There has also been an increasing interest in extending time-geographic methods for the study of how information and communications technologies (ICT) influence human activity-travel patterns in recent years. These studies seek to describe and analyze the impact of Internet and mobile phone use on people's behavior—for example, will e-shopping or e-banking reduce people's trips to shops or banks in the physical world? These studies also try to understand how ICT use may affect people's space-time constraints—for example, will people undertake more social and recreational activities by using the time they saved through using ICT (like e-shopping)? Both the substantive and methodological development in analytical methods for representing space-time paths and understanding human movement in the urban context will further enhance our knowledge in this important area of research.

References

Adams, P. C. 1995. A reconsideration of personal boundaries in space-time. *Annals of the Association of American Geographers* 85:267–285.

Battelle. 1997. *Lexington Area Travel Data Collection Test: Final Report.* Battelle Memorial Institute, Columbus, Ohio.

Bhat, C. R. and S. K. Singh. 2000. A comprehensive daily activity-travel generation model system for workers. *Transportation Research A* 34 (1):1–22.

Burns, L. D. 1979. *Transportation, Temporal, and Spatial Components of Accessibility.* Lexington, Mass.: Lexington Books.

Carlstein, T., D. Parkes and N. Thrift. 1978. *Timing Space and Spacing Time II: Human Activity and Time Geography.* London: Arnold.

Chapin, F. S. Jr. 1974. *Human Activity Patterns in the City.* New York: John Wiley & Sons.

Couclelis, H. 1998. Geocomputation in context. In *Geocomputation: A Primer,* edited by P. Longley, S. M. Brooks, R. McDonnell and B. MacMillan. New York: John Wiley & Sons.

Cullen, I., V. Godson and S. Major. 1972. The structure of activity patterns. In *Patterns and Processes in Urban and Regional Systems.* Edited by A. G. Wilson, 281–296. London: Pion.

Forer, P. 1998. Geometric approaches to the nexus of time, space, and microprocess: Implementing a practical model for mundane socio-spatial systems. In *Spatial and Temporal Reasoning in Geographic Information Systems.* Edited by M. J. Egenhofer and R. G. Golledge. Oxford: Oxford University Press.

Golob, T. F. 1985. Analyzing activity pattern data using qualitative multivariate statistical methods. In *Measuring the Unmeasurable.* Edited by P. Nijkamp, H. Leitner and N. Wrigley, 339–356. Boston: Martinus Nijhoff.

Golob, T. F. and M. G. McNally. 1997. A model of activity participation and travel interactions between household heads. *Transportation Research B* 31 (3):177–194.

Goulias, K. G. 1999. Longitudinal analysis of activity and travel pattern dynamics using generalized mixed Markov latent class models. *Transportation Research* B 33 (8):535–558.

Hägerstrand, T. 1970. What about people in regional science? *Papers of Regional Science Association* 24:7–21.

Hornsby, K. and E. Egenhofer. 2002. Modeling moving objects over multiple granularities. *Annals of Mathematics and Artificial Intelligence* 36 (1-2):177–194.

Janelle, D. G. 1973. Measuring human extensibility in a shrinking world. *Journal of Geography* 72:8–15.

Kim, H.-M. and M.-P. Kwan. 2003. Space-time accessibility measures: A geocomputational algorithm with a focus on the feasible opportunity set and possible activity duration. *Journal of Geographical Systems* 5:71–91.

Kwan, M.-P. 1998. Space-time and integral measures of individual accessibility: A comparative analysis using a point-based framework. *Geographical Analysis* 30:191–216.

———. 1999a. Gender, the home-work link, and space-time patterns of non-employment activities. *Economic Geography* 75:370–394.

———. 1999b. Gender and individual access to urban opportunities: A study using space-time measures. *The Professional Geographer* 51:210–227.

———. 2000a. Gender differences in space-time constraints. *Area* 32:145–156.

———. 2000b. Interactive geovisualization of activity-travel patterns using three-dimensional geographical information systems: A methodological exploration with a large data set. *Transportation Research C* 8:185–203.

———. 2000c. *Evaluating Gender Differences in Individual Accessibility: A Study Using Trip Data Collected by the Global Positioning System.* A report to the Federal Highway Administration (FHWA), US Department of Transportation, 400 Seventh Streets, S.W., Washington, D.C. 20590.

————. 2000d. Human extensibility and individual hybrid-accessibility in space-time: A multi-scale representation using GIS. In *Information, Place, and Cyberspace: Issues in Accessibility.* Edited by D. G. Janelle and D. C. Hodge, 241–256. Berlin: Springer-Verlag.

————. 2001a. Analysis of LBS-derived data using GIS-based 3D geovisualization. Paper presented at the *Specialist Meeting on Location-Based Services,* Center for Spatially Integrated Social Science (CSISS), University of California, Santa Barbara, 14–15 December 2001.

————. 2001b. Cyberspatial cognition and individual access to information: The behavioral foundation of cybergeography. *Environment and Planning* B 28:21–37.

————. 2004. GIS methods in time-geographic research: Geocomputation and geovisualization of human activity patterns. *Geografiska Annaler B* 86 (4):267–280.

Kwan, M.-P. and G. Ding. 2008. Geo-Narrative: Extending geographic information systems for narrative analysis in qualitative and mixed method research. *The Professional Geographer,* forthcoming.

Kwan, M.-P. and X.-D. Hong. 1998. Network-based constraints-oriented choice set formation using GIS. *Geographical Systems* 5:139–162.

Kwan, M.-P. and J. Lee. 2004. Geovisualization of human activity patterns using 3D GIS: A time-geographic approach. In *Spatially Integrated Social Science.* Edited by M. F. Goodchild and D. G. Janelle. New York: Oxford University Press.

Kwan, M.-P. and J. Weber. 2003. Individual accessibility revisited: Implications for geographical analysis in the twenty-first century. *Geographical Analysis* 35:341–353.

Laws, G. 1997. Women's life courses, spatial mobility, and state policies. In *Thresholds in Feminist Geography: Difference, Methodology, Representation.* Edited by J. P. Jones III, H. J. Nast and S. M. Roberts. New York: Rowman and Littlefield.

Lenntorp, B. 1976. *Paths in Time-Space Environments: A Time Geographic Study of Movement Possibilities of Individuals.* Lund Studies in Geography B: Human Geography, Gleerup, Lund, Sweden.

————. 1978. A time-geographic simulation model of individual activity programs. In *Timing Space and Spacing Time Volume 2: Human Activity and Time Geography.* Edited by T. Carlstein, D. Parkes and N. Thrift. London: Edward Arnold.

Longley, P. 1998. Foundations. In *Geocomputation: A Primer.* Edited by P. Longley, S. M. Brooks, R. McDonnell and B. MacMillan. New York: John Wiley & Sons.

Lowe, J. W. 2003. Special handling of spatio-temporal data. *Geospatial Solutions* 13 (11):42–45.

Ma, J. and K. G. Goulias. 1997a. An analysis of activity and travel patterns in the Puget Sound transportation panel. In *Activity-based Approaches To Travel Analysis.* Edited by D. Ettema and H. Timmermans, 189–207. Tarrytown, NY: Elsevier Science.

————. 1997b. A dynamic analysis of person and household activity and travel patterns using data from the first two waves in the Puget Sound Transportation Panel. *Transportation* 24 (3):309–331.

Mårtensson, S. 1977. Childhood interaction and temporal organization. *Economic Geography* 53:99–125.

McCormack, E. 1999. Using a GIS to enhance the value of travel diaries. *ITE Journal* 69 (1):38–43.

Miller, H. J. 1999. Measuring space-time accessibility benefits within transportation networks: Basic theory and computational procedures. *Geographical Analysis* 31:187–212.

Murakami, E. and D. Wagner. 1999. Can using global positioning system (GPS) improve trip reporting? *Transportation Research* C 7:149–165.

Odland, J. 1998. Longitudinal analysis of migration and mobility spatial behavior in explicitly temporal contexts. In *Spatial and Temporal Reasoning in Geographic Information Systems.* Edited by M. J. Egenhofer and R. G. Golledge. Oxford: University of Oxford Press.

Ohmori, N., Y. Muromachi, N. Harata and K. Ohto. 1999. A study on accessibility and going-out behavior of aged people considering daily activity pattern. *Journal of the Eastern Asia Society for Transportation Studies* 3:139–153.

O'Sullivan, D., A. Morrison and J. Shearer. 2000. Using desktop GIS for the investigation of accessibility by public transport: An isochrone approach. *International Journal of Geographical Information Science* 14: 85–104.

Parkes, D. N. and N. Thrift. 1975. Timing space and spacing time. *Environment and Planning A* 7:651–670.

Recker, W. W., M. G. McNally and G. S. Root. 1987. An empirical analysis of urban activity patterns. *Geographical Analysis* 19 (2):166–181.

Ren, F. and M.-P. Kwan. 2007. Geovisualization of human hybrid activity-travel patterns. *Transactions in GIS* 11 (5):721–744.

Rollinson, P. 1998. The everyday geography of the homeless in Kansas City. *Geografiska Annaler B* 80:101–115.

Shoval, N. and M. Isaacson. 2007. Sequence alignment as a method for human activity analysis in space and time. *Annals of the Association of American Geographers* 97 (2):282–297.

Takeda, Y. 1998. Space-time prisms of nursery school users and location-allocation modeling. *Geographical Sciences* (Chiri-kagaku, in Japanese) 53:206–216.

Thrift, N. 1977. *An Introduction to Time Geography.* Geo Abstracts, University of East Anglia, Norwich.

Tivers, J. 1985. *Women Attached: The Daily Lives of Women with Young Children.* London: Croom Helm.

Weber, J. 2003. Individual accessibility and distance from major employment centers: An examination using space-time measures. *Journal of Geographical Systems* 5:51–70.

Weber, J. and M.-P. Kwan. 2002. Bringing time back in: A study on the influence of travel time variations and facility opening hours on individual accessibility. *The Professional Geographer* 54:226–240.

———. 2003. Evaluating the effects of geographic contexts on individual accessibility: A multilevel approach. *Urban Geography* 24 (8):647–671.

Yu, H. 2006. Spatial-temporal GIS design for exploring interactions of human activities. *Cartography and Geographic Information Science* 33:3–19.

SECTION 5
Analysis and Modeling

CHAPTER 26

Spatial Modeling of Ecological Processes within a Conceptual Framework

Kevin M. Johnston

26.1 Introduction

Ecological models can only represent a specified aspect of an ecological process. To create an ecological model, the modeler should have a fair understanding of the components that are universal to all ecological processes—an ecological conceptual framework. By generalizing an ecological process and then identifying that process within the ecological conceptual framework, the ecological model can explore the assumptions, questions and decisions set up within the model.

Historically space has been incorporated into ecological models from a top-down approach. That is, initially space was used to limit the extent of the ecological process to make it more manageable. A geographic extent was defined and the interactions within that space were described or quantified. As modeling progressed, space was incorporated into ecological models more for explaining the resulting patterns from an ecological process. It then transitioned further toward integrating space into the mechanisms creating the processes.

Space is not used as a constraint that bounds an ecological process. In the *ecological conceptual framework* presented here, space is an integral part of the mechanisms creating the processes and patterns. In this framework, physical and mental requirements and behavior integrate in a spatial context to influence the decision making of the individuals creating the processes that produce the ecological patterns. Spatial boundaries and extents for the ecological process are emergent properties of the interaction of the mechanisms of the process.

In this chapter I present an ecological conceptual framework as a tool to help gain a better understanding of ecological processes. This conceptual framework may be used to unify the many specific, although disparate, spatially oriented ecological models. Every ecological model addresses some aspect of this conceptual framework. Knowing what aspect a model addresses in this conceptual framework may allow the modeler or the decision maker using the model to understand an ecological process in its entirety: the assumptions being made, what questions can be asked, where uncertainty exists in the modeling process, and what decisions can be made from the results. Since this ecological conceptual framework is based on space, this chapter will explore how space has been and is integrated into ecological models. Because Geographic Information Systems (GIS) are key tools used to model space, this chapter will identify how and where GIS has been used to model ecological processes, and what types of problems need spatially explicit modeling tools. Several hundred existing spatially explicit ecological models that have used GIS are presented in the context of the aspect they address in this ecological conceptual framework. Four case studies explore in detail how certain GIS models have addressed the different aspects of this conceptual framework.

26.2 The Integration of Space into Ecological Theory and Models

Space has been an integral part of ecological theory and models. Early on the ecological relationships of plants were mapped by botanists accompanying the early explorers. Space is an important concept when understanding the term 'ecology,' since ecology has a long history of studying and examining the distribution and abundance of species in space. Initially ecological modeling with space has been treated from a top-down approach by defining a geographic extent and then describing or quantifying the interactions within that space.

There have been attempts to establish reasonable geographic extents or boundaries to make the conceptualization of ecological process more manageable. One of the first was Grinnell who in 1917 developed the concept of a niche defining the geographic extent as a subdivision of an environment occupied by a species. The spatial extent of the niche for a particular creature is identified by the structural and functional limitations of that creature.

Tansley in 1935 not only provided a definition for deriving the spatial extent for an ecological process but also acknowledged space as an integral part of the ecological process, by using the term ecosystem to refer to, "the whole system ... including not only the organism-complex, but also the whole complex of physical factors forming what we call the environment." He goes on to say, "We cannot separate them (the organisms) from their special environment with which they form one physical system.... It is the system so formed which [provides] the basic units of nature on the face of the earth."

After World War II, systems theory took hold in ecological models (O'Neill 2001). In systems theory usually a geographic extent defines the boundaries of the system applying a minimum bounding rectangle to encompass the group of elements that will be acting as a system. Systems theory has been challenged due to the idea no ecological process is closed and that external mass disturbances were treated as exceptions to the system and not part of them (O'Neill 2001).

Eventually space became an implicit mechanism behind the functionality of an ecological process. The concept of space was generalized in ecological models through assumptions as demonstrated by the mathematical formulas of population growth and species interactions of Lotka-Volterra (Lotka 1956; Kingsland 1995), Schoener (1974, 1976, 1983), Tilman (1980, 1985) and many more. These models generalized space in the model assumptions by assuming each modeling member has equal access to all locations and that they are modeling the average individual—the so-called mean or mass action approximation.

Space increasingly became more central to ecological theory and models as an explanatory variable rather than a mechanistic component as demonstrated by the island biogeographic theory (Preston 1962a, 1962b; MacArthur and Wilson 1963; MacArthur and Wilson 1967). Generally the theory states, 1) the number of species on an island increases with the area of the island (species-area relationship), 2) there is more diversity of species on islands closer to a mainland and the number of species on an island decreases as the remoteness or isolation of the island increases (the distance effect), 3) an island reaches equilibrium when the immigration and extinction rates are equal (theory of equilibrium), and 4) immigration rates vary according to the distance from a mainland while extinction rates vary according to area and distance from mainland. These spatial relationships set the basis for many of the current conservation biology and reserve design models. Not only was space used to define the boundaries of the process, in this case the islands, but distance was one of the explanatory variables behind the patterns.

Carrying space even more central to its theme, the field of landscape ecology focuses on three characteristics of the landscape (Forman and Godron 1986):

1) structure: the spatial relationships among the distinctive ecosystems or elements present and the species in relation to the sizes, shapes, numbers, kinds and configurations of the ecosystem,

2) function: the interactions among the spatial elements, and

3) change: the alteration in the structure and function of the ecological mosaic over time.

Landscape ecology divides the landscape of a given spatial extent into patches, corridors, and the matrix, and explores how elements move between them. Like the earlier models, landscape ecology uses space to define the boundaries of the ecological process but differs in that it quantifies the spatial relationships of the elements in the landscape. However, the mechanisms that produce these spatial patterns are only implied.

Space is the basis for metapopulation theory (Hanski and Gilpin 1991; Hanski 1991). Prior to metapopulation theory, a population was comprised of a stand-alone group of individuals interacting. A metapopulation describes a population of populations. That is, metapopulation theory studies the dynamics of populations, and the extinction and establishments of new populations. As one population dies off in one location then individuals from other populations can immigrate in and establish a new population. The driving force behind the interaction between populations is the spatial arrangements of where the populations exist within a given spatial extent and the functional connectivity between them. The use of space is moving more along the continuum from an explanatory variable to a central component of the functioning of the ecological process.

Source-sink population theory (Pullian 1986) explores the fact that in some populations a large number of the individuals live in "sink" habitats that cannot balance local mortality, but the populations persist in these habitats because there is a continued immigration from more-productive "source" habitats nearby. The spatial relationships of the source and sink habitats is important in order to maintain a constant immigration. In the source-sink population theory, what the elements are doing and where they are doing it makes space even more central to the functioning of the ecological process.

Optimal foraging theory accounts for space by determining which patch an animal is likely to feed within a habitat of patches. Optimum foraging theory states that an optimal forager will maximize its fitness (Ritchie 1988, 1990, 1991). Stephens and Krebs (1986) discuss how a forager will move from one patch to another patch depending on the forager's characteristics and the energy return from a patch. A risk-prone forager will move into another patch if the energy return is potentially greater even though that return is unknown. In contrast a risk-averse forager will move into a patch it knows has lesser return but the forager is more certain that the energy return is found within that patch. Charnov (1976) presents the marginal value theorem for optimal foraging which states that a forager should move out of a patch into another if the marginal rate of forage from that patch is less than the average rate for the entire habitat. In the marginal value theorem, the forager will subtract the travel time to move from one patch to another when determining the marginal rate. As with the source-sink theory, in optimum foraging theory, where and what the individuals are doing are more central to the functioning of the process. By incorporating travel time, Charnov integrates space even more as a mechanism behind an ecological process. Still, generally in these models spatial limits are artificially imposed to bound the process.

These are some ecological theories or models that incorporate space as an integral component. In most of these theories, the spatial extent is first established followed by descriptions within the specified space. Throughout history, space becomes more central to models by moving on a continuum as an explanatory variable to a mechanism behind the functioning of the process. But these models fall short of fully conceptualizing space as an integrated mechanism within ecological processes.

What I propose in this ecological conceptual framework is how to fully incorporate space as an integral part of the mechanisms producing the processes that create the patterns. Space is not an explanatory variable in this conceptual framework but is integrated with physical and mental requirements and behavior and behavior strategies, creating a spatial context that produces predictable processes and patterns. In this conceptual framework the spatial extent of an ecological process is an emergent property of the actions by individuals driven by physiology and behavior.

26.3 The Ecological Conceptual Framework

The ecological conceptual framework proposed here attempts to identify the essence of any ecological process. This conceptual framework does not rely on an abstraction of the reality of an ecological process. It attempts to break down an ecological process based on how it works in the real world. If the essence of an ecological process can be understood and principles can be developed, then the same modeling principles can be applied to any ecological process. This conceptual framework can guide the model developer by focusing the scope of the model to the aspect of the ecological process being modeled. The following proposed ecological conceptual framework is a composite of many ideas ranging from Aristotle's theory of understanding (Moravcsik 1975), Maslow's hierarchy of needs (Maslow 1943), systems ecology, an aggregation of existing literature, and through direct observation.

An ecological process (in this case the world) is made up of individuals. Examples of individuals in this process include white-tailed deer, humans, beech trees, mountain laurel, hummingbirds and grass hoppers. The individuals are driven by physical requirements: hunger, thirst, heat, or cold. The individuals can become hungry, thirsty, hot, or cold. Within the category of physical requirements is the necessity to survive.

The physical requirement spawns a response. In certain individuals it is a "mental" response whether conscious or unconscious; "I need to eat." "I should drink." "I must run and hide". (Note: whether the response is conscious or occurs in some central processing location is a philosophical debate beyond the scope of this chapter; we are only interested in responses.) There are many responses including no action such as, "I am not hungry, therefore I do not need to eat." From the suite of responses, the individual makes a decision or multiple decisions. Just because the individual formulates a response does not mean it will act on it. For instance, the individual may not eat even though it is hungry because eating may put it into a position of danger. There are usually many decisions being contemplated based on the mental responses. But before the individual makes a decision or decisions it at some level will weigh its state, outside influences, the physical environment, and its capability. It will also anticipate at some level the actions and reactions of other individuals, future conditions (e.g., the weather) and its own requirements (e.g., does it need to store food for the winter). The decision made in any particular moment is a tradeoff between the individual's objective to meet its basic physical requirements (e.g., hunger, thirst, cold) and once its basic requirements are met it may attempt to meet its more complex requirements such as the need for social interaction, reproduction, comfort and play.

Based on these decisions, the individual performs an action. It may eat, drink, move its leaves toward the sun, or any number of other actions. The decision process can be shortened to a reaction, as would be the case if a predator attacks. In such a case the physical requirement would be survival, the mental requirement would be survival, the decision for a white-tailed deer might be run, and the action would be run.

The accumulation of all the individual actions and reactions create a process. A process is comprised of all the individual decisions from all the individuals. Given the same conditions (e.g., climate, physical environment and time of year), the types of individuals, and the underlying motivations for decision making, a predictable outcome should result. If the outcome is predictable, patterns may result.

A pattern is an aggregation of resources, individuals, or the physical environment in time or space that group together, and not by chance. Sometimes the patterns are obvious, other times the pattern may be obscure, thus difficult to identify and quantify.

Patterns can be specific to a particular physical location with a given set of conditions. But the goal of certain ecologists is to generalize the patterns to all locations with similar conditions to create a law or a general principle. Examples of ecological laws or general principles include succession, island biogeographic theory (Preston 1962a, 1962b; MacArthur and Wilson 1963; MacArthur and Wilson 1967) and the concept of a niche (Grinnell 1917).

Essentially, this proposed ecological conceptual framework for an ecological process consists of a series of individuals, each making individual decisions for the improvement of itself. Through their collective decisions in similar conditions and similar physical environments, they create processes that produce predictable patterns. The hope of some ecologists is to generalize those processes and patterns to include even more conditions and locations to produce laws and general principles to describe the underlying causality of the natural world.

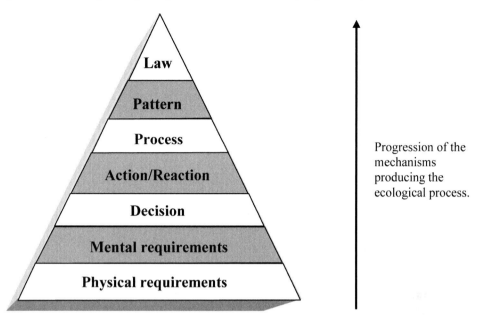

Figure 26-1 A graphical depiction of the ecological conceptual framework. See included DVD for color version.

26.3.1 Boundaries of Ecological Processes

Where an ecological process begins and ends either in time or space is difficult to define. In this proposed conceptual framework, the temporal and spatial boundaries are determined by the interaction of the defined individuals and the individuals that they come in contact with, thus influencing the decision making, thus creating the process. The starting time and place in which the individuals interact are generally defined by the problem being addressed. All ecological models need some spatial and temporal boundaries, however all boundary definitions are flawed.

26.3.2 Individuals Within Ecological Processes

When creating an ecological model, the individual must be defined either implicitly or explicitly. Thus far in this ecological conceptual framework, individuals have been living objects. The word individual can be expanded to include non-living objects such as water droplets, fire, soil and rocks. In this conceptual framework, individuals or objects are anything that makes decisions. The word decision here is used generally, "it did this" and "not this" and is not based in all cases on the mental process. The individual generates an action based on a reason. For instance, the oil in an oil spill moves with the currents.

Evolutionary biologists may say actions are based on the selection process whereby individuals with the strongest genes will pass the favorable traits onto their offspring. However, within this proposed ecological conceptual framework there are two plausible responses. First, individuals are making individual decisions to improve their situation. The individuals making the best decisions will probably have a greater chance for survival and reproduction thus passing the genes along that allow for more strategic decision making. As a result, the offspring will use these genes to make better decisions. Or second, a more esoteric response, might be the gene is the individual and it is making its own decisions based on its survival and the animal is a composite of many individual genes.

Manual of Geographic Information Systems

26.3.3 Physical Requirements

The physical process is generally driven by the basic biophysical requirements for survival and comfort. The state of the individual (e.g., "I have not eaten in a long time") will dictate the emphasis and urgency of the requirement or no requirement.

The physical requirements of non-living individuals or objects will usually revolve around the physical properties surrounding the objects. For instance fire can continue if it moves down wind and into areas containing fuel. Rock erosion will occur given certain physical properties like wind and water.

26.3.4 Mental Requirements

Mental requirements are the conscious realization of the physical requirements of living individuals but also include extraneous requirements that individuals choose to meet once the physical requirements are met (e.g., social interactions and play). This later process is more often found among complex individuals.

26.3.5 Decisions

Decisions may be conscious or not and are based on the tradeoffs between the mental options and consequences of each decision. Sometimes the tradeoffs are not weighed but are imposed by the physical characteristics of the object or of the conditions surrounding the object. For example, a plant that spreads by wind may only vegetate downwind from itself with the seeds only germinating in favorable conditions.

Decisions are also based on the anticipated action of other individuals and what the physical conditions will be at the time of the action. There is a degree of uncertainty in predicting most decisions.

Some decisions are based on the strategy of the individual. For example, optimum foraging (Ritchie 1988, 1990, 1991) and patch selection (Stephens and Krebs 1986; Charnov 1976) are based on different strategies for a grazer to maximize its energy intake while grazing in various patches. A herding animal choosing to stay in a herd is based on a survival strategy knowing the herd will reduce the individual's predation risk.

26.3.6 Action / Reaction

An action or reaction is the realization of a decision. The action or reaction may or may not be the result of a conscious decision.

26.3.7 Process

The process is the aggregation of individual actions or reactions. The herding process is created by the individuals trying to reduce the predation risk. The "V" created by migrating geese is created by individual geese trying to reduce their energy output when flying long distances. Streams are created by thousands and thousands of water droplets taking the least path of resistance.

26.3.8 Patterns

Process and patterns are the emergent properties of the aggregation of the individual decision making. This is not to say that all decision rules are currently understood or due to their complexity will ever be quantified, but within this ecological conceptual framework, if the correct individuals and objects can be identified and the full interactions and decision rules understood, anything can be modeled, with varying levels of difficulty.

Patterns can be observed from many processes working in similar ways in similar conditions. If the conditions for the individuals change, different physical influences, mental responses, decision strategies, actions and processes might result. Similarly, if different individuals or objects are in similar conditions, again different physical influences, mental responses, decision strategies, actions and processes might result. By changing the conditions the individuals are in or changing the individuals in specific conditions, predictable patterns may not be produced.

Knowledge helps identify patterns. The more that is known about a process, the more the patterns become evident. Many of the patterns that we continue to discover have always been here, yet were simply unrecognized. However, the identification of a pattern can be subjective to the modeler. There may be cultural, gender and religious bias in identifying and defining patterns. The tools that are available to quantify patterns may influence the definition of a pattern.

The strength of the pattern is a measure of how discernable the aggregation of resources is from random. Not all patterns should be considered obvious within a given spatial or temporal extent. For example, the rise in the mean temperature of the Earth in any particular year may not be an abnormal event when considering the temperature over thousands of years. Or the number of individuals of a certain species may appear to be declining when looking at a small geographic extent but when looking at the regional extent, the number of individuals maybe increasing.

In any particular location, there are usually many interacting processes that produce a pattern. Each of these processes is constantly changing based on the changing physical conditions, inherent characteristics of the individuals making the decisions that create the pattern, on the changing decisions of the individuals based on the changing conditions produced by the changing decisions of individuals creating other processes, and external forces causing great changes in the conditions to the area (e.g., a storm, a cold winter). The consistent thread is that processes are continuously changing which result in continuously changing patterns. Therefore a pattern is only valid for a slice of time in a particular slice of conditions. How long a pattern continues, or how robust a pattern metric is depends on how quickly the decisions of the individuals producing the interacting processes result in a change in the pattern. A pattern of evolutionary migration will probably be longer-lasting than one quantifying the swarming of an insect.

26.3.9 Scale

In this proposed ecological conceptual framework, spatial scale is relative to the decision making of the individuals being modeled. For example, an ant will make decisions based on grains of sand while an elephant will probably not consider the sand but will make decisions on coarser objects, such as trees and mountains.

Temporal scale is also determined by the individuals being modeled. A desert tortoise may make decisions based on water availability when precipitation may only occur sometime within a three year period. The time scale need not be based on any human interval and the time interval can vary relative to events.

As scale becomes finer, the model must be able to account for the additional complexity (Levin 1992). Spatial and temporal scales and their extents are only arbitrary boundaries. Most ecological processes do not have beginnings or endings.

26.3.10 Implicit and Explicit Change

The individuals react to the features and objects around them but the individuals can also produce change explicitly or implicitly. Change is explicit when the beaver builds a dam or a water droplet becomes part of a flood. Change is implicit as seen in nesting bird preference. The attributes at a location may make a site more preferred than others, such as proximity to water. The nesting preference for each nest may increase for certain species as more birds build nests in the area because there is a reduced risk for predation, however some saturation point is usually reached. There is a feedback with the inherent characteristics of the location and with the changing decisions of the individuals. The individual's decisions are not independent.

26.3.11 Decision Making, Error and Uncertainty

There is some level of uncertainty in all models (Burrough et al. 1996; Burrough and McDonnell 1998; Zhang and Goodchild 2002; Longley et al. 2005). This uncertainty can come from an incomplete understanding of the phenomena being modeled, error in the input data, or uncertainty in the model parameters. Some phenomena are not understood well enough so they appear as random events such as locations for lightning strikes. However, with a more sophisticated model and information, it may be concluded that they are not random events. Even with this uncertainty decisions must be made.

Error analysis and sensitivity analysis can help the decision maker using the model to understand uncertainty. Error analysis explores and quantifies the effect of error on the results of a model. Sensitivity analysis systematically changes a parameter slightly in a model and explores how much the change affects the outcome. With knowledge of the quality of the input data and understanding the effects of the parameters, the decision maker using the model can identify the risk that is involved in any decision made from the model results. See case study 4 later in this chapter for additional discussion of how uncertainty can be addressed in models.

26.4 Representing Spatial Relationships in Current Ecological Models

Since space is central to the ecological conceptual framework, tools to model space are necessary to realize ecological processes through the conceptual framework. Historically modelers have either generalized space or explicitly calculated space (usually distances) with mathematical formulas. With the increased power of computers, spatial models have become more complex. With the development of GIS and other spatially oriented software many ecological theories can be more rigorously tested and tested in a more spatially explicit manner.

Generally, space is represented in ecological models in these ways:

1) mathematically, either by direct measurement of distances between features or through optimization models such as linear and multi-criteria programming,
2) through maps and GIS,
3) in charts and graphs (e.g., semivariogram),
4) by other means: Urban and Keitt (2001) used graph theory to represent the functional relationships between spatial features. Graph theory models entities through nodes that are connected by lines. The nodes can be patches and the lines can be corridors. If one corridor is removed, graph theory can identify which patches will be affected. And Williams et al. (2005) proposed dispersal chains to identify if a particular population of a species can survive with a changing landscape due to global change. Dispersal chains are a series of bars, one for each time slice. The dispersal chain has rules; for example, a species can only survive into the next chain or time step if their existing habitat can support them or habitats immediately next to them can (these dispersal rules can change based on the animal species). The species can survive to the final time step if it can move successfully between dispersal chains.

26.5 The Application of GIS to Ecological Models

Because GIS is one of the most-used spatial modeling tools, used in a wide variety of applications for modeling ecological processes, the technology can be important in realizing models through the ecological conceptual framework. GIS has been used extensively in forestry, hydrology, agriculture, land use change and wildlife management. Tables 26-1 through 26-4 present a sampling of how GIS was used in modeling mammals, birds and insects, fish and amphibians, and plants and communities, respectively.

Table 26-1 A sampling of mammal studies utilizing GIS in spatial models.

Wildlife species	Researchers	Application
Alpine Ibex	Hirzel et al. 2002	Created prediction maps for Alpine Ibex population size, density and production.
Bighorn sheep	McKinney et al. 2003	Explored how escape terrain influenced
Black bear	Bowmann et al. 2004	From surveys, determined human attitudes toward bears to identify best restoration locations.
	Mitchell and Powell 2003	Examined bear response to timber harvests and how their preference changed with regeneration.
Bobcat	Riley et al. 2003	Examined home range size in relation to exposure to human activity (included coyotes).
Brown bear (in Spain)	Naves et al. 2003	Identified suitability for bears relative to mortality and reproduction.
Chipmunk	Bender and Fahrig 2005	Looked at the influence of the matrix relative to chipmunk movement between patches.
Cougar	Anderson and Lindzey 2003	Determined predation rates from GPS cluster locations.
	Dickson and Beier 2002	Calculated home range and analyzed its composition, in particular to human influences.
Coyote	Grinder and Krausman 2001	Studied use home range around urban areas.
Deer	Bowyer et al. 2002	Explored how sexual segregation affects the spatial distribution throughout the year.
	Bruinderink et al. 2003	Examined connectivity of landscape network.
	Kie et al. 2002	Studied how heterogeneity of the landscape increased or decreased home range.
	Rothley 2001	Tested HSI before and after timber cuts.
	Conner and Miller 2004	Explored the spread of disease in mule deer.
Elk	Peinetti et al. 2002	Studied change in willows and elk winter range.
Flying squirrel	Kurttila and Pukkala 2003	Examined economic land holding and spatial ecological goals and their influence on squirrels.
Grey kangaroo	Selkirk and Bishop 2002	Enhanced home range calculations for kangaroos.
Grey squirrel	Lutz et al. 2001	Predicted grey squirrel expansion in Italy.
Grizzly bear	Apps et al. 2004	Predicted bear distribution and abundance.
	Johnson et al. 2004	Identified high risk areas for bear mortality, highest where human interaction.
Moose	Girard et al. 2002	Looked at sampling size on home range calc.

(continued on next page)

Mustelid	Gough and Rushton 2000	Review of GIS modeling of mustelids.
Ocelot	Harveson et al. 2004	Examined soils as indicator for vegetation type to restore the habitat for ocelots.
Panda	Loucks et al. 2003	Identified habitat for the conservation of pandas.
Red squirrel	Verbeylen et al. 2003	Explored the influence of the matrix on the squirrel distribution.
Rodents	Williams et al. 2005	Examined dispersal capabilities of rodents in South Africa with climate change.
Sumatran rhino	Regan et al. 2005	Examined how uncertainty of information may affect management options.
Swift fox	Sargeant et al. 2005	Used image restoration to fill absent data to help predict the distribution of swift fox.
Thomson's gaz	Fryxell et al. 2004	Explored different foraging strategies and movement rules and compared to field data.
Wolf	Carroll, Phillips et al. 2003	Studied the viability of a reintroduction proposal.
	Treves et al. 2004	Explored wolf predation habits on livestock.
	Whittington et al. 2005	Quantified response to roads and development.

Table 26-2 A sampling of birds and insects studies utilizing GIS in spatial models.

Wildlife species	Researchers	Application
Birds		
Sharp-tailed grouse	Akçakaya et al. 2004	Used landscape and metapopulation models to evaluate forest plans' effects on grouse viability.
Red grouse	Palmer and Bacon 2000	Studied habitat composition and boundaries as they relate to vegetation of territories.
Bell's sage sparrow	Akçakaya et al. 2005	Observed how different fire regimes affect sparrow.
Red-cockaded woodpecker	Cox et al. 2001	Examined clusters of activity within tree cavities.
Glanville fritillary	Drechsler et al. 2003	Examined the effects of different scenarios on metapopulations.
White throated sparrow	Formica et al. 2004	Identified how habitat quality and social structure of neighbors determines habitat choice.
Southwestern willow Flycatcher	Hatten and Paradzick 2003	Created prediction maps of flycatcher distribution.
Pintail	Ji and Jeske 2000	Studied the effects of environmental and land use impacts on the distribution of wintering locations.
	Podruzny et al. 2002	Examined decline of pintail relative to wetlands and ponds as they relate to agriculture.
Ovenbird	Larson et al. 2004	Linked population model with landscape model to examine effects of cutting scenarios.

Marbled murret	Meyer et al. 2002	Recommend preserve habitat for wide-range of species at multiple scales. Murret found most often in old growth.
Mexican Spotted Owl	Moou and DeAngelis 2003	Three models of complexity. Concluded spatially explicit models are less affected by uncertainty.
Spotted and Barred Owls	Peterson and Robins 2003	Looked at how Barred owl invasion affect the spotted owl since they share same niche.
Cerulean warbler	Thogmartin et al. 2004	Predict abundance from Bayesian approach.
Coal tit, golden plover, and snipe	Tucker et al. 1997	Determined suitability from Bayesian rules.
Waterfowl (mallard)	Royle and Dubovsky 2001	Modeled spatial variability in band-recovery data to produce spatial estimations of recovery rates.
Song birds	Dettmers and Bart 1999	Created habitat models from presence data using GIS operations.
Insects		
Speckled wood butterfly	Chardon et al. 2003	Explored the effects of the matrix composition on the movement of the butterfly between patches.
Butterflies (*Plebejus argus*)	Gutierrez et al. 2005	Analyzed *Lasius niger* (ant) habitat use to predict the use of a correlated butterfly species.
Cricket	Kidd and Ritchie 2000	Modeled intra-species variation and adaptation of crickets.
Beetle	Morales and Ellner 2002	Created movement model to incorporate behavior. The model is more realistic than random.
Cactus bug	Schooley and Wiens 2005	Examined how suitable patches and the connecting matrix affects the distribution of bugs.
Grasshopper	Skinner et al. 2000	Explored correlations between grasshopper species, available water capacity and soil permeability.

Table 26-3 A sampling of fish and amphibian studies utilizing GIS in spatial models.

Wildlife species	Researchers	Application
Aquatic fish and mammals		
Trout	Clark et al. 2001	Examined how climate change will affect trout.
	Railsback and Harvey 2002	Studied three foraging theories for habitat selection.
	Railsback et al. 2003	Examined the assumptions of habitat models on a virtual trout population (e.g., high quality means high density).
Manatees	Flamm et al. 2001	Mapped the distribution of manatees using a variable-shaped filter. The filter grows based on the habitat below.
Bass	Vander Zanden et al. 2004	Identified the lakes that would be most likely be invaded by bass.
Amphibians		
Salamander	Gustafson et al. 2001	Examined alternative forest management scenarios and their effect on the distribution of salamanders.
Lizards	Kearney and Porter 2004	Created a mechanistic approach to define the fundamental niche based on physiological measurements and biophysical models to determine distribution and abundance of lizards.
Tortoise	Kristan and Boarman 2003	Studied how predation by ravens affects the distribution of tortoise.
Amphibians	Rustigian et al. 2003	Examined how alternative land use scenarios affected amphibians in agricultural wetlands.

These different ecological models exhibit commonalities in their applications of GIS:
1) GIS-produced maps for visual display that show the study area and context.
2) GIS used as an integrating environment, processing data from a variety of sources.
3) Data creation through the GIS editing or analysis functions that can be exported to outside modeling packages.
4) To extract specific data to run analysis with a statistical software package or to use in an outside modeling package. The GIS is also used to identify how and what data should be extracted (e.g., only forested areas or locations in a certain watershed will be extracted).
5) To display the results from outside modeling software.
6) For analysis using numerous tools to quantify the spatial relationships of the features in the study area.
7) For the modeling of processes.

26.6 Spatially Explicit Ecological Models

This proposed ecological conceptual framework is based on objects making decisions and moving in space. Models that do not incorporate space explicitly focus on an abstraction of reality. Thus, to realize space for models based on the framework, only spatially explicit models and modeling tools are relevant.

Spatially explicit prediction models incorporate space as an integrated or essential part of the process and decision making. Berger et al. (2002) present four tests to determine if a model is spatially explicit. A model is spatially explicit if:

1) its results are affected by randomly moving the objects that participate in the model.
2) location is included in the representation of the system being modeled.
3) spatial concepts such as location or distance appear directly in the model.
4) the spatial forms of inputs and outputs are different. A spatially explicit model modifies the landscape on which it operates.

The ecological processes that require spatially explicit modeling include 1) ecological processes that are inherently spatial, 2) land management, 3) when the modeler is exploring a theory that is spatially oriented, and 4) when designing field experiments. A GIS alone or coupled with other software is a productive environment to model these spatially explicit processes.

Table 26-4 A sampling of invasive species, hotspot analysis, and wetland interactions studies utilizing GIS in spatial models.

Area of concern	Researchers	Application
Invasive species	DirnBöck et al. 2003	Examined the effects of invasive plants in Robinson Crusoe Island in Chile.
	MacIsaac et al. 2004	Studied the effects of invasive plants into lakes.
	Peterson 2003	Reviewed tools to study effects of invasive plants.
	Arriaga et al. 2004	Identified where buffel grass can invade.
	Chakraborti et al. 2002	Analyzed factors affecting the spread of Zebra Mussels in Lake Huron.
	Haltuch and Berkman 2000	Studied the spread of Dreissenid mussels.
Hotspot analysis	Gjerde et al. 2004	Identified that certain species are not represented in hotspots.
	Grand et al. 2004	Identified hotspots in pitch pine scrub oak in Massachusetts.
	Luoto et al. 2002	Identified hotspots for plant species richness.
Wetlands	Cedfeldt et al. 2000	Identified the functionally significant wetlands.
	Palik et al. 2003	Identified unmapped wetlands, particularly seasonal ones, using a hierarchy classification.
	Liu and Cameron 2001	Modeled patterns of wetlands in Texas.
	McCauley and Jenkins 05	Quantified wetland characteristics in Illinois.
	Kelly 2001	Explored the impact of permits on ecological processes.

26.6.1 GIS Applications of Ecological Processes That Are Inherently Spatial

The following is a sampling in which GIS has been used to model applications and ecological processes that are by nature spatially oriented and spatially explicit:

To define distributions or potential distributions of species as a primary focus or an intermediate step to an additional analysis. The distributions can be determined from field data, expert opinion, or from a number of statistical and modeling tools used to quantify the distribution patterns (see Section 26.7.3, Models Quantifying Patterns of the Conceptual Ecological Framework).

To describe a specified spatial area or shape and its characteristics is often the primary purpose of a GIS for certain studies. The geographic space can be defined as habitat patches (McKinney et al. 2003), clusters of activity (Cox et al. 2001), existing human settlement and fields (Fritz et al. 2003), blowdowns (Lindemann and Baker 2001), vegetation type (Peinetti et al. 2002) or home range (Dickson and Beier 2002), or the geographic space can be artificially defined by the GIS, such as through concentric circles (Meyer et al. 2002).

Examples of characteristics being quantified as attributes within the geographic shapes can include the quantity of roads (Crist et al. 2005), suitability of land use type for restoration (Lee et al. 2002), elevation ranges (Loucks et al. 2003), amount of escape terrain (McKinney et al. 2003), human activity (Grinder and Krausman 2001), or any other number of landscape features that have some relationship to the process being analyzed.

To analyze home ranges to define the activity extents of territorial animals. There are two main approaches for home range calculations. The first assumes that the home range can be defined by some geometrically defined shape (e.g., an ellipse) or polygon derived to be the minimum bounding polygon for a set of field observations—the convex-hull polygon. The second approach derives the home range based on the density of the field observations (Hooge and Eichenlaub 1997; Selkirk and Bishop 2002). There are a variety of methods in each approach. The calculations for home range definitions are based on the spatial distribution of the observations. Several software packages are used as standalone or extensions to GIS software to perform such calculations (see Section 26.9, Ecological Spatial Modeling Software).

There are several examples of home range analysis applied in GIS. Dickson and Beier (2002) defined and quantified the home range of cougars relative to habitat fragmentation, roads and other anthropogenic influences. Formica et al. (2004) identified that the habitat selection of white and brown-throated sparrows is based not solely on quality, but also on the social structure of its neighbors. Harveson et al. (2004) used home range definitions to quantify the existing use of soils as indicators for identifying the preferred locations for ocelots. Kie et al. (2002) looked at how heterogeneity of landscape increased or decreased the home range size. And Riley et al. (2003) showed that bobcat and coyote home range size was positively correlated with urban associations suggesting human dominated areas are less suitable.

Studies specific to improving home range calculations include Girard et al. (2002) who examined how sampling size influenced home range calculations. Kenward et al. (2001) proposed a nearest-neighbor clustering alteration to improve the linkage method of home range calculations. Selkirk and Bishop (2002) not only improved on calculations but used 3-dimensional (3D) analysis and orthophotographs as a backdrop to improve the understanding of the home ranges.

To model reserve design, metapopulation, corridors and source sink populations that are by nature spatial problems since they examine the spatial relationships and interconnections between habitat patches. Reserve design and regional conservation plans may be used to derive favorable options for the configuration of existing patches (Carroll, Noss et al. 2003; Rothley et al. 2004) or to analyze the structure of the landscape network (Bruinderink et. al. 2003). Reserve design models can incorporate threat into the design (Lawler et al. 2003), account for sensitivity of the parameters (Warman et al. 2004), or adjust for diversity patterns and compositional turnover (Wiersma and Urban 2005).

Metapopulation analysis explores how different populations of a species might interact. The interaction between populations occurs through some spatial movement between the patches (i.e., corridors). The area through which the species travels, the matrix, also influences movement (Chardon et al. 2003; Verbeylen et al 2003) with behavior possibly influencing the functional connectivity of the movement (Belisle 2005).

In a source-sink model, the source locations are defined as where the species can live and reproduce, whereas in the sink locations the animals can survive but mortality exceeds reproduction. The spatial arrangement of the patches and their attributes determine what is a source, a sink, and the migration between them. Source-sink models are particularly well suited for GIS (Naves et al. 2003).

To assess fire impact on the landscape is a spatial problem since fires move through space. GIS models have been used to examine the spatial effects of fire suppression (Duncan and Schmalzer 2004), identify locations susceptible to fire based on regeneration scenarios (Gustafson et al. 2004) and fire suppression scenarios (Sturtevant et al. 2004), analyze the fire regime over time with vegetation type (Wells et al. 2004), and to identify the areas to perform prescribed burns (Hiers et al. 2003).

To model natural disturbance. The magnitude of damage from a natural disaster is usually dictated by where, how hard, and the amount of area affected making it a spatially explicit process. Boose et al. (2004) examined the long-term ecological effects of the spatial patterns of hurricane damage in Puerto Rico. Kallimanis et al. (2005) demonstrated that increasing the spatial aggregation of disturbance generally increases extinction risk while random disturbance decreases it. Kramer et al. (2001) predicted spatial patterns of windthrow based on slope, elevation, soil stability and exposure to prevailing storm winds. And Lindemann and Baker (2001) showed that the spatial characteristics of windthrows are highly variable and that it would be inappropriate to try to mimic them.

To access change in landscape patterns. Landscape change is a spatially explicit problem since it quantifies how space changes through time. Alados et al. (2004) examined how land cover has changed in the Mediterranean from 1957 to 1994. Axelsson et al. (2002) studied the change in mixed deciduous forest in Sweden between 1866 and 1999 and concluded that large changes occurred due to complex interactions between fire disturbance, fire suppression, logging and silviculture. Weaver and Petera (2004) demonstrated that applying forest transition models that do not incorporate spatial dependence created more-fragmented landscapes than those that incorporate it (which more accurately represent reality). Metternicht (2001) explored how the salinity distribution on the landscape changed in South America. Verburg and Veldkamp (2004) studied how the forests of the Philippines will change with different development scenarios.

26.6.2 Land Management Requiring Spatially Explicit Models

Land management as an ecological process may require spatially explicit models. Spatial models are used by many land managers because they are responsible for a specific park, habitat, town, or conservation area. The managers generally wish to manage for what they have now and project for the future. Management applications include wildlife, forestry, exotic species control, reserves and sanctuaries, urban parks, town planning and agriculture. Many managers will create various alternative scenarios of possible outcomes (with or without the GIS). Some managers include the stakeholders in the design process to create alternative scenarios (Berger and Bolte 2004; Hiers et al. 2003; Hulse et al. 2004), while others use GIS to evaluate and possibly rate alternative scenarios based on the management objectives.

26.6.3 Spatially Oriented Theoretical Models

In this set of models, space is at the center of a concept or theory that the modeler is exploring. Many times these models are applied to small, simulated landscapes to explore the spatial concept. Some of the ecological processes studied in these simulated landscapes include foraging behavior, and movement and dispersal activity. (For a sampling of spatially oriented theoretical studies see Section 26.7.2, Models Addressing Process, Decisions, and Actions of the Conceptual Ecological Framework.)

26.6.4 Using Spatially Explicit Models for Designing Field Experiments

The location of data collection for some field studies can optimize the data gathering process in order to obtain the most representative samples possible and to reduce bias. Kadmon et al. (2003) quantified the degree of sampling bias with respect to climate conditions which negatively affected model predictive accuracy. Kadmon et al. (2004) studied the distribution of 129 species of woody plants and concluded they suffered from roadside collection bias and the impact of the bias depended on the magnitude of climatic bias of the geographic distribution of the road network. Legendre et al. (2004) quantified the effects of spatial autocorrelation on Type I error and explored ways to control the effect of spatial autocorrelation in data collection.

26.7 Existing Spatially Explicit Ecological Models in Context of the Ecological Conceptual Framework

The following sections group a wide variety of existing ecological models according to the aspect of the ecological conceptual framework they address. The models are aggregated into those that address 1) physical and mental requirements, 2) process, decisions and actions, 3) quantifying patterns, and 4) laws and general principles.

26.7.1 Models Addressing the Physical and Mental Requirements of the Conceptual Ecological Framework

There are few spatially explicit models that incorporate the physiological and mental requirements as the driving force for defining the spatial arrangements in ecological processes. Kearney and Porter (2004) looked at physiology to define the fundamental niche of lizards, and Johnston and Schmitz (1997) explored how global warming may affect several species' physiological requirements (see Case Study 1 later in this chapter).

26.7.2 Models Addressing Process, Decisions, and Actions of the Conceptual Ecological Framework

Generally these models explore the decision making of individuals and the resulting actions that produce the process. Many times these models are theoretical in nature and are initially applied to small simulated landscapes. From these simple models, various decision strategies are applied to explore the effects on the simulated landscape. Occasionally the principles are applied to actual processes or species.

Basset et al. (2002) proposed on a 6x6 raster the optimal foraging of individuals and when they would leave one patch for another. They used energetics and population stability in the decision making and proposed a potential tradeoff between short-term individual fitness and long-term population stability. Fryxell et al. (2004) simulated 12 alternative foraging objectives and movement rules to replicate field data on gazelles. They suggested energetics as an appropriate parameter. Railsback and Harvey (2002) used three foraging theories for habitat selection of trout—maximum growth, survival and expected maturity—to simulate actual data.

Dispersion by definition is movement. Kallimanis et al. (2005) used a simulated landscape to create various disturbance regimes to determine how disturbance affects extinction. They concluded that the farther the species can disperse the less likely they will become extinct. Molofsky et al. (2002) use cellular automata to explore the effects of different scales on dispersion. Sondgerath and Schroder (2002) demonstrated that the spatial configuration of habitat quality affects the spatial spread of populations in a heterogeneous landscape and that species with limited dispersal ability require stepping stones to reach isolated patches. Bender and Fahrig (2005) showed not only does distance facilitate movement between patches but the matrix they travel through may also influence the movement. Morales and Ellner (2002) demonstrated on a hypothetical landscape that it is behavior, not spatial structure, that drives the movement of beetles. And Wiegand et al. (2005) explored on a 50 x 50 grid of varying quality of habitat types how various degrees of fragmentation affect population response. They conclude that to predict the effects of fragmentation on a population requires a good understanding of the biology and habitat use which is complicated by the uniqueness of the species and the landscapes they occupy.

Individual-Based Models (IBM) (DeAngelis and Gross 1992) and Agent-base Models (ABM) often apply decision rules to individuals based on their state that initiates an action. From the aggregation of the individual actions, patterns are created. (See Case Study 3 later in this chapter for further information.)

26.7.3 Models Quantifying Patterns of the Conceptual Ecological Framework

The majority of ecological models using GIS fall into the category of quantifying patterns. These models are based on the premise that the elements are arranged non-randomly to create the spatial configuration of the landscape. Models quantifying patterns generally deviate from the ecological conceptual framework in that they take a top-down approach to space and impose a spatial extent. Within that extent they identify and quantify the patterns. Implicit in these models is a biological or physical explanation that produces the pattern; causality can only be deduced from these models. Generally the focus of these models is to implement tools that can accurately quantify the pattern rather than conceptualize and explore the interaction of the elements creating the pattern. Most models that explore patterns generalize and model the average case. However, it may be the unusual individuals and objects that define the process and its future direction. Generally models quantifying patterns make predictions from the patterns and attempt to replicate the pattern, not the process that created the pattern. The various types of models that quantify patterns 1) develop associations from experts, 2) derive patterns from the attributes at locations, 3) quantify patterns through metrics, and 4) optimize using patterns. They are explained below.

26.7.3.1 Developing Associations from Experts

Habitat Suitability Index (HSI) and weighted suitability models weight each location in a study area based on the preference of a species for the attributes located there. The ranking and weights assigned to each significant attribute may be derived from expert opinion, field research and existing literature. HSI and weighted suitability models are descriptive models since they weight or describe each location by what is located there. These models generally do not account for space explicitly (for example, the suitability of one location does not influence the suitability of its neighboring locations).

HSI and suitability models have been used for many types of applications including wildlife (Loucks et al. 2003; McKinney et al. 2003; Kurttila and Pukkala 2003), plants (Burnside et al. 2002) and shell fish (Chakraborti et al. 2002). Larson et al. (2004) applied different versions of HSI models to the output from landscape simulations to explore various alternatives.

Even though HSI and suitability models are commonly used, some caution should be exercised when interpreting the results. Johnson and Gillingham (2004) demonstrated that the sensitivity of the parameters and uncertainty can influence the reliability of predictive distributions from the models. And Rothley (2001) created HSI models for deer before and after a timber cut. She used not only density as a measure of preference but also foraging intensity and pellet count. She concluded that HSI models are unreliable for prediction of deer distribution.

HSI and suitability models attempt to identify or create the pattern of distribution based on the attributes located at the site. Generally there is no interaction between the attributes with each contributing independently to the preference (Malczewski 1999). The implied causality is the species is reacting to the attributes. Underlying these models is the assumption that the more positive attributes associated with a location the more preferred it is. These models are highly sensitive to rankings and weights (Malczewski 1999).

26.7.3.2 Deriving Patterns from the Attributes at Locations

The following set of models use statistical relationships to quantify the patterns of where, in what quantity, or how much preference the phenomenon has for the attributes at each location. Unlike the HSI and suitability models above that rely on expert opinion, the following models (regression, factor analysis, discriminant analysis, classification, Bayesian, neural networks and other methods) attempt to have the phenomena quantify their preference or magnitude for each location based on their existing distribution.

26.7.3.2.1 Regression

Regression analysis is generally a classical statistical analysis based on the attributes at each independent observation. Spatial data challenge the independent observation assumption since spatial data are usually spatially autocorrelated. The dependent variable represents the magnitude of the phenomenon being estimated (e.g., quantity of biomass) and independent variables define the attributes that influence or determine the magnitude of the dependent variable. The output of regression analysis includes one coefficient per independent variable; the coefficient indicates the influence of the independent variable on the dependent variable.

A prediction surface can be created by multiplying the value for each cell for each independent variable, by the corresponding coefficient. When creating the preference surface, if there is a magnitude of how much the species prefers the location, a multiple regression can be used (Johnston 1992; Haltuch and Berkman 2000) and if presence and absence is known a logistics regression is used (Johnston 1992; Grand et al. 2004; Hatten and Paradzick 2003; Kramer et al. 2001; Whittington et al. 2005). Instead of using species data, Bowmann et al. (2004) used survey data of people's attitudes toward bear reintroduction to create a surface of locations where bears will be most favorably received based on human demographics. DirnBöck et al. (2003) used logistic regression to identify the distribution of the potential worst plant invaders on Robinson Crusoe island in Chile. Kristan and Boarman (2003) developed a Poisson regression for raven distribution and a logistic regression for risk when mapping raven density and its effects on tortoise distribution. And Naves et al. (2003) created four regression models for bear suitability, a general model, one for natural landscapes, one for human and one for reproductive. From the models they identified source, refuge, sink and matrix habitats.

Chardon et al. (2003) used logit regression to compare the results of a Euclidean versus a cost distance analysis defining connectivity for the speckled wood butterfly. Regression analysis has also been used to compare prediction models to the actual distribution (Gustafson et al. 2001), to explain the factors of patch abundance and occupancy (Schooley and Wiens 2005), to predict predation and kill type based on time spent at locations collected from GPS points (Anderson and Lindzey 2003), and to develop a resource-selection function relating wolf distribution (Carroll, Phillips et al. 2003).

Logistics regression requires presence and absence information. Although alluring for modeling ecological processes, in particular animal observations, logistics regression is commonly used when the phenomenon is present at an observation; however, for all other locations the phenomenon usually has not been observed, which does not necessarily mean it is absent. In most of the models in this section, regression analysis was used to quantify an existing pattern from independent observations, continue the pattern, and infer the underlying process.

26.7.3.2.2 *Factor Analysis*

Factor analysis is used to explain the variability among a number of observable random variables in terms of a smaller number of unobservable random variables called factors. The observable random variables are modeled as linear combinations of the factors, plus "error" terms. There is no distinction made between dependent and independent variables and the factor analysis is based on the similarities between the variables. Factor analysis addresses the issue of absence data since it only requires presence data as input (Hirzel et al. 2002).

Hirtzel et al. (2002) used factor analysis to determine the attributes that ibex prefer by comparing those cells where the ibex has been observed to all other cells in the study area. Like in regression analysis, prediction maps can be created from the analysis.

26.7.3.2.3 *Discriminant Analysis*

Discriminant analysis, like regression analysis, is applicable when there is one dependent variable but multiple independent variables. Unlike regression and analysis of variance, the dependent variable must be categorical. It is similar to factor analysis in that both look for the underlying dimensions, but differs from factor analysis in that the underlying dimensions are based on differences rather than similarities. Discriminant analysis also differs from factor analysis in that a distinction must be made between the dependent variable and the independent variables. Cox et al. (2001) used discriminant analysis to identify clusters of red-cockaded woodpecker use when identifying the preferred habitat for the woodpecker.

26.7.3.2.4 *Classification*

Various classification techniques are used in multivariate statistical analysis in a GIS (e.g., maximum likelihood classification) and for determining land use types in a remote sensing package. The goal is the same in a classification (as with most of the models presented in this section) in that they aggregate locations into a number of categories based on the similarities of the attributes at the locations.

Classification methods such as numeric classification (Leathwick et al. 2003), Classification and Regression Trees (CART) and logistic regression (Line et al. 2003), and image restoration (Sargeant et al. 2005) have been used in classifying ecological processes. Fuzzy logic has been used to examine the effects of crisp boundaries in the classification results (Guneralp et al. 2003).

26.7.3.2.5 *Bayesian*

The assumption behind a Bayesian model is that a joint distribution of a collection of random samples can be divided into a series of conditional models (Wikle 2003). It is easier to specify the conditional distributions, and the product of a series of simple conditions can lead to a complicated joint distribution (Wikle 2003).

Bayesian techniques have been used to predict the breeding distribution of birds in the northeastern USA (Tucker et al. 1997), to predict the abundance of house finches (Wikle 2003) and the distribution of ground flora (Hooten et al. 2003). Bayes and Poisson regression have been used to account for the observer effects and spatial correlation between counts in survey data (Thogmartin et al. 2004).

26.7.3.2.6 *Neural Networks*

Neural networks are based on a conceptual framework similar to the human brain. There are many interconnecting processes (or neurons) that interact to solve a specific problem. Like the human brain the neural network can learn and is proficient at deriving patterns from imprecise or complex information (Stergiou and Siganos 2006).

Vander Zanden et al. (2004) selected neural networks to model the vulnerability of lakes to the invasive small-mouth bass over the parametric approaches listed above because, 1) neural networks can model nonlinear relationships, 2) they accept either continuous or discrete data, 3) they require no assumptions about the distribution of the independent variables, and 4) they have been demonstrated to predict better for nonlinear relationships over logistic regression and discriminant analysis.

26.7.3.2.7 Other Pattern Quantifying Methods: General Linear Model, Analysis of Variance and Detrended Correspondence Analysis (DCA)

The general linear model (GLM) is the general form for a series of statistical models. Analysis of variance and ordinary linear regression are special cases of general linear models. If there is only one dependent variable or one column in the Y matrix then the model becomes a multiple regression. GLM's have been used to describe the spatial patterns of *Lasius niger* (an ant) in order to predict the distribution of the butterfly *Plebejus argus* (Gutierrez et al. 2005). Luoto et al. (2004) predicted bird species richness from remote sensing and topographic data using GLM.

Analysis of variance was used to evaluate the effects of timber harvest and regeneration type on HSI (Mitchell and Powell 2003).

Skinner et al. (2000) describe Detrended Correspondence Analysis (DCA) as a nonlinear, weighted averaging method that can be used for measuring the variance among species and site data (e.g., soil type, elevation) in a single analysis. The species and sites are ordered along axes according to their similarities to each other. Each axis is a gradient in an ecological factor. The eigenvalues on each axis indicate how similar or dissimilar two sites are from each other. Skinner et al. (2000) used DCA to test the correlations among grasshopper species.

Most of the models described in this section for quantifying spatial patterns do not take into account spatial relationships but rather independently quantify the attributes at the location. Using GIS spatial operations (e.g., Euclidean distance) some spatial relationships are included in the independent variables. However, the analysis generally does not adjust its parameters by location based on what surrounds the location.

26.7.3.3 Quantifying Patterns Through Metrics

These models quantify the non-random spatial distribution of the elements in the landscape that produce patterns. Usually quantifying the pattern is not the end goal but trying to determine causality from the patterns is not well understood (Wagner and Fortin 2005). Crist et al. (2005) used five indices to measure the connectivity and isolation of roadless areas. Tinker et al. (2003) used metrics to evaluate the structural effects of small and large fires compared to landscape structure pre- and post-clearcutting. Kurttila and Pukkala (2003) used metrics to evaluate the fragmentation of the landscape for flying squirrels. And Lindemann and Baker (2001) used metrics to quantify windthrow characteristics such as size, perimeter and distance to nearest patch.

Problems can arise when applying metrics. Wagner and Fortin (2005) summarize four main problems: 1) there is no single index to capture landscape structure, 2) metrics are highly sensitive to scale, 3) organisms can respond to the landscape in a non-linear way, and 4) metrics assume the property to be nominal or binary. There are additional problems with applying metrics. Most patterns are the result of several processes creating the heterogeneity, thus it is difficult to attribute any one pattern to a single process (Wagner and Fortin 2005; Li and Wu 2004). The scale and the error in the classification of the data can cause error in metrics (Arnot et al. 2004). Metrics can be sensitive to the resolution and the spatial extent in which they are measured (Baldwin et al. 2004). Depending on the data set used (e.g., the scale of the road data set from which the metrics are being taken) can alter the metric (Hawbaker and Radeloff 2004). Indices can be inconsistent with different landscape structures and dispersal behaviors (Tischendorf 2001). Bender et al. (2003) demonstrated that metrics did not adequately predict immigration when patch size and shape varied. Finally, Li and Wu (2004) stated that there is a conceptual flaw in landscape pattern analysis because of unwarranted relationships between pattern and process, ecological irrelevance of landscape indices, and confusion between the scales of observation and analysis.

26.7.3.4 Optimizing Using Patterns

Optimization models attempt to optimize (maximize or minimize) some criterion based on a series of constraints. There are two main areas where optimization models have been implemented to model ecological processes. The first is quantifying the configuration of spatial patterns. Identifying the best patterns in a landscape based on a specified criterion is referred to as spatial optimization (Duh and Brown 2005).

The second area where optimization models are used is based on Maynard Smith's (1978) theory of behavioral optimization. The theory assumes that most species make optimal behavioral choices such as when they forage, to maximize their fitness. Optimization models in this context explore optimal behavioral choices.

Three tools that can be used for optimization of patterns are genetic algorithms, simulated annealing and linear programming.

26.7.3.4.1 Genetic Algorithm

A genetic algorithm is comprised of an objective function to quantify an alternative and a means to represent a solution or alternative. The algorithm begins by randomly combining a series of the components (chromosomes) to create a series of possible solutions (or individuals). Each solution is evaluated. The more fit solutions, defined by the objective function, are allowed to reproduce. These individuals will create new offspring (individuals) and the offspring will be comprised of a different combination of chromosomes (components) many of which are coming from the parents. Which chromosomes are passed to the child is determined by methods such as crossover and mutation. A new population of individuals (or solutions) is created and another generation is produced from the most fit individuals. This process can be continued for generations; it concludes when any number of objectives are met such as a minimum criterion is obtained or the defined number of generations is reached. The genetic algorithm can converge toward a local optimum if the problem is complex. It is difficult to formulate the representation space when exploring an objective function defined by a landscape metric (Duh and Brown 2005).

Genetic algorithm can also be used to explore behavior and habitat preference. From field observations a genetic algorithm can be used to define the criteria a species prefers and these criteria can be extrapolated to all locations in the GIS to create prediction maps. The Genetic Algorithm for Rule Set Prediction (GARP) software was used to create surfaces to predict where buffel grass was to potentially invade (Arriaga et al. 2004), generate ecological niche models from museum records (Illoldi-Rangel et al. 2004) and observations (Peterson 2001), and to define the native distribution of Barred and Spotted Owl to look at the overlap of the distributions to predict potential invasion of the Barred Owl into the Spotted Owl distribution (Peterson and Robins 2003).

26.7.3.4.2 Simulated Annealing

Simulated annealing comes from the process of heating and cooling metals to create a strong material (Cook and Auster 2005). Simulated annealing is an iterative process that attempts to find the optimal (or near optimal) solution to an objective function. Like with the metals, the algorithm goes through a cooling process to examine different alternatives. For example, simulated annealing can be used to identify the optimal configuration of landscape patches for a species for a reserve design given that only a limited number of patches can be selected with an objective function based on a measure of connectivity. The algorithm begins by randomly selecting the specified number of patches. When the annealing begins, the temperature is hot and one patch might be removed from the alternative and another (randomly selected) replaces it. The first alternative will be compared based on the objective function to the second. As the annealing cools down, the replacement process will not be random but will only select an alternative replacement that improves the objective function (Duh and Brown 2005). Because of the randomness of the iteration process, the annealing should not get caught in local optima when searching for alternatives (Cook and Auster 2005; Duh and Brown 2005).

Cook and Auster (2005) evaluated the potential for simulated annealing to identify the essential fish habitat for multiple species in four ecological regions in the eastern continental shelf of the USA.

26.7.3.4.3 Linear Programming

Some optimization models use space in the objective function and others include space in their constraints. Maximizing the connectivity of a series of patches and minimizing the amount of land to allocate is an example where space is used in the objective function. An example of a spatial constraint would be to constrain the distance between two subsequent cuts to reduce road construction costs in a forestry cutting model.

Arthur et al. (2004) used integer programming (linear programming with only integer variables) to maximize the expected number of the species and maximize the likelihood that a subset of endangered species is represented in a conservation plan. Haight et al. (2005) created a site selection model using integer programming with two objectives, maximize the expected number of species represented in protected sites and maximize the expected number of people with access to the sites.

Multi-criteria decision analysis (MCDA) identifies the feasible options when multiple objective functions from multiple criteria are simultaneously evaluated. Drechsler et al. (2003) used MCDA to rank different realizations from a metapopulation model. Rothley et al. (2004) used MCDA for reserve design but cautioned that selection criteria and the reserve-network objectives may be inconsistent.

26.7.4 Models Addressing the Laws and General Principles of the Conceptual Ecological Framework

Since there are so few laws and general principles in ecology there are rarely ecological models that are designed to model them.

26.8 Case Studies of Ecological Models as They Relate to the Ecological Conceptual Framework

In this section four case studies are presented to provide details of how models have addressed the different aspects of the ecological conceptual framework proposed in this chapter. The first case study (Johnston and Schmitz 1997) takes a more traditional approach to ecological modeling by defining a spatial extent, the continent of the USA, and quantifies the existing species distribution patterns within it. The criteria creating these patterns are extrapolated and applied to a new realization of the USA after global warming. The mechanism creating the patterns is only implied. However, unlike traditional models, the model incorporates the lower levels of the ecological conceptual framework pyramid by exploring the direct physical effects of the increase in temperature due to global warming on several wildlife species. The second case study, (Johnston, Schmitz et al. in review) focuses on modeling the decision making of moose driven by biophysical requirements to create processes and then patterns. The study demonstrates how to take the physical requirements and space within the conceptual framework and incorporate them into the mechanisms producing the processes and patterns. The third case study (Johnston, Collier et al. in review) presents an agent-based model that creates processes and patterns from the decisions of individual cougars. The cougar decisions are based on their physical and mental requirements. This case study is the closest to realizing the bottom-up approach to ecological modeling presented in the conceptual framework. The final case study (Johnston et al. 2005) explores how uncertainty in random events, error analysis and sensitivity analysis can be incorporated into the modeler's decision-making process. Uncertainty exists at each level of the ecological conceptual framework pyramid; thus, the techniques presented in case study four are applicable to many ecological models.

26.8.1 Case Study 1 - Quantifying Patterns and Inferring Causality: Assessing the Sensitivity of Selected Species to Simulated Doubling of Atmospheric CO_2

This case study performs a regression analysis on the existing distributions (patterns) of four wildlife species to determine how their distribution might change with the doubling of CO_2. A top-down definition of the spatial boundary is imposed, the continental USA, but within that extent, the spatial distribution of the species is an emergent property from the analysis. Biophysical constraints are also analyzed in the study to determine the effects of global warming on the physiology of the species. For a full discussion of this case study see Johnston and Schmitz (1997).

The study simulated the sensitivity of four mammal species within the continental USA (elk, *Cervus Canadensis*; white-tailed deer, *Odocoileus virginianus*; Columbian ground squirrel, *Spermophilus columbianus*; and chipmunk, *Tanias striatus*) to the effect of anticipated levels of global climate change brought about by a doubling of CO_2. Sensitivities to the direct effects of climate change were evaluated using a climate-space approach to delineate the range of thermal conditions tolerable by each species. Sensitivities to indirect effects were evaluated by quantifying the association of each species to the current vegetation distribution within the continental USA and using this association to assess whether wildlife species distributions might shift in response to vegetation shifts under climate change. Results of the direct effects of climate change indicate that altered thermal conditions alone should have little or no effect on the wildlife species distributions as physiological tolerance to heat load would allow them to survive. Analyses of the indirect effects of vegetation change indicate that deer and chipmunks should retain their current distributions and possibly expand westward in the USA. For elk and ground squirrels, there is a possibility that their current distributions would shrink and there is little possibility that each species would spread to new regions. This study emphasizes that the distributions of the four mammalian species are likely to be influenced more by vegetation changes than by thermal conditions. Future efforts to understand the effects of global change on wildlife species should focus on animal-habitat and climate-vegetation linkages.

26.8.1.1 Data Used in the Assessments

Evaluations of climate change were based on data generated by the Vegetation Ecosystem Modeling and Analysis Project (VEMAP) (Kittle et al. 1995; VEMAP Members 1995). VEMAP's aim is to model the sensitivity of terrestrial ecosystems in the continental USA to altered climate brought about by a doubling of atmospheric CO_2. Eight different climate models were used to predict the effects of doubling the atmospheric CO_2.

26.8.1.2 Assessing the Sensitivity of Wildlife to Climate Change

To model the biophysical effects of the doubling of CO_2 on these four species, the Porter and Gates' (1969) 'climate-space' approach was used to delineate the range of thermal environmental conditions that these animals can tolerate and still survive. (See also Gates 1980.) The climate space is bounded by extreme values of tolerable temperature and radiation. The climate-space model and data used to parameterize the model are presented in Johnston and Schmitz (1997).

The climate-space diagrams for each species are presented with current temperature and solar radiation in Figure 26-2. The lines represent the minimum night-time (top line) and maximum daytime (bottom line) temperatures and radiation that can be tolerated when the animal has attained the minimum (left line) and maximum (right line) allowable body temperatures.

Each species was evaluated to determine whether it could withstand the changes in temperature and absorbed radiation by comparing the climate values predicted under a doubling of CO_2 with the boundary conditions of the climate space. Example data for one of the worst-case scenarios (generated by the UKMO [United Kingdom Meteorological Office] model) are presented in Figures 26-3 and 26-4. For each species, the temperature and absorbed radiation values shifted farther toward the north-east boundary of the climate space (Figure 26-3), indicating a warming trend. Only white-tailed deer encountered locations in which they could not tolerate the level of change. However, this was only at one location in the Gila River region of Arizona where white-tailed deer populations are normally at low densities.

26.8.1.3 Assessing the Sensitivity to Indirect Effects of Climate Change on Wildlife

The association between the current distribution of a wildlife species and the current vegetation types was quantified within each wildlife distribution. Changes in wildlife distributions were projected as vegetation changes in response to a doubling of atmospheric CO_2.

Projections of vegetation change were used in response to climate change generated by the MAPSS model (Neilson 1995) for the vegetation data. MAPSS was used in assessments conducted by VEMAP (VEMAP Members 1995); it predicts the presence of the dominant vegetation type at a geographical location on a comparatively fine scale of resolution, 10 x 10 km grid within the continental USA (Neilson 1995).

The researchers quantified the association between the presence or absence of a wildlife species at each geographical location and vegetation types at those locations using logistic regression. Areas with presence of an animal species were coded as a 1, while areas of absence received a 0 score. Habitat types were coded categorically from 1 to 16 to correspond with the 16 dominant vegetation types in the continental USA.

Logistic regression was used to predict the probability that each wildlife species was present at different geographical locations within the continental USA based on the vegetation attributes at those locations under current climate and under a doubling of CO_2.

Comparison of the differences between current wildlife distributions and distributions under a doubling of CO_2 are presented in Figure 26-4 for the UKMO GCM scenario. The categories in the illustrations are refinements of four basic possibilities: 1) present currently and under a doubling of CO_2 (blues and gray); 2) present currently, absent under a doubling of CO_2 (orange and yellow), 3) absent currently and under a doubling of CO_2 (white), and 4) absent currently and present under a doubling of CO_2 (red).

Johnston and Schmitz (1997) suggest that white-tailed deer could expand their geographical range westward and should not be lost from most of their current geographical range (Figure 26-4a). Chipmunks could continue to inhabit the entire continent (Figure 26-4b). Both elk and ground squirrels have a greater sensitivity to changing habitat, as shown by the yellow and orange (loss) and red (gain) in Figures 26-4c and 4d. This emphasizes that predictions for range expansion by chipmunks and loss of elk must be interpreted cautiously because there is a high degree of uncertainty that their current distributions are determined solely by available vegetation types within their geographical ranges.

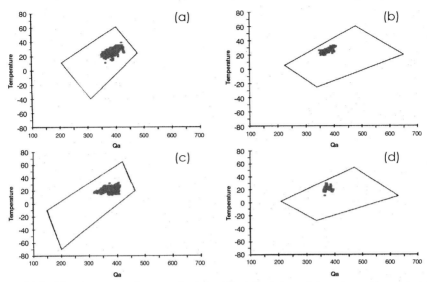

Figure 26-2 Climate-space diagrams for (a) white-tailed deer, (b) eastern chipmunk, (c) elk and (d) ground squirrel. Lines on the graphs bound a region representing combinations of temperature (°C) and absorbed solar radiation (Qa, W m-2) that each animal species can tolerate without incurring lethal physiological stress. Combinations of current (base) temperature and absorbed radiation from all locations within each species' geographical range are presented relative to the species' climate space. See included DVD for color version.

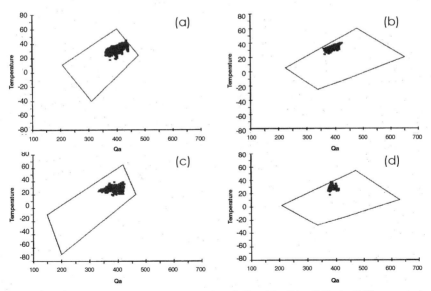

Figure 26-3 Temperature (°C) and absorbed radiation (Q_a, W m^{-2}) that wildlife species may experience under a doubling of atmospheric CO_2 relative to the boundaries for each species' climate-space. The values cover all locations within each species' current geographical range. The data represent one of the worst possible scenarios and were generated by VEMAP using atmospheric conditions predicted by the United Kingdom Meteorological Office (UKMO) GCM model. Data are for (a) white-tailed deer, (b) eastern chipmunk, (c) elk and (d) ground squirrel. See included DVD for color version.

Manual of Geographic Information Systems

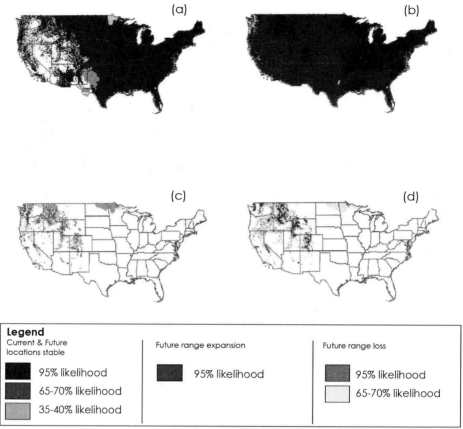

Figure 26-4 Effects of doubling of atmospheric CO_2 on the distribution of (a) white-tailed deer, (b) eastern chipmunk, (c) elk and (d) ground squirrel. The assessment is based on differences between current predicted wildlife species distributions and predicted distributions under climate change. This is accomplished by overlaying maps of current wildlife species distributions, predicted using current vegetation values substituted into the corresponding species' logistic regression model, and wildlife distributions under climate change, predicted by substituting new vegetation values generated by the MAPSS model into the same logistic regression model for each species. See included DVD for color version.

26.8.2 Case Study 2 - Modeling the Biophysical to Identify Process and Patterns: Landscape Level Fitness Assessment of Wildlife Habitat Quality

This case study explores the patterns produced as a result of a species selecting activities to balance its physiological heat loss and gain within a 24-hour day. The study deviates from the ecological conceptual framework since space is imposed, defined by a study area, and the size of individual home ranges is constant. For a full discussion of this case study see Johnston, Schmitz et al. (in review).

This study is an alternative habitat assessment procedure that incorporates animal behavioral ecology at the landscape level. This assessment procedure uses optimization theory to predict habitat quality based on the fitness consequences of selecting different habitat types on the landscape level. This approach recognizes that the mere existence of suitable habitat may not be sufficient to assess habitat quality. Rather, given restrictions due to animal home range size, the modeler must also consider the spatial arrangement of habitat types on the landscape to derive a predictive measure of suitability. Below, is an outline of the assessment procedure applied to the hypothetical example of habitat selection by moose (*Alces alces*) in a boreal forest ecosystem. The habitat assessment model uses a state-dependent habitat selection based on thermal characteristics and food resource availability, and assigns a fitness value to any combination of habitat types that are juxtaposed in a given location on a landscape.

26.8.2.1 Habitat Assessment Procedure

Conceptually, this model places an artificial moose on a computational landscape that is identical to a specified study area. Characteristics of the study area are coded in data layers of a GIS. The model then queries relevant attributes within a specified home range to determine if the biophysical conditions of habitat types satisfy all constraints such that the thermal intake for the moose balances to zero, the moose gut capacity constraints are met, and the moose obtains the necessary nutrition. This process continues for each location in the study site.

Behavioral decisions made by animals are subject to natural selection and animals behave in ways that tend to maximize fitness (Stephens and Krebs 1986; Krebs and Davies 1981; Mangel and Clark 1989; Mangel and Ludwig 1992). Fitness is defined as the ability of an animal to survive and reproduce within a specific location based on optimal feeding time budgets and habitat selection given the available habitat types. At a minimum, data on habitat types at different geographic locations within the landscape, data on the thermal properties within each habitat type (e.g., air temperature, ground-vegetation temperature, solar insolation, wind speed) and data on food abundance for the species being examined within the different habitat types were required. The habitat assessment procedure then converts a map of these habitat attributes into a map of fitness attributes. It is noteworthy that this habitat assessment predicts fitness for specific combinations of habitat types at each geographic location within the landscape.

26.8.2.2 Fitness Assessment for Different Habitat Associations

The fitness assessment procedure quantifies fitness by linking an empirically validated dynamic optimization model of habitat selection based on foraging ecology (Belovsky 1981; Schmitz 1991) with habitat types in a spatially explicit context. The optimization model is based on the fundamental biological property that animals must, during the course of any daily activity, balance their thermal budgets by regulating their body temperature within certain physiological tolerance limits. It is well known that animals regulate their body temperatures behaviorally by seeking habitats and exhibiting behaviors that offset heat losses or gains incurred during activities in habitats in previous time periods. The particular choices during any time period will depend on the physiological state of previous time periods.

The fitness assessment model assumes that an animal attempts to maximize its net nutrient gain during a given period of time and that the activity budget during that time is primarily geared toward achieving this foraging goal. This assumption is borne out in analyses of the foraging ecology of a variety of species (Belovsky and Schmitz 1994; others). Other demands, such as predator avoidance or competition may modify how animals will achieve both their foraging goals and the level of fitness they could achieve within a location. These factors can be added to an optimization model once the basic algorithm is in place.

The fitness assessment model assumes that each animal is limited by three constraints. First, the net gain or loss of body heat must be zero at the end of each time period over which the habitat-choice–feeding-time-budget is being optimized. Otherwise, the animal will become physiologically stressed or die. Next, an animal may not eat more in each time period than is permitted by its gut capacity. Finally, between time intervals, animals may not permit their body temperatures to exceed a specified upper and lower level to minimize the risk of thermal stress due to unexpected activity (e.g., escaping predation, sudden exposure to deleterious climate).

Fitness and survival for each combination of habitats is estimated using the following generalized algorithm. First the potential behaviors that an animal may exhibit (e.g., resting, feeding, walking) need to be identified. The choice of the number of behaviors requires some judgment about which are most important to achieving the optimization goal, rather than consider the entire behavioral repertoire. There should be enough behaviors to capture the desired process, but the number of behaviors should be limited to improve processing speed. Next, the heat gain or loss (HS_{ij}) for each potential behavior (i) in each potential habitat type (j) needs to be estimated. Values for heat gain or loss are calculated using an empirically calibrated model of heat flow (Porter and Gates 1969; Gates 1980; Belovsky 1981).

Three main activities were considered for each hour of the day; feeding, resting and walking. Three activity-combinations were also calculated within the model to refine the decision-making of the animal: 1) walk 25% rest 75%, 2) walk 50% rest 50%, and 3) walk 75% rest 25%. Additionally six habitat types typical of a boreal forest ecosystem were considered: hardwoods, conifer, a mixture of hardwoods and softwoods, aquatic, open, and open in regeneration. For each activity in each habitat a heat flux value (HS) is assigned. The heat flux value identifies how much heat will be gained or lost by the moose when performing a given activity in a habitat. To solve for the feeding-habitat choice time budget the model uses a dynamic optimization algorithm (Belovsky 1981).

The goal of the original foraging model was to determine the animal's activity for each hour of the day; therefore, the model assumed the animal can live at the location. However, it is important to determine not only what the animal is doing and where for each hour of the day, but if the animal can even live at any particular location or not. The assumption, if the animal cannot balance its HS values within the 24-hour period then it can not survive at the location, was expressed in this model.

However, the model still did not indicate if the animal received enough nutrition to survive at the location. For instance, the moose may be able to survive at the site only if it rests every hour, therefore no nutrition will be gained. The dry mass intake that would be gained if the animal fed in the habitat for the designated time period was estimated. Since the activity-budget model indicates where and what the animal will be doing each hour, it is easy to calculate the dry mass intake for the animal for a single day. Assuming nutritional intake positively relates to fitness in large ungulates such as moose and deer (Schmitz 1992), higher nutritional intake are assumed to result in higher fitness.

The spatial carrying capacity assessment divides the landscape into non-overlapping home ranges. Within each home range, the available habitats and their quantity are identified. A fitness value is assigned to each home range based on the fitness assessment determined above and the cropping rate of the animal in the habitat.

The spatial carrying capacity assessment begins on an identified base plan of the study area. The base plan identifies the spatial configuration of six habitats in a study area, hardwoods, softwoods, hardwood/softwood mix, aquatic, open, and open after regeneration. A proposed cut is considered open. A home range size is specified.

For each home range, the spatial algorithm identifies the combination of habitats available to the animal on day one. The animal will gain the fitness for that day as determined in the fitness assessment described above. The available habitats are reduced by the measured cropping rate of the animal based on the time spent in that particular habitat (the time spent in each habitat was determined in the fitness assessment above). This process continues for each day of the season. If the animal consumes the available quantity of a particular habitat, the fitness of the new available combination of habitats is used in subsequent days. A running sum of the fitness gained per day is assigned to each home range.

26.8.2.3 The Planning Management Environment

The study site is Crown (government-owned) land located in northern Saskatchewan, Canada and managed by the Mistik Timber company. The habitat assessment model was initially run on the existing landscape to determine the carrying capacity for moose within the area. Several possible forest cutting scenarios were created. On each scenario the habitat assessment model was run to quantify the impact the scenario would have on the carrying capacity for moose.

Scenario two proposes six clear cuts totaling 6,500 hectares (see Figure 26-5). The clear cuts were designed in conjunction with the total fitness map derived from the base landscape. The cuts are located in areas that cannot support moose or that have low fitness. The scenario is somewhat counter-intuitive since the total fitness actually increases slightly by 2,916 fitness units. When examining the difference map between the base fitness and the scenario 2 fitness, certain home ranges actually gain fitness (see Figure 26-6), while others slightly lose some. A home range can gain fitness because when certain combination of habitats are available, the animal may feed the entire feeding in one particular habitat, for instance, 3 hours in hardwoods. However, when fewer combinations are available, the model predicts that the animal will, for example, feed four hours in the deciduous. This counter-intuitive behavior can be attributed to either an error in the logic of the algorithm or that the moose does not optimize as well in certain combinations.

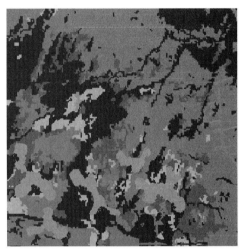

Figure 26-5 Management scenario two. The proposed cuts are in tan. See included DVD for color version.

Figure 26-6 The difference between the fitness derived from the initial base landscape and the fitness derived from the proposed cuts in scenario two. Net fitness gains and losses are shown in greens and reds respectively. The brighter the green, the more fitness is gained. Home ranges colored olive green had no change between scenarios. The brighter the red, the greater was the loss of fitness in scenario two compared to the base landscape. See included DVD for color version.

Manual of Geographic Information Systems

26.8.3 Case study 3 - Modeling the Physical and Mental to Identify Process and Patterns: Exploring the Causality of Cougar Movement Through Agent-Based Modeling

The third case study is the closest to capturing the essence of the ecological conceptual framework. Patterns were created from individuals cougars making decisions based on physiology requirements and behavioral strategies. The spatial boundaries of the individuals and their ecological processes are defined by the decision making of the individuals. For a full discussion of this case study see Johnston, Collier et al. (in review).

Agent-based modeling (ABM) is an alternative modeling approach to represent processes that are derived from the aggregation of individual decisions from each time step. The intent of the ABM is not to quantify or describe existing patterns but to explore the causality of how these processes and patterns are created. Agent-based modeling attempts to replicate patterns and processes that are derived from the aggregation of individual decisions from each time step.

This case study presents the creation of an agent-based model for the movement of cougars around Flagstaff, Arizona using the Agent Analyst software extension to ArcGIS from Environmental Systems Research Institute (ESRI). The Agent Analyst is a midlevel integration between Repast (a toolkit for ABM) and ArcGIS.

26.8.3.1 How an Agent-Based Model Works

In its simplest form, an ABM works as follows. First, the agents have to be identified. These agents can be animals, terrorists, land parcels, or anything that "makes a decision" or performs an action. "Making a decision" is being used in the most general meaning, the agent does "this" as opposed to "that." Whether the decision is conscious or how it is made is beyond the scope of this chapter.

The agents do things; they perform an action or no action. The animal agent can run, walk, or sleep. A terrorist can attack. The land parcel can change from agricultural to residential.

What each individual agent does is based on 1) its state, which can be physical or mental or any other measure influencing its decision making (e.g., how hungry is the animal), 2) its interactions with other agents (e.g., an animal may avoid a predator), and 3) its interactions with the external world. Three such external interactions can be from 3a) global factors—for example, many animals will seek shelter if a storm is imminent, 3b) environmental landscape factors—for example, the animal will slow down as it ascends a steep slope, and 3c) external societal factors—for example, whether a land parcel should change from agriculture to residential will be based on the economic value of residential relative to agriculture.

Time is explicit in the simulation of an ABM. Each decision is made in some specified time step. The length of each time step is controlled through a scheduler. The duration for the time step should be determined based on the decision making and characteristics of the agents being modeled. For example, if the ABM is exploring the movement of an animal, then the time step will be much smaller (e.g., every hour) as opposed to possibly using a much longer time interval when modeling the change in land use types based on an economic model.

26.8.3.2 Some General Cougar Biology

Cougars are opportunistic. When wandering, even if they are not hungry, if they happen to run into a prey, they will pursue it.

When a cougar is hungry and is hunting, whether a cougar captures a prey is dependent on the available prey, the probability of catching the prey, and how hungry the cougar is. Not only is it important to have prey available, the cougar must be able to catch it. Whether the cougar can catch the prey is based on its hunting advantage. The more rugged the landscape the more hunting advantage the cougar will have since it can pounce on its prey. The hunger level of the cougar will dictate how determined it will be at pursuing the prey.

The size of the prey will determine how long the cougar will take to consume it. The larger the prey, the more time the cougar will take to eat it; larger prey can take up to days to consume. When the cougar has made a kill it will eat some of the kill, possibly bury it, and wander off a short distance to digest, then return to consume more. When wandering from the prey, the activity level for the cougar is low.

A male must first sense a female before he can pursue mating. Quite possibly he can sense a female if she is within 3 kilometers of him. The closer she is, the greater possibility he will sense her. Once he detects her, he then will need to find her.

Cougars are territorial. If the male cougar is not hungry and does not sense a female, he will usually wander within his home range.

26.8.3.3 The Model Implementation

The data for the rules for the model agents came from literature, cougar experts at the USGS in Flagstaff, Arizona, and from collared data.

There are two main agents in the model, male cougars and female cougars. To simplify the cougar activities, the male cougars are driven by only a few behaviors. Male cougars either eat, wander, look for females, or avoid being killed. Female cougars have similar behaviors but instead of pursuing male cougars they try to attract them and they also raise young.

The overall model is based on energetics. What drives the decision making of a cougar is the amount of energy is in its system. Another way of looking at it is, the decision making for any time step is influenced by how hungry the cougar is. If the cougar is very hungry, it will intently hunt. A cougar can gain energy from hunting and can lose energy through activity.

The model is built on a random walk. The cougar in a time step will move (or not) into a neighboring cell location based on considerations of hunting, wandering and mating with the decision being driven by the state of the cougar in the time step, its enegertic state.

26.8.3.4 Attractors

A cougar has long term intent in its movement, so a traditional random walk is not adequate. To incorporate long term structure or intent to the decision making of the cougar, the concept of attractors influences the random walk. The attractors allow the cougars to have "memory" and "intelligence." Four main attractors (or repellants) drive the long-term decision making of the cougar: home range, good habitat, a kill, and a mate. The home-range and the habitat attractors influence movement for each times step, while the kill and mating attractors are temporary and influence decision making when present.

The home range attractor insures that the cougar stays within its home range. Cougars usually are familiar with the territory in their home range. They know where there is good habitat for hunting and security. The most ideal habitats within the home range become attractors. When moving through their territory, especially if just wandering, the cougars will move toward the good habitat within their home range. The habitat attractors keep the cougars moving with intent within their home range.

When a cougar makes a kill, the location of the kill becomes a temporary strong attractor. The kill site remains a strong attractor until the cougar consumes the prey. Once the male detects the female, depending on his energetic state and what other attractors are influencing his decisions, he may go looking for the female with the intent to mate. Once the male cougar finds her, she becomes a strong attractor for 12 hours.

26.8.3.5 The Model Iterations

For each time step, for each agent, the model conceptually goes through the following steps to determine what the agent should do during the time step.
1) Identifies the energetic state (i.e., How hungry are you?)
2) Weights each location around the agent for:
    a. Staying within the home range (home range attractor)

  b. Moving toward a habitat attractor

  c. Staying secure within cover

3) Checks for other attractors

  a. Is prey being consumed (kill attractor)

  b. For a male cougar, was a female detected or found in a previous time step (female attractor)

4) Evaluates the tradeoff between the above objectives

5) Makes a move

6) Asks if a kill was made during the movement? If so, what kind of prey was killed?

Weights are assigned to each of the 8 cellular neighbors based on each attractor identified in steps 2 and 3. The higher the weight the greater the probability the cougar agent will have to move into that cell based on the criteria. The weights for each attractor and security layer are added together and the agent chooses one of the cells into which to move. The neighboring cells with the highest weights have a greater chance of being selected. The weights will vary based on the energetic state of the agent. That is, if the cougar is in a low energetic state (is hungry) more weight will be placed on the habitat attractor.

26.8.3.6 Making a Kill

Since the cougar is an opportunistic hunter, for each time step the cougar has a probability of killing a prey. The probability is based on the hunting advantage (determined by the roughness of the terrain), the availability of the prey, and the energetic level of the cougar at the time step. The hungrier the cougar (the lower the energetic state) the more intently it will hunt (increasing the probability of a kill) resulting in higher success.

Once the cougar agent makes a kill, the individual prey species distribution layers are queried to see what potential prey could have been killed. The more the prey is captured in the wild, the higher that prey has for being selected (if available) by the cougar agent.

The type of prey killed by the cougar agent will determine how long the cougar should stay with the prey to consume it. To keep the cougar agent with the kill, the kill attractor will be given a high weight until the prey is consumed.

Several simulations were run to see different possible scenarios. Below is one simulation (Figure 26-7).

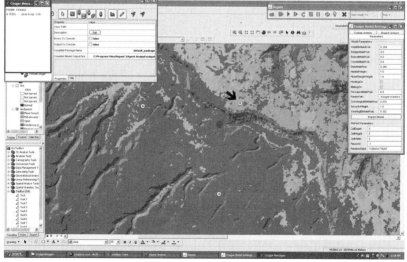

Figure 26-7 Early in a simulation the male has detected a female within 3 kilometers of its location (to the northeast). The female cougars are identified as yellow circles and the male in red. He started to move toward her since his energetic level was within an acceptable level. On his way toward the female the male cougar encountered a prey and killed it. The prey was a young elk. The male cougar will now stay with the prey until it is consumed. See included DVD for color version.

26.8.4 Case Study 4 - Ecological Process and Uncertainty: Process, Simulation, Error and Sensitivity Modeling Integrated in a Modeling Environment

The fourth case study explores methods to address uncertainty in models as they relate to the ecological conceptual framework. For a full discussion of this study see Johnston et al. 2005.

With every model there is some degree of uncertainty. Uncertainty can stem from error in—or not having complete understanding of—the phenomena being modeled, error in the input data, and uncertainty in the model parameters. These uncertainties can be explored through:

- Modeling stochastic events through random variables,
- Performing error analysis by adding random values to the input data and model parameters,
- Performing sensitivity analysis by exploring the influence of parameters within a model.

26.8.4.1 Modeling Stochastic Events

Modeling stochastic events using random variables allows you to model events that appear to be random; random variables let you account for the uncertainty in your knowledge of the events. Using a fire growth model as an example, fire spotting appears to be random. Fire spotting occurs when a piece of the fire breaks off and jumps some distance from the main body. When the spot occurs and where the spot will land appear to be random.

In Figure 26-8 a spot that randomly occurred will be modeled as a conditional branch to the fire model and whether the branch is to be implemented or not in a time step will be based on the evaluation of a randomly calculated value from a Gamma distribution (since a spot should not occur every time step).

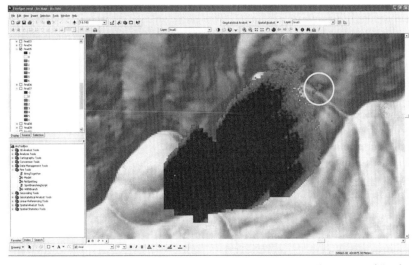

Figure 26-8 During time step 35, a spot occurred in the northeast leading edge of the fire (circled in yellow). Notice how the spot jumped the stream. See included DVD for color version.

26.8.4.2 Performing Error Analysis

Error creates uncertainty for the modeler with respect to the model input, parameters, and its interactions (Burrough and McDonnell 1998; Eastman 2001; Zhang and Goodchild 2002; Longley et al. 2005). There are two main ways to explore how error influences the output in a model: 1) through simulations, for example, Monte Carlo simulations (Malczewski 1999; Goodchild et al. 1999; Aerts et al. 2003) and 2) through a structured mathematical approach (Heuvelink 1998; Zhang and Goodchild 2002).

The basic concept of using simulations for error analysis is based on running the model with a possible combination of input data and parameters to produce a plausible scenario of your model with your data. It is unlikely that any of the simulations will produce the actual combination of input values and parameters to represent reality, especially if your model is complex. You then run the model again with another possible combination of input and parameters. And you continue this process for 50, 100, 1000 or more times depending on the data, the phenomenon being modeled, and the error in the model to derive many scenarios of possible outputs. By combining all the scenarios, you derive probabilities of the occurrence of the phenomenon at each location.

For example, let us explore how error in the input elevation or digital elevation model (DEM) surface might affect delineating a stream network. A series of tools is used to delineate streams from a DEM. Conceptually, the tools drop water onto the DEM surface and track where the drops flow and how they accumulate as they flow. If a location has enough water passing through it, that location will become part of the stream network. Determining how water flows from one cell into a neighboring cell is dependent on the heights of the cells around it. The water will travel to the cell with the steepest descent. But if the input DEM has an accuracy of plus or minus 2 feet, this can have an effect on which cell provides the steepest descent. In one scenario of possible elevation values, the water may travel one way, and in another scenario of possible elevation values the drop may travel a slightly different way. By combining the possible scenarios the water can travel, you will create a probability surface of finding a stream in each location.

The model in Figure 26-9 was run 100 times varying the input elevation data plus or minus 2 feet using a normal distribution with a mean of zero and a standard deviation of 1.5. The 100 possible realizations of the elevation surface were combined to produce a probability surface identifying for each cell how many runs out of 100 realized a stream passing through the cell. It may be deduced that the more times a stream passed through a cell, the higher the probability that the incidence would occur.

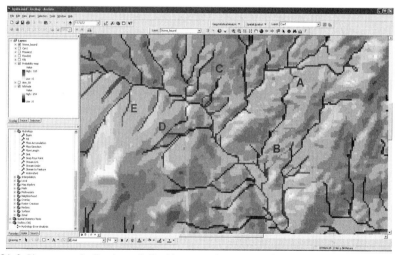

Figure 26-9 Blue areas indicate cells that have a stream passing through them most of the time. Stream occurrences are ramped from yellow to blue, with the lowest occurrence being yellow. A few points of interest:
- At "A" (to the left of the letter), "B" (to the lower right of the letter), and "C" (to the left of the letter) some cell locations followed down different watersheds.
- At "D" (to the lower right of the letter) uncertainty occurs because during multiple scenarios water traveled down different paths.
- At "E" there is a fair amount of noise surrounding the dominant path of the stream.
Notice that most of the locations that have greatest uncertainty occur in the areas with less relief. The error analysis identifies those locations having the greatest uncertainty in the model and may be considered for more detailed data collection or ground truthing. See included DVD for color version.

26.8.4.3 Performing Sensitivity Analysis

Sensitivity analysis explores the influence of a parameter on the model output. During sensitivity analysis, you systematically make small changes in a parameter and explore the influence the parameter change has on the output. If there is little effect on the output, the output is not sensitive to that parameter change. This process continues by increasing the change on the parameter or applying similar changes in other parameters of interest. If a particular change or series of changes causes great change in the output, the model is sensitive to the parameter; that parameter has great influence in the output. If there is little change in the output with the systematic changes in the parameters, the model is considered robust. If the model is sensitive to a particular parameter and there is great uncertainty of the actual value of the parameter, caution may be exercised before making a decision based on the output from the model.

One goal of a housing suitability model might be to identify the most desired locations to build a house based on specified criteria. Figure 26-10 is an example of a housing suitability model, where sensitivity analysis was used to explore how the output housing preference locations changed with slight changes in the input parameters.

Sensitivity analysis was performed by changing each of the input criteria in 5% increments. Only the top three most preferred classes were altered for each criterion, since we were only interested in identifying the best locations for a house.

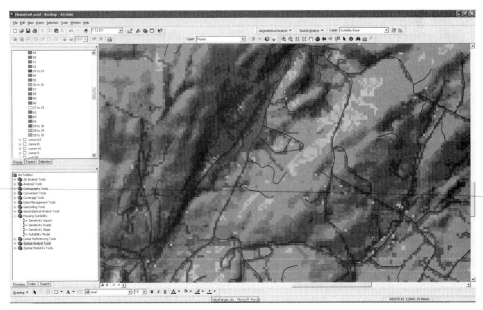

Figure 26-10 The housing suitability model after changing the aspect by 5%. Dark blue indicates areas that were most preferred on the base suitability and remain the most preferred after the parameter change. Light blue indicates values that were second-most preferred and remain second-most preferred after the parameter change. The few red locations indicate where the second-most preferred on the base became the most preferred after the parameter change. The orange cells are values that were third-most preferred on the base that became the most preferred after the parameter changes. The two yellow classes moved from values of third- and fourth-most preferred to second-most preferred after the parameter change. In particular, those areas that became red or orange should be further considered for the housing site. Note that in the upper right of the image two new isolated yellow locations appear, opening an entirely new area for investigation because they are not just expansions of existing preferred locations. See included DVD for color version.

26.9 Ecological Spatial Modeling Software

There are additional ecological software packages that allow the modeler to incorporate space explicitly into an ecological model that may be linked or not to a GIS. A description of the capabilities of several of the more commonly used packages found in the literature is provided in Table 26-5.

26.9.1 Metapopulation and Forest Succession Models

RAMAS metapopulation model and the LANDIS landscape dynamic forest succession model were used to explore the effects of different forest management plans on sharp-tailed grouse (Akçakaya et al. 2004), to identify how altering the fire regime would affect the Bell's sage sparrow (Akçakaya et al. 2005), and to understand how tree harvesting would influence ovenbirds (Larson et al. 2004). In each of these three studies the researchers found that by linking landscape simulation models with metapopulation models they were able to more accurately predict the dynamics in their ecological process.

26.9.2 Population Models

PATCH, a population viability analysis model, was used by Carroll, Noss et al. (2003) to examine the viability of populations of reintroduced wolves under various human growth scenarios. Rustigian et al. (2003) used PATCH to examine the viability of amphibians in agricultural wetlands under different future land use scenarios. Schumaker et al. (2004) used PATCH to examine the wildlife responses to different landscape changes in Oregon's Willamette River Basin. In all cases, PATCH accentuated the GIS by quantifying the population viability of species from different spatial scenarios.

26.9.3 Forest Growth Models

Forest growth models can be integrated with a GIS to predict future scenarios that the GIS alone could not produce. Ranson et al. (2001) used Zelig, a forest growth model, to predict the growth and development of forests in Maine. Ehman et al. (2002) applied a forest growth model to the Great Lakes region of the USA, and determined that there will be a decrease in total basal area of northern conifers and northern deciduous with a doubling of CO_2. Gustafson et al. (2001) used HARVEST, a timber-harvest simulator, to predict the distribution of salamanders based on the forest structure of various forest management scenarios over a 150-year period.

26.9.4 Other Software Models Related to Ecological Processes

Iverson et al. (2004) used DISTRIB to assess potential suitable habitats under a CO_2 increase. They then used SHIFT to determine if the species will migrate under the habitat change scenarios from DISTRIB.

Other software complementary to GIS include Agent-based modeling software such as Agent Analyst, Repast, and Swarm; system thinking and population dynamics software such as Stella; habitat analysis software such as NestCalc, and the many general statistical software packages.

Table 26-5 Ecological spatial modeling software complementary to a GIS. This list favors wildlife modeling and does not address other related areas such as hydrology, land planning, land use change, and fire movement and protection.

Software	Description	Studies using software
Metapopulation and Forest Succession		
RAMAS	Links a GIS to metapopulation dynamics model for population viability analysis and extinction assessment (Applied Biomathematics 2005).	Akçakaya et al. 2004 Akçakaya et al. 2005 Larson et al. 2004
LANDIS	The Landscape Disturbance and Succession Model models forest succession, disturbance and seed dispersal over large landscapes (NCRS 2003a).	Gustafson et al. 2004 Sturtevant et al. 2004
Population		
PATCH	Program to Assist in Tracking Critical Habitat performs a population viability analysis. Tracks survival and reproduction of territorial females (Schumaker 1998).	Carroll, Noss et al. 2003 Rustigian et al. 2003 Schumaker et al. 2004
Forest Growth		
Zelig	An individual tree simulator for tree growth and mortality (Urban 1990).	Ranson et al. 2001
HARVEST	Evaluates consequences of timber management (NCRS 2003b).	Gustafson et al. 2001
DISTRIB	Assesses potential suitable habitat under equilibrium of 2 times CO_2 (Iverson et al. 2004).	Iverson et al. 2004
SHIFT	A cellular automata to estimate migration of species (Iverson et al. 2004).	Iverson et al. 2004
JABOWA II	Simulates forest growth (Botkin 1993).	Ehman et al. 2002
Landscape Ecology, Reserve Design, and Conservation Planning		
LARCH	Explores connectivity of ecological networks.	Bruinderink et al. 2003
SITES	A reserve selection algorithm.	Carroll, Noss et al. 2003
CPLAN	Used for conservation planning to identify conservation sites (National Parks and Wildlife Service 1999).	Warman et al. 2004
VISTA	Conservation planning software (Barker 2004).	Barker 2004
Landscape Metrics		
Fragstats	Computes a wide range of landscape metrics on categorical map patterns (McGarigal et al. 2002).	Dickson and Beier 2002 Kie et al. 2002 Kurttila and Pukkala 2003 Thogmartin et al. 2004 Tinker et. al 2003 Tischendorf 2001
APACK	Calculates 25 landscape metrics on raster data (FLEL 2006).	Verburg and Veldkamp 2004
Leap II	Explores fragmentation, edge content, spatial geometry, and connectivity (Forest Landscape Ecology Program 2005).	Weaver and Petera 2004
Home Range and Animal Movement		
Animal Movement	Functions to analyze animal movement; includes home range analysis and movement tracking; is an extension to ArcGIS (Hooge and Eichenlaub 2000).	Dickson and Beier 2002 Harveson et al. 2004

(continued on next page)

Manual of Geographic Information Systems

Ranges 6	Analysis of locations. Capabilities include home range delineation and analysis of dispersal (Anatrack Ltd. 2006).	Girard et al. 2002 Grinder and Krausman 2001
CALHOME	Calculates home range by various methods (Kie 2004).	Kie et al. 2002
KernelHR	Calculates home range by kernel estimates (Seaman 2004).	Mitchell and Powel 2003
Genetic Algorithm		
GARP	Genetic Algorithm for Rule-set Prediction has inferential tools that work in an iterative, artificial-based approach (Stockwell and Peters 1999, KUCR 2002).	Arriaga et al. 2004 Illoldi-Rangel et al. 2004 Peterson and Robins 2003

26.10 Conclusion

The proposed ecological conceptual framework explores the motivating factors resulting in patterns that are perceived in the landscape. This ecological conceptual framework can provide insight into ecological processes and can be useful in guiding the decision-maker using the model in making better-informed decisions. When applying this ecological conceptual framework, instead of defining a spatial extent and then describing the processes within it, space is fully integrated into the physiological and behavioral decisions. As a result, spatial extents become an emergent property of the decision making.

However, the ecological conceptual framework also uncovers at least three questions and possible deficiencies in our current ecological modeling approaches. First, this ecological conceptual framework argues that change appears to be inherent in ecological processes. Thus instead of measuring static, quantifiable spatial and temporal patterns in ecological processes, perhaps the goal in modeling ecological processes should be to describe patterns and processes as trajectories or motions of change. Examples of such attempts include Galton (2000), who quantifies the change in the geometry of shapes changing from one form to another. Kim and Cova (2005, 2007) used a computer graphics technique called tweening to interpolate the shape transformation between discrete time steps or snapshots of a moving fire. Laube et al. (2005) developed the Relative Motion (REMO) software to compare the motion attributes of point objects in space and time and relate the motion of an object to the motion of all other objects. Each of these studies attempts to describe the changing processes and resulting patterns as a motion.

Possibly a new language needs to evolve to describe the motion of change in ecological processes. Yattaw (1999) presents a language for describing changing shapes in time and space. She characterizes time by the progression of the event which includes direction, duration and frequency, and recognizes that most phenomena cannot change locations without moving across contiguous space. She classifies the movement of areal, linear and point data according to the movement type: continuous, cyclical and intermittent movement. Most descriptions of patterns and trajectories depend on the process of moving toward a goal, usually a stable state. Possibly there is no goal. Perhaps the goal for the individuals in an ecological process is to react to change; change may be the pattern.

The second question this ecological conceptual framework addresses is motivation. If altruistic actions do in fact exist in complex living individuals, then the ecological conceptual framework proposed here must be altered. However, possibly when the motivations of altruistic actions are fully understood, the actions might prove to be self-serving, thus making them consistent within this framework.

Third, many attempts have been made to define laws and general principles from the descriptions of patterns. But it has been proposed here, that the driving forces behind the observed patterns are individuals making decisions based on their physical or mental requirements, decision strategies, and the actions and reactions. Possibly laws and general principles should be established from these base mechanisms. If general building blocks of laws and general principles can be created for physical and mental requirements, decisions and decision strategies, and behaviors which result in actions and reactions, then these can be used to develop models that could possibly be applied to any location with any given physical conditions and comprised of any set of individuals.

Other disciplines such as behavioral ecology appear to take a bottom-up approach by observing the behavior of the individuals and then producing generalities from these behaviors. Even though the individual is being observed, these studies generally quantify behavioral patterns rather than spatial patterns. The mechanisms creating the behaviors and resulting patterns are implied and are not generally modeled.

Implementing models based on the ecological conceptual framework requires a change in many of the existing ecological approaches. First, the modeler must see, explore, understand, and experience the world and the underlying decision making from the perspective of the species or the phenomenon being modeled. Second, the modeler must conceptualize and generalize the mechanisms driving the decision making (e.g., the physical and mental requirements and the decision-making strategies). Third, what type and the manner in which data are collected needs to change. Data collection should depict and quantify decisions based on physical and mental requirements as well as on behavioral and decision strategies. Fourth, new tools need to be created for implementing and evaluating the output from simulation models. Researchers need to continue to explore and improve the tools to quantify patterns so that we have tools to use for validating and quantifying the results of simulated data relative to data collected in the field. With a deeper understanding of the mechanisms behind ecological processes, hopefully we will make more-informed decisions concerning our limited resources and vanishing wildlife.

Acknowledgements

I would like to thank Michelle Gudorf for her contributions to the chapter through discussions, ideas and comments on the text. And to Oswald Schmitz for his contributions to the case studies and for his input into this chapter.

References

Aerts, J.C.J.H., M. F. Goodchild and G.B.M. Heuvelink. 2003. Accounting for spatial uncertainty in optimization with spatial decision support systems. *Transactions in GIS* 7(2):211–230.

Akçakaya, H. R., V. C. Radeloff, D. J. Mladenoff and H. S. He. 2004. Integrating landscape and metapopulation modeling approaches: viability of the sharp-tailed grouse in a dynamic landscape. *Conservation Biology* 18(2):526–537.

Akçakaya, H. R., J. Franklin, A.D. Syphard and J. R. Stephenson. 2005. Viability of Bell's Sage Sparrow (*Amphispiza belli* ssp. *belli*): Altered fire regimes. *Ecological Applications* 15(2):521–531. <http://www.ramas.com/SageSparrow.pdf> Accessed 25 June 2007.

Alados, C. L., Y. Pueyo, O. Barrantes, J. Escos, L. Giner and A. B. Robles. 2004. Variations in landscape patterns and vegetation cover between 1957 and 1994 in a semiarid Mediterranean ecosystem. *Landscape Ecology* 19:543–559.

Anatrack Ltd. 2006. Ranges – Analysis and Tracking. <http://www.anatrack.com/> Accessed 3 July 2007.

Anderson Jr., C. R. and F. G. Lindzey. 2003. Estimating cougar predation rates from GPS location clusters. *Journal of Wildlife Management* 67(2):307–316.

Applied Biomathematics. 2005. RAMAS Ecological Software. <http://www.ramas.com> Accessed 2 July 2007.

Apps, C. D., B. N. McLellan, J. G. Woods and M. F. Proctor. 2004. Estimating grizzly bear distribution and abundance relative to habitat and human influence. *Journal of Wildlife Management* 68(1):138–152.

Arnot, C. P., F. Fisher, R. Wadsworth and J. Wellens. 2004. Landscape metrics with ecotones: Patterns under uncertainty. *Landscape Ecology* 19:181–195.

Arriaga, L., A. E. Castellanos V, E. Moreno and J. Alarcón. 2004. Potential ecological distribution of alien invasive species and risk assessment: a case study of buffel grass in arid regions of Mexico. *Conservation Biology* 18(6):1504–1514.

Arthur, J. L., J. D. Camm, R. G. Haight, C. A. Montgomery and S. Polasky. 2004. Weighting conservation objectives: maximum expected coverage versus endangered species protection. *Ecological Applications* 14(5):1936–1945.

Axelsson, A., L. Ostlund and E. Hellberg. 2002. Changes in mixed deciduous forests of boreal Sweden 1866-1999 based on interpretation of historical records. *Landscape Ecology* 17:403–418.

Baldwin, D.J.B., K. Weaver, F. Scknekenburger and A. H. Perera. 2004. Sensitivity of landscape pattern indices to input data characteristics on real landscapes: implications for their use in natural disturbance emulation. *Landscape Ecology* 19:255–271.

Barker, K. 2004. NatureServe Vista Software for Biodiversity Planning. *Proceedings for the 2004 ESRI User Conference*, San Diego, Calif. <http://gis2.esri.com/library/userconf/proc04/docs/pap2185.pdf> Accessed 2 July 2007.

Basset, A., M. Fedele and D. L. DeAngelis. 2002. Optimal exploitation of spatially distributed trophic resources and population stability. *Ecological Modeling* 151:245–260.

Belisle, M. 2005. Measuring landscape connectivity: the challenge of behavioral landscape ecology. *Ecology* 86(8):1988–1995.

Belovsky, G. E. 1981. Optimal activity times and habitat choice of moose. *Oecologia* 48:22–30.

Belovsky, G. E. and O. J. Schmitz. 1994. Plant defenses and optimal foraging by mammalian herbivores. *Journal of Mammalogy* 75(4):816–832.

Bender, D. J., L. Tishendorf and L. Fahrig. 2003. Using patch isolation metrics to predict animal movement in binary landscapes. *Landscape Ecology* 18:17–39.

Bender, D. J. and L. Fahrig. 2005. Matrix structure obscures the relationship between interpatch movement and patch size and isolation. *Ecology* 86(4):1023–1033.

Berger, P. A. and J. P. Bolte. 2004. Evaluating the impact of policy options on agricultural landscapes: an alternative-futures approach. *Ecological Applications* 14(2):342–354.

Berger, T., M. Goodchild, M. A. Janssen, S. M. Manson, R. Najlis and D. C. Parker. 2002. Methodological considerations for agent-based modeling of land-use and land-cover change. In *Agent-Based Models of Land-Use and Land-Cover Change*, 7–25. LUCC Report Series No. 6, LUCC International Project office.

Boose, E. R., M. I. Serrano and D. F. Foster. 2004. Landscape and regional impacts of hurricanes in Puerto Rico. *Ecological Monographs* 74(2):335–352.

Botkin, D. B. 1993. *Forest Dynamics: An Ecological Model.* New York: Oxford University Press.

Bowman, J. L., B. D. Leopold, F. J. Vilella and D. A. Gill. 2004. A spatially explicit model, derived from demographic variables, to predict attitudes toward black bear restoration. *Journal of Wildlife Management* 68(2):223–232.

Bowyer, T. R., K. M. Stewart, S. A. Wolfe, G. M. Blundell, K. L. Lehmkuhl, P. J. Joy, T. J. McDonough and J. G. Kie. 2002. Assessing sexual segregation in deer. *Journal of Wildlife Management* 66(2):536–544.

Bruinderink, G. G., T. Van Der Sluis, D. Lammertsma, P. Opdam and R. Pouwels. 2003. Designing a coherent ecological network for large mammals in northwestern Europe. *Conservation Biology* 17(2):549–557.

Burnside, N. G., R. F. Smith and S. Waite. 2002. Habitat suitability modelling for calcareous grassland restoration on the South Downs, United Kingdom. *Journal of Environmental Management* 65:209–221.

Burrough, P. A., R. van Rijn and M. Rikken. 1996. Spatial data quality and error analysis issues: GIS functions and environmental modeling. In *GIS and Environmental Modeling: Progress and Research Issues*, edited by M. F. Goodchild, L.T. Steyaert, B.O. Parks, C. Johnston, D. Maidment, M. Crane and S. Glendinning, 29–34. Fort Collins, Colo.: GIS World Books.

Burrough, P. A. and R. A. McDonnell. 1998. *Principles of Geographical Information Systems*, 2nd ed. Oxford, UK: Oxford University Press.

Carroll, C., R. F. Noss, P. C. Paquet and N. H. Schumaker. 2003. Use of population viability analysis and reserve selection algorithms in regional conservation plans. *Ecological Applications* 13(6):1773–1789.

Carroll, C., M. K. Phillips, N. H. Schumaker and D. W. Smith. 2003. Impacts of landscape change on wolf restoration success: planning a reintroduction program based on static and dynamic spatial models. *Conservation Biology* 17(2):536–548.

Cedfeldt, P. T., M. C. Watzin and B. D. Richardson. 2000. Using GIS to identify functionally significant wetlands in the northeastern United States. *Environmental Management* 26(1):13–24.

Chakraborti, R. K., J. Kaur and J. V. DePinto. 2002. Analysis of factors affecting zebra mussel (*Dreissena polymorpha*) growth in Saginaw Bay: a GIS-based modeling approach. *J. Great Lakes Res.* 28(3):396–410. <http://sgnis.org/publicat/papers/chakkaur.pdf> Accessed 22 June 2007.

Chardon, I. P., F. Adriaensen and E. Matthysen. 2003. Incorporating landscape elements into a connectivity measure: a case study for the speckled wood butterfly (*Pararge aegeria* L.). *Landscape Ecology* 18:561–573.

Charnov, E. L. 1976. Optimal foraging, the marginal value theorem. *Theoretical Population Biology* 9:129–136.

Clark, M. E., K. A. Rose, D. A. Levine and W. W. Hargrove. 2001. Predicting climate change effects on Appalachian trout: combining GIS and individual-based modeling. *Ecological Applications* 11(1):162–178.

Conner, M. M. and M. W. Miller. 2004. Movement patterns and spatial epidemiology of a prion disease in mule deer population units. *Ecological Applications* 14(6):1870–1881.

Cook, R. R. and P. J. Auster. 2005. Use of simulated annealing for identifying essential fish habitat in a multispecies context. *Conservation Biology* 19(3):876–886.

Cox, J. A., W. W. Baker and R. T. Engstrom. 2001. Red-cockaded woodpeckers in the Red Hills region: a GIS-based assessment. *Wildlife Society Bulletin* 29(4)1278–1288.

Crist, M. R., B. Wilmer and G. H. Aplet. 2005. Assessing the value of roadless areas in a conservation reserve strategy: biodiversity and landscape connectivity in the Northern Rockies. *Journal of Applied Ecology* 42:181–191.

DeAngelis, D. L. and L. J. Gross. 1992. *Individual-Based Models and Approaches in Ecology*. New York: Chapman and Hall.

Dettmers, R. and H. Bart. 1999. A GIS modeling method applied to predicting forest songbird habitat. *Ecological Applications* 9(1):152–163.

Dickson, B. G. and P. Beier. 2002. Home-range and habitat selection by adult cougars in Southern California, *Journal of Wildlife Management*, 66(4):1235–1245.

DirnBöck, T., J. Greimler, P. Lopex and T. F. Stuessy. 2003. Predicting future threats to the native vegetation of Robinson Crusoe Island, Juan Fernandez Archipelago, Chile. *Conservation Biology* 17(6):1650–1659.

Drechsler, M., K. Frank, I. Hanski, R. B. O'Hara and C. Wissel. 2003. Ranking metapopulation extinction risk: from patterns in data to conservation management decisions. *Ecological Applications* 13(4):990–998.

Duh, J. and D. G. Brown. 2005. Generating prescribed patterns in landscape models. In *GIS, Spatial Analysis, and Modeling*, edited by D.J. Maguire, M. Batty and M.F. Goodchild, 423–444. Redlands, Calif.: ESRI Press.

Duncan, B. W. and P. A. Schmalzer. 2004. Anthropogenic influences on potential fire spread in a pyrogenic ecosystem of Florida, USA. *Landscape Ecology* 19:153–165.

Eastman, R. 2001. Uncertainty management in GIS: decision support tools for effective use of spatial data. In *Spatial Uncertainty in Ecology: Implications for Remote Sensing and GIS Applications*, edited by C. T. Hunsaker, M. F. Goodchild, M .A. Friedl and T. J. Case, 379–390. New York: Springer.

Ehman, J. L., W. Fan, J. C. Randolph, J. Southworth and N. T. Welch. 2002. An integrated GIS and modeling approach for assessing the transient response of forests of the Southern Great Lakes Region to a double CO_2 dlimate. *Forest Ecology and Management* 155:237–255.

Flamm, R. O., L. I. Ward and B. L. Weigle. 2001. Applying a variable-shape spatial filter to map relative abundance of manatees (*Trichechus manatus latirostris*). *Landscape Ecology* 16:279–288.

FLEL (Forest Landscape Ecology Lab). 2006. APACK: an analysis package for rapid calculation of landscape metrics on large scale data sets. Madison: University of Wisconsin-Madison Department of Forestry. <http://landscape.forest.wisc.edu/projects/apack/> Accessed 2 July 2007.

Forest Landscape Ecology Program. 2005. Leap II. Sault Ste. Marie, Ontario, Canada: Forest Landscape Ecology Program, Ontario Forest Research Institute.<http://www.ai-geostats.org/index.php?id=102> Accessed 3 July 2007.

Forman, R.T.T. and M. Godron. 1986. *Landscape Ecology*. New York: John Wiley & Sons.

Formica, V. A., R. A. Gonset, S. Ramsay and E. M. Tuttle. 2004. Spatial dynamics of alternative reproductive strategies: the role of neighbors. *Ecology* 85(4):1125–1136.

Fritz, H., S. Said, P. Renaud, S. Mutake, C. Coid and F. Monicat. 2003. The effects of agricultural fields and human settlements on the use of rivers by wildlife in the mid-Zambezi Valley, Zimbabwe. *Landscape Ecology* 18:293–302.

Fryxell, J. M., J. F. Wilmshurst and A.R.E. Sinclair. 2004. Predictive models of movement by Serengeti grazers. *Ecology* 85(9):2429-2435.

Galton, A. 2000. *Qualitative Spatial Change*. Oxford: Oxford University Press.

Gates, D. M. 1980. *Biophysical Ecology*. New York: Springer-Verlag.

Girard, I., J. Ouellet, R. Courtois, C. Dussault and L. Breton. 2002. Effects of sampling efforts based on GPS telemetry on home-range size estimations. *Journal of Wildlife Management* 66(4):1290–1300.

Gjerde, I., M. Saetersdal, J. Rolstad, H. H. Blom and K. O. Storaunet. 2004. Fine-scale diversity and rarity hotspots in northern forests. *Conservation Biology* 18(4):1032–1042.

Goodchild, M. F., A. Shortridge and P. Fohl. 1999. Encapsulating simulation models with geospatial data sets. In *Spatial Accuracy Assessment: Land Information Uncertainty in Natural Resources*, edited by K. Lowell and A. Jaton, 123–130. Ann Arbor, Mich.: Ann Arbor Press.

Gough, M. C. and S. P. Rushton. 2000. The application of GIS-modeling to mustelid landscape ecology. *Mammal Rev.* 30(3-4):197–216.

Grand, J., J. Buonaccorsi, S. A. Cushman, C. R. Griffin and M. C. Neel. 2004. A multiscale landscape approach to predict bird and moth rarity hotspots in a threatened pitch pine-scrub oak community. *Conservation Biology* 18(4):1063–1077.

Grinder, M. I. and P. R. Krausman. 2001. Home range, habitat use, and nocturnal activity of coyotes in an urban environment. *Journal of Wildlife Management* 65(4):887–898.

Grinnell, J. 1917. The niche-relations of the California Thrasher. Auk 34:427-433.

Guneralp, B., G. Mendoza, G. Gertner and A. Anderson. 2003. Spatial simulation and fuzzy threshold analyses for allocation restoration areas. *Transaction in GIS* 7(3):325–343.

Gustafson, E. J., N. L. Murphy and T. R. Crow. 2001. Using a GIS model to assess terrestrial salamander response to alternative forest management plans. *Journal of Environmental Management* 63:281–292.

Gustafson, E. J., P. A. Zollner, B. R. Sturtevant, H. S. He and D. J. Mladenoff. 2004. Influence of forest management alternatives and land type on susceptibility to fire in northern Wisconsin, USA. *Landscape Ecology* 19:327–341.

Gutierrez, D., P. Fernandez, A. S. Seymour and D. Jordano. 2005. Habitat distribution models: are mutualist distributions good predictors of their associates? *Ecological Applications* 15(1):3–18.

Haight, R. G., S. A. Snyder and C. S. Revelle. 2005. Metropolitan open-space protection with uncertain site availability. *Conservation Biology* 19(2):327–337.

Haltuch, M. A. and P.A. Berkman. 2000. Geographic information system (GIS) analysis of ecosystem invasion: exotic mussels of Lake Erie. *American Society of Limnology and Oceanography, Inc.* 45(8):1778–1787.

Hanski, I. and M. Gilpin. 1991. Metapopulation dynamics: brief history and conceptual domain. *Biol. J. Linn. Soc.* 42:3–16.

Hanski, I. 1991. Single species metapopulation dynamics: concepts models and observations. *Biol. J. Linn. Soc.* 42:17–38.

Harveson, P. M., M. E. Tewes, G. L. Anderson and L. L. Laack. 2004. Habitat use by ocelots in South Texas: implications for restoration. *Wildlife Society Bulletin* 32(3):948–954.

Hatten, J. R. and C. E. Paradzick. 2003. A multiscaled model of Southwestern Willow Flycatcher breeding habitat. *Journal of Wildlife Management* 67(4):774–787.

Hawbaker, T. J. and V. C. Radeloff. 2004. Roads and landscape pattern in Northern Wisconsin based on a comparison of four road data sources. *Conservation Biology* 18(5):1233–1244.

Heuvelink, G.B.M. 1998. *Error Propagation in Environmental Modelling*. London: Taylor and Francis Ltd.

Hiers, J. K., S. C. Laine, J. J. Bachant, J. H. Furnam, W. W. Green Jr. and V. Compton. 2003. Simple spatial modeling tool for prioritizing prescribed burning activities at the landscape scale. *Conservation Biology* 17(6):1571–1578.

Hirzel, A. H., J. Halusser, D. Chessel and N. Perrin. 2002. Ecological-niche factor analysis: how to compute habitat-suitability maps without absence data? *Ecology* 83(7):2027–2036.

Hooge, P. N. and B. Eichenlaub. 2000. Animal movement extension to ArcView, version 2.0. Alaska Science Center - Biological Science Office. Anchorage, Alaska: US Geological Survey.

Hooten, M. B., D. R. Larsen and C. K. Wikle. 2003. Predicting the spatial distribution of ground flora on large domains using a hierarchical Bayesian model. *Landscape Ecology*. 18:487–502.

Hulse, D. W., A. Branscomb and S. G. Payne. 2004. Envisioning alternatives: using citizen guidance to map future land and water use. *Ecological Applications* 14(2):325–341.

Illoldi-Rangel, P., V. Sanchez-Cordero and A. T. Peterson. 2004. Predicting distributions of Mexican mammals using ecological niche modeling. *Journal of Mammalogy* 85(4):658–662.

Iverson, L. R., M. W. Schwartz and A. M. Prasad. 2004. Potential colonization of newly available tree-species habitat under climate change: an analysis for five eastern US species. *Landscape Ecology* 19:787–799.

Ji, W. and C. Jeske. 2000. Spatial modeling of the geographic distribution of wildlife populations: a case study in the Lower Mississippi River region. *Ecological Modeling* 132:95–104.

Johnson, C. J., M. S. Boyce, C. C. Schwartz and M. A. Haroldson. 2004. Modeling survival: application of the Andersen-Gill model to Yellowstone grizzly bears. *Journal of Wildlife Management* 68(4):966–978.

Johnson, C. J. and M. P. Gillingham. 2004. Mapping uncertainty: sensitivity of wildlife habitat ratings to expert opinion. *Journal of Applied Ecology* 41:1032–1041.

Johnston, K. M. 1992. Using statistical regression analysis to build three prototype GIS wildlife models. In *Technical Papers Vol. 1* of the 7th International GIS/LIS Conference, San Jose, Calif., November 1992. Bethesda, Md.: ASPRS.

Johnston, K. M. and O. J. Schmitz. 1997. Wildlife and climate change: assessing the sensitivity of selected species to simulated doubling of atmospheric CO_2. *Global Change Biology* 3:531–544.

Johnston, K. M., S. Kopp and C. Tucker. 2005. Process, simulation, error, and sensitivity modeling in an integrated environment. *Proceedings of the 8th International Conference on GeoComputation*, held 31 July-3 August 2005 in Ann Arbor, Michigan.

Johnston, K. M., N. Collier, M. North and T. Arundel. In review. Modeling cougar movement through and agent-based model.

Johnston, K. M., O. J. Schmitz, K. D. Rothley and R. Mendelsoh. In review. Landscape level fitness-based assessment of wildlife habitat quality.

Kadmon, R., O. Farber and A. Danin. 2003. A systematic analysis of factors affecting the performance of climatic envelope models. *Ecological Applications* 13(3):853–867.

———. 2004. Effect of Roadside bias on the accuracy of predictive maps produced by bioclimatic models. *Ecological Applications* 14(2):401–413.

Kallimanis, A. S., W. E. Kunin, J. M. Halley and S. P. Sgardelis. 2005. Metapopulation extinction risk under spatially autocorrelated disturbance. *Conservation Biology* 19(2):534–546.

Kearney, M. and W. P. Porter. 2004. Mapping the fundamental niche: physiology, climate, and the distribution of a nocturnal lizard. *Ecology* 85(11):3119–3131.

Kelly, N. M. 2001. Changes to the landscape pattern of coastal North Carolina wetlands under the Clean Water Act, 1984-1992. *Landscape Ecology* 16:3–16.

Kenward, R. E., R. T. Clarke, K. H. Hodder and S. S. Walls. 2001. Density and linkage estimators of home range: nearest-neighbor clustering defines multinuclear cores. *Ecology* 83(7):1905–1920.

Kidd, D. M. and M. G. Ritchie. 2000. Inferring the patterns and causes of geographic variation in *Ephippiger ephippiger* (Orthoptera: Tettigoniidae) using geographic information systems (GIS). *Biological Journal of the Linnean Society of London* 71:269–295.

Kie, J. G. 2004. Program CALHOME: A Home Range Analysis Program. Champaign, Ill.: Illinois Natural History Survey. <http://nhsbig.inhs.uiuc.edu/wes/calhome_info.html> Accessed 3 July 2007.

Kie, J. G., T. Bowyer, M. C. Nicholson, B. B. Boroski and E. R. Loft. 2002. Landscape heterogeneity at differing scales: effects on spatial distribution of mule deer. *Ecology* 83(2):530–544.

Kim, T. H. and T. J. Cova. 2005. Tweening grammars: deformation rules for representing change between discrete geographic entities, Proceedings of the 8th International Conference on GeoComputation, 31 July-3 August 2005, Ann Arbor, Mich. <http://www.geocomputation.org/2005/Kim.pdf> Accessed 3 July 2007.

————. 2007. Tweening grammars: deformation rules for representing change between discrete geographic entities. *Computers, Environment & Urban Systems*, 31(3): 317-336.

Kingsland, S. E. 1995. *Modeling Nature*. Chicago: The University of Chicago Press.

Kittel T.G.F, N. A. Rosenbloom, T. H. Painter, D. S. Schimel and VEMAP Modeling Participants, 1995. The VEMAP integrated database for modeling United States ecosystem/vegetation sensitivity to climate change. *Journal of Biogeography* 22(4-5): 857–862.

Kramer, M. G., A. J. Hansen, M. L. Taper and E. J. Kissinger. 2001. Abiotic controls on long-term windthrow disturbance and temperate rain forest dynamics in southeast Alaska. *Ecology* 82(10):2749–2768.

Krebs, J. R. and N. B. Davies. 1981. *An Introduction to Behavioural Ecology*. Boston, Mass.: Blackwell Scientific Publications.

Kristan III, W. B. and W. I. Boarman. 2003. Spatial pattern of risk of common raven predation on desert tortoises. *Ecology* 84(9):2432–2443.

KUCR (University of Kansas Center for Research, Inc.). 2002. DesktopGarp. <http://nhm.ku.edu/desktopgarp/> Accessed 3 July 2007.

Kurttila, M. and T. Pukkala. 2003. Combining holding-level economic goals with spatial landscape-level goals in the planning of multiple ownership forestry. *Landscape Ecology* 18:529–541.

Larson, M. A., F. R. Thompson III, J. J. Millspaugh, W. D. Dijak and S. R. Shifley. 2004. Linking population viability, habitat suitability, and landscape simulation models for conservation planning. *Ecological Modelling* 180:103–118.

Laube, P., D. Imfeld and R. Weibel. 2005. Discovering relative motion patterns in groups of moving point objects. *International Journal of Geographic Information Science* 19(6):639–668.

Lawler, J. J., D. White and L. L. Master. 2003. Integrating representation and vulnerability: two approaches for prioritizing areas for conservation. *Ecological Applications* 13(6):1762–1772.

Leathwick, J. R., J. M. Overton and M. Mcleod. 2003. An environmental domain classification of New Zealand and its use as a tool for biodiversity management. *Conservation Biology* 17(6):1612–1623.

Lee, J. T., N. Bailey and S. Thompson. 2002. Using geographical information systems to identify and target sites for creation and restoration of native woodlands: a case study of the Chiltern Hills, UK. *Journal of Environmental Management* 64:25–34.

Legendre, P., M.R.T. Dale, M. Fortin, P. Casgrain and J. Gurevitch. 2004. Effects of spatial structures on the results of field experiments. *Ecology* 85(12):3202–3214.

Levin, S. A. 1992 The problem of pattern and scale in ecology. *Ecology* 73:1943–1983.

Li, H. and J. Wu. 2004. Use and misuse of landscape indices. *Landscape Ecology* 19:389–399.

Lindemann, J. D. and W. L. Baker. 2001. Attributes of blowdown patches from a severe wind event in the Southern Rocky Mountains, USA. *Landscape Ecology* 16:313–325.

Line, J. D., D. L. Azuma and A. Mosses. 2003. Modeling the spatial dynamic distribution of humans in Oregon (USA) Coast Range. *Landscape Ecology* 18:347–361.

Liu, A. J. and G. N. Cameron. 2001. Analysis of landscape patterns in coastal wetlands of Galveston Bay, Texas (USA). *Landscape Ecology* 16:581–595.

Longley, P. A., M. F. Goodchild, D. J. Maguire and D. W. Rhind. 2005. Uncertainty. In *Geographic Information Systems and Science*, 2nd ed., 127–153. New York: John Wiley & Sons.

Lotka, A.J. 1956. *Elements of Mathematical Biology*. New York: Dover Press.

Loucks, C. J., L. Zhi, E. Dinerstein, W. Dajun, F. Dali and W. Hao. 2003. The giant pandas of the Qinling Mountains, China: a case study in designing conservation landscapes for elevation migrants. *Conservation Biology* 17(2):558–565.

Luoto, M., T. Toivonen and R. K. Heikkinen. 2002. Prediction of total and rare plant species richness in agricultural landscapes from satellite images and topographic data. *Landscape Ecology* 17:195–217.

Luoto, M., R. Virkkala, R. K. Heikkinen and K. Rainio. 2004. Predicting bird species richness using remote sensing in boreal agricultural-forest mosaics. *Ecological Applications* 14(6):1946–1962.

Lutz, P.W.W., S. P. Rushton, L. A. Wauters, S. Bertolino, I. Currado, P. Mazzoglio and M.D.F. Shirley. 2001. Predicting grey squirrel expansion in North Italy: a spatially explicit modelling approach. *Landscape Ecology* 16:407–420.

MacArthur, R. H. and E. O. Wilson. 1963. An equilibrium theory of insular zoogeography. *Evolution* 17:373–387.

———. 1967. *The Theory of Island Biogeography*. Princeton, New Jersey: Princeton University Press.

MacIsaac, H. J., J.V.M. Borrely, J. R. Muirhead and P. A. Graniero. 2004. Backcasting and forecasting biological invasions of inland lakes. *Ecological Applications* 14(3):773–783.

Malczewski, J. 1999. *GIS and Multicriteria Decision Analysis*. New York: John Wiley & Sons.

Mangel, M. and C. W. Clark. 1989. *Dynamic Modeling in Behavioral Ecology*, Monographs in Behavior and Ecology. Princeton, New Jersey: Princeton University Press.

Mangel, M. and D. Ludwig. 1992. Definition and evaluation of the fitness of behavioral and developmental programs. *Annual Review of Ecology and Systematics* 23:507–536. Palo Alto, Calif.: Annual Reviews Inc.

Maslow, A. 1943. *Motivation and Personality*, 2nd ed. New York: Harper and Row.

Maynard Smith, J. 1978. Optimization Theory in Evolution. *Am. Rev. Ecol. Syst.* 9:31–56.

McCauley, L. A. and D. G. Jenkins. 2005. GIS-based estimates of former and current depressional wetlands in an agricultural landscape. *Ecological Applications* 15(4):1199–1208.

McGarigal, K., S. A. Cushman, M. C. Neel and E. Ene. 2002. FRAGSTATS: Spatial Pattern Analysis Program for Categorical Maps. Computer software program produced by the authors at the University of Massachusetts, Amherst. <www.umass.edu/landeco/research/fragstats/fragstats.html> Accessed 29 June 2007.

McKinney, T., S. R. Boe and J. C. deVos Jr. 2003. GIS-based evaluation of escape terrain and desert bighorn sheep populations in Arizona. *Wildlife Society Bulletin* 31(4):1229–1236.

Metternicht, G. 2001. Assessing temporal and spatial changes of salinity using fuzzy logic, remote sensing, and GIS. Foundations of an expert system. *Ecological Modelling* 144(2):163–179.

Meyer, C. B., S. L. Miller and C. J. Ralph. 2002. Multi-scale landscape and seascape patterns associated with marbled murrelet nesting areas on the U.S. West Coast. *Landscape Ecology* 17:95–115.

Mitchell, M. S. and R. A. Powell. 2003. Response of black bears to forest management in the Southern Appalachian Mountains. *Journal of Wildlife Management* 67(4):692–705.

Molofsky, J., J. D. Bever, J. Antonovics and T. J. Newman. 2002. Negative frequency dependence and the importance of spatial scale. *Ecology* 83(1):21–27.

Moou, W. M. and D. L. DeAngelis. 2003. Uncertainty in spatially explicit animal dispersal models. *Ecological Applications* 13(3):794–805.

Morales, J. M. and S. P. Ellner. 2002. Scaling up animal movements in heterogeneous landscapes: the importance of behavior. *Ecology* 83(8):2240–2247.

Moravcsik, J. M. 1975. *Aitia* as generative factor in Aristotle's philosophy. *Dialogue* 14:622–636.

National Parks and Wildlife Service (NPWS). 1999. C-Plan: conservation planning software. User manual. Version 2.2. Armidale, New South Wales, Australia: NPWS.

Naves, J., T. Wiegand, E. Revilla and M. Delibes. 2003. Endangered species constrained by natural and human factors: the case of brown bears in Northern Spain. *Conservation Biology* 17(5):1276–1289.

NCRS (North Central Research Station). 2003a. LANDIS Landscape Disturbance and Succession model. Rhinelander, Wis.: USDA Forest Service NCRS. <http://www.ncrs.fs.fed.us/4153/landis/default.asp> Accessed 3 July 2007.

NCRS (North Central Research Station). 2003b. Harvest - Version 6.1. Rhinelander, Wis.: USDA Forest Service NCRS. <http://www.ncrs.fs.fed.us/4153/Harvest/v61/default.asp> Accessed 3 July 2007.

Neilson, T. P. 1995. A model for prediction continental-scale vegetation distribution and water balance. *Ecological Applications* 5:362–385.

O'Neill, R.V. 2001. Is it time to bury the ecosystem concept? (With full military honors, of course!). *Ecology* 83(2):3275–3284.

Palik, B. J., R. Buech and L. Egeland. 2003. Using an ecological land hierarchy to predict seasonal-wetland abundance in upland forests. *Ecological Applications* 13(4):1153–1163.

Palmer, S.C.F. and P. J. Bacon. 2000. A GIS approach to identifying territorial resource competition. *Ecography* 23:513–524.

Peinetti, H. R., M. A. Kalkhan and M. B. Coughenour. 2002. Long-term changes in willow spatial distribution on the elk winter range of Rocky Mountain National Park (USA). *Landscape Ecology* 17:341–354.

Peterson, A. T. 2001. Predicting species' geographic distributions based on ecological niche modeling. *The Condor* 103:599–605.

———. 2003. Predicting the geography of species' invasions via ecological niche modeling. *Quarterly Review of Biology* 78(4):419–434.

Peterson, A. T. and C. R. Robins. 2003. Using ecological-niche modeling to predict barred owl invasions with implications for spotted owl conservation. *Conservation Biology* 17(4):1161–1165.

Podruzny, K. M., J. H. Devries, L. M. Armstrong and J. J. Rotella. 2002. Long-term response of northern pintails to changes in wetlands and agriculture in the Canadian Prairie Pothole region. *Journal of Wildlife Management* 66(4):993–1010.

Porter W. P. and D. M. Gates. 1969. Thermodynamic equilibria of animals with the environment. *Ecological Monographs* 39:245–270.

Preston, F. W. 1962a. The canonical distribution of commonness and rarity: part I. *Ecology* 43:185–215.

———. 1962b. The canonical distribution of commonness and rarity: part II. *Ecology* 43:410–432.

Pullian, H. R. 1986. Sources, sinks, and population regulation. *The American Naturalist* 132(5):652–661.

Railsback, S. F. and B. C. Harvey. 2002. Analysis of habitat-selection rules using an individual-based model. *Ecology* 83(7):1817–1830.

Railsback, S. F., H. B. Stauffer and B. C. Harvey. 2003. What can habitat preference models tell us? Tests using a virtual trout population. *Ecological Applications* 13(6):1580–1594.

Ranson, K. J., G. Sun, R. G. Knox, E. T. Levine, J. F. Weishampel and S. T. Fifer. 2001. Northern forest ecosystem dynamics using coupled models and remote sensing. *Remote Sens. Environ.* 75:291–302.

Regan, H. M., Y. Ben-Haim, B. Langford, W. Langford, W. G. Wilson, P. Lundberg, S. J. Andelman and M. A. Burgman. 2005. Robust decision-making under severe uncertainty for conservation management. *Ecological Applications* 15(4):1471–1477.

Riley, S. P., R. M. Sauvajot, T. K. Fuller, E. C. York, D. A. Kamradt, C. Bromley and R. K. Wayne. 2003. Effects of urbanization and habitat fragmentation on bobcats and coyotes in Southern California. *Conservation Biology* 17(2):566–576.

Ritchie, M. E. 1988. Individual variation in the ability of Columbian ground squirrels to select an optimal diet. *Evolutionary Ecology* 2:232–252.

———. 1990. Optimal foraging and fitness in Columbian ground squirrels. *Oecologia* 82:56–67.

———. 1991. Inheritance of optimal foraging behaviour in Columbian ground squirrels. *Evolutionary Ecology* 5:146–159.

Rothley, K. D. 2001. Manipulative, multi-standard test of a white-tailed deer habitat suitability model. *Journal of Wildlife Management* 65(4):953–963.

Rothley, K. D., C. N. Berger, C. Gonzalez, E. M. Webster and D. I. Rubenstein. 2004. Combining strategies to select reserves in fragmented landscapes. *Conservation Biology* 18(4):1121–1131.

Royle, J. A. and J. A. Dubovsky. 2001. Modeling spatial variation in water-fowl band-recovery data. *Journal of Wildlife Management* 65(4):726–737.

Rustigian, H. L., M. V. Santelmann and N. H. Schumaker. 2003. Assessing the potential impacts of alternative landscape designs on amphibian population dynamics. *Landscape Ecology* 18:65–81.

Sargeant, G. A., M. A. Savado, C. C. Slivinski and D. H. Johnson. 2005. Markov Chain Monte Carlo estimation of species distributions: a case study of the swift fox in Western Kansas. *Journal of Wildlife Management* 69(2):483–497.

Schmitz, O. J. 1991. Thermal constraints and optimization of winter feeding and habitat choice by white tailed deer. *Holarctic Ecology* 14:104–111.

———. 1992. Optimal diet selection by white-tailed deer: balancing reproduction with starvation risk. *Evolutionary Ecology* 6:125–141.

Schoener, T. W. 1974. Competition and the form of habitat shift. *Theor. Pop. Biol.* 6:265–307.

———. 1976. Alternatives to Lotka-Volterra Competition: models of intermediate complexity. *Theor. Pop. Biol.* 10:309–333.

———. 1983. Simple models of optimal feeding-territory size: a reconciliation. *The American Naturalist* 121(5):608–629.

Schooley, R. L. and J. A. Wiens. 2005. Spatial ecology of cactus bugs: area constraints and patch connectivity. *Ecology* 86(6):1627–1639.

Schumaker, N. H. 1998. A user's guide to the PATCH model. EPA/600 R-98/135. Corvallis, Oregon: US Environmental Protection Agency Western Ecology Division.

Schumaker, N. H., T. Ernst, D. White, J. Baker and P. Haggerty. 2004. Projecting wildlife responses to alternative future landscapes in Oregon's Willamette Basin. *Ecological Applications* 14(2):381–400.

Seaman, D. E. 2004. KernelHR 4.27. Champaign, Ill.: Illinois Natural History Survey. <http://nhsbig.inhs.uiuc.edu/wes/khr427.html> Accessed 3 July 2007.

Selkirk, S. W. and I. D. Bishop. 2002. Improving and extending home range and habitat analysis by integration with a geographic information system. *Transactions in GIS* 6(2):151–159.

Skinner, K. M., W. P. Kemp and J. P. Wilson. 2000. GIS-based indicators of Montana grasshopper communities. *Transactions in GIS* 4(2):113–128.

Sondgerath, D. and B. Schroder. 2002. Population dynamics and habitat connectivity affecting the spatial spread of populations – a simulation study. *Landscape Ecology* 17:57–70.

Stephens, D. W. and J. R. Krebs. 1986. *Foraging Theory, Monographs in Behavior and Ecology.* Princeton, New Jersey: Princeton University Press.

Stergiou, C. and D. Siganos. 2006. Neural Networks. London: Imperial College London. <http://www.doc.ic.ac.uk/~nd/surprise_96/journal/vol4/cs11/report.html> Accessed 2 July 2007.

Stockwell, D.R.B. and D. Peters. 1999. The GARP modeling system: problems and solutions to automated spatial prediction. *International Journal of Geographical Information Science* 13:143–158.

Sturtevant, B. R., P. A. Zollner, E. J. Gustafson and D. T. Cleland. 2004. Human influence on the abundance and connectivity of high-risk fuels in mixed forests of northern Wisconsin, USA. *Landscape Ecology* 19:235–253.

Tansley, A. G. 1935. The use and abuse of vegetational concepts and terms. *Ecology* 16:284–307.

Thogmartin, W. E., J. R. Sauer and M. G. Knutson. 2004. A hierarchical spatial model of avian abundance with application to cerulean warblers. *Ecological Applications* 14(6):1766–1779. <http://www.umesc.usgs.gov/documents/publications/2004/thog-martin_ecolappl_2004.pdf> Accessed 25 June 2007.

Tilman, D. 1980. Resources: a graphical-mechanistic approach to competition and preda-tion. *Am. Nat.* 116:362–393.

———. 1985. The resource-ratio hypothesis of plant succession. *Am. Nat.* 125:827–852.

Tinker, D. B., W. H. Romme and D. G. Despain. 2003. Historic range of variability in landscape structure in subalpine forests of the greater Yellowstone area, USA. *Landscape Ecology* 18:427–439.

Tischendorf, L. 2001. Can landscape indices predict ecological processes consistently? *Landscape Ecology* 16:235–254.

Treves, A., L. Naughton-Treves, E. K. Harper, D. J. Mladenoff, R. A. Rose, T. A. Sickley and A. P. Wydeven. 2004. Predicting human-carnivore conflict: a spatial model derived from 25 years of data on wolf predation on livestock. *Conservation Biology* 18(1):114–125.

Tucker, K., S. P. Rushton, R. A. Sanderson, E. B. Martin and J. Blaiklock. 1997. Modeling bird distributions – a combined GIS and Bayesian rule-based approach. *Landscape Ecology* 12(2):77–93.

Urban, D. L. 1990. *A Versatile Model to Simulate Forest Pattern: A Users Guide to Zelig.* Environmental Sciences Department. Charlottesville: University of Virginia.

Urban, D. and T. Keitt. 2001. Landscape connectivity: a graph-theoretic perspective. *Ecology* 82(5):1205–1218.

Vander Zanden, M. J., J. D. Olden, J. H. Thorne and N. E. Mandrak. 2004. Predicting occurrences and impacts of smallmouth bass introductions in north temperate lakes. *Ecological Applications* 14(1):132–148.

VEMAP Members [J. Borchers, J. Chaney, H. Fisher, S. Fox, A. Haxeltine, A. Janetos, D. W. Kicklighter, T.G.F. Kittel, A. D. Mcguire, R. Mckeown, J. M. Melilo, R. Neilson, R. Nemani, D. S. Ojima, T. Painter, Y. Pan, W. J. Parton, L. Pierce, L. Pitelka, C. Prentice, B. Tizzo, N. A. Rosenbloom, S. Tunning, D. S. Schimel, S. Sitch, T. Smith, I. Woodward] (1995) Vegetation/ecosystem modeling and analysis project (VEMAP): comparing biogeography and biogeochemistry models in a continental-scale study of terrestrial ecosystem responses to climate change and CO_2 doubling. *Global Biochemical Cycles* 9:407–437.

Verbeylen, G., L. De Bruyn, F. Adriaensen and E. Matthysen. 2003. Does matrix resistance influence red squirrel (*Sciurus vulgaris* L. 1758) distribution in an urban landscape? *Landscape Ecology* 18(8):791–805.

Verburg, P. H. and A. Veldkamp. 2004. Projecting land use transitions at forest fringes in the Philippines at two spatial scales. *Landscape Ecology* 19:77–98.

Wagner, H. H. and M. J. Fortin. 2005. Spatial analysis of landscapes: concepts and statistics. *Ecology* 86(8):1975–1987.

Warman, L. D., A.R.E. Sinclair, G.G.E. Scudder, B. Klinnkenberg and R. L. Pressey. 2004. Sensitivity of systematic reserve selection to decisions about scale, biological data, and targets: case study from southern British Columbia. *Conservation Biology* 18(3):655–666.

Weaver, K. and A. H. Petera. 2004. Modeling land cover transitions: a solution to the problem of spatial dependence in data. *Landscape Ecology* 19:273–289.

Wells, M. L., J. F. O'Leary, J. Franklin, J. Michaelsen and D. E. McKinsey. 2004. Variation in a regional fire regime related to vegetation type in San Diego County, California (USA). *Landscape Ecology* 19:139–152.

Whittington, J., C. C. St. Clair and G. Mercer. 2005. Spatial responses of wolves to roads and trails in mountain valleys. *Ecological Applications* 15(2):543–553.

Wiegand, T., E. Revilla and K. A. Moloney. 2005. Effects of habitat loss and fragmentation on population dynamics. *Conservation Biology* 19(1):108–121.

Wiersma, Y. F. and D. L. Urban. 2005. Beta diversity and nature reserve system design in the Yukon, Canada. *Conservation Biology* 19(4):1262–1272.

Wikle, C. K. 2003. Hierarchical Bayesian models for predicting the spread of ecological processes. *Ecology* 84(6):1382–1394.

Williams, P., L. Hannah, S. Andelman, G. Midgley, M. Araújo, G. Hughes, L. Manne, E. Martinez-Meyer and R. Pearson. 2005. Planning for climate change: identifying minimum-dispersal corridors for the Cape Proteacease. *Conservation Biology* 19(4):1063–1074.

Yattaw, N. J. 1999. Conceptualizing space and time: a classification of geographic movement. *Cartography and Geographic Information Science* 26(2):85–98.

Zhang, J. and M. F. Goodchild. 2002. *Uncertainty in Geographic Information*. New York: Taylor and Francis.

CHAPTER 27

Geographic Data Mining: An Introduction

Giorgos Mountrakis

27.1 Introduction

Recent developments in data acquisition methods and sensors have supported an explosion of geospatial information. A variety of geographic phenomena are observed in continuously improved spatial and temporal resolutions. As datasets have grown in size and complexity there has been a move away from manual data analysis towards automated techniques using more complex, sophisticated tools. Initially, methods and models targeted capturing of geospatial information (e.g., remote sensing and GPS sensors). However, as information increased the next step was to store efficiently and make available these geospatial datasets (e.g., data warehouses). Parallel to these efforts, the need has increased to investigate intelligent ways for:

   i) Identification of hidden relationships within geospatial collections, and

   ii) Retrieval of geospatial information.

The first relates to **geographic data mining** (GDM), development of methods/rules that provide insight into geospatial data that is not directly available. For example, establish a relationship between spatial proximity to water and real estate prices.

The latter is often expressed as **geographic information retrieval** (GIR), methodologies used to extract information from a geographic database. Typical geospatial queries would fall under the GIR category—for example, retrieve all houses sold this year within a 1-km buffer from water. The key difference is that GDM creates new information/knowledge, while GIR attempts to find the most relevant (to a user request) existing information from a database, most often without further processing.

As models become more complex a crisp categorization into either GDM or GIR may not be feasible. For example, content retrieval from image databases can be seen as a GIR task but hidden information may be identified in the process, giving the retrieval characteristics of GDM. Think of a query for a house that is within area S and that at time T looked like image pattern P. A typical GIR query would filter images within a specific spatio-temporal region. An image extraction algorithm would follow to perform a pattern match (GDM), therefore transforming the raw pixels into previously unknown knowledge (e.g., that house exists on that image).

The focus of this chapter is to introduce non-experts to geographic data mining. Some issues on intelligent geographic information retrieval have been previously discussed (Mountrakis et al. 2004). This chapter begins with a short introduction to data mining and its multi-disciplinary history. Major tasks performed within data mining are introduced. A discussion is presented on the progress of data mining of geographic datasets. The chapter concludes with popular machine-learning methodologies with potential application to geographic data mining. This overview is largely based on three excellent books (Dunham 2002; Kantardzic 2002; Miller and Han 2001).

27.2 Data Mining and Knowledge Discovery in Databases

Database content has grown significantly in science, government and business. Automated methods have naturally emerged as manual analysis is not feasible for such large information volumes. These methods are the subject of the rapidly growing field of **knowledge discovery in databases** (KDD). The terms KDD and data mining are often used interchangeably.

Dunham (2002) suggests the following distinction:
- Knowledge discovery in databases is the process of finding useful information and patterns in data.
- Data mining is the use of algorithms to extract the information and patterns derived by the KDD process.

According to Fayyad et al. (1996) the KDD process is composed of the following five steps (Figure 27-1):

i) *Selection*: In this step data used in the data mining process are collected or merged from a variety of different and heterogeneous sources.

ii) *Preprocessing*: The purpose of this step is to correct potential discrepancies in the data. Examples of targeted errors include incorrect, noisy or missing data, and data inconsistency due to multiple sources using different data types and metrics.

iii) *Transformation*: Here, a common processing data format is established. If applicable, dimensionality reduction techniques are used to enhance expressiveness and minimize redundancy.

iv) *Data Mining*: This is the core step where data mining algorithms are employed.

v) *Interpretation/evaluation*: At this last step statistical and visualization methods are used to understand and explain the data mining results.

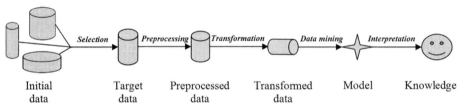

Figure 27-1 Knowledge discovery process.

27.3 Data Mining Multi-disciplinary History

The current evolution of data mining (DM) algorithms has resulted from multi-disciplinary research. Disciplines such as databases, statistics and artificial intelligence have unified their algorithmic approaches for data mining purposes. However, the wide variety of data mining problems combined with different backgrounds and expertise can also lead to different perspectives. It is not uncommon to find that similar problems or solutions are described differently. Nonetheless, the multi-disciplinary character of DM has initiated discussions and created algorithms than reach beyond DM applications. Geographic data mining (GDM) is a new and exciting field that focuses on knowledge extraction from geographic data. GDM— as we will discuss more fully later—imposes unique challenges but also has distinct benefits.

Table 27-1 (Dunham 2002) shows theoretical and algorithmic developments in the areas of Artificial Intelligence (AI), Databases (DB) and Statistics (Stat) as applied to data mining. For an extended review of the statistical methods developed over the past 40 years and their contribution to KDD the reader is advised to read Elder and Pregibon (1996).

Table 27-1 Time line of data mining development (Dunham 2002).

Time	Area	Contribution	Reference
Late 1700s	Stat	Bayes theorem of probability	Bayes (1763)
Early 1900s	Stat	Regression analysis	
Early 1920s	Stat	Maximum likelihood estimate	Fisher (1921)
Early 1940s	AI	Neural networks	McCulloch and Pitts (1943)
Early 1950s		Nearest neighbor	Fix and Hodges (1951)
Early 1950s		Single link	Florek et al. (1951)
Late 1950s	AI	Perceptron	Rosenblatt (1958)
Late 1950s	Stat	Resampling, bias reduction, jackknife estimating	
Early 1960s	AI	Machine Learning (ML) started	Feigenbaum and Feldman (1963)
Early 1960s	DB	Batch reports	
Mid 1960s		Decision trees	Hunt et al. (1966)
Mid 1960s	Stat	Linear models for classification	Nilsson (1965)
	Stat	Exploratory data analysis	
Late 1960s	DB	Relational data model	Codd (1970)
Mid 1970s	AI	Genetic algorithms	Holland (1975)
Late 1970s	Stat	Estimation with incomplete data (EM algorithm)	Dempster et al. (1977)
Late 1970s	Stat	K-means clustering	
Early 1980s	AI	Kohonen self-organizing map	Kohonen (1982)
Mid 1980s	AI	Decision tree algorithms	Quinlan (1986)
Early 1990s	DB	Association rule algorithms	
		Web and search engines	
1990s	DB	Data warehousing	
1990s	DB	Online analytic processing (OLAP)	

27.4 Data Mining Tasks

Data mining is one of the fastest-growing fields in the computer industry. As we mentioned, a wide range of methods can be applied to a variety of problems. These methods involve a collaborative effort between humans and computers, where human experts provide the necessary knowledge in describing problems and goals and computers act in a supportive role with their computational capabilities. In practice, the two "high-level" primary tasks of data mining are prediction and description (Kantardzic 2002). Prediction focuses on calculation of unknown or future values of existing variables. Description focuses on finding human-interpretable patterns or relationships in the data. Thus we can categorize data mining activities into one of the two categories:

1. Predictive data mining which produces the underlying model described by the given dataset, or
2. Descriptive data mining which produces new, non-trivial information based on the available dataset.

Several task groupings have been proposed in the literature, especially in books and introductory tutorials on data mining. There is a significant overlap between different task groupings, and sometimes the distinction between them is based solely on terminology. For our review we use the rather comprehensive task representation of Figure 27-2 and an explanation as presented by Dunham (2002).

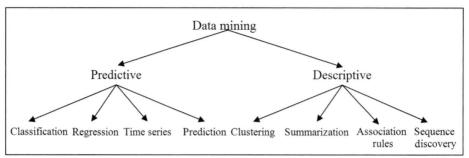

Figure 27-2 Data mining primary goals and tasks.

27.4.1 Predictive Tasks

The following tasks are categorized as predictive based on their functionality: classification, regression, time series analysis and prediction. Boundaries along these tasks are not crisp, as one task might borrow techniques developed for another.

Classification is learning a function that maps a data item into one of several classes (Hand 1981; Weiss and Kulikowski 1991; McLachlan 1992). It is also known as supervised learning because classes are identified before examining the data. Pattern recognition is a classification type where an input pattern is classified into one out of several classes based on its similarity to these predefined classes. For example, in face recognition a feature vector is produced describing facial characteristics (distance between eyes, size and shape of mouth, shape of head, etc.). This is then compared to the entries in a database to see if there is a successful match.

The modeling of continuous outputs is the focus of *regression*. Parametric regression assumes that the target data fit into some known type of function (e.g., linear, logistic, etc.) while non-parametric regression introduces complex non-linear models (e.g., neural networks). An example of regression is the calculation of recovery probability of a patient (output) based on a set of health diagnostics (inputs).

Time series analysis focuses on attribute changes over time. Three general operation types can be identified, namely:

i) distance measures evaluating similarity between different time series,
ii) time series examination to determine its behavior based on its structure, and
iii) prediction of future values based on historical time series.

Prediction forecasts future states based on existing data. It has many real-world data mining applications such as flooding and earthquake prediction. Even though prediction has close ties to classification, time series analysis and regression methods, it should be recognized as a distinct task since it includes methods not used by the other three tasks.

27.4.2 Descriptive Tasks

Descriptive tasks investigate data properties and their internal relationships (instead of predicting new properties). Clustering, summarization, association rules and sequence discovery are usually viewed as descriptive in nature.

Clustering is a common descriptive task where we identify groupings (clusters) based on predefined attributes within the data. The clusters may be mutually exclusive and exhaustive or contain a richer representation such as hierarchical or overlapping categories. A clustering example in KDD would be the identification of sub-populations of consumers in marketing applications.

Summarization involves methods for finding a compact description for a subset (or all) of data. It is also called characterization or generalization as the goal is to detect representative information from the data. Summarization techniques are often applied to interactive exploratory data analysis and automated report generation. A simple example would be the representation of fields based on their mean and standard deviation.

Association rules refer to the data mining task of identifying hidden relationships among data. It is also called link analysis, affinity analysis or dependency modeling. Here we try to find a model that detects dependencies between variables in the dataset (or parts of it). A frequent application of this task involves its use in the retail sales community, for example to identify items that are purchased together (e.g., milk and cereal).

Sequence discovery is used to determine temporal ordering in data. It is a special case of association rules where the relationship is based on time. An example would be the discovery of sequence within which goods are purchased (e.g., people who purchase CD players may purchase Audio CDs within one week).

27.5 Geographic Data Mining

New applications of data mining techniques continue to drive research in the field. To date, most data mining research concentrates on relational and transactional data. Despite the importance and proliferation of geospatial information, work in this field has only emerged recently (Gunopulos 2001). Nonetheless, temporal and spatial data mining continue to grow rapidly as exciting subfields of data mining. There are many reasons for this, including the following (Roddick et al. 2001):

- Constantly increasing acquisition of geospatial information requiring analysis.
- Improved data availability through the Internet and as a result of electronic commerce and inter-enterprise applications.
- Recognition of the value and scientific/commercial advantage that geographic data mining provides.
- The realization that temporal and spatial data are special and need to be explicitly accommodated.

In the next section the distinction of geospatial data from other data is presented. This distinction is seen from the data mining perspective and it is important as it propagates to algorithmic goals and requirements. An organization of current geospatial mining tasks is also provided.

27.5.1 Special Characteristics of Geographic Data Mining

The recent geographic data explosion is not different from other areas that have seen increased digital information (e.g., marketing, biology, astronomy). But is there a difference between geographic data mining and data mining in other fields? Several papers were recently published addressing this issue (Buttenfield et al. 2000; Miller and Jiawei 2001; Gahegan 2001) and are used in the following review.

Many of the challenging issues arise from the fact that geography is an integrative discipline. Geographic data are incorporated in fields as diverse as social science and engineering. The interdisciplinary acquisition, processing and modeling of geographic information poses or magnifies the following obstacles:

- **Data volume.** Improvements in geographic information acquisition methods, such as newly developed sensors, are increasing data in non-linear fashion. Many of the large databases containing consumer, medical and financial transactions nowadays contain spatial and temporal attributes and therefore are waiting to be converted to geographic knowledge (Miller and Jiawei 2001). The terabyte proportions of some geographic datasets mandate significant changes to traditional retrieval methods in order to provide efficient and scalable solutions.

- **Data format.** Because of monetary, privacy or processing constraints data is often manipulated or combined in uncooperative ways. A typical example is census socioeconomic data, where its spatially aggregated format makes integration with other sources difficult (e.g., remotely sensed imagery).

- **Data dependencies.** Spatial processes are clearly interrelated which makes it difficult to isolate specific study areas. Geographically-explicit data are often subjected to spatial dependency and spatial heterogeneity (Buttenfield et al. 2001). The former refers to values being related in (proximal) locations. The latter refers to the non-stationarity of most geographic processes. This is a typical problem encountered in spatial statistics where global models cannot adequately model localized variations. Ignoring these properties affects geographic data mining results (Chawla et al. 2001). Spatial and temporal codependence that occurs across a variety of scales and from a variety of causes can also have a significant affect (Roddick and Lees 2001). For example, the cyclic frequency of many geographical systems (daily, seasonal, annual, circulatory, El-Nino, sunspot) imposes a strong signal on data that overshadows more localized variance (Gahegan 2001). For a further discussion on this challenging issue the reader may check several existing publications (e.g., Anselin 1995; Brunsdon et al. 1996; Fotheringham et al. 1997; Getis and Ord 1992, 1996).

- **Data representation and manipulation.** The complexity of spatio-temporal objects and patterns imposes additional constraints. In most non-geographic domains, data objects can be meaningfully represented discretely within the information space without losing important properties (Buttenfield et al. 2001). This does not necessarily propagate to geographic objects: size, shape and boundaries can affect geographic processes (e.g., geographic objects cannot always be reduced to points or simple line features without information loss). Relationships such as distance, direction and connectivity become more complex with dimensional objects (Egenhofer and Herring 1994; Okabe and Miller 1996; Peuquet and Zhang 1987). Operations on these objects over time are also complex and information-bearing (Egenhofer and Hornsby 2000). Temporal scales and granularities can also be intricate, preventing a simple "dimensioning up" of space to include time (Roddick and Lees 2001). Developing tools for data mining on spatio-temporal objects is an important challenge.

- **Geographic conceptualization.** A significant difficulty with geographic knowledge discovery activities is the underlying complex conceptualization (Gahegan 2001). There is no universally accepted conceptual model for geography (e.g., Goodchild 1992) either by scientists or commercial products. This leads to three distinct problems (Buttenfield et al. 2001):

  i) Comparison or combination of data is not always trivial.
  ii) Formal geographical knowledge is not readily available, making its application to the process of knowledge discovery difficult.
  iii) When new knowledge is uncovered, its formalization is difficult.

27.5.2 Tasks Within Geographic Data Mining

Over the past decade there has been a substantial increase in temporal, spatial and spatio-temporal data mining applications and a variety of papers have been published. In this section we make use of the data mining task categorization previously discussed. We propagate these categories to geographic data mining. Our intent is to provide a useful guide rather than an exhaustive classification. An extensive list of geographic data mining papers is available in (Roddick et al. 2001).

Table 27-2 Organization and examples of geographic data mining tasks.

Data Mining Tasks	Geographic Data Mining Tasks	
Aspatial Data	**Spatial Data**	**Spatio-temporal Data**
Predictive (tasks using known results on additional data)		
a. Classification	☑ Compare spatial data to a given pattern and classify them (e.g., land use classification of remote sensing data)	☑ Compare spatio-temporal data to a given pattern and classify them (e.g., moving target identification from a video)
b. Regression	☑ Use spatial data to create geographically weighted regression models (e.g., rainfall)	☑ Use spatio-temporal data in regression models (e.g., traffic routes)
c. Time series analysis	☑ Not applicable without spatial component	☑ Not applicable without spatial component
d. Prediction	☑ Not applicable without temporal component	☑ Forecasting future states of a spatio-temporal model (e.g., weather prediction)
Descriptive (tasks investigating relationships within existing data)		
a. Clustering	☑ Identify spatial groupings (e.g., crime groupings)	☑ Identify spatio-temporal groupings (e.g., deer mortality)
b. Summarization	☑ Aggregation of spatial data (e.g., census household data)	☑ Generalization of spatio-temporal data (e.g., traffic flow)
c. Association rules	☑ Identification of spatial dependencies (e.g., snowfall amount and lake proximity)	☑ Identification of spatio-temporal dependencies (e.g., human health and hazardous exposures)
d. Sequence discovery	☑ Not applicable without temporal component	☑ Spatio-temporal sorting (e.g., bank robberies)

As we mentioned, the key difference between predictive and descriptive GDM tasks is that in the predictive case knowledge is used to design the models, while in descriptive tasks knowledge is extracted after model implementation. Frequently, methods are used interchangeably, for example a typical descriptive method such as clustering can be used in a remote sensing classification. Similarly, association rules are generally considered descriptive; however, such rules can be used to assist in building a geographically weighted regression model, a predictive task. Our task distinction is motivated by differences in algorithmic implementations. We should note that the presented dataset examples can be used in multiple tasks depending on modeling specifics.

A general review with data mining tutorials can be found in Moore (2006). Spatial data mining literature collections are available at Ester et al. (1997), Shekhar and Chawla (2003), NASA (2004) and TKK (2005). For a detailed collection of applications the reader can review Miller and Han (2001). Finally, one of the first works in GDM, a generalization-based spatial data mining method, is by Lu et al. (1993) while one of the first software packages targeting specifically GDM (spatial association rules) is demonstrated by Rodman et al. (2006).

27.6 Machine Learning for Geographic Data Mining

Machine learning is a popular combination of computer science and statistics that has led to a variety of algorithms solving a diverse set of problems. These algorithms vary in their goals, training datasets, learning strategies and representation of data (Kantardzic 2002). Machine learning methods have been used extensively in data mining. Here we introduce some representative machine learning methods that can be applied to GDM. Popular machine learning methods examined include decision trees, genetic algorithms, case-based reasoning, neural networks and fuzzy logic. If the reader is interested in an in-depth analysis of regression models a valuable starting book would be Hastie et al. (2001).

27.6.1 Decision Trees

The decision tree is a widely used method that adopts a top-down strategy to identify a solution in a part of the input space. It is similar to a flow-chart type structure, where each internal node performs a test on an attribute, each branch corresponds to the outcome of each test (true/false), and each leaf node represents a single class (Han and Kamber 2001). In order to classify a new sample, the attribute values of the sample are propagated through the decision tree with basic IF statements. A path is created as the sample values progress from the root to the leaf node. Each node corresponds to a single class, the class prediction of that sample.

Decision trees are mostly popular in classification problems. They offer many advantages because they are easy to use and interpret. They also scale well in large data samples because the tree size is independent of database size (Dunham 2002). On the other hand there are several disadvantages, the most important being their inability to easily support continuous data, although some efforts exist to overcome this limitation (e.g., Cubist software by RuleQuest Research). In order to do so each attribute has to be discretized to several subclasses. Furthermore, handling missing data is not straightforward and overfitting may occur if tree pruning is not implemented. Finally, another important drawback is the lack of direct support of identified correlations/dependencies among attributes.

Some well-known examples of decision trees include the ID3 algorithm (Quinlan 1986), the C4.5 (Quinlan 1993), which does support continuous data, and PRISM (Cendrowska 1987) which allows individual testing of each attribute to support an attribute importance ranking. Another tree type is the scalable parallelizable induction of decision trees algorithm, also known as SPRINT (Shafer et al. 1996).

27.6.2 Genetic Algorithms

Genetic algorithms are optimization methods that are based on concepts such as natural selection and evolutionary processes. The original idea for genetic algorithms was developed in 1975 by John Holland, a biologist performing adaptive learning on natural genetic systems. In general, genetic learning follows these steps (Han and Kamber 2001). Randomly generated rules are used to create an initial population. A string of bits is assigned to each rule. A new population is formed based on the "best" rules in the current population. Typically, a rule is evaluated from its classification accuracy on a set of training samples. A new population is created by applying genetic operators known as crossover and mutation. In crossover, substrings from pairs of rules are replaced with new pairs of rules. In mutation, random bits in a rule's string are inverted.

Genetic algorithms have been used in data mining for classification, clustering and generating association rules as well as other optimization problems. They are popular because they do not depend on derivatives, are easily parallelizable, and are applicable to both continuous and discrete data. However, they suffer some shortcomings as it is difficult to i) identify the best fitness function and ways to perform crossover and mutation, and ii) understand the provided results and perform an error assessment since the algorithm operates in a different space than the space of the problem. For more information on genetic algorithms a comprehensive book is available by Goldberg (1989).

27.6.3 Case-based Reasoning

Case-based reasoning is a popular method with origins in the cognitive science community. Case-based reasoning methods do not measure similarity between cases numerically. Instead they form a model in memory of the relationships based on examples. These relationships may either be induced or supplied by expert users. New examples are presented and their relationship to the ones in the memory is identified. The underlying idea with this method is that humans try to recall past cases when solving new problems. Thus case-based reasoning is considered a plausible model for knowledge discovery (Bareiss and Porter 1988). Significant examples include CYRUS (Kolodner 1984), UNIMEM (Lebowitz 1987) and PROTOS (Bareiss and Porter 1988). The first two assume generalization hierarchies from examples while the third maintains a complex set of user-supplied associations continually refined as new examples are added. While case-based reasoning approaches have not been extended to GDM, it would be interesting to investigate possible applications in the future.

27.6.4 Neural Networks and Fuzzy Logic

Neural networks and fuzzy systems have been successfully implemented in information analysis. They both offer certain advantages: neural networks incorporate learning capabilities in their process; fuzzy inference systems provide a structured knowledge representation and are easy to interpret and analyze.

Artificial neural networks try to mimic the organization of the human brain. Initially, research in this area was driven by neurobiological interests. Modern interest from the data mining perspective considers the development of architectures and learning algorithms that are applied in information processing tasks. In the literature, a large amount of information exists on the network types, learning methodologies and applications of neural networks. Good starting books would be Haykin (1994) and Bishop (1995). Here a brief description is provided of the most popular architecture, the feed-forward multi-layer. Neural networks consist of a number of independent, simple processors called neurons (nodes). Neurons are linked to each other through weighted connections called synaptic weights. After the network design is chosen a dataset is presented to the network and the learning process begins. The goal is to optimize the network behavior (fit) by adjusting the weights appropriately so the network output is close to the provided output. In other words the network creates a mapping of the input data to the desired output using the presented examples. The most important advantage of neural networks is that they are universal approximators, meaning they have the ability to approximate any arbitrary function (Haykin 1994). They do not need a mathematical model describing the problem and no prior knowledge is necessary. On the other hand neural networks are often considered as black boxes due to the difficulty in interpreting their internal structure; for example, it is difficult to identify specific rules within a neural network. The learning process can take a long time and the success is not guaranteed.

Fuzzy set theory (Zadeh 1965) has become a popular method for dealing with complexity, uncertainty and imprecision in various systems. Fuzzy sets may be represented by a mathematical formulation known as the membership function. This function gives a degree of membership within a fuzzy set. Interpretations of membership degrees include similarity, preference and uncertainty (Dubois et al. 1996). They show how similar an object is to a prototype or they model uncertainty by using imprecise terms. There is extensive literature on fuzzy sets applications, such as visual retrieval systems (Santini and Jain 1999) and fuzzy integrals for similarity approximation (Ishii and Wang 1998), both closely related to GDM.

The discussion above on neural networks and fuzzy methods leads to an intuitive combination of the approaches. The fuzzy system can be used to represent knowledge and the neural network techniques can provide a learning mechanism to determine membership values. The drawbacks of both individual methods, the black box behavior and the non-adaptable membership functions, could be addressed. The combination constitutes an interpretable model, with learning and prior knowledge incorporation capabilities (Klose et al. 2001). Therefore, neuro-fuzzy methods are especially suited for applications where user interaction in model design or interpretation is desired. Further discussion and an overview of current approaches can be found in Lin and Lee (1996), Nauck et al. (1997), Klose et al. (2001), Liu and Miyamoto (2000) and Tettamanzi and Tomassini (2001).

27.7 Summary

In this chapter we presented an overview of data mining and identified possible requirements to handle spatial and spatio-temporal data. As geographic information continues to increase so will the need to convert raw data into knowledge through intelligent models. The unique characteristics and challenges presented by geospatial data suggest that direct transition of traditional data mining methods to the geospatial case is not straightforward. However, design of new models and knowledge validation procedures are beginning to emerge. The potential scientific and economic benefits are already positioning geographic data mining as an important research area.

References

Anselin, L. 1995. Local indicators of spatial association. *Geographical Analysis* 27:93–115.

Bareiss, E. R. and B. W. Porter. 1988. PROTOS: an exemplar-based learning apprentice. *International Journal of Man-Machine Studies* 29:549–561.

Bayes, T. 1763. An essay towards solving a problem in the doctrine of chances. *Philosophical Transactions of the Royal Society of London* 53:370–418.

Bishop, C. M. 1995. *Neural Networks for Pattern Recognition*. Oxford, UK: Oxford University Press.

Brunsdon, C., A. S. Fotheringham and M. E. Charlton. 1996. Geographically weighted regression: A method for exploring spatial nonstationarity. *Geographical Analysis* 28:281–298.

Buttenfield, B., M. Gahegan, H. Miller and M. Yuan. 2000. Geospatial Data Mining and Knowledge Discovery. UCGIS White Paper on Emergent Research Themes. Alexandria, Va.: University Consortium for Geographic Information Science. <http://www.ucgis.org/priorities/research/research_white/2000%20Papers/emerging/gkd.pdf> Accessed 7 July 2007.

Cendrowska, J. 1987. PRISM: an algorithm for inducing modular rules. *International Journal of Man-Machine Studies* 27:349–370.

Chawla S., S. Shekhar, W. L. Wu and U. Ozesmi. 2001. Modeling spatial dependencies for mining geospatial data: An introduction. In *Geographic Data Mining and Knowledge Discovery*, edited by H. J. Miller and J. Han, 131–159. London: Taylor and Francis.

Codd, E. F. 1970. A relational model of data for large shared data banks. *Communications of ACM* 13(6):377–387.

Dempster, A., N. Laird and D. Rubin. 1977. Maximum likelihood from incomplete data via the EM algorithm. *Journal of the Royal Statistical Society* 39:1–38.

Dubois, D., H. Prade and R. R. Yager. 1996. Information engineering and fuzzy logic. Pages 1525–1531 in *5th IEEE International Conference on Fuzzy Systems, Volume 3*, held New Orleans, Louisiana, 8-11 September 1996. New York: IEEE.

Dunham, M. H.. 2002. *Data Mining: Introductory and Advanced Topics*. Upper Saddle River, New Jersey: Prentice Hall.

Egenhofer, M. J. and J. R. Herring. 1994. Categorizing binary topological relations between regions, lines and points in geographic databases. In *The 9-intersection: Formalism and Its Use for Natural-language Spatial Predicates*, edited by M. Egenhofer, D. M. Mark and J. R. Herring, 1–28. NCGIA Technical Report 94-1. Santa Barbara, Calif.: National Center for Geographic Information and Analysis.

Egenhofer, M. J. and K. Hornsby. 2000. Identity-based change: a foundation for spatio-temporal knowledge representation. *International Journal of Geographical Information Science* 14:207–224.

Elder, J. and D. Pregibon. 1996. A statistical perspective on KDD. In *Advances in Knowledge Discovery and Data Mining*, edited by U. Fayyad, G. Piatetsky-Shapiro, P. Smyth and R. Uthurusamy, 83-116. Cambridge, Mass.: AAAI/MIT Press.

Ester, M., H.-P. Kriegel and J. Sander. 1997. Spatial data mining: A database approach. Pages 47–66 in *Proceedings Advances in Spatial Databases, 5th International Symposium, SSD'97* held in Berlin, Germany, 15-18 July 1997. Edited by M. Scholl and A. Voisard. Lecture Notes in Computer Science 1262. Berlin: Springer.

Fayyad, U. M., G. Piatetsky-Shapiro and P. Smyth. 1996. From data mining to knowledge discovery: An overview. In *Advances in Knowledge Discovery and Data Mining*, edited by U. M. Fayyad, G. Piatetsky-Shapiro, P. Smyth and R. Uthurusamy, 1–34. Cambridge, Mass.: AAAI Press/MIT Press.

Feigenbaum, E. and J. Feldman. 1963. *Computers and Thought*. New York: McGraw-Hill.

Fisher, R. A. 1921. On the probable error of a coefficient of correlation deduced from a small sample. *Metron International Journal of Statistics* 1(4):3–32.

Fix, E. and J. L. Hodges. 1951. Discriminating analysis: non-parametric distribution. Technical Report 21-49-004(4). Randolph Field, Texas: USAF School of Aviation Medicine.

Florek, K., J. Lukaszewicz, J. Perka, H. Steinhaus and S. Zubrzycki. 1951. Taksonomia wroclawska. *Przeglad Antropologiczny* 17(4):93–207.

Fotheringham, A. S., M. Charlton and C. Brunsdon. 1997. Two techniques for exploring non-stationarity in geographical data. *Geographical Systems* 4:59–82.

Gahegan, M. 2001. Data mining and knowledge discovery in the geographical domain. White Paper submitted to National Academies Computer Science and Telecommunications Board: Intersection of Geospatial Information and Information Technology. <http://www7.nationalacademies.org/cstb/wp_geo_gahegan.pdf> Accessed 6 July 2007.

Getis, A. and J. K. Ord. 1992. The analysis of spatial association by use of distance statistics. *Geographical Analysis* 24:189–206.

———. 1996. Local spatial statistics: An overview. In *Spatial Analysis: Modelling in a GIS Environment*, edited by P. Longley and M. Batty, 261–277. Cambridge, UK: GeoInformation International.

Goldberg, D. E. 1989. *Genetic Algorithms in Search, Optimization, and Machine Learning*. Reading, Mass.: Addison-Wesley.

Goodchild, M. F. 1992. Geographical data modeling. *Computers and Geosciences* 18(4):401–408.

Gunopulos, D. 2001. Data mining techniques for geospatial applications. White Paper submitted to National Academies Computer Science and Telecommunications Board. <http://www7.nationalacademies.org/cstb/wp_geo_gunopulos.pdf > Accessed 6 July 2007.

Han, J. and M. Kamber. 2001. *Data Mining: Concepts and Techniques*. San Francisco, Calif.: Morgan Kaufmann.

Hand, D. J. 1981. *Discrimination and Classification*. New York: Wiley.

Hastie, T., R. Tibshirani and J. Friedman. 2001. *The Elements of Statistical Learning: Data Mining, Inference, and Prediction*. New York: Springer.

Haykin, S. 1994. *Neural Networks*. Upper Saddle River, New Jersey: Prentice Hall.

Holland, J. H. 1975. *Adaptation in Natural and Artificial Systems*. Ann Arbor: University of Michigan Press.

Hunt, E. B., J. Marin and P. Stone. 1966. *Experiments in Induction*. New York: Academic Press.

Ishii, N. and Y. Wang. 1998. Learning feature weights for similarity measures using genetic algorithms. Pages 27–33 in *IEEE Intl Joint Symposia on Intelligence and Systems*, held in Rockville, Maryland, 21-23 May 1998. Los Alamitos, Calif.: IEEE Computer Society.

Kantardzic, M. 2002. *Data Mining: Concepts, Models, Methods, and Algorithms*. New York: Wiley-IEEE Press.

Klose, A., A. Nürnberger, D. Nauck and R. Kruse. 2001. Data Mining with Neuro-Fuzzy Models. In *Data Mining and Computational Intelligence*, edited by A. Kandel, H. Bunke, and M. Last, 1–36. Berlin: Physica-Verlag.

Kohonen, T. 1982. Self-organized formation of topologically correct feature maps. *Biological Cybernetics* 43:59–69.

Kolodner, J. L. 1984. *Retrieval and Organizational Strategies in Conceptual Memory: A Computer Model*. Hillsdale, N.J.: Lawrence Erlbaum Associates.

Lebowitz, M. 1987. Experiments with universal concept formation: UNIMEM. *Machine Learning* 2:103–138.

Lin, C. T. and C. C. Lee. 1996. *Neural Fuzzy Systems. A Neuro-Fuzzy Synergism to Intelligent Systems*. Upper Saddle River, New Jersey: Prentice Hall.

Liu, Z. Q. and S. Miyamoto. 2000. *Soft Computing and Human-Centered Machines*. Tokyo: Springer.

Lu, W., J. Han and B. C. Ooi. 1993. Knowledge discovery in large spatial databases. Pages 275–289 in *Proc. Far East Workshop Geographic Information Systems* (FEGIS'93), held in Singapore, June 1993.

McCulloch, W. S. and W. Pitts. 1943. A logical calculus of the ideas immanent in nervous activity. *Bulletin of Mathematical Biophysics* 5:115–133.

McLachlan, G. J. 1992. *Discriminant Analysis and Statistical Pattern Recognition*. New York: John Wiley & Sons.

Miller, H. J. and Han, J., eds. 2001. *Geographic Data Mining and Knowledge Discovery*. London: Taylor and Francis.

Moore, A. 2006. Statistical Data Mining Tutorials. <http://www.autonlab.org/tutorials/> Accessed 6 July 2007.

Mountrakis, G., P. Agouris and A. Stefanidis. 2004. Similarity learning in GIS: An overview of definitions, prerequisites and challenges. In *Spatial Databases: Technologies, Techniques and Trends*, edited by M. Vassilakopoulos, A. Papadopoulos and Y. Manolopoulos, 294–321. Hershey, Penn.: Idea Group Inc.

NASA. 2004. Intelligent Data Understanding – Research Task List. <http://is.arc.nasa.gov/IDU/tasks.html> Accessed 6 July 2007.

Nauck, D., F. Klawonn, and R. Kruse. 1997. *Foundations of Neuro-Fuzzy Systems*. Chichester, UK: Wiley.

Nilsson, N. J. 1965. *Learning Machines: Foundations of Trainable Pattern-Classifying Systems*. New York: McGraw-Hill.

Okabe, A. and H. J. Miller. 1996. Exact computational methods for calculating distances between objects in a cartographic database. *Cartography and Geographic Information Systems* 23:180–195.

Peuquet, D. J. and C.-X. Zhang. 1987. An algorithm to determine the directional relationship between arbitrarily-shaped polygons in the plane. *Pattern Recognition* 20:65–74.

Quinlan, J. R.. 1986. Induction of decision trees. *Machine Learning* 1:81–106.

———. 1993. *C4.5: Program for Machine Learning*. San Mateo, Calif.: Morgan Kaufmann.

Roddick, J. F. and B. Lees. 2001. Paradigms for spatial and spatio-temporal data mining. In Geographic Data Mining and Knowledge Discovery, edited by H. J. Miller and J. Han, 33–50. London: Taylor and Francis.

Roddick, J., K. Hornsby, and M. Spiliopoulou. 2001. An updated bibliography of temporal, spatial and spatio-temporal data mining research. Pages 147–163 in *Proceedings of the First International Workshop on Temporal, Spatial, and Spatio-temporal Data Mining*, TSDM2000, held in Lyon, France, 12 September 2000. Edited by J. Roddick and K. Hornsby. Lecture Notes in Artificial Intelligence Volume 2007. Berlin: Springer.

Rodman, L. C., J. Jackson, R. Huizar III and R. K. Meentemeyer. 2006. An association rule discovery system for geographic data. Pages 3478–3481 in *Proceedings of the 2006 IEEE International Geoscience and Remote Sensing Symposium*, IGARSS 2006, held in Denver, Colo., Jul. 31- Aug. 4, 2006. New York: IEEE Xplore.

Rosenblatt, F. 1958. The Perceptron: A probabilistic model for information storage and organization in the brain. *Psychological Review* 65:386–408.

Santini, S. and R. Jain. 1999. Similarity measures. *IEEE Transactions on Pattern Analysis and Machine Learning* 21(9):871–883.

Shafer, J. C., R. Agrawal and M. Mehta. 1996. SPRINT: A scalable parallel classifier for data mining. Pages 544–555 in *Very Large Data Bases*, Proc. of the 22th Int'l Conference on Very Large Data Bases, held in Mumbai (Bombay), India, Sept. 3-6 1996. Edited by T. M. Vijayaraman, A. P. Buchmann, C. Mohan and N. L. Sarda. San Francisco: Morgan Kaufmann.

Shekhar, S. and S. Chawla. 2003. *Spatial Databases: A Tour*. Upper Saddle River, New Jersey: Prentice Hall.

Tettamanzi, A. and M. Tomassini. 2001. *Soft Computing: Integrating Evolutionary, Neural, and Fuzzy Systems*. Heidelberg: Springer-Verlag.

TKK (Helsinki University of Technology). 2005. Spatial Data Mining - Literature. Laboratory of Geoinformation and Positioning Technology. <http://www.tkk.fi/Units/Cartography/research/sdm/literature.html> Accessed 6 July 2007.

Weiss, S. M. and C. A. Kulikowshi. 1991. *Computer Systems that Learn: Classification and Prediction Methods from Statistics, Neural Nets, Machine Learning and Expert Systems*. San Mateo, Calif.: Morgan Kaufmann.

Zadeh, L. A. 1965. Fuzzy sets. *Information and Control* 8:338–353.

CHAPTER 28
Geostatistics and GIS

Peter M. Atkinson and *Christopher D. Lloyd*

28.1 Introduction

Geostatistics is a set of techniques for handling spatial data based on the random function (RF) model (Journel and Huijbregts 1978; Chilès and Delfiner 1999). Geostatistical techniques include the spatial prediction tool known as kriging, spatial simulation, regularization and spatial optimization. The dependence of these techniques on the RF model characterizes them as geostatistics. Short introductions to geostatistics are provided by Oliver and Webster (1990) and Burrough and McDonnell (1998).

A RF $Z(\mathbf{x})$ is a random variable (RV) Z that varies as a function of location $\mathbf{x}$. Hence the term random *function*. A RV is just a stochastic or random process. A simple example of a discrete RV is rolling a die; clearly a stochastic process. The die has six possible outcomes: a number between 1 and 6. In combination, these possible outcomes define the distribution function of the die. Each number between 1 and 6 has an equal chance of being rolled; i.e., a probability of 1 in 6. A roll of the die leads to a particular outcome, called a realization. For continuous variables the discrete definition of the distribution function is replaced with a continuous function: either the probability density function (pdf) or cumulative distribution function (cdf). The cdf defines the probability of the outcome being less than a selected value (Goovaerts 1997). The cdf may be estimated by fitting a model to an empirical distribution. Each candidate model is defined mathematically by an equation involving a few parameters (coefficients to be estimated). For example, the Gaussian model involves two parameters; the mean and variance. See Isaaks and Srivastava (1989) for a discussion of RVs in a geostatistical context.

Now let us consider again the RF model in which a RV is allowed to vary as a function of $\mathbf{x}$. The question is how should the *function* be defined? In the simplest case, every position $\mathbf{x}$ in space has its own cdf and each is independent of any other cdf. While simple to understand, this model requires a huge number of parameters; one set for each possible location. Further, it does not accord with common sense; experience tells us that places close together are likely to have similar characteristics. For these reasons, we place some restrictions on the RF model. The most common set of restrictions are referred to as stationarity constraints, meaning that certain parameters are held constant from place to place. In the strictest sense, the mean and variance parameters can be held constant for all locations $\mathbf{x}$. However, fixing both parameters turns out to be unhelpful too because in this RF model each point is independent and identically distributed (iid), meaning that spatial inference is severely limited.

In geostatistics, it is common to define a stationary mean parameter. Various alternatives are commonly applied in which the mean is allowed to vary in specific ways (see Goovaerts 1997, and Section 28.3.2 below). For our purposes, the stationary mean provides a basic starting point. A second restriction, which is central to geostatistics, is to replace the stationary variance with a stationary spatial covariance function (second-order stationarity) or variogram (intrinsic stationarity, a weaker form of stationarity). Although the spatial covariance is often used in geostatistical theory, the variogram is by far the most commonly used in practice and so we refer to it from here on. The variogram defines the relations between points and, thus, facilitates spatial statistical inference.

The mean and variogram are, thus, the parameters of the RF model that need to be estimated. They replace the mean and variance of the RV model. It is possible to estimate these parameters, in particular the variogram, by first estimating an empirical variogram from data and then fitting a mathematical model to it. Various important considerations must be taken into account during this process (see McBratney and Webster 1986) and some of these are discussed below. Once the parameters are estimated (either with or without the uncertainty of estimation accounted for) the RF is defined and geostatistical operations can proceed.

Since geostatistics is founded on the RF model, then immediately a constraint is placed on what geostatistics can and cannot do. Let us draw a distinction first between the RF and object-based views of the world. Here the term RF is used to describe, in general, a stochastic model of continuous variation over a continuous geographical space, acknowledging that in geostatistics we are interested in that part of the set in which a spatial covariance or equivalent is defined. In a GIS context, these views are commonly associated with the raster data model (RF and less commonly object-based) and vector data model (object-based and less commonly RF) (Burrough and McDonnell 1998).

Geostatistics does not apply to objects and, thus, is unlikely to find application to vector data. A separate, but related, body of statistics known as mathematical morphology (Serra 1982) has been developed to deal with objects. Within the RF view, alternative classes of models may be defined. Geostatistics is concerned with a RF in which a spatial covariance or variogram is used to define the relations between Z at different locations in space. Alternative classes of RF model are the spatially autoregressive process (SAP) and Markov RF (MRF) which are popular in statistics and geography as tools for spatial regression and classification. Although a spatial covariance emerges implicitly from the recursive fitting of a SAP, the spatial covariance is not defined *a priori* and hence the SAP falls outside the traditional field of geostatistics.

Geostatistical operations include spatial prediction, spatial simulation, regularization and spatial optimization. Let us first consider spatial prediction or kriging. The objective is to predict the value $z(\mathbf{x}_0)$ of some unobserved location $\mathbf{x}_0$ given a sample of data $z(\mathbf{x}_i)$, $i = 1,2,\ldots,n$ usually defined on points or supports (the space on which each observation is defined) that can be treated as quasi-points. The RF model helps because to predict $z(\mathbf{x}_0)$ sensibly some understanding or model of the underlying processes or form is required. Since in environmental science (in the broadest sense) process knowledge is usually severely limited the RF model provides a plausible alternative. It does not mean that we believe that environmental phenomena are generated stochastically. Rather it represents our limited understanding; it is the best that we can do in the circumstances.

The RF model is useful because—given a spatial sample of data—the relations between the data $z(\mathbf{x}_i), i = 1,2,\ldots,n$ and the value to be predicted $z(\mathbf{x}_0)$ are estimated by the variogram. Generally, the closer $z(\mathbf{x}_0)$ is to a given datum the more similar the two values are likely to be, and the variogram quantifies this spatial dependence. Moreover, in geostatistics the relations between the sample data themselves are accounted for so that a cluster of data points will contribute less to the prediction than a dispersed set (Journel and Huijbregts 1978). This means that the cdf of the predicted value (i.e., the set of possible values from which we must draw one realization) can be conditioned on the sample data. In particular, the variance of the conditional cdf (ccdf) is likely to be less than that of the original cdf. In general terms, this means that the range of possible values for the unknown value is restricted to be close to the neighboring data by an amount determined by the spatial proximity of the prediction to the neighbors.

In kriging, the most likely value is selected from the ccdf, that is, the one that has the greatest probability. However, if the most likely value is selected for all locations for which predictions are required then the resulting map will be smoothed, meaning that the variance in the predicted set will be less than in the original set of data. Smoothing means that the predicted map cannot be correct by definition; the spatial character of the map is guaranteed to be different to that of the original data. Further, smoothing can have serious consequences where the objective is to compare the predicted variable with other data layers (as might be the case within a GIS). An alternative, called conditional simulation, is to draw a value stochastically from the ccdf in such a way that the original variogram (i.e., representing the spatial character of the variable) is retained in the predicted map. The RF model allows the definition of the ccdf required for this operation. While only half as precise as kriging on a point by point basis, conditional simulation does lead to a spatial distribution that might exist in reality (a "possible reality").

Other geostatistical operations include regularization (changing the support on which the variogram is defined) (Atkinson and Tate 2000) and techniques for spatial optimization such as spatial simulated annealing (van Groenigen 1999). All geostatistical operations make use of the RF model for statistical inference. The next section describes the process of fitting the RF model parameterized by a spatial covariance or variogram, while Section 28.3 describes in detail the basic geostatistical operations. Section 28.4 considers more advanced operations and Section 28.5 discusses the outlook for the adoption of geostatistics by the GIS community.

28.2 Characterizing Spatial Variation

28.2.1 Estimating the Experimental Variogram

Much of the effort and time associated with geostatistical analysis is expended in analysis of the spatial structure of a variable. One simple way of examining spatial structure is through computing the variogram cloud. The variogram cloud is a plot of the semivariances for paired data against the distances separating the paired data points in a given direction. The semivariance is estimated as half the squared difference between values at two locations. The variogram cloud shows how dissimilar paired data points are as a function of their separation distance and direction (termed spatial lag, **h**). If data are spatially structured then pairs separated by small lags will tend to be less dissimilar than pairs separated by large lags.

A core idea in geostatistics is that the spatial structure in a variable should be characterized and used for spatial prediction and simulation. The objective of geostatistical prediction is to find weights to assign to observations located around the prediction location. If information is available on how dissimilar two observations are likely to be for a given lag then this information can be used to assign these weights. The most commonly used approach utilizes the estimated variogram. The experimental variogram is estimated by calculating the squared differences between all the available paired observations and obtaining half the average for all observations separated by a given lag (or within a lag tolerance where the observations are not on a regular grid). So, while the variogram cloud provides semivariances as a function of a set of actual lags the experimental variogram provides only a set of average semivariances at a set of discrete lags. Examination of the variogram cloud provides a means of identifying heterogeneities in spatial variation of a variable (Webster and Oliver 2000) that is obscured through the summation over lags that occurs with the experimental variogram. Therefore, examination of the variogram cloud is a sensible step prior to estimation of the experimental variogram.

Figure 28-1 gives a simple example of a transect along which observations have been made at regular intervals. Lags (**h**) of 1 and 2 units are indicated. So in this case, half the average squared difference between observations separated by a lag of 1 unit is calculated and the process is repeated for a lag of 2 units and so on. The variogram can be estimated for different directions to enable the identification of directional variation (termed anisotropy).

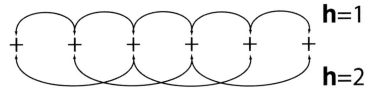

Figure 28-1 Observations (+) made along a transect, with lag (h) of 1 and 2 units indicated.

The experimental variogram for lag **h**, $\hat{\gamma}(\mathbf{h})$, can be estimated from $p(\mathbf{h})$ (the number of paired observations for lag **h**), $z(\mathbf{x}_\alpha)$, $z(\mathbf{x}_\alpha + \mathbf{h})$, $\alpha = 1,2,\dots p(\mathbf{h})$ using:

$$\hat{\gamma}(\mathbf{h}) = \frac{1}{2p(\mathbf{h})} \sum_{\alpha=1}^{p(\mathbf{h})} \left\{ z(\mathbf{x}_\alpha) - z(\mathbf{x}_\alpha + \mathbf{h}) \right\}^2 \qquad (28\text{-}1)$$

In the situation where a variable is preferentially sampled in areas with large or small values of the property of interest, the histogram will be unrepresentative and often a declustering algorithm is used in estimation of the histogram. For example, values in areas or cells with more data may be given smaller weights than values in sparsely sampled areas (Deutsch and Journel 1998). Preferential sampling of a variable also impacts on the form of the experimental variogram. Richmond (2002) shows that clustering can, in some cases, alter drastically the form of the variogram. Two methods of declustering for weighting paired data in estimation of the experimental variogram are given by Richmond (2002).

In the presence of large-scale, low-frequency variation (e.g., that would be fitted well by a trend model), the form of the variogram will be affected. If the variogram increases more rapidly than a quadratic polynomial for large lags then a RF which is non-stationary in the mean should be adopted (Armstrong 1998). This topic is explored in greater depth in Section 28.3.2.

28.2.2 Fitting a Variogram Model

A mathematical model may be fitted to the experimental variogram and the coefficients of this model can be used for a range of geostatistical operations such as spatial prediction (kriging) and conditional simulation (defined below). A model is usually selected from one of a set of so-called authorized models. McBratney and Webster (1986) provide a review of some of the most widely used authorized models. There are two principal classes of variogram model. Transitive (bounded) models have a sill (finite variance), and indicate a second order stationary process. Unbounded models do not reach an upper bound; they are intrinsically stationary only (McBratney and Webster 1986). Figure 28-2 shows the parameters of a bounded variogram model (the spherical model as defined below). The nugget effect, c_0, represents unresolved variation (a mixture of spatial variation at a finer scale than the sample spacing and measurement error). The sill, sometime referred to as the structured component, c, represents the spatially correlated variation. The total sill, $c_0 + c$, is the *a priori* variance. The range, a, represents the scale of spatial variation. For example, if a measured property varies markedly over quite small distances then the property can be said to exhibit short-range spatial variation while if the measured property is quite similar over much of the region and varies markedly only at the extremes of the region (that is, at large separation distances) then the property can be said to exhibit long-range spatial variation.

Some of the most commonly used authorized models are detailed below. The nugget effect model, defined above, is given by:

$$\gamma(h) = \begin{cases} 0 & \text{for } h = 0 \\ c_0 & \text{for } |h| > 0 \end{cases} \qquad (28\text{-}2)$$

Three of the most frequently used bounded models are the spherical model, the exponential model and the Gaussian model and these are defined in turn. The exponential model is given by:

$$\gamma(h) = c \cdot \left[1 - \exp\left(-\frac{h}{d} \right) \right]. \qquad (28\text{-}3)$$

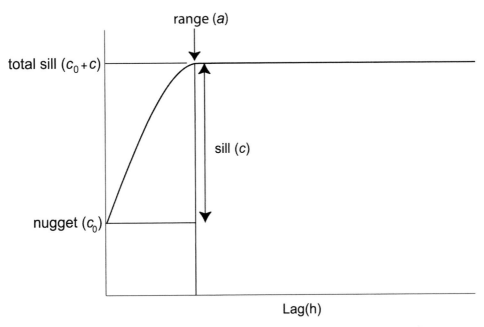

Figure 28-2 The components of a bounded variogram model with a nugget effect.

where c is the sill of the exponential model and d is the non-linear distance parameter. The exponential model reaches the sill asymptotically and the practical range is $3d$ (i.e., the separation at which approximately 95% of the sill is reached).

$$\gamma(h) = \begin{cases} c \cdot [1.5\frac{h}{a} - 0.5(\frac{h}{a})^3] & \text{if } h \leq a \\ c & \text{if } h > a \end{cases} \qquad (28\text{-}4)$$

where a is the non-linear parameter known as the range.

The Gaussian model is given by:

$$\gamma(h) = c \cdot \left[1 - \exp\left(-\frac{h^2}{d^2} \right) \right]. \qquad (28\text{-}5)$$

The Gaussian model does not reach a sill at a finite distance and the practical range is $d\sqrt{3}$ (Journel and Huijbregts 1978). Variograms with parabolic behavior at the origin, as represented by the Gaussian model here, are indicative of very regular spatial variation (Journel and Huijbregts 1978). Authorized models may be used in positive linear combination where a single model is insufficient to represent well the form of the variogram.

Where no sill is reached and the variogram model is unbounded the most widely used model is the power model:

$$\gamma(h) = m \cdot h^{\omega} \qquad (28\text{-}6)$$

where ω is a power $0<\omega<2$ and m is a positive slope (Deutsch and Journel 1998). The linear model is a special case of the power model.

Manual of Geographic Information Systems

One of the advantages of kriging is that it is often fairly straightforward to model aniso-tropic structure using the variogram. Two primary forms of anisotropy have been outlined in the geostatistical literature. If the sills for all directions are not significantly different and the same structural components (for example, spherical or Gaussian) are used then anisotropy can be accounted for by a linear transformation of the co-ordinates: this is called geometric or affine anisotropy (Webster and Oliver 1990). Where the sill changes with direction but the range is similar for all directions the anisotropy is called zonal (Isaaks and Srivastava 1989). However, the modeling of zonal anisotropy is much more problematic than the modeling of geometric anisotropy. In practice, a mixture of geometric and zonal anisotropy has been found to be common (Isaaks and Srivastava 1989).

There is a variety of different approaches to the fitting of models to variograms. Some geostatisticians prefer fitting variogram models 'by eye' on the grounds that it enables one to use personal experience and to account for features or variation that may be difficult to quantify (Christakos 1984; Journel and Huijbregts 1978). Weighted least squares (WLS) has been proposed as a suitable means of fitting models to variograms (Cressie 1985; Pardo-Igúzquiza 1999) and the approach has been used by many geostatisticians. The technique is preferred to unweighted ordinary least squares (OLS) because in WLS the weights can be made proportional to the number of pairs at each lag (Cressie 1985). Thus, lags with many pairs have greater influence in the fitting of a model. The use of generalized least squares (GLS) has also been demonstrated in a geostatistical context (Cressie 1985; McBratney and Webster 1986). Use of Maximum Likelihood (ML) estimation (McBratney and Webster 1986) has become widespread amongst geostatisticians and has been used for WLS. The success (or goodness) of the fit of models to the variogram, and of the relative improvement or otherwise in using different numbers of parameters, may be compared through the exami-nation of the sum of squares of the residuals or through the use of the Akaike Information Criterion (McBratney and Webster 1986; Webster and McBratney 1989).

Figure 28-3 shows an experimental variogram estimated from precipitation data acquired in Great Britain in January 1999. The data are described by Lloyd (2002). The variogram was fitted with a nugget and two spherical components and the values of the coefficients are indicated on the figure (where nug. indicates nugget and sph. indicates spherical). Autho-rized models are often used in combination in this way to model nested spatial structures. In Figure 28-4, the directional variogram, estimated from the same data, is shown. It indicates that the scale of spatial variation is similar in all directions while the magnitude of the variation (the semivariance) is clearly different for different directions.

28.2.3 Non-stationary Variograms

In cases where the variogram does not represent well the spatial variation across the whole of the region of interest, some approach may be necessary to account for the change in spatial variation locally. In the geostatistical literature, several approaches are presented for estimation of non-stationary variograms. These vary from approaches that estimate and model automatically the variogram in a moving window to approaches that transform the data so that the transformed data have a stationary variogram. Reviews of some methods are provided by Sampson et al. (2001) and Schabenberger and Gotway (2005). Estimation of the variogram where the mean is modeled as nonstationary is discussed in the following section.

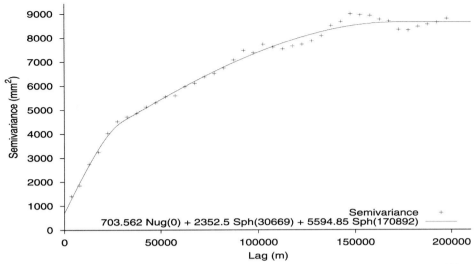

Figure 28-3 Omnidirectional variogram estimated from precipitation data for January 1999. Based on Lloyd 2002.

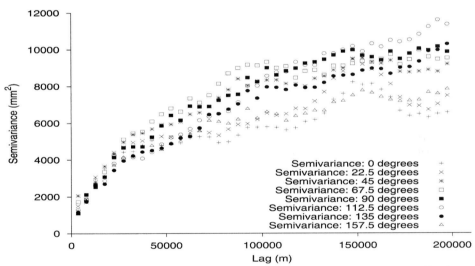

Figure 28-4 Directional variogram estimated from precipitation data for January 1999. Directions are decimal degrees clockwise from North. Based on Lloyd 2002.

28.3 Spatial Prediction

28.3.1 Simple and Ordinary Kriging

There are many varieties of kriging. Its simplest form is called simple kriging (SK). To use SK it is necessary to know the mean of the property of interest and this must be modeled as constant across the region of interest. In practice, this model is rarely suitable. The most widely used variant of kriging, ordinary kriging (OK), allows the mean to vary spatially: the mean is modeled as constant within each prediction neighborhood only. For each point to be predicted a new neighborhood is defined and so effectively the mean is allowed to vary locally.

OK predictions are weighted averages of the n available data. The OK weights define the Best Linear Unbiased Predictor (BLUP). The OK prediction, $\hat{z}_{OK}(\mathbf{x}_0)$, is defined as:

$$\hat{z}_{OK}(\mathbf{x}_0) = \sum_{\alpha=1}^{n} \lambda_{\alpha}^{OK} z(\mathbf{x}_{\alpha}) \tag{28-7}$$

with the constraint that the weights, λ_{α}^{OK}, sum to 1 to ensure an unbiased prediction:

$$\sum_{\alpha=1}^{n} \lambda_{\alpha}^{OK} = 1 \tag{28-8}$$

So, the objective of the kriging system is to find appropriate weights by which the available observations will be multiplied before summing them to obtain the predicted value. These weights are determined using the coefficients of a model fitted to the variogram (or another function such as the covariance function).

The kriging prediction error must have an expected value of 0:

$$E\{\hat{Z}_{OK}(\mathbf{x}_0) - Z(\mathbf{x}_0)\} = 0 \tag{28-9}$$

The kriging (or prediction) variance, $\hat{\sigma}_{OK}$, is expressed as:

$$\begin{aligned}
\hat{\sigma}_{OK}^2(\mathbf{x}_0) &= E[\{\hat{Z}_{OK}(\mathbf{x}_0) - Z(\mathbf{x}_0)\}^2] \\
&= 2\sum_{\alpha=1}^{n} \lambda_{\alpha}^{OK} \gamma(\mathbf{x}_{\alpha} - \mathbf{x}_0) - \sum_{\alpha=1}^{n}\sum_{\beta=1}^{n} \lambda_{\alpha}^{OK} \lambda_{\beta}^{OK} \gamma(\mathbf{x}_{\alpha} - \mathbf{x}_{\beta})
\end{aligned} \tag{28-10}$$

That is, we seek the values of $\lambda_1, \ldots, \lambda_n$ (the weights) that minimize this expression with the constraint that the weights sum to one (Equation 28-8). This minimization is achieved through Lagrange multipliers. The conditions for the minimization are given by the OK system comprising $n + 1$ equations and $n + 1$ unknowns:

$$\begin{cases}
\sum_{\beta=1}^{n} \lambda_{\beta}^{OK} \gamma(\mathbf{x}_{\alpha} - \mathbf{x}_{\beta}) + \psi_{OK} = \gamma(\mathbf{x}_{\alpha} - \mathbf{x}_0) & \alpha = 1, \ldots, n \\
\sum_{\beta=1}^{n} \lambda_{\beta}^{OK} = 1
\end{cases} \tag{28-11}$$

where ψ_{OK} is a Lagrange muliplier. Knowing ψ_{OK}, the prediction variance of OK can be given as:

$$\hat{\sigma}_{OK}^2 = \sum_{\alpha=1}^{n} \lambda_{\alpha}^{OK} \gamma(\mathbf{x}_{\alpha} - \mathbf{x}_0) + \psi_{OK} \qquad (28\text{-}12)$$

The kriging variance is a measure of confidence in predictions and is a function of the form of the variogram, the sample configuration and the sample support (Journel and Huijbregts 1978). The kriging variance is not conditional on the data values locally and this has led some researchers to use alternative approaches such as conditional simulation (discussed in the next section) to build models of spatial uncertainty (Goovaerts 1997).

There are two varieties of OK: punctual OK and block OK. With punctual OK the predictions cover the same area (of the support, $\mathbf{v}$) as the observations. In block OK, the predictions are made to a larger support than the observations. With punctual OK the data are honored. That is, they are retained in the output map. Block OK predictions are averages over areas (that is, the support has increased). Thus, at $\mathbf{x}_0$ the prediction is not the same as an observation and does not need to honor it.

The choice of variogram model affects the kriging weights and, therefore, the predictions. However, if the form of two models is similar at the origin of the variogram then the two sets of results may be similar (Armstrong 1998). The choice of nugget effect may have marked implications for both the predictions and the kriging variance. As the nugget effect is increased, the predictions become closer to the global average (Isaaks and Srivastava 1989).

A map of precipitation in Britain in January 1999 generated using OK is shown in Figure 28-5. It was generated using the variogram model given in Figure 28-3; the 16 nearest neighbors to each grid cell were used in the prediction process. The map is very smooth in appearance; this is a common feature of maps derived using OK. In Figure 28-6, the kriging standard error (square root of the kriging variance) is shown. It is notably large in, for example, south west Wales and parts of the Highlands of Scotland and the Western Isles since there are fewer rain gauges in those locations. One way in which such maps are sometimes used is to identify areas with kriging standard errors of greater than some fixed amount. Addition of observations at such locations is likely to be appropriate to increase the precision of maps derived from the data.

28.3.2 Fitting a Trend and Universal Kriging

OK is robust but, in some cases, an even more general form of kriging may be appropriate. In cases where the mean of the variable changes markedly over small distances, a non-stationary model of the mean may provide more accurate spatial predictions. While the mean varies from place to place with ordinary kriging it does not vary within the search window but several approaches exist which provide a non-stationary mean. The most widely used approach is called kriging with a trend model (KT; sometimes termed universal kriging). In KT, the mean is commonly modeled using a polynomial. The principal problem with KT is that the underlying trend-free variogram must be estimated yet the local trend (or drift) is estimated as a part of the KT procedure which itself requires the variogram. Various approaches for estimating the trend-free variogram are described in the literature. In such cases, the trend could be modeled as a low-order polynomial and analysis can proceed with the residuals from the trend. Another approach to estimating the trend-free variogram is discussed in the following section. An alternative approach is Intrinsic Random Functions of Order k kriging whereby the generalized covariance is used in place of the variogram (Chilès and Delfiner 1999).

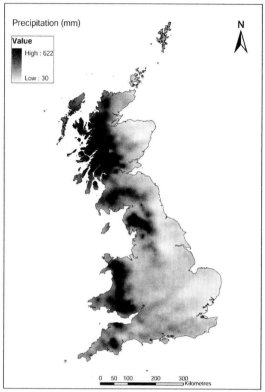

Figure 28-5 Map of precipitation in January 1999 generated using ordinary kriging. See included DVD for color version.

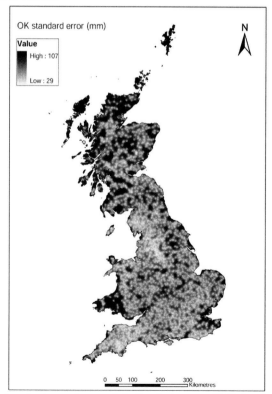

Figure 28-6 Kriging standard error for precipitation in January 1999. See included DVD for color version.

28.3.3 Cokriging, KED and SKlm

Where a secondary variable is available that is cross-correlated with the primary variable both variables may be used simultaneously in prediction using cokriging. To apply cokriging, the variograms (that is, autovariograms) of both variables and the cross-variogram are required. The operation of cokriging is based on the linear model of coregionalization (see Webster and Oliver 2000). For cokriging to be beneficial, the secondary variable should be cheaper to obtain or more readily available to make the most of the technique. If the variables are clearly linearly related then cokriging may estimate more accurately than, for example, OK.

There are various other widely used approaches that make use of secondary variables. If we have some variable that is linearly related to the primary variable and varies smoothly (i.e., there are no marked local changes in values) it could be used to inform spatial prediction of values of the primary variable. Two such approaches are described below.

Simple kriging is the most basic form of kriging. With SK, the mean is assumed to be constant (there is no systematic change in the mean of the property across the region of study) and known. If the mean is not constant, but we can estimate the mean at locations in the domain of interest, then this locally varying mean can be used to inform estimation. That is, the local mean can be estimated prior to kriging. The locally-varying mean (lm) can be estimated in various different ways. One approach is to use regression to estimate the value of the primary variable at (i) all observation locations and (ii) all locations where SKlm predictions will be made. The variogram is then estimated using the residuals from the regression predictions at the data locations. SKlm is conducted using the residuals and the trend is added back after the prediction process is complete.

An alternative approach is kriging with an external drift model (KED). In KED, the secondary data act as a shape function (the external trend) and the function describes the average shape of the primary variable (Wackernagel 2003). The local mean of the primary variable is derived as a part of the kriging procedure using the secondary information and SK is carried out on the residuals from the local mean. The approach differs from SKlm in that the local mean is estimated as part of the kriging procedure and not before it, as is the case with SKlm (Goovaerts 1997). Lloyd (2002) illustrates the use of KED in mapping monthly precipitation, with elevation used as the external trend.

As noted above, a major problem with KT and KED is that the underlying (trend-free) variogram is assumed known. That is, if the mean changes from place to place the variogram estimated from the raw data will be biased, so it is necessary to remove the local mean and estimate the variogram of the residuals. Since the trend (that is, local mean) is estimated as a part of the KED (and KT) system, which requires the variogram model coefficients as inputs, we are faced with a circular problem. A potential solution is to infer the trend-free variogram from paired data that are largely unaffected by any trend (Goovaerts 1997; Wackernagel 2003). Hudson and Wackernagel (1994), in an application concerned with mapping mean monthly temperature in Scotland, achieved this by estimating directional variograms and retaining the variogram for the direction that showed least evidence of trend. That is, temperature values systematically increase or decrease in one direction (there is a trend in the values), but values of temperature are more constant in the perpendicular direction. In such cases, the concern is to characterize spatial variation in the direction for which values of temperature are constant. Hudson and Wackernagel (1994) assumed that the trend-free variogram was isotropic and the variogram for the direction selected was used for kriging.

28.4 Further Geostatistical Operations

28.4.1 Conditional Simulation

Kriging predictions are weighted moving averages of the available sample data. Kriging is, therefore, a smoothing interpolator. Conditional simulation (also called stochastic imaging) is not subject to the smoothing associated with kriging (conceptually, the variation lost by kriging due to smoothing is added back) as predictions are drawn from equally probable joint realizations of the RVs which make up a RF model (Deutsch and Journel 1998). That is, simulated values are not the expected values (i.e., the mean) but are values drawn randomly from the ccdf: a function of the available observations and the modeled spatial variation (Dungan 1999). The simulation is considered "conditional" if the simulated values honor the observations at their locations (Deutsch and Journel 1998). As noted above, simulated realizations represent a possible reality whereas kriging does not. Simulation allows the generation of many different possible realizations that may be used as a guide to potential errors in the construction of a map (Journel 1996) and multiple realizations encapsulate the uncertainty in spatial prediction.

Arguably, the most widely used form of conditional simulation is sequential Gaussian simulation (SGS). With sequential simulation, simulated values are conditional on the original data and previously simulated values (Deutsch and Journel 1998). In SGS the ccdfs are all assumed to be Gaussian. The SGS algorithm follows several steps (Goovaerts 1997; Deutsch 2002) as detailed below:

1. Apply a standard normal transform to the data.
2. Go to the location $\mathbf{x}_1$.
3. Use SK (note OK is often used instead; see Deutsch and Journel 1998 about this issue), conditional on the original data, $z(\mathbf{x}_\alpha)$, to make a prediction. The SK prediction and the kriging variance are parameters (the mean and variance) of a Gaussian ccdf:

$$F(\mathbf{x}_1 : z | (n)) = \text{Prob}\{Z(\mathbf{x}_1) \le z | (n)\}$$

4. Using Monte Carlo simulation, draw a random residual, $z^l(\mathbf{x}_1)$, from the ccdf.
5. Add the SK prediction and the residual which gives the simulated value; the simulated value is added to the data set.
6. Visit all locations in random order and predict using SK conditional on the n original data and the i-1 values, $z^l(\mathbf{x}_i)$, simulated at the previously visited locations $\mathbf{x}_j$, j=1,…, i-1 to model the ccdf:

$$F(\mathbf{x}_i ; z | (n + i - 1)) = \text{Prob}\{Z(\mathbf{x}_i) \le z | (n + i - 1)\}$$

7. Follow the procedure in steps 4 and 5 until all locations have been visited.
8. Back transform the data values and simulated values.

By using different random number seeds the order of visiting locations is varied and, therefore, multiple realizations can be obtained. In other words, since the simulated values are added to the data set, the values available for use in simulation are partly dependent on the locations at which simulations have already been made and, because of this, the values simulated at any one location vary as the available data vary. SGS is discussed in detail in several texts (for example, Goovaerts 1997; Deutsch and Journel 1998; Chilès and Delfiner 1999; Deutsch 2002).

28.4.2 Regularization

One advantage of GIS is that a large range of data formats can be handled in a single system. The proper handling of scale is important. In particular, it is important to define a model of the effect of the support—the size, geometry and orientation of the space on which an observation is defined—on the spatial variation contained in observed variables. An example of the support is the pixel in a remotely sensed image. Scaling the data can be problematic. For example, if the objective is to increase the support and data are defined at a sparsely distributed set of locations (e.g., a set of soil cores as part of a soil survey) then upscaling (increasing the support) is difficult. Block Kriging can be used to predict over larger supports, but the predicted variable is likely to be smoothed as a result of the interpolation process. Downscaling the data is even more problematic and depends on a good model of the point process or "underlying" punctual spatial variation. Examples of downscaling methods for continuous variables include the area-to-point Kriging (Kyriakidis 2004) and downscaling cokriging (Pardo-Iguzquiza et al. 2006) methods. An example of a downscaling method for categorical variables (classification of remotely sensed imagery) is the super-resolution approach (Tatem et al. 2001).

Geostatistics provides a means of scaling the actual RF model. This amounts to scaling or regularizing the actual variogram defined as a parameter of the RF model. The equations are given in several articles (see Atkinson and Tate 2000 for an introduction). It may be useful to explain conceptually why this is possible. The variogram represents the expected dissimilarity between positions at a given separation lag. Let us assume the simplest case of point observation, with an average distance between points of around 10 m. If the point support is increased to some positive finite area, say 2.5 m, then the variation within the support is integrated or averaged out leaving only the variation between supports. Thus, the variance decreases, but the new variance σ^2_{group} is not predicted adequately by the classical equation

$$\sigma^2_{\text{group}} = \frac{\sigma^2}{n} \tag{28-13}$$

which assumes independence between the n observations in the original set. The amount by which the variance decreases is less than predicted by Equation 28-13 because of spatial dependence. Critically, the variance lost is less than predicted because of local dependence between points (i.e., within the support). Values are likely to be similar within the support and, thus, the removed variance is likely to be less than if the points were independent. The exact decrease in variance can be predicted by regularizing the variogram. In fact, the outcome of regularizing the variogram is a complete variogram defined on the new support.

28.4.3 Optimizing Sampling Design

Kriging predicts with minimum prediction error or kriging error, σ_K, and also estimates this kriging error for every predicted value. The kriging error depends only on the geometry of the domain or support $\mathbf{V}$ to be predicted, the distances between $\mathbf{V}$ and the $n(\mathbf{x}_0)$ data points $\mathbf{x}_\alpha$, the geometry of the $n(\mathbf{x}_0)$ data, and finally the variogram (Journel and Huijbregts 1978). The values of the sample observations locally have no influence. Thus, if the variogram is known, the kriging error can be predicted for any proposed sampling strategy prior to the actual survey. Kriging is, therefore, an ideal tool for designing optimal sampling strategies. The optimal sampling density for a given sampling scheme can be designed by solving the kriging equations for several sampling densities and plotting the maximum kriging variance, $\hat{\sigma}_{Kmax}$, against sample spacing (McBratney et al. 1981; McBratney and Webster 1981). If the budget for the survey is limited then so too is the maximum precision attainable. If the survey is not limited by funding and the investigators can define a maximum tolerable prediction error, then the optimal sampling strategy is the one that just achieves the desired precision. Greater precision would be wasteful. The optimal strategy is found by reading the required sample spacing from the plot of $\hat{\sigma}_{Kmax}$ against sample spacing. Figure 28-7a shows an experimental variogram estimated from elevation data from an area in Britain and Figure 28-7b shows the corresponding plot of maximum kriging error against sample spacing. The sample spacing required to achieve a maximum kriging standard error of 4 m (using a neighborhood of 16 observations) is indicated by a solid line and it gives a grid with a spacing of some 13.5 m.

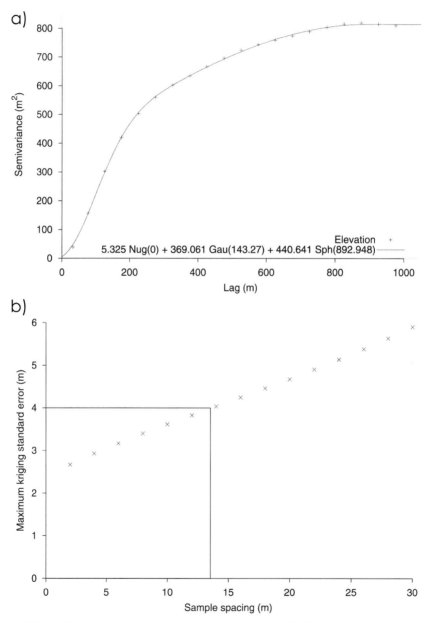

Figure 28-7 (a) Experimental variogram of elevation data (with fitted model coefficients indicated) and (b) corresponding plot of sample spacing against maximum kriging variance for a kriging neighborhood of 16 observations. A maximum kriging variance of 4 m corresponds to a sample spacing of approximately 13.5 m.

28.5 Discussion

28.5.1 Geostatistics Within Existing GIS

We now consider the past, present and future adoption of geostatistics by the GIS community. In the past, GISystems included only limited geostatistical functionality. Over time, some geostatistical functions have been added, presumably in recognition of the theoretical superiority of operations such as kriging over standard interpolation algorithms such as inverse distance weighting (IDW) that had been available in GIS as standard long before. There are further examples of the tools of geostatistics superseding those of GIS. In many cases, overlay is used in situations where regression would provide a more empirically grounded and statistically rigorous alternative. Regression, which is aspatial, often should be replaced by geostatistical techniques such as cokriging, SKlm or KED that make use of the spatial information available. Unfortunately, in addition to providing only limited geostatistical functionality, early versions of some GIS software provided, in some cases, misleading options (e.g., variogram models that were not permissible in 2-D).

Moving to the present, both ArcInfo and IDRISI Kilimanjaro provide a relatively wide range of geostatistical functions. For IDRISI, the software is based on the Gstat software (Pebesma and Wesseling 1998) which is widely used within the geostatistical community. So we may conclude that ample opportunities exist for geostatistical operations to be adopted by the GIS community. If more extensive functionality is required then the GIS user can export their data in the required format and use geostatistical software such as Gstat, GSLIB (Deutsch and Journel 1998) and ISATIS which are widely available.

As suggested by Atkinson (2005), it is questionable whether GIS users are able to devote the time required to learn the fundamental principles and practice of geostatistics (which should surely be a pre-requisite to use of the geostatistical tools available). GIS users have to organize and manipulate (often massive) spatial data sets which encompass a wide range of data models. The operations available in a GIS are arguably more varied than the set of geostatistical operations, and a GIS project requires training and expertise in the same way that geostatistics does. The question is, should we expect a GIS user to be an expert in geostatistics as well? This question becomes more acute if one considers the impressive advances that are being made in geostatistics all the time (see Atkinson 2005 for a discussion). Such current state-of-the-art methods are unlikely to be adopted by the GIS community; there is always a substantial lag in time between development and wide adoption. Indeed, it is unclear whether even long-standing advanced geostatistical concepts—e.g., IRF-k's (intrinsic random functions of order k), surface deformation—will ever be adopted by the GIS community.

Generally, GIS operations such as overlay and buffering are not subject to the same statistical constraints as geostatistics (although recognition of uncertainty has grown in GIS in recent years). Therefore, a concern exists over the potential for mis-use of geostatistics by GIS users who are accustomed to relative freedom in their choice of operations and model parameters. This difference in approach can be seen in the wider context of computer science (i.e., technological research methodology or engineering, see Curran 1987) versus statistics (and mathematics). Geostatistics has its roots in mining engineering. There remains an art and sense of adventure within GIS which has arguably been gradually stripped away from geostatistics, to some extent by the involvement of statisticians in the field. This statistical rigor is welcome and necessary, but it does lay traps for unsuspecting GIS users who may wish to play with geostatistical tools.

Atkinson (2005) argued for greater collaboration between organizations with mapping needs, GIS operators and geostatisticians (i.e., the people developing the techniques). The benefits should be clear from the above arguments: opportunities to use up-to-the-minute, advanced algorithms and less potential for mis-use. Ultimately, the question of who should "do" geostatistics (i.e., the GIS user or the trained geostatistician) may be ill-posed. In practice, the relevant population contains a multivariate spectrum of expertise with much overlap between any detectable clusters.

28.5.2 Geostatistics as Complementary to GIS

As clarified in the introduction, geostatistics is not useful for all problems. In particular, the object-based view of the world is beyond the scope of geostatistics. What makes this point important is that much of the real world is best conceptualized using the object-based view. There are some phenomena which fall obviously into this category. These are mostly human-created functional features at either the within-building scale (e.g., telephone, computer, chair, desk, piano) or between-building scale (e.g., house, car, road, pavement, bridge, shop and so on) (Atkinson and Tate 2000). However, the object-based view is more pervasive than this. Consider a remotely sensed scene which has been imaged at a particular spatial resolution. The image may at first seem to lend itself to being treated as a realization of a RF model. However, the scene itself may be an area comprised of agricultural fields between which there exist sharp discontinuities in terms of remotely sensed brightness (Strahler et al. 1986; Hay et al. 2005). The object-based view has a role here, and it would be folly to apply the RF model directly without considering these objects. Similar arguments apply to other common scenes. For example, the scene may be a forested area in which humans have shaped the boundaries between individual forest stands. Urban areas are even more influenced by humans and, consequently, appropriately modeled using the object-based view.

The point is not that geostatistics is redundant in the above scenarios, but rather that more can be achieved if the RF model is nested within a Boolean-type model of the objects in the scene. Consider, for example, the agricultural scene above. One possibility is to use vector data to represent the field boundaries and then adopt the RF model for variation within the objects. Berberoglu et al. (2000) and Lloyd et al. (2004) adopted a similar strategy for texture classification applied to a Mediterranean agricultural landscape. If vector data are not available *a priori* then an alternative is to segment the image prior to application of geostatistics. Many algorithms are available for segmentation including the popular eCognition software (Definiens, http://www.definiens-imaging.com/). GIS software has a key role to play in object-based processing as complementary to geostatistics.

Given the above discussion, the tools of geostatistics can be seen a complementary to those of GIS.

28.6 Conclusion

Geostatistics should be seen as a set of tools that can be used to perform a wide range of spatial analysis tasks based on the RF model. The specific RF model used in geostatistics includes a spatial covariance function or variogram (or equivalent) and this distinguishes geostatistics from other approaches that depend on a RF view such as the SAP and MRF. Geostatistical techniques include kriging, conditional simulation, regularization and spatial optimization. These techniques have great potential in spatial analysis and sit well alongside traditional GIS functionality, particularly where a nested object-based and RF model provides an appropriate model of reality.

Acknowledgements

We thank Dr. Jennifer McKinley and Prof. Pierre Goovaerts for commenting on the manuscript. We thank the Editor for her patience while the chapter was completed.

References

Atkinson, P. M. 2005. Spatial prediction and surface modelling. *Geographical Analysis* 36:113–123.

Atkinson, P. M. and N. J. Tate. 2000. Spatial scale problems and geostatistical solutions: a review. *Professional Geographer* 52:607–623.

Armstrong, M. 1998. *Basic Linear Geostatistics*. Berlin: Springer.

Berberoglu, S., C. D. Lloyd, P. M. Atkinson and P. J. Curran. 2000. The integration of spectral and textural information using neural networks for land cover mapping in the Mediterranean. *Computers and Geosciences* 26:385–396.

Burrough, P. A. and R. A. McDonnell. 1998. *Principles of Geographical Information Systems*. Oxford: Oxford University Press.

Chilès, J.-P. and P. Delfiner. 1999. *Geostatistics: Modeling Spatial Uncertainty*. New York: Wiley.

Christakos, G. 1984. On the problem of permissible covariance and variogram models. *Water Resources Research* 20:251–265.

Cressie, N.A.C. 1985. Fitting variogram models by weighted least squares. *Mathematical Geology* 17:563–586.

Curran, P. J. 1987. Remote sensing methodologies and geography. *International Journal of Remote Sensing* 8:1255–1275.

Deutsch, C. V. 2002. *Geostatistical Reservoir Modelling*. New York: Oxford University Press.

Deutsch, C. V. and A. G. Journel. 1998. *GSLIB: Geostatistical Software and User's Guide, 2nd ed.* New York: Oxford University Press.

Dungan, J. L. 1999. Conditional simulation. In Spatial Statistics for Remote Sensing, edited by A. Stein, F. van der Meer and B. Gorte, 135–152. Dordrecht, Netherlands: Kluwer Academic Publishers.

Goovaerts, P. 1997. *Geostatistics for Natural Resources Evaluation*. New York: Oxford University Press.

Hay, G. J., G. Castilla, M. A. Wulder and J. R. Ruiz. 2005. An automated object-based approach for the multiscale image segmentation of forest scenes. *International Journal of Applied Earth Observation and Geoinformation* 7:339-359.

Hudson, G. and H. Wackernagel. 1994. Mapping temperature using kriging with external drift: theory and an example from Scotland. *International Journal of Climatology* 14:77–91.

Isaaks, E. H.and R. M. Srivastava. 1989. *An Introduction to Applied Geostatistics*. New York: Oxford University Press.

Journel, A. G. 1996. Modelling uncertainty and spatial dependence: stochastic imaging. *International Journal of Geographical Information Systems* 10:517–522.

Journel, A. G. and C. J. Huijbregts. 1978. *Mining Geostatistics*. London: Academic Press.

Kyriakidis P. C. 2004. A geostatistical framework for area-to-point spatial interpolation. *Geographical Analysis* 36:259–289.

Lloyd, C. D. 2002. Increasing the accuracy of predictions of monthly precipitation in Great Britain using kriging with an external drift. In *Uncertainty in Remote Sensing and GIS*, edited by G. M. Foody and P. M. Atkinson, 243–267. Chichester, UK: John Wiley and Sons.

Lloyd, C. D., S. Berberoglu, P. J. Curran and P. M. Atkinson. 2004. Per-field mapping of Mediterranean land cover: A comparison of texture measures. *International Journal of Remote Sensing* 15:3943–3965.

McBratney, A. B. and R. Webster. 1981. The design of optimal sampling schemes for local estimation and mapping of regionalised variables. II. Program and examples. *Computers and Geosciences* 7:335–365.

———. 1986. Choosing functions for semi-variograms of soil properties and fitting them to sampling estimates. *Journal of Soil Science* 37:617–639.

McBratney, A. B., R. Webster and T. M. Burgess. 1981. The design of optimal sampling schemes for local estimation and mapping of regionalised variables. I. Theory and method. *Computers and Geosciences* 7:331–334.

Oliver, M. A. and R. Webster. 1990. Kriging: a method of interpolation for geographical information systems. *International Journal of Geographical Information Systems* 4:313–332.

Pardo-Igúzquiza, E. 1999. VARFIT: a Fortran-77 program for fitting variogram models by weighted least squares. *Computers and Geosciences* 25:251–261.

Pardo-Iguzquiza, E., M. Chico-Olmo and P. M. Atkinson. 2006. Downscaling cokriging for image sharpening. *Remote Sensing of Environment* 102:86–98.

Pebesma, E. J. and C. G. Wesseling. 1998. Gstat, a program for geostatistical modelling, prediction and simulation. *Computers and Geosciences* 24:17–31.

Richmond, A. 2002. Two-point declustering for weighting data pairs in experimental variogram calculations. *Computers and Geosciences* 28:231–241.

Sampson, P. D., D. Damien and P. Guttorp. 2001. Advances in modelling and inference for environmental processes with nonstationary spatial covariance. In *GeoENV III: Geostatistics for Environmental Applications*, edited by P. Monestiez, D. Allard and R. Froidevaux, 17–32. Dordrecht, Netherlands: Kluwer Academic Publishers.

Schabenberger, O. and C. A. Gotway. 2005. *Statistical Methods for Spatial Data Analysis*. Boca Raton, Fla.: Chapman and Hall/CRC.

Serra J. 1982. *Image Analysis and Mathematical Morphology*. London: Academic Press.

Strahler, A. H., C. E. Woodcock and J. A. Smith. 1986. On the nature of models in remote sensing. *Remote Sensing of Environment* 20:121–139.

Tatem, A. J., H. G. Lewis, P. M. Atkinson and M. S. Nixon. 2001. Super-resolution target identification from remotely sensed images using a Hopfield neural network. *IEEE Transactions on Geoscience and Remote Sensing* 39:781–796.

Van Groenigen, J.-W. 1999. *Constrained Optimization of Spatial Sampling: A Geostatistical Approach*. Enschede, Netherlands: ITC Publication Series.

Wackernagel, H. 2003. *Multivariate Geostatistics. An Introduction with Applications, 3rd ed.* Berlin: Springer.

Webster, R. and M. A. Oliver. 1990. *Statistical Methods in Soil and Land Resource Survey*. Oxford: Oxford University Press.

———. 2000. *Geostatistics for Environmental Scientists*. Chichester, UK: John Wiley and Sons.

Webster, R. and A. B. McBratney. 1989. On the Akaike information criterion for choosing models for variograms of soil properties. *Journal of Soil Science* 40:493–496.

CHAPTER 29
GIS Modeling and Analysis

Joseph K. Berry

29.1 Introduction

Although GIS technology is just a few decades old, its analytical approaches have evolved as much as its mapping capabilities and practical expressions. In the 1960s analytical software development primarily occurred on campuses and its products were relegated to library shelves. These formative years provided the basic organization for both data structures and processing structures found in a modern GIS. A raging debate centered on "vector vs. raster" formats and efficient algorithms for processing—technical considerations with minimal resonance outside of the small (but growing) group of innovators.

The early 1970s saw *Computer Mapping* automate the cartographic process. The points, lines and areas defining geographic features on a map are represented as organized sets of X,Y coordinates. In turn these data form input to a pen plotter that can rapidly update and redraw the connections at a variety of scales and projections. The map image, itself, is the focus of this processing.

The early 1980s exploited the change in the format and the computer environment of mapped data. *Spatial Database Management Systems* were developed that link computer mapping techniques to traditional database capabilities. The demand for spatially and thematically linked data focused attention on data issues. The result was an integrated processing environment addressing a wide variety of mapped data and digital map products.

During the 1990s a resurgence of attention was focused on analytical operations and a comprehensive theory of spatial analysis began to emerge. This "map-ematical" processing involves spatial statistics and spatial analysis. *Spatial statistics* has been used by geophysicists and climatologists since the 1950s to characterize the geographic distribution, or pattern, of mapped data. The statistics describe the spatial variation in the data, rather than assuming a typical response occurs everywhere within a project area.

Spatial analysis, on the other hand, expresses a geographic relationship as a series of map analysis steps, leading to a solution map in a manner analogous to basic algebra. Most of the traditional mathematical capabilities, plus an extensive set of advanced map analysis operations, are available in contemporary GIS software. You can add, subtract, multiply and divide maps; apply exponential, root, log and cosine functions to maps; and differentiate and even integrate maps. After all, maps in a GIS are just an organized set of numbers. However, in map analysis, the spatial coincidence and juxtaposition of values among and within mapped data create new operations, such as effective distance, optimal path routing, visual exposure density, landscape diversity, shape and pattern.

GIS modeling encompasses the varied applications of the concepts, procedures and approaches ingrained in spatial analysis and statistics. This chapter investigates a generalized framework supporting GIS modeling and analysis within a grid-based GIS environment.

Several sections and chapters in this manual address specific analytical and statistical techniques in more detail, as well as comprehensively describing additional modeling applications. In addition, the online book, *Beyond Mapping: Compilation of Beyond Mapping columns appearing in GeoWorld magazine* provides more detailed discussion of the material presented in this chapter (see Author's Note).

29.1.1 Mapping to Analysis of Mapped Data

The evolution (or is it a revolution?) of GIS technology has certainly taken it well beyond the traditional roles of mapping. For thousands of years maps were graphic representations of physical features primarily for the purpose of navigation. With the advent of geotechnology, maps have changed form, to digital representations that are linked to databases and a host of new processing and analytical capabilities.

Figure 29-1 identifies two key trends in the movement from mapping to map analysis. *Traditional GIS* treats geographic space in a manner similar to our paper map legacy. Points, lines and polygons are used to define discrete spatial objects, such as houses, streams and lakes. In turn, these objects are linked to attributes in a database; the attributes describe the objects' characteristics and conditions. The result is a tremendously useful system enabling users to make complex geo-queries of the information and then map the results.

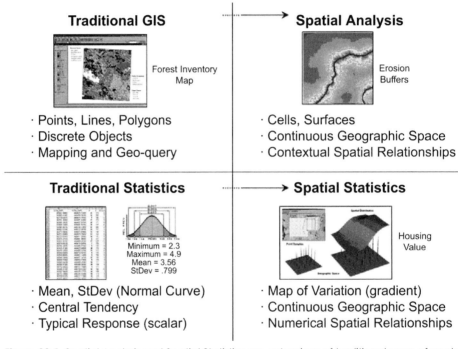

Figure 29-1 Spatial Analysis and Spatial Statistics are extensions of traditional ways of analyzing mapped data. See included DVD for color version.

Spatial Analysis extends the basic set of discrete map features of points, lines and polygons to map "surfaces" that represent continuous geographic space as a set of contiguous grid cell values. The consistency of this grid-based structuring provides the foothold for a wealth of new analytical tools for characterizing "contextual spatial relationships," such as identifying the visual exposure of an entire road network.

In addition, it provides a mathematical/statistical framework by numerically representing geographic space. *Traditional Statistics* is inherently non-spatial as it seeks to represent a data set by its typical response regardless of spatial patterns. The mean, standard deviation and other statistics are computed to describe the central tendency of the data in abstract numerical space without regard to the relative positioning of the data in real-world geographic space.

Spatial Statistics, on the other hand, extends traditional statistics on two fronts. First, it seeks to map the variation in a data set to show where unusual responses occur, instead of focusing on a single typical response. Secondly, it can uncover "numerical spatial relationships" within and among mapped data layers, such as generating a prediction map identifying where likely customers are within a city based on existing sales and demographic information.

29.1.2 Vector-based Mapping Versus Grid-based Analysis

The close conceptual link of vector-based desktop mapping to manual mapping and traditional database management has fueled its rapid adoption. In many ways, a database is just picture waiting to happen. The direct link between attributes described as database records and their spatial characterization is easy to conceptualize. Geo-query enables one to click on a map to pop-up the attribute record for a location or to search a database and then plot all of the records meeting the query. Increasing data availability and Internet access, coupled with decreasing desktop mapping system costs and complexity, make the adoption of spatial database technology a practical reality.

Maps in their traditional form of points, lines and polygons identifying discrete spatial objects align with manual mapping concepts and experiences. Grid-based maps, on the other hand, represent a different paradigm of geographic space that opens entirely new ways to address complex issues. Whereas traditional vector maps emphasize *precise placement of physical features, grid maps seek to analytically characterize continuous geographic space in both real and cognitive terms.*

29.2 Fundamental Map Analysis Approaches

The tools for mapping database attributes can be extended to analysis of spatial relationships within and among mapped data layers. Two broad classes of capabilities form this extension—spatial statistics and spatial analysis.

29.2.1 Spatial Statistics

Spatial statistics can be grouped into two broad camps—surface modeling and spatial data mining. *Surface modeling* involves the translation of discrete point data into a continuous surface that represents the geographic distribution of the data. Traditional non-spatial statistics involves an analogous process when a numerical distribution (e.g., standard normal curve) is used to generalize the central tendency of a data set. The derived average and standard deviation reflects the typical response and provides a measure of how typical it is. This characterization seeks to explain data variation in terms of the numerical distribution of measurements without reference to the data's geographic distribution and patterns.

In fact, an underlying assumption in most traditional statistical analyses is that the data is randomly or uniformly distributed in geographic space. If the data exhibits a geographic pattern (termed spatial autocorrelation) many of the non-spatial analysis techniques are less valid. Spatial statistics, on the other hand, utilizes inherent geographic patterns to further explain the variation in a set of sample data.

The numerous techniques for characterizing the spatial distribution inherent in a data set can be categorized into four basic approaches:
- *Point Density* mapping, that aggregates the number of points within a specified distance (number per acre);
- *Spatial Interpolation*, that weight-averages measurements within a localized area (e.g., Kriging);
- *Map Generalization*, that fits a functional form to the entire data set (e.g., polynomial surface fitting);
- *Geometric Facets*, that construct a map surface by tessellation (e.g., fitting a Triangular Irregular Network of facets to the sample data).

For example, consider Figure 29-2 showing a point density map derived from a map identifying housing locations. The project area is divided into an analysis frame of 30-meter grid cells (100 columns x 100 rows = 10,000 grid cells). The number of houses for each grid space is identified in the left portion of the figure as colored dots in the two-dimensional (2D) map and "spikes" in the three-dimensional (3D) map.

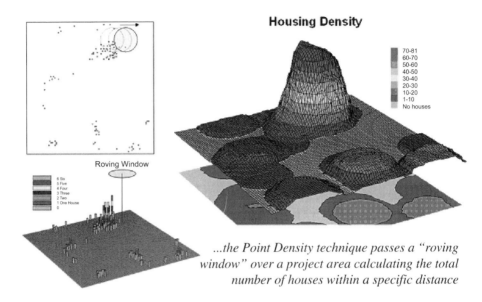

...the Point Density technique passes a "roving window" over a project area calculating the total number of houses within a specific distance

Figure 29-2 Calculating the total number of houses within a specified distance of each map location generates a housing density surface. See included DVD for color version.

A neighborhood summary operation is used to pass a "roving window" over the project area calculating the total number of houses within a quarter-mile of each map location. The result is a continuous map surface indicating the relative density of houses—'peaks' where there are a lot of nearby houses and 'valleys' where there are few or none. In essence, the map surface quantifies what your eye sees in the spiked map—some locations with lots of houses and others with very few.

While surface modeling is used to derive continuous surfaces, spatial data mining seeks to uncover numerical relationships within and among mapped data. Some of the techniques include coincidence summary, proximal alignment, statistical tests, percent difference, surface configuration, level-slicing, and clustering that is used in comparing maps and assessing similarities in data patterns.

Another group of spatial data mining techniques focuses on developing predictive models. For example, one of the earliest uses of predictive modeling was in extending a test market project for a phone company (Figure 29-3). Customers' addresses were used to "geo-code" map coordinates for sales of a new product that enabled distinctly different rings to be assigned to a single phone line—one for the kids and one for the parents. Like pushpins on a map, the pattern of sales throughout the test market area emerged with some areas doing very well, while other areas' sales were few and far between.

The demographic data for the city were analyzed to calculate a prediction equation between product sales and census block data. The prediction equation derived from the test market sales was applied to another city by evaluating exiting demographics to "solve the equation" for a predicted sales map. In turn the predicted map was combined with a wire-exchange map to identify switching facilities that required upgrading before release of the product in the new city.

Analyzing Spatial Relationships
The Spatial Data Mining Process

Geo-Registered Maps are Used to Uncover and Apply Spatial Relationships

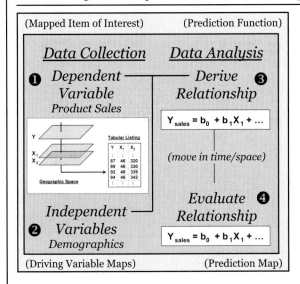

(Mapped Item of Interest)   (Prediction Function)

Maps of the item of interest ❶ and related variables ❷ are encoded then analyzed to derive a "map-ematical" relationship ❸ that can be used to predict the item at another place or time ❹.

For example, test market sales of a product ❶ can be related to demographic characteristics ❷ then the derived relationship ❸ moved to another city to generate a map of predicted sales ❹.

Figure 29-3 Spatial Data Mining techniques can be used to derive predictive models of the relationships among mapped data. See included DVD for color version.

29.2.2 Spatial Analysis

Whereas spatial data mining responds to 'numerical' relationships in mapped data, spatial analysis investigates the 'contextual' relationships. Tools such as slope/aspect, buffers, effective proximity, optimal path, visual exposure and shape analysis fall into this class of spatial operators. Rather than statistical analysis of mapped data, these techniques examine geographic patterns, vicinity characteristics and connectivity among features.

One of the most frequently used map analysis techniques is Suitability Modeling. These applications seek to map the relative appropriateness of map locations for particular uses. For example, a map of the best locations for a campground might be modeled by preferences for being on gentle slopes, near roads, near water, good views of water and a southerly aspect.

These spatial criteria can be organized into a flowchart of processing (see Figure 29-4). Note that the rows of the flowchart identify decision criteria with boxes representing maps and lines representing map analysis operations. For example, the top row evaluates the preference to locate the campground on gentle slopes.

The first step calculates a slope map from the base map of elevation by entering the command—

SLOPE Elevation Fitted FOR Slopemap

The derived slope values are then reclassified to identify locations of acceptable slopes by assigning 1 to slopes from 0 to 20 percent—

RENUMBER Slopemap ASSIGNING 1 TO 0 THRU 20
ASSIGNING 0 TO 20 THRU 1000 FOR OK_slope

—and assigning 0 to unacceptably steep slopes that are greater than 20 percent.

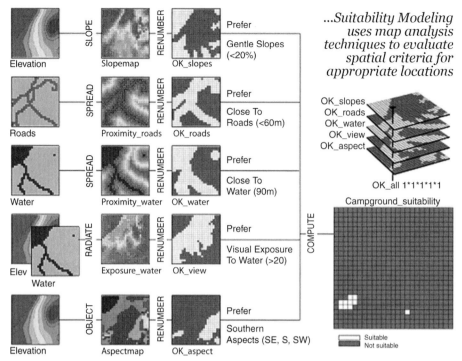

Figure 29-4 Map Analysis techniques can be used to identify suitable places for a management activity. See included DVD for color version.

In a similar manner, the other criteria are evaluated and represented by maps with a value of 1 assigned to acceptable areas (white) and 0 assigned to unacceptable areas (dark grey). The individual preference maps are combined by entering the command—

> CALCULATE OK_slope * OK_road * OK_water * OK_view * OK_aspect
> FOR Campground_suitability

The map analysis procedure depicted in Figure 29-4 simply substitutes values of 1 and 0 for suitable and non-suitable areas. The multiplication of the digital preference maps simulates stacking manual map overlays on a light-table. A location that computes to 1 ($1*1*1*1*1= 1$) corresponds to acceptable. Any numeric pattern with the value 0 will result in a product of 0 indicating an unacceptable area.

While some map analysis techniques are rooted in manual map processing, most are departures from traditional procedures. For example, the calculation of slope is much more exacting than ocular estimates of contour line spacing. And the calculation of visual exposure to water presents a completely new and useful concept in natural resources planning. Other procedures, such as optimal path routing, landscape fragmentation indices and variable-width buffers, offer a valuable new toolbox of analytical capabilities.

29.3 Data Structure Implications

Points, lines and polygons have long been used to depict map features. With the stroke of a pen a cartographer could outline a continent, delineate a road or identify a specific building's location. With the advent of the computer, manual drafting of these data has been replaced by stored coordinates and the cold steel of the plotter.

In digital form these spatial data have been linked to attribute tables that describe characteristics and conditions of the map features. Desktop mapping exploits this linkage to provide tremendously useful database management procedures, such as address matching and driving directions from your house to a store along an organized set of city street line segments. Vector-based data forms the foundation of these processing techniques and directly builds on our historical perspective of maps and map analysis.

Grid-based data, on the other hand, is a relatively new way to describe geographic space and its relationships. At the heart of this digital format is a new map feature that extends traditional points, lines and polygons (discrete objects) to continuous map surfaces.

The rolling hills and valleys in our everyday world comprise a good example of a geographic surface. The elevation values constantly change as you move from one place to another forming a continuous spatial gradient. The left side of Figure 29-5 shows the grid data structure and a sub-set of values used to depict the terrain surface shown on the right side. Note that the shaded zones ignore the subtle elevation differences within the contour polygons (vector representation).

Grid data are stored as a continuous organized set of values in a matrix that is geo-registered over the terrain. Each grid cell identifies a specific location and contains a map value representing its average elevation. For example, the grid cell in the lower-right corner of the map is 1800 feet above sea level (falling within the 1700 to 1900 contour interval). The relative heights of surrounding elevation values characterize the subtle changes of the undulating terrain of the area.

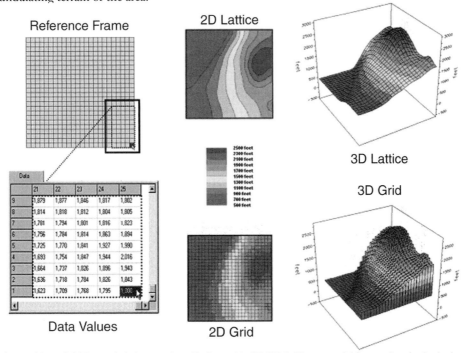

Figure 29-5 Grid-based data can be displayed in 2D/3D lattice or grid forms. See included DVD for color version.

29.3.1 Grid Data Organization

Map features in a vector-based mapping system identify discrete, irregular spatial objects with sharp abrupt boundaries. Other data types—raster images and raster grids—treat space in an entirely different manner forming a spatially continuous data structure.

For example, a *raster image* is composed of thousands of 'pixels' (picture elements) that are analogous to the dots on a computer screen. In a geo-registered black-and-white (B&W) aerial photo, the dots are assigned a grayscale color from black (no reflected light) to white (lots of reflected light). The eye interprets the patterns of gray as forming the forests, fields, buildings and roads of the actual landscape. As aerial imaging has become digital, it has also more often come to be color. While such raster images contain tremendous amounts of information that are easily "seen" and can be computer classified using remote sensing software, the data values representing color intensity (electromagnetic energy) are all positive integers, limiting their support of the full suite of map analysis operations involving relationships within and among maps.

Raster grids, on the other hand, contain a robust range of values and organizational structure amenable to map analysis and modeling. As depicted on the left side of Figure 29-6, this organization enables the computer to identify any or all of the data for a particular location by simply accessing the values for a given column/row position (spatial coincidence used in point-by-point overlay operations). Similarly, the immediate or extended neighborhood around a point can be readily accessed by selecting the values at neighboring column/row positions (zonal groupings used in region-wide overlay operations). The relative proximity of one location to any other location is calculated by considering the respective column/row positions of two or more locations (proximal relationships used in distance and connectivity operations).

There are two fundamental approaches to storing grid-based data—individual "flat" files and "multiple-grid" tables (right side of Figure 29-6). Flat files store map values as one long list, most often starting with the upper-left cell, then sequenced left to right along rows ordered from top to bottom. Multi-grid tables have a similar ordering of values but contain the data for many maps as separate data fields in a single table.

Generally speaking the flat file organization is best for applications that create and delete a lot of maps during processing as table maintenance can affect performance. However, a multi-gird table structure has inherent efficiencies useful in relatively non-dynamic applications. In either case, the implicit ordering of the grid cells over continuous geographic space provides the topological structure required for advanced map analysis.

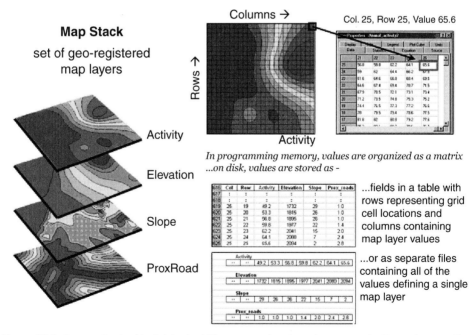

Figure 29-6 A map stack of individual grid layers can be stored as separate files or in a multi-grid table. See included DVD for color version.

29.3.2 Grid Data Types

Understanding that a digital map is first and foremost an organized set of numbers is fundamental to analyzing mapped data. The locations of map features are translated into computer form as organized sets of X,Y coordinates (vector) or grid cells (raster). Considerable attention is given data structure considerations and their relative advantages in storage efficiency and system performance.

However, this geo-centric view rarely explains the full nature of digital maps. For example, consider the numbers themselves that comprise the X,Y coordinates—how does number type and size affect precision? A general feel for the precision ingrained in a "single precision floating point" representation of Latitude/Longitude in decimal degrees is—

> 1.31477E+08 ft = equatorial circumference of the Earth
> 1.31477E+08 ft / 360 degrees = 365,214 ft per degree of Longitude

A single precision number carries six decimal places, so—

> 365,214 ft/degree * 0.000001= .365214 ft *12 in/ft = 4.38257 inch precision

In analyzing mapped data, however, the characteristics of the attribute values are just as critical as precision in positioning. While textual descriptions can be stored with map features they can only be used in geo-query. Figure 29-7 lists the data types by two important categories—numeric and geographic. *Nominal* attribute values do not imply ordering, even when the name of the value is a number. A nominal attribute value of 3 isn't bigger, tastier or smellier than a 1, it is just not a 1. In the figure these data are schematically represented as scattered and independent pieces of wood.

Numeric and Geographic Data Types

...all digital maps are composed of organized sets of numbers– the data type determines what "map-ematical" processing can be done with the numbers on a map, or stack of map layers

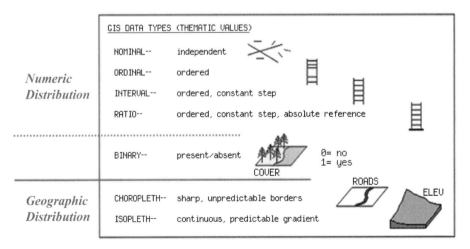

Figure 29-7 Map values are characterized from two broad perspectives—numeric and geographic—then further refined by specific data types. See included DVD for color version.

Ordinal numbers, on the other hand, imply a definite ordering and can be conceptualized as a ladder, however with varying spaces between rungs. The numbers form a progression, such as smallest to largest, but there isn't a consistent step. For example you might rank five different soil types by their relative crop productivity (1= worst to 5= best) but it doesn't mean that soil type 5 is exactly five times more productive than soil type 1.

When a constant step is applied, *Interval* numbers result. For example, a 60° Fahrenheit spring day is consistently/incrementally warmer than a 30° winter day. In this case one "degree" forms a consistent reference step analogous to a typical ladder with uniform spacing between rungs.

A *ratio* number introduces yet another condition—an absolute reference—that provides a consistent footing or starting point for the ladder; for example all molecular movement ceases at temperature measurement zero Kelvin. A final type of numeric data is termed *Binary*. In this instance the value range is constrained to just two states, such as forested/non-forested or suitable/not-suitable.

So what does all of this have to do with analyzing digital maps? The type of number dictates the variety of analytical procedures that can be applied. Nominal data, for example, do not support direct mathematical or statistical analysis. Ordinal data support only a limited set of statistical procedures, such as maximum and minimum. These two data types are often referred to as Qualitative Data. Interval and ratio data, on the other hand, support a full set mathematics and statistics and are considered Quantitative Data. Binary maps support special mathematical operators, such as AND and OR.

The geographic characteristics of the numbers are less familiar. From this perspective there are two types of numbers. *Choropleth* numbers form sharp and unpredictable boundaries in space such as the values on a road or cover type map. *Isopleth* numbers, on the other hand, form continuous and often predictable gradients in geographic space, such as the values on an elevation or temperature surface.

Figure 29-8 puts it all together. Discrete maps identify mapped data with independent numbers (nominal) forming sharp abrupt boundaries (choropleth), such as a cover type map. Continuous maps contain a range of values (ratio) that form spatial gradients (isopleth), such as an elevation surface.

The clean dichotomy of discrete/continuous is muddled by cross-over data such as speed limits (ratio) assigned to the features on a road map (choropleth). Understanding the data type, both numerical and geographic, is critical to applying appropriate analytical procedures and construction of sound GIS models.

29.3.3 Grid Data Display

Two basic approaches can be used to display grid data— grid and lattice. The *Grid* display form uses cells to convey surface configuration. The 2D version simply fills each cell with the contour interval color, while the 3D version pushes up each cell to its relative height. The *Lattice* display form uses lines to convey surface configuration. The contour lines in the 2D version identify the breakpoints for equal intervals of increasing elevation. In the 3D version the intersections of the lines are "pushed-up" to the relative height of the elevation value stored for each location.

Figure 29-9 shows how 3D plots are generated. Placing the viewpoint at different look-angles and distances creates different perspectives of the reference frame. For a 3D grid display, entire cells are pushed to the relative height of their map values. The grid cells retain their projected shape forming blocky extruded columns. The 3D lattice display pushes up each intersection node to its relative height. In doing so the four lines connected to it are stretched proportionally. The result is a smooth wire-frame that expands and contracts with the rolling hills and valleys.

Generally speaking, lattice displays create more pleasing maps and knock-your-socks-off graphics when you spin and twist the plots. However, grid displays provide a more honest

Discrete versus Continuous Data

Covertype map- *values are independent and represent discrete categories (independent numbers); map values form sharp abrupt boundaries in geographic space (abrupt boundaries)*

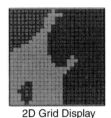

2D Grid Display

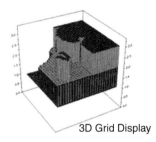

3D Grid Display

	Discrete	Continuous
Numeric distribution	independent numbers	range of values
Geographic distribution	abrupt boundaries	spatial gradient
	Discrete	Continuous

Elevation map- *values form a continuous range with an absolute reference (range of values); map values form a continuous gradient in geographic space (spatial gradient)*

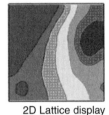

2D Lattice display

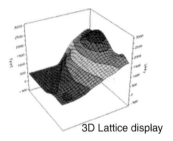

3D Lattice display

Figure 29-8 Discrete and Continuous map types combine the numeric and geographic characteristics of mapped data. See included DVD for color version.

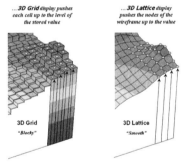

Figure 29-9 3D display "pushes-up" the grid or lattice reference frame to the relative height of the stored map values. See included DVD for color version.

picture of the underlying mapped data—a chunky matrix of stored values. In either case, one must recognize that a 3D display is not the sole province of elevation data. Often a 3-dimensional plot of data such as effective proximity is extremely useful in understanding the subtle differences in distances.

29.3.4 Visualizing Grid Values

In a GIS, map display is controlled by a set of user-defined tools—not the cartographer/publisher team that produced hardcopy maps just a couple of decades ago. The upside is a tremendous amount of flexibility in customizing map display (potential for tailoring). The downside is a tremendous amount of flexibility in customizing map display (potential for abuse).

Manual of Geographic Information Systems

The display tools are both a boon and a bane as they require minimal skills to use but considerable thought and experience to use correctly. The interplay among map projection, scale, resolution, shading and symbols can dramatically change a map's appearance and thereby the information it graphically conveys to the viewer.

While this is true for the points, lines and areas comprising traditional maps, the potential for cartographic effects is even more pronounced for contour maps of surface data. For example, consider the mapped data of animal activity levels from 0.0 to 85.2 animals in a 24-hour period shown in Figure 29-10. The map on the left uses an *Equal Ranges* display with contours derived by dividing the data range into nine equal steps. The flat area at the foot of the hill skews the data distribution toward lower values. The result is significantly more map locations contained in the lower contour intervals—first interval from 0 to 9.5 = 39% of the map area. The spatial effect is portrayed by the radically different areal extent of the contours.

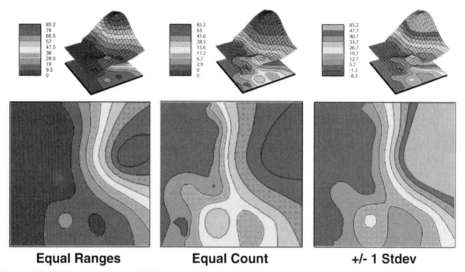

Equal Ranges   **Equal Count**   **+/- 1 Stdev**

Figure 29-10 Comparison of different 2D contour displays using Equal ranges, Equal Count and +/-1 Standard deviation contouring techniques. See included DVD for color version.

The middle map in the figure shows the same data displayed as *Equal Counts* with contours that divide the data range into intervals that represent equal amounts of the total map area. Notice the unequal spacing of the breakpoints in the data range but the balanced area of the color bands—the opposite effect as equal ranges.

The map on the right depicts yet another procedure for assigning contour breaks. This approach divides the data into groups based on the calculated mean and *Standard Deviation*. The standard deviation is added to the mean to identify the breakpoint for the upper contour interval and subtracted to set the lower interval. In this case, the lower breakpoint calculated is below the actual minimum so no values are assigned to the first interval (highly skewed data). In statistical terms the low and high contours are termed the "tails" of the distribution and locate data values that are outside the bulk of the data—identifying "unusually" lower and higher values than you normally might expect. The other five contour intervals in the middle are formed by equal ranges within the lower and upper contours. The result is a map display that highlights areas of unusually low and high values and shows the bulk of the data as a gradient of increasing values.

The bottom line is that the same surface data generated dramatically different map products. All three displays contain nine intervals but the breakpoints were assigned to the contours employing radically different approaches that generated fundamentally different map displays.

So which display is correct? Actually all three displays are proper, they just reflect different perspectives of the same data distribution—a bit of the art in the art and science of GIS. The translation of continuous geographic data into discrete 2D maps invariably conjures up artful interpretation. A good rule of thumb is to be skeptical of the lines on any map that portrays continuous phenomena, such as elevation, proximity, buffers, density, visual exposure or activity. Visualizing and analyzing these data are usually best when the full information content of map surfaces is employed.

29.4 Spatial Statistics Techniques

As outlined in Section 29.2.1, *Spatial Statistics* can be grouped into two broad camps—surface modeling and spatial data mining. *Surface Modeling* involves the translation of discrete point data into a continuous surface that represents the geographic distribution of the data. *Spatial Data Mining*, on the other hand, seeks to uncover numerical relationships within and among sets of mapped data.

29.4.1 Surface Modeling

The conversion of a set of point samples into its implied geographic distribution involves several considerations—an understanding of the procedures themselves, the underlying assumptions, techniques for benchmarking the derived map surfaces and methods for assessing the results and characterizing accuracy.

29.4.1.1 Point Samples to Map Surfaces

Soil sampling has long been at the core of agricultural research and practice. Traditionally point-sampled data were analyzed by non-spatial statistics to identify the typical nutrient level throughout an entire field. Considerable effort was expended to determine the best single estimate and assess just how good the average estimate was in typifying a field.

However non-spatial techniques fail to make use of the geographic patterns inherent in the data to refine the estimate—the typical level is assumed everywhere the same within a field. The computed standard deviation indicates just how good this assumption is—the larger the standard deviation the less valid is the assumption of "…everywhere the same."

Surface Modeling utilizes the spatial patterns in a data set to generate localized estimates throughout a field. Conceptually it maps the variance by using geographic position to help explain the differences in the sample values. In practice, it simply fits a continuous surface to the point data spikes as depicted in Figure 29-11.

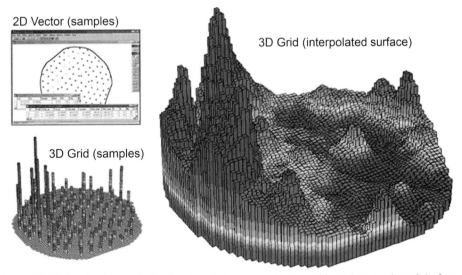

Figure 29-11 Spatial interpolation involves fitting a continuous surface to sample points. See included DVD for color version.

Manual of Geographic Information Systems

While the extension from non-spatial to spatial statistics is quite a theoretical leap, the practical steps are relatively easy. The left side of the figure shows 2D and 3D point maps of phosphorous soil samples collected throughout the field. This highlights the primary difference from traditional soil sampling—each sample must be geo-referenced as it is collected. In addition, the sampling pattern and intensity are often different than traditional grid sampling to maximize spatial information within the data collected.

The surface map on the right side of the figure depicts the continuous spatial distribution derived from the point data. Note that the high spikes in the left portion of the field and the relatively low measurements in the center are translated into the peaks and valleys of the surface map.

When mapped, the traditional, non-spatial approach forms a flat plane (average phosphorous level) aligned within the bright yellow zone. Its "…everywhere the same" assumption fails to recognize the patterns of larger levels and smaller levels captured in the surface map of the data's geographic distribution. A fertilization plan for phosphorous based on the average level—22 parts per million (ppm)—would be ideal for very few locations and be inappropriate for most of the field as the sample data vary from 5 to 102 ppm phosphorous.

29.4.1.2 Spatial Autocorrelation

Spatial Interpolation's basic concept involves *Spatial Autocorrelation*, referring to the degree of similarity among neighboring points (e.g., soil nutrient samples). If the neighboring points exhibit a lot of similarity, termed spatial dependence, they ought to create a good map. If they are spatially independent, then expect a map of pure, dense gibberish. So how can we measure whether "what happens at one location depends on what is happening around it?"

Common sense leads us to believe more similarity exists among the neighboring soil samples (lines in the left side of Figure 29-12) than among sample points farther away. Computing the differences in the values between each sample point and its closest neighbor provides a test of the assertion as nearby differences should be less than the overall difference among the values of all sample locations.

If the differences in neighboring values are a lot smaller than the overall variation, then a high degree of positive spatial dependency is indicated. If they are about the same or if the neighbors variation is larger (indicating a rare checkerboard-like condition), then the assumption of spatial dependence fails. If the dependency test fails, it means an interpolated map likely is just colorful gibberish.

The difference test, however, is limited, as it merely assesses the closest neighbor, regardless of its distance. A *Variogram* (right side of Figure 29-12) is a plot of the similarity among

SAMPLE POINTS   **VARIOGRAM PLOT**

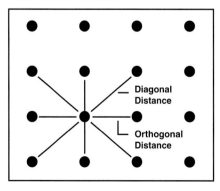 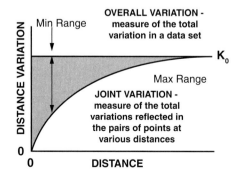

Figure 29-12 Variogram plot depicts the relationship between distance and measurement similarity (spatial autocorrelation).

values based on the distance between them. Instead of simply testing whether close things are related, it shows how the degree of dependency relates to varying distances between locations. The origin of the plot at 0,0 is a unique case where the distance between samples is zero and there is no dissimilarity (data variation = 0) because a location is exactly the same as itself.

As the distance between points increases, subsets of the data are scrutinized for their dependency. The shaded portion in the idealized plot shows how quickly the spatial dependency among points deteriorates with distance. The maximum range (Max Range) position identifies the distance between points beyond which the data values are considered spatially independent. This tells us that using data values beyond this distance for interpolation actually can mess up the interpolation.

The minimum range (Min Range) position identifies the smallest distance contained in the actual data set and is determined by the sampling design used to collect the data. If a large portion of the shaded area falls below this distance, it tells you there is insufficient spatial dependency in the data set to warrant interpolation. If you proceed with the interpolation, a nifty colorful map will be generated, but likely of questionable accuracy. Worse yet, if the sample data plots as a straight horizontal line or circle, no spatial dependency exists and the map will be of no value.

Analysis of the degree of spatial autocorrelation in a set of point samples is mandatory before spatially interpolating any data. This step is not required to mechanically perform the analysis as the procedure will always generate a map. However, it is the initial step in determining if the map generated is likely to be a good one.

29.4.1.3 Benchmarking Interpolation Approaches

For some, the previous discussion on generating maps from soil samples might have been too simplistic—enter a few things then click on a data file, and in a few moments you have a soil nutrient surface. Actually, it is that easy to create one. The harder part is figuring out if the map generated makes sense and whether it is something you ought to use for subsequent analysis and important management decisions.

The following discussion investigates the relative amounts of spatial information provided by comparing a whole-field average to interpolated map surfaces generated from the same data set. The top-left portion in Figure 29-13 shows the map of the average phosphorous level in the field. It forms a flat surface because there isn't any information about spatial variability in an average value.

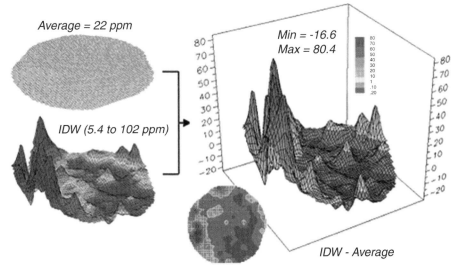

Figure 29-13 Spatial comparison of a whole-field average and an IDW interpolated map. See included DVD for color version.

The non-spatial estimate simply adds up all of the sample measurements and divides by the number of samples to get 22 ppm. Since the procedure didn't consider the relative position of the different samples, the variations in the measurements cannot be mapped. The assumption is that the average is everywhere, plus or minus the standard deviation. But the result offers no spatial guidance as to where phosphorous levels might be higher or lower than the average.

The spatially based estimates are shown in the interpolated map surface below the average plane. As described in the previous section (Section 29.4.1.2), spatial interpolation looks at the relative positioning of the soil samples as well as the measured phosphorous levels. In this instance the big bumps were influenced by high measurements in that vicinity while the low areas indicate surrounding low values.

The map surface in the right portion of Figure 29-13 compares the two maps simply by subtracting them. The color ramp was chosen to emphasize the differences between the whole-field average estimates and the interpolated ones. The center yellow band indicates the average level while the progression of green tones locates areas where the interpolated map estimated that there was more phosphorous than the whole field average. The higher locations identify where the average value is less than the interpolated ones. The lower locations identify the opposite condition where the average value is more than the interpolated ones. Note the dramatic differences between the two maps.

Now turn your attention to Figure 29-14 that compares maps derived by two different interpolation techniques—IDW (inverse distance-weighted) and Kriging. Note the similarity in the peaks and valleys of the two surfaces. While subtle differences are visible the general trends in the spatial distribution of the data are identical.

The difference map on the right confirms the coincident trends. The broad band of yellow identifies areas that are +/- 1 ppm. The brown color identifies areas that are within 10 ppm with the IDW surface estimates a bit more than the Kriging ones. Applying the same assumption, that a difference of +/- 10 ppm is negligible in a fertilization program, the maps are effectively identical.

So what's the bottom line? That there often are substantial differences between a whole field average and any interpolated surface. It suggests that finding the best interpolation technique isn't as important as using an interpolated surface in preference to the whole field average. This general observation holds for most mapped data exhibiting spatial autocorrelation.

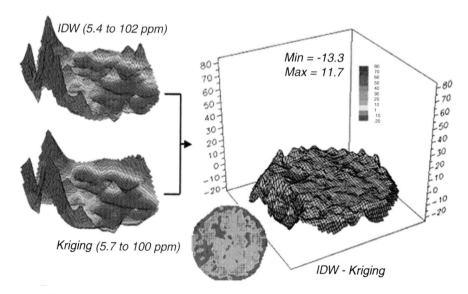

Figure 29-14 Spatial comparison of IDW and Kriging interpolated maps. See included DVD for color version.

29.4.1.4 Assessing Interpolation Results

The previous discussion compared the assumption of the field average with map surfaces generated by two different interpolation techniques for phosphorous levels throughout a field. While the differences between the average and the derived surfaces were considerable (from -20 to +80 ppm), there was relatively little difference between the two interpolated surfaces (+/- 10 ppm).

But which surface best characterizes the spatial distribution of the sampled data? The answer to this question lies in *Residual Analysis*—a technique that investigates the differences between estimated and measured values throughout a field. Common sense requires that one should not simply accept an interpolated map without assessing its accuracy. Ideally, one designs an appropriate sampling pattern and then randomly locates a number of test points to evaluate interpolation performance.

So which surface, IDW or Kriging, did a better job in estimating the measured phosphorous levels for a test set of measurements? The table in Figure 29-15 reports the results for twelve randomly positioned test samples. The first column identifies the sample ID and the second column reports the actual measured value for that location.

Column C simply depicts estimating the whole-field average (21.6) at each of the test locations. Column D computes the difference of the estimated value minus actual measured value for the test set—formally termed the *residual*. For example, the first test point (ID#59) estimated the average of 21.6 but was actually measured as 20.0, so the residual is (21.6 – 20.0 = 1.6 ppm) …very close. However, test point #109 is way off (21.6 – 103.0 = –81.4 ppm) … that is, the estimate is only 1/5th of the measured value, a serious under-estimation error.

The residuals for the IDW and Kriging maps are similarly calculated to form columns F and H, respectively. First note that the residuals for the whole-field average are generally larger than either those for the IDW or Kriging estimates. Next note that the residual patterns between the IDW and Kriging are very similar—when one is way off, so is the other and usually by about the same amount. A notable exception is for test point #91 where Kriging dramatically over-estimates.

The rows at the bottom of the table summarize the residual analysis results. The *Residual sum* row characterizes any bias in the estimates—a negative value indicates a tendency to underestimate with the magnitude of the value indicating how much. The –92.8 value for the whole-field average indicates a relatively strong bias to underestimate.

...Residual Analysis is used to evaluate interpolation performance (IDW at .18 appears best)

	A	B	C	D	E	F	G	H
1	Test Set (randomly selected)							
2	Id	P_actual	Average	Avg-Actual	P_IDW	IDW-Actual	P_Krig	Krig-Actual
3	59	20.0	21.6	1.6	18.80	-1.20	17.80	-2.20
4	2	15.0	21.6	6.6	11.70	-3.30	13.50	-1.50
5	25	5.0	21.6	16.6	5.39	0.39	5.69	0.69
6	63	27.0	21.6	-5.4	28.80	1.80	27.40	0.40
7	18	9.0	21.6	12.6	9.68	0.68	7.90	-1.10
8	34	14.0	21.6	7.6	12.80	-1.20	13.30	-0.70
9	22	10.0	21.6	11.6	10.00	0.00	10.20	0.20
10	91	20.0	21.6	1.6	31.60	11.60	42.30	22.30
11	109	103.0	21.6	-81.4	91.10	-11.90	90.70	-12.30
12	46	12.0	21.6	9.6	16.80	4.80	15.80	3.80
13	89	93.0	21.6	-71.4	67.60	-25.40	68.60	-24.40
14	50	24.0	21.6	-2.4	24.60	0.60	27.40	3.40
15								
16	Average values =	29.3	21.6		27.41		28.38	
17	Residual sum =			-92.8		-23.10		-11.40
18	Average error =			19		5.24		6.08
19	Normalized error =			0.65		0.18		0.21

Figure 29-15 A residual analysis table identifies the relative performance of average, IDW and Kriging estimates.

Manual of Geographic Information Systems

The *Average error* row reports how typically far off the estimates were. The 19.0 ppm average error for the whole-field average is three times worse than Kriging's estimated error (6.08) and nearly four times worse than IDW's (5.24).

Comparing the figures to the assumption that +/-10 ppm is negligible in a fertilization program it is readily apparent that the whole-field estimate is inappropriate to use and that the accuracy differences between IDW and Kriging are minor.

The *Normalized error* row calculates the average error as a proportion of the average value for the test set of samples (5.24/29.3 = 0.18 for IDW). This index is the most useful as it enables the comparison of the relative map accuracies between different maps. Generally speaking, maps with normalized errors of more than .30 are suspect and one might not want to make important decisions using them.

The bottom line is that Residual Analysis is an important tool when spatially interpolating data. Without an understanding of the relative accuracy and interpolation error of the base maps, one can't be sure of any modeling results using the data. The investment in a few extra sampling points for testing and residual analysis of these data provides a sound foundation for site-specific management. Without it, the process can become one of blind faith and wishful thinking.

29.4.2 Spatial Data Mining

Spatial data mining involves procedures for uncovering numerical relationships within and among sets of mapped data. The underlying concept links a map's geographic distribution to its corresponding numeric distribution through the coordinates and map values stored at each location. This 'data space' and 'geographic space' linkage provides a framework for calculating map similarity, identifying data zones, mapping data clusters, deriving prediction maps and refining analysis techniques.

29.4.2.1 Calculating Map Similarity

While visual analysis of a set of maps might identify broad relationships, it takes quantitative map analysis to handle a detailed scrutiny. Consider the three maps shown in Figure 29-16— what areas identify similar patterns? If you focus your attention on a location in the lower right portion how similar is the data pattern to all of the other locations in the field?

The answers to these questions are much too complex for visual analysis and certainly beyond the geo-query and display procedures of standard desktop mapping packages. While the data in the example show the relative amounts of phosphorous (P), potassium (K) and nitrogen (N) throughout a field, it could as easily be demographic data representing income, education and property values; or sales data tracking three different products; or public health maps representing different disease incidences; or crime statistics representing different types of felonies or misdemeanors.

Regardless of the data and application arena, a multivariate procedure for assessing similarity often is used to analyze the relationships. In visual analysis you move your eye among the maps to summarize the color assignments at different locations. The difficulty in this approach is two-fold—remembering the color patterns and calculating the difference. The map analysis procedure does the same thing except it uses map values in place of the colors. In addition, the computer doesn't tire as easily and completes the comparison for all of the locations throughout the map window (3289 in this example) in a couple of seconds.

The upper-left portion of Figure 29-17 illustrates capturing the data patterns of two locations for comparison. The "data spear" at map location column 45, row 18 (45c, 18r) identifies the P-level as 11.0 ppm, the K-level as 177.0 and N-level as 32.9. This step is analogous to your eye noting a color pattern of dark-red, dark-orange and light-green. The other location for comparison (32c, 62r) has a data pattern of P = 53.2, K = 412.0 and N = 27.9; or as your eye sees it, a color pattern of dark-green, dark-green and yellow.

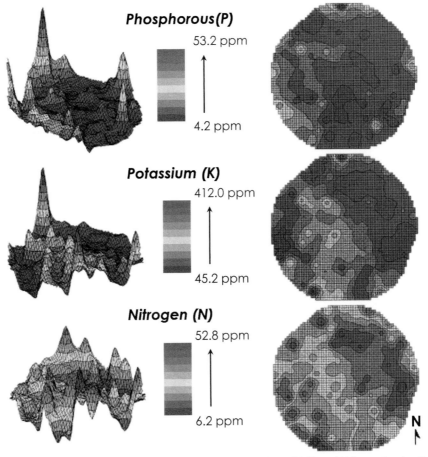

Figure 29-16 Map surfaces identifying the spatial distribution of P,K and N throughout a field. See included DVD for color version.

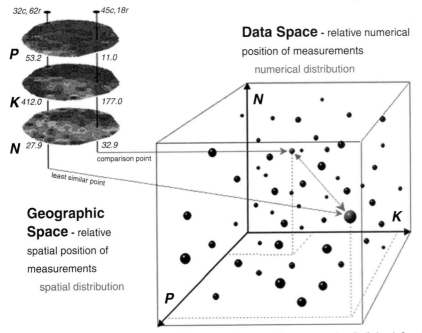

Figure 29-17 Geographic space and data space can be conceptually linked. See included DVD for color version.

Manual of Geographic Information Systems

The right side of the figure conceptually depicts how the computer calculates a similarity value for the two response patterns. The realization that mapped data can be expressed in both geographic space and data space is a key element to understanding the procedure.

Geographic space uses coordinates, such as latitude and longitude, to locate things in the real world—such as the southeast and extreme north points identified in the example. The geographic expression of the complete set of measurements depicts their spatial distribution in familiar map form.

Data space, on the other hand, is a bit less familiar but can be conceptualized as a box with balls floating within it. In the example, the three axes defining the extent of the box correspond to the P, K and N levels measured in the field. The floating balls represent grid cells defining the geographic space—one for each grid cell. The coordinates locating the floating balls extend from the data axes—11.0, 177.0 and 32.9 for the comparison point. The other point has considerably higher values in P and K with slightly lower N (53.2, 412.0, 27.9) so it plots at a different location in data space.

The bottom line is that the position of any point in data space identifies its numerical pattern—low, low, low is in the back-left corner, while high, high, high is in the upper-right corner. Points that plot in data space close to each other are similar; those that plot farther away are less similar.

In the example, the floating ball in the foreground is the farthest one (least similar) from the comparison point's data pattern. This distance becomes the reference for 'most different' and sets the bottom value of the similarity scale (0%). A point with an identical data pattern plots at exactly the same position in data space resulting in a data distance of 0; that equates to the highest similarity value (100%).

The similarity map shown in Figure 29-18 applies the similarity scale to the data distances calculated between the comparison point and all of the other points in data space. The green tones indicate field locations with fairly similar P, K and N levels. The red tones indicate dissimilar areas. It is interesting to note that most of the very similar locations are in the left portion of the field.

Map Similarity can be an invaluable tool for investigating spatial patterns in any complex set of mapped data. Humans are unable to conceptualize more than three variables (the data space box); however a similarity index can handle any number of input maps. In addition, the different layers can be weighted to reflect relative importance in determining overall similarity.

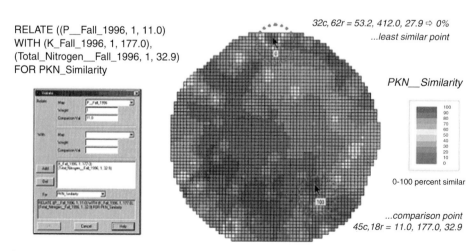

Figure 29-18 A similarity map identifies how related the data patterns are for all other locations to the pattern of a given comparison location. See included DVD for color version.

In effect, a similarity map replaces a lot of laser-pointer waving and subjective suggestions of how similar/dissimilar locations are with a concrete, quantitative measurement for each map location.

29.4.2.2 Identifying Data Zones

The preceding section introduced the concept of 'data distance' as a means to measure similarity within a map. One simply mouse-clicks a location and all of the other locations are assigned a similarity value from 0 (zero percent similar) to 100 (identical) based on a set of specified maps. The similarity value replaces difficult visual interpretation of map displays with an exact quantitative measure at each location.

Figure 29-19 depicts level slicing for areas that are unusually high in P, K and N. In this instance the data pattern coincidence is a box in 3-dimensional data space.

A mathematical "trick" was employed to get the map solution shown in the figure. On the individual maps, areas high in phosphorous were set to P = 1, areas high in potassium were set to K = 2 and areas high in nitrogen were set to N = 4; then the maps were added together. The result is a range of coincidence values from zero (0 + 0 + 0 = 0; gray = no high concentrations) to seven (1 + 2 + 4 = 7; high P, high K and high N, shown in red). The map values between these extremes identify the individual map layers having high measurements. For example, the yellow areas with the value 3 have high P and K but not N (1 + 2 + 0= 3). If four or more binary maps are to be combined, the areas of interest are assigned progressive power-of-two values (…8, 16, 32, etc.)—the sum will always uniquely identify the combinations.

While *Level Slicing* is not a sophisticated classifier, it illustrates the useful link between data space and geographic space. This fundamental concept forms the basis for most geo-statistical analysis including map clustering and regression.

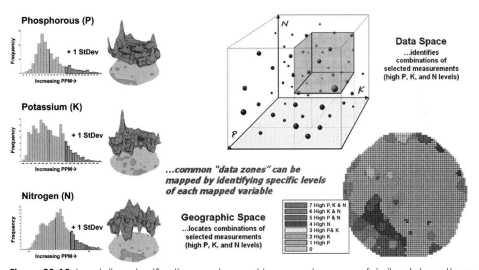

Figure 29-19 Level-slice classification can be used to map sub-groups of similar data patterns. See included DVD for color version.

29.4.2.3 Mapping Data Clusters

While both Map Similarity and Level Slicing techniques are useful in examining spatial relationships, they require the user to specify data analysis parameters. But what if you don't know what level slice intervals to use or which locations in the field warrant map similarity investigation? Can the computer on its own identify groups of similar data? How would such a classification work? How well would it work?

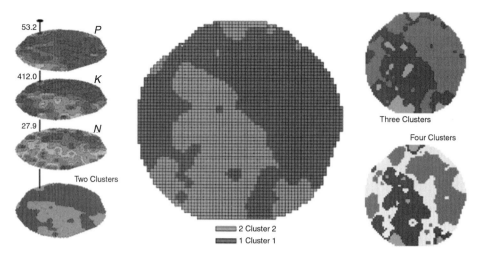

Figure 29-20 Map clustering identifies inherent groupings of data patterns in geographic space. See included DVD for color version.

Figure 29-20 shows some examples derived from *Map Clustering*. The map stack on the left shows the input maps used for the cluster analysis. The maps are the same P, K, and N maps identifying phosphorous, potassium and nitrogen levels used in the previous discussions in this section. However, keep in mind that the input maps could be crime, pollution or sales data—any set of application-related data. Clustering simply looks at the numerical pattern at each map location and sorts the patterns into discrete groups regardless of the nature of the data or its application.

The map in the center of the figure shows the results of classifying the P, K and N map stack into two clusters. The data pattern for each cell location is used to partition the field into two groups that meet the criteria as being 1) *as different as possible between groups* and 2) *as similar as possible within a group.*

The two smaller maps at the right show the division of the data set into three and four clusters. In all three of the cluster maps red is assigned to the cluster with relatively low responses and green to the one with relatively high responses. Note the encroachment on these marginal groups by the added clusters that are formed by data patterns at the boundaries.

The mechanics of generating cluster maps are quite simple: specify the input maps and the number of clusters you want then miraculously a map appears with discrete data groupings. So how is this miracle performed? What happens inside cluster's black box?

The schematic in Figure 29-21 depicts the process. The floating balls identify the data patterns for each map location (geographic space) plotted against the P, K and N axes (data space). For example, the large ball appearing closest to you depicts a location with high values on all three input maps. The tiny ball in the opposite corner (near the plot origin) depicts a map location with small map values. It seems sensible that these two extreme responses would belong to different data groupings.

While the specific algorithm used in clustering is beyond the scope of this chapter, it suffices to note that 'data distances' between the floating balls are used to identify cluster membership—groups of floating balls that are relatively far from other groups and relatively close to each other form separate data clusters. In this example, the red balls identify relatively low responses while green ones have relatively high responses. The geographic pattern of the classification is shown in the map in the lower right portion of the figure.

Identifying groups of neighboring data points to form clusters can be tricky business. Ideally, the clusters will form distinct clouds in data space. But that rarely happens and the clustering technique has to enforce decision rules that slice a boundary between nearly identical responses. Also, extended techniques can be used to impose weighted boundaries based on data trends or expert knowledge. Treatment of categorical data and leveraging spatial autocorrelation are other considerations.

So how do know if the clustering results are acceptable? Most statisticians would respond, "…you can't tell for sure." While there are some elaborate procedures focusing on the cluster assignments at the boundaries, the most frequently used benchmarks rely on standard statistical indices.

Figure 29-22 shows the performance table and box-and-whisker plots for the map containing two clusters. The average, standard deviation, minimum and maximum values within each cluster are calculated. Ideally the averages would be radically different and the standard deviations small—large difference between groups and small differences within groups.

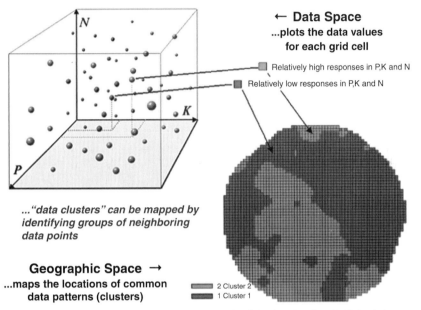

Figure 29-21 Data patterns for map locations are depicted as floating balls in data space. See included DVD for color version.

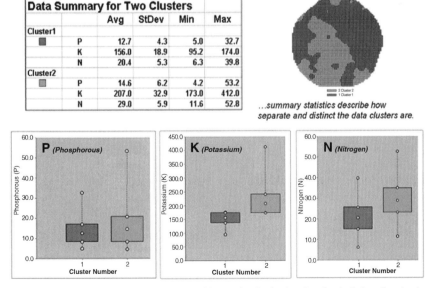

Figure 29-22 Clustering results can be roughly evaluated using basic statistics. See included DVD for color version.

Manual of Geographic Information Systems

Box-and-whisker plots enable a visual assessment of the differences. The box is centered on the average (position) and extends above and below one standard deviation (width) with the whiskers drawn to the minimum and maximum values to provide a visual sense of the data range. When the diagrams for the two clusters overlap, as they do for the phosphorous responses, it suggests that the clusters are not distinct along this data axis.

The separation between the boxes for the K and N axes suggests greater distinction between the clusters. Given the results a practical user would likely accept the classification results. And statisticians hopefully will accept in advance apologies for such a conceptual and terse treatment of a complex spatial statistics topic.

29.4.2.4 Deriving Prediction Maps

For years non-spatial statistics has been predicting things by analyzing a sample set of data for a numerical relationship (equation) then applying the relationship to another set of data. The drawbacks are that the nonspatial approach doesn't account for geographic relationships and the result is just a table of numbers. Extending predictive analysis to mapped data seems logical; after all, maps are just organized sets of numbers. And GIS enables us to link the numerical and geographic distributions of the data.

To illustrate the data mining procedure, the approach can be applied to the same field that has been the focus for the previous discussion. The top portion of Figure 29-23 shows the yield pattern of corn for the field varying from a low of 39 bushels per acre (bu/ac) (red) to a high of 279 (green). The corn yield map is termed the *dependent map variable* and identifies the phenomenon to be predicted.

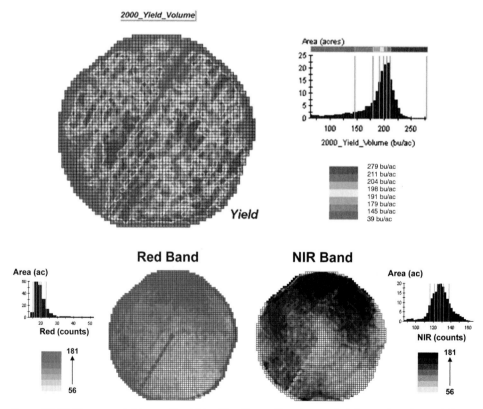

Figure 29-23 The corn yield map (top) identifies the pattern to predict; the red and near-infrared maps (bottom) are used to build the spatial relationship. See included DVD for color version.

The *independent map variables* depicted in the bottom portion of the figure are used to uncover the spatial relationship used for prediction—the *prediction equation*. In this instance, digital aerial imagery will be used to explain the corn yield patterns. The map on the left indicates the relative reflectance of red light off the plant canopy while the map on the right shows the near-infrared (NIR) response (a form of light just beyond what we can see).

While it is difficult to visually assess the subtle relationships between corn yield and the red and near-infrared images, the computer "sees" the relationship quantitatively. Each grid location in the analysis frame has a value for each of the map layers— 3,287 values defining each geo-registered map covering the 189-acre field.

For example, top portion of Figure 29-24 identifies that the example location has a 'joint' condition of red band equals 14.7 and yield equals 218. The lines parallel to axes in the scatter plot on the right identify the precise position of the pair of map values—X = 14.7 and Y = 218. Similarly, the near-infrared and yield values for the same location are shown in the bottom portion of the figure.

The set of dots in both of the scatter plots represents all of the data pairs for each grid location. The slanted lines through the dots represent the prediction equations derived through regression analysis. While the mathematics is a bit complex, the effect is to identify a line that 'best fits the data'—just as many data points are above as below the regression line.

In a sense, the line identifies the average yield for each step along the X-axis for the red and near-infrared bands and a reasonable guess of the corn yield for each level of spectral response. That's how a regression prediction is used—a value for the red band (or near-infrared band) in another field is entered and the equation for the line calculates a predicted corn yield. Repeating the calculation for all of the locations in the field generates a prediction map of yield from remotely sensed data.

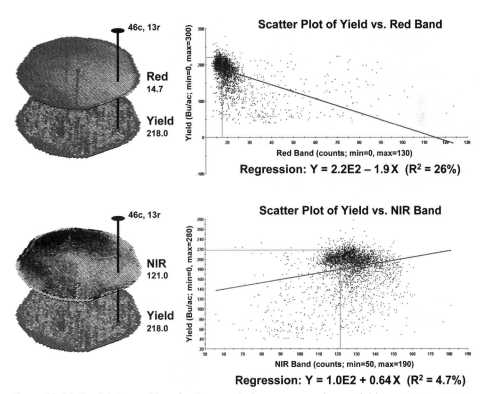

Figure 29-24 The joint conditions for the spectral response and corn yield maps are summarized in the scatter plots shown on the right. See included DVD for color version.

Manual of Geographic Information Systems

A major problem is that the R-squared statistic (R^2 in computer notation) summarizing the residuals for both of the prediction equations is fairly small (R^2 = 26% and 4.7% respectively) which suggests that the prediction lines do not fit the data very well. One way to improve the predictive model might be to combine the information in both of the images. The Normalized Density Vegetation Index (NDVI) does just that by calculating a new value that indicates plant density and vigor—NDVI = ((NIR – Red) / (NIR + Red)).

Figure 29-25 shows the process for calculating NDVI for the sample grid location—((121-14.7) / (121 + 14.7)) = 106.3 / 135.7 = .783. The scatter plot on the right shows the yield versus NDVI plot and regression line for all of the field locations. Note that the R^2 value is higher at 30% indicating that the combined index is a better predictor of yield.

The bottom portion of the figure evaluates the prediction equation's performance over the field. The two smaller maps show the actual yield (left) and predicted yield (right). As you would expect the prediction map doesn't contain the extreme high and low values actually measured.

The larger map on the right calculates the error of the estimates by simply subtracting the actual measurement from the predicted value at each map location. The error map suggests that overall the yield estimates are not too bad—average error is 2.62 bu/ac over estimate and 67% of the field is within +/- 20 bu/ac. Also note the geographic pattern of the errors: most of the over-estimates occur along the edge of the field, while most of the under estimates are scattered along northeast-southwest strips.

Evaluating a prediction equation on the data that generated it is not validation; however the procedure provides at least some empirical verification of the technique. It suggests hope that with some refinement the prediction model might be useful in predicting yield from remotely sensed data well before harvest.

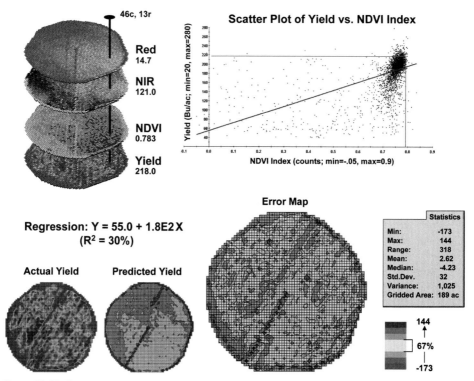

Figure 29-25 The red and NIR maps are combined for NDVI value that is a better predictor of yield. See included DVD for color version.

29.4.2.5 Stratifying Maps for Better Predictions

The preceding section described a procedure for predictive analysis of mapped data. While the underlying theory, concerns and considerations can be quite complex, the procedure itself is quite simple. The grid-based processing preconditions the maps so each location (grid cell) contains the appropriate data. The 'shish kebabs' of numbers for each location within a stack of maps are analyzed for a prediction equation that summarizes the relationships.

The left side of Figure 29-26 shows the evaluation procedure (including the error map) for regression analysis used to relate a map of NDVI to a map of corn yield for a farmer's field. One way to improve the predictions is to stratify the data set by breaking it into groups of similar characteristics. The idea is that prediction equations tailored to each stratum will result in better predictions than a single equation for an entire area. The technique is commonly used in non-spatial statistics where a data set might be grouped by age, income, and/or education prior to analysis. Additional factors for stratifying, such as neighboring conditions, data clustering and/or proximity can be used as well.

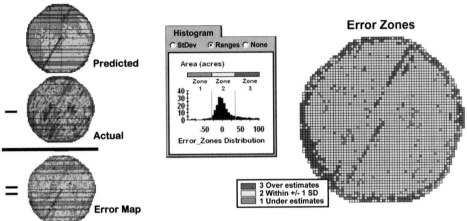

Figure 29-26 A project area can be stratified based on prediction errors. See included DVD for color version.

While there are numerous alternatives for stratifying, subdividing the error map will serve to illustrate the conceptual approach. The histogram in the center of Figure 29-26 shows the distribution of values on the Error Map. The vertical bars identify the breakpoints at +/- 1 standard deviation and divide the map values into three strata—zone 1 of unusually high under-estimates (red), zone 2 of typical error (yellow) and zone 3 of unusually high over-estimates (green). The map on the right of the figure maps the three strata throughout the field.

The rationale behind the stratification is that the whole-field prediction equation works fairly well for zone 2 but not so well for zones 1 and 3. The assumption is that conditions within zone 1 make the equation under estimate while conditions within zone 3 cause it to over estimate. If the assumption holds one would expect a tailored equation for each zone would be better at predicting corn yield than a single overall equation.

Figure 29-27 summarizes the results of deriving and applying a set of three prediction equations. The left side of the figure illustrates the procedure. The Error Zones map is used as a template to identify the NDVI and Yield values used to calculate three separate prediction equations. For each map location, the algorithm first checks the value on the Error Zones map then sends the data to the appropriate group for analysis. Once the data have been grouped a regression equation is generated for each zone.

The R^2 statistic for all three equations (.68, .60, and .42 for zones 1, 2 and 3 respectively) suggests that the equations fit the data fairly well and ought to be good predictors. The right side of Figure 29-27 shows a composite prediction map generated by applying the equations to the NDVI data respecting the zones identified on the template map.

Manual of Geographic Information Systems

The left side of Figure 29-28 provides a visual comparison between the actual yield and predicted maps. The stratified prediction shows detailed estimates that more closely align with the actual yield pattern than the 'whole-field' derived prediction map using a single equation. The error map for the stratified prediction shows that eighty percent of the estimates are within +/- 20 bushels per acre. The average error is only 4 bu/ac and has maximum under/over estimates of –81.2 and 113, respectively. All in all, fairly good yield estimates based on remote sensing data collected nearly a month before the field was harvested.

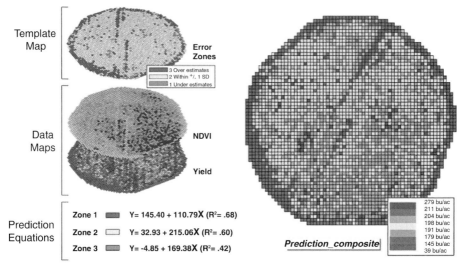

Figure 29-27 After stratification, prediction equations can be derived for each element. See included DVD for color version.

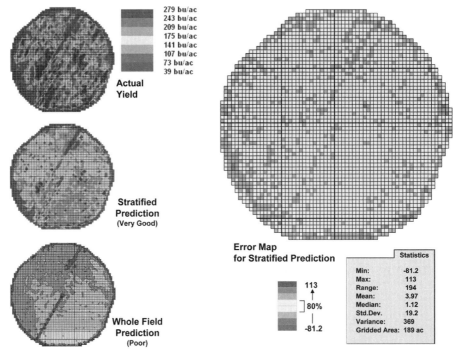

Figure 29-28 Stratified and whole-field predictions can be compared using statistical techniques. See included DVD for color version.

A couple of things should be noted from this example of spatial data mining. First, that there is a myriad of other ways to stratify mapped data—1) Geographic Zones, such as proximity to the field edge; 2) Dependent Map Zones, such as areas of low, medium and high yield; 3) Data Zones, such as areas of similar soil nutrient levels; and 4) Correlated Map Zones, such as micro terrain features identifying small ridges and depressions. The process of identifying useful and consistent stratification schemes is an emerging research frontier in the spatial sciences.

Second, the error map is a key part in evaluating and refining prediction equations. This point is particularly important if the equations are to be extended in space and time. The technique of using the same data set to develop and evaluate the prediction equations isn't always adequate. The results need to be tried at other locations and dates to verify performance. While spatial data mining methodology might be at hand, good science is imperative.

Finally, one needs to recognize that spatial data mining is not restricted to precision agriculture but has potential for analyzing relationships within almost any set of mapped data. For example, prediction models can be developed for geo-coded sales from demographic data or timber production estimates from soil/terrain patterns. The bottom line is that maps are increasingly seen as organized sets of data that can be map-ematically analyzed for spatial relationships—we have only scratched the surface.

29.5 Spatial Analysis Techniques

While map analysis tools might at first seem uncomfortable they simply are extensions of traditional analysis procedures brought on by the digital nature of modern maps. The previous Section 29.4 described a conceptual framework and some example procedures that extend traditional statistics to a spatial statistics that investigates numerical relationships within and among mapped data layers.

Similarly, a mathematical framework can be used to organize spatial analysis operations. Like basic math, this approach uses sequential processing of mathematical operations to perform a wide variety of complex map analyses. By controlling the order that the operations are executed, and using a common database to store the intermediate results, a mathematical-like processing structure is developed.

This 'map algebra' is similar to traditional algebra where basic operations, such as addition, subtraction and exponentiation, are logically sequenced for specific variables to form equations—however, in map algebra the variables represent entire maps consisting of thousands of individual grid values. Most of the traditional mathematical capabilities, plus an extensive set of advanced map processing operations, comprise the map analysis toolbox.

As with matrix algebra (a mathematics operating on sets of numbers) new operations emerge that are based on the nature of the data. Matrix algebra's transposition, inversion and diagonalization are examples of the extended set of techniques in matrix algebra.

In grid-based map analysis, the spatial coincidence and juxtaposition of values among and within maps create new analytical operations, such as coincidence, proximity, visual exposure and optimal routes. These operators are accessed through general purpose map analysis software available in most GIS systems. While the specific command syntax and mechanics differs among software packages, the basic analytical capabilities and spatial reasoning skills used in GIS modeling and analysis form a common foundation.

There are two fundamental conditions required by any spatial analysis package—a consistent data structure and an *iterative processing environment*. The earlier Section 29.3 described the characteristics of the grid-based data structure by introducing the concepts of an analysis frame, map stack, data types and display forms. The traditional discrete set of map features (points, lines and polygons) were extended to map surfaces that characterize geographic space as a continuum of uniformly-spaced grid cells. This structure forms a framework for the map-ematics underlying GIS modeling and analysis.

Manual of Geographic Information Systems

The second condition of map analysis provides an iterative processing environment by logically sequencing map analysis operations. This involves:
- retrieval of one or more map layers from the database,
- processing that data as specified by the user,
- creation of a new map containing the processing results, and
- storage of the new map for subsequent processing.

Each new map derived as processing continues aligns with the analysis frame so it is automatically geo-registered to the other maps in the database. The values comprising the derived maps are a function of the processing specified for the input maps. This cyclical processing provides an extremely flexible structure similar to "evaluating nested parentheses" in traditional math. Within this structure, one first defines the values for each variable and then solves the equation by performing the mathematical operations on those numbers in the order prescribed by the equation.

This same basic mathematical structure provides the framework for computer-assisted map analysis. The only difference is that the variables are represented by mapped data composed of thousands of organized values. Figure 29-29 shows a solution for calculating the percent change in animal activity.

The processing steps shown in the figure are identical to the algebraic formula for percent change except the calculations are performed for each grid cell in the study area and the result is a map that identifies the percent change at each location. Map analysis identifies what kind of change (thematic attribute) occurred where (spatial attribute). The characterization of "what and where" provides information needed for continued GIS modeling, such as determining if areas of large increases in animal activity are correlated with particular cover types or near areas of low human activity.

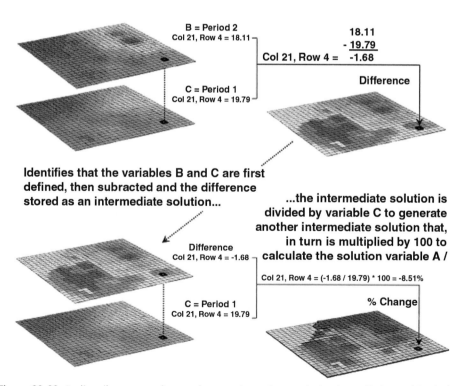

Figure 29-29 An iterative processing environment, analogous to basic math, is used to derive new map variables. See included DVD for color version.

29.5.1 Spatial Analysis Framework

Within this iterative processing structure, four fundamental classes of map analysis operations can be identified. These include:

- **Reclassifying Maps** – involving the reassignment of the values of an existing map as a function of its initial value, position, size, shape or contiguity of the spatial configuration associated with each map category.
- **Overlaying Maps** – resulting in the creation of a new map where the value assigned to every location is computed as a function of the independent values associated with that location on two or more maps.
- **Measuring Distance and Connectivity** – involving the creation of a new map expressing the distance and route between locations as straight-line length (simple proximity) or as a function of absolute or relative barriers (effective proximity).
- **Summarizing Neighbors** – resulting in the creation of a new map based on the consideration of values within the general vicinity of target locations.

Reclassification operations merely repackage existing information on a single map. Overlay operations, on the other hand, involve two or more maps and result in the delineation of new boundaries. Distance and connectivity operations are more advanced techniques that generate entirely new information by characterizing the relative positioning of map features. Neighborhood operations summarize the conditions occurring in the general vicinity of a location.

The reclassifying and overlaying operations based on point processing are the backbone of current GIS applications, allowing rapid updating and examination of mapped data. However, other than the significant advantage of speed and ability to handle tremendous volumes of data, these capabilities are similar to those of manual map processing. Map-wide overlays, distance and neighborhood operations, on the other hand, identify more advanced analytic capabilities and most often do not have paper-map legacy procedures.

The mathematical structure and classification scheme of Reclassify, Overlay, Distance and Neighbors form a conceptual framework that is easily adapted to modeling spatial relationships in both physical and abstract systems. A major advantage is flexibility. For example, a model for siting a new highway could be developed as a series of processing steps. The analysis likely would consider economic and social concerns (e.g., proximity to high housing density, visual exposure to houses), as well as purely engineering ones (e.g., steep slopes, water bodies). The combined expression of both physical and non-physical concerns within a quantified spatial context is a major benefit.

However, the ability to simulate various scenarios (e.g., steepness is twice as important as visual exposure and proximity to housing is four times more important than all other considerations) provides an opportunity to fully integrate spatial information into the decision-making process. By noting how often and where the proposed route changes as successive runs are made under varying assumptions, information is gained on the unique sensitivity to siting a highway in a particular locale.

In addition to flexibility, there are several other advantages in developing a generalized analytical structure for map analysis. The systematic rigor of a mathematical approach forces both theorist and user to carefully consider the nature of the data being processed. Also it provides a comprehensive format for learning that is independent of specific disciplines or applications. Furthermore the flowchart of processing succinctly describes the components and weightings capsulated in an analysis.

This communication enables decision-makers to more fully understand the analytic process and actually interact with weightings, incomplete considerations and/or erroneous assumptions. These comments, in most cases, can be incorporated and new results generated in a timely manner. From a decision-maker's point of view, traditional manual techniques for analyzing maps are a distinct and separate task from the decision itself. They require considerable time to perform and many of the considerations are subjective in their evaluation.

In the old environment, decision-makers attempt to interpret results, bounded by vague assumptions and system expressions of the technician. Computer-assisted map analysis, on the other hand, engages decision-makers in the analytic process. In a sense, it both documents the thought process and encourages interaction—sort of like a "spatial spreadsheet."

29.5.2 Reclassifying Maps

The first, and in many ways the most fundamental, class of analytical operations involves the reclassification of map categories. Each operation involves the creation of a new map by assigning thematic values to the categories of an existing map. These values may be assigned as a function of the *initial value*, *position*, *contiguity*, *size*, or *shape* of the spatial configuration of the individual categories. Each of the reclassification operations involves the simple repackaging of information on a single map, and results in no new boundary delineation. Such operations can be thought of as the purposeful "re-coloring of maps."

Figure 29-30 shows the result of reclassifying a map as a function of its initial thematic values. For display, a unique color is associated with each value. In the figure, the Cover Type map has categories of Open Water, Meadow and Forest. These features are stored as thematic values 1, 2 and 3, respectively, and displayed as separate colors. A binary map that isolates the Open Water locations can be created by simply assigning 0 to the areas of Meadow and Forest. While the operation seems trivial by itself, it has map analysis implications far beyond simply re-coloring the map categories, as will soon be apparent.

A similar reclassification operation might involve the ranking or weighing of qualitative map categories to generate a new map with quantitative values. For example, a map of soil types might be assigned values that indicate the relative suitability of each soil type for residential development.

Quantitative values may also be reclassified to yield new quantitative values. This might involve a specified reordering of map categories (e.g., given a map of soil moisture content, generate a map of suitability levels for plant growth). Or, it could involve the application of a generalized reclassifying function, such as "level slicing," which splits a continuous range of map category values into discrete intervals (e.g., derivation of a contour map of just ten contour intervals from an elevation surface composed of thousands of specific elevation values).

Other quantitative reclassification functions include a variety of arithmetic operations involving map category values and a specified or computed constant. Among these operations are addition, subtraction, multiplication, division, exponentiation, maximization, minimization, normalization and other scalar mathematical and statistical operators. For example, an elevation surface expressed in feet might be converted to meters by multiplying each map value by the appropriate conversion factor of 3.28083 feet per meter.

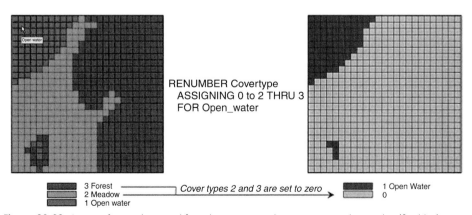

Figure 29-30 Areas of meadow and forest on a cover type map can be reclassified to isolate large areas of open water. See included DVD for color version.

Reclassification operations can also relate to location, as well as purely thematic attributes associated with a map. One such characteristic is position. An overlay category represented by a single location, for example, might be reclassified according to its latitude and longitude. Similarly, a line segment or area feature could be reassigned values indicating its center or general orientation.

A related operation, termed parceling, characterizes category contiguity. This procedure identifies individual clumps of one or more cells having the same numerical value and spatially contiguous (e.g., generation of a map identifying each lake as a unique value from a generalized map of water representing all lakes as a single category).

Another location characteristic is size. In the case of map categories associated with linear features or point locations, overall length or number of points might be used as the basis for reclassifying those categories. Similarly, an overlay category associated with a planar area might be reclassified according to its total acreage or the length of its perimeter. For example, a map of water types might be reassigned values to indicate the area of individual lakes or the length of stream channels. The same sort of technique might also be used to deal with volume. Given a map of depth to bottom for a group of lakes, for example, each lake might be assigned a value indicating total water volume based on the area of each depth category.

Figure 29-31 identifies a similar processing sequence using the information derived in the previous Figure 29-30. While your eye sees two distinct blobs of water on the open water map the computer only 'sees' distinctions by different map category values. Since both water bodies are assigned the same value of 1 there isn't a categorical distinction and the computer cannot easily differentiate.

The *Clump* operation is used to identify the contiguous features as separate values—clump #1, #2 and #3. The *Size* operation is used to calculate the size of each clump—clump #1 = 78 hectares, clump #2 = 543 ha and clump #3 = 4 ha. The final step uses the *Renumber* operation to isolate the large water body in the northwest portion of the project area.

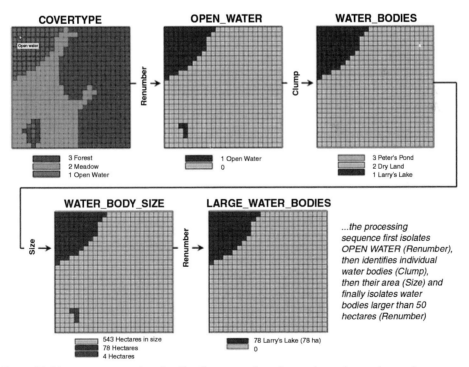

Figure 29-31 A sequence of reclassification operations (renumber, clump, size and renumber) can be used to isolate large water bodies from a cover type map. See included DVD for color version.

Manual of Geographic Information Systems

In addition to the initial value, position, contiguity, and size of features, shape characteristics also can be used as the basis for reclassifying map categories. Shape characteristics associated with linear forms identify the patterns formed by multiple line segments (e.g., dendritic stream pattern). The primary shape characteristics associated with polygonal forms include feature integrity, boundary convexity, and nature of edge.

Feature integrity relates to "intact-ness" of an area. A category that is broken into numerous 'fragments' and/or contains several interior 'holes' is said to have less spatial integrity than ones without such violations. Feature integrity can be summarized as the Euler Number that is computed as the number of holes within a feature less one short of the number of fragments which make up the entire feature. An Euler Number of zero indicates features that are spatially balanced, whereas larger negative or positive numbers indicate less spatial integrity.

Convexity and edge are other shape indices that relate to the configuration of boundaries of polygonal features. Convexity is the measure of the extent to which an area is enclosed by its background, relative to the extent to which the area encloses this background. The Convexity Index for a feature is computed by the ratio of its perimeter to its area. The most regular configuration is that of a circle which is totally convex and, therefore, not enclosed by the background at any point along its boundary.

Comparison of a feature's computed convexity to a circle of the same area, results in a standard measure of boundary regularity. The nature of the boundary at each point can be used for a detailed description of boundary configuration. At some locations the boundary might be an entirely concave intrusion, whereas others might be entirely convex protrusions. Depending on the "degree of edginess," each point can be assigned a value indicating the actual boundary convexity at that location.

This explicit use of cartographic shape as an analytic parameter is unfamiliar to most GIS users. However, a non quantitative consideration of shape is implicit in any visual assessment of mapped data. Particularly promising is the potential for applying quantitative shape analysis techniques in the areas of digital image classification and wildlife habitat modeling. A map of forest stands, for example, might be reclassified such that each stand is characterized according to the relative amount of forest edge with respect to total acreage and the frequency of interior forest canopy gaps. Those stands with a large proportion of edge and a high frequency of gaps will generally indicate better wildlife habitat for many species.

29.5.3 Overlaying Maps

The general class of overlay operations can be characterized as "light table gymnastics." These involve the creation of a new map where the value assigned to every point, or set of points, is a function of the independent values associated with that location on two or more existing map layers. In location specific overlaying, the value assigned is a function of the point by point coincidence of the existing maps. In category wide composites, values are assigned to entire thematic regions as a function of the values on other overlays that are associated with the categories. Whereas the first overlay approach conceptually involves the vertical spearing of a set of map layers, the latter approach uses one map to identify boundaries by which information is extracted from other maps.

Figure 29-32 shows an example of location specific overlaying. Here, maps of cover type and topographic slope classes are combined to create a new map identifying the particular cover/slope combination at each map location. A specific function used to compute new category values from those of existing maps being overlaid can vary according to the nature of the data being processed and the specific use of that data within a modeling context. Environmental analyses typically involve the manipulation of quantitative values to generate new values that are likewise quantitative in nature. Among these are the basic arithmetic operations such as addition, subtraction, multiplication, division, roots and exponentials.

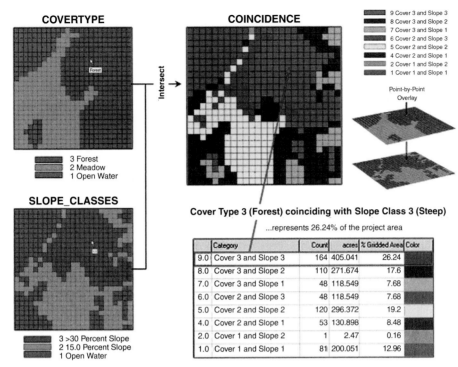

COVERTYPE

3 Forest
2 Meadow
1 Open Water

SLOPE_CLASSES

3 >30 Percent Slope
2 15.0 Percent Slope
1 Open Water

Intersect →

COINCIDENCE

9 Cover 3 and Slope 3
8 Cover 3 and Slope 2
7 Cover 3 and Slope 1
6 Cover 2 and Slope 3
5 Cover 2 and Slope 2
4 Cover 2 and Slope 1
2 Cover 1 and Slope 2
1 Cover 1 and Slope 1

Point-by-Point
Overlay

Cover Type 3 (Forest) coinciding with Slope Class 3 (Steep)

...represents 26.24% of the project area

	Category	Count	acres	% Gridded Area	Color
9.0	Cover 3 and Slope 3	164	405.041	26.24	
8.0	Cover 3 and Slope 2	110	271.674	17.6	
7.0	Cover 3 and Slope 1	48	118.549	7.68	
6.0	Cover 2 and Slope 3	48	118.549	7.68	
5.0	Cover 2 and Slope 2	120	296.372	19.2	
4.0	Cover 2 and Slope 1	53	130.898	8.48	
2.0	Cover 1 and Slope 2	1	2.47	0.16	
1.0	Cover 1 and Slope 1	81	200.051	12.96	

Figure 29-32 Point-by point overlaying operations summarize the coincidence of two or more maps, such as assigning a unique value identifying the cover type and slope class conditions at each location. See included DVD for color version.

Functions that relate to simple statistical parameters such as maximum, minimum, median, mode, majority, standard deviation or weighted average also can be applied. The type of data being manipulated dictates the appropriateness of the mathematical or statistical procedure used. For example, the addition of qualitative maps such as soils and land use would result in mathematically meaningless sums, since their thematic values have no numerical relationship. Other map overlay techniques include several that might be used to process either quantitative or qualitative data and generate values which can likewise take either form. Among these are masking, comparison, calculation of diversity, and permutations of map categories.

More complex statistical techniques can be applied in this manner, assuming that the inherent interdependence among spatial observations can be taken into account. This approach treats each map as a variable, each point as a case, and each value as an observation. A predictive statistical model, such as regression, can then be calculated for each location, resulting in a spatially continuous surface of predicted values. The mapped predictions contain additional information over traditional non spatial procedures, such as direct consideration of coincidence among regression variables and the ability to spatially locate areas of a given level of prediction. Sections 29.4.2.4 and 29.4.2.5 discussed considerations involved in spatial data mining derived by statistically overlaying mapped data.

An entirely different approach to overlaying maps involves category wide summarization of values. Rather than combining information on a point by point basis, this approach summarizes the spatial coincidence of entire categories shown on one map with the values contained on another map(s). Figure 29-33 contains an example of a category-wide overlay operation. In this example, the categories of the cover type map are used to define an area over which the coincidental values of the slope map are averaged. The computed values of average slope within each category area are then assigned to each of the cover type categories.

Manual of Geographic Information Systems

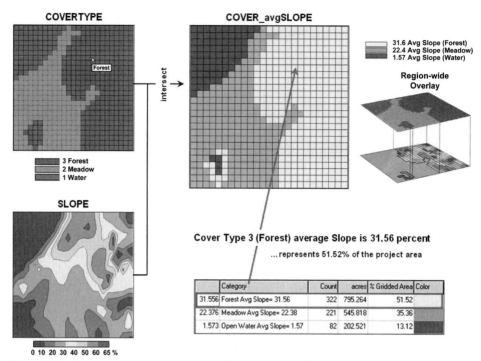

Figure 29-33 Category-wide overlay operations summarize the spatial coincidence of map categories, such as generating the average slope for each cover type category. See included DVD for color version.

Summary statistics which can be used in this way include the total, average, maximum, minimum, median, mode, or minority value; the standard deviation, variance, or diversity of values; and the correlation, deviation, or uniqueness of particular value combinations. For example, a map indicating the proportion of undeveloped land within each of several counties could be generated by superimposing a map of county boundaries on a map of land use and computing the ratio of undeveloped land to the total land area for each county. Or a map of zip code boundaries could be superimposed over maps of demographic data to determine the average income, average age, and dominant ethnic group within each zip code.

As with location specific overlay techniques, data types must be consistent with the summary procedure used. Also of concern is the order of data processing. Operations such as addition and multiplication are independent of the order of processing. Other operations, such as subtraction and division, however, yield different results depending on the order in which a group of numbers is processed. This latter type of operation, termed non commutative, cannot be used for category wide summaries.

29.5.4 Establishing Distance and Connectivity

Measuring distance is one of the most basic map analysis techniques. Historically, distance is defined as the *shortest straight line* between *two points*. While this three-part definition is both easily conceptualized and implemented with a ruler, it is frequently insufficient for decision-making. A straight-line route might indicate the distance "as the crow flies," but offer little information for the walking crow or other flightless creature. It is equally important to most travelers to have the measurement of distance expressed in more relevant terms, such as time or cost.

Proximity establishes the distance to all locations surrounding a point— *the set of shortest straight lines among groups of points*. Rather than sequentially computing the distance between pairs of locations, concentric equidistance zones are established around a location or set of locations (Figure 29-34). This procedure is similar to the wave pattern generated when a rock is thrown into a still pond. Each ring indicates one unit farther away—increasing distance as the wave moves away. Another way to conceptualize the process is nailing one end of a ruler at a point and spinning it around. The result is a series of data zones emanating from a location and aligning with the ruler's tick marks.

However, nothing says proximity must be measured from a single point. A more complex proximity map would be generated if, for example, all locations with houses (set of points) are simultaneously considered target locations (left side of Figure 29-35).

Proximity is defined as the set of shortest straight lines between a location and all other points in a project area

...Pythagorean Theorum procedure is replaced

...by the "Splash Algorithm" that is analogous to tossing a rock into a pond and counting the ripples (grid spaces) to determine distance to everywhere - like nailing a ruler at a point and spinning it

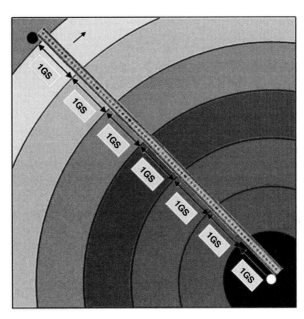

Figure 29-34 Proximity identifies the set of shortest straight-lines among groups of points (distance zones). See included DVD for color version.

Simple Proximity Surfaces

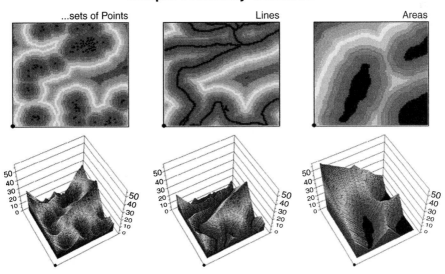

Figure 29-35 Proximity surfaces can be generated for groups of points, lines or polygons identifying the shortest distance from all location to the closest occurrence. See included DVD for color version.

Manual of Geographic Information Systems

In effect, the procedure is like throwing a handful of rocks into pond. Each set of concentric rings grows until the wave fronts from other locations meet; then they stop. The result is a map indicating the shortest straight-line distance to the nearest target area (house) for each non-target area. In the figure, the red tones indicate locations that are close to a house, while the green tones identify areas that are far from a house.

In a similar fashion, a proximity map to roads is generated by establishing data zones emanating from the road network—sort of like tossing a wire frame into a pond to generate a concentric pattern of ripples (middle portion of Figure 29-35). The same result is generated for a set of areal features, such as sensitive habitat parcels (right side of Figure 29-35).

It is important to note that proximity is not the same as a buffer. A buffer is a discrete spatial object that identifies areas that are within a specified distance of a map feature; all locations within a buffer are considered the same. Proximity is a continuous surface that identifies the distance to a map feature(s) for every location in a project area. It forms a gradient of distances away composed of many map values; not a single spatial object with one characteristic distance away.

The 3D plots of the proximity surfaces in Figure 29-35 show detailed gradient data and are termed accumulated surfaces. They contain increasing distance values from the target point-, line- or area-locations displayed as colors from red (close) to green (far). The starting features are the lowest locations (black= 0) with hillsides of increasing distance and forming ridges that are equidistant from starting locations.

In many applications, however, the shortest route between two locations might not always be a straight line (or even a slightly wiggling set of grid steps). And even if it is straight, its geographic length may not always reflect a traditional measure of distance. Rather, distance in these applications is best defined in terms of "movement" expressed as travel-time, cost, or energy that is consumed at rates that vary over time and space. Distance-modifying effects involve weights and/or barriers—concepts that imply the relative ease of movement through geographic space might not always constant.

Effective proximity responds to intervening conditions or barriers. There are two types of barriers that are identified by their effects—absolute and relative. *Absolute barriers* are those completely restricting movement and therefore imply an infinite distance between the points they separate. A river might be regarded as an absolute barrier to a non-swimmer. To a swimmer or a boater, however, the same river might be regarded as a *relative barrier* identifying areas that are passable, but only at a cost which can be equated to an increase in geographical distance. For example, it might take five times longer to row a hundred meters than to walk that same distance.

In the conceptual framework of tossing a rock into a pond, the waves can crash and dissipate against a jetty extending into the pond (absolute barrier; no movement through the grid spaces). Or they can proceed, but at a reduced wavelength through an oil slick (relative barrier; higher cost of movement through the grid spaces). The waves move both around the jetty and through the oil slick with the ones reaching each location first identifying *the set of shortest, but not necessarily straight, lines among groups of points*.

The shortest routes respecting these barriers are often twisted paths around and through the barriers. The GIS database enables the user to locate and calibrate the barriers; the wave-like analytic procedure enables the computer to keep track of the complex interactions of the waves and the barriers. For example, Figure 29-36 shows the effective proximity surfaces for the same set of starter locations shown in the previous Figure 29-35 expressed as simple proximity.

The point features in the left inset respond to treating flowing water as an absolute barrier to movement. Note that the distance to the nearest house is very large in the center-right portion of the project area (green) although there is a large cluster of houses just to the north. Since the water feature can't be crossed, the closest houses are a long distance to the south.

Effective Proximity Surfaces

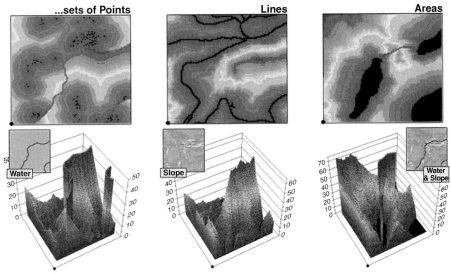

Figure 29-36 Effective proximity surfaces consider the characteristics and conditions of movement throughout a project area. See included DVD for color version.

Terrain steepness is used in the middle inset to illustrate the effects of a relative barrier. Increasing slope is coded into a friction map of increasing impedance values that make movement through steep grid cells effectively farther away than movement through gently sloped locations. Both absolute and relative barriers are applied in determining effective proximity-sensitive areas in the right inset.

Compare these results in figures 29-35 and 29-36 and note the dramatic differences between the concept of distance "as the crow flies" (simple proximity) and "as the crow walks" (effective proximity). In many practical applications, the assumption that all movement occurs in straight lines disregards reality. When traveling by trains, planes, automobiles, and feet there are plenty of bends, twists, accelerations and decelerations due to characteristics (weights) and conditions (barriers) of the movement.

Figure 29-37 illustrates how the splash algorithm propagates distance waves to generate an effective proximity surface. The Friction Map locates the absolute (blue/water) and relative (light blue = gentle/easy through red = steep/hard) barriers. As the distance wave encounters the barriers their effects on movement are incorporated and distort the symmetric pattern of simple proximity waves. The result identifies the "shortest, but not necessarily straight" distance connecting the starting location with all other locations in a project area.

Note that the absolute barrier locations (blue) are set to infinitely far away and appear as pillars in the 3-D display of the final proximity surface. As with simple proximity, the effective distance values form a bowl-like surface with the starting location at the lowest point (zero away from itself) and then ever-increasing distances away (upward slope). With effective proximity, however, the bowl is not symmetrical and is warped with bumps and ridges that reflect intervening conditions— the greater the impedance the greater the upward slope of the bowl. In addition, there can never be a depression as that would indicate a location that is closer to the starting location than everywhere around it. Such a situation would violate the ever-increasing concentric rings theory and is impossible.

The past four sections comprised a series focused on how simple distance is extended to effective proximity and movement in a modern GIS. Considerable emphasis was given to the calculations involving a propagating wave of increasing distance (algorithm) instead of

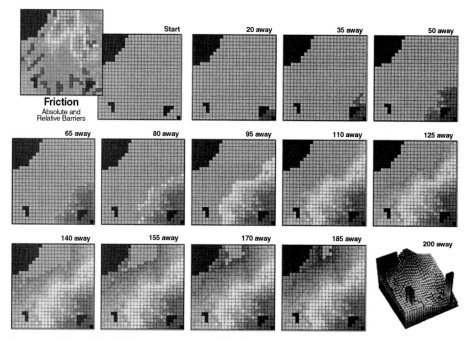

Figure 29-37 Effective Distance waves are distorted as they encounter absolute and relative barriers, advancing faster under easy conditions and slower in difficult areas. See included DVD for color version.

our more familiar procedures of measuring with a ruler (manual) or solving the Pythagorean Theorem (mathematical).

While the computations of simple and effective proximity might be unfamiliar and appear complex, once programmed they are easily and quickly performed by modern computers. In addition, there is a rapidly growing wealth of digital data describing conditions that impact movement in the real world. It seems that all is in place for a radical rethinking and expression of distance—computers, programs and data are poised.

However, what seems to be the major hurdle for adoption of this new way of spatial thinking lies in the experience base of potential users. Our paper map legacy suggests that the shortest straight line between two points is the only way to investigate spatial context relationships and anything else is wrong (or at least uncomfortable).

This restricted perspective has lead most contemporary GIS applications to employ simple distance and buffers. While simply automating traditional manual procedures might be comfortable, it fails to address the reality of complex spatial problems or fully engage the potential of GIS technology.

The first portion of Figure 29-38 identifies the basic operations described in the previous sections. These procedures have set the stage for even more advanced distance operations, as outlined in the lower portion of the figure. For example, consider a *Guiding Surface* which can be used to constrain movement up, down or across a surface: the algorithm can check an elevation surface and only proceed to downhill locations from a feature such as roads to identify areas potentially affected by the wash of surface chemicals applied.

The simplest *Directional Effect* involves compass directions, such as establishing proximity only in the direction of a prevailing wind. A more complex directional effect is consideration of the movement with respect to an elevation surface—a steep uphill movement might be considered a higher friction value than movement across a slope or downhill. This consideration involves a dynamic barrier that the algorithm must evaluate for each point along the wave front as it propagates.

Manual of Geographic Information Systems

Basic Distance Operations —
✓ **Simple Proximity** as the crow flies (straight lines)
✓ **Effective Proximity** as the crow walks (not necessarily straight respecting absolute and relative barriers)
✓ **Weighted Proximity** recognizes differences in mover characteristics

Advanced Distance Operations —
✓ **Guiding Surface** restricted movement (Up/Downhill)
✓ **Directional Effects** (bearing; up/across slope)
✓ **Accumulation Effects** (wear and tear)
✓ **Momentum Effects** (acceleration/deceleration with movement)
✓ **Stepped Movement** (go until specified location then restart)
✓ **Back Azimuth** (direction of travel)
✓ **1st and 2nd Derivative** (speed and change in speed)

Figure 29-38 The basic set of distance operations can be extended by considering the dynamic nature of the implied movement.

Accumulation Effects account for wear and tear as movement continues. For example, a hiker might easily proceed through a fairly steep uphill slope at the start of a hike but balk and pitch a tent at the same slope encountered ten hours into a hike. In this case, the algorithm "carries" an equation that increases the static/dynamic friction values as the movement wave front progresses. A natural application is to enter the gas tank size and average mileage of your car so the algorithm would automatically suggest refilling stops along a proposed route.

A related consideration, *Momentum Effects*, tracks the total effective distance but in this instance it calculates the net effect of uphill/downhill conditions that are encountered. It is similar to a marble rolling over an undulating surface—it picks up speed on the downhill stretches and slows down on the uphill ones.

The remaining three advanced operations interact with the accumulation surface derived by the wave front's movement. Recall that this surface is analogous to a football stadium with each tier of seats being assigned a distance value indicating increasing distance from the field. In practice, an accumulation surface is a twisted bowl whose rim is always increasing asymmetrically: the different rates reflect the differences in the spatial patterns of relative and absolute barriers.

Stepped Movement allows the proximity wave to grow until it reaches a specified location, and then restart at that location until another specified location and so on. This generates a series of effective proximity facets from the closest to the farthest location. The steepest downhill path over each facet, as you might recall, identifies the optimal path for that segment. The set of segments for all of the facets forms the optimal path network connecting the specified points.

The direction of optimal travel through any location in a project area can be derived by calculating the *Back Azimuth* of the location on the accumulation surface. Recall that the wave front potentially can step to any of its eight neighboring cells and keeps track of the one with the least friction to movement. The aspect of the steepest downhill step (N, NE, E, SE, S, SW, W or NW) at any location on the accumulation surface therefore indicates the direction of the best path through that location. In practice there are two directions—one in and one out for each location.

An even more bizarre extension is the interpretation of the *1st and 2nd Derivative* of an accumulation surface. The 1st derivative (rise over run) identifies the change in accumulated value (friction value) per unit of geographic change (cell size). On a travel-time surface, the result is the speed of optimal travel (best path) across the cell. The second derivative generates values whether the movement at each location is accelerating or decelerating.

Chances are these extensions to distance operations seem a bit confusing, uncomfortable, esoteric and bordering on heresy. While the old "straight line" procedure from our paper map legacy may be straightforward, it fails to recognize the reality that most things rarely move in straight lines, through constant conditions.

Effective distance recognizes the complexity of realistic movement by utilizing a procedure of propagating proximity waves that interact with a map indicating relative ease of movement. By assigning values to relative and absolute barriers to travel, a user enables the algorithm to consider locations to favor or avoid as movement proceeds. The basic distance operations assume static conditions, whereas the advanced ones account for dynamic conditions that vary with the nature of the movement.

So what's the "take home" from the discussions involving effective distance? Two points seem to define the bottom line. First, the digital map is revolutionizing how we perceive distance, as well as how we calculate it. It is the first radical change since Pythagoras came up with his theorem a couple of thousand years ago. Secondly, the ability to quantify effective distance isn't limited by computational power or available data; rather it is most limited by difficulties in understanding and accepting the concept.

29.5.5 Summarizing Neighbors

Analysis of spatially defined neighborhoods involves summarizing the context of surrounding locations. Four steps are involved in neighborhood analysis—1) define the neighborhood, 2) identify map values within the neighborhood, 3) summarize the values and 4) assign the summary statistic to the focus location. The process is then repeated for every location in a project area.

Summarizing Neighbors techniques fall into two broad classes of analysis—*Characterizing Surface Configuration and Summarizing Map Values* (see Figure 29-39). It is important to note that all neighborhood analyses involve mathematical or statistical summary of values on an existing map that occur within a roving window. As the window is moved throughout a project area, the summary value is stored for the grid location at the center of the window resulting in a new map layer reflecting neighboring characteristics or conditions.

Approach	Description	Example Techniques
Characterizing Surface Configuration	The map values in the roving window are assumed to form a portion of a continuous map surface and a new map value characterizing the configuration of the implied gradient is stored for the grid location at the center of the roving window	Slope, Aspect, Profile
Summarizing Map Values	The map values in the roving window are treated numerically and a new map value summarizing the neighboring values is stored for the grid location at the center of the roving window	Total, Average, StDev, CoffVar, Maximum, Minimum, Median, majority, Minority, Diversity, Deviation, Proportion, Custom Filters, Spatial Interpolation

Figure 29-39 The two fundamental classes of neighborhood analysis operations involve Characterizing Surface Configuration and Summarizing Map Values.

The difference between the two classes is in the treatment of the values—implied surface configuration or direct numerical summary. For example, Figure 29-40 shows a small portion of a typical elevation data set, with each cell containing a value representing its overall elevation. In the highlighted 3x3 window there are eight individual slopes, as shown in the calculations on the right side of the figure. The steepest slope in the window is 52% formed by the center and the NW neighboring cell. The minimum slope is 11% in the NE direction.

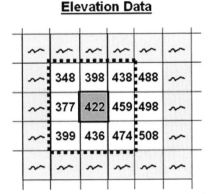

Elevation Data

Eight Individual Slopes

N = (398 - 422) / (1.00 *100) = 24%
NE = (438 - 422) / (1.41 *100) = 11%
E = (459 - 422) / (1.00 *100) = 37%
SE = (474 - 422) / (1.41 *100) = 37%
S = (436 - 422) / (1.00 *100) = 14%
SW = (399 - 422) / (1.41 *100) = 16%
W = (377 - 422) / (1.00 *100) = 45%
NW = (348 - 422) / (1.41 *100) = 52%

Maximum = 52%   Median = 30%
Minimum = 11%   Average = 29%

Figure 29-40 At a location, the eight individual slopes can be calculated for a 3x3 window and then summarized for the maximum, minimum, median and average slope. See included DVD for color version.

But what about the general slope throughout the entire 3x3 analysis window? One estimate is 29%, the arithmetic average of the eight individual slopes. Another general characterization could be 30%, the median of slope values. But let's stretch the thinking a bit more. Imagine that the nine elevation values become balls floating above their respective locations, as shown in Figure 29-41. Mentally insert a plane and shift it about until it is positioned to minimize the overall distances from the plane to the balls. The result is a "best-fitted plane" summarizing the overall slope in the 3x3 window.

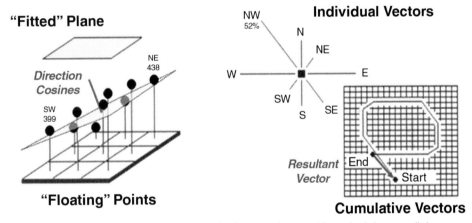

Figure 29-41 Best-fitted plane and vector algebra can be used to calculate overall slope. See included DVD for color version.

The algorithm is similar to fitting a regression line to a set of data points in two-dimensional space. However in this case, it's a plane in three-dimensional space. There is an intimidating set of equations involved, with a lot of Greek letters and subscripts to "minimize the sum of the squared deviations" from the plane to the points. Solid geometry calculations, based on the plane's "direction cosines," are used to determine the slope (and aspect) of the plane.

Another procedure for fitting a plane to the elevation data uses vector algebra, as illustrated in the right portion of Figure 29-41. In concept, the mathematics draws each of the eight slopes as a line in the proper direction and relative length of the slope value (individual vectors). Now comes the fun part. Starting with the NW line, successively connect the lines as shown in the figure (cumulative vectors). The civil engineer will recognize this procedure as similar to the latitude and departure sums in "closing a survey transect." The length of the "resultant vector" is the slope (and direction is the aspect) of the fitted plane.

Manual of Geographic Information Systems

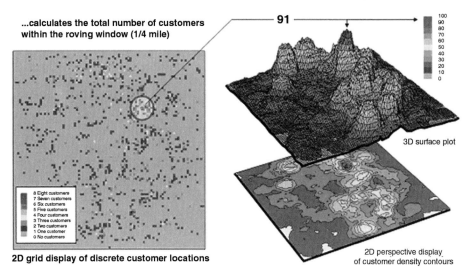

Figure 29-42 Approach used in deriving a Customer Density surface from a map of customer locations. See included DVD for color version.

There is a lot more to neighborhood analysis than just characterizing the lumps and bumps of the terrain. Figure 29-42 shows a direct numerical summary identifying the number of customers within a quarter of a mile of every location within a project area.

The procedure uses a roving window to collect neighboring map values and compute the total number of customers in the neighborhood. In this example, the window is positioned at a location that computes a total of 91 customers within quarter-mile.

Note that the input data is a discrete placement of customers while the output is a continuous surface showing the gradient of customer density. While the example location does not even have a single customer, it has an extremely high customer density because there are a lot of customers surrounding it.

The map displays on the right of the figure show the results of the processing for the entire area. A traditional vector GIS forces the result into a set of 2D contour intervals stored as discrete polygon spatial objects—1 10 customer range, 10 20, 20 30, etc. The 3D surface plot, on the other hand, shows all of the calculated spatial detail—mountains of high customer density and valleys of low density. An importance difference is that the vector representation aggregates the results, whereas the grid representation contains all of the detailed information.

Figure 29-43 illustrates how the information was derived. The upper-right map is a display of the discrete customer locations of the neighborhood of values surrounding the "focal" cell. The large graphic on the right shows this same information with the actual map values superimposed. Actually, the values are from a worksheet in a database, with the column and row totals indicated along the right and bottom margins. The row (and column) sum identifies the total number of customers within the window—91 total customers within a quarter-mile radius.

This value is assigned to the focal cell location as depicted in the lower-left map. Now imagine moving the worksheet window to focus on the next cell on the right, determining the total number of customers and assigning the result—then on to the next location, and the next, and the next, etc. The process is repeated for every location in the project area to derive the customer density surface.

The processing summarizes the map values occurring within a location's neighborhood (roving window). In this case the resultant value was the sum of all the values. But summaries other than *Total* can be used—*Average, Standard Deviation, Coefficient of Variation, Maximum, Minimum,*

Median, Majority, Minority, Diversity, Deviation, Proportion, Custom Filters, and *Spatial Interpolation.* The next section of this chapter will focus on how these spatial techniques (Reclassify, Overlay, Distance and Neighbors) can be used to derive valuable insight into the conditions and characteristics through GIS modeling of relationships within and among map layers.

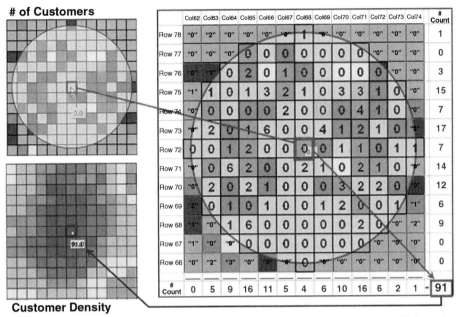

Figure 29-43 Calculations involved in deriving customer density. See included DVD for color version.

29.6 GIS Modeling Frameworks

Three elements are commonly recognized as essential to GIS—data, operations and applications. To use the technology a set of digital maps, an analytic engine to process the maps, and interesting problems to solve are needed. However, not all users have the same view of the relative importance of the three elements. Some have a ***data-centric*** perspective, as they prepare individual data layers and/or assemble the comprehensive databases forming the cornerstone of GIS. Others are ***operations-centric*** and are locked in on refining and expanding the GIS toolbox of processing and display capabilities. A third group is ***applications-centric*** and sees the portentous details of data and operations as merely impediments to solving real-world problems. Such is the occasionally fractious fraternity of GIS.

In the early years, data and software development dominated the developing field. As GIS has matured, the focus has extended to innovative ways of addressing complex spatial problems beyond simply mapping and geo-query. As a result, attention is increasingly directed toward the assumptions and linkages embedded in GIS models that weave data layers into logical expressions of spatial interrelationships that solve pressing problems. From this perspective there are three dominant GIS modeling frameworks—Suitability, Decision Support and Statistical modeling.

29.6.1 Suitability Modeling

A simple habitat model can be developed using only reclassify and overlay operations. For example, a Hugag is a curious mythical beast with strong preferences for terrain configuration: prefers low elevations, prefers gentle slopes, and prefers southerly aspects.

29.6.1.1 Binary Model

A binary habitat model of Hugag preferences is the simplest to conceptualize and implement. It is analogous to the manual procedures for map analysis popularized in the landmark book *Design with Nature*, by Ian L. McHarg, first published in 1969. This seminal work was the ancestor of modern map analysis: it described an overlay procedure involving paper maps, transparent sheets and pens.

For example, if avoiding steep slopes was an important decision criterion, a draftsperson would tape a transparent sheet over a topographic map, delineate areas of steep slopes (contour lines close together) and fill-in the precipitous areas with an opaque color. The process is repeated for other criteria, such as the Hugag's preference to avoid areas that are northerly-oriented and at high altitudes. The annotated transparencies then are aligned on a light-table and the transparent areas showing through identify acceptable habitat for the animal.

An analogous procedure can be implemented in a computer by using the value 0 to represent the unacceptable areas (opaque) and 1 to represent acceptable habitat (clear). As shown in Figure 29-44, an *Elevation* map is used to derive a map of terrain steepness (*Slope_map*) and orientation (*Aspect_map*). A value of 0 is assigned to locations Hugags want to avoid—

- Greater than 1800 feet elevation = 0   …*too high*
- Greater than 30% slope = 0   …*too steep*
- North, northeast and northwest = 0   …*too northerly*

All other locations are assigned a value of 1 to indicate acceptable areas. The individual binary habitat maps are shown in the displays on the right side of the figure. The dark red portions identify unacceptable areas that are analogous to McHarg's opaque areas delineated on otherwise clear transparencies.

A *Binary Suitability* map of Hugag habitat is generated by multiplying the three individual binary preference maps. If a zero is encountered on any of the map layers, the solution is sent to zero (bad habitat). For the example location on the right side of the figure, the preference string of values is 1 * 1 * 0 = 0 (Bad). Only locations with 1 * 1 * 1 = 1 (Good) identify areas without any limiting factors—good elevations, good slopes and good orientation. These areas are analogous to the clear areas showing through the stack of transparencies on a light-table.

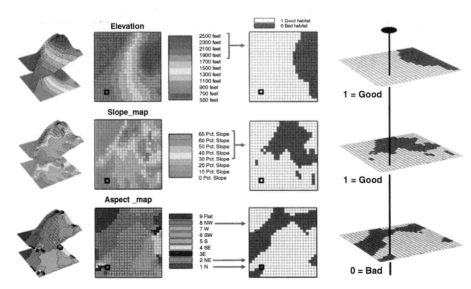

Figure 29-44 Binary maps representing Hugag habitat preferences are coded as 1= good and 0= bad. See included DVD for color version.

While this procedure mimics manual map processing, it is limited in the information it generates. The solution is binary and only differentiates acceptable and unacceptable locations. But an area that is totally bad (0 * 0 * 0 = 0) is significantly different from one that is just limited by one factor (1 * 1 * 0 = 0) as two factors are acceptable, thus making it nearly good.

29.6.1.2 Ranking Model

The right side of Figure 29-45 shows a *Ranking Suitability* map of Hugag habitat. In this instance the individual binary maps are simply added together for a count of the number of acceptable locations. Note that the areas of perfectly acceptable habitat (light grey) on both the binary and ranking suitability maps have the same geographic pattern. However, the unacceptable area on the ranking suitability map contains values 0 through 2 indicating how many acceptable factors occur at each location. The zero value for the area in the northeastern portion of the map identifies very bad conditions (0 + 0 + 0= 0). The example location, on the other hand, is nearly good (1 + 1 + 0= 2).

The ability to report the degree of suitability stimulated a lot of research into using additive colors for the manual process. Physics suggests that if blue, green and red are used as semi-opaque colors on the transparencies, the resulting color combinations should be those indicated in the left side of the table in Figure 29-46. However, in practice the technique collapses to an indistinguishable brownish-purple glob of colors when three or more map layers are overlaid.

Computer overlay, on the other hand, can accurately differentiate all possible combinations as shown on the right side of the figure. The trick is characterizing the unacceptable areas on each map layer as a binary progression of values—1, 2, 4, 8, 16, 32, etc. In the example in the figure, 1 is assigned to areas that are too high, 2 assigned to areas that are too steep and 4 assigned to areas that are too northerly oriented. When adding a binary progression of values, each combination of values results in a unique sum. The result is termed a *Ranking Combination* suitability map (right side of Figure 29-46).

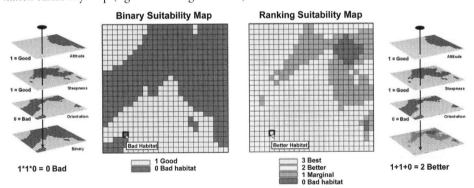

Figure 29-45 The binary habitat maps are multiplied together to create a Binary Suitability map (good or bad) or added together to create a Ranking Suitability map (bad, marginal, better or best). See included DVD for color version.

Comparison of Manual and Computer Overlay Techniques			
Manual Overlay		**Computer Overlay**	
Layer Colors	**Composite Color**	**Limiting Factor**	**Map Value**
Clear, clear, clear	Clear	--	0 + 0 + 0 = 0
Blue, clear, clear	Blue	High	1 + 0 + 0 = 1
Clear, green, clear	Green	Steep	0 + 2 + 0 = 2
Blue, green, clear	Cyan	High, steep	1 + 2 + 0 = 3
Clear, clear, red	Red	Northerly	0 + 0 + 4 = 4
Blue, clear, red	Indigo	High, northerly	1 + 0 + 4 = 5
Clear, green, red	Yellow	Steep, northerly	0 + 2 + 4 = 6
Blue, green, red	Black	All three (H,S,N)	1 + 2 + 4 = 7

Figure 29-46 The sum of a binary progression of values on individual map layers results in a unique value for each combination. See included DVD for color version.

Manual of Geographic Information Systems

In the three-map example, all possible combinations are contained in the number range 0 (best) to 7 (bad) as indicated. A value of 4 can only result from a map layer sequence of 0 + 0 + 4 that corresponds to OK Elevation, OK Steepness and Bad Aspect. If a fourth map were added to the stack, its individual binary map value for unacceptable areas would be 8 and the resulting range of values for the four map-stack would be from 0 (best) to 15 (bad) with a unique value for each combination of habitat conditions.

29.6.1.3 Rating Model

The *binary*, *ranking* and *ranking combinations* approaches to suitability mapping share a common assumption—that each habitat criterion can be discretely classified as either acceptable or unacceptable. However, suppose Hugags are complex animals and actually perceive a continuum of preference. For example, their distain for high elevations isn't a sharp boundary at 1800 feet. Rather they might strongly prefer low land areas, venture sometimes into transition elevations but absolutely detest the higher altitudes.

Figure 29-47 shows a *Rating Suitability* map of Hugag habitat. In this case the three criteria maps are graded on a scale of 1 (bad) to 9 (best). For example, the elevation-based conditions are calibrated as—

- 1 (bad)   =   > 1800 feet
- 3 (marginal)   =   1400-1800 feet
- 5 (OK)   =   1250-1400 feet
- 7 (better)   =   900-1250 feet
- 9 (best)   =   0-900 feet

The other two criteria of slope and orientation are similarly graded on the same scale as shown in the figure.

This process is analogous to a professor grading student exams for an overall course grade. Each map layer is like a test, each grid cell is like a student and each map value at a location is like the grade for one test. To determine the overall grade for a semester a professor averages the individual test scores. An overall habitat score is calculated in a similar manner—take the average of the three calibrated map layers.

The location in the example is assigned an average habitat value of 5.67 that is somewhere between "OK" and "Better" habitat conditions on the 1 (bad) to 9 (best) suitability scale. The rating was derived by averaging the three calibrated values of—

9   …9 assigned "Best" elevation condition of between 0 and 900 feet

7   …7 assigned "Better" steepness condition of between 5 and 15 %

1   …1 (northwest) assigned "Bad" aspect condition

17 / 3 = 5.67 average habitat rating   …slightly better than OK

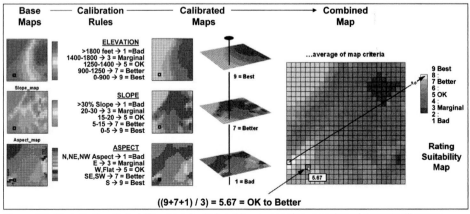

Figure 29-47 A Rating Suitability map is derived by averaging a series of "graded" maps representing individual habitat criteria. See included DVD for color version.

Note the increased information provided by a rating suitability map. All three of the extended techniques (ranking, ranking combination and rating) consistently identify the completely bad area in the northeast and southeast. What changes with the different techniques is the information about subtle differences in the marginal through better areas. The rating suitability map contains the most information as it uses a consistent scale and reports habitat "goodness" values to the decimal point.

That brings the discussion full circle and reinforces the concept of a map-ematical framework for analyzing spatial relationships. Generally speaking, map analysis procedures such as Suitability Modeling take full advantage of the digital nature of maps to provide more information than traditional manual techniques and thus support better decision-making.

29.6.2 Decision Support Modeling

In the past, siting electric transmission lines required thousands of hours around paper maps, sketching hundreds of possible paths, and then assessing their feasibility by 'eyeballing the best route.' The tools of the trade were a straight edge and professional experience. This manual approach capitalizes on expert interpretation and judgment, but it is often criticized as a closed process that lacks a defensible procedure and fails to engage the perspectives of external stakeholders in what constitutes a preferred route.

29.6.2.1 Routing Procedure

The *Least Cost Path* (LCP) procedure for identifying an optimal route based on user-defined criteria has been used extensively in GIS applications for siting linear features and corridors. Whether applications involve movement of elk herds or herds of shoppers, or locating highways, pipelines or electric transmission lines, the procedure is fundamentally the same— 1) develop a discrete cost surface that indicates the relative preference for routing at every location in a project area, 2) generate an accumulated cost surface characterizing the optimal connectivity from a starting location (point, line or area) to all other locations based on the intervening relative preferences, and 3) identify the path of least resistance (steepest downhill path) from a desired end location along the accumulated surface.

Figure 29-48 schematically shows a flowchart of the GIS-based routing procedure for a hypothetical example of siting an electric transmission line that avoids areas that
- have high housing density,
- are far from roads,
- are near or within sensitive areas and
- have high visual exposure to houses.

These four criteria are shown as rows in the left portion of the flowchart in the figure. The *Base Maps* are field collected data such as elevation, sensitive areas, roads and houses. *Derived Maps* use computer processing to calculate information that is too difficult or even impossible to collect, such as visual exposure, proximity and density. The discrete *Preference Maps* translate this information into decision criteria. The calibration forms maps that are scaled from 1 (most preferred—favor siting, gray areas) to 9 (least preferred—avoid siting, red areas) for each of the decision criteria.

The individual cost maps are combined into a single map by averaging the individual layers. For example, if a grid location is rated 1.0 in each of the four cost maps, its average is 1.0 indicating an area strongly preferred for siting. As the average increases for other locations it increasingly encourages routing away from them. If there are areas that are impossible or illegal to cross these locations are identified with a "null value" that instructs the computer to never traverse these locations under any circumstances.

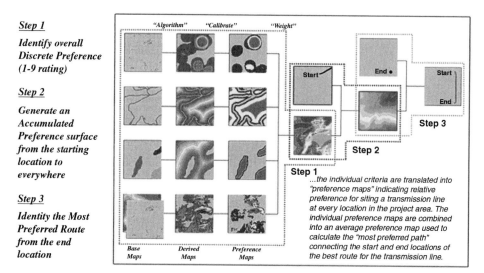

Step 1

Identify overall Discrete Preference (1-9 rating)

Step 2

Generate an Accumulated Preference surface from the starting location to everywhere

Step 3

Identity the Most Preferred Route from the end location

...the individual criteria are translated into "preference maps" indicating relative preference for siting a transmission line at every location in the project area. The individual preference maps are combined into an average preference map used to calculate the "most preferred path" connecting the start and end locations of the best route for the transmission line.

Figure 29-48 GIS-based routing uses three steps to establish a discrete map of the relative preference for siting at each location, generate an accumulated preference surface from a starting location(s) and derive the optimal route from an end point as the path of least resistance guided by the surface. See included DVD for color version.

29.6.2.2 Identifying Corridors

The technique generates accumulation surfaces from both the Start and End locations of the proposed power line. For any given location in the project area one surface identifies the accumulated cost of the best route to the start and the other surface identifies the accumulated cost of the best route to the end. Adding the two surfaces together identifies the total cost of forcing a route through every location in the project area.

The series of lowest values on the total accumulation surface (valley bottom) identifies the best route. The valley walls depict increasingly less optimal routes. The red areas in Figure 29-49 identify all of the locations that are within five percent of the optimal path. The green areas are 5% to 10% more costly than the optimal path.

The corridors are useful in delineating boundaries for detailed data collection, such as high resolution aerial photography and ownership records. The detailed data within the macro-corridor are helpful in making slight adjustments in centerline design and in generating and assessing alternative routes.

29.6.2.3 Calibrating Routing Criteria

Implementation of the LCP routing procedure provides able room for interpretation and relative preferences. For example, one of the criteria in the routing model seeks to avoid locations having high visual exposure to houses. But what constitutes high visual exposure—5 or 50 houses visually impacted? Are there various levels of increasingly high exposure that correspond to decreasing preference? Is "avoiding high visual exposure" more or less important than "avoiding locations near sensitive areas." If so, how much more (or less) important?

The answers to these questions are what tailor a model to the specific circumstances of its application and the understanding and values of the decision participants. The tailoring involves two related categories of parameterization—calibration and weighting.

Calibration refers to establishing a consistent scale from 1 (most preferred) to 9 (least preferred) for rating each map layer used in the solution. Figure 29-50 shows the result for the four decision criteria used in the routing example.

The *Delphi Process*, developed in the 1950s by the Rand Corporation, is designed to achieve consensus among a group of experts. It involves directed group interaction consisting

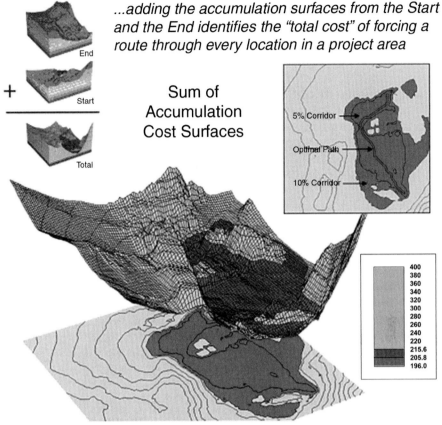

...adding the accumulation surfaces from the Start and the End identifies the "total cost" of forcing a route through every location in a project area

Figure 29-49 The sum of accumulated surfaces is used to identify siting corridors as low points on the total accumulated surface. See included DVD for color version.

Model calibration refers to establishing a consistent scale from 1 (most preferred) to 9 (least preferred) for rating each map layer...

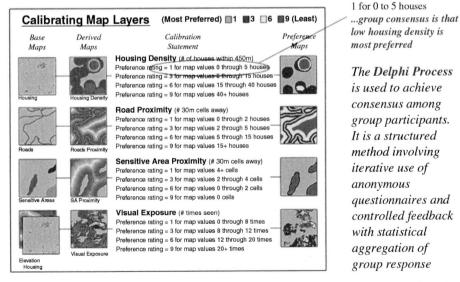

Figure 29-50 The Delphi Process uses structured group interaction to establish a consistent rating for each map layer. See included DVD for color version.

Manual of Geographic Information Systems

of at least three rounds. The first round is completely unstructured, asking participants to express any opinions they have on calibrating the map layers in question. In the next round the participants complete a questionnaire designed to rank the criteria from 1 to 9. In the third round participants re-rank the criteria based on a statistical summary of the question-naires. "Outlier" opinions are discussed and consensus sought.

The development and summary of the questionnaire is critical to Delphi. In the case of continuous maps, participants are asked to indicate cut-off values for the nine rating steps. For example, a cutoff of 4 (implying 0-4 houses) might be recorded by a respondent for Housing Density preference level 1 (most preferred); a cut-off of 12 (implying a range of 4-12) for preference level 2; and so forth. For discrete maps, responses from 1 to 9 are assigned to each category value. The same preference value can be assigned to more than one category; however, there has to be at least one condition rated 1 and another rated 9. In both continuous and discrete map calibration, the median, mean, standard deviation and coefficient of variation for group responses are computed for each question and used to assess group consensus and guide follow-up discussion.

29.6.2.4 Weighting Criteria Maps

Weighting of the map layers is achieved using a portion of the *Analytical Hierarchy Process (AHP)* developed in the early 1980s as a systematic method for comparing decision criteria. The procedure involves mathematically summarizing paired comparisons of the relative importance of the map layers. The result is a set of map layer weights that serves as input to a GIS model.

In the routing example, there are four map layers that define the six direct comparison statements identified (# pairs = (N * (N – 1) / 2) = 4 * 3 / 2= 6 statements) as shown in Figure 29-51. Members of the group independently order the statements so they are true, then record the relative level of importance implied in each statement. The importance scale is from 1 (equally important) to 9 (extremely more important).

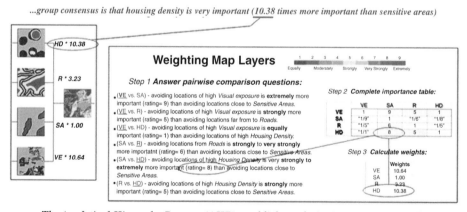

Model weighting establishes the relative importance among map layers (model criteria) on a multiplicative scale...

...group consensus is that housing density is very important (10.38 times more important than sensitive areas)

The Analytical Hierarchy Process (AHP) establishes relative importance overall by mathematically summarizing paired comparisons of map layers' importance.

Figure 29-51 The Analytical Hierarchy Process uses direct comparison of map layers to derive their relative importance. See included DVD for color version.

This information is entered into the importance table a row at a time. For example, the first statement in the figure views avoiding locations of high Visual Exposure (VE) as extremely more important (importance level = 9) than avoiding locations close to Sensitive Areas (SA). The response is entered into table position row 2, column 3 as shown. The reciprocal of the statement is entered into its mirrored position at row 3, column 2. Note that the last weighting statement is reversed so its importance value is recorded at row 5, column 4 and its reciprocal recorded at row 4, column 5.

Once the importance table is completed, the map layer weights are calculated. The procedure first calculates the sum of the columns in the matrix, and then divides each entry by its column sum to normalize the responses. The row sum of the normalized responses derives the relative weights that, in turn, are divided by minimum weight to express them as a multiplicative scale.

The relative weights for a group of participants are translated to a common scale then averaged before expressing them as a multiplicative scale. Alternate routes are generated by evaluating the model using weights derived from different group perspectives.

29.6.2.5 Transmission Line Siting Experience

Figure 29-52 shows the results of applying different calibration and weighting information to derive alternative routes for a routing application in central Georgia. Four routes and corridors were generated emphasizing different perspectives—*Built* environment (community concerns), *Natural* environment (environmental), *Engineering* (constructability) and the *Simple* un-weighted average of all three group perspectives.

While all four of the routes use the same criteria layers, the differences in emphasis for certain layers generate different routes/corridors that directly reflect differences in stakeholder perspective. Note the similarities and differences among the Built, Natural, Engineering and un-weighted routes. The bottom line is that the procedure identified constructible alternative routes that can be easily communicated and discussed.

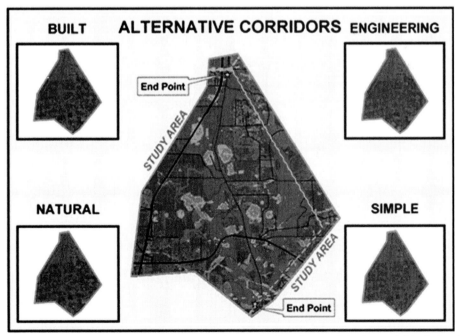

Figure 29-52 Alternate routes are generated by evaluating the model using weights derived from different group perspectives. See included DVD for color version. (Courtesy of Photo Science and Georgia Transmission Corporation)

The final route is developed by an experienced transmission line siting team who combine alternative route segments for a preferred route. Engineers make slight centerline realignments responding to the detailed field surveys along the preferred route, and then design the final pole placements and construction estimates for the final route.

The ability to infuse different perspectives into the routing process is critical in gaining stakeholder involvement and identifying siting sensitivity. It acts at the front end of the routing process to explicitly identify routing corridors that contain constructible routes reflecting different perspectives that guide the deliberations of the siting engineer. Also, the explicit nature of the methodology tends to de-mystify the routing process by clearly identifying the criteria and how they are evaluated.

In addition, the participatory process 1) encourages interaction among various perspectives, 2) provides a clear and structured procedure for comparing decision elements, 3) involves quantitative summary of group interaction and dialog, 4) identifies the degree of group consensus for each decision element, 5) documents the range of interpretations, values and considerations surrounding decision criteria, and 6) generates consistent, objective and defendable parameterization of GIS models.

The use of Delphi and AHP techniques provides a structure for infusing diverse perspectives into calibrating and weighting criteria maps used in GIS-based routing. Traditional siting techniques rely on unstructured discussions that often seem to mystify the routing process by not clearly identifying the criteria used or how they were derived and evaluated. The approach described in this section is objective, consistent and comprehensive and encourages multiple perspectives for generating alternative routes as well as thoroughly documenting the decision process itself. The general approach is readily applicable to other siting applications of linear features, such as pipelines and roads.

29.6.3 Statistical Modeling

Site-specific management, often referred to as Precision Farming or Precision Agriculture, is about doing the right thing, in the right way, at the right place and time. It involves assessing and reacting to field variability and tailoring management actions, such as fertilization levels, seeding rates and variety selection, to match changing field conditions. It assumes that managing field variability leads to cost savings, production increases and better stewardship of the land. Site-specific management isn't just a bunch of pretty maps, but a set of new procedures that link mapped variables to appropriate management actions.

Several of the procedures, such as map similarity, level-slicing, clustering and regression, used in precision agriculture are discussed thoroughly in Section 29.4 above, Spatial Statistics Techniques. The following discussion outlines how the analytical techniques can be used to relate crop yield to driving variables like soil nutrient levels that support management action.

29.6.3.1 Elements of Precision Agriculture

To date, much of the analysis of yield maps has been visual interpretation. By viewing a map, all sorts of potential relationships between yield variability and field conditions spring to mind. These 'visceral visions' and explanations can be drawn through the farmer's knowledge of the field—"…this area of low yield seems to align with that slight depression," or "…maybe that's where all those weeds were," or "…wasn't that where the seeder broke down last spring?"

Data visualization can be extended by GIS analysis that directly links yield maps to field conditions. This processing involves three levels—cognition, analysis and synthesis. At the *cognition* level (termed desktop mapping) computer maps of variables, such as crop yield and soil nutrients, are generated. These graphical descriptions form the foundation of site-specific management. The *analysis level* uses the GIS's analytical toolbox to discover relationships among the mapped variables. This step is analogous to a farmer's visceral visions of relationships, but uses the computer to establish mathematical and statistical connections. To many farmers this step is an uncomfortable leap of scientific faith from pretty maps to pure, dense technical gibberish.

However, map analysis greatly extends data visualization and can more precisely identify areas of statistically high yield and correlate them to a complex array of mapped field conditions. The *synthesis level* of processing uses GIS modeling to translate the newly discovered relationships into management actions (prescriptions). The result is the prescription map needed by intelligent implements in guiding variable rate control of field inputs. Admittedly, the juvenile science of site-specific management is a bit imprecise, and raises several technical issues. In less than two decades, however, the approach has placed millions of acres worldwide under site-specific management, as well as completely altering equipment with GPS (global positioning system) and variable-rate hardware.

The precision agriculture process can be viewed as four steps: Data Logging, Point Sampling, Data Analysis and Prescription Modeling as depicted in Figure 29-53. *Data logging* continuously monitors measurements, such as crop yield, as a tractor moves through a field. *Point sampling*, on the other hand, uses a set of dispersed samples to characterize field conditions, such as phosphorous, potassium and nitrogen levels. The data derived by the two approaches are radically different in nature—a "direct census" of yield consisting of thousands of on-the fly samples versus a "statistical estimate" of the geographic distribution of soil nutrients based on a handful of soil samples.

In data logging, issues of accurate measurement, such as GPS positioning and material flow adjustments, are major concerns. Most systems query the GPS and yield monitor every second, which at 4 mph translates into about 6 feet. With differential positioning the coordinates are accurate to about a meter. However the paired yield measurement is for a location well behind the harvester, as it takes several seconds for material to pass from the point of harvest to the yield monitor. To complicate matters, the mass flow and speed of the harvester are constantly changing when different terrain and crop conditions are encountered. The precise placement of GPS/Yield records are not reliant as much on the accuracy of the GPS receiver as on "smart" yield mapping software.

The Precision Ag Process (Fertility example)

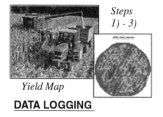

Steps 1) - 3)

Yield Map
DATA LOGGING

As a combine moves through a field it (Step 1) uses GPS to check its location then (2) checks the yield at that location to (3) create a continuous map of the yield variation every few feet. Soil samples are (4) collected and interpolated into continuous maps of nutrient levels. The yield map and nutrient maps are (5) analyzed to (6) derive a "Prescription Map" that is used to (7) adjust fertilization levels every few feet in the field (variable rate application)

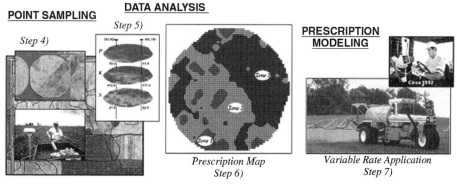

POINT SAMPLING

Step 4)

DATA ANALYSIS

Step 5)

PRESCRIPTION MODELING

Prescription Map
Step 6)

Variable Rate Application
Step 7)

Figure 29-53 The Precision Agriculture process analysis involves Data Logging, Point Sampling, Data Analysis and Prescription Mapping. See included DVD for color version.

In point sampling, issues of surface modeling (estimating between sample points) are of concern, such as what sampling frequency/pattern and interpolation technique to use. The cost of soil lab analysis dictates "smart sampling" techniques based on terrain and previous data be used to balance spatial variability with a farmer's budget. In addition, techniques for evaluating alternative interpolation techniques and selecting the "best" map using residual analysis are available in some of the soil mapping systems.

In both data logging and point sampling, a critical concern is the resolution of the analysis grid used to geographically summarize the data. Like a stockbroker's analysis of financial markets, the fluctuations of individual trades must be "smoothed" to produce useful trends. If the analysis grid is too coarse, information is lost in the aggregation over large grid spaces; if too small, spurious measurement and positioning errors dominate the information.

The technical issues surrounding mapped *data analysis* involve the validity of applying traditional statistical techniques to spatial data. For example, regression analysis of field plots has been used for years to derive crop production functions, such as corn yield (dependent variable) versus potassium levels (independent variable). In a GIS, you can use regression to derive a production function relating mapped variables, such as the links among a map of corn yield and maps of soil nutrients—like analyzing thousands of sample plots. However, technical concerns, such as variable independence and autocorrelation, have yet to be thoroughly investigated. Statistical measures assessing results of the analysis, such as a spatially responsive correlation coefficient, await discovery and acceptance by the statistical community, let alone the farm community.

In theory, *prescription modeling* moves the derived relationships in space or time to determine the optimal actions, such as the blend of phosphorous, potassium and nitrogen to be applied at each location in the field. In current practice, these translations are based on existing science and experience without a direct link to data analysis of on-farm data. For example, a prescription map for fertilization is constructed by noting the existing nutrient levels (condition) then assigning a blend of additional nutrients (action) tailored to each location forming an if-(condition)-then-(action) set of rules. The issues surrounding spatial modeling are similar to data analysis and involve the validity of using traditional "goal seeking" techniques, such as linear programming, neural network or induction modeling, to calculate maps of the optimal actions.

29.6.3.2 The Big Picture

The precision agriculture process is a special case of spatial data mining described in Sections 29.2.1 and 29.4.2. It involves deriving statistical models that spatially relate map variables to generate predictions that guide management actions. In the case discussed earlier the analysis involved a prediction model relating product sales to demographic data. In the precision agriculture example, corn yield is substituted for sales and nutrient levels are substituted for demographics.

The big picture however, is the same—relating a dependent map variable (sales or yield) to independent map variables (demographics or nutrients) that are easily obtained and are thought to drive the relationship. What separates the two applications is their practical implementation. Sales mapping utilizes existing databases to derive the maps and GIS to form the solution.

Precision agriculture, on the other hand, is a much more comprehensive solution. It requires the seamless integration of three technologies—*global positioning system* (GPS), *geographic information systems* (GIS) and *intelligent devices and implements* (IDI) for on-the-fly data collection (monitors) and variable-rate application (controls) as depicted in the left-side of Figure 29-54.

Modern GPS receivers are able to establish positions within a field to within a few feet. When connected to a *data collection device*, such as a yield/moisture monitor, these data can

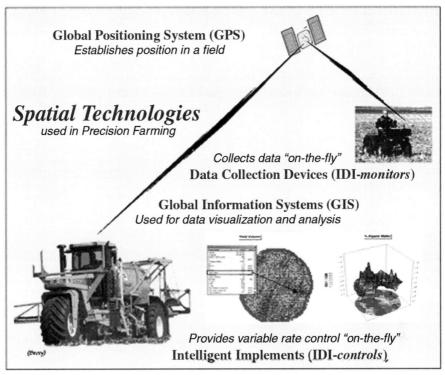

Figure 29-54 Precision Agriculture involves applying emerging spatial technologies of GPS, GIS and IDI. See included DVD for color version.

be assigned geographic coordinates. The GIS is used to extend map visualization of yield to analysis of the relationships among yield variability and field conditions.

Once established, these relationships are used to derive a prescription map of management actions required for each location in a field. The final element, *variable rate implements*, notes a tractor's position through GPS, continuously locates it on the prescription map, and then varies the application rate of field inputs, such as fertilizer blend or seed spacing, in accordance with the instructions on the prescription map.

Site-specific management through statistical modeling extends our traditional understanding of farm fields from "where is what" to analytical renderings of "so what" by relating variations in crop yield to field conditions, such as soil nutrient levels, available moisture and other driving variables. Once these relationships are established, they can be used to insure the right thing is done, in the right way, at the right place and time. Common sense leads us to believe the efficiencies in managing field variability outweigh the costs of the new technology. However, the enthusiasm for site-specific management must be dampened by reality consisting of at least two parts: empirical verification and personal comfort.

To date, studies have not conclusively established that site-specific management is economically justifiable in all cases. In addition, the technological capabilities appear to be somewhat ahead of scientific understanding and a great deal of spatial research lies ahead. In the information age, a farmer's ability to react to the inherent variability within a field might determine survival and growth of tomorrow's farms.

From the big picture perspective however, precision agriculture is pushing the envelope of GIS modeling and analysis as well as linking it to robotics. This is quite a feat for a discipline that had minimal use of mapping just a decade ago.

29.7 Conclusions

Research, decision-making and policy development in the management of land have always required information as their cornerstone. Early information systems relied on physical storage of data and manual processing. With the advent of the computer, most of these data and procedures have been automated during the past three decades. As a result, land-based information processing has increasingly become more quantitative. Systems analysis techniques developed links between descriptive data of the landscape and the mix of management actions that maximizes a set of objectives. This mathematical approach to land management has been both stimulated and facilitated by modern information systems technology. The digital nature of mapped data in these systems provides a wealth of new analysis operations and an unprecedented ability to model complex spatial issues. The full impact of the new data form and analytical capabilities is yet to be determined.

Effective map analysis applications have relatively little to do with data and everything to do with understanding, creativity and perspective. It is a common observation of the Information Age that the amount of knowledge doubles every fourteen months or so. It is believed, with the advent of the information super highway, this period will likely shorten. But does more information directly translate into better decisions? Does the Internet enhance information exchange or overwhelm it? Does the quality of information correlate with the quantity of information? Does the rapid boil of information improve or scorch the broth of decisions?

Geotechnology is a major contributor to the tsunami of information, as terabytes of mapped data are feverishly released on an unsuspecting (and seemingly ungrateful) public. From a GIS-centric perspective, the delivery of accurate base data is enough. However, the full impact of the technology is in the translation of "*where* is *what*, to *why* and *so what*." The effects of information rapid transit on our changing perceptions of the world around us involve a new expression of the philosophers' view of the stages of enlightenment—data, information, knowledge and wisdom. The terms are often used interchangeably, but they are distinct from one another in some subtle and not-so-subtle ways.

The first is data, the "factoids" of our Information Age. *Data* are bits of information, typically—but not exclusively—in a numeric form, such as cardinal numbers, percentages, statistics, etc. It is exceedingly obvious that data volumes are increasing at an incredible rate. Coupled with the barrage of data, is a requirement for the literate citizen of the future to have a firm understanding of averages, percentages, and to a certain extent, statistics. More and more, these types of data dominate the media and are the primary means used to characterize public opinion, report trends and persuade specific actions.

The second term, information, is closely related to data. The difference is that we tend to view information as more word-based and/or graphic than numeric. *Information* is data with explanation. Most of what is taught in school is information. Because it includes all that is chronicled, the amount of information available to the average citizen substantially increases each day. The power of technology to link us to information is phenomenal. As proof, simply "surf" the exploding number of "home pages" on the Internet.

The philosophers' third category is *knowledge*, which can be viewed as information within a context. Data and information that are used to explain a phenomenon become knowledge. It probably does not double at fast rates, but that really has more to do with the learner and processing techniques than with what is available. In other words, data and information become knowledge once they are processed and applied.

The last category, *wisdom*, certainly does not double at a rapid rate. It is the application of all three previous categories, and some intangible additions. Wisdom is rare and timeless, and is important because it is rare and timeless. We seldom encounter new wisdom in the popular media, nor do we expect a deluge of newly derived wisdom to spring forth from our computer monitors each time we log on.

Knowledge and wisdom, like gold, must be aggressively processed from tons of near-worthless overburden. Simply increasing data and information does not assure the increasing amounts of the knowledge and wisdom we need to solve pressing environmental and resource problems. Increasing the processing "throughput" by efficiency gains and new approaches might.

How does this philosophical diatribe relate to geotechnology and map analysis? What is GIS's role within the framework? What does it deliver—data, information, knowledge or wisdom? Actually, if GIS is appropriately presented, nurtured and applied, it can affect all four. That is, provided the technology's role is recognized as an additional link that the philosophers failed to note.

Understanding sits at the juncture between the data/information and knowledge/wisdom stages of enlightenment. *Understanding* involves the honest dialogue among various interpretations of data and information in an attempt to reach common knowledge and wisdom. Note that understanding is not a "thing," but a process. It is how concrete facts are translated into the slippery slope of beliefs. It involves the clash of values, tempered by judgment based on the exchange of experience. Technology, and in particular GIS modeling and analysis, has a vital role to play in this process. It is not sufficient to deliver spatial data and information; a methodology for translating them into knowledge and wisdom is needed.

Tomorrow's GIS builds on the cognitive basis, as well as the spatial databases and analytical operations, of the technology. This new view pushes GIS beyond data mapping, management and modeling, to spatial reasoning and dialog focusing on the communication of ideas. In a sense, modeling and analysis extends the GIS toolbox to a social "sandbox," where alternative perspectives are constructed and discussed, and a common understanding is distilled.

This step needs to fully engage the end-user in GIS itself, not just its encoded and derived products. It requires a democratization of geotechnology that goes beyond a graphical user interface and cute icons. It obligates the GIS community to explain concepts in lay terms and provide access to the community's onceptual expressions of geographic space. In turn, it requires the user community to embrace the new modeling and analysis approaches to spatial reasoning and dialogue. GIS has an opportunity to empower people with new decision-making tools, not simply entrap them in a new technology and an accelerating avalanche of data. The mapping, management and modeling of spatial data is necessary—but not sufficient—for effective solutions. Like the automobile and indoor plumbing, GIS will not be an important technology until it fades into the fabric of the decision-making process and is taken for granted. Modeling and analysis for understanding spatial relationships needs to become as second nature and comfortable as the traditional paper map.

Author's Note

See *Beyond Mapping: Compilation of Beyond Mapping columns appearing in GeoWorld magazine*, an online compilation of Joseph K. Berry's *Beyond Mapping* columns published in *GeoWorld* magazine from 1989 through 2007. It is intended to be used as a self-instructional text or in support of formal academic courses for study of grid-based map analysis and GIS modeling.

<http://innovativegis.com/basis/MapAnalysis/> *Map Analysis* online book. Accessed 5 August 2007.

<http://innovativegis.com/basis/> instructional and other materials. Accessed 5 August 2007.

CHAPTER 30
Spatial Analysis and Modeling

Paul A. Longley

30.1 A Brief Primer to Spatial Analysis and Modeling

Analysis is deemed to be 'spatial' if the results of an investigation are likely to depend upon how the bounding frame of a representation is delineated. Any spatial representation is a selective abstraction, a simplification, or a 'model' of reality—what we choose, or are forced, to leave out can be as important as the aspects of reality that we choose to retain. In practice, the ways in which the bounding extent and content of a representation are defined are likely to be interdependent outcomes of choice, convention and chance, and thus the terms 'spatial analysis' and 'spatial modeling' are frequently taken as near synonyms. We will adopt this practice in much of what follows.

Some 40 years ago, Chorley and Haggett (1967) distinguished between iconic and symbolic models. *Iconic models* can be thought of as 'scaled-down versions of (relevant aspects of) the real thing,' sometimes with working parts which are similarly scaled down. Examples include the architect's block model, in which buildings are made from balsa wood or cardboard, the fluvial geomorphologist's flume, and the paper map. *Analog models* can be thought of as a special case of iconic models. Here the functioning of the model (in addition to, or instead of, its physical form) is in analogy to some other system, for example:

- with the economic geographer's Varignon Frame the 'pull' of a production center is represented by the physical weights exerted using pulleys (Longley et al. 2005, 345–6);
- in simulations of the interactions within a local economy using electrical networks, the components of the electrical network represent different flows, resistances and potentials in the economy.

In contrast, *symbolic models* are based on logical (mathematical or statistical) relationships, where relevant attributes of the system in question are modeled or simulated. The core ideas of spatial analysis relate to symbolic, rather than analog or iconic, models. Digital mapping is symbolic because it embodies explicit topological relations, and the property of scale has no physical meaning within the confines of the computer.

For over 40 years, digital representation using computers has made it possible for us to specify symbolic relationships and, more recently, has made it possible for us to create plausible visual representations of the physical entities that are symbolized. This is achieved using computer graphics which, through GIS, allow us to go much further than paper mapping, through symbolic modeling of spatial relationships. In practice, this entails explicit specification of adjacency and topology, which makes digital mapping considerably more powerful than its iconic forbear. Specification of symbolic relationships is not restricted to two dimensions, and extending geographic representation from the icon of the map to the icon of the building or landscape presents us with the opportunity of extending symbolic modeling from 2 dimensions to 3 dimensions (3D).

Taken together, the development of GIS-enabled spatial analysis and modeling links powerful ideas embodied in software to excellence of visual communication, in ways that are of wide applicability in our quest to understand the functioning of environmental and human systems. GIS allows us to manage the granularity of space and time, in ways that facilitate generalization whilst also preserving relevant unique attributes of 'places' and particular time periods. This can realize the goal of bringing together the idiographic tradition in geography (which emphasizes the uniqueness of places) with the nomothetic (which emphasizes the generality of processes), in a science that is relevant to practical problem-solving. These qualities make GIS-based spatial analysis and modeling of interest and applicability well beyond the academic domain.

The deployment of the methods and techniques of spatial analysis and modeling changed profoundly following the innovation of the Internet as society's medium for information exchange. The client-server architectures that developed in the 1990s have been augmented by data inputs and feeds to the smallest of handheld devices, and the innovation of 'sensor webs' opens up new prospects for spatial analysis based upon real measurement and monitoring in real time. Yet despite these seismic shifts, many of the core themes to spatial analysis and modeling remain of enduring importance. Of particular importance are issues of scale, of representation of human behavior, of uncertainty in representation, of visualization and user interaction, of representation of dynamics and spatial process, and of policy application through questions of planning and design. These are core to the various research agendas of GIScience, described in more detail by Longley et al. (Chapter 3 of this volume). As such, spatial analysis and modeling should be seen as core to the GIScience agenda.

30.2 GIS and Spatial Analysis: Retrospective

GIS has diverse roots in many spatial sciences, ranging from geography and the Earth sciences to architecture, urban planning and ecology. Other entries in this Manual document how GIS emerged as a cognate area and as a significant domain of commercial activity in the late 1970s and 1980s, after declining costs of computer hardware had made processing affordable by a wide base of users. The reference to the classic work of Chorley and Haggett in the previous section illustrates that spatial analysis has a longer history as an area of activity. Some of these origins can be traced to the development of quantitative methods in geography, macroeconomics and the Earth sciences, as well as the discipline of regional science. Spatial statistics and locational analysis dominated the quantitative geography of the 1960s, with the development and application of techniques that together laid the foundations of a new scientific geography (Johnston 2005). These developments culminated in the reader 'Spatial Analysis,' edited by Brian Berry and Duane Marble and published in 1968, and the tradition retains a core following in geography today.

Many of the basic operations of GIS—such as distance measurement, area calculation, or map overlay—are not viewed as spatial analysis per se, although software that provides these functions is now the environment of choice for much spatial analysis. Since its inception as a field in the 1970s, GIS can also be thought of as providing a series of parallel, yet more application-centered, developments in spatial statistics and locational analysis. Despite the practical applications focus, this has led to the incorporation of some of the more general spatial analysis techniques into GIS software, and its adoption by a wider user community. By the mid 1990s, it is reasonable to suggest that GIS had broadened its remit sufficiently to embrace spatial analysis, and the research communities in modeling and simulation were developing alongside GIS in a more complementary fashion. Today, the cutting edge of GIS extends across many fields of spatial representation in modeling, policy and design, hence the aforementioned importance of spatial analysis to the GIScience agenda.

Spatial analysis, then, can be thought of as extending GIS from a restricted range of 'off the shelf' representational forms to customized depictions of the world and critical interpretation of the assumptions that are invoked to represent it. Like GISystems and GIScience, the core goal of spatial analysis can be thought of as moving beyond 'what is?' representation of how the world looks towards 'what if?' scenario-building based upon how the world works. The embedding of spatial analysis in a GIS environment can be seen as having moved the spatial analysis community beyond preoccupation with technique to embrace ideas about policy and application. These concerns are evident in a recent compendium of spatial analysis applications (Longley and Batty 2003), which also contains a broad discussion of the remit of spatial analysis. Viewed from this perspective, the standard functionality of GIS provides a framework for data manipulation and visualization which, when harnessed to spatial analysis techniques, also provides powerful ways of generating and evaluating strategic policies and plans. Some spatial analysis requires very intensive use of computer processing power, and has benefited considerably from the continuing precipitous falls in the cost of computing, as well as the development of new networked hardware and software architectures.

A number of other changes, and the challenges they present, are worth enumerating here: they are discussed at greater length in Longley and Batty (2003). First, GIS in its various guises now pervades our lives, and this has increased the challenges to effective user interaction. This provides both a motivation for spatial analysis research into the ways in which the medium of GIS may overwhelm the message of spatial analysis in the eyes of most users, as well as a need for spatial analysts to understand the importance of good user interface design. Second, modularity has become the dominant procedure for adding new spatial analysis functions to GIS software. In practice, software linkage still occurs predominantly on stand-alone machines, often in a high-performance computing environment. However, grid processing architectures are increasingly mooted as providing solutions to problems that require intensive processing of diverse data sets scattered across remote locations—and which often require different access protocols. Third, developments in 3D representation have provided a cutting edge to research, in applications as diverse as Web-enabled photogrammetry or location-based services (Brimicombe and Li 2008). Fourth, the miniaturization of computer devices, allied to ongoing developments in computer networking, is taking some of academic geography back to its roots in field measurement and environmental cognition, and this poses important new challenges for relaying the results of spatial analysis applications to the field in real time. The networking of hardware installations and supplementation of laboratory installations with small mobile devices opens up a vast array of new spatial analysis applications. This is part of an ongoing and massive decentralization of computing, where data, software and even applications themselves are being distributed across networks and spatial analysis can benefit from the real time feeds of sensor webs and other field observations.

30.3 Prospects: Spatial Analysis in a Networked World

The networking of spatial analysis and modeling is having profound and ongoing implications for the ways in which we communicate, process and visualize spatial data across a range of scales. Batty (2006) has considered these developments in the context of a world that is moving from a paradigm in which centralized control from the top down, to a world characterized by a much more decentralized, bottom-up approach to organization. The impact of this is evident from ways in which we now communicate information and expertise that was once the preserve of select institutions, but which is now readily available from online sources (e.g., de Smith et al. 2007, UCL undated) and easily distributed in real time. This decentralization has contributed to our ability to develop much richer, more realistic and more applicable forms of representation and hence knowledge—often at fine geographical temporal levels of granularity that were unimaginable even a decade ago. This wave of change towards more decentralized thinking is also leading to rethinking of past practices with regard to the conduct of scientific research. Top-down strategies are now widely regarded as being rather insensitive ways of understanding and hence changing our systems of interest. They are also viewed as being suspect in local democracies.

What is clear from this brief overview of spatial analysis and modeling is that this tide of decentralization has some way to run and that by 2020 the world is likely to be a much more decentralized place than it is today. As regards spatial analysis, this is likely to lead to new theoretical imperatives concerning our approach to science in general and GIScience in particular; new practical problems that will affect the way in which we build applications and the end users for whom we will build them; and new ethical concerns for GIStudies which will determine the way we respond to problems and influence the GIScience that we wish to pursue.

References

Batty, M. 2006. Globalisation, scale and interaction in spatial modelling. *Environment and Planning B* 33:637–8.

Berry, B.J.L. and D. F. Marble, eds. 1968. *Spatial Analysis: A Reader in Statistical Geography.* Englewood Cliffs, N.J.: Prentice Hall.

Brimicombe, A. and C. Li. 2008. *Location-based Services and Geo-information Engineering.* Chichester, UK: Wiley.

Chorley, R. and P. Haggett, eds. 1967. *Models in Geography.* London: Methuen.

de Smith, M. J., M. F. Goodchild and P. A. Longley. 2007. Geospatial Analysis: Web site, PDF and Book. <www.spatialanalysisonline.com> Accessed 1 August 2007.

Johnston, R.J. 2005. Geography and GIS. In *Geographical Information Systems: Principles, Techniques, Management and Applications* (abridged edition), edited by P. A. Longley, M. F. Goodchild, D. J. Maguire and D. W. Rhind, 39–47. Hoboken, N.J.: Wiley.

Longley, P. A. and M. Batty, eds. 2003. *Advanced Spatial Analysis: The CASA Book of GIS.* Redlands, Calif.: ESRI Press.

Longley, P. A., M. F. Goodchild, D. J. Maguire and D. W. Rhind. 2005. *Geographic Information Systems and Science,* 2nd edition. Chichester, UK: Wiley.

UCL (University College London). Undated. Spatial-Literacy.org. <www.spatial-literacy.org> Accessed 1 August 2007.

CHAPTER 31
Cellular Automata and GIS for Urban Planning

Anthony Gar-On Yeh and *Xia Li*

31.1 Introduction

Cities are open, nonlinear, dynamic and complex systems. Since conventional models based on strict mathematical equations have problems modeling the complex behaviors of cities, recent studies have used cellular automata (CA) techniques for urban simulation (Batty and Xie 1994; White and Engelen 1993; Wu and Webster 1998; Li and Yeh 2000). CA are computational models that can simulate complex systems of spatio-temporal features—including processes of reproduction, self-organization and evolution of systems—in physics, chemistry and biology. CA can model complex natural phenomena in a way that is conceptually clearer, more accurate and more complete than conventional mathematical systems (Itami 1994). Originating from computer sciences, CA have now been used in many research fields, mainly for simulating natural phenomena. CA can be dated back to von Neumann's efforts at understanding the logic of self-reproduction in the 1940s (Burks 1970). He used CA to demonstrate that universal machines could simulate themselves and, if they could do this, there lay the logic for their self-reproduction (Batty and Xie 1994). An example of CA simulation is the well-known *Game of Life*, created by the British mathematician John Horton Conway in 1970, which shows that simple local rules give rise to complex global patterns (Gardner 1970, 1971). CA can be used to simulate the unexpected behaviors of complex systems which cannot be represented by concrete equations.

Researchers have proposed a variety of urban CA models to deal with complex urban systems. Many *ad hoc* modifications to the general CA formalism have been introduced by urban modelers for creating greater realism. This is because urban systems usually have some unique features that are distinct from other systems. Urban systems are strongly influenced by social factors and human interventions. Many urban phenomena cannot be simply explained by local interactions. For example, transport improvements or zoning policies can drastically change the path of urban development. There are constraints from external or exogenous forces that should be reflected in CA transition rules. Urban CA models have become considerably more complex when they are applied to real cities, especially for different activities or land use types.

CA are suitable for the simulation of both artificial cities and real cities. Couclelis (1985) has demonstrated that CA can generate very complex spatial patterns from simple rule sets. However, her work is not intended to produce a realistic representation of urban growth. White and Engelen (1993, 1997) developed a CA model to investigate general features of urban structure—the fractal or bifractal properties of cities and their evolution. In contrast, Clarke and Gaydos (1998) created a CA model for the realistic simulation and prediction of the urban growth in two urban areas: the San Francisco Bay region in California and the Washington / Baltimore corridor in the Eastern United States. The model is calibrated by using historical digital maps to ensure that the simulation is close to reality.

CA modeling techniques also have strong implications for planning purposes by using constraints. Embedding constraints into CA can regulate simulation processes so that various plans can be obtained. It is convenient to explore various urban forms by using different sets of parameters, transition rules and model structures (Yeh and Li 2001a). Specified CA models have been designed to deal with a variety of ecological and environmental issues. For example, Li and Yeh (2001) have proposed a CA model using satellite images, which they used to evaluate zoning options to protect good-quality agricultural land. Clarke et al. (1994) present a CA model to simulate the propagation and extinction of wildfire. Couclelis (1988) also provides a very simple CA model for rodent population dynamics that can generate a wide variety of different spatio-temporal structures corresponding to different forms of equilibrium.

One of the unique features of urban CA models is that they are usually integrated with geographical information systems (GIS). The integration of CA with GIS can allow urban modeling to generate more-realistic simulation results. GIS plays a key role in urban simulation because it provides a rich source of spatial information and a convenient environment for spatial data handling. Indeed, over the last twenty years, GIS had profound impacts in a variety of disciplines that are related to spatial information.

31.2 The Development of Cellular Automata

Cellular automata (CA) are, in essence, a kind of spatial dynamic simulation modeling based on discrete temporal and spatial concepts. CA originated from biological and calculating science and were firstly developed by Stanislaw Marcin Ulam in the 1940s (White and Engelen 1993); the model was soon used by John von Neumann to investigate the evolution process of self-reproduction system. The most famous early CA model is Conway's *Game of Life* mentioned above (Gardner 1970, 1971).

CA models can simulate both natural and man-made systems in physics, chemistry, biology and geography. These models are attractive because they can generate very complex behaviors and global structures by using very simple local rules. Conventional mathematical equations can hardly be defined to create such models because of the complexity and uncertainty of natural systems.

Wolfram's studies considered influences on the later development of CA models (Wolfram 1984). He generalizes five fundamental characteristics of CA:

1. CA models consist of a discrete lattice of sites;
2. They evolve in discrete time steps;
3. Each cell takes on a finite set of possible values;
4. The value of each cell evolves according to the same deterministic rules;
5. The rules for the evolution of a cell depend only on a local neighborhood of sites around it.

In addition, Wolfram (1984) lists a series of advantages can be identified for CA in modeling physical systems:

- The correspondence between physical and computation processes are clear;
- CA models can produce more comprehensive results just by using simpler rules than complex mathematical equations;
- They can be modeled by computers without loss of precision;
- They can simulate the actions of any possible physical systems;
- These models are irreducible (Itami 1994).

One-dimensional (1D) cellular automata drew great attention in the early stages of CA development and various in-depth studies were reported to describe the detailed behavior of these models. A 1D CA is strictly defined as consisting of a line of sites, with each site carrying a state value of 0 or 1. A neighborhood function is then used to update the state value ai at each position i in discrete time:

$$a_i^{t+1} = \phi[a_{i-r}^t, a_{i-r+1}^t, ..., a_{i+r}^t] \tag{31-1}$$

where a_i is the state of cell i, r is related to the size of neighborhood (e.g., $r = 1$ or 2), and ϕ is the neighborhood function.

Table 31-1 One of the possible transition rules for a one-dimensional CA model (ai = 0 or 1; r = 1).

t	111	110	101	100	011	010	011	000
t+1	0	1	0	0	1	1	0	0

Transition rules should be provided for CA simulation. When $r = 1$ and $a_i = 0$ (dead) or 1 (alive), there are $2^8=256$ sets of possible transition rules for this type of CA model. Table 31-1 is one set of these possible transition rules. For example, an alive cell (state 1) which has two alive neighbors (111) will become dead, but it can remain alive if it is surrounded by one alive cell and one dead cell (011 or 110). Wolfram (1984) discovered that these transition rules could yield very complex spatial patterns. However, these patterns can be generalized into four classes of behavior which have emerged from thousands of simulations for one-dimensional cellular automata. They are:

1. Spatially homogenous state;
2. Sequence of simple stable or periodic structures;
3. Chaotic aperiodic behavior;
4. Complicated localized structures with some propagating.

Wolfram found that all CA within each class, regardless of the details of their construction and evolution rules, exhibit qualitatively similar behavior. Starting from all possible initial configurations, CA may generate only special organized configuration, and self-organization may occur. Such universality produces general results of these classes applicable to a wide range of systems modeled by CA.

The simulation of population dynamics is a good demonstration of CA's capabilities in modeling complex natural systems. Couclelis (1988) successfully generated a wide variety of different spatio-temporal structures of rodent population by using a very simple one-dimensional cellular automaton. Her research clearly indicates that the whole range of complex and apparently bizarre population dynamics can be easily reproduced by the simple cellular automaton. Some vole populations cycle regularly for a few years and then, for no apparent reason, switch to a phase of random fluctuations. Elsewhere, the same species—under virtually identical environmental conditions—exhibits regular cycling in part of its range and irregular fluctuations in another. Her CA model can well explain these irregularities. The simulation is just based on a very simple kind of transition rule—a 'totalistic' rule. It means that the next state of a cell is a function of the simple sum of the present values of the cell itself and its neighbors.

Itami (1994) shows that the complex dynamics of rodent populations also can be simulated by a two-dimensional CA model. The two-dimensional CA model was run as a four-state system using totalistic rules. These states are: state 0 (vacant), state 1 (low density), state 2 (medium density), and state 3 (high density). The transition is based on the summation of neighborhood values. The hypothesis is that vole populations do best at low (but not too low) densities and drop off at very low and increasingly high densities. The simulation results compares well with those of Couclelis' one-dimensional model. Moreover, the two-dimensional model has characteristics that are not evident in the one-dimensional model, such as symmetry, information transfer, edge effect, local cyclical density patterns and equilibrium in chaos.

31.3 Urban Cellular Automata

The application of two-dimensional (2D) CA for urban simulation is straightforward since urban growth is analyzed primarily as growth in area—two-dimensional space—not in volume. CA have great potential for simulating urban growth and exploring alternative development forms by using predefined rules. In the last two decades, urban models based on CA techniques were reported with interesting outcomes (Deadman et al. 1993; Batty and Xie

1994; Batty and Xie 1997; Couclelis 1997; White and Engelen 1997; Wu and Webster 1998; Li and Yeh 2000).

Urban CA models differ from typical CA models as described by Wolfram (1984). Some of the strict conditions attached to conventional CA models have to be relaxed to meet the specific requirements of urban simulation. For example, typical CA models have a very limited total number of cells and a small number of temporal iterations (Batty and Xie 1994; Wu and Webster 1998). Urban CA models involve the use of a large number of cells and iterations. They also adopt heterogeneous cellular space which is different from that of conventional CA models. Thus, there are significant differences between urban CA models and other conventional CA models that have been developed for physics, artificial life, chemistry and biology.

Actually, Hägerstrand's (1967) spatial diffusion model could be regarded as an early CA-like model for geography because he used neighborhood effects. Hägerstrand developed diffusion models through 'Monte Carlo' simulation techniques. His models were specifically for human migration based on historical population records and using action-at-distance through gravitation effects. The models relied on microscopic behavior to describe macroscopic behavior of the system by using simple predefined rules.

Tobler (1979) was perhaps the first to recognize the advantage of CA models in solving geographical problems (White and Engelen 1993). In his cellular space model, the state of a cell is determined by the states of a set of 'neighbor' cells according to some uniform location-independent rules. The basic principle of such types of models is to use a cell-space representation to realize spatial dynamics.

Couclelis (1985, 1988, 1989) subsequently carried out some pioneering research on urban simulation using CA. Her studies attempted to explore the links with the theory of complex systems and examine the possible uses in an urban planning context (White and Engelen 1993). She showed that CA might be used as an analog or metaphor to study how different varieties of urban dynamics might arise.

Batty and his colleagues (Batty and Xie 1994; Batty and Xie 1997; Batty et al. 1999) also have carried out interesting research on urban CA models. In their early studies (Batty et al. 1989), a closely related technique—diffusion limited aggregation (DLA)—was used to model the growth of built-up areas. DLA models can generate complex forms by using a simple process like CA models. They also developed a general class of CA models which emerged through insights originating in computation and biology (Batty and Xie 1994). Their models are very similar to the Game of Life because each cell can only take on one of two states (dead or alive). However, these models have different features in three ways. First, they are nondeterministic because births and deaths at time t are computed stochastically. Second, a system-wide survival rate is used to control the whole pattern of actual survival rate. Third, a threefold hierarchy of neighborhood assessment is used to locally decide an actual birth. The simulation demonstrates that micro processes can lead to aggregate development patterns.

CA models can be used for testing hypotheses, simulating urban forms and dynamics, and generating alternative land use plans. However, most CA models to-date have been developed for hypothetical applications (Couclelis 1997; Batty et al. 1999). They were focused on testing ideas without providing enough details for realistic representation (White et al. 1997). Most urban CA models are primarily focused on testing urban theories by exploring the mechanisms of urban growth. They are designed to investigate basic questions of urban forms and evolution of urban systems (White and Engelen 1993; Couclelis 1997). These models are useful to explore how local actions give rise to global patterns.

CA models are able to generate cellular cities that have features very similar to those of real cities (White and Engelen 1993; White et al. 1997). Recently, much effort has been made to generate greater realism in urban CA simulation, to generate detailed and complex urban patterns. There are many aspects to this endeavor, including attention to fractal structures, higher spatial resolution, heterogeneous space, urban structure and sustainability. Fractal structures have been considered as the most important feature of urban geometry and fractal dimension can be used to evaluate the validity of urban simulation models (Wu 1998). It has been shown that stochastic disturbance variables can be incorporated into CA models in generating fractal patterns (White and Engelen 1993). Batty and Xie (1994) have illustrated how CA can be used to simulate suburban expansion of the town of Amherst in metropolitan Buffalo, New York. White et al. (1997) also employed CA to simulate the land use pattern of Cincinnati, Ohio. Clarke and Gaydos (1998) applied CA models to simulate and predict urban development in the San Francisco Bay region in California and the Washington/ Baltimore corridor in the Eastern United States.

In addition to simulating existing urban forms, CA models also can provide procedures for the design of optimal forms (Batty 1997). They may become powerful planning tools when integrated with GIS which can supply physical, social and economic data for the simulation. Recent CA models are usually linked to GIS by using a heterogeneous array of cells, so that the transition rules are in a sense site-specific. Remote sensing and GIS are integrated with CA in providing detailed land use information and other characteristics of cities for realistic urban simulation (Li and Yeh 2000).

Recently, attempts have been made to develop a kind of CA model which can be used as a planning tool for urban planning. Planning objectives are translated into transition rules that are the basis of CA simulation (Li and Yeh 2000; Ward et al. 2000; Yeh and Li 2001a). Urban planning usually involves the comparison between a set of planning scenarios and development options before making a plan. CA models can produce various development options that are dependent on the structures of models and inputs of data. The basic strategy is to properly define the structures of CA models that can incorporate planning objectives in the simulation.

Li and Yeh (2000) have used constrained CA and GIS to plan for sustainable urban development which aims at minimizing agricultural land loss and promoting compact development. Various urban forms which are associated with different development and energy 'costs' can also be explored using constrained CA models for testing different planning options (Yeh and Li 2001a). Ward et al. (2000) also developed a constrained CA model which has been applied to an area in Australia—Gold Coast, a rapidly urbanizing region of coastal eastern Australia. They demonstrate that CA models can simulate planned, realistic development by incorporating sustainability in the simulation. Their study shows that economic, physical and institutional control factors can be incorporated to modify, constrain and prohibit urban growth.

31.4 Configuration of Urban Cellular Automata

An urban CA model has four major elements: cells, states, neighborhood and transition rules. First, urban CA simulation operates on a lattice of cells in two-dimensional space. In most situations, the simulation space is divided into uniformly sized and regularly spaced cells. Second, only one state among a set of possible states is assigned to each cell at time t, although a 'gray' or 'fuzzy' state can sometimes be used (Li and Yeh 2000). In many urban CA models, there are only two common binary states – urbanized or not. Third, the configuration of neighborhood is to address the influences of neighboring cells in determining the conversion of states. There are usually two typical neighborhoods—the von Neumann neighborhood which consists of the four cells adjoining the central cell, and the Moore neighborhood which is composed of the eight adjacent cells (Figure 31-1). However, other configurations of neighborhood have been proposed to fit urban environments, such as a circular neighborhood (White and Engelen 1993; Li and Yeh 2000) and some action-at-distance-windows (Batty and Xie 1994). The cells with closer distance to the central cell will have greater neighboring effects on the conversion of states at the central cell. Fourth, transition rules which are usually expressed by some neighborhood functions are essential to urban simulation and iterations are required for the accomplishment of the simulation. Sometimes, stochastic variables are embedded in transition rules to address the influences of unknown factors.

	Neighbor Cell {x,y+1}	
Neighbor Cell {x-1,y}	Central Cell {x,y}	Neighbor Cell {x+1,y}
	Neighbor Cell {x,y-1}	

Neighbor Cell {x-1,y+1}	Neighbor Cell {x,y+1}	Neighbor Cell {x+1,y+1}
Neighbor Cell {x-1,y}	Central Cell {x,y}	Neighbor Cell {x+1,y}
Neighbor Cell {x-1,y-1}	Neighbor Cell {x,y-1}	Neighbor Cell {x+1,y-1}

a. von Neumann Neighborhood   b. Moore Neighborhood Area

Figure 31-1 von Neumann and Moore Neighborhoods.

Some studies prefer to use probabilities instead of deterministic rules for urban simulation because of the inherent uncertainties in urban systems. There are many substantial differences between urban CA transition rules and traditional CA transition rules. Unlike traditional CA models, urban CA models have not adopted strict transitional rules; a variety of urban CA transition rules have been proposed to satisfy various users' preferences. An essential part of CA simulation, therefore, is to define transition rules that are proposed for different applications.

A general transition rule function can be expressed as follows:

$$S_{ij}^{t+1} = f_N(S_{ij}^t)$$ (31-2)

where S^{t+1} is a set of possible states at location ij, N is the neighborhood of all cells providing input values to the function f, and f is the transition function that defines the change of the state S from time t to $t+1$.

Operational transition rules have to be provided for the implementation of CA. The most famous set of transition rules is related to the aforementioned *Game of Life* (Gardner 1970, 1971). The rules of the game are extremely simple. If an inactive cell has exactly 3 live neighbors, it becomes active (born). If an active cell is surrounded by 2 or 3 neighbors, it remains alive. An active cell will die under the situations of isolation (surrounded by fewer than 2 neighbors) or overcrowding (surrounded by more than 3 neighbors). Such a simple set of rules can yield patterns of surprising complexity which can be stabilized after many interactions.

In contrast to the rigid transition rules of traditional CA models, the determination of transition rules for urban CA models is quite relaxed. These transition rules are usually represented by transition probability or transition potentials. A simplified urban CA model can be defined using the following rule-based structure (Batty 1997):

IF any cell {x±1, y±1} is already developed

THEN $P_d\{x,y\} = \Sigma_{ij\sigma\Omega} P_d\{i,j\}/8$

&

IF $P_d\{x,y\}$ > some threshold value

THEN cell{x,y} is developed with some other probability $\rho\{x,y\}$

where $P_d\{x,y\}$ is the urban development probability for cell {x,y} and $\Sigma_{ij} P_d \{i,j\}$ is the sum of urban development probability for all cells from the Moore neighborhood Ω including the cell {x,y} itself.

Transition potentials also can be calculated by the combination of a series of factors. White and Engelen (1997) calculate transition potentials according to three factors: 1) the intrinsic suitability of the cell itself; 2) the aggregate effect of the cells in the neighborhood; and 3) a stochastic perturbation. The equation is given as follows:

$$P_Z = S_Z N_Z + \varepsilon_Z \quad \text{for all z,} \tag{31-3}$$

$$Nz = \sum_{d,i} I_{d,i} \, W_{z,y,d} \tag{31-4}$$

where

P_z is the potential for transition to state z;

S_z is the suitability of the cell for activity z; $S_z \varepsilon[0,1]$;

N_z is the neighborhood effect for activity z;

$W_{z,y,d}$ is the weight applied for activity z to cells in state y in distance zone d;

i is the index of cells in distance zone d;

ε_z is a stochastic disturbance term; and

$$I_{d,i} = \begin{cases} 1, & \text{if cell } i \text{ in distance zone } d \text{ is in state } y, \\ 0, & \text{otherwise,} \end{cases}$$

Clarke et al. (1997) use five factors to control the behavior of urban simulation: *DIFFUSION, BREED, SPREAD, SLOPE_RESISTANCE* and *ROAD_GRAVITY*. Transition rules based on these five factors affect the acceptance level of randomly drawn numbers, are set by the user for every model run and are varied as part of the calibration process.

In another approach, Wu and Webster (1998) present an integrated CA and multi-criteria evaluation (MCE) method to estimate the probability of urban transition in a nondeterministic CA. MCE is used to capture the different blends of government and private developer preferences that govern various development regimes. The development probability p_{ij} is determined by a combined evaluation score r_{ij} and nonlinear transformation is used to discriminate the simulation patterns. The equation is expressed by:

$$p_{ij} = \phi\,(r_{ij}) = \exp\left[\alpha(\frac{r_{ij}}{r_{max}} - 1\,)\right] \tag{31-5}$$

where:

α is a dispersion parameter ranging from 0 to 1;

r_{ij} is the combined evaluation score at location ij;

r_{max} is the maximum value of r_{ij}.

The composite evaluation score is calculated from the following linear equation:

$$r_{ij} = (\beta_1 \text{ CENTER} + \beta_2 \text{ INDUSTRL} + \beta_3 \text{ NEWRAILS} + \beta_4 \text{ HIGHWAY} + \beta_5 \text{ NEIGHBOR}) \text{ RESTRICT} \qquad (31\text{-}6)$$

where $\beta_1, \ldots, \beta_5$ are weighting parameters acquired from the Analytic Hierarchy Process (AHP) analysis of MCE; CENTER, INDUSTRL, NEWRAILS, HIGHWAY and NEIGHBOR are development factors.

'Gray value' can be defined to represent the 'fuzzy' probability of conversion in transition rules (Li and Yeh 2000; Yeh and Li 2001a). Since land use conversion can be considered to take place by a gradual course, a 'gray value' can be used to indicate the degree of urbanized process for a cell, as opposed to the conventional concept of purely urbanized or not. The 'gray value' is calculated by iterations:

$$G_{xy}^{t+1} = G_{xy}^{t} + \Delta G_{xy}^{t} \qquad (31\text{-}7)$$

where G is the 'gray value' for development which falls within the range of 0–1; and xy is the location of the cell. A cell will be urbanized when the 'gray value' reaches 1. ΔG^t is the gain of the 'gray value' at each loop. ΔG^t can be defined using the neighborhood function which is the basis of CA simulations. According to the neighborhood function, the conversion probability at a cell depends on the states of its neighboring cells. There is a higher chance of conversion at a cell if it is surrounded by more converted cells. The increase of 'gray value' should be determined by the amount of developed cells in the neighborhood.

ΔG^t can be simply defined by the following neighborhood function:

$$\Delta G_{ij}^{t} = f_N(q_{ij})$$
$$= \frac{q_{ij}}{\pi \xi^2} \qquad (31\text{-}8)$$

where q is the total amount of developed cells in the neighborhood; ξ is the radius of the circular neighborhood. A circular neighborhood is used because it has no bias in any direction (Li and Yeh 2000).

Another set of urban CA models is developed based on a different approach, the concept of life cycle (Batty and Xie 1994; Batty et al. 1999). Activities in urban systems are considered to follow life cycles which apply both to the physical stock that is developed, as well as to the activities that occupy the stock at different locations. In this type of CA model, urban activities can be defined by new, mature and declining housing, industry and commercial land uses.

Transition rules have become even more complex when CA models are integrated with economic theories and conventional urban models. Urban CA models usually involve not only simple local rules, but also other economic and social factors (Semboloni 1997). Examples of this mix include the use of social actors, Lowry models (Lowry 1964; Webber 1984) and system dynamics in the definition of CA transition rules. Market mechanisms governing competition of land uses should be emphasized in this type of models. Semboloni (1997) presents a CA model which is closely related to Lowry models, but is much more complex than simple CA models. White and Engelen (1997) also propose a cellular automaton in conjunction with a macro-scale model to represent non-local dynamics of population, economy and natural environments. Webster and Wu (1999a, 1999b) infuse many behavioral rules into urban CA models. The transition rules are defined by mixing economic equilibrium and CA paradigms. They are based on behavioral economic models rather than pure heuristics and development potential is measured in monetary value (profit).

The question arises of how to maintain the original simplicity of CA models when they are integrated with other conventional urban models. They may collapse into other non-CA models if the emphasis of local rules is lost. There also are questions of how to define model structures and parameter values for these complex models. It is almost impossible to calibrate CA models and determine parameter values when these models become too complex and use too many variables. Calibration also cannot be carried out by conventional statistical methods to obtain parameter values, such as logistic or multi-logistic regression models. These models are only valid for the simple linear relationships between independent variables and dependent variables.

It is not easy to define model structures and determine parameter values in urban CA models. Studies in calibration of urban CA models also are limited because of their complexities. Section 31.6 will discuss different new methods in calibrating urban CA models using neural networks and data mining techniques.

31.5 Incorporating GIS and Remote Sensing for Data Inputs and Constraints

Studies have shown that the integration of CA models with GIS can yield more-plausible simulation results (Li and Yeh 2000; Yeh and Li 2001a). CA and GIS strongly complement each other in terms of their spatial and temporal modeling capabilities. First, CA can serve as the analytical engine for GIS because CA can significantly enhance GIS to perform spatial dynamic modeling. Although GIS have been popularly applied in many spatial analysis and spatial decision making situations, they have poor performance for many operators and poor ability to handle dynamic spatial models (Wagner 1997). CA with its temporal modeling abilities can enrich existing GIS functions. Spatial and sectoral interactions for modeling cannot be easily solved by the functionality of current GIS software (Batty et al. 1999). The simulation of urban systems needs to run models indefinitely, or for as long as the user requires, while conventional GISs have limitations dealing with temporal processes. CA models have been proven to implement spatial and temporal interactions rather easily because they are quite adept at handling complex systems. A class of urban models has been developed by the integration of CA and GIS for more realistic simulation and better simulation performance (Wu and Webster 1998; Batty et al. 1999; Li and Yeh 2000).

Second, high resolution, location-specific information can be retrieved from GIS to meet the data realism requirement for CA models. GIS provides a rich source of spatial information that can be used as various kinds of constraints for CA modeling. For example, the constraints of resources and environment can be conveniently measured in GIS and the results imported into CA models. The integration can allow the simulation to explore a large set of spatial variables and readily provide the information of the relationships between urban growth and these variables. Operational urban models are often linked to land use, transport and other economic and environmental factors. GIS are most appropriate to provide such type of information.

CA can alleviate limitations in urban simulation by providing a high resolution of spatial reality. The basic unit of CA is the cell which can be defined at a very fine resolution. The principle of urban CA is that local decisions give rise to global patterns and thus generate very complex features of cities. CA models have been increasingly used in urban simulation recently because of their powerful modeling capabilities and high degree of reality. Many researchers indicate that CA can simulate complex systems by using some simple local rules. Cities become computable in various ways within the generic framework of CA models (Batty and Xie 1994; Batty 1997). The framework can simulate a range of urban morphologies from strict determinism to complete randomness and from complete predictability to complete unpredictability.

The employment of CA in urban simulations often entails substantial departures from the original formal structure of CA described by von Neumann, Ulam, Conway, and Wolfram

(Torrens and O'Sullivan 2001). Standard CA deal only with neighborhood effects. There are usually two typical neighborhoods—the von Neumann neighborhood which consists of the four cells adjoining the central cell, and the Moore neighborhood which is composed of the eight adjacent cells. The transition rules of standard CA usually are defined within a homogeneous cell space where cells have no inherent qualities that can affect the transition rules. A given configuration of cells in the neighborhood of a cell will result in the same state transition regardless of the location of the cell on the grid (White and Engelen 1997). No constraints are applicable for standard CA during the simulation.

These strict assumptions of standard CA need to be relaxed so that the simulation can fit the experience of real cities. For example, the adoption of action-at-a-distance in urban simulation is much different from the von Neumann or Moore neighborhoods in traditional CA models. Exogenous forces can also exert influences on urban growth. White and Engelen (1997) demonstrate that the aggregate demand for land for each activity will influence the simulation process. All kinds of constraints from local, regional and global areas can be prepared from GIS and incorporated in urban simulation (Li and Yeh 2000). Another significant adaptation is to incorporate some randomness into transition rules (White and Engelen 1993). The simulation results are not deterministic for this type of CA models.

Over the last two decades, satellite remote sensing has provided a rich source of ground information for various disciplines that can be treated as empirical data for CA simulation. The recent availability of high-resolution image data provides useful spatial and temporal information that can be input to urban CA models. Land use information is a very important input to urban models that can be effectively acquired by the classification of satellite remote sensing data. Since remote sensing data are in raster format, they are easier to use for CA models than any other sources of data and their use ensures more realistic CA simulation.

31.6 Calibrating Urban Cellular Automata

A critical issue in CA simulation is to provide proper parameter values or weights so that realistic results can be generated. Real cities are complex dynamic systems that involve the use of many spatial variables, each of which makes a contribution to the CA simulation. The influence of a spatial variable is determined by its associated parameter or weight in the simulation with a larger parameter value usually indicating that it is more important than other variables. There are usually many parameter values to be defined in a CA model and the results of CA simulation are very sensitive to these parameters (Wu 2000).

There are very limited studies on the calibration of geographical CA models. Validation of most of the existing urban CA models is based on the so-called 'trial and error' approach with a visual assessment of model results (Clarke et al. 1997; White et al. 1997; Ward et al. 2000). There are some other attempts to develop more elaborate methods to tackle the problems of uncertainties in defining transition rules and parameter values. Wu and Webster (1998) use multi-criteria evaluation (MCE) to heuristically define the parameter values for CA simulation. Calibration also can largely rely on repetitive runs of the same model with different combinations of parameter values (Wu 2000). Wu (2002) provides a method to estimate the global development probability by using a logistic regression model. The initial global probability is calibrated according to historical land use data and the meanings of the coefficients are easier to understand in the logistic regression equation.

Clarke and Gaydos (1998) also provided a more elaborate calibration method by statistically testing the observed against the expected. The method determines which set of parameter values can lead the model to produce the best fit. The set of parameter values with the best fit are then used for prediction. However, there are numerous possible combinations of parameter values. Their experiments have tried more than 3,000 combinations which need a high-end workstation to run several hundreds of hours for the calibration. A thorough search procedure, therefore, is difficult to develop for this calibration.

Another method is to use neural networks to deal with the complicated calibration issue in CA urban simulation (Li and Yeh 2001a). Neural networks are simple and convenient because complex, nonlinear relationships can be modeled by simple network structures. The parameter values can be automatically obtained by a back-propagation calibration procedure. The method is also more robust due to the well-developed procedure of back-propagation training. Moreover, the model is able to deal with complex interactions among variables which are not required to be independent of each other because of using a neural network. The model structure overall is much simpler and more stable compared with traditional CA models.

Knowledge discovery or machine learning techniques also can be used to reconstruct the transition rules of geographical CA. The process for acquiring domain knowledge, however, is tedious and time-consuming. Although experts are capable of using their knowledge to solve problems, they cannot guarantee that the knowledge is explicitly expressed in a systematic, correct and complete form. A well-known problem when creating expert systems is often called the 'knowledge acquisition bottleneck' (Huang and Jensen 1997). It is found that explicit transition rules of CA can be automatically reconstructed through the rule induction procedure of data mining and these explicit transition rules are much more intuitive to decision-makers (Li and Yeh 2004). The transition rules can be obtained by applying data mining techniques to GIS and remote sensing data. The proposed method can reduce the uncertainties in defining transition rules and helps to generate more reliable simulation results.

31.7 A Constrained CA Model for Land Use Planning

As stated above, transition rules in urban CA modelling decide whether a cell will change its state from one to another and the transition rules according to Equation 31-8 only address the influences of the states (developed cells) in the neighborhood. The evolution of real cities, however, is influenced by a series of complicated factors which can be obtained at various local, regional and global levels. The neighborhood function also cannot address the issue of urban structures and environmental problems. Some kinds of constraints, therefore, should be used to regulate the simulation to improve modeling accuracy. Without constraints, urban simulation will generate patterns as usual based on historical trends. Constraints should be added into urban CA models to reflect environmental and sustainable development consider-ations since they are important factors for the formation of idealized urban patterns.

By taking environmental and other constraints into considerations, Equation 31-8 can be revised into a generic constrained CA model as follows:

$$\Delta G'_{ij} = f_N(q_{ij}) \times \prod_{m=1}^{M} \delta_{mij}$$

$$= \frac{q_{ij}}{\pi \xi^2} \times \prod_{m=1}^{M} \delta_{mij} \tag{31-9}$$

where q is the total amount of developed cells in the neighborhood; Π is the radius of the circular neighborhood; δ_{mij} is the function to represent various types of constraints in which the values should be normalized within the range from 0 to1. It can be regarded as a scaling factor to readjust the increase of the 'gray value.'

A stochastic disturbance term (γ) is added to the model to represent unknown errors which are frequently exhibited in many complex systems. This can allow generated patterns that are more similar to realistic development patterns. Incorporating the random variable in Equation 31-9, the final equation for calculating the increase of the 'gray value' is then given by:

$$\Delta G_{ij}^{'t} = (1 + (-1n\,\gamma)^{\alpha}) \times f_{N}(q_{ij}) \times \prod_{m=1}^{M} \delta_{mij}$$

$$= (1 + (-1n\,\gamma)^{\alpha}) \times \frac{q_{ij}}{\pi\xi^{2}} \times \prod_{m=1}^{M} \delta_{mij}$$

(31-10)

The generic constrained CA model developed above takes into account not only the influences of neighboring states, but also a series of economic and environmental constraints. These constraints may include environmental suitability, urban forms and development density. The following examples use the city of Dongguan in southern China as an example to show how the generic constrained CA model can be used to generate CA planning models that take the environment, urban form and density into consideration, respectively, and how these factors can be combined into one constrained model. The CA models were programmed using the Arc Macro Language (AML) within a GIS package, ARC/INFO GRID.

Dongguan, a very fast-growing city covering 2,465 km² in the Pearl River Delta of southern China, underwent urban development at a tremendous speed and scale in the 1990s (Yeh and Li 1997; Yeh and Li 1999). It includes a city proper and 29 towns. Remote sensing and GIS data were used to provide the basic information for the simulation. The 1988 and 1993 Landsat Thematic Mapper (TM) images were classified to retrieve land use and land use change information (Li and Yeh 1998). A GIS database was built to contain the information on land use, transportation, population and administrative boundaries. The database was converted into a raster format for the simulation. The basic unit is a cell which has an area of 50×50 m on the ground. The initial map for the simulation was from the land use classification of the 1988 satellite TM image and the models attempt to generate land development options for 1988-93. The actual urban areas (built-up areas and development sites) in 1993 obtained from the classification of the 1993 satellite TM image were used as a baseline for evaluating the results of the simulation.

31.7.1 Model 1: Environmentally Constrained CA Model

The aim of the first model is to incorporate environmental constraints in urban simulation. Due to growing concern for environmental issues related to urban growth in the world, urban development should be determined not only by pure economic factors, but also by environmental constraints. Environmental consciousness reflected in CA models will result in more idealized urban development patterns. It is convenient to obtain environmental constraints and embed them in urban simulation based on the integration of CA and GIS technologies.

Environmental constraints are used to indicate whether a piece of land should be protected from development with regard to environmental considerations. The generic constrained CA model in Equation 31-10 can be modified into an environmentally constrained CA model by incorporating environmental constraints as follows:

$$\Delta G_{ij}^{'t} = (1 + (-1n\,\gamma)^{\alpha}) \times \frac{q_{ij}}{\pi\xi^{2}} \times \delta_{ENVij}$$

(31-11)

where δ_{ENVij} is the function of environmental constraints.

The constraint function δ_{ENVij} in Equation 31-11 is related to a number of environmental factors that can be defined by using GIS data. This model emphasizes the protection of strategic agricultural land and other important ecological areas in urban planning. The score of the constraint function can be calculated by combining an agricultural suitability score with other environmental scores for protecting resources and the environment. Agricultural suitability, which reflects the potential of agricultural production, can be used to address the need to reserve important agricultural land. Higher costs are then associated with encroachment on sites of good-quality agricultural land. Other environmental scores also can be calculated to address the disturbance of development in protected areas (e.g., river basins for supplying drinking water) and ecologically sensitive areas (e.g., wetlands and mangroves). Development in the neighborhood of these types of land use can bring about environmental degradations and ecological disturbances. The environmental scores can be defined based on the buffer distances to these sensitive areas using GIS functions. The influences should be in the forms of distance decay functions.

Multi-criteria evaluation (MCE) techniques can be employed to calculate the total combined constraint score for various environmental factors. Before the calculation, it is necessary to standardize the score for each factor because these factors may be measured at different scales. A typical method of standardization is to use the minimum and maximum values as scaling points for a simple linear transformation (Voogd 1983). However, CA simulation based on a linear transformation cannot generate typical development patterns. Other types of nonlinear transformation can provide more-plausible results by achieving greater discrimination between cells (Wu and Webster 1998; Li and Yeh 2000). The nonlinear transformation can be defined in an ad hoc way because a unique transformation does not exist. The transformation can be in exponential (Wu and Webster 1998), logistic (Wu 1998) or power forms (Li and Yeh 2000). Usually, the values of the adjusted scores should fall within the range of 0 to 1 for comparison.

The calculation of δ_{ENVij} can be accomplished by using the following expression:

$$\delta_{ENVij} = \sum_{\Theta=1}^{N} w_{\Theta} \, (1\text{-}ENV_{\Theta ij})^k \tag{31-12}$$

where $ENV_{\Theta ij}$ is the score of the Θth environmental factor, w_{Θ} is the weight and k is the parameter for the nonlinear transformation (Li and Yeh 2000). Each factor should be normalized within the range of 0 to 1. A higher value of k will ensure that the environmentally sensitive land can be protected strictly, but the simulated patterns may be fragmented (Li and Yeh 2000).

The combination of Equations 31-11 and 31-12 yields the environmentally constrained CA model:

$$\Delta G_{ij}^{'t} = (1 + (\text{-}\ln\gamma)^{\alpha}) \times \frac{q_{ij}}{\pi\xi^2} \times \sum_{\Theta=1}^{N} w_{\Theta} \, (1\text{-}ENV_{\Theta ij})^k \tag{31-13}$$

The protection of important agricultural land and other ecologically valuable land and water resources has become a major issue in the Pearl River Delta since the economic reform in 1978 (Yeh and Li 1999). The region used to be one of the most important agricultural production bases in China. However, economic development in the region has triggered severe agricultural land loss and depletion of other ecological land, such as wetlands. The rate of agricultural land loss is astonishing in the Pearl River Delta in the 90s because of the rapid urbanization process. An extreme case of land loss has been witnessed in the Shenzhen metropolitan region which has the closest proximity to Hong Kong. The entire agricultural land base has been almost destroyed by massive land development in the early 1990s according to the analysis of satellite images (Li and Yeh 1998).

Figure 31-2 is the simulation results obtained by applying the environmentally constrained CA model as described in Equation 31-12. In this study, environmental constraints of agricultural suitability and preservation of natural resources were incorporated in the model for the protection of cropland, forest and wetland by allocating land development in other less sensitive areas. Agricultural suitability, which is one of the major environmental considerations, is calculated from the slope and soil maps in the GIS. The locations of forest and wetland, which also are obtained from the GIS, are used as environmental constraints of the model (Li and Yeh 2000). The environmental constraint is essential for the model to find

a) Loss of good agricultural land (k=0, non-constrained)

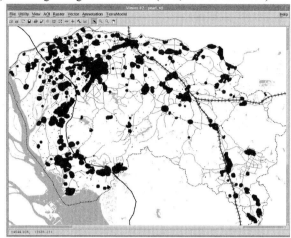

b) Controlling urban development (k=1, general control)

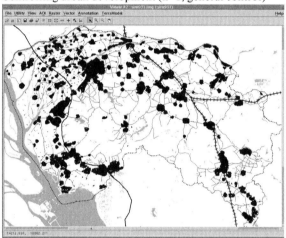

c) Strictly controlling urban development (k=3, strictly control)

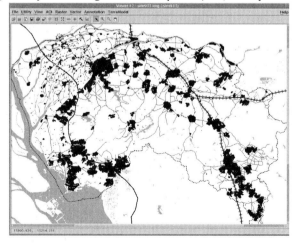

Figure 31-2 Land development scenario for agricultural and ecological protection in Dong-guan, China in 1988-93 from the constrained CA planning model.

the solution in minimizing the impacts of urban development on agricultural and ecological conservation. A nonlinear transformation ($k=3$) was used for strict protection of resources. In most situations, environmental resources exist in heterogeneous patterns across space and a GIS is essential for providing real data to the modeling process. In this study area, the most fertile agricultural land is concentrated in the alluvial plain in the northwest part near the city proper. This model can preserve productive agricultural land and ecological land by incorporating a series of environmental constraints.

31.7.2 Model 2: Urban-Form-Constrained CA Model

The planning of urban form is central to the promotion of sustainable development (Breheny 1996). There have been many debates on how to confine urban sprawl and conserve agricultural land resources (Bryant et al. 1982; Gierman 1977; Ewing 1997; Daniels 1997). An essential part of plan-making is to develop simulation models for generating regional and sub regional growth options. The California Urban Future (CUF) Model is one such model which is valuable for presenting and comparing the details of different development scenarios (Landis 1995). The CUF model was the first large-scale metropolitan simulation model to use a GIS for data integration and spatial analysis. Clarke et al. (1997) also provide a CA model that can simulate urban dispersion using self-modifying rules. The model has been applied to the simulation of urban growth in the San Francisco Bay area and the behavior of the system is mainly controlled by five factors, which can be calibrated by historical data. The model is used to generate three predictions of urban growth in the region—uncontrolled rapid growth, sustained slow growth and the growth which stabilizes at a desirable or sustainable level. Deadman et al. (1993) applied a CA model to predict the patterns related to the spread of rural residential development. They demonstrate the potential of the model to be run 'into the future' to predict the outcome of policy decisions.

There are many advantages in using CA and GIS to simulate possible urban forms for the planning of sustainable urban development. Urban CA models can help planners to explore various options and evaluate possible environmental impacts. When urban form is used as the main constraint, the generic constrained CA model in Equation 31-10 can then be modified into an urban-form-constrained CA model:

$$\Delta G'_{ij} = (1 + (-\ln\gamma)^\alpha) \times \frac{q_{ij}}{\pi\xi^2} \times \delta_{ij}(FORM) \tag{31-14}$$

The function, $\delta_{ij}(FORM)$, is determined by the relationships between urban growth and urban centers. Land development can be concentrated around the main center for promoting the growth of large cities, or can be shifted to around sub-centers for promoting polycentric growth. There are many possible development patterns which are the main concerns of urban planning. Land development and urban forms, for example, are closely related to the efficient use of energy, capital and land resources (Banister et al. 1997; Burchell et al. 1998). The constraint is decided by location factors in terms of the distances to urban centers. Urban centers play an important role in urban growth as they provide the support for the requirements of energy, materials, capital and techniques for development. The influence of urban centers can be measured by a distance decay function. The classical measure of urban structure is the density gradient from the Central Business District (CBD) (Muth 1969). Although the density gradient is related to the monocentric concept of urban form, nevertheless it gives us an index of the degree of decentralization. Two distances can be defined to capture the hierarchy of urban structures that consist of a major center and many sub-centers. The constraint score, which indicates the attractiveness of urban centers, can be expressed by the following function:

$$\delta_{ij}(FORM) = \exp(-\frac{\sqrt{w_R^2 d_{Rij}^2 + w_r^2 d_{rij}^2}}{\sqrt{w_R^2 + w_r^2}}) \qquad (31\text{-}15)$$

where d_R is the distance from a cell to the main center and d_r is the distance from cell ij to its closest sub-center. w_R and w_r are the weights for the two distance variables, respectively.

The ratio of w_R/w_r determines what kind of urban forms can be generated at the macro-level, with a higher value of w_R/w_r giving more weight to the main center. In contrast, a lower value of w_R/w_r puts more weight to sub-centers. The growth rate of the 'gray value' at a location is affected by the constraint. Different types of urban forms can emerge by simply changing the value of w_R/w_r. A higher value of w_R/w_r results in monocentric development, whereas a lower value leads to polycentric development. Finally, the combination of Equations 31-14 and 31-15 yields the urban-form-constrained CA model:

$$\Delta G_{ij}^{'t} = (1 + (-\ln\gamma)^\alpha) \times \frac{q_{ij}}{\pi\xi^2} \times \exp(-\frac{\sqrt{w_R^2 d_{Rij}^2 + w_r^2 d_{rij}^2}}{\sqrt{w_R^2 + w_r^2}}) \qquad (31\text{-}16)$$

Preparing development scenarios is one of the routine tasks in many planning departments. It is a tedious job because a large set of spatial data should be processed to reflect the complexity of urban systems. Urban CA models can be automatically used to simplify the job by generating development scenarios. Furthermore, a large set of spatial data can be easily handled when CA are integrated with GIS. It is convenient to explore the different types of urban forms by adjusting the weights which are associated with different spatial variables.

Five types of urban forms were simulated using the model represented by Equation 31-16. They include compact-monocentric, compact-polycentric, dispersed, highly dispersed and very highly dispersed development forms. Table 31-2 lists the parameter values used in the model. Other mixed urban forms can be easily generated using the same method by changing the parameters. Figure 31-3 shows one of the simulated scenarios, the compact-monocentric development, to demonstrate the capability of the CA model in generating development options. The simulation was to emphasize the role of the main center in supporting urban growth. The growth rate was affected by the distance to the city proper rather than by the distances to sub-centers. Higher growth rates were only allowed to take place around the city proper according to the criterion and there was very limited growth around those sub-centers that were farther away from the city proper. The model results showed a large part of the projected development took place within a small area around the city proper (the north-west part) and the pattern is much more compact, compared with the actual dispersed development pattern obtained by using satellite images.

Urban forms have direct impacts on urban sustainability because the forms are associated with various types of costs. There are two major types of development costs: 1) for the connection of various types of infrastructure between new development sites and their closest networks within zones; and 2) for the improvements of networks to accommodate the flows of population and goods between zones. Urban forms affect the activities related to inter-zonal infrastructure requirements and between-zone interactions. The costs of these activities for various development forms can be estimated from GIS analysis (Yeh and Li 2001a). The actual development pattern is associated with a larger amount of costs compared with the simulated development patterns. It is easy to identify a better development alternative based on the assessment. This provides important implications for urban planners who are required to choose a suitable development plan among various development scenarios.

Table 31-2 Model parameters for simulating different types of urban forms.

Urban Forms and Developments	Dispersion Factor	Urban Form
1.Compact-Monocentric Development	= 0	$W_R = 1; W_r = 0$
2. Compact-Polycentric Development	= 0	$W_R = 0; W_r = 1$
3. Dispersed Development	= 1	[factor not incorporated]
4. Highly Dispersed Development	= 5	[factor not incorporated]
5. Very Highly Dispersed Development	= 10	[factor not incorporated]

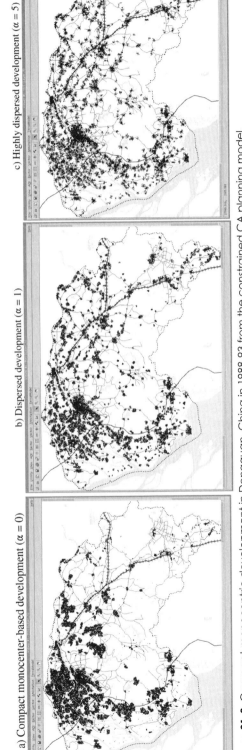

a) Compact monocenter-based development ($\alpha = 0$)

b) Dispersed development ($\alpha = 1$)

c) Highly dispersed development ($\alpha = 5$)

Figure 31-3 Compact-monocentric development in Dongguan, China in 1988-93 from the constrained CA planning model.

31.7.3 Model 3: Development-Density-Constrained CA Model

Development density is another important factor for compact development that, unfortunately, has not been well integrated in the process of general GIS site selection and urban CA simulation. Development density can be represented by the total number of people that can be accommodated by a developed cell. High development density can significantly reduce the expenditures for providing infrastructure and public service and the costs of consuming energy and resources.

Rising development densities of cities have many benefits for sustainable development (Newman and Kenworthy 1988; Pushkarev and Zupan 1977). Per capita infrastructure costs almost certainly fall as densities rise, although extremely high densities may cause an increase in costs (Ewing 1997). Urban sprawl, on the other hand, will lead to the encroachment on much more land than compact development. For example, an assessment of two development plans for the state of New Jersey indicates that the compact plan can reduce land consumption as much as 60% (CUPR 1992; Ewing 1997). An urban-sprawl plan will result in the loss of five times as much environmentally sensitive land and 60% more farmland than compact development. A solution to contain urban sprawl is to increase development density properly. This can reduce a series of costs for land development and increase the benefits for environmental protection.

This third model includes the factor of density in urban simulation. The essential part of simulation is to determine the increase of 'gray value' for a cell based on neighborhood functions. The increase of 'gray value' is proportional to the total population in the neighborhood. When development density is used as the main constraint, the generic constrained CA model in Equation 31-10 can be modified into a development-density-constrained CA model:

$$\Delta G_{ij}^{'t} = (1 + (-\ln \gamma)^{\alpha}) \times \Omega_N (Den_{ij})$$

$$\tag{31-17}$$

$$= (1 + (-\ln \gamma)^{\alpha}) \times \frac{\sum\limits_{ij \in \Omega_N} Den_{ij}}{Den_{max} \pi \xi^2}$$

where Den_{ij} is development density, f_N is the set of developed cells in the Moore neighborhood N, ξ is the radius of the circular neighborhood, and Den_{max} is the maximum value of the development density. A circular neighborhood is used to calculate the total population.

When a cell is selected for development, a development density should be assigned to the cell according to local experience or historical data. Development density in terms of population should be dependent on the distance to urban centers. Density decay functions used to determine the development density of a developed cell assume that development density (population density) declines in an inversely exponential manner. The notion that population density declines from centers has been previously discussed in many studies (Clark 1951; Thrall 1988; Papageorgiou 1971). The function is generally given (Clark 1951) as:

$$Den_{ij}^0 = A \ \exp (-\beta l_{ij})$$

$$\tag{31-18}$$

where Den_{ij}^0 is the assigned development density, l_{ij} is the distance to a center, and A and β are the parameters of the density decay function. The function has been repeatedly tested for the past 150 years and examined to be statistically significant (Papageorgiou 1971). Especially, for most large cities, the negative exponential model seems to describe the real-world observations as a first approximation (King and Golledge 1978).

Incorporating development density in urban simulation is a relatively new approach for urban CA models. Although an important factor in urban planning that plays a crucial rule in influencing urban forms, development density has not been incorporated in most urban models. The relationship between development density and urban forms is apparent. As development density rises, cities will become more compact. This study attempts to generate different scenarios of development densities to provide more-detailed information for urban planning. We use the model to generate different combinations of development densities, which are compared with the actual development density. Figure 31-4 is an example of simulating development density around the city proper of Dongguan, China, by incorporating the density gradient into the CA model according to Equation 31-16. The parameters of A and β of the density decay function were set to 80 and 0.005, respectively.

Urban development inevitably consumes the land previously used for farming, forestry or wetland. However, the encroachment on agricultural land can be minimized if the development density is raised properly. More options will then be available for the protection of strategic agricultural land, wetland and forest when there is a large reserve of land resources. In most situations, raising development density will reduce the costs of infrastructure construction and infrastructure maintenance, and also the costs related to energy consumption.

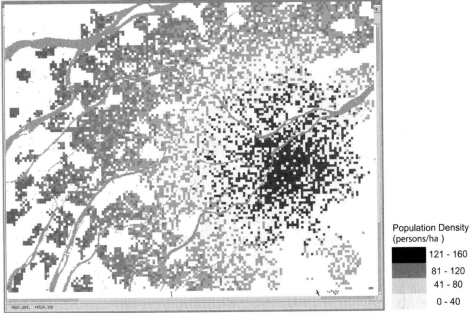

Population Density
(persons/ha)

▮ 121 - 160

▮ 81 - 120

▮ 41 - 80

▯ 0 - 40

Figure 31-4 Simulating development density from the constrained CA planning model.

31.7.4 Model 4: A Combined CA Model with the Constraints of Environmental Factors, Urban Forms and Development Densities

This model combines the constraints of environmental factors, urban forms and development densities together in urban simulation. It enables planners to explore the impacts of all three constraints or some combination of them together in one model. In this model, the increase of 'gray value' is proportional to development density in the neighborhood and also is subject to the factors of environmental suitability and urban forms.

By incorporating the factors of density, urban form and environmental suitability in the generic constrained CA model, Equation 31-10 becomes as follows:

$$\Delta G_{ij}^{'t} = (1 + (-\ln \gamma)^{\alpha}) \times f_N(Den_{ij}) \times \delta_{ij}(FORM) \times \delta_{mij}(ENV)$$

(31-19)

$$= (1 + (-\ln \gamma)^{\alpha}) \times \frac{\sum\limits_{ij \in \Omega_n} Den_{ij}}{Den_{max} \pi \xi^2} \times \exp(-\frac{\sqrt{w_R^2 d_{Rij}^2 + w_r^2 d_{rij}^2}}{\sqrt{w_R^2 + w_r^2}}) \times \sum\limits_{m=1}^{N} w_m (1-ENV_{mij})^k$$

where Ω_N is the set of developed cells in the Moore neighborhood N, $\delta_{mij}(ENV)$ is the environmental constraint function for the mth criterion, ξ is the radius of the circular neighborhood, and Den_{max} is the maximum value of the development density.

The combined model as described in Equation 31-19 was used to produce a more plausible urban simulation. This combined model simultaneously takes into account the three factors in sustainable land development—environmental suitability, urban form and development density. These factors can be handled by the constrained CA through a step-by-step approach. They are important in regulating urban simulation for generating different land development patterns according to different planning objectives. Without these constraints, cities will prevail along the trajectory of historical trends, which may be economically oriented and resource consuming. This will lead to increases in environmental costs and degradation of the environment. Constraints in CA models are important for achieving sustainable development objectives, especially for the protection of resources and environment. An example is to restrict land development in environmentally sensitive areas in urban simulation. Constraint scores are defined to indicate the degree that is needed to control land development. Constraint scores can be estimated from land evaluation (McRae and Burnham 1981).

Many types of constraints can be defined to reflect different planning objectives for urban simulation. The environmental constraints consider the factors of protecting agricultural land and other ecological land such as forest and wetland. Buffer analysis from GIS was carried out to define the constraint score. There are higher values of ENV_{mij} for environmental sensitive areas. This can guide urban development away from environmentally sensitive areas. Table 31-3 lists the parameter values for producing various development options according to these constraints. Figure 31-5 only shows one of the simulated options—the simulation of polycentric, high-density and environmental-based development. This shows the utility of this combined constrained CA model generated by enhancing the generic constrained CA model.

Table 31-3 Parameters for generating development options with the constraints of environmental factors, urban forms and development densities.

Development Scenarios	Environment Factors	Urban Forms		Density Functions	
	k	w_R	w_r	A (persons/ha)	
1. Monocentric-based					
1A) High Density & Fast Density Decay	3	1	0	80 (city proper); 60 (towns)	0.005
1B) High Density & Slow Density Decay	3	1	0	80 (city proper); 60 (towns)	0.001
1C) Low Density & Fast Density Decay	3	1	0	40 (city proper); 30 (towns)	0.005
1D) Low Density & Slow Density Decay	3	1	0	40 (city proper); 30 (towns)	0.001
1. Polycentric-based					
2A) High Density & Fast Density Decay	3	0	1	80 (city proper); 60 (towns)	0.005
2B) High Density & Slow Density Decay	3	0	1	80 (city proper); 60 (towns)	0.001
2C) Low Density & Fast Density Decay	3	0	1	40 (city proper); 30 (towns)	0.005
2D) Low Density & Slow Density Decay	3	0	1	40 (city proper); 30 (towns)	0.001

The simulation results can be evaluated by using GIS overlay analysis to identify which types of development can have better performance in terms of resource savings. The baseline scenario is the actual land development in 1988-93 which is considered as very low-density and highly dispersed development. The comparison can help planners to evaluate the advantages and disadvantages of different planning options according to different planning objectives.

The simulation results can be evaluated by using GIS overlay analysis to identify which types of development can have better performance in terms of resource savings. The baseline scenario is the actual land development in 1988-93 which is considered as very low-density and highly dispersed development. The comparison can help planners to evaluate the advantages and disadvantages of different planning options according to different planning objectives.

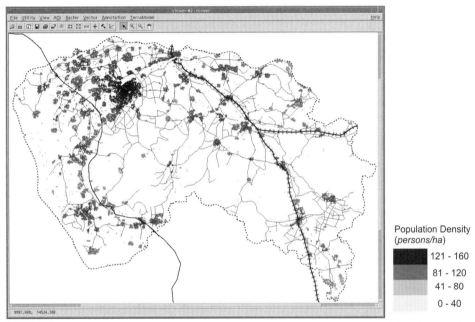

Figure 31-5 Simulation of polycentric, high-density and environmental-based development.

31.8 The Applications of CA Models in Urban Planning

Above simulation results indicate that CA models can have many applications in urban planning. The first category is the baseline-growth simulation models, which often hypothesize a stable urban development trend and then simulate urban growth based on the trajectory of historical development. The second category of models aims to provide planning evaluation and analysis. Planners can use such kinds of models to generate the optimal urban development scenarios and then compare them with the past or the current urban development situations. This can help to identify some unreasonable urban development, which should be avoided in future decision making. The third kind of models aims to generate various development alternatives or options for assisting the planning processes. Generally these models will incorporate different planning objectives or regulations as the constraints and correspondingly generate the development scenarios. Based on these different scenarios, decision makers can compare these different urban planning layouts and select the most suitable one.

31.8.1 Application 1: Baseline Growth Simulation and Prediction

One of the basic applications of these planning-CA models is to simulate baseline urban growth, in which modelers can use CA models to predict possible future urban development results based on the past development trend and see what will happen if cities develop without any constraints. Planners can compare these baseline growth scenarios (no constraints) with some well-planned urban development layouts and then find out whether such a growth trend is reasonable (Figure 31-6).

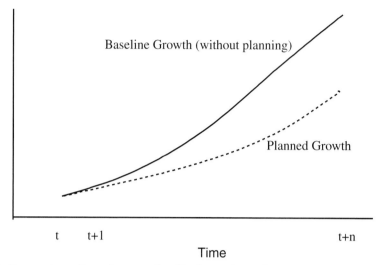

Figure 31-6 Comparison of baseline growth with planned growth.

In this study, possible urban development results of Dongguan, China in 1993, 1997 and 2005 were simulated, respectively, based on the historical urban growth trend. It is found that the conversion from agricultural land to urban land will be very fast if the city keeps on the growth trend of 1988-1993 (Figure 31-7). In 1993, according to the simulation, substantial agricultural land has been encroached upon by urban development. In 2005, this situation was worse, with the urban construction area enlarged by three times compared with that of 1988. Clearly, if the government does not control urban development, such rapid transition will bring considerable impacts on both the urban environments and natural resources. By simulating and predicting the future land-use changes with CA models, we may provide important references for planners and decision makers in the urban planning process.

31.8.2 Application 2: An Evaluation and Analysis Tool

The second application of CA models in urban planning is to evaluate the underlying rationale of existing development situations. CA models can be used to simulate constrained urban growth based on various constraints, e.g., agricultural suitability. The simulated results can be compared with the existing land use situations for evaluating the rationale of current land use plans. This is useful for identifying land use problems which should be avoided in the future. Such kind of application can be demonstrated in Figure 31-8 showing the real land development situation of Dongguan, China during 1988-1993, in which the urban land development is characterized by disorder, dispersion, low development density and low efficiency. Figure 31-8b shows a simulated urban development result generated by urban CA models; it is compact, orderly and efficient. Considerable agricultural land loss could be prevented if CA model-based planning were to be adopted. Thus, by comparing the real urban development results and simulated scenarios, we can easily identify the existing land use problems. According to the simulation results, the city should take a more compact and rational development form to avoid the disorder of urban sprawl.

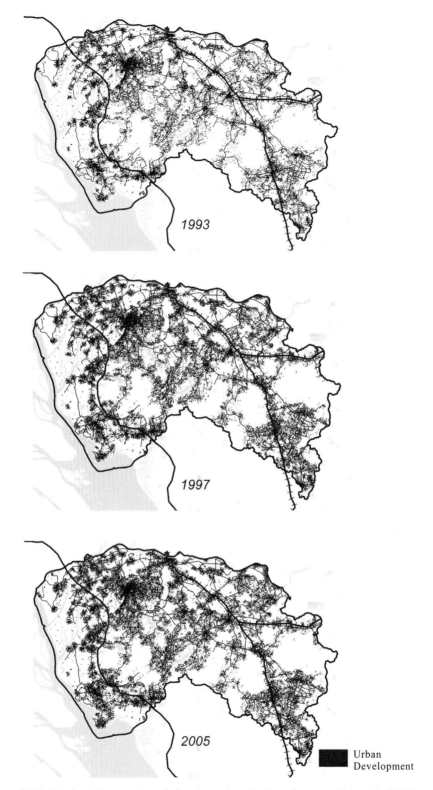

Figure 31-7 CA simulation and prediction based on the baseline growth trend of 1988.

a) Actual urban development

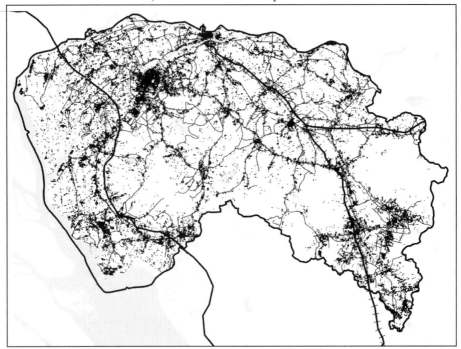

b) Optimal urban development

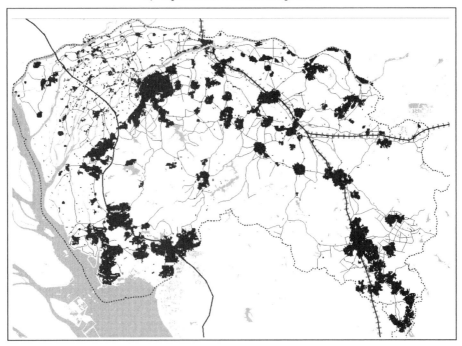

Figure 31-8 The comparison of actual urban development from Landsat TM (a) and optimal urban development generated by CA model (b).

A quantitative measure can be provided to evaluate the impacts of each development option. For example, we can use the index of suitability loss to quantify the impact of land use conversion. When agricultural land use has been converted into urban land use, its suitability for agricultural production will be lost. Besides considering the amount of land loss, another concern is the loss of good-quality agricultural land. Therefore, agricultural suitability loss is an important indicator that should be considered in the evaluation of urban development. The total loss of suitability for agricultural activities during land development can be calculated by the following formula:

$$S_{loss} = \sum_i \sum_j S(i,j) \qquad (31\text{-}20)$$

Here S_{loss} is the total agricultural loss, $S(i,j)$ is the agricultural suitability score for agricultural type j in location i, where agricultural land was encroached for urban development. Table 31-4 shows the comparison of agricultural loss of real urban development and the simulated urban development generated by urban CA models, during the period of 1988-1993. It is clear in the optimal development situation, that not only the urban form will be more compact, but the suitability loss is much lower than that of real development. Thus, if the urban development was based on the simulated scenario, many land use problems can be avoided.

Table 31-4 Comparison of the land use efficiency of different urban development strategies (Same land consumption, Dongguan 1988-1993).

	Compactness Index	Suitability Loss	Standardized Suitability Loss
Actual Development	1.78	5,537	6.2
Sustainable Compact Development	9.79	3,878	4.8

Source: Yeh and Li (2001b)

Further, constrained CA models also can be used to generate a compact and orderly urban form, which can greatly save investment in urban infrastructure in terms of the supply of water, electricity, gas, telecommunication and the construction of roads. Table 31-5 shows a rough comparison of urban infrastructure investment of actual and optimal urban development of Dongguan 1988-1993. Compared with the optimal development, actual urban development causes wasteful investment in urban infrastructure. The investment can be very substantially saved if the urban development is based on the simulated option from the CA model.

Table 31-5 Comparison of Infrastructure Costs Between Actual Development and Sustainable Compact Development (in million US$).

Items	Unit Price (US$/km)	Actual Development (in million US$)	Sustainable Compact Development (in million US$)
Electricity	770,000	528.31	339.31
Water	130,000	89.20	57.29
Gas	520,000	356.78	229.15
Telecommunication	520,000	356.78	229.15
Roads	2,580,000	1,770.18	1,136.92
TOTAL		3,101.25	1,991.82

Source: Yeh and Li (2001b)

Manual of Geographic Information Systems

31.8.3 Application 3: Simulating Development Alternative for Assisting the Planning Process

The third application of CA models for urban planning is to simulate different urban development forms according to planning objectives and regulations. These simulation results can provide useful references for urban planners and CA models can be used to simulate alternative development plans conveniently by modifying planning objectives and regulations.

Modelers can select different constraints related to environmental factors, urban forms, development density and combined factors according to different planning objectives. They can easily modify simulation outputs by adjusting different control parameters. For example, by adjusting the k value in Equation 31-13, planners can generate many different urban forms in terms of *non-constrained development, general constrained development and strictly constrained development* (see Figure 31-2).

The dispersion index value (α) property in Equation 31-16 also can be modified to generate *compact monocentric urban development* or *dispersed polycentric urban development* (see Figure 31-3). Modelers can simulate different urban forms by adjusting the density value and its decay speed. All these factors can be incorporated in the models by using combined constraints (see Figure 31-5). Therefore, these constrained CA models are flexible and transparent in creating development alternatives as useful planning tools.

31.9 Conclusion

CA has become a useful tool for modeling urban spatial dynamics and encouraging results have been documented. These models are often used to simulate urban growth based on the trajectory of historical growth, also called 'baseline development' simulation. However, urban CA models can simulate not only 'projected' growth, but also various possible development scenarios by incorporating constraints and planning objectives. Such CA models are usually implemented by the use of GIS. Integrated with GIS, the constrained CA models can benefit by using social, economic, physical and environmental data contained in GIS databases. The rapid development of GIS helps to foster the application of CA in urban simulation. Some research indicates that cell-based GIS may indeed serve as a useful tool for implementing CA models for the purposes of geographic analysis. CA has attracted growing attention in urban simulation because of its capability in spatial modeling that is not fully developed in GIS.

By adjusting parameters values, CA models are able to reflect the priorities of various environmental, social and economic factors. The users can conveniently explore the possible outcomes corresponding to a specific land use policy. Planning objectives also can be incorporated in the simulation process. For example, a larger weight value for a variable often means that it has a greater importance in affecting urban growth. Transportation-oriented growth patterns also can be generated by adjusting the weight value of transportation factors. Planners can compare the urban growth generated by the baseline-growth CA models and the constrained CA models, and then identify the existing land use problems. CA can be used as a planning experimental tool to test different planning theories and various hypotheses.

Since urban CA models are still under development, they are not without limitations. There are problems on how to define model structures and elicit transition rules. There are no agreements on how to select model variables and parameter values. There also are many ways of incorporating planning objectives and policies in simulation processes. Meanwhile, the inherent data error and the modeling uncertainties also will affect the application of CA models. The issues of data errors and model uncertainties have been well addressed in GIS literature. Although there are many studies on data errors and error propagation in GIS analysis, very little research has been carried out to examine these issues in urban CA simulation. Furthermore, CA models have apparent limitations in incorporating the decisions and behaviors of governments and investors in shaping urban growth. The influences of human factors are difficult to implement in traditional CA models. Recently, there are increasing studies on using agent-based systems to simulate complex behaviors of various players in urban systems. The interactions between these micro-agents can give rise to global spatial patterns of cities. The integration of CA and agent-based systems may be a solution to these problems that cannot be explained by using CA models alone.

References

Banister, D., S. Watson and C. Wood. 1997. Sustainable cities: transport, energy, and urban form. *Environment and Planning B: Planning and Design* 24:125–143.

Batty, M. 1997. Cellular automata and urban form: a primer. *Journal of the American Planning Association* 63(2):266–274.

Batty, M. and Y. Xie. 1994. From cells to cities. *Environment and Planning B: Planning and Design* 21:531–548.

———. 1997. Possible urban automata. *Environment and Planning B: Planning and Design* 24:175–192.

Batty, M., P. Longley and S. Fotheringham. 1989. Urban growth and form: scaling, fractal geometry, and diffusion-limited aggregation. *Environment and Planning A: Environment and Planning* 21:1447–1472.

Batty, M., Y. Xie and Z. L. Sun. 1999. Modeling urban dynamics through GIS-based cellular automata. *Computers, Environment and Urban Systems* 23:205–233.

Breheny, M. 1996. Centrists, de-centrists and compromisers: views on the future of urban form. In *The Compact City: A Sustainable Urban Form?* Edited by M. Jenks, E. Burton and K. Williams, 13–35. London: E&FN SPON.

Bryant, C. R., L. H. Russwurm and A. G. McLellan. 1982. *The City's Countryside: Land and Its Management in the Rural-urban Fringe*. New York: Longman Group Ltd.

Burchell, R. W., N. A. Shad, D. Listokin, H. Phillips, A. Downs, S. Seskin, J. S. Davis, T. Moore, D. Helton and M. Gall. 1998. T*he Costs of Sprawl – Revisited*, Washington, D.C.: National Academy Press.

Burks, A. W., ed. 1970. *Essays on Cellular Automata*. Urbana: University of Illinois Press.

Clark, C. 1951. Urban population densities. *Journal of Royal Statistical Society, Series A* 114:490–496.

Clarke, K. C., J. A. Brass and P. J. Riggan. 1994. A cellular automata model of wildfire propagation and extinction. *Photogrammetric Engineering & Remote Sensing* 60:1355–1367.

Clarke, K. C. and L. J. Gaydos. 1998. Loose-coupling a cellular automata model and GIS: long-term urban growth prediction for San Francisco and Washington/Baltimore. *International Journal of Geographical Information Science* 12(7):699–714.

Clarke, K. C., L. Gaydos and S. Hoppen. 1997. A self-modifying cellular automaton model of historical urbanization in the San Francisco Bay area. *Environment and Planning B: Planning and Design* 24:247–261.

Couclelis, H. 1985. Cellular worlds: a framework for modelling micro-macro dynamics. *Environment and Planning A: Environment and Planning* 17:585–596.

———. 1988. Of mice and men: what rodent populations can teach us about complex spatial dynamics. *Environment and Planning A: Environment and Planning* 20:99–109.

———. 1989. Macrostructure and micro-behaviour in a metropolitan area. *Environment and Planning B: Planning and Design* 16:141–54.

———. 1997. From cellular automata to urban models: new principles for model development and implementation. *Environment and Planning B: Planning and Design* 24:165–174.

CUPR (Center for Urban Policy Research). 1992. *Impact Assessment of the New Jersey Interim State Development Plan*. Trenton, N.J.: New Jersey Office of State Planning.

Daniels, T. L. 1997. Where does cluster zoning fit in farmland protection? *Journal of the American Planning Association* 63(1):129–137.

Deadman, P. D., R. D. Brown and H. R. Gimblett. 1993. Modelling rural residential settlement patterns with cellular automata. *Journal of Environmental Management* 37:147–160.

Ewing, R. 1997. Is Los Angeles-style sprawl desirable? *Journal of the American Planning Association* 63(1):107–126.

Gardner, M. 1970. The fantastic combinations of John Conway's new solitaire game "Life". *Scientific American* 223(4):120–123.

———. 1971. Mathematical games: on cellular automata, self-reproduction, the Garden of Eden, and the game 'life'. *Scientific American* 224(2):112–117.

Gierman, D. M. 1977. *Rural to urban land conversion*. Occasional Paper 16, Lands Directorate. Ottawa: Environment Canada.

Hägerstrand, T., 1967. *Innovation Diffusion as a Spatial Process*. Chicago: University of Chicago Press.

Huang, X. Q. and J. R. Jensen. 1997. A machine-learning approach to automated knowledge-base building for remote sensing image analysis with GIS data. *Photogrammetric Engineering and Remote Sensing* 63(10):1185–1194.

Itami, R. M. 1994. Simulating spatial dynamics: cellular automata theory. *Landscape and Urban Planning* 30:27–47.

King, L. and R. G. Golledge. 1978. *Cities, Space, and Behavior: the Elements of Urban Geography*. Englewood Cliffs, New Jersey: Prentice-Hall, Inc.

Landis, J. D. 1995. Imagining land use futures: applying the California urban futures model. *Journal of American Planning Association* 6(4):438–457.

Li, X. and A.G.O. Yeh. 1998. Principal component analysis of stacked multi-temporal images for monitoring of rapid urban expansion in the Pearl River Delta. *International Journal of Remote Sensing* 19(8):1501–1518.

———. 2000. Modelling sustainable urban development by the integration of constrained cellular automata and GIS. *International Journal of Geographical Information Science* 14(2):131–152.

———. 2001a. Calibration of cellular automata by using neural networks for the simulation of complex urban systems. *Environment and Planning A: Environment and Planning* 33:1445–1462.

———. 2001b. Zoning land for agricultural protection by the integration of remote sensing, GIS and cellular automata. *Photogrammetric Engineering & Remote Sensing* 67(4):471–477.

———. 2004. Data mining of cellular automata's transition rules. *International Journal of Geographical Information Science* 18(8):723–744.

Lowry, I. S. 1964. *A Model of Metropolis*. Santa Monica, Calif.: Rand Corp.

McRae, S. G. and C. P. Burnham. 1981. *Land Evaluation*. Oxford: Clarendon Press.

Muth, R. F. 1969. *Cities and Housing*. Chicago: University of Chicago Press.

Newman, P.W.G. and J. R. Kenworthy. 1988. The transport energy trade-off: fuel-efficient traffic versus fuel-efficient cities. *Transport Research* A 3:163–174.

Pushkarev, B. S. and J. M. Zupan. 1977. *Public Transportation and Land Use Policy*. Bloomington, Ind.: Indiana University Press.

Papageorgiou, G. J. 1971. A theoretical evaluation of the existing population density gradient function. *Economic Geography* 47:21–26.

Semboloni, F. 1997. An urban and regional model based on cellular automata. *Environment and Planning B: Planning and Design* 24:589–612.

Thrall, G. I. 1988. Statistical and theoretical issues in verifying the population density function. *Urban Geography* 9(5):518–537.

Tobler, W. R. 1979. Cellular geography. In *Philosophy in Geography*, edited by S. Gale and G. Olsson, 279–386. Dordrecht, Netherlands: D. Reidel Publishing Co.

Torrens, P. M. and D. O'Sullivan. 2001. Cellular automata and urban simulation: Where do we go from here? *Environment and Planning B: Planning and Design* 28:163–168.

Voogd, H. 1983. Multi-criteria Evaluation for Urban and Regional Planning. London: Pion.

Wagner, D. F. 1997. Cellular automata and geographic information systems. *Environment and Planning B: Planning and Design* 24:219–234.

Ward, D. P., A. T. Murray and S. R. Phinn. 2000. A stochastically constrained cellular model of urban growth. *Computers, Environment and Urban Systems* 24:539–558.

Webber, M. J. 1984. *Explanation, Prediction and Planning*: the Lowry Model. London: Pion, 214 pp.

Webster, C. J. and F. Wu. 1999a. Regulation, land use mix and urban performance. Part 1: Theory. *Environment and Planning A: Environment and Planning* 31:1433–1442.

———. 1999b. Regulation, land use mix and urban performance. Part 2: Simulation. *Environment and Planning A: Environment and Planning* 31:1529–1545.

White, R. and G. Engelen. 1993. Cellular automata and fractal urban form: a cellular modelling approach to the evolution of urban land-use patterns. *Environment and Planning A: Environment and Planning* 25:1175–1199.

———. 1997. Cellular automata as the basis of integrated dynamic regional modelling. *Environment and Planning B: Planning and Design* 24:235–246.

White, R., G. Engelen and I. Uijee. 1997. The use of constrained cellular automata for high-resolution modeling of urban land-use dynamics. *Environment and Planning B: Planning and Design* 24:323–343.

Wolfram, S. 1984. Cellular automata as models of complexity. Nature 31(4):419–424.

Wu, F. 1998. An experiment on the general poly-centricity of urban growth in a cellular automatic city. *Environment and Planning B: Planning and Design* 25:103–126.

———. 2000. A parameterized urban cellular model combining spontaneous and self-organizing growth. In *GIS and Geo-computation*, edited by P. Atkinson and D. Martin, 73–85. New York: Taylor and Francis.

———. 2002. Calibration of stochastic cellular automata: the application to rural-urban land conversions. *International Journal of Geographical Information Sciences* 16(8):795–818.

Wu, F. and C. J. Webster. 1998. Simulation of land development through the integration of cellular automata and multi-criteria evaluation. *Environment and Planning B: Planning and Design* 25:103–126.

Yeh, A.G.O. and X. Li. 1997. An integrated remote sensing and GIS approach in the monitoring and evaluation of rapid urban growth for sustainable development in the Pearl Rive Delta, China. *International Planning Studies* 2(2):193–210.

———. 1999. Economic development and agricultural land loss in the Pearl River Delta, China. *Habitat International* 23(3):373–390.

———. 2001a. A constrained CA model for the simulation and planning of sustainable urban forms using GIS. *Environment and Planning B: Planning and Design* 28:733–753.

———. 2001b. The need and challenges for compact development in the fast growing areas in China - the Pearl River Delta. In *Compact City: Sustainable Urban Form for Developing Countries*, edited by M. Jenks and R. Burgess, 73–90. London: SPON Press.

CHAPTER 32

Using a Cellular Automaton Model and GIS to Simulate the Spatial Consequences of Different Growth Scenarios in the Atlanta Metropolitan Area

C. P. Lo and *Xiaojun Yang*

32.1 Introduction

The concept of cellular automata (originally known as "cellular spaces") was originally introduced by JOhn von Neumann and Stanislaw Marcin Ulam in the 1960s to model biological self-reproduction. They are mathematical idealizations of physical systems in which space and time are discrete (Wolfram 1994, p. 5). A cellular automaton consists of a regular uniform lattice of cells of infinite extent, each of which contains a discrete variable. The values of the variables determine the state of a cellular automaton. A cellular automaton evolves in discrete time steps. The value of the variable in each cell is affected by the values of variables in neighborhood cells on the previous time step. The values of these variables are continuously updated according to a definite set of local transition rules. For two-dimensional cellular automata, the two most commonly defined neighborhood cells are the five-cell von Neumann (four corner cells plus cell itself) or the nine-cell Moore neighborhood (eight surrounding cells plus cell itself). Despite its simple construction, cellular automata are dynamical systems with complex self-organizing behavior (Wolfram 1994, p. 115). Hence, they can be associated with emergence and complex adaptive systems, characterized by phase shifts, self-organization, self-similarity and fractal dimensions (Torrens 2000). Urban systems have been found to exhibit similar characteristics, and hence cellular automata are ideally suited to model the complexity of urban systems (Clarke and Gaydos 1998). Because the model is cell-based, it is compatible with the image data collected by remote sensing and easily handled by raster GIS in data extraction and data-layer overlay in the process of model construction and calibration. By combining with GIS, different scenarios of urban development strategies can be easily tested with the cellular automata simulation.

It is therefore not surprising that cellular automata have become a popular tool for urban growth simulation in the past decade, as demonstrated by the work of Batty and Xie (1994), Batty et al. (1999), Couclelis (1997), Clarke et al. (1997), Clarke and Gaydos (1998), Silva and Clarke (2002), and Cheng and Masser (2003, 2004). The British journal, *Environment and Planning B*, has published at least two special issues on applying cellular automata to urban modeling, in 1997 (volume 24) and 2001 (volume 28), respectively edited by Batty et al. (1997) and Torrens and O'Sullivan (2001). In recent years, cellular automata modeling for urban planning has taken into account the spatial configuration of the real world and the external forces or constraints that will affect development. These can be achieved by changing the von Neumann and Moore neighborhood configurations of cells, and by incorporating constraints in the local transition rules in producing constrained cellular automata models (White et al. 1997; Sui and Zeng, 2001; Li and Yeh 2000, 2002; see also the previous chapter by Yeh and Li in this GIS Manual).

The purpose of this chapter is to demonstrate an application of a cellular automata model to simulate the spatial consequences of different growth scenarios of a major American city—Atlanta Metropolitan Area, Georgia, using GIS, remote sensing and spatial analysis (Barredo et al. 2003). This is the loose coupling approach (Clarke and Gaydos 1998). The cellular automata model selected for this purpose is the SLEUTH Urban Growth Model (Clarke 2000; USGS Undated). SLEUTH derives its name from the six types of data inputs: **S**lope, **L**and cover, **E**xclusion, **U**rban extent, **T**ransportation, and **H**illshade. This model considers four types of growth behavior: spontaneous growth, diffusive growth and creation of new spreading centers, organic growth, and road-influenced growth. These four growth types are applied sequentially during each growth cycle (a year), and are controlled by five growth coefficients: diffusion, breed, spread, slope resistance, and road gravity (Table 32-1) (Clarke and Gaydos 1998; Clarke 2000; Silva and Clarke 2002). The excluded layer (exclusion) is defined by users to exclude areas that should not be developed.

Table 32-1 Growth cycle, growth type, and controlling coefficients used in the SLEUTH cellular automata simulation model (Sources: Clarke and Gaydos 1998; Jantz et al. 2003)

Growth Cycle	Growth Type	Controlling Coefficients	Description
1	spontaneous	diffusion	A randomly chosen cell falls in a suitable location for urbanization
2	diffusive	breed	Cells flat enough for urban development, even if not near an established urban area
3	organic (edge)	spread	Outward growth from existing urban centers, representing tendency of all urban areas to expand
4	road-influenced	road gravity	Urbanized cells to develop along transportation network
Throughout	slope resistance	slope	Effects of slope on urban development
Throughout	excluded layer	user-defined	User specifies area not to be developed

Another characteristic of this model is that it possesses a functionality known as self-modification (Clarke et al. 1997), which allows the growth coefficients to change if an unusually high or low growth rate—above or below a threshold—is encountered. When the rate of growth is above the threshold, the controlling coefficients (diffusion, spread, and breed) are multiplied by a factor greater than one (simulating a boom cycle). When the rate of growth is below the threshold, the controlling coefficients (diffusion, spread, and breed) are multiplied by a factor less than one (simulating a bust cycle). Self-modification produces the S-curve growth rate of urban use instead of the linear or exponential growth without the self-modification (Clarke and Gaydos 1998).

A number of reasons lead to the choice of the SLEUTH model for the current research. First, the model is scale independent, so that it can encompass local, regional, and continental scale processes in a single context. It is dynamic and future oriented, conforming to the essential requirement of urban growth simulation in this project. The behavior rules guiding urban growth in the model consider not only the spatial properties of neighboring cells but also existing urban spatial extent, transportation and terrain slope. The transportation and terrain conditions were found to be significant factors driving land use and land cover changes in the study area (Yang 2002; Lo and Yang 2002). These behavior rules therefore have realistically accounted for the driving forces in the formation of edge cities in a postmodern metropolis. The model can also modify itself if extensive growth or stagnation leads to aberrations from the linear normal growth development. This provides a way of feedback, which may be meaningful for ensuring reasonable prediction. Second, the model can be verified through rigorous past-to-present calibration using historical data. The model incorporates some rigid statistical measures for characterizing historical fit in the phase of model calibration. These contrast greatly to many other models, which are largely game-like simulators without a component of rigid validity. Third, the model can be used to simulate urban growth under different conditions by altering the growth control coefficients. This can be useful not only for guiding future urban planning practices but also for studying metropolitan dynamics. Last, in contrast to many other models that were developed by different research teams with essentially no reuses, this model allows itself to be applied to other regions with different data sets. This can save a great deal of labor and time in model design and programming, allowing researchers to concentrate on model calibration and validation as well as scenario designing and simulation.

The SLEUTH model has been tested to demonstrate its usefulness for long-term urban growth prediction for some American cities, including San Francisco and the Washington/Baltimore area (Clarke and Gaydos 1998), Santa Barbara (Herold et al. 2001), Houston (Oguz et al. 2004), and Sioux Falls (Goldstein 2004). The model has also been applied to European cities (Lisbon and Porto)(Silva and Clarke 2002), Sydney, Australia (Liu and Phinn 2004) and in South America and Africa. A listing of cities around the world to which the model has been applied is found on the website about Project Gigalopolis (<http://www.ncgia.ucsb.edu/projects/gig/v2/About/abApps.htm>, accessed 14 August 2007). More recently, Jantz et al. (2003) have successfully employed the model to simulate the impacts of future policy scenarios on urban land use in the Baltimore-Washington metropolitan area.

32.2 Atlanta as a Rapidly Suburbanizing City

For the past three decades, Atlanta has been one of America's fastest-growing metropolises, population having increased 27%, 33% and 39% respectively during the 1970-80, 1980-90 and 1990-2000 periods (Research Atlanta, Inc. 1993; SSDAN Undated). The city has expanded greatly as suburbanization consumes large areas of forest and open land adjacent to the city, pushing the peri-urban fringe farther and farther away from the original urban boundary. Because of the significant physical growth, Atlanta's urban spatial structure has changed dramatically (Yang 2002; Yang and Lo 2002). Urban geographers have recognized Atlanta as one of the few typical postmodern metropolises in North America (Hall 2001). Research on Atlanta has led to the formulation of a new urban model, i.e., the urban realms model, which describes the appearance of a multiple-nuclei city in contrast to the conventional single-centered city (Hartshorn and Muller 1989; Fujü and Hartshorn 1995). Atlanta's outward expansion has created the problem of urban sprawl, which has affected the quality of life and health of the people. It has also intensified economic and racial polarization (Bullard et al. 2000). Clearly, future growth of the city will be of great concern, and planners need to know how to control and manage the growth of the city. Using cellular automata modeling is therefore most appropriate to test out different scenarios of development—in particular, the spatial consequences and impacts on the population and the environment.

The actual modeled area is a rectangle covering thirteen counties in the Atlanta Metropolitan Statistical Area (AMSA) (Figure 32-1). This area includes ten counties under the Atlanta Regional Commission (ARC) as well as three additional counties—Coweta, Forsyth, and Paulding—which have exhibited similar growth pattern to the ARC counties (Yang 2002). The City of Atlanta is located in the center of the study area, mainly in Fulton county bordering DeKalb county. The total area is approximately 16,284 km^2.

Physiographically, the Atlanta area is situated mainly in the foothills of the southern Appalachians in northern Georgia at an elevation of 300 to 350 m above mean sea level. The northwestern part (approximately 18% of the total area) is in the Appalachian mountain. It has an even terrain that slopes downward toward the east and south. The climate is generally characterized as mild. The Chattahoochee River traverses the study area from northeast to southwest.

32.3 Data Layers Preparation

The five data layers required for cellular automata modeling using SLEUTH were obtained with the aid of satellite remote sensing through image classification and GIS for digitizing, co-registering, buffering and topological overlay analysis. The five layers are urban extent, road, excluded area for development, slope and shaded relief (Figure 32-2).

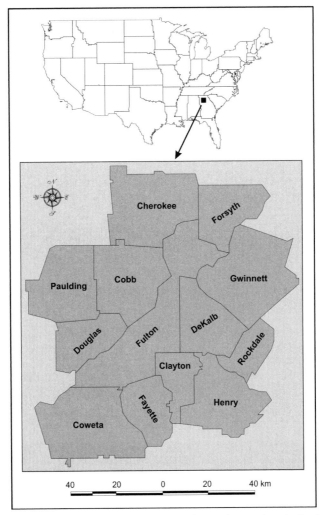

Figure 32-1 Location of the study area. The modeled area is a rectangle that covers 13 counties in the Atlanta metropolitan area, Georgia. This includes 10 counties under the Atlanta Regional Commission (ARC) and three additional counties that show similar growth patterns. The modeled area is approximately 16,284 km².

32.3.1 Urban Extent

Urban extent is actually urban built-up land, which includes all types of urban uses. In conjunction with the NASA-funded project ATLANTA (ATlanta Land-use ANalysis: Temperature and Air-quality), five land use/cover maps were produced from Landsat Multispectral Scanner (MSS) and Thematic Mapper (TM) satellite images for 1973, 1979, 1987, 1993, and 1999, using a hybrid approach combining unsupervised classification and knowledge-based spatial reclassification (Yang 2002; Yang and Lo 2002) (Figure 32-3). The producer's and user's accuracies for urban classes are well above 80 percent with an overall classification accuracy of 90 percent. The urban use areas for each of these years were extracted from the five land use/cover maps to form five layers of urban extent. These five layers are therefore historic data showing the spatial expansion of Atlanta between 1973 and 1999. The first four layers (1973, 1979, 1987 and 1993) were used primarily for model calibration while the 1999 layer was employed for both model calibration and future growth simulation. These layers were converted into binary images with 1 for urban and 0 for non-urban.

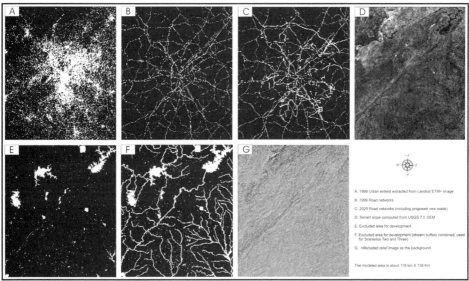

Figure 32-2 Examples of model input data layers. Basically, all of these data layers are unsigned 8-bit GIF images with the same dimension.

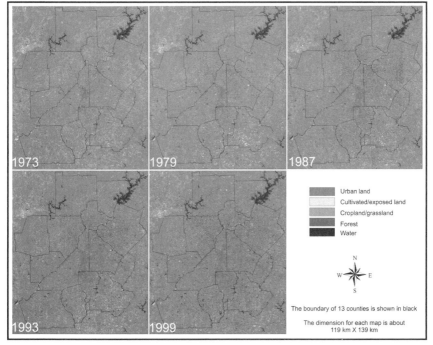

Figure 32-3 Historical urban development (land use/cover changes) in Atlanta, 1973-1999. These maps were produced with a method described by Yang and Lo (2002). See included DVD for color version.

Manual of Geographic Information Systems

32.3.2 Road

Road contains not only road networks but also node points and large shopping malls. A driving force analysis conducted by the authors indicates that highways, nodes, and shopping malls generally promote urban development in Atlanta (Lo and Yang 2002). For convenience, however, this theme is still called "road."

The major highways were extracted from the February 1998 version of Automotive Navigation Data (AND) global highway database (<http://www.and.com>, accessed 14 August 2007), and then updated with satellite images to form highway layers for three years: 1973, 1987 and 1999. A weight system ranking these highways' relative importance was developed with highway traffic measures and highway classification (Table 32-2). Major node points are either (major) highway exits, junctions, or towns crossed by one or more major highways. They are of strategic significance for commercial or industrial development. The major node points were extracted from the 1998 AND global road database. Similarly, a weight system was assigned to create radius-variable buffer rings around these node points (Table 32-2). Three layers of large mall polygons were extracted from the 1973, 1987 and 1999 Landsat images. The relative size of these shopping malls was used to assign weights for creating radius-variable buffer rings (Table 32-2).

The layers of highways, nodes, and shopping malls in the same year were combined to form a single 'road' layer. In this way, three 'road' layers were produced, one each for 1973, 1987 and 1999. The 'road' layer for the year of 2025 was produced by overlaying the 1999 roads with the improved roadways and new roadways according to the 2025 Regional Transportation Plan (Atlanta Regional Commission 2000). The transportation plan has been recently updated to Mobility 2030 (Atlanta Regional Commission 2007).

32.3.3 Excluded Area for Development

Two layers of excluded areas were created. The first layer is a binary image, consisting of the water features extracted from the 1973 Landsat MSS image and the public lands. The latter includes national parks/refuge and wilderness areas, archaeological sites/areas, historic sites, off-road vehicle sites/areas, wild and scenic areas, state parks, USDA land, wildlife management areas, and parks. These areas were not allowed for urban development. This layer was mainly used for the model calibration.

For the future growth prediction, a second layer was built, with probabilities of exclusion included. All excluded areas in the first layer were still preserved and assigned a value of 100. Additionally, this layer contains three levels of buffer zones around major streams in the study area. Specifically, for the Chattahoochee River, the buffer zone within 50 meters was assigned a value of 100, meaning that this area was not allowed for urban development at all; the buffer zone between 50 and 100 meters was assigned a value of 60, indicating a 60% probability of exclusion; and the buffer zone between 100 and 200 meters was assigned a value of 20, indicating a 20% probability of exclusion. For other large streams, the three buffer zones and their probabilities of exclusion were 0-30m, 100%; 30-60m, 60%; and 60-120m, 20%.

32.3.4 Slope and Shaded Relief Image

In order to produce terrain slope and shaded relief images, a seamless DEM image was constructed by mosaicking 159 USGS 7.5' DEMs covering the entire modeled area. Then, a terrain slope image was computed by using a standard algorithm. The slope is represented in percentages. A layer of the hill-shaded image was computed from the DEM too. This image shows the topographic relief in the study area. It was used as a background image for visualization purposes only.

Table 32-2 Buffer radius and weight assigned in the production of the "roads" layers.

No	Highways				Node Points				Shopping Malls	
	Class*	Description	Buffer Radius (m)	Weight assigned**	Class***	Description	Buffer Radius (m)	Weight assigned	area (100,000 m²)	Weight assigned
1	I285	Interstate Highway	60	200	1	junction (exit)	60	30	<1	255
2	I20/75/85	Interstate Highway	60	150						
3	I575/675/975	Other Interstate Highways	60	150	2	intersection	135	55	1-2	175
4	GA400/19	State Highway	60	150						
5	Type 1	Other Motor Way	60	150						
6	Type 2	Federal Highway Dual Carriage Way	45	100	14	town >20,000	90	40	2-4	150
7	Type 3	Federal Highway	30	50						
8	Type 4	Regional Road	30	50	15	town > 5,000	60	30	<4	100
9	Type 5	Local Road	30	50						
10	D11	Digitized Major Highway	30	50	16	town > 1,000	30	20		
11	D22	Digitized Highway	30	50						
12	Improved	Highways Expanded by 2025	60	220						
13	New	New Roads by 2025	60	255						

* Highway Classes Type 1 through Type 5 are based on the AND global road database (www.and.com). Classes D11 and D22 are two types of highways that can be observed clearly from satellite images. The layers were created in this project using on-screen digitizing. D11 is for digitized major highways whose reflectance is quite strong from images; D22 is for other digitized highways. Classes Improved and New (No. 12 and No. 13) are based on the 2025 Regional Transportation Plan proposed by the Atlanta Regional Commission (2000).

** All images are unsigned 8-bit data. The weight range is from 0 to 255.

*** The Node Point Classes are based on the AND global database. The population of towns where highways cross is considered in the database.

To comply with the model data input requirements, all of the input data layers need to be standardized in data format, resolution and dimension. In this project, each layer was resampled into three levels of spatial resolution, namely 60 m, 120 m and 240 m. All data layers were converted into 8-bit GIF (Graphics Interchange Format) files and named with the convention stipulated by the model.

32.4 Model Calibration

The purpose of model calibration was to determine the best values for the five controlling coefficients that can effectively simulate growth during the 1973-1999 period (Table 32-1):
- diffusion—overall dispersiveness of growth,
- breed—likelihood of new settlements being generated,
- spread—growth outward from existing spreading centers,
- slope—likelihood of settlements extending up steeper terrain and
- road gravity—attraction of urbanization near road networks.

As explained above, these coefficients determine the behavior of cellular automaton. The model calibration will also be used to determine the controlling coefficients for different scenarios of growth in the future from 1999 to 2025.

The calibration makes use of a brute-force Monte Carlo method. The user sets a range of values for each controlling coefficient, and the model will try all possible combinations to come up with the best coefficient values within the range specified. The simulated growth is then compared with the actual growth based on the historic data, and 13 statistical measures are computed to evaluate the goodness of fit between the two (Table 32-3). Six of these measures are least squares regression scores, comparing the modeled urban pixels (popR2; urbR2), urban edge count (edgR2), urban clustering (cstR2), average urban cluster size (McstR2), and average slope of urbanized cells with the actual (slpR2). The most useful measure is the compare (comp) statistic, which is a ratio of the number of modeled final urban pixels to the number of actual urban pixels. Lee and Sallee shape index (Lee and Sallee 1970), which measures the spatial fit between the modeled and the actual urban extent for the controlling years, is also useful. The possible range of values specified for each controlling coefficient is 0 to 100, yielding 101^5 (more than 10 billion) possible combinations. Ideally, each combination should be assessed. Given the computational resources available at the time this research project was conducted (a Sun Ultra Model 1, with 143 MHz CPU and 64 Mb RAM), this would take years to complete according to an earlier test. In recognition of the constraints of time and computational resources, the calibration was broken down into three phases as described below, and only data layers with a spatial resolution of 240 meters were used in the calibration. In each phase of calibration, both the spatial scale and the range of values for the controlling coefficients are narrowed down to find the set of values that best simulates the historic urban growth.

32.4.1 Phase One: Coarse Calibration

The goal here was to narrow down the range for each controlling coefficient. The entire range of possible values (0-100) was tested for each controlling coefficient, with 25 as the step (Table 32-4). This resulted in 55 or 3,125 combinations. Given the time constraint, a small value (4) was assigned for the number of Monte Carlo computations. Based on the weighted sums of all the statistical measures, three best combinations were identified (Table 32-5a). A closer look at these combinations led to the observation that the values of the first four coefficients (diffusion, breed, spread, and slope resistance) were highly consistent but the last one, road gravity, had values over the whole range.

Table 32-3 The statistical measures used to test the historical fit in the model calibration (Source: USGS Undated).

No.	Abbreviation	Definition
1	**comp**	Comparison of modeled final urban pixels to actual urban counts for the control years. If modeled counts are larger than the actual counts, divide the latter by the former; otherwise, divide the former by the latter
2	**popR2**	Least squares regression score for modeled urban pixels compared to actual urban pixels for the control years
3	**edgR2**	Least squares regression score for modeled urban edge count compared to actual urban edge count for the control years
4	**cstR2**	Least squares regression score for modeled urban clustering compared to actual urban clustering for the control years
5	**McstR2**	Least squares regression score for modeled average urban cluster size compared to actual average urban cluster size for the control years
6	**lees1**	a shape index, a measurement of spatial fit between the model's growth and the actual urban extent for the control years
7	**slpR2**	Least squares regression of average slope for modeled urbanized cells compared to average slope of actual urban cells for the control years
8	**urbR2**	Least squares regression of percent of available pixels urbanized compared to actual urbanized pixels for the control years
9	**xmuR2**	Least squares regression of average x_values for modeled urbanized cells compared to average x_values of actual known urban cells for the control years
10	**ymuR2**	Least squares regression of average y_values for modeled urbanized cells compared to average y_values of known urban cells for the control years
11	**SD_R2**	Averaged standard deviation in x and y
12	**LUcmp**	A proportion of goodness of fit across landuse classes
13	**composite**	All the above scores (1-12) multiplied together

32.4.2 Phase Two: Fine Calibration

The goal of the fine calibration was to narrow down the ranges of the controlling coefficients to approximately 10 or fewer, based on the values obtained from the coarse calibration. In doing so, these coefficients were held to 2-4 steps, and increments of 5-10 were used (see Table 32-4). Because fewer steps were used, the resultant combinations of different coefficient values should decrease substantially. This means that the entire computation time should decrease proportionally. Thus, more times of Monte Carlo computations would be possible. The number of iterations was 900, which was 2,225 fewer than the number at the first phase. But the unit computation time increased as a slightly larger number of Monte Carlo computations (6) was used at this stage. Using the weighted sums, the best three combinations were identified (Table 32-5b). Based on these combinations, the values of both spread coefficient and slope resistance were still very consistent. Moreover, the other three coefficients showed much narrower ranges. These results are also quite encouraging.

32.4.3 Phase Three: Final Calibration

Based on the values of the controlling coefficients obtained from the fine calibration, the final calibration was to determine the best combination. To this end, the coefficients were held to two steps and the increments of 3 and 5 were used (see Table 32-4). These limits substantially reduced the number of combinations when compared to the fine calibration stage, making more iterations of the Monte Carlo computations affordable. Given these conditions, the number of iterations was 243. Using the weighted sum of statistical measures, the best combination was identified, which had the following starting values: diffusion—55, breed—8, spread—25, slope resistance—53 and road gravity—100, indicating that they became more sensitive to the local conditions (Table 32-5c).

Manual of Geographic Information Systems

Table 32-4 Calibration runs: input data, calibration files, number of Monte Carlo iterations, computation time, and outputs.

Item		Calibration Runs			Parameter Averaging
		Coarse	Fine	Final	
Input Data	resolution (m)	240			
	urban extent (year)	1973, 1979, 1987, 1993, 1999			
	'roads' (year)	1973, 1987, 1999			
	excluded area	stream buffered zones not considered			
	slope	same (only one layer is available)			
	hillshaded relief	same (only one layer is available)			
Calibration File	seed*	2840	2840	2840	2840
	number_of_times**	4	6	10	100
	diffusion_coeff_start	0	40	52	55
	diffusion_coeff_step	25	5	3	1
	diffusion_coeff_stop	100	60	58	55
	breed_coeff_start	0	0	2	8
	breed_coeff_step	25	5	3	1
	breed_coeff_stop	100	10	8	8
	spread_coeff_start	0	15	22	25
	spread_coeff_step	25	5	3	1
	spread_coeff_stop	100	35	28	25
	slope_resistance_start	0	40	47	53
	slope_resistance_step	25	5	3	1
	slope_resistance_stop	100	55	53	53
	road_gravity_start	0	80	90	100
	road_gravity_step	25	10	5	1
	road_gravity_stop	100	100	100	100
Number of iteration(s)		3125	900	243	1
Computation time		14445'58"	5760'14"	1112'03"	104'14"

*These are about 1 percent of the total number of pixels.
**This number is for the Monte Carlo iterations.

32.4.4 1973-1999 Simulation

One more step was to determine the starting values to use for future growth simulation. The best values identified in the final calibration were selected to be the starting values. Because of the self-modification functionality incorporated in the model, these starting values tend to be altered when a run is completed. Thus, a coefficient may have different starting and finishing values for each run. The finishing values are better for future simulations (Clarke 2000). The best values derived from the final calibration were used as the starting values in simulating urban growth between 1973 and 1999. Since only one combination was available for the computation, more times of Monte Carlo computations were affordable. The finishing values of the controlling coefficients were further averaged using the MEAN utility provided by the model. At the end, the final values of the controlling coefficients were: diffusion—71, breed—10, spread—32, slope resistance—73, and road gravity—100. Clearly, there is an overall increase in the values of these controlling coefficients with the exception of the

Table 32-5 Controlling coefficient values from coarse, fine, and final calibrations.

(a) Coarse Calibration

Weighted Total	dif	brd	spd	slp	rdg	comp	popR2	edgR2	cstR2	McstR2	leesl	slpR2	urbR2	xmuR2	ymuR2	SD_R2	LUcmp
500.1310	50	1	25	50	100	0.9937	0.9456	0.9989	0.8942	0.8859	0.4265	0.9623	0.9456	0.4615	0.6232	0.8665	0.6673
499.9140	50	1	25	50	50	0.9923	0.9470	0.9976	0.8947	0.8988	0.4259	0.9862	0.9470	0.0843	0.6352	0.8683	0.6659
499.6435	50	1	25	50	1	0.9924	0.9473	0.9977	0.9108	0.8754	0.4259	0.9430	0.9473	0.3520	0.5947	0.8690	0.6666

(b) Fine Calibration

Weighted Total	dif	brd	spd	slp	rdg	comp	popR2	edgR2	cstR2	McstR2	leesl	slpR2	urbR2	xmuR2	ymuR2	SD_R2	LUcmp
502.2294	55	5	25	50	100	0.9978	0.9442	0.9980	0.9108	0.9109	0.4266	0.9789	0.9442	0.3319	0.5993	0.8647	0.6646
500.2560	50	10	25	50	100	0.9972	0.9442	0.9987	0.9059	0.8943	0.4261	0.9927	0.9442	0.0287	0.5859	0.8646	0.6653
500.0975	50	1	25	50	80	0.9939	0.9465	0.9983	0.9058	0.8859	0.4261	0.9959	0.9465	0.0214	0.6374	0.8677	0.6652

(c) Final Calibration

Weighted Total	dif	brd	spd	slp	rdg	comp	popR2	edgR2	cstR2	McstR2	leesl	slpR2	urbR2	xmuR2	ymuR2	SD_R2	LUcmp
500.5085	55	8	25	53	100	0.9878	0.9448	0.9988	0.9122	0.9109	0.4261	0.9829	0.9448	0.0784	0.6180	0.8653	0.6659
499.6955	52	8	25	50	100	0.9991	0.9459	0.9988	0.9126	0.8962	0.4258	0.9613	0.9459	0.0458	0.6313	0.8670	0.6645
499.6045	58	5	25	50	90	0.9980	0.9451	0.9986	0.9020	0.9109	0.4262	0.9546	0.9451	0.0731	0.6349	0.8659	0.6650

Note: dif = diffusion, brd = breed, spd = spread, slp = slope, rdg = road gravity
Explanations of the abbreviations of the12 statistical measures can be found in Table 32-3.

road gravity coefficient, which is unchanged and remains high. It should also be noted that diffusion, slope resistance and road gravity appeared to be the three major growth coefficients in Atlanta, thus suggesting the importance of spontaneous growth as well as the influences of roads and slopes to urban development.

Using the final values for the controlling coefficients given above, a past to present simulation (1973-1999) was conducted. The simulated results were compared with the actual urban extent extracted from Landsat images for the corresponding years, thus providing visual verification of the accuracy of the model calibration (Figure 32-4). Five hundred randomly sampled points per year were used in the assessment. The overall accuracy ranges from 52.4% to 63.8%. This result should be very encouraging, given the development status of dynamic modeling technologies (Wang and Zhang 2001). Animated movies were also generated that allowed the general trend of urban development in Atlanta to be verified. (The movies are included on the DVD accompanying this manual. Refer to Chapter 63 for additional information.)

It should be pointed out that the model calibration was carried out with the 240-m resolution dataset only. An earlier test estimated that the time for completing the first stage of calibration using the 120-m resolution data set would be about 32,500 hours or 135 days given the computer resources available. Higher spatial resolution datasets should reflect better the local conditions than coarser ones. For practical reason, the two higher resolution data sets were not used in the model calibration. This issue will be further discussed in the last section.

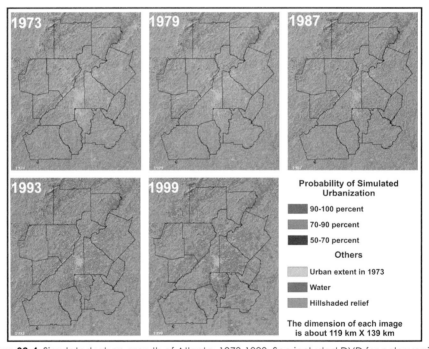

Figure 32-4 Simulated urban growth of Atlanta, 1973-1999. See included DVD for color version.

32.5 Scenario Design and Simulation Beyond 1999

The SLEUTH model allows simulation of the future growth of Atlanta beyond 1999. In view of the urban development problems faced by Atlanta as exhibited during the period 1973-1999—depletion of forest, air pollution, traffic congestion and urban sprawl—the following three possible planning scenarios were designed.

32.5.1 Scenario One

The first scenario assumes the factors for the growth of Atlanta remain unchanged. As discussed before, Atlanta has been well-known as a "postmodern" city. "Postmodern" is a term firstly used by Soja (1989) to describe cities that have undergone restructuring in the United States after the rise of post-Fordist industrial organization, which is characterized by a flexible subcontracted production system based on small-size and small-batch units. Under the above assumptions, the city of Atlanta would continue to grow. Thus, this scenario may be viewed as the continuation of the 1973-1999 trend. It provides therefore a benchmark for comparison with alternative growth strategies. To implement this scenario in model simulation, the final values of the control coefficients obtained from the model calibration were used as the starting values. These values were determined with the historical urban extent data as the reference. The 1999 urban extent data were used in the simulation and other conditions and input data set are shown in Table 32-6.

Table 32-6 The conditions applied for each simulation.

Items		Past-to-Present Simulation	Future Simulations		
			Scenario 1	Scenario 2	Scenario 3
Time Span		1974-1999	2000-2050		
Input Data	Resolution (m)	240			
	urban extent (year)	1973	1999		
	'roads'	1973, 1987, and 1999	1999	1999, 2025	
	excluded areas	stream buffered zones not considered	stream buffered zones considered		
	slope	same (only one layer can be chosen)			
	hillshaded relief	same (only one layer can be chosen)			
Self Modification Constraints*	critical_high	1.500			
	critical_low	0.050			
	boom	1.010			
	bust	0.090			
	critical_slope	21			10
Control Coefficients**	diffusion	71/92	71/88	71/88	100/100
	breed	10/13	10/12	10/12	100/100
	spread	32/41	32/40	32/40	15/15
	slope resistance	73/95	73/100	73/100	10/40
	road gravity	100/100	100/100	100/100	200/200***
Number of Monte Carlo Computations		100			
Random Samples		2840			

* The definitions are: critical_high : when the growth rate exceeds this value, self_modification increases the control parameter values; critical_low: when the growth rate falls below this value, self_modification increases the control parameter values; boom: value of the multiplier (greater than one) by which parameter values are increased when the growth rate exceeds critical_high; bust: value of the multiplier (less than one) by which parameter values are decreased when the growth rate falls below critical_low, and Critical_slope: average slope at which system increases spread.

** Both starting and ending values are given. It should be noted that the ending values were the averaged values after 100 times of Monte Carlo computations.

*** Program code was changed to allow up to 200 for road gravity.

32.5.2 Scenario Two

The second scenario considers future road development and environmental protection while other growth conditions used in the first scenario remain unchanged (Table 32-6). This reflects an alternative growth strategy in which environmental protection is emphasized so that the city is more livable. In order to address the problems of air quality and traffic congestion in Atlanta (Bullard et al. 2000), the Atlanta Regional Commission (ARC) adopted an air-quality-conforming 25-year Regional Transportation Plan (RTP) on 22 March 2000 (Atlanta Regional Commission 2000). This plan proposed some roadway improvements and construction including the Northern Arc, the so-called "second loop" highway in Atlanta. With this updated information, a new "roads" layer was prepared for the year of 2025 (as explained in Section 32.3.2 above).

On the other hand, environmental protection is another important concern for future urban development planning in Atlanta. Water conservation is a critical issue to ensure clean water supply throughout the region. The protection of lakes, rivers, and streams is therefore critical towards this goal. To implement this idea in the urban modeling, buffered zones were created along major streams and lakes. These buffered zones cover areas of 120 to 200 m wide from the center lines (or banks) of these streams (or lakes). These areas contain not only fresh water but also floodplains and wetlands, which are of great ecological value. Different probabilities of exclusion for urban development were assigned for these buffered zones (as explained in Section 32.3.3 above). Then, these buffered zones were combined with the existing layer of excluded areas to create a new file for future urban simulation. In addition to the conditions used for the first scenario, the second scenario used one more "roads" layer (2025) and an updated layer of excluded areas (see Table 32-2).

32.5.3 Scenario Three

The last scenario considers a hybrid growth strategy in which both conventional suburban development and alternative growth efforts are addressed. Sprawling development over two decades has produced rapid and profound changes in Atlanta (Bullard et al. 2000). Because this pattern of development is not environmentally, economically or socially sustainable, a number of alternatives are being developed towards a better management of urban growth (Gillham 2002). These alternatives have a profound root in new urbanism, a notion of returning to traditional neighborhood patterns for restoring functional, sustainable communities (Hamer 2000). Among those alternatives, smart growth has been considered the most effective way to combat suburban sprawl (Daniels 2001). The goal of smart growth is to create more compact development that is cheaper to service, consumes less land, and generates less traffic. Smart growth efforts are being adopted in Atlanta and their impacts upon urban development will be felt soon. Nevertheless, suburban development will be more likely to dominate over quite a long period of time although its speed may slow down with the enforcement of other alternatives.

To implement this idea in model simulation, the starting values for the five growth control coefficients used in the first two scenarios need to be changed in order to slow down the growth rate and to alter the growth pattern. As discussed earlier, the SLEUTH model considers four types of urban growth. An examination of the simulation results for the first two scenarios indicates that an overwhelming portion of the projected urban growth is accounted for by organic growth (controlled by the spread coefficient), which describes the expansion of existing cities into their surroundings. The other three types of growth (i.e., spontaneous, diffusive and road-influenced) contribute a very small portion to the growth (Table 32-7). This third scenario was intended to suppress the organic growth as an effective way to slow down the growth rate. Meanwhile, this scenario also explored how urbanization could take place in undeveloped areas. Atlanta's urban development has been basically a form of suburbanization characterized primarily by rapid residential development, which tended to take place away from existing large urban facilities in Atlanta (Lo and Yang 2002). Given these findings, the current scenario promotes spontaneous growth and diffusive growth (the development of urban settlements in undeveloped areas) as well as the road-influenced growth. The conditions and controlling coefficients used in this scenario can be seen from Table 32-6, where the spread coefficient was lowered to 15 while diffusion and breed coefficients were raised to 100. The road gravity coefficient was raised to 200. It should be noted that the proposed transportation improvements and the new additions as well as environmental conservation introduced in the second scenario are still valid here.

Table 32-7 Selected statistical measures for the stop year (2050) in relation to different simulation runs under the conditions given in Table 32-6.

Statistical Measures*	Past-to-Present Simulation	Future Simulations		
		Scenario 1	Scenario 2	Scenario 3
average slope	4.87	8.32	8.08	4.46
sng	900.38	529.14	514.72	885.60
sdc	191.84	101.31	97.37	1494.15
og	44189.96	143207.96	138478.80	79644.26
rt	26.09	12.02	11.55	16.85

* The definitions of these statistical measures: average_slope: average slope of urbanized size (in percent); sng:cumulative number of urbanized pixels by Spontaneous Neighborhood Growth; sdc: cumulative number of urbanized pixels by Diffusive Growth and Creation of a New Growth Center; og: cumulative number of urbanized pixels by Organic Growth; and rt: cumulative number of urbanized pixels by Road Influenced Growth. Please note that all values given were the averaged values after 100 runs of Monte Carlo computations.

Although the above three scenarios are different in policies and environmental conditions, there are several commonalities. The time span is the same, which is from 2000 to 2050. Because of the limitation in computation resources, only the data set with 240-m spatial resolution was used. The two input data layers of slope and hillshaded relief were used without change for all the runs. The number of times of Monte Carlo computations was 100 and the random samples were 2,840, or about 1% of the total pixels available.

32.5.4 Assessment of Spatial Consequences

Procedures were developed to assess the spatial distribution of future urban growth and to investigate the emerging urban spatial structure. These involved the use of visualization, statistical analysis, spatial modeling, and landscape metrics. First, each simulation's graphic outputs were aligned to form a time series of images by which the progressive urban growth was assessed with respect to different development scenarios (Figure 32-5).

Second, basic statistics were computed to show the area and percent of projected urban land at different time periods (Table 32-8).

Third, GIS spatial modeling was used to generate three maps showing the spatial distribution of projected urbanization for different periods under three different scenarios (Figure 32-6). Specifically, these maps were produced with the conditional run logical function powered by Imagine's Spatial Modeler (ERDAS 2001), in which the 1999 urban extent showed up fully while only the projected addition in the next period was shown. For example, the cyan-colored patches on the map represent the projected net addition of urban land between 2010 and 2020. The county boundary was overlaid on those maps for further characterization.

Last, a few selected landscape metrics were computed to characterize the emerging urban spatial structure: number of urban clusters (NC), mean urban cluster area (MCA), inter-

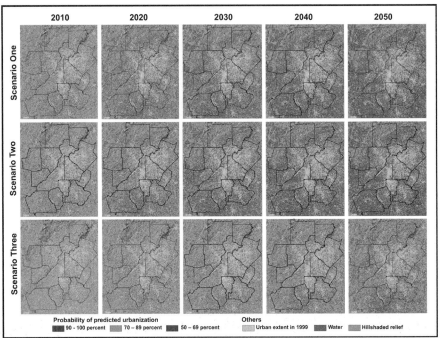

Figure 32-5 Simulation of the spatial consequences of future urban growth of Atlanta under three different scenarios. The county boundary is overlaid for location purposes. See included DVD for color version.

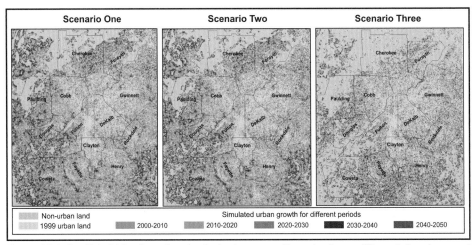

Figure 32-6 Spatial distribution of projected urban growth under three different development scenarios. See included DVD for color version.

spersion juxtaposition index (IJI), and aggregation index (AI) (Table 32-9). The first two indices are explicitly defined by their names. The third index, IJI, which was originally proposed by McGarigal and Marks (1995), was used to measure the extent to which urban patch types are interspersed. Higher values result from landscapes in which the urban patch types are well interspersed while lower values characterize landscapes in which the urban patch types are poorly interspersed. Aggregation index was used to measure the degree by which urban patches are aggregated. Detailed definitions and discussions of these metrics can be found in McGarigal and Marks (1995).

Table 32-8 Statistics of simulation results for different development scenarios.

		2010 Area (hectare)	2010 Percent*	2020 Area (hectare)	2020 Percent*	2030 Area (hectare)	2030 Percent*	2040 Area (hectare)	2040 Percent*	2050 Area (hectare)	2050 Percent*
Scenario One (Probability)	50-59**	41985	2.57	29889	1.83	21767	1.33	15978	0.98	12390	0.76
	60-69	51817	3.17	40602	2.48	29716	1.82	22067	1.35	17447	1.07
	70-79	60653	3.71	58343	3.57	43419	2.65	33615	2.06	26173	1.60
	80-89	61269	3.75	94654	5.79	76326	4.67	58291	3.56	46276	2.83
	90-100	27740	1.70	263820	16.13	469428	28.70	603821	36.92	691281	42.26
	Total urban area***	736595	45.03	980433	59.94	1133787	69.32	1226903	75.01	1286692	78.67
Scenario Two (Probability)	50-59	40383	2.47	28898	1.77	21796	1.33	16030	0.98	12920	0.79
	60-69	49409	3.02	39565	2.42	29169	1.78	23259	1.42	17856	1.09
	70-79	56385	3.45	56690	3.47	43373	2.65	34197	2.09	27780	1.70
	80-89	57531	3.52	90346	5.52	74632	4.56	59052	3.61	47508	2.90
	90-100	25770	1.58	247611	15.14	437990	26.78	565822	34.59	652994	39.92
	Total urban area	722609	44.18	956235	58.46	1100091	67.26	1191491	72.84	1252184	76.56
Scenario Three (Probability)	50-59	29843	1.82	37123	2.27	31444	1.92	26346	1.61	22239	1.36
	60-69	22038	1.35	44444	2.72	41507	2.54	35983	2.20	31110	1.90
	70-79	7188	0.44	49144	3.00	54950	3.36	50970	3.12	45942	2.81
	80-89	559	0.03	43338	2.65	75635	4.62	80185	4.90	74920	4.58
	90-100	0	0.00	11048	0.68	75848	4.64	160865	9.83	238798	14.60
	Total urban area	552758	33.79	678223	41.46	772514	47.23	847480	51.81	906134	55.40

* It is computed by using urban area divided by the total modeled area (1,635,656 hectares).
** This is the probability of predicted urbanization.
*** It contains 1999 urban area (493,131 hectares).

32.5.5 Interpretations

The progressive urban development as projected into the future 51 years under the different scenarios can be perceived quite well from Figure 32-5. The graphical outputs of the first two scenarios are quite similar. By evaluating these graphical outputs carefully, it is found that a Los Angeles-like metropolis characterized by huge urban agglomerations would emerge by 2030, if the current development conditions prevail. The vegetation area and open space in the 13 metro counties (excluding the northwestern mountainous area) will be very limited. If future road development and environmental protection are considered, the projected urban-

Table 32-9 Spatio-temporal pattern of predicted urbanization under different scenarios.

Scenario	Cluster Type*	Number of Clusters (NC)			Mean Cluster Area (MCA; in hectare)			Interspersion Juxtaposition Index (IJI)			Aggregation Index (AI)		
		2010	2030	2050	2010	2030	2050	2010	2030	2050	2010	2030	2050
One	1	9218	9218	9217	53.50	53.50	53.50	96.95	97.89	95.37	48.64	48.64	48.64
	2	5651	2951	1749	7.43	7.38	7.08	66.69	97.59	95.11	6.57	6.70	6.32
	3	6644	3830	2375	7.80	7.76	7.35	68.34	96.51	91.72	7.70	7.89	5.54
	4	7191	5259	3292	8.43	8.26	7.95	66.83	90.78	84.75	9.54	9.15	8.65
	5	6745	7774	5164	9.08	9.82	8.96	57.82	78.55	71.25	11.43	13.47	11.95
	6	3732	4754	2777	7.43	98.74	248.93	47.34	43.19	34.36	7.42	49.39	60.60
	Landscape	39181	33786	24574	18.80	33.56	52.36	78.79	53.27	36.68	35.49	43.20	51.94
Two	1	9218	9218	9217	53.50	53.50	53.50	77.15	89.77	59.76	48.64	48.64	48.64
	2	5487	3001	1820	7.36	7.26	7.10	65.34	78.51	93.31	6.47	6.54	5.35
	3	6375	3860	2432	7.75	7.56	7.34	66.46	79.91	92.16	7.55	7.16	6.96
	4	6805	5350	3519	8.29	8.11	7.89	65.25	78.12	87.83	9.20	8.97	8.48
	5	6499	7728	5287	8.85	9.66	8.99	57.54	78.70	72.46	11.05	13.04	11.35
	6	3506	5163	3089	9.08	84.83	211.39	45.26	43.73	34.80	6.96	47.81	58.93
	Landscape	37890	34320	25364	19.07	32.05	49.37	78.34	54.22	37.97	35.92	42.39	50.66
Three	1	9218	9218	9217	53.50	53.50	53.50	41.49	21.98	20.38	48.64	48.64	48.64
	2	4195	4297	3115	7.11	7.32	7.14	22.87	78.51	93.31	5.48	6.44	5.73
	3	3206	5520	4257	6.87	7.52	7.31	26.56	79.91	92.16	5.02	7.32	6.57
	4	1173	6626	5761	6.13	8.29	7.97	28.11	78.12	87.83	2.10	9.28	8.32
	5	97	7749	7926	5.76	9.76	9.45	28.68	68.43	78.07	0.00	12.93	12.90
	6		6819	7705		11.12	30.99		45.93	44.69		16.47	34.45
	Landscape	17889	40229	37981	30.90	19.20	23.86	54.39	54.22	37.97	43.91	35.25	37.40

* "Cluster Type" refers to the type of class in the simulated image. Type 1 is the 1999 urban extent. Types 2–6 are for projected urbanized classes at different levels of probability (50–59%, 60–69%, 70–79%, 80–89% and 90–100%). Landscape is the entire urbanized area, the total of Cluster Types 1–6.

ization still shows a similar trend (Scenario Two). In contrast, the simulated urbanization under the third scenario appears to be relatively constrained, indicating that the effort of slowing down urbanization through restricting organic growth has been quite successful.

The basic statistics shown in Table 32-8 reveal much more information. Under the first scenario, the total urban area for 2050 would be 1,286,692 ha. The total net increase in urban area with at least 50% probability would be 793,567 ha, or 43.6 ha per day on average, representing an increase of 160% between 1999 and 2050. As a result of such a dramatic growth, urban land would occupy approximately 78.67% of the total modeled land by 2050. The averaged slope steepness for urban land would increase from 4.87% in 1999 to 8.32% in 2050 (Table 32-7), suggesting that even steep slope land would be converted into urban use. For the second scenario, the projected urban land for 2050 would be 1,252,184 ha (Table 32-8). The net increase in urban land from 1999 to 2050 would be 759,053 ha, or about 40.8 ha per day, showing an increase of 154% for the entire period. By 2050, urban settlements would occupy 76.56% of the entire area. The mean slope steepness for urban land would increase from 4.87% in 1999 to 8.08% in 2050. A closer look at the statistical data for the first two scenarios leads to the observation that the net urban area increase as projected has been overwhelmingly concentrated for the period of 1999-2030 (Table 32-8). In this sense, the first two scenarios are quite similar. This further confirms the conclusion based on the graphical outputs that a huge metropolis would take shape by 2030. In contrast, under the third scenario, by 2050, the total urban area would be 906,134 ha, or approximately 55.40% of the entire modeled area. The total net urban area increase would be 413,003 ha, or 22.2 ha per day, indicating an increase of 84% between 1999 and 2050 (Table 32-8). Clearly, the magnitude of urban growth as projected under the third scenario has been substantially suppressed. The mean slope steepness for urban land would decrease from 4.87% in 1999 to 4.46% in 2050, implying that only land with more gentle gradients would be converted into urban uses.

The spatial distribution of simulated urbanization under these three scenarios can be discerned from Figure 32-6. For the first two scenarios, the projected urban additions for the period of 1999-2010 are largely adjacent to the 1999 urban pixels, which can be viewed as a continuation of urbanization in the form of organic growth. This is confirmed by the statistics given in Table 32-7, which show that more than 99 percent of the net urban growth under the first two scenarios is accounted for by organic growth. For the same period, the simulated urban additions are mostly dispersed over several inner counties such as Clayton, Cobb, DeKalb, and northern Fulton, as well as a few exterior counties, such as eastern Coweta, Fayette, southern Forsyth, Gwinnett, Henry, and Rockdale, consistent with Atlanta's reputation for urban sprawl. Some new additions are also found over the eastern and southern parts, where the terrain is relatively flat and thus slope resistance is quite low. The projected urban additions during 2010-2030 are largely distributed over places far away from the 1999 urban land. A large share of these new additions are spread over some exterior counties such as southeastern Cherokee, Coweta, western Douglas, northern Forsyth, and southern Fulton. Many projected additions are also found in western, northwestern and southeastern parts. Some large urban clusters can be clearly recognized. The projected urban additions after 2030 are predominantly scattered over the western and southeastern parts. Under the third scenario, the projected urbanization for the period of 1999-2010 has been very limited. Most of the new additions are for the period of 2010-2030, represented by blue and green pixels in Figure 32-6 (left). They mainly spread over a few exterior counties such as western Cobb, eastern Coweta, northern Douglas, Fayette, Henry, and Rockdale. Numerous large urban clusters can be clearly recognized, particularly in southern and western parts.

Based on the landscape metrics shown in Table 32-9, the emerging urban spatial structure can be further characterized. For the first two scenarios, the total number of urban clusters tends to decrease but the mean urban clusters tend to increase in size for the entire period.

The urban cluster classes (with different levels of urbanization probability) tend to be less interspersed but more aggregated. This indicates that smaller urban settlements would grow outward and join together to form much larger ones, thus favoring the formation of more edge cities in the metropolis. Under the third scenario, the total number of urban clusters will increase from 9,218 in 1999 to 40,229 in 2030, and then will decrease to 37,981 in 2050. The mean size of urban clusters decreases from 53.50 ha in 1999 to 19.20 ha in 2030, and then increases to 23.86 ha in 2050. The urban cluster classes tend to be more interspersed before 2030 but this trend reverses after 2030. The urban cluster classes tend to be less aggregated before 2030 but this trend reverses after 2030. This indicates that many smaller dispersed urban settlements will emerge before 2030. These scattered clusters tend to merge together and form larger settlements after 2030.

Based on the above comparisons, it is clear that the results from the first two scenarios are quite different compared to those from the third scenario. The first two scenarios are very similar in appearance, making visual discrimination difficult. But the second scenario would preserve some buffer zones along large rivers, streams and lakes quite well. These preserved zones, although relatively small in area (about 27,358 ha), contain the most important fresh water supply, wetlands and floodplains for the metropolis. The first two scenarios illustrate that unchecked urban sprawl would consume almost the entire vegetation and open space in the metro area, with an exception in the northwestern mountainous area. The dramatic growth in urban land as projected under these two scenarios would change the city's spatial form substantially with numerous edge cities scattered over a huge area. This would greatly deteriorate the quality of life in Atlanta. In contrast, the third scenario allows much more green areas and open space, including buffer zones of large streams and lakes. The urban growth rate would be reduced by approximately 50 percent when compared to the first two scenarios. Therefore, the last scenario should be the most desirable one for the future growth planning in Atlanta.

Finally, for effective visualization and communication of trends, animation movies were produced for each of the three scenarios with the use of a professional software package. The movies are included on the DVD accompanying this manual. Refer to Chapter 63 for additional information.

32.6 Discussion and Conclusions

The simulation should be useful for characterizing postmodern urbanization in Atlanta. On the other hand, the future growth simulations further examine metropolitan dynamics under different conditions. The SLEUTH model has been successfully used in conjunction with GIS in a loose coupling fashion to simulate the future urban growth in Atlanta from 1999 to 2050. The model has several strengths.

- First, the SLEUTH model's urban growth transition rules are established with a set of socio-economic and biophysical factors such as urban extent, location conditions and terrain slope. In a postmodern city such as Atlanta, an overwhelming share of the urban land is residential which is assumed to have a linear relationship with population distribution. Thus, the model considers population distribution in at least an indirect way. Different location conditions, such as road networks, business centers, urban centers, etc., can be considered as gravity with various weights assigned according to their relative significance.
- Second, certain policy conditions can be incorporated in the model. Environmentally protected buffer zones can be quantified with various weights for different levels of exclusion. Urbanization rate can be manipulated by modifying one or more growth control coefficients.

- Third, the model's calibration functionality permits a valid linkage between simulation and ground truth, thus allowing more realistic predictions to be obtained. In addition, the model's source code is open, and thus it is possible for users to update the model or add new components for improving the model's performance.

However, there are certain limitations of the SLEUTH model that should be noted. First, the urban extent is under-predicted. A major cause is that the model has not considered some other factors such as human behavior, taxes, income or race. These factors could be significant under certain circumstances. Another reason is that the model emphasizes linear growth although—in the real world—non-linear urban growth has been quite common. Second, for application in rapidly suburbanizing cities such as Atlanta, the model's transition rules need to be adjusted. The urban growth model overwhelmingly favored the so-called organic growth, or expansion from established urban cells to their surroundings (Table 32-7). This growth pattern is generally true for the development of high-density urban uses such as commercial, industrial, and large transportation facilities. But for low-density urban uses dominated by residential, new developments tend to move away from existing urban facilities in search of a better living environment (i.e., suburbanization), and thus, undeveloped areas (notably forests and croplands) are gaining popularity in Atlanta (Lo and Yang 2002). In this sense, the other three types of urban growth—spontaneous, diffusive, and road-influenced—should be more strongly represented. The third scenario of this project addressed this issue. The results indicate that the other three types of growth, despite an increase in absolute development area, are still inadequately represented. Last, the model used in this project demanded intensive computational resources, which has prevented the model calibration and simulation from using the other sets of higher-resolution data. Higher-resolution calibration and simulation should not only give more accurate results (Silva and Clarke 2002) but also be more useful for analyzing emergent urban spatial patterns. These limitations as a whole may affect the model's performance when applied to the study area.

By using Atlanta as the study site, the following conclusions at the technological, theoretical, and application levels can be made. At the technological level, the study has demonstrated the usefulness of cellular automaton modeling and geographic information system technologies for urban planning. From a user's perspective, this study has demonstrated the effectiveness of the model as a tool to imagine, test and choose between different development scenarios. These scenarios represent different growth strategies that can be adopted by planners. Three major scenarios have been designed and successfully simulated by manipulating the input data layers, self-modification constraints, and growth control coefficients. The results are quite encouraging, although more-accurate simulations could be achieved if more growth constraints were considered and the model's transition rules were modified. On the other hand, the model uses a loose coupling approach to integrate with GIS and remote sensing for the preparation of input data layers for registration and overlay, model calibration and verification, emergent urban spatial pattern analysis, growth impact assessment, and scientific visualization (such as map display and animation).

At the theoretical level, this study has examined the emergent spatial structure and patterns as related to different planning scenarios for Atlanta, one of the few recognized postmodern metropolises in North America. The model's past-to-present simulation shows a clear polycentric trend in the evolution of the urban spatial form, which is compatible with the findings obtained by urban geographers from the social and economic perspectives (e.g. Hartshorn and Muller 1989; Fujü and Hartshorn 1995). These simulations suggest that smaller urban clusters tend to grow outward and join together to form much larger clusters in the form of organic growth; thus, many edge cities would be formed throughout the entire metropolis. Eventually, these edge cities would coalesce to form an even larger city, thus changing Atlanta's spatial form substantially in the future.

At the application level, the SLEUTH cellular automaton model allows different planning strategies to be tested, as demonstrated in this study of Atlanta, by changing the growth control coefficients and the exclusion layer. The three scenarios were designed with different environmental and policy conditions in mind. The first scenario simulated the continued growth trend if the urban sprawl is allowed to continue. The second scenario projected the growth trend if future road development and environmental protection are considered. The third scenario simulated the development trend when the growth rate is slowed down and the growth pattern is altered. Despite some degree of uncertainty, the three scenarios of future urban growth simulation predict the general trends under different conditions very nicely. The results from the first two scenarios indicate that Atlanta will lose a lot of green vegetation area by circa 2030 if the current rate and pattern of urban growth do not change. In contrast, the result from the third scenario shows that much more green area and open space, including buffer zones of large streams and lakes, could be preserved. Accordingly, the last scenario should be the most desirable for the future urban growth of Atlanta. This third scenario is most similar to a smart growth strategy with emphasis on environmental protection so that the livability of the city of Atlanta will be maintained for the future generations.

These simulations will also allow projection of the urban heat island effect in the city of Atlanta and evaluation of its health implications (Lo and Quattrochi 2003). Urban growth has resulted in the replacement of soil and vegetation with impervious urban materials, such as concrete, asphalt and buildings, which change the albedo and runoff characteristics, giving rise to warmer temperatures in the city than its peripheral rural area, a phenomenon known as the urban heat island. The higher temperatures in the city have adversely affected air quality because of the production of ground level ozone from volatile organic compounds (VOCs) in the presence of nitrogen oxides (NOx) and sunlight by the photochemical smog mechanism (Cardelino and Chameides 1990). Ground level ozone is a public health hazard that can cause respiratory and cardiovascular illness. In addition, the urban heat island effect could also contribute to human deaths during high heat events, as seen in Chicago, Illinois in July 1995. Therefore, urban growth poses a public health threat. Research has shown that spatial structure and urban morphology can affect the intensity of the urban heat island effect (Nichol 1996). Armed with the SLEUTH cellular automaton model, urban planners can test various strategies designed for the development of the city to determine what would be the best spatial pattern to achieve for a healthy livable environment. Greenness (vegetative fraction), albedo and soil moisture properties can be measured for each scenario, which can serve as input into the regional climate and air quality models so that the adverse environmental impact of the city's growth can be minimized (Solecki and Oliveri 2004). In conclusion, despite the limitations noted above, the SLEUTH cellular automaton model is a useful planning tool that couples well with GIS for simulating and evaluating the impacts of future policy scenarios.

References

Atlanta Regional Commission (ARC). 2000. The 2025 Regional Transportation Plan. <http://www.atlantaregional.com> Accessed 25 October 2005.

———. 2007. Mobility 2030. <http://www.atlantaregional.com> Accessed 17 August 2007.

Barredo, J. I., M. Kasanko, N. McCormick and C. Lavalle. 2003. Modelling dynamic spatial processes: simulation of urban future scenarios through cellular automata. *Landscape and Urban Planning* 64:145–160.

Batty, M., H. Couclelis and M. Eichen. 1997. Urban systems as cellular automata. *Environment and Planning B: Planning and Design* 24:159–164.

Batty, M. and Y. Xie. 1994. From cells to cities. *Environment and Planning B: Planning and Design* 21:531–548.

Batty, M., Y. Xie and Z. Sun. 1999. Modeling urban dynamics through GIS-based cellular automata. *Computers, Environment and Urban Systems* 23: 205–233.

Bullard, R. D., G. S. Johnson and A. O. Torres, eds. 2000. *Sprawl City: Race, Politics, and Planning in Atlanta*. Washington, D.C.: Island Press. 236 pp.

Cardelino, C. A. and W. L. Chameides. 1990. Natural hydrocarbons, urbanization, and urban ozone. *Journal of Geophysical Research* 95(D9):13971–13979.

Cheng, J. and I. Masser. 2003. Urban growth modeling: a case study of Wuhan city, PR China. *Landscape and Urban Planning* 62:199–217.

———. 2004. Understanding spatial and temporal processes of urban growth: celluar automata modeling. *Environment and Planning B: Planning and Design* 31:167–194.

Clarke, K. C. 2000. SLEUTH: Land Cover Transition Model (Version 3.0). <http://www.ncgia.ucsb.edu/projects/gig/v2/Dnload/download.htm> Accessed 13 August 2007.

Clarke, K. C., S. Hoppen and L. Gaydos. 1997. A self-modifying cellular automaton model of historical urbanization in the San Francisco Bay area. *Environment and Planning B: Planning and Design* 24:247–261.

Clarke, K. C. and L. Gaydos. 1998. Loose-coupling a cellular automaton model and GIS: long-term urban growth prediction for San Francisco and Washington/Baltimore. *International Journal of Geographical Information Science* 12:699–714.

Couclelis, H. 1997. From cellular automata to urban models: new principles for model development and implementation. *Environment and Planning B: Planning and Design* 24:165–174.

Daniels, T. 2001. Smart growth: a new American approach to regional planning. *Planning Practice and Research* 16:271–279.

Dietzel, C. and K. C. Clarke. 2007. Toward optimal calibration of the SLEUTH land use change model. *Transactions in GIS* 11:29–45.

ERDAS. 2001. *ERDAS Field Guide*, 5th edition. Atlanta, Georgia: ERDAS, Inc.

Fujü, T. and T. Hartshorn. 1995. The changing metropolitan structure of Atlanta, Georgia: locations of functions and regional structure in a multinucleated urban area. *Urban Geography* 16:680–707.

Gillham, O. 2002. *The Limitless City: A Primer on the Urban Sprawl Debate*. Washington, D.C.: Island Press. 309 pp.

Goldstein, N. C. 2004. Brains versus brawn--comparative strategies for the calibration of a cellular automata-based urban growth model. In *GeoDynamics*, edited by P. Atkinson, G. Foody, S. Darby and F. Wu, 249–272. Boca Raton, Fla.: CRC Press.

Hall, T. 2001. *Urban Geography*, 2nd edition. London: Routledge. 209 pp.

Hamer, D. 2000. Learning from the past: historical districts and the new urbanism in the United States. *Planning Perspectives*. 15:107–122.

Hartshorn, T. and P. Muller. 1989. Suburban downtown and the transformation of metropolitan Atlanta's business landscape. *Urban Geography* 10:375–395.

Herold, M., G. Menz and K. C. Clarke. 2001. Remote sensing and urban growth models–demands and perspectives. *Proceedings of the Second Symposium of Remote Sensing of Urban Areas*, held in Regensburg, Germany, 22-23 June 2001. Edited by C. Juergens. Unpaginated CD-ROM. <http://www.geogr.uni-jena.de/~c5hema/pub/herold_menz_clarke.pdf> Accessed 13 August 2007.

Jantz. C. A., S. J. Goetz and M. K. Shelley. 2003. Using the SLEUTH urban growth model to simulate the impacts of future policy scenarios on urban land use in the Baltimore-Washington metropolitan area. *Environment and Planning B: Planning and Design* 30:251–271.

Lee, D.R. and G. T. Sallee. 1970. A method of measuring shape. *The Geographical Review* 60:555–563.

Li, X. and A. Yeh. 2000. Modelling sustainable urban development by the integration of constrained cellular automata and GIS. *International Journal of Geographical Information Science* 14:131–152.

Li, X. and A. Yeh. 2002. Urban simulation using principal components analysis and cellular automata for land-use planning. *Photogrammetric Engineering and Remote Sensing* 68:341–351.

Liu, Y. and S. R. Phinn. 2004. Mapping the urban development of Sydney (1971-1996) with cellular automata in a GIS environment. *Journal of Spatial Science* 49:57–74.

Lo, C. P. and D. A. Quattrochi. 2003. Land-use and land-cover change, urban heat island phenomenon, and health implications: a remote sensing approach. *Photogrammetric Engineering and Remote Sensing* 69:1053–1063.

Lo, C. P. and X. Yang. 2002. Drivers of land-use/land-cover changes and dynamic modeling for the Atlanta, George Metropolitan Area. *Photogrammetric Engineering and Remote Sensing* 68:1073–1082.

McGarigal, K. and B. Marks. 1995. FRAGSTATS: Spatial Pattern Analysis Program for Quantifying Landscape Structure. General Technical Report: PNW-GTR-351. Portland, Oregon: USDA Forest Service, Pacific Northwest Research Station.

Nichol, J. E. 1996. High-resolution surface temperature patterns related to urban morphology in a tropical city: a satellite-based study. *Journal of Applied Meteorology* 35:135–146.

Oguz, H., A. Klein and R. Srinivasan. 2004. Modeling urban growth and landuse and land-cover change in the Houston metropolitan area from 2002 to 2030. In *Proceedings of the ASPRS 2004 Fall Conference*, held in Kansas City, Missouri, 12-16 September 2004. CD-ROM. Bethesda, Md.: ASPRS.

Research Atlanta, Inc. 1993. *The Dynamics of Change: An Analysis of Growth in Metropolitan Atlanta over the Past Two Decades*. Atlanta: Policy Research Center, Georgia State University.

Silva, E. A. and K. C. Clarke. 2002. Calibration of the SLEUTH urban growth model for Lisbon and Porto, Portugal. *Computers, Environment and Urban System* 26:525–552.

Soja, E. 1989. *Postmodern Geographies: The Reassertion of Space in Critical Social Theory*. London: Verso. 266 pp.

Solecki, W. D. and C. Oliveri. 2004. Downscaling climate change scenarios in an urban land use change model. *Journal of Environmental Management* 72:105–115.

SSDAN (Social Science Data Analysis Network). Undated. CensusScope: Your Portal to Census 2000 Data. <http://www.censusscope.org/> Accessed 14 August 2007.

Sui, D. Z. and H. Zeng. 2001. Modeling the dynamics of landscape structure in Asia's emerging desakota regions: a case study in Shenzhen. *Landscape and Urban Planning* 53:37–52.

Torrens, P. M. 2000. How cellular models of urban systems work: 1. Theory. Centre for Advanced Spatial Analysis (CASA) Working Paper Series, Paper 28. London: University College London. <http://www.casa.ucl.ac.uk/working_papers/paper28.pdf> Accessed 13 August 2007.

Torrens, P. M. and D. O'Sullivan. 2001. Cellular automata and urban simulation: where do we go from here. *Environment and Planning B: Planning and Design* 28:163–168.

USGS. Undated. Project Gigalopolis: urban and land cover modeling. <http://www.ncgia.ucsb.edu/projects/gig/> Accessed 17 August 2007.

Wang, Y. and X. Zhang. 2001. A dynamic modeling approach to simulating socioeconomic effects on landscape changes. *Ecological Modelling* 140:141–162.

White, R., G. Engelen and I. Uijee. 1997. The use of constrained cellular automata for high-resolution modeling of urban land-use dynamics. *Environment and Planning B: Planning and Design* 24:323–343.

Wolfram, S. 1994. *Cellular Automata and Complexity: Collected Papers*. Reading, Mass.: Addison-Wesley Publishing Company. 596 pp.

Yang, X. 2002. Satellite monitoring of urban spatial growth in the Atlanta metropolitan region. *Photogrammetric Engineering and Remote Sensing* 68:725–734.

Yang, X. and C. P. Lo. 2002. Using a time series of normalized satellite imagery to detect land use/cover change in the Atlanta, Georgia metropolitan area. *International Journal of Remote Sensing* 23:1775–1798.

CHAPTER 33
Artificial Neural Networks for Urban Modeling

Xiaojun Yang

33.1 Introduction

Urban growth modeling and simulation started from the 1950s, reached the height of fashion in the late 1960s and early 1970s, and then fell out of favor in most of the 1970s and 1980s (Wilson 2000). The vigorous resurgence of urban model-based analysis since the early 1990s has been largely prompted by the advancements in spatial data acquisition, computer simulation techniques and geographic information systems (GIS). Over the past years, various predictive models have been developed as stand-alone packages or as subcomponents that are linked with different software packages from GIS, visualization or urban planning. They can be categorized as either stochastic, such as logit, neural networks, and cellular automata, or processes-based, such as dynamic ecosystem models. General reviews of these models are given elsewhere (e.g., Sui 1998; EPA 2000).

The purpose of this chapter is to examine the utility of artificial neural networks for urban predictive modeling. The material is organized into five components. First, a brief overview of cities as complex spatial systems is given, and the need for applying artificial neural networks to urban modeling is highlighted. Second, the fundamentals of artificial neural networks are discussed with the focus on the multilayer perceptron network because of its robustness and popularity, forming the necessary basis for further discussion. Third, several recent studies that applied neural networks to the problem of urban modeling are reviewed. Based on this status, a rigorous framework for neural network applications to urban modeling is proposed. This framework emphasizes the adoption of a systematic approach to optimizing model development that considers problem conceptualization, data preprocessing, architecture design, network training and model validation in a sequential mode. Finally, several areas are identified for further research in order to improve the success of neural networks applications to urban modeling.

33.2 Cities as Complex Systems and Neural Networks

The introduction of artificial neural networks into the problem of urban predictive modeling has largely been inspired by the emergence of the science of complexity and the increased awareness of cities as complex spatial systems (Allen 1998). The term "complex system" refers to a system of many parts that are mathematically coupled in a nonlinear fashion, and are physically associated with active exchanges of matter, energy and information between parts and their environment. There are two major features associated with a complex system. First, the relationships among parts are nonlinear. This means that change in one side is not proportional to change on the other, and thus a small perturbation may cause a large effect. In fact, nonlinear changes, including accelerating, abrupt, and essentially irreversible changes, are quite common in cities (Portugali 2000). As a result, a city as a whole is not necessarily equal to the sum of its parts. Furthermore, most of these nonlinear changes are not expected and are unpredictable, and some can be very large in magnitude and have substantial impacts on other systems. Nevertheless, it is the non-linearities that form the basis of what is interesting in complex system behaviors (Wilson 2000). On the other hand, the relationships among parts contain feedback loops. Cities consist of a largish number of interconnected parts that are organized into different levels in a hierarchical system where higher levels constrain and regulate the lower levels to various degrees, depending on the time constraint of the behavior. Most processes are constrained within a small range around a certain optimal level under certain social and environmental conditions. Any deviation of the optimal value of the controlled parameters can result from the changes in internal and external environments, thus adding to the complexity of cities.

Because of the above complex behaviors, modeling urban growth has been quite challenging. This is particularly true for the regression-based urban modeling approach. Because of its statistical robustness and easy implementation, logistic regression has been widely used to model urban growth (e.g., Landis 1995; Landis and Zhang 1997; Allen and Lu 2003; Cheng and Masser 2003). Nevertheless, this approach has some limitations when applied to the urban environment. First, logistic regression does not require linear relationships between the independent variables and the dependent variable, but it does assume a linear relationship between the logit of the independent variables and the dependent variable. In this sense, logistic regression models are intrinsically linear. When this assumption is violated, logistic regression will lack power and generate unexpected errors. This suggests that the rules governing regression models may be too restrictive, making it difficult to utilize them in the urban environment. Second, logistic regression is very sensitive to the variables considered. Any omission of relevant variables or inclusion of irrelevant variables would boost the level of errors. This implies the necessity for rigid data preprocessing. Last, logistic regression is not capable of modeling the properties based on feedbacks in the complex urban environment.

Many efforts have been made to tackle the problems associated with complex systems over the past decades. Some employ analogy or metaphor, such as neural network computing that combines human intelligence with computational power. A neural network is a mathematical model of theorized human brain activity, attempting to parallel and simulate the powerful capabilities for knowledge acquisition, recall, synthesis and problem solving. It originated from the concept of artificial neurons introduced by McCulloch and Pitts in 1943. Over the past six decades, neural networks have gone through three phases: 1) the preliminary development of artificial neurons; 2) the rediscovery and popularization of the back-propagation training algorithm; and 3) the implementation of artificial neural networks using dedicated hardware packages. Theoretically, neural networks are highly robust regardless of data distribution, and can handle incomplete, noisy and ambiguous data. They are well-suited to modeling complex, nonlinear phenomena ranging from financial management and hydrological modeling to natural hazard prediction. Since the late 1990s, neural networks have been applied to the problem of urban growth simulations, and some of these studies have been quite encouraging (e.g., Weisner and Cowen 1997; Pijanowski et al. 2002; Yeh and Li 2003; Guan and Wang 2005; Pijanowski et al. 2005).

33.3 Basics of Artificial Neural Networks

The basic structure of an artificial neural network involves a network of many interconnected neurons. These neurons are very simple processing elements that individually handle pieces of a big problem. A processing element computes an output using an activation function that considers the weighted sum of all its inputs. These activation functions can have many different types, but the logistic sigmoid function is probably most commonly used:

$$f(x) = \frac{1}{1+e^{-x}} \tag{33-1}$$

where *f(x)* is the output of a processing element and x represents the weighted sum of inputs to a processing element.

Apparently, the principles of computation at the processing element level are quite simple. The power of neural computation is built upon the use of distributed, adaptive and nonlinear computing. The distributed computing environment is realized through the massive interconnected processing elements that share the load of the overall processing task. The adaptive property is embedded within the network by adjusting the weights that interconnect the processing elements during the training phase. The use of an activation function in each processing element introduces the nonlinear behavior to the network.

There are many different types of neural networks, but most can fall into one of the paradigms listed here: multilayer perceptrons, generalized feed-forward networks, modular feed-forward networks, Jordan and Elman networks, principal component analysis networks, radial basis function networks, generalized regression-probabilistic networks, self-organizing feature maps, time-lagged recurrent networks, fully recurrent networks, the Co-Active Neuro-Fuzzy Inference System (CANFIS) model and the Support Vector Machine (SVM). Each paradigm has advantages and disadvantages depending upon specific applications. A detailed discussion about each of these paradigms is beyond the scope of this section, and interested readers are directed towards texts such as Bishop (1995) and Principe et al. (2000). This section focuses on the multilayer perceptron network because it has been fully documented and widely used (Bishop 1995). Indeed, all the urban modeling applications reported recently involved the use of this paradigm. This type of network is quite easy to use and can approximate virtually any function.

Figure 33-1 illustrates a simple multilayer perceptron neural network with a 4×5×4×1 structure. This is a typical feed-forward network that allows the connections between neurons to flow in one direction. Information flow starts from the neurons in the input layer and then moves along weighted links to neurons in the hidden layers for processing. The weights are normally determined through training. Each neuron contains a nonlinear activation function that combines information from all neurons in the preceding layers. The output layer is a complex function of inputs and internal network transformations.

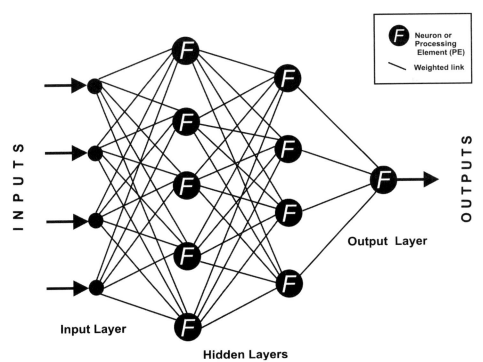

Figure 33-1 A simple Multilayer Perceptron (MLP) neural network with a 4x5x4x1 structure.

Manual of Geographic Information Systems

The topology of a neural network is critical for neural computing to solve a predefined problem with acceptable training time and performance. For any neural computing, training time is always the biggest bottleneck and thus, in an actual application, a great deal of effort is needed to make training effective and affordable. It has been shown that training time grows exponentially as the degrees of freedom in a network increase. The latter is determined by the complexity of the network topology that is ultimately dependent upon the combination of hidden layers and neurons. A trade-off is needed to balance the processing purpose of the hidden layers and the required training time. A network without any hidden layer is only able to solve a linear problem. To tackle a nonlinear problem, a reasonable number of hidden layers is needed. A network with one hidden layer has the power to approximate any function providing that the number of neurons and the training time are not constrained (Hornik 1993). But in practice, many functions are difficult to approximate with one hidden layer, and thus Flood and Kartam (1994) suggested using two hidden layers as a starting point. It should be noted that adding more hidden layers tends to increase the discriminant power of the network, which can attenuate the propagation of errors generated at the output back through the network to train the weights, but this slows down the training process.

The number of neurons for the input and output layers can be defined according to the research problem identified in an actual application. The critical aspect is related to the choice of the number of neurons in hidden layers and hence the number of connection weights. If there are too few neurons in hidden layers, the network may be unable to approximate very complex functions because of insufficient degrees of freedom. On the other hand, if there are too many neurons, the network will have a large number of degrees of freedom that may lead to overtraining and hence poor performance in generalization (Rojas 1996). Thus, it is crucial to find the 'optimum' number of neurons in hidden layers that adequately capture the relationship in the training data.

Traditionally, this optimization has been realized by trial and error. Recently, several systematic approaches have been proposed, including the use of pruning and constructive algorithms. A pruning algorithm approaches the problem of optimizing the number of hidden layer neurons by removing or disabling unnecessary weights or neurons from a large network that is initially constructed to capture the input-output relationship. A review of different pruning algorithms was given by Reed (1993). Conversely, a constructive algorithm approaches the solution by adding neurons or weights (connections) to a simple network (normally with one hidden layer) for improving the performance. Some different constructive algorithms have been reviewed by Kwok and Yeung (1997). It should be noted that these systematic methods are not without problems (see Doering et al. 1997). In practice, the trial-and-error approach still remains a popular solution to determining the appropriate number of neurons in hidden layers.

Training is a learning process by which the connection weights are adjusted until the network is optimal. This involves the use of training samples, an error measure and a learning algorithm. Training samples are presented to the network with input and output data over many iterations. They should not only be large in size, but also be representative of the entire data set to ensure the generalization ability of the trained network. There are several different error measures including the mean squared error (MSE), the mean squared relative error (MSRE), the coefficient of efficiency (CE), and the coefficient of determination ($r2$) (Dawson and Wilby 2001). The mean squared error (MSE) has been most commonly used to construct an error surface (Warner and Misra 1996). The overall goal of training is to optimize the error function through a learning algorithm that is based on either a local or global method. Local methods adjust weights of the network based on its localized input signals and localized first- or second-derivative of the error function. They are computationally effective for changing the weights in a feed-forward network but are susceptible to local minima in the error surface. Global methods have the ability to escape local minima in the error surface and are thus able to find optimal weight configurations (Maier and Dandy 2000).

By far the most widely used algorithm for optimizing feed-forward neural networks is error back-propagation (Rumelhart et al. 1986). This is a first-order local approach based on the method of steepest descent in which the descent direction is equal to the negative of the gradient of the error. In other words, when the back-propagation algorithm is used, the weights are being updated in the direction opposite to the largest local gradient. This idea is relatively simple, but its search for the optimal weight can become caught in local minima, thus resulting in suboptimal solutions. This vulnerability increases when the step size taken in weight space is too small and increasing the step size can help escape local error minima. When the step size becomes too large, however, training can fall into oscillatory traps (Rojas, 1996). If this happens, the algorithm will diverge and the error will increase rather than decrease. Apparently, it is difficult to find a step size that can balance high learning speed and minimization of the risk of divergence. Recently, several algorithms have been introduced to help adapt step sizes during training (see Maier and Dandy 2000). In practice, however, a trial-and-error approach has often been used to optimize step size. In doing so, a large step size is normally used at the beginning, and an appropriate value can be found by decreasing the size until the network becomes stable. Another sensitive issue in back-propagation training is the choice of initial weights. In the absence of any *a priori* knowledge, random values should be used for initial weights.

The stop criteria for learning are very important. Training can be stopped when the total number of iterations specified or a targeted value of error is reached, or when the training is at the point of diminishing returns. It should be noted that using a low error level is not always a safe way to stop the training because of possible overtraining or overfitting. When this happens, the network memorizes the training patterns, thus losing the ability to generalize. A highly recommended method for stopping the training is through cross validation (Amari et al. 1997). In this method, an independent data set is required for test purposes, and a close monitoring of the error in the training set and the test set is needed. Once the error in the test set increases, the training should be stopped since the point of best generalization has been reached.

33.4 Neural Network Applications for Urban Modeling

The application of artificial neural networks for urban predictive modeling is a quite new but rapidly expanding area of research. Neural networks have been used to compute development probability by integrating a set of predictive variables as the core of a land transformation model (e.g. Pijanowski et al. 2002; Pijanowski et al. 2005) or a cellular automata-based model (e.g., Yeh and Li 2003; Guan and Wang 2005). All of the applications documented so far involve the use of the multilayer perceptron network, a grid-based modeling framework, and a GIS that was loosely or tightly integrated with the network for input data preparation, modeling validation and analysis.

The land transformation model developed by Pijanowski et al. (2002) was based on these ten predictive variables: agricultural density, highway distance, inland lake distance, lakeshore distance, river distance, county road distance, residential street distance, urban distance, recreation distance, and quality of views. These variables were integrated using a feed-forward neural network with 1×1×1 structure through the Stuttgart Neural Network Simulator that was developed at the University of Stuttgart in Germany. The input layer had ten neurons corresponding to the ten predictive variables, the hidden layer had the identical number of neurons as in the input layer and the output layer had a single neuron representing the development likelihood. The entire learning process was done using a classic back-propagation algorithm, while strategies adopted to avoid overtraining or overfitting included the use of a subset of data from every other cell that were then randomly ordered. For validating the prediction accuracy, model results were overlaid with the land use changes derived primarily from remotely sensed data. The connection weights learned from a small area (a county) during the training were extended to simulation of the entire study area (a watershed), and the prediction accuracy was at 46%. It should be noted that the future predictions were not directly projected with neural network modeling but through a land demand approach. In doing so, the amount of land that was expected to transition into urban over a given time period was determined by using population projection estimates. Then, future urbanization projections were made by selecting the appropriate number of cells in priority order according to the development likelihood determined through neural computing.

Using the same network architecture, Pijanowski et al. (2005) further tested the ability of neural networks to generalize across two large metropolitan areas. For this purpose, four neural network models were built: one for each of the two study areas, one by using the training weights from one area and applying them to the other, and one by using a small set (1%) sampled from a highly urbanized area. The performance of these models was assessed by using Kappa and landscape pattern metrics (i.e., number of patches, mean patch area, patch area standard deviation, landscape shape index, landscape fractal dimension, and patch cohesion index). The simulation results varied greatly by the number of training cycles, and these models mostly performed well on spatial patterns using landscape metrics (if judged by location using Kappa they did not perform well). These researchers suggested that landscape metrics are good to judge model performance and Kappa may not be reliable for situations where the magnitude of change is fairly low.

In recent years, cellular automata modeling has rapidly gained popularity among geographers and urban planners as a promising tool for urban simulation (e.g., Yeh and Li 2001; Herold et al. 2003; Deal and Schunk 2004). But this type of model has been constrained by the extremely time-consuming calibration process for determining the most appropriate set of parameters from numerous possible combinations. This process may be unaffordable for finer-resolution data with moderate computational resources (Yang and Lo 2003). As a solution to this bottleneck, neural networks have been used to find the suitable parameters or weights for urban simulation by learning from historical land use data (e.g., Yeh and Li 2003; Guan and Wang 2005).

The neural network used by Yeh and Li (2003) was a feed-forward type with $1\times1\times1$ structure available through the THINK PRO software package. The input layer had seven neurons that represent the seven site variables including distance to the major urban areas (city proper), distance to suburban areas (town), distance to the closest road, distance to the closest expressway, distance to the closest railway, neighborhood development quantity and agricultural suitability. The hidden layer had seven neurons, while the output layer had only one neuron to represent the development probability. In total 1,000 training points were randomly sampled from the entire data set— the overlaying of the historical urban growth extracted from remotely sensed data with the GIS-derived site attributes. In addition to the training data set, a test set was prepared to validate the training performance. The training was conducted with a classic back-propagation learning algorithm. With the weights determined from neural training, a simulation run was conducted for the period of 1988 to 1993, resulting in an overall accuracy of 79%. For addressing several planning objectives linked with different development conditions, several new data sets were created by modifying the original training data set using several subjective rules. Then, each new data set was used in network training to obtain the best weights that were further used for future development simulation.

Another similar research project was conducted recently by Guan and Wang (2005) using the Matlab Neural Network Kit. They used a slightly more sophisticated feed-forward network that had two hidden layers in addition to one input layer and one output layer. The input layer had eight neurons corresponding to the eight predictive variables considered in their study: elevation, slope, distance to the city center, distance to town centers, distance to railways, distance to highways, distance to the freeways, and number of urbanized cells in an 11×11 neighborhood of each cell. The expected output was based on a time series of historical urban land use data derived from remotely sensed data. Training data were extracted from the entire data set through a stratified random sampling method. The first hidden layer had six neurons, and the second layer had one neuron. The learning algorithm used was an improved version of the back-propagation algorithm—namely, the scaled conjugate gradient method—because of its better performance in function fitting and pattern classification. The stopping criteria used were predefined error levels or a maximum number of epochs. The model output was the urbanization probability for each cell. The past trend simulation was found to be quite good, as indicated by the Lee-Sallee index (Lee and Sallee 1970) of 0.83 and the correlation index of 0.90. It should be noted that the future predictions under three different planning scenarios were not directly projected using neural computing, but through a similar method used by Pijanowski et al. (2002). In doing so, future populations were projected over several time periods using the Tietenberg model (Tietenberg 2006) and logistic regression. Then, the total number of new urban cells needed for each scenario was computed. This number of cells was selected from the non-urban cells of the past trend simulation in descending order of the development probability for each scenario.

33.5 Towards a Systematic Modeling Framework

Based on the applications described in Section 33.4, the prospect of artificial neural networks for urban growth prediction and forecasting seems to be quite promising. On the other hand, the capability of neural networks tends to be oversold as an all-inclusive 'black box' that is capable of formulating an optimal solution to any problem regardless of network architecture, urban system conceptualization or data quality. As a result, this field has been characterized by inconsistent research design and poor modeling practice. Nevertheless, recent studies highlight the need to adopt a systematic approach for effective neural network model development considering problem conceptualization, data preprocessing, network architecture design, training methods and model validation in a sequential mode (e.g., Maier and Dandy 2000; Principe et al. 2000; Dawson and Wilby 2001).

Clearly, there is no rigorous framework for neural network applications to urban growth modeling. Therefore, a system approach is proposed here (Figure 33-2). This approach consists of several core components, and the component of research problem formulation is placed at the center because of its crucial role in an urban modeling project. This is actually the phase of conceptualization through which an appropriate theoretical framework needs to be established. During this phase, the variables and conditions that appear relevant to the research problem being analyzed need to be carefully identified. Specifically, this component involves the identification of appropriate biophysical and socio-economic variables that may contain a large number of distance measurements and will serve as independent variables in the neural network model. These variables must be defined in accordance with an appropriate observational scale designed in an urban modeling project. In addition, the expected prediction accuracy and modeling time frame must be clearly defined as they are crucial for the neural network training.

Once the research problem has been defined, the next step is to collect the data. This may be the most time-consuming part in any urban modeling project. At this stage, every effort should be made to acquire sufficient data that cover the conditions that the network may encounter later. In other words, it is not only necessary to collect large data sets, but also representative data sets. Information content and data quality are essential goals in data acquisition. All data sets should be in digital format. There are some examples that provide mature guidance for spatial data acquisition, and interested readers are referred to Chen and Lee (2001) and Lo and Yeung (2002).

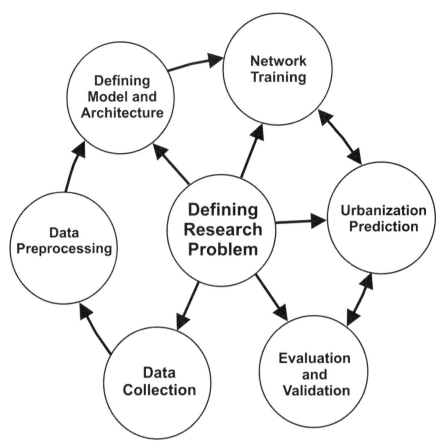

Figure 33-2 A systematic approach to urban predictive modeling by using neural networks. The component of research problem formulation is placed at the center because of its crucial role in an urban modeling project. Data collection and preprocessing are normally quite time-consuming in such a project but the training phase is the most computationally intensive. Note that the component of probing is not represented here because it is not required. However probing does provide access to all internal variables such as weights, errors and sensitivities, which can help improve the process of neural network design.

Data preprocessing is necessary as it can have a significant effect on neural model performance. The first step for data preprocessing is to divide the available data into calibration, test and validation subsets. The calibration subset is for network training and should take up the large share of the entire data set. The test subset is used to assess the performance of the model at the stage of training for cross-validation purposes. The validation subset is used to evaluate the model against independent data. Both the test and validation sets should be much smaller in size when compared to the calibration subset. Each subset should be representative of the same population. If the available data are quite limited, the division of data may be difficult, and some other methods should be attempted to maximize utilization of the available data. Examples of these methods include the holdout method (Masters 1993) and the method described by Maier and Dandy (1998).

Once the entire data set is appropriately divided, the next step in data preprocessing is to normalize the data so that each variable can receive equal attention during the training process. This is particularly relevant for population, income and distance variables because of their large variances in general. These variables are often included in an urban model. Whenever possible, deterministic components that account for trends and heteroscedasticity in the data should be removed to emphasize the dominant signal. The predictors (i.e., independent variables) need to be carefully determined, and only significant variables should be considered for further processing. If too many variables are available, a data reduction procedure (e.g., principal component analysis) should be attempted to help identify a smaller group of significant predictors that can capture the major properties of urbanization.

Before the actual training and prediction, it is essential to define the appropriate neural network type, architecture and internal parameters. Begin with a multilayer perceptron neural network and a back-propagation learning algorithm as the benchmark to evaluate any other network types and learning methods. Specify the appropriate number of hidden layers and nodes unless a pruning algorithm or cascade correlation is used. Begin with two hidden layers as a starting point. Choose either logistic sigmoid or hyperbolic tangent function as the activation function. Also choose appropriate values for learning parameters. It should be noted that, as discussed in Section 33.3, a number of trial-and-error experiments may be needed in order to optimize the network architecture. The initial weights should be randomly chosen. Use the calibration data set in the training, and a separate test set for cross-validation in order to determine when to terminate the training process.

Once the training is completed, save the weights and architecture of the neural model, and run the simulation to compute the urbanization probability. The model's performance is assessed by using the independent validation data for calculation of an error matrix or landscape metrics (see Section 33.4). Future urban development can be estimated either through modifying the independent variables or using future urban land demand based on population projections (see Section 33.4).

33.6 Future Research Directions

There are a few areas where further research is needed. First, there are many arbitrary decisions involved in the construction of a neural network model, and therefore, there is a need to develop guidance that helps identify the circumstances under which particular approaches should be adopted and how to optimize the parameters that control them. For this purpose, more empirical, inter-model comparisons and rigorous assessment of neural network performance with different inputs, architectures, and internal parameters are needed. Second, data preprocessing is an area where little guidance can be found. There are many theoretical assumptions that have not been confirmed by empirical trials and it is not clear how different preprocessing methods can affect the model outcome. Further investigation is needed to explore the impact of data quality and different methods in data division, data standardization, or data reduction upon the urban growth prediction. Third, continuing research is needed to develop effective strategies and probing tools for mining the knowledge contained in the connection weights of trained neural network models for urbanization prediction. This can help uncover the 'black-box' construction of the neural network, thus facilitating the understanding of physical meanings of spatial factors and their contribution to urban development. This should help improve the success of neural network applications to urban modeling. Last, further research is needed to develop improved methods that are based on neural network models to predict the future urban development under different planning scenarios.

References

Allen, J. and K. Lu.2003. Modeling and prediction of future urban growth in the Charleston region of South Carolina: a GIS-based integrated approach. *Conservation Ecology* 8(2) <http://www.ecologyandsociety.org/vol8/iss2/art2/print.pdf> Accessed 29 August 2007.

Allen, P. M. 1998. *Cities and Regions as Self-Organizing Systems: Models of Complexity*. Amsterdam: Gordon&Breach. 296 pp.

Amari, S., N. Murata, K. R. Muller, M. Finke and H. H. Yang. 1997. Asymptotic statistical theory of overtraining and cross-validation. *IEEE Transactions on Neural Networks* 8(5):985–996.

Bishop, C. 1995. *Neural Networks for Pattern Recognition.* Oxford: University Press. 504 pp.

Chen, Y. Q. and Y. C. Lee, editors. 2001. *Geographical Data Acquisition.* New York: Springer. 265 pp.

Cheng, H. Q. and I. Masser. 2003. Urban growth pattern modeling: a case study of Wuhan city, PR China. *Landscape and Urban Planning* 62(4):199–217.

Dawson, C. W. and R. L. Wilby. 2001. Hydrological modelling using artificial neural networks. *Progress in Physical Geography* 25(1):80–108.

Deal, B. and D. Schunk. 2004. Spatial dynamic modeling and urban land use transformation: a simulation approach to assessing the costs of urban sprawl. *Ecological Economics* 51(1-2):79–95.

Doering, A., M. Galicki and H. Witte. 1997. Structure optimization of neural networks with the A*-algorithm. *IEEE Transactions On Neural Networks* 8(6):1434–1445.

EPA (Environmental Protection Agency). 2000. *Projecting Land-Use Change: A Summary of Models for Assessing the Effects of Community Growth and Change on Land-Use Patterns.* EPA/600/R-00/098. Cincinnati, Ohio: U.S. Environmental Protection Agency, Office of Research and Development. 260 pp.

Flood, I. and N. Kartam. 1994. Neural networks in civil engineering.2. Systems and application. *Journal of Computing in Civil Engineering* 8(2):149–162.

Guan, Q. and L. Wang. 2005. An artificial-neural-network-based constrained CA model for simulating urban growth and its application. AutoCarto 2005 Technical Papers, conference held in Las Vegas, Nevada, 18-23 March 2005.

Herold, M., N. C. Goldstein and K. C. Clarke. 2003. The spatiotemporal form of urban growth: Measurement, analysis and modeling. *Remote Sensing of Environment* 86(3):286–302.

Hornik, K. 1993. Some new results on neural-network approximation. *Neural Networks* 6(8):1069–1072.

Kwok, T. Y. and D. Y. Yeung. 1997. Constructive algorithms for structure learning in feed-forward neural networks for regression problems. *IEEE Transactions on Neural Networks* 8(3):630–645.

Landis, J. 1995. Imaging land use futures: Applying the California Urban Futures Model. *Journal of the American Planning Association* 61(1):438–457.

Landis, J. and M. Zhang. 1997. Modeling urban land use change: the next generation of the California Urban Futures Model. <http://www.ncgia.ucsb.edu/conf/landuse97/papers/landis_john/paper.html> Accessed 28 August 2007.

Lee, D. R. and G. T. Sallee. 1970. A method of measuring shape. *The Geographical Review* 60:555–563.

Lo, C. P. and K. W. Yeung. 2002. *Concepts and Techniques of Geographic Information Systems.* Upper Saddle River, New Jersey: Prentice Hall. 492 pp.

Maier, H. R. and G. C. Dandy. 1998. Understanding the behaviour and optimising the performance of back-propagation neural networks: An empirical study. *Environmental Modelling & Software* 13(2):179–191.

———. 2000. Neural networks for the prediction and forecasting of water resources variables: a review of modeling issues and applications. *Environmental Modelling & Software* 15:101–124.

Masters, T. 1993. *Practical Neural Network Recipes in C++.* San Diego, Calif.: Academic Press.

McCulloch, W. S. and W. Pitts. 1943. A logical calculus of the ideas imminent in nervous activity. *Bulletin of Mathematical Biophysics* 5:115–33.

Pijanowski, B.C., D. Brown, B. Shellito and G. Manik. 2002. Using neural networks and GIS to forecast land use changes: A land transformation model. *Computers, Environment and Urban Systems* 26:553–575.

Pijanowski, B. C., S. Pithadia, B. A. Shellito and K. Alexandridis. 2005. Calibrating a neural network-based urban change model for two metropolitan areas of the Upper Midwest of the United States. *International Journal of Geographical Information Science* 19(2):197–215.

Portugali, J. 2000. *Self Organization and the City.* New York: Springer. 352 pp.

Principe, J. C., N. R. Euliano and W. C. Lefebvre. 2000. *Neural and Adaptive Systems: Fundamentals Through Simulations.* New York: John Wiley & Sons. 565 pp.

Reed, R. 1993. Pruning algorithms - a survey. *IEEE Transactions On Neural Networks* 4(5):740–747.

Rojas, R. 1996. *Neural Networks: A Systematic Introduction.* Springer-Verlag, Berlin. 502 pp.

Rumelhart, D. E., G. E. Hinton, R. J. Williams. 1986. Learning internal representations by error propagation. In *Parallel Distributed Processing*, edited by D. E. Rumelhart and J. L. McClelland. Cambridge, Mass.: MIT Press.

Sui, D. Z. 1998. GIS-based urban modelling: practices, problems, and prospects. *International Journal of Geographical Information Science* 12(7):651–671.

Tietenberg, T. 2006. *Environmental and Natural Resource Economics*, 7th edition. Reading, Mass.: Addison-Wesley, 655 pp.

Warner, B. and M. Misra. 1996. Understanding neural networks as statistical tools. *American Statistician* 50(4):284–293.

Weisner, C. and D. Cowen. 1997. Modeling urban dynamics with artificial neural networks and GIS. Volume 5, pages 66–75 in *Proceedings, AUTOCARTO 13, ACSM/ASPRS '97*, held in Seattle, Washington, April 1997. Bethesda, Md.: ASPRS.

Wilson, A. G. 2000. *Complex Spatial Systems: The Modeling Foundations of Urban and Regional Analysis.* New Jersey: Prentice Hall. 174 pp.

Yang, X. and C. P. Lo. 2003. Modelling urban growth and landscape changes in the Atlanta metropolitan area. *International Journal of Geographical Information Science* 17(5):463–488.

Yeh, A.G.O. and X. Li. 2001. A constrained CA model for the simulation and planning of sustainable urban forms by using GIS. *Environment and Planning B-Planning & Design* 28(5):733–753.

———. 2003. Simulation of development alternatives using neural networks, cellular automata, and GIS for urban planning. *Photogrammetric Engineering and Remote Sensing* 69(9):1043–1052.

CHAPTER 34
Transportation Spatial Indicators: Relating the Transportation Network to the Land

Raymond D. Watts and *Giorgos Mountrakis*

34.1 Introduction

Before there were roads, there was space. That space is now fragmented by roads; in GIS terms, the space is broken into polygons. Road-bounded polygons, however, often have dead end roads within them, which partially dissect them. This chapter introduces strategies for the analysis of this complex dissection of space. Central to the analysis are distance transformations, particularly *distance to the nearest road* (DTR) and its areal integration, *roadless volume* (RV).

In areas of dense population or high economic activity there is little space between roads; in areas of sparse population and economic activity there is more intervening space. As time goes on new roads are built, extending from high road density areas into lower density areas—what we might call *network extension*. At the same time, however, additional roads are built within high density areas—what is often called *infilling*. The distinction between extension and infilling is, however, a matter of the scale of inspection; in both cases, roads penetrate from the earlier road network into roadless space. It is simple to make the intuitive statement that roadless space is reduced or lost, but more difficult to quantify the reduction. The central contribution of this chapter is to illustrate the quantification of these intuitive notions of roadless space.

Space between roads is at least as interesting to study as the road network itself. With appropriate spatial metrics, we can ask questions such as these:

- How much roadless space is there in, say, a particular county or in the conterminous 48 states of the United States?
- When a new road is built, how much space is lost?
- What is the difference between building a road adjacent to an antecedent road versus building one that penetrates to the core of an otherwise roadless area?

Much of our work has employed the simplest of transformations: distance to the nearest road (DTR)—a transformation that is available in most GIS systems. When DTR is integrated over an area it becomes an explicit measure of the space between roads for that area, which we call roadless volume (RV). We also illustrate a related—but much more difficult to calculate—transformation into estimated off road travel time.

Our focus on the space between roads is motivated by more than analysis strategy. It is in roadless space that ecosystem services are rendered. The reduction of roadless space diminishes the capability of the land to provide these services. It is not our purpose to specifically evaluate the correlation between roadless volume and ecosystem services—such correlations are unique to each geographic setting and ecosystem service—but rather to illustrate tools that might be used for future analyses of such problems. The interested reader can explore recent literature in the field of road ecology (e.g., Forman and Alexander 1998; Forman and Deblinger 2000; Gelbard and Belnap 2003; Havlick 2002; Trombulak and Frissell 2000) to gain an understanding of the myriad landscape phenomena that are influenced by the presence of roads.

34.1.1 Distance to the Nearest Road—Creating a US National Road Indicator Dataset

A regular grid of DTR is one of the simplest ways to describe the relationship of roads to their encompassing roadless spaces, to compare this spatial juxtaposition from place to place, and potentially to monitor over time. DTR is straightforward to calculate using standard spatial analysis tools. If, however, the reader is interested—as we have been—in evaluating DTR at high resolution over large areas, then the analysis must be done in segments. We here describe our approach to segmented evaluation of DTR at 30 m resolution for the United States.

The starting point for developing a national DTR dataset is a national road vector dataset. The first data resource that claimed to describe roads across the United States was the 1990 TIGER road dataset assembled by the US Census Bureau with support from the US Geological Survey (USGS). Much of the data for the 1990 TIGER files derived from laser scans of lines on 1:100,000-scale USGS quadrangle maps; this scanning was done in the late 1980s. The source maps in some cases had publication dates twenty or thirty years earlier, and they were based on aerial photography made years before publication. Thus, the road distribution represented in the 1990 TIGER files is an amalgam of conditions roughly from 1950 to 1990. Without source date attributes, there is no immediate way for users to evaluate vintage of road features.

The Census Bureau made an updated version of TIGER for the 2000 census (US Census Bureau 2005); road vectors for the Census 2000 TIGER data were released in 1999. During the 1990s various commercial firms also worked to improve upon the 1990 TIGER data, some to support web mapping applications. Address ranges are a key ingredient for general purpose web mapping services, because address ranges attached to road segments allow users to search for specific addresses and be given a map showing the approximate, interpolated location. The Bureau of Transportation Statistics (BTS) of the US Department of Transportation and USGS negotiated purchase of one of these commercial datasets stripped of its address ranges. The acquired dataset is complete with road attributes, following the scheme of 1990 TIGER data, and names. It is now in the public domain and can be obtained from a USGS web site (USGS 2002). The company that provided the data was Geographic Data Technology, or GDT (which was purchased by Tele Atlas mid-2004). The dataset goes by several names: *BTS roads*, *GDT roads*, and *Dynamap 1000* (its commercial designation)*.

Because so many road effects relate to natural resources, and those resources respond to land cover as well as roads, we built a national DTR dataset structured to match the National Land Cover Dataset (NLCD), which is in the Albers Equal Area projection† with cell size of 30 m. The size of this dataset, approximately 10 billion cells, precluded its assembly in a single piece owing to 32-bit address range limits. Because the BTS road vector dataset was available by state, we developed the DTR data also by state. Larger states were themselves too large to process as single segments, so these were divided and their sections treated as if they were states.

State-by-state processing requires care because neighboring states are not roadless. Thus, our state-by-state workflow consists of the following steps:

1. Conflate road vectors from adjoining states, and in the case of border states conflate roads from adjacent areas of Canada and México.
2. Clip the multi state dataset to a polygon buffered outward from the central state boundary by 30 km (or farther over water, as explained below).
3. Convert road vectors to a 30 m grid.

---

* Commercial names are used for descriptive purposes only and do not imply endorsement by the authors or their institutions.
† North American Datum of 1983, Central meridian -96°, Standard parallels 29.5° and 45.5°, Latitude of origin 23°, Units meters, False northing zero, False easting zero.

4. Calculate Euclidean distance to nearest road (DTR).

5. Clip DTR to the central state boundary buffered outward by 60 m.

The buffering in the last step is necessary to prevent occurrence of *nodata* cells along borders when the state DTR grids are merged. When generating the DTR grid for each state we align its corner on an integer multiple of 30 m in both easting and northing in the Albers coordinate system. Alaska and Hawaii are processed similarly, but using different projections more appropriate to their locations. DTR values are calculated for both inland and offshore waters; over water calculations are generally terminated at 100 km except where greater distance is needed to determine DTR for islands.

Buffering and clipping operations are so routine in GIS work that they need little explanation. Distance calculations on a grid are done less frequently and deserve amplification. The procedure is to generate a grid with road-containing cells set to zero (or any other constant) and roadless cells set to a value of *nodata*. The proximity calculation then uses the data cells as the origin (zero-distance locus) and calculates distances outward from there. Users should consult their GIS software documentation to determine the mechanics of setting up and performing the calculations.

We refer to the national 30-m distance to road dataset as NORM ED, the National Overview Road Metrics – Euclidean Distance dataset. NORM ED reflects the shortcomings of the BTS dataset: lack of currency, missing roads, timespan of original source materials, and so on. Nevertheless, it has proven sufficient for depicting and understanding large scale patterns of roadless space for the nation. Areas of low DTR generally are found in cities (Figure 34-1), but there are notable exceptions, such as in the oil and gas fields of western Texas. There is a correlation of DTR with climate; much of the Interior West has both greater DTR and lower mean annual precipitation. Topographic inhibition of road building is apparent in both the high mountain ranges of the West and the lower Appalachian and Adirondack mountains of the East. Water and swamps restrain road building, as evidenced along the Gulf Coast and the Canadian border west of the Great Lakes.

Full resolution NORM ED data are useful in analyses that relate to the National Land Cover Dataset, with which it is registered. With this combination of datasets it is possible, for example, to answer such questions as "Is average distance to road greater for grasslands or

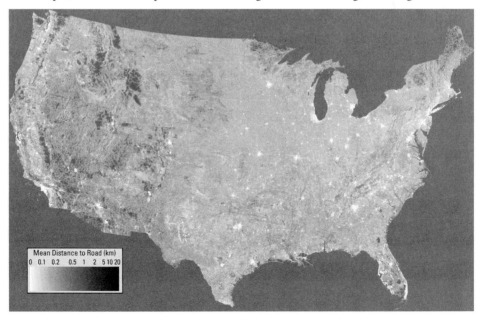

Figure 34-1 Map of the Conterminous United States depicting distance to the nearest road (DTR) averaged over 1 km squares.

Manual of Geographic Information Systems

for forests?" or "Is average distance to road greater in Eastern or Western forests?" For many other analyses lower resolution data suffice. For such analyses we have floating-point statistics datasets at 1,020-m resolution made from 34x34 clusters of 30-m cells; available statistics are minimum, mean, and maximum DTR values, as well as the count of valid 30-m DTR data cells. The 1,020-m statistical grids for the conterminous states are 4,800 columns by 3,000 rows in size.

34.1.2 Roadless Volume — An Improved Metric of Roadless Space

Roadless volume (RV) is the integral of DTR values over an area, and is our fundamental measure of space between roads. RV can be understood in the following terms: replace elevation with DTR to create a synthetic topography; the volume beneath the synthetic topography is the roadless volume. A square road pattern 1 km (or mi) on a side produces a pyramid with a height of ½ km (mi) and a volume of 1/6 km³ (mi³). Although RV provides a quantitative integrated metric, it is also helpful to visualize the synthetic topographic surface in order to see where volume contributions occur. RV is amplified with an example application in Section 34.2.2. The roadless volume of the conterminous 48 states (plus the District of Columbia) of the United States is 2.1 million km³ (Watts et al. 2007).

Much of the effectiveness of the RV metric comes from its geometric behavior. A road penetrating to the core of an otherwise roadless area reduces RV much more than a road of the same length that runs alongside perimeter roads. In the case of the square pyramid, a road to its center reduces its apex pseudo elevation to zero; a road running close to its edge shaves a sliver from a face. The volume reduction is far greater in the former case than the latter. In this way, RV directly measures the invasion of space by roads.

34.2 Analyses Based Exclusively on Distance to Road

We describe three applications derived only from DTR data: (1) identification of the 200 most remote places in the conterminous US; (2) analysis of regional changes in roadless volume; and (3) application of DTR to assessment of potential biases in environmental sampling along roads.

34.2.1 The 200 Most Remote Points in the Conterminous US

Using the 1,020-m resolution version of NORM ED we can identify points most remote from roads in the conterminous United States. Distance from nearest road is a simple measure of remoteness; more comprehensive measures might account for the difficulty of crossing roadless areas to reach remote points. Nevertheless, we believe that the analysis presented here is the most comprehensive assessment to date of remote areas in the US.

Processing begins by identifying local maximum values in the 1,020-m resolution dataset of maximum 30 m values, *maxDTR* (described in Section 34.1.1). This is done by calculating maxima of 3x3-cell sliding windows and then testing whether the central cell has the same value as the maximum. If the central cell is maximal, its value is retained; otherwise it is set to zero. If a cell cluster has no data, the result is *nodata*.

Every road polygon has at least one maximum DTR point, and may contain more if the polygon has a complex shape or dead end roads. The number of local maxima in NORM ED exceeds 314,000, but we are interested only in those with the largest values. We sort the data values, search the sorted table for the distance value that corresponds to the 201st largest maximum, which is 18,432 m, and write a text file containing the 201 maximal distances.

We want to produce a map that colors all cells with DTR values greater than the 201st local maximum, according to the cells' ranks in remoteness. The processing logic is most easily illustrated starting with the most distant and second most distant points. We color all cells in *maxDTR* that exceed the DTR value of the second most remote point, thus marking a cluster of cells around and including the most remote point. We now color, in a different shade, all uncolored cells with DTR values greater than the third most remote point, which adds an annulus in the new color around the first cluster and creates a new cluster around the second most distant point. This process continues up to the 201st most remote point, at which stage we have added colors to the 200 most remote areas. This process is readily implemented as a classification procedure, with the DTR values at successively remote points serving as break points in the classification. The table to specify this reclassification is easily built in a spreadsheet program or with simple programming code, using the table of 201 maximal distances produced earlier. Figure 34-2 shows the results of the calculations for the conterminous 48 states and, in higher resolution, the parts of the nation where the 200 points most remote from roads occur.

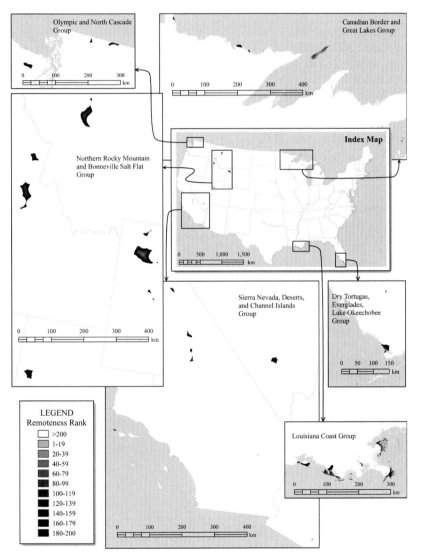

Figure 34-2 Map illustrating the spatial distribution of the 200 points most remote from roads in the conterminous United States. Each ranked point is a local maximum of distance to road (DTR) and is surrounded by concentric shaded fringes of unranked points that are not local maxima. The shades in these fringes correspond to the ranks of distant maximum DTR points of like DTR value.

Comprehensive tabulation and reporting of these results will be done elsewhere, but some highlights follow. Water and swamps are the strongest predictors of remoteness. The most remote point in the conterminous states is in the Dry Tortuga Islands west of the Florida Keys, 115.6 km from the nearest road. Other remote areas occur along the Louisiana coast, in the Everglades, the Channel Islands of California, and islands in the Great Lakes. The extensive network of lakes in Minnesota produces remoteness along the Canadian border. Of the landlocked remote areas, the largest is near the southeast corner of Yellowstone National Park in Wyoming (maximum DTR = 34,944 m or 21.7 mi).

34.2.2 A Regional Study of Change in Roadless Volume

Shortcomings of national road vector datasets for the United States have been discussed above, and consistent national multi-temporal road vector data are not currently available. To illustrate change assessment, we switch to the scale of a large landscape or small region—the northern Front Range area of Colorado. We use USGS road network datasets developed from comparable 1-m resolution vertical images made in 1937, 1957, 1977 and 1997 (Langer et al. 2000). There are few places in the United States where such consistent multi-temporal data are available. We calculated DTR on a 30-m grid; values near the edges of our grid are erroneous because, for lack of data, we treated external areas as roadless.

Figure 34-3 shows the result of substituting DTR values for elevation and rendering the resulting synthetic terrain with hill shading. Summits are local maxima in DTR values, and roads are all at zero pseudo elevation. The volume filled by the synthetic terrain is the roadless volume (RV). Units of RV are m^3 or, more conveniently for an area of this size, km^3. It is easy to see from Figure 34-3 that RV is useful as a metric for the loss of roadless space. With multiple dates of data it is possible to animate change of RV; an example video, an animation of Figure 34-3, is included on the DVD that accompanies this volume.

Roadless volume is calculated by adding DTR cell values over an area of interest and multiplying the sum by the area of one cell. Roadless volume is proportional to average DTR, but there is added value in thinking of RV as the integration of the synthetic terrain and in visualizing its spatial distribution. Figure 34-3 illustrates that the overall 44% loss of roadless volume along the Front Range occurred in large volume fragments near Denver and in a more widespread loss of small volume fragments farther north. Volume loss was minor in an area south of Greeley.

Roadless volume can be used to evaluate scenarios in a planning process. More roadless volume is lost when a new road penetrates an area of high DTR values, than is lost when another new road of the same length is built close to existing roads. Conversely, the removal or closure of a relatively isolated road results in greater gains of roadless volume than the removal of the same length of road that lies close to other roads.

Extensions are possible. For example, it is clear that sprawl not only impacts urban and suburban conditions and dwellers—by increasing travel times, adversely impacting air quality, and so on—but also impacts surrounding rural areas by diminishing their intrinsic rurality and converting them to a more developed condition. Many urban sprawl metrics have been developed primarily from an urban-centered perspective (Galster et al. 2001; Hasse 2004); these measure connectivity to earlier development and to urban social amenities. What is lacking is consideration of the penetration of roadless space, which we believe is an essential characteristic of rural areas. Sprawl metrics that account for diminished rurality may be more comprehensive or balanced than those that ignore this change.

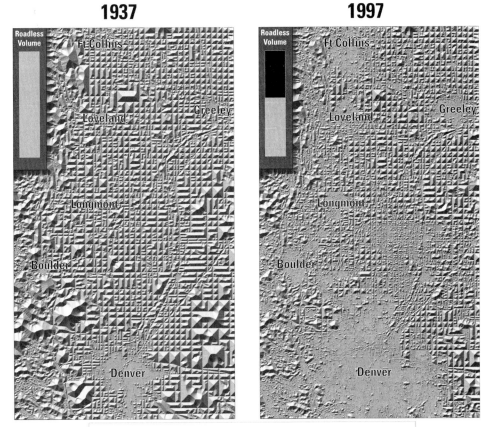

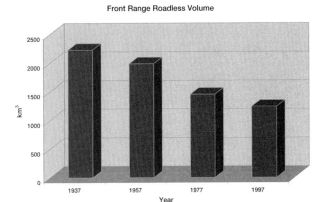

Figure 34-3 The indicator *roadless* volume is the volume filled by the synthetic terrain when distance to road (DTR) is substituted for elevation. This illustration shows hill shaded synthetic topography of the urbanizing Front Range of Colorado for 1937 and 1997. The volume of the synthetic topography was 2,230 km^3 in 1937 and 1,241 km^3 in 1997; the loss—a reduction of 44% over the 60-year period—is represented by the shrinkage of the gray bar between the 1937 and 1997 panels.

34.2.3 Sampling Along Roads: Assessing Possible Biases in the Breeding Bird Survey

DTR analysis can sometimes help other disciplines to untangle multi-dimensional problems. An interesting example comes from the field of avian biology. The Breeding Bird Survey (BBS) is an annual effort in the United States and Canada to observe trends in bird populations. Observers follow a protocol that requires visual and auditory observations at ½-mile intervals along more than 2000 road routes, each 24.5 mi. in length. Thus, the BBS observations are intrinsically biased by roads because the observations occur strictly on roads.

Biologists rightly look for the relationship of BBS protocols to bird biology. Some bird species are influenced by roads, traffic, human presence, and road-associated features such as suspended power and communication lines. Some effects are direct, others indirect—for example, the population of ground-dwelling birds may be depleted by the availability of elevated perches for raptors and increased efficiency of predation. From a biological perspective there is little doubt that conditions sampled on roads differ from conditions on the whole landscape.

As spatial analysts, we can ask some slightly different, yet informative, questions. We illustrate by addressing the following question: are the roads used in the BBS typical of roads in the region in terms of their relationship to surrounding roadless space? Biased sampling would occur, for example, if BBS routes were preferentially laid out along the borders of large roadless areas, leaving areas of higher road density under-represented in the sample.

Because the BBS routes are on roads, their values of DTR are all zero and provide no information. One can invent any number of buffering schemes and examine DTR statistics within buffers, but most such schemes suffer from an arbitrary choice of buffer distance and from buffer overlaps resulting from sinuosity of some routes. A landscape-oriented approach is more productive. We illustrate with data from South Dakota, focusing more on spatial analysis strategy than on GIS processing details.

Our approach is to divide the landscape into sections, each associated with a BBS route. The simplest way to do this is to use distance calculations with source allocation. In this analysis, the GIS determines Euclidean distances and simultaneously tags each cell with a value taken from the closest source cell (closest cell on the locus of zero distance, which is on a BBS route). When source cells on BBS routes are given values equal to their route numbers, the result is a grid with all cells tagged with the number of the closest route. The grid can be converted into polygons that circumscribe routes, with a polygon attribute equal to the route number (Figure 34-4). Polygon boundaries lie at equal distance from two or more routes. Because states around South Dakota have their own BBS routes, some of the adjoining states' routes must be included in order to properly determine the limits of the areas associated with South Dakota's internal routes; polygons associated with these external routes are evident in Figure 34-4.

The next step is to isolate each route's area and divide it into the half that is closest to the route and the half that is farthest; we call these the near and far zones (these are not explicitly depicted in Figure 34-4). We then assemble the distance to road values in the near and far zones and compare their statistics (Figure 34-5).

There are statistical complications because distributions of DTR often show no central tendency. It is easy to demonstrate for simple geometries—circles and squares, for example—that DTR distributions decrease monotonically with increasing DTR. We circumvent this difficulty by replacing zero DTR values (roads themselves) with 15 m, which is a unique and appropriately low value when DTR is computed on a 30 m grid. We then take the log of DTR and do statistical analysis in the log DTR domain, where distributions are approximately normal, although skewed to low values (Figure 34-5). Results are summarized by assigning to each route's area, in either polygon or grid format, the difference in mean log

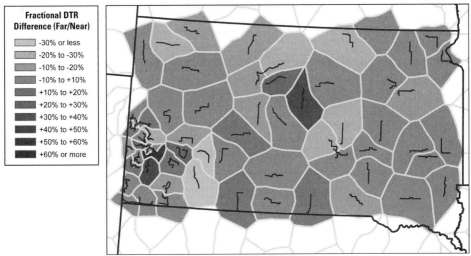

Figure 34-4 Distance analysis defines areas associated with Breeding Bird Survey routes (dark lines) in South Dakota and adjoining states. Area boundaries (light gray) are equidistant from two or more routes. Each area bounded by these equidistant lines is further divided for analysis into the half that lies closest to the route (the near zone) and the half that lies farthest (the far zone); this internal division is not depicted. Shading of each area reflects the DTR discrepancy between its near and far zones. The darkest shade, for example, indicates that mean DTR in the far zone exceeds the mean for both zones by a factor of at least 1.6.

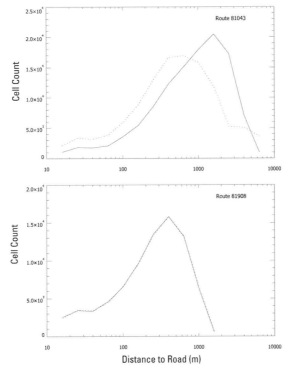

Figure 34-5 Graphs show distributions of log DTR for near zones (solid lines) and far zones (dashed lines) of two Breeding Bird Survey routes in South Dakota. Differences between near and far zone are minimal for some routes (81908), and substantial for others (81043). For routes with identical near and far zone statistics, the relationship of the sample roads to their surrounding space would change minimally if the sample route were altered. Routes with significantly different statistics would sample a different roadless space environment if routes were moved to alternate roads.

Manual of Geographic Information Systems

DTR values between the far and near zones. A far/near DTR ratio is obtained by calculating $10^{(\text{difference of mean log DTR})}$. Classification and route-area shading in Figure 34-4 is based on the deviation of this ratio from unity.

For routes where the near and far zones have similar statistics, there is no apparent spatial road bias; where they differ, there may be a road bias. In some places the bias may be unavoidable, for example where a large roadless area occurs, and there is no possibility to relocate the route to better sample it. Figure 34-6, a scatter plot of mean log DTR values in the near and far zones for South Dakota's 61 BBS routes, demonstrates that there are as many instances of DTR in the near zone exceeding DTR in the far zone as the opposite. Where DTR is greater in the near zone than in the far zone, it is possible, in principle, to balance the DTR statistics by moving the route into areas with more roads. Such a change would, however, disrupt temporal continuity in the BBS surveys and would be ill-advised unless it were demonstrated that a spatial bias is likely to cause a bias in population estimates. In our example, spatial analysis serves to stratify the problem into spatial and biological components, and the spatial analysis suggests that most BBS routes appropriately sample their portions of the landscape. For the few that do not, biological analysis is needed in order to understand the relevance of the spatial bias.

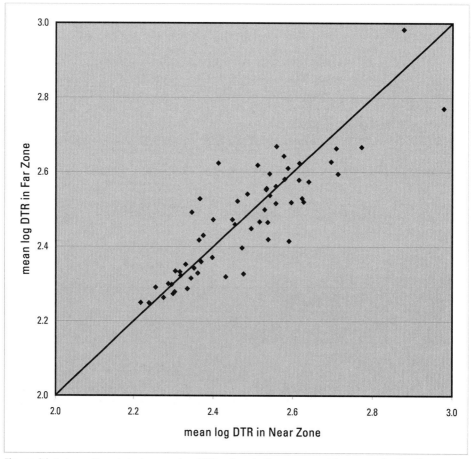

Figure 34-6 A scatter plot of mean log DTR values in the near and far zones for South Dakota's 61 Breeding Bird Survey routes shows that there is no significant overall bias. The line represents equal mean DTR in the near and far zones. Routes with points below the line could potentially be relocated into areas with more roads in order to reduce individual route bias; bias reduction is more difficult for routes with points above the line.

The statistical methods that we have illustrated are by design nearly as simple as they can be. There are many more-sophisticated tests that might, among other things, give explicit measures of confidence in observed differences. Our objective here has been to illustrate spatial processing strategies, not to do definitive statistical analysis.

34.3 Combined Analyses

In many situations distance alone poorly reflects process realities, and it is clear that distance must be combined with other descriptors to better estimate effects that are only partly determined by roads. A powerful generalization of Euclidean distance is cost-weighted distance. Below we illustrate cost weighting for estimation of travel times across a landscape, with part of the travel on road and the remainder on foot. Estimated travel times are a simple measure of relative accessibility; long travel times imply relative freedom from human activity. In a second example we investigate the spatial association of constructed impervious surface with DTR to illustrate that such profound human disturbances are indeed closely associated with roads.

34.3.1 Modeling Remoteness as Travel Time

Humans move away from roads in several ways: on foot, horses, or mountain bikes, and on or in motorized vehicles. The ease or difficulty of reaching any point on the landscape is determined by the limitations that topography and surface conditions, including vegetation, impose on each transportation mode. In many places there are additional controls, such as fenced private lands that must be circumvented rather than passed through.

Here we extend the analysis of the transportation network to the landscape itself. Time is a metric that applies to all modes of transportation, and is a powerful determinant in human decisions about where and how to travel. Energy expenditure would be another useful metric, but loses much of its meaning when applied to travel on motor vehicles or horseback. With a temporal metric, we can evaluate mixed travel modes. For simplicity of illustration, we analyze mixed motor vehicle and foot travel, restricting vehicular travel to roads. This model can be elaborated with travel speed information for many other modes of transport, including both land (mountain bike, all terrain vehicle, etc.) and water (canoe, kayak, motor boat, etc.) modes.

Paved roads in the area that we analyzed could be driven at speeds so great that the travel time from one point to another on the paved road network was minimal. Thus, the goal of our model was to estimate the time required to reach any point on the landscape from a starting point on a paved road. The locus of zero travel time is the paved portion of the road network.

We assume that the traveler uses the most time-efficient route. Thus, the starting point is not arbitrary but is dictated by the destination—which is confusing, because the paved-road starting point for any destination point is not known before the fact. Although this is not how the problem is solved, stating it in reciprocal terms makes it easy to understand: given an off-road starting point, what is the route that results in the shortest travel time to any paved road (using speeds appropriate for travel in the opposite direction)?

The problem is solved by estimating travel times away from all paved roads; in other words, it is not solved point-by-point for each destination, but rather is solved globally. This is equivalent to documenting the time of arrival of a front that starts on the margins of all paved roads and moves across the landscape at the traveler's speed as determined by local conditions. Because we allow multiple modes of travel, frontal speed requires special treatment for unpaved roads, trails, and other features that affect choice of mode and speed. Front propagation problems are best solved using fast marching methods (Sethian 1999), but network solutions (Dijkstra 1959) are adequate approximations given all the other unknown factors and approximations that influence overland travel speed.

The model is calculated on a grid by assigning travel speeds (m/min) to each grid cell, taking the reciprocal to get the time cost per unit distance traveled (min/m), and then calculating the cost (time in minutes) to traverse the network from paved roads. A complete model would employ speeds that depend on travel direction (the direction of frontal propagation), but we opted for the simplicity of assigning an isotropic speed to each cell. The greatest travel time error induced by the isotropy assumption occurs for foot travel on steep slopes. Van Wagtendonk and Benedict (1980) studied trail walking speeds, demonstrating that fastest travel occurs on nearly level trails. Exponential reductions in speed occur with increasing slope, but with distinctly different exponential coefficients for uphill and downhill travel. Their formula for trail walking speed is

$$v = 55.5 \exp(-(0.03 + 0.06\,\delta)\,s)\ \text{m/min}, \tag{34-1}$$

where s is slope in m/m and δ is 1 for uphill travel and 0 for downhill travel. We use the average of the uphill and downhill coefficients,

$$v = 55.5 \exp(-0.06\,s)\ \text{m/min}, \tag{34-2}$$

as a simplified isotropic estimate, and therefore underestimate downhill speeds and overestimate uphill speeds. We believe that the error so introduced is minor compared to errors in factors such as land cover penalties (see following paragraph). Fritz et al. (2000) introduced a similar modeling method based on a mountaineering approach formula called Naismith's rule; their model does not adjust for land cover.

In the course of a number of field studies we have driven all unpaved roads in the study area and have hiked to several off road destinations. Thus, we are able to estimate driving speeds on the unpaved roads and walking speed penalties for several land cover types: deciduous (aspen) forests, evergreen (spruce-fir and lodgepole pine) forests, wetland shrubs (willows), dryland shrubs, meadows, tundra, and so on. We expressed the penalties as fractional speed reductions compared to estimated bare ground speed in the same locations.

Steps in making the reciprocal-speed time_per_m map are these:

1. Calculate slopes from a 30 m digital elevation model.
2. Estimate bare ground walking speed (m/min) from slope using the isotropic speed formula given above.
3. Simplify National Land Cover Dataset (NLCD) 30 m land cover classification to the four classes considered to have distinctly different travel impacts: coniferous forest, deciduous forest, shrubland or meadow, and wetlands.
4. Make a trail grid and set land cover (from step 3) to bare ground for trail cells.
5. Make a fractional speed multiplier grid (floating point numbers ranging from 0.0 to 1.0) from land cover adjusted for trails (from step 4).
6. Apply land cover speed penalty factors (from step 5) to bare ground speeds (from step 2).
7. Make a road grid and assign driving speeds (m/min) to unpaved road cells.
8. Impose driving speeds (from step 7) onto road cells, leaving other cells (from step 6) unchanged.
9. Take the reciprocal of speed (from step 8) to get time cost per meter of travel, time_per_m.

Time calculations were done as follows:

1. Make a grid paved_rds with paved road cells set to zero and all others nodata.
2. Calculate the time cost to traverse the network using cost values from time_per_m.

The ArcInfo Grid command to do the time calculations is:

```
time_grid = costdistance( paved_rds, time_per_m )
```

The result is estimated optimum travel time in minutes from a paved road, for each cell in the model.

Figure 34-7 shows selected input data and results of the computations. US Highway 191 runs along the main Gallatin River in the northeast corner of the study area and is the only paved road; this is the locus of zero travel time. The network of unpaved roads in the valley of Taylor Fork shows as light gray tones in the speed image; driving speed on these roads varies according to surface and maintenance. Trails also appear on the speed image in medium gray. Differences between trail speeds and adjacent off-trail speeds reflect the lack of land-cover speed penalties on trails (slopes are generally comparable on- and off-trail). Off the road network, it is easy to see the influence of steep slopes by comparing the shaded relief image with the travel speed image, particularly along the western edge of the study area.

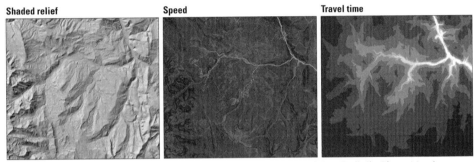

Figure 34-7 Three images show the valley of the Taylor Fork of the Gallatin River, Montana; each image is 25 km in width. (The shaded relief image is traditional elevation, not the synthetic topography discussed in the section on RV above.) The travel speed image is shaded logarithmically; the influences of slope, trails (medium gray and linear), and unpaved roads (light gray and linear) are obvious. Travel times are also shaded logarithmically; white areas can be reached from paved roads within 5 minutes; the darkest gray areas take more than four hours to reach.

Calculated travel times are dominated by the transportation network. The main east-west road in the Taylor Fork valley provides access to most of the valley floor within 30 minutes. In the roadless north-south valley that lies to the east, in the southeast quadrant of the images, travel over the same distance takes four hours. The four-hour contour, the transition to the darkest shade of gray, has special meaning in backcountry settings such as these because further travel exceeds an eight hour round trip, and thus is likely to imply an overnight stay.

These calculations were made as experiments in the analysis of grizzly bear habitat. Many habitat models for large mammals use either distance to road or road density as explanatory variables for habitat use (Mace et al. 1999). Some of these models indicate that animals avoid roads. We postulate that the avoidance may not be of roads themselves, but rather of the presence of humans, which is closely associated with roads. Human activities are not, however, confined to roads, and our accessibility modeling provides information about the likelihood of spatially distributed human presence influenced by the use of roads. Our model is a good deal more sophisticated than simple Euclidean distance, but it does not account for other important factors such as line of sight to places most frequented by humans.

Distance and remoteness models allow analysts and decision-makers to examine implications of decision scenarios. The reader can readily imagine how easy it is to simulate the closure of a road, which effectively turns it into a trail, in our model. After such a manipulation, the travel time map reflects the increased difficulty in reaching affected parts of the landscape. Similarly, proposed new roads and trails can be artificially inserted and the resulting areas of altered access can be identified. Such analysis is particularly valuable in areas of sensitive habitat; our study area is a small section of designated habitat for grizzly bear, which has been a protected species in the conterminous United States.

In much the same way that DTR can be integrated into roadless volume, remoteness measured in travel time can be integrated over an area. A simple way to do this is to multiply the grid of travel times (in minutes) by the nominal walking speed on flat bare ground, 55.5 m/min. The result is an equivalent flat-ground walking distance to paved road for each point on the landscape. As with roadless volume, this measure can replace elevation to create a synthetic terrain, and the volume of the synthetic terrain is an integrated measure, in m^3 or km^3, of remoteness.

34.3.2 Spatial Association of Roads and Impervious Surfaces

The previous section focused on places where the landscape surrounding roads is little modified by humans; this section focuses on the areas of human modification. Without roads, little large-scale or long-term modification can be done. With roads, construction equipment and materials can be brought and applied to the nearby landscape. Among the most profound of changes is construction of impervious surface area (ISA), which significantly affects downstream hydrographs (Berry and Horton 1974), local climate (Changnon 1992), ecological integrity of stream biota (Kennen 1999), and numerous terrestrial conditions and resources (Trombulak and Frissell 2000).

Over the past 25 years, researchers have tested numerous methods for extracting urban land cover from remote sensing data. Complex urban landscapes present significant challenges because broadly available 30-m and coarser satellite multispectral data mix multiple land covers in single pixels. Progress has been made, however, by focusing on the extraction of ISA fractional area (Wang and Zhang 2004; Wu and Murray 2003), and this work has been complemented with studies using high-resolution images and field checking (Slonecker and Tilley 2004). ISA mapping accuracy is increasing through the use of advanced processing methods such as neural networks and spectral unmixing.

We have recently tested adaptive algorithm approaches that apply simple algorithms in parts of the input parameter space where they deliver accurate results, and more complicated algorithms in areas where simple processes fail. Our estimates of classification error, and therefore our algorithmic segmentation, are based on training data interpreted from 1-m resolution images. This approach is amenable to addition of independent data (expansion of the input parameter space) because data are used only where they are needed. Rather than using fractional ISA coverage, we used binary (ISA present or absent) data for our preliminary analysis. Our study area was Las Vegas, Nevada, covering an area of 2,800 km^2. What follows is an assessment of spatial correlation of DTR and ISA, where ISA was the result of our 90%-accurate extraction method.

Our hypothesis is straightforward: proximity to roads increases likelihood of occurrence of ISA. Figure 34-8 illustrates the cumulative percentage of ISA occurrence as it increases with distance to road. 75% of ISA detected on the scene falls within 100 m of a road, and approximately 85% with 300 m.

Using Figure 34-8 we can easily identify the influence of the road network on ISA development, which brings us to a natural question: if a statistical model can be established correlating DTR and ISA, why not incorporate such a model within the currently used methodology for ISA detection, which is exclusively dependent on multispectral information? By including additional knowledge, increased accuracy may be achieved, especially in areas where materials such as roof tops and soil types have almost identical spectral responses. A pixel classified spectrally as ISA because of its similarity to typical rooftops, but 2 km from the nearest road, might receive a more accurate interpretation as roof-similar soil given the low probability of roof occurrence at this distance from roads. There is also potential to develop inverse interpretations: if a reliable ISA method exists based on multispectral analysis, then spatial filters and pattern recognition algorithms could estimate road proximity expressed as pseudo-DTR. This estimate could also be applied in Bayesian fashion to further refine spectral ISA determinations.

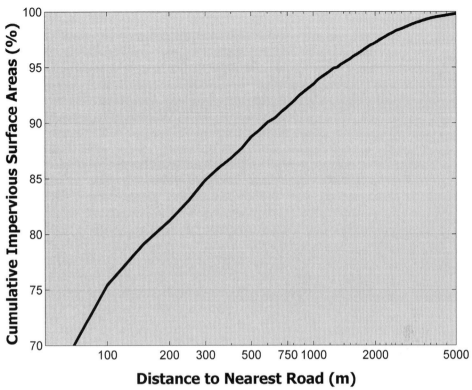

Figure 34-8 Graph showing the spatial association of impervious surfaces with roads. 75% of impervious surfaces in Las Vegas, Nevada, occur within 100 m of roads and 95% within 1.2 km.

34.4 Discussion and Conclusions

We have demonstrated several transformations that usefully extend vector road data across the landscape. Euclidean distance to road (DTR) is the simplest of these, and can be calculated by most GIS software packages. DTR serves not only as a direct measure of open space but also as a proximity device for associating landscape areas with road segments, and setting up for subsequent comparisons. More complex transformations, such as estimated travel time, are also accessible, but require additional input data. In the industrialized world, roads are the spatial template for many human activities, so road proximity helps in the mapping and analysis of these activities. Using Euclidean distance and more complex transformations of vector road data, spatial road-association information can be combined with remote sensing information on a pixel-by-pixel basis. By adding context, this addition may help to resolve ambiguities in land use, land cover, and other pixel-based interpretations, including those of constructed impervious surface area.

Roads will be built. As the road network grows in extent and density, society would benefit from adoption of road-related indicators; roadless volume (RV) has been illustrated as a useful summary spatial indicator. The Euclidean distance and travel-time metrics presented here, when integrated over areas and compared across time or space, can help policy-makers, decision-makers, and the public to understand the geographic distribution and pace of human manipulation of natural conditions.

Acknowledgments

The NORM ED dataset is the foundation for much of our analysis; the GIS work for NORM ED was done by Roger Compton, John McCammon, Carl Rich, Stewart Wright, and Tom Owens of USGS. Colin Homer, Mike Crane, George Xian, and Cory McMahon of the NLCD team at USGS Earth Resources Observation and Science (EROS, formerly EROS Data Center) provided invaluable ISA calibration datasets for our ISA detection algorithm. David Greenlee at EROS first called our attention to the opportunities implicit in the BTS dataset and has supported the development of NORM ED in many ways. Doug Muchoney, Jonathan Smith, and the Geographic Analysis and Monitoring program of USGS provided support for this work, including the National Research Council Postdoctoral Fellowship held by one of the authors (GM). Douglas Ouren of USGS has worked constantly with us on remoteness modeling and other road-related habitat analyses.

References

Berry, B.J.L. and F. E. Horton. 1974. *Urban Environmental Management: Planning for Pollution Control*. Englewood Cliffs, N.J.: Prentice-Hall. 425 pp.

Changnon, S. A. 1992. Inadvertent weather modification in urban areas: lessons for global climate change. *Bulletin - American Meteorological Society* 73(5):619–627.

Dijkstra, E. W. 1959. A note on two problems in connection with graphs. *Numerische Math.* 1:269–271.

Forman, R.T.T. and L. E. Alexander. 1998. Roads and their major ecological effects. *Annu. Rev. Ecol. Syst.* 29:207–231.

Forman, R.T.T. and R. D. Deblinger. 2000. The ecological road-effect zone of a Massachusetts (U.S.A.) suburban highway. *Conserv. Biol.* 14(1):36–46.

Fritz, S., S. Carver and L. See. 2000. New GIS approaches to wild land mapping in Europe. In *Proceedings, Conference on Wilderness Science in a Time of Change*. USDA Forest Service, Ogden, Utah. USDA Forest Service Proceedings RMRS-P-15-VOL-2. 2000. <http://www.fs.fed.us/rm/pubs/rmrs_p015_2/rmrs_p015_2_120_127.pdf> Accessed 30 August 2007.

Galster, G., R. Hanson, M. R. Ratcliffe, H. Wolman, S. Coleman and J. Freihage. 2001. Wrestling sprawl to the ground: defining and measuring an elusive concept. *Housing Policy Debate* 12(4):681–717. <http://www.fanniemaefoundation.org/programs/hpd/pdf/HPD_1204_galster.pdf> Accessed 30 August 2007.

Gelbard, J. L. and J. Belnap. 2003. Roads as conduits for exotic plant invasions in a semiarid landscape. *Conserv. Biol.* 17(2):420–432.

Hasse, J. 2004. A geospatial approach to measuring new development tracts for characteristics of sprawl. *Landscape Journal* 23(1):52–67.

Havlick, D. G. 2002. No Place Distant : Roads and Motorized Recreation on America's Public Lands. Washington, D.C.: Island Press. 297 pp.

Kennen, J. G. 1999. Relation of macroinvertebrate community impairment to catchment characteristics in New Jersey streams. *J. Am. Water Resources Assn.* 35(4):939–955.

Langer, W. H., N. S. Fishman, D. H. Knepper, Jr., D. A. Lindsey, C. S. Mladinich, L. D. Nealey, S. G. Robson, J. E. Roelle and D. R. Wilburn. 2000. The Front Range Infrastructure Resources Project – An Overview. <http://rockyweb.cr.usgs.gov/frontrange/overview.htm> Accessed 21 September 2007.

Mace, R. D., J. S. Waller, T. L. Manley, K. Ake and W. T. Wittinger. 1999. Landscape evaluation of grizzly bear habitat in western Montana. *Conservation Biology* 13(2):367–377.

Sethian, J. A. 1999. Level set methods and fast marching methods : evolving interfaces in computational geometry, fluid mechanics, computer vision, and materials science. Cambridge, UK: Cambridge University Press. 378 pp.

Slonecker, E. T. and J. S. Tilley. 2004. An evaluation of the individual components and accuracies associated with the determination of impervious area. *GIScience and Remote Sensing* 41(2):165–184.

Trombulak, S. C. and C. A. Frissell. 2000. Review of ecological effects of roads on terrestrial and aquatic communities. *Conserv. Biol.* 14(1):18–30.

US Census Bureau. 2005. TIGER®, TIGER/Line® and TIGER-Related Products. <http://www.census.gov/geo/www/tiger/index.html> Accessed 30 August 2007.

USGS. 2002. BTS road vectors, Seamless Data Distribution System <http://seamless.usgs.gov> Accessed 30 August 2007.

van Wagtendonk, J. W. and J. M. Benedict. 1980. Travel time variation on backcountry trails. *J. Leisure Research* 12(2):99–106.

Wang, Y. and S. Zhang. 2004. A SPLIT model for extraction of subpixel impervious surface information. *Photogrammetric Engineering and Remote Sensing* 70(7):821–828.

Watts, R. D., R. W. Compton, J. H. McCammon, C. L. Rich, S. M. Wright, T. Owens and D. S. Ouren. 2007. Roadless space of the conterminous United States. *Science* 316(5825):736–738.

Wu, C. and A. T. Murray. 2003. Estimating impervious surface distribution by spectral mixture analysis. *Remote Sensing of Environment* 84(4):493–505.

CHAPTER 35

The Role of Remote Sensing and GIS for Wildland Fire Hazard Assessment

James E. Vogelmann, Donald O. Ohlen, Zhi-liang Zhu, Stephen M. Howard and *Matt G. Rollins*

35.1 Introduction

Wildland fire plays a natural role in many of the Earth's landscapes, and benefits the environment in several ways. Historically, wildland fire has performed an important set of ecological functions, including the maintenance of forest structural and compositional heterogeneity, biodiversity, and soils (Keane et al. 2002; Agee 2000). Some historical wildfires have been exceptionally large when allowed to burn unchecked. In North America, for example, the National Interagency Fire Center (NIFC Undated) lists seven separate fires that each burned 400,000 hectares (1,000,000 acres) or more between 1825 and the present, including the 1988 Yellowstone fire, which burned 634,000 hectares (1,585,000 acres). Wildfires occur naturally in many landscapes throughout North America (Figure 35-1), Australia, South America, Asia, Africa and southern Europe (Justice et al. 2003; Mistry 1998; Soja et al. 2004; Russel-Smith et al. 2003; Csiszar et al. 2005; Hoelzemann et al. 2004).

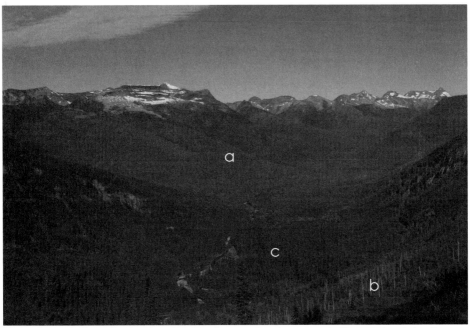

Figure 35-1 Northern Rocky Mountain landscape showing several different stand structures modified by recent fire history of the region. Brown areas in the background (a) are related to the 2003 Trapper fire in Glacier National Park in Montana. The dead trees with smaller living trees in the lower right (b) indicate an area affected by a fire in 1967. The oldest and most mature forest stands, which have not been recently affected by fire, can be seen as dark green areas in the valley (c). See included DVD for color version.

In the western United States, well-intentioned fire suppression policies implemented throughout most of the 20th century resulted in gradual increases in surface and crown fuel loadings in many of our forests (Figure 35-2). These policies helped create conditions that today can result in large, severe and catastrophic wildland fires (Peterson et al. 2005; Allen et al. 2002). The problem of fire fuels buildup has become an especially thorny issue for land use managers. The scientific community and much of the general public recognize the value of naturally occurring wildfires for maintaining and restoring valuable ecosystem goods and services. Nonetheless, many people still choose to live in these fire-prone areas in spite of the known risks. Suppression of wildfires encourages the buildup of fire fuels, and increases the likelihood of even more severe fires in the future, but removal of fire fuels substantially alters the very landscape characteristics that residents find so appealing. During the last several decades, the population has increased in the fire-prone ecosystems of the western United States, greatly expanding the wildland urban interfaces (WUI) where wildfire can have detrimental effects on communities. Many recent wildfires throughout the region have resulted in substantial loss of life and property. Not surprisingly, there have been societal pressures to "do something" about the problem and mitigate the adverse effects of burning where wildfire is a common and natural occurrence.

Figure 35-2 A representative fuel loading for a higher elevation conifer forest within the western United States. See included DVD for color version.

In response to recent wildfires, a number of public policies and initiatives have been implemented in the United States that have a major impact on the nation's management of fire and wildlands. These include the National Fire Plan (NFP 2007) and the Healthy Forest Initiative (NFP 2007). Both the National Fire Plan and the Healthy Forest Initiative were initiated after several seasons of severe wildland fires, and are intended to ensure appropriate responses to wildland fires and to help mitigate the effects of the fires on local communities in the future. The plans address firefighting, rehabilitation, fuels reduction, community assistance, and accountability. As part of the National Fire Plan, the USDA Forest Service and the Department of the Interior were charged with providing technical, financial and resource guidance and support for wildland fire management across the United States. The Healthy Forests Initiative was intended to use the best science available to help restore United States public lands to healthy conditions.

The implementation of wildland fire policies can be controversial, but the need for high quality spatial information is undisputed. Detailed knowledge about the status and conditions of our natural resources in fire-prone ecosystems helps local land managers to initiate and implement fire hazard analyses for their communities. Such information also provides broad overviews to regional and national managers, and gives them the building blocks to develop long-term plans, including resource allocation strategies.

Access to current and accurate spatial information in a GIS is especially useful to a manager dealing with fire-related issues. Much of the spatial information relevant to fire management issues is reasonably stable through time, such as roads, topography, political boundaries and waterways, but some of the requisite information changes relatively frequently and requires updates on a regular basis. Examples of frequently changing information include vegetation type, structure and condition for both the overstory and understory of forests. Much of this can be derived from remote sensing data in conjunction with other spatial information, along with detailed field data. The integration of remote sensing information into a GIS is especially powerful for natural resource managers, because it affords the opportunity of combining the static and non-static data layers in ways that are optimal for the project at hand. The incorporation of remotely sensed data into a GIS has become more and more common within the fire community (Chuvieco 2003; Keane et al. 2001; Schmidt et al. 2002; Perry 1998).

We are currently conducting two major, interrelated national wildland fire projects that rely heavily on remote sensing data and GIS technology. These are the (a) Monitoring Trends in Burn Severity and (b) LANDFIRE projects. Our goal in this chapter is to summarize both projects, and to describe the relevant geographic information systems (GIS) applications of each to the resource manager. Both projects are large, multi-agency efforts and will take several years to complete. As of 2006, the projects are in the early phases of implementation, and data for specific regions are just now becoming available. Ultimately, LANDFIRE and Burn Severity products will be available for the entire United States.

35.2 Monitoring Trends in Burn Severity

35.2.1 General Overview of Burn Severity and Wildland Fire

The burn scars from wildland fires are readily discernible on satellite images, especially in the Earth's drier landscapes (Figure 35-3). Within an image, there may be many individual burn scars, representing a range of sizes, shapes and patterns. Burn scars can have variable spectral characteristics. In many cases, the burn scars are visible in the imagery even when the fires that caused them occurred many years before image acquisition. The spatial and spectral patterns of the burn scars can provide much site-specific information about actual vegetation structure and composition. In order to maximize the usefulness of this information, however, a solid understanding of the linkages between ground-based site characteristics and satellite-based image data is needed.

Wildland fires are influenced by several natural physical factors. Fuel type and condition, weather and topography all play major roles in determining how fires behave. Fire fuels that contain fine structure provide a flash ignition source and the capacity to carry a fire line at a rapid pace. Conversely, fuel that contains coarse woody material can sustain a fire for longer periods of time. The vertical profile of fuels influences whether a fire maintains surface burn characteristics or is carried into the crown of the overstory vegetation. Weather elements, including atmospheric moisture, temperature and wind, are important factors in fire behavior. Sustained drought or low relative humidity can determine live and dead fuel moisture content, with drier fuel more volatile to ignition and burning. Wind increases the amount of oxygen to the fire and influences the direction and speed that a fire advances.

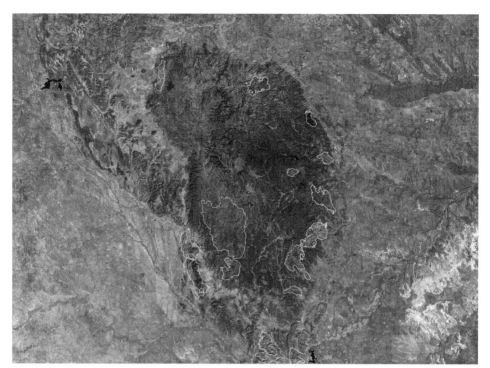

Figure 35-3 Mosaic of four Landsat Thematic Mapper (TM) scenes covering the Black Hills of South Dakota and Wyoming. Perimeters illustrating locations of fires since the 1980s are white. See included DVD for color version.

Topographic elements of slope and aspect influence the rate and direction of fire spread, as fire generally advances at a more rapid rate on upslope areas. This is because the fuels are being preheated on the upslope; the upslope fuels are closer to the flame front, and rising heat from the fire creates upslope winds that help drive the fire line. Aspect also affects wildland fire behavior; southern and southwestern slopes are typically warmer and dryer than other aspects. This in turn relates to the type and condition of the available fuels in these areas. Collectively, these physical fire elements create a great deal of variance in fire behavior and therefore result in variance in spatial burn patterns. Many of the factors that affect wildland fire behavior can be readily stored in a GIS and used by natural resource managers to evaluate and predict wildland fire conditions and effects.

Vegetation response to fire can vary greatly within a fire boundary, depending on many factors, including exposure to the heating factor of the fire line, differences in plant species, and pre- and post-fire weather. The heating factor includes both the highest temperature reached and the duration of time exposed to a particular temperature. Higher temperatures and longer durations may result both in the death of plants and the destruction of seeds. Many plant species that grow in fire-prone areas have adaptations for reproducing and regenerating post-fire to ensure survival. Some tree species adapt by having thicker bark as insulation against the heat, while other species have seed cones that only open after exposure to heat, and still other plant species depend upon the ability to produce new growth by re-sprouting from a root system or tubers. Weather determines the condition of the vegetation before the fire; if a plant was water-stressed before exposure to the fire, it will likely have more difficulty surviving. Post-fire weather conditions determine moisture conditions before or during the next growing season. Fires also affect plant species composition: in the post-fire competition for resources, perennial plants have the advantage of an already established root system. After a wildland fire occurs, it may take a number of years to fully understand the total ecological effects on vegetation.

Burn severity as used here can be defined as the degree of environmental change caused by the fire. The ecological effects of fire range from low to high, and depend upon the interactions of many different physical and biological factors. An example of a "low impact" fire would be one that moves through the understory of a mature and reasonably open "park-like" forest, where the overstory vegetation is relatively unaffected, and the understory vegetation burns, but comes back relatively quickly (i.e., within a year or two). An example of a "high impact" fire would be the type of fire that has been prevalent throughout much of the western United States in recent years, where fuel loading is high because of fire suppression activities, and all vegetation, including the overstory, is significantly (and maybe even totally) burned. This type of burning can be followed by slow re-vegetation and high erosion. It is difficult to characterize and quantify the impact of fire to ecosystems, but we can use fuel consumption and the response of vegetation to the fire as relative measures that indicate burn severity. Mapping the spatial patterns of burn characteristics is necessary to understand landscape dynamics and to provide information to land managers and scientists. Remotely sensed data processed in a GIS can be used to measure the gradient of change resulting from fire.

The impacts of fire are not limited to destruction of vegetation. With the inception of a "wildland urban interface," where homes are built close to forests, fire's impacts have expanded to include a social and economic component (Radeloff et al. 2005). Between 2000 and 2004, fire destroyed on average over 1,000 homes each year.

After a fire has been extinguished, an assessment of burn severity characteristics can be completed and then used to identify potential impacts, such as increases in flooding or mudslides. Identifying fire-impacted areas—where vegetation has been removed by fire, erodable soils are on steep slopes, and areas that are upslope from human habitation—is critical for planning any rehabilitation and restoration. Remote sensing and GIS methods are useful for these assessments.

35.2.2 Monitoring Trends in Burn Severity—Background

Historically, fire crews have sketched boundaries of burn scars and fire severity on topographic maps based on ground observations or from aircraft. Both methods are costly and inefficient, often providing simple line drawings that show little detail. With the advent of satellite imagery, fire perimeters and severity may be mapped at finer detail. For example, a 1,000 hectare (2,500 acre) fire is composed of over 14,000 individual picture elements in a Landsat Thematic Mapper (TM) or Enhanced Thematic Mapper Plus (ETM+) satellite image; each one represents some level of fire severity. For the satellite image to be meaningful, however, the data that the satellite acquires must be correlated with observed conditions on the ground (i.e., field reference information). The fire effects crew typically visits each wildland fire burn and assesses fire severity information, and this field reference information is then used to "ground-truth" the image. The images can be used to update the vegetation and fire history layers in a geographic information system.

Yellowstone National Park has sporadic records of fires dating back to the 1880s. After 1930, records were consistently kept, although differences in filing and storage over the years prevented their general use. Beginning in 2000, the Fire Management Office researched and collected all the information from the archives and created a systematic database consisting of narrative and spatial data. Fire perimeters were created in the park's geographic information database for fires larger than 40 hectares (100 acres). Smaller fires were mapped as points. All related information for each fire was transcribed into a searchable database.

The National Park Service has expanded upon these initial Yellowstone National Park burn mapping activities and currently maintains the Fire Effects Monitoring Program (NIFC Undated) to monitor the effects of natural and prescribed wildland fires in all parks having fire management activities. Yellowstone National Park has now had an official fire effects monitoring program since 1998. The fire effects monitoring crew collects information on the long-term effects of fire and fire management activities. The crew monitors fuel loads, plant populations, tree regeneration, exotic species and other aspects of the park's ecosystems. Monitoring ensures that management objectives are met and that adverse effects are minimized.

35.2.3 Current Burn Mapping Activities

Remote sensing and GIS technologies have become increasingly valuable tools for assessing the impacts of fire on the landscape. Historically, fire severity characteristics have been mapped and evaluated by ground observation or through aircraft reconnaissance, but these practices become especially difficult for large, spatially complex fires. Satellite imagery can provide accurate post-fire assessments with sufficient precision for scientists and land managers, and imagery represents a viable solution for mapping wildland fire locations and conditions (Diaz-Delgado et al. 2004; Ruiz-Gallardo et al. 2004; Epting et al. 2005; Brewer et al. 2005; Lopez Garcia and Caselles 1991). Many platforms are available for the collection of remote sensing data: airplanes, helicopters, unmanned aerial vehicles (UAVs) and satellites. Helicopters, airplanes and UAVs are typically used to gather imagery to support on-going fire fighting activities, while pre- and post-fire aerial photography and satellite images are used to plan mitigation efforts and to assess impacts.

Satellite data are good sources of fire-related information. For burn severity work, our primary sources of satellite data are the Landsat TM and ETM+ sensors. We rely heavily on TM and ETM+ because (1) the spatial resolution (30-m pixels) is appropriate for burn mapping, (2) the spectral bands of the sensor, which include bands in the visible, near infrared and middle infrared portion of the electromagnetic spectrum, are appropriate for detecting, mapping and monitoring burn areas, and (3) there is a large archive of TM and ETM+ data sets, acquired from 1984 to the present, from which to select imagery. When Landsat TM and ETM+ imagery is not available, data from other satellite sensors, such as SPOT and ResourceSat, can be used as alternatives.

A number of multi-band indices are useful for characterizing various vegetation conditions. For example, the Normalized Difference Vegetation Index (NDVI) is useful for assessing the amount and vigor of green growing vegetation. NDVI combines spectral data from the red and near infrared bands using this formula:

$$\text{NDVI} = (\text{Infrared Band} - \text{Red Band}) / (\text{Infrared Band} + \text{Red Band}) \qquad (35\text{-}1)$$

or, using Landsat TM/ETM+ band nomenclature:

$$\text{NDVI} = (\text{Band 4} - \text{Band 3}) / (\text{Band 4} + \text{Band 3}). \qquad (35\text{-}2)$$

A similar index, the Normalized Burn Ratio (NBR) is particularly useful for mapping areas burned by wildland fires and for assessing fire's effects on the landscape (Key and Benson 2006). The NBR is formulated similarly to the NDVI but instead substitutes a mid-infrared band (TM/ETM+ Band 7) for the red band (Band 3):

$$\text{NBR} = (\text{Band 4} - \text{Band 7}) / (\text{Band 4} + \text{Band 7}) \qquad (35\text{-}3)$$

Recently burned areas appear very dark in the NBR images. Normal vegetation appears much brighter (Figure 35-4).

Fire can dramatically change the land surface. To measure land surface change and monitor the recovery of vegetation, we can generate an NBR image for the area prior to the fire and compare it with an NBR image generated after the fire. This is the basis of the Differenced Normalized Burn Ratio, or dNBR (Figure 35-5), which can be used to map the extent of the fire and estimate its severity. In practice, there are several times when a dNBR analysis of

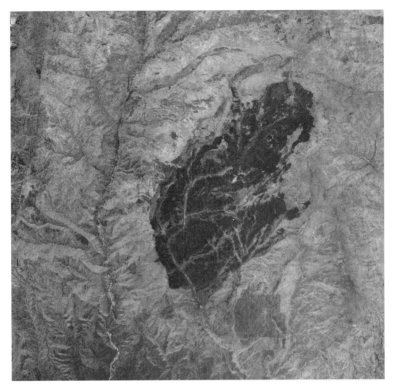

Figure 35-4 Normalized Burn Ratio (NBR) data and ground photo acquired for the 2000 Viveash fire in the Santa Fe National Forest of New Mexico. The Viveash fire started on May 15, 2000 and burned an estimated 9,360 hectares (23,400 acres). The Landsat 5 TM data set was acquired on 5 June 2001. The ground photo from the Viveash burn scar shows a typical mosaic pattern of burn characteristics. See included DVD for color version.

burn severity may be conducted. We conduct dNBR analysis at three levels of assessment: rapid, initial and extended.

A rapid assessment is conducted as soon as possible after the fire is contained. Rapid assessment may be hindered by the still-burning fire and the imagery may be partially obscured by smoke and haze. The rapid assessment is used by Burned Area Emergency Response (BAER) teams to help formulate emergency plans to control erosion and to protect lives, property and resources. The BAER teams require the dNBR data as quickly as possible to help identify the areas most affected by the fire and to guide mitigation efforts.

Initial assessment is less urgent than rapid assessment, and may be conducted up to several months after the fire has been extinguished. This type of assessment is typically done for grass fires, because grasses can recover fairly quickly and the fire scar may not be easily detectable on satellite imagery after a few months. The initial assessment can thus be used to assess vegetation recovery rates.

Manual of Geographic Information Systems

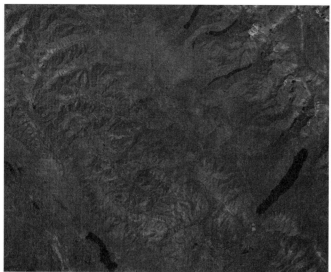

Figure 35-5a Subset of a Landsat 7 Enhanced Thematic Mapper Plus (ETM+) data set, acquired on 7 July 2001, for a region of Glacier National Park. Dark green areas depict dense coniferous forest; light green areas depict zones dominated by other types of vegetation, such as deciduous forest. A few small burned areas (maroon) can be seen, but most of the image is dominated by healthy forest. Light blue areas in the upper right represent extant patches of snow. See included DVD for color version.

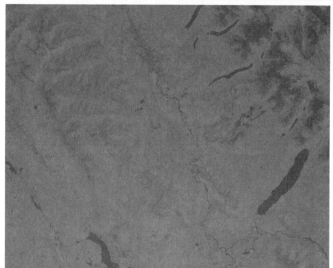

Figure 35-5b Normalized Burn Ratio calculated from the Landsat ETM+ image shown in Figure 5a. Dark tones represent areas of water or sparsely vegetated alpine areas and rocks, whereas lighter tones represent normal "healthy" forest conditions. Some variation in tone is to be expected to occur in healthy forests, which can be see throughout the image.

Figure 35-5c A subset of a Landsat 7 ETM+ scene acquired on 10 July 2002, covering the same region as shown in Figure 5a. The area impacted by the Moose Fire, which burned approximately 26,670 hectares (66,680 acres) of Glacier National Park in August, 2001, is depicted in maroon. See included DVD for color version.

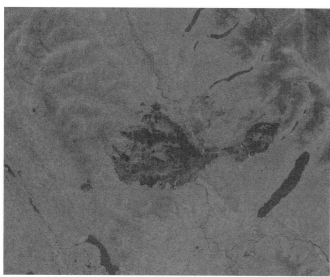

Figure 35-5d Normalized Burn Ratio data set calculated from the 10 July 2002 Landsat ETM+ image shown in Figure 5c. Dark tones in the center of the image represent previously forested areas that now have very low Normalized Burn Ratio (NBR) values due to wildland fire. These areas were much brighter in the pre-burn image (compare with Figure 5b). Other dark tones represent areas of low vegetation cover, including lakes and sparsely vegetated high elevation alpine areas and rocks (also present in Figure 5b). Lighter tones represent healthy forests (i.e., not recently burned).

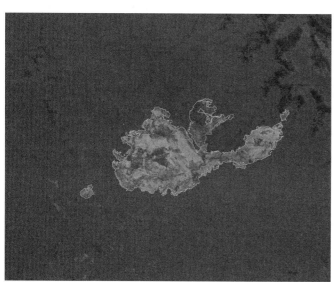

Figure 35-5e The Differenced Normalized Burn Ratio data set for the Moose fire. The fire is outlined with an image-interpreted fire perimeter. There are different shades of gray within the perimeter: darker gray areas are where burn severity was relatively low; lighter gray areas are where burn severity was higher. See included DVD for color version.

An extended assessment is the most common type of dNBR analysis. The post-fire image is acquired at the peak of greenness during the following growing season, whereas the pre-fire image is acquired during a previous peak of green season. The rationale of the 'extended assessment' is that delayed mortality and vegetation survival and recovery cannot be detected directly after the fire, but instead become apparent during the next growing season.

Field studies are ongoing, but preliminary results have shown the dNBR to be well correlated to the severity of impacts noted at sample plots located in burned areas. The dNBR images can then be color-coded to show the variations of burn severity within the fire perimeter. Burn perimeters are also easily delineated with standard GIS tools. Once the perimeter and burn severity of a fire have been delineated in a GIS (Figure 5e), the total acres of high, medium and low burn severity can readily be tabulated. If a vegetation map is available, the types and areas of vegetation burned at different severity levels can be assessed, and the vegetation map can be updated based upon the size and severity of the fire. Future vegetation types and conditions can be projected with successional models.

Areas of very low burn severity can be more difficult to discern from satellite imagery than from the ground. Likewise, ground fires under a closed forest canopy are difficult to map with satellite imagery. Despite these potential problems, land managers have embraced the burn mapping methodology, and are currently using the satellite-derived burn severity data sets to manage their natural resources. Since 2000, the United States National Park Service (NPS) and United States Geological Survey (USGS) have used the dNBR methodology to map burn severity for over 260 fires, each at least 500 acres in size, that occurred on National Park Service lands.

The use of Landsat imagery enables scientists to assess the burn severity of historical fires (Figure 35-6). The archive of Landsat imagery extends back to the early 1970s, but the NBR requires information from the middle infrared region of the spectrum (band 7). Because Landsats 1, 2, and 3 did not acquire middle infrared data, NBR and dNBR assessments cannot be made using satellite data prior to 1982, when Landsat 4 was launched. Other fire-related trend information, however, can be gleaned from early Landsat data. Using the Landsat archive, the NPS and USGS have compiled a Landsat-based 'fire atlas' for a number of National Parks in the United States, including Glacier, Mesa Verde, Sequoia-Kings Canyon, Yellowstone-Grand Tetons, Yosemite, Grand Canyon and Yukon-Charlie.

The fire atlas not only enables the assessment of the burn severity of a particular fire, but also includes scenes acquired many years afterwards. These allow resource managers to monitor and assess the recovery of the burned landscape and the effectiveness of rehabilitation efforts. The imagery also aids fire planning efforts, such as fuel reduction and tree thinning to protect infrastructure, safety zones, and escape routes. The success of the NPS/USGS burn severity mapping program and the usefulness of the fire atlas were the impetus for the start of a five year program to map all the fires meeting certain size criteria that have occurred in the United States since 1984. The USGS and USDA Forest Service are collaborating to process at least 8,000 Landsat scenes to generate burn severity information for all historical fires larger than 200 hectares (500 acres) in the east, and 400 hectares (1,000 acres) elsewhere. This information will allow researchers to study the fire history of the United States at local, regional and national scales, and will help monitor fire trends and assess the effectiveness of programs to reduce fire occurrence and severity.

Figure 35-6 This series of images represents the current collection of Landsat data for the Mesa Verde National Park burn severity fire atlas. Reddish tones show areas of recent burns. The 1970s data are Landsat Multispectral Scanner (MSS). The remaining satellite images are Landsat TM and ETM+. The last image is shaded relief derived from a digital elevation model (DEM) with 10-meter resolution. See included DVD for color version.

35.3 LANDFIRE

35.3.1 General Overview of LANDFIRE

The Monitoring Trends in Burn Severity project maps wildland fires and determines the ecological effects of these fires across the United States. The LANDFIRE project has a different goal: to assess the current potential for fire in the Nation's wildlands (LANDFIRE 2007). The intent of LANDFIRE is to provide high resolution spatial data to help evaluate fire hazard status and to plan responses to wildland fires. The project will provide consistent and comprehensive maps and predictive models that describe vegetation, fire and fuels characteristics across the United States. This will include identifying areas that are at risk because of the accumulation of hazardous fuel, prioritizing fuel reduction projects, modeling real-time fire behavior to support decisions being made in the field during fires, and modeling potential fire behavior and effects to strategically plan projects for hazardous fuel reduction and restoration of ecosystem integrity. Spatial data layers being developed include existing vegetation type, vegetation structure, biophysical gradients, biophysical settings, fire regime condition class, and fire fuels.

The LANDFIRE project, which began in 2003, is a five-year joint effort of the USDA Forest Service, the Department of the Interior, and The Nature Conservancy. LANDFIRE data will be available first for the western United States, followed by the eastern and central United States, Alaska and Hawaii.

35.3.2 LANDFIRE—A Brief History

Fire managers have for many years recognized the need to reduce excessive fuel accumulations in order to lessen the threat of catastrophic wildfires (USDA Forest Service 2000), but the development of fire management plans has been hampered by a lack of national-level spatial data about wildland fire and fuel conditions (GAO 2002). This recognition has led to a series of investigations (Schmidt et al. 2002) in which 1-km resolution spatial data sets were produced for the conterminous United States in support of national-level fire planning risk assessments. Data layers included potential natural vegetation, current cover type, historical natural fire regimes, and current fire regime condition class (FRCC). FRCC depicts the degree of departure from historical fire regimes, and offers an interesting spatial perspective on where key ecosystem characteristics have been significantly altered. The process used to generate these data layers included the integration of biophysical data, remote sensing data, disturbance, and succession information in a GIS, with input from regional ecologists and fire managers. A significant percentage of the United States is in FRCC Class 3, which depicts the greatest departures from expected historical ecological conditions.

Although these coarse-scale products were an important first step toward providing land managers with spatial information to support fire risk assessments, data with greater spatial detail were needed to better meet the needs of the fire management community. This led to the current LANDFIRE effort, in which most of the same principles employed for the coarse-resolution analyses are now applied to generate finer-resolution spatial data sets.

The LANDFIRE project offers a suite of seamless data products that wildland fire managers need in order to identify lands and communities with hazardous fuel build-up and/or extreme departure from historical conditions. These data also aid in setting priorities for ecosystem restoration and hazardous fuel reduction treatments to protect ecosystems, property and people. However, the generation of fine-scale spatial fire products for the entire United States is no simple task. Before LANDFIRE began, a "LANDFIRE Prototype" project was conducted (Rollins and Frame 2006), with the goal of determining the best methods for generating wall-to-wall fine-scale fire risk information for the entire United States. Two pilot areas were selected for analyses: (1) an area of over 2.8 million hectares (7 million acres) in central Utah that included the Wasatch and Uinta Ranges, and (2) an area of over 4 million hectares (10 million acres) in western Montana and north central Idaho in the north-central Rocky Mountains. After the LANDFIRE prototype was completed, the project shifted towards national implementation, in which LANDFIRE data sets are being generated for the entire United States.

35.3.3 Requisite Field Information

Ground reference data are important to any remote sensing-based land cover mapping and modeling effort (Congalton and Biging 1992). Accordingly, LANDFIRE allocates much time and effort to the acquisition of high quality field plot data. These data are maintained in the LANDFIRE Reference Database that is used during several phases of the study (Caratti 2006). The database supplies field or ground-truth data for the vegetation mapping and modeling processes, and for assessing general accuracy of the final products. The LANDFIRE reference database integrates existing field data from other projects. Gaps in existing data are filled through coordinated field campaigns in target areas.

Each field plot has a series of attributes, including a plot identification number, data source, reference coordinates, acquisition date, vegetation type, life form category (i.e., tree, shrub or herbaceous), percent canopy cover, dominant species, height, and fuels information. For many plots, on-site photographs are available as well.

Data from many sources are incorporated into the reference database. The bulk of the points come from several national-scale efforts, including the Forest Inventory and Analysis National Program (FIA 2007), the GAP Analysis Program (GAP Undated), and data collected by the Student Conservation Association (SCA 2007) specifically for LANDFIRE. In general, the various field data sets were collected for different applications, by different groups, and during different years. As a result, plot data vary in quality. Each field point used for mapping and modeling has to pass a series of quality assurance/quality control (QA/QC) steps to ensure that the plots to be used in the analyses are reasonable. In the LANDFIRE QA/QC process, plots are overlaid on satellite imagery and visually assessed as to whether the points make logical sense. For example, a ponderosa pine plot that falls on a lake or a road is clearly in error and is therefore removed from the analysis. Plot data are also compared with other sources of information, and if inconsistencies are found, the plots are discarded. Some plots are located in areas that have recently changed, such as a forest plot located in an area that has recently burned. In these cases, the vegetation information in the reference database does not accurately reflect the current conditions, and these plots are also discarded. The QA/QC approach is intended to provide the best possible field data set. The field data are in turn used to develop the most accurate map products possible. Typically, many thousands of field plots are used to generate maps for a given region (Figure 35-7).

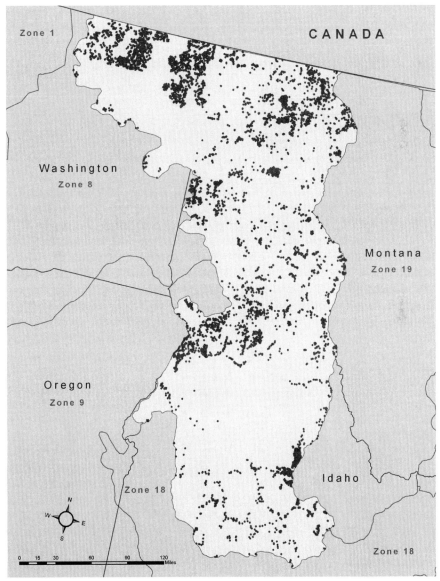

Figure 35-7 Location of field plots used for classification of the Northern Rocky Mountain map zone. Plot locations exclude Forest Inventory and Analysis (FIA) plots (not shown because of issues of confidentiality). See included DVD for color version.

Manual of Geographic Information Systems

35.3.4 Spatial Data Layers being Generated by LANDFIRE

The LANDFIRE project is developing maps and data sets on a map-zone-by-map-zone basis (Figure 35-8). These mapping zones are the same as those used by the National Land Cover Data 2000 project (Homer et al. 2004), and are based on the principle that the efficiency and accuracy of land cover classification is improved when it is based on individual landscape units that are relatively homogeneous with respect to landform conditions, vegetation parameters, spectral reflectance and other variables. Zonal data sets are merged and made available to users in an online, seamless data distribution system (LANDFIRE 2007). LANDFIRE now generates many spatial data layers at 30-m resolution. These fall under several major subheadings, and are described in greater detail below. Examples of selected data layers are shown in Figures 35-9, 35-10 and 35-11. A description of the methods used to develop these data layers is beyond the scope of this chapter; for more information, please see Rollins and Frame (2006) and LANDFIRE (2007).

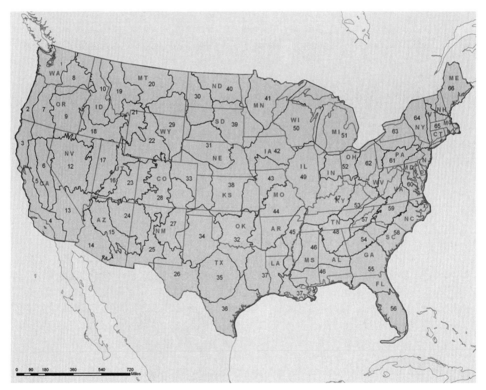

Figure 35-8 Map zone boundaries used for LANDFIRE project. Mapping is being done on a map-zone-by-map-zone basis, starting with the western part of the country. See included DVD for color version.

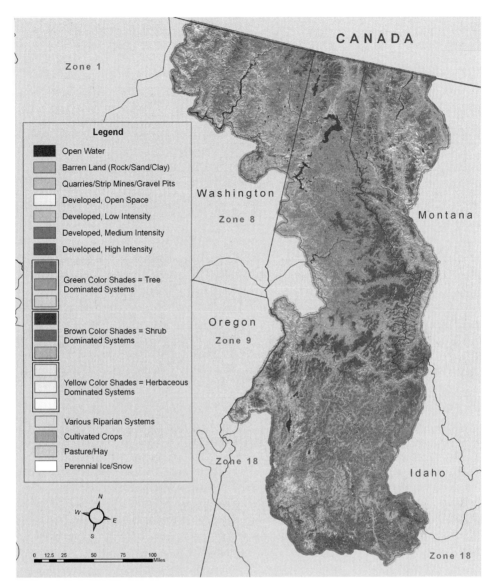

Figure 35-9 Existing vegetation map of the Northern Rocky Mountains map zone. For each map zone, there are typically too many natural vegetation classes to portray effectively on a single map. In this rendition, tree classes are in shades of green, shrub classes in shades of brown, and herbaceous classes in shades of yellow. A typical LANDFIRE map zone might have 30-35 types of natural vegetation. See included DVD for color version.

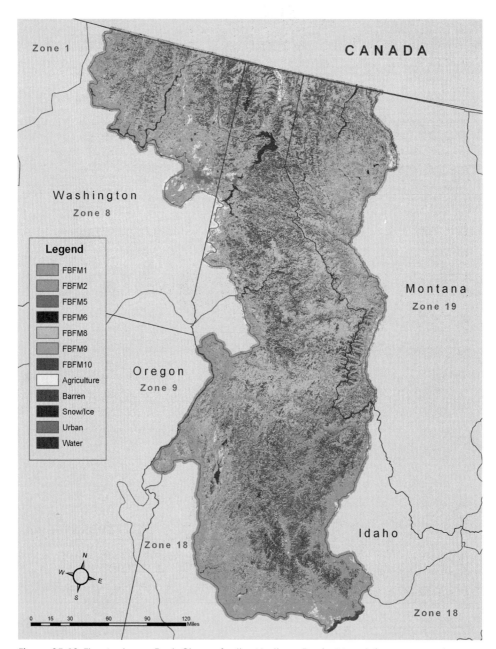

Figure 35-10 The Anderson Fuels Classes for the Northern Rocky Mountain map zone. In general, higher class numbers represent denser and more closed conditions (e.g., dense forests) with higher fuel loadings than low class numbers (e.g., sagebrush-dominated shrubland). See included DVD for color version.

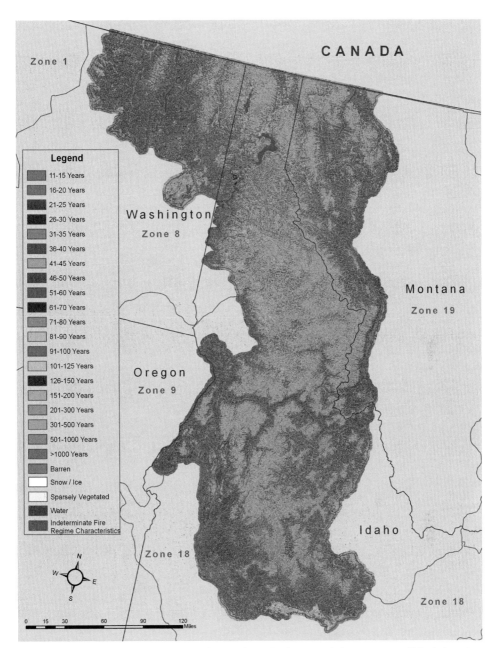

Figure 35-11 Mean fire return interval for Northern Rocky Mountain map zone. This data set indicates where fire events are expected to be frequent under historical conditions. For this zone, the regions of highest fire frequency are located at lower elevations, which historically have been dominated by relatively sparse forests and shrublands. See included DVD for color version.

35.3.4.1 Existing Vegetation Mapping Data Layers

LANDFIRE generates three different data layers of existing vegetation: existing vegetation type (EVT), percent canopy cover and canopy height. All three data layers are needed to develop fire fuels models and FRCC maps.

The classification legend used for EVT is the Ecological Systems Classification developed by NatureServe (2007). This is a mid-scale classification that was developed largely for conservation and management planning. Ecological systems represent recurring groups of biological communities that are found in similar physical environments. The classes can be identified by researchers in the field, and mapped from remote sensing data. Approximately 600 ecological system classes have been identified for the United States (Comer et al. 2003), although many of the rarer classes will not be mapped by LANDFIRE. An example of an EVT data layer for the Northern Rocky Mountains map zone is shown in Figure 35-9.

35.3.4.2 Potential Vegetation Data Layers

Two different types of potential vegetation data layers are being produced: the LANDFIRE environmental site potential (ESP) and biophysical settings (BpS). The ESP layer represents the vegetation that could be supported at a given site based on the biophysical environment. Similar to EVT, the ESP data layers are based on ecological systems (Comer et al. 2003). The ESP layers are generated using a predictive modeling approach that relates spatially explicit layers, representing biophysical gradients and topography, to field training sites. The BpS layer represents the vegetation that may have been dominant on the landscape prior to Euro-American settlement. It is essentially a refinement of the ESP layer that incorporates current knowledge about ecological processes. These data layers have the same overall appearance as the EVT data, but are generated using complex vegetation models without satellite imagery.

35.3.4.3 Fire Behavior Data Layers

A series of data layers is being developed and used to model fire behavior. These include a series of fire fuel models, such as the Anderson Fire Behavior Fuel models (1982) and the Scott and Burgan Fire Behavior Fuel models (2005). Fire behavior fuel models represent distinct distributions of fuel loading found among live and dead surface fuel components, size classes and fuel types. These can be used as inputs to models of surface fire behavior, and can be important for predicting where wildland fires will likely spread and at what rates the fires will move. In addition to the fuels models, other data layers pertinent to fire behavior modeling are being developed, including forest canopy bulk density and forest canopy base height. These are generally analyzed in conjunction with digital elevation model data, slope and aspect data, and weather data. Figure 35-10 shows the Anderson Fire Behavior Fuel models for the Northern Rocky Mountains map zone.

35.3.4.4 Fire Regime Data Layers

Included in the suite of fire regime data layers are FRCC (Hann and Bunnell 2001; Hardy et al. 2001), fire regime groups, mean fire return interval, and succession classes. Of these, the FRCC data set is arguably the most important data layer, and perhaps the most controversial one. FRCC represents a synthesis of many of the other LANDFIRE data sets. It represents in three levels the degree to which current vegetation has departed from the historical vegetation reference conditions: low departure from the normal condition (FRCC 1), moderate departure (FRCC 2) and high departure (FRCC 3). LANDFIRE simulates vegetation reference conditions using the vegetation and disturbance dynamics model LANDSUM (Keane et al. 2002), and current vegetation conditions are derived from the Landsat-derived vegetation data layers described above. FRCC data sets, as well as other spatial LANDFIRE data sets, are currently free and downloadable from http://landfire.cr.usgs.gov/viewer/. Figure 35-11 shows the mean fire return interval data set generated for the Northern Rocky

Mountains map zone. Mean fire return interval, which portrays the average time between fires for different land cover types, is a precursor data set for developing FRCC. For more information regarding FRCC and the precursor data layers for developing FRCC, see Schmidt et al., (2002) and Rollins and Frame (2006).

35.4 Conclusions

The various fire data products described in this chapter are intended to help identify areas across the nation that are at risk to the accumulation of wildland fuel. When used appropriately, the data sets will help prioritize national wildland fuel reduction projects, provide information for modeling of real-time fire behavior, and support tactical decisions to increase wildland firefighting capacity and safety. The data sets will also support regional modeling of potential fire behavior and fire effects, which will aid plans to reduce hazardous fuels and restore ecosystem integrity.

The LANDFIRE program is the first project to offer seamless data on fire fuels and condition for all lands, regardless of ownership. As fire sciences and geospatial techniques advance, so will the ability of LANDFIRE to offer improved data layers for fire and fuel management. Meanwhile, the Monitoring Trends in Burn Severity project provides important information about the location and severity of large fires in the United States, and on recent trends in fire behavior. This information, when combined with the LANDFIRE data, will provide critical information to fire and fuel managers that can be used to predict fire behavior and risk.

The spatial data layers being generated have relevance beyond the fire community. We expect that researchers and land use managers will use the data sets in many other applications, such as conservation planning and biodiversity assessments, assessing the effects of climate change, and modeling the carbon cycle. LANDFIRE and MTBS data will also provide a spatial framework for many other basic ecological investigations. With new sources of imagery and improvements in our ability to analyze multiple sources of data, the role of remote sensing and GIS will steadily expand for fire-related assessments. The GIS data layers now being developed are essential for making informed decisions about our natural resources, and the projected enhancements in these data sets will only improve our decision making process.

Acknowledgements

A portion of this work was conducted by SAIC staff, performed under US Geological Survey contract 03CRCN0001 at the US Geological Survey Center for Earth Resources Observation and Science, Sioux Falls, South Dakota 57198. The authors would like to thank Brian Tolk for his assistance with the illustrations.

References

Agee, J. K. 2000. Wilderness fire science: a state of the knowledge review. Pages 5–22 in Wilderness Science in a Time of Change Conference—Volume 5: Wilderness Ecosystems, Threats, and Management. Compiled by D. N. Cole, S. F. McCool, W. T. Borrie, J. O'Loughlin. Missoula, Montana, 23–27 May 1999. Proc. RMRS-P-15-VOL-5. Ogden, Utah: US Department of Agriculture, Forest Service, Rocky Mountain Research Station:.

Allen, C. D., M. Savage, D. A. Falk, K. F. Suckling, T. W. Swetnam, T. Schulke, P. B. Stacey, P. Morgan, M. Hoffman and J. T. Klingel. 2002. Ecological restoration of southwestern ponderosa pine ecosystems: a broad perspective. *Ecological Applications* 12:1418–1433.

Anderson, H. 1982. Aids to determining fuel models for estimating fire behavior. General Technical Report GTR INT-122. USDA Forest Service, Intermountain Forest and Range Experiment Station, Ogden, Utah.

Brewer, C. K., J. C. Whinne, R. L. Redmond, D. W. Opitz and M. V. Mangrich. 2005. Classifying and mapping wildfire severity: a comparison of methods. *Photogrammetric Engineering and Remote Sensing* 71 (11):1311–1320.

Caratti, J. F. 2006. The LANDFIRE reference database. In: The LANDFIRE Prototype Project: Nationally Consistent and Locally Relevant Geospatial Data for Wildland Fire Management. General Technical Report RMRS-GTR-175. Fort Collins, Colo.: US Department of Agriculture, Forest Service, Rocky Mountain Research Station.

Chuvieco, E., editor. 2003. *Wildland Fire Danger Estimation and Mapping: the Role of Remote Sensing Data.* (Series in Remote Sensing Vol.4). Singapore: World Scientific Publishing Co., Pt. Ltd. 264 pp.

Comer, P., D. Faber-Langendoen, R. Evans, S. Gawler, C. Josse, G. Kittel, S. Menard, M. Pyne, M. Reid, K. Schulz, K. Snow and J. Teague. 2003. Ecological systems of the United States: a working classification of U.S. terrestrial systems. Arlington, Va.: Nature-Serve. 75 pp.

Congalton, R. G. and G. S. Biging. 1992. A pilot study evaluating ground reference data collection efforts for use in forest inventory. *Photogrammetric Engineering and Remote Sensing* 58:1669–1671.

Csiszar, I., L. Denis, L. Giglio, C. O. Justice and J. Hewson. 2005. Global fire activity from two years of MODIS data. *International Journal of Wildland Fire* 14:117–130.

Diaz-Delgado, R., F. Lloret and X. Pons. 2004. Spatial patterns of fire occurrence in Catalonia, NE, Spain. *Landscape Ecology* 19:731–745.

Epting, J., D. Verbyla and B. Sorbel. 2005. Evaluation of remotely sensed indices for assessing burn severity in interior Alaska using Landsat TM and ETM+. *Remote Sensing of Environment* 96 (3-4):328–339.

FIA (Forest Inventory and Analysis). 2007. Forest Inventory and Analysis National Program. <http://fia.fs.fed.us/> Accessed 10 October 2007.

GAO. 2002. Wildland Fire Management. Report to congressional requesters, GAO-02-158. Washington, DC: United States General Accounting Office. 32 pp.

GAP. Undated. The GAP Analysis Program. < http://gapanalysis.nbii.gov/portal/server.pt> Accessed 10 October 2007.

Hann, W. J. and D. L. Bunnell. 2001. Fire and land management planning and implementation across multiple scales. *International Journal of Wildland Fire* 10:389–403.

Hardy, C. C., K. M. Schmidt, J. M. Menakis and N. R.Sampson. 2001. Spatial data for national fire planning and fuel management. *International Journal of Wildland Fire* 10:353–372.

Hoelzemann, J. J., M. G. Schultz, G. P. Brasseur and C. Granier. 2004. Global wildland fire emission model (GWEM): evaluating the use of global area burnt satellite data. *Journal of Geophysical Research* 109:D14S04.

Homer, C., C. Huang, L. Yang, B. Wylie and M. Coan. 2004. Development of a 2001 national land-cover database for the United States. *Photogrammetric Engineering and Remote Sensing* 70:829–840.

Justice, C. O., R. Smith, A. M. Gill and I. Csiszar. 2003. A review of current space-based fire monitoring in Australia and the GOFC/GOLD program for international coordination. *International Journal of Wildland Fire* 12:247–258.

Keane, R. E., R. Burgan and J. van Wagtendonk. 2001. Mapping wildland fuels for fire management across multiple scales: integrating remote sensing, GIS and biophysical modeling. *International Journal of Wildland Fire* 10:301–319.

Keane, R. E., R. Parsons and P. Hessburg. 2002. Estimating historical range and variation of landscape patch dynamics: limitations of the simulation approach. *Ecological Modelling* 151:29–49.

Key, C. H. and N. C. Benson. 2006. Landscape assessment: ground measure of severity, the Composite Burn Index; and remote sensing of severity, the Normalized Burn Ratio. In *FIREMON: Fire Effects Monitoring and Inventory System*, by D. C. Lutes, R. E. Keane, J. F. Caratti, C. H. Key, N. C. Benson, S. Sutherland and L. J. Gangi. USDA Forest Service. Ogden, Utah: Rocky Mountain Research Station. Gen. Tech. Rep. RMRS-GTR-164-CD: LA1-51.

LANDFIRE. 2007. LANDFIRE. <http://www.landfire.gov> Accessed 10 October 2007.

Lopez Garcia, M. J. and V. Caselles. 1991. Mapping burns and natural reforestation using Thematic Mapper data. *Geocarto International* 1:31–37.

Mistry, J. 1998. Fire in the cerrado (savannas) of Brazil: an ecological review. *Progress in Physical Geography* 22:425–448.

NatureServe. 2007. A network connecting science with conservation. <http://www.nature-serve.org> Accessed 10 October 2007.

NFP (National Fire Plan). 2007. Healthy forests and rangelands. <http://www.forestsan-drangelands.gov/> Accessed 10 October 2007.

NIFC (National Interagency Fire Center). Undated. Welcome to the nation's logistical support center. <http://www.nifc.gov/> Accessed 10 October 2007.

Perry, G.L.W. 1998. Current approaches to modeling the spread of wildland fire: a review. *Progress in Physical Geography* 22:222–245.

Peterson, D. L., M. C. Johnson, J. K. Agee, T. B. Jain, D. McKenzie and E. D. Reinhardt. 2005. Forest structure and fire hazard in dry forests of the western United States. General Technical Report, US Department of Agriculture Forest Service, Pacific Northwest Research Station. PNW-GTR-628. 38 pp.

Radeloff, V. C., R. B. Hammer, S. I. Stewart, J. S. Fried, S. S. Holcomb and J. F. McKeefry. 2005. The wildland-urban interface in the United States. *Ecological Applications* 15 (3):799–805.

Rollins, M. G. and C. K. Frame, editors. 2006. The LANDFIRE Prototype Project: nationally consistent and locally relevant geospatial data for wildland fire management. General Technical Report RMRS-GTR-175. Fort Collins, Colo.: US Department of Agriculture, Forest Service, Rocky Mountain Research Station.

Ruiz-Gallardo, J. R., S. Castano and A. Calera. 2004. Application of remote sensing and GIS to locate priority intervention areas after wildland fires in Mediterranean systems: a case study from South-Eastern Spain. *International Journal of Wildland Fire* 13 (3):241–252.

Russel-Smith, J., C. Yates, A. Edwards, G. E. Allan, G. D. Cook, P. Cooke, R. Craig, B. Heath and R. Smith. 2003. Contemporary fire regimes of northern Australia, 1997-2001: change since Aboriginal occupancy, challenges for sustainable management. *International Journal of Wildland Fire* 12:283–297.

SCA (The Student Conservation Association, Inc.). 2007. <http://www.thesca.org/> Accessed 10 October 2007.

Scott, J. H. and R. E. Burgan. 2005. Standard fire behavior fuel models: a comprehensive set for use with Rothermel's surface fire spread model. General Technical Report RMRS-GTR-153. Fort Collins, Colo.: US Department of Agriculture, Forest Service, Rocky Mountain Research Station. 72 pp.

Schmidt, K. M., J. P. Menakis, C. C. Hardy, W.J. Hann and D. L. Bunnell. 2002. Development of coarse-scale spatial data for wildland fire and fuel management. General Technical Report RMRS-GTR-87. Fort Collins, Colo.: US Department of Agriculture, Forest Service, Rocky Mountain Research Station. 41 p. + CD.

Soja, A. J., A. I. Sukhinin, D. R. Cahoon, H. H. Shugart and P. W. Stackhouse. 2004. AVHRRR-derived fire frequency, distribution and area burned in Siberia. *International Journal of Remote Sensing* 25:1939–1960.

US Department of Agriculture, Forest Service. 2000. Protecting people and sustaining resources in fire-adapted ecosystems: a cohesive strategy. The Forest Service management response to the General Accounting Office Report GAO/RCED-99-65. April 13, 2000. 89 pp.

SECTION 6

Blending Technologies:
Remote Sensing, GPS and Visualization

CHAPTER 36

Integrating Remote Sensing and GIS: From Overlays to GEOBIA and Geo-visualization

Marguerite Madden, Thomas Jordan,
Minho Kim, Hunter Allen and *Bo Xu*

36.1 Introduction

By the mid 1980s, it was evident that two emerging geospatial technologies, remote sensing and geographic information systems (GIS), were on a collision course towards integration. Developments a decade earlier in computers and scanners had led to new opportunities for monitoring and managing natural resources using digital imagery acquired from airborne and spaceborne platforms. Equally important was a growing environmental awareness often attributed to the 1962 publication of Rachel Carson's book, *Silent Spring*, warning about ecological hazards of chemical pesticides and impacts of human activities (Carson 1962). Widespread interest in monitoring natural resources, especially in forestry, geology and hydrology, coincided with the birth of the US land remote sensing program in the early 1970s and nationally organized programs for the acquisition of aerial photographs in the 1980s. The availability of synoptic, systematic and repetitive images of Earth resources for the first time, coupled with new computerized image processing capabilities and the desire for automated thematic classification spurred advances in remote sensing.

On a parallel track, growing familiarity of researchers with computers, programming and digital data in the 1960s led to the development of computer mapping software and early GIS programs operating on mainframes. The 1970s saw GIS analysis functionality increase and line printer maps as output, while the 1980s witnessed the explosion of GIS applications on newly purchased personal computers, with digital maps displayed on 16-color monitors and large format pen plotters. It wasn't long before image processing of remotely sensed data and GIS were being performed on the same computer, albeit using separate software accessing data in different formats. Increased interest in the 1990s for integrated remote sensing and GIS capabilities, especially the overlay of predominantly vector GIS data layers on raster images for use in natural resource management, soon led to the supplementation of GIS display and analysis functions to image processing software and the reciprocal addition of image display and manipulation functions in GIS software. The dividing line between the two geospatial technologies began to blur in the 2000s with advances in raster-vector data conversion, on-the-fly ground coordinate and datum transformation, integrated analysis and improved data visualization. Two-dimensional (2D) displays of GIS data on images quickly advanced to two-and-a-half-dimensional (2.5D) drapes of vector and image data on digital elevation models (DEMs), true stereo three-dimensional (3D) perspective views of GIS and image data and temporal considerations with four-dimensional (4D) animations of time series data and GIS-data-driven virtual reality using very high resolution (VHR) imagery.

The advent of VHR imagery led to a paradigm shift in image classification of remote sensing data. Prior to the launch of the first commercial high resolution remote sensing satellite, *Ikonos*, by Space Imaging in 1999, pixel-based thematic classification algorithms and procedures were adequate for successfully classifying medium-resolution satellite data on the order of 10-m SPOT and 30-m Landsat Thematic Mapper (TM) imagery for broad-scale natural resource monitoring and management. Mapping resources at a finer scale could be accomplished by manual interpretation of aerial photographs to create detailed, yet labor-intensive, data sets for making management decisions. The recent availability of satellite images with pixels sizes less than 5 m, typically 1-m panchromatic and 4-m multispectral images, along with airborne scanners, photogrammetric digital cameras and high-resolution quality scanners for the conversion of hardcopy photographs to digital format, produced a plethora of image data for which traditional pixel-based classification techniques were no longer adequate. Pervasive use of Global Positioning System (GPS) and Inertial Measurement Unit (IMU) technologies to capture the geographic location and attitude of scanners and cameras at the instant of exposure, thus relieving the requirement for numerous accurate ground control points, also alleviated a major barrier to remote sensing and GIS integration. Increased production of DEMs meant images could be orthorectified and made readily available to users following acquisition. Increased availability of DEMs also facilitated orthorectification of images for ready input to existing GIS databases. Users of GIS no longer required advanced knowledge of photogrammetry or image processing to take advantage of image data content for GIS analyses.

Global adoption of the World Wide Web (www) in the 2000s has brought another paradigm shift in data access. Anyone with an internet connection and a computer now has access to extensive remote sensing and GIS data organized as national spatial data infrastructures (NSDIs), served by regional/statewide clearinghouses, searchable via virtual globes and connected for analysis over distributed cyber infrastructures and GRID architectures. Users no longer work alone in isolated computer labs with open manuals. They are members of globally and virtually connected communities sharing information, applications and products through social forums and domain-specific wikis. Researchers eager to populate new object-based geodatabases, as well as update existing GIS data sets with information on the current status of features (i.e., objects), share new methods for image analysis and thematic classification. Geographic object-based image analysis (GEOBIA), for example, is a recent sub-discipline of geographic information science (GIScience) that addresses limitations of traditional pixel-based classification by grouping pixels into meaningful objects (i.e., segmenting the image). GEOBIA uses powerful contextual analyses to classify objects based on their spectral, spatial and temporal characteristics and spatial correlation with ancillary data from GIS databases (Blaschke et al. 2008, Hay and Castilla 2008). Output from GEOBIA is designed for direct input to vector-based GIS and is expected to be a definitive integrator of the remote sensing and GIS communities.

This chapter will highlight some of the critical research and innovations of the past four decades that led to the integration of remote sensing and GIS technologies for applications in natural resource management. It will also address developments in GEOBIA and geo-visualization that have advanced the use and analysis of geospatial data by geospatially aware managers and decision makers. Case studies of the southeastern US will demonstrate these trends in technology used for national monitoring and inventory by state and federal agencies tasked with resource preservation.

36.2 Historical Background of Remote Sensing

In the US, remote sensing studies for resource management started in the 1930s, with the use of aerial photographs and manual interpretation to map resource distributions (Spurr and Brown 1946; Colwell 1950; Moessner 1960) with many applications focused on forest damage and decline (Haack 1962; Ciesla et al. 1967; Murtha 1972; Friedland et al. 1984) and wetlands (Seher and Tueller 1973; Bogucki et al. 1980; Welch et al. 1988, 1992). Photointerpreters traditionally used basic elements of manual interpretation such as size, shape, tone, texture and association to delineate and classify landforms, hydrologic networks, exposed soils/geology, wetlands, prairies and forests for natural resource inventories (Avery 1962; Teng et al. 1997). Manual interpretation is a labor-intensive and cost-demanding procedure that requires interpreters to have a high level of experience, knowledge and expertise (Heller and Ulliman 1983; Welch et al. 1995). Interpreters often used stereoscopes to view stereo pairs of air photos in 3D and delineate features on plastic overlays registered to the photographs (Dale et al. 1986; Welch et al. 1988). The stereoscopic view of the stereo pair permits a realistic understanding of the terrain, hydrology, relative moisture conditions and relative elevations of features in the landscape (Welch et al. 2002b). It also provides the image analyst with information on texture, relative height and 3D shape that is often critical for feature detection and identification. Transferring information interpreted from the photos to orthorectified maps corrected to eliminate errors of distortion and tip, tilt and relief displacement, however, was often challenging to resource managers who did not have access to expensive photogrammetric equipment.

Resource managers conducting remote sensing applications in the 1960s through the 1990s with hardcopy photographs and maps used mechanical instruments such as Kail and Kargl reflecting projectors, vertical Sketchmasters by the Keuffel & Esser Company, and Bausch & Lomb zoom transfer scopes (ZTS) based on the *camera lucida* principle to superimpose the delineated polygons onto a base map (Paine and Kiser 2003). The resulting hardcopy resource maps provided managers with valuable information on the distribution of forests, wetlands and water bodies, but in order to summarize descriptive statistics, resource analysts resorted to the use of dot grids and planimeters to manually measure the areas of individual polygons. Once resources were mapped for a particular area using multiple dates of aerial photographs, change maps could be generated from a time series by registering maps of different dates to one another on a light table and manually delineating changes on a transparent overlay (Remillard and Welch 1992). In spite of these extremely time-consuming procedures that were prone to user errors, resource managers benefited greatly from the ability to manually interpret air photos and map natural resources for critical data required for management decisions.

The world's first multispectral satellite devoted to remote sensing of land resources, Earth Resources Technology Satellite (ERTS-1) later renamed Landsat-1, was successfully launched on July 23, 1972 by the US National Aeronautics and Space Administration (NASA), and with it began a new era of spaceborne digital image data collection (Lillesand et al. 2007). Computer capabilities to manipulate and utilize these data also evolved rapidly during this period. In 1970, the first dynamic random access memory (DRAM) chip had been released by a newly formed company named Intel and by 1972 "it was the best selling semiconductor memory chip in the world" (Bellis 2008). The following year saw a number of computing "firsts" including: 1) the first advertisement for a microprocessor, the Intel 4004, in *Electronic News*; 2) advertisement of the first personal computer, the Kenbak-1, in *Scientific American*; 3) the invention of the 8-inch floppy disk by IBM; and 4) the first network email (sent by researcher Ray Tomlinson over a military network called ARPAnet—credited as the precursor of the internet) (Computer History Museum 2006). The simultaneous development of computer technology, environmental awareness and availability of Earth imagery in digital format created conditions ripe for the birth of the new discipline of remote sensing.

Since the US launches of Landsats-1, -2 and -3 carrying 80-m Multispectral Scanner (MSS) sensors and Landsats-4, -5 and -7 with the 30-m TM and Enhanced Thematic Mapper Plus (ETM+) sensors, automated pixel-based methodologies have been used extensively with satellite imagery for resource management. Initial applications such as forest, crop and aquatic vegetation classifications (Dodge and Bryant 1976; Hoffer et al. 1978; Jensen et al. 1978, 1980) evolved into broad-scale operational resource monitoring programs (Loveland et al. 2002; Pearlstine et al. 2002; GAP 2008). Many resource applications also have objectives that require high spatial resolution images (< 5-m pixel resolution) to identify details such as forest crown morphology, tree heights, individual tree species and communities or associations of species. Attempts have been made to employ automatic methodologies to develop natural resource inventories with VHR satellite imagery; such imagery is similar to aerial photographs in spatial resolution and information content (Ehlers

2004). However, conventional pixel-based approaches are often found to have limitations with VHR imagery due to high spectral variation of individual ground features and lack of contextual consideration (Marceau et al. 1990; Schiewe et al. 2001; Brandtberg and Warner 2006). Remote sensing researchers looked for ways to utilize the spectral information of image data in combination with contextual information and ancillary data from existing GIS databases to improve classification results (Lu and Weng 2007). There was, therefore, a growing interest in approaches that integrated remote sensing and GIS.

36.3 Historical Background of GIS

With the advent of computers came the realization that digital data could be used to represent phenomena in space either implicitly by the location of values in a 2D array (i.e., a raster data file) or explicitly by associating a geographic location with a value, either by Cartesian x and y coordinates (i.e., vector data) or surveyors' metes and bounds (i.e., coordinate geometry). The "value" with "place" could represent a thematic class (e.g., 1 represents forest and 2 represents water), a ranked order (e.g., 1 is best and 2 is second-best) or a measurement of continuous data (e.g., 106 represents an elevation of 106 m). With computer programming, digital files representing spatial data, and maps displayed on line printer paper and PC screens, came the realization that computer "maps" could also be analyzed and manipulated. These were the components of a GIS.

The origins of GIS involve multiple researchers and practitioners working in different places and being exposed to emerging computer technology during the 1960s. Foresman (1998) provides the history of GIS from the perspectives of the pioneers while Chrisman (2006) describes the important role of the Harvard Laboratory for Computer Graphics and Spatial Analysis in the origins of GIS.

> *There are many stories about the origins of geographic information systems technology. A few of them are true. But no matter which story you hear, if you probe a little bit, you will find a connection to the Harvard Laboratory for Computer Graphics and Spatial Analysis, where, beginning in 1965, planners, geographers, cartographers, mathematicians, computer scientists, artists, and many others converged to rethink thematic mapping, spatial analysis, and what is now called GIS* (Chrisman 2006, p. 3).

According to Chrisman (2006), Howard Fisher of Harvard attended a workshop in 1963 on computer mapping of census tracts. The workshop was organized by Edgar Horwood, a professor of Urban Planning and Civil Engineering from the University of Washington. Fisher immediately recognized the value of computer mapping for planning and landscape architecture, which led him to apply for a grant from the Ford Foundation to develop a computer-mapping software program called Synagraphic Mapping System (SYMAP) (Fisher 1982). [Fisher coined the term "synagraphic" from Greek roots of "together" and "graphic" to mean 'seeing things together' (Chrisman 2006).] That Ford Foundation grant resulted in Fisher founding the Harvard Laboratory for Computer Graphics and Spatial Analysis in Cambridge, Massachusetts.

Fisher's program extended Horwood's limited graphic output capability by assuming that any computer had a card reader to input data and a line printer with 130 parallel slugs hitting the paper through a carbon ribbon, each the same width and height to create 10 characters per inch. Since these printers also had overprinting capability, the paper could be stopped from advancing and any character could be repeatedly struck up to four times. Very dark areas could, therefore, be created by overprinting O, X, A and V, while other shades of light and dark gray could be created with different combinations of letters and numbers of overstrikes. The position of the character on the page depicted its location in geographic space and together the characters produced a shaded choropleth map he called "conformant" because the map symbols conformed to the object. Although the output looked inherently like a grid, SYMAP actually used a vector model that foreshadowed modern GIS, "a collection of objects – points, lines, and areas – in planar coordinate space with thematic values attached to them" (Chrisman 2006). Designed for use by planners, cartographic considerations of symbolism, legends, maps of different scales, titles and text were important from the beginning.

The addition of improvements in methods for mathematical interpolation—now called Inverse Distance Weighting—by computer programmer Donald Shepard, brought spatial analysis capability to SYMAP (Chrisman 2006). This development was linked to a funded project from the US Public Health Service from 1967-1969 that combined point samples of temporal air pollution measures with demographic variables. Shepard used his interpolation algorithm in an undergraduate thesis to consider environmental consequences of different fuels burned by power plants at different locations. A student in the Harvard Masters of Landscape Architecture program, Jack Dangermond, worked on the project using SYMAP and testing the next-generation plotter-based version of SYMAP to produce visualizations of air pollution levels for different times of the day, and with and without barriers to air movements. Dangermond and his wife, Laura, would go on to found a consulting firm in 1973 called the Environmental Systems Research Institute (ESRI) of Redlands, California.

In 1967, SYMAP was used by the Harvard Departments of City Planning and Landscape Architecture to conduct a GIS-type regional study led by Carl Steinitz for the Delmarva Peninsula covering Delaware, eastern Maryland and a small portion of Virginia (Steinitz 1967 and Tomlinson 1968 as cited in Chrisman 2006). Base layers of topographic maps, soils, land use interpretation from aerial photographs and county-level census statistics were coded as 2-by-2-mile grid cells, while county boundaries, shorelines and roads were entered as vector objects. Land use was indicated as the percentage of a cell that was forest or agriculture and topographic relief was indicated by average elevation. Interpolation of the grid cell values as points by SYMAP produced smoother map output and the suitability for a number of different uses was determined by digital map overlay of weighted factors, following Ian McHarg's concept of physical map overlay analysis (McHarg 1969). Since each SYMAP overlay consisted of an attribute value representing a grid cell center, Steinitz and his student assistant, David Sinton, questioned the need for the vector model and proposed a simple grid cell data structure that would be easier to conceptualize and manage for the analysis. Sinton pulled out the map production capability from SYMAP Version 4 and added a grid-based input system to create GRID. Rewritten for Landscape Architecture projects in the 1970s as IMGRID, the software became the "springboard for early commercial GIS enterprises" including ESRI, Intergraph, Synercom, Earth Resources Data Analysis System, Inc. (ERDAS) and the MAP Analysis software written by Dana Tomlin, thus strongly influencing grid-based spatial analysis in GIS (Chrisman 2006).

For many years SYMAP was the dominant thematic-mapping computer program in the world. This innovative program was the precursor to today's GIS software packages and its use of both vector and raster data formats laid the foundation for integrated remote sensing and GIS.

36.4 Integrated Remote Sensing and GIS

Remote sensing has traditionally supplied spatial information for GIS databases, analysis, manipulation and display. Aerial photographs, Earth observation electro-optical satellite images, radar and more recently, lidar, have been used extensively to develop geo-databases appropriate for resource management, environmental planning and policy decision making. Numerous examples of resource mapping from aerial photographs and Landsat/SPOT image data have been reported since the 1980s (Jensen et al. 1980; Scarpace and Quirk 1980;

You-Ching 1980; Remillard and Welch 1992; Welch et al. 1992, 1995, 1999; Rutchey and Vilcheck 1994; Heipke et al. 2000 to name a few). As multiple dates of resource databases were created, GIS software was used to perform change analysis, predictive modeling and suitability assessment (Lillesand et al. 1981; Loveland and Johnson 1983; Lyon 1983; Remillard and Welch 1993; Hilton 1996; Narumalani et al. 1997). In addition to the use of remote sensing to collect data for GIS databases, Mesev and Walrath (2007) consider other "time-honoured ways in which GIS and remote sensing have been integrated" to include GIS data used as ancillary information for image processing (Hutchinson 1982; Foody 1988; Janssen et al. 1990; Mesev 1998) and combined analytical functions such as spatial queries and overlay of attributes from both GIS and remote sensing data.

Ehlers et al. (1989) proposed three stages in the evolution towards complete and total integration of remote sensing and GIS. Criteria for determining the stage of integration include the degree of interaction between vector and data models, data exchange, geometric registration, cartographic representations, common user interface and geographic abstraction. The three stages are:

- Stage 1 – Database development separately and equally—from remote sensing (raster data model) and GIS (predominantly vector data model)—with simultaneous display by vector-on-raster overlay. Analysis in Stage 1 is limited to update of GIS databases with information from classified remotely sensed images or use of GIS data to facilitate image geo-registration.
- Stage 2 – Continued use of separate remote sensing and GIS databases, but with a shared user interface. Data are converted through vectorization or rasterization to a common data model for simultaneous display and operational use of spatial and temporal attributes.
- Stage 3 – Total integration achieved when remote sensing and GIS become one indistinguishable system with vector and raster data models "handled interchangeably through data uniformity across object-based (GIS data) and field-based (remotely sensed data) geographic representation" (Mesev and Walrath 2007).

The National Center for Geographic Information and Analysis (NCGIA), a consortium consisting of the University of California, Santa Barbara (USCB), State University of New York (SUNY) Buffalo and the University of Maine at Orono, was created in 1988 by an award from the National Science Foundation to conduct research, education and outreach to remove impediments to the broad application of GIS and geographic analysis (Star et al. 1991). The goal of Research Initiative No. 12 (RI-12) of the NCGIA list of priority research topics was the integration of remote sensing and GIS. A special issue of *Photogrammetric Engineering and Remote Sensing* in 1991 was devoted to NCGIA RI-12 and addressed integrated GIS (IGIS) in terms of integrated data availability and access by users (Ehlers et al. 1991), scale dependence and data transformation (Davis et al. 1991), required computer technology (Faust et al. 1991), institutional barriers to implementation (Lauer et al. 1991), and error sources and decision confidence (Lunetta et al. 1991). Research since 1991 addressing NCGIA RI-12, coupled with advances in computer technology, increased availability of multi-resolution remote sensing data and demands for easy-to-use interfaces (for data search, download, manipulation, integration, analysis and display) have resulted in near IGIS functionality in the major current software packages. Almost all GIS programs today offer display and query of digital images with some image processing functionality and hybrid processing of raster and vector data (Ehlers 1997, 2000). Some issues still remaining include: 1) the need for uncertainty descriptors for integrated remote sensing and GIS data of multiple scales and accuracies, as well as system-independent taxonomy of GIS-image analysis functions (Gahegan and Ehlers 1991; Ehlers 2007); 2) issues of scale variation when upscaling and downscaling observed data, sampling processes, measurement and data transformation (Atkinson and Tate 2000; Atkinson, 2007); and 3) desired fusion of remote sensing and GIS data prevented by lack of consistent data standards and differences in scale, legends and accuracy (Gamba and Dell'Acuqa 2007).

36.4.1 Geographic Object-Based Image Analysis (GEOBIA) for Integrating Remote Sensing and GIS

Progress towards true integration of remote sensing and GIS can be made if knowledge can be extracted from images and readily accessed by GIS functions and/or knowledge from GIS data can be used to analyze images. Both are possible through object-based image analysis (OBIA), a paradigm in image processing that deals with increasingly complex imagery to model reality and extract geospatial information compatible with GIS (Blaschke et al. 2008). Lang (2008) states the guiding principle of OBIA is as "clear as it is ambitious: to represent complex scene content in such a way that the imaged reality is best described and a maximum of the respective content is understood, extracted and conveyed to users". Object-based analysis of Earth remote sensing imagery is now being referred to as Geographic Object-Based Image Analysis (GEOBIA).

> Geographic Object-Based Image Analysis (GEOBIA) *is a sub-discipline of* Geographic Information Science (GIScience) *devoted to developing automated methods to partition remote sensing imagery into meaningful image-objects, and assessing their characteristics through spatial, spectral and temporal scales, so as to generate new geographic information in GIS-ready format...Since GEOBIA relies on RS (remote sensing) data, and generates GIS (Geographic Information Systems) ready output, it represents a critical bridge between the (often disparate) raster domain of RS, and the (predominantly) vector domain of GIS. The 'bridge' linking both sides of these domains is the generation of* polygons *(i.e.,* classified image-objects*) representing geographic objects* (Hay and Castilla 2008, p. 77).

The process of GEOBIA requires image segmentation, the use of contextual and ancillary data to group similar image pixels into segments that can be related to real objects via object-based classification techniques (Blaschke et al. 2000; Hay et al. 2001, 2003). Partitioning image pixels into objects follows the way humans manually interpret aerial photographs and conceptually organize and understand objects in a landscape, the objects can be characterized by useful features such as shape, texture and context, and they are less sensitive to the modifiable areal unit problem (MAUP) since they represent meaningful geographic entities rather than arbitrary spatial units (i.e., pixels). (Hay et al. 1996, 2002, 2008).

According to Hay and Castilla (2008), there are many reasons that GEOBIA can be considered the link between remote sensing and GIS.

- Image objects are more readily integrated into a vector GIS than pixel-based classification results.
- Object-oriented concepts and approaches have been adopted by modern GIS software and they can be adapted to GEOBIA.
- There is a growing community of remote sensing and GIS practitioners who currently use image segmentation for different geospatial applications, so as GEOBIA matures new opportunities for research and commercial software with GEOBIA processing capabilities are expected.
- Adopting existing open GIS standards for programming such as Open Source GIS (http://opensourcegis.org) and guidelines of the Open Geospatial Consortium (OGC) (http://www.opengeospatial.org) will allow GEOBIA to be integrated among different platforms and data types and promote web-based applications of GEOBIA.

36.4.2 Integrating Remote Sensing and GIS Using GEOBIA and Geo-visualization

The Center for Remote Sensing and Mapping Science (CRMS), Geography Department, University of Georgia (http://www.crms.uga.edu) has worked cooperatively with resource management agencies for over 25 years to develop digital databases for the preservation of natural and cultural resources on state and federal lands (Welch et al. 1988, 1992; Welch and Remillard 1994; Remillard and Welch 1992, 1993; Madden 2004a). Vegetation community distributions, off-road vehicle/airboat trails and land cover/land use (LULC) were mapped in 21 National Park Service (NPS) units of the southeastern US as part of the NPS National Vegetation Inventory (NVI). Over 10,000 km[2] of NPS units were mapped in south Florida including Everglades National Park, Big Cypress National Preserve and Biscayne National Park (Welch et al. 1995, 1999, 2002a; Welch and Remillard 1996; Madden et al. 1999; Hirano et al. 2003). Overstory and understory databases were developed for over 2,000 km[2] in Great Smoky Mountains National Park (Jordan 2002, 2004; Welch et al. 2002b; Madden et al. 2004) and 17 additional national park units including the Blue Ridge Parkway, Big South Fork National River and Recreational Area, and Cumberland Gap National Historical Park (Jordan and Madden 2008) (Figures 36-1 and 36-2, Table 36-1). These vegetation databases and association-level classification systems contribute to NPS management objectives such as modeling fire fuel, eradicating exotic plants, assessing human impacts of off-road vehicles/airboats and predicting the effects of exotic insect invasions on forest community structure (Madden 2004b; Madden et al. 2006).

Vegetation databases—originally created from manually interpreted large-scale color infrared (CIR) aerial photographs integrating photogrammetry, Global Positioning System (GPS), image processing and GIS technologies—were effective and valuable, yet costly and labor intensive to create. Efficient methods are now required to meet two objectives: 1) update the NVI databases in the future to monitor changes over time; and 2) preserve the expert knowledge of CRMS botanists and photo interpreters who are now approaching retirement age. Both of these objectives were addressed using GEOBIA to delineate and

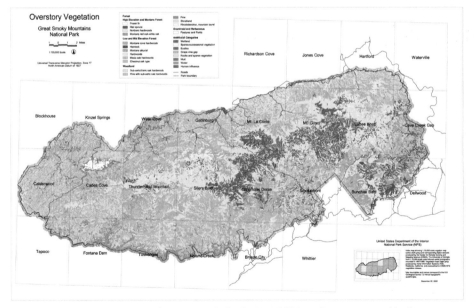

Figure 36-1 Hardcopy vegetation maps plotted at 1:15,000 scale correspond to the area covered by 25 individual USGS 7.5-minute topographic quadrangles in Great Smoky Mountains National Park, as outlined on this generalized overview map of overstory vegetation. The black box indicates the area enlarged in Figure 36-2. See included DVD for color version.

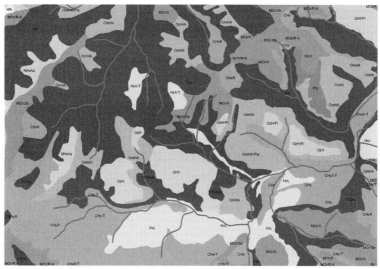

Figure 36-2 A detailed enlargement of a 1-km² portion of Cataloochee Valley in the Great Smoky Mountains National Park vegetation database with vegetation association labels listed in Table 36-1. See included DVD for color version.

classify vegetation communities and land use. Noted issues in assessing the effectiveness of GEOBIA techniques for NVI mapping include image segmentation quality and contextual fuzzy classification using VHR image data. The accuracies of results were compared to the NPS NVI vegetation databases that were field-verified to an overall accuracy 80.4 % with a Kappa Index of 0.80 by NPS (Jenkins 2007).

Table 36-1 Selected forest classes in the Great Smoky Mountains National Park vegetation database (Jackson et al. 2002).

Association Code	Association Description	Veg Type	Vegetation Description
CHx	Cove Hardwoods	MF	Mixed Forest
CHxR	Cove Hardwoods Rich Type	MF	Mixed Forest
Hth	Heath Shrubs	SB	Shrub
HxL	Southern Appalachian Early Successional Hardwoods Liriodendron Type	DF	Deciduous Forest
HxA	Southern Appalachian Early Successional Hardwoods Acid Type	DF	Deciduous Forest
MOr/R-K MOr/R/K	Montane Northern Red Oak with Rhododendron and Kalmia	DF	Deciduous Forest
MOr/Sb	Montane Northern Red Oak with Deciduous Shrubs and Herbaceous	DF	Deciduous Forest
MOz	Montane Xeric Northern Red Oak-Chestnut Oak-Woodland	MW	Mixed Woodland
NHxAz	Southern Appalachian Mixed Hardwood Forest, Acid Type	DF	Deciduous Forest
OmHA	Submesic to Mesic Oak/Hardwoods Acid Type	DF	Deciduous Forest
OmHr	Submesic to Mesic Oak/Hardwoods Red Oak-Red Maple-Mixed Hardwoods	DF	Deciduous Forest
OzH	Chestnut Oak/Hardwoods Xeric Woodland	MW	Mixed Woodland
T/NHxAz T/NHxR	Eastern Hemlock (T) with Northern Hardwood Acid Type or Rich Type	DF	Deciduous Forest

Manual of Geographic Information Systems

The selection of initial parameters when using GEOBIA software such as Definiens Developer 7.0 (Definiens AG) was found to be critical for deriving information from 4-m multispectral Ikonos images (GeoEye, Inc.) with reasonable accuracy. The scale parameter of segmentation, for example, is set by the user and without guidance is often varied by trial and error to produce different results that affect classification accuracy. Using an *a priori* approach to determine the optimal scale parameter when performing the segmentation procedures, Kim et al. (2008) were able to classify four forest types (deciduous broadleaf, evergreen coniferous, deciduous dominant mixed and evergreen dominant mixed), in Guildford Courthouse National Military Park in Greensboro, North Carolina, to an accuracy of 79% and 0.65 Kappa. Adding object-specific correlation, entropy and gray-level co-occurrence matrix (GLCM) texture information to the Guildford Courthouse GEOBIA segmentation and classification improved overall classification accuracies for forest types to 83% with 0.71 Kappa (Kim et al. *in press*). Texture measures have previously been shown to improve image segmentation and classification in forest and landform studies (Hay et al. 1996; Ryherd and Woodcock 1996; Kayitakire et al. 2006).

Ancillary data such as topographic elevation, aspect, and slope, in combination with texture measures and VHR spectral information, should further improve the results of GEOBIA segmentation and classification of targeted vegetation (Parker 1982; Treitz and Howarth 2000). In fact, topographic variables have aided in performing object-based vegetation classification (Domaç et al. 2006; Chastain et al. 2008) and have been adopted in GEOBIA to perform automatic landform unit classifications (Dragut and Blaschke 2006). Information from GIS databases also has been used in vegetation and land cover mapping (Sader et al. 1995; Debeir et al. 2002).

Topographic information and GIS data were added to the GEOBIA analysis of forests in Great Smoky Mountains National Park to improve classification accuracies. Xu (2007) combined ancillary information on terrain characteristics (slope, aspect, elevation, slope position) with texture and spectral information from orthoimages derived from the same CIR aerial photographs used to create the CRMS-NPS vegetation database (Jordan 2002). Five forest types—deciduous broadleaf, coniferous evergreen, deciduous dominant mixed, grass and shrub—were mapped to an overall accuracy of 90% and Kappa 0.46. At the next level of increasing detail, the alliance level, 7 forest classes—hemlock, cove hardwoods, mixed hardwoods, montane oak, northern hardwoods, heathbalds and pasture—were mapped to an overall accuracy of 73% with 0.42 Kappa.

A multiscale GEOBIA approach—with Definiens Developer Version 7.0. and 4-m Ikonos spectral bands, texture and ancillary data, in the Smokemont area of southeastern Great Smoky Mountains National Park—was used to map five vegetation types: deciduous broadleaf forest (DF), evergreen coniferous forest (EF), deciduous and evergreen mixed forest (MF), shrub (SB) and grass (GR). A proximity layer to stream channels was added to the GEOBIA analysis with topographic variables and spectral bands to improve the segmentation quality. Figure 36-3a illustrates stream channels overlaid on the manual interpretation and Figure 36-3b the proximity layer derived with Euclidean distance. The figures show the contextual relationship between stream channels and forest stands, a relationship which may be utilized in GEOBIA segmentation and classification. A series of image segmentations with the proximity layer as well as topographic variables and spectral bands resulted in classification accuracies that were similar to those from spectral-only GEOBIA. However, the highest results were obtained at a scale parameter of 50: 75.0 % overall accuracy and 0.54 Kappa. In this scenario, segmentation scales were subdivided in steps of 1 around scale 50 and produced classification accuracies as shown in Figure 36-4. With fine scales, maximum accuracies were achieved at scale 48 with an overall accuracy of 76.6 % and a Kappa of 0.57. Segmentation with a proximity-to-stream-channels layer, topographic variables and spectral bands yielded enhanced classification results (76.6 % accuracy and 0.56 Kappa) over spectral-only segmentation (67.1 % accuracy and 0.44 Kappa) – a gain of 9.5 % in overall accuracy and 0.13 in Kappa (Figure 36-5). It is hoped that progress on the success of this transition from manual to object-based automated techniques will provide resource managers with geospatial tools and spatial information needed to make management decisions.

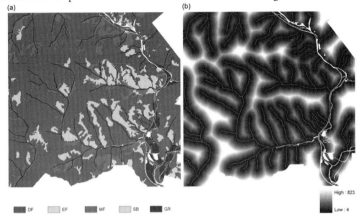

Figure 36-3 A portion of the Great Smoky Mountains National Park CRMS-NPS vegetation database created by manual interpretation with stream channels (black) (a) and Euclidean distance (in meters) layer with stream channels (red) (b). Vegetation is generalized as Deciduous Forest (DF), Evergreen Forest (EF), Mixed Evergreen and Deciduous Forest (MF), Shrub (SB) and Grass (GR). See included DVD for color version.

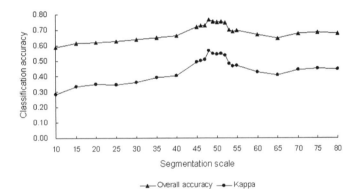

Figure 36-4 Classification accuracies and Kappa statistic of GEOBIA based on Euclidean distance to stream channels, topographic variables and spectral bands across segmentation scales ranging from 10 to 80.

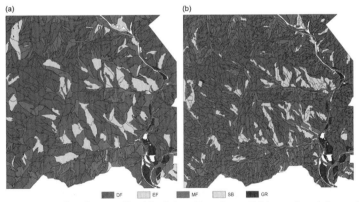

Figure 36-5 GEOBIA classification results of Definiens Developer 7.0 based on (a) spectral bands and ancillary topographic variables using a scale parameter of 75 and (b) Euclidean distance to stream channels, topographic variables and spectral bands at a scale parameter of 48. Vegetation is generalized as Deciduous Forest (DF), Evergreen Forest (EF), Mixed Evergreen and Deciduous Forest (MF), Shrub (SB) and Grass (GR). See included DVD for color version.

Manual of Geographic Information Systems

36.4.3 Geo-visualization of Image and GIS Data for Resource Assessment

Throughout the process to create digital vegetation databases for NPS units in the southeastern US, geo-visualization techniques have been used to aid in the extraction and assessment of vegetation patterns, quality control evaluation and communication of information to managers and users of park resources. The techniques traditionally used to map the vegetation communities relied on stereo interpretation of large-scale CIR aerial photographs. Interpreters preferred to use analog methods—delineating vegetation boundaries on plastic overlays registered to the film transparencies while viewing the photos in stereo with a mirror stereoscope—over heads-up on-screen digital procedures. This allowed interpreters to view relatively large areas of the terrain in stereo and in color for best determination of site conditions, (i.e., moisture, hydrology and relative soil richness) and identification of individual species and vegetation community types. Digital manipulation of the scanned and orthorectified air photos for 3D visualization, however, provides interpreters with additional information that can be used in the interpretation process. Drapes of the orthorectified photos onto DEMs or Triangulated Irregular Networks (TINs) of the terrain aided in visualizing vegetation patterns related to topographic factors influencing habitat conditions (Madden 2004b) (Figure 36-6 and 36-7). The 2.5D drapes of vegetation maps on the terrain were used to assist in quality control checks of the geometric integrity of the completed vegetation databases and to convey information on development trends in watersheds surrounding parks to assess potential impacts on park resources (Madden et al. 2006). Animations also were constructed of vegetation mapped from historical aerial photographs to depict changes in vegetation patterns over time (Madden and Giraldo 2005).

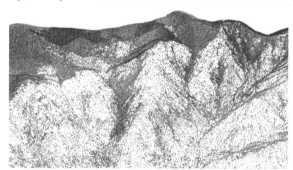

Figure 36-6 Lidar bare earth data of an area corresponding to a 1-km² portion of Cataloochee Valley in Great Smoky Mountains National Park. Lidar was provided by the North Carolina Floodplain Mapping Program for Swain County and displayed using ESRI ArcScene 9.3. Point density is approximately 5-10 m. See included DVD for color version.

Figure 36-7 The 1999 false color USGS DOQQ image is draped over a TIN created from the lidar bare earth points for an area corresponding to a 1-km² portion of Cataloochee Valley in the Great Smoky Mountains National Park. See included DVD for color version.

Photorealistic, 3D perspective views of landscapes—a recent advancement in geo-visualization—create a virtual reality of park resources used to portray the projected results of land use operations, management plans and proposed development (Muhar 2001; Madden et al. 2006). Allen and Madden (in press) used photorealistic visualization techniques to better understand the implications of extensive dieback of eastern hemlock (*Tsuga canadensis* L.) in Great Smoky Mountains National Park due to the invasion of the exotic hemlock woolly adelgid (*Adelges tsugae* Annand). According to Soehn et al. (2005), the hemlock woolly adelgid entered the park in 2002, and spreading populations now pose a threat to hemlocks throughout the park. Resource managers are treating infected trees by applying insecticidal soaps, oils and biological control agents in an attempt to prevent complete devastation of hemlock and they have solicited input from the public and interested agencies concerning treatment alternatives. Geo-visualization techniques are well-suited in this situation for education and outreach to a broad audience (Gardiner 2006).

GEOBIA and manual classification methods were combined to map current hemlock dieback from 4-m multispectral Ikonos imagery (Kim et al. 2008). The vector files of dieback areas and 1997 hemlock distributions from the pre-invasion CRMS-NPS vegetation database were input to Visual Nature Studio (VNS) for integration with overlays of the Ikonos imagery and TINs (3D Nature 2008). Next, the vector files were linked to 2D texture maps of tree bark characteristics and 3D objects of vegetation tree models created in Xfrog organic modeling software to provide local scale and ground perspective visualizations of potential changes within stands of hemlock (Greenworks Organic Software 2008). Figures 36-8 and 36-9 depict a VNS geo-visualization corresponding to a 1-km² portion of the Cataloochee Valley in eastern Great Smoky Mountains National Park; in this visualization hemlock, eastern white pine *(Pinus strobus)* and dog hobble *(Leucothoe fontanesiana)* shrub objects created in Xfrog were placed in the 3D lidar landscape according to a density derived from the CRMS-NPS vegetation database. These visualizations can help park managers convey information about potential impacts of the dieback to park visitors and assist in targeting critical areas for treatment and restoration.

Figure 36-8 A geo-visualization created in Visual Nature Studio corresponding to a 1-km² portion of Cataloochee Valley in the Great Smoky Mountains National Park and depicting hemlock, white pine and dog hobble shrub objects created in Xfrog and placed in the 3D lidar landscape according to a density derived from the CRMS-NPS vegetation database. See included DVD for color version.

Figure 36-9 A virtual reality ground-perspective view created in Visual Nature Studio of hemlock, white pine and dog hobble shrubs in Great Smoky Mountains National Park. See included DVD for color version.

In summary, GEOBIA and geo-visualization techniques are enhancements to a long history of integrated remote sensing and GIS. The plethora of geospatial data available today is reaching the desktops of researchers and resource managers faced with the preservation of vital resources in an ever-changing environment. In order to provide societal benefits from geospatial data, decision makers increasingly require fully integrated tools. Although the concepts of integrated GIS have been with us for decades, it now time for full implementation.

Acknowledgements

This study was sponsored by the US Department of Interior, National Park Service, Cooperative Agreement Numbers. 1443-CA-5460-98-019 and H5028-01-0651. The authors wish to express their appreciation for the devoted efforts of the staff at the Center for Remote Sensing and Mapping Science, Department of Geography, The University of Georgia, NatureServe and national park units of the Appalachian Highlands and Cumberland/Piedmont Networks.

References

3D Nature, LLC. 2008. What is Visual Nature Studio? <http://3dnature.com/vnsinfo.html> Accessed 9 December 2008.

Allen, H. and M. Madden. In press. Geovisualization of forest dynamics: Hemlock woolly adelgid damage in Great Smoky Mountains National Park. *Geospatial Today*.

Atkinson, P. M. 2007. The importance of scale in remote sensing and GIS and its implications for data integration. In *Integration of GIS and Remote Sensing*, edited by V. Mesev, 69–91. West Sussex, England: John Wiley & Sons Ltd.

Atkinson, P. M. and N. J. Tate. 2000. Spatial scale problems and geostatistical solutions: A review. *Professional Geographer* 52:607–623.

Avery, T. E. 1962. *Interpretation of Aerial Photograph*, 2nd edition. Minneapolis, Minn.: Burgess Publishing Co. 324 pp.

Bellis, M. 2008. The World's First Available DRAM Chip. *Inventors of the Modern Computer: The Invention of the Intel 1103*. <http://inventors.about.com/library/weekly/aa100898.htm> Accessed 7 November 2008.

Blaschke, T., S. Lang, E. Lorup, J. Strobl and P. Zeil. 2000. Object-oriented image processing in an integrated GIS/remote sensing environment and perspectives for environmental applications. In *Environmental Information for Planning, Politics and the Public,* edited by A. Cremers and K. Greve, 555–570. Marburg: Metropolis Verlag.

Blaschke, T., S. Lang and G. J. Hay, eds. 2008. Object-Based Image Analysis: Spatial Concepts for Knowledge-Driven Remote Sensing Applications. Berlin Heidelberg: Springer-Verlag. 817 pp.

Bogucki, D. J., G. K. Gruendling and M. Madden. 1980. Remote sensing to monitor water chestnut growth in Lake Champlain. *Journal of Soil and Water Conservation* 35 (2):79–81.

Brandtberg T. and T. Warner. 2006. High resolution remote sensing. In *Computer Applications in Sustainable Forest Management,* edited by G. Shao and K. M. Reynolds, 19–41. Dordrecht, Netherlands: Springer-Verlag.

Carson, R. 1962. *Silent Spring.* Boston, Massachusetts: Houghton Mifflin Co.

Chastain, R. A., A. S. Matthew, H. S. He, and D. R. Larsen. 2008. Mapping vegetation communities using statistical data fusion in the Ozark National Scenic Riverways, Missouri, USA. *Photogrammetric Engineering and Remote Sensing* 74 (2):247–264.

Chrisman, N. 2006. *Charting the Unknown: How Computer Mapping at Harvard became GIS.* Redlands, California: ESRI Press. 218 pp.

Ciesla, W. M., J. C. Bell, Jr. and J. W. Curtin. 1967. Color photos and the southern pine beetle. *Photogrammetric Engineering* 33:883–888.

Colwell, R. N. 1950. New technique for interpreting aerial color photography. *Journal of Forestry* 48:204–205.

Computer History Museum. 2006. Timeline of computer history, 1971. <http://www.computerhistory.org/timeline/?year=1971> Accessed 7 November 2008.

Dale, P.E.R., K. Hulsman and A. L. Chandica. 1986. Seasonal consistency of salt-marsh vegetation classes classified from large-scale color infrared aerial photographs. *Photogrammetric Engineering and Remote Sensing* 52 (2):243–250

Davis, F. W., D. A. Quattrochi, M. K. Ridd, N. S-M. Lam, S. J. Walsh, J. C. Michaelsen, J. Franklin, D. A. Stow, C. J. Johannsen and C. A. Johnston. 1991. Environmental analysis using integrated GIS and remotely-sensed data: Some research needs and priorities. *Photogrammetric Engineering and Remote Sensing* 57 (6):689–697.

Debeir, O., I.V. den Steen, P. Latinne, P. V. Ham and E. Wolff. 2002. Textural and contextual land-cover classification using single and multiple classifier systems. *Photogrammetric Engineering and Remote Sensing* 68 (6):597–605.

Dodge, A. G. Jr. and E. S. Bryant. 1976. Forest type mapping with satellite data. *Journal of Forestry* 74 (8):526–531.

Domaç, A., M. L. Suzen and C. Bilgin. 2006. Integration of environmental variables with satellite images in regional scale vegetation classification. *International Journal of Remote Sensing* 27 (7):1329–1350.

Dragut, L. and T. Blaschke. 2006. Automated classification of landform elements using object-based image analysis. *Geomorphology* 81:330–344.

Ehlers, M. 1997. Rectification and registration. In *Integration of Remote Sensing and GIS,* edited by J. L. Star, J. E. Estes and K. C. McGwire, 13–36. New York, New York: Cambridge University Press.

———. 2000. Integrated GIS- from data integration to integrated analysis. *Surveying World* 9:30–33.

———. 2004. Remote sensing for GIS applications: new sensors and analysis methods. Pages 1-13 In *Proceedings of SPIE Remote Sensing for Environmental Monitoring, GIS Applications, and Geology III,* held in Bellingham, Washington. Edited by M. Ehlers, H. J. Kaufmann and U. Michel. Vol. 5239. Bellingham, Washington: SPIE.

————. 2007. Integration taxonomy and uncertainty. In *Integration of GIS and Remote Sensing,* edited by V. Mesev, 17–42. West Sussex, England: John Wiley & Sons Ltd.

Ehlers, M., G. Edwards and Y. Bédard. 1989. Integration of remote sensing with geographic information systems: A necessary evolution. *Photogrammetric Engineering and Remote Sensing* 55 (11):1619–1627.

Ehlers, M., D. Greenlee, T. Smith and J. Star. 1991. Integration of remote sensing and GIS: data and data access. *Photogrammetric Engineering and Remote Sensing* 57 (6):669–675.

Faust, N. L., W. H. Anderson and J. L. Star. 1991. Geographic information systems and remote sensing future computing environment. *Photogrammetric Engineering and Remote Sensing* 57 (6):655–668.

Fisher, H. T. 1982. *Mapping Information: The Graphic Display of Quantitative Information.* Cambridge, Massachusetts: Abt Books.

Foody, G. M. 1988. Incorporating remotely sensed data into a GIS: The problem of classification evaluation. *Geocarto International* 3:13–16.

Foresman, T. W., ed. 1998. *History of Geographic Information Systems: Perspectives from the Pioneers.* Upper Saddle River, New Jersey: Prentice Hall PTR. 397 pp.

Friedland, A. J., R. A. Gregory, L. Karenlampi and A. H. Johnson. 1984. Winter damage to foliage as a factor in red spruce decline. *Canadian Journal of Forest Research* 14:963–965.

Gahegan, M. and M. Ehlers. 1991. A framework for the modeling of uncertainty between remote sensing and geographic information systems. *ISPRS Journal of Photogrammetry and Remote Sensing* 55:176–188.

Gamba, P. and F. Dell'Acuqa. 2007. Data fusion related to GIS and remote sensing. In *Integration of GIS and Remote Sensing,* edited by V. Mesev, 43–67. West Sussex, England: John Wiley & Sons Ltd.

GAP. 2008. The Gap Analysis Project. <http://gapanalysis.nbii.gov/portal/community/GAP_Analysis_Program/Communities/GAP_Home/GAP_Home/> Accessed: 31 October 2008.

Gardiner, N. 2006. High definition geovisualization: Earth and biodiversity sciences for informal audiences. In *Geographic Hypermedia,* edited by E. Stefanakis, M. P. Peterson, C. Armenakis and V. Delis, 423–446. Berlin Heidelberg: Springer.

Greenworks Organic Software. 2008. Xfrog, Greenworks Organic Software. <http://www.xfrogdownloads.com/greenwebNew/products/productStart.htm> Accessed 9 December 2008.

Haack, P. M. 1962. Evaluating color, infrared, and panchromatic aerial photos for the forest survey of interior Alaska. *Photogrammetric Engineering* 28:592–598.

Hay, G. J. and G. Castilla. 2008. Geographic object-based image analysis (GEOBIA): A new name for a new discipline. In *Object-Based Image Analysis - Spatial concepts for knowledge-driven remote sensing applications*, edited by T. Blaschke, S. Lang, and G. J. Hay, 75–89. Berlin: Springer-Verlag.

Hay, G. J., K. O. Niemann and G. F. McLean. 1996. An object-specific image-texture analysis of H-resolution forest imagery. *Remote Sensing of Environment* 55:108–122.

Hay, G. J., D. J. Marceau, A. Bouchard and P. Dube. 2001. A multiscale framework for landscape analysis: Object-specific upscaling. *Landscape Ecology* 16:471–490.

Hay, G. J., P. Dube, A. Bouchard and D. J. Marceau. 2002. A scale-space primer for exploring and quantifying complex landscapes. *Ecological Modelling* 153 (1-2):27–49.

Hay, G. J., T. Blaschke, D. J. Marceau and A. Bouchard. 2003. A comparison of three image-object methods for the multiscale analysis of landscape structure. *ISPRS Journal of Photogrammetry and Remote Sensing* 57:327–345.

Hay, G. J., T. Blaschke and D. J. Marceau, eds. 2008. GEOBIA 2008 – Pixels, Objects, Intelligence, Proceedings of the GEOgraphic Object Based Image Analysis for the 21st Century Conference held on August 5-8, 2008 at the University of Calgary, Calgary, Alberta, Canada, ISPRS Archives, vol. XXXVIII-4/C1. 373 pp.

Heipke, C., K. Pakzad and B. M. Straub. 2000. Image analysis for GIS data acquisition. *Photogrammetric Record* 16 (96):963–985.

Heller, R. C. and J. J. Ulliman, eds.1983. Forest resources assessments. In *Manual of Remote Sensing,* 2nd Edition, edited by R. N. Colwell, Editor-in-Chief, vol. II, 2229–2324. Falls Church, Virginia: American Society of Photogrammetry.

Hilton, J. C. 1996. GIS and remote sensing integration for environmental applications. *International Journal of Geographic Information Systems* 10:877–890.

Hirano, A., M. Madden and R. Welch. 2003. Hyperspectral image data for mapping wetland vegetation. *Wetlands* 23 (2):436–448.

Hoffer, R. M., S. C. Noyer and R. P. Mroczynski. 1978. A comparison of Landsat and forest survey estimates of forest cover. Pages 221–231 in *Proceedings of the Fall Technical Meeting of the American Society of Photogrammetry* held in Albuquerque, New Mexico. Falls Church, Va.: American Society of Photogrammetry.

Hutchinson, C. F. 1982. Techniques for combining Landsat and ancillary data for digital classification improvement. *Photogrammetric Engineering and Remote Sensing* 48 (1):123–130.

Jackson, P., R. White and M. Madden. 2002. *Mapping Vegetation Classification System for Great Smoky Mountains National Park.* Center for Remote Sensing and Mapping Science, Department of Geography, The University of Georgia. 7 pp.

Janssen, L.L.F., M. N. Jaarsma and E.T.M. van der Linden. 1990. Integrating topographic data with remote sensing for land cover classification. *Photogrammetric Engineering and Remote Sensing* 56 (11): 1503–1506.

Jenkins, M. 2007. *Thematic Accuracy Assessment: Great Smoky Mountains National Park Vegetation Map.* Gatlinburg, Tennessee: National Park Service, Great Smoky Mountains National Park. 26 pp.

Jensen, J. R., J. E. Estes, and L. R. Tinney. 1978. Evaluation of high altitude photography and Landsat imagery for digital crop identification. *Photogrammetric Engineering and Remote Sensing* 44 (6):723–733.

————. 1980. Remote sensing techniques for kelp surveys. *Photogrammetric Engineering and Remote Sensing* 46 (6):43–755.

Jordan, T. R. 2002. *Softcopy Photogrammetric Techniques for Mapping Mountainous Terrain: Great Smoky Mountains National Park.* Doctoral Dissertation, Department of Geography, The University of Georgia, Athens, Georgia. 193 pp.

————. 2004. Control extension and orthorectification procedures for compiling vegetation databases of National Parks in the Southeastern United States. In *International Archives of Photogrammetry and Remote Sensing,* edited by M. O. Altan, vol. 35, Part 4B:422–428. Istanbul, Turkey: International Society for Photogrammetry and Remote Sensing.

Jordan, T. R. and M. Madden. 2008. *Digital Vegetation Maps for National Park Service Cumberland-Piedmont Inventory and Monitoring Network.* Final Report to the U.S. Department of Interior, National Park Service, Cooperative Agreement Number H5028-01-0651, Center for Remote Sensing and Mapping Science, The University of Georgia, Athens, Georgia. 105 pp.

Kayitakire, F., C. Hamel and P. Defourny. 2006. Retrieving forest structure variables based on image texture analysis and IKONOS-2 imagery. *Remote Sensing of Environment* 102:390–401.

Kim, M., M. Madden and T. Warner. 2008. Estimation of optimal image object size for the segmentation of forest stands with multispectral IKONOS imagery. In *Object-Based Image Analysis - Spatial concepts for knowledge-driven remote sensing applications,* edited by T. Blaschke, S. Lang and G. J. Hay, 291–307. Berlin: Springer-Verlag.

————. In press. Forest type mapping using object-specific texture measures from multispectral IKONOS imagery: segmentation quality and image classification issues. *Photogrammetric Engineering and Remote Sensing.*

Lang, S. 2008. Object-based image analysis for remote sensing applications: Modeling reality – dealing with complexity. In *Object-Based Image Analysis: Spatial Concepts for Knowledge-Driven Remote Sensing Applications,* edited by Th. Blaschke, S. Lang and G. J. Hays, 3–27. Berlin Heidelberg: Springer-Verlag.

Lauer, D. T., J. E. Estes, J. R. Jensen and D. D. Greenlee. 1991. Institutional issues affecting the integration and use of remotely sensed data and geographic information systems. *Photogrammetric Engineering and Remote Sensing* 57 (6):647–654.

Lillesand, T. M., D. E. Meisner, D. W. French and W. L. Johnson. 1981. Evaluation of digital photographic enhancement for Dutch Elm disease. *Photogrammetric Engineering and Remote Sensing* 47 (11):1581–1592.

Lillesand, T. M., R. W. Kiefer and J. W. Chipman. 2007. *Remote Sensing and Image Interpretation*, 6th edition. New York: John Wiley & Sons. 768 pp.

Loveland, T. R. and G. E. Johnson. 1983. The role of remotely sensed and other spatial data for predictive modeling: The Umatilla, Oregon Example. *Photogrammetric Engineering and Remote Sensing* 49 (8):1183–1192.

Loveland, T. R., T. L. Sohl, S. V. Stehman, A. L. Gallant, K. L. Sayler and D. E. Napton. 2002. A strategy for estimating the rates of recent United States land-cover changes. *Photogrammetric Engineering and Remote Sensing* 68 (10):1091–1099.

Lu, D. and Q. Weng. 2007. Survey of image classification methods and techniques for improving classification performance. *International Journal of Remote Sensing* 28 (5):823–870.

Lunetta, R. S., R. G. Congalton, L. K. Fenstermaker, J. R. Jensen, K. C. McGwire and L. R. Tinney. 1991. System data integration: Error sources and research issues. *Photogrammetric Engineering and Remote Sensing* 57 (6):677–687.

Lyon, J. G. 1983. Landsat derived land cover classifications for locating potential Kestrel nesting habitat. *Photogrammetric Engineering and Remote Sensing* 49 (2):245–250.

Madden, M. 2004a. Remote sensing and GIS methodologies for vegetation mapping of invasive exotics. *Weed Technology* 18:1457–1463.

———. 2004b. Vegetation modeling, analysis and visualization in U.S. National Parks. In *International Archives of Photogrammetry and Remote Sensing,* edited by M. O. Altan, vol. 35, Part 4B:1287–1293. Istanbul, Turkey: International Society for Photogrammetry and Remote Sensing.

Madden, M. and M. Giraldo. 2005. Landscape modeling and geovisualization. *Geospatial Today* 3 (7):14–20.

Madden, M., D. Jones and L. Vilchek. 1999. Photointerpretation key for the Everglades Vegetation Classification System. *Photogrammetric Engineering and Remote Sensing* 65 (2):171–177.

Madden, M., R. Welch, T. Jordan, P. Jackson, R. Seavey and J. Seavey. 2004. *Digital Vegetation Maps for the Great Smoky Mountains National Park*. Final Report to the U.S. Dept. of Interior, National Park Service, Cooperative Agreement Number 1443–CA–5460–98–019, Center for Remote Sensing and Mapping Science, The University of Georgia, Athens, Georgia. 112 pp.

Madden, M., T. R. Jordan and J. Dolezal. 2006. Geovisualization of vegetation patterns in National Parks of the Southeast. In *Geographic Hypermedia: Concepts and Systems,* edited by E. Stefanakis, M. P. Peterson, C. Armenakis and V. Delis, 329–344. New York: Springer-Verlag.

Marceau, D. J., P. J. Howarth, J.M.M. Dubois and D. J. Gratton.1990. Evaluation of grey-level co-occurrence matrix method for land-cover classification using SPOT imagery. *IEEE Transactions on Geoscience and Remote Sensing* 28 (4):513–519.

McHarg, I. 1969. *Design with Nature*. Garden City, New Jersey: Natural History Press.

Mesev, V. 1998. Use of census data in urban image classification. *Photogrammetric Engineering and Remote Sensing* 64:431–438.

Mesev, V. and A. Walrath. 2007. GIS and remote sensing integration: In search of a definition. In *Integration of GIS and Remote Sensing*, edited by V. Mesev, 1–16. West Sussex, England: John Wiley & Sons.

Moessner, K. E. 1960. *Basic Techniques in Forest Photo Interpretation*. Forest Service Training Handbook, US Forest Service, Inter-Mountain Forest and Range Experiment Station, Ogden, Utah. 73 pp.

Muhar, A. 2001. Three-dimensional modelling and visualisation of vegetation for landscape simulation. *Landscape and Urban Planning* 54:5–19.

Murtha, P. A.1972. *A Guide to Air Photo Interpretation of Forest Damage in Canada*. Environment Canada, Canadian Forest Service Publication No. 1292. 62 pp.

Narumalani, S., J. R. Jensen, J. Althausen, S. Burkhalter and H. E. Mackey. 1997. Aquatic macrophyte modeling using GIS and logistic multiple regression. *Photogrammetric Engineering and Remote Sensing* 63 (1):41–49.

Paine, D. P. and J. D. Kiser. 2003. *Aerial Photography and Image Interpretation*, 2nd ed. New York: John Wiley & Sons. 632 pp.

Parker, A. J. 1982. The topographic relative moisture index: an approach to soil-moisture assessment in mountain terrain. *Physical Geography* 3 (2):160–168.

Pearlstine, L., S. Smith, L. Brandt, C. Allen, W. Kitchens and J. Stenberg. 2002. Assessing state-wide biodiversity in the Florida Gap Analysis Project. *Journal of Environmental Management* 66 (2):127–144.

Remillard, M. and R. Welch. 1992. GIS technologies for aquatic macrophyte studies: I. Database development and changes in the aquatic environment. *Landscape Ecology* 7 (3):151–162.

———. 1993. GIS technologies for aquatic macrophyte studies: II. Modeling applications. *Landscape Ecology* 8 (3):163–175.

Rutchey, K. and L. Vilcheck. 1994. Development of an Everglades vegetation map using a SPOT image and the global positioning system. *Photogrammetric Engineering and Remote Sensing* 60 (6):767–775.

Ryherd, S. and C. Woodcock. 1996. Combining spectral and texture data in the segmentation of remotely sensed images. *Photogrammetric Engineering and Remote Sensing* 62 (2):181–194.

Sader, S. A., D. Ahl and W. Liou. 1995. Accuracy of Landsat-TM and GIS rule-based methods for forest wetland classification in Maine. *Remote Sensing of Environment* 53:133–144.

Scarpace, F. L. and B. K. Quirk. 1980. Land cover classification using digital processing of aerial imagery. *Photogrammetric Engineering and Remote Sensing* 46 (8):1059–1065.

Schiewe, J., L. Tufte and M. Ehlers. 2001. Potential and problems of multi-scale segmentation methods in remote sensing. *GIS-Zeitschrift für Geoinformationssysteme* 6:34–39.

Seher, J. S. and P. T. Tueller. 1973. Color aerial photos for marshland. *Photogrammetric Engineering* 39 (5): 489–499.

Soehn, D., G. Taylor, T. Remaley and K. Johnson. 2005. *Draft Environmental Assesment of Hemlock Woolly Adelgid Control Strategies in the Great Smoky Mountains National Park*. U.S. Department of the Interior National Park Service (NPS). 64 pp. <http://www.nps.gov/grsm/parkmgmt/upload/Hemlock-Woolly-Adelgid-Control-EA%5B1%5D.pdf> Accessed 10 December 2008.

Spurr, S. H. and C. T. Brown, Jr. 1946. Specifications for aerial photographs used in forest management. *Photogrammetric Engineering* 12:131–141.

Star, J. L., J. E. Estes and F. Davis. 1991. Improved integration of remote sensing and geographic information systems: A background to NCGIA Initiative 12. *Photogrammetric Engineering and Remote Sensing* 57 (6):643–645.

Steinitz, C. 1967. Computer mapping and the regional landscape. Unpublished manuscript, Laboratory for Computer Graphics, Harvard Graduate School of Design.

Teng, W. T., E. R. Loew, D. I. Ross, V. G. Zsilinsky, C. P. Lo, W. R. Philipson, W. D. Philpot and S. A. Morain. 1997. Fundamentals of photographic interpretation. In *Manual of Photographic Interpretation,* edited by W. R. Philipson, 49–110. Bethesda, Maryland: American Society for Photogrammetry and Remote Sensing.

Tomlinson, R. F. 1968. A geographic information system for regional planning. In *Land Evaluation*, edited by G. A. Stewart, 200–210. Melbourne: Macmillan.

Treitz, P. and P. Howarth. 2000. Integrating spectral, spatial, and terrain variables for forest ecosystem classification. *Photogrammetric Engineering and Remote Sensing* 66 (3):305–317.

Welch, R. and M. Remillard. 1994. Integration of GPS, digital image processing and GIS for resource mapping applications. *International Archives of Photogrammetry and Remote Sensing*, 30 (Part 4):10–14.

———. 1996. GPS, photogrammetry and GIS for resource mapping applications. In *Digital Photogrammetry: An Addendum to the Manual of Photogrammetry*, 183–194. Bethesda, Maryland: American Society for Photogrammetry and Remote Sensing.

Welch, R., M. Remillard and R. Slack. 1988. Remote sensing and geographic information system techniques for aquatic resource evaluation. *Photogrammetric Engineering and Remote Sensing* 54 (2):177–185.

Welch, R., M. Remillard and J. Alberts. 1992. Integration of GPS, remote sensing and GIS techniques for coastal resource management. *Photogrammetric Engineering and Remote Sensing* 58 (11):1571–1578.

Welch, R., M. Remillard and R. Doren. 1995. GIS database development for South Florida's National Parks and Preserves. *Photogrammetric Engineering and Remote Sensing* 61 (11):1371–1381.

Welch, R., M. Madden and R. Doren. 1999. Mapping the Everglades. *Photogrammetric Engineering and Remote Sensing* 65 (2):163–170.

———. 2002a. Maps and GIS databases for environmental studies of the Everglades. Chapter 9. In *The Everglades, Florida Bay and Coral Reefs of the Florida Keys: An Ecosystem Sourcebook,* edited by J. Porter and K. Porter, 259–279. Boca Raton, Florida: CRC Press.

Welch, R., M. Madden and T. Jordan. 2002b. Photogrammetric and GIS techniques for the development of vegetation databases of mountainous areas: Great Smoky Mountains National Park. *ISPRS Journal of Photogrammetry and Remote Sensing* 57 (1-2):53–68.

Xu, Bo. 2007. *Vegetation Mapping using Object-based Image Analysis in Great Smoky Mountains National Park.* Masters of Science Thesis, Department of Geography, University of Georgia. 110 pp.

You-Ching, F. 1980. Aerial photo and Landsat image use in forest inventory in China. *Photogrammetric Engineering and Remote Sensing* 46 (11):1421–1424.

CHAPTER 37

High Resolution Image Data and GIS

Manfred Ehlers, Karsten Jacobsen and *Jochen Schiewe*

37.1 Introduction

The number of remote sensing programs and systems has increased dramatically since the turn of the century, serving the needs of geographic information systems (GIS) with high demands for geospatial data. New technologies such as coupled global positioning systems (GPS) and inertial navigation systems (INS) allow airborne sensors to produce digital data of excellent geometric accuracy and challenge standard large format aerial cameras. Multi-source remote sensing systems are creating data at higher spatial and temporal resolution than have been collected at any other time on Earth. GIS technology allows the efficient storage and management of spatial data sets in digital formats. Remote sensing systems, in return, acquire current, accurate and synoptic data that can be used to update GIS databases. In combination with the appropriate data transfer and interoperability standards that are currently being developed the technology is being put in place that will eventually allow standardized exchange, processing and dissemination of geospatial information.

In this overall context this chapter will describe the latest developments of remote sensing systems (Section 37.2) and respective processing methods–for the derivation of both geometric and thematic information (Sections 37.3 and 37.4). Finally, conceptual issues and examples for integrating remote sensing and GIS will be demonstrated (Section 37.5).

37.2 Systems

37.2.1 Taxonomy of Remote Sensing Systems

Remote sensing can be broadly defined as the art, science and technology of obtaining reliable information about physical objects and the environment, through the process of recording, measuring and interpreting imagery and digital representation of energy patterns derived from noncontact sensors (Colwell 1997). A number of other definitions are also being used that make more specific references to the types of sensors used to record information and the wavelengths of electromagnetic radiation that are employed for transmitting this information to the respective sensors (see, for example, Jensen 2000).

Remote sensing has come a long way from its origins, aerial photography and image interpretation. Electromagnetic wavelengths used for remote sensing have extended from the visible light to the near infrared, thermal infrared and the microwave domain. Remotely sensed information is recorded by digital sensors onboard of satellite platforms. The sensors can passively record emitted or reflected radiation from the Earth surface or act as their own energy source in an active mode, such as radar and lidar systems. Images can be acquired as panchromatic (1-band), multispectral or even hyperspectral data depending on the type of sensor employed. One might also discuss the question of remoteness and look at the recording platform for sensor differentiation. Images can be acquired by satellites or space stations and downlinked to ground receiving stations. Higher spatial resolution is usually associated with airborne sensors. For ground truthing, also ground based (stationary) remote sensing platforms (e.g., boomtrucks) are used for measurements that still qualify as remote sensing. Which way is chosen to categorize remote sensing system depends mainly on the application range and the background of the scientists involved. Common categories include:

- Recording Platform
- Recording Mode
- Recording Medium
- Spectral Range
- Spectral Resolution
- Radiometric Depth
- Spatial Resolution

There are, however, no established rules concerning how sensors are classified based on the above parameters. Ehlers (2002 and 2004a) used the classification scheme as shown in Table 37-1 to propose a taxonomy for remote sensing systems.

Table 37-1 Characterization of remote sensing systems (after Ehlers 2004a).

Taxonomy of Remote Sensing Systems					
Recording Platform	Satellite/Shuttle		Aircraft/Balloon		Stationary
Recording Mode	Passive (Visible, Near Infrared, Thermal Infrared, Thermal Microwave)		Active (Laser, Radar)		
Recording Medium	Analog (Film Camera, Video)		Digital (Whiskbroom, Line Array, 2D CCD)		
Spectral Coverage	Visible/ Ultraviolet	Reflected Infrared	Thermal Infrared	Microwave	
Spectral Resolution	Panchromatic 1 Band	Multispectral 2–20 Bands	Hyperspectral 20–250 Bands	Ultraspectral > 250 Bands	
Radiometric Resolution	Low (< 6 bit)	Medium (6–8 bit)	High (8–12 bit)	Very High (> 12 bit)	
Spatial Ground Resolution	Very Low > 250 m	Low 50–250 m	Medium 10–50 m	High 4–10 m / Very High 1–4 m	Ultra High < 1m

37.2.2 Technical Background

Optical imaging systems record the energy reflected or emitted from the object via a lens or mirror system to a recording device. One of three recording devices is used: the classical film, charge coupled devices (CCDs) or complementary metal-oxide semiconductor (CMOS) sensors or single or systems of diodes. Corresponding to the recording device the energy can be captured in a plane, a line or as individual pixels. The sensor is moving during imaging, determining the scene geometry. With film, CCD- or CMOS-arrays, perspective geometry can be achieved (Figure 37-1 left). With the CCD-line or film-slit in the moving direction of the sensor, panoramic geometry exists (Figure 37-1 center) and with the arrangement across the moving direction a CCD-line or film-slit geometry exists (Figure 37-1 right). All of the described systems are used from aircraft and satellite platforms. Cameras based on diodes are not of geometric high resolution or available for civilian use, so they are not discussed here (Jacobsen 2008a).

For metric purposes the image geometry has to be reconstructed. This is not a problem for perspective images. For line sensors, the movement of the sensor must be smooth and well predictable, which it is on satellites in space, or it has to be measured. In aerial application inertial measurement units (IMUs), based on GPS-positioning and a system of gyros, are used. The drift of the gyros has to be determined by the positioning system, which is possible with some additional special flight data. Such flight data cannot be measured in space, so imaging satellites are equipped with star sensors for the update of the gyros.

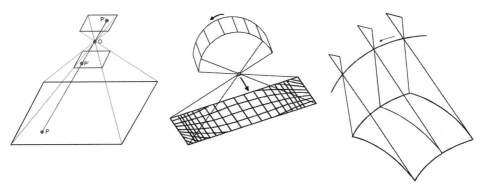

Figure 37-1 Image geometries: left, perspective geometry; center, panoramic geometry; right, line scanner geometry (push-broom).

Slit-cameras with film do not play an important role in remote sensing, so the line cameras are equipped with CCD-lines. Up to now, CMOS-sensors are not used for standard remote sensing applications; so all digital line sensors are using CCD-lines. The free electrons generated in the CCD-elements by the incoming light energy have to be shifted through the CCD-line to the read-out register. For a longer CCD-line, not enough time may be available for the read out before the platform has moved into position for the next line of pixels, for this reason often a combination of shorter CCD-lines is used. So for example, the European satellite, *SPOT*, has for the HRV, HRVIR and HRG sensors a combination of 3 panchromatic CCD-lines similar to the Indian *IRS-1C / 1D* PAN-sensor (Jacobsen 1997) and also the US *Ikonos* (GeoEye, Inc.). The US DigitalGlobe's *QuickBird* uses a combination of 6 CCD-elements (Figure 37-2).

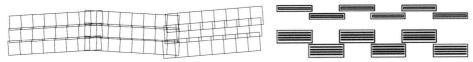

Figure 37-2 Calibration situation of IRS-1C sub-scenes, left. Arrangement of CCD-lines in focal plane of QuickBird, right, with panchromatic (above) and multispectral (below).

Corresponding to the optical combination of the CCD-lines, there may be an offset between the individual CCD-lines in the imaging plane, causing a time delay of some lines in the finally merged image (Figure 37-3a). A geometric exact combination of the sub-images having different imaging instants is only possible for a defined height level in the object space (Figure 37-3b). If the object has a varying height, by theory a mismatch cannot be avoided. For *IRS-1C* and *IRS-1D*, a mismatch of 1 pixel is caused by objects having a height difference of 450 m against the reference height; for QuickBird, the panchromatic CCD-lines are very close together so that a mismatch of one pixel is caused by a height difference of 2.8 km; that is, the mismatch is unimportant under operational conditions. Only in pan-sharpened images can the time delay between the panchromatic and the color channels be seen at moving objects. For example in Ikonos images the color follows the gray values of moving cars (Jacobsen 2007a).

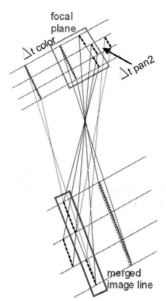

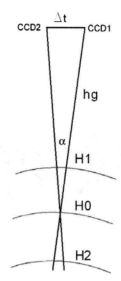

Figure 37-3a Merged image line from different image instants. See included DVD for color version.

Figure 37-3b Mismatch of merged CCD-lines.

For the very high resolution space sensors, the time interval for imaging one pixel is very short. For example, the imaging of QuickBird's 62-cm ground sampling distance (GSD) – the distance between the centers of the neighboring pixels on the ground – has only 0.09 ms available. This is not enough for a sufficient image quality, so some of the very high resolution imaging satellites are equipped with time delay and integration (TDI) sensors. They do have a small CCD-array. The free electrons in the CCD-element generated by the energy reflected by an object are shifted with the speed of the image motion to the neighboring CCD-element; there more charge is generated and again free electrons are shifted to the next CCD-element and so on. Ikonos and QuickBird usually are using 13 CCD-elements for the summation of the energy (see Figure 37-4). TDI sensors also are used by the digital aerial cameras, Z/I Imaging Digital Metric Camera (DMC) and Microsoft Vexcel Imaging UltraCamD and UltraCamX as electronic forward motion compensation.

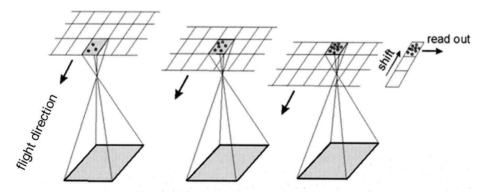

Figure 37-4 Time delay and integration sensor (TDI). See included DVD for color version.

The very high resolution optical satellite systems not equipped with TDI have to slow down the image motion by a permanent rotation of the view direction (Figure 37-5). QuickBird is equipped with TDI-sensors, but the sampling rate is limited to 6500 lines/second, so QuickBird has to use a slow-down-factor of 1.5 (see Figure 37-5).

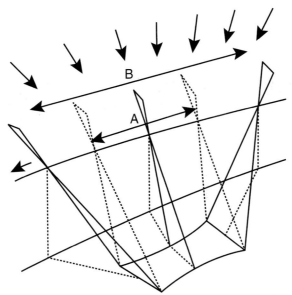

Figure 37-5 Asynchronous imaging mode - enlargement of integration time with slow-down-factor B/A by continuous change of view direction.

The slow-down-factor between the time interval really used and the imaging time corresponding to constant view direction in relation to the orbit is dependant upon the ground pixel size. The ground pixel size is not identical to the ground sampling distance because of possible over-sampling and the use of staggered CCD-lines. Staggered CCD-lines are 2 neighboring CCD-lines shifted in the line direction half a pixel against each other. By an over-sampling of 2 – one object pixel has the width of 2 GSD – the effective resolution is improved. The effective resolution, in this case, does not theoretically correspond to the GSD, but rather to the object pixel size. *SPOT 5* and *OrbView-3* use staggered CCD-lines. The HRG-sensors of *SPOT 5* have an object pixel size of 5 m and for the "supermode" 2.5 m GSD. The effective resolution checked by test targets is in the range of 3 m. For *OrbView-3* the slow-down-factor is smaller than for *EROS-A* because of the staggered CCD-lines.

The area covered by one pixel depends upon the nadir angle. The field of view (FOV) for a pixel is a fixed angle projected to the ground. One pixel covers in the view direction the pixel size in the nadir view divided by the cosine square of the incidence angle, while across the view direction the pixel size in the nadir view is divided by the cosine of the incidence angle. The incidence angle is the nadir angle from the ground point to the sensor which is larger than the nadir angle in the orbit because of the Earth curvature. The GSD in the orbit direction is determined by the sampling rate. The actual object pixel size causes an over-sampling only in the orbit direction.

The newer satellite systems can rapidly and precisely change the view direction. With *SPOT 5* this can be done only by rotating the mirror, thus determining the view direction. The influence of the Earth's rotation can then be removed by the so-called yaw correction. Other new satellites are equipped with reaction wheels or control moment gyros. Systems with fast rotating gyros can slow down or accelerate the gyros and cause a moment to the satellite, so the whole satellite rotates only based on electrical energy from the solar paddles. Control moment gyros do allow a faster rotation than the reaction wheels.

Manual of Geographic Information Systems

The replacement of photographic material (film) by digital images is leading, in most cases, to an improvement of the image quality. Film quality is influenced by the film grain and very often scratches on the photos negatively influence automatic image matching. In addition, digital sensors are more light sensitive, allowing imaging under less than optimal conditions. The high resolution space sensors and also the large size digital aerial cameras, usually have a combination of a higher resolution panchromatic band with lower resolution multispectral bands. Although by definition the panchromatic band should cover the visible range without spectral separation, in no case does the so-called panchromatic band follow this definition. Instead, the sensitivity is usually reduced for the short wavelength (blue) and extended to the near infrared. Because the short wavelength is strongly influenced by the atmospheric scatter effect, the blue band is not available for all sensors. On the other hand, the extension of the panchromatic band to the near infrared improves object contrast for vegetation and leads to better image quality. Figure 37-6 shows the system sensitivity for Ikonos and QuickBird, which are equipped with the same Kodak CCD-elements. The effective sensitivity is a combination of the overall effect of the sun's energy depending upon the wavelength, the transmission of the atmosphere, the filters and the CCD-sensitivity. The imaging systems with mid-infrared bands (SWIR) do compensate for the lower sun energy in this spectral range with larger GSD.

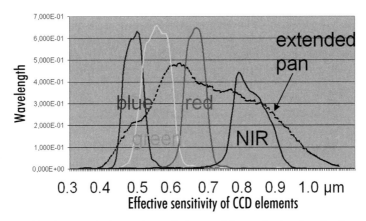

Figure 37-6 Effective spectral quantum efficiency of the optical system for Ikonos and QuickBird. See included DVD for color version.

Because of the usually higher resolution, the geometric processing—such as automatic image matching—will be made in most cases with the panchromatic band. This has disadvantages in forest areas for systems limited with their panchromatic range close to the visible range. In forest areas the near infrared has a quite better contrast. The spectral range of the panchromatic channel of SPOT is limited to 0.51 up to 0.73µm while Ikonos and QuickBird extends the range to 1.2µm causing in dark forest areas a quite better gray value distribution than for SPOT (Figure 37-7).

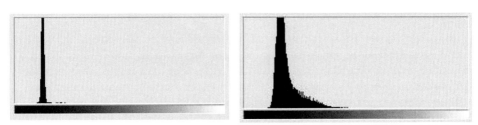

Figure 37-7 Gray value histograms of the same forest area. Left: SPOT 5 panchromatic. Right: Ikonos

Very often the pixel size of the multispectral image is 4 times larger in linear dimension (for example, 4 m × 4 m compared to 1 m × 1 m) than the panchromatic pixel size. This is an acceptable relationship for the generation of pan-sharpened images. For the human eye, it is not necessary for the color information to have the same resolution as the gray value information – since the human eye is more sensitive for gray values than for color information.

37.2.3 Satellite System Overview

High resolution satellite sensors with stereo capability such as ASTER (Advanced Spaceborne Thermal Emission and Reflection Radiometer), *Cartosat 1*, *ALOS (Advanced Land Observing Satellite)* PRISM (Panchromatic Remote-sensing Instrument for Stereo Mapping) and *SPOT 5* HRS have 2 or 3 sensors viewing backwards and forwards or to the nadir. This has the advantage of producing a stereoscopic coverage within a few seconds, while the standard sensors on *SPOT, IRS-1C* and *-1D, Resourcesat, KOMPSAT 1* and *CBERS* (China-Brazil Earth-Resources Satellite, ZiYuan) can only generate stereo models by viewing from neighboring orbits (Table 37-2, see next page). Under optimal conditions, the second image can be taken one day later. If the weather conditions do not allow this, however, a longer time interval may occur leading to problems in image-matching caused by any changes of the imaged object. The acquisition of stereoscopic coverage within the same orbit is thus advantageous. The design of the MOMS (Modular Optoelectronic Multispectral Stereo) sensor has responded to this problem by viewing forward, backward and to the nadir while *SPOT 5* has an additional HRS (high resolution stereo) camera viewing forward and backward. PRISM on *ALOS* has sensors with multiple views. The satellites with flexible view direction can generate stereoscopic coverage within the same orbit, but the required rotation to the second view direction takes time and reduces the possibility of taking other images from the same orbit. Stereoscopic imaging from the same orbit, therefore, is not an economic use of the capacity and the satellite vendors must reduce this stereo configuration to exceptions.

Regarding the high spatial resolution "spy" imagery, only the high resolution photos taken with the *CORONA 4* series and the combination of the Russian TK350 with the KVR1000 play a role in GIS. The CORONA images (available for a handling fee) are especially used as a base for change detection and archaeological purposes. In addition, they are still useful for the generation of detailed height models.

Unfortunately, actual images are not available from all satellite systems listed in Table 37-2. Some of the satellites and/or systems are now inactive (or even nonexistent) such as the first SPOT satellites, JERS-1, MOMS, CBERS-1 and OrbView-3, while others are not in an image distribution system, for instance KOMPSAT-1 and TES (Technology Experiment Satellite). ASTER scenes are often used for remote sensing and GIS purposes because of the regular imaging program, the good distribution system, the low price of a little over 60 USD for a scene combination and the stereoscopic coverage. The satellite images from the commercial companies, GeoEye (formerly OrbImage / SpaceImaging), Imagesat and DigitalGlobe, are well distributed and simple to order, but still relatively expensive. In a similar manner, SPOT Image and India are widely distributing their products. With the exception of the former MOMS, used on Space Shuttle and MIR-Priroda, all listed systems do have a sun-synchronous orbit and with the exception of KOMPSAT-1, the imaging is in the descending orbit.

Table 37-2 Larger optical high resolution space systems (satellites > 250kg).

Satellite, sensor, country, (company)	Launch	GSD [m] pan / MS	Swath [km]	Remarks
SPOT 1, HRV, France	1986	10 / 20	60	±27° across orbit
SPOT 2, HRV, France	1990	10 / 20	60	±27°
SPOT 4, HRV, France	1998	10 / 20	60	±27°
SPOT 5, HRG, France SPOT 5, HRS, France	2002	5 / 10 (2.5) 5*10 pan	60 120	±27° (staggered) +23°, −23° in orbit
JERS-1, Japan	1992	OPS 18 pan	75	+ SAR
Space Shuttle, MOMS 02, Germany	1993	4.5 / 13.5	37 / 78	nadir, + 21.5°, −21.5° in orbit direction
MIR-Priroda, MOMS-2P, Germany	1996	6 / 18	48 / 100	same as MOMS 02
IRS-1C, PAN, India	1995	5.7 / 23	70 / 142	±26° across orbit
IRS-1D, PAN, India	1997	5.7 / 23	same as IRS-1C	
IRS P6 Resourcesat, LISS-IV, India	2003	5.7 MS	24 / 70	±26° across orbit
KOMPSAT-1, South Korea	1999	6.6 pan	17	±45° across orbit
CBERS-1, CCD, China + Brazil	1999	20 MS	113	±31° across orbit
CBERS-2, CCD, China + Brazil	2003	same as CBERS-1		
Terra, ASTER, USA / Japan	1999	15, 30, 90 all MS	60	nadir, + 24° backwards
IKONOS-2 USA SpaceImaging	1999	0.82 / 3.24	11	free view direction, TDI
EROS A1, Israel, Imagesat	2000	1.8 pan	12.6	free view direction
TES, India	2001	1 pan	15	free view direction
QuickBird-2, USA DigitalGlobe	2002	0.62 / 2.48	17	free view direction, TDI
OrbView-3, USA GeoEye	2003	1 / 4	8	free view direction, TDI
FORMOSAT-2, Taiwan	2004	2 / 8	24	free view direction, TDI
IRS-P5 Cartosat-1, India	2005	2.5 pan	30	−5°, +26° , 2 cameras in orbit direction
ALOS, PRISM, Japan	2006	2.5 pan	35/70	−24°, 0°, 24°
RESURS DK-1	2006	1 / 3	28	free view direction
KOMPSAT-2, S. Korea	2006	1 pan	14	free view direction
EROS B-1, Israel	2006	0.7 pan	14	free view direction
Cartosat-2A, India	2007	1 pan	10	free view direction
CBERS 2B, HRC/CCD China/Brazil	2007	2.5 / 20	27 / 120	free view direction
WorldView-1, USA DigitalGlobe	2007	0.46 pan	16.4	free view direction, TDI
GeoEye-1, USA, GeoEye Inc	2008	0.41 / 1.65	15.2	free view direction, TDI
THEOS, Thailand	2008	2.0 / 15.0	22 / 90	free view direction

Note: GSD = ground sampling distance, pan = panchromatic, MS = multispectral,
TDI = time delay and integration

FORMOSAT-2, previously named *ROCSAT-2*, was made by EADS Astrium for Taiwan. Indeed, today it is possible to order the whole systems with satellites and ground stations including image processing. The launches since longer time are still in strong international competition and it is absolutely not a problem to launch satellites.

Table 37-3 Comparison of very high resolution satellites.

	IKONOS	**QuickBird**	**OrbView-3**	**WorldView-1**
GSD in nadir, pan / ms	0.82m / 3.28m	0.61m / 2.44m	1m / 4m	0.46m / –
Number of pixels pan	13,816	27,552	8,000	42,434
Swath width in nadir	11 km	16.8 km	8 km	16.4 km
Flying height	681 km	450km	470 km	494 km
Agility – time to slew 300km	25 sec	62 sec	31 sec	10 sec
Collection rate	2365 km²/min	2666 km²/min	1483 km²/min	6888 km²/min

Table 37-3 shows a comparison of the characteristics of very-high-resolution commercial satellite images. In the early years of Ikonos, images were distributed only with 1-m GSD, but today they also are available with the full resolution of 0.82-m GSD in the nadir view (McGill 2005). One of the differences between the image types is the swath width. Although QuickBird has the largest swath width, it also has some capacity limitations in the orbit direction caused by the required slow-down-factor of 1.5 (Section 37.2.2). Nevertheless, the collection rate for QuickBird is larger than for Ikonos and OrbView-3. For all three sensors the orientation accuracy projected to the ground, without control points, is ~15 m CE90 (circular error at the 90% confidence level), corresponding to standard deviation for X and Y of 7 m.

The required hardware components for the satellites are becoming smaller and smaller, so today it is no longer necessary to have very heavy satellites for reconnaissance purposes. A weight of 100kg to 300kg should be enough for satellites with ground resolutions suitable for topographic mapping purposes. Such satellites also can be equipped with off-the-shelf components for cost reduction. Several small Earth observation satellites have been launched during the last years (Table 37-4). With the exception of *KITSAT* and *SunSAT*, the small satellites listed in this table are produced or partially produced by Surrey Satellite Technologies (SSTL) in the UK. These satellites can be and have been ordered as complete solutions including the ground stations with the required equipment. Most of the listed countries are cooperating in the disaster monitoring constellation – in the case of natural disasters, the affected area is being imaged within 24 hours by at least one of the satellites. *TopSat*, *BLMIT-1*, *UOSAT 12* and *BilSat 1* images have a GSD in the range required for topographic mapping, but they do not have a regular image distribution system (Jacobsen 2007c).

Table 37-4 Small optical space systems (satellites < 250kg).

System, country, (company)	Launch	GSD [m] pan / MS	Swath [km]	Remarks
UOSAT 12 UK	1999	10 / 20	10 / 30	CCD arrays
KITSAT 3, South Korea	1999	15 MS	50	
SunSAT, South Africa,, SunSpace	2000	15	52	
Alsat 1, Algeria	2002	32 MS	600	DMC
BilSat 1, Turkey	2003	12 / 28	24 / 53	DMC, CCD arrays
BNSCSat, UK	2003	32 MS	600	DMC
NigeriaSat, Nigeria	2003	32 MS	640	DMC
DMC+4 (BLMIT-1), China	2005	4 / 32	24 / 600	DMC
TopSat, UK	2005	2.5 / 5	10 / 15	free view direction, TDI
RapidEye, Germany, RapidEye AG	2008	– / 6.5	77	System of 5 satellites

Several high-resolution optical satellite systems have been announced (Table 37-5). The proposed launch time in most cases was delayed. In addition, some systems may fail, others may be cancelled and additional systems may come. In general, the number of high- and very-high-resolution systems will be enlarged very soon. The tendency is toward higher resolution and lower weight; that translates in most cases to a lower price. In addition, the tendency of international cooperation for the component assembly is growing. It is no longer necessary to develop all the components in one's own country and components (or even the whole systems) can be bought for a reasonable price.

Table 37-5 Announced larger optical space sensors.

System	Launch	GSD [m] pan / MS	Swath [km]	Remarks
EROS C, Israel	2009	0.7 / 2.8	11	free view direction, TDI
CBERS-3, China, Brazil	2008	5 / 20	60 / 120	+/-32° across
WorldView-2 DigitalGlobe, US	2008	0.5 / 2	16.4	free view direction, TDI, 8 bands
GeoEye-1, US	2008	0.41 / 1.64	15	free view direction, TDI
THEOS, Thailand	2008	2 / 15		free view direction, TDI
Pleiades 1 , France	2009	0.7 / 2.8	20	free view direction, TDI
Pleiades 2, France	2009	same as Pleiades 1		

Most of the systems, however, will have a flexible view direction also allowing the generation of stereo models within the same orbit. Three systems will have a GSD below 1m and TDI will be used in most of the very-high-resolution systems. The French *Pleiades* will be the follow-on system for *SPOT* with higher resolution and lower weight; thus also for lower cost. *THEOS* will be made by EADS Astrium for Thailand. GeoEye made a proposal for *GeoEye-2* having 0.25 m GSD.

WorldView-1, WorldView-2 and *GeoEye-1* are based on contracts within the NextView program of the US military. Not only these satellites, but most of the high resolution observation satellites, are based on dual use – the highest percentage of the required funds coming from the military and the remaining free capacity available for commercial applications. *WorldView-1, WorldView-2* and *GeoEye-1* will lead to 0.5-m GSD. In addition to the bands used by QuickBird, four new spectral bands are included in WorldView-2 (Figure 37-8). The main reason for the higher spectral resolution is based on military requirements, but this also is useful for civilian applications. The coastal band with its possibility of water penetration may be especially useful in shallow water.

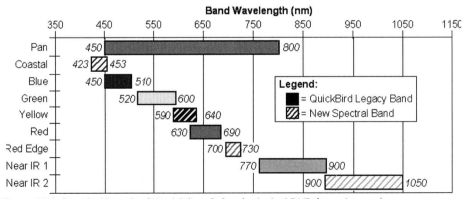

Figure 37-8 Spectral bands of WorldView-2. See included DVD for color version.

WorldView-1 has and *WorldView-2* will have control-moment gyros, with agility strongly improved over that of *QuickBird*. They will be able to slew (turn in order to point the sensor at an area of interest) over a distance of 300 km within 10 or 9 seconds. Also the positioning accuracy without control points shall be improved to 7–10 m CE90 (equivalent to 3–4 m standard deviation of X and Y).

Several small satellites with medium- to high-resolution optical sensors also were recently launched or will be launched. In this group *RapidEye* is one of the first totally commercial applications without financial support by dual military and civilian use. The 5 low-cost satellites comprising the system were constructed by SSTL and are operated by RapidEye, a spin-off of the German Aerospace Centre DLR, in cooperation with MacDonald Dettwiler, Canada. The main application is in the field of precision farming.

With the high altitude long endurance (HALE) unmanned aerial vehicles (UAVs), an alternative between space and airborne systems will come. The HALE UAV uses solar energy and may stay in the air for a month. With *Pathfinder-Plus,* NASA has already made a successful test. The Flemish Institute for Technological Research (VITO) in Belgium had the first test flight of its HALE UAV *Pegasus* in 2006. The sun-powered *Pegasus* is designed for continuous operation over several months in up to 55° northern latitude from March to September. It operates at a height of 20,000 m, overnight it will go down to 16,000 m and rise again the next morning. These altitudes are above aeronautic control, thereby avoiding safety problems. In a partially autonomous flight it can be directed to the area for imaging. Starting in 2007, it carried a digital camera with 4 spectral bands and 12,000 pixels having a GSD of 20 cm. In the future, this shall be extended to 10 spectral bands and 30,000 pixels and SAR and lidar may be included.

37.2.4 Airborne System Overview

Complementary to images of satellite systems, those of conventional **aerial photographic cameras** are a valuable input for GIS applications with high demands concerning geometric details, in particular for topographic mapping at large scales. Obviously, the usage of photographic imagery in a digital GIS environment demands an analog-to-digital conversion or scanning. Besides the additional efforts of film handling and scanning, photographic imagery is restricted concerning the separability of spectral channels. An overview of photographic systems is not the subject of this section and for this we refer to the ASPRS *Manual of Photogrammetry* by McClone et al. (2004).

The above-mentioned drawbacks have led to the development and use of **digital imaging systems** since the end of the 1990s. Advanced technologies such as global positioning systems (GPSs) and inertial measurement units (IMUs) coupled in navigation systems, as well as improved digital sensor technologies, have overcome the strongest impediment of aircraft scanners—the lack of geometric stability. This new generation of digital airborne remote sensing systems offers a broader field of topographic and thematic applications at large scales. This is not only due to the digital form of the data, enabling a direct and faster data flow, but also to the ability for acquiring multispectral data at a reasonable level of geometric accuracy and an improved radiometric resolution.

Comparing the **technical properties** of these digital systems with their film-based counterparts in more detail can be summarized as follows:

- The major difference is concerned with the *spectral resolution and range*: Digital camera systems have a multispectral ability, leading to the measurement of absolute radiances with separable, non-overlapping and rather narrow bands in the visible and near infrared range of the electromagnetic spectrum. The design of the spectral bands is very similar to that of the digital camera's satellite-borne counterparts (e.g., Ikonos, QuickBird).
- Empirical tests state that the dynamic range of 12 bits in digital imagery (the *radiometric resolution*) is superior to the estimated 8 bits in film and that "radiometric noise" due to

the film granularity is a major drawback of film recordings (Leberl and Gruber 2003). Consequently, digital systems show a comparatively better signal-to-noise ratio.

- A comparison of the *geometric resolution* between digital and film-based systems is, strictly speaking, not possible. Depending on the (fixed) focal length and the flying altitude, one obtains ground sampling values (down to 5 cm) which are in the same order or even superior as those obtained with film-based cameras. Due to the wider angle (and the lower flying altitude) one suffers less atmospheric interference, but also more shadowing and obstruction effects (at least across the direction of flight if line scanners are used).

- With most digital systems we have (still) to cope with a significantly smaller *coverage* in the across-flight direction, in contrast to film cameras.

- With respect to *stereoscopic data acquisition,* digital systems are using two different approaches. One option is to employ area CCD arrays, which is similar to the image frame principle of conventional film cameras and which produces stereoscopy through overlapping scenes. The other technical realization is to use line CCD arrays that apply the triplet stereo principle. For a discussion on these approaches refer to Spiller (2000) and Fricker et al. (2000).

Table 37-6 gives an overview of the parameters of selected digital systems which are presently available on the market. For further systems – such as STARIMAGER (Tsuno et al. 2004), DiMAC (Digital Modular Aerial Camera; GIM International 2004), JAS-150 (Jena Airborne Scanner; Georgi et al. 2005) or DigiCAM (Grimm and Kremer 2005) – we refer to the referenced literature. The change from analog to digital cameras has happened very fast (Jacobsen 2007b).

Table 37-6 Selected digital airborne camera systems (based on company information).

Company	DLR	Leica Geosystems	Microsoft Photogrammetry	Microsoft Photogrammetry	Z/I Imaging
Sensor	HRSC-AX	ADS 40	UltraCam-D	UltraCamX	DMC
Sensor type	system of CCD-lines	system of CCD-lines	CCD array	CCD array	CCD array
Year of introduction	2000	2000	2003	2006	2002
Focal length	151 mm	62.7 mm	100 mm (28 mm multispectral)	104 mm (28mm multispectral)	120 mm (25 mm multispectral)
Total field of view	29°	62.5°	55° × 37°	55° × 37°	74° × 44°
Number of CCD lines	9	7	11500 × 7500 (pan) 2672 (ms)	9420 (pan)	7680 (pan) 2 000 (ms)
Sensors per CCD line	12 172	2 × 12 000 (pan) 12 000 (ms)	11 500 (pan) 4008 (ms)	14430 (pan)	13 824 (pan) 3 000 (ms)
Sensor size	6.5 μm	6.5 μm	9 μm	7.2 μm	12 μm
Radiometric resolution	12 bit	12 bit	12 bit	12 bit	12 bit
Spectral resolution (nm)	520–760 (pan) 450–510 (blue) 530–576 (green) 642–682 (red) 770–814 (NIR)	465–680 (pan) 428–492 (blue) 533–587 (green) 608–662 (red) 703–757 (NIR) or 833–887 (NIR opt.)	4 bands: blue, green, red, infrared 1 band pan	4 bands: blue, green, red, infrared 1 band pan	400–580 (pan) 400–580 (blue) 500–650 (green) 590–675 (red) 675–850 (NIR)
Read-out frequency	1640 lines/sec	800 lines/sec	> 1 images/sec	1 image/sec	0.5 images/sec

In addition to stereo imaging for acquiring 3D information, airborne laser scanning or light detection and ranging (lidar) systems can be used. Such systems also have reached maturity within recent years due to the enhanced performance of GPS/IMU solutions for capturing position and orientation data of the associated moving platforms. Today's standard systems are able to capture multiple reflections—either as discrete pulses (first and last echo data) or even as the complete waveform—from which digital surface models (DSMs) can be derived. In addition, some systems also are able to record the strength of the reflected laser pulses (generally, in the near infrared portion of the spectrum), leading to **lidar intensity images**. However, in contrast to the well-known behavior of passive systems, one has to consider:

- superimposition with other illumination sources, leading to significant noise;
- the decrease of intensity with the increase of the traveling distance of the laser beam;
- a rather small ratio of reflected to emitted energy; and
- the existence of disturbing multiple reflections.

Indeed, intensity images have an additional value for orientation and rough visual interpretation purposes, but due to the mentioned limitations they only have rather small potential with regard to interpretability compared to the imagery of passive sensors as presented earlier in this section.

For many GIS applications, images and 3D information (as well as lidar intensity imagery in some cases) are used. The option to simultaneously acquire image and lidar data leads to high potential for **multi-sensor systems**. Such systems are able to deliver more accurate and reliable elevation data compared to stereo image matching solutions, and, on the other hand, by-pass the disadvantage of the "blind" laser scanning information. Examples of such systems (for example, FALCON by TopoSys, or the combination of ALS50 and ADS40 by Leica Geosystems) are described by Schiewe and Ehlers (2005).

37.3 Geometric Processing Methods

37.3.1 Overview of Methods

The geometric processing of images critical for use in a GIS is based on the relation between image and ground coordinates. In order to derive the three dimensions of the object coordinate system from two-dimensional images, a geometrically correct object reconstruction has to be based on at least two images taken from different locations. Alternatively, if only one image is available, a height model is required for the reconstruction of the horizontal ground coordinates X and Y. Without a height model, only an approximate solution is possible based on just one image.

With known sensor geometry and orientation from image positions, object coordinates can be determined. As general sensor geometry we do have perspective geometry, panoramic geometry and CCD-line geometry (see Figure 37-1). The perspective geometry is dominating in aerial applications, while some small satellites have CCD-arrays. For aerial applications, the dominating method of orientation is the common adjustment of all images with their image coordinates to the base of ground control points by bundle block adjustment (Kraus 1993, Volume 1, Chapter 5.3 Bundle Block Adjustment). Since the year 2000, the use of direct and integrated sensor orientation based on relative kinematic GPS-positioning, together with inertial measurement unit (IMU) data, has been growing. The exact image geometry can be determined by self-calibration with additional parameters. Self-calibration by additional parameters also can be used for the reconstruction of the imaging geometry of the dynamic panoramic and CCD-line cameras, having by theory a different projection center coordinates and attitudes for every CCD-line and scan-lines of panoramic images. In aerial applications, control points have to be used for the reconstruction of the imaging geometry. The number of required control points depends upon the positional and attitude information of the sensor (Jacobsen 2007a).

For space applications, the geometry of the dominant CCD-line scanner images can be reconstructed with just a few control points if the available general satellite orbit (ellipse and orbit inclination) is used in addition to the usually given view direction in relation to the orbit. The direct reconstruction of the imaging geometry is only possible with original space images, usually corrected by the inner sensor geometry. The user does not have to be concerned with the merging of sub-images from the individual CCD-lines. The original images are named by *SPOT, IRS, KOMPSAT* and some other satellite operators as level 1A and for *QuickBird* and *WorldView-1* as Basic Imagery. Original images are not available for *Ikonos*. For *Ikonos*, and also from most of the sensors, images projected to a plane with constant height are distributed as *Ikonos* Geo, level 1B or for *QuickBird* and *WorldView-1*, OrthoReady (OR) Standard. See Figure 37-9 for the geometry of imaging products.

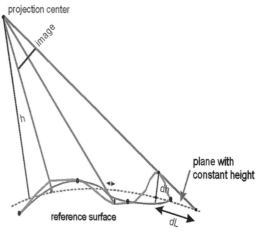

Figure 37-9 Imaging products: original images = level 1A; images projected to plane with constant height = level 1B. See included DVD for color version.

Also available for *QuickBird* are rough orthoimages based on the DEM GTOPO30 designated as Standard Imagery. The level 1B images – including Ikonos Geo and QuickBird OR Standard – are based on the direct sensor orientation of the satellite. For mapping purposes, the actual ground height has to be used for the terrain relief correction shown as dL in Figure 37-9. For precise positioning, a correction by control points is required.

The information of the direct sensor orientation is not available for all sensors. When the sensor geometry has not been published (for *Ikonos*, for example) it can be reconstructed based on the view direction from the scene center to the satellite and the general satellite orbit. Another possibility for the reconstruction of the relation between image and ground coordinates is to use the rational polynomial coefficients (RPCs), also named rational polynomial functions (Equation 37-1).

$$xij = \frac{Pi1(X, Y, Z)j}{Pi2(X, Y, Z)j} \tag{37-1a}$$

$$yij = \frac{Pi3(X, Y, Z)j}{Pi4(X, Y, Z)j} \tag{37-1b}$$

$$\begin{aligned} Pn(X,Y,Z)j = {} & a_1 + a_2{*}Y + a_3{*}X + a_4{*}Z + a_5{*}Y{*}X + a_6{*}Y{*}Z \\ & + a_7{*}X{*}Z + a_8{*}Y^2 + a_9{*}X^2 + a_{10}{*}Z^2 + a_{11}Y{*}X{*}Z + a_{12}{*}Y^3 + \\ & a_{13}{*}Y{*}X^2 + a_{14}{*}Y{*}Z^2 + a_{15}{*}Y^2{*}X + a_{16}{*}X + a_{17}{*}X{*}Z^2 + \\ & a_{18}{*}Y^2{*}Z + a_{19}{*}X^2{*}Z + a_{20}{*}Z^3 \end{aligned} \tag{37-1c}$$

where xij, yij =scene coordinates and X,Y = geographic object coordinates, Z = height, P = polynomials.

Based on the direct sensor orientation and the inner satellite geometry, the relationship between the scene and the ground positions can be expressed by the ratio of two polynomials (Equation 37-1). Since third-order polynomials are usually used, 80 coefficients are required to express the imaging geometry. For precise positioning, a "bias correction" based on control points also is required.

37.3.2 Image Orientation and Registration

The orientation of images projected to a plane with constant height (level 1B, Ikonos Geo, QuickBird ORStandard) can be based on correct mathematical models, but also on approximations. If RPCs are available, they can be used together with control points as **bias corrected RPCs**.

For the scene center or the first line, the direction to the satellite is available in the image header data. This direction can be intersected with the orbit of the satellite as defined by its published Kepler elements. Depending upon the location of an image point, the location of the corresponding projection center on the satellite orbit together with the view direction can be computed. The view direction from any ground point to the corresponding projection center can be reconstructed. This method of the **reconstruction of the imaging geometry** requires, therefore, the same number of control points as the sensor-oriented RPC-solution. That means this method can also be used without control points if the direct sensor orientation is accepted as accurate enough, or this method requires the same additional transformation of the computed object points to the control points as the sensor-oriented RPCs if additional accuracy is required.

The **three-dimensional affine transformation** does not use available sensor orientation information. The 8 unknowns for the transformation of the object point coordinates to the image coordinates have to be computed based on control points located not in the same plane (Equation 37-2). At least 4 well-distributed control points are required. The computed unknowns should be checked for high correlation values between the unknowns – large values indicate numerical problems which cannot be seen at the residuals of the control points, but may cause large geometric problems for extrapolations outside the three-dimensional area of control points. Three dimensional means also the height, so problems with the location of a mountain top may be caused if the control points are only located in the valleys. A simple significance check of the parameters, e.g., by a Student test, is not sufficient. The 3D-affinity transformation is based on a parallel projection which is approximately given in the orbit direction, but not in the direction of the CCD-line. The transformation can be improved by a correction term for the correct geometric relation of the satellite images.

$$\text{xij} = a_1 + a_2 * X + a_3 * Y + a_4 * Z \qquad \text{yij} = a_5 + a_6 * X + a_7 * Y + a_8 * Z \qquad (37\text{-}2)$$

In this 3D-affine transformation xij, yij = image coordinates; X, Y, Z = object coordinates; $a1, \ldots, a6$ = unknown transformation parameters.

Direct Linear Transformation (DLT): Like the 3D affine transformation, the direct linear transformation (Equation 37-3) does not use available sensor orientation information. The 11 unknowns for the transformation of the object point coordinates to the image coordinates have to be determined with at least 6 control points. The small field of view for high resolution satellite images, together with the limited object height distribution in relation to the satellite flying height, causes considerably more numerical problems than for the 3D affine transformation. The DLT is based on a perspective image geometry which is available only in the direction of the CCD line. Although some commercial software packages offer this method for satellite orientation, its use is not justified, since other solutions have fewer unknowns.

$$xij = \frac{L1*X + L2*Y + L3*Z + L4}{L9*X + L10*Y + L11*Z + 1}$$ (37-3a)

$$yij = \frac{L5*X + L6*Y + L7*Z + L8}{L9*X + L10*Y + L11*Z + 1}$$ (37-3b)

Terrain-dependent RPCs: The relationship of scene to object coordinates can be approximated by a limited number of polynomial coefficients shown in Equation 37-1 based on control points. The number of chosen unknowns is quite dependent upon the number and three-dimensional distribution of the control points. The success of using terrain-dependent RPCs cannot be checked just by the residuals at the control points. Some commercial programs offering this method do not use any statistical checks for high correlations of the unknowns, making the correct handling extremely difficult. A selection of the unknowns may lead to the three-dimensional affine transformation.

The different orientation methods have been compared with very high resolution satellite images (Jacobsen 2007a). The orientation process includes mainly two steps, the terrain relief correction (Equation 37-4) and the orientation in relation to the control points. The images projected to a plane with constant height do have location differences (DL) compared to an orthoimage, depending upon the height differences (Δh) and the incidence angle v (Equation 37-4).

$$DL = \Delta h * \tan$$ (37-4)

After terrain relief correction (correction by DL—Equation 37-4), the scene has to be related to the control points. Based on the high quality direct sensor orientation of very-high-resolution satellites, the differences of the terrain-relief-corrected scene against the control points are small. For some satellites, just a shift in X and Y is sufficient, while for some others a 2D affine transformation or even more parameters may be required.

The terrain dependent RPCs—the rational polynomial coefficients determined by control points – have been shown to be a very unsafe solution. Because of the higher number of unknowns, the discrepancies at the control points are small, but the discrepancies at independent check points may be very large (Figure 37-10). In one test (Jacobsen 2007a) with less-than-optimal distribution of control points, the discrepancies at check points were in the range of 50 m even if it was not an extrapolation beyond the 2-dimensional range of the control points. Since typically-used commercial programs do not indicate these problems, this method of orientation should generally be avoided.

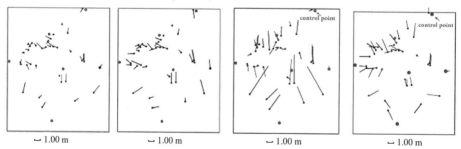

Figure 37-10 Shows the discrepancies at independent check points of the orientation of a QuickBird scene based on 6 control points using different orientation methods (Jacobsen 2006). After terrain relief correction, an affine transformation to the control points was required for the RPCs (based on the direct sensor orientation) and also the geometric reconstruction. Just with a shift to the control points, the RPC-solution has root mean square discrepancies for 2.09 m in X and 1.04 m in Y and corresponding large discrepancies at the check points. This was similar for the geometric reconstruction with the Hannover program CORIKON. See included DVD for color version.

Table 37-7 Orientation of QuickBird scene Zonguldak based on 6 and all 41 control points [discrepancies in (m)].

	sensor oriented RPCs		geometric reconstruction		3D-affine transformation		DLT	
	SX	SY	SX	SY	SX	SY	SX	SY
6 control points – discrepancies at control points	0.45	0.62	0.84	1.12	0.48	0.18	0.02	0.08
6 control points – discrepancies at check points	0.53	0.75	0.84	1.01	0.96	2.22	1.08	1.20
41 control points – discrepancies at control points	0.40	0.59	0.76	0.62	1.04	1.04	1.36	1.50

The best results have been achieved with sensor-oriented RPCs, followed by the results of geometric reconstruction (Table 37-7). The discrepancies generated by the 3D affine transformation are clearly larger. The Hannover program TRAN3D does perform statistical checks for high correlations of the unknowns, and it warned of high correlation values for the DLT-method based on 6 control points. The unrealistic discrepancies at the control points of 2 cm and 8 cm are caused by the poor over-determination—6 control points are giving 12 observations and the DLT-method has 11 unknowns. By this reason, the small discrepancies at the check points were a surprise. But this is just a random result; if just 1 control point is exchanged with a neighboring point, the discrepancies at the check points are in the range of 4 m (Figure 37-11). If all control points are used, it is possible to include more unknowns in the geometric reconstruction. With 2 orientation parameters and 2 additional parameters for the orientation dependent upon the location in the scene, the geometric reconstruction decreased to RMSX=0.48 m and RMSY=0.46 m.

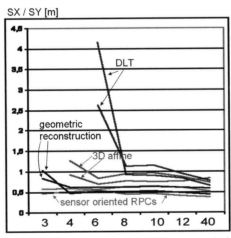

Figure 37-11 QuickBird data of Zonguldak: Results at independent check points for the different orientation methods as a function of the number of control points - only the result achieved with 32 control points shows the residuals at control points

A similar test has been made with Ikonos images of the same area, Zonguldak (Figure 37-12), with some of the same control points. In general the internal accuracy of Ikonos images seems to be more stable than for QuickBird. After geometric reconstruction a shift to the control points was sufficient. The transformation of the terrain-relief-corrected locations to the control points gave better results at the check points when just a shift was used instead of an affine transformation. Depending upon the number of control points, up to 10% larger discrepancies at the check points have been achieved with an affine transformation. So with just 2 unknowns, the Ikonos orientation can lead to optimal results.

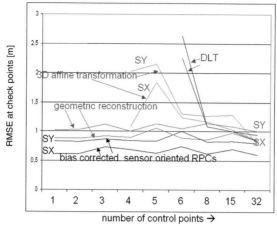

Figure 37-12 Ikonos data of Zonguldak: Results at independent check points for the different orientation methods as a function of the number of control points—only the result achieved with 32 control points shows the residuals at control points. See included DVD for color version.

The results achieved with an Ikonos scene shows the same trend as with QuickBird (Jacobsen 2007a). The best results are from the sensor-oriented RPCs based on the direct sensor orientation; next best results are from geometric reconstruction. By theory the 3D affine transformation requires at least 4 control points and DLT at least 6 control points. With just one control point the RPC solution and also the geometric reconstruction are directly dependent upon the chosen control point. The missing reliability does not allow a check of the solution; so at least 2 control points should be used. The location of the control points within the scene is not important. This is quite different for the 3D affine transformation and for DLT. Both orientation methods do not use any information about the scene orientation and for this reason a good three-dimensional distribution of the control points is absolutely necessary. A 3D affine transformation using 4 control points, well-distributed in X and Y, resulted in RMSX=1.9 m and RMSY=18.5 m at check points because the 4 control points described a surface that was nearly a tilted plane. This was indicated by the Hanover program, TRAN3D, with a warning that the program found extreme correlations between the unknowns. It should be noted that commercial programs usually do not use such statistical tests and the result may look very nice even if it is poor. The same is true with the DLT where the unnecessarily high number of orientation unknowns causes even more problems.

Orientation of original images (level 1A): Space images are distributed as close-to-original images, just corrected by inner sensor orientation (level 1A), and as images projected to a surface with constant height above the geoid (level 1B). The names level 1A and 1B originally came from SPOT Image. Other organizations use different names, so Ikonos Geo is a level 1B-type image, QuickBird Basic is a level 1A-type image, while QuickBird OR Standard belongs to level 1B.

The solutions for handling original images are well known since the start of *SPOT 1*. In the Hannover program BLASPO, the image geometry is reconstructed based on the given view direction, the general satellite orbit and few control points. Based on control points the attitude and the satellite height are improved. The X- and Y-locations are fixed because they are nearly linear dependent upon the view direction. In addition, two parameters for image affinity and angular affinity are required. For these 6 unknowns, 3 control points are necessary. More parameters can be introduced if geometric problems exist. Only for scenes with totally unknown orientation the full sensor orientation with 6 orientation elements (3 positions plus 3 attitude unknowns) has to be adjusted together with necessary additional parameters. This requires a good vertical distribution of control points; so for flat areas the full orientation cannot be computed. Other solutions are based on the satellite ephemeris, available in the header files together with the view direction. These methods also have to be improved by the same number of control points. Some solutions do not use available orientation information (see 3D affine transformation, direct linear transformation (DLT) and terrain dependent RPCs above). As with the solution for level 1B-data, more control points with a good three-dimensional distribution are required if the existing sensor orientation information is ignored.

Table 37-8 Orientation accuracy of different original (level 1A) space images by University of Hannover (Jacobsen 2008b).

	RMSX / RMSY [m]	RMSx″ / RMSy″ [GSD]
TK 350, Zonguldak	*8.3*	*(0,8)*
KVR 1000, level 1A, Duisburg	*3.3*	*(1.6)*
KFA 3000, Vienna	*2.5*	*(2.0)*
ASTER, Zonguldak	10.8	0.7
KOMPSAT-1, Zonguldak	8.5	1.3
SPOT, level 1A, Hannover	4.6	0.5
SPOT 5, level 1A , Zonguldak	5.1	1.0
SPOT HRS, Bavaria	6.1	0.7 / 1.1
IRS-1C, level 1A, Hannover	5.1	0.9
QuickBird Basic, Atlantic City	0.60	1.0
OrbView-3 Basic, Zonguldak	1.3	1.3
Cartosat-1, Warsaw	1.5	0.6
WorldView-1, Istanbul	0.45	0.9

The quality of orientation by geometric reconstruction of different types of original images—computed with the Hannover programs BLUH and BLASPO—is shown in Table 37-8. In general, it is possible to reach accuracy in the range of the ground sampling distance (GSD) or even smaller (i.e., better). In all cases having a standard deviation exceeding 1.0 GSD, this is caused by limited accuracy of the ground control points. A QuickBird scene from Atlantic City has been handled also with 3D affine transformation leading to poor results of root mean square (RMS) for the X-component of 12.6 m and RMSY = 8.7 m. The DLT with 11 unknowns gave the values RMSX = 8.9 m and RMSY = 7.8 m. The area is flat and so the determination of the unknowns for the 3D-handling is poor. All Student test values (size of parameter divided by its own standard deviation) are below 1.0, even for 174 control points, well distributed in X and Y. The correlation of the unknowns caused a warning to be issued by the used Hannover program TRAN3D. In the area of Phoenix, however, the orientation of a QuickBird scene by 3D affine transformation led to RMSX = 1.4 m and RMSY = 2.1 m, and the DLT to RMSX = 1.2 m and RMSY = 0.9 m. This is a poor result in relation to the exact reconstruction of the geometry which can be done with a smaller number of unknowns and without problems of large correlations. Large correlation values do indicate problems of point determination outside the 3D-area of control points. It can be summarized that the orientation of the QuickBird scenes with 3D affine transformation and DLT did not lead to satisfying results. Also the orientation of SPOT and SPOT HRS images with these methods led to 50% up to 100% higher standard deviations in relation to the orientation with geometric reconstruction.

37.3.3 Analysis of DSMs Generated by SRTM Data

The generation of orthoimages requires a digital elevation model (DEM) with the height information of the bare ground or a digital surface model (DSM) with the height information of the visible surface. Only a few stereo combinations of high resolution optical satellite images are available, so other height information is required. Based on the Shuttle Radar Topography Mission (SRTM) in February 2000 for the area from latitude 56° south up to 60.25° north, NASA generated DSMs , available for download free of charge. During the 11-day mission, two different radar systems were used in the Space Shuttle – the US C-band with 5.6-cm wavelength and the German-Italian X-band with 3-cm wavelength. In addition to the active antenna systems in the Shuttle, passive antennas were available at a 60-m-long arm outside the Shuttle. With the combination of the active and the passive antenna by interferometric synthetic aperture radar (InSAR), the visible surface on the ground could be determined in three dimensions leading to DSMs showing the height of the visible surface as the upper part of vegetation and buildings. The C-band as well as the X-band radar cannot penetrate the vegetation, so a DEM with the height of the bare ground only can be achieved by filtering the SRTM height model for points not belonging to the bare ground (Passini et al. 2002).

The US C-band was operated with a scan-SAR mode having a swath width of 225 km, while the X-band was limited to a swath width of 45 km. For this reason, the X-band data have large gaps between the strips, while the C-band has nearly complete coverage; 94.6% of the mapped area is covered at least twice and approximately 50% at least three times. Only in very steep areas, in dry sand deserts and water surfaces are gaps totaling 0.15% caused by radar layover. In the 1° × 1° area around Mount Everest, 9% of the possible points are missing.

Outside the USA, the C-band height model is limited to a point spacing of 3 arcsec, corresponding to 92 m at the equator. The X-band data can be bought from the German Aerospace Centre DLR with a spacing of 1 arcsec, corresponding to 31m at the equator. The accuracy loss caused by the point spacing of a height model depends upon the roughness of the terrain (i.e., the change of the terrain inclination). In very rough mountainous areas, for example, the loss of accuracy by interpolation over 92 m can result in RMS values above 10 m, while it is negligible for flat areas.

Like a height model generated by automatic image matching, the height model based on C-band and X-band InSAR corresponds to the visible surface. Such a DSM can be reduced by automatic filtering to a DEM if enough height points are available for the bare ground and if the accuracy of the height values is below the terrain roughness. This is not usually a problem in open and built-up areas, but it is difficult in closed forest areas where no point is located on the bare ground. In such forested areas, the DSM still can be improved by filtering, especially at the border, but not in the center. In mountainous areas where the height accuracy is in the range of the terrain roughness, a filtering of a DSM to a DEM is not possible.

The characteristics of the SRTM C-band DSM can be analyzed very well in the flat area of the city of Bangkok where the elevation of the SRTM height model reaches 44 m. The bare ground of the Bangkok area is extremely flat with heights not exceeding 4 m. After filtering the SRTM C-band DSM the largest elevation in the filtered height model is only 6 m—this is a realistic value for some artificially raised ground levels (Jacobsen 2005). By filtering in the city area, 59% of the height points are eliminated—and this is realistic.

The vertical orientation accuracy of SRTM-DSMs is in the range of ± 3 to 4 m RMSE. This can be improved with control values. The horizontal orientation accuracy is in the range of 30 m RMSE. Problems such as the horizontal orientation mismatch are caused by a merge with existing height models. Often the geometric reference and datum are not known and horizontal shifts of 200 m are common. For this reason, for an analysis and a merging of height data, the horizontal shift of the DEMs or DSMs has to be determined in advance. This should be done by adjustment of all available data to avoid local problems. The functional model of such an adjustment is

$$DZ = \tan \alpha \, DL \qquad\qquad (37\text{-}5)$$

where α is the terrain inclination and DL is the horizontal shift.

The analysis of a SRTM C-band DSM in a rolling area in Bavaria indicates the characteristics of the height model (Table 37-9). The accuracy reached in forest areas cannot be compared with the open areas and the accuracy shows a clear dependency on the terrain inclination. A DEM could be achieved by filtering in the open area, but not in the forest areas (Passini and Jacobsen 2007).

Table 37-9 Analyses of SRTM C-band height model against lidar reference–test area Gars, Bavaria.

	RMSZ [m]	bias [m]	RMSZ F(inclination) [m]
open areas, no filtering	5.44	−2.33	4.37 + 2.5 * tan α
forest, no filtering	16.46	−13.84	14.98 + 3.8 * tan α
open areas filtered	4.03	−2.15	3.45 + 1.9 * tan α
forest filtered	11.77	−9.07	10.76 + 3.5 * tan α

The accuracy achieved in rolling areas of Bavaria has been confirmed in other areas (Passini and Jacobsen 2007) where qualified reference data were available (Table 37-10). In mountainous parts, the dependency upon the terrain inclination is much stronger than in more smooth areas. In addition, a small dependency upon the aspects (the inclination as a function of direction) can be seen. If the systematic height errors (bias) are removed, the variation of the RMS height error is reduced by 2.5-3.7 m for open and flat areas. Similar tests have been made with the SRTM X-band data. It shows the same characteristics and the same accuracy, but it has the advantage of a smaller spacing.

Table 37-10 Accuracy of SRTM C-band DEMs in different test areas (only open areas).

	RMSZ [m]	bias [m]	RMSZ F(slope)
Arizona, flat to mountainous	3.9	1.3	2.9 + 22.5 * tan α
Williamsburg NJ, flat	4.7	−3.2	4.7 + 2.4 * tan α
Atlantic City NJ, flat	4.7	−3.6	4.9 + 7.6 * tan α
Bavaria, rolling	4.6	−1.1	2.7 + 8.8 * tan α
Bavaria, mountainous	8.0	−2.4	4.4 + 33.4 * tan α

The comparison of contour lines (Figure 37-13) in the mountainous area of Zonguldak shows the loss of morphologic details of the SRTM C-band height model with the spacing of 3 arcseconds against the 1 arcsecond spacing of the X-band data and the reference height model with a spacing of 40 m. Even the quite less accurate ASTER height model provides more morphologic details (Büyüksalih and Jacobsen 2007).

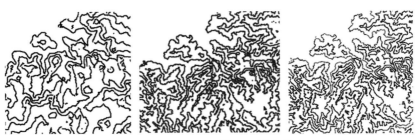

Figure 37-13 Contour lines in mountainous area of Zonguldak, Turkey, contour interval = 50 m. Left, SRTM C-band DEM; center, SRTM X-band DEM; right, reference DEM.

Manual of Geographic Information Systems

The accuracy of the SRTM C-band DSM is sufficient for orthoimages using very high resolution space images if the nadir angle is not too large. For orthoimages, a standard deviation of 2 GSD is sufficient and so even for QuickBird orthoimages a standard deviation of 1.2 m can be accepted. The dominating error component of the elevation model to the orthoimage corresponds to the standard deviation of the height model multiplied by the tangent of the incidence angle. Thus for QuickBird in open and flat areas, the SRTM DSM can be used for the generation of orthoimages with images having an incidence angle not exceeding 16°. For orthoimages having 1-m pixel size, use of the SRTM DSM can be extended to 26° incidence angle. The lower accuracy in mountainous areas usually can be accepted because accuracy requirements are usually lower in such areas.

37.4 Thematic Processing Methods

37.4.1 Overview of Methods

The purpose of thematic processing of remotely sensed data—in an optional combination with other spatial or non-spatial data—is to generate and present application-specific information and knowledge. From a methodological point of view we can model the inherent interpretation process as outlined in Figure 37-14. This interpretation process can be seen as containing procedures in both top-down (model-driven) and bottom-up (data-driven) directions. Such a hybrid process is analogous to cognitive perception, with the sequences and weights depending on the application. Furthermore, image interpretation is not simply a linear process, but also includes feedback mechanisms at various stages. This workflow is the basis for both (semi-) automatic and visual interpretation approaches.

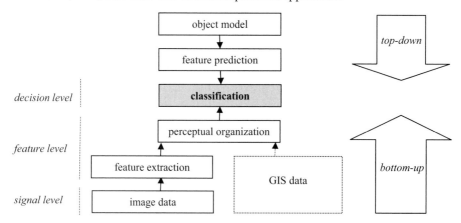

Figure 37-14 Generalized, bi-directional workflow for the interpretation of image data in an optional combination with GIS data

Starting from the given data sources, attributes (features) belonging to graphical primitives (feature carriers, i.e., points (or pixels), lines, regions or solids) are extracted. Features associated with image data can be: tone, color, shape, size, height, texture, shadow, pattern, context and neighborhood. In order to close the application-dependent gap between those image features and the related object characteristics which are described in a knowledge representation (object model) a perceptual organization of the features becomes necessary. This can be done at different levels and with different methods, for example based on statistical cluster algorithms or by grouping lines and areas into meaningful real-world structures using geometrical properties, similar spectral characteristics, etc. Finally the classification, i.e. the assignment of an instance of the application-dependent object set to the feature carrier under consideration, has to take place. The corresponding decision is based on one of many well-known methods which will be briefly summarized in Section 37.4.4.

Recent developments in advanced spaceborne and airborne sensors as described above have been a reaction to the limitations of the earlier available data sources as input to the interpretation process. With the advanced resolutions (spatially, spectrally, radiometrically and temporally) and the improved information content, the potential for thematic applications (like mapping urban, forestry or transportation structures or processes at large scales) has been significantly increased. However, with the variety and the advanced features of these sensors, the user community faces new problems in its thematic processing:

- The increasing multi-dimensionality of remotely sensed scenes as described above imposes the necessity for a fusion at different levels as a pre-processing step for both visual and automatic classification approaches. The multi-dimensionality can accrue, for example, from multispectral channels (with different geometrical resolution) of one sensor, from scenes of different sensors or from the increasing availability of GIS data in the public and private domains. Hence, in the following section we will describe necessary fusion methods on signal, feature and decision-based levels (see also Figure 37-14) with respective examples.

- The high spatial resolution of the advanced sensors increases the spectral within-field variability—in contrast to the integration effect of earlier sensors—and therefore may decrease the classification accuracy of traditional per-pixel based methods (like the maximum-likelihood method). In this context, segmentation algorithms already have been recognized as a valuable and complementary approach that—similar to human operators—create regions instead of points or pixels as feature carriers; these feature carriers are then introduced into the perceptual organization and classification stage. Section 37.4.3 will elaborate on these region-building methods.

37.4.2 Image Fusion

In a special issue of the *International Journal of Geographical Information Science* (IJGIS) on data fusion, Edwards and Jeansoulin (2004) state, "data fusion is a complex process with a wide range of issues that must be addressed. In addition, data fusion exists in different forms in different scientific communities. Hence, for example, the term is used by the image community to embrace the problem of sensor fusion, where images from different sensors are combined. The term is also used by the database community for parts of the interoperability problem. The logic community uses the term for knowledge fusion." (Edwards and Jeansoulin 2004, p. 303).

Consequently, it comes as no surprise that several definitions for data fusion can be found in the literature. Pohl and van Genderen (1998) proposed "image fusion is the combination of two or more different images to form a new image by using a certain algorithm." (p. 825) Mangolini (1994) (coref. Wald 2002, p. 40) extended data fusion to information in general and he also refers to quality. He defined data fusion as "a set of methods, tools and means using data coming from various sources of different nature, in order to increase the quality (in a broad sense) of the requested information." Wald (1999) defined "data fusion as a formal framework in which are expressed means and tools for the alliance of data originating from different sources. It aims at obtaining information of greater quality; the exact definition of 'greater quality' will depend upon the application." (p. 1191)

In the imaging community, fusion techniques are used to merge panchromatic image information of high spatial resolution into multispectral images of lower spatial resolution. These techniques are designed to produce images which present the 'best of both worlds': high spatial resolution combined with high spectral resolution, both of which are often desired for the extraction of information for integrating with or updating GIS data layers.

37.4.2.1 Fusion Techniques

Fusion techniques for remotely sensed data can be classified into three levels: pixel level (iconic), feature level (symbolic) and knowledge or decision level (Pohl and van Genderen 1998). Of highest relevance for remote sensing data are techniques for iconic image fusion to merge panchromatic images of high spatial resolution with multispectral data of lower resolution (see Cliche et al. 1985; Welch and Ehlers 1987; Zhang 2002). They also offer the widest range of theoretical background whereas symbolic and decision based fusion methods are usually related to specific application examples. However, existing techniques hardly satisfy conditions for successful fusion of the new generation high-resolution satellite images such as Ikonos, Landsat-7, SPOT-5 and QuickBird or ultra high resolution airborne data (Zhang 2002). All of the new generation satellite and almost all airborne sensors provide high-resolution information only in their panchromatic mode, whereas the multispectral images are of lower spatial resolution. The ratios between high resolution panchromatic and low resolution multispectral images typically vary between 1:2 and 1:8 (Ehlers 2004b). To produce high resolution multispectral data sets, the panchromatic information has to be merged with the multispectral images.

In general, image fusion techniques can be grouped into three classes: (1) color-related techniques, (2) statistical methods, and (3) numerical methods. The first comprises the color composition of three image channels in the red-green-blue (RGB) color space as well as more sophisticated color transformations such as intensity-hue-saturation (IHS) or the hue-saturation-value (HSV) transforms. Statistical approaches are developed on the basis of band statistics including correlation and filters. Techniques such as principal-component-analysis (PCA) and regression belong to this group. The numerical methods employ arithmetic operations such as image multiplication and summation. A sophisticated numerical approach uses wavelets in a multi-resolution environment. The most significant problem with image fusion techniques, however, is the color distortion of the fused image.

37.4.2.2 Color-Related Techniques

Merging various combinations of image bands from different dates or sensors to create color composites can aid the interpreter in discriminating the various ground features present. Examples of successfully used color composites of optical and microwave data are described, for example, by Vornberger and Bindschadler (1992), Marek and Schmidt (1994), and Pohl et al. (1996). Reports about multi-sensor optical composites can be found in Welch et al. (1985), and Chavez (1986). Multi-sensor SAR fusion by RGB is reported by Marek and Schmidt (1994) and Welch and Ehlers (1988). In many cases, the RGB technique is applied in combination with another image fusion procedure.

The IHS color transformation can effectively separate a standard RGB image into spatial (I) and spectral (H, S) information. This separation property may be utilized for the fusion of multi-resolution images. The basic concept of IHS fusion is: (1) transform a color image composite from RGB space into IHS space, (2) replace the I (intensity) component by a panchromatic image with a higher resolution, and (3) reverse the transformation of the replaced components from IHS space back to the original RGB space to obtain a fused image. The IHS technique is one of the most widespread image fusion methods in the remote sensing community and has been employed as a standard procedure in many commercial packages (Ehlers 2004b).

37.4.2.3 Statistical Techniques

The principal component analysis (PCA) is a general statistical technique that transforms a multivariate data set with correlated variables into one with uncorrelated variables. The variables are usually arranged in descending order of their eigenvalues, meaning that the highest amount of information is contained in the first principal component. These new variables are obtained as linear combinations of the original variables. PCA has been widely used in image encoding, image data compression, image enhancement and image fusion. The basic steps in PCA fusion are: 1) perform a principal component transformation to convert a set of multispectral bands (three or more bands) into a set of principal components; 2) replace one principal component, usually the first component, by a panchromatic image with higher resolution; and 3) perform a reverse principal component transformation to convert the replaced components back to the original image space so as to produce a fused multi-spectral image. Previous works about PCA for data fusion include Richards (1984), Singh and Harrison (1985), Fung and LeDrew (1987), Chavez et al. (1991), Shettigara (1992), Jutz and Chorowicz (1993), and Yesou et al. (1993a and 1993b). Related statistical methods include the regression variable substitution (RVS) and the canonical variate substitution.

37.4.2.4 Numerical Methods

The possibilities of combining image data using multiplication or summation are manifold. For example, to enhance the contrast, adding and multiplication of images are useful. The choice of weighting and scaling factors may improve the resulting images. Details can be found in Cliche et al. (1985), Price (1987), Welch and Ehlers (1987), Carper et al. (1990), Ehlers (1991), Munechika et al. (1993), and Pellemans et al. (1993). Difference or ratio images are very suitable for change detection (Mouat et al. 1993). The ratio method is even more useful because of its capability to emphasize slight signature variations (Singh 1989). The Brovey Transform, a special combination of arithmetic combinations including ratio, is a formula that normalizes multispectral bands used for RGB display, and multiplies the result by any other desired higher resolution image to add the intensity or brightness component to the image (Jurio and van Zuidam 1998).

As a powerful mathematical tool, wavelet transform has been developed in the field of signal and image processing. More recently, wavelets have started playing a role in image fusion. The original principle of the wavelet image fusion is to get the best resolution without altering the spectral contents of the image. More clearly, this principle is based on multi-resolution analyses provided by the wavelet transforms (Mallat 1989). Wavelet transforms are capable of decomposing a digital image into a set of multi-resolution images, accompanied by wavelet coefficients for each resolution level. The wavelet coefficients for each level contain the spatial differences between two successive resolution levels. The general steps of wavelet-based fusion are: 1) decompose a high resolution panchromatic image into a set of low resolution panchromatic images with wavelet coefficients for each level; 2) replace a low resolution panchromatic band with a multispectral band at the same resolution level; and 3) perform a reverse wavelet transform to convert the decomposed and replaced panchromatic set back to the original panchromatic resolution level. The replacement and reverse transform is done three times, once for each multispectral band. Previous works about wavelet transform for data fusion can be found in Ranchin and Wald (1993), Li et al. (1995), Yocky (1996), Zhou et al. (1998), Nunez et al. (1999), Ranchin and Wald (2000), Shi, Zhu et al. (2003) and Ling et al. (2007).

37.4.2.5 FFT Based Filtered IHS Fusion (Ehlers Fusion)

The most significant problem with image fusion techniques is the color distortion of the fused image, especially for multi-sensor and multi-temporal image fusion (Zhang 2002; Ehlers and Klonus 2004). The principal idea behind an image fusion that preserves spectral characteristics is this: the high resolution image must sharpen the multispectral image without adding new information to the spectral components. As a basic image fusion technique, we will make use of the IHS transform. This technique will be extended to include more than the standard 3 bands (red, green, blue color transform) from color theory. In addition, filter functions for the multispectral and panchromatic images have to be developed. The filters have to be designed in a way that the effect of color change from the high resolution component is minimized.

The ideal fusion function would add the high resolution spatial components of the panchromatic image (i.e., edges, object changes) but disregard its actual gray values. For a thorough analysis of the information distribution along the spatial frequencies of an image, use is made of Fourier transform (FT) theory (Gonzalez and Woods 2001). An overview flowchart of the method is presented in Figure 37-15 (see Ehlers and Klonus 2004 for a complete description).

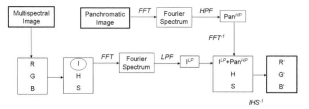

Figure 37-15 FFT-based filter fusion using a standard IHS transform. Three selected bands (RGB) of the low resolution multispectral image are transformed into the IHS domain. Intensity component and the high resolution panchromatic image are transformed into the Fourier domain using a two dimensional Fast Fourier Transform (FFT). The power spectrum of both images is used to design the appropriate low pass (LP) filter for the intensity component and high pass (HP) filter for the high resolution panchromatic image. An inverse FFT transforms both components back into the spatial domain. The low pass filtered intensity (ILP) and the high pass filtered panchromatic band (PHP) are added and matched to the original intensity histogram. At the end, an inverse IHS transform converts the fused image back into the RGB domain (after Ehlers 2004b).

The Ehlers fusion was applied to the fusion of multi-temporal Landsat and SPOT images of an urban area in Germany. The panchromatic SPOT image was acquired on 16 March 2003 with a pixel size of 5 m (Figure 37-16). The Landsat ETM scene from 26 June 2001 with an original GSD of 30 m was registered to the SPOT image and resampled to 5 m using a bicubic convolution technique (Figure 37-17).

Figure 37-16 Panchromatic SPOT image of 16 March 2003 (512 × 512 subset). The image shows an area of the northern part of the city of Aachen, Germany.

Figure 37-17 Landsat ETM image of 26 June 2001 registered to the SPOT scene and resampled to 5-m pixel size. Shown is a false-color display of bands 4 (red), 5 (green) and 7 (blue). See included DVD for color version.

Bands 4, 5, and 7 of the Landsat ETM image were transformed into the IHS domain. The SPOT panchromatic band and the Landsat intensity component were transformed into the Fourier domain and filtered according to Fig. 37-15. Figures 37-18 and 37-19 show the comparison of the intensity component before and after fusion. Figure 37-20 presents the final result after the Ehlers fusion. It is evident that besides the pan-sharpening effects of the fusion process, the original colors are extremely well preserved.

Figure 37-18 Intensity component of the Landsat ETM image.

Figure 37-19 Intensity component after low-pass filtering and the integration of the high-pass filtered panchromatic SPOT image.

Figure 37-20 Final results of the Ehlers fusion. The colors (see Figure 37-17) are preserved. See included DVD for color version.

With fusion techniques that preserve spectral characteristics, the IHS transform can be extended to any number of bands by performing a set of transforms and creating a final multi-band image composite (Ehlers 2004b). For visual comparison, the results of standard fusion techniques are presented in Figures 37-21 to 37-23. Interestingly, the Principal Component (PC) merge which usually shows good results for standard false color infrared bands (i.e., bands 4, 3, and 2 for Landsat ETM) produces very poor results for this all-infrared color composite (Figure 37-23). All standard techniques show significant color differences to the original Landsat image.

Figure 37-21 Fused Landsat/SPOT image after Brovey transform. See included DVD for color version.

Figure 37-22 Fused Landsat/SPOT image after multiplicative merge. See included DVD for color version.

Figure 37-23 Fused Landsat/SPOT image after Principal Component (PC) merge. See included DVD for color version.

Manual of Geographic Information Systems

The positive results for the new technique have been confirmed for a number of image data sets. Even for multi-sensor and multi-temporal image fusion, the FFT-based technique preserves the spectral characteristics of the multispectral images while keeping the spatial resolution of the panchromatic images. This is reflected by the correlation coefficient for the multispectral bands before and after fusion. The Ehlers fusion achieved a correlation coefficient of 0.994 which is far superior to all the other methods (Table 37-11). Since the filter function for the high resolution image can be adjusted to the size of the geo-objects, the Ehlers fusion technique can also be used for an optimum spatial enhancement of selected geo-objects (e.g., houses, parcels or field boundaries) (Ehlers 2005).

Table 37-11 The correlation coefficients between the multispectral bands of the original and the pansharpened image prove the superiority of the Ehlers fusion method.

Pansharpening Method	Correlation Coefficient with Original Bands
IHS	0.762
Brovey	0.816
Principal Component	0.850
Multiplicative	0.932
Ehlers	0.994

To investigate the quality of the fused images, they were tested using automated maximum likelihood classification techniques. Twelve selected classes were identified for a study site and sampled for training and test areas. They were based on the spectral signatures of the original Landsat images and the training areas were used without modifications for the fused images. Accuracies were checked using the same test areas for all images. The results confirmed the superiority of the Ehlers fusion. It proved to be the only fused image which displayed an increase in classification accuracy, whereas all other fusion techniques yielded a decrease in classification accuracy (see Table 37-12 for results). Figures 37-24 and 37-25 show the results of the maximum likelihood classification for the original Landsat image and the merged Landsat/SPOT image after Ehlers fusion.

Figure 37-24 Classified Landsat ETM image (12 classes). See included DVD for color version.

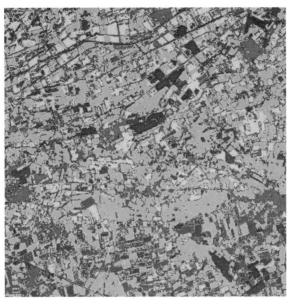

Figure 37-25 Classified merged SPOT/Landsat ETM image after Ehlers fusion. Note that training and test areas are the same as for the original Landsat image. See included DVD for color version.

Table 37-12 Classification accuracies for selected fusion techniques

Image Data	Classification Accuracy (min-max)	Overall Classification Accuracy	Kappa Coefficient
Original Landsat Data	40%–100%	87%	0.86
Landsat/SPOT IHS Fusion	20%–95%	74%	0.71
Landsat/SPOT Brovey Fusion	40%–100%	77%	0.74
Landsat/SPOT PC Fusion	25%–100%	73%	0.70
Landsat/SPOT Multiplicative Fusion	25%–100%	79%	0.76
Landsat/SPOT Ehlers Fusion	70%–100%	90%	0.89

37.4.2.6 Decision Based Fusion: Settlement Analysis

The advantages of iconic image fusion are that rich theories exist to discuss appropriate techniques and characteristics. Also, pan-sharpened images produce a better visual appearance by combining images with the multispectral information from the lower resolution image. Unfortunately, for many fusion techniques we experience more or less significant color shifts which, in most cases, impede a subsequent automated analysis. Even with a fusion technique that preserves the original spectral characteristics, automated techniques may not produce the desired results because of the high resolution of the fused data sets. For this purpose, feature-based or decision-based fusion techniques are employed that are usually based on empirical or heuristic rules. Because a general theory is lacking for these types of fusion, we present a case study for a decision-based fusion.

The basis for an automated detection of settlement areas was a number of selected satellites of high and medium resolution. The high resolution satellite data sets were panchromatic images from SPOT-5 with 5-m resolution and KOMPSAT with 6.5-m resolution. Medium resolution multispectral data were obtained from Landsat ETM and ASTER data sets with 30- and 15-m resolution, respectively (see Figure 37-26). Table 37-13 provides an overview of the selected satellite data.

Figure 37-26 Study site near the village of Vettweiß north of Cologne, Germany: panchromatic KOMPSAT (left) and multispectral (bands 3, 2, 1) ASTER image (right). See included DVD for color version.

Table 37-13 Selected satellites for settlement analysis.

System	Landsat 7	SPOT 5	Terra	KOMPSAT 1
Sensor	ETM+	HRG	Aster	EOC
Recording date	6/26/2001	3/16/2003	8/3/2003	5/20/2004
Geometric resolution in m	30 m	5 m	15 m	6.6 m
Spectral resolution	multispectral (6 bands)	panchromatic	multispectral (4 bands)	panchromatic
Scene size in km²	180 × 180	60 × 60	60 × 60	17 × 17
Approximate cost in $	600	2750	45	110

Contrary to the iconic image fusion techniques described above, the images were rectified to ground coordinates, but otherwise left in their original format. Parameters such as texture and shape parameters were extracted from the high-resolution panchromatic data, and vegetation information from the multispectral images. Using an adaptive threshold procedure, the information from the image data sets was fused and resulted in a binary mask for the areas 'settlements candidates' and 'definitely no settlements' (Figure 37-27). This process was repeated at a hierarchy of different-sized segments ranging from coarse to fine with a set of different threshold parameters at each level (Figure 37-28). At each step, the next level analysis was only performed in areas that were identified as settlement candidates. More details can be found in Ehlers et al. (2005).

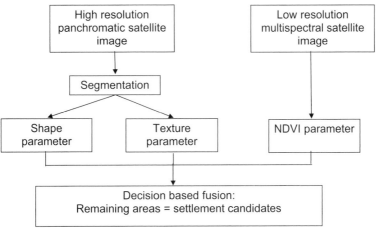

Figure 37-27 Decision based fusion process. Texture and shape parameters are calculated from the high resolution panchromatic data, whereas the multispectral data are used to calculate vegetation indices.

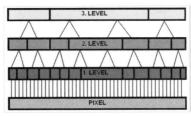

Figure 37-28 Hierarchical network diagram for the fusion process. At each level, the fusion process is performed with adapted threshold parameters. The process is performed with a segment-oriented image processing system (after Ehlers et al. 2005). See included DVD for color version.

The process was performed at a three-level hierarchy resulting in improved areas for settlement candidates at each step (see Figures 37-29 to 37-31). For the final results, areas such as parks and lakes that were surrounded by settlement areas were included by neighborhood analysis; small features with settlement characteristics (farms, etc.) that were outside of settlement areas could be excluded by perimeter/area analysis (Ehlers et al. 2005). The final result is shown in Figure 37-32.

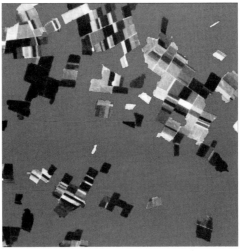

Figure 37-29 Settlement candidates (red) after level 1 processing. The area that is rejected as potential settlement is on the KOMPSAT satellite image. See included DVD for color version.

Manual of Geographic Information Systems

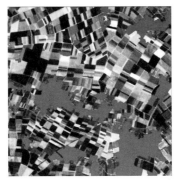

Figure 37-30 Settlement candidates (red) after level 2 processing. See included DVD for color version.

Figure 37-31 Settlement candidates (red) after level 3 processing. See included DVD for color version.

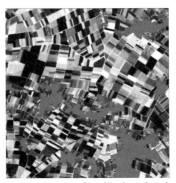

Figure 37-32 Final selected settlement area after filtering. See included DVD for color version.

For an analysis of the final accuracy, settlement areas were manually digitized and compared to the ones selected by the automated hierarchical processing at each level (Table 37-14). For both combinations, results were virtually identical and exceeded 90% at the final level. The method will be further tested with very high resolution satellite images such as Ikonos and QuickBird.

Table 37-14 Accuracies for the detection of settlement areas.
Both satellite combinations show almost identical accuracy values.

Hierarchical Level	KOMPSAT / Aster	SPOT-5 / Landsat ETM
3	18.6%	13.6%
2	64.4%	70.0%
1	86.7%	86.9%
Final	90.4%	90.3%

37.4.3 Segmentation

The high spatial resolution of the advanced sensors as described in the previous sections increases the spectral within-field variability—in contrast to the integration effect of earlier sensors. As Figure 37-33 demonstrates, topographic objects under consideration (here: forest or buildings) are represented by their spatially, spectrally and ontologically distinct components. Therefore the classification accuracy of traditional (semi-)automatic methods on a per-pixel basis (like the maximum-likelihood method) decreases as objects are represented by more and more pixels. While a visual interpreter has no problem with the aggregation of the components into meaningful objects, specific and efficient analysis techniques need to be developed for (semi-)automatic processing.

Figure 37-33 Representation of topographical objects by a digital camera system (FALCON II) with high spatial resolution (image courtesy of TopoSys). See included DVD for color version.

In this context, a simple aggregation of pixels (e.g., by applying a quadratic average filter) could remove undesired details. However, the degree of aggregation varies from one object to another or even within one object class, and the resulting quadratic or rectangular "super pixels" do not correspond to actual real-world objects in general.

As an alternative, segmentation can be applied. Segmentation is the process of completely partitioning a scene (i.e., a remote sensing image) into non-overlapping and irregular regions (segments) in scene space (i.e., image space) based on various features. Algorithms have been developed within pattern recognition and computer vision since the 1980s with successful applications in disciplines such as medicine and telecommunication engineering. In the meantime, they also have been recognized as a valuable and complementary approach for image interpretation, in particular for processing scenes with high spatial resolution. The conceptual idea is—similar to the work of human interpreters—to create regions instead of points or pixels as feature carriers which are then introduced into the classification stage where each of these regions shall correspond exactly to one and only one object class (object-oriented approach).

37.4.3.1 Concepts

The basic **principle** of segmentation algorithms is the merging of (image) elements based on homogeneity parameters or on the differences from neighboring regions (heterogeneity). Thus, segmentation methods follow the two strongly correlated *principles of neighborhood* and *value similarity*.

We can distinguish the following **strategies** for partitioning a scene into regions:
- point-based
- edge-based
- region-based
- combined

In the following discussion, only the basic principles of those strategies will be explained, for detailed studies refer to the literature (e.g., Haralick and Shapiro 1985; Sonka et al. 1998).

Point-based approaches search for homogeneous elements within the entire scene by applying global threshold operations which combine such data points that show an equal or at least a similar signal or feature value. This threshold can divide the feature space into two or more parts (binarization or generation of equidensities, respectively). The choice of threshold values can be performed statically or dynamically based on histogram information. Because this grouping has not considered the principle of neighborhood so far, a connection analysis in scene space is performed in a second step. Here, spatially connected elements (components) of equal value (e.g., gray value "1") are grouped to one region (*component labeling*). It has to be noted that point-based approaches are less suitable for the evaluation of remotely sensed data due to varying reflection values for a certain object placed at different locations within the real world and the sensed scene.

Edge-based approaches describe the segments by their outlines. These are generated through an edge detection (e.g., a Sobel filtering) followed by a contour-generating algorithm. Optionally, the transition from the outlines to the interior region can be achieved by contour-filling methods like the watershed algorithm (Sonka et al. 1998). The main disadvantage of edge-based approaches is that the edge and also the contour image are strongly affected by noise, with artificial objects being generally not as noisy as wooded regions or other natural features. More noise may lead to an unacceptable over-segmentation.

Region-based approaches start in the scene space where the available elements (pixels or already existing regions) are tested for similarity against other elements (see Section 37.4.3.2). Concerning the definition of the initial segmentation the procedures of *region growing* (bottom-up, i.e., starting with a seed pixel) and *region splitting* (top-down, i.e., starting with the entire scene) are distinguished. One disadvantage of the splitting method is that it tends to result in an over-segmentation because a splitting always produces a fixed number of sub-regions (normally 4) although two or three of them might actually be homogeneous with respect to each other. As a consequence, one can apply a method combination which leads to the *split-and-merge* algorithm that after every split tests whether neighboring regions are so similar that they should be re-merged again.

37.4.3.2 Homogeneity Criteria

The central underlying idea of segmentation is the *principle of value similarity* (or homogeneity) between two elements which are generally adjacent to each other (*principle of neighborhood*). Given the two elements A and B (i.e., pixels or regions) one possibility for deriving *a homogeneity measure* (Δh) is to compare a certain feature of A and B (e.g., the gray value) through its Euclidian distance:

$$\Delta h = +\sqrt{(f_A - f_B)^2} \tag{37-6}$$

where f_A is a measured value of one particular feature for element (pixel or region) A and f_B is a measured value of that particular feature for element (pixel or region) B.

In addition, it is also possible to simultaneously consider multiple features f_i (i=1, …, n) of A and B, with the option to introduce individual weights (g_i). Hence, a homogeneity measure Δh is obtained by a feature based fusion (refer to Section 37.4.2) as follows:

$$\Delta h = +\sqrt{\sum_{i=1}^{n} g_i \cdot (f_{A,i} - f_{B,i})^2} \tag{37-7}$$

As an additional alternative, the homogeneity measures can be computed before and after an eventual merge of the elements A and B. With the obtained measure, Δh, and a given threshold h_T it can be decided whether the elements A and B have to be merged to a larger segment or not (*single linkage*). If inquiry aims at the extension of an already existing region, normally the average values of this segment are taken into account (*centroid linkage*). The choice of the threshold value h_T controls the size and number of segments and with that, the level of generalization of the segmentation process. Hence, it is possible to repeat the segmentation with varying thresholds leading to a *multi-scale representation* and the possibility to introduce many segmentation layers into the follow-up classification; such a process can produce multiple results which can be compared in order to come to an optimum (see Figure 37-34).

Figure 37-34 Multi-scale representation of a remotely sensed scene by multiple segmentation operations based on different thresholds. See included DVD for color version.

The merging algorithm also can consider further constraints concerning neighborhood and similarity. In the simplest case element A accepts B if the homogeneity measure is below the given threshold (*fitting*). In contrast, A may accept only that neighboring element B which fulfils the homogeneity criterion best (*best fitting*). Furthermore an element C is connected to A (which is similar enough to B) only if B and C as well as A and C are similar enough (*local mutual best fitting*).

Due to different application demands and the multi-dimensionality of geo-data segmentation, approaches for the evaluation of remotely sensed data have to be rather complex systems (e.g., the system eCognition, introduced in 1999 by Definiens-Imaging, Baatz and Schäpe 2000) that should
- handle various input data simultaneously (*multi-source aspect*),
- integrate a couple of segmentation strategies serving all object types which shall be extracted (*multi-method aspect*),
- create various levels of generalization at the same time—due to the fact that different objects are represented best at different scales (*multi-scale aspect*).

37.4.3.3 Evaluation of Segmentation

In general the described segmentation methods do not yield a perfect partition of the scene, but often produce either too many and small regions (*over-segmentation*) or too few and large segments (*under-segmentation*). The first effect is normally a minor problem because in the following classification step neighboring segments can be assigned to the same category and merged.

Applying segmentation methods to remotely sensed data, we can observe that over- and under-segmentation can even occur simultaneously within a single scene depending on the heterogeneity of objects under consideration. As Figure 37-35 shows, natural objects tend to be more strongly partitioned than regular artificial objects. Furthermore, different levels of generalization are desired depending on the specific applications (e.g., evaluation scales). For instance, some applications may demand the delineation of individual trees while other applications need larger wooded areas.

Manual of Geographic Information Systems

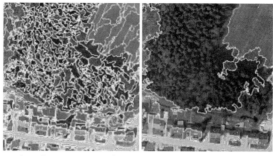

Figure 37-35 Over- and under-segmentation within a single scene, depending on segmentation scale. See included DVD for color version.

Methods for the evaluation of segmentation results are discussed for example by Hoover et al. (1996), Zhang (1996) or Levine and Nazif (1985). In the last case, the authors also present developments for a dynamic determination of the segmentation quality by continuously computing homogeneity measures of all intermediate regions. However, it has be noted that presently the most reliable evaluation method is still a visual interpretation that has to consider the exact geometrical position of the segment borders as well as the membership of one and only one object class to a single region. With visual inspection, the generalization level as well as the homogeneity features and parameters are controlled in a rather subjective manner.

37.4.4 Classification

Image fusion at the iconic level produces images of enhanced quality which may be interpreted by human analysts or subjected to automated information extraction. For an automated analysis, statistical techniques such as minimum distance or maximum likelihood determine the probability that a certain pixel belongs to a given land use/land cover class. These pixel-based methods are often augmented by auxiliary information such as GIS data, parcel boundaries or texture parameters.

High resolution sensors especially, however, pose new challenges for automated interpretation. The homogenizing effects of comparably large pixel sizes are no longer valid. With these ultra/very high resolution sensors, simple pixel-based analyses are no longer applicable because of the difficulty of classifying high resolution data where each pixel is related not to the character of an object or an area as a whole, but to components of it (Blaschke and Strobl 2001). Instead, we have image pixels that might belong to the same class but exhibit totally different reflectance values. For example, a high resolution image of the class 'house' can be represented by hundreds of pixels that might belong to undesired subclasses of 'window', 'chimney', 'sunlit roof', 'shadowed roof', 'front lawn', or 'driveway' (see Figure 37-36).

Figure 37-36 High resolution image of a house with several identifiable subclasses of different spectral reflectance (causing high in-class variance). See included DVD for color version.

It is well known that spectral classification of higher resolution data does not automatically lead to more-detailed classification results (see, for example, Metternicht 1999; Petit and Lambin 2001). Further, using just multispectral information for classifications does not lead to accurate interpretation results because the differentiation between object classes is performed not only with the help of spectral information, but also with spatial (contextual) information of the image data. For example, using only multispectral information, different objects like roofs and streets might not be separated into two object-classes because they are built with the same material (Hoffmann et al. 2000). Consequently, new intelligent techniques are being developed that make use of GIS integration, multi-sensor approaches and context-based interpretation schemes (see, for example, Ehlers 2000; Schiewe 2003). Otherwise, the last step of an all-digital image acquisition and handling process has to consist of manual on-screen digitizing.

Case Study: Biotope Type Monitoring from Ultra-high Resolution Data

Fairway expansion projects on rivers in northwestern Germany due to shipping requirements mandated that continuous environmental monitoring be conducted after the end of the expansion. Emphasis was placed on monitoring the tidally influenced riverside biotopes. Changes in composition and size of these biotopes should be documented over the long term to assess the impacts of hydraulic engineering measures. The term "biotope type" is being used as a more specific one than the term "land cover class". In contrast to "land cover class", "biotope type" not only involves similar structures in vegetation or other surfaces, but also includes abiotic parameters such as altitude or relative distance to water.

The High Resolution Stereo Camera Airborne (HRSC-AX) of the German Aerospace Center (DLR) was employed for the data acquisition (see Table 37-15). An integrative monitoring concept was developed using a combination of GIS, image analysis and modeling software (Ehlers et al. 2003a and b). The selected approach seemed to guarantee the best possible accuracy compared to traditional mapping and surveying methods that are always combined with extensive fieldwork. In the several projects different HRSC sensors/data sets were applied (see Table 37-15). The data were recorded from flying heights between 3,000 m and 6,000 m and delivered with 15-cm to 60-cm ground pixel resolution. Absolute accuracies were given as ± 20 to 30 cm in horizontal and ± 30 to 50 cm in vertical direction, respectively.

Table 37-15 *Sensors and data sets.*

	Pilot Project River Elbe (1999)	**River Elbe (2000a)** (area north of Hamburg)	**River Elbe (2000b)** (area south of Hamburg)	**River Elbe (2002)**	**River Weser (2002)**
Sensor	**HRSC-A**	**HRSC-A**	**HRSC-AX**	**HRSC-AX**	**HRSC-AX**
Data set provided by	DLR	DLR	DLR	Terra Imaging/ ISTAR	Terra Imaging/ ISTAR
Flight Height	3000 m	6000 m	6000 m	6000 m	6000 m
Ground Pixel Resolution	15 cm 50 cm (DSM)	30 cm 50 cm (DSM)	60 cm 100 cm (DSM)	25 cm 100 cm (DSM)	32 cm 100 cm (DSM)
Radiometric Resolution	8 bit	8 bit	12 bit	12 bit	12 bit

The study sites are located along the tidally influenced areas of the rivers Elbe and Weser in northwestern Germany close to the large cities Hamburg and Bremen (see Figure 37-37). In both river areas, reeds and some relics of willow forests are of major interest for nature conservation. For the sites, complete HRSC data sets (i.e., panchromatic and multispectral images, and digital surface model data) were used. An example for various band combinations is presented in Figure 37-38.

Manual of Geographic Information Systems

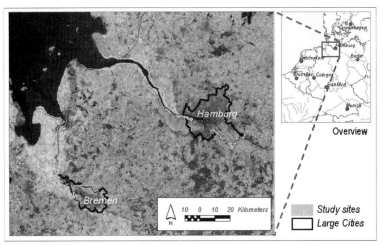

Figure 37-37 Study sites in Germany. See included DVD for color version.

Figure 37-38 Various views recorded with only one HRSC-AX camera system: digital surface model (DSM) with orange = high elevation, yellow and green = medium elevation and blue = low elevation; color infrared (CIR); panchromatic; true color (RGB). See included DVD for color version.

Methodology and Results: For an automated and reproducible classification process, a hierarchical procedure was developed. It consisted of an index-based segmentation and pre-classification procedure followed by a stepwise hierarchical classification process (see Figure 37-39). The first step (Level 1) is the computation of ancillary information such as texture and vegetation indices. The computation of vegetation indices is a standard procedure for satellite remote sensing applications. Indices like the Normalized Difference Vegetation Index (NDVI) and its derivatives are commonly used for separating vegetation from bare soil, as well as for estimating quality and vitality of vegetation stands (Jensen 2000, pp. 361 ff).

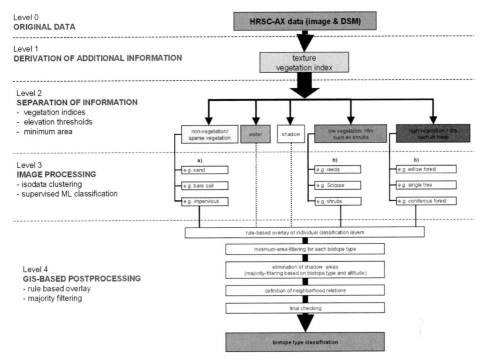

Figure 37-39 Hierarchical classification process for high resolution image data.

In the field of airborne remote sensing, texture and vegetation indices actually have only rarely been used, mostly due to the limited availability of an appropriate database. Scanned aerial photos have generally proven to be inappropriate due to their heterogeneous radiometric properties and their evident inconsistencies even within single scenes. Even with the first models of operational digital scanners for airborne applications, computations of vegetation indices were limited. For example, the HRSC-A offered an inappropriate spectral resolution (i.e., no true red band) for vegetation mapping tasks. The new HRSC-AX sensor displays better-suited multispectral bands for terrestrial applications and higher radiometric resolution (see Table 37-15). The excellent quality of the HRSC-AX data enables the computation vegetation indices. However, the applicability of each sensor should be examined thoroughly because its spectral bands and scales differ from those of standard satellite systems.

During our research, numerous indices have been tested for the HRSC image data. For the HRSC-A, best results were obtained with a combination of the near infrared, the panchromatic and a calculated virtual red band (Ehlers et al. 2003a). For the HRSC-AX, standard NDVI methods were sufficient. It has to be noted that the use of vegetation indices combined with a hierarchical classification procedure improves the classification process with respect to speed and accuracy.

The vegetation indices are used in the next step (Level 2) to identify and separate four coarse classes (non-vegetation/sparse vegetation; vegetation; water; shadow) (Figure 37-38). With the additional incorporation of height information from the digital surface model (DSM) provided by the stereo sensor HRSC, the vegetation class is further divided into high vegetation (e.g., trees) and low vegetation (e.g., shrubs, grass). This can be achieved by incorporating all data in an integrated GIS/image processing environment using the appropriate GIS or image analysis procedures. Thus, biotope types that do not show a difference in their multispectral reflectance characteristics but are of different height can easily be separated. This 'separation of information' step, therefore, permits the detail and accuracy of the classification to be improved. The resulting segments are already pre-classified into the semantic layers 'non-vegetation/sparse vegetation', 'water', 'shadows', 'low vegetation' and 'high vegetation'. For each segment and each object class a minimum size is defined. Smaller segments are eliminated using standard GIS operations. In the next step (Level 3), the separated layers (for an example see Figure 37-40) were treated with appropriate classification algorithms (i.e., isodata clustering for the non-vegetation/sparse vegetation layer and supervised classification for the herbaceous vegetation layers). With this approach, the level of detail in the biotope type classification could be significantly improved and it was possible to identify more than 20 different classes.

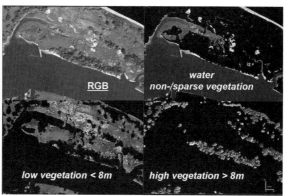

Figure 37-40 Separation of information into different independent layers. See included DVD for color version.

Finally, a GIS-based postprocessing is involved to produce the final classification result (Level 4). GIS operations such as majority filtering, logical overlay, definition of neighborhood relations and minimum area functions were used to estimate appropriate classes for shadow areas and to combine the individual information layers. The final output was a GIS layer with 21 biotope types for the study sites. The differences between the final classification results and maps created by fieldwork and photointerpretation are presented in Figure 37-41 for a small test site located on an island in the Elbe River. The richness of detail of the classification results corresponds well with the structures in the original image. The visual interpretation result already shows the generalization that was performed by the human operator. The older reference map shows the subset mapped with only a few polygons whereas the new classification result consists of more and better-fitting polygons, described by over 1400 vertices. Even single trees, shrubs or open forest areas smaller than 100 square meters can be detected over large areas.

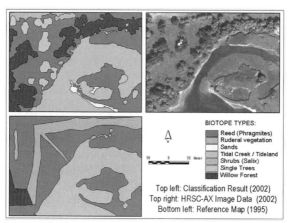

Figure 37-41 Comparison of reference map and result after hierarchical classification. See included DVD for color version.

Results of accuracy checking and change analysis proved the validity of this procedure. For the classes of particular ecological interest (e.g., ruderal vegetation growing in waste areas, aquatic bulrush (Scirpus spp.), willow trees (*Salix* spp.)) mapping accuracy exceeded 95% (Ehlers et al. 2006). The average overall accuracy was found to be better than 85%. By using GIS overlay techniques, areas of change could be identified and loss of critical vegetation quantified (Figure 37-42).

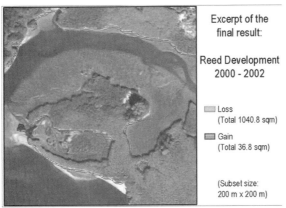

Figure 37-42 Change analysis for a selected test site. For a small 200 m × 200 m area, a net loss of more than 1000 m² of reed vegetation could be detected. See included DVD for color version.

37.5 Integration of GIS and Remote Sensing

Geographic information systems (GIS) are increasingly seen as an integral part of the modern information and communication society. Improved methods for data access and integration have accelerated this process. Scientific advances also have paved the way for GIS as a catalyst for a new evolving discipline geoinformatics. One of the problems in applying geospatial technology has been the currency, quality, accessibility and completeness of geo-information (GI). Remotely sensed image data, especially from satellites, can be used to generate current, accurate and synoptic information about all parts of the Earth as a basis for geo-scientific analyses in GIS. Consequently, almost all major GIS software packages offer now at least the possibility to display and query digital images as part of their GIS database. With the advent of the new satellites of 1-m resolution or even better, we will see another push for the integration of remote sensing images into GIS.

The advantages of the integration of GIS and remote sensing have been demonstrated in a large number of application-oriented projects (see, for example, Star et al. 1997). However, the merging of remote sensing (and its associated image analysis) and GIS has often resulted in the creation of just another 'dumb' GIS layer with pictorial information. Integration is restricted to a mere georeferencing and image overlay (more accurately called image underlay). A complete analysis from a remotely sensed image to a geo-object can be performed only by manual interpretation. GIS and remote sensing information are usually processed independently from each other. The ideal goal should be that GIS objects can be extracted from a remote sensing image to update the GIS database. In return, GIS 'intelligence' (e.g., object and analysis models) should be used to automate this object extraction process (see Figure 37-43). However, the current status can still be described primarily as data exchange between a GIS and an image analysis system or an add-on of some image processing functionality to a separate GIS. Images are seen as another GIS layer, integration consists more or less of a georeferencing and overlay process.

Ehlers and colleagues presented as early as 1989 a concept for a totally integrated system for remote sensing and GIS. They differentiated between three integration levels: 1) two separate systems with a data interface; 2) two principally separate systems with a common user interface; and 3) a totally integrated system (Ehlers et al. 1989). Most of today's GISs offer hybrid processing, i.e., the analysis of raster and vector data. They also have image display capabilities or image analysis add-ons which offer some "level 2" (common user interface) functionality (Bill 1999; Ehlers 2000). However, geospatial information is usually processed in either raster or vector form and has to be converted into the desired processing or output format. A truly integrated processing option (without prior conversion) does not exist. Similarly, a truly integrated processing option does not exist for integrated remote sensing/GIS analyses. The requirements for totally integrated systems are usually defined on an ad hoc basis, driven by project demands or the data sources to be incorporated (Ehlers et al. 1994; Johnston et al. 1997). What is needed is an analysis of the necessary processing components of such an integrated system. The data integration approach has to be replaced by an analysis integration approach. This implies that we need a taxonomy of system-independent analysis functions.

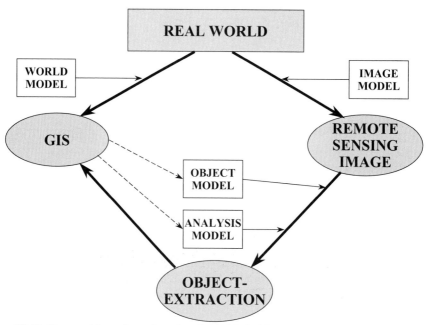

Figure 37-43 Concept for automatic extraction of GIS objects from remote sensing imagery.

37.5.1 Taxonomy Issues

37.5.1.1 Taxonomy of GIS Operators

If one looks into the functionality of current GISs, it is immediately evident that GIS operations are usually based on the underlying system and its associated data structure. A general description of GIS functions could offer a system-independent view. GIS functions are, however, predominantly concerned with low level functions (see, for example, Laurini and Thompson 1992; Worboys 2004). A GIS user, on the other hand, wants to perform a spatial analysis or a comparison of two possible locations for a specific development. He or she is normally faced with a system that offers a huge number of functions which depend upon system and data structures and have to be put together in a specified order to perform the desired analysis. A taxonomy of universal high-level GIS operations that are independent of the system and the data structure is still lacking. A notable exception is the cartographic modeling (*Map Algebra*) approach of Berry (1987) and Tomlin (1990). However, this approach is still restricted to raster-based systems.

Tomlin structures his cartographic modeling functions into four classes with about 40 subfunctions. These functions are sufficient to perform almost every possible high-level GIS analysis. The strength of his approach is the mathematical rigidity which is incorporated in a computer programming type GIS language. Structuring the Map Algebra commands in a procedure allows the composition of very complex GIS analyses. The basic function classes of Tomlin's Map Algebra are:

- LocalFunctions (e.g., point operations, overlay, recoding)
- FocalFunctions (e.g., neighborhood operations, buffering, distance calculation)
- ZonalFunctions (e.g., attribute operations, intersections)
- IncrementalFunctions (e.g., nearest neighbor, connectivity, slope, aspect)

Although these functions are system independent and form the basis of many raster GIS packages in current GIS software, they are still data-structure dependent, i.e., designed for raster GIS. A step further towards a universal GIS language is the approach of Albrecht (1996). Twenty data-structure- and system-independent high-level GIS functions are grouped into 6 classes (Table 37-16). The user communicates with the system through a flowchart tool similar to those in modern GIS and image analysis packages. The difference is that the GIS functions are independent of the underlying system.

Table 37-16 Universal High Level GIS Operators (modified after Albrecht 1996).

Function Group	(Sub)Functions			
Search	Thematic Search	Spatial Search	Interpolation	(Re)Classification
Location Analysis	Buffer	Corridor	Overlay	Thiessen/Voronoi
Terrain Analysis	Slope/Aspect	Catchment/Basins	Drainage/Network	Viewshed Analysis
Distribution/Neighborhood	Cost/Diffusion/Spread	Proximity	Nearest Neighbor	
Spatial Analysis	Pattern/Dispersion	Centrality/Connectivity	Shape	Multivariate Analysis
Measurements	Distance	Area		

37.5.1.2 Taxonomy of Image Analysis Operators in Remote Sensing

Digital image processing started in the early '70s and is viewed as a young but established discipline. It was influenced by its one-dimensional counterpart, signal processing, by photography and optics, and by the scientific and technological developments in electrical engineering and computer science. Its interdisciplinary heritage is clearly visible in the very different descriptions of image processing functionality which can be seen in standard textbooks such as Pratt (1992) or Sonka et al. (1998).

Even in a well-defined application area like remote sensing, we experience very diverse approaches toward image analysis taxonomies. Textbook authors typically do not present a systematic taxonomy of image analysis functions. Nevertheless, textbooks mix hardware, sensors, systems and operations, or present structures that are inconsistent with a rational image-processing taxonomy (Ehlers 2000). This inconsistency when dealing with image analysis taxonomies is an impediment for the development of a stronger theoretical background for the design and implementation of integrated GIS. Without such a theoretical basis, however, the only way to GIS/remote sensing integration seems to be a project driven *ad hoc* approach with limited usefulness and applicability.

37.5.1.3 An Integrated Taxonomy

To set up a taxonomy of data-structure- and system-independent high-level GIS/image analysis functions, one has to start either from the remote sensing or the GIS side. The 20 universal operators from Table 37-16 represent currently the only formalized taxonomy that meets the requirements stated above. Although the grouping can be debated (are *Buffer* and *Corridor* really different functions? does *Interpolation* belong to the *Search* or rather to the *Spatial Analysis* group?), the 20 operators are used as starting point for an iterative approach. Based on typical remote sensing analyses, four groups with 17 image processing functions were selected to be added to the 20 universal GIS operators (Table 37-17). The derivation of these functions is a first step and is based on an in-depth analysis of remote sensing literature and intensive project experiences (Richards and Jia 1999; Ehlers and Schiewe 1999; Ehlers 2000; Schiewe and Ehlers 2003; Jensen 2005).

Table 37-17 Universal image processing functions for integrated GIS (after Ehlers 2000).

Function Group	(Sub)Functions			
Preprocessing	Parametric Radiometric Sensor Correction	Parametric Geometric Sensor Correction		
Geometric Registration	Deterministic Techniques	Statistical Techniques (Interpolation)	Automated Techniques (Matching)	Error Assessment
3D Image Analysis	Ortho Image Generation	DEM Extraction		
Atmospheric Correction	Deterministic Approaches	Histogram Based Manipulations (Point Operations)	Filtering	Image Enhancement
Feature/Object Extraction	Unsupervised Techniques	Supervised Techniques	Model-Based Techniques	Error Assessment

It has to be noted, however, that the operators presented in Tables 37-16 and 37-17 are not sufficient to define and describe the complete functionality of integrated GIS. Still required is a thorough analysis of hybrid processing capabilities, i.e. functions that allow a joint analysis of remote sensing and GIS information. It remains to be investigated how polymorphic techniques can be used to extend the capacities of the universal high-level GIS/image processing functions. The operator *Overlay*, for example, should be able to process image-image, GIS-image, and GIS-GIS overlays without a different name for every function option. First results of such polymorphisms were investigated, for example, by Jung (2004). Additional functions have to be developed, on the other hand, that extend the capabilities of integrated GIS beyond the sum of the single components. 3D urban information systems created from GIS and remote sensing can be seen as an example for these extensions.

37.5.2 Uncertainty Issues

37.5.2.1 Uncertainty in GIS

The advantages of an integrated geo-processing framework have been proven by many examples. It is, however, also evident that the issues of accuracy and errors within this integration process have to be addressed. Good science requires statements of accuracy, by which the reliability of results can be understood and communicated. Where accuracy is known objectively then it can be expressed as error, where it is not, the term uncertainty applies (Hunter and Goodchild 1993). Thus uncertainty covers a broader range of doubt or inconsistency, and in the context of this chapter, includes error as a component. The understanding of uncertainty as it exists in geographic data remains a problem that is only partly solved (see, for example, Story and Congalton 1986; Goodchild and Gopal 1989; Veregin 1995; Ruiz 1997; Worboys 1998; Gahegan and Ehlers 2000; Zhang and Goodchild 2002). However, without quantification, the reliability of any results produced remains problematic to assess and difficult to communicate to the user. GISs provide a whole series of tools with which data can be manipulated, without offering any control over misuse. To that instance, Openshaw et al. (1991) state:

> *"A GIS gives the user complete freedom to combine, overlay and analyze data from many different sources, regardless of scale, accuracy, resolution and quality of the original map documents and without any regard for the accuracy characteristics of the data themselves."* (p. 79)

This is a serious issue; without quantification of uncertainty, the results themselves may only be considered as qualitative information, and this greatly devalues their merit in both a scientific and a practical sense. To compound the problem, in the fusion of activities from remote sensing and GIS, an integrated approach to managing geographic information is required. This must necessarily support many different types of data (Ehlers et al. 1991), gathered according to different models of geographic space (Goodchild 1998), each possessing different types of inherent errors and uncertainties (Chrisman 1991). As well as providing individual support for these different models of space, it is necessary to explicitly include methods to keep track of uncertainty, as data is changed from low-level forms (such as remotely-sensed image data) to the higher-level abstractions required by digital cartography and GIS (such as objects and themes).

Whether a particular data set can be considered suitable for a given task depends on many different criteria, and despite the fact that various aspects of uncertainty can be measured objectively, their importance will be largely determined by the task. The overall goal when modeling uncertainty is therefore threefold: 1) to produce a statement of uncertainty to be associated with each data set so that an objective statement of reliability may be reported; 2) to develop methods to propagate uncertainty as the data are processed and transformed; and 3) to ultimately determine the suitability of a data set for a given task ('fitness for use'). Another goal is to communicate uncertainty information to the user (e.g., Hunter and Goodchild 1996).

A useful framework for handling uncertainty, recognizing the separate error components of value, space, time, consistency and completeness, was proposed by Sinton (1978) and later embellished by Chrisman (1991). However, uncertainty in geographic data can be described in a variety of alternative ways, such as those provided by Bédard (1987), Miller et al. (1989) and Veregin (1989). Although different, these approaches all have a number of aspects in common, including the observation that uncertainty itself occurs at different levels of abstraction. For example, positional and temporal error describe uncertainty in a metric sense within a spatio-temporal framework, whereas completeness and consistency represent more abstract concepts describing coverage and reliability, and are consequently more problematic to describe.

Uncertainty in its many forms has been on the research agenda of the GIS and remote sensing community for at least two decades, gaining much of its early momentum from the very first research initiative of the US National Center for Geographic Information and Analysis (NCGIA) (Goodchild and Gopal 1989). Work to date on uncertainty addresses the inherent errors present within specific types of data structure (e.g., raster or vector) or data models (e.g., field or object). The effects of error propagation and analysis within these various paradigms has been studied by Veregin (1989, 1995); Openshaw et al. (1991); Goodchild et al. (1992), Heuvelink and Burrough (1993); Ehlers and Shi (1997); Leung and Yan (1998); Shi (1998); Arbia et al. (1999); Zhang and Kirby (2000); Zhang and Stuart

(2001); Shi, Cheung et al. (2003). In a recent compendium on uncertainty in geographic information, Zhang and Goodchild (2002) investigate methods for uncertainty assessment for continuous variables (fields), categorical variables (classes), and objects. Despite the progress made to date, they conclude that "academics, technologists, government information agencies, the general public and the commercial sector must work together to take advantage of the benefits of geographical information in new applications, while being fully informed of the nature and implications of the associated uncertainties. Scientists and workers lead the leap forward." (p. 324) This does not sound like a problem solved.

37.5.2.2 Uncertainty in the Integration of GIS and Remote Sensing

Even if we restrict uncertainty description to one specific problem—the integration of remote sensing and GIS—a generally applicable method for quantifying uncertainty does not exist. In 1990, the error analysis research group of the NCGIA Initiative 12 (integration of remote sensing and GIS) identified the research on uncertainty as one of the major challenges in the integration of these two technologies (Lunetta et al. 1991). Remote sensing scientists have always had the need to quantify errors that were associated with the processing of remotely sensed data. Most efforts have gone into the error analysis of rectification and registration processes and of information extraction or multispectral classification techniques (see, for example, Ehlers 1997 and Congalton and Green 1999).

Only scant attention has so far been given to the problem of modeling uncertainty as the data are transformed through different models of geographic space, a notable exception being the work of Lunetta et al. (1991). A typical path taken by data captured by satellite, then abstracted into a suitable form for GIS is shown in Figure 37-44, and involves four such models. Continuously varying fields are quantized by the remote sensing device into image form, then classified and finally transformed into discrete mapping objects. The overall object extraction process is sometimes referred to as semantic abstraction (Waterfeld and Schek 1992) due to the increasing semantic content of the data as it is manipulated into forms that are easier for people to work with.

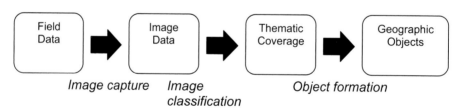

Figure 37-44 Continuum of Abstraction from Field Model to Object Model (after Gahegan and Ehlers 2000).

When transforming data between different conceptual models of geographic space, the uncertainty characteristics in the data may change, in that techniques used to transform the data also alter the inherent uncertainty and, in addition, may introduce further uncertainty of their own. Furthermore, many of the abstraction techniques employed combined data with different uncertainty characteristics. Consequently, two interrelated problems must be addressed, namely:

- How do the uncertainty characteristics of data change as data are transformed between models?
- How do the transformation methods used affect and combine the uncertainty present in the data?

One of the consequences of the traditional separation of GIS and remote sensing activities into distinct communities and separate software environments is that there is an artificial barrier between the two disciplines. Thus, the integration of these two branches of science is to some extent an artificial problem. As a result, there is no easy flow of metadata between systems, interoperability is often restricted to the exchange of image files or object geometry and the problem of managing uncertainty is compounded. The four stages shown in Figure 37-44 represent the four models of geographic space, namely field, image, thematic, and object (or feature) models and are typical (though not exhaustive) of those used in the integration of GIS and remote sensing activities. These models represent the conceptual properties of the data only and can be considered here as independent from any particular data structure that might be used to encode and organize the data.

Gahegan and Ehlers (2000) developed an integrated error simulation model for the transition from field (raw remote sensing) data to geo-objects. The description of uncertainty followed that proposed by Sinton (1978). It covers the sources of error as they occur in remote sensing and GIS integration (although other approaches may be equally valid). Uncertainty is restricted to the following properties: 1) value (including measurement and label errors); 2) spatial; 3) temporal; 4) consistency; and 5) completeness. Of these, measurement and label errors as well as uncertainties in space and time can apply either individually to a single datum or to any set of data. The latter two properties of consistency and completeness can only apply to a defined data set since they are comparative (either internally amongst data or to some external framework). Their findings are summarized in Table 37-18.

Table 37-18 Types of uncertainty and their sources in four models of geographic space [(α) = data or value, (β) = space, (χ) = time, (δ) = consistency, (ε) = completeness)] (from Gahegan and Ehlers 2000).

	FIELD	IMAGE	THEMATIC	OBJECT
α	Measurement error and precision.	Quantization of value in terms of spectral bands and dynamic range	Labeling uncertainty (classification error)	Identity error (incorrect assignment of object type), object definition uncertainty.
β	Locational error and precision	Registration error, sampling precision.	Combination effects when data represented by different spatial properties are combined.	Object shape error, topological inconsistency, 'split and merge' errors.
χ	Temporal error and precision	(Temporal error and precision are usually negligible for image data.)	Combination effects when data representing different times is combined.	Combination effects when data representing different times is combined.
δ	Samples / readings collected or measured in an identical manner.	Image is captured identically for each pixel, but medium between satellite and ground is not consistent; inconsistent sensing, light falloff; shadows.	Classifier strategies are usually consistent in their treatment of a data set.	Methods for object formation may be consistent, but often are not. Depends on extraction strategy.
ε	Sampling strategy covers space, time and attribute domains adequately.	Image is complete, but parts of ground may be obscured (clouds, trees).	Completeness depends on the classification strategy. (Is all the data set classified or are only some classes extracted?).	Depends on extraction strategy. Spatial and topological inconsistencies may arise as a result of object formation.

Other uncertainty issues in the integration of the two spatial technologies can be related to scale and representation of the data (Bruegger 1995; Guptill and Morrison 1995) or the provision of lineage information (e.g., Lanter 1991).

New research on uncertainty deals with the development of advanced processing techniques for the information extraction from remotely sensed images. The inclusion of contextual information (textures, neighborhood), object-or segment-based analysis techniques together with the application of fuzzy set theory and artificial intelligence, challenge the standard image processing strategies (Wang 1993; Ryherd and Woodcock 1996; Lucieer and Stein 2002; Ibrahim et al. 2005). In a recent special issue of the International Journal of Remote Sensing on "Uncertainties in Integrated Remote Sensing and GIS", the editors conclude: "Within the framework of uncertainties in integrated remote sensing and GIS, we can describe the uncertainties in terms of positional accuracy, attribute and thematic accuracy, temporal accuracy, logical consistency, and completeness. In this special issue, we mainly address the modeling of uncertainty in terms of attribute and positional accuracy. Relatively less attention is paid to the issue of the completeness of temporal uncertainties. Modelling uncertainties in newly emerging data sets, such as laser scanning data, high-resolution satellite images, InSar, and high spectral satellite images will be an area for future research" (see Shi et al. 2005, page 2914).

References

Albrecht, J. 1996. Universal Analytical GIS Operations. Ph.D. Thesis, ISPA-Mitteilungen 23, University of Vechta, Germany.

Arbia, G., D. Griffith and R. Haining. 1999. Error propagation modelling in raster GIS: adding and rationing operations. *Cartography and Geographic Information Science* 26:297–315.

Baatz, M., and A. Schäpe. 2000. Multiresolution segmentation – an optimization approach for high quality multi-scale image segmentation. In *Angewandte Geographische Informations-verarbeitung XII,* edited by J. Strobel, T. Blaschke and G. Griesebner, 24–29. Heidelberg: Wichmann.

Blaschke, T. and J. Strobl. 2001. What's wrong with pixels? Some recent developments interfacing remote sensing and GIS. *GeoBIT/GIS* 6:12–17.

Bédard, Y. 1987. Uncertainties in land information databases. Pages 175–184 in *Proceedings of Auto-Carto 8: , Proceedings of the Eighth International Symposium on Computer-Assisted Cartography,* held in Baltimore, Md. 30 March-2 April 1987. Bethesda, Md.: ASPRS.

Berry, J. K. 1987. Fundamental operations in computer-assisted map analysis. *International Journal of Geographic Information Systems* 1 (2):119–136.

Bill, R. 1999. GIS-produkte am markt – stand und entwicklungstendenzen. *Zeitschrift für Vermessungswesen* 6:195–199.

Bruegger, B. P. 1995. Theory for the integration of scale and representation formats: major concepts and practical implications. In *Spatial Information Theory. Lecture Notes in Computer Science, 988,* edited by A. U. Frank and W. Kuhn, 297–310. Berlin: Springer.

Büyüksalih, G. and K. Jacobsen. 2007. Comparison of DEM generation by very high resolution optical satellites. Pages 627–637 in *New Developments and Challenges in Remote Sensing. Proceedings of the 26th Annual Symposium of the European Association of Remote Sensing Laboratories (EARSeL),* held in Warsaw, Poland, 29 May – 2 June 2006, edited by Z. Bochenek. Rotterdam: Millpress.

Carper, W. J., T. M. Lillesand and R. W. Kiefer. 1990. The use of Intensity-Hue-Saturation transformations for merging SPOT panchromatic and multispectral image data. *Photogrammetric Engineering and Remote Sensing* 56:459–467.

Chavez, P. S. 1986. Digital merging of Landsat TM and digitalized NHAP data for 1:24 000 scale image mapping. *Photogrammetric Engineering and Remote Sensing* 52:1637–1646.

Chavez, P. S., S. C. Sides and J. A. Anderson. 1991. Comparison of three different methods to merge multiresolution and multispectral data: TM & SPOT pan. *Photogrammetric Engineering and Remote Sensing* 57:295–303.

Chrisman, N. R. 1991. The error component in spatial data. In *Geographical Information Systems, Vol. 1: Principles,* edited by D. J. Maguire, M. F. Goodchild and D. W. Rhind, 165–174. Essex: Longman Scientific & Technical.

Cliche, G., F. Bonn and P. Teillet. 1985. Integration of the SPOT pan channel into its multispectral mode for image sharpness enhancement. *Photogrammetric Engineering and Remote Sensing* 51:311–316.

Colwell, R. N. 1997. History and place of photographic interpretation. In *Manual of Photographic Interpretation,* 2nd edition, edited by W. R. Philipson, 33–48. Bethesda, Md.: American Society for Photogrammetry and Remote Sensing (ASPRS).

Congalton, R. G. and K. Green. 1999. *Assessing the Accuracy of Remotely Sensed Data: Principles and Practices.* Boca Raton, Fla.: Lewis Publishers.

Edwards, G. and R. Jeansoulin. 2004. Data fusion - from a logic perspective with a view to implementation. Guest Editorial. *International Journal of Geographical Information Science* 18 (4):303– 307.

Ehlers, M. 1991. Multisensor image fusion techniques in remote sensing. *ISPRS Journal of Photogrammetry and Remote Sensing* 46:19–30.

———. 1997. Rectification and registration. In *Integration of Remote Sensing and GIS,* edited by J. L. Star, J. E. Estes and K. C. McGwire, 13–36. New York: Cambridge University Press.

———. 2000. Integrated GIS – from data integration to integrated analysis. *Surveying World* 9:30–33.

———. 2002. Fernerkundung für GIS-anwender: sensoren und methoden zwischen anspruch und wirklichkeit. In *Fernerkundung und GIS: Neue Sensoren – innovative Methoden* edited by T. Blaschke, 10–23. Heidelberg: Wichmann Verlag.

———. 2004a. Remote sensing for GIS applications: new sensors and analysis methods. In *Remote Sensing for Environmental Monitoring, GIS Applications, and Geology III, Proceedings of SPIE Vol. 5239,* edited by M. Ehlers, H. J. Kaufmann and U. Michel, 1–13. Bellingham, Wash.: SPIE.

———. 2004b. Spectral characteristics preserving image fusion based on Fourier domain filtering. In *Remote Sensing for Environmental Monitoring, GIS Applications, and Geology IV, Proceedings of SPIE Vol. 5574,* edited by M. Ehlers, F. Posa, H. J. Kaufmann, U. Michel, and G. De Carolis, 1-13. Bellingham, Wash.: SPIE.

———. 2005. Urban remote sensing: new developments and trends. In *Proceedings, 5th International Symposium Remote Sensing of Urban Areas (URS 2005),* held in Tempe, Ariz., USA, 14-16 March 2005 (CD proceedings), 6 pp.

Ehlers, M. and W. Z. Shi. 1997. Error modelling for integrated GIS. *Cartographica* 33:11–21.

Ehlers, M. and J. Schiewe, editors. 1999. *Geoinformatik 99: Ausgewählte Themen der Forschungsgruppe GIS/Fernerkundung,* Materialien Umweltwissenschaften Vechta, 5. University of Vechta, Germany.

Ehlers, M. and S. Klonus. 2004. Erhalt der spektralen charakteristika bei der bildfusion durch FFT basierte filterung. *Photogrammetrie-Fernerkundung-Geoinformation (PFG)* 2004 (6):495–506.

Ehlers, M., G. Edwards and Y. Bédard. 1989. Integration of remote sensing with GIS: a necessary evolution. *Photogrammetric Engineering and Remote Sensing* 55:1619–1627.

Ehlers, M., D. D. Greenlee, J. L. Star and T. R. Smith. 1991. Integration of remote sensing and GIS: data and data access. *Photogrammetric Engineering and Remote Sensing* 57:669–675.

Ehlers, M., D. R. Steiner and J. B. Johnston, editors. 1994. *Requirements for Integrated Geographic Information Systems*. Ann Arbor, Mich.: Environmental Research Institute of Michigan.

Ehlers, M., M. Gähler and R. Janowsky. 2003a. Automated analysis of ultra high resolution remote sensing data for biotope type mapping: new possibilities and challenges. *ISPRS Journal of Photogrammetry and Remote Sensing* 57:315–326.

Ehlers, M., R. Janowsky and M. Gähler. 2003b. Ultra high resolution remote sensing for environmental monitoring. *Earth Observation Magazine* 12 (9):27–32.

Ehlers, M., U. Michel, G. Bohmann and D. Tomowski. 2005. Entscheidungsbasierte datenfusion von multisensoralen fernerkundungsdaten zur erkennung von siedlungsgebieten. Pages 209-216 in *Vorträge der 25. wissenschaftlich-technische Jahrestagung der DGPF*. Publikationen der Deutschen Gesellschaft für Photogrammetrie, Fernerkundung und Geoinformation, Nr. 14.

Ehlers, M., M. Gähler and R. Janowsky. 2006. Automated techniques for environmental monitoring and change analyses for ultra high resolution remote sensing data. *Photogrammetric Engineering and Remote Sensing* 72 (7):835–844.

Fricker, P., R. Sandau, U. Tempelmann and S. Walker. 2000. ADS 40 – why LH systems took the three-line road. *GIM International* 7:45–47.

Fung, T. and E. LeDrew. 1987. Application of principal component analysis to change detection. *Photogrammetric Engineering and Remote Sensing* 53:1649–1658.

Gahegan, M. and M. Ehlers. 2000. A framework for modeling of uncertainty in an integrated Geographic Information System. *ISPRS Journal of Photogrammetry and Remote Sensing* 55:176–188.

Georgi, C., R. Stognienko, S. Knuth and G. Albe. 2005. JAS: the next generation digital aerial scanner. Pages 147–154 in *Photogrammetric Week '05*, edited by D. Fritsch. Heidelberg: Wichmann Verlag.

GIM International. 2004. Product News: New large-format DC concept. *GIM International* 18 (9):49.

Gonzalez, R. C. and R. E. Woods. 2001. *Digital Image Processing*. Upper Saddle River, N.J.: Prentice Hall.

Goodchild, M. F. 1998. Different data sources and diverse data structures: metadata and other solutions. In *Geocomputation: A Primer*, edited by P. A. Longley, S. M. Brooks, R. McDonnell and W. Macmillan, 61–74. London: John Wiley & Sons, Inc.

Goodchild, M. F. and S. Gopal, editors. 1989. *The Accuracy of Spatial Databases*. London, UK: Taylor & Francis.

Goodchild, M. F., S. Guoqing and Y. Shiren. 1992. Development and test of an error model for categorical data. *International Journal of Geographical Information Systems* 6:87–104.

Grimm, C. and J. Kremer. 2005. DigiCAM and LiteMapper – versatile tools for industrial projects. In *Photogrammetric Week '05*, edited by D. Fritsch, 207–216. Heidelberg: Wichmann Verlag.

Guptill, S. C. and J. L. Morrison. 1995. *Elements of Spatial Data Quality*. New York, N.Y.: Elsevier Science.

Haralick, R. and L. G. Shapiro. 1985. Image segmentation techniques. *Computer Vision, Graphics and Image Processing* 12:100–132.

Heuvelink G.B.M. and P. A. Burrough. 1993. Error propagation in cartographic modelling using boolean logic and continuous classification. *International Journal of Geographical Information Systems* 7:231–246.

Hoffmann, A., J. W. van der Vegt and F. Lehmann. 2000. Towards automated map updating: is it feasible with new digital data-acquisition and processing techniques? Pages 295-302 in *International Archives of Photogrammetry and Remote Sensing (IAPRS), Proceedings XIX ISPRS Congress*, 33(B2), Amsterdam, The Netherlands, 16-22 July 2000.

Hoover, A., J.-B. Gillian, X. Jiang, P. J. Flynn, H. Bunke, D. Goldgof, K. Bowyer, D. Eggert, A. Fitzgibbon and R. Fisher. 1996. An experimental comparison of range image segmentation algorithms. *IEEE Transactions on pattern analysis and machine intelligence* 18 (7):673–689.

Hunter G. and M. F. Goodchild. 1993. Mapping uncertainty in spatial databases, putting theory into practice. *Journal of Urban and Regional Information Systems Association* 5:55–62.

Hunter G. and M. F. Goodchild. 1996. Communicating uncertainty in spatial databases. *Transactions in GIS* 1:13–24.

Ibrahim, M. A., M. K. Arora and S. K. Ghosh. 2005. Estimating and accommodating uncertainty through the soft classification of remote sensing data. *International Journal of Remote Sensing* 26:2995–3007.

Jacobsen, K. 1997. Calibration of IRS-1C PAN-camera. Joint Workshop: Sensors and Mapping from Space. Hannover, 1997 < http://www.ipi.uni-hannover.de/uploads/tx_tkpublikationen/IRS1CKa.pdf > Accessed 20 November 2008.

———. 2005. Analysis of SRTM elevation models. EARSeL 3D-Remote Sensing Workshop, held in Porto, Portugal, 10-11 June 2005. <http://www.ipi.uni-hannover.de/uploads/tx_tkpublikationen/ASEjac.pdf> Accessed 22 December 2008.

———. 2006. Pros and cons of the orientation of very high resolution optical space images. Pages 41–47 in *IntArchPhRS. Band XXXVI 1/WG I/5.* Paris, 2006, 7 S.

———. 2007a. Orientation of high resolution optical space images. In *ASPRS Tampa 2007 Proceedings* held in Tampa, Fla. 7-11 May 2007. Bethesda, Md.: ASPRS.

———. 2007b. Geometry of digital frame cameras. In *ASPRS Tampa 2007 Proceedings* held in Tampa, Fla. 7-11 May 2007. Bethesda, Md.: ASPRS.

———.2007c. 3D-remote sensing, status report. Pages 591–599 in *Proceedings of the 27th Annual Symposium of the European Association of Remote Sensing Laboratories (EARSeL),* held in Bolzano, Italy, 4-7 June 2007, edited by M. A. Gomarasca. Rotterdam, The Netherlands: Millpress.

———. 2008a. Geometric modelling of linear CCDs and panoramic imagers. Pages 145–155 in *2008 ISPRS Congress Book,* edited by Z. Li, J. Chen and E. Baltsavias. London: Taylor & Francis.

———. 2008b. Satellite image orientation. ISPRS Congress, Beijing 2008. *IntArchPhRS.* Vol XXXVIII, Part B1 (WG I/5), pages 703–709.

Jensen, J. R. 2000. *Remote Sensing of the Environment: An Earth Resource Perspective.* Upper Saddle River, N.J.: Prentice Hall. 544 pp.

———. 2005. *Introductory Digital Image Processing,* 2nd edition. Upper Saddle River, N.J.: Prentice Hall.

Johnston, J. B., M. Ehlers, D. R. Steiner and M. A. Gomarasca, eds. 1997. *New Developments in Geographic Information Systems.* Ann Arbor, Mich.: Environmental Research Institute of Michigan.

Jung, S. 2004. HYBRIS: Hybride räumliche Analyse Methoden als Grundlage für ein integriertes GIS. Ph.D. Thesis, University of Vechta, Germany (CD Publication).

Jurio, E. M. and R. A. van Zuidam. 1998. Remote sensing, synergism and geographical information system for desertification analysis: an example from northwest Patagonia, Argentina. *ITC Journal* 3-4:209–217.

Jutz, S. L. and J. Chorowicz. 1993. Geological mapping and detection of oblique extension structures in the Kenyan Rift Valley with a SPOT/Landsat-TM datamerge. *International Journal of Remote Sensing* 14:1677–1688.

Kraus, K. 1993. *Photogrammetry, Volume 1, Fundamentals and Standard Processes.* Bonn: Dümmler. ISBN 3-427-78684-6.

Lanter, D. P. 1991. Design of a lineage-based meta-data base for GIS. *Cartography and Geographic Information Systems* 18:255–261.

Laurini, R. and D. Thompson. 1992. *Fundamentals of Spatial Information Systems.* London: Academic Press.

Leberl, F. and M. Gruber. 2003. Flying the new large format digital aerial camera ultracam. Pages 67–76 in *Photogrammetric Week '03.* Edited by D. Fritsch. Heidelberg: Wichmann.

Leung, Y. and J. Yan. 1998. A locational error model for spatial features. *International Journal of Geographical Information Science* 12:607–620.

Levine, M. D. and A. M. Nazif. 1985. Dynamic measurement of computer generated image segmentations. *IEEE Transactions on Pattern Analysis and Machine Intelligence* 7 (2):155–164.

Li, H., B. S. Manjunath and S. K. Mitra. 1995. Multisensor image fusion using the wavelet transform. *Graphical Models and Image Processing* 57 (3):235–245.

Ling, Y., M. Ehlers, E. L. Usery and M. Madden. 2007. FFT-enhanced IHS transform method for fusing high-resolution satellite images. *ISPRS Journal of Photogrammetry and Remote Sensing* 61 (6):381–392.

Lucieer, A. and A. Stein. 2002. Existential uncertainty of spatial objects segmented from satellite sensor imagery. *IEEE Transactions on Geoscience and Remote Sensing* 40:2518–2521.

Lunetta R. S., R. G. Congalton, L. K. Fenstermaker, J. R. Jensen, K. C. McGwire and L. R. Tinney. 1991. Remote sensing and geographic information system data integration: error sources and research issues. *Photogrammetric Engineering and Remote Sensing* 57:677–687.

Mallat, S. G. 1989. A theory for multiresolution signal decomposition: the wavelet model. *IEEE Transactions on Pattern Analysis and Machine Intelligence* 11:674–693.

Mangolini, M. 1994. Apport de la Fusion D'images Satellitaires Multicapteurs au Niveau Pixel en Télédétection et Photointerprétation. Ph.D. Thesis, Université Nice - Sophia Antipolis, France, 174 pp.

Marek, K. H. and K. Schmidt. 1994. Preliminary results of the comparative analysis of ERS-1 and ALMAZ-1 SAR data. *ISPRS Journal of Photogrammetry and Remote Sensing* 49:12–18.

McClone, C., E. Mikhail and J. Bethel, editors. 2004. *Manual of Photogrammetry,* 5th edition. Bethesda, Md.: American Society for Photogrammetry and Remote Sensing (ASPRS). 1000 pp.

McGill, M. 2005. A US perspective on phase-2 of high resolution satellite remote sensing. Paper presented at Eurimage Meeting, held in Rome, Italy in 2005.

Metternicht, G. 1999. Change detection assessment using fuzzy sets and remotely sensed data: an application of topographic map revision. *ISPRS Journal of Photogrammetry and Remote Sensing* 54 (4):221–233.

Miller, R., H. Karimi, and M. Feuchtwanger. 1989. Uncertainty and its management in geographical information systems. Pages 252-259 in *Proceedings CISM'89.* Ottawa, Canada: Canadian Institute of Surveying and Mapping.

Mouat, D. A., G. G. Mahin and J. Lancaster. 1993. Remote sensing techniques in the analysis of change detection. *Geocarto International* 2:39–50.

Munechika, C. K., J. S. Warnick, C. Salvaggio and J. R. Schott. 1993. Resolution enhancement of multispectral image data to improve classification accuracy. *Photogrammetric Engineering and Remote Sensing* 59:67–72.

Nunez, E., X. Otazu, O. Fors, A., Prades, V. Pala and R. Arbiol. 1999. Multiresolution-based image fusion with adaptive wavelet decomposition. *IEEE Transactions on Geoscience and Remote Sensing* 37(3):1204–1211.

Openshaw, S., M. Charlton and S. Carver. 1991. Error propagation: a Monte Carlo simulation. In *Handling Geographic Information,* edited by I. Masser and M. Blakemore, 78–101. Essex: Longman Scientific & Technical.

Passini, R. and K. Jacobsen. 2007. Accuracy analysis of SRTM height models. In *ASPRS Tampa 2007 Proceedings* held in Tampa, Fla. May 7–11 2007. Bethesda, Md.: ASPRS.

Passini, R., D. Betzner and K. Jacobsen. 2002. Filtering of digital elevation models. In *Proceedings, XXII FIG International Congress ACSM-ASPRS Conference and Technology Exhibition 2002*. Held in Washington, DC, 19-26 April 2002. Bethesda, Md.: ASPRS. CD-ROM.

Pellemans, A.H.J.M., R.W.L. Jordans and R. Allewijn. 1993. Merging multispectral and panchromatic SPOT images with respect to the radiometric properties of the sensor. *Photogrammetric Engineering and Remote Sensing* 59:81–87.

Petit, C. C. and E. F. Lambin. 2001. Integration of multi-source remote sensing data for land cover change detection. *International Journal of Geographical Information Science* 15 (8):785–803.

Pohl, C. and J. L. van Genderen. 1998. Multisensor image fusion in remote sensing: concepts, methods and applications. *International Journal of Remote Sensing* 19:823–854.

Pohl, C., Y. Wang and B. N. Koopmans. 1996. The 1995 flood in the Netherlands from space. *ITC Journal* 1996-4:414–415.

Pratt, W. K. 1992. *Digital Image Processing,* 3rd edition. New York: Wiley.

Price, J. C. 1987 Combining panchromatic and multispectral imagery from dual resolution satellite instruments. *Remote Sensing of Environment* 21:119–128.

Ranchin, T. and L. Wald. 1993. The wavelet transform for the analysis of remotely sensed images. *International Journal of Remote Sensing* 14:615–619.

———. 2000. Fusion of high spatial and spectral resolution images: the arsis concept and its implementation. *Photogrammetric. Engineering and Remote Sensing* 66:49–61.

Richards, J. A. 1984. Thematic mapping from multitemporal image data using the principal component transformation. *Remote Sensing of Environment* 16:35–46.

Richards, J. A. and X. Jia. 1999. *Remote Sensing Digital Image Analysis.* Berlin: Springer.

Ruiz, M.O. 1997. A causal analysis of viewshed error. *Transactions in GIS* 2:85–94.

Ryherd, S. and C. Woodcock. 1996. Combining spectral and textural data in the segmentation of remotely sensed images. *Photogrammetric Engineering and Remote Sensing* 62:181–194.

Schiewe, J. 2003. Integration of multi-sensor data for landscape modeling using region-based approach. *ISPRS Journal of Photogrammetry and Remote Sensing* 57 (5-6):371–379.

Schiewe, J. and M. Ehlers, editors. 2003. *Geoinformatik 03: Ausgewählte Themen der Forschungsgruppe GIS/Fernerkundung.* Materialien Umweltwissenschaften Vechta (MUWV) 17. Vechta, Germany.

Schiewe, J. and M. Ehlers. 2005. A novel method for generating 3D city models from high resolution and multi-sensoral remote sensing data. *International Journal for Remote Sensing* 26 (4):683–698.

Shettigara, V. K. 1992. A generalized component substitution technique for spatial enhancement of multispectral images using a higher resolution data set. *Photogrammetric Engineering and Remote Sensing* 58:561–567.

Shi, W. Z. 1998. A generic statistical approach for modelling errors of geometric features in GIS. *International Journal of Geographical Information Science* 12:131–143.

Shi, W. Z., C. K. Cheung and C. Q. Zhu. 2003. Modelling error propagation in vector-based GIS. *International Journal of Geographical Information Science* 17:251–271.

Shi, W. Z., C. Q. Zhu, C. Y. Zhu and X. M. Yang. 2003. Multi-band wavelet for fusing SPOT panchromatic and multispectral images. *Photogrammetric Engineering and Remote Sensing* 69 (5):513–520. <http://www.asprs.org/publications/pers/scans/2003journal/may/2003_may_513-520.pdf> Accessed 22 December 2008.

Shi, W. Z., M. Ehlers and M. Molenaar, editors. 2005. Uncertainties in integrated remote sensing and GIS. *International Journal of Remote Sensing* (special issue) 26:2909–3120.

Singh, A. 1989. Digital change detection techniques using remotely-sensed data. International Journal of Remote Sensing 10:9891003.

Singh, A. and A. Harrison. 1985. Standardized principal components. *International Journal of Remote Sensing* 6:883–396.

Sinton, D. 1978. The inherent structure of information as a constraint to analysis: mapped thematic data as a case Study. In *Harvard Papers on Geographic Information Systems, 6,* edited by G. Dutton. Reading, Mass.: Addison-Wesley.

Sonka, M., V. Hlavac and R. Boyle. 1998. *Image Processing, Analysis, and Machine Vision,* 2nd edition. Pacific Grove, Calif.: PWS/Brooks and Cole Publishing.

Spiller, R. 2000. DMC – Why Z/I imaging preferred the matrix approach. *GIM International* 7:66–68.

Star, J. L., J. E. Estes and K. C. McGwire, editors. 1997. *Integration of Remote Sensing and GIS.* New York: Cambridge University Press.

Story, M. and R. G. Congalton. 1986. Accuracy assessment: a user's perspective. *Photogrammetric Engineering and Remote Sensing* 52:397–399.

Tomlin, D. 1990. *GIS and Cartographic Modeling.* Englewood Cliffs, N.J.: Prentice-Hall.

Tsuno K., A. Gruen, L. Zhang, S. Murai and R. Shibasaki. 2004. STARIMAGER – a new airborne three-line scanner for large-scale applications. *International Archives of Photogrammetry, Remote Sensing and Spatial Information Sciences* (Congress Istanbul). 35 (B1):226–234.

Veregin, H. 1989. Error modelling for the map overlay operation. In *Accuracy of Spatial Databases,* edited by M. F. Goodchild and S. Gopal, 3–19. London: Taylor & Francis.

———. 1995. Developing and testing of an error propagation model for GIS overlay operations. *International Journal of Geographical Information Systems* 9:595–619.

Vornberger, P. L. and R. A. Bindschadler. 1992. Multi-spectral analysis of ice sheets using co-registered SAR and TM imagery. *International Journal of Remote Sensing* 13:637–645.

Wald, L. 1999. Definitions and terms of references in data fusion. *International Archives of Photogrammetry and Remote Sensing* 32, part 7-4-3 W6. Valladolid, Spain.

Wald, L. 2002. *Data Fusion - Definitions and Architectures.* Paris: Les Presses de l'École des Mines. 198 pp.

Wang, F. 1993. A knowledge-based vision system for detecting land changes at urban fringes. *IEEE Transactions on Geoscience and Remote Sensing* 31:136–145.

Waterfeld, W. and H. J. Schek. 1992. The DASBDS Geokernel - an extensible database system for GIS. In *Three-Dimensional Modelling with Geoscientific Information Systems,* edited by A. K. Turner, 45–55. Dordrecht, The Netherlands: Kluwer Academic Publishers.

Welch, R., T. R. Jordan and M. Ehlers. 1985. Comparative evaluations of geodetic accuracy and cartographic potential of Landsat-4/5 TM image data. *Photogrammetric Engineering and Remote Sensing* 51:1249–1362.

Welch, R. and M. Ehlers. 1987. Merging multiresolution SPOT HRV and Landsat TM data. *Photogrammetric Engineering and Remote Sensing* 53:301–303.

———. 1988. Cartographic feature extraction from integrated SIR-B and Landsat TM images. *International Journal of Remote Sensing* 9:873–889.

Worboys, M. F. 1998. Computation with imprecise geospatial data. *Computers, Environment and Urban Systems* 22:85–106.

———. 2004. *GIS: A Computing Perspective,* 2nd edition. Keele, UK: Taylor & Francis.

Yesou, H., Y. Besnus and J. Rolet. 1993a. Extraction of spectral information from Landsat TM data and merger with SPOT panchromatic imagery -- a contribution to the study of geological structures. *ISPRS Journal of Photogrammetry and Remote Sensing* 48:23–36.

Yesou, H., Y. Besnus, J. Rolet and J. C. Pion. 1993b. Merging Seasat and SPOT imagery for the study of geologic structures in a temperate agricultural region. *Remote Sensing of Environment* 43:265–280.

Yocky, D. 1996. Multiresolution wavelet decomposition image merger of Landsat Thematic Mapper and SPOT panchromatic data. *Photogrammetric Engineering and Remote Sensing* 62:1067–1074.

Zhang, Y. J. 1996. A survey on evaluation methods for image segmentation. *Pattern Recognition* 29 (8):1335–1346.

Zhang, Y. 2002. Automatic image fusion: a new sharpening technique for Ikonos multispectral images. *GIM International* 16 (5):54–57.

Zhang, J. and M. F. Goodchild. 2002. *Uncertainties in Geographical Information.* London: Taylor & Francis.

Zhang, J. and R. P. Kirby. 2000. A geostatistical approach to modeling positional errors in vector data. *Transactions in GIS* 4:145–159.

Zhang, J. and N. Stuart. 2001. Fuzzy methods for categorical mapping with image-based land cover data. *International Journal of Geographical Information Science* 15:175–195.

Zhou, J., D. L. Civco, and J. A. Silander. 1998. A wavelet transform method to merge Landsat TM and SPOT panchromatic data. *International Journal of Remote Sensing.* 19 (4):743–75.

CHAPTER 38
Location-Based Services

Michael F. Goodchild

38.1 Introduction

Over the past ten years location-based services (LBS) have been variously touted as the next generation of GIS; as the killer application that will bring the GIS industry to its next level of prosperity; and as a growing threat to human privacy. This chapter attempts to provide a basic introduction to LBS, and a discussion of some of the major issues surrounding them. It begins with basic definitions, and a short history of LBS. This is followed by an overview of the services currently provided by this form of geospatial technology, or likely to be provided in the near future. A major section of the chapter then describes the various methods for determining the location of a device, and significant gaps in current coverage. The penultimate section introduces some of the major social issues associated with LBS, and the final section discusses some broader implications and current research directions.

A location-based service can be defined as an information service provided by a device that knows where it is, and modifies the information it provides accordingly (Küpper 2005). This is a very broad definition, and not surprisingly there are some trivial fits. A human, for example, is capable of knowing his or her location, and of providing information based on that knowledge, but one would not describe a human as "a device" – instead, "a device" is clearly intended to imply something digital, such as a cell phone, laptop computer, desktop computer, ATM (automated teller machine), or PDA (personal digital assistant). Moreover human knowledge of location is often informal (e.g., I am "at the airport"), rather than the formal and precise coordinates demanded by GIS technology.

Early computing devices were designed for the solution of numerical problems such as the equations governing nuclear reactions, for breaking secret codes, and for handling and analyzing massive amounts of data. None of these applications is in any way dependent on knowing the location of the computing device. Location was and is important in GIS, but while it is important in these computing applications to know the locations and attributes of features on the Earth's surface, the locations of the user and the computing device are largely irrelevant. Indeed, much is still made of the ability of GIS to analyze conditions in parts of the world that are remote from the user, and may never have been visited by the user.

While LBS are defined by their ability to provide information, the recipient of that information may not be the person using the device, and may not be in the same location. To be precise, then, the user for the purposes of this discussion will be the person co-located with the device, while the term recipient will be used in those cases where the information provided by the device is not directed to the user, at least in the first instance. For example, when a cell phone user initiates a 911 emergency call, the service provider will attempt to determine the location of the device by one of a number of techniques described later in the chapter, and will forward that information to the emergency dispatcher. Indeed, in this scenario the user may never know his or her location, but that information is enormously valuable to the dispatcher. US law explicitly authorizes this release of location information by cell phone providers for emergency purposes.

The first truly mobile computing devices began to appear in the 1980s (in my own case, in the form of a Hyperion portable computer with two 5.25 inch floppies, an alphanumeric screen, an Intel 8088 processor and no hard drive). Since then mobility has become an increasingly important factor in computing, and today we expect to find far more powerful performance available in devices that can fit in a pocket. Mobility has in turn opened up the possibility of providing information about the user's location in space and time, about the user's surroundings, and about features in the environment that are nearby but beyond the user's own sensory perception. This information may be useful to the user, and it may also be useful to someone else, such as the user's employer, bank, parents or government. The next section reviews examples of LBS, and their value to users and other recipients.

38.2 Why Is Location Important, and Who Wants to Know?

GPS (Global Positioning System) provides the simplest and most obvious form of LBS, allowing a user to determine position on the Earth's surface, in the precise coordinates of a formally specified location, to within meters and in some cases centimeters (Kennedy 2002; Leick 2004). The US Global Positioning System was originally developed for military purposes, with only a degraded signal available to civilians. But a change of policy in 1990 opened full accuracy to all, and today GPS is used worldwide as a free, ubiquitous means of determining position. It is based on comparing the timing of signals arriving from that subset of a constellation of 24 satellites that is above the user's horizon. A comparable Russian system, Global Orbiting Navigation Satellite System (GLONASS), is in operation, and a European system known as Galileo will be operational within a few years.

A GPS position alone is not of much value—exact position relative to the Equator and the Greenwich Meridian is not in itself very useful. Position can be determined relative to other features if the GPS signal can be integrated with a digital map, or if a paper map is available and suitably marked with a coordinate graticule. But the real power of GPS comes from its integration into other functions, including those of a GIS with a suitable database. Today, simple GPS devices are available that contain digital maps, databases of features such as hotels, retail stores and restaurants, and GPS has been integrated into the in-vehicle navigation systems that are increasingly common either as installed features or after-market additions. Such systems are often described as providing an *augmented reality*, by providing the user with information that is beyond his or her sensory perception, in contrast to the *virtual reality* of traditional GIS.

GPS integrated with a database in a mobile device such as a PDA (personal digital assistant) or cell phone can provide a powerful form of augmented reality. But an isolated database is inherently static, since it can be updated only from physical media. For example, many in-vehicle navigation systems must be updated by the cumbersome process of downloading from a CD via a personal computer and a USB connection. A more dynamic approach results when the mobile device is linked wirelessly to the Internet, allowing more-or-less continuous updating. Information on minute-by-minute changes in traffic congestion, the positions of trains and buses, and the real-time locations of flights can be continuously fed from servers, providing truly dynamic LBS.

In many cities it is now possible to obtain real-time feeds of the actual locations of buses, and to service queries such as "When will the next Number 24 bus arrive at this stop?", where "this stop" is determined from the user's current location as determined by the mobile device's GPS (e.g., NextBus, Inc. 2007). Several Web services offer real-time data on traffic densities on major roads (e.g., Microsoft 2007), and several airlines allow users to query the locations of flights, returning the results in map form (RLM Software Undated).

Similar technology has been exploited to provide valuable services for emergency response, military operations, and construction. An emergency responder, for instance, may find it difficult to determine his or her current location, particularly if street signs and house numbers have been obliterated, or if the street pattern itself has become distorted, or if visibility is obscured by smoke or dust (NRC 2007). In such situations it is useful to have a mobile device equipped with GPS and a map database, and if possible to receive real-time feeds of other data sets, including recent post-event imagery. In the construction industry, LBS offers the possibility of on-site access to databases showing the underground locations of infrastructure, allowing workers to dig without risk of interrupting services.

Researchers also have recognized the potential of augmented reality in field work. At Columbia University, Steven Feiner and his group have experimented with head-mounted devices that superimpose historic scenes on the user's field of view, based on GPS positions and sensors that determine the user's head orientation (Feiner et al. Undated). At the University of California, Santa Barbara, Reginald Golledge and his group have developed prototypes of a personal navigation system for the visually impaired (UCSB Undated), while Project Battuta, a collaborative project with Iowa State University, has developed LBS to support researchers working in the field, providing information about the user's position and the locations of sample sites, and performing real-time spatial analysis of observations as they are gathered by the user (Nusser et al. Undated), and based on wearable devices.

Besides these rather utilitarian applications, LBS are also proving popular in social and recreational activities. Several cell phone operators now provide variations on the theme of mapping friends. For example, the Find Friends service offered in some markets by AT&T prior to its merger with Cingular (Hanrahan 2003) allows participants to display maps on their cell phones of the locations of friends who are also participants. Geocaching (Groundspeak Inc. 2007) is a popular and rapidly growing sport that involves players in using GPS and digital maps to find caches where they can exchange interesting objects. A number of *location-based games* have been developed that engage one or more users in fictitious scenarios. In Undercover 2 (YDreams 2007), for example, players are tracked using cell phones, and maps are displayed on user devices showing the locations of each player.

In all of these examples it is the user who receives the information provided by the LBS. Reference has already been made to 911 calls, when it is the dispatcher rather than the caller who receives the information about the caller's location. However, there are many other examples of LBS applications for which the recipient is someone other than the user; in many of these the user may not be aware that the information is being provided, and may even be disturbed to know that his or her locational privacy is being compromised in this way.

The term *tracking* is often used to refer to real-time acquisition of location information by a third party, often by wireless communication of GPS measurements. Such tracking is common in the transportation industry, and car rental agencies have been known to track their fleets to monitor violations of rental agreements, for example by speeding or by traveling into explicitly prohibited areas such as other countries. Tracking is used by probation officers to monitor compliance, by vehicle owners to detect theft, by researchers and wildlife managers to monitor the movement of animals and birds, and by pet owners and even parents to monitor their charges. The OnStar system (OnStar Corp. 2007) installed in many vehicles initiates tracking when activated by the user, transmitting the user's current location to a dispatcher who can respond to queries, for example for driving directions. The system is automatically activated in the event of airbag deployment, prompting the dispatcher to activate appropriate emergency services.

Tracking devices monitor the location of the device on a fixed interval of time. Often the major constraint on the frequency of sampling and the length of time over which data can be collected is the capacity of batteries, and this is particularly problematic when devices have to be carried by small mammals or birds. But when battery power is not a constraint, for example when tracking devices are carried on vehicles, it is common for sampling intervals to be as short as 1 second. With such rates it is possible to detect stops, starts, and changes of speed, and thus to infer many potentially useful properties. For example, changes of speed can indicate the presence of traffic accidents long before the event is known through more conventional means, such as cell phone calls from drivers.

Other techniques make it possible for third parties to determine the locations of events, and to associate them with individuals. Although no mobile devices are involved, every use by a purchaser of a credit or debit or store affinity card creates a record that is easily referenced in space and time. Card companies use this information to detect fraud, as for example when a card is used for several large purchases in quick succession in an unusual location. Many of us have had the embarrassing experience of having a card refused at a restaurant, store, or hotel that is not within our normal geographic range. Other devices record the passage and identity of vehicles on toll highways, and video cameras increasingly record the presence of individuals in public places. Automated license-plate recognition is now routine, and facial recognition is increasingly effective.

Finally, information on the user's location is of substantial commercial value to the operator of a search service, such as Google or Yahoo!, since it can be used to prioritize hits in geographic proximity to the user. Such prioritization is of value to vendors, who are consequently willing to pay the search service operator accordingly. Although geographic location is not normally known precisely, it can be inferred approximately from the user's IP address, as described in the following section.

38.3 Location-Determining Technologies

Mention has already been made of GPS and its analogs, which today constitute the primary means by which a mobile device is able to determine its geographic location. Miniaturization of GPS receivers has made it possible to embed them in cell phones, PCMCIA cards (devices to connect peripherals to computers, so-called because they use the standard developed by the Personal Computer Memory Card Industry Association) , and even wristwatches such as the Garmin Forerunner 201 (Garmin Ltd. 2007). Battery power is currently the major constraint to further miniaturization, however, and the operating times of some devices on a single charge can be disappointingly short.

GPS requires the line-of-sight presence of at least three satellites for horizontal positioning, and at least four for additional vertical positioning. Unfortunately this means that GPS signal is frequently lost by mobile devices, owing to tree cover, steep slopes, and buildings—and is never available within buildings. It is estimated that while an average vehicle driving in the San Francisco area has signal available 80% of the time, in Hong Kong with its urban canyons the proportion drops to 20%. For pedestrians and for vehicles parked in structures the percentages are substantially lower.

Many options for filling these coverage gaps in positioning technology have been explored, especially within buildings, with varying degrees of success. Beacons can be installed at fixed positions, radiating signals in various parts of the spectrum that can be used by mobile devices to determine position (WiFi signals offer one of the more successful options; e.g., Ekahau, Inc. (2007)). At this time, however, no one system appears to be ideal, and to offer smooth interoperability and transition with GPS positioning outdoors. Meanwhile research projects continue to develop prototypes of integrated indoor/outdoor positioning (e.g., CRCSI Undated).

Although GPS provides the positioning capability of many cell phones, other and in some respects simpler techniques are also used. Accurate timing of signals from multiple towers, accurate timing by multiple towers of the signal from a phone, and estimation of distance by measuring signal attenuation, are all used to provide positional information, with accuracies ranging from hundreds to tens of meters (Goodchild 2007). In addition the identity of the cell containing the phone is sufficient in some areas to provide positioning to a few hundred meters, particularly with the micro-cell technology now being implemented in some urban environments. Note that these technologies can provide limited coverage beyond the range of GPS signals.

Several other techniques are capable of providing at least approximate locations. For point-of-sale systems and ATMs, geographic location is determined when the system is installed, and is used to georeference any transactions recorded by the system. Mention was made earlier of the potential for approximate determination of location from the IP address of the user, based on the publicly available registration information associated with the address. Several companies specialize in inferring geographic locations from IP addresses (e.g., Digital Envoy 2007).

Finally, there has been much interest recently in the potential of RFID (Radio-Frequency Identification) as a technique for determining position (Finkenzeller 2003). RFID is perhaps best thought of as the radio equivalent of a barcode, its major advantage being that no optical device such as a laser needs to be physically pointed at the tag. Instead, tags are detected and read by remote radio-frequency transmitters, which receive and interpret the returned signal. RFID tags are sometimes incorporated into standard optical barcodes, but can also be concealed in articles and made difficult to remove. They are cheap to produce, and easily miniaturized.

RFID tags are now being used by major retailers to keep track of the production, shipment, storage and sale of goods (Williams 2004). They are being used to identify and track the components of a building during construction, to manage farm animals, and to track containers through ports. A new version of RFID allows tracking of objects enclosed in metal containers (Navigational Sciences 2007). Several recent news stories have focused on the potential for RFID tracking of individuals, including one in which a school used RFID tags to track its students entering and leaving school property (Leff 2005), and one in which a night club used RFID to tag its members (Losowsky 2004). Another recent news story described the Geo-Spatial Web (Ashbrook 2006), a future world in which many types of objects would be "intelligent"—capable of returning their identity, location, and other properties in response to a radio signal, just as air traffic controllers today can track the identity and properties of aircraft.

38.4 Social Issues

It will be apparent already that many forms of LBS raise serious and significant issues of an ethical nature. Many of the applications allow third parties to gain access to information about the location of the user, whether or not such access has been explicitly authorized. Users of credit cards, for example, in effect surrender certain aspects of their locational privacy in return for the convenience of the card, and some degree of assurance that tracking of location will protect them from fraud. US legislation in the form of the Wireless Communication and Public Safety Act of 1999 protects the users of cell phones from unauthorized release of location information by the cell phone operator, allowing such release only in carefully defined circumstances:

> *(4) to provide call location information concerning the user of a commercial mobile service (as such term is defined in section 332(d)):*
>
> *(A) to a public safety answering point, emergency medical service provider or emergency dispatch provider, public safety, fire service, or law enforcement official, or hospital emergency or trauma care facility, in order to respond to the user's call for emergency services;*
>
> *(B) to inform the user's legal guardian or members of the user's immediate family of the user's location in an emergency situation that involves the risk of death or serious physical harm; or*

*(C) to providers of information or database management services solely for purposes
of assisting in the delivery of emergency services in response to an emergency.'*

The State of California has aggressively pursued rental car operators who track renters in ways that can be regarded as intrusive (San Jose Business Journal 2004).

Jonathan F. Raper (CSISS 2002) defines two distinct contexts of locational privacy. The public persona governs those circumstances in which the individual is willing to surrender locational privacy in return for certain benefits—the convenience of a credit card, or the discounts offered by a store affinity card. The private persona governs those circumstances in which the individual demands to keep location private, perhaps because harm or embarrassment might result if location were revealed.

Dobson has defined and written extensively on the concept of *geoslavery* (Dobson and Fisher 2003), drawing an analogy between the effects of surveillance and tracking on individual freedom today, and the effects of shackles, branding, and skin color on the freedom of individuals in past centuries. Public debates over the tension between individual freedoms and the need for security have grown in intensity since the events of 11 September 2001, and will likely continue to grow.

A quite different set of social issues arises from the ability of LBS to modify other aspects of human behavior. For example, in the past it has been important for retailers, particularly of fast foods, to locate in the most visible urban sites, notably on retail strips and on street corners. But in a world of augmented reality, when the senses are aided by an array of information sources including LBS, it is no longer necessary to be able to see a vendor. In principle, the effects of widespread adoption of LBS on the urban retail landscape could be as profound as those of the Interstate Highway network beginning in the 1950s. At a more immediate level, LBS has the ability to assist drivers in scheduling daily tasks, choosing routes and sequences that avoid the worst congestion (e.g., Garling et al. 1998). Some daily activities, such as the journey to work or to drop children at school, are fixed in time, but others, such as shopping, dry cleaning, or visiting the gym, can be rescheduled in response to real-time and forecasted levels of congestion.

38.5 The Broader Context

The growth of LBS in recent years has stimulated a challenging research agenda. A 2001 specialist meeting organized by the Center for Spatially Integrated Social Science at the University of California, Santa Barbara, in conjunction with the University Consortium for Geographic Information Science, identified a series of short- and long-term research issues ranging from three-dimensional positioning indoors and outdoors, to the social impacts of widespread LBS use (CSISS 2002). Some of the highlights of the research agenda were:

- There is a need for methods of analysis that infer activity types from the form of tracking data, and other potentially available ancillary data such as the characteristics or socioeconomic status of the person being tracked. We need a library of track types, an understanding of the distinguishability of each common form of mobile human activity, and an understanding of the importance of sampling rate and spatial accuracy in each case. Alternatively, we need to define a new set of activity types that are both detectable in tracks and of interest in social science, given the potential abundance of such data from LBS.

- Data mining techniques are already being used to discover misuse of credit cards, based in part on abnormal behavior in space and time. Other data mining tools could be developed for purposes more closely related to social science, including the detection of types of behavior from tracking data.

- There is a pressing need for new theory, to frame the new analytic tools and models. New theory should deal with the types of behavior revealed by tracking, and with the impacts of LBS on behavior of consumers and entrepreneurs.

- LBS have the potential to affect profoundly the spatial organization of society, and the behavior of individuals and groups within it. How rapidly will this occur? Which social groups will lead the process and which will lag?
- Will LBS adoption lead to fundamental change in retailing structure? Will retail outlets become more dispersed or more concentrated in space, and what impact will LBS have on micro scale location strategies (e.g., within malls, or along streets)?
- What factors determine how people react to the intrusiveness of LBS, and their willingness to trade privacy? Are there differences due to gender, age, or ethnicity?

At a more technical level, LBS raise issues of user interface design, computational geometry and network architectures. The small space available on the typical mobile device presents challenges for the cartographer, who is used to much larger display areas (e.g., Dillemuth 2005; Reichenbacher 2004). Mobile devices must often be used in conditions that make it difficult to see the device screen (smoke, bright sunlight). Because viewing a small screen and interacting with a keyboard can be distracting, it is desirable that drivers interact with mobile devices using sound and speech, raising additional questions about user interface design. There is a growing literature on the computational problems posed by large numbers of mobile devices querying large spatial databases over wireless connections with limited bandwidth (e.g., Kulik and Tanin 2006). More broadly, LBS can be seen as part of a growing research interest in sensor networks (e.g., Lesser et al. 2003), and to pose particular problems due to the mobile nature of the sensors.

References

Ashbrook, T. 2006. On Point : The New Sense of the Web. <http://www.onpointradio. org/shows/2006/01/20060103_b_main.asp> Accessed 22 October 2007.

CRCSI (Cooperative Research Centre - Spatial Information). Undated. Project: 1.3 Integrated Positioning & Geo-referencing Platform. <http://www.spatialinfocrc. org/pages/project.aspx?projectid=65> Accessed 22 October 2007.

CSISS (Center for Spatially Integrated Social Science). 2002. Specialist Meeting on Location-Based Services, Final Report. Santa Barbara, Calif.: Center for Spatially Integrated Social Science. <http://www.csiss.org/events/meetings/location-based/goodchild_lbs.htm> Accessed 23 October 2007.

Digital Envoy. 2007. Digital Envoy, IP Intelligence and Authentication Solutions. <http:// www.digitalenvoy.net/> Accessed 22 October 2007.

Dillemuth, J. 2005. Map design for mobile display. Proceedings, AutoCarto 2005. <http://media.igert.ucsb.edu/pubdls/DillemuthAutoCarto2005.pdf>. Accessed 25 October 2007.

Dobson, J. E. and P. F. Fisher. 2003. Geoslavery. *IEEE Technology and Society Magazine* 22(1):47–52.

Ekahau, Inc. 2007. Wireless network planning and verification: Ekahau Wi-Fi Tools. <http://www.ekahau.com/> Accessed 22 October 2007.

Feiner, S., T. Höllerer, E. Gagas, D. Hallaway, T. Terauchi, S. Güven and B. MacIntyre. Undated. MARS - Mobile Augmented Reality Systems. <http://www1.cs.columbia. edu/graphics/projects/mars/> Accessed 22 October 2007.

Finkenzeller, K. 2003. *RFID Handbook: Fundamentals and Applications in Contactless Smart Cards and Identification*. Tr. R. Waddington. Hoboken, N.J.: Wiley.

Garling T., T. Kalen, J. Romanus, M. Selart and B. Vilhelmson. 1998. Computer simulation of household activity scheduling. *Environment and Planning A* 30:665–679.

Garmin Ltd. 2007. Forerunner® 201. <http://www.garmin.com/products/forerunner201/> Accessed 22 October 2007.

Goodchild, M. 2007. Geography 176B: Technical Issues in Geographic Information Systems. <http://www.geog.ucsb.edu/~good/176b/n14.html> Accessed 22 October 2007.

Groundspeak Inc. 2007. Geocaching - The Official Global GPS Cache Hunt Site. <http://www.geocaching.com> Accessed 22 October 2007.

Hanrahan, T. 2003. New cellphone services help find friends and places to go. *The Wall Street Journal Online*, 22 May 2003. <http://online.wsj.com/article/SB105355745962282600.html> Accessed 22 October 2007.

Kennedy, M. 2002. *The Global Positioning System and GIS: An Introduction.* 2nd edition. New York: Taylor and Francis.

Kulik, L. and E. Tanin. 2006. Incremental rank updates for moving query points. Pages 251–268 in *Geographic Information Science*, edited by M. Raubal, H. J. Miller, A. U. Frank and M. F. Goodchild.. Fourth International Conference, GIScience 2006, Münster, Germany, September, Proceedings. Lecture Notes in Computer Science 4197. New York: Springer.

Küpper, A. 2005. *Location-Based Services: Fundamentals and Operation.* New York: Wiley.

Leff, L. 2005. Students ordered to wear tracking tags. Associated Press, 9 February 2005. <http://www.msnbc.msn.com/id/6942751/> Accessed 22 October 2007.

Leick, A. 2004. *GPS Satellite Surveying.* 2nd edition. Hoboken, NJ: Wiley.

Lesser, V., C. L. Ortiz and M. Tambe, editors. 2003. *Distributed Sensor Networks: A Multiagent Perspective.* New York: Springer.

Losowsky, A. 2004. I've got you under my skin. *The Guardian*, 10 June 2004. <http://technology.guardian.co.uk/online/story/0,3605,1234827,00.html> Accessed 22 October 2007.

Microsoft. 2007. Live Search Maps. <http://maps.live.com/> Accessed 22 October 2007.

National Research Council. 2007. *Successful Response Starts with a Map: Improving Geospatial Support for Disaster Management.* Washington, DC: National Academy Press. <http://www.nap.edu/catalog.php?record_id=11793> Accessed 22 October 2007.

Navigational Sciences. 2007. Navigational Sciences, Inc. - Home. <http://www.navsci.com> Accessed 22 October 2007.

NextBus, Inc. 2007. How NextBus works. <http://www.nextbus.com/corporate/works/index.htm> Accessed 22 October 2007.

Nusser, S., L. Miller, M Goodchild and K. Clarke. Undated. Collecting & Using Geospatial Data in the Field. <http://dg.statlab.iastate.edu/dg/> Accessed 22 October 2007.

OnStar Corp. 2007. OnStar Explained. <http://www.onstar.com/us_english/jsp/explore/index.jsp> Accessed 22 October 2007.

Reichenbacher, T. 2004. Mobile Cartography: Adaptive Visualisation of Geographic Information on Mobile Devices. Ph.D. Dissertation, Technical University of Munich, Munich, Germany.

RLM Software. Undated. FlightView Flight Tracker. <http://www.flightview.com/> Accessed 22 October 2007.

San Jose Business Journal. 2004. Rental car firm cited for secretly tracking customers, then imposing surcharges. *San Jose Business Journal*, 9 November 2004. <http://www.bizjournals.com/sanjose/stories/2004/11/08/daily16.html> Accessed 22 October 2007.

UCSB. Undated. UCSB Personal Guidance System (PGS). <http://www.geog.ucsb.edu/pgs/main.htm> Accessed 22 October 2007.

Williams, D. H. 2004. The strategic implications of Wal-Mart's RFID mandate. Directions Magazine 29 July 2004. <http://www.directionsmag.com/article.php?article_id=629&trv=1> Accessed 22 October 2007.

YDreams. 2007. Undercover 2: Merc Wars. <http://www.ydreams.com/ydreams_2005/index.php?page=193> Accessed 22 October 2007.

CHAPTER 39
Location-Based Services Using WiFi Positioning on a Wireless Campus

Barend Köbben

39.1 The Wireless Campus at the University of Twente

In June 2003 the "Wireless Campus" was inaugurated at the University of Twente (UT), allowing cable–free internet access to staff and students anywhere on campus. University of Twente is a young university in the Eastern part of The Netherlands. It employs 2,500 people and has over 6,000 students. On its campus, the university has 2,000 student rooms. The university campus is situated between the cities of Enschede and Hengelo, near the Dutch-German border. Spread over the 140-hectare campus 650 individual wireless network access points have been installed, making it Europe's largest uniform wireless hotspot. Anyone with a PC, laptop, PDA or other WiFi (wireless fidelity)-enabled device can access the university's network and the internet from any building, the campus park and other facilities without cabling.

University of Twente's Wireless Campus aims at a broad range of research and applications of wireless and mobile telecommunication. The UT wants to use the WLAN (Wireless Local-Area Network) in cooperation with the adjacent Business and Science Park and is busy covering this by access points as well. Furthermore, a project has just started in cooperation with Enschede Municipality to install further access points to cover the downtown area of Enschede.

The wireless network facility was made possible with financial support of the Dutch Ministry of Economic Affairs and has been built in cooperation with IBM Netherlands and Cisco Systems. It consisted in the first instance mainly of access points that use the 802.11b wireless networking standard, offering a data transfer speed of 11 megabits per second for most users. There is an ongoing effort to upgrade the entire network to the 802.11g standard, providing data at speeds up to 56 megabits per second.

Research projects investigate the technology and the applications of wireless and mobile communication in several ways, mostly in cooperation with industrial and other knowledge partners. The Wireless Campus has become a 'test bed' for wireless and mobile applications. The major part of this research takes place at the Centre for Telematics and Information Technology (CTIT) and the research institute MESA+. Both are key research institutes of the University of Twente. MESA+ is an institute that conducts research in the fields of nanotechnology, microsystems, materials science and microelectronics. CTIT is an institute that conducts research into the design of advanced ICT (Information, Communications and Technology) systems and their application in a variety of application domains. Its Computer Architecture Design and Test for Embedded Systems Group became interested in using the WiFi technology in the wider framework of the SmartSurroundings research program. This program is

> *investigating a new paradigm for bringing the flexibility of information technology to bear in every aspect of daily life. It foresees that people will be surrounded by deeply embedded and flexibly networked systems …. This presents a paradigm shift from personal computing to ubiquitous computing, …. Relevant knowledge areas include embedded systems, computer architecture, wireless communication, distributed computing, data and knowledge modelling, application platforms, human-computer interaction, industrial design, as well as application research in different settings and sectors"* (Havinga et al. 2004, page 64).

An important part of such systems is establishing the position of persons, services and devices, and one of the possible strategies to achieve that is to use the WiFi network.

39.2 Positioning Using WiFi Technology

Using WiFi technology is just one of the many ways available of using wireless networks for positioning of (mobile) users. An overview of wireless location papers, a website maintained by Youssef (2005), currently distinguishes some 10 categories, among them mobile phone network– and GPS–based techniques. For an overview of many of these techniques, see Hightower and Borriello (2001).

There are various reasons to choose WiFi-based localization over other methods. The most prominent possibly is the fact that it is an economical solution. Because the wireless network infrastructure already exists, localization can be done by software-only methods without adding any additional hardware. This is also true of cell-phone-network-based solutions, but these currently offer only poor accuracy. Secondly, compared to other indoor techniques such as InfraRed, Bluetooth or RFID, the range covered by WiFi is significantly larger. And although coverage is not as ubiquitous as mobile phone networks or GPS, WiFi-based WLANs are being installed at an ever-increasing rate all over the world in public places like airports, conference centers and shopping malls, as well as in universities, hotels, offices and such.

39.2.1 Positioning Methodologies

There are different basic methods of using WiFi signals for determining the location of users (Muthukrishnan, Meratnia et al. 2005), shown in figure 39-1. Hardware-based systems use additional hardware on top of the existing infrastructure to determine characteristics such as Time of Arrival or Angle of Arrival of the signal received from known fixed locations of the WLAN Access Points (APs). Triangulation and other geodetic techniques can then be used to calculate positions. Apart from the disadvantage of the additional hardware need, there is also the problem that the signals are reflected from various objects, especially indoors. Because of this multipath environment, these techniques are complex and reliable results are difficult to obtain.

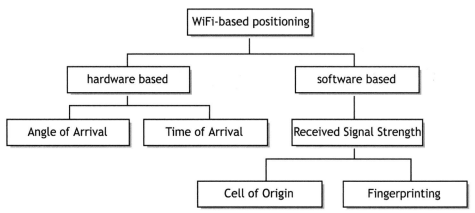

Figure 39-1 Simplified taxonomy of WiFi-based positioning systems, after Muthukrishnan, Meratnia et al. (2005).

The *software-based* systems mostly use the *Received Signal Strength* (RSS). The big advantage of RSS-based techniques is that we can use the existing infrastructure to deploy a positioning system without additional devices, other then the standard WLAN network card in the computer or PDA.

A very simple method is to determine which AP's signal is the strongest one received and then assume the location to be in the area that is covered by this AP's signal. This *Cell of Origin* method, that is similarly used in mobile phone networks, will generally result in very coarse-grained location information.

More-accurate methods use so-called *location fingerprinting* schemes, where selected characteristics of the signal that are location dependent are stored in a database. These fingerprints are then matched to the characteristics actually measured at the current location of a receiver. Unfortunately, the RSS is a highly variable parameter and issues related to positioning systems based on RSS fingerprinting are not yet understood well. The fingerprint information can be obtained in two ways. Firstly there are *Radio Map*-based methods. They involve location fingerprinting done at as many locations as possible, building up a fine-grained RSS map. When a device requests a location, it sends the signal strengths from all access points it can detect to the database, that finds the closest match and returns that as the probable location. The main drawback is the necessary calibration of signal strength as a function of a particular location, and even particular time (since the radio wave properties in an indoor environment can vary greatly depending on the number of people inside the building). There is a trade-off between the amount of effort put into the calibration (it requires lots of time and work and should be performed repeatedly) and the accuracy obtained. Little research as of yet has addressed the issue of optimizing the calibration effort. Secondly, *Model Based* approaches are based on automatically generated fingerprint information based on models of the signal propagation and detailed information about the geometry and topology of the environment, including the materials used in walls, ceiling, etcetera, as these greatly influence the multipath characteristics of the signal.

In Muthukrishnan, Meratria et al. (2005), the authors defined evaluation criteria for WiFi-based location systems that can be used as guidelines to compare and evaluate several indoor location/positioning systems.

1. *Accuracy and Precision* of estimated location are the key metrics for evaluating a localization technique. Accuracy is defined as the deviation of the estimated position from the true position and is denoted by an accuracy value and precision value (e.g. 15 cm accuracy over 95% of the time). The precision indicates how often we expect to get at least the given accuracy. The accuracy of a positioning system is often used to determine whether the chosen system is applicable for a specific use.

2. *Calibration* is also very important. The uncalibrated readings are highly erroneous and device calibration (the process of forcing a device to conform to a given input/output mapping) is needed. Often there is a trade-off between the accuracy and the calibration effort.

3. *Responsiveness* is defined as how quickly the system delivers the location information. It is an important parameter, especially when dealing with mobility. However, this parameter is mostly ignored in the description of the existing systems.

4. *Scalability* is a significant parameter, as the proposed design should be scalable for large networks. If an approach is calibration intensive then eventually it is not a scalable solution.

5. *Self-organization* is of great importance, as it is infeasible to manually configure the location determination processes for a large number of devices in random configurations with random environmental characteristics.

6. *Cost* is also a crucial issue. It includes the cost of installation, deployment, infrastructure and maintenance.

7. *Power Consumption* is of great concern when running the system in a real environment, especially for mobile devices.

8. *Privacy* includes major concerns: Using localization, it is very easy to create a Big Brother infrastructure that can track users' movements and facilitate the deduction of

patterns of behavior. However, this issue is being generally overlooked in the design of systems and considered as an after thought only. Centralized systems are particularly weak with regard to privacy.

39.2.2 Localization Algorithm in the Prototype

The positioning component currently employed in the project described here is based upon an earlier pilot project called "FriendFinder" (Bockting et al. 2004), done in 2004 for two specific buildings on the University campus. In this pilot a prototype client-server architecture was built, where the client program on the mobile device determined its location by detecting the Access Points (APs) in range and comparing them with data about the APs in a server-side database. This database stored the location in XYZ of the APs, their BSSIDs (Basic Service Set Identifier, the unique identifier of an AP), and the strength of their antenna output (in mW). The application first buffered the RSS measurements because not all APs are detected in any single scan. Then it detected probable faulty measurements and deleted them. The accepted measurements were then put through a filter that calculated their centroid. Now the client had a first estimate of its position. Further filtering took place, and the final estimate was determined by so-called "iterative multilateration." In this technique a client's position, with its estimated inaccuracy, is used by other clients as a reference frame. In that way all nodes use each other's information to jointly improve the accuracy of the positioning. An important part is played by further filters that implement a learning effect from the stored positioning history of the application to further improve the accuracy.

Tests have shown that the FriendFinder pilot achieved an average positioning accuracy just under 5 meters, for non-moving devices. In the current project, the positioning component is part of a wider PhD research (Muthukrishnan, Lijding et al. 2005) that aims at significantly increasing this accuracy. To achieve this, the research investigates the possible approaches, the filters and methods used and the positioning algorithms themselves. An area of further research will be the self-learning abilities of the system, which should make the positioning more accurate over time. The Model Based approach, mentioned in 39.2.1, will be employed to achieve calibration-free localization, preserving quality and accuracy. This is ongoing research and the positioning system used for the current prototype "Flavour" (see 39.4.1) will be updated with the emerging knowledge and methods.

At present, the localization works as follows: The WiFi device inside the laptop or PDA periodically scans its environment to discover WLAN Access Points (APs) in the vicinity. The location of the APs in a 3D coordinate system is maintained in a geodatabase (see 39.3). During the access point scanning phase, the BSSID address of the access points and their Recorded Signal Strengths (RSS) are determined and stored. At any unknown location 'n' in the conference venue, the variation of the signal strength will be:

$$\text{at each } n \text{ RSS varies as } 0 >= \text{RSS} <= \text{MAX}$$

However, the signal strengths are usually contaminated by noise. In order to have a better estimation of the actual location, an *exponential moving average filter* is employed to smooth the signal strength. The formula used in doing so, in which $\alpha = 0.125$ and RSS denotes the observed signal strength is:

$$\text{Current RSS} = \alpha * (1 - \text{Current RSS}) + \alpha \, (\text{previous RSS})$$

The accuracy of this method can be improved by introducing previous traces (history) of the users. The knowledge of users' earlier determined locations can be used to predict their movement in a certain area. For example, assume that there are already 'p' predetermined positions prior to the user's new location 'q', then there is a high probability that the new location 'q' is near the vicinity of the predetermined positions 'p'. Due to fluctuations in the signal strength, caused for example by reflections off walls, or by obscuring of the antenna by the device user, the system might return a false location. The algorithm computes the difference in the distance between the last determined locations and the currently calculated one. If the difference is larger than a certain threshold then it is can be assumed the location will be in error. This system can be greatly improved by introducing the internal geometry of the building into the equation. If the system would be able to have a notion of the distances as well as topological relations within the environment, major improvements in the final localization accuracy can be expected.

39.3 Mapping the Access Points

The WiFi-based positioning methods described above are highly dependent on an initial mapping stage, in which the coordinates of all the access points in a 3D coordinate system had to be recorded in a database. For the FriendFinder project mentioned above, only a limited number of the Access Points (APs) had been used. As no geo-information experts were involved at that stage, their positioning was done in a rather improvised way. The height of the APs especially was a problem: it was determined only by estimate and with respect to the building's ground floor height. In this limited project that was not a big problem, as only one building was involved, but for the larger project the elevation differences between the buildings (more than 5 meters, which is a lot for The Netherlands!) had to be taken into account.

The 650 individual wireless network APs that have been installed were only indicated on paper maps, one map per floor, of the individual buildings of the University. The base maps are print-outs from CAD–drawings (blueprints) maintained by the Facility Management Services; they have a high level of detail, but they are not georeferenced and thus have a local, arbitrary, coordinate system that's basically just 'paper coordinates.. Furthermore, the location of the APs had been indicated on these maps haphazardly by hand-drawn symbols at the time of installation of the devices.

Therefore the first task has been the digital mapping of the AP locations in a geodatabase. In order to do this, it was decided to digitize all locations using GIS software and digitally georeferenced versions of the CAD drawings. The georeferencing was achieved by trans-formation of the CAD drawings, using control points from an overview map of the whole campus that is available in the Dutch national coordinate system "Rijks Driehoeksstelsel." It was possible, when using simple first order transformations, to achieve Root Mean Square Errors of less than 0.1 m.

For all buildings a base elevation was determined in meters above NAP (the Dutch vertical datum) by combining the campus map with the Actual Height Model of the Netherlands, a detailed elevation model of the whole country made by airborne laser altimetry, which has a point density minimum of 1 per 16 m2 and a systematic error of 5 cm maximum (Rijkswaterstaat Undated). In order to get precise location measurements, it was necessary to physically visit all APs and use a laser measurement device to determine the relative location of the AP antenna with respect to the elements of the building present in the CAD drawings (walls, floors, windows). By combining these relative measurements with the georeferenced maps a precise 3D location has been determined and put into a geodatabase. The added bonus is that all APs have been checked and additional attributes were gathered, such as antenna type, antenna connection length (for estimating signal loss), etc.

To efficiently store and use spatial data in a database, it has to support geographic data types and methods, preferably using the standards of the OGC (Open Geospatial Consortium 2007). Among other things, the OGC has set the Simple Features SQL Specifi-cation that provides for publishing, storage, access, and simple operations on spatial features (point, line, polygon, multi-point, etcetera) through an SQL interface. There are several commercial database systems with OGC-compliant spatial extensions, of which Oracle is probably the most prominent, but there are also open source alternatives: PostgreSQL, a database system with PostGIS as a spatial extension and MySQL. MySQL's recent versions include OGC-compliant spatial extensions, although not implementing the full set of OGC specifications. The reason for the choice of MySQL in this project was the simple 'light-weight' character of the software, as compared to the complicated though more fully-featured PostGIS. By adhering strictly to the OGC standards it should be straightforward to change or even mix database platforms in the future.

39.4 Wireless Campus Location Based Services

There has recently been a lot of industry and research activity in the realm of "Location Based Services" (LBS), which has been defined in Urquhart et al. (2004, page 70) as *"wireless services that use the location of a (portable) device to deliver applications which exploit pertinent geospatial information about a user's surrounding environment, their proximity to other entities in space (eg. people, places) and/or distant entities (eg. destinations)"*. The purpose of the project described here is not the development of 'the' or even 'a' Wireless Campus LBS, but rather to investigate and set up the infrastructure necessary to be the basis for LBS's. It combines input from several research projects with the practical application of new as well as established techniques to provide useful services for the UT campus population. The research mentioned has a wider scope then just this project: the Wireless Campus LBS is intended *to serve as a testbed* for the research as well as *to benefit from* the outcomes of the research.

39.4.1 First Tests: FLAVOUR

The first use case test of the Wireless Campus LBS was to provide the participants of a conference held at the UT grounds in the summer of 2005 with an LBS to help them navigate the conference locations and locate fellow attendees. This conference, SVG Open 2005, the 4th Annual Conference on Scalable Vector Graphics (SVGopen.org), was deemed to be a good testbed as it drew a crowd of some 180 people from 20 countries all over the world, from a very wide field of applications: electronic arts & media, geospatial sciences, information technologies, computer sciences, software developers, Web application designers, etc. They share an interest in Scalable Vector Graphics (SVG), the W3C open standard enabling high-quality, dynamic, interactive, styleable and scalable graphics to be delivered over the Web using XML. Most of them are technology-oriented and there is a high degree of interest in, and ownership of, mobile devices.

The application built for testing by the participants has been called FLAVOUR (Friendly Location-aware conference Assistant with priVacy Observant architectURe). Services offered by FLAVOUR can be categorized into:
- *Pull services*, in which location of attendees plays an important role as the attendee's request will be replied by the system on the basis of their whereabouts. Examples of pull services offered are:
  – Finding fellow attendees;
  – Locating resources available in the infrastructure such as printers, copiers, coffee machines etc.
- *Push services*, in which individual and bulk messages are sent to the attendees. This enables the attendees to:
  – Be notified about important events by conference organizers;
  – Communicate with their contacts, i.e., colleagues, friends, etc.

The system architecture, described in more detail in (Muthukrishnan, Meratnia et al. 2005), is based on a Location Manager, which provides services using the Jini platform (Sun Microsystems 2007). Jini is a Java-based open architecture that enables developers to create network-centric services. Each Location Manager registers with the Jini Lookup Service to offer the location of the user it represents. Interested users can look up the service and subscribe to the location of a given conference participant. This is done using a publish–subscribe mechanism. The Location Manager uses a privacy policy to decide if a client is allowed to subscribe to the location of its owner (publisher). It also publishes to all the subscribers relevant changes in the location of its owner.

The Jini architecture also provides other kinds of services, such as a message board to which every conference participant can subscribe. The message board can be used by the conference organization to publish changes in the schedule, information related to the social events, etc. Participants can also use the message board to make announcements to the other participants, as for example asking about lost objects, or to chat.

The graphical depiction of the maps and the location of the users is done in SVG, providing vector graphics in high graphical quality with a small memory– and file–footprint. The system also provides the user with an estimation of the current positioning accuracy. A screen dump of the user interface can be seen in Figure 39-2.

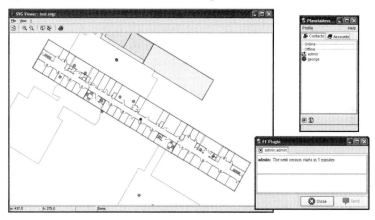

Figure 39-2 Screen dump of the FLAVOUR user interface.

The tests at SVG Open 2005 were relatively successful. Most conference participants experimented with the localization features of the system. The messaging and friend-finder functions were used to a lesser extent. Various extensive interviews have been held with test persons and also written feedback was collected. The localization functionality worked quite reliably, although the accuracy varied quite a bit over the different conference locations. In the computer science building the results were clearly better then in the main conference halls. The tests still have to be analyzed further, but the most obvious reasons are the non-optimal configuration of access points and the fact that the database of these access points still was incomplete at the time of testing.

39.5 Outlook

The implementation of the Wireless Campus LBS described here has only just started. But as it builds on the solid foundations of the well-established infrastructure of the Campus-wide WLAN at the University of Twente, and has had a successful pilot in the FLAVOUR tests at SVG Open 2005, we expect that it will be put into use and expanding relatively quickly in the coming years. On the client-side of the system, ongoing research at ITC on data dissemination for LBS and mobile applications (Köbben 2004) will be concentrating on the Wireless Campus LBS as a testbed for adaptive, task-oriented delivery of mapping information to mobile users using *cartographically aware database objects*. Cartography and GIS more and more involve the use of spatial database technology. In the database world, there is a growing focus on *context awareness* of database objects. One of the important context parameters is location awareness, which is important for all spatial applications. Our goal is to extend context awareness with the idea of database objects that are *cartographically aware*. As a simple example one might think of a location map in the Wireless Campus LBS where the spatial distance of the objects to the focus of the map (the user's position) influences their representation: Interior walls only show themselves in buildings that the user is in, room numbers only if the user is within reading distance. But more complicated systems could be thought up, which might be especially useful for generalization techniques that are hard to achieve in traditional layer-based systems. It is our intention to use the Wireless Campus LBS to test these concepts in practice and provide a proof-of-concept application.

Probably the most exciting aspect of the project is the fact that it provides the opportunity for a very diverse group of people from quite different disciplines to contribute to a technical infrastructure that can serve as a testbed for their respective research projects, and at the same time has the potential to become a useful everyday feature for mobile users at the University Campus.

Acknowledgements

The Wireless Campus LBS project started in early 2005. It's an informal co-operation between people at the University of Twente (UT) Computer Architecture Design and Test for Embedded Systems group (Arthur van Bunningen, Kavitha Muthukrishnan, Nirvana Meratnia, Georgi Koprinkov), the UT department of Information Technology, Library & Education (Sander Smit, Jeroen van Ingen Schenau) and the International Institute for Geo-Information Science and Earth Observation (Barend Köbben).

References

Bockting, S., N. Hoogma, V. Y. Nguyen, M. J. Ooms and G. D. Oostra. 2004. FriendFinder Project 2003/2004, Localisatie met Wireless LAN. MSc project. Enschede, The Netherlands: University of Twente, 54 pp. <http://www.sanderbockting.nl/documents/FriendFinder.pdf> Accessed 23 October 2007.

Havinga, P., P. Jansen, M. Lijding and H. Scholten. 2004. Smart Surroundings. Pages 64–69 in *Proceedings of the 5th PROGRESS Symposium on Embedded Systems*, held in Nieuwegein, The Netherlands, October 2004. ISBN 90-73461-41-3.

Hightower, J. and G. Borriello. 2001. A Survey and Taxonomy of Location Systems for Ubiquitous Computing. *Computer* 34(8):57–66.

Köbben, B. 2004. RIMapper - a test bed for online Risk Indicator Maps using data-driven SVG visualisation. Pages 189–195 in *Proceedings of 2nd Symposium on Location Based Services and TeleCartography*, held in Vienna, Austria 28 – 29 January 2004. Edited by G. Gartner. Vienna: Institute of Cartography and Geo-Media Techniques. <http://kartoweb.itc.nl/kobben/publications/RIMapper_paper_Vienna2004.pdf> Accessed 23 October 2007.

Muthukrishnan, K., M. E. Lijding and P. Havinga. 2005. Towards Smart Surroundings: Enabling Techniques and Technologies for Localization. Proceedings of LOCA2005 – co-located with the 3rd International Conference on Pervasive Computing., held in Munich Germany May 2005. Munich: Springer Verlag, 11 pp.

Muthukrishnan, K., N. Meratnia and M. Lijding. 2005. FLAVOUR- Friendly Location-aware conference Aid with priVacy Observant architectURe. CTIT Technical report TR-CTIT-05-28. Enschede, The Netherlands: University of Twente CTIT, 16 pp.

Open Geospatial Consortium. 2007. Welcome to the OGC Website. <http://www.opengeospatial.org/> Accessed 23 October 2007.

Rijkswaterstaat. Undated. AHN: Actual Height model of the Netherlands. (Actueel Hoogtebestand Nederland.) <http://www.ahn.nl/english.php> Accessed 23 October 2007.

Sun Microsystems. 2007. Jini Network Technology. <http://www.sun.com/software/jini/> Accessed 23 October 2007.

SVGopen.org. Undated. SVG Open 2005 Conference and Exhibition - Home. <http://www. svgopen.org/2005/> Accessed 23 October 2007.

Urquhart, K., S. Miller and W. Cartwright. 2004. A user-centred research approach to designing useful geospatial representations for LBS. Pages 69–78 in *Proceedings of 2nd Symposium on Location Based Services and TeleCartography*, held in Vienna, Austria 28 – 29 January 2004. Edited by G. Gartner. Vienna: Institute of Cartography and Geo-Media Techniques.

Youssef, M. A. 2005. Location Determination Papers. <http://www.cs.umd.edu/~moustafa/location_papers.htm> Accessed 23 October 2007.

CHAPTER 40
Multimedia Decision-Support Tools

William Cartwright

40.1 Introduction

Community collaborative decision-making has traditionally used conventional mapping and presentation tools for conducting consultative processes. These are generally undertaken as a group exercise facilitated by an expert moderator who is conversant with the community decision-making process and the tools, including geospatial tools, necessary to adequately address an issue and to provide essential information.

This chapter focuses on a project that uses 3D delivered via the World Wide Web (Web) to supplement these processes with a Web-delivered interactive 3D tool. The tool provides a method whereby users are able to dynamically interact with the virtual world that is built on a 'footprint' generated from a GIS database. The tool was developed using Virtual Reality Modeling Language (VRML) and designed to be delivered via the Web, and used at home or at Internet cafés and thus easily accessible by the general community. It could also be used at meetings, where the tool would be 'driven' by an experienced operator to support collaborative meetings and discussions. The focus of the research was to evaluate qualitative components of urban scenarios and how an environment is depicted.

To test the effectiveness of the product it was evaluated in two stages:
1. An initial qualitative evaluation with an expert group of users from the general community; and
2. Testing with both the general community and professionals to better understand how the 'geographical dirtiness' of the Virtual Environment changes the perception of a space.

The first part of this chapter provides a brief background of the use of multimedia to support planning decisions and to broadcast general geographic information to professionals and citizens. This is followed by a description of a prototype developed to evaluate the effectiveness of a 3D multimedia Web-delivered product for decision-support in an urban planning context. Lastly the chapter provides the results of evaluations of the prototype.

40.2 Using Multimedia Tools for Decision Support

Over the last two decades there has been considerable interest in the development of multimedia-based tools for Spatial Decision Support Systems (SDSS) for use by professionals in making decisions or in communicating the implications of those decisions at public meetings. In fact multimedia itself was first developed at MIT (Negroponte 1995) to solve a spatial representation problem. When multimedia was new, Jungert—looking at how GIS might be enhanced by Rich Media—contended that, "In the future, GIS must allow inclusion of a variety of new information types" (Jungert 1990, p. 190). Included among the new types of information, he cited maps, associated media (on secondary devices like videodiscs and CD-ROM) and structural data. Multimedia databases would, according to Jungert, contain both conventional maps and multimedia images (and associated methods for handling the attributes of multimedia information). He drew two conclusions about future new information types to be included in GIS: 1. Other (non-map) data would be needed—remote sensing imagery, sketches and sensor signals; and 2. Novel methods for multimedia interaction needed to be invented. Rivamonte (1992) saw multimedia being used in several GIS applications: 3D modeling, vector/raster draping, drive/walk-throughs, fly-overs, video conferencing, video logging and animation. She saw the future of multimedia in the areas of interactive video, holographic data storage, the use of Photo CDs and multi-function drives.

The following sections provide 'snapshots' of how various multimedia tools have provided innovative decision-support solutions. They look at how many media types were combined—audio, video and text—and then moves to video, videodisc, hypertext, multimedia and multimedia on the Web.

The 1992 Property Revaluation Project that was developed for Franklin Township, New Jersey, stored a collection of videos (Barlaz and Gottsegan 1992; Barlaz 1993). A still video camera captured 35,000 pictures of properties, and related assessment data was transferred from the city's main assessment system. The system could handle custom reports and the images displayed appeared to the user as photographic quality images.

A hypermedia-based tool for SDSS was a prototype used to support the siting of the Wilson Bridge Crossing in the US. This domain-specific, hypermedia-based, SDSS consisted of a Graphic User interface (GUI) and a hypermedia engine. 'Tools' of the system included GIS, CAD, spreadsheets, drawing packages and a word processor. The media stored and displayed consisted of images, audio and text; as well as an 'audit trail' to support decision traceability and comparison of alternative design decisions. Hypermedia assisted in assigning weights, addressing utility curves and performing sensitive analyses. This technology provided users with control over the sequence and content of the material used to support the decision-making process.

The city of St. Louis developed two videodisc-based systems: *Show Me St. Louis*, in the mid-1980s, and *Riverfront 2000*, in the late-1980s (Shiffer 1991). Later, the information system was redeveloped as a purely digital system, and provided as an information kiosk. It incorporated many GIS-like features whereby maps were linked to pertinent data and information. Later, St. Louis provided information on-line using the Web (Poset and Kindleberger 1996).

Astori et al. (1993) defined a *Sistemi Informativi Territoriali (SIT)* as a number of elements that gather, exchange, elaborate and store data. It was a hypermedia tool for urban planning that provided a package that was easily accessible to institutional and private users for the monitoring and comparison of urban conditions with planned zonings. Astori et al. outlined 'Hyperterritory' as a methodological platform for the achievement of Sit. It provided a network of informative systems marked both per theme and per area and it made use of Hypertext to represent territory—for this project the urban area of Cesena, Italy.

Multimedia-assisted public information systems were developed as either complete packages or as prototypes for larger proposals. Multimedia Australia developed an information kiosk (which used touch-screen, digital and video technology) for the City of Woodville, South Australia. This provided access to community information files and the ability to view a continual story about the City in terms of accommodation, housing, employment and training and the general workings of the council (City of Woodville 1993).

Multimedia GIS products were developed to assist environmental management projects. For example: *Expo '98* environmental studies prototype (Câmara et al. 1992; Fonseca et al. 1992, 1993), a PC-based GIS (Hoey et al. 1991); the *Wildlife at Risk* prototype (Armenakis 1993); Georeferenced Pavement Information System (Sekioka et al. 1992); The *Scolt Multimedia Project* (SMP) (Raper et al. 1992); and Fire Simulation Model (Fonseca et al. 1992).

But perhaps one of the most innovative applications of multimedia tools to spatial decision was the National Capital Planning Commission, Washington D.C., Collaborative Planning System developed by the Planning Support Systems Group at the Department of Urban Strategies and Planning at MIT. Researchers produced a multimedia environmental impact assessment system for use with major developments. The system was based on Shiffer's Collaborative Planning System (CPS) (Shiffer 1993). It was implemented to demonstrate the capabilities to present proposals and conclusions, to undertake geographically/visually/textually related searches and to visualize large amounts of data such as traffic projections or shifts in demographics by using multiple representation aids. Displays included maps (with overlays), descriptive video images, sounds and text. It used networked computers, a video projector, an infrared pointing device and microphones for user interaction. Information was stored as both digital and analog (video tape for audio and visual "note-taking" on field trips). Communication was real-time and asynchronous (for generating shadow diagrams etc.). Planning-related support in the system was for land use analysis, traffic analysis, and assessment of visual elements and illustration of proposed changes to visual elements.

More recently the Web has become an extremely efficient channel for transferring and viewing data because of its visual capabilities and the relatively advanced hypermedia and on-line geographical information tools developed. This project uses the Web to facilitate the delivery of tools to professionals and the general public. It combines the Web with non-immersive Virtual Reality (VR) and delivers product via 'standard' computers and browsers. This project was developed in the spirit of what Jungert and Rivamonte envisaged, but its scope is somewhat modest compared to Shiffer's efforts. However, it does explore the use of interactive multimedia to support decision-making by the provision of interactive 3D tools.

40.3 The Project

A project was undertaken to develop Web-delivered 3D community decision-making tools to support community decision-making. A prototype product was produced and tested with residents in an inner municipality in Melbourne, Australia, and with a group of professional planners. It was designed to be screened in community forums or used online (at home or in Internet cafés) to provide citizens and professionals with tools for visualizing the locality.

It addressed the use of New Media, and particularly multimedia, delivered via the Web, as a potential tool for improving Public Participatory Planning Support Systems (PPPSS). It also evaluated the effectiveness of the use of the tool, with users from unknown and diverse backgrounds with potentially different skills for using geographical visualization tools and for developing mental maps. A VRML world was built as a workhorse to facilitate the exploration of these areas of interest.

The project simulated part of an Australian inner city area using Web-delivered 3D tools. Of great interest was how successful these simulations were, especially when delivered through the sometimes restricted 'pipe' of the Internet. As well, there was interest to discover how users 'move' through this virtual world and how complex, or 'dirty' (dirtied with elements like street furniture, power lines, traffic, graffiti, etc.) a virtual world needs to be before it could be considered to be 'real.'

The approach had three underlying needs:
- It would be delivered as non-immersive 3D;
- It must use Open Source and a widely accepted format; and
- It had to be deliverable off hard disk, CD-ROM or via the Web.

From the users' perspective a number of criteria had to be met. It had to:
- Work with users with different, non-geospatial backgrounds;
- Work with different skills bases;
- Work with different age groups (with the potential of having all age groups accessing the product);
- Work with different language skills; and
- Be able to be delivered on-line—at home and via the municipality's Web site or be accessible on computers at local Internet cafés.

Virtual Reality Modeling Language (VRML) was chosen as a development tool as it allowed open, extensible formats to be used and the 'built' worlds could be constructed in Web browsers that included a VRML plug-in. (VRML was used rather than X3D as there was no need to link to databases in this initial development phase). VRML is extensible, interpreted language and it became an industry-standard scene description language.

40.3.1 Building the Initial Model

The first part of model development was to construct the actual VRML world. All buildings in the study area were surveyed to ascertain position, use and building height. Also, each building façade was photographed for use in 'stitching' the images onto the sides of VRML primitive shapes. As this area is busy day and night every day of the week, there was not one instance when the streets were free of objects—static and moving. Therefore considerable time was spent in 'cleaning' the images using a graphics-editing program. "Cleaning" involved removing trees, cars, signage, and any street furniture that, if not removed, would appear as a 2D image in the 3D world, thus degrading the perception of the Virtual World. (However, as explained later in the chapter, these building facades were later 'dirtied' to produce models for the second stage of the project.). To remove all elements like street signs was an impossible task, so these were retained in most cases. (See the parking signs outside the shops in Figure 40-1.) The photograph in Figure 40-1a illustrates this point—all of the items in front of the buildings were removed to ensure a clear image. Figure 40-1b shows how the buildings appeared in the VRML package.

Figure 40-1a 'Raw' image taken in the retail shopping strip.

Figure 40-1b The same shop in the VRML world. Note the parking signs that still appear as 2D.

(See included DVD for color versions)

40.3.2 Addressing Problems Associated with Working with Virtual Worlds

When users use traditional paper maps there is no guarantee that the 'reality' they build as a mental map is the reality that the map designer thinks they will build. According to Cartwright (1994), ordinary psychological principles do not exist, or they operate differently, in the virtual world. Also, 'new realities' can emerge from maps that the designer did not consider could be (Liben 2003). The more senses that are involved at once, the more immersed one may become in virtual reality and / or cyberspace, and the harder it may become to distinguish the real world from the artificial (Cartwright 1994). But perhaps the biggest problem associated with the development of such products is the distortion of reality. This may be related to the 'viewing media,' and this is one aspect of the delivery (Web) process map authors have little control over. To this can be added the uncertainty, or, better still, lack of knowledge about whether different viewpoints, and therefore different realities, can be associated with the use of different viewing 'portholes.'

To be able to fully appreciate the model of the social and physical space within which particular users will work, they need to have access to the tools with which to construct their own 'virtual world.' Once constructed, users can immerse themselves (virtually) into a resource that presents the best real-world picture.

There is interest in using the expertise of many professions to understand how these geographical visualization tools 'work.' Fabrikant and Buttenfield (2001, p. 264) noted:

> Research questions such as "How do people learn about geographical information?" or "How do people develop concepts and reason about geographical space?" beg for an interdisciplinary approach, drawing upon expertise in cognitive psychology, geographic information science, cartography, urban and environmental planning and cognitive science.

40.3.3 Evaluation of the Tools Developed in the Project

We assume that when we provide users with a product that relies on their understanding of geography to 'work,' we cannot assume that they really 'know' the geography of the area being studied. Perhaps they assume that they know the geography, whereas they only have a naïve version of reality—Naïve Geography. Naïve Geography was defined by Egenhofer and Mark (1995) as "the body of knowledge that people have about the surrounding geographic world" —the primary theories of space, entities and processes, and as "… the field of study that is concerned with formal models of the common-sense geographic world" (p. 1). The term describes a formal model of common-sense geography (Mark and Egenhofer 1996). This would form the basis for developing intuitive and 'easy-to-use' Geographic Information Systems. It "… captures and reflects the way humans think and reason about geographic space and time. Naive stands for instinctive or spontaneous" (Egenhofer and Mark 1995, p. 4).

So, if we build what can be called naïve geographical representations of real spaces, and, if users only have a naïve viewpoint of the geography of the study area, is the tool effective? Or, is it doubly ineffective due to the double imposition of naïve geographical understanding and naïve presentations (that is, presentations built on simplistic geographical models—e.g., flat earth and 'clean' environments, especially clean inner urban environments). Therefore, we cannot take for granted that the user group will automatically be able to make best use of our tools and hence the need to evaluate the product.

40.3.3.1 Stage 1 Evaluation Results

An alpha prototype online model was usability tested at a special workshop for local community members who had past experience in consultations on planning issues. They were asked to explore the models and signal difficulties in its online use. The participants were asked to complete a questionnaire to assess the potential of the tools and suggest further developments. The Reeves and Harmon (1993) User Interface Rating Tool for Interactive Multimedia was used as a 'model' for constructing the questionnaire. Candidates were asked to rate the product according to 9 criteria:

1. Navigation. Referred to the ease with which users can move through the product. They were asked to consider the general navigation through the product, especially the 3D interface, and how you navigate to other parts of the product. They were also asked to consider whether this type of 3D interface either assists or hampers their comprehension about where they are in the package.
2. Cognitive load. Consideration of how much harder, or easier, the product was to use compared to the 2D products. They had to consider whether they needed to work harder mentally, or whether it was easier, or more intuitive when using this 3D product compared to products that they used in the past.

3. Mapping. Related to how the program tracks as they use the program, and how it provides feedback. This can be graphically, or through the use of some other medium. When considering this section they were asked to reflect upon whether this 3D assists in understanding the spatial and locational aspects of the area more.

4. Screen design. The actual design components, colors, text, symbols etc. They were asked to not focus on the overall 'look' of the product (which is covered in point 8 below).

5. Knowledge space compatibility. Focused on the type of information provided in the product. This prototype 'test' product was designed as a decision-making support tool. Comments asked for here were related to whether the prototype provides the needed tools for making decisions.

6. Information presentation. Considered whether the information presented was in an understandable form. They were asked whether they understood the information contained in the product, and if they learned something about the area that would not have been possible with conventional 2D products.

7. Media integration. The 'putting together' of the different media types. Usually, when multimedia products are evaluated they will contain many media types, thus presenting a 'Rich Media' product. As this product was a combination of VRML and Web information, candidates were asked to consider both of these elements.

8. Aesthetics. Candidates considered here the 'look' of the product, and its form.

9. Overall functionality. The perceived utility of the product. As this product aims at providing geographical information in a 'Virtual Environment' format, this was the main criterion users were asked to consider.

Reeves and Harmon (1993) recommend that novice users are generally not good candidates for using this type of form. They consider experienced users of the type of program being rated, or experienced designers of interactive products, to be the best candidates for providing feedback. An initial 'filtering' of candidates was made to ensure that only expert user/producers completed the assessment. Candidates were asked to rate the product in each of the nine evaluation areas, from 1 (Difficult) to 10 (Easy). General comments about the product also were solicited. During general discussion after the formal input via the questionnaire, general comments were also provided. Positive and negative comments are shown in Table 40-1.

Table 40-1 Comments from the focus group.

Positive comments	Negative comments
• Useful in planning decisions • Useful adjunct to 'on-site' inspections. • Don't necessarily need to 'dress' the bulk of the buildings (ie add image of building to VRML primitive shapes). There was some disagreement with this concept by some members of the group. • Would result in better decision-making	• Would be improved if shadows were added • Images need to look 'more VR' • Vegetation in the sandbox is hard to visualize. • Users were wary that images might be manipulated to create a certain image. • Didn't like scenario's limiting factors–i.e., it provided only a corridor to work within. Needs purpose to drive system.

In summary, the test group generally liked the concept, but the actual product needed to be refined as per the feedback and use of 3D improved the interpretation of the area being studied. There was some comment on the use of navigation tools associated with the actual VRML browser plug-in and this was addressed in product development between this and the next stage of development. There were some comments on the need to provide high levels of detail for all buildings. This was also addressed in developing prototypes for Stage 2.

40.3.3.2 Stage 2 of the Evaluation – How Naive Can the Virtual World Be?

This Stage evaluated how the 'geographical dirtiness' of the Virtual Environment changes the perception of a space. As the product was to be delivered on-line, the developers had no information about whether users of the product had an understanding of the 'real' reality of the area. There was no knowledge about whether the models developed should show a 'clean' neighborhood, devoid of street furniture, people, cars, graffiti, etc., or whether a model that included the activities and elements in that environment would be better. The model view in Figure 40-1b above shows this clean virtual world. A walk through the area shows the real reality—one of a densely populated inner city area, with associated things like graffiti on buildings. To solve this problem more detail can be added, but this is a costly exercise in time. Therefore, this stage of the research had also to determine how 'dirty' a model needed to be to depict an inner city area effectively. And, did the level of geographic dirtiness required to 'paint' the best picture change from inexpert user to expert user?

In this stage of the evaluation, the main task was to understand how complex a computer graphics 3D-environment really needed to be to support community discussion of urban planning developments. The approach to this stage of the evaluation was designed to complete two tasks:

- To determine the test candidates' understanding of the geography of the test neighborhood; and
- To determine how 'dirty' (complex/detailed) the computer visualization needed to be so as to provide a more usable product.

What was wanted was to better understand how the 'geographic dirtiness' (complexity) of the virtual environment changes the perceptions of users of a virtual space. Basically, our question was: how much detail (or dirtiness) is necessary for a visualization to 'work'?

40.3.3.2.1 Testing

Users were first asked to grade themselves as belonging to one of five typical user types:

- Grade 1: they had a simple understanding of the geography of the area—they knew the general pattern of the streets.
- Grade 2: they knew the major landmarks in the area—the major buildings. They did not have as much knowledge of the side streets.
- Grade 3: They knew all of the major buildings in all of the streets.
- Grade 4: They knew all of the intermediate buildings as well as the major buildings. They had a good understanding of the geography, but they did not know the exact details of every building and every street.
- Grade 5: They had a comprehensive knowledge of the area—they knew details of all buildings, including what they are used for, and every street and laneway.

Users were then asked to view five visualizations, each with an increasing level of geographic dirtiness, and then to complete a feedback document. Candidates were asked to consider whether the amount of detail provided is sufficient for you to understand the general geography of the area, and did it provide sufficient information for a community discussion on new developments in the area? The following sections provide brief snapshots and descriptions of the five scenarios built for testing. Level 1 was the simplest VRML world and Level 5 the most complex (and most 'populated').

Level 1 (Figure 40-2) was a simple visualization of the part of the main road in the study area. This was a 3D world that only contained transparent 'shells' of the buildings in the study area. It was considered to be a basic level of information for community discussion of urban developments in the area.

Level 2 (Figure 40-3) had roads added and the surrounding buildings were color-coded to indicate the current building use. A sky 'dome' was added, which provided the means for the model to appear to continue above the buildings, whereas the Level 1 model only provided a flat dark blue backdrop.

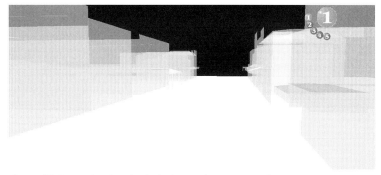

Figure 40-2 Level 1. See included DVD for color version.

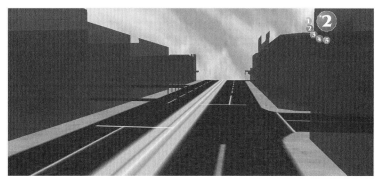

Figure 40-3 Level 2. See included DVD for color version.

Level 3 (Figure 40-4) had all of the buildings in full detail. Road detail remained as it appeared in Level 2.

Level 4 (Figure 40-5) included signage suspended beneath verandas, above verandas and as sandwich boards on the street. Road detail was enhanced with different textures on footpaths. Street furniture also was included—seats, bicycle stands, rubbish bins, parking signs, etc. Light poles and overhead tram power cables were added.

Level 5 (Figure 40-6) had cars, trams and people 'populating' the main road. This model was further enhanced with synthetic shadows.

For this evaluation, two user groups were canvassed for their opinions: a community group and professional planners. Each group met on separate occasions. The community group was drawn from the local area and the professional group came from the Department of Primary Industry, Victoria, and particularly from the group that deals with planning issues, and planners attending the Australian Planning Institute 2005 conference. The results reported in this section have grouped the professional planners as one composite group.

Candidates were asked to view each section independent of the following, more complex scenes. They then rated the statements, listed below, for accuracy in describing how they felt about each stage's visualization.

- There are adequate landmarks to assist me in orienting myself;
- The amount of detail is sufficient;
- Having the buildings shown as outlines is sufficient;
- I need more detail for this 3D model to provide me with an adequate mental representation of the area; and

Figure 40-4 Level 3.

Figure 40-5 Level 4.

Figure 40-6 Level 5.

- The addition of extra information would make it easier for me to build a mental image of the area.

The community group was asked one additional question: "Is it obvious where this area is located in my neighborhood?"

After rating each of the statements the candidates were shown the next level of information and they addressed the same questions. At the end of this formal processing a 'free comment' session was held, where candidates were encouraged to make more general comments on the various stages. The following sections report initially on results from the community group evaluation, and then provide results from the professional user group.

40.3.3.2.2 Results – Community Group: Levels of Complexity

As the tool was designed to work for both community members and professional planners, it was important to ascertain what the community members needed in a virtual landscape so it could be used as an effective tool (Tables 40-2 to 40-6). Comments that were seen to be important for developing a final, Stage 3 evaluation tool are highlighted in gray.

Level 1:

Table 40-2 Comments from the community group on Level 1.

Person	Comments
1	• Too hard to recognise which is the building next door • Not enough differentiation between buildings • Have trouble identifying the streets
2	• Would be very limiting in terms of what this could be useful for • Bit like a maze
3	• I would have put "5" for questions 1-4 (on questionnaire) if the street names had been marked
4	• It is too difficult to get orientation reference points quickly when viewing at this level • The transparency is confusing—the distinction between buildings is too fuzzy
5	• Works best when viewed from above i.e. when used like a 2D map/aerial view. Down at street level it is too difficult to pick out features

Level 2:

Table 40-3 Comments from the community group on Level 2.

Person	Comments
1	• Aerial view assists in orientation but once I was back on street level had difficulty establishing locations. • Multiple views a plus
2	• Great map/model for zoning purposes • Aerial good
3	• Color coding makes little difference
4	• Slightly easier to orient oneself in space than Level 1 but still not enough detail • Transparency is more acceptable when color coded
5	• Much better than Level 1 but best when viewed from above—at ground level seems a bit confusing

Level 3:

Table 40-4 Comments from the community group on Level 3.

Person	Comments
1	• Once again, aerial or semi-aerial view assists in orientation when looking at the whole area face on. • Much easier to locate area when presented with detail, such as recognisable buildings
2	• Depends on the purpose of the exercise and what you hope to achieve from the model. – Home use – plenty of detail required but ok here – Work – could use more detail in Sydney Rd
3	
4	• Any further detail – eg not such a sharp cut off at boundary—would help marginally by allowing me to "place" the section in the overall landscape, but more detail is not really necessary. • The sky is important
5	• The first level to make sense at ground level, but is no more useful than Level 2 when looking from above/aerial.

Level 4:

Table 40-5 Comments from the community group on Level 4.

Person	Comments
1	• Probably you don't need this much detail if looking at major changes to the neighborhood
2	• Great to add traffic congestion, people. Planning/development; painting facades/roofs all excellent
3	• NO COMMENT
4	• For some purposes (eg detailed urban design strategies) this level of detail would be helpful • But for many planning purposes the extra detail is not necessary or even helpful
5	• Signs and other detail are a great gimmick but maybe a bit of overkill • Could be very useful for some applications but previous levels of detail is sufficient for most visualisation needs associated with planning decisions (except maybe more detailed things) • Trams/powerlines are good though as such a dominant feature of Sydney Rd

Level 5:

Table 40-6 Comments from the community group on Level 5.

Person	Comments
1	• The figures of people are distracting • Vistas at the end of the road make the picture look more realistic, but I'm not sure they are necessary • This gives a more realistic view of Sydney Rd and surrounds
2	• Extras are great but shadows generally not necessary. People not really necessary. • Picture of street at the end is excellent
3	
4	• Shadows not necessary • Simple representations of people adequate, but don't need to make them so colorful or dominant.
5	• This is highly detailed – fun and interesting but unnecessary for a lot of planning purposes. • Addition of cars/people helps to give the "vibe" of Sydney Rd – clutter, bustle, chaos confusing. • Feels too much like computer game rather than helpful tool

General Comments:

The final stage of the evaluation asked two questions:

Q1. Leaving the models unchanged, which level of information did you prefer?
   A. Levels 3, 4, or 5 were seen to be useful.
Q2. If a basic model only could be provided, which level of information could you work with and achieve a useful mental image of the study area?
   A. 1, 2, 3, 4 and 5

After further discussing the models it was generally agreed "less is more!" Therefore it was ascertained that the simplest model possible needed to be built, but this model must contain certain design elements. From the formal questions and later discussion, design guidelines were assembled (see list below). These were used, along with the complementary information from the professional user group to determine the contents of and adequate levels of detail for a virtual landscape that is: 1) Usable and useful and 2) A model that can be built and maintained without heavy development and maintenance costs.

A model that would 'work' would contain the following elements:
- Transparency must include color coding.
- It would contain street signs.
- An aerial view must be included for orientation.
- Detail is only necessary at street level—it is not needed at the aerial view. This could be realized by using the 'level of detail' in VRML coding, whereby detail is 'added' when the user moves within a certain proximity to the model.
- Street furniture is not really needed for considering general planning aspects.
- The addition of extra elements in the model "makes it look like Sydney Rd" (people, cars, horizon), but people in the model can be distracting. Therefore this is a consideration that needs to be further investigated, so as to ascertain the usefulness of adding these items.
- The 'visual returns' from adding shadows are minimal and some candidates saw them as a distraction.
- 'End of the world' cannot be shown and adding images at the end of the streets improved the image. (I call this the 'Truman Show' effect, after the movie where the main actor, Truman, lived his whole life on a massive movie set—the world did end, and this was shown at the movie climax.). (See the image at the rearmost end of the street in Figure 40-7, where an image taken at the end of the prototype area was rendered in image manipulation software to have the same appearance as the model. This was then inserted into the VRML world.)
- The use of the sky on the inside of the enclosing sphere is necessary.

40.3.3.2.3 Results – Professional Group: Levels of Complexity

The professional group was shown each of the five levels in turn. After a 'guided tour' they were asked to rate the virtual world level of detail by indicating whether they agreed or disagreed with the statements provided on an evaluation form. The sections below tabulate the results from each evaluation candidate. This has been done statement by statement, with the rating for each level of detail shown in subsequent table rows. In Tables 40-7 through 40-11 below, a 1 represents the lowest or most negative impression—the candidates disagree with the statement. A grading of 3 indicates that the candidate generally agreed with the statement. A value of 5 is the highest value and this indicates that the candidate agrees with the statement.

So as to be able to consider whether the responses were in general agreement, the table cells have been coded with different levels of tone: white means a respondent indicated that a level was inadequate, medium gray means the respondent indicated a visualization was generally adequate, and black cells indicate that the candidates indicated that a level was appropriate and they could use it in their professional activities:

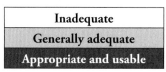

When looking at the tables the black infill indicates that this level of information 'works' for:
1. Orientation
2. Providing enough detail
3. An adequate mental representation of the study area.
4. That the information level is sufficient to build a mental image of the area.
5. The addition of extra elements does not make the image too complex. (Refers to level 5 only.)

Question 1
There are adequate landmarks to assist me in orienting myself.

Table 40-7 Results from questions related to orientation.

LEVEL	PERSON									
	1	2	3	4	5	6	7	8	9	10
1	1	2	2	4	2	3	3	3	2	1
2	2	3	2	4	2	3	3	3	2	1
3	3	5	4	5	4	4	5	5	4	4
4	3	5	5	5	4	5	5	5	5	5
5	4	5	5	5	5	5	4	5	5	5

Looking at Table 40-7 it is obvious that, in general, the professional candidates could work effectively with the landmarks in Levels 3 – 5. For all candidates Level 5 contained the best landmark information.

Question 2

The amount of detail is sufficient.

Table 40-8 Results from questions related to whether the amount of detail was sufficient.

LEVEL	PERSON									
	1	2	3	4	5	6	7	8	9	10
1	2	2	2	3	1	1	3	3	1	1
2	3	2	2	3	2	3	4	3	1	1
3	3	5	5	4	3	4	5	5	2	4
4	4	5	5	4	4	5	4	5	4	5
5	5	5	5	5	5	5	4	5	4	4

Levels 4 and 5 have sufficient levels of detail, but Level 3 appears to be adequate.

Question 3

The third question addressed was "Having the buildings shown as an outline is sufficient." This question was asked to the planners at DPI but, upon reflection, the question was considered to be quite similar to the question that followed this one: "I need more detail to provide me with an adequate mental representation of the area." Thus the question was not used in the second 'professional' evaluation. Therefore, the answers given by the first professional group were removed from consideration.

Question 4

I need more detail to provide me with an adequate mental representation of the area.

Table 40-9 Results from questions related to improving the ability to make an adequate mental representation of the area.

LEVEL	PERSON									
	1	2	3	4	5	6	7	8	9	10
1	4	4	5	4	5	3	4	4	5	4
2	4	4	5	4	4	3	4	3	4	4
3	3	1	3	2	4	2	2	1	4	4
4	3	1	2	2	–	1	2	1	4	1
5	2	1	1	–	1	1	2	1	2	1

Levels 1 and 2 contain insufficient detail to be considered useful. The other three levels have enough detail.

Question 5

The addition of extra information would make it easier for me to build a mental image of the area.

Table 40-10 Results from questions related to whether the addition of extra information would make it easier to visualize the study area.

LEVEL	PERSON									
	1	2	3	4	5	6	7	8	9	10
1	4	4	5	4	5	4	4	5	5	4
2	4	4	5	4	4	4	4	5	5	4
3	3	1	3	2	4	2	3	5	5	4
4	3	1	2	2	5	1	3	–	4	1
5	2	1	3	–	3	1	4	1	2	1

Levels 1 and 2 need extra detail added. Level 3 is just adequate.

Question 6

The addition of extra elements—people, cars, trams, street furniture, etc.—makes the image too complex. (i.e., it has a negative effect, rather than improving the model. Note—this question was only associated with Level 5.)

Table 40-11 Results from questions related to whether the addition of extra 'clutter' information actually improves the visualization.

LEVEL	PERSON									
	1	2	3	4	5	6	7	8	9	10
5	2	2	3	–	–	3	2	5	2	5

The candidates were divided here. Some liked more detail and others were non-plussed about the need for extra detail.

The questionnaires also asked for additional comments about how each level could be improved. Obviously, with an unlimited budget and with no time constraints, all desired elements for each level could be 'built.' But, a compromise was needed so as to be able to specify what could really be built for professional users of such tools. These comments are listed in the following sections:

Level 1
- Level of detail "just" OK for building a mental image but not sufficient for most, if not all, planning purposes;
- Seeing buildings through other buildings is confusing; and
- The level of detail required should depend strongly on the purpose of the study.

Level 2
- Needs a legend so user can tell what building type it is;
- The color-coded land use is more useful;
- The road texture made a big difference to being able to place the geography; and
- Can only the major landmarks be detailed?

Level 3

Vegetation and facades were included in the model.

Users were asked by the facilitator their thoughts on a Level 2.5 image. (This) would be very useful, with landmarks, etc. included. (Combining photos of facades with level 2 color scheme). This led to some further discussion on the attributes of a simplified Level 3 model. Here comments were included:

"If had a few key buildings at Level 3, could have rest at Level 2".
"If knew one building, I know what's next door"

Level 4

- Useful having street signs to help navigate/educate user where they are;
- Extra detail supplied in this level useful in asset management and traffic control tasks;
- Maybe too much detail; and
- For other types of planning studies this level might be sufficient.

Level 5

- Huge jump from 3 to 5;
- The cars and traffic are a good addition—potentially quite useful for planning;
- Unsure if the people are necessary;
- The photos for the end of the streets ["end of the world"] are important;
- Don't think shadows are that important; and
- Perhaps all objects (including trees) should be represented at the same level of complexity.

40.3.3.2.4 General Comments

As per the general comments asked of the community group, the same was done for both professional groups. These comments are shown below:

Q1. Leaving the models unchanged, which level of information did you prefer?
A. Levels 3, 4, or 5.
Q2. If a basic model only could be provided, which level of information could you work with and achieve a useful mental image of the study area?
A. 2, 3, 4 and 5. (Again, there was a range of preferences. After discussion related to the possibility of building Level 2.5, there was interest in such a level being developed.)

Other comments related to using a Level 2 model with landmarks. Also, it was thought that the color coding of land use should probably have 2 tiers: 1. general industrial, residential etc and 2. restaurant, office etc. And, one respondent commented that the design guidelines might need to consider the area being studied—e.g., within 10 km of CBD and in an Activity Centre we need probably a Level 4/5. In a new area (e.g., fringe) probably a Level 3 would suffice. The comment that "people not necessary" again was given.

40.3.3.3 Discussion from Stage 2 Evaluation

From the evaluations completed with the community and professional groups a wealth of information has been assembled that can be used for developing guidelines for building a world that satisfies the needs of both user groups, but also is 'buildable' with modest inputs of time and data maintenance. It was decided to base the model for final testing on the Level 2 world, and to enhance this by the following additions:

- Color code land use;
- Include a legend;
- Add street signs for navigation and location awareness;
- Provide an aerial viewpoint for orientation;
- Use 'level of detail' to reduce detail at the aerial view, but have increased detail at street level;
- Avoid 'end of the world' models, and use images at the end of each street. (Perhaps include collision detection that disallows users penetrating these images);
- Include the sky image;
- Add appropriate landmarks;
- Include different road textures; and
- Add some street furniture for specific use.

These general guidelines were used to build the final model for evaluation. This model was called Level 2.5. This model forms the basis for further testing.

The Level 2.5 model was designed around the guidelines developed during Stage 3 of the research, outlined in the previous section. It is 'built' on the Level 2 world, with landmark buildings, street signs, road surfaces, selected street furniture (power poles and lines) and 'end of the world' images (with appropriate collision detection added). As well, a legend has been added. The Level 2.5 model is depicted in Figure 40-7.

Figure 40-7 Level 2.5. See included DVD for color version.

40.4 Conclusion and Future Developments

This chapter has described a product developed for community decision-making support and the results from testing the tool in community collaborative decision-making simulations. It described the evaluation conducted and it described the results from two evaluation stages. Evaluations indicated that both community members and professional users deemed that 3D multimedia tools were useful adjuncts to what they usually employ. Preferences for the amount of detail desired in virtual landscapes were found to vary from user to user. However, a 'compromise' model, one that includes basic planning and building information, supported by more detailed landmark building and adjunct information would provide a useful and usable tool.

Acknowledgements

The evaluation prototype development was supported by a VRII grant from RMIT University. The author acknowledges the production work done on the project by Dane McGreevy, Adam Andrenko and Dominik Ertl. Anitra Nelson and Andrea Babon provided testing support.

References

Armenakis, C. 1993. Hypermedia: an information management approach for geographic data. Pages 19–28 in *GIS/LIS'93 Proceedings, Volume I*. ACSM-ASPRS-URISA-AM/FM, held Minneapolis, Minnesota. Bethesda, Maryland: ASPRS.

Astori, E., G. Beato, L. Di Prinzio and P. Pitarello. 1993. Hypermedia tool for urban planning. Pages 1222–1229 in *Proceedings of Fourth European Conference and Exhibition on Geographical Information Systems*, EGIS '93, Volume 2. Genoa: EGIS Foundation.

Barlaz, E. C. 1993. Objectives analysis chops cost of property revaluation photo imaging system. *GIS World* May:50.

Barlaz, E. C. and J. Gottsegan. 1992. Implementing photographic imaging systems in local government. Pages 221–228 in *URISA 1992 Annual Conference Proceedings*. Edited by W. Mumbleau and M. J. Salling. Washington, D. C.: URISA.

Câmara, A., F. C. Ferreira, D. P. Loucks and M. J. Seixas. 1992. Multidimensional simulation applied to water resources management. *Water Resources Research* 26(9):1877.

Cartwright, G. F. 1994. Virtual or Real? The mind in cyberspace. *The Futurist* March–April:22–26.

City of Woodville. 1993. Promotional Information Sheet.

Egenhofer, M. J. and D. M. Mark. 1995. Naive geography. In *Spatial Information Theory: A Theoretical Basis for GIS*, edited by A. U. Frank and W. Kuhn, 1–15. Lecture Notes in Computer Sciences No. 988. ,Berlin: Springer-Verlag.

Fabrikant, S. I. and B. P. Buttenfield. 2001. Formalizing semantic spaces for information access. *Annals of the Association of American Geographers* 91(2):263–280.

Fonseca, A., C. Gouveia, A. S. Câmara and F. C. Ferreira. 1992. Functions for a multimedia GIS. Pages 1095–1101 in *Proceedings of EGIS '92, Volume 2*. Munich: EGIS Foundation.

Fonseca, A., C. Gouveia, J. Raper, F. Ferreira and A. S. Câmara. 1993. Adding video and sound to GIS. Pages 187-193 in *Proceedings of EGIS '93, Volume 1*. Genoa: EGIS Foundation.

Hoey, M., D. Whelan and S. Folving. 1991. A PC based geographical information system that integrates image data, map data and attribute data. Pages 469–476 in *Proceedings of EGIS '91*. Brussels: EGIS Foundation.

Jungert, E. 1990. Is multi-media the next step in GIS? Pages 190–193 in *Proceedings Third Scandinavian Research Conference on Geographical Information Systems, Volume 2*. Held in Denmark, November 14 - 16.

Liben, L. S. 2003. Thinking through maps. Pages 45–67 in *Spatial schemas and abstract thought*. M. Gattis, editor. Cambridge, Mass.: MIT Press/A Bradford Book.

Mark, D. M. and M. J. Egenhofer. 1996. Common-sense geography: foundations for intuitive Geographic Information Systems. Pages 935–941 in *GIS/LIS '96 Proceedings*. Bethesda, Md.: ASPRS.

Negroponte, N. 1995. Affordable computing. *Wired* July:192.

Poset, J. and C. P. Kindleberger. 1996. City governments go on line. *GIS World* June:50–52.

Raper, J., T. Connolly and D. Livingstone. 1992. Embedding spatial analysis in multimedia courseware. Pages 1232–1237 in *Proceedings of EGIS '92, Volume 2*. Munich: EGIS Foundation.

Reeves, T. C. and S. W. Harmon. 1993. User Interface Rating Tool for Interactive Multimedia. <http://it.coe.uga.edu/~treeves/edit8350/UIRF.html> Accessed 12 November 2007.

Rivamonte, L. 1992. A multimedia primer for GIS? Pages 188–199 in *URISA 1992 Annual Conference Proceedings, Volume 2*. Edited by W. Mumbleau and M. J. Salling. Washington, D. C.:URISA.

Sekioka, K., W. Taniguro, T. Minamisawa and K. Soma. 1992. Pavement management implementation with video-based GIS. Pages 38–47 in *URISA 1992 Annual Conference Proceedings*. Edited by W. Mumbleau and M. J. Salling. Washington, D. C.: URISA.

Shiffer, M. J. 1993. Augmenting geographic information with collaborative multimedia technologies. Pages 367–376 in *Auto Carto 11 Proceedings, Volume 1*. Meeting held in Minneapolis, Minnesota. Bethesda, Md.:ASPRS.

Shiffer, M. J. 1991. A hypermedia implementation of a collaborative planning system. Pages 121–135 in *URISA Annual Conference Proceedings, Volume 3*. Edited by Lyna L. Wiggans. San Francisco: URISA.

CHAPTER 41

Visualizing Rural Landscapes from GIS Databases in Real-time – A Comparison of Software and Some Future Prospects

Katy Appleton and *Andrew Lovett*

41.1 Introduction

Landscape visualization is an increasingly accepted method for use in participatory exercises within land-use planning and other environmental decision-making processes. Its primary benefit is to increase the accessibility of complex spatial and environmental information for non-expert users (Tufte 1990; Bishop 1994), thus increasing both satisfaction with the process and the quality of decisions made. The history of visualization can be traced from early drawings and models through augmented photographs to computer-based methods (Zube et al. 1987). Developments in the technology employed in computerized visualization have historically been driven by the desires of users such as the military, government departments (e.g., the US Forest Service), the entertainment industry, engineering, and oil and gas interests; from small beginnings in the 1960s, development in visualization has continued at an increasing speed, especially in the last 10 years (Ervin and Hasbrouck 2001). Today, numerous computer-based visualization techniques are now available, encompassing everything from near-photorealistic still images to less-detailed navigable models with interactive and interrogable features.

Real-time visualization has long been seen as the logical 'next step' from still imagery and animations. To give viewers control over the location and direction of views and allow them to explore a landscape in their own way by 'flying' or 'walking,' should mean that every viewer can find the scenes that are most meaningful personally, and thereby gain greater understanding of the displayed landscape. Adding the ability to interrogate and interact with objects also allows information beyond the visual to be presented. Urban environments have benefited from real-time visuals early on, due to the ease with which urban geography and geometry can be simplified to geometric primitives. Rural environments, with their organic and complex shapes, however, have historically been more difficult and it is relatively recently that software began to be able to depict such landscapes in real-time (Appleton at al. 2002; Doyle et al. 2002).

Software to create such visualizations is not necessarily new—tools to convert GIS data into virtual reality modeling language (VRML, originally virtual reality markup language) have existed since the mid 1990s (see Lovett et al. 2002)—but due to the variety and sophistication level of programs available today, a comparison of some of the alternatives is now warranted. Given some of the disparities in system cost, output file size and, most importantly, the appearance of and interaction with the resulting visualization, there is a clear need for some kind of guidance aimed at those users who are venturing into real-time landscape visualization for the first time. This chapter compares the approaches of three current software packages—ESRI ArcGIS®, Terrex TerraVista™ and 3D Nature Visual Nature Studio™—that are capable of constructing real-time visualizations using features from a GIS database. Comparison is from the user's point of view (that is, the person compiling the visualization), including consideration of the resulting images. Three primary issues are considered: data needs, ease of use, and quality of output. A framework has been devised to compare more objectively the features and limitations of the software packages, as well as examples of the output they generate.

In addition to the three packages being compared, other software tools are mentioned, and data sources are identified. The URLs for details of the data and software are included in the final section of this chapter, "Links to Software and Data Mentioned in the Text."

41.2 Data

The land use database used to examine the different visualization approaches covered 1 square kilometer and used polygon-based MasterMap® data from the UK national mapping agency, Ordnance Survey (OS), and 10-cm color aerial photographs from Bluesky), both covering part of the University of East Anglia campus. This data set was chosen primarily because it already existed in a relatively detailed state, having been initially set up to examine development proposals in the area, and also as a prototype for a larger model of the whole campus. From the OS data's raw state, polygons had been given their correct land use classification (as the default MasterMap classification is rather too general), and new point data were created to show the locations of bushes, hedges and single trees using the aerial photography. Buildings had been given simple classifications as to their general color and estimated height in order to display them as masses. The river outline had also been extracted and modified to form a continuous polygon instead of being interrupted by bridges. Elevation data at 5-m resolution were taken from NextMAP Britain obtained through Bluesky. All land use and elevation data were in British National Grid coordinates. The project also used non-spatial data including 3D models of an electricity pylon, a sample campus building, and the bridge over the river. Tree and bush images and models were primarily taken from the libraries provided with each piece of software, although comments are made later on the process of adding new objects to these libraries.

41.3 Real-time Landscape Visualization Software

Various options are available to those wishing to use GIS information to create landscape models for real-time viewing. They range from high-end products designed for the type of large-area models required in flight simulation and other military applications, to others which offer the kind of fine control applicable to smaller, more-detailed areas. The degree of GIS integration may be very high if the software is part of a commercial GIS package, or more moderate in the case of stand-alone packages, but in general the situation has greatly improved in the past few years (cf. Appleton et al. 2002). Output format and viewing requirements are the other major determinants of a technique's suitability for a particular purpose, with sophisticated viewing hardware and software (e.g., multi-projector systems) usually requiring particular file formats in order to take full advantage of display capabilities. By contrast, desktop-based equipment is often able to employ a variety of simpler viewers, some of which are free of charge and/or may allow for online viewing.

This research examines three currently-available packages which cover the range of possibilities addressed above. TerraVista is designed for large-scale models, offers a high degree of control over the landscape, and outputs to OpenFlight format only. Visual Nature Studio with the SceneExpress™ extension offers a very high degree of control over the landscape model, and outputs to numerous real-time formats including VRML, the manufacturers' own format NatureView Express™, and, to a limited extent, OpenFlight. Finally, ArcScene is a module of ArcGIS 8 onwards. Building on the capabilities of its predecessor, ArcView® 3D Analyst™, ArcScene™ offers moderate control over the landscape model and has viewing capabilities within the GIS and outputs to VRML format only. All three packages accept a wide range of GIS data formats and projections. Other packages are available including Sitebuilder 3D™ for ArcGIS, Creator Terrain Studio™, *Fledermaus*, VirtualGIS for ERDAS® Imagine, Blueberry3D®, the upcoming Lenné3D®, and the possibilities offered by open-source toolkits such as OpenSceneGraph© and the Virtual Terrain Project© (see "Links to Software and Data

Mentioned in the Text" at the end of the chapter for details of all software mentioned). For various reasons including cost and compatibility with existing systems and experience within the research group, these other pieces of software were not assessed in this exercise.

When investigating the strengths and weaknesses of the three software packages, the first consideration was the technical aspects relevant to the user, as follows:

- ease of import of GIS data, including data types and file formats
- ability to manipulate data and assign visual properties to features
- ease of use of landscape objects, e.g., trees and buildings, and the range of objects supplied
- time taken to generate output
- size and format of output file
- ease of navigating the scene

Other requirements for each package also were considered:

- price
- hardware required
- complexity, and learning aids such as tutorials and courses.

Summary comparisons are presented in Tables 41-1 and 41-2, with additional description and discussion in the sections below.

Table 41-1 Summary of resource requirements for the three packages.

Software	Viewing software required (PC)	Minimum system requirements	Price Range*	Complexity and documentation	Training and support
Visual Nature Studio + Scene Express from 3D Nature	VRML viewer, NatureView Express, Virtual Terrain Project, Audition™ or TerraVue™	1 GHz P2, 256 Mb RAM, 1Gb HD, OpenGL® video card, Windows® 95** or later	Low (VNS2 + Scene Express)	High complexity. Small printed manual; comprehensive documentation in online help.	Basic tutorials included (HTML and video) with others available. 3DNature offers multi-day residential training (US). Independent consultants offer customized training globally, plus UK residential course. Individual support: email free, phone subject to fees. Free online support forum.
TerraVista from Terrex	Audition or TerraVue	2.5GHz P4, 1 GB RAM HD not specified OpenGL video card Windows® XP/2000**	High	High complexity. Printed "getting started" guide rather small; online help useful but not 100% comprehensive.	Basic tutorial useful but short. Terrex offers multi-day residential training (US). Few consultants offer training (US only). Individual support by email, subject to maintenance fees.
ArcGIS 9: ArcView + 3D Analyst (ArcScene™ application) from ESRI	VRML viewer or ArcScene itself	1GHz Pentium® Xeon® 512 MB RAM 700Mb HD OpenGL video card Windows® XP Pro**	Medium (ArcView + 3D Analyst)	Moderate complexity (reduced if users already familiar with ESRI products). Extensive printed documentation and online help.	"Getting Started" guide includes tutorials. ESRI offers various courses and customized training. Various consultants offer customized training. Free online support forum. Individual support subject to maintenance fees.

*Price ranges are Low (< 5,000 USD), Medium (> 5,000 and <10,000 USD) and High (> 20,000 USD).
***Windows® is a registered trademark of Microsoft® Corporation in the United States and other countries.*

Manual of Geographic Information Systems

Table 41-2 Summary of usage factors for the three packages.

Software	Importing, creating, editing data	Ease of model set-up	Control over model content	Ability to accept imagery* and objects
Visual Nature Studio + Scene Express	Import: Excellent. Many file types and projections supported. Add/Edit: Good. Simple vector operations are relatively easy, others are best accomplished in a GIS.	Good. Can be a lengthy and complicated process for the first model but re-usable components make subsequent models easier. Scenario feature makes presenting multiple options easier.	Excellent. Large degree of control over all landscape elements. Extensive texture engine can control many environmental attributes besides color. Many visual parameters can be based on database attributes. Terrain can be modified locally, e.g. making embankments or ponds.	Images: Excellent. Many formats accepted. Transparency can be defined without an alpha channel. Objects: Excellent/Good. Import shows no problems but appearance can be problematic (too dark).
TerraVista	Import: Excellent. Many file types and projections supported. Add/Edit: Adequate. Vectors can be added but editing is laborious.	Adequate. Pre-set feature types may not be relevant. Many parameters attached to each feature but only a few are relevant in most cases. Cannot save and re-use components.	Good. Photo-based ground textures only, leading to visible tiling. Landscape features such as sky and water less controllable. Terrain locally modifiable to a limited extent. Terrain area may be larger than imported data, out of user control.	Images: Good. Limited formats if transparency required. Some transparency issues. Setting parameters for new images (e.g., real world size) takes time. Objects: Adequate, due to very limited formats.
ArcScene	Import: Excellent. Many file types and projections supported. Add/Edit: Excellent. Full GIS capabilities available through underlying program.	Good. Features more limited but existing options easy to use. Requirement for trees to be attached to points is tedious.	Adequate. Photo-based ground textures only, very memory intensive. Objects must be positioned individually. No local terrain modification.	Images: Good. Limited formats if transparency required (some issues evident), but adding items to scenes is straightforward. Objects: Good, due to relatively limited formats.

* Imagery may be geo-specific textures, geo-typical and object-based textures, or foliage images (not applicable to ArcScene – it can accept billboard trees only as OpenFlight models).

41.4 Discussion

41.4.1 Resource Requirements

While it is fair to say that the general trend for landscape modeling software has been for it to become more affordable over the last few years, the packages used in this work show that there is, nonetheless, a wide range of prices. In terms of initial outlay, support, and training,

TerraVista is clearly the most expensive piece of software considered here, which will be a major disincentive to many users. That said, its output is the most suited to the dedicated virtual reality display systems which are also available, and therefore, it is more likely to be considered as part of a high-specification visualization facility than for use with standard PCs. The overall pricing of ArcScene reflects the underlying ESRI components needed, and it may well be that potential users already have the basic GIS software and would only need to buy ArcScene. Visual Nature Studio is a standalone package like TerraVista, albeit at a fraction of the price, and while the Scene Express extension is available at additional cost, the overall price is still the lowest of the three. For all packages, free viewers are available for at least one of their output formats (see below).

The basic hardware requirements given by each manufacturer are broadly similar and are not particularly high at first glance. However, processing power and memory are the primary determinant of the size of landscape model that can be created, and the level of detail that can be shown, and so generally the more resources available, the better. The database used in this investigation was not particularly large, and all three packages coped well with it as the model was created. The only noticeable effects were in Visual Nature Studio, which features real-time previews during the set-up stage. Here, it was beneficial to program response times to turn off these previews when they were not needed. Otherwise, effects were more obvious when interacting with the models produced.

41.4.2 Learning Process

With each package having a different user interface, it is a subjective exercise to try and compare them in terms of ease of use and speed of learning. It should be stated at this point that the authors are more familiar with Visual Nature Studio than the other two packages, but efforts have been made to consider all packages from a new user's viewpoint.

ArcScene (Figure 41-1) benefits from being an integral part of a widely-used GIS package, and users already familiar with ArcGIS would learn to use it relatively quickly. Setting up 3D symbols (such as trees and ground textures) for the database features is done via the same basic process as creating any other sort of symbology, with a few additional parameters. Visual Nature Studio and TerraVista both have much more complex user interfaces, but this reflects their versatility. The complexity in Visual Nature Studio and TerraVista is down to slightly different factors in each—Visual Nature Studio offers a high degree of control over all aspects of the visual environment, whereas TerraVista's controls are oriented somewhat more towards management of the resulting real-time database. These differences, in turn, stem from the origins of the software, with Visual Nature Studio originating as a renderer of still images and animations and TerraVista always having been a real-time terrain modeler, and the two packages do take quite different approaches to landscape set-up. Visual Nature Studio (Figure 41-2) employs a graphical user interface (GUI) with a plethora of editor windows, check-boxes, radio buttons, numerical entry fields, graphs and other input devices. TerraVista's approach (Figure 41-3) is a hierarchical structure whereby the components of the landscape are defined and applied via textual database entries. This contrast between text-based and more visual control methods is quite marked, and it was felt that the latter was slightly more user-friendly in that the range of possible values for any one setting was usually clearer. Visual Nature Studio further benefits from an online help system where every editor window has an entry discussing the effects of each parameter, compared to that in TerraVista where only some parameters are described.

Training is available in some form for all three packages, but the price and accessibility vary widely. As might be expected from a widely established company like ESRI, training on its products is available through its sales offices as well as from independent consultants; both are able to tailor the training to individual needs, although at a cost. TerraVista and Visual Nature Studio are both from much smaller software companies, and as such the training is more limited with manufacturer-led courses only available in the USA, and running once or

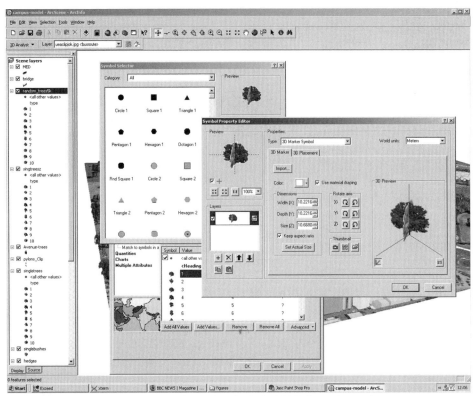

Figure 41-1 The ArcScene user interface. See included DVD for color version.

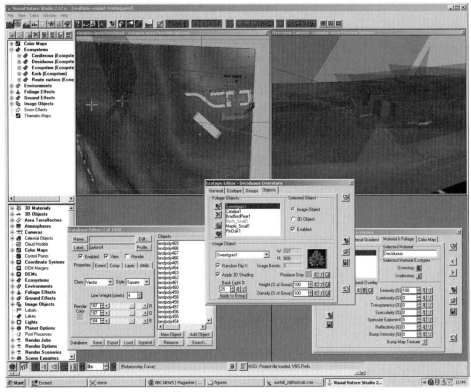

Figure 41-2 The Visual Nature Studio user interface. See included DVD for color version.

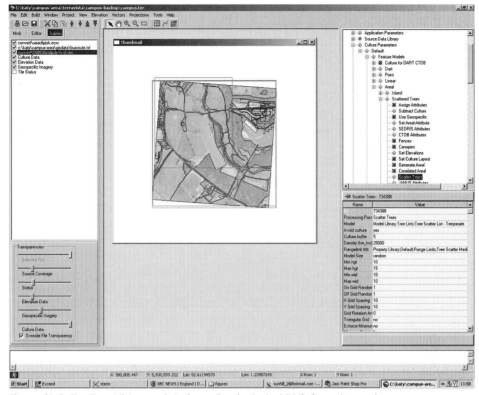

Figure 41-3 The TerraVista user interface. See included DVD for color version.

twice a year. A European training session for Visual Nature Studio has recently become established by an independent company, running approximately yearly, but nothing similar exists for TerraVista—training outside Terrex facilities takes place on-demand and at a premium. All packages also have self-paced tutorials available to varying degrees, and ranging from comprehensive video-based instruction and online tutorials (Visual Nature Studio) through detailed printed exercises (ArcScene) to a relatively short "getting started" guide (TerraVista). There are noticeably more tutorials for Visual Nature Studio than the other two, due at least in part to the existence of an active online user community, supported by the manufacturer, with members willing to share their expertise. This is also the case for ESRI products, albeit with the focus more on answering questions than providing tutorials, but with TerraVista aimed primarily at military users it is perhaps not surprising that there is little public discussion of the software and the projects for which it is used.

41.4.3 GIS Data Preparation and Import

Across the field of landscape visualization, the ability to import GIS data of various formats and projections has greatly improved since previous software assessments (Appleton et al. 2002). Test data in ESRI shapefile and ASCII grid format and using the British National Grid projection were loaded into all three packages with no problems.

In terms of attribute coding, data preparation considerations depend to some extent on whether other landscape models have been constructed in the past, from which components or settings can be re-used within the software. In this case, it is usually desirable to code the land cover/land use attributes to match those referenced in the previous project. Data standardization is in its infancy, but in the case of TerraVista, a certain amount of support exists for the SEDRIS (Synthetic Environment Data Representation and Interchange Specification) data specification model, and 3DNature offers a pre-built template Visual Nature Studio project to work with the US Geological Survey National Land Cover Data set (USGS 2007). No such support is known to be available within ESRI's tools, although it would certainly be possible to put together a legend utilizing a collection of textures and/or 3D symbols relevant to a particular data or coding format.

Some thought also must be given to the way in which features are stored in the database, as there are various ways in which they may be represented in the output model—similar issues were found by Appleton et al. (2002). For example, a simple way of representing buildings is via an extrusion of their footprint (possible in Visual Nature Studio and ArcScene), whereas more complex and realistic structures can be shown using a 3D model created in an external package attached to a point, which may or may not exist as a feature in the original database. Linear features, such as roads and rivers, may also be represented as a single line (usually the centerline), or their real-life outline. Mismatches can occur with centerline-driven features which in real life vary in width along their length, possibly leading to gaps or overlaps with adjacent land cover. Indeed, none of the software can vary a linear feature's width so precisely as to avoid this. However, using an area-based representation usually cannot show directional features such as markings on a road in the same way that linear representations can. A similar consideration relates to hedges—these features can be represented by rows of bush-type vegetation attached to points, or extruded lines with appropriate texturing. Both will likely require at least a little pre-processing of the base data to add extra points: in the former case, one point is required for each bush model; in the latter, points should be sufficiently dense that the hedge can follow terrain elevation variations along its length. Multiple bushes may look odd if viewed in a format where single-billboard 2D vegetation rotates to face the viewer, as the hedge's apparent width will change according to viewing angle. The extruded 'wall' type feature is likely to be much more efficient to display in real-time, although 2D shapes can disappear when seen from one end or directly above. Display can, therefore, be enhanced by buffering the line in the GIS in order to obtain a 3D, as opposed to 2D, object when extruded, but again this adds to the processing load at the time of viewing.

Finally, perhaps the most significant consideration in this category is the representation of trees within woodland. Unlike both TerraVista and Visual Nature Studio, ArcScene does not offer a "scatter" function to randomly place trees or any other point symbols within a defined area. If detailed tree locations are not available, the user must create random points to represent them—either by hand, or using an extension such as that offered by Hawths Analysis Tools for spatial ecologists (SpatialEcology.com Undated). If the trees are to show variation, such as in species or height, then attributes must also be added for ArcScene use, whereas TerraVista and Visual Nature Studio both offer the user ways to add random variation to the models used.

41.4.4 Setting Up the Model's Appearance

In terms of the GIS base data, landscape models contain three types of feature: point, line and area. At its most basic level, the process of assigning visual properties to these features is akin to creating a legend, and the three packages allow different degrees of control over the features' visual properties. Point features in the GIS database are used to position single 2D or 3D features such as buildings and vegetation models; 'scatter' functions are discussed below. Linear features can be given a width in all packages, and appearance can be varied across that width if using TerraVista or Visual Nature Studio. For areal features, ArcScene allows plain color, patterned or image-derived textures; Visual Nature Studio allows all of these plus procedural textures (see below); and TerraVista relies on image-based textures alone. In all cases, image-based textures may be either geo-specific imagery or geo-typical textures derived from photographs.

In any software, repeated photo-based textures tend to exhibit noticeable tiling, depending on their real-world size, although the effect can be reduced a little by skilled editing of the source images to make them more homogenous. Visual Nature Studio is unique in containing a texture editor—this is a powerful tool allowing the user to combine and control a variety of computer-generated patterns in order to mimic a real-life texture without having to photograph it, and without visible seams or repeats when the texture is applied to a surface. It also allows the capability to blend tiled photographs through its texture editor, greatly reducing the artifacts mentioned above. Within Visual Nature Studio, textures can be used not only on the ground surface but on any object (in TerraVista and ArcScene the object must be imported with its textures applied) and also used to drive such parameters as the distribution of vegetation (e.g., natural clumps or cultivated rows) and the appearance of clouds. This gives an important advantage when trying to represent the natural variation of rural landscapes.

All three packages allow some additional set-up beyond the basic point, line and area features. Simple 3D objects may be created by the extrusion of linear or areal features, for example to create fences or block buildings, although TerraVista only allows this to be done for lines. More complex entities should be created externally and imported to be attached to point features. In Visual Nature Studio, additional visual properties can be set up for water, and light, shade, sky and atmospheric effects also can be configured; not all of these translate well to real-time, since factors such as reflective water, detailed waves and volumetric clouds all add to the processing load. Such capabilities are largely absent from TerraVista as the landscapes it creates are more often used within dedicated display software such as Mantis™ or Vega Prime™, which offer control over this sort of feature at run-time.

TerraVista and Visual Nature Studio have different approaches to landscape model set-up in terms of the user interface. Visual Nature Studio's use of low-detail preview windows and the ability to do a quick still render of a landscape make it much easier to 'tweak' settings during the creation process, whereas there is no way to monitor TerraVista models without performing a build. Changes made to ArcScene models take place in real-time, although some changes to models with considerable foliage or extruded buildings can take several seconds to update. Another noticeable difference is the ability of Visual Nature Studio to use template projects and saved 'components' to speed up the workflow when working with similar data sets and land cover types; these can be placed on the manufacturer's website for download by other users. Similarly, once an ArcScene model's symbology has been set up, the resulting set of parameters can be saved with the data set as a Layer (.lyr) file and used with similar data sets in other projects. This kind of facility is a noticeable absence in TerraVista.

As discussed in the previous section, the data set used to create a model may not describe the landscape at the desired level of detail, and, in particular, groups of scattered landscape objects may be described as areal features (e.g., woodland or village) rather than individuals (e.g., trees or buildings). "Scatter" functions in TerraVista and Visual Nature Studio allow point features to be spread over a defined area, and this process is a significant part of the set-up of most models. Of the three programs, Visual Nature Studio offers the most control over object scatter: either foliage images or 3D objects can be spread in Visual Nature Studio with user-defined height, density and distribution (e.g., to allow for clumps or rows), and the ability to specify the proportional mix of tree or object types used, as well as density and height for each type. TerraVista requires "scatter lists" to be created, but without the finer control over the object mix and placement. It does, however, offer the ultra-low detail option of using a raised canopy surface (following the terrain surface at a specified elevation above it) edged with a vertical wall, both with appropriate textures applied so as to mimic the top and edge of a woodland. This works well for distant features, but is less convincing at close range, and not appropriate for models where users may wish to enter a woodland area.

Lack of detail in the database also may lead to the need for localized terrain modification if the digital elevation model (DEM) is of insufficient resolution. This technique, where available, allows defined areas to be raised or lowered to represent features such as ponds and embankments which are not reflected in the DEM. ArcScene does not offer this facility, except via manual editing of the elevation model. TerraVista offers some functionality in this area with its "complex lake" and "complex river" options, whereby relative elevations and surface appearances can be defined for the bottom, sides, banks and surface of a water body (or other features such as earthworks if the geometry is inverted and textures appropriately defined). Visual Nature Studio offers a large degree of control, allowing both areal and linear modifications with user-defined edge profiles, with the possibility of non-uniform changes through the use of generic or specific textures. Water features are applied separately from terrain modifications.

41.4.5 Landscape Objects

All the packages considered here come with a selection of objects for inclusion in landscape models, and all can accept imported objects with varying degrees of limitation. For example, ArcScene accepts OpenFlight (.flt), 3Dstudio (.3ds) and VRML models; TerraVista only accepts .flt; and Visual Nature Studio accepts .3ds, Lightwave (.lwo), Object Wavefront (.obj) and DXF formats. One advantage of .flt models is that they can be assigned differing levels of detail (LoD) for efficient display at increasing observer distance; however, they can only be created in a relatively small number of (usually expensive) programs such as Creator. While some packages offer conversion (e.g., Okino PolyTrans®), the LoD functionality is not available. Simpler and cheaper programs, including some free software such as Wings3D, can be used to create .3ds or .obj models and assign materials to the various surfaces. An important consideration is the ability to construct a model in such a way as to keep the polygon count as low as possible, by merging coincident surfaces and minimizing interior geometry, and possibly by applying photographic or other textures to give the illusion of surface detail without increasing the model's complexity. Complexity can be a problem if using models from one of the many websites offering free or paid-for objects, as they are usually not created with real-time use in mind, but rather to be as realistic as possible.

For 2D images, as used for textures and foliage objects, there are fundamental differences between the packages. TerraVista accepts bitmap (.bmp) .gif, .jpg, .rgb/rgba and .tif files for textures (although their dimensions must be a power of two) but requires .rgba (SGI RGB with Alpha channel to dictate transparency) format for foliage objects. While higher-end graphics packages support this format, the creation of alpha channels did cause some problems when processed in Corel Paint® Shop Pro® v8, leaving a white fringe around the trees when viewed in real-time. Visual Nature Studio does not require the use of alpha channels, instead allowing the user to designate a color (pure black by default) which the software will treat as transparent for foliage images; foliage images be imported from .bmp, .jpg, .iff, .rgb, .tif and other formats. The same formats are available for georeferenced ground imagery and photographic surface textures (whose dimensions are not constrained in any way). In contrast to the other two programs, ArcScene has no facility for 2D landscape objects, and trees must be imported as .flt models even if they are simple billboards or crossboards; however, geo-specific ground textures such as aerial photographs can be loaded in the same way as in other ArcGIS components, and a wide range of image formats including .bmp, .gif, .jpg, .rgb and .tif are supported for texturing surfaces such as the ground and extruded objects.

Trees can be created in 3D or 2D format in a number of specialist packages, including XFrog™, Onyx Tree Pro™, and RealNat®. The ability to vary certain parameters and output several variations of the same tree addresses the problem of vegetation looking too uniform in a simulation. That said, efficient real-time visualization often involves the re-use of tree models via "instancing," whereby a model is defined once and used many times, and so a large number of different models may not be helpful. Fully three-dimensional trees will most likely be too complex for efficient display in real-time visualizations intended for desktop viewing, but there are various options for getting 2D image or 3D double-billboard (or crossboard) versions of trees from these packages, including direct export from the software or rendering in an external package.

Visual Nature Studio gives an additional option for landscape objects: it allows the creation of labels to be displayed within the terrain. Essentially 2D billboards, their size and appearance can be defined, and they can take their position from point data, with their content coming from the points' attributes. This allows labeling of locations or features within a scene and can be useful for viewer orientation. Similar objects in the other two pieces of software would have to be created externally and then imported.

41.4.6 Beyond Navigation: Including Animated and Interactive Objects

Animated objects (e.g., moving vehicles or rotating wind turbines) may enhance the realism of a 3D scene, and are possible in VRML and OpenFlight scenes but not in NatureView Express at present. Implementation is not possible from within the authoring software tested here, but could be done using additional coding in VRML. Animated objects within OpenFlight require 'Degrees of Freedom Nodes' to be defined with separate software, and can generally only be displayed within more sophisticated viewing systems. As such, animated objects are beyond the scope of this comparison except to note that the capability exists.

All the formats used here allow for user interaction with elements of the landscape, although the process for OpenFlight output is complex and additional software is required. 3D objects can be assigned hyperlink-type properties in VRML and NatureView Express, causing a file of some sort to be displayed when they are clicked. This could be text, an image, a sound, a video clip, or a web page containing some combination of these media. The audience can therefore be exposed to information beyond the purely visual when exploring a scene. VRML standards allow for a variety of other actions to be associated with nodes, with the additional capability to trigger actions as the viewer gets close to a particular point. Both VRML and NatureView Express models can include objects that link to custom-written scripts, thereby giving immense possibilities for further interaction and viewer feedback. However, in both VRML and NatureView Express, none of these actions is currently able to be automatically written from any of the software tested here, and must be coded in by other means. For VRML, further examples of non-geographical authoring software are available and these would be of great use in enhancing the content of a scene produced with Visual Nature Studio or ArcScene. No other authoring software exists for the NatureView Express format, but like VRML, the code is hand-editable (although edited files must be re-validated using Visual Nature Studio before they can be viewed).

41.4.7 Generating the Real-time Output

Once the scene has been set up within the modeling software, the model must be built or rendered. All the packages allow certain features or classes of element to be disabled so as not to appear in the final model: ArcScene can be set to ignore disabled themes when exporting, and Visual Nature Studio and TerraVista both allow the user to disable individual components or classes of component, either within the project's global settings (TerraVista and Visual Nature Studio) or in the export control (Visual Nature Studio only). Such capabilities mean that test models can be rendered and viewed quickly, e.g., to check the appearance of just one sort of feature. When rendering multiple models for testing or other purposes, it is useful to be able to separate them, and Visual Nature Studio and ArcScene both allow users to specify a destination file for the output. On the other hand, TerraVista creates a "flight" directory in the project folder and does not give the option to specify an alternative.

Visual Nature Studio is unique in that it also allows real-time models to be rendered out at user-specified resolutions, whereas in TerraVista and ArcScene, terrain and drape image resolutions default to those of the imported data (5 m and 0.1 m, respectively, in this comparison). In the Visual Nature Studio exports, terrain and texture dimensions were kept as similar to the original as possible for comparative purposes, with terrain at ~5 m (excluding VRML) and drape images at 0.1 m. VRML terrain is limited to 16,000 heights per tile (Web3D Consortium 1997), so the single-tile terrain must be limited to this in ArcScene, giving ~8-m resolution; Visual Nature Studio allows multiple terrain tiles to be output in VRML format, and so a terrain resolution of ~5 m was achieved using 4 tiles of 104 x 104 cells. NatureView Express terrain is not limited in size but the texture dimensions (number of pixels) must be a power of two; 8096 x 8096, for example, gave a resolution of approximately 0.12 m.

ArcScene offers little control over the export process itself, offering the single menu option "File > Export > 3D"—although all components of the scene apparently should export, only the terrain, texture and extruded buildings were present in the VRML output; point marker symbols, i.e., vegetation and externally-created 3D objects, were missing.

Summary comparisons of navigation aspects of the packages are presented in Table 41-3.

Table 41-3 Summary comparison of the viewing software used.

Software	Price Range*	Ease of navigation	Range of navigation features	Other features
Cortona (VRML plugin for Internet Explorer®)	Free	Adequate	Very good. Limited choice of navigation modes, plus pan and tilt, goto and reset position. Pre-set viewpoints can be toured manually. Control frustrating for large models due to slow response.	VRML allows objects to have associated hyperlinks to any media the web browser can handle.
Blaxxun Contact 5 (VRML plugin for Internet Explorer)	Free (non-commercial use)	Good	Very good. Limited choice of navigation modes, plus pan and tilt, goto and reset position. Pre-set viewpoints can be toured manually. Control frustrating for large models due to slow response.	VRML allows objects to have associated hyperlinks to any media the web browser can handle.
NatureView Express	Free	Excellent	Excellent. Variety of navigation modes to choose from, plus stop, goto and reset position. Pre-set viewpoints can be toured manually or automatically, and may be animated.	Objects can be made clickable, linking to text, html, sounds, or other models. Logo or location map can be added.
Audition	Free	Adequate	Adequate. Single navigation mode. Cannot climb/descend while moving forwards/backwards.	Ability to view wire-frame model.
ArcScene	Low	Good	Good. Choice of navigation modes plus goto.	Not standalone, requires ArcGIS/ArcScene. Using too many textures causes some to be blank.
TerraVue (TerraViz if purchased separately from TerraVista)	Low	Good	Adequate. Single navigation mode. Cannot climb/descend while moving forwards/backwards.	

*Price ranges are Low (< 5,000 USD), Medium (> 5,000 and <10,000 USD) and High (> 20,000 USD).
**Windows is a registered trademark of Microsoft® Corporation in the United States and other countries.*

41.4.8 Database Management Issues

A primary factor relevant to real-time landscape models is the way in which database content is optimized for fast and efficient display, usually by turning off parts of the model which are not in view, and using low-detail substitutes for areas which are visible but distant (LoD controls). This capability is a combined function of the authoring program, the output format and, to a more limited extent, the viewing software, and there are significant differences between the packages assessed.

VRML allows both terrain information and objects to be assigned a LoD, albeit with activation based merely on observer distance and not view direction or terrain visibility (McCann 2004). Visual Nature Studio allows the user to specify levels of detail for VRML terrains and the observer distances at which they will be activated; landscape objects can be turned on and off in this way, but no substitution of lower-detail models is possible. ArcScene does not allow for LoD control in its export of VRML models. NatureView Express also allows objects and foliage to be turned on and off, but not substituted for low-detail versions; terrain is limited to a single model.

The background of TerraVista and the OpenFlight format in terms of development for large databases becomes clear in this respect, as this consideration can effectively be addressed in reverse, at least for terrain. When setting up a new model, a polygon budget can be specified – this is the number of terrain polygons to be displayed at the time of viewing, and the resolution is then managed to stick to this polygon budget. Some trial and error may be needed at first, but suitable values can be determined. Levels of detail can also be set for 3D objects brought into the model, since the .flt format allows for this, but the billboarded trees are a fixed feature within the landscape. The facility already mentioned whereby woodland areas can be represented as a 'canopy' plus 'edges' is the alternative offered, and while it certainly reduces the processing load significantly, the level of realism is adversely affected. Visual Nature Studio's output of OpenFlight information is limited, with difficulties having been encountered during coding despite OpenFlight's intended status as an open and documented format for real-time 3D models (3DNature, personal communication).

41.4.9 Viewing the Real-time Output

All output was viewed on an AMD Athlon™ 64 2.8 GHz PC with a 256 Mb NVIDIA® GeForce® 6800 GT graphics card and 1 Gb RAM. Time to generate output files, and sizes of files generated, are summarized in Table 41-4.

It is clear that the VRML display method as used here is not particularly efficient and navigation is slow for the comprehensive output from Visual Nature Studio (Figure 41-4). ArcScene's output (Figure 41-5) displayed quicker due to the lack of vegetation. With appropriate software, the code as output from ArcScene and Visual Nature Studio could probably be optimized for faster display, for example with dedicated optimization software such as Chisel; alternatively, since it is a markup language and can easily be read and edited, sufficient expertise could also allow authors to optimize the code manually or with another authoring program. However, for those wishing to produce usable real-time output without the need for further processing, VRML does not seem to be the best way to do so. In particular, ArcScene's output, in lacking vegetation and 3D object buildings, is severely limited. (It is worth noting that the scene can be viewed with all elements within ArcScene itself, but this is not helpful for distributing the information beyond the GIS community.)

Table 41-4 Generation time and file size for the outputs.

File type	Origin	Time (0.1m)	Time (1m)	File size[1] (0.1m)	File size[1] (1m)
OpenFlight	TerraVista	2 min.	2 min.	85 Mb	85 Mb
OpenFlight	VNS	7 hrs.	2 min.	85 Mb[2]	16 Mb[2]
VRML	ArcScene	10 sec.	10 sec.	24 Mb[3]	22 Mb[3]
VRML	VNS	4.5 hrs.	1 min. 45 sec.	14 Mb	33 Mb
NatureView Express	VNS	7 hrs.	2 min.	104 Mb	8 Mb

[1] including any subdirectories, image files, etc.
[2] terrain, texture and buildings only
[3] some landscape elements apparently not exported (see text); this may affect file size

Manual of Geographic Information Systems

Figure 41-4 VRML output from Visual Nature Studio displayed in Cortona. See included DVD for color version.

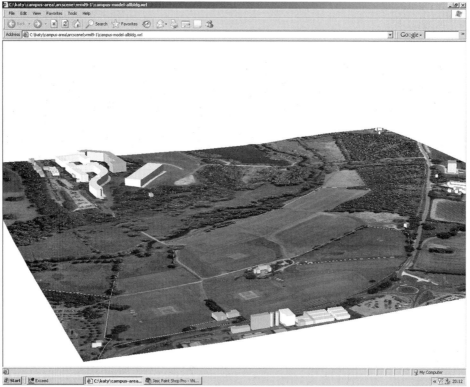

Figure 41-5 VRML output from ArcScene displayed in Blaxxun Contact. See included DVD for color version.

Positive points for VRML include the fact that it is the only format used here to be accessible online, and the range of plug-ins available to view it either online or locally. Of the two viewing plug-ins tested here, Blaxxun Contact seems to have more intuitive controls than Cortona®, but the functionality is largely similar. Both viewers have a variety of features including numerous navigation options (such as fly, walk, study), 'go-to' functions, and the ability to select pre-set viewpoints or visit them as part of a tour. The capability for interactivity and hyperlinking is also an important benefit of VRML, although it was not used here.

OpenFlight (OF) is clearly a powerful format, but Visual Nature Studio output capabilities in this format were limited by the exclusion of vegetation (Figure 41-6). This was queried, and it was stated that the OF format is not well-documented and different viewing software appears to have different expectations regarding file structure and content; at present, there is little commercial incentive to refine the output further (3DNature, personal communication). In TerraVue™, the output is acceptable and interaction is fast as would be expected. Output from TerraVista (Figure 41-7) suffers a little from the inability to confine the output area to the specific boundary of the data imported, leading to confusing "blank" areas. In addition, the terrain texture derived from 0.1-m imagery must be down-sampled to maintain response speed when viewing. Some trees display white fringing, but this seems to be a by-product of using PaintShop Pro to create the necessary alpha channels, and can be avoided by using other software. The Audition™ viewer was not able to display the final output from either piece of software, although other models have been viewed in Audition in the past. The cause of this problem is not known; it has been reported and is under investigation. Both OpenFlight viewers tested here offer only adequate navigation controls in comparison to the other viewing software, with no choice over the navigation mode and no ability to use pre-set viewpoints. No capacity for interaction with objects is included in these viewers.

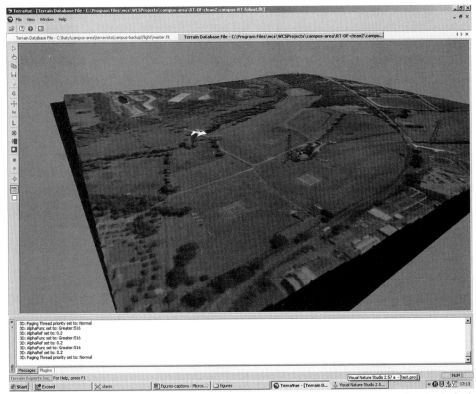

Figure 41-6 OpenFlight output from Visual Nature Studio displayed in TerraVue. See included DVD for color version.

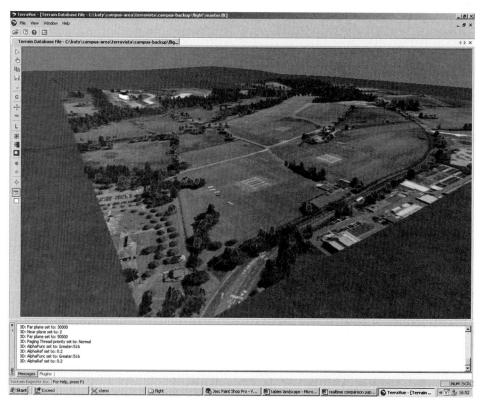

Figure 41-7 OpenFlight output from TerraVista displayed in TerraVue. See included DVD for color version.

NatureView Express viewing (Figure 41-8) benefits from the largest choice of navigation modes, meaning most users should find one they are comfortable with, although the variety may initially be confusing. To remain compliant with Mac control systems, the right mouse button is not used in navigation, which leads to a smaller variety of movements available at any one time without changing navigation mode. 'Go-to' and viewpoints are available, with the additional capability for animated viewpoints; this can be an important benefit for users who feel uncomfortable with interactive navigation. Speed and smoothness are as good as the OpenFlight viewers and better than the VRML plug-ins for fully comprehensive scenes, and display of 3D objects is generally good, albeit with very dark shadows on unlit faces.

Uniquely, NatureView Express allows the control of landscape display in terms of enabling and disabling classes of objects so that, for example, place-name labels could be turned on initially and removed once the user is oriented. Disabling objects will of course increase the model's responsiveness. It is not possible to disable only some of a class of object, e.g., only the proposed buildings in a scene. The viewer also has some customizable options including the ability to display a logo or other image and the opportunity to include a small orientation map (a miniature of the ground texture with a location/direction arrow).

In addition to standard desktop display, OpenFlight output is an industry standard format for display on dedicated virtual reality systems running software such as Vega Prime and Mantis. Such display generally requires some additional work to set up the run-time environment and make best use of the lighting and other environmental factors, as well as animated and interactive models. (Evaluation of such display is beyond the scope of this work.) With no additional processing, NatureView Express output can be viewed on Elumens spherical section screens by setting an option when starting the software. NatureView Express is also available in an Extended Developer Edition, allowing for

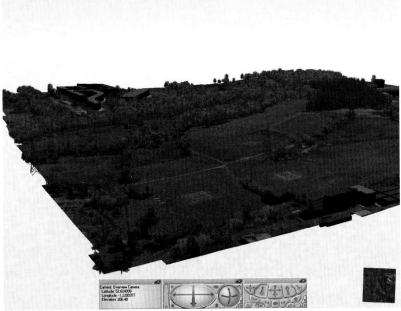

Figure 41-8 NatureView output from Visual Nature Studio displayed in NatureView Express. See included DVD for color version.

customization of the display and its capabilities according to the requirements of a particular project; this is at extra cost and requires programming experience. A variety of other VRML viewers is available beyond those used here, including open-source software such as FreeWRL which could be modified according to project needs to provide more display, navigation and interaction options.

41.5 Conclusions

The software packages assessed here provide a range of options for GIS users wishing to begin visualizing landscape information in real time and, inevitably, each has strengths and weaknesses. The familiarity of ArcScene's ESRI interface will be an important benefit for many GIS users, but unfortunately the output it produces for viewing outside the software is inadequate for landscape visualization in that it lacks vegetation and any other elements applied as point markers. The VRML export format also has limitations in that scenes featuring all the correct elements, such as output from Visual Nature Studio, are slow to display, and while it may be possible to optimize the code for faster interaction, many visualization users will not have the time or expertise to do so. It is true that VRML is the only method offering online viewing, but the file sizes associated with landscapes as opposed to smaller and simpler models mean that network speeds, bandwidth and the viewer's likely computer specification are all potentially problematic.

NatureView Express and OpenFlight output formats seem to be more promising than VRML, but again there are limitations. However, each is suitable for a different type of use. The two OpenFlight-capable packages assessed here offer either comprehensive output at very high cost, or limited output at more moderate cost. Both pieces of software are complex to learn, although Visual Nature Studio benefits from greater community support to help overcome this, and its complexity does allow greater control over the environment. However, these factors cannot compensate for the lack of vegetation or the large file sizes for OpenFlight output, and this along with better control over display efficiency (i.e., polygon budget) means that TerraVista has to be the choice for this type of model. Its main drawbacks are the relatively specific requirements for the dimensions and format of foliage images (and geo-typical textures if used), and the inability to extrude footprint vectors to quickly create simple building masses—often a useful option for structures which are not of great importance to the visualization's primary purpose. Overall, its cost means that it would be best suited to sophisticated display systems that really require OpenFlight, since an equivalent or, in some respects, better solution for desktop viewing is offered by the NatureView Express format.

The main limitation of NatureView Express seems to be the lack of options for controlling the display of a scene to maximize its efficiency, including the use of models at different levels of detail. In terms of viewing speed the output compares well with OpenFlight, and in terms of appearance and the ability to interact with objects and link to auxiliary information it equals or betters VRML, although animated objects and automatically-triggered events are not possible. The viewer is simple to use, freely available, and offers more navigation options than the others tested here.

Regardless of output format, Visual Nature Studio also has certain important advantages over the other packages used, in both the visual quality of the output and the workflow associated with creating it. Its capabilities for controlling the appearance of the ground surface are beyond those offered in either of the other two packages due to its texture editor—an invaluable asset if aerial photography is not available, or undesirable due to low resolution or the unavoidable presence of high-level detail such as treetops and roofs on the ground surface. It also can output to other media, which the other packages cannot—although only real-time output has been examined here, Visual Nature Studio's history is as a still-image renderer, allowing the user to put in some additional set-up time and output more detailed still images or compiled animations from the same data and working project. Finally, the core ability to save and re-use collections of visual properties cannot be overstated in terms of allowing the user to save progressively more time on each project created.

In summary, of the three options compared here, the combination of Visual Nature Studio and Scene Express, outputting to the NatureView Express format, offers the best overall solution for real-time visualizations which will be viewed only on a desktop PC, when considering quality and content of output along with software costs and viewing options. Due to the complex nature and high cost of packages in this field, it was not possible to compare a wider variety of programs, thus the scope of this chapter is limited in that solutions for high-end visualization systems have not been addressed. Although this review could not be comprehensive, it is hoped that the material presented here will prove useful to GIS users wishing to enter into the field of real-time landscape visualization.

41.6 Future Prospects

The field of real-time visualization is undergoing constant change as technological advances lead to new hardware and software, and bring previously expensive solutions within the reach of academic, government and other organizations with moderate budgets. Although it is increasingly used in public participation exercises for planning and other environmental decision-making purposes, its precise effects on the outcomes of such exercises are not yet clear (Bishop 2005), and this will be an area of important research in the next few years. Improvements come within two general categories: display capabilities, including greater realism and viewing technologies such as multi-channel and stereo viewing; and non-display capabilities, relating to content and interaction.

New-generation graphics cards, such as the Matrox Parhelia, also offer the possibility of driving multi-channel and other wide-screen systems from a relatively standard PC as opposed to specialized display hardware (Matrox 2007). Realism and other display factors are largely a function of the available processing power—as processor speeds increase and graphics cards improve, the elements of a scene can be reproduced more faithfully while retaining an acceptable degree of interactivity. Algorithm-based modeling, especially of vegetation, also offers further possibilities for increasing display efficiency (Deussen et al. 2005). However, the capability for more detailed display then raises more questions about the adequacy of the input data in terms of content and accuracy (Lange 2005), and real-time visualizations are subject to the same concerns as more established methods about transparency and ethics in the production process (Sheppard 2001); there may even be more danger of the medium overpowering the message. Beyond visual representations of a landscape come the possibilities for modifying it in real-time, for example through the manipulation of scenario parameters, interactive movement and alteration of elements within the scene, or communication and discussion with other audience members (Bishop 2005). Technology initially developed for computer games often can be used as a base for implementing these and other extensions to existing visualization methods (Stock et al. 2005; Herwig et al. 2005).

Importantly, it is these same improvements to the capabilities and accessibility of computer technology that play a significant part in increasing the familiarity of the next generation of users with the technology employed in landscape visualization. It also perhaps raises their expectations in terms of the visual quality and responsiveness of the models used. It is important that the limitations of both the available data and the technology employed are recognized, and that the most appropriate resources are used for the purpose at hand. To this end, there is important research remaining to be done in the field of real-time landscape visualization.

References

Appleton, K., A. Lovett, G. Sünnenberg and T. Dockerty. 2002. Visualizing rural landscapes from GIS databases: a comparison of approaches, options and problems. *Computers, Environment and Urban Systems* 26 (2–3):141–162.

Bishop, I. 1994. The role of visual realism in communicating and understanding spatial change and process. In *Visualization in Geographic Information Systems*, edited by D. Unwin and H. Hearnshaw 60–64. London: Bellhaven Press.

Bishop, I. 2005. Visualization for participation: the advantages of real-time? In *Trends in Real-Time Landscape Visualization and Participation*, edited by E. Buhmann, P. Paar, I. Bishop and E. Lange, 2–15. Heidelberg: Wichmann.

Deussen, O., C. Colditz, L. Coconu and H.-C. Hege. 2005. Efficient modeling and rendering of landscapes. In *Visualization in Landscape and Environmental Planning*, edited by I. Bishop and E. Lange, 56–61. London: Taylor and Francis.

Doyle, S., M. Dodge and A. Smith. 2002. The potential of web-based mapping and virtual reality technologies for modelling urban environments. *Computers, Environment and Urban Systems* 22 (2):137–155.

Ervin, S., and H. Hasbrouck. 2001. *Landscape Modeling: Digital Techniques for Landscape Visualization*. New York: McGraw-Hill.

Herwig, A., E. Kretzler and P. Paar. 2005. Using games software for interactive landscape visualization. In *Visualization in Landscape and Environmental Planning*, edited by I. Bishop and E. Lange, 62–67. London: Taylor and Francis.

Lange, E. 2005. Issues and questions for research in communicating with the public through visualizations. In *Trends in Real-Time Landscape Visualization and Participation*, edited by E. Buhmann, P. Paar, I. Bishop and E. Lange, 16–25. Heidelberg: Wichmann.

Lovett, A., R. Kennaway, G. Sünnenberg, D. Cobb, P. Dolman, T. O'Riordan and D. Arnold. 2002. Visualizing sustainable rural landscapes. In *Virtual Reality in Geography*, edited by P. Fisher and D. Unwin, 102–130. London: Taylor and Francis.

McCann, M. P. 2004. Using GeoVRML for 3D oceanographic data visualizations. *Proceedings of the Ninth International Conference on 3D Web Technology*, Monterey, California, pp. 15–21.

Matrox. 2007. Matrox Graphics – CAD and GIS – Products – Graphics Cards – Parhelia Series. Matrox Graphics Inc., Quebec, Canada. <http://www.matrox.com/graphics/en/cadgis/products/home.php> Accessed 19 November 2007.

Sheppard, S. 2001. Guidance for crystal ball gazers: Developing a code of ethics for landscape visualization. *Landscape and Urban Planning* 54 (1-4):183–199.

SpatialEcology.com. Undated. Hawth's Analysis Tools for ArcGIS. <http://www.spatialecology.com/htools/tooldesc.php> Accessed 19 November 2007.

Stock, C., I. Bishop and A. O'Connor. 2005. Generating virtual environments by linking spatial data processing with a game engine. *Trends in Real-Time Landscape Visualization and Participation*, edited by E. Buhmann, P. Paar, I. Bishop and E. Lange, 324–329. Heidelberg: Wichmann.

Tufte, E. 1990. *Envisioning Information*. Cheshire, Conn.: Graphics Press.

USGS (US Geological Survey). 2007. Get Land Cover Data. <http://landcover.usgs.gov/landcoverdata.php> Accessed 19 November 2007.

Web3D Consortium. 1997. The Virtual Reality Modeling Language: 7 Conformance and minimum support requirements. VRML97, ISO/IEC 14772-1:1997 <http://www.web3d.org/x3d/specifications/vrml/ISO-IEC-14772-VRML97/part1/conformance.html> Accessed 19 November 2007.

Zube, E. H., D. E. Simcox and C. S. Law. 1987. Perceptual landscape simulations: history and prospect. Landscape Journal 6:62–80.

Links to Software and Data Mentioned in the Text

All URLs below were accessed 19 November 2007.

Data

Bluesky <http://www.bluesky-world.com/>
NEXTMap Britain <http://www.bluesky-world.com/dem-nextmap.html>
Ordnance Survey MasterMap <http://www.ordnancesurvey.co.uk/osmastermap/>

Authoring Software Used

ArcScene, ArcGIS 9.0, ArcView 3.x (ESRI, Inc.) <http://www.esri.com>
TerraVista (Terrain Experts/TERREX, now Presagis, Inc.) <http://www.presagis.com>
Visual Nature Studio, Scene Express (3DNature, LLC) <http://www.3dnature.com>

Viewing Software Used

Audition (Quantum 3D) <http://www.quantum3d.com/support/downloads/audition/index.htm>
Blaxxun Contact 5 <http://www.blaxxun.com/home/>
Cortona VRML Client <http://www.parallelgraphics.com/products/cortona/>
NatureView Express (3DNature, LLC) <http://3dnature.com/nv.html>

Other Software Mentioned

Blueberry3D (Bionatics EUROPE and Bionatics North America) <http://www.bionatics.com/Blueberry3D.php>

Chisel (created and placed in the public domain by Trapezium Development LLC) <http://create.ife.no/vr/tools/chisel/install.htm>

Creator (MultiGen-Paradigm, Inc., now Presagis, Inc.) <http://www.multigen.com/products/database/creator/index.shtml>

Creator Terrain Studio (MultiGen-Paradigm, Inc., now Presagis, Inc.) <http://www.multigen.com/products/database/creator/modules/mod_terrain_studio.shtml>

Fledermaus (IVS-3D) <http://www.ivs3d.com/products/fledermaus/>

FreeWRL <http://freewrl.sourceforge.net>

Lenné3D project <http://www.lenne3d.de/>

Mantis <http://www.quantum3d.com/products/Software/mantis.html>

Onyx Tree Pro <http://www.onyxtree.com>

OpenSceneGraph <http://www.openscenegraph.org/>

Paint Shop Pro (Corel) <http://www.paintshoppro.com>

Polytrans (Okino) <http://www.okino.com/conv/conv.htm>

RealNAT <http://www.bionatics.com/Site/product/index.php3?Community=3>

Sitebuilder 3D (a component of CommunityViz® by Placeways, LLC) <http://www.placeways.com/communityviz/?p=site>

VirtualGIS for ERDAS IMAGINE (Leica Geosystems AG) <http://gi.leica-geosystems.com/LGISub1x39x0.aspx>

Virtual Terrain Project <http://www.vterrain.org/>

Vega Prime <http://www.multigen.com/products/runtime/vega_prime/index.shtml>

Wings 3D <http://www.planit3d.com/source/software_files/3dmod.html >

XFrog <http://www.xfrog.com>

CHAPTER 42

Interactive Maps for Exploring Spatial Data

Robert Edsall, Gennady Andrienko,
Natalia Andrienko and *Barbara Buttenfield*

42.1 Introduction

A hallmark of modern geographic information system (GIS) software is its capability for user interaction. Interactivity—referring here to the myriad ways that a system, or data represented in a system, can change according to user input—is now ubiquitous enough in computerized information systems that it is often taken for granted. However, interactivity should be examined critically in the context of maps and geographic representations in order to understand how far GIS has come in facilitating data analysis, and what might still be developed in order to examine data sets that are presently difficult to examine given the state-of-the-art of GIS. This chapter explores the design and use of interactive maps for spatial data exploration and analysis, paying particular attention to applications incorporating approaches from cartography, statistics and computer science. We also consider the role of the interactive capabilities and potential of GIS for the burgeoning field of visual analytics, and look ahead to the possibilities of incorporating interaction into future designs of GIS, in particular those that can handle spatio-temporal data. Present GISs, with their unique and highly interactive interfaces, are well-designed for many spatial data exploration tasks. However, developers and users should consider implementing modes of interaction not presently enabled in GIS that would facilitate a wider range of geographic visualization and analysis. We thus advocate the increase of interactive capabilities for map products created in a GIS environment based on the improvements for data exploration that could be gained.

42.2 Interaction and Geo-visualization

Traditional maps are designed by cartographers to communicate information to a map user or group of map users. The incorporation of interactivity into maps allows for this same communication between cartographer and map user, but it also affords a dialogue of sorts between the representation and the user, as the map user becomes an important agent in the creation and re-presentation of the information. In the days before users were given opportunities to interact with maps, the process of cartography was a one-way path that led from "reality" through several intermediate steps—the cartographer's filtering of that reality (tempered by his or her own intentional or unintentional bias), selection, generalization, symbolization, the user's perception of the map, and a (potential) alteration of the user's mental model of the represented phenomenon (Figure 42-1) (Robinson et al. 1995; MacEachren 1995). The provision of interactivity adds an important feedback to this process: though the user cannot alter reality itself, every other step of the cartographic process can be subject to alteration with an interactive map. User interaction can alter the themes and base map information that are displayed, the scale and aggregation of the data, the level of detail, the type of map, the classification, the color scheme, the viewing angle, the highlighted elements—interactive maps afford a user an infinite number of representation possibilities, each with the potential to alter mental models and to construct knowledge in a unique way (Figure 42-2).

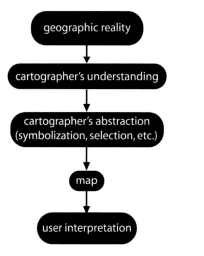

Figure 42-1 Typical model of cartographic communication.

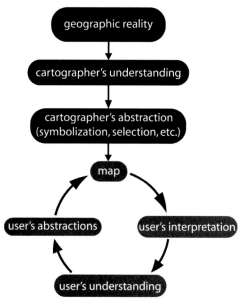

Figure 42-2 Interaction allows a user of a map to incorporate his/her own abstractions in order to refine mental models and explore data. Interactive maps enable this feedback loop.

42.2.1 Geo-visualization and GIS

A high degree of user interactivity opens a new world of opportunities and map use options. One of the most important is the ability to use the map as a tool for exploration and analysis, and not simply as a tool for presentation (Dykes 1997; Andrienko and Andrienko 1999; Edsall 2001; Crampton 2002). Over the past two decades, as GISs have evolved, theories within geographic visualization (or geo-visualization) have been developed to support the creation and use of maps and other associated representations for interactive analysis of complex spatial data sets. This approach in geographic information science generally involves a small group of expert users (often only one) that examines a phenomenon or set of phenomena through dynamic maps on a computer (MacEachren and Kraak 1997). Geo-visualization is characterized by highly interactive representations designed for use by individuals, expert in the understanding of the mapped phenomenon, to reveal previously undetected features of a data set. Recent research has indicated that, with innovative interaction techniques, geo-visualization can be facilitated even in collaborative situations, where multiple users, sometimes in different locations, can construct shared understanding of spatial phenomena (Rauschert et al. 2002; MacEachren 2005). Geo-visualization itself is not a set of tools and representations but is a process where human and computer interact continually to transform, select, and map data to a visual form in the quest for patterns and relationships (MacEachren, Wachowicz et al. 1999; MacEachren et al. 2004). With this new kind of map use, cartographers have become not only producers of well-designed maps for the communication of concepts, but also providers of opportunities (through user interfaces, modeling algorithms and visualization environments) for the exploration of data.

The connections between this shift in map use and GIS are clear. Geographic information systems, perhaps to a fault from a cartographic design point of view, were not developed for the presentation of information in the form of map output, but rather for the interactive analysis of spatially referenced information. A vast majority of maps created in GIS are ephemeral, existing for only the amount of time they are useful for an analyst. As such, they are meant to be seen by a single researcher or research group during the exploration of a geographic problem.

Exploratory geo-visualization often happens with a GIS at the center, but it can also occur with special-purpose software specifically designed to, for example, handle very large data sets through aggregation and/or other generalization forms (Fredrikson et al. 1999), represent many attribute variables at once through novel representation techniques (DiBiase et al. 1994; Erbacher et al. 1995; Dai and Hardisty 2002; Edsall 2003; MacEachren et al. 2003), simulate three-dimensional space with virtual reality motion and depth cues (Forsberg et al. 2006; MacEachren, Edsall et al. 1999; Fisher and Unwin 2002), or view the temporal nature of geographic data through map animations (DiBiase et al. 1992; Dykes 1997; Peterson 1999; Harrower 2004; Blok 2006). These applications, central to visual geographic analysis, are relatively difficult with "off-the-shelf" commercial GIS packages.

Often, the very personal activity of data exploration using maps involves the examination of those variables for features that might only be interesting to a single analyst with highly individualized mental models of the problem at hand. The uniqueness and specificity of data exploration tasks makes the creation of guidelines for representing and interacting with large disparate data sets with multiple variables relatively elusive (Haug et al. 2001). One way to begin a discussion about a framework for designing interaction with maps for exploratory visualization is to define a set of abstract roles of interaction to support data exploration.

42.2.2 The Roles of Interaction in Visual Analytics and Information Exploration

"Interaction is the fuel for analytical discourse." (Thomas and Cook 2005, p. 9). Visual representations put the extraordinary power of the human mind to work in concert with sophisticated computational tools. Recently, the field of *visual analytics* has focused attention on the tremendous power of combining the cognitive resources of both scientists and domain experts with the interactive high-performance resources of interactive visual computing (Thomas and Cook 2005). At this point, work in this field has been primarily oriented toward developing research agendas and building bridges across disciplines that have previously worked separately on visual analytical theory and methodology. One of the major focus areas of visual analytics is "visual representation and interaction techniques," and, while stressing the vital role of interaction in exploring information, the primary agenda document acknowledges that theories of how user interaction enables analytical thinking are few. It calls for the development of "a new science of interactions that supports the analytical reasoning process. This interaction science must provide a taxonomy of interaction techniques ranging from the low-level interactions to more complex interaction techniques and must address the challenge to scale across different types of display environments and tasks" (Thomas and Cook 2005, p. 9).

We believe that interaction techniques designed for data exploration have several basic roles, regardless of whether the data being represented are spatial. In upcoming sections, we describe specific modes of interaction that perform these roles.

First, interaction allows the environment to *compensate for the indispensable deficiencies* arising from representing information on a computer display. The depiction of data on a two-dimensional screen with limited resolution and size restricts the amount and form of visible information. In addition, regardless of the limitations of the screen, representing data also involves necessary biases associated with generalization and symbolization choices. Interacting with a computer environment (specifically, in ways detailed in Section 42.4.1), allows a user to access additional information and to use several different representation forms for multiple perspectives.

Second, interaction helps to *discover unobvious patterns* in data. Some patterns and relationships are only visible after much trial-and-error of manipulating the representation. Such manipulations might include rotations of scatter plots, transformations of axes or reordering of a set of images. We describe interaction modes best-utilized for this role in Section 42.4.2. Using interactive capabilities for this purpose is useful and relevant for data sets of any size.

However, interaction is imperative, not simply relevant, to *explore particularly large databases*, defined both in terms of the number of data records and in terms of the number of attributes. Even millions of pixels may be insufficient to represent terabytes of information. In addition, the perceptual and cognitive capabilities of a human impose their serious limitations on the amount and density of the information a display may contain irrespective of its size and resolution. Only with interaction can all the data become accessible to an analyst. There are a number of interaction modes that are specifically designed to handle very large databases, which we categorize (in detail in Section 42.4.3) into two major types: *roll-up/drill-down* and *slice-and-dice,* which refer, respectively, to varying the levels of abstraction of data to give an overall impression, and the division of data into manageable portions. With the increasing prevalence of such data sets in geographic information science (GIScience), these modes will become more and more important in the design of interactive information systems.

We argue below that there are aspects of geographic inquiry that make visualization of *spatial* data distinctive. However, we also can borrow from (and support) research in related disciplines that are not particularly concerned with spatial data for some guidance in developing interaction for data exploration and knowledge discovery.

42.3 Designing Interaction for Exploration and Knowledge Discovery: Lessons from Outside Geography

The GIScience community can look outside its traditions for novel methodologies in interactive and/or visual analysis of large databases. These techniques are research priorities in the communities of statisticians, computer scientists, psychologists and graphic designers involved in exploratory data analysis (EDA), information visualization (InfoVis), and knowledge discovery in databases (KDD).

42.3.1 Exploratory Data Analysis

Statistics over the last three decades has witnessed the development of techniques that counterbalance a long-term bias in statistical research towards developing mathematical methods for hypothesis testing. Tukey (1977) saw exploratory data analysis as a return to the original goals of statistics—detecting and describing patterns, trends and relationships in data (both spatial and non-spatial). In this philosophy, EDA is about hypothesis generation rather than hypothesis testing. Though it need not be strictly visual, EDA is strongly associated with the use of graphical representations of data:

> Most EDA techniques are graphical in nature with a few quantitative techniques. The reason for the heavy reliance on graphics is that by its very nature the main role of EDA is to open-mindedly explore, and graphics gives the analysts unparalleled power to do so, enticing the data to reveal its structural secrets, and being always ready to gain some new, often unsuspected, insight into the data. In combination with the natural pattern-recognition capabilities that we all possess, graphics provides, of course, unparalleled power to carry this out.
>
> (NIST/SEMATECH 2006, Section 1.1.1)

Visual representations developed in EDA are often designed to complement confirmatory techniques, depicting data sets that can be gigantic, complex, multivariate, temporal, and/or spatial. The development of graphics for EDA has occurred somewhat atheoretically (with the possible exception of work by Wilkinson (1999)), with a large number of (very creative and useful) tools and techniques developed for specific applications or problems (Keim and Kriegel 1994; Rao and Card 1995; Hummel 1996; Spence and Tweedy 1998; and many others). Recognizing the importance of spatially referenced information, statisticians have bolstered their visualization systems with mapping techniques borrowed from cartography and GIScience (Cook et al. 1997; Swayne et al. 1997; Unwin and Hofmann 1998; Carr et al. 2002).

The standard EDA "toolbox" contains representations designed for non-spatial data such as histograms, scatterplots and boxplots (Becker et al. 1988; MacDougall 1992). Adding dynamic features to these EDA tools have made them useful for data exploration. These dynamic features include both animation and interaction. Much of the work in EDA deals with developing ways of exploring complex numerical data. Many analogous techniques and innovative interaction modes have been developed for the exploration of abstract and/or non-numerical data. Such exploration falls into the related realm of *information visualization*.

42.3.2 Information Visualization

Information visualization (InfoVis) is concerned with the mapping of abstract data (like library card catalogs, web pages, financial transactions and mathematical functions, for example), and the design of such mapping has taken more creativity and originality than analogous mapping of physical data, which may already exist in some spatial context (like the atmosphere near the ozone hole, the interior of a human body, or the inter-molecular structure of a protein). This is a challenge that has led to novel and important research into using and interacting with the space of a graphic presentation to display complex information efficiently and clearly (Card et al. 1999).

The process of creating an information visualization representation from raw data, according to Card et al. (1999), is a "pipeline" of transformations and mappings, each stage of which, in a well-designed information visualization environment, benefits from human interaction. The general stages in the pipeline are *data transformations, visual mapping* and *view transformations.*

In the data transformation stage, raw data are translated into information that can be used in visualization by first deriving structure or value from them. This translation is accomplished by creating interactive tables of information (GIS practitioners would consider these highly interactive "attribute tables") that can then be sorted, summarized statistically, classified, aggregated, or otherwise manipulated to give the information additional meaning to the researcher.

Visual mapping requires the abstraction and mapping of information not just to space, but also to color, position, size, visual hierarchy (connections between visual elements, like "branches" of a "tree") and value. These visual variables are those that were described and implemented by Bertin (1983), and have, of course, been the subject of a large proportion of literature in statistical graphics and cartography. Skupin and Fabrikant (2003) argue that GIScience can contribute theoretical depth to this step, calling the visual mapping of abstract data "spatialization." Tweedie et al. (1996) and Wise et al. (1995) are among several authors who emphasize user interaction in this step, enabling different sorts of visual representations to be created on the fly to conform to user needs.

Finally, "view transformations" refer to various abilities to alter parameters of the view itself, as opposed to the methods for mapping the data onto the view. Such transformations, according to Card et al. (1999) include adjusting viewpoint, altering distortion, and revealing additional information with an interaction such as a mouse-over, in which information is revealed when the mouse cursor hovers over specific text or graphics. View transformations also occur as a result of so-called *direct manipulation* interaction (Shneiderman 1997; North and Shneiderman 1999), through which the user may use the representation itself, and not buttons, menu items or other elements external to the display window, to alter what is represented. We shall argue that direct manipulation interfaces are particularly important in spatial data exploration; interacting with the space of the representation itself allows users to think spatially when exploring spatial data.

In the next section, we demonstrate how interaction modes developed in EDA and InfoVis can be applied to spatial data analysis, and how these modes correspond to and complement existing methods for interacting with spatial data in GISs and interactive cartographic environments.

42.3.3 Knowledge Discovery in Databases

Though many EDA techniques, particularly those adapted by Tukey (1977), are visual, visualization is just one way of exploring statistical information. Visualization, for example, is one important component of analyzing very large databases. However, it is typically used in conjunction with other non-visual computational techniques that have been developed under the research umbrella of "knowledge discovery in databases" (KDD), of which the better-known procedure of "data mining" is a part (KDD consists of several generic steps: data preparation, data mining, interpretation of the results and reporting; see Miller and Han (2001)). The application of such computational data mining techniques with subsequent visualization of their results allows a human analyst to gain additional knowledge that cannot be (easily) gained directly from the viewing and manipulation of the original data. In modern KDD software, data mining results are represented in interactive displays. The interactivity is essential since the outcomes of data mining are often very voluminous and/or have a complex structure, which does not permit putting all information in a single static picture.

Some work has been recently done on designing interactive data mining techniques, where the user can guide the discovery process. However, as it is admitted in the KDD literature, guided knowledge discovery through interactive data mining is still in its infancy (Ceglar et al. 2003), perhaps because the visualization and graphical interfaces do not belong to the primary competences (and interests) of the mathematicians and statisticians designing and elaborating data mining methods. A closer cooperation with researchers in visualization, in particular, geo-visualization, would be of clear benefit for both sides.

Thus, a very effective and powerful exploratory environment for large spatial databases may be built by combining data mining with interactive maps (MachEachren et al. 1999; Andrienko et al. 2001a; Guo 2003; Koua and Kraak 2005). In such an environment, the analyst can first explore the data visually using interactive map displays. From this interaction, an analyst can gain the general idea of the data and uncover significant features that require a closer look. This directs the further investigation, including the choice of appropriate computational methods. Then, to interpret and make use of the results, the analyst will again need map displays which adequately represent the spatial aspect of the data.

Besides the obvious idea that KDD can aptly complement interactive maps in exploration of spatial data, it may be useful for designers of interactive mapping environments to adopt the view of data exploration and knowledge discovery as an essentially iterative process rather than a direct course from data to knowledge. An analyst often needs to return to one of the previous steps (data selection, data pre-processing, formulation of the question to be answered, selection of a data mining method, or setting method parameters) either for refining the results obtained or for testing their sensitivity to the selections and settings made in order to avoid possible biases and artifacts. This view is also relevant to the exploration of spatial data with the use of maps, which basically involves the same steps as KDD except that maps are used instead of data mining methods. In this process, the standard cartographers' attitude to a map as a final product is no more appropriate. A designer of a mapping environment intended for exploration of spatial data should care about supporting all the steps of the exploration process, and enabling the iterative feedback loop. According to the steps involved, a minimum set of required interactive operations might include data selection (in particular, attributes, but also samples of data records), various data transformations (e.g.

from absolute values to relative or per unit values), selection of mapping techniques, and variation of their parameters.

In upcoming sections, we review specific methods from EDA, InfoVis and KDD that have proven (or would be) useful in representing and interacting with spatial and spatio-temporal data, and discuss in what ways the present capabilities of GIS complement those methods.

42.4 Exploring Spatial Information with Interactive Graphics: Justifications and Existing Techniques in GIS

In this section, we examine interaction forms present in existing GISs and classify them according to the roles for interacting with geographic information described in Section 42.2. Instrumental to the development (and wide usability and acceptance) of GIS—and indeed most software—was the introduction and adaptation of: 1) the spatial and user-friendly graphical user interfaces (GUIs) of operating systems; and 2) interaction devices—in particular, the mouse. The GUI was a non-trivial advancement of GIS because it took advantage of the well-established cartographic techniques of using the two dimensions of a map (in a GIS, the computer display) to represent two dimensions (x and y) of the surface of the Earth. Command line interaction, which might be useful and acceptable in the analysis of non-spatial data, removes a key cognitive connection between a user and spatial data: the map.

The map works uniquely as a visual representation of geographic data because it is *isomorphic* with the represented phenomenon (Arnheim 1997). The concept of isomorphism states that any representation on any level of abstraction needs to meet one condition: it must be structurally similar (isomorphic) to the pertinent features of the phenomenon being represented. Hence, effective exploration of the geographical space is impossible without an isomorphic representation of its pertinent features and relationships. From all existing types of display, only maps can satisfy this requirement.

Most of the interaction forms that make interactive maps effective for spatial data exploration take advantage of the isomorphism of maps on a computer display, and many other interaction modes (now present only in concept or in prototype systems) can be imagined that would increase the usefulness of interactive maps for exploration tasks.

Here, we follow the lead of EDA specialists Buja et al. (1996) and InfoVis pioneers Card et al. (1999) in organizing interaction modes according to task types. The roles we see for interaction with maps for exploring geographic information, as introduced in Section 42.2 above, include: 1) compensating for the indispensable deficiencies in the display; 2) revealing unobvious patterns in data; and 3) exploring large and complex databases.

42.4.1 Compensating for Indispensable Deficiencies in the Display

Because the cognition and understanding of maps on a computer display are limited by factors such as the display screen's resolution, color depth and two-dimensionality, as well as the visual perceptual limitations of the user, environments for spatial data exploration require interaction to compensate for these limitations. Present GIS and interactive cartographic environments feature methods by which these limitations can be (at least partially) overcome. Some of these are borrowed from (or replicated in) EDA and InfoVis environments. They include:

- *Zooming*. Presuming that there is sufficient detail in the raw (input) data, interactively adjusting the display scale restricts the extent of the view and can reveal features undetectable at smaller scales. This is, of course, common in interactive maps, and is used in more abstract representations such as scatter plots.

- *Panning, re-centering.* Upon restriction of the display extent such that data are hidden off-screen, capabilities become necessary to reveal the hidden information. In GIS, panning is enabled with click-drag actions, interface "arrow" buttons, or inset navigation maps, which show the display extent relative to the overall study area. Popular web-based interactive mapping systems have incorporated *animation* of these pan actions, representing a definite improvement over panning in discrete jumps with buttons or other navigation devices, and allowing for a more natural continuous scanning of our environment.

- *Re-projecting.* The two-dimensionality of computer displays (and static graphics such as paper maps) force a projection of multi-dimensional information. These "projections" include familiar map projections (at which GIS are adept) and also the projection of abstract data clouds (e.g. scatter plots of more than two variables; see Asimov 1985). Scatter plot matrices are multiple projections of data clouds, and are implemented and featured in many geo-visualization environments including those built with the GeoVISTA Studio toolkit (Gahegan et al. 2002) and particularly the work of Dai and Hardisty (2002) (Figure 42-3). In GIS, map projections can be altered through dialog boxes, pull-down menus, and form fill-in interaction, but using the mouse to *rotate* a three-dimensional scene also serves to re-project the representation. Rotation has long been identified as essential for the recognition of patterns and features in an abstract data representation, revealing features (such as in a three-dimensional data cloud) that would otherwise be obscured from view (Becker et al. 1988; Hurley and Buja 1990).

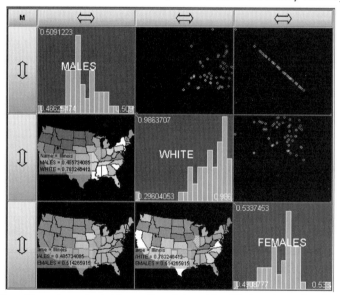

Figure 42-3 Conditioned manipulable scatter plot matrix of Dai and Hardisty (2002). The scatter plot matrix represents a form of projection of a multidimensional attribute data set; in this case, a three-dimensional data cloud is projected in three different ways (with corresponding choropleth maps). The matrix could contain two-dimensional projections of an indefinitely high number of dimensions. "Strumming" and "accessing exact values" are also illustrated: mouse-over actions immediately reveal details and are linked to corresponding glyphs in other windows. See included DVD for color version. Source: Dai and Hardisty (2002).

- *Accessing exact data.* Typically in maps and other graphics, data items are encoded by symbolic representations. However, the extraction of exact data from the graphical feature (particularly if it has been simplified, symbolized or classified) is difficult for a user. An interactive map can provide easy access to the exact data standing behind graphical features. This is a typical operation in a GIS, for example, as data can be shown in a special window when the user points to an element in a map.

- *Focusing* involves increasing the legibility and degree to which display elements may be differentiated (so-called *display expressiveness*) of a subset of data. Even high-definition displays are limited in legible sizes and differences in symbols. Through the selection and re-symbolization of a subset of data, focusing reveals detail within the subset and de-emphasizes the differences in the data set as a whole. Owing to the conceptual similarity to display zooming, which enlarges a selected part of a display area in order to improve its legibility, focusing may also be called "data zooming." A specific case of focusing is removal of outliers, i.e. extremely high or extremely low values standing far apart from the bulk of the data, from the display.

42.4.2 Revealing Unobvious Patterns in Data

The revelation of previously unrecognized features or patterns in a data set is one of the basic purposes of cartography in general. However, comprehensive data exploration is best supported by interactive methods to alter the methods of abstraction, giving unique and potentially important perspectives on the information. Some of the ways these multiple perspectives can be created in present GISs include:

- *Altering representation type.* Though cartography has long-standing guidelines regarding the appropriateness of symbolization schemes for different types of data, symbolization for data exploration must be flexible enough to support non-traditional strategies in order to reveal patterns. Switching between a choropleth and proportional symbol map, for example, may prompt different mental models of the information and facilitate insight.
- *Altering symbolization.* Within a chosen representation type, data exploration can be supported by simple interactions that allow the alteration of symbolization choices such as classification schemes, color schemes, interpolation type, and contour intervals. Figure 42-4 shows an example of this in the CommonGIS environment of Andrienko et al. (2001b). Again, traditional cartographic guidelines governing such choices for data communication to wide audiences may not be ideal for private data exploration.

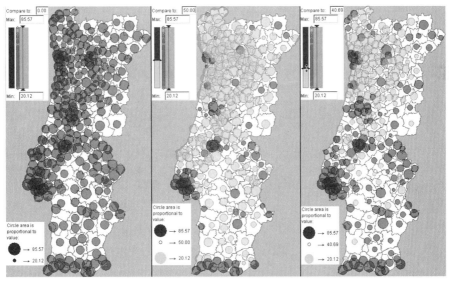

Figure 42-4 Altering symbolization (here, in the CommonGIS environment of Andrienko et al. (2001b)) facilitates revealing unobvious patterns in data. Left: the percentages of working people employed in services are represented by graduated circles whose sizes are proportional to the values they represent. Center: the symbolization has been changed so that two colors are used to represent the values below and above an interactively specified reference value, 50%. Salient spatial patterns are immediately noticeable. Right: the reference value can be dynamically changed, e.g., by moving a slider (top left of the map). In response, the map is automatically redrawn and can expose new interesting patterns.

- *Posing queries.* Maintaining the isomorphism of maps is also reflected in the spatial nature of many queries in GIS, particularly the ability to make selections with a tool that draws a box (or, more generally, a "lasso") directly on the map. Selecting (by drawing a box or lasso, or by pointing and clicking) points, lines or polygons according to their spatial arrangement, rather than a value of a thematic variable, gives privilege to the spatial nature of GIS queries.
- *Transforming data.* Manipulation of the data represented in a map often allows an analyst to see what was previously not evident. One of the most commonly used transformations is the derivation of relative data from absolute data, for example, proportions of parts in a whole, densities or amounts per capita. Statistical standardization is often applied when several attributes need to be jointly explored and compared. Computing of changes is used when data refer to several time moments (Figure 42-5). Interpolation and smoothing are applied to data representing spatially continuous phenomena, which are typically specified in raster (grid) form.

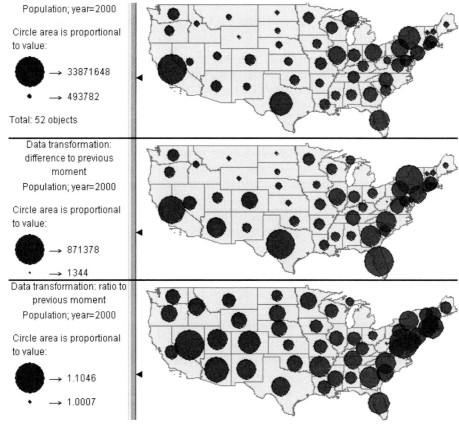

Figure 42-5 Three screenshots of the same map demonstrate an interactive application of data transformation (also from the CommonGIS environment of Andrienko et al. (2001b)). Top: the circles represent the absolute population numbers in 2000. Middle: the absolute population numbers in 2000 have been transformed to differences in comparison to 1999; the circles now represent the differences. Bottom: the circles represent the relative changes (ratios) with respect to 1999. Note the explanations in the legend on the left of the map.

42.4.3 Exploration of Voluminous and/or Multidimensional Databases

Spatial data can be particularly voluminous, and also can have dozens or hundreds of thematic variables associated with each record. Interaction is imperative for the exploration of such databases, assisting in several related but separate conceptual operations. *Roll-up* is a means of simplifying by means of aggregation, generalization and abstraction. By means of this "rolling up," the analyst can get an overview of the entire data set and build a simplified mental model of it. Verification of this model is accomplished by *drill-down,* an approach by which an analyst can view portions of the data closer to its raw (not abstracted) form. These two operations are often used in concert with one another. Another approach facilitated by interaction is *slice-and-dice,* which divides the data into manageable portions and allows separate exploration of these portions. The partial mental models obtained in this way must then be united in an overall model of the entire data set. Figure 42-6 illustrates the concepts of rolling up, drilling down, and slicing and dicing in CommonGIS.

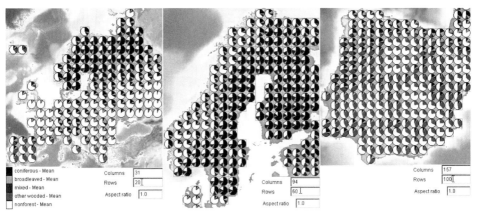

Figure 42-6 An illustration of the concepts of roll-up, drill-down and slice-and-dice. The original data about the proportions of different types of forests and non-forest land are specified as 5 high-resolution rasters, where each raster contains 40 million values. In order to get an overview of the distribution of forests over Europe, the data have been aggregated by cells of a coarse grid (roll-up). The aggregated values (specifically, the mean values in the cells) have been represented using pie charts. The analyst can select parts of the territory (slice-and-dice) and decrease the level of aggregation for these parts (drill-down) in order to examine the data in more detail. Irrespectively of the chosen level of aggregation, the interactive map preserves the representation type and symbolization parameters, which are re-applied to the data after each re-aggregation operation.

Interaction can support roll-up and drill-down in natural and convenient ways. For example, a GIS or other interactive map can be set to intelligently zoom, increase and decrease the level of detail when the user changes scale. Interaction can support slice-and-dice by enabling an analyst to divide a data set into portions and consider, for example, the south of a country separately from the north or urban areas separately from countryside. More subtly, interaction also is important in the synthetic activity of building an overall model from partial ones, which requires intensive comparisons between portions to extract similarities, note differences and establish relationships.

There are other interaction forms of present GIS and cartographic exploration systems that seem particularly well suited to the role of handling large complex databases. Among these are:

- *Toggling visibility* of loaded themes. Though a fundamental operation in GIS, the ability to select themes on a map is an extraordinarily powerful interaction mode, allowing for comparison of a subset of dimensions from a large multidimensional data set. This partially replaces the necessary bias of non-interactive maps created by the cartographer's selection of particular themes.

• *Brushing* and *linking*. In most GISs, a subset of records of a large data set is selectable through a spatial query (e.g., the lasso tool on a map, described above) or through an attribute query (selecting records that are similar along one attribute dimension after sorting an attribute table). This brushing operation is often used in combination with linking, through which selected records in one view (e.g. the map) are also selected (and manipulated) in other representations (e.g. tables, charts, other maps). A recent effective application of these techniques is the Exploratory Spatio-Temporal Analysis Toolkit (ESTAT) environment (Figure 42-7) of Robinson et al. (2005).

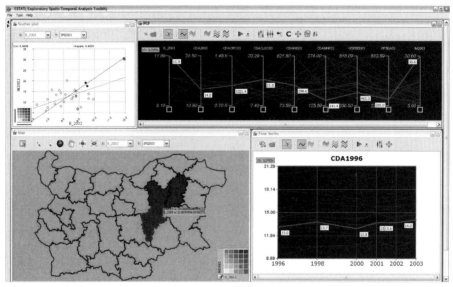

Figure 42-7 Exploration of spatial data through brushing (Robinson et al. 2005). A user selects a single enumeration unit (on map in lower left, or any of the windows) and all correspond- ing glyphs and traces are highlighted in other windows. Brushing can also occur with multiple spatial units to uncover regional patterns in attribute or temporal spaces, or with multiple at- tribute or temporal units to uncover corresponding patterns in the spatial representation. See included DVD for color version. Source: Robinson et al. (2005)

• *Conditioning*. Conditioning (or *filtering*) typically uses slider bars or other graphical interface tools to perform attribute queries on individual variables, displaying, for example, only those points between two values of one important variable (Carr et al. 2005). Conditioning reduces the amount of data displayed by displaying only those records that meet certain criteria (Figure 42-8).

42.5 Supporting Exploratory Spatio-temporal Visualization with GIS

In this section, we imagine an interface for geographic information that includes windows that each affords access to one (or more) of three aspects of geographic data: the spatial component, the temporal component and the attribute(s) component (following the TRIAD model of Peuquet and Qian 1996). Many interaction forms can be utilized in any of these interface windows of a GIS. Brushing, for example, can be implemented in the spatial representation (the map display), the attribute representation (the database table, or a graphic version of the attributes such as a scatter plot or parallel coordinate plot), or the temporal representation (if one would come to exist as standard in GIS—presently, timelines and other temporal query and display widgets are not standard). We organize the following ideas for interaction with spatio-temporal information through a GIS according to these three

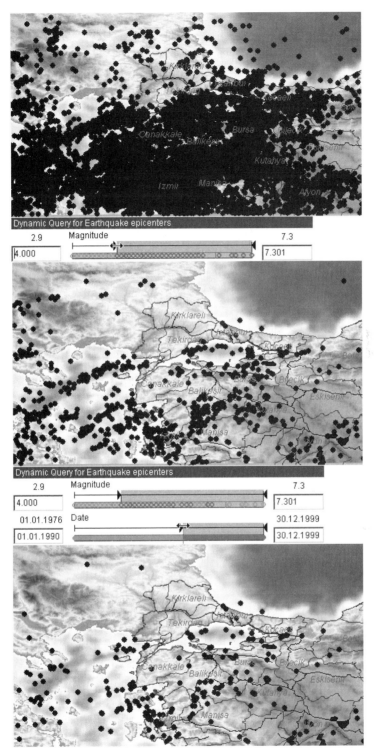

Figure 42-8 Application of a conditioning tool to a map in CommonGIS. Top: epicenters of 10,560 earthquakes that occurred in the area of Marmara (Western Turkey) during the period from 1976 to 1999. Middle: according to an interactively specified filter condition, only earthquakes with magnitudes 4.0 or higher are shown. Bottom: another filter condition has been added to show the epicenters of the earthquakes with the magnitude 4 or more that occurred since the beginning of 1990.

Manual of Geographic Information Systems

dimensions (space, time and attribute); however, many of these concepts below can be implemented in parallel ways to other displays.

42.5.1 Extending the Forms of Interaction with Spatial and Attribute Dimensions in GIS

Present GISs are designed to facilitate the analysis of the spatial and attribute dimensions of geographic data, and the discovery of connections between these two dimensions. A typical GIS query, for example, often requires an attribute "sort" (display only those units in a certain range of some variable) and a spatial "sort" (display only those units a certain distance from a selected polygon). We believe that this emphasis can inspire new interaction methods that would facilitate the exploration of these two dimensions of geographic data. Below is a non-exhaustive but hopefully provocative survey of interaction forms that might assist in the understanding of large, multivariate, disparate and/or complex geographic information.

- *Strumming*. Many of the representations that GIS might borrow from the EDA toolbox consist of relatively confusing and complicated displays. A scatter plot matrix, for example, is enhanced when a mouse-over of one point leads to the highlighting of all corresponding points in the matrix (Cleveland 1985, pp. 210–218). Dai and Hardisty (2002), among others, have adapted a type of mouse-over capability to maps and other corresponding statistical displays (see Figure 42-3). By simply passing a cursor over an object (e.g., an enumeration unit on a map, a point on a scatter plot, or a row of a table) or pixel (e.g., in a raster image), a user may highlight that unit of the display. Parallel coordinate plots (Inselberg 1985; Inselberg 1998; Edsall 2003) are vastly improved when the entire trace (representing the multivariate signature of one record) is highlighted upon the "strumming"–like a guitar string–of one segment of the trace. This strumming capability alone is not particularly useful, but when corresponding symbols, either in the same display or in other representations, are highlighted automatically with the strumming, complex displays can be simplified and the construction of knowledge is facilitated. Conceptually similar to linking (discussed as a present capability of GIS above), strumming, unlike linking, also can be utilized within a single, complex data display, and works with a simple mouse-over, reducing the need for fine-scale dexterity with the mouse.

- *Lensing*. Adopting a popular mode of interaction from InfoVis, GIS could use lensing, which simulates a fish-eye lens or other physical magnifying tools to highlight a region while deemphasizing unlensed regions (Leung and Apperley 1993; Stone et al. 1994). This adds detail to a region of interest, distorting its surroundings and reducing their visual importance but allowing them to remain visible. This is conceptually similar to zooming, which necessarily hides parts of the map or display that are not within the restricted extent of the view, but allows for (highly generalized) visualization of the parts of the data set that are not the focus of attention.

- *Manipulating symbolization with data-rich interactive tools*. Exploratory tasks are facilitated when a user is able to visualize as much information as possible. Classification and other methods of symbolization are more informative if they are associated with representations of distributions (even in the legend itself), such as histograms, parallel coordinate plots or scatter plots. In the context of map animations, Peterson (1999) called these "active legends." Recent prototype environments (Andrienko and Andrienko 1999; Andrienko et al. 2001b; Edsall 2003; Haug et al. 1997) allow for dynamic reclassification and color scheme manipulation by interacting directly with such representations via movable elements like slider bars, rather than via text fill-in boxes or command-line interactions. Interactive classification and representation changes are core components of the GeoVISTA Studio toolkit (Gahegan et al. 2002; see Figure 42-7).

Adding interactivity to the legend allows for on-the-fly adjustments to the classification scheme to better align the symbolization to the mental model of the user than the default schemes in GIS (equal interval, quantile, natural breaks, etc.).

42.5.2 Extending the Forms of Interaction with Temporal Information in GIS

The representation of, and interaction with, temporal aspects of geographic data, both in GIS databases and GIS displays, have been the focus of researchers who acknowledge the need for analysis and representation of process rather than stasis in geographic inquiry (Langran 1992; Peuquet 1996; Wachowicz 2002; Andrienko and Andrienko 2006). In its representations and graphical user interface, GIS presently gives privilege to the spatial and attribute relationships among data. Advances must be made in the database representation of spatio-temporal data; such advances will lead to and require innovations in the visual display (and its associated interactive capabilities) of the temporal geographic data. If, as Frank (1993) succinctly states, "the interface is the system" to users of GIS, simply adding interface tools that enable interaction with the temporal dimension of geographic information might make investigation of time in geography more central and fundamental to users of GIS. Some of the ways that users might be able to interact with temporal geographic data include:

- *Constructing and controlling animation.* Animations are the temporal analogy of maps—while maps represent physical space as display space, animations represent physical time as display time (DiBiase et al. 1992). As such, animations are isomorphic to their represented phenomenon, and logically should be the first choice to represent temporal information. GIS interfaces that might support animation should borrow concepts from animation software in order to make animation construction and viewing intuitive. Exploratory tasks with temporal data might be facilitated by reordering the frames according to a variable other than time, similar to "mapping" geographic data onto an attribute-based representation such as a histogram.
- *Temporal panning.* Buttons that mimic VCR controls, including start, stop, pause, fast forward, and rewind, are logical to include on any moving representation in order to reduce the cognitive disadvantages of animation (including "disappearance," in that users are asked to remember the state of a phenomenon from several frames before in order to compare it to the current frame; see Harrower (2003) for a survey of animation design choices). Some animated maps include a useful "rock" function that allows a user to move back and forth from one frame to the next and back again, over and over, in order to detect changes between a pair of moments in time. Temporal panning is applicable not only to animated maps but also to maps representing temporal variation by means of diagrams or symbols (Figure 42-9).
- *Temporal zooming.* More difficult to implement but likely necessary for creative exploration would be a temporal zoom tool. These tools could take several forms. First, to preserve the time-to-time isomorphism, a "zoom-in" on the timeline tool would increase the temporal scale. We define such a scale in the same sense as a map—1:86,400 would imply that one second of animation time would be equivalent to 86,400 seconds (one day) of real world time. A temporal zoom might increase this scale such that one second of animation would be equivalent, say, to 21,600 seconds, or six hours. If the data supported it, this zoom would, in effect, slow down the animation to highlight a segment of the animation of particular interest. The inverse also could occur, of course—a "zoom out" would devote fewer frames per unit real time (this opens new questions, not in the scope of this chapter, about methods of temporal generalization and extrapolation).

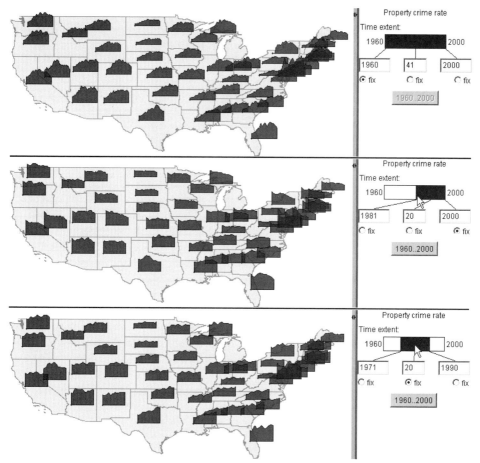

Figure 42-9 Top: the symbols on the map show the dynamics of the property crime rates in the states of the USA over the period of 41 years from 1960 to 2000. Middle: the user has interactively reduced the represented time interval to 20 years from 1981 to 2000. Bottom: the user shifts the 20-year interval along the time axis, which results in the map being dynamically redrawn.

- *Temporal querying.* Standard "timelines" form the most typical representation for time passing on an animation. However, this interface metaphor (time "moving" from left to right) may limit the creative exploration of information, particularly those temporal data that are periodic or cyclic in nature (Edsall and Sidney 2005). Ideal exploratory interaction techniques would allow for the selection and/or ordering of frames of an animation differently from simple linear time. This linear-cyclic distinction might be manifested in a GUI through a temporal querying tool, a graphical representation of time that can alternate between a "time line" and a "time wheel" at the user's request (Edsall and Peuquet 1997; Edsall 2001). In the case of the wheel interface tool, the user might specify the period of the cycle represented and choose to query only those dates that correspond to a specified duration within the cycle. For example, suppose a researcher were interested in the variation of rainfall each monsoon season over several years. He or she would customize the time wheel to a yearly period and then select the days, weeks, or months of interest to limit the investigation.
- *Temporal data transformation:* A map can be supplied with interactive controls allowing the user to apply various data transformations, for example, from the original values to the changes (differences or ratios) with respect to the previous moment or a selected moment (Oberholzer and Hurni 2000; Andrienko et al. 2001c). The map immedi-

ately would react to the actions of the user by re-applying the current representation technique and symbolization parameters to the transformed data. Besides computing changes, useful transformations for time-dependent numeric data might include temporal smoothing, which hides minor fluctuations and exposes major trends, and comparison to the running mean or median for the whole territory or to the local mean or median in each place. A possible user interface for interactive transformations of time-dependent data is demonstrated in Figure 42-10.

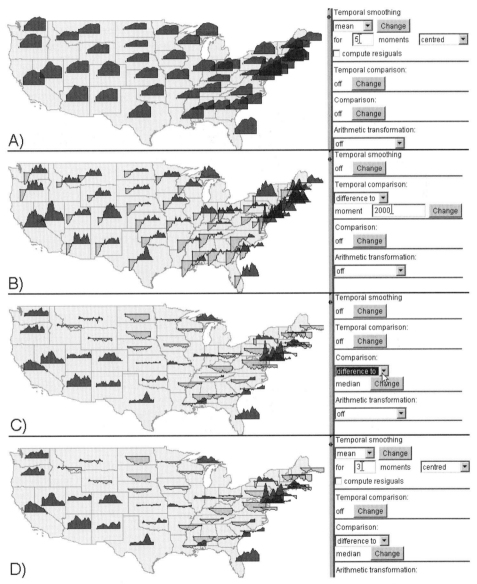

Figure 42-10 Various transformations can be interactively applied to time-series data represented on a map in CommonGIS. A) Temporal smoothing (compare to the map in Fig.9 top). B) Comparison to a selected time moment. C) Comparison to the country's running median, i.e., the values in each year are compared to the country's median in this year. D) It is possible to combine two or more transformations, for example, smoothing with comparison to the running median. In maps B, C, and D, the darker shade corresponds to positive differences and the lighter shade to negative differences.

Manual of Geographic Information Systems

42.6 Summary

In this chapter, we have described and examined a multitude of different modes of inter-activity that either exist presently in modern GIS or might benefit GIS if they were widely implemented. The focus here has been on the interaction with maps and graphics with the goal of exploring spatial (and spatio-temporal) data; this contrasts somewhat with typical roles of GIS which include confirmatory analysis and data communication. While user interfaces of present GISs would almost never be described as "intuitive," they do provide a remarkable range of interaction possibilities that position GIS well as an environment for exploratory tasks. With some additional techniques and theories of interaction borrowed from other disciplines, particularly human-computer interaction, information visualization and knowledge discovery in databases, GIS could evolve into a more natural exploratory tool. In doing so, it would support tasks such as compensating for limitations in the display, discovering unobvious patterns in data and examining particularly large spatial databases about which little is known going into the analysis.

We have chosen in this chapter to limit our survey to those techniques and systems designed for interaction of a single human analyst with a single desktop computer. It is unrealistic, however, to assume that data exploration and expert-driven analysis will occur solely in this mode. Particularly in the realm of visual analytics, the combined efforts of many analysts working collaboratively often lead to the greatest insight (MacEachren 2005). Mouse-based interaction will not be sufficient for exploratory analysis in group work, and interfaces need to be multi-modal, responding to not only mouse actions but also, possibly, gestures and voice commands (Rauschert et al. 2002). Additionally, theoretical and empirical examinations of the influence on the globalization of visualization may be necessary to maximize the capabilities of interactive maps for spatial data exploration; with the internet, the ability to carry out international collaboration with users of diverse cultural backgrounds will prove to be powerful. However, it may be necessary to examine differences in symbolic conventions and social interaction customs in order to design representations and interac-tions to involve diverse users (Marcus 2001; Shen et al. 2006; Edsall 2007).

In this chapter, we hope to persuade users and designers of GIS to appreciate the great power of interactivity in present systems. Interaction through GUIs is isomorphic with—and therefore highly appropriate for—the spatial analysis priorities of GIS, enabling the space-based queries and visualization that today's GIS users take for granted. GIS would benefit from expanding those capabilities in order to interact with other dimensions of GIS data, through, for example, spatialization of abstract and/or attribute data, or animation and/or transformation of temporal data. Research in related fields such as geo-visualization and information visualization can be utilized for designing such interaction. In this chapter, thus, we also hope to inspire users and designers to consider the potential of GIS for spatio-temporal data exploration through innovative representations and interactions that represent natural extensions of existing systems.

References

Andrienko, G. and N. Andrienko. 1999. Interactive maps for visual data exploration. *International Journal of Geographic Information Science* 13 (4):355–374.

Andrienko, N. and G. Andrienko. 2006. *Exploratory Analysis of Spatial and Temporal Data.* Berlin: Springer. 715 pp.

Andrienko, N., G. Andrienko, A. Savinov, H. Voss and D. Wettschereck. 2001a. Exploratory analysis of spatial data using interactive maps and data mining. *Cartography and Geographic Information Science* 28 (3):151–165.

Andrienko, G., N. Andrienko and A. Savinov. 2001b. Choropleth maps: classification revisited. Pages 1209–1219 in *Proceedings, 20th International Cartographic Conference,* Beijing, China, 6-10 August 2001. Beijing: State Bureau for Surveying and Mapping.

Andrienko, N., G. Andrienko and P. Gatalsky. 2001c. Exploring changes in census time series with interactive dynamic maps and graphics. *Computational Statistics* 16 (3):417–433.

Arnheim, R. 1997. *Visual Thinking.* Berkeley: University of California Press. 352 pp.

Asimov, D. 1985. The grand tour: a tool for viewing multidimensional data. *Siam Journal on Scientific and Statistical Computing* 6 (1):128–143.

Becker, R. A., W. S. Cleveland and A. R. Wilks. 1988. Dynamic graphics for data analysis. In *Dynamic Graphics for Statistics,* edited by W. S. Cleveland and M. E. McGill, 1–49. Belmont, California: Wadsworth & Brooks.

Bertin, J. 1983. *Semiology of Graphics: Diagrams, Networks, Maps.* Madison, Wisconsin: University of Wisconsin Press. 415 pp.

Blok, C. A. 2006. Interactive animation to visually explore time series of satellite imagery. In *Visual Information And Information Systems,* edited by S. Bres and R. Laurini, 71–82. Lecture Notes In Computer Science 3736. Berlin: Springer.

Buja, A., D. Cook and D. Swayne. 1996. Interactive high-dimensional data visualization. *Journal of Computational and Graphical Statistics* 5 (1):78–99.

Card, S., J. Mackinlay and B. Shneiderman. 1999. Information visualization. In *Readings in Information Visualization: Using Vision to Think,* edited by S. Card, J. Mackinlay and B. Shneiderman, 1–34. San Francisco: Morgan Kaufman.

Carr, D. B., D. MacPherson, D. White and A. M. MacEachren. 2005. Conditioned choropleth maps and hypothesis generation. *Annals of the Association of American Geographers* 95 (1):32–53.

Carr, D. B., Y. Zhang and Y. Li. 2002. Dynamically conditioned choropleth maps: shareware for hypothesis generation and education. *Statistical Computing & Statistical Graphics Newsletter* 13 (2):2–7.

Ceglar, A., J. F. Roddick and P. Calder. 2003. Guiding knowledge discovery through interactive data mining. In *Managing data mining technologies in organizations: techniques and applications,* edited by P. C. Pendharkar, 45–87. Hershey, Penn.: Idea Group Publishing.

Cleveland, W. S. 1985. *The Elements of Graphing Data.* Monterey, California: Wadsworth Advanced Books and Software. 323 pp.

Cook D., J. Symanzik, J. J. Majure and N. Cressie. 1997. Dynamic graphics in a GIS: More examples using linked software. *Computers and Geosciences* 23 (4):371–385.

Crampton, J. 2002. Interactivity types in geographic visualization. *Cartography and Geographic Information Science* 29 (2):85–98.

Dai, X. and F. Hardisty. 2002. Conditioned and Manipulable Matrix for Visual Exploration. Pages 489–492 in *Proceedings, 2002 National Conference for Digital Government Research,* Los Angeles, California, 20-22 May 2002. <http://www.diggov.org/library/library/pdf/dai.pdf> Accessed 26 November 2007.

DiBiase D., A. M. MacEachren, J. B. Krygier and C. Reeves. 1992. Animation and the role of map design in scientific visualization. *Cartography and Geographic Information Systems* 19 (4):201–214, 265–266.

DiBiase, D., C. Reeves, A. M. MacEachren, J. Krygier, J. Sloan and M. Detweiller. 1994. Multivariate display of geographic data: applications in Earth systems science. In *Visualization in Modern Cartography,* edited by A. M. MacEachren and D. R. Fraser Taylor, 287–312. New York: Elsevier.

Dykes, J. A. 1997. Exploring spatial data representation with dynamic graphics. *Computers and Geosciences* 23:345–370.

Edsall, R. M. 2001. *Interacting with space and time: designing dynamic geo-visualization environments.* Unpublished dissertation. University Park, Penn.: Department of Geography, Pennsylvania State University.

Edsall, R. M. 2003. Design and usability of an enhanced geographic information system for exploration of multivariate health statistics. *The Professional Geographer* 55:146–160.

Edsall, R. M. 2007. Cultural factors in digital cartographic design: implications for communication to diverse users. *Cartography and Geographic Information Science* 34 (2):121–128.

Edsall, R. M. and D. Peuquet. 1997. A graphical user interface for the incorporation of time into GIS. Pages 182–189 in *Proceedings of the 1997 ACSM/ASPRS Annual Convention and Exhibition, Volume 2.* Seattle, Washington, April 1997. Bethesda, Md.: ASPRS.

Edsall, R. M. and L. R. Sidney. 2005. Applications of a cognitively informed framework for the design of interactive spatiotemporal representations. In *Exploring Geo-visualization,* edited by J. Dykes, A. MacEachren, and M.-J. Kraak 577–590. Oxford: Elsevier.

Erbacher R., G. Grinstein, H. Levkowitz, L. Masterman, R. Pickett and S. Smith. 1995. Exploratory visualization research at the University of Massachusetts at Lowell. *Computers and Graphics Journal, Special Issue on Visual Computing* 19(1):131–139.

Fisher, P., and D. Unwin, editors. 2002. *Virtual Reality in Geography.* New York: Taylor & Francis. 404 pp.

Forsberg, A., Prabhat, G. Haley, A. Bragdon, J. Levy, C. I. Fassett, D. Shean, J. W. Head III, S. Milkovich and M. Duchaineau. 2006. Adviser: immersive field work for planetary geoscientists. *IEEE Computer Graphics and Applications* 26(4):46–54.

Frank, A. 1993. The use of geographical information systems: the user interface is the system. In *Human Factors in Geographical Information Systems,* edited by D. Medyckyj-Scott and H. Hearnshaw, 15–31. Hoboken, New Jersey: Wiley & Sons.

Fredrikson, A., C. North, C. Plaisant and B. Shneiderman. 1999. Temporal, geographic and categorical aggregations viewed through coordinated displays: a case study with highway incident data. Pages 26–34 in *Proceedings of Workshop on New Paradigms in Information Visualization and Manipulation (NPIVM'99).* New York: ACM Press.

Gahegan, M., M. Takatsuka, M. Wheeler and F. Hardisty. 2002. Introducing GeoVISTA Studio: an integrated suite of visualization and computational methods for exploration and knowledge construction in geography. *Computers, Environment and Urban Systems* 26:267–292.

Guo, D. 2003. Coordinating computational and visual approaches for interactive feature selection and multivariate clustering. *Information Visualization* 2 (4):232–246.

Harrower, M. 2003. Tips for designing effective animated maps. *Cartographic Perspectives* 44:63–65.

Harrower, M. 2004. A look at the history and future of animated maps. *Cartographica* 39 (3):33–42.

Haug, D. B., A. M. MacEachren, F. P. Boscoe, D. Brown, M. Marra, C. Polsky and J. Beedasy. 1997. Implementing exploratory spatial data analysis methods for multivariate health statistics. Pages 190–198 in *GIS/LIS '97 Annual Conference and Exposition Proceedings,* Cincinnati, Ohio, October 1997.

Haug, D., A. M. MacEachren and F. Hardisty 2001. The challenge of analyzing geo-visualization tool use: taking a visual approach. Pages 3119–3128 in *Proceedings, 20th International Cartographic Conference.* Beijing, China, 6-10 August 2001. Beijing: State Bureau for Surveying and Mapping.

Hummel, J. 1996. Linked bar charts: analysing categorical data graphically. *Computational Statistics* 11 (1):23–33.

Hurley, C. and A. Buja. 1990. Analyzing high-dimensional data with motion graphics. *SIAM Journal of Statistical Computing* 11 (6):1193–1211.

Inselberg, A. 1985. The plane with parallel coordinates. *The Visual Computer* 1:69–97.

Inselberg, A. 1998. Visual data mining with parallel coordinates. *Computational Statistics* 13 (1):47–63.

Keim D., and H.-P. Kriegel. 1994. VisDB: database exploration using multidimensional visualization. *IEEE Computer Graphics and Applications* 14 (5):40–49.

Koua, E. L. and Kraak, M.-J. 2005. Evaluating Self-organizing Maps for Geo-visualization. In *Exploring Geo-visualization,* edited by J. Dykes, A. MacEachren, and M.-J. Kraak, 627–643. Oxford: Elsevier.

Langran, G. 1992. *Time in Geographic Information Systems.* London: Taylor & Francis. 180 pp.

Leung, Y. K. and M. D. Apperley. 1993. A review and taxonomy of distortion-orientation presentation techniques. *ACM Transactions on Computer-Human Interaction* 1 (2):126–160.

MacDougall, E. B. 1992. Exploratory analysis, dynamic statistical visualization, and geographic information systems. *Cartography and Geographic Information Systems* 19 (4):237–246.

MacEachren, A. M. 1995. *How Maps Work.* New York: Guilford. 513 pp.

MacEachren, A. M. 2005. Moving geo-visualization toward support for group work. In *Exploring Geo-visualization,* edited by J. Dykes, A. MacEachren, and M.-J. Kraak, 445–462. Oxford: Elsevier.

MacEachren, A. M., X. Dai, F. Hardisty, D. Guo and E. Lengerich 2003. Exploring High-D Spaces with Multiform Matricies and Small Multiples. In *Proceedings of the International Symposium on Information Visualization,* Seattle, Washington, USA, 19-21 October 2003. <http://www.geovista.psu.edu/publications/2003/MacEachren_highD_ISIV.pdf> Accessed 26 November 2007.

MacEachren, A. M., R. Edsall, D. Haug, R. Baxter, G. Otto, R. Masters, S. Fuhrmann and L. Qian. 1999. Virtual Environments for Geographic Visualization: Potential and Challenges. Pages 35–40 in *Proceedings of the ACM Workshop on New Paradigms for Information Visualization and Manipulation,* Kansas City, Kansas, November, 1999. New York: ACM Press.

MacEachren, A. M., M. Gahegan, W. Pike, I. Brewer, G. Cai, E. Lengerich and F. Hardisty. 2004. Geo-visualization for knowledge construction and decision-support. *Computer Graphics and Applications* 24 (1): 13–17.

MacEachren, A. M. and M.-J. Kraak. 1997. Exploratory cartographic visualization: advancing the agenda. *Computers and Geosciences* 23 (4): 335–344.

MacEachren, A. M., M. Wachowicz, R. Edsall, D. Haug and R. Masters. 1999. Constructing knowledge from multivariate spatiotemporal data: integrating geo-visualization (GVis) with knowledge discovery in databases (KDD). *International Journal of Geographic Information Science* 13 (4):311–334.

Marcus, A. 2001. International and Intercultural User Interfaces. In *User Interfaces For All: Concepts, Methods, and Tools,* edited by C. Stephanidis, 47–63. Mahwah, New Jersey: Lawrence Erlbaum.

Miller, H. J. and J. Han. 2001. Geographic data mining and knowledge discovery: an overview. In *Geographic Data Mining and Knowledge Discovery,* edited by H. J. Miller and J. Han, 3–32. London: Taylor & Francis.

NIST/SEMATECH 2006. NIST/SEMATECH *e-Handbook of Statistical Methods. Chapter 1: Exploratory Data Analysis,* <http://www.itl.nist.gov/div898/handbook/> Accessed 26 November 2007.

North, C. and B. Shneiderman. 1999. Snap-together visualization: coordinating multiple views to explore information. Report. College Park, Maryland: University of Maryland Human Computer Interaction Laboratory.

Oberholzer, C. and L. Hurni. 2000. Visualization of change in the interactive multimedia atlas of Switzerland. *Computers and Geosciences* 26 (1):423–435.

Peterson, M. P. 1999. Active legends for interactive cartographic animation. *International Journal of Geographic Information Science* 13 (4):375–383.

Peuquet, D. J. 1996. It's about time: a conceptual framework for the representation of temporal dynamics in geographic information systems. *Annals of the Association of American Geographers* 84 (3):441–461.

Peuquet, D. J. and L. Qian. 1996. An integrated database design for temporal GIS. Pages 1–11 in *Advances in GIS Research II: Proceedings 7th International Symposium on Spatial Data Handling,* Volume I, Session 2.

Rao, R. and S. K. Card. 1995. Exploring large tables with the table lens. Pages 403–404 in *Proceedings of ACM Conference on Human Factors in Computing Systems (CHI'95),* Denver, Colo., May 1995.

Rauschert, I., S. Fuhrmann, I. Brewer and R. Sharma. 2002. Approaching a new multimodal GIS-interface. Pages 145–148 in *Proceedings, GIScience 2002,* Boulder, Colo., September 2002.

Robinson, A. C., J. Chen, G. Lengerich, H. Meyer and A. M. MacEachren. 2005. Combining usability techniques to design geo-visualization tools for epidemiology. *Cartography and Geographic Information Science* 32 (4):243–255.

Robinson, A. H., J. L. Morrison, P. C. Muehrcke, A. J. Kimerling and S. C. Guptill. 1995. *Elements of Cartography,* 6th ed. Hoboken, New Jersey: Wiley & Sons. 688 pp.

Shen, S.-T., M. Wooley and S. Prior. 2006. Towards culture-centered design. *Interacting with Computers* 18:820–852.

Shneiderman, B. 1997. Direct manipulation for comprehensible, predictable, and controllable user interfaces. Pages 33–39 in *Proceedings of the ACM International Workshop on Intelligent User Interfaces '97*, Orlando, Florida, January 1997.

Skupin, A. and S. Fabrikant. 2003. Spatialization methods: a cartographic research agenda for non-geographic information visualization. *Cartography and Geographic Information Science* 30 (2):95–115.

Spence, R. and L. Tweedy. 1998. The attribute explorer: information synthesis via exploration. *Interacting with Computers* 11:137–146.

Stone, M., K. Fishkin and E. Bier. 1994. The movable filter as a user interface tool. Pages 306–312 in *Proceedings of the Human Factors in Computing Systems (CHI '94)*, Boston, Mass., May 1994.

Swayne, D., D. Cook and A. Buja. 1997. XGobi: Interactive dynamic data visualization in the X window system. *Journal of Computational and Graphical Statistics* 7 (1):113–130.

Thomas, J. J. and K. A. Cook, editors. 2005. *Illuminating the Path. The Research and Development Agenda for Visual Analytics.* New York: IEEE Computer Society.

Tukey, J. W. 1977. *Exploratory Data Analysis.* Reading, Mass.: Addison-Wesley. 688 pp.

Tweedie, L., R. Spence, H. Dawkes and H. Su. 1996. Externalizing abstract mathematical models. Pages 406–412 in *Proceedings of the ACM SIGCHI Conference on Human Factors in Computing Systems (CHI 96)*, Vancouver, BC, April 1996.

Unwin, A. R., and H. Hofmann. 1998. New interactive graphics tools for exploratory analysis of spatial data. In *Innovations in GIS*, Vol. 5, edited by S. Carver, 46–55. London: Taylor & Francis.

Wachowicz, M. 2002. *Object-Oriented Design for Temporal GIS.* London: Taylor & Francis. 136 pp.

Wilkinson, L. 1999. *The Grammar of Graphics.* New York: Springer-Verlag. 694 pp.

Wise, J., J. Thomas, K. Pennock, D. Lantrip, M. Pottier, A. Schur and V. Crow. 1995. Visualizing the non-visual: spatial analysis and interaction with information from text documents. Pages 51–58 in *Proceedings of IEEE 1995 Symposium on Information Visualization,* Atlanta, Georgia, October 1995. New York: IEEE.

CHAPTER 43
Real-time Visualization Techniques for Natural Hazards Assessment

John J. Kosovich, Jill J. Cress, Drew T. Probst and *Thomas P. DiNardo*

43.1 Introduction

Clear and concise visualization of data and information are important when there is a great amount of data to be interpreted in a short period of time. Consumers of information have been using graphics, images and photos as a method to make an immediate assessment of available information for various situations for a long time. The advancement of technology and the consumer need for this information in ever shorter periods of time have resulted in the blending of visualization techniques and dissemination technology to provide consumers with information to make assessments of information in near real-time. Whether the information gathered is simple or complex, the availability of the information to the customer is what makes the data valuable.

43.2 General Visualization Techniques

Hazards assessment can benefit from several different forms of visualization. The simplest form is an informational map using common data readily available from online or archive sources. Much of the data shown in the following images is publicly available online from local, state and federal sources (Figure 43-1).

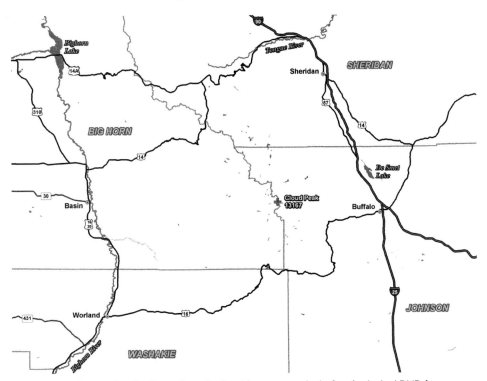

Figure 43-1 Simple visualization using planimetric map symbols. See included DVD for color version.

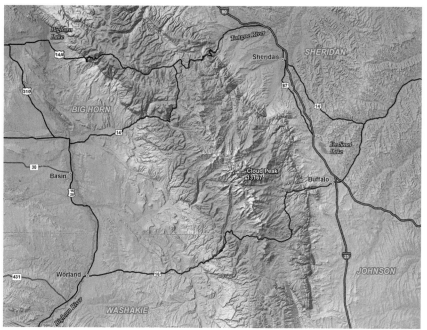

Figure 43-2 Enhanced visualization using shaded relief to show topography. See included DVD for color version.

Such maps are greatly enhanced with the addition of the vertical component of surface elevation in the form of contours or shaded relief (Figure 43-2).

Additional techniques, such as displaying the map as an oblique perspective or an anaglyph (Figure 43-3) using synthetic parallax from the surface elevation data, can be applied to further enhance the visual information provided by the data content.

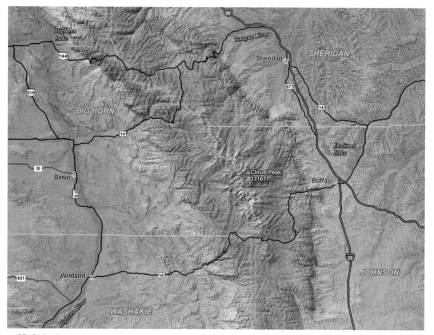

Figure 43-3 Further enhancement using anaglyphic offset (use red-blue glasses to see 3D effect). See included DVD for color version.

43.3 Combining Data

Often the most effective visualization technique is also one of the simplest: combining, or fusing (Figures 43-4 and 43-5), two or more datasets which, when separate, provide adequate but unlinked information, but when combined allow the user to see the information in a whole different light. Such techniques allow those working on hazard issues to more rapidly and completely understand the story behind the data.

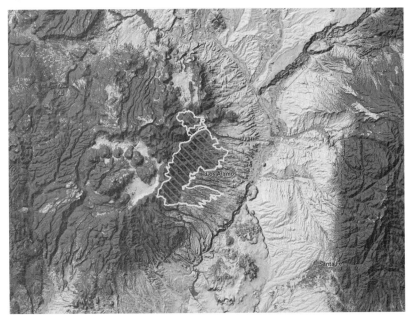

Figure 43-4 Fusion of USGS elevation and land-cover data overlain by the outline of the May, 2000, Cerro Grande, New Mexico, wildfire which threatened the town of Los Alamos, New Mexico. See included DVD for color version.

Figure 43-5 3D perspective view of the May, 2000, Cerro Grande, New Mexico, wildfire (large red area) draped on USGS elevation and land-cover data. See included DVD for color version.

Manual of Geographic Information Systems

43.4 IFSAR Example

Data are often specially acquired during or after a hazardous event. A specific example is the Interferometric Synthetic Aperture Radar (IFSAR) elevation and radar imagery data (Figure 43-6) collected several months after Colorado's Hayman wildfire, which in June 2002 burned nearly 140,000 acres and is, six years later, still the largest fire in the state's history. Visualization of the data (Figures 43-6, 43-7 and 43-8) helped affected agencies and the public to better understand and analyze the post-fire situation.

Figure 43-6 Fusion of IFSAR elevation and IFSAR ortho-imagery near the Hayman fire area, Colorado. See included DVD for color version.

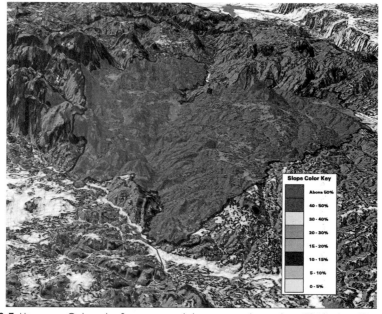

Figure 43-7 Hayman, Colorado, fire area and slope map draped on IFSAR elevation. See included DVD for color version.

Figure 43-8 Hayman, Colorado, fire area and IFSAR magnitude image, colored by elevation range, draped on IFSAR elevation, showing close proximity of the fire to metropolitan Denver area and reservoirs. See included DVD for color version.

43.5 Lidar Visualization

Identification and assessment of potential hazards, including homeland security issues, form a developing market for visualization tools and methods. Landslide, wildfire, and flood potential and mitigation studies all use digital elevation model (DEM) data to provide topographic and geomorphologic input. Newer technology known as lidar (light detection and ranging) uses laser pulses and precise Global Positioning System (GPS) and inertial measurement/navigation unit (IMU) systems to provide very accurate terrain and intensity-image data (Figure 43-9). Visual presentation of these data (Figure 43-10) to researchers, planners, and emergency responders is an important part of the process.

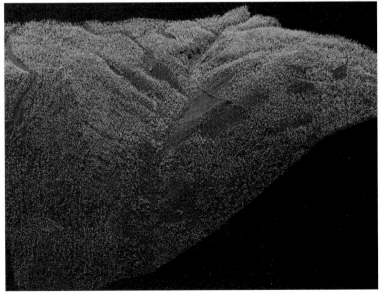

Figure 43-9 Lidar points colored by return type (brown = ground, green = trees). See included DVD for color version.

Figure 43-10 Lidar points, colored by height, over Las Vegas, Nevada. See included DVD for color version.

43.6 Data Resolution

Data resolution is important to hazards and other geo-scientific studies in which visualization is a large part. As more complex and rigorous viewing methods are applied to lower resolution data, the need for higher levels of detail in the data becomes apparent. An example is the derivation of hill slope from an elevation model. Figure 43-11 shows the real-world footprints of various cell sizes on a hilly landscape. The larger 30- and 10-meter cells of a DEM will attenuate the topography so that smaller variations are lost in the data, whereas a finer cell size will preserve more of the true surface form. Figures 43-12 and 43-13 show slope categories derived from a 10-meter and a 2-meter DEM, respectively. Researchers are more apt to glean beneficial details from the finer resolution product, although the coarser resolution data is certainly useful at a macro scale.

Figure 43-11 Visualization of DEM/imagery cell size footprint (in meters) in the field. See included DVD for color version.

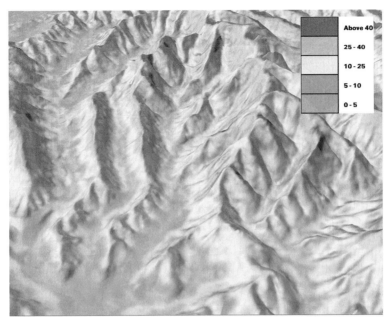

Figure 43-12 Slope derived from a 10-meter DEM. See same area in Figure 43-15. See included DVD for color version.

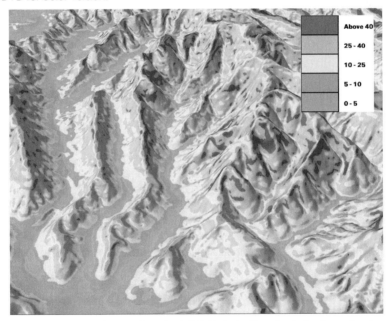

Figure 43-13 Slope derived from a 2-meter DEM. Note better discrimination of categories compared to 10-meter slope map (Figure 43-12). See included DVD for color version.

43.7 Visualizing Natural Hazards Information in the Natural Hazards Support System (NHSS)

The US Geological Survey (USGS) Natural Hazards Support System (NHSS) helps users visually monitor and analyze natural hazard events, including earthquakes, hurricanes, severe weather, floods, wildfires, and tsunamis. NHSS provides this capability by combining a wide-range of geospatial reference data with dynamic near real-time natural hazards information and presenting it visually via the internet (http://nhss.cr.usgs.gov).

NHSS allows users to visually monitor a single significant natural hazard event and all of its underlying impacts. For example, in July 2005 when Hurricane Dennis neared the Florida coast, NHSS included dynamic information on the location of the hurricane, and access to current tide conditions, stream gage readings and weather information.

NHSS also allows users to easily see the geospatial relationships of different natural hazard events in the same area, which can significantly improve analysis of potential impacts. Southern California, for instance, is periodically subject to multiple natural hazards at the same time, including gale warnings, earthquakes, and wildfires. The combination of information available at NHSS can be used to analyze the potential for mudslides in the area if the gale warning produces heavy rains over either the earthquake or wildfire locations.

NHSS visually displays the current event and provides a hyperlink to the primary agency for more detailed information for each of the following:

- Global earthquakes (USGS National Earthquake Information Center)
- Volcanoes (Global Volcanism Program)
- North American weather watches/warnings (National Oceanic and Atmospheric Administration)
- Global hurricane tracking points (National Hurricane Center)
- North American wildfires (National Interagency Fire Center)
- Global tide monitoring buoys (NOAA's National Data Buoy Center and other sources)
- Stream Gages (USGS)
- Remote Automated Weather Stations (RAWS)

In summary, NHSS provides an overview of current natural hazard information, geospatial data and detailed information directly from expert sources. This method of data presentation allows users to visually track and analyze numerous natural hazard events across the country and around the world.

43.8 Sumatra-Andaman Islands Earthquakes and Tsunami Visualization

43.8.1 Overview

On 26 December 2004, the Sumatra-Andaman Islands earthquakes generated a massive tsunami that was among the deadliest in modern history. There were a total of fifty-one earthquakes ranging between 5.2 and 9.0, reported by the US Geological Survey, occurring in the region on that day. The 9.0 earthquake was the 2nd largest ever recorded on a seismograph, and lasted for a period of nearly ten minutes. The entire Earth's surface is estimated to have moved vertically by up to 1 cm. The large quake caused the fault-line to drop 10 feet, producing a 30 meter high wave. The ensuing tsunami destroyed an enormous number of structures, and caused 240,000+ deaths. The waves even caused destruction and deaths 5,000 miles away in Port Elizabeth, South Africa. The total estimated energy of the tsunami waves equaled five megatons of TNT, more than twice the explosive energy produced during World War II, including the two atomic bombs dropped on Japan.

43.8.2 The Demand for Near-Real-Time Natural Hazards Visualization

At the time of this natural disaster, there were no Indian Ocean tsunami warning systems in place to give early warning to populated areas in the path of the waves. There was high demand for all types of visual media (Figures 43-14 through 43-17). The Rocky Mountain Geographic Science Center (RMGSC) produced and freely distributed informational posters of the disaster and created a dedicated web page on its web mapping applications website,

Figure 43-14 a) Thailand's coast before the earthquake, and b) the same area three days after the tsunami hit. Sandy beaches and green vegetation are gone. Courtesy: Space Imaging. See included DVD for color version.

displaying both 2D and 3D terrain models and animated (before and after) visualizations of the Sumatra earthquakes and tsunami. RMGSC also collected and augmented satellite imagery to produce animations, of various file sizes (for users with different bandwidth),of the Banda Aceh Province and Khao Lak regions to visually convey the amount of structural devastation produced by the waves on populated coastal areas. These animations were presented in January of 2005 to Congress by speakers from the US Department of Interior at an event for the evaluation of US Government funding programs for the relief of natural disasters.

Manual of Geographic Information Systems

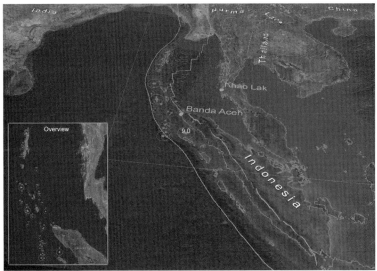

Figure 43-15 3-D terrain view of the multiple 5.2+ earthquake locations occurring on 2/26/04. See included DVD for color version.

Figure 43-16 Banda Aceh before the tsunami, observed from QuickBird. Courtesy: DigitalGlobe. See included DVD for color version.

Figure 43-17 Banda Aceh after the tsunami, also observed from QuickBird. Compare to Figure 43-16. Courtesy: DigitalGlobe. See included DVD for color version.

43.9 Wildland Fire Visualization

43.9.1 Overview

In the year 2000, more than 79,000 US fires had burned an estimated 6,838,748 acres along with hundreds of structures and valuable natural resources. Long-term weather forecasts indicated that the hot, dry conditions throughout the west would continue until fall weather brought enough rain to put out the larger fires. Across the west, priorities were set by geographic fire coordination centers for deployment of fire fighting resources based on human safety.

43.9.2 Determining Needs

Determining these priorities required more information than printed maps and situation reports could provide. Fire managers requested a real-time application that provided geospatial information on the status, location, and proximity of wildfires in relation to terrain slope, property and infrastructure.

43.9.3 Plan of Action

A coordinated multi-government-agency effort for combining information was devised, resulting in a web-based, one-stop wildland fire information viewer named GeoMAC or Geospatial Multi-Agency Coordination. The GeoMAC website application (Figure 43-18) was designed and developed by the USGS's Rocky Mountain Geographic Science Center (RMGSC). The application developers used eXtensible Markup Language (XML) to combine all the wildland fire data from the National Incident Fire Center (NIFC) and other fire reporting agencies, resulting in a visual representation of active fire locations in a dynamically driven mapping application (Figure 43-19). Now wildland fire mitigation managers

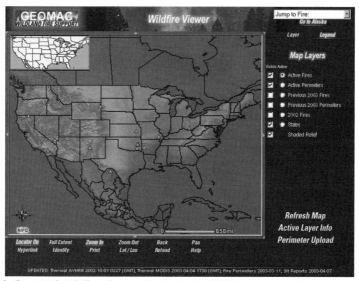

Figure 43-18 GeoMAC Wildfire Viewer. See included DVD for color version.

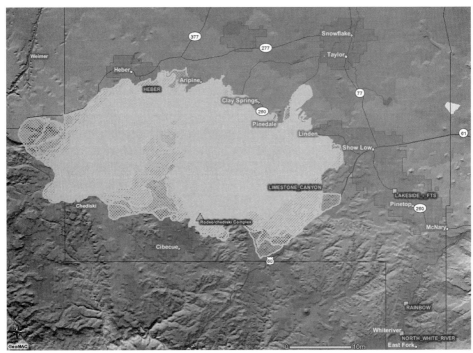

Figure 43-19 The rapid perimeter growth of the Rodeo Chediski-wildland fire impacted residential communities in Arizona are displayed in a lighter shade (pink on the DVD). See included DVD for color version.

and the public can view wildland fire locations anywhere in the conterminous US and Alaska with near real-time results, right from their web browser.

43.9.4 Other Visualization Projects

RMGSC provides wildland fire geospatial information in various forms. There is a need to know at what rate wildland fires spread and behave, so containment can be performed safely and with the least loss of life and property. RMGSC has provided wildland fire mitigation maps for FEMA, developed slide presentations for Air Force 1 Press Corps briefings, and furnished wildland fire perimeter animations for Discovery Channel High Definition television.

43.10 Future Direction

Providing clear and concise, usable information to customers in ever shorter periods of time will no longer be the exception, but will be the expectation. In emergency response, it is imperative that information be immediately accessible to allow for assessment and action for the situation. The continuing impacts of natural hazards on life, resources and property mandate quick response to minimize loss. The timing of the information is critical, as is the type of information available. The challenge for data providers will be to collect large amounts of information from a variety of sources and blend the data into a product that provides a great deal of information in a clear, usable format for the user. This will require systems that can store and process very large volumes of data, and tools and applications that will allow non-GIS experts to read and manipulate the information with relative speed and ease.

CHAPTER 44
Real-time Visualization Techniques

Brian N. Davis and *Brian G. Maddox*

44.1 Introduction

Real-time visualization implies a non-static display—interaction with data rather than passive viewing, or automated playback of animations. Animations are pre-recorded—rendered before viewing—and then played back from a single view, or on a pre-recorded path through the data or a model. Animations, therefore, are a passive experience for a user, much like watching a movie.

A real-time environment is an active experience for the user because scenes are rendered "on the fly" so the user can move and look freely within the modeled data. However, numerous obstacles must be overcome to enable a realistic "virtual environment." Some technological hurdles have been removed in the past decade, while at the same time an improved understanding of the human components of virtual environments is providing insight into the future technology developments necessary to make virtual reality a reality.

Real-time visualization techniques attempt to address the influence of human factors to allow the human brain to better process the data from virtual environments by presenting data in a visual form that the human brain has evolved to process. We are only beginning to be able to visualize data in three dimensions in "real-time" (or four dimensions when adding time-steps), if that data are streaming from the source, such as a satellite, a storage server over high-speed networks, or innumerable field-based instruments. However, progress is being made towards this goal, and real-time visualization techniques can now be applied to problems from the real-world, using environments from a virtual world.

44.2 Virtual Environments

44.2.1 History of Virtual Reality

Virtual Reality (VR) is defined as "A computer simulation of a real or imaginary system that enables a user to perform operations on the simulated system and shows the effects in real time" (American Heritage Dictionary 2006). It has emerged in recent years as one of the latest and greatest buzzwords. It's touted in the news as being a breakthrough that will allow anything from better medical treatment to training combat troops of the future. Even popular culture has gotten on the bandwagon with several movies being made that prominently feature VR as a central plotline.

VR, however, is neither a new technology nor even a new idea. VR actually became an idea back in the 1950s, an era when computer technology was still very primitive and computers the size of small buildings were less powerful than the wristwatches of today. Most computer technology during this time was used simply to perform complex mathematical calculations. They were difficult to "program," and the output devices were primitive at best.

Douglas Engelbart, however, had an idea in the late 1950s that computers could be connected to lighted screens, allowing the users to interact and actually see information displayed. At that time, however, many people ignored his ideas as the technological climate at the time saw computers as only being useful for performing calculations. Of course, computers at that time really were only capable of being "big calculators," and computer science was still in its infancy because the technology really had only gotten its start a decade earlier.

The 1960s marked a change in how computers were viewed and became the era when Engelbart's ideas were realized. Technology advanced and computers started to become smaller and more powerful. As computers became smaller, work was done on improving user interfaces. Instead of hard coding wires, operators had better and faster methods of data entry into a computer. Technology was even created to allow people to draw images directly on the screen. Ivan Sutherland created the concept of virtual reality in 1965 when he proposed what he called the "Ultimate Display" (Lanier 2001). His display would allow the user to experience a computer-generated environment that would be perceived as reality.

One of the main drivers to use computers in a visual manner was the Cold War. The military funded creation of real-time radar systems to process and display large amounts of data. ARPA, the Advanced Research Projects Agency, also came into being and funded research into projects such as computer modeling of airflow data. This decade marked large increases in funding and research into advancing technology and computer science.

The 1970s and 1980s marked an explosion in use of computers for graphical display and virtual reality, both by the military and the public sector. The military began experimenting with virtual reality to run flight simulators. Although the graphics were primitive, they were good enough to train pilots to see what it was really like flying an aircraft. The movie industry started to make use of computers for generating graphics for movies. The computer gaming industry took off and provided products ranging from flight simulators to adventure games. It was during this time period that people really began to envision using computers to recreate environments.

In the 1980s to 1990s, significant accomplishments were made in scientific visualization with computers, which was now attainable because desktop computers became much more powerful and much less expensive. This allowed more people to afford computational technology that previously was only available through super- or mini-computers. The visual display capabilities of computers of the era also increased greatly, partly due to advances in technology and partly due to the demands of the computer gaming industry. Computers enabled much more information to be produced than before, so scientists needed a way to visualize this information in ways more sophisticated than simple numbers on a printout.

From the 1990s to today, VR research has experienced continued rapid growth. Technology in general has been expanding at a rapid pace, providing better display capabilities and even better computer processing capability. Research has gone into things from head-mounted displays to rooms such as a CAVE™ (CAVE Automatic Virtual Environment) (Board of Trustees 2005) where multiple people can walk in and visualize data in three and four dimensions. VR research has also begun to incorporate ideas from fields such as human factors to better present information to a user.

Today, VR is a wide-ranging field encompassing many different ideas. VR can be something as simple as a computer video game that presents a virtual environment for someone to run through, or as complex as a goggle system that tricks both the visual and audio processing portions of the brain into thinking it is in another environment. This wide range of applications makes VR much more useful for human-computer interaction because some applications may require a fully immersive environment while others might only need to portray a three dimensional (3D) on-screen display.

44.2.2 Immersive Visualization

Immersive visualization is an application of virtual reality that seeks to take geospatial data and recreate an environment in such a way as to make the brain believe that it is actually there. This type of VR is best suited for a goggle-type setup, where the goggles track the user's head movements and the computer can change what is being displayed so that it looks natural as it moves with the user. Audio cues can also be included in this type of immersion so the audio sense can also be used to process information. Immersive VR allows users to recreate a remote environment and experience it from a distance.

There are good reasons for presenting information to the user so that it recreates the natural environment that we all experience every day. Modern technology is placing increased demands on the information processing capabilities of data users. Projects that make use of geospatial data are becoming more complex as a greater amount of information becomes available. All of this information can place such a burden on the user that understanding is impossible due to the limitations of human information processing capabilities. This is especially problematic in situations where geographic data must be analyzed. It can also be problematic for quality control tasks of geospatial datasets. Lack of understanding of the data can cause a critical reduction of knowledge based on the information.

Immersive VR is a technique that can allow people to deal with and understand data better than traditional GIS applications allow. Traditional techniques of overlaying multiple data layers and projecting the result to a 2-dimensional (2D) screen can result in a loss of perceived information. In other cases, the representation of geographic features by 2D means cannot truly show the magnitude of features in relation to each another. Geospatial data consists of imagery, elevation, vector, and other types of information. One of the faults with working in a 2D environment is not determined as much by technology as by the capabilities of the human brain. Though the brain is capable of processing feats that a computer will never match, it also has limitations that a computer does not. The problem is that the human brain has a hard time dealing with large amounts of information at the same time.

Immersive VR tries to deal with this problem by merging the data together and presenting it to the user in "true scale," where the environment created is presented as if the user is standing there. With geospatial data, this would involve imagery being draped over elevation data with some vector data overlaid when necessary. This aids the human information processing that our brains have evolved over a long period of time to be able to deal with the environment around us. An artificial 2D display forces the brain to process data differently, when multiple data layers are still treated separately. Recreating the environment, however, merges data together to be more compatible with our natural abilities, and allow humans to deal with geospatial data quickly and efficiently.

Human factors is a field that studies how humans process information and seeks to redesign things to make them easier to process. The human factors elements will be examined in more detail in Section 44.5, but generally speaking, immersive VR aids in information processing because the human brain is limited in its information processing capabilities. Short-term, or working memory, is limited by the small number of independent concepts that it can hold at any given time. With traditional GIS applications, each data layer is dealt with separately and from an artificial point of view. These layers can actually exceed the number of concepts that working memory is capable of handling. By combining them together and presenting them to the user in a means that mimics the natural environment, the number of independent concepts is actually small enough to fit into working memory at the same time. This aids greatly in information processing because the user can deal with more data simultaneously and can better understand the information.

There are problems, however, with converting geospatial data to recreate the natural environment. Most of the data is fairly coarse. For example, the majority of USGS Digital Elevation Models (DEMs) have a resolution of 30 meters per pixel. This means that each pixel represents a 30×30-meter spot on the ground. While this resolution is fine for on-screen 2D display, it is almost unusable for immersive VR as 30×30-meter data would create a very blocky elevated environment. Some work has been done in trying to interpolate this data to a finer resolution but has not been successful to date because the difference between needed and available resolutions is too great. In the case of DEM data, interpolation methods such as Fourier Transforms can be used, but the overall environment would then look too smooth.

Another problem is with imagery. For example, much of USGS Digital Orthophoto (DOQ) data is gray-scale. The remainder is either false color or infrared. Any of these color types make the environment look unusual as the "ground" does not look as people expect. This type of imagery is easier to interpolate, but ends up looking washed out when interpolated to the high resolution necessary for immersive VR. High-resolution orthoimagery can help slightly, but still poses a problem because detailed features such as vehicles would look fuzzy to the viewer.

There is a difference here between immersing someone in an environment and how GIS applications treat 3D data today. In these applications, it is common to display imagery and elevation data in a false perspective view. In this type of display, the plane of display is angled and the "back" of the data is shrunk while the front is enlarged in order to display depth. Figure 44-1 illustrates this type of visualization.

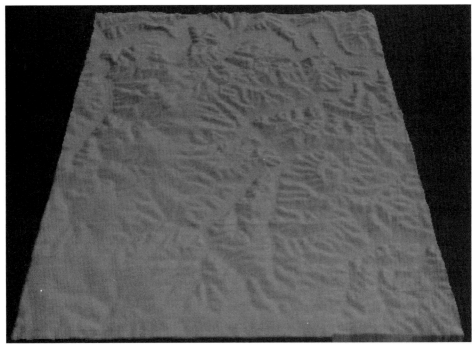

Figure 44-1 Typical perspective view of elevation data (visualization by Brian Maddox). See included DVD for color version.

This type of view is often given a vertical exaggeration so that differences in elevation are more visible. However, as can be seen from Figure 44-1, this is still a display that puts the viewer at a virtual distance from the data. Subtle details of the data are not visible from this type of view, and once multiple data layers are draped over it, visibility will be even less effective.

In addition to imagery, audio cues can also be used in the presentation of geospatial data in VR. As an example, a tone could be modulated with elevation based on the user's current virtual location. Audio could also be used to indicate when the user moves over a specific feature such as a road. Audio signals are useful because the human brain has a separate processing center for audio. Audio can therefore be used as another cue in allowing the user to experience and interact with the data being presented.

There are several ways for a user to interact with such an environment. The traditional keyboard and mouse model does not work well in this situation as it takes away from the "immersiveness" of the environment and introduces distractions. One of the longest-researched methods for interacting with a virtual environment is through the use of special sensor gloves. These gloves contain sensors that can track how the fingers are bent, motion, position, and direction of movement. Integrating this tracking into the virtual world allows a user to do things such as point in a direction of desired movement or even pick up and examine an object in the VR environment. This presents the user with a natural way to interact and does not require a distracting switch to a mouse-keyboard combination.

44.3 Evolution of Virtual Reality and Immersive Environments

Most human understanding is attained through visual interpretation of our surroundings. The CAVE technology described above, initiated at the University of Illinois at Chicago, was developed to implement virtual reality environments attempting to mimic the inputs to human visual analysis, interpretation, and perception. "CAVE," the name selected for the virtual reality technology, is both a recursive acronym (Cave Automatic Virtual Environment) and a reference to "The Simile of the Cave" found in Plato's Republic, in which the philosopher explores the ideas of perception, reality, and illusion. Plato used the analogy of a person facing the back of a cave alive with shadows that provide the only basis for ideas of what real objects are (Cruz-Neira et al. 1992). CAVEs were developed for not only displaying data in 3D, but also for analyzing and interacting with the data in real time. The leading edge technology required, such as head-tracking devices, sensor gloves, and haptic feedback mechanisms, were becoming available, but costs were prohibitive. Descendants of the CAVE, such as Immersadesk, attempted to reduce some of these limitations, but for an audience of only one or two, and the presentation of only one wall of a CAVE, the complexity and cost were still concerns.

Another approach attempted to replace supercomputers from Silicon Graphics Inc. (SGI), expensive projectors and room-sized venues with desktop SGI systems using the same shuttered goggles with a double refresh-rate monitor, combined with an RF transmitter synching the goggles and the monitor. Though a lower-cost solution, this was still expensive for the limited budgets of universities and smaller labs. Also, this approach limited audience size, and was also limited to essentially the same core applications developed for larger SGI systems. These systems provided very limited GIS capability—the ability to interactively add layers, geo-reference information and real-time data. In addition, applications were not portable between differing operating systems, and therefore limited collaboration. However, even though the cost or complexity could still be barriers, technologies such as these, the Immersadesk, and CAVEs helped to continue the momentum of some real-time visualization aspects. The "tech" was cool, demonstrated potential, and more people wanted it.

44.3.1 The GeoWall Consortium

More-recent technology developments have introduced an entry point into VR that does not require leading-edge, expensive technology. The GeoWall Consortium (Leigh et al. 2001) was founded to promote the use of low-cost, commodity hardware and software components for an adequate virtual reality experience at a fraction of the cost. The consortium formed in a spirit of open software, data, and information to provide a cost-effective way to broaden participation in the visualization community and develop new stereo and 3D data interaction (Steinwand and Davis 2003). The prototype system, coined "GeoWall," though only mimicking one wall of a CAVE, and therefore only partially immersive, enabled room-sized venues, and interactive collaboration, and made applications hardware and software platform-independent.

This technology transitioned approaches to VR technology from "build it and they will come," to "make it cheap and easy to build, and anyone can afford the financial and time investments necessary to determine a way to make use of it in their own domain."

Such visualization systems have remarkable potential for any educational or scientific discipline that deals with complex spatial relationships. Initially, these systems can increase the ability to visualize and interpret spatial relationships of maps and images, and ultimately the ability to interactively manipulate complex 3D, and 4-dimensional (4D) time-dependent mathematical models—all at an order of magnitude less cost than previous generation technologies.

Beyond the traditional domains of scientific visualization, such as astronomy and physics, now disciplines traditionally averse to technology such as archeology can afford and make use of 3D visualization. For example, a GeoWall system can be used to "virtually" look at dig sites in the classroom, before actually beginning the dig, thereby getting a better idea how to approach the excavation of the site with reduced possibility of damaging priceless artifacts.

Even farther away from the technology mainstream are fields such as stereo photography, which has been around since the 1870s, but until now did not widely make use of computer technology. Now a GeoWall system can allow a history department professor to more accurately analyze the content and cultural significance of historical stereo photography (Kolbe and Davis 2003).

Several factors have contributed to the genesis and appeal of GeoWall systems.

44.3.1.1 Commodity Computers

The computers used can now be ordinary desktop Personal Computers (PCs) running the Windows® operating system, or MAC OS/X℠, or Linux®. Whatever operating system and computer you use to do your GIS work, other desktop applications, or just office automation work could now be used as a 3D display system. It doesn't matter whether your platform is Intel®, or Apple®, or AMD® (Advanced Micro Devices).

44.3.1.2 Commodity Graphics Cards

As graphics cards and the chip technology they are built upon became more powerful and less expensive, 3D Open Graphics Library (OpenGL) applications performance requirements could for the first time be satisfied with a desktop computer containing a commodity graphics card. Because the chipsets of these cards used the same technology as the exploding personal video game industry, performance of these chips and cards accelerated rapidly while prices plummeted at a similar rate.

44.3.1.3 Linux®

The first GeoWall application was the result of porting the CAVE library of software from proprietary SGI platforms to a platform-independent version for PCs running the freely available Linux® operating system. As soon as "CAVE" software was free, and the hardware required to run it was low-cost, or potentially already on anyone's desktop, the time and money of a shoestring budget could be spent on applications development, instead of expensive hardware.

44.3.1.4 OS X

All this was happening about the time that Apple® was transitioning from OS 9 to OS X, which is based on the FreeBSD version of Linux®. Therefore, OS X ports and development were fairly straightforward, and followed soon after the generic Linux® port. This made a wide variety of visualization applications available to the portion of the academic and research community that preferred Apple's Macintosh.

44.3.1.5 Windows®

As soon as Windows® developers ported applications to the Microsoft operating system, using the same low-cost commodity hardware components, anyone that owned a Windows® computer could now do 3D visualization. Initial caution to acceptance of GeoWall technology was mostly due to users without an IT support staff, or one hesitant to absorb the learning curve and support of Linux®. As soon as these users could employ the computers already supported through their normal mode of operation, GeoWalls became widely attractive.

44.3.1.6 Applications Software

Data could now be viewed in stereo/3D in native formats with multiple software applications, or as before, in native/proprietary formats, if desired. Now users and consumers have multiple options.

44.3.1.7 Commercial GIS Software

Commercial software vendors were motivated to provide stereo 3D compatibility, after witnessing the momentum of the open-source aspect of GeoWall technology, and deducing the potential. Several GIS applications are now compatible with low-cost GeoWall systems, representing data in stereo 3D, no longer just 2D representations of perspective views. Although not totally immersive like the parent CAVE technology, GeoWall implementations of stereo viewing software are dramatically better than 2D visualization which was being labeled and marketed as 3D. These GIS applications, allow real-time interaction and data integration.

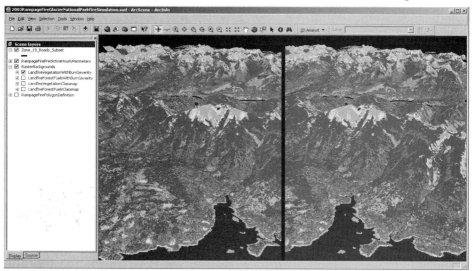

Figure 44-2 Portrayal of the 4D progression of a predictive forest-fire model simulation with background high-level classification maps draped on topography, and integrated with other GIS layers, all of which are viewable in stereo (visualization by Brian Davis). See included DVD for color version.

As the technology became less expensive, yet interchangeable, installations of GeoWall systems in university labs sky-rocketed after 2002 (Johnson et al. 2003). Instead of a high-powered, well-connected lab with unlimited funding, all one needed to "do 3D" was a lab of any kind. Some of the components were probably already on hand. The only additions needed were perhaps a new graphics card, a projector or two, some paper-framed glasses, and maybe some super glue or duct tape.

These low-cost, commodity component systems for 3D visualization were perhaps most impressive to those technologists responsible for developing and using the parent CAVE technology. Experts were impressed with the impact achieved with such a minimal financial investment. This contrasted the slightly greater impact created by CAVE technology for an order of magnitude more expense. GeoWall technology is now advancing VR to the point of enabling 4D GIS for the general public (Figure 44-2).

These improved, more realistic representations enable additional capability to interact with data. Volume rendering, 3D display, and GIS software have existed for more than a decade, but for the first time these capabilities can all be integrated into a single, low-cost, familiar, desktop environment (Davis 2004a and Davis 2004b). Earth scientists require interactive access to elevation, satellite, and a myriad of other data sources. A GeoWall system running a GIS-like application in 3D and 4D now allows a scientist to analyze these data interactively. In the spirit of openness advocated by the GeoWall Consortium, GIS users are now motivated and encouraged to use this technology in ways not conceived by the founders of the consortium. That is the point. 3D display technology is now affordable, so uses never before practical or possible now are.

44.4 Future Visualization Technology Research

While systems like GeoWall are pushing VR technology downward to lower costs and outward toward more GIS users, research into new and exciting visualization technology is still continuing at the high-end. Such research will investigate the next generation of visualization technology needed to address new and daunting questions. For example, what if your visualization problem is not inherently a 3D problem, but a large display real estate problem? What if you just have more data than you are able to view all at once, but you would sure like to?

Scalable display systems will be required to enable the visualization and analysis of the large volumes of data from emerging data sources such as satellites, airborne sensors, and numerous remote sensing devices that provide live feeds of data. These data source technologies are being deployed and producing data at unprecedented rates. Continuing the trend to take advantage of the commodity computing and related technologies, scalable display system will serve geoscience applications that require greater display resolution and visualization capacity. A so-called "GeoWall-2" display system consists of Liquid-Crystal Display (LCD) panels tiled in an array, each driven by a single node of a cluster computer. These display systems are scalable in that smaller or larger versions can be built by adjusting the number of LCDs and computers. Applications of the GeoWall-2 include the visualization of large remote sensing data, volume rendering imagery, mapping, seismic interpretation, museum exhibits and any applications that require a large collaborative screen area (Leigh et al. 2003). These future applications will create the capability to interactively roam through terabytes of distributed, georeferenced data ranging from satellite imagery to aerial photography at detailed resolutions (Krishnaprasad et al. 2003).

The CAVE was instrumental in transforming the VR world from using heavy head-mounted displays to room-sized projection-based displays. Since then, over a hundred CAVE and similar devices have been deployed around the world. More recently, GeoWalls leveraged the increasing capabilities of commodity components to lower the cost and increase the availability of visualization tools, leading to the building and deployment of GeoWall systems too numerous to count. Now, continued leveraging of trends in commodity computing, graphics, networking, and display technology are leading to the development of a new generation of network-enabled, scalable visualization and collaboration systems that will remove the barriers of distance. These systems are proving useful in application areas such as geoscience, bioinformatics and homeland security (Leigh et al. 2004).

Future visualization systems will allow interactive network access, data interaction, and visual analysis via optical-switched networks such as LambdaRail (Smarr 2003). This capability will address the visualization problem of always needing more: access to more data than will fit into the memory of a desktop computer, or a server, or a cluster of computers; and more data access that is faster, better, cheaper, and interactive. The data volumes of this class of visualization application preclude any viewer or user from replicating the source data bases. These huge storage requirements cannot be efficiently duplicated. Ultimately, users just want to query and view data interactively to deduce results. They are not interested in building the infrastructure necessary to create a duplicate data archive.

The OptIPuter project is a National Science Foundation (NSF) funded high-speed networking project to interconnect distributed storage, computing and visualization resources using photonic networks in an attempt to do just that—enable interactive analysis of larger than ever data volumes. The main goal of the project is to exploit the trend that network capacity is increasing at a rate far exceeding processor speed gains, while at the same time plummeting in cost. The objective is high-speed, zero-latency visualization, i.e., no visible lag between a request for data and its display. For example, not only do users desire access to vast data stores, but some applications need to be continually supplied new data to be meaningful. Another example is applications that must in real-time visualize live feeds of data from sensors, cameras and satellites. Zero latency in data delivery will be required, possibly spanning global distances, and at high bandwidths, when the data requirements are vast and the display technology is scalable to the application (Brown 2003). These technologies and applications may not appear on the desktop anytime soon. However, remember CAVEs were introduced almost ten years before the more affordable GeoWall systems.

Figure 44-3 One-foot resolution aerial photography of New Orleans after the 2005 Hurricane Katrina flooding, displayed on an NSF OptIPuter LambaVision 105.6 mega-pixel display (photograph by Luc Renambot). See included DVD for color version.

44.5 Human Factors

Some questions may arise with the various virtual reality systems presented here, such as "Why go through all the trouble?" and "Isn't an on-screen display enough?" The answer to these questions involves an understanding of human factors and how the human brain processes information.

The human factors field has included many studies on the capabilities of the human brain. Some of these studies have dealt with the amount of information that the brain is able to simultaneously process. What these studies have found is that the human brain can only handle a very small amount of information at any given time. Studies conducted by Mowbray and Gebhard (1961) have found that people are limited in their abilities to process items such as different audible tones. They found people were only able to identify five tones of different pitches if they had to hear them independently of the others (Sanders and McCormick 1993:54). When testing on a relative basis—making comparisons—the test subjects were able to identify around 1800 tone pairs. This is an important part of information processing, as humans need to be able to compare and contrast different pieces of information with each other.

These studies have found the limitations are not caused by any problems with the body's sensory organs. In fact, organs such as the eyes and the ears are quite capable of overloading the brain with sensory information. The problem has also been determined not to be the brain's ability to receive this information and act on it. The problem is with human memory and its relationship to how the brain processes information. Specifically, the problem lies with the limitations of the type of memory known as short-term or working memory.

The human brain contains three types of memory: sensory storage; working; and long term (Sanders and McCormick 1993, page 65). Sensory storage briefly retains the input from a sensory organ just long enough for it to be processed. Working memory is where the brain stores things that it is currently processing. Working memory is also a temporary storage area, although it can hold items much longer than sensory memory can. Long-term memory is where the brain will permanently store things for later use.

In order for the brain to perform any type of processing, the items to process must make their way into working memory. This can take place either by copying from sensory or from long-term memory. Information stored in working memory can be encoded via visual, phonetic, or semantic methods (Sanders and McCormick 1993, page 66). Once in working memory, the brain can then manipulate the information and send it back to long-term memory when necessary. If the brain must keep information in working memory for an extended period of time, it must focus attention on the concepts and keep them active to avoid decay. This process is also known as rehearsal (Sanders and McCormick 1993, page 67).

Working memory's problem is that it can only store a finite number of independent concepts at any given time. In 1956, Dr. George A. Miller found that on average, human working memory could only store five to nine independent concepts at any time. This is known in the human factors field as the "magical 7±2 number" (Miller 1956). The problem this creates is that for any new concept to be processed by the brain, it must enter into working memory by displacing something that is already there. Combine this problem with the working memory's short storage time and one can begin to understand the problems inherent with human information processing.

While it may appear miraculous that humans can perform complex tasks at all, there are ways that the brain can get around its memory limitations. One of the ways that the brain can manipulate larger amounts of information is to group them together into what Miller (1956) called a chunk. For example, consider a string of letters DOGBIRDTREE. By themselves, it could be hard to remember long strings such as this. But, by grouping it into parts such as DOG.BIRD.TREE, the brain is able to manipulate more information through the use of chunking. Chunking recodes the information into a different form by placing more bits into each unit (Miller 1956).

Recoding information into a different, more easily processed form is also an important part of human information processing. The human brain has evolved over thousands years to process some types of information more efficiently than others. We live in a 3D world that is alive with many different colors and is filled with many different sounds. The human brain has evolved accordingly, and has developed sophisticated techniques for processing this type of information, such as excellent depth perception of visual and auditory information. In the above DOGBIRDTREE example, when the phrase is broken up, it is not stored as the actual letters D.O.G. Instead, it is encoded either visually or phonetically into the image of a dog or the sound that the word dog makes when spoken aloud (Sanders and McCormick 1993, page 66).

Recoding and chunking are used to process information about our environment. When the brain tries to identify something in the environment, it does not compare it as a whole object. Instead, it breaks the object up into different parts and uses those to identify the object. Consider the example of identifying the face of a friend. The face as a whole is not stored in working memory and acted upon. The face as an entity is composed of many different components such as the eyes, nose, ears, and color intensities. This is too complex an object to store as a single concept in working memory. Instead, the face is recoded and broken down into components in working memory. These components are then searched in a database-like manner to determine to whom the face belongs (Hancock et al. 1996). In fact, research into the Principal Components Analysis (PCA) method has found that depth is an important component in feature identification.

Studies have shown that people can better process 3D information when actually presented on a 3D display than when projected onto a 2D plane. Ellis et al. (1987) performed tests on pilots to judge the differences between displaying aircraft positional information in a 3D fashion and in the traditional projected 2D form. The goal was to test how quickly and in what manner pilots would respond between the differences in display methods. What they found was that with the 3D displays, the pilots on average had a three to six second faster decision time than with the 2D displays (Sanders and McCormick 1993, page 154). Because people live in 3D environments, the brain is better able to process positional information when it is presented in a 3D manner. In the case of the pilots, those using the flat 2D displays would make mostly banking maneuvers because the displays left out the Z, or altitude, component of the positional information and treated the world as a flat X-Y plane. Those pilots using the 3D displays were able to execute changes in altitude in addition to left/right banking.

The implications for geographic information—positional information on the Earth's surface—become obvious. As the above study demonstrated, when positional information is presented in a more natural 3D form, the brain can better process it because it more closely matches what people experience on an everyday basis. Taking geographic information and projecting it to a 2D plane forces the brain to deal with something familiar in an unfamiliar manner. The slower response times of the pilots given 2D environments demonstrate the brain does have difficulty dealing with positional information when in this form. Some of the causes include the brain trying to recode the data but lacking the information to do so, or that it could be trying to break the data into chunks and has trouble dealing with the unfamiliar representation.

Projecting information to a 2D plane is accomplished through discarding some of the positional information. In the case of the Ellis et al. (1987) study, as it is with current geographic information systems, the height is discarded. Reducing the amount of information available also reduces the brain's understanding of the information, which leads to a reduction in the number of decisions that can be made. The pilots who had 2D displays made decisions that focused on moving left or right since their brains were not considering height. The pilots with the 3D displays were able to process the information in a more natural manner and had the additional information of height, which led them to consider changes in altitude.

These studies can be compared to how geographic information systems operate. Layers of data are frequently stacked on top of each other. These layers can also be combined with elevation data and imagery for potential interaction. Traditional methods of manipulating the information are through means of an overhead view, where everything is projected to a 2D plane. Elevation data is typically shaded with intensity values based on height to give the appearance of depth. As was the case with the pilots, this type of view is unnatural to humans, and can result in substantial loss of data interpretation. This can affect the knowledge that can be learned from this information, as was shown in the Ellis et al. (1987) study. Quality assurance methods can also be limited because people may be unable to accurately relate geographic features to each other due to the data loss. This may cause errors in data to go unnoticed in a quality check.

Modern advances in the computer sciences and computer technology can finally tie all of this together to allow users to process larger amounts of geospatial information. The goal of VR is to allow a user to experience telepresence, which is "the experience of presence in an environment by means of a communication medium" (Steuer 1992). Understanding human perception and how technology can be applied to improve it can enable us to take the large amounts of geospatial data being created and allow users to actually navigate, manipulate and understand it.

Other scientific fields have been using these virtual environments with great success for several years now. The medical community, for example, has been using VR to visualize the vast amount of information that is produced by systems such as magnetic resonance imaging (MRI) machines. These types of systems generally produce highly detailed 3D information about areas of the body. Neurosurgery has been making use of 3D environmental interaction in the planning of medical procedures, and has been identified by the NSF OptIPuter project as a field to address with improved VR technology. Neurosurgeons work and think in 3D terms, and providing the imaging information in simulated 3D form allows them to see the data in those same terms (Hinckley et. al 2001). Now, the geosciences are beginning to use virtual GIS environments to aid in interpreting and understanding natural processes. Geology departments at leading universities have employed 3D display systems to improve by 20% undergraduate students' performance in geology field exercises, with relatively little previous field experience (Kelly and Riggs 2005). Studies are underway to determine the effectiveness of visualization systems, and how much they can advance the introduction of more complex geologic concepts. University astronomy departments are measuring the effectiveness of 3D display systems on undergraduate students' ability to understand the origins of lunar phases (Turner et al. 2004), and other physical phenomena requiring inherently 3D spatial reasoning skills.

44.6 Conclusion

For the first time in the history of Virtual Reality, we are starting to see measurable results demonstrating that VR and immersive environments can aid our understanding of actual environments. Continuing research to improve the technology of virtual environments, combined with a continued improvement in the understanding of the human factors that influence our interpretation of virtual environments, will not only continue to make our virtual worlds better: applying real-time visualization techniques to real-world scientific problems will help to make our real world a better place to live.

References

American Heritage® Dictionary of the English Language, fourth edition. 2006. Boston: Houghton Mifflin Company. <http://dictionary.reference.com/browse/virtual reality> Accessed 30 November 2007.

Board of Trustees of the University of Illinois, The. 2005. Virtual Reality: History. <http://archive.ncsa.uiuc.edu/Cyberia/VETopLevels/VR.History.html> Accessed 27 November 2007.

Brown, M., guest editor. 2003. Blueprint for the future of high-performance networking. *Communications of the Association of Computing Machinery* 46 (11):30–77.

Cruz-Neira, C., D. Sandin, T. DeFanti, R. Kenyon and J. Hart. 1992. The CAVE: audio visual experience automatic virtual environment. *Communications of the ACM* 35 (6):65–72.

Davis, B. 2004a. Virtual Reality Meets GIS: 3D on the Wall. *ArcNews* 26 (2):33. <http://www.esri.com/news/arcnews/summer04articles/virtual-reality.html> Accessed 27 November 2007.

———, 2004b. Affordable System for Viewing Spatial Data in Stereo. ArcUser 7 (3):48–49. <http://www.esri.com/news/arcuser/0704/files/geowall.pdf> Accessed 27 November 2007.

Ellis, S. R., M. W. McGreevy and R. J. Hitchcock. 1987. Perspective traffic display format and airline pilot traffic avoidance. *Human Factors* 29 (4):371–382

Hancock, P., A. Burton and V. Bruce. 1996. Face processing: human perception and principal components analysis. *Memory and Cognition* 24:26–40.

Hinckley, K., R. Pausch, J. Goble and N. Kassell. 2001. A Three Dimensional User Interface for Neurosurgical Visualization. <http://www.cs.cmu.edu/~stage3/publications/94/conferences/MedImg/paper.html>   Accessed 30 November 2007.

Johnson, A., P. Morin, P. van Keken. 2003. The GeoWall in the Earth Sciences Classroom, Special Session of the *American Geophysical Union 2003 Fall Meeting*, 8-12 December 2003, unpaginated CD-ROM.

Kelly, M. and N. Riggs. 2006. Use of a virtual environment in the GeoWall to increase student confidence and performance during field mapping. *Journal of Geoscience Education* 54 (2):158–164.

Kolbe, R. and B. Davis. 2003. The Digital Landscape of Lewis and Clark: Using Low-cost Digital 3D Technology for Interactive Display of Current and Historical Landscapes. *Thirty-fourth Annual Dakota Conference on Northern Plains History, Literature, Art, and Archaeology – The Lewis and Clark Expedition: Then and Now*, 30-31 May 2003, Sioux Falls, South Dakota, p. 61.

Krishnaprasad, N., V. Vishwanath, S. Venkataraman, A. Rao, L. Renambot, J. Leigh, A. Johnson and B. Davis. 2004. JuxtaView – A tool for interactive visualization of large imagery on scalable tiled displays. In unpaginated proceedings of the 2004 IEEE International Conference on Cluster Computing, San Diego, 20-23 September 2004. CD-ROM.

Lanier, J. 2001. Virtually there. *Scientific American* April 2001.

Leigh, J., P. Morin, P. van Keken. Undated. The Geowall Consortium. <http://GeoWall.org> Accessed 27 November 2007.

Leigh, J., P. Morin, A. Johnson, T. DeFanti, M. Brown, D. Sandin, F. Rack, F. Vernon, J. Orcutt, B. Davis, P. van Keken and L. Smarr. 2003. GeoWall-2: a scalable display system for the geosciences. In unpaginated proceedings of American Geophysical Union 2003 Fall Meeting, San Francisco, 8-12 December 2003. CD-ROM.

Leigh, J., L. Renambot, A. Johnson, M. Brown, D. Sandin, T. DeFanti, M. Ellisman, J. Orcutt, L. Smarr, B. Davis, P. Morin, E. Ito and F. Rack. 2004. Challenges in ultra-high-resolution visualization and collaboration. In unpaginated proceedings of *U.S. Display Consortium High Information Content Displays Symposium*, Arlington, Va., 14-15 November 2004. CD-ROM.

Miller, G. 1956. The magical number seven, plus or minus two: some limits on our capacity for processing information. *The Psychological Review* 63:81–97.

Mowbray, G. H. and J. W. Gebhard. 1961. Man's senses as informational channels. In Human Factors in the Design and Use of Control Systems, edited by H. W. Sinaiko, 115–149. New York: Dover.

Sanders, M. and E. McCormick. 1993. Information input. In *Human Factors in Engineering and Design*, 7th ed. McGraw-Hill, Inc., 54–154.

Smarr, L. 2003. OptIPuter. University of California, San Diego, San Diego. <http://OptIPuter.net> Accessed 30 November 2007.

Steinwand, D. and B. Davis. 2003. GeoWall: Investigations into Low-Cost Stereo Display Technologies. *USGS Open-File Report 03-198*, 9 September 2003, pp. 1-26.

Steuer, J. 1992. Defining virtual reality: dimensions determining telepresence. *Journal of Communications* 42 (4):73–93.

Turner, N. E., R. E. Lopez, D. S. Corralez, C. L. Gray and E. J. Mitchell. 2004. Effectiveness of GeoWall Technology in Conceptualizing Lunar Phases. In *Cosmos in the Classroom 2004: A Hands-on Symposium on Teaching Astronomy*, held at Tufts University, Massachusetts, July 2004. Edited by A. Fraknoi and W. Waller. San Francisco: Astronomical Society of the Pacific.

CHAPTER 45

Virtual GIS:
Efficient Presentation of 3D City Models

Norbert Haala and *Martin Kada*

45.1 Introduction

While the use of GIS was originally restricted to the processing, analysis and presentation of two-dimensional (2D) geospatial data, 2.5-dimensional (2.5D) data such as Digital Terrain Models (DTMs) soon became available. Meanwhile, especially for applications in urban areas, complex three-dimensional (3D) objects, including buildings, roads and bridge structures, also were integrated and used mainly for various presentation purposes. The provision of such visualization components within 3D virtual reality GIS was facilitated considerably by the rapid developments in the field of computer graphics. Tasks like the real-time visualization of complex 3D scenes can meanwhile be realized by standard hardware and software components. As a result, components for the presentation of structured 3D geodata are provided for an increasing number of applications. Photorealistic visualization of urban environments is, therefore, available in the context of urban planning, tourism and personal navigation systems, as well as entertainment such as games based on real locations.

In order to generate realistic visualizations of urban landscapes, suitable 3D city models have to be supplied. These data sets represent the relief of a city with a DTM and the sites with 3D building models as the main objects. Frequently, traffic infrastructure of street and railway networks, including tunnels and bridges, also are indicated. Water bodies, vegetation structures, and so-called city furniture, including street and traffic lights or benches, are other GIS objects of interest. In addition to the provision of tools for efficient data collection, the need for standardization has to be met. By these means, interoperability and integration of 3D city models from different data sources are feasible. For this purpose, the open data model, CityGML, was developed by a consortium of several dozen German public and private entities (Kolbe and Gröger 2005). In this data model, the different levels of detail of city models can be represented by four main categories. According to this categorization, LOD 0 (level of detail) relates to a regional model, which is given by a 2.5D DTM, and LOD 1 refers to a city or site model defined by a block model for the buildings without roof structures. In LOD 2 buildings are already available with differentiated and textured roof structures. Detailed architecture models have been generated in LOD 3, while the interior of the building is integrated to generate walkable models (i.e., models the user can virtually walk through) in LOD 4. This definition conforms to the amount of detail that is required for the different visualization categories in urban planning. These categories are separated into visualizations of a complete city by a bird's eye view at a height of the virtual observer more than 200 m above the scene, presentations of smaller districts at oblique views and distances from 200 m to 5 m, and pedestrian perspectives for the visualization of places and single buildings (Danahy 1999).

While 3D virtual city models have become a standard product integrated in a growing number of applications, their efficient collection is also feasible by a number of algorithms. A good overview on the state-of-the-art is given in Baltsavias et al. (2001). Usually the area covering sets of 3D building models is collected by photogrammetric 3D measurement from airborne stereo imagery or lidar. By these means the footprints and the roof shapes of all buildings can be supplied at sufficient detail and accuracy and are then used to generate the required 3D building models in a subsequent step. However, the viewpoint restrictions of airborne platforms frequently limit the amount of detail that can be made available for the façades of the buildings. For this reason visualizations of building models collected from airborne data are mainly useful for overviews from elevated viewpoints. This is demonstrated in Figure 45-1 (top). In contrast, a number of applications in the context of urban planning or 3D navigation require visualizations with a very high degree of realism for pedestrian or only moderately elevated viewpoints. An example for this type of visualization is depicted in the bottom left and right of Figure 45-1.

Figure 45-1 Different levels of detail for visualization of urban landscapes. See included DVD for color version.

As demonstrated in Figure 45-1, different visualization scenarios require the modification and further processing of the basic 3D city models as provided from airborne data collection. In order to achieve a sufficient degree of realism for pedestrian viewpoints, as required for a number of applications in the context of urban planning or 3D navigation, data from terrestrial platforms have to be integrated. One example is image texture, which is frequently mapped against the façades of the buildings to improve their visual appearance. This can be realized efficiently by programmable graphics hardware (Section 45.2.1). Techniques originally developed within computer graphics also can be used for geometric refinement of building faces based on data from terrestrial laser scanning (Section 45.2.2).

While realistic visualizations from pedestrian perspectives require a refinement of the original building models, their simplification can be necessary for presentation at small scales. Originally such a generalization of city models was used to reduce the amount of data to be processed during real-time visualization of large urban areas. Even more important, simplified 3D building models also can support the visual interpretability of the generated urban scenes. For example, in navigation applications important landmark buildings can be emphasized and presented with a great amount of geometric detail, while unimportant buildings may be considerably simplified. This is especially important for mobile applications to be realized by handheld devices with small displays. Such a generalization process of building models will be discussed in Section 45.3.

45.2 Improving the Virtual Realism of Building Façades

The integration of image information during surface rendering can help to improve the visual realism of objects that are represented by relatively simple geometric models. For this reason, either artificial texture or real-world imagery is frequently mapped to the building façades and roofs of virtual city models. Despite techniques such as synthetic image generation based on grammars, which have been proposed for procedural façade texturing (Wonka et al. 2003), most applications aiming for virtual realism are based on real-world imagery. From these images, the façade textures are extracted and associated with the visible object surfaces. If the interior and exterior orientations of the respective images are available, the texture can be directly mapped to the corresponding surface patches of the georeferenced 3D building models. This process of texture extraction and placement can be realized very efficiently by making use of the functionality of 3D graphics hardware, as discussed in Section 45.2.1. Within this step tasks such as the elimination of image distortions or the integration of multiple images can be solved on-the-fly.

Since texture from real-world imagery modifies the appearance of the rendered surfaces, it can substitute for geometric modeling, at least to a certain degree. However, even though real-world imagery implies a detailed representation, the object geometry is still defined by the original 3D model. Thus, the façades of the buildings are represented by planar surfaces, as they were captured originally from airborne data. While this approximation is acceptable for visualizations from orthogonal views, misplaced protrusions will disturb the visual impression for oblique views due to the discrepancies between geometric model and reality. Real-world surface texture also represents the object's illumination at the time of image acquisition, while applications in architecture and city planning frequently require the simulation of different illumination conditions. Thus, if the surface geometry is not represented well, the rendered scene as it is generated by a shading algorithm will appear unrealistic. For this reason, large-scale visualizations of 3D building models require an increased amount of geometric detail. This can be realized by integration of data from terrestrial laser scanning (Section 45.2.2).

45.2.1 Texture Mapping Using Programmable Graphics Hardware

Texture mapping can be realized easily, if correspondences between the given 3D building model and the respective images of the façades are available. The required corresponding primitives can, for example, be defined by a human operator using a suitable graphics user interface (GUI). Alternatively, the co-registration of image and model data can be determined directly if the exterior orientation of the camera is available. The required orientation parameters can be determined by direct georeferencing, which is then refined by an automatic matching of the image and the 3D model of the building (Haala and Böhm 2003). The resulting correspondences can then be used to improve the directly measured exterior orientation by a spatial resection. Further refinement is feasible by a modified spatial resection using corresponding linear features, which can be extracted very accurately by standard image processing algorithms (Klinec 2004).

Manual of Geographic Information Systems

As depicted in Figure 45-2, once the camera parameters and the exterior orientation are available, the 3D object coordinates of the georeferenced building model can be mapped to the texture image. Based on corresponding object and image coordinates, the presentation of the textured model is feasible using a standard virtual reality modeling language (VRML) viewer. Nevertheless, since these viewers only allow for simple transformations during texture mapping, the quality of visual realism is limited. As an example, complex geometric image transformations to model a perspective rectification or lens distortion are not available. These effects usually have to be eliminated before texture mapping by generating "ideal" images, which are then used as an input for the standard viewer. However, this generation of additional data requires substantial effort for preprocessing and storage. Additionally, if the geometry of the 3D model is, for example, modified by interactive processing, the whole process has to be repeated. Thus, a solution allowing for on-the-fly processing is highly preferable.

Figure 45-2 Mapping of building model and image. See included DVD for color version.

The extraction of façade textures from digital images mainly requires the transformation of vertices and the processing of pixel data. In principle, these computations can be highly parallelized for increased performance. However, a software solution is generally not adequate since the main central processing unit (CPU) does not exploit this parallelism very effectively. In contrast, graphics processing units (GPUs) that are integrated in today's commodity PC graphics cards are optimized for this kind of data processing. As an example, the GPU is used to generate the rasterized image based on the vector representation of the 3D scene. During this step, the 3D vertices are transformed repeatedly until the screen coordinates are available, while the output of one transformation is the input of the next one. In order to allow for interactive speeds during real-time rendering, the required geometric transformations are separated into finer steps, which are then executed by the GPU in a parallel or time-sliced fashion. Parallel processing also can be realized for tasks like clipping to remove non-visible parts or pixel shading. For this reason, today's commodity 3D graphics hardware features a number of parallel processing pipelines, which can be used independently and thus allow an enormous increase in computational speed.

Meanwhile, 3D graphics hardware also has evolved from a fixed function to a programmable GPU design. Thus, in addition to standard rendering tasks, 3D graphics hardware can now be utilized for various fields of applications. Since even complex algorithms can be implemented using high-level programming languages, texture mapping can be realized very efficiently by exploiting the functionality of 3D graphics hardware. For this purpose, the GPU is used both for visualization and direct texture extraction of the façade texture from the original images. By these means, the generation and storage of intermediate images can be avoided. Since pixels only have to be interpolated once within the whole mapping process, the quality of the resulting façade texture is increased. If input photographs and extraction parameters are modified interactively, the user can observe the resulting textured building model in real time. Finally, self-occlusions of the model are detected automatically, so that several images can be fused to provide the final texture images (Kada et al. 2005).

In our approach, so-called pixel shaders are used to exert control over (projective) texture lookups, a depth buffer algorithm and the on-the-fly removal of lens distortions for calibrated cameras. These shaders are small programs that are executed on the 3D graphics card. They can be conceived of as functions that are called within the GPU at specific points during the generation of the image. Two types of shaders exist—vertex shaders replace the transformation module in the geometry stage and pixel shaders replace the processing of individual pixels in the rasterizer stage of the GPU. Nowadays, shaders can be developed using High-Level Shader Language (HLSL developed by Microsoft) or C for graphics (Cg developed by NVIDIA). Both are based on the programming language C and offer the flexibility and performance of an assembly language but with the expressiveness and ease-of-use of a high-level language.

The complete process of texture extraction is depicted in Figure 45-3. Based on the original input image (top left), different steps such as texture extraction, texture placement, detection of self-occlusions, image fusion and removal of lens distortion have to be performed. In the first step, the object geometry is mapped to the original image of the building façade to extract the required texture, which is then stored in the frame buffer. If a calibrated camera is used, the lens distortion is corrected on-the-fly in the pixel shader. This distortion is represented by the parameter set introduced by Brown (1971) and denotes the transition of pixels from the distorted to the idealized image. The resulting texture maps are then placed on the corresponding polygons by computing two-dimensional texture coordinates (s,t) for polygon vertices. This transformation is based on the linear combination

$$s = A_1 x + B_1 y + C_1 z + D_1$$
$$t = A_2 x + B_2 y + C_2 z + D_2$$

(45-1)

of vertex coordinates   as described in OpenGL (Shreiner and OpenGL Architecture Review Board 1999). The parameters A, B, C and D can be interpreted as the definition of the respective planes in parameter form. The normal vector components A, B and C of the two planes are defined by the vector from the bottom left vertex to the bottom right vertex of the bounding box and from the bottom left vertex to the top left vertex, accordingly. The values for D are simply computed by inserting the bottom left vertex into the equation. The result of the linear combination of the polygon vertices are then in the range 0 to 1, as required. The result of this step is depicted in Figure 45-3 (top right). As evident in this example, the visual impression is disturbed due to the occlusions, which have not yet been eliminated. Within the GPU, occluded areas are determined based on the so-called depth buffer algorithm, which can be modified for our purposes.

In our approach, occluded areas are determined using the depth buffer algorithm. First, the depth value of the closest polygon is determined for each pixel in the photograph and stored in a depth texture. This can be done simply by rendering all polygons with the hardware depth buffer functionality enabled and by copying the resulting depth buffer into a 32-bit floating-point texture. In our application a more efficient approach is used by calculating the depth value in a pixel shader and direct rendering into the depth texture. Modern 3D graphics processors support the floating-point texture formats even as render targets. During texture extraction, the depth value is read out in the pixel shader using the same texture coordinates as for the color lookup. After the perspective divide is applied to the texture coordinates, the z-component holds the depth value for the current polygon. A comparison of these two depth values then determines if the pixel in the color texture belongs to the polygon. If, for example, the value from the depth texture is lower than the computed value, then the polygon is occluded at this pixel by another polygon. Figure 45-3 (bottom left) shows the results with occluded pixel values blackened out. To suppress artifacts caused by precision errors, the depth test is done by applying a small depth bias in the depth test.

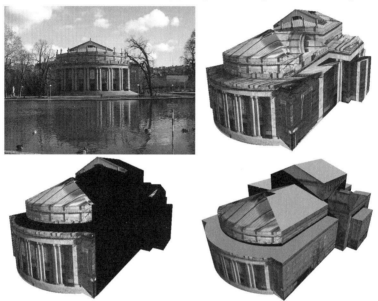

Figure 45-3 Results of texture mapping using programmable graphics hardware. See included DVD for color version.

The problem of image fusion using the hardware is how to get the GPU to decide from which image a pixel should be taken. The solution is to process all images and make the hardware accept or reject pixel values by using the depth, stencil or alpha test. Even though the approach is brute force, it is still very efficient with the hardware support. The presented per pixel approach merges the final façade texture by using the color value of only the closest, non-occluded pixel found in all images. The required depth test is similar to the approach for occlusion detection and again implemented on the hardware. The final texture mapping from three images with selected occlusion-free pixels is depicted in Figure 45-3 (bottom right).

45.2.2 Geometric Refinement of Building Façades Using Terrestrial Laser Scanning

Texture mapping efficiently increases the realism for visualizations of virtual city models. However, since texture maps only modify the color values of the rendered pixel for the respective surfaces, the underlying object geometry is still defined by the coarse 3D model. Such an approximation of the true façade geometry by planar polygons can be sufficient for moderate 3D structures and orthogonal views. However, due to the difference between the available model and the true geometry as represented by the texture image, protrusions like balconies or window ledges will disturb the visual impression, especially for oblique views. Additionally, texture maps from real images depend on the illumination at the time of image acquisition, while the 3D model can be shaded using light sources at arbitrary locations and directions. Thus, large differences between the illumination of the 3D model and the texture images will considerably deteriorate the quality of the generated visualization.

One well-known standard technique to improve the shading of object surfaces is so-called "bump maps." This approach exploits the fact that fine surface details influence the perceived intensity primarily due to their effect on the orientation of the surface rather than their effect on the position of the surface (Blinn 1978). Bump mapping simulates the bumps or wrinkles in a surface by recording the effect of such fine details on the orientation of the surface. These orientation changes are represented by a map, which contains the respective surface normal vectors. During shading these modified normal vectors are then used instead of the original surface orientation. By these means, an explicit geometric modeling of small details can be avoided.

Such bump maps can be provided for virtual city models based on terrestrial laser scanning. As depicted in Figure 45-4, these systems provide densely sampled point clouds consisting of several million points. Since these point clouds are usually measured from multiple viewpoints, a transformation to a suitable reference coordinate system is required as a first processing step. Frequently, this is realized by signalized targets, which are provided in the selected reference frame. Alternatively, low-cost components like navigation-grade GPS and a digital compass can provide an approximate position and orientation. This coarse georeferencing is then improved by an automatic alignment of the laser point clouds to the 3D city model using standard approaches like the Iterative Closest Point approach (Böhm and Haala 2005). Figure 45-4 depicts an overlay of the available 3D building models from an aerial data collection, which is already given in the required reference frame and the point clouds from terrestrial laser scanning after this georeferencing process.

Terrestrial laser scanning from multiple stations provides unordered sets of 3D point clouds, which are then frequently triangulated for surface generation. While this can be a tedious task for complex 3D object geometries, the point processing can be facilitated considerably if the approximate object geometry is already available from the 3D building model. As a first processing step, relevant laser measurements can be extracted from the point cloud for each façade by a simple buffer operation. The selected 3D points are then transformed to a local coordinate system as defined by the façade plane. After mapping of the 3D points to this reference plane, further processing can be simplified to a 2.5D problem. While assuming that the refined geometry of the façade can be described sufficiently by a relief, which is centered on the respective 3D polygon of the coarse 3D building models, the difference between the measured 3D laser points and the given façade polygon are interpolated to a regular grid. The resulting height field is depicted in Figure 45-5.

Figure 45-4 Point cloud from terrestrial laser scanning aligned with a virtual city model. See included DVD for color version.

Figure 45-5 Height field of a single façade.

Manual of Geographic Information Systems

Based on this information, the modification of surface normals as required for bump mapping can be easily computed. Figure 45-6 demonstrates surface shading for two different light sources using the bump map derived from the available height field. For demonstration purposes, the respective illumination conditions are visible on the sphere depicted in the upper right corner.

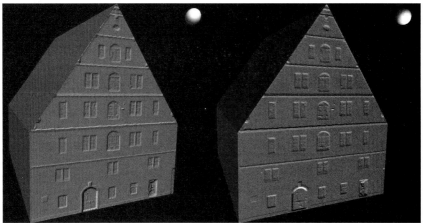

Figure 45-6 Shading of the bump map as generated from terrestrial laser scanning.

Bump mapping is a standard in computer graphics that is implemented on nearly all computer graphics cards and can be used for high-performance rendering and real-time animation. However, this method only modifies the basic shading while effects of the refined object geometry on shadows, occlusions or object silhouettes are not considered. For this purpose, small-scale detail is added to the respective surfaces by displacement mapping (Cook 1984). Unlike bump mapping, which affects only the surface normals, displacement mapping actually modifies the positions of the surface elements. Traditionally, displacement mapping is implemented by subdividing the original surface into a large number of smaller polygons. The vertices of these polygons are then shifted in normal direction. In our application this information is directly provided from the height map (see Figure 45-5). An exemplary visualization of the building façade by displacement mapping is given in Figure 45-7. While adaptive remeshing methods have been proposed to reduce the number of subdivisions during tessellation, the number of generated triangles is still large and remains difficult to process in real time without development of hardware support. Nevertheless, displacement mapping actually modifies the object geometry, thus the amount of geometric detail is increased and allows for a more realistic visualization.

Figure 45-7 Visualization of building façade using displacement mapping.

45.3 Generalization of City Models

Large-scale visualization of urban scenes from pedestrian or moderately elevated viewpoints requires supplementary data from airborne data collection be integrated with 3D city models. An additional processing of these 3D city models is also necessary for small-scale visualizations of larger areas. In such a scenario the amount of information to be displayed has to be limited by geometrically simplified representations as they are provided by a generalization step. Within this step, important information can be preserved and enhanced, while unimportant data is eliminated. In addition to a potential acceleration of the visualization process, which can be required for real-time rendering during interactive manipulation, analysis and exploration of 3D city models, this step is especially necessary if more abstract presentations are aspired. One example is the so-called non-photorealistic rendering of 3D city models for applications in city development planning and city information systems (Buchholz et al. 2005). This approach aims to effectively present thematic information to support the exploration and analysis of the urban data. For this purpose simplified object shapes are generated similar to architectural drawings and sketches (Döllner and Walther 2003). Such sketches usually take a relatively small number of important lines, which are, however, sufficient for human viewers to quickly comprehend the scene. Thus, one key task in non-photorealistic rendering is the determination of visually important edges (Sousa and Prusinkiewicz 2003), which can be provided from a suitable generalization of the original 3D building models.

In principle, the generalization of 3D objects can be realized by surface simplification processes. Within these algorithms, the number of polygonal or triangular meshes used to represent the 3D object geometry is gradually reduced. During this process a suitable error metric is applied in order to preserve the original shape of the input model as well as possible. A good overview on such algorithms originally developed in computer graphics is given by Heckbert and Garland (1997), De-Floriani and Puppo (1995) and Puppo and Scopigno (1997). Also, as a result of the geometric error metric, the algorithms show best results for highly complex objects made from hundreds of thousands to millions of primitives. In contrast to arbitrary objects that can be processed by these approaches, single 3D building models are of comparatively low complexity. Typically, they consist of at most a few hundred polygons. However, each individual building model, and also several objects, that makes up a building block, exhibits special characteristics that need to be preserved during simplification. These are, for example, right angles that can often be found in the building architecture. A geometric error metric as used in surface simplification does not account for such characteristics. In addition, because the standard simplification operators were developed for smooth surfaces rather then for angular 3D shapes, their application to a 3D building model often results in a skewed or tilted model. In principle, the situation is similar to the automatic generalization of 2D building ground plans that cannot be solved by the application of standard line simplification algorithms, such as that described by Douglas and Peucker (1973). Consequently, a number of specialized generalization solutions have been proposed in the past since the early work of Staufenbiel (1973). Sester (2000) simplifies the shape of 2D ground plans by applying a set of simple rules followed by a least squares adjustment.

An analogous approach for the generalization of 3D building models has been presented by Kada (2002). The polygonal building models are iteratively simplified by combining a number of edge collapse operations into a single step. Building regularities that need to be detected prior to the simplification are maintained in the process. Another work on 3D generalization is based on scale-space theory (Forberg 2004). Orthogonal building structures are simplified by shifting parallel facets that have been found to be under a certain distance until they merge. A squaring operation is used for non-orthogonal structures such as roofs. Thiemann and Sester (2004) present a segmentation algorithm that is adapted from the work of Ribelles et al. (2001). The resulting partitioning of the 3D building model is transformed into a Constructive Solid Geometry (CSG) tree. Fragments of the building can then be emphasized, aggregated or eliminated depending on their importance. All these methods are very dependent on the quality of the input models or rely on specific building characteristics. If, for example, angles are close but not exactly orthogonal, building regularities may not be found (Kada 2002), shifted facets may never merge (Forberg 2004) or the segmentation may produce too many fragments (Thiemann and Sester 2004). Furthermore, the proposed simplification operators are not general enough and only work on a subset of structural elements that can be found in 3D building models.

45.3.1 Building Generalization Using Approximating Planes

One option to overcome these shortcomings is to simplify the respective 3D building models using approximating planes. An approximating plane is the result of averaging a set of polygons that have been identified to belong to the same façade, including protrusions and other small structural elements.

The shape of the generalized building model is highly dependent on the geometric properties of the approximating planes that are used in its generation. Because each subdivision adds complexity to the final model, it is favorable to use the smallest number of planes that optimally approximate the original shape. Each approximating plane represents a set of polygons that belong to the same building façade. These are primarily the polygons with the same orientation. Matching façade polygons to corresponding planes is often ambiguous for complex models even if human interaction is involved. We, therefore, kept the algorithm for finding the approximating planes rather simple, accepting that a small number of polygons are erroneously linked to planes that represent other façades. This effect can usually be ignored, however, as it only affects small polygons that are usually of limited significance to the overall shape of the building.

We use an iterative approach for finding the set of approximating planes. Each iteration results in the plane of greatest importance, which is measured by the total area of included polygons that are parallel to the approximating plane. At the beginning of an iteration step, a buffer defined by two delimiting, parallel planes is created for each polygon. The initially planar buffers are then merged pair-wise to create larger buffers until the delimiting planes reach a maximum threshold distance. The iteration step stops when no more buffers can be merged and the averaged plane equation of the buffer with the largest total area is returned. The polygons inside the buffer are discarded from further processing. By repeating this process, the set of approximating planes is found in descending order of importance. See Figure 45-8 (a-f) for example planes determined for a 3D building model.

In order to preserve right angles, and also to endorse parallelism in the building model, the approximating planes are analyzed in a last step. If the angle of the normal vectors from two or more planes is found to be below a certain threshold, these planes are made rectangular or parallel, respectively.

45.3.2 Half-space Modeling of Simplified Buildings

In the modeling step, the approximating planes are then used to subdivide an initially infinite space into smaller subspaces. This process is very similar to half-space modeling where a cuboid is, for example, modeled by trimming the infinite space using six planar half-space primitives. Depending on the definition, the object is then either on the positive or negative side of all six planes. Because our approximating planes have no preferred orientation, a recursive subdivision is not possible. Consequently, the algorithm divides all subspaces that intersect a plane by brute force, creating a large number of fragments in the process. These fragments define solids in 3D space that must be further differentiated in building and non-building objects. Non-building fragments are discarded subsequently. Up to this point, the roof polygons have been neglected by the algorithm, leaving flat-top building fragments as an intermediate result. Because the structure of the roof can be very complex, we remodel it individually per fragment using approximating planes averaged from the roof polygons. The resulting objects are referred to as building primitives. The generalized building model is then generated by merging these building primitives.

After the approximating planes equations have all been determined, they are then used to generate a fragmentation of the building model. For this purpose, an infinite 3D space is subdivided brute force by the planes that serve the purpose of half-space primitives (Figure 45-9). In practice an infinite space is unsuitable, so a solid two times the size of the building's bounding box is used as a substitute. Because the equations of the approximating planes have no horizontal component, the infinite space is at this point only divided in two dimensions. The resulting solids are, therefore, convex 2D ground polygons swept along the vertical direction.

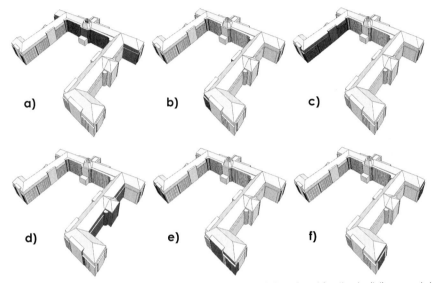

Figure 45-8 Thirteen approximating planes have been determined for the building model. (a-f) shows six of these planes and their corresponding polygons (highlighted). (e) Some of the polygons, which are given by the original building model, have been erroneously assigned to an approximating plane (b, c+d). See included DVD for color version.

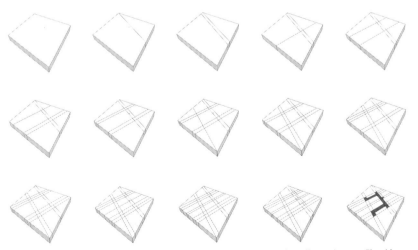

Figure 45-9 Brute force subdivision of infinite space by approximating planes. Final image depicts the differentiation in building (highlighted) and non-building fragments. See included DVD for color version.

Manual of Geographic Information Systems

As previously mentioned, the dividing process makes it impossible to directly identify the building fragments. Therefore, solids must be differentiated in building and non-building fragments in a subsequent step as it is depicted in the final image of Figure 45-9. For each fragment, a percentage value is calculated that denotes the fraction of building to non-building space. Because the space division was basically done in 2D, we use the area of the original building ground plan inside the fragment divided by the ground area of the fragments itself. All solids with a percentage value under a given threshold value are then denoted as non-building fragments and discarded from further processing. A suitable threshold value is dependent on the buffer size used to determine the approximating planes.

At this point, the merging of the building fragments would result in a flat-roof building model with a generalized building ground plan as is visible in Figure 45-10. It should be noted, however, that the process so far is not only a 2D ground plan generalization, because the fragmentation is essential for generating the generalized roof structure. The roof structure can be very complex for general 3D building models. However, the previous generalization step simplifies this task as the roof can now be generated for the individual building fragments (Figure 45-11). It must be ensured, however, that the roof polygons of neighboring fragments still fit against each other.

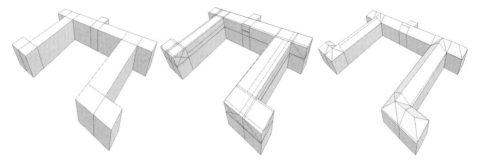

Figure 45-10 Fragments divided by roof planes.   **Figure 45-11** 3D building model with simplified roof.

In order to generate a simplified set of roof planes again, approximating planes are generated from the original model. However, in this step the original roof planes are used. Then a subset is created for each fragment that includes planes that have polygons intersecting the respective fragment. The geometric properties of the planes are not changed in this process. Fragments are then divided by their individual set of approximating planes as described above (see Figure 45-10). The resulting fragments are now real 3D solids, so the differentiation in building and non-building fragments is done in 3D space. A percentage value is calculated this time that denotes the volume of the original building model inside the respective fragment. However, the non-building fragments are not entirely discarded, but rather saved for further processing.

The generalization algorithm has been implemented and tested on a number of 3D building models. It shows good results for models ranging from rather simple to very complex shapes. A selection of examples is depicted in Figure 45-12. In all cases, the complexity of the objects could be highly reduced without destroying the overall appearance of the building. However, there are circumstances where incorrect symmetries are generated during the determination of the approximating planes. Because the erroneous symmetries do not usually affect the topology of the resulting building model, this is corrigible during a geometric postprocessing.

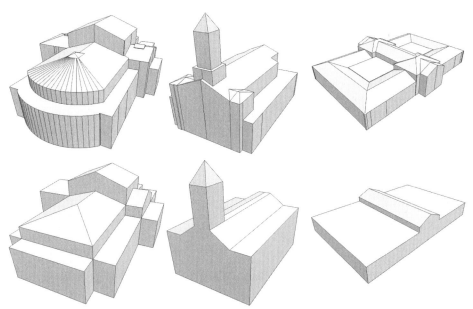

Figure 45-12 Generalized 3D building models in their original (top) and simplified (bottom) shape.

45.4 Discussion

3D city models have become important standard databases that are frequently used for a number of planning purposes, for example, by the generation of visualizations and virtual fly-throughs. Up to now, these data sets were mainly captured by order of national mapping agencies or city administrations. While 3D urban landscape visualization has become a standard component within a number of GIS applications, similar tools like interactive 3D city maps are still emerging in the private market. This development benefits from the fact that the presentation of large 3D city models is now feasible using standard software and hardware and, thus, is appropriate without additional effort for private users. The underlying tremendous improvement of visualization tools is still driven by the requirements of the computer game industry, which is the most important market for applications based on real-time visualization.

Up to now, the achievement of visual realism is the main goal during visualization of 3D city models. Especially for large scales, this is only feasible if a sufficient amount of detail is provided by suitable data collection. Thus, in order to enhance the 3D building geometries to look like they usually do when supplied from airborne sensor data, images or lidar points collected from terrestrial platforms increasingly are used. As it has been demonstrated, the integration and combined evaluation of these different data sets can be supported considerably by the application of techniques, which are well known in computer graphics.

While many applications aspire to generate realistic visualization from 3D city models, this realism can be unnecessary or even a hindrance for some other tasks. As an example, photo-realistic presentations of urban areas for navigation purposes can be adequate for pedestrian users of a tourist information system, while the same amount of visual information will distract a user of a car navigation system during driving. In such scenarios the visualization has to be simplified, e.g., by the limitation to landmark buildings that are really necessary for orientation. Such abstractions, by the selection of important information while simplifying other objects by suitable generalization operations, also are required for the visualization of thematic information. Similar requirements have to be fulfilled for map-based mobile services, which are realized using mobile handheld devices with limited display sizes. Thus, in addition to a further improvement of the data collection process, adequate techniques are required to select and present the visual information from 3D city models that are really suitable for a specific task.

References

Baltsavias, E., A. Grün and L. van Gool. 2001. *Automatic Extraction of Man-Made Objects from Aerial and Space Images (III)*. Lisse, The Netherlands: Swets & Zeitlinger B.V.

Blinn, J. F. 1978. Simulation of wrinkled surfaces. *ACM SIGGRAPH Computer Graphics* 12 (3):286–292.

Böhm, J. and N. Haala. 2005. Efficient integration of aerial and terrestrial laser data for virtual city modeling using LASERMAPS. *IAPRS* 36, Part 3/W19, ISPRS Workshop on Laser scanning 2005, 192–197.

Brown, D. C. 1971. Close-range camera calibration. *Photogrammetric Engineering* 37 (8):855–866.

Buchholz, H., J. Döllner, M. Nienhaus and F. Kirsch. 2005. Real-time non-photorealistic rendering of 3D city models. *Proceedings of the First International Workshop on Next Generation 3D City Models*, held in Bonn, June 2005.

Cook, R. L. 1984. Shade trees. *ACM SIGGRAPH Computer Graphics* 18 (3):223–231.

Danahy, J. 1999. Visualization data needs in urban environmental planning and design. *Photogrammetric Week* 99:351–365.

De-Floriani, L. and E. Puppo. 1995. Hierarchical triangulation for multiresolution surface description. *ACM Transactions in Graphics* 14 (4):363–411.

Döllner, J. and M. Walther. 2003. Real-time expressive rendering of city models. Pages 245–250 in *Seventh International Conference on Information Visualization, Proceedings IEEE 2003 Information Visualization*, held in London, July 2003.

Douglas, D. and T. Peucker. 1973. Algorithms for the reduction of the number of points required to represent a digitized line or its caricature. *The Canadian Cartographer* 10 (2):112–122.

Forberg, A. 2004. Generalization of 3D building data based on scale-space approach. Pages 194–199 in *International Archives of Photogrammetry and Remote Sensing (IAPRS)*, Vol. 35, Part B, Istanbul, Turkey.

Haala, N. and J. Böhm. 2003. A multi-sensor system for positioning in urban environments. *ISPRS Journal of Photogrammetry and Remote Sensing* 58 (1-2):31–42.

Heckbert, P. S. and M. Garland. 1997. Survey of polygonal surface simplification algorithms. Pages 8–38 in *Course Notes Multiresolution Surface Modeling*, SIGGRAPH '97, Los Angeles, Calif.

Kada, M. 2002. Automatic generalization of 3D building models. *International Archives of Photogrammetry and Remote Sensing (IAPRS)*, Vol. 34, Part 4, on CD.

Kada, M., D. Klinec and N. Haala. 2005. Façade texturing for rendering 3D city models. Pages 78–85 in *Technical Papers of the ASPRS Conference 2005*, held in Baltimore, Maryland.

Klinec, D. 2004. A model based approach for orientation in urban environments. Pages 903–908 in *International Archives of Photogrammetry and Remote Sensing (IAPRS)*, Vol. 35, Part B, Istanbul, Turkey.

Kolbe, T. H. and G. Gröger. 2005. CityGML—interoperable access to 3D city models. *First International Symposium on Geo-Information for Disaster Management*, held in Delft, The Netherlands.

Puppo, E. and R. Scopigno. 1997. Simplification, LOD and multiresolution—principles and applications. In *Eurographics '97 Tutorial Notes*, held in Budapest, Hungary.

Ribelles, J., P. Heckbert, M. Garland, T. Stahovich and V. Srivastava. 2001. Finding and removing features from polyhedra. In *American Association of Mechanical Engineers (ASME) Design Automation Conference*, held in Pittsburgh, Pa.

Sester, M. 2000. Generalization based on least squares adjustment. Pages 931–938 in *International Archives of Photogrammetry and Remote Sensing (IAPRS)*, Vol. 33, Part B4, held in Amsterdam, Netherlands.

Shreiner, D. and OpenGL Architecture Review Board. 1999. *Open GL Reference Manual*. Addison-Wesley Professional.

Sousa, M. C. and P. Prusinkiewicz. 2003. A few good lines: Suggestive drawing of 3D models. *Comput. Graph. Forum* 22 (3): 381–390.

Staufenbiel, W. 1973. Zur Automation der Generalisierung topographischer Karten mit besonderer Berücksichtigung großmaßstäbiger Gebäudedarstellungen. (in German) (Automatic Generalization of Topographic Maps with Regard to Building Representations at Large Scale) Ph.D. thesis, Universität Hannover, Fachrichtung Vermessungswesen. (University of Hanover, Field of Study: Surveying).

Thiemann, F. and M. Sester. 2004. Segmentation of buildings for 3D-generalization. *Proceedings of the ICA Workshop on Generalization and Multiple Representation*, Leicester, UK.

Wonka, P., M. Wimmer, F. Sillion and W. Ribarsk. 2003. Instant architecture. *ACM Transactions on Graphics (TOG) Special Issue: Proceedings of ACM SIGGRAPH 2003* 22 (3):669–677.

CHAPTER 46
Three-dimensional GIS Data Acquisition

Yongwei Sheng

46.1 Introduction

Like any other information system, a geographic information system (GIS) is meaningless without data. Uniquely, data in a GIS are spatial data, commonly characterized by their reference to specific geographic locations. The terms GIS data and spatial data are used interchangeably in the following discussion. Spatial data consist of information about the surface and near-surface of the Earth (Goodchild et al. 2002). Data collected on or under the ground, in the air, or in the ocean can be broadly considered to be spatial data as long as they have a spatial component. The history of spatial data acquisition is in many ways the history of mapmaking, which can be dated back thousands of years. In a modern GIS, spatial data are managed in spatial databases. The creation of a spatial database is a long and complex process that involves many stages, such as data acquisition, interpretation, analysis, generalization, and quality control. Data acquisition is quoted as the most fundamental requirement for GIS-based applications, superseding hardware and software purchasing issues (Goodchild 1996). When data are not readily available, spatial data collection is the most expensive part of establishing a GIS (Thapa and Burtch 1991). Spatial data acquisition has been raised as a critical topic in the UCGIS (University Consortium on GIS) research agenda (Jensen et al. 2005). Much attention has been paid to two-dimensional (2D) spatial data, which are the dominant data sources in current GIS systems (Chen and Lee 2001). The UCGIS research agenda on spatial data acquisition has also emphasized 2D data acquisition and harmony among various GIS data sets. This chapter focuses on three-dimensional (3D) spatial data and reviews the state-of-the-art technologies for 3D spatial data acquisition.

We live in a three-dimensional world. Modern Earth sciences require increasingly quantitative and accurate representation and characterization of Mother Nature within the 3D environment. Three-dimensional data are particularly important in the fields of geology, oceanography, meteorology and atmospheric sciences, water resources, environmental assessments, and petroleum. Many spatial phenomena that appear on the Earth have a 3D distribution, but are often handled and mapped two-dimensionally for simplicity, omitting the third dimension that is less accessible. This simplification may lead to loss of critical information since 2D maps provide only one view (usually the top view) of the more complicated 3D objects. This is not a serious problem for many applications because the horizontal dimensions (i.e., X and Y) of the phenomena are usually several orders of magnitude larger than the vertical dimension (i.e., Z) (Kraak 1989). For example, a river that runs hundreds of kilometers may have a variation of its vertical dimension in the range of only several hundred meters. It is usually an acceptable simplification to map the river as a 2D geographic object in many applications. However, this representation becomes insufficient in applications such as river dynamics studies. One has to consider the elevation drop between the upper and the lower reaches of the river and treat it as a 3D object, even though the horizontal dimensions are much larger than the vertical. In addition, a 3D treatment is often necessary when we examine the phenomenon at a local scale, e.g., a waterfall within a small segment of a river, whose vertical variation is much larger than the horizontal dimensions.

Data acquisition pertaining to GIS has been broadly classified into two categories—primary and secondary (Thapa and Burtch 1991). While primary data acquisition produces new data from the ground up, secondary data acquisition refers to generating data from existing sources, which does not produce new information. The widely used map digitizing in GIS that converts paper maps into digital GIS data layers is a typical example of secondary data acquisition. Unlike 2D geographic data, which are often available from other sources such as existing maps, 3D spatial data acquisition is predominantly primary since very few existing maps, charts or graphs contain 3D information.

The technologies employed in 3D data acquisition are diversified and vary with the application fields. For example, photogrammetry is the operational method for terrain surface data and feature data collection, while the petroleum industry widely uses seismic surveys to acquire 3D subsurface data to investigate underground conditions. Three-dimensional spatial data describe a wide variety of objects on or near the Earth's surface, ranging from a single surface point or the surface itself, to more complicated caverns underground. This chapter first classifies 3D spatial data into three broad categories—surface feature data, surface data, and subsurface data—and reviews 3D data acquisition technologies employed in each of these categories.

46.2 Categorization of 3D GIS Data Acquisition Techniques

Three-dimensional spatial data describe many different kinds of 3D geo-objects, such as terrain surface and features on the surface (e.g., trees and buildings), underground objects such as caverns and faults, upwelled cold water bodies in the ocean, and tornadoes in the air. These geo-objects relate to various subjects of the Earth sciences. Three-dimensional spatial data can be as simple as indicating the (X, Y, Z) coordinates of a well or as complicated as describing an irregular-shaped cavern. The fundamental atom of spatial information is the tuple $<S, V>$, where S typically refers to a location in space, and V defines attributes of that location (Zhang and Goodchild 2002). In 3D spatial data acquisition, the tuple becomes (X, Y, Z, V), where (X, Y, Z) denotes the coordinates in the 3D Cartesian system. The fundamental task in 3D data acquisition is to acquire a series of (X, Y, Z) coordinates of the feature under investigation.

Remote sensing-based methods have been effective for spatial data collection over large areas. Owing to their efficiency, airborne and spaceborne remote sensing technologies have been widely used in 3D spatial data acquisition. Since these remote sensing sensors have limited ability in penetrating water or ground, they are restricted to acquiring data for the Earth's surface and features on the surface, and other technologies are employed to obtain 3D data under the ground and water. Based on the technologies employed, we can broadly categorize 3D data into surface feature data, surface data, and subsurface data.

Three-dimensional surface features can be any features on the Earth's surface, ranging from a simple 3D point feature (e.g., a mountain peak), or a 3D line feature (e.g., a mountain trail), to a 3D polygon feature (i.e., a property on rolling hills). They can also be features standing on top of the terrain surface such as trees, buildings, etc. These features can be collected through field surveying or 3D remote sensing.

Surface data are a description of the 3D surface of the Earth. They are of great interest in a variety of fields. Highway planning needs surface data to design routes at least cost; the military needs surface data to guide missiles. The most popular representation of the terrain surface is a digital elevation model (DEM), which could be either grid-based (a grid of regularly spaced cells with known elevation values) or TIN-based. TINs, or triangulated irregular networks, represent the surface using triangles formed by irregularly spaced elevation points (Hutchinson and Gallant 1999). It is common for the data acquisition process to form a TIN-based DEM using a set of elevation points, which is then converted to a grid-based DEM through interpolating the TIN model. A DEM is actually a 2.5-dimensional expression in which the third dimension of space is carried along as the thematic attribute at (X, Y) locations. Under such a representation, the same (X, Y)

location is not allowed to have multiple Z values. This, however, is sufficient to describe the terrain surface for most areas. Though surface data can be arduously collected through field surveying, nowadays acquisition predominantly uses 3D remote sensing technologies. Surface data are one of the most fundamental data layers in GIS and can serve as the reference for converting 2D feature data into 3D data, as described in the mono-plotting discussion (Section 46.3.2).

Subsurface data discussed here refers to the data that describe phenomena under the ground or in the ocean (i.e., under the surface of the water). Since subsurface features are hidden below the surface, they are collected by means other than surface remote sensing. Underground data are collected through coring in the field or by geophysical tools, such as ground penetrating radar (GPR) and 3D seismic reflection. Underwater data are usually collected through cruise probing or by remotely operated vehicles (ROV) (Su and Sheng 1999). It must be noted that 3D subsurface data acquisition is often very expensive when compared with surface feature data and surface data acquisition.

46.3 3D Surface Feature Data Collection Technologies

Large amounts of 3D GIS data are collected for features on the Earth, for example, the *(X, Y, Z)* coordinates of a well and a 3D geometric model of a building. Among various 3D data acquisition tasks, the most fundamental is to determine the *(X, Y, Z)* coordinates for a feature point. A real advantage for surface feature data collection is that these features are often accessible in the field and visible on high-resolution remotely sensed imagery. Thus, they can be collected from field measurements or via remote sensors. Field surveying, photogrammetry and remote sensing are considered primary 3D spatial data collection methods. Field surveying remains a crucial method for spatial data collection, but it is labor-intensive and time-consuming and therefore not suitable for large-area mapping. On the other hand, airborne or spaceborne remote sensing techniques provide great efficiency in collecting such data over large areas and even at global scales, but often need ground control and field validation.

46.3.1 Field Surveying

The traditional method of spatial data collection is field surveying, in which the surveyor needs to gain physical access to the features to be measured. With the help of existing survey networks of control points, land surveying employs geometric principles to determine the *(X, Y, Z)* ground coordinates of a feature point through measurements of angles, distances, and height differences from the previously established control points. The *(X, Y)* coordinates are determined by traversing, triangulation, and/or trilateration methods in a horizontal survey, while the Z coordinate is obtained through differential or trigonometric leveling methods in a vertical survey. Total station instruments are standard tools for angular and distance measurement in field surveying (Wolf and Dewitt 2000). As described, this *in situ* data collection technique is reliable due to the physical contact with the target; however, it is labor-intensive and time-consuming.

A major advancement in surveying is the operational use of the global positioning system (GPS), a network of 24 transmitting satellites orbiting the Earth at an altitude of ~20,000 km and ground-based receivers, which enables accurate 3D positions to be determined anywhere on the Earth (Lange and Gilbert 1999; Wolf and Dewitt 2000). It is common for a single GPS receiver to establish the *(X, Y, Z)* coordinates for a point at submeter accuracy. Advanced differential positioning GPS, which involves a base GPS station at a known location and a roving GPS receiver to capture unknown positions, can remove systematic errors inherent in the GPS signal and achieve centimeter level accuracy. GPS field surveying is a simple process. To acquire the *(X, Y, Z)* coordinates of a target, the surveyor places the GPS receiver (i.e., the antenna) on the target to be measured and allows some time to collect signals from visible transmitting satellites. The GPS is replacing traditional land surveying systems as an efficient ground data collection tool, with improved precision and reduced cost and labor intensity. When integrated with a handheld GIS data collection tool, the GPS-collected spatial data can be directly loaded into a GIS system.

In spite of the efficiency gained from GPS tools, field surveying is still impractical and prohibitive for large-volume spatial data collection over large areas. However, this does not reduce its significance in data collection for GIS and mapping. Field surveying remains an irreplaceable tool for ground control and field validation for remote sensing-based data acquisition.

46.3.2 Image-based 3D Feature Data Collection

Compared to field surveying, remote sensing techniques (including photogrammetry) are more suitable for large-area mapping and serve as the primary technology for GIS data acquisition and update (Heipke et al. 2000). Advances in remote sensing and photogrammetry have revolutionized spatial data acquisition. Remotely sensed imagery has become a major and dominant source of spatial information for the GIS community. However, a remotely sensed image is a 2D representation of the 3D world, and spatial data collected from the image are 2D. Remote sensing imagery can be employed to extract 3D information of the features through stereoscopic techniques, which require stereo pairs of high-resolution images. The images can be collected by spaceborne, mobile-based sensors (Li et al. 1996), or, more popularly, with airborne cameras. With the recent developments in remote sensing technology, spaceborne sensors are providing a stable and cost-effective source of high-resolution imagery with stereoscopic capabilities, thus allowing the 3D objects to be collected more economically. Independent of the image source, the principles used in image-based 3D spatial data collection are largely the same, i.e., the photogrammetric principles.

Photogrammetry is the discipline of taking measurements from photographs and serves as the foundation of the image-based spatial data production industry. By employing photogrammetry, spatial data can be compiled more economically than through field surveying methods, while achieving comparable or even better accuracy. Photogrammetry is a mainstream technique for 3D feature data acquisition.

The fundamental task in photogrammetry-based 3D spatial data collection is to derive a feature's 3D coordinates from photo (or image) coordinates. The relationship between the feature's object space coordinates and photo coordinates is governed by the well-known photogrammetric collinearity equations. If we assume that the camera properties, location (X_s, Y_s, Z_s) and orientation (ϕ, ω, κ) are known, and the lens distortions are negligible, the feature's object space coordinates (X, Y, Z) are to be derived from the 2D photo coordinates (x, y) through the inverse collinearity equations (Equation 46-1)

$$\begin{cases} X = X_s + (Z - Z_s) \cdot \dfrac{m_{11}(x - x_0) + m_{21}(y - y_0) + m_{31}(-f)}{m_{13}(x - x_0) + m_{23}(y - y_0) + m_{33}(-f)} \\[2em] Y = Y_s + (Z - Z_s) \cdot \dfrac{m_{12}(x - x_0) + m_{22}(y - y_0) + m_{32}(-f)}{m_{13}(x - x_0) + m_{23}(y - y_0) + m_{33}(-f)} \end{cases} \qquad (46\text{-}1)$$

where, f is the camera focal length; x_0 and y_0 are the photo principal point offsets; and m_{11}, m_{12}, ..., and m_{33} are the elements of the rotation matrix, which is determined from ϕ, ω and κ.

Determining the feature coordinates (X, Y, Z) from photo coordinates (x, y) using Equation 46-1 is not a straightforward process, since two equations are not sufficient for three unknowns (i.e., X, Y and Z). Additional information is therefore needed. In photogrammetry, 3D feature coordinates can be obtained from a stereo pair of photos through the process of stereo-plotting, or from a single photo with the assistance of a surface model through the process of mono-plotting.

Photogrammetric stereo plotters are traditionally the instruments used to transfer aerial photogrammetric information to planimetric and topographic maps. With recent developments in digital photogrammetry, 3D spatial data can be collected more efficiently in digital GIS formats. The core process in digital stereo-plotting is space intersection to determine the 3D coordinates from two or more images. The photo coordinates of the same feature point are measured on multiple photos. Each pair of photo coordinates provides a pair of two equations in Equation 46-1; therefore multi-photo measurements lead to four or more equations with three unknowns (i.e., X, Y, Z), which can be solved optimally using the least squares solution.

Based on the above-described process of determining the 3D coordinates for a single point, more complicated 3D features can be collected. Figure 46-1 shows 3D features collected in the University of California, Berkeley campus from stereo pairs of 1:2,400 scale photographs (Sheng 2000; Gong et al. 2002). This serves as a good example of 3D feature collection since various types of features were collected, including a smooth ground surface model, road features, building features, and two types of trees (conifers in light green and hardwoods in dark green). They are all collected from the photos using digital stereo-plotting techniques. The ground surface model is interpolated from a number of critical ground points collected. The building models are interactively extracted. The roads are collected as 3D polygon features. The trees are modeled using a geometric tree model with parameters including treetop 3D coordinates, crown shape and dimensions, which are measured from the photos. All the features are integrated together in ArcView® 3D Analyst for a 3D visualization of the scene reconstructed from the photographs (Figure 46-1).

Figure 46-1 Visualization of acquired 3D spatial data of various features. See included DVD for color version.

Stereo plotting is not always applicable since images sometimes are not available in stereo pairs. It is possible to collect 3D features from a single photo when a surface model is available. Three-dimensional coordinates of a point can be established through the process of mono-plotting, whose core is a single-ray back-projection algorithm (Sheng 2004). The mono-plotting approach is particularly useful to collect features on the surface. Digital surface/elevation models are being acquired more independently of photographs using other technologies such as lidar and radar interferometry, and are often readily available as a fundamental layer in GIS. Mono-plotting becomes more important in GIS for geographical information acquisition. With the support of surface models, the digital mono-plotting process extracts accurate 3D information directly from an unrectified photograph to be used in a GIS. This process does not necessitate the use of sophisticated or expensive equipment, which is an appealing advantage. When a reliable surface model is available, mono-plotting is a simple, fast, economical and accurate way to acquire 3D surface feature data (Baltsavias 1996).

Manual of Geographic Information Systems

46.4 3D Surface Reconstruction Technologies

Surface data are an essential type of spatial information describing 3D surface shapes. Digital elevation models (DEMs), which are a 3D description of the bare Earth surface, are the most popular surface data and provide important 3D information widely used in GIS. DEMs were rated as one of the three foundation spatial databases in the United States National Spatial Data Infrastructure (NSDI) by the National Research Council (NRC 1995). Photogrammetric methods were the only operational ways to produce DEMs prior to the 1990s. With recent technology advancements, radar interferometry and lidar (light detection and ranging) have become practical and are rapidly gaining ground in this field. This section reviews these three dominant technologies for surface reconstruction.

46.4.1 Photogrammetry

As an operational process, national topographic maps and DEMs are produced using photogrammetric principles. In the past, production was done on stereo plotters. Using a photo pair, the operator determined a large number of crucial points and break lines at a feature boundary or where the terrain formed sudden changes, such as hill tops, valleys, or mountain ranges. Though more efficient than field surveying, this is still a tedious process. With the development of photogrammetry, image-based DEM generation has become a semi-automatic process using digital photogrammetric workstations. A cloud of critical points can be automatically generated through image matching between the photos in a stereo pair. The automatically produced critical points usually need to be edited, and break lines often need to be interactively added to produce a good-quality DEM.

Ground objects such as buildings and trees appear as 3D features on high-resolution imagery. As resolution has increased to submeter levels, image-based building modeling and canopy surface reconstruction has attracted attention. Unlike DEMs, digital surface models (DSMs) depict both the terrain and the object surfaces, such as trees and buildings. DSM generation is a challenge to the current photogrammetric systems since the object surface is not as smooth as the terrain surface. By introducing geometric tree models, Sheng et al. (2003a) have employed a model-based image matching technique to successfully reconstruct canopy surface for a redwood tree stand in the campus of University of California, Berkeley, and the surface model has been used to produce a true orthoimage for the stand (Sheng et al. 2003b). Figure 46-2 illustrates a 3D view of the obtained surface model with the orthoimage draped on top. Most trees and buildings stand up in the 3D view, demonstrating the success of the model-based DSM generation.

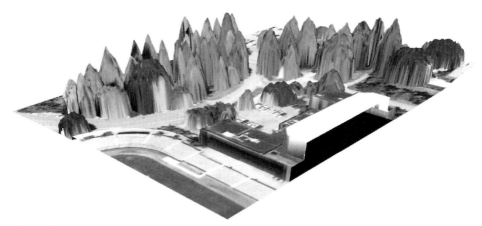

Figure 46-2 3D view of canopy surface model with orthoimage draped.

46.4.2 Radar Interferometry

Surface can also be reconstructed using a stereo pair of radar images, which is known as radargrammetry (Leberl 1990). This is achieved by stereoscopic image matching, which can be considered an extension of photogrammetry using radar imagery (Toutin and Gray 2000). Considered to be one of the most exciting advancements in radar technologies, radar interferometry is a surface reconstruction and surface deformation monitoring technique that has matured recently. Subtle elevation differences can be recorded by phase coherence of the interferometric radar signals received by either two antennae at different locations or with the same antenna at two different times. A comprehensive review on radar interferometry is available in Madsen and Zebker (1998). This technology is being used to produce high-resolution DEMs on a global scale. Employing interferometry principles, the Shuttle Radar Topography Mission (SRTM) in February 2000 acquired seamless DEM data with 80% global coverage at a resolution of 3 arc seconds (nominally 92 m on the ground in equatorial regions), with the absolute horizontal and vertical accuracy of ~10 m (Rabus et al. 2003; Sheng and Alsdorf 2005).

46.4.3 Lidar (Light Detection and Ranging)

As its name indicates, lidar is an active range detection technology similar to radar, but it operates using laser light (usually in near-infrared wavelengths) instead. The principle is quite simple: the range is measured in units of time, which can be converted to distance because the speed of light is constant and known. Lidar mapping is made possible by the development of GPS and inertial navigation technologies, which instantaneously record sensor position and orientation. There are two families of lidar technologies—small- and large-footprint lidar. Large-footprint lidar sensors collect continuous return energy (waveform) within a large sampling area (i.e., footprint), which is usually from several meters to hundreds of meters in diameter. Small-footprint lidar sensors detect the return signal as discrete echoes from a small footprint usually 10-30 cm in diameter. Large-footprint lidar is attractive in vegetation structure studies. Small-footprint lidar is described in this chapter as a surface construction technology, due to its capability to produce accurate high-resolution DEM/DSM.

It is a straightforward process for the lidar sensor to determine the 3D *(X, Y, Z)* coordinates of a target on the ground with known sensor position and orientation. As illustrated in Figure 46-3, the sensor emits a laser pulse and detects a returned echo from a target after time *t*. The onboard GPS receiver records the position *(Xs, Ys, Zs)* of the lidar sensor when the echo is received, and the inertial mapping unit (IMU) records the direction of the returned echo in two angles—the off-vertical-line angle *θ*, and the horizontal azimuth angle *α* from the North. Through simple geometric derivations, the coordinates *(X, Y, Z)* of the target can be determined using Equation 46-2, where *c* is laser light traveling speed.

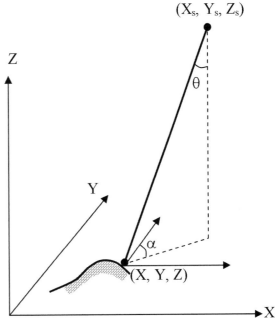

Figure 46-3 Illustration of target position determination using lidar.

$$\begin{cases} X = X_S - \dfrac{t \cdot c}{2} \cdot \sin\theta \cdot \sin\alpha \\[2mm] Y = Y_S - \dfrac{t \cdot c}{2} \cdot \sin\theta \cdot \cos\alpha \\[2mm] Z = Z_S - \dfrac{t \cdot c}{2} \cdot \cos\theta \end{cases} \qquad (46\text{-}2)$$

Based on this principle, a lidar sensor equipped with a scanning mechanism can collect clouds of massive surface points at a sampling rate as high as 100 kHz with a submeter ground posting, with their precise (X, Y, Z) coordinates determined. The points themselves are important 3D spatial data. For example, they precisely locate power line towers and may even outline the shape of power lines. By applying interpolation techniques to these dense irregularly distributed lidar points, a high-resolution DEM and/or DSM can be derived.

Many lidar sensors have the capability to simultaneously record multiple return signals. The first and the last returns are most important, especially for vegetation studies, since the first return bounced back from the canopy recovers the canopy surface model, while the last return can be used to generate the bare Earth DEM. Figure 46-4 shows a case study of surface reconstruction using lidar mapping. As shown in the aerial photograph (Figure 46-4a), this is a typical farm site with dominant farmland, patches of trees, and scattered farmhouses. Figures 46-4b and 4c are the hillshaded surface models reconstructed from the first return and the last return, respectively. The farmland and buildings are solid objects and appear identical in the first and the last return surface models, while the two surface models exhibit a huge discrepancy in forested areas. When a laser pulse interacts with a tree (i.e., a semi-transparent object), part of the signal is reflected by the outer leaves of the tree as the first echo, and the transmitted signal continues to travel in the canopy and may hit a solid target such as a tree branch/trunk and the ground to form the last return. The first return directly produces the canopy surface, and the last return can be used to produce the ground DEM with a filtering process to remove the forest returns. The two surface models have great applications in forestry, e.g., extraction of forest biomass information.

These three surface reconstruction technologies have their own advantages and disadvantages. Where radar signal has limited ability to penetrate canopy, radar interferometry can only produce a surface model rather than a terrain elevation model in vegetated areas. Photogrammetric methods can produce terrain/surface models with human interaction, but the core image-matching process is vulnerable to ambiguities and lack of features. Manual editing is necessary in this technology. Lidar technology can produce accurate (~15 cm in Z) high-resolution (e.g., 1 m) surface and elevation models in a highly automated way. However, filtering surface objects to produce terrain models still remains a challenging problem (Sithole and Vosselman 2004). In addition, lidar only provides massive data points rather than imagery. Without image data as reference, lidar technology alone is not favorable for 3D feature collection, and is not able to produce break lines in surface model generation. It is generally agreed that a comprehensive system is desirable that integrates a lidar sensor with an imaging sensor for improved performance (Tao 2006).

In addition to the above technologies, sonar (sound navigation and ranging) technology is widely used in lake/sea floor mapping to produce bathymetric data. Being an active sensor, sonar emits sound waves to penetrate deep water to collect depth data from the water surface (Aronoff 2005).

46.5 3D Subsurface Data Acquisition Techniques

The main uses of GIS have dealt with the Earth's surface and the objects on the surface. If objects are below the surface, they are projected to the surface, allowing them to be analyzed two dimensionally. However, this 2D approach does not always meet the needs since the vertical

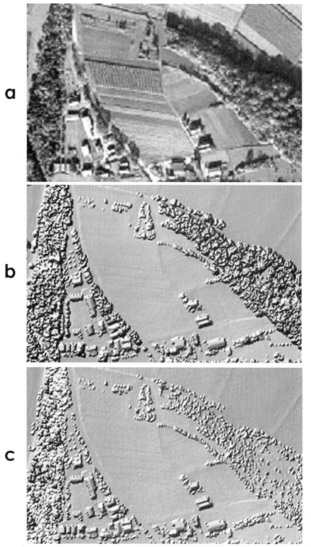

Figure 46-4 Lidar surface reconstruction. (a) Aerial image of a farm site. (b) First return surface model. (c) Last return surface model. Data available through ISPRS lidar filtering test. Source: Sithole and Vosselman 2004.

dimension is missing (Smith and Paradis 1989). This is particularly true in the fields of geology, geophysics, mining, oceanography, ground-water hydrology, hazardous contaminations, and petroleum. Many phenomena in these fields are 3D in nature and require quantitative properties of the subsurface environment, and thus involve 3D subsurface data collection.

Three-dimensional subsurface data are more challenging to acquire than surface feature data since the subsurface features are hidden. They may be collected from a variety of sources including boreholes, well cores, ground penetrating radar (GPR) surveys, and seismic reflection surveys (Youngmann 1989). Boreholes and well cores are collected from intensive drilling operations. Cores are point observations of a variety of geophysical, petrochemical and geological attributes at various depths and provide the ground truth for GPR and seismic surveys. Drilling cores can be either vertical or non-vertical (Figure 46-5). For vertical cores, all measurements at various depths share the same (X, Y) coordinates, and only depth and properties need to be recorded. However, the (X, Y, Z) coordinates and the properties at various depths need to be recorded for non-vertical cores.

Manual of Geographic Information Systems

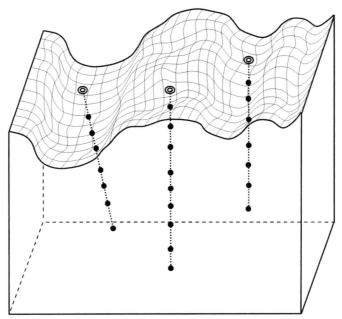

Figure 46-5 Vertical and non-vertical cores.

Ideally with a sufficiently large number of cores, we can model the scattered or randomly located coring data into a 3D cube (i.e., a uniform or regular data structure) through the process of interpolation and extrapolation to estimate the properties at locations without observation. The 3D cube is always desirable to interpret and locate horizons, faults, lithofacies and objects in the form of 3D surfaces and 3D objects.

Since space is continuous, ideal spatial data acquisition would require an infinite number of samples, which is hardly possible in practice. This is particularly true for subsurface spatial data acquisition. Since the subsurface features are hidden under the surface, subsurface coring data are very costly. The cost can prevent a sufficient sampling being acquired to resolve all uncertainties. It is always necessary to limit sampling only to critical locations. However, the site cannot be previewed before sampling since the shape and properties of the features are under investigation and to be determined through the sampling. The sampling scheme and sampling density are therefore difficult to determine in advance (Raper 1989). An iterative hierarchical sampling strategy is most suitable. First, initial sparse samples are collected using a suitable sampling scheme with empirical knowledge, if any, followed by a variability assessment of the 3D feature. In the next step, sampling is densified in areas of greatest importance to describe the subsurface features. Thus, the data collection can be done both economically and efficiently. Even though the coring samples are normally sparser than desired, they are the most detailed source, providing attribute data and serving as a calibration and validation tool for GPR and seismic surveys.

Ground penetrating radar (GPR) techniques collect subsurface data over a broader area without coring or destruction. In principle, GPR provides images of subsurface profiles through velocity sounding. Radar signal travels at varying speeds in subsurface layers of different materials or conditions, and results in a visible boundary between layers in the generated GPR profile image. The radar antenna equipped in the GPR instrument operates at a wide range of frequency from 10 to 1000 MHz. Lower frequency antennae can yield greater depth of penetration (usually up to 30 m) but at lower resolution (Jol and Bristow 2003). GPR in recent years has proven its value in various shallow subsurface investigations, for example, in detecting peatland depth in West Siberia (Sheng et al. 2004).

When profiles are collected in closely spaced trace lines, a 3D cube of subsurface data can be constructed that is more informative than 2D profiles and normally requires very dense cores to produce. The 15 m deep 3D cube in Figure 46-6 was generated by parallel GPR profiles at 1 m spacing in a gravel pit (Heinz and Aigner 2003). The grid has a length of 50 m and a width of 35 m and provides information up to 15 m in depth. The cube provides an insight into the layers of depositional elements. Various layers and structures (on the right side of the figure) can be interpreted and reconstructed from the cube.

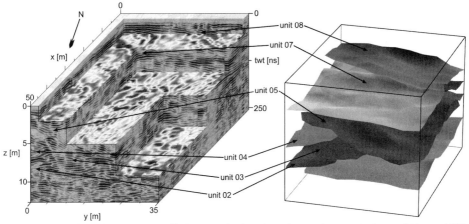

Figure 46-6 3D GPR data cube and interpreted subsurface layers. Source: Heinz and Aigner 2003. See included DVD for color version.

With their capability to sense deeper subsurface than a GPR survey, 3D seismic reflection techniques are the most exciting geophysical methods for imaging the underground and acquiring information about subsurface features. They are extensively used in mineral and petroleum exploration (Bacon et al. 2003). Seismic images are produced by generating, recording, and analyzing sound waves that travel through the Earth. The technique uses artificial wave sources (i.e., explosions, vibrating trucks, etc.) for generating short but intense seismic sound waves that travel through the ground or the water and recording the reflected sound bounced back from the underground by a series of detectors on the surface. The measured depth (up to kilometers) depends on the power of the wave source and the underground conditions. The collected data are processed as a 3D cube like the GPR-generated data shown in Figure 46-6. Since boundaries between different subsurface layers often reflect seismic waves, variations and discontinuities in the cube are usually indicative of subsurface feature boundaries. Three-dimensional subsurface feature information can then be obtained through subsequent analysis.

46.6 The Role of GPS in 3D Spatial Data Acquisition

The Global Positioning System (GPS) has greatly enhanced our ability to acquire 3D spatial data. When discussing 3D spatial data acquisition, we cannot overemphasize the influence of GPS. GPS is involved in any type of data collection and impacts almost every aspect of 3D data acquisition. As a direct field surveying tool, GPS can measure accurate *(X, Y, Z)* coordinates of features to be loaded into a GIS. GPS is helpful even in subsurface data collection. With GPS, it becomes extremely easy to determine the precise *(X, Y, Z)* coordinates of the coring sites. GPS also helps to position the GPR and seismic surveys. More importantly, the critical role of GPS in 3D spatial data acquisition is in assisting remote sensing-based acquisition technologies, which are considered the standard method of spatial data collection.

GPS is one of the key technologies in the advent of lidar remote sensing. Lidar technology would have been impossible without GPS. Laser ranging technology was invented many years ago, but lidar remote sensing has been available only recently. One of the main obstacles was that the position of the flying sensor could not be precisely determined. As indicated in Equation 46-2, the coordinates of ground targets cannot be derived without sensor position data. By measuring the sensor's position instantaneously, onboard GPS makes lidar data acquisition possible.

In addition, GPS assists image-based spatial data acquisition in many ways. GPS may be used to collect ground control points for georeferencing remotely sensed images to reduce image distortion and improve their positioning accuracy, allowing reliable spatial data to be collected. Three-dimensional data collection from images requires the position and orientation of the images, which are traditionally established using ground control. Onboard GPS, by providing the sensor's position, significantly reduces the number of ground control points needed, saving a large amount of fieldwork. Ground truthing or field validation for remote sensing is made easier and more efficient by GPS's navigation function. The coordinates of any area of interest in the image can be loaded into a GPS receiver to guide the field inspector directly to that location. This also allows the inspector to locate any pixel of the image (even if it is not a feature point) in the field.

46.7 Discussion and Conclusion

Data acquisition very often accounts for the most significant costs in a GIS project. Three-dimensional GIS data acquisition is even more costly, especially subsurface data collection. This chapter has given a review on 3D spatial data acquisition technologies. Unlike 2D GIS data, which are often available from digitizing existing maps, 3D GIS data are collected predominantly using primary data acquisition approaches, mainly through field surveying and 3D remote sensing. Though field surveying is labor-intensive and time-consuming, it is not replaceable and serves as a ground validation tool to remote sensing-based GIS data acquisition. As the primary spatial data acquisition technologies, photogrammetry and remote sensing are cost-effective data production methods over large areas. Owing to the technology advancement in remote sensing and the rapid increase of computing resources, mainly in the past ten years or so, our ability to acquire 3D spatial data has been remarkably improved. Three-dimensional photogrammetric data collection has evolved from stereoplotters to digital photogrammetry, allowing 3D data to be captured with automation and in a digital form that can be imported directly into a GIS. The recently available radar interferometry and lidar technologies have revolutionized 3D surface reconstruction. In addition to providing a direct data collection tool in the field, GPS is the most critical supporting technology in 3D spatial data acquisition. It not only makes image-based 3D data collection more productive but also makes lidar data collection possible and practical.

It is expected that remote sensing-based acquisition technologies will continue to grow and dominate the field. With the increasing availability of high-resolution stereoscopic satellite imagery, 3D spatial data are being acquired increasingly by spaceborne sensors. Though many data collection tasks require human-machine interaction, future data acquisition technologies will allow a higher level of automation. For example, the newly emerged lidar technology produces surface models in a more automatic way than the traditional photogrammetric methods. Digital photogrammetric workstations are building automated tools for feature extraction and collection. Integrating the currently available technologies is a trend in 3D data acquisition. For example, an integrated sensor system, consisting of both an imaging sensor and a lidar sensor, is able to simultaneously capture stereoscopic imagery and lidar data of the same area. Such an integrated system has a better capability to collect 3D data. The combination of lidar points and break lines extracted from images can produce high-quality surface models in an easy way. In addition, this integration also promotes a better identification and extraction of 3D features. In summary, our ability to collect 3D surface features and 3D surface data is improving substantially. However, subsurface data acquisition still remains a challenge and requires further development in affordable active-sensing technologies for underground investigation over large areas.

This chapter focuses on the technological aspect of 3D spatial data acquisition without touching upon other relevant issues. Though data quality and data uncertainty are not discussed in this chapter, they are critical issues in any data acquisition. These issues are particularly challenging in 3D GIS data acquisition and warrant further research.

References

Aronoff, S. 2005. *Remote Sensing for GIS Managers*. Redlands, Calif.: ESRI Press.

Bacon, M., R. Simm and T. Redshaw. 2003. *3-D Seismic Interpretation*. Cambridge, UK: Cambridge University Press.

Baltsavias, E. P. 1996. Digital ortho-images – a powerful tool for the extraction of spatial- and geo-information. *ISPRS Journal of Photogrammetry and Remote Sensing* 51 (1):63–77.

Chen, Y. Q. and Y. C. Lee. 2001. *Geographic Data Acquisition*. New York: Springer.

Gong, P., Y. Sheng and G. S. Biging. 2002. 3D model-based tree measurement from high-resolution aerial imagery. *Photogrammetric Engineering & Remote Sensing* 68 (11):1203–1212.

Goodchild, M. F. 1996. The spatial data infrastructure of environmental modeling. In *GIS and Environmental Modeling: Progress and Research Issues*, edited by M. F. Goodchild, L. T. Steyaert, B. O. Parks, C. Johnston, D. Maidment, M. Crane and S. Glendinning, 61–74. New York: John Wiley & Sons.

Goodchild, M. F., P. F. Fisher and W. Shi. 2002. Preface to *Spatial Data Quality*, edited by W. Shi, P. F. Fisher and M. F. Goodchild, xv–xviii. New York: Taylor & Francis.

Heinz, J. and T. Aigner. 2003. Three-dimensional GPR analysis of various Quaternary gravel-bed braided river deposits. In *Ground Penetrating Radar in Sediments*, Geological Society Special Publications 211, edited by C. S. Bristow and H. M. Jol, 99–110. London: Geological Society.

Heipke, C., K. Pakzad and B. M. Straub. 2000. Image analysis for GIS data acquisition. *Photogrammetric Record* 16 (96):963–985.

Hutchinson, M. F. and J. G. Gallant. 1999. Representation of terrain. In *Geographical Information Systems*. Vol. 1, *Principles and Technical Issues*, 2nd ed, edited by P. A. Longley, M. F. Goodchild, D. J. Maguire and D. Rhind, 105–124. New York: John Wiley & Sons.

Jensen, J., A. Saalfeld, F. Broome, D. Cowen, K. Price, D. Ramsey, L. Lapine and E. L. Usery. 2005. Spatial data acquisition and integration. In *A Research Agenda for Geographic Information Science*, edited by R. B. McMaster and E. L. Usery, 17–60. Boca Raton, Fla.: CRC Press.

Jol, H. M. and C. S. Bristow. 2003. GPR in sediments: advice on data collection, basic processing and interpretation, a good practice guide. In *Ground Penetrating Radar in Sediments*, Geological Society Special Publications 211, edited by C. S. Bristow and H. M. Jol, 9–27. London: Geological Society.

Kraak, M. J. 1989. Computer-assisted cartographical 3D imagining techniques. In *Three Dimensional Applications in Geographic Information Systems*, edited by J. Raper, 99–113. London: Taylor & Francis.

Lange, A. F. and C. Gilbert. 1999. Using GPS for GIS data capture. In *Geographical Information Systems*. Vol. 1, *Principles and Technical Issues*, 2nd ed, edited by P. A. Longley, M. F. Goodchild, D. J. Maguire and D. Rhind, 467–476. New York: John Wiley & Sons.

Leberl, F. 1990. *Radargrammetric Image Processing*. Norwood, Mass.: Artech House.

Li, R., M. A. Chapman, L. Qian, Y. Xin and C. Tao. 1996. Mobile mapping for 3D GIS data acquisition. *International Archives of Photogrammetry and Remote Sensing*, Vol. XXXI, part B2, 232–237.

Madsen, S. N. and H. A. Zebker. 1998. Imaging radar interferometry. In *Manual of Remote Sensing*, 3rd ed, Vol. 2, *Principles & Applications of Imaging Radar*, edited by F. M. Henderson and A. J. Lewis, 359–380. New York: John Wiley & Sons.

National Research Council. 1995. *A Data Foundation for the National Spatial Data Infrastructure*. Mapping Science Committee. Washington, DC: National Academy Press.

Rabus, B., M. Eineder, A. Roth and R. Bamler. 2003. The shuttle radar topography mission – a new class of digital elevation models acquired by spaceborne radar. *ISPRS Journal Photogrammetry and Remote Sensing* 57 (4):241–262.

Raper, J. 1989. The 3-dimensional geoscientific mapping and modeling system: a conceptual design. In *Three Dimensional Applications in Geographic Information Systems*, edited by J. Raper, 11–19. London: Taylor & Francis.

Sheng, Y. 2000. Model-Based Conifer Crown Surface Reconstruction from Multi-Ocular High-Resolution Aerial Imagery, Ph.D. diss., University of California, Berkeley.

———. 2004. Comparative evaluation of iterative and non-iterative methods to ground coordinate determination from single aerial images. *Computers & Geosciences* 30 (3):267–279.

Sheng, Y. and D. Alsdorf. 2005. Automated ortho-rectification of Amazon basin-wide SAR mosaics using SRTM DEM data. *IEEE Transactions on Geoscience and Remote Sensing* 43 (8):1929–1940.

Sheng, Y., P. Gong and G. S. Biging. 2003a. Model-based conifer canopy surface reconstruction from photographic imagery: overcoming the occlusion, foreshortening and edge effects. *Photogrammetric Engineering & Remote Sensing* 69 (3):249–258.

———. 2003b. True orthoimage production for forested areas from large-scale aerial photographs. *Photogrammetric Engineering & Remote Sensing* 69 (3):259–266.

Sheng, Y., L. C. Smith, G. MacDonald, K. V. Kremenetski, K. E. Frey, A. A. Velichko, M. Lee, D. W. Beilman and P. Dubinin. 2004. A high-resolution GIS-based inventory of the west Siberian peat carbon pool. *Global Biogeochemical Cycles* 18 (30):GB3004: 1–14 (doi:10.1029/2003GB002190).

Sithole, G. and G. Vosselman. 2004. Experimental comparison of filter algorithms for bare-Earth extraction from airborne laser scanning point clouds. *ISPRS Journal of Photogrammetry and Remote Sensing* 59 (1-2):85–101.

Smith, D. R. and A. R. Paradis. 1989. Three-dimensional GIS for earth sciences. In *Three Dimensional Applications in Geographic Information Systems*, edited by J. Raper, 11–19. London: Taylor & Francis.

Su, Y. and Y. Sheng. 1999. Visualizing upwelling at Monterey Bay in an integrated environment of GIS and scientific visualization. *Marine Geodesy* 22 (2):93–104.

Tao, C. V. 2006. 3D data acquisition and object reconstruction for AEC/CAD. In *Large-Scale 3D Data Integration*, edited by S. Zlatanova and D. Prosperi, 39–56. Boca Raton, Fla.: CRC Press.

Thapa, K. and R. C. Burtch. 1991. Primary and secondary methods of data collection in GIS/LIS. *Surveying and Land Information Systems* 51 (3):162–170.

Toutin, T. and L. Gray. 2000. State-of-the-art of elevation extraction from satellite SAR data. *ISPRS Journal of Photogrammetry and Remote Sensing* 55 (1):13–33.

Wolf, P. and B. A. Dewitt. 2000. *Elements of Photogrammetry with Applications in GIS*. 3rd ed. New York: McGraw Hill.

Youngmann, C. 1989. Spatial data structures for modeling subsurface features. *Three Dimensional Applications in Geographic Information Systems*, edited by J. Raper, 129–136. London: Taylor & Francis.

Zhang, J. and M. F. Goodchild. 2002. *Uncertainty in Geographical Information*. New York: Taylor & Francis.

SECTION 7
GIS and the World Wide Web

CHAPTER 47

US National Databases: *The National Map*

Mark L. DeMulder and *Gail A. Wendt*

The need for a national mapping program for the United States has never been stronger. Policy and decisions regarding economic development and environmental management rely on spatially oriented information to discern patterns and to provide a visual context. As land use changes and population demographics shift correspondingly, a national perspective is imperative to monitor the altered landscape. As a Nation, we are increasingly dependent on the natural resources that come from that landscape—energy resources, minerals, food and fiber products. The litany of natural disasters with which we have had to contend has had a marked impact on the landscape as well, from floods and hurricane effects to wildfires and the proliferation of non-native species that invade mutable ecosystems. It is to these impacts, alterations, and implications for the national landscape that the mapping mission of the US Geological Survey (USGS), a bureau of the US Department of the Interior, is directed. Issues of national and homeland security that affect society as a whole add urgency to the requirement for immediately available and complete geospatial information, organized and accessible in a national database.

The USGS, founded in 1879, has met the civilian mapping needs of the US Government since the early 1880s when the second USGS Director, John Wesley Powell, began the national topographic mapping program. Topographic maps have become the most visible and best-known product of the USGS. Americans have relied on the accuracy and quality of these products for generations, building "brand" recognition and market loyalty. As technology and user needs change, however, so too must the Nation's mapping program. Coupled with this are differing expectations from the public for the resources to and the services from government programs. In 2001, the USGS set a visionary course that is energizing its mapping activities in exciting and forward-looking ways and will continue to have impact well into this new century. In this vision, *The National Map* (www.nationalmap. gov) and its rational beginning and its promise for the future are presented as a critical national database to be sustained as an enduring resource for the Nation. A brief look at the history upon which *The National Map* is built (USGS 2001) provides context for the emergence of this program as the future for national geospatial assets.

The mission of the USGS was a natural outgrowth from and the formal institution of the four Great Western Surveys conducted after the Civil War between 1867 and 1879. The original charge of that mission embodies a unique combination of responsibilities: "classification of the public lands, and examination of the geological structure, mineral resources, and products of the national domain." Early USGS topographic mapping efforts "of the national domain" were focused on supporting the science of the bureau, then predominantly geology and hydrology. John Wesley Powell, Director from 1881–1894, refocused the mapping program to create a national map that "once constructed should be enduring, that the expense of frequent resurveys may be avoided." (Rabbitt 1980). By the early and mid-twentieth century, the topographic mapping program that Powell initiated had evolved and grown significantly. The program called for complete national coverage at a scale of 1:24,000. It depended on very extensive field crews made up of full-time USGS employees who, for months at a time, traveled their territory conducting survey work and field checking information such as whether streams were intermittent or perennial and the use of buildings (e.g., churches, schools, and post offices), and mapping trails, springs, and other important features. This was information that normally could not be interpreted from aerial photographs. These field crews would return periodically to one of four mapping centers that were geographically distributed across the continental United States to support their operations. These were the Eastern Mapping Center in Reston, Virginia; the Mid-Continent Mapping Center in Rolla, Missouri; the Rocky Mountain Mapping Center in Lakewood, Colorado; and the Western Mapping Center in Menlo Park, California. The geographic distribution of the centers allowed crews to operate most efficiently during the field season and then return to their home center for the winter months.

The last mapping field crews stopped working in the late 1980s and early 1990s. At that time, the USGS had implemented a limited revision program that depended largely on the interpretation of aerial photography and on the collection of other information from ancillary sources. In the early 1990s the USGS marked a significant milestone in Powell's original vision and in its mapping mission—complete, once-over coverage of the United States—in the standard 7.5 minute or 1:24,000 topographic map series. This monumental effort resulted in more than 55,000 topographic quadrangle maps, which had required more than 33 million person-hours to construct. These maps remain the most complete topographic map coverage of the United States in the public domain. The early 1990s also marked a period when other governmental and many private sector organizations were beginning to use geographic information systems (GIS), automated mapping, Global Positioning Systems (GPS), and orthoimagery to conduct mapping operations in their jurisdictions. The USGS came to rely heavily on partnerships with those organizations to share mapping information in order to maintain national coverage for topographic products and to remain the pre-eminent leader for a nationally consistent representation of the Nation's landscape. As the national geospatial archivist, it also became critical for the USGS to identify any gaps in partner-generated data and to develop strategies to fill those gaps in order to ensure that the resulting data set is truly "national" in its scope, coverage, and integrity.

A perfect storm of forces have coalesced in the last decade or so that not only support but virtually mandate the partnership-based business model of *The National Map*. The advent of the internet and the World Wide Web, on which mapping tools and software were widely available, broadened access and applications for topographic mapping. The decision in 2000 by President Clinton to end "selective availability," or degradation, of GPS signals enabled civilian users to pinpoint locations far more accurately than previously. Federal, state and local government agencies now had affordable mapping technologies available to them with unprecedented capabilities and opportunities. The broad recognition that geographic information is critical to promote economic development, improve stewardship of natural resources, and protect and manage the environment led to the embracing of a coordinated National Spatial Data Infrastructure (NSDI) that would support public and private sector applications of geospatial data in such areas as transportation, community development, agriculture, emergency response, environmental management, and information technology. The NSDI encompasses the policies, standards, and procedures for organizations to cooperatively produce and share geographic data. It became the mandate of the Federal Geographic Data Committee (FGDC) to develop the NSDI in cooperation with organizations from state, local and tribal governments, the academic community, and the private sector.

As the era of field-based crews has waned and these new forces, mandates, and opportunities have come into play, the USGS has increasingly relied on advances in technology, on the delivery and interactive capabilities of the World Wide Web, on its role to coordinate and lead rather than to produce, and on the resources and contributions of partners in ways that would have been unimaginable when the four USGS mapping centers were founded.

In November 2001, the USGS re-committed to Powell's vision of an enduring national map for the United States, built on up-to-date, accurate, and integrated digital geospatial data. That vision is being made a reality through the partnership model of *The National Map*, with all levels of government, the private sector, academia, citizen volunteers, and others sharing in the creation of a national map that not only will endure but will also capitalize on the initial investments of partners and provide opportunities to leverage collectively those independent contributions. This, too, fulfills Powell's vision of reducing the expense of a national map and creates an investment model that is in keeping with the evolving notion of a focused and responsive federal government. These broad-ranging and invested partnerships, under USGS leadership, are the threads from which *The National Map* is woven. The greatest challenge for the USGS in achieving *The National Map* as a national database is to establish and sustain these vital long-term partnerships as vibrant collaborators in a shared vision. Today, *The National Map* has more than 5,230 sources providing the content, most of which comes from registered partners who provide accurate and current content for their particular geographic area.

The National Map is comprised of these principal layers:
- Orthorectified imagery, both medium- and high-resolution
- Surface elevation data and, increasingly, bathymetry
- Vector feature data for hydrography, transportation, structures and publicly owned lands
- Geographic names
- Land cover data

Woven together, these principal layers are the foundation for an enduring portrayal of the Nation's landscape and the essential threads for creating cartographic renderings in a host of products, whether manipulated and integrated online or printed out in graphic, paper format.

In 2003, the National Academy of Science's Mapping Science Committee published a report, sponsored by the National Research Council, on *The National Map*, entitled *Weaving a National Map, Review of the U.S. Geological Survey Concept of The National Map* (National Research Council 2002). The committee critically analyzed the concept's intentions and determined that "the USGS…has made a bona fide effort to confront its future head-on with *The National Map* vision. If successful, the program will have great benefits to the nation. *The National Map* vision of the USGS is ambitious, challenging, and worthwhile."(page 2) The committee also emphasized that additional planning was required for this ambitious effort to be realized. Responding to this advice, the USGS developed and published a detailed implementation plan in October 2003 (USGS 2003). The USGS is continuing to implement and evolve that plan, which calls for the following actions:
- Establishing partnerships with federal, state and local governments, the private sector, and public organizations that will provide public domain data to build *The National Map*.
- Assisting partners to inventory existing data sets, document them using FGDC metadata standards, publish them in the NSDI clearinghouse and post them to the Geospatial One-Stop portal.
- Providing technical assistance to help develop and implement commercial translation tools for data that do not meet NSDI standards.
- Developing and implementing data filtering, generalization, and validation tools needed to obtain a nationally consistent representation from the composite of contributions from numerous partners.
- Determining where there are data gaps and developing cooperative strategies to fill in these gaps.
- Providing technical assistance to partners to help them implement NSDI standards.
- Ensuring unrestricted access to public domain data by operating an Internet-accessible distributed data network.

- Providing visualization and analysis tools to ensure users can display and interpret the basic geographic data of *The National Map*.
- Implementing the capability to provide derivative products from *The National Map* to meet the needs of users in non-digital environments, such as emergency first responders for whom computer access is not always possible and for which quick notations on a paper map are critical to split-second decisions that save lives and property. Future derivative products are on the technology horizon in wireless applications that can be deployed in the field on personal digital assistants (PDAs) and perhaps on yet-to-be envisioned devices.

At this point in its implementation (2005), *The National Map* is a distributed network of databases that provide a consistent mapping framework for the country. It serves as an information foundation for integrating, sharing and using other spatial data easily, ranging from real-time stream-flow-monitoring data to critical national infrastructure, to demographic data and economic statistics, to weather and hazards data.

The partnerships that underlie *The National Map* are providing the foundation and the essential threads that will result in well-woven and spatially strong national coverage that will be complete and up to date. While *The National Map* is clearly focused on delivery and access through the Web, the USGS recognizes that there is still, and will continue to be for the foreseeable future, a demand for paper topographic map products. *The National Map* supports this need through product generation capabilities that the USGS is developing to enable users to quickly and easily generate maps for printing from *The National Map* data.

In August 2005, the USGS took an additional step to organize its geospatial activities by bringing together topographic mapping, partnerships, the FGDC function, the Geospatial One-Stop portal, and related offices and facilities under the National Geospatial Programs Office (NGPO). Three fundamental contributions that *The National Map* brings to the NGPO are:

i) Improving the currentness, information content, and completeness of geospatial data available in the public domain by building distributed databases through partnerships with state and local governments that collect and maintain high-resolution, current data;

ii) Ensuring the availability of nationally consistent and integrated geospatial data by leading implementation by partners of international and national standards sponsored by the FGDC and NSDI communities;

iii) Maintaining and promoting access to a long-term archive of geospatial data that describe the Earth's surface and are essential to understanding fundamental Earth processes and to serving the collaborative research needs of USGS science programs as part of an integrated information environment.

The goals for *The National Map* would be unattainable without the active participation of federal, state, and local government agencies, the private sector, volunteers, and non-governmental and professional organizations as invested and involved business partners. To build and sustain these partnerships, the USGS has established a network of employees who are strategically distributed across the country and whose responsibility it is to establish, implement, and promote partnerships. Because of the critical role of this network of liaisons to the success of *The National Map*, the USGS is focusing increased resources on this effort, including deploying additional people in key spots across the country to add breadth and depth to the partnership focus.

In an action related to the creation of the NGPO, in 2005 the USGS consolidated its mapping activities with the establishment of the National Geospatial Technical Operations Center (NGTOC). The NGTOC is a center of government excellence for geospatial information, technology, and innovation for key partners and stakeholders. The NGTOC provides geospatial technical expertise, services, and innovative solutions in support of the NGPO in its development of *The National Map* and in the implementation of key components of the

NSDI. The center consolidates the four USGS mapping centers described earlier into a single unified organization.

To carry out the operational components of *The National Map*, the NGTOC:

- Provides IT infrastructure (servers, portals, expertise) to support the NGPO
- Manages multipurpose geospatial data products and services contracts
- Performs integration and quality assurance of geospatial data provided by contractors and partners
- Supports geospatial standards and data model development, testing, and implementation
- Provides geospatial data systems integration and technical support for a national geospatial enterprise architecture, as a component of the Federal Enterprise Architecture
- Provides for archive of and access to digital geospatial data holdings, including dissemination to the public incorporating an e-commerce solution
- Generates map products consistent with the role of government
- Provides technical assistance on geospatial data issues to the USGS and its partners
- Provides education and training related to the program activities of the NGPO
- Conducts and sponsors research in geospatial information technology best practices and innovative solutions as a center of excellence
- Provides incentives to partners in developing technological solutions for geospatial activities

The development of the full capacity of the NGTOC is a key step in modernizing the USGS mapping program. It represents a fundamental shift in the business model that supports USGS mapping efforts, from relying on USGS employees for data collection and maintenance to capitalizing on the capabilities of partners and the private sector for many of these services and many of the other functions necessary to assure the availability of current, complete and consistent topographic data in the public domain.

The USGS is committed to providing *The National Map* for use by governments at all levels, the research and science community, and the public, ranging from digital data to be used in a GIS application to high-quality paper maps. John Wesley Powell got it right when he determined that for the USGS to carry out its science mission it would need to understand the landscape of the Nation and to have—at its own disposal and that of countless other users—a national data set that was enduring. The "national domain" of the USGS mission has changed in the ensuing years. Its motto today is *science for a changing world*, and in *The National Map*, the Nation has a national mapping program that meets the geospatial needs of that changing world.

References

National Research Council. 2002. *Weaving a National Map: Review of the U.S. Geological Survey Concept of The National Map.* Washington, DC: National Academy Press.

Rabbitt, Mary C. 1980. *Minerals, Lands, and Geology for the Common Defence and General Welfare,* Vol. 2, 1879-1904, A History of Geology in Relation to the Development of Public Land, Federal Science and Mapping Policy, and the Development of Mineral Resources in the United States During the First 25 Years of the U.S. Geological Survey. Washington, DC: US Government Printing Office.

US Geological Survey. 2001. *The National Map: Topographic Mapping for the 21st Century.* Reston, Virginia: USGS.

US Geological Survey. 2003. *Implementation Plan for The National Map v. 1.0.* Reston, Virginia: USGS.

CHAPTER 48

The Integration of Internet GIS and Wireless Mobile GIS

Ming-Hsiang Tsou

48.1 Introduction

The paradigm of geographic information systems (GISs) is shifting from a closed, centralized GIS architecture, to an open, distributed geographic information services (GIServices) framework. With advances in computer networking and wireless communication technology, GIS is moving toward an integration of GIServices and spatial analysis functions via the Internet and wireless communication. Wireless mobile GIS and Internet GIS can be combined together to provide ubiquitous GIS power for various users at different locations with different tasks, such as urban planners, cartographers, first responders, bus drivers, realtors, and the general public.

The integration of GIServices (including Internet GIS and wireless mobile GIS) can synergize information-sharing and facilitate the diffusion of GIS technologies into broader applications and potential users. Online GIServices will encourage multidisciplinary cooperation between the GIS community and other communities, such as the computer science community, the education community, and the geoscience communication community (Plewe 1997; Tsou and Buttenfield 2002; Peng and Tsou 2003). For example, Figure 48-1 illustrates a web-based GIS education project developed for the teacher and students at the Hoover High School in San Diego. The map shows the location of fast food restaurants and the overweight population nearby the high school from the census data set. High school students and teachers can use the online mapping tools to create a community-based project regarding the obesity problem. One unique potential of this application is that the high school students might utilize handheld Global Positioning System (GPS) units to record the locations of fast food restaurants on the streets and then combine these points into the web maps (Figure 48-1). This web GIS education program was developed by San Diego State University and was funded by the National Science Foundation–Advanced Technology Education (NSF–ATE) program (project website: http://geoinfo.sdsu.edu/hightech). Besides this education example, other disciplines and communities can also adopt integrated Internet GIS and wireless mobile GIServices for providing better services and functions, such as disaster management, homeland security, tourism, and natural resource conservation.

Recent developments in advanced network computing technologies, such as web services, grid computing, and semantic web provide a promising future for the next generation of Internet GIS applications. Traditional GIS, designed as isolated islands, will become increasingly less attractive, and may disappear altogether. The reusable and interoperable open and distributed GIServices will broaden geographic information uses into an increasingly wide range of online geospatial applications. New wireless communication technology, such as Bluetooth, Wi-Fi, 4G cellular systems, and WiMAX, also will transform GIS into wireless mobile GIS, which can provide ubiquitous access of geospatial information from portable devices or mobile phones from anywhere.

This section will highlight the importance of combining wired Internet GIS and wireless mobile GIS technology that can provide complementary GIServices for each other. Without the linkage to mobile GIS devices, most Internet GIS applications will lack ground-truth information for real-time or near real-time data update and GPS tracking functions. Without the powerful databases and large-size remotely sensed imagery archived in the Internet

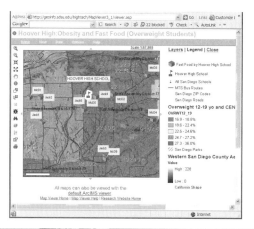

Figure 48-1 The top image combines Internet GIS and wireless mobile GIS for a high school GIS education project (http://geoinfo.sdsu.edu/hightech/mapviewer.htm). The locations of fast food chains are tracked by using mobile GIS devices and GPS (a Trimble GeoExplorer XM unit), image on left. The image on the right shows an alternative method of mobile GIS by using a Pocket PC and a cellular phone. See included DVD for color version.

GIS servers, mobile GIS applications will become very limited by the incomplete data sets available in restricted computer hardware. In the future, a comprehensive GIService will rely on both wireless mobile GIS and wired Internet GIS. The following sections will highlight the major characteristics of Internet GIS, components of wireless mobile GIS and the major challenges in integrating the two technologies.

48.2 Characteristics of Internet GIS

There are many different Internet GIS applications, ranging from data clearinghouses, web mapping, data portals, web-based decision support systems and GPS tracking, to digital Earth virtual globes. In general, we can categorize various Internet GIServices into three types: data sharing, information sharing and knowledge sharing (Tsou 2004a) (Figure 48-2).

The first type of Internet GIServices is for data sharing, which combines the functions of online data archive and data search services. Two typical applications are online data warehouses (or data archive centers) and online data clearinghouses. An online data warehouse is for archiving, accessing and downloading both GIS databases and/or remotely sensed imagery. A web-based data clearinghouse can help users to search and index the contents of metadata, and then access the actual data through the descriptions of metadata.

Three Types of Internet GIS	
Data *Sharing (Exchanging)*	***Applications*** Online data warehouses (data archive) Online data clearinghouse (metadata)
Information *Sharing (Map Sharing)*	***Applications*** Web-based map display Navigation services
Knowledge *Sharing (Spatial Analysis and Modeling)*	***Applications*** Online GIS models Web-based spatial analysis tools

Figure 48-2 Three types of Internet GIS (modified from Tsou 2004a).

The second type is for information sharing and map sharing. Multiple interactive map servers and mobile navigation services are the typical applications. Web-based mapping functions include the display, zoom-in/out and query of spatial information. The major requirement of information sharing services is to provide effective web-based display mechanisms and client/server communication protocol.

The third type of Internet GIS focuses on the sharing of knowledge and GIS models. This is the most challenging task for the development of Internet GIS and only a few applications are available today. The goal is to provide online GIS modeling and spatial analysis functions without running GIS engines or software packages locally. Some Internet GIS applications utilize Java language or other distributed component technologies (like .NET or web services) to develop online GIS model functions. The implementation of these web-based software components can provide ubiquitous access for all different types of GIS applications.

More recently, a new concept of Internet GIS has started to emerge called "GIS portals." A web-based GIS portal can provide all three types of Internet GIS services together, including data sharing, map display and some spatial analysis functions via web services. This new trend will integrate various GIS functions, maps, and data servers into a systematic framework via a single entry point rather than create scattered Internet GIS applications.

Another important change in Internet GIS is the dramatic growth of its users. According to recent research from ComScore Network (http://www.ebrandz.com/newsletter/2005/July/1july_31july_article1.htm), online map users are a huge market for business applications. In May 2005, Time Warner (MapQuest.com) estimated 43.7 million US visitors, Yahoo!Maps (maps.yahoo.com) 20.2 million users, Google Maps (maps.google.com) 6.1 million and Microsoft's MSN MapPoint (mappoint.msn.com) 4.68 million visitors. This is a huge market for online mapping services compared to traditional GIS users. It is also very interesting to see the new online mapping providers, such as Google, Microsoft, Yahoo, and Amazon.com join the market of online mapping and provide more diversified geospatial information services to the public.

One key issue of the development of Internet GIS is to provide the most updated data online. Therefore, we will need to utilize wireless mobile GIS to provide real-time or near-real-time information feedback to the Internet GIS servers. The following section will highlight the major components of mobile GIS.

48.3 Components of Wireless Mobile GIS

Mobile GIS is an integrated software/hardware framework for the access of geospatial data and services through mobile devices via wired or wireless networks (Tsou 2004b). There are two major application areas of mobile GIS: field-based GIS and location-based services (LBS). Field-based GIS focuses on the GIS data collection, validation and update in the field, such as adding or editing map features or changing the attribute tables in an existing GIS data set. Location-based services focus on business-oriented location management functions, such as navigation, street routing, finding a specific location, tracking a vehicle, etc. (Jagoe 2002, OGC 2003). The major differences between the field-based GIS and LBS are the data-editing capabilities. Most field-based GIS applications need to edit or change the original GIS data or modify feature attributes. LBS rarely changes original GIS data sets but rather uses them as background or reference maps for navigation or tracking purposes. Most field-based GIS software packages are cross-platform and independent of hardware devices. On the other hand, LBS technologies focus on creating commercial value from locational information. Each mobile phone system has its own proprietary operating system that is very difficult to customize.

The architecture of mobile GIS is very similar to the Internet GIS. It follows the concepts of client/server architecture as in traditional Internet GIS applications. Client-side mobile GIS components are the end-user hardware devices that display maps or provide analytical results of GIS operations. Server-side components provide comprehensive geospatial data and perform GIS operations based on a request from the client-side components. Between the client and server, there are various types of communication networks (such as hard-wired cable connections or wireless communications) to facilitate the exchanges of geodata and services. Figure 48-3 illustrates the six basic components of mobile GIS: 1) positioning systems; 2) mobile GIS receivers; 3) mobile GIS software; 4) data synchronization and wireless communication; 5) geospatial data; and 6) GIS content servers (Tsou 2004b).

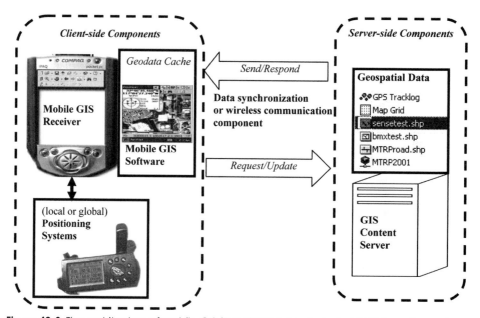

Figure 48-3 The architecture of mobile GIS (Tsou 2004b). See included DVD for color version.

Mobile GIS can provide geospatial information and GPS coordinates for field-based personnel conducting remote field (*in situ*) GIS tasks. To enable comprehensive mobile GIS, wireless communication is essential for connecting mobile GIS devices and GIS content servers. Recent progress in broadband wireless technology is the major momentum for the integration of mobile GIS and Internet GIS. The wireless service coverage and the bandwidth (speed) are the two key issues for wireless communication. There are many different wireless technologies, ranging from a walkie-talkie to high-speed WiMAX, to satellite phone systems. Based on the speed of data transfer, current wireless technologies can be categorized into two groups: narrowband wireless systems and broadband wireless systems. To communicate between mobile GIS and Internet GIS, broadband wireless technology is a better choice because most geospatial information and remote sensing data are very large and complicated, which require broadband wireless communication. The size of wireless coverage area is also an important criterion for mobile GIS applications.

Generally, there are three different types of wireless communication systems: ad-hoc systems, cellular phone systems and Wi-Fi/WiMAX data network systems. The ad-hoc wireless systems are custom designed for specific applications, such as direct satellite phone systems, General Mobile Radio Service (GMRS) for walkie-talkie devices, or ham radio communication. Usually, these systems are narrowband and localized for a small group of special users and require specialized user licenses. 3G and 4G mobile cellular phone communication systems can provide a good coverage area with decent-bandwidth communication. Cellular phone systems can allow other wireless devices, such as PDA and Pocket PC, to receive multimedia services (such as streaming audio and video on the devices). Most mobile GIS devices can utilize 3G or 4G cellular communication systems for collaborative work with Internet or web-based GIS applications. However, the major disadvantage of cellular phone systems is the limit of bandwidth between 300Kbps to 3Mbps. Wi-Fi/WiMAX data network systems are another promising category for broadband wireless mobile GIS communication (Intel 2004). Currently, the most common wireless LAN infrastructure is the IEEE 802.11 (or Wi-Fi) technology. Many computers, PDAs, printers, etc., have begun to adopt Wi-Fi—or IEEE 802.11—as their major communication channels. The problem of Wi-Fi for wireless mobile GIS is its short coverage area from a Wireless Access Point (WAP), usually within 100 m only. WiMAX will be a better choice for mobile GIS because it can provide a large 4–6 mile coverage range. WiMAX is an emerging IEEE 802.16 standard (available in late 2006) for broadband Wireless Wide Area Network (WWAN) or Metropolitan Area Network (MAN) applications. Its communication signals can cover a large area range (up to 20 miles for the long-distance setting) with the speed of 30–75 Mbps. With such range and high throughput, WiMAX is capable of delivering comprehensive data sets or large imagery files from Internet GIS servers to wireless mobile GIS units. The next section will discuss the major issue for the integration of wireless mobile GIS and Internet GIS.

48.4 The Integration of Wireless Mobile GIS and Internet GIS

Different types of GIS tasks and applications will require different methods to combine wireless mobile GIS and Internet GIS. Various integration techniques will be adopted for the linkage between wireless mobile GIS and Internet GIS in order to fit the needs of different users, different requirements, and different data formats. The following discussion will focus on three aspects of the system integration methods: 1) asynchronous vs. synchronous connections; 2) thin-client model vs. thick-client model; and 3) loosely connected methods vs. tightly integrated methods.

48.4.1 Asynchronous versus Synchronous Connections

The priority of data update needs is the key to deciding whether or not an Internet GIS server requires synchronous connections to wireless mobile GIS. If a GIS task requires real-time information update, such as GPS tracking and emergency response, the Internet GIS connection should be synchronized. A synchronous connection will allow wireless mobile GIS to provide real-time data update back to one or multiple Internet GIS servers (adding new points and polygons, tracking the locations from GPS, modifying attributes, etc.) Also, Internet GIS servers can provide updated information in real time and distribute to multiple mobile GIS units immediately via wireless communications. However, it is difficult to create the two-way synchronous communication because of the heterogeneous operating systems between Internet GIS servers and wireless mobile GIS units. Many mobile GIS applications only provide one-way synchronization (sending data from Internet GIS to wireless mobile GIS units).

On the other hand, if a GIS task does not rely on the critical information updated from the field, such as urban planning, and facility and utility management tasks, asynchronous methods might be more appropriate. An asynchronous communication model can allow mobile units to work independently and save the updated information temporally in the local cache. Mobile GIS users then bring the units back to specific locations or devices to upload the new data sets back to an Internet GIS server at the end of the task (per day or per week). Usually, the cost of establishing synchronized connections is much higher than for asynchronous methods. Two-way synchronized connections also are more expensive than one-way synchronized connections.

48.4.2 Thick-Client Model versus Thin-Client Model

The second consideration is the choice of thick-client model or thin-client model. The terms thick client and thin client are defined by the computer networking community. In networking terminology, the thick-client model is defined as having major operations and calculations executed on the client side. On the other hand, a thin-client model may require that selected operations run on the server side. Therefore, the thick-client model for wireless mobile GIS indicates that the client side (mobile GIS units) will have powerful computing capability with high-speed CPU and large-size memories. Users can process advanced GIS functions locally on handheld devices or portable PCs. The thin-client mobile GIS model indicates that the client-side mobile GIS is only a terminal for Internet GIS servers. In this case, the client units will need to send GIS operations and tasks back to Internet GIS servers, then get the results from the GIS servers. The balance of functionality and performance between wireless mobile GIS clients and Internet GIS servers will be a critical issue for the success of integration.

48.4.3 Loosely Connected versus Tightly Integrated Systems

The third issue is the implementation of a relationship between wireless mobile GIS and Internet GIS—loosely connected systems vs. tightly integrated systems. Currently, most GIS projects and applications are loosely connecting mobile GIS and Internet GIS via public communication channels or asynchronous communication methods. Loosely connected mobile GIS units can be easily switched to connect different Internet GIS servers. There is no mandatory regulation between client units and servers. Flexibility is the major advantage for loosely connected mobile GIS frameworks. However, the loosely connected systems are less reliable compared to tightly integrated systems because the client units are not guaranteed to access Internet GIS servers during the operations. If the GIS servers shut down accidentally or the public communication network has problems, the whole system will be out of service. On the other hand, tightly integrated wireless mobile GIS are usually more reliable, but also more expensive to set up. Most tightly integrated systems are using preparatory and customized products with private communication channels for specific tasks or projects, such as E-911 systems or flood control water gauge censor systems.

In general, every GIS project requires a different strategy and method for successful implementation due to the nature of GIS tasks, geodata format and size, map display requirements, etc. There currently is no perfect framework for connecting wireless mobile GIS and the Internet GIS server. Some GIS task requirements and implementation methods will rely on the progress of new technologies in web applications. The next section will highlight some potential new technologies for bridging wireless mobile GIS and Internet GIS applications.

48.5 New Potential Technologies for Bridging Wireless Mobile GIS and Internet GIS

The progress of Internet GIS relies greatly on new web technologies. Recently, many technology breakthroughs have been achieved in online graphic display techniques and web services. For example, Scalable Vector Graphics (SVG), mobile web services, and Ajax (Asynchronous JavaScript and XML) are three promising future technologies for the integration of mobile GIS and Internet GIS. SVG is an XML-based, two-dimensional vector graphics media format specified by the W3C in 2001 (version 1.0) and in 2003 (version 1.1) (W3C 2003). There are three types of SVG profiles: SVG Full, SVG Basic, and SVG Tiny. SVG Full is suitable for desktop or laptop PCs. SVG Basic (smaller than SVG Full) is designed for Pocket PC or PDAs. SVG Tiny is designed for mobile phones. The advantage of mobile SVG (Basic and Tiny) compared to other graphic formats is that it can provide a compact, multimedia-enabled vector display format. Therefore, SVG images are scalable and dynamic, and can be used for both the regular Internet mapping display on PCs and the small screen display of various mobile GIS devices, such as PocketPC and cellular phones.

Mobile web services are another promising new technology that is an extension of general web services built upon XML; Simple Object Access Protocol (SOAP); Universal Description, Discovery, and Integration (UDDI); and Web Services Description Language (WSDL). Mobile web services can combine multiple functions and customizable information provided by web service providers for different mobile applications and users. The advantage of adopting web services for mobile GIS applications is that web services can provide a flexible combination of multiple web computing techniques with modern enterprise GIS architecture. The contents of mobile web services include short messaging services (SMS), multimedia messaging services (MMS) and location-based services (LBS). Most mobile web services significantly rely on server-side computing power. For mobile GIS, web services works like the thin-client model, with more flexible choices of GIS functions provided by remote web servers.

The development of Ajax (Asynchronous JavaScript and XML) is also a promising web technology (Garrett 2005). Traditional Internet GIS applications and web-based mapping tools always suffer from the slow response and the lack of high-resolution images resulting from the limitation of image data sizes and the client/server communications. The new Ajax technologies can significantly improve the performance and response times of Internet GIS applications. [maps.search.ch] and [maps.google.com] are the two early examples of Ajax Internet GIS applications. One unique advantage of Ajax is the key word "Asynchronous." According to the first Ajax paper (written by Jesse James Garrett in 2005), "An Ajax application eliminates the start-stop-start-stop nature of interaction on the web by introducing an intermediary—an Ajax engine—between the user and the server…. The Ajax engine allows the user's interaction with the application to happen asynchronously—independent of communication with the server. So the user is never staring at a blank browser window and

classic web application model (synchronous)

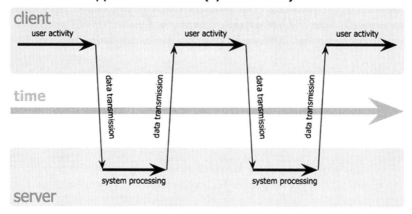

Ajax web application model (asynchronous)

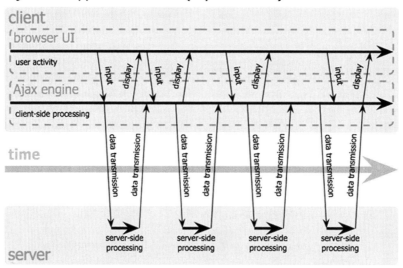

Jesse James Garrett / adaptivepath.com

Figure 48-4 The synchronous interaction pattern of a traditional web application (top) compared with the asynchronous pattern of an Ajax application (bottom) (Garrett 2005).

an hourglass icon, waiting around for the server to do something." (Garrett 2005). Figure 48-4 illustrates the synchronous interaction pattern of a traditional web application (top) compared with the asynchronous pattern of an Ajax application.

For the Internet mapping applications, Ajax can create temporary image cache functions in the background when the web browser displays the image or maps. While the user watches the downloaded imagery, the server keeps sending additional images or maps into the cache area (Figure 48-5). Therefore, when the user moves the displayed maps or imagery, the cached image will be displayed at the user's end more quickly. The key concept in adopting Ajax for Internet mapping is to be able to predict the users' behaviors (Zoom-in, Zoom-out, and Pan) and then pre-cache the required images and actions before the users actually perform them. In the future, we will see more intelligent web technologies applied in Internet and mobile GIS for creating more responsive, user-friendly and interactive GIServices.

The recent development of grid technology (Foster et al. 2001; Armstrong et al. 2005) might provide a possible framework for the deployment of dynamic Internet GIServices with

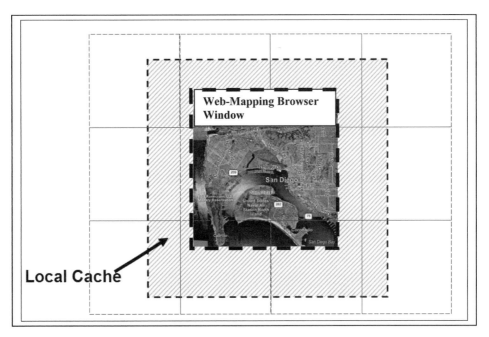

Figure 48-5 A local cache of geospatial data (bigger than the browser window) by using Ajax technology. See included DVD for color version.

its powerful computational and resource management capabilities. In the computer science community, the focus of grid technology is to resolve low-level grid computing technique issues (e.g., communication, protocols and resource management) and to build a collaborative network grid architecture (e.g., Globus Toolkit). However, the deployment of Internet GIS needs a high-level application framework rather than low-level grid computing architecture.

Semantic web (Berners-Lee et al. 2001) is another important development for the future of distributed GIServices because semantic web can facilitate web-based data sharing within a global network system. The technology can provide better definition of web data and services, thus broad-scale data sharing and reuse could become possible. A working group within W3C (World Wide Web Consortium) has defined related standards and languages for the real applications of semantic web technologies (www.w3.org). RDF (Resource Description Framework) was designed to organize web information into triple terms for easier data retrieval (http://www.w3.org/RDF/). To better handle terms and relations in semantic webs, a Web Ontology Language (OWL) is also proposed to define terminology used for specific contexts and properties in terms of classes and relations (http://www.w3.org/TR/owl-guide/).

One very interesting recent development is the emergence of Digital Earth viewers or digital globes by using 3D representation and visualization tools, such as Google Earth (http://earth.google.com) and NASA World Wind (http://worldwind.arc.nasa.gov/). The framework of Digital Earth provides a very promising direction for integrating multiple types of geospatial information, including remote sensing images, GIS data, GPS tracking path, etc. Open source GIS software is another promising direction in the development of Internet GIS and mobile GIS. More and more GIS professionals are developing open-source Internet GIS servers and toolkits because the open-source framework can provide a more flexible way to customize functions and tools (such as "Mapserver" by the University of Minnesota, "GeoTools" by a group of GIS programmers, and "PostGIS" by Refractions Research, Inc.). GIS professionals are increasingly choosing an open-source GIS development environment rather than proprietary, black-box and vendor-based GIS packages (Ramsey 2005).

Manual of Geographic Information Systems

48.6 Conclusion

The integration of Internet GIS servers and wireless mobile GIS units will facilitate the adoption of geospatial technology in many fields, such as education, homeland security, disaster management and public transportation. Ideally, multiple map layers and GIS functions from heterogeneous Internet GIS servers could be connected dynamically to hundreds of wireless mobile GIS units remotely. To accomplish such tasks, tremendous amounts of geospatial data and computational power would be requested and transferred across the public Internet or private communication networks. Therefore, the GIS community needs to establish a comprehensive cyber-infrastructure to organize all available heterogeneous GIServices and web servers and to streamline various data flow and operation procedures across the networks. Such GIService cyber-infrastructure needs to become dynamically adjustable in order to accommodate complicated network environments (dial-up modem, cable modem service, ADSL, T1/T3, Wi-Fi, WiMAX, cellular networks, etc.) This will be a great challenge for the GIS community to establish a comprehensive cyber-infrastructure for integrating Internet GIS and wireless mobile GIS applications.

One possible approach is to develop web-based GIS portals, which can utilize grid computing technology and semantic web technology to establish a high-level service-oriented cyber-infrastructure. Users can use the portal to analyze geospatial problems, to find out possible useful mapping services, to request geospatial data and to aggregate multiple GIS analysis results. The intelligence of web portals can be introduced by the semantic web technology that organizes geospatial ontologies to facilitate the interoperability between users and the machines. The computing power of web portals will be supported by the grid computing technology that can gather hundreds of computers to provide more effective online GIS functions and spatial analysis.

This chapter illustrates the major approaches of integrating Internet GIS and wireless mobile GIS and highlights the importance of combining the two frameworks. In the future, a comprehensive, distributed GIService will rely on both wireless mobile GIS and wired Internet GIS. As time wears on, more progress and changes will be made in Internet GIS and mobile GIS, leading to infinite possibilities and great potentials. In the next decade, for example, everyone might be able to utilize their mobile phones to download a $5 e-coupon for a coffee shop in the next street block, or track their children's location remotely, or find any available parking lots near a shopping mall. To summarize, the integration of Internet GIS and wireless mobile GIS will attract more people to use GIS and the new cyber-infrastructure will transform the way people live, work and behave. The general public will start to use novel ways to collect, analyze and distribute information—geographically.

Acknowledgements

The author wishes to acknowledge and express the appreciation of funds received from the National Science Foundation–Advanced Technology Education (NSF–ATE) program. This paper forms a portion of the NSF–ATE project, "A Scalable Skills Certification Program in GIS" (NSF-ATE DUE 0401990). Additional acknowledgement is extended to Dr. Carl Eckberg at Computer Science Department, San Diego State University, and Anthony Howser, Gagan Arora, and Kimberly Dodson, three graduate students at San Diego State University, for their efforts in support of the NSF–ATE research project.

References

Armstrong, M. P., M. K. Cowles and S. Wang. 2005. Using a computational grid for geographic information analysis: A reconnaissance. *The Professional Geographer* 57 (3):365–375. <http://www.blackwell-synergy.com/doi/pdf/10.1111/j.0033-0124.2005.00484.x> Accessed 5 February 2008.

Berners-Lee, T., J. Hendler and O. Lassila. 2001. The semantic web. *Scientific American* 284 (5):34–43.

Foster, I., C. Kesselman and S. Tuecke. 2001. The anatomy of the grid: Enabling scalable virtual organizations. *International J. Supercomputer Applications* 15(3).

Garrett, J. J. 2005. Ajax: A New Approach to Web Applications. White paper. <http://www.adaptivepath.com/publications/essays/archives/000385.php> Accessed 25 January 2008.

Intel. 2004. Understanding WiMAX and 3G for Portable/Mobile Broadband Wireless. White paper. <http://www.itr-rescue.org/bin/pubdocs/mtg-weekly/9-16-05%20Intel_WiMAX_White_Paper%20(Hassib).pdf> Accessed 19 July 2008.

Jagoe, A. 2002. *Mobile Location Services: The Definitive Guide.* Upper Saddle River, N.J.: Prentice Hall.

OGC (Open GIS Consortium). 2003. *OpenGIS Location Services (OpenLS): Core Services (Part 1–Part 5).* Version 0.5.0. OGC-03-006r1. Wayland, Mass.: Open GIS Consortium, Inc.

Peng, Z. R. and M. H. Tsou. 2003. *Internet GIS: Distributed Geographic Information Services for the Internet and Wireless Networks.* New York: John Wiley & Sons.

Plewe, B. 1997. *GIS Online: Information Retrieval, Mapping, and the Internet.* Santa Fe, N.M.: OnWord Press.

Ramsey, P. 2005. The State of Open Source GIS. White paper. Refractions Research Inc. <www.refractions.net/white_papers/> Accessed 25 January 2008.

Tsou, M. H. 2004a. Integrating web-based GIS and online remote sensing facilities for environmental monitoring and management (special issue on the Potential of Web-based GIS). *Journal of Geographical Systems* 6 (2):155–174.

———. 2004b. Integrated mobile GIS and wireless Internet map servers for environmental monitoring and management (special issue on Mobile Mapping and Geographic Information Systems). *Cartography and Geographic Information Science* 31 (3):153–165.

Tsou, M. H. and B. P. Buttenfield. 2002. A dynamic architecture for distributing geographic information services. *Transactions in GIS* 6 (4):355–381.

W3C. 2003. Scalable Vector Graphics (SVG) 1.1 Specification. <http://www.w3.org/TR/SVG/> Accessed 25 January 2008.

CHAPTER 49

Principles and Techniques of Interactive Web Cartography and Internet GIS

Maged N. Kamel Boulos

49.1 Introduction

Maps published in cyberspace (the Internet) could be either maps covering topics related to cyberspace itself, e.g., cybermaps of Web resources and non-geographic information spaces—see examples in Boulos (2003a) and at http://www.cybergeography.org/atlas/[1], or maps using cyberspace, usually the Web, as a publishing and dissemination medium, e.g., MARA/ARMA (Mapping Malaria Risk in Africa/*Atlas du Risque de la Malaria en Afrique*) maps of malaria risk in Africa—see http://www.mara.org.za/. The same principles and opportunities of Web cartography apply to both types. However, the main focus of this text is on the latter type.

With the advent of Internet GISs (Geographic Information Systems) and mapping, GIS information products (GIS output) have become accessible to a much wider audience of decision makers who have no GIS software skills. The only software expertise they are required to have is the ability to use any standard Web browser. Internet GIS applications include:

- Routing/driving directions and similar applications that help with identifying/locating real-world services according to some user-defined criteria, e.g., the nearest or cheapest service. Examples of this type of application include healthcare facility finders and restaurant locators;
- Those applications that help with indexing and accessing/navigating data sets based on location, for example, weather maps (e.g., http://uk.weather.com/maps/ and the US Environmental Protection Agency [EPA] daily updated interactive air quality maps http://airnow.gov/), pollen maps for allergic people (e.g., http://uk.weather.com/outlook/health/allergies/precipmaps?mdefmaptype=grass), and maps of US presidential election results by state (e.g., http://www-personal.umich.edu/~mejn/election/);
- Those applications enabling the integration and interactive visualization of any/many geographically-differentiated indicators/outcomes, services, and statistics of interest through thematic mapping and other visualization techniques (e.g., the US Centers for Disease Control and Prevention—CDC Oral Health Maps http://apps.nccd.cdc.gov/gisdoh/). The online interfaces of these applications sometimes feature advanced tools for graphically interrogating and manipulating the underlying data, and visually and instantly observing the effects on the resultant online maps, e.g., applications allowing users to experiment with different "what if" scenarios for the optimal siting of a new healthcare or commercial facility; and
- Applications and maps used for real-time or near-real-time monitoring and surveillance (environmental, outbreak and disease surveillance, including early detection of bioterrorist attacks; see, for example, the RODS—Real-time Outbreak and Disease Surveillance system: http://rods.health.pitt.edu/NRDM.htm and http://openrods.sourceforge.net/).

---

1. All hyperlinks in this chapter were accessed 11 January 2008.

Web-based maps offer the following extra features not found in conventional paper-based maps: (1) real-time or near-real-time map updates based on the latest data sets, (2) interactivity—desktop GIS-like functionality, e.g., drill-down and zooming, map querying, measuring distances on the map, and switching map layers on and off, and (3) availability to larger audiences/wider and more rapid dissemination of information compared to other publishing media (Boulos 2004b). It should be noted that different security measures are also available to prevent unauthorized access to those maps and applications dealing with confidential or sensitive data (usually running over secured networks, intranets and extranets).

Web-based interactive geographical map interfaces offer an intuitive way of indexing, accessing, mining, and understanding large data sets describing geographically-differentiated phenomena, and can act as an enhanced alternative or supplement to purely textual online interfaces. Such online interactive maps, when based on good science, can help us understand and instantly spot the relationships, patterns and trends buried in the original data sets, and also enable quick and effective visual comparisons to be made between different geographical areas and also over time when data sets and maps for successive periods of time are available (Boulos 2004c; Boulos et al. 2005a).

The process of turning raw tabular data into much more useful and accessible visual information in the form of interactive Web maps is much needed to support and empower organizations, planners and decision makers in their planning and management tasks, and even members of the general public.

Some of the best examples of Web-based maps were produced during the 2002–2003 SARS (Severe Acute Respiratory Syndrome) outbreak, which is considered the first major new infectious disease of the 21st century and the Internet age that took full advantage of the opportunities for rapid spread along international air routes. A review by this author (Boulos 2004b) describes several geographic mapping efforts of SARS on the Internet that employed a variety of techniques like thematic mapping and choropleth rendering, graduated circles, graduated pie charts, buffering, overlay analysis, point-in-polygon analysis, and animation. The aim of these mapping services was to improve global vigilance and awareness at all levels, educate the public (especially travelers to potentially at-risk areas), and assist public health authorities in analyzing the temporospatial trends and patterns of SARS in order to make well-informed decisions when designing and following up epidemic control strategies, or issuing and updating travel advisories.

The online Bradford Community Statistics Project (http://www.communitystats.org.uk/) provides another good example of public-participation GIS projects, with the aim in this case to empower residents in developing their own policy initiative and funding proposals. Other examples that aim at empowering members of the general public include Neighborhood Information Systems like UpMyStreet (http://www.upmystreet.com/) in the UK, and the US EPA WME (Window to My Environment http://www.epa.gov/enviro/wme/). The latter example (WME) is designed to provide public accessibility to a wide range of federal, state, and local geospatial data about environmental conditions and features in any US location.

Table 49-1 lists some other examples of freely available online maps, in addition to the many examples mentioned elsewhere throughout this text.

49.1.1 A Simple Technical Classification of Web Maps

Web maps are sometimes classified into static maps and dynamic maps according to whether or not they are animated (Kraak and Brown 2001). Each group is further subdivided into view-only maps and interactive maps (Figure 49-1). Non-animated (static) interactive Web maps, also called clickable maps, imagemaps or hypermaps, can be served either as client-side imagemaps or as server-side imagemaps, depending on where mouse-click coordinates are resolved.

Interacting with a map can stimulate a user's (visual) thinking and encourage exploration (Kraak and Brown 2001; Kraak 2004). Clicking an object on such maps can lead to other

Table 49-1 Assorted examples of freely available online maps, providing a glimpse at the wide range of applications using such maps, the various degrees of functionality they offer, and the many methods that are available today for implementing them.

Web map title/short description	Internet address
Dartmouth Atlas of Health Care series – Adobe Reader for PDF documents required	http://www.dartmouthatlas.org/ and this example: http://www.dartmouthatlas.org/atlases/99Atlas.pdf
Interactive injury maps (US Centers for Disease Control and Prevention – CDC)	http://www.cdc.gov/ncipc/maps/default.htm
Kids' Well-being Indicators Clearinghouse (New York State, US) – Macromedia Flash required	http://www.nyskwic.org/index.cfm and http://www.nyskwic.org/access_data/map_select.cfm
Leeds Health Atlas	http://www.leeds.nhs.uk/professional/healthatlas/
London Air Quality Network – LAQN London urban air pollution maps (updated daily)	http://www.erg.kcl.ac.uk/london/asp/home.asp
Maps of NHS (National Health Service) Hospitals in England	For example, http://www.nhs.uk/ServiceDirectories/Pages/Hospital.aspx?id=RK953
Plymouth Informed (Plymouth, Devon, UK)	http://www.plymouth-informed.org.uk/
Social Explorer (US Demography) – Macromedia Flash required	http://www.socialexplorer.com/maps/home.asp
The Community Health Information Profile (CHIP) for Manchester, Salford and Trafford, UK	http://www.healthprofile.org.uk/
The Health Atlas for the East of England – Scalable Vector Graphics – SVG support, e.g., Adobe SVG Viewer required	http://www.erpho.org.uk/extras/geowise-svg/svgIntro.asp
Interactive Atlases of Coronary Heart Disease and Stroke Mortality in England – Scalable Vector Graphics –SVG support, e.g., Adobe SVG Viewer required	http://www.sepho.org.uk/extras/interactiveAtlas/reports/atlasIndex.aspx
The Multi-Agency Internet Geographic Information Service (see Theseira 2002)	Replaced by PHASE (Public Health and Statistics Exchange) http://phase.esriuk.com/
TOXMAP – environmental health e-maps from the US National Library of Medicine	http://toxmap.nlm.nih.gov
US CDC Heart Disease and Stroke Maps	http://www.cdc.gov/dhdsp/library/maps/index.htm
US CDC Interactive Atlas of Reproductive Health	http://apps.nccd.cdc.gov/gisdrh/
US spoken languages maps	http://www.mla.org/census_main
West Nile Virus Maps (US Geological Survey and CDC)	http://westnilemaps.usgs.gov/
World Health Organization –WHO Global Health Atlas	http://www.who.int/GlobalAtlas/
WHO Atlas of Health in Europe – Adobe Reader for PDF documents required	http://www.euro.who.int/document/e79876.pdf

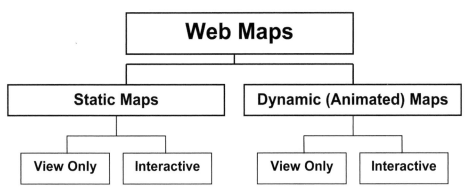

Figure 49-1 Technical classification of Web maps. Adapted with permission from Kraak and Brown (2001).

Manual of Geographic Information Systems

Web resources, including other Web maps, and can even trigger a query against an underlying database and display the results. It is possible to put all kinds of additional information behind the map image, thus reducing map clutter and size. Mouse events such as mouse-over (e.g., ToolTips or MapTips for map feature labeling) and mouse clicking of map objects can be associated with this extra information. Interactivity could also mean users have the option to define map contents by switching layers on and off. The map interface itself can be made interactive by providing the user with control options like panning, zooming in and out, and a smaller interactive overview map to highlight/select the area covered by the currently displayed map tile in relation to a bigger map (Kraak and Brown 2001).

Interactivity features can vary widely depending on the methods and software used to produce the online interactive version of a desktop GIS map (Boulos et al. 2002). Some online interactive maps also offer map feature search—see example in Boulos (2004c) and at http://healthcybermap.org/PCT/ratings/ (click the "binocular" icon on the left), and the ability to control class intervals (thresholding of hues) in choropleth maps and see the results of this instantly on the map. The Scalable Vector Graphics (SVG) choropleth map example described in Boulos et al. (2005a) and available to browse at http://healthcybermap.org/PCT/STDs/ allows users to "Choose classification method" to select how data are mapped (with equal ranges, equal counts or by highlighting the highest values), select "Number of colors" to control how many classes or ranges are used in the choropleth map (two to five ranges), and even pick from the "Color of theme" list a color theme to be used in the main map (orange, green, or blue).

Figure 49-2 provides a good example of some of these interactivity options and features. Note the different map interface buttons on the left for zooming, panning, displaying attribute information (the blue "i"—Identify button) and other functions. A help window is displayed by clicking the button with "?"—the question mark symbol. Also note the overview map with a red positional square on the top right; this helps users know where they are within the larger map that cannot be displayed in full detail in one screen. The overview map is also clickable and can be used to select a different area for viewing.

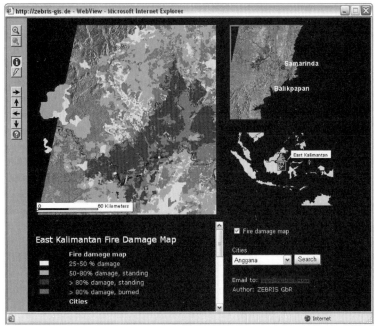

Figure 49-2 Example of an interactive Web map: Fires in Indonesia (Integrated Forest Fire Management, Samarinda, Indonesia - available from http://zebris-gis.de/index.php?id=48&L=1). See included DVD for color version.

49.2 Cartographic Grammar and Guidelines for Web Maps

Cartographic methods and techniques are a kind of visual grammar. They allow for the optimal design and production of effective maps (including Web maps) that are well suited to the application at hand (Brewer 2005). The ultimate goal is always to visually reveal the spatial objects, relations and patterns in the underlying data, so that the map user can visually locate spatial objects, while the color and shape of symbols representing them on the map inform him or her about their characteristics.

49.2.1 Bertin's Visual Variables

To find the proper symbology for a map, a cartographic data analysis must be performed to assess the characteristics of the data and find out how they can be visualized. In his *"Sémiologie Graphique,"* Bertin (1967—cited in Kraak and Brown 2001) distinguished six categories, which he called the visual variables. They are size, texture (grain), orientation, shape (form), value and color (Figure 49-3). These visual variables can be used to make one symbol (point, line or area symbol—see below) look different than another one. The choice of the correct variables for the data at hand is of utmost importance in the cartographic visualization and communication process; misuse of variables can create misleading results, cause loss of information, and convey a wrong impression.

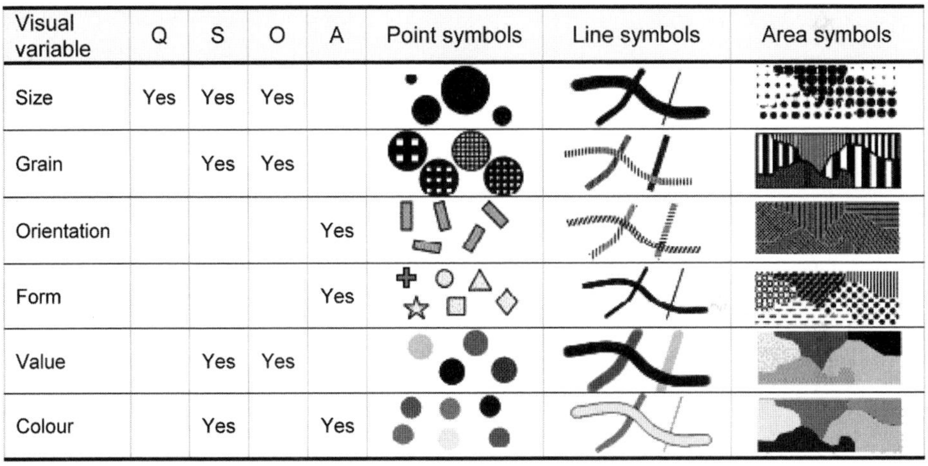

Visual variable	Q	S	O	A	Point symbols	Line symbols	Area symbols
Size	Yes	Yes	Yes				
Grain		Yes	Yes				
Orientation				Yes			
Form				Yes			
Value		Yes	Yes				
Colour		Yes	Yes				

Figure 49-3 Bertin's visual variables—Q = Quantitative; S = Selective; O = Ordered; A = Associative. Adapted with permission from Kraak and Brown (2001). See included DVD for color version.

The grouping in these six categories is based on the perceptual behavior(s) that each category stimulates in the map user. These behaviors are quantitative, selective, ordered and associative. For example, differences in values (gray values or lighter/darker shades of the same color) convey differences in order, or relative quantity, e.g., maps of population or resource density. This variable (value) forms the basis of choropleth maps. The variables color, shape and orientation have associative properties, as they can give the impression of difference among things and are the basis of chorochromatic or mosaic maps used to represent data such as land use (Kraak and Brown 2001).

Contour maps and climate maps are probably the most well known representatives of the isoline maps (Kraak and Brown 2001). Each line in an isoline map has the same value, e.g., of terrain height or temperature.

Color schemes can convey different semantics. The study by Boulos and Phillipps (2004) and the companion online Web map demonstrator available at http://healthcybermap.org/PCT/dentists/ provide a good example of the use of color as a visual variable in a "traffic light" scheme. The simplest "traffic light" maps use only one tint/shade of each of the three basic hues (red, yellow and green) to represent three "traffic light" classes, but if required, more than one tint/shade of red and/or yellow and/or green can be used to represent more transition points/level ranges or classes of mapped variables. Blue is also sometimes inserted between yellow and green. The colors are used to distinguish between, and compare at a glance, map areas associated with: (1) a range of decisions or plans/actions of varying quantity, quality or priority; or (2) high, medium, and low level ranges of variables that have practical implications, e.g., prompting different courses of action or different degrees of endorsement of the same action depending on level range. Depending on the mapped topic and the quality to be conveyed, e.g., capacity, potentiality, severity, vulnerability, risk, suitability/readiness for some action, or urgency/priority, the red and green ends of the "traffic light" map spectrum can assume different meanings.

49.2.2 Cartographic Symbol Design for Web Maps

49.2.2.1 Point Symbols

Point symbols are mainly used to represent spatial features occupying a very small area on the map at a given scale. They also can be used to provide shorthand information, e.g., a set of symbols next to a village on a tourist map to point to tourist facilities like restaurants, hotels, hospitals, post offices, etc. Three categories of point symbols exist: pictorial (e.g., airplane, tree, etc.), geometric (polygonal symbols, squares, circles, etc.), and alphanumeric (letters and numbers). On interactive Web maps a point symbol can also function as a Web object, i.e., as an area that can respond to mouse events like mouse-over and mouse clicks, triggering, for example, JavaScript functions and hyperlinks.

Pictorial point symbols are usually very easily understood by inexperienced map-readers, sometimes even without the use of a legend. The main problem in designing these symbols for Web maps is that the essential characteristics of the features they represent must be visualized without much clutter within a small screen area with a limited number of pixels (cf. designing icons for Microsoft® Windows). The symbols may also need to be larger than on equivalent paper maps to aid legibility and must be less complex regarding detail and possibly also colors to suit the lower resolution constraints of Web maps. However, this should not be considered a drawback since Web objects can be used to access the second level of information content that can cover a smaller area in more detail. For best results, a clear metaphorical relation should exist between the symbol and the user's real world knowledge.

Geometric or abstract symbols do not attempt to resemble the real feature represented. On different maps the same symbol can have a different meaning. Therefore geometric symbols should always be explained in a legend (Kraak and Brown 2001). Alphanumerical point symbols also require a legend.

49.2.2.2 Line Symbols

Line symbols on topographic maps can represent features like roads, railways and contours. In thematic maps, lines can show the position of geological fault lines, ocean currents, and trade flows. However, the thin and elongated shape of line symbols makes them very difficult to handle as interactive Web objects, especially when they are highly curved (Kraak and Brown 2001).

49.2.2.3 Area Symbols

Area symbols are used to represent area-based information. The graphic variables typically used in designing area symbols for maps are color, value, texture, shape and orientation. These can create complex area patterns. The proper use of variables can promote the semantic meaning of an area symbol, e.g., green tree symbols filling an area to represent forest area. It can also decrease the possibility of confusion among adjacent area symbols, e.g., by assigning a different color to each country on a map of the world and using a blue color for seas and oceans (Kraak and Brown 2001). In Web maps, area symbols can also function as clickable Web objects.

49.2.3 Features Labeling and Typographic Variables

Text labels on Web maps cannot be omitted, as text can express information like geographic names that are not possible to express using any other symbol. Map typography and map symbol design cannot be separated from each other. Inappropriate application of typographic variables may affect the legibility of the text on a map and/or clash with other graphic variables.

Typographic variables include (Kraak and Brown 2001):

- type size: expressed in points;
- shape: refers to variations in font or type face, e.g., serif types like Times or sans serif types like Helvetica, and sometimes lower and UPPER case of the same font are considered shape variations; and
- orientation: refers to upright or italic variations within one font. Value refers to light, medium and bold font variations, or some gray value.

Text placement is also important; it can be horizontal, inclined, or curved along a path. Users of Web maps cannot rotate it to read very inclined or upside down text as they do with paper maps. Other important issues include text-background relation, e.g., any outline around text to make it clearer, and any antialiasing (font smoothing) used and its amount. Antialiasing might not be a good choice for very small font sizes and together with many other options available for typing text on a map, like cast shadows, they tend to increase image file size (for maps served in raster formats). Also important is character kerning (space between letters).

Considering text on Web maps, one can distinguish two types (Kraak and Brown 2001):

- text applied outside the map face, such as in the legend. This can be saved as graphic in raster format or saved as real text on a Web page (to be displayed in a separate frame, with the usual precautions to avoid/care for different fonts on users' machines, and control text flow, letter spacing and space between lines); and
- text within the map face, e.g., for geographical names. This is usually saved as part of the map in raster format to be sure no text changes will take place at the user's side. MapTips used to label map features (on mouse-over) also belong to this type.

The density of text (the amount of textual information on the map) is also very important and an overcrowded map can become illegible. Text in this case can be made available as selective MapTips only appearing on mouse hovering or as message boxes available by clicking hotspots on the map (Kraak and Brown 2001).

49.2.4 Contrast and Visual Hierarchy

Contrast will increase the communicative role of the map since it will create a kind of visual hierarchy or figure-ground relation in map contents, since usually not all map information is of equal importance (Kraak and Brown 2001; Skupin 2000). The visual hierarchy of Web map information content deserves more attention compared to maps in general. Three distinct levels exist (Kraak and Brown 2001):

- *Primary content level.* This is formed by the main theme of the map. Interactive Web objects such as hotspots, mouse-overs, etc., can also be considered to form part of this level. These objects trigger specific events resulting in the supply of main theme information.
- *Secondary content level.* This refers to the (often topographic) base map information, but also to pop-up menus and windows, and any movies, tables, sounds, etc., supplying information on the main map theme.

- *Supportive content level.* This includes the legend, other marginal information like grid, and any information that is not directly related to the main theme of the map, e.g., map interface help. A legend is necessary to understand how the topic is represented, e.g., a key to the color tints used in a choropleth map and the class ranges they represent. If the legend and the map are saved as one image, the legend will disappear when the viewer zooms in to part of the map (or scrolls away from the legend corner if the map does not fit the whole window). The legend in Web maps is thus best treated as a separate image so that it can be kept in view in a separate frame.

49.2.5 Scale, Generalization and Isomorphism

The visualization of geographic and non-geographic information spaces on a two-dimensional screen can be severely impeded as the volume and complexity of the respective data set grows. Skupin (2000) cites methods such as windowing, fish-eye views, hierarchical display, and tree condensing as examples of methods used to reduce the complexity of information visualizations.

Cartographers are capable of creating maps where geographic meaning is preserved throughout the scales, despite the large number of features involved. The processes of abstraction that achieve such scale-dependent representation are collectively referred to as cartographic generalization. Generalization can be based on controlling the number of classes into which features are grouped, or on simplification of form (geometric generalization). At the root of the complexity problem lies a conflict between the number of visualized objects, the size of symbols representing them (and their labels) and the size of the display surface. Cartographic generalization is deeply tied to the notion of scale, which is defined as the ratio between the size of a feature on the map to its size in the real world. Digital cartography has further expanded the classical concept of scale. Now the accuracy with which a given map scale represents the location and details of features is known as resolution. Computers can enlarge the scale of any map (by zooming in), but no additional (true) detail will be gained and map accuracy will not change unless a new higher resolution map is loaded replacing the first (Skupin 2000).

Conventional maps are also isomorphic, i.e., identical to or similar to the domains they represent, though they usually present a selective view of reality, only showing a subset of the features in their domains. During the process of creating this scaled-down view of reality, the cartographer has to select those features that will become amplified, while discarding the rest of the information in the domain according to the purpose of the map (Lie 1991).

However, sometimes isomorphism is abandoned all together for the sake of clarity. A good example of non-isomorphic maps is the very popular London Underground map, which Henry C. Beck first designed in 1931 (latest version available at http://www.tfl.gov.uk/assets/downloads/Standard-Tube-map.pdf). The map is not drawn to scale and therefore is not 100% isomorphic. If enlarged to the actual size of London it would diverge significantly from the actual geography of the city, a fact many tourists soon discover when they try to use the Underground diagram as a walking map. Isomorphism is sacrificed to give downtown areas better coverage. The density of information is roughly the same all over the diagram (distances between stations are nearly equal), while downtown areas in reality have a much higher density of Underground installations. The result is a product that is highly functional for its purpose and is a good model, which some cybermaps have followed (Kahn and Lenk 2001; Lie 1991).

49.2.6 Download Time

If the information takes too long to download (large map image/data set files or Web browser plug-ins are required to display the maps), users will lose interest and abort the process (Dodge 2000; Kraak and Brown 2001). The ideal Web map should not be too large in both file size and image size, to cope with the limited size of display screens. This means graphic and information density in Web maps should be kept low (Kraak and Brown 2001). The Web offers many techniques for adding interactivity to maps (see Section 49.1.1), which can be used to make smaller, "smarter" maps that are faster to download and provide additional functionality.

With Web GIS there is always a trade-off between the sophistication of user tools and response times. Applications using the Internet are constrained by connection download speed. The smaller the amounts of data being transmitted are and the simpler the client user interface is, the faster the application will be. Web GIS applications differ in the way this balance is approached. Some applications use a very simplistic user front-end and display the results of the server-side process by delivering a simple raster image (thin client). Such applications tend to be fast and robust and will work within a standard browser. Other applications require a client-side plug-in, Java applet or ActiveX control to be downloaded to give richer functionality to the user (fat client). Larger amounts of data can be downloaded from the server to the plug-in, which can make the application run more slowly, but has the benefit of providing the user with a more sophisticated set of tools and greater interaction with the data. The choice whether to go for a thin-client solution or a fat-client one should be made based on the specific requirements of the application at hand, and the likely expertise and needs of the target user community (Ordnance Survey 2003).

49.2.7 Screen Resolution, Color Palette and Web Graphic Formats

The screen resolution and color palette settings of display adapters can limit the amount of map detail that can be displayed on a computer/device monitor (compared to a map of similar size printed on paper). Mobile Internet devices carry with them even more challenges to the Web cartographer because of their much smaller (and sometimes monochrome) display screens.

Moreover, although Web graphics, including maps, are stored in platform-independent formats like GIF (CompuServe Graphical Interchange Format), JPEG (Joint Photographic Experts Group) and PNG (Portable Network Graphic), the same graphic or map still might not appear exactly the same to every user. This is because of differences in users' devices, browsers and operating systems, which handle colors in different ways, and in the quality (e.g., resolution) of their graphic cards and display screens. Besides, users are able to manually adjust their displays for resolution, contrast, brightness and color balance. Fortunately, there are established ways of dealing with most of these different output conditions.

Web map designers should adopt a cautious approach by assuming the minimum configuration and lowest settings on users' machines, e.g., 8-bit/256-color depth (though this is becoming much less common, with most new machines now having their display set to 24- or 32-bit/True Color). The famous 216-color Web- or browser-safe palette fits well into this configuration and should be used for Web images saved in a palletized format like GIF. The 216 colors in this palette are guaranteed to be non-dithered—smooth, solid colors—on any configuration, unless combinations of them are used to represent other colors not originally in the palette. This does not mean, however, that these 216 colors will always appear exactly the same on any system, since much depends on the calibration of the monitor and other factors as mentioned above (Kraak and Brown 2001). For many applications, this is not a big problem.

GIF is a palletized raster (bitmapped) format, i.e., images saved in this format are limited to a palette with a maximum of 256 colors. GIF is also a lossless compression standard (cf. JPEG in next paragraph). GIF is more suitable for line art images and images with solid colors. GIF images can have transparent backgrounds if a color from the underlying palette has been defined as transparent. Animated GIF maps can be seen as the view-only version of the dynamic maps.

Another common raster format, JPEG, does not work with a palette; it compresses images based on color and intensity (usually lossy compression). Even if a map is designed to contain only Web-safe colors, some of them might shift slightly during JPEG compression, possibly leading to dithered results on a 256-color display configuration. For maps making considerable use of color blends, e.g., contour maps, JPEG is however the best compression algorithm.

Web map designers should test both formats—GIF and JPEG—on the maps they intend to produce, varying the various parameters available, including JPEG compression/quality settings and GIF palette sizes. GIF palettes can be adaptive, i.e., include only the colors that need to appear in an image. The aim is to find which setting gives good results, usually assuming a 256-color configuration, while maintaining a reasonably small file size (Kraak and Brown 2001).

PNG (http://www.w3.org/Graphics/PNG/) is an emerging royalty-free raster Web graphic format that promises:

- to put an end to differences in color display on different platforms because of color control through gamma settings (see http://www.cgsd.com/papers/gamma.web.html);
- to deliver real transparency by specifying an alpha channel so that a transparent layer really blends into whatever color is underneath;
- to achieve smaller file sizes compared to similar GIF images; and
- to be always editable with the possibility to resave without loss of information.

The latest Web browser versions are natively supporting the PNG format, but sometimes not all of its features (Kraak and Brown 2001). PNG is the format used, for example, by Google Maps (http://maps.google.com/), though it also uses JPEG to display satellite imagery.

Wavelets image compression is a relatively new technique that also promises to achieve high quality images with smaller file sizes. However, a special browser plug-in is required to display images and maps saved in this format.

49.2.7.1 Vector Graphic Formats

The vector graphics approach to online maps offers scalable, editable, interactive and more bandwidth-efficient Web maps. Three formats will be briefly discussed here, namely the Vector Markup Language (VML—http://www.w3.org/TR/NOTE-VML), the Scalable Vector Graphics format (SVG—http://www.w3.org/Graphics/SVG/) and Macromedia® Flash (http://www.adobe.com/products/flash/).

VML is an application of XML (eXtensible Markup Language) that defines a format for the encoding of vector information together with additional markup to describe how that information may be displayed and edited. VML is built into the latest versions of Microsoft® Internet Explorer Web browser, with no plug-in required. VML maps can be easily copied and pasted into Microsoft® Office applications.

SVG is a non-proprietary language for describing rich, stylable two-dimensional graphics and graphical applications in XML (eXtensible Markup Language). SVG is fully endorsed by the W3C (World Wide Web Consortium), and is rapidly becoming a popular choice for delivering interactive Web maps, being designed to work effectively across platforms, output resolutions, color spaces, and a range of available bandwidths. It offers a rich modern graphics format providing the ability for better map display, and advanced graphical features such as transparency, arbitrary geometry, filter effects (shadows, lighting effects, etc.), scripting, and animation (Harwell 2004). All these features have made SVG a direct competitor to the proprietary Macromedia® Flash format (Held et al. 2004), and the debate is continuing regarding which of the two formats is the best (Strømberg 2005).

Vector-based images (describing shapes and paths), such as those in SVG and SWF (Macromedia® Shockwave/Flash File) formats, will keep their sharp character when enlarged, while raster-based images that store information about each and every pixel in the image, such as those saved in GIF (Graphics Interchange Format) or JPEG (Joint Photographic Experts Group) formats, will show jagged edges (Boulos et al. 2005a).

A free SVG Web browser plug-in is available from Adobe for different platforms (Adobe SVG Viewer—http://www.adobe.com/svg/viewer/install/main.html), in the same way the free Adobe Reader software is available for rendering PDF (Portable Document Format) files. Users will have to wait for a short time while the SVG browser plug-in of a few megabytes in size is downloading (unless their Web browser natively supports SVG), but this needs to be done only once.

The Office for National Statistics' England and Wales 2001 Census Key Statistics maps (freely available at http://www.statistics.gov.uk/census2001/censusmaps/index_new.html) and their UK 2001 Area Classification for Health Areas maps (freely available at http://www.statistics.gov.uk/about/methodology_by_theme/area_classification/ha/maps.asp) provide two excellent examples of interactive SVG Web maps, besides the other SVG map examples mentioned in Table 49-1, Figure 49-4, and elsewhere in this paper. Interested readers are also referred to Carto.net, which provides some excellent papers about SVG (see http://www.carto.net/papers/svg/), as well as many additional examples of interactive SVG maps that are freely available to browse (see http://www.carto.net/papers/svg/samples/ and http://www.carto.net/papers/svg/samples/wien.shtml).

Commercial tools exist for producing interactive SVG and Flash maps from desktop GIS projects. For example, this author used GeoReveal from Graphical Data Capture Ltd. (http://www.graphdata.co.uk/products_and_services/public_access/georeveal/) to produce the SVG maps at http://healthcybermap.org/PCT/STDs/ (Boulos et al. 2005a). Wizard-driven tools like GeoReveal have made it very easy to transform complex raw data into valuable decision support information products (interactive Web maps) in very little time and without requiring much expertise. Using GeoReveal Enterprise Server, GeoReveal projects can also be connected to any database capable of serving data via XML to deliver live interactive mapping.

Another market player, GeoWise Ltd., provides InstantAtlas (http://www.instantatlas.com/), an online service that allows users to quickly create interactive atlas applications that use SVG to display spatial and statistical data in map, chart and tabular formats. These atlases can then be published on the user's own Web site without the need for any specialized software (Figure 49-4).

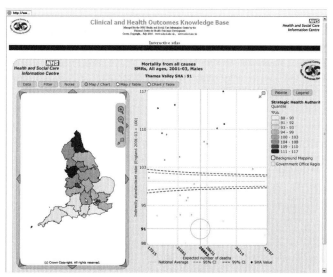

Figure 49-4 Screenshot of England's SHA (Strategic Health Authority) Funnel Plot atlas (available from http://www.nchod.nhs.uk/). This online atlas was produced using InstantAtlas from GeoWise Ltd. See included DVD for color version.

Manual of Geographic Information Systems

InstantAtlas features a simple technical architecture. Users use a combination of its Publisher and Templates to produce stand-alone dynamic SVG reports. Each stand-alone dynamic report comprises an SVG file (with extension .SVG or .SVGZ), a set of associated JavaScript files (with extension .JS), a cascading style sheet file (with extension .CSS), an XML data file (with extension .XML), and an XML configuration file (with extension .XML). The files can remain separate and reside in the same Web server/site directory, or they can all be bundled into the SVG file to simplify distribution.

The .SVG/.SVGZ is the main file that contains the layout of the page together with the vector boundary data for mapping. The .JS set of files contains the functions that each page delivers. The .CSS file is a style sheet that defines the fonts and colors used in each stand-alone dynamic report. The .XML data file contains all the attribute data linked to the maps by means of an ID code, while the .XML configuration file is a text file containing parameters that define the layout and display of the associated stand-alone dynamic report. The .XML data file can be modified without changing any of the other files. This allows a user to add new attribute data to a stand-alone dynamic report. The .XML data file can also be delivered as an XML stream. This allows an organization to link a stand-alone dynamic report to a back-end database or data store. The streaming function is handled by the Adobe SVG plug-in.

Other SVG/Flash mapping tools available today for publishing maps created in desktop GIS include GéoClip (http://www.geoclip.net/an/), Corda OptiMap (http://www.corda.com/products/optimap/ and examples directory http://www.corda.com/examples/go/), SVGMapMaker for MapInfo (http://www.tetrad.com/software/svgmapmaker/), MapViewSVG (http://www.uismedia.de/mapview/eng/), and SVGMapper (http://www.svgmapper.com/). The latter two tools (MapViewSVG and SVGMapper) are specific to ESRI ArcView GIS (Boulos et al. 2005a). Beacon Dodsworth's MapVision (http://www.beacon-dodsworth.co.uk/products/web-mapping/) supports Web map creation in VML.

49.2.8 Other Online Map Usability and Accessibility Issues

Usability testing of online interactive maps prior to their final release is of paramount importance and must be carried out with representatives of the target audience of the maps under testing to make sure the map message is properly communicated in an unambiguous way, and the interactive map interface is usable and accessible by its respective users (Tiits 2003, 2004)—see also the usability issues covered in Boulos (2003a). For example, some map color schemes are not color-blind friendly, e.g., the aforementioned red-yellow-green "traffic light" scheme, which means color-blind online map users should ideally be provided with alternative color scheme options. Color schemes can be tested for color-blind friendliness using the Vischeck simulator (http://www.vischeck.com/), a tool based on SCIELAB from the Wandell Laboratory at Stanford University, US (see http://white.stanford.edu/~brian/scielab/scielab.html).

The online map example described in Boulos (2004c) used Cynthia Brewer's ColorBrewer (http://www.colorbrewer.org/) to select a suitable color scheme. The chosen scheme is color-blind friendly, black-and-white photocopy friendly (for printed output), LCD (Liquid-Crystal Display) projector friendly, laptop (LCD) friendly, CRT (Cathode-Ray Tube) screen friendly, and color printing friendly—all at the same time (Figure 49-5). ColorBrewer is a research-based, free-to-use online tool available from Pennsylvania State University Web site and designed to help people select good color schemes for maps and other graphics (Brewer 2005; Harrower and Brewer 2003; Olson and Brewer 1997).

Audio-tactile (haptic) and screen reader maps have also been proposed as another option for the visually impaired (Figure 49-6) (Jacobson 1998; Natural Resources Canada 2003; Siekierska and Müller 2003).

Another important usability/accessibility issue is the need to care for the various devices available today to access the Internet like WebTV, mobile phones, PDAs (Personal Digital Assistants) and Pocket PCs, in addition to standard PCs and laptops. These devices do not all have the same input device and display capabilities. For example, there is a very limited support of imagemaps (imagemaps or hypermaps are graphics with clickable hotspots—see Section 49.3.1) in WebTV compared to a standard Web browser running on a desktop PC. Only rectangular hotspots are supported in WebTV (see http://web. archive.org/web/20070608205749/http://developer.msntv.com/Develop/seamimgmap.asp); complex shapes, e.g., the outline of Canada on the map shown in Figure 49-7 below, are not supported. Therefore, equivalent textual interfaces (e.g., the country drop-down list shown in the WebTV figure below) have to be provided as an alternative way to access the same material accessible via the hypermaps in a standard Web browser.

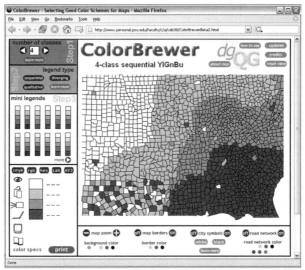

Figure 49-5 Screenshot of ColorBrewer online tool (http://www.colorbrewer.org/) showing the color-blind friendly yellow-green-blue quadri-color scheme chosen for the maps described in Boulos (2004c). The corresponding Hue-Saturation-Value numerical triplets for the four colors in the selected scheme are also shown, ready for use in ArcView 3.x. The online interactive maps based on this scheme are available at http://healthcybermap.org/PCT/ratings/. See included DVD for color version.

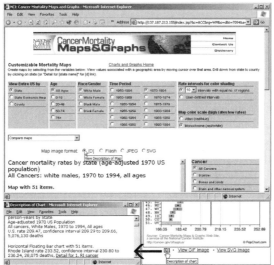

Figure 49-6 Screenshot of the customizable cancer mortality maps and graphs developed by the US National Cancer Institute (NCI—http://www3.cancer.gov/atlasplus/index.html). These maps (upper part of screenshot) and the associated charts and graphs (lower part of screenshot) are compliant with Section 508 of the US Rehabilitation Act. This means they can be accessed by the blind or visually impaired through screen readers that read the text description file ([D] link) accompanying each map, graph or chart. See included DVD for color version.

Manual of Geographic Information Systems

Figure 49-7 An imagemap and the alternative textual drop-down list in WebTV viewer (this viewer can be downloaded from http://web.archive.org/web/20070118150906/http://developer.msntv.com/Tools/WebTVVwr.asp). See included DVD for color version.

49.2.9 Maps Must be Underpinned by Good Science

GIS integration of complex data into visually easy-to-understand pictures can sometimes be a setup for misunderstanding and misuse. It is therefore necessary that we design maps in such a way as to avoid drawing false conclusions from them (Boulos 2004a).

According to Monmonier (1996), it is not just easy but also essential to lie with maps. The cartographer's paradox is that to avoid hiding critical information in a fog of detail, the map must offer a selective, incomplete view of reality. Map users always need to be alert for "lies" that can range from legitimate and appropriate selective suppression of some details to help the user focus on what needs to be seen, to more serious distortions in which the visual image suggests conclusions that would not be supported by careful scientific analysis.

Sound scientific and statistical principles and methods should provide the foundation for all data analyses to be displayed on maps (Boulos 2004a). For example, when classifying data to create "traffic light" maps like those described in Boulos and Phillipps (2004), we need to decide on the number of classes to use (the minimum is three, corresponding to the three basic "traffic light" hues: red, yellow and green; more than three classes can be defined and associated with additional shades/tints of the three basic hues). We also need to carefully choose the endpoints (cut-off points or hue thresholds) for our class intervals. The effects of changing the number of classes and hue thresholds of a "traffic light" map can be dramatic, and may even convey very different interpretations of the same underlying data. In deciding on the number of classes to use in a "traffic light" map and the associated cut-off points for class intervals, we should strive to neither raise false alarms nor overlook/miss areas of concern on the resultant map.

As stated above (Figure 49-1), animated (dynamic) time series maps can also be created to communicate findings (for some examples of animated time series maps, see http://circ.rupri.org/animation/). A very important point to note for animated maps is that the same classification method and values should be used for each map in the series to avoid conveying any false impressions to map users.

49.3 Techniques for Implementing Web Map Interactivity

49.3.1 Basic Imagemaps

Images referenced in an HTML (HyperText Markup Language) file can be made to have "sensitive areas" or hotspots, defined by their bounding coordinates. Hotspots can be specified as rectangles, circles, or polygons. Clicking within a hotspot will usually cause the browser to launch an Internet location, e.g., another image or document. Different hotspots can be tied to different links. In this way, a single image can provide multiple hyperlink destinations, each associated with a defined region or regions of the image.

49.3.1.1 Server-side Imagemaps

The ismap attribute for the HTML img tag can be used to turn an image into a graphically active element, so that clicking different regions on the same image causes the server to take different actions. The bounding coordinates of hotspots are not stored in the HTML file that calls the map image, but in a separate map file (a text file) stored on the Web server. This map file also includes the corresponding actions to be taken when these hotspots are clicked.

When a user clicks on a server-side hypermap, the browser sends the mouse coordinates to Imagemap, a program on the server that looks in the corresponding map file for the action associated with those coordinates. Imagemap first determines which hotspot among the defined hotspots these coordinates fall within. It then reads the action associated with this hotspot, and returns a server redirect message back to the browser telling it which Internet location or document it should access.

Imagemap is a CGI program that runs on the Web server. CGI stands for Common Gateway Interface. It is an interface definition that allows an HTML application to invoke a program on the server, like Imagemap, and pass arguments to it. CGI programs can be programmed in many languages like C, Pascal and PERL (Practical Extraction and Report Language).

In the following example, a GIF image, mymap.gif, is declared active using the ismap attribute:

```
<a href="http://www.myserver.com/cgi-bin/imagemap.exe/maps/mymap.map">
<img src="mymap.gif" ismap>
</a>
```

Note the path to the map file (mymap.map) associated with mymap.gif. When a user clicks over the image, the browser sends the mouse-click coordinates to the server, with respect to the image origin (0,0—top left corner of the image); the co-ordinates are appended to the map file address after a question mark "?" symbol as follows:

```
http://www.myserver.com/cgi-bin/imagemap.exe/maps/mymap.map?239,80
```

It is noteworthy that Microsoft® ASP pages (Active Server Pages) and PHP pages (PHP Hypertext Preprocessor), which are sometimes used in online interactive map sites, employ a similar technique to pass arguments to the server.

The main disadvantage of server-side imagemaps is that they create extra data traffic between the server and its clients. That is why client-side clickable maps have evolved as a replacement to server-side maps (Graham 1998; Kraak and Brown 2001; McCauley et al. 1996).

49.3.1.2 Client-side Imagemaps

Client-side imagemaps store hotspot coordinates and associated hyperlink information in the same HTML document in which the image is referenced, not in a separate map file on the server. When the user clicks a hotspot in the image, the associated hyperlink location is determined by the Web browser software (from the underlying HTML code) and the user is transferred directly to that location. This makes client-side imagemaps faster than server-side imagemaps and reduces server load (Crossley and Boston 1995; Kraak and Brown 2001).

Manual of Geographic Information Systems

Many imagemap editors exist that allow users to define hotspots on an image as well as the actions associated with these hotspots, and then automatically generate the necessary client-side HTML code (or server-side map file if needed), e.g., Mapedit (http://www.boutell.com/mapedit/). The client-side HTML code below defines the coordinates of a rectangular hotspot on a JPEG image, myimage.jpg, near its lower right corner. The description "The Roman Baths, Bath, UK" should appear when the user moves the mouse cursor over the defined hotspot. Clicking the hotspot will take the user to the Roman Baths Web site (http://www.romanbaths.co.uk/). Nothing will happen if the image area outside the hotspot is clicked.

```
<img src="myimage.jpg" usemap="#mymap" width="445" height="523" border="0">
<map name="mymap">
<area shape="rect" alt=" The Roman Baths, Bath, UK" coords="237,311,443,467"
href="http://www.romanbaths.co.uk/">
<area shape="default" nohref>
</map>
```

Client-side imagemaps provide better accessibility compared to server-side imagemaps because authors are able to assign appropriate text to each imagemap hotspot by including the alt attribute and area description inside each <area> tag. This feature means that someone using a screen reader can easily identify and activate regions of the map (Federal IT Accessibility Initiative 2005; United States Access Board 2001).

ZEBRIS WebView (http://www.zebris.com/english/main_webview.htm), the Internet extension to ESRI ArcView GIS and ArcGIS, is one example of a tool that can be used to generate client-side imagemaps from ArcView themes. (The example shown in Figure 49-2 above was generated using WebView.) WebView adds extra interactive functionality to these maps using JavaScript. JavaScript is embedded directly in HTML pages (as readable text) and is interpreted by the browser completely at runtime. JavaScript statements can respond to user events such as mouse-over and mouse-clicks. Web map ToolTips can also be implemented in Javascript. VBScript is another popular scripting language by Microsoft® and an alternative to JavaScript. More information is also provided below on WebView and other related tools (under Section 49.3.3).

Client-side interactive maps can also be created in SVG, Macromedia® Flash (e.g., http://kartoweb.itc.nl/webcartography/webmaps/static/si-example4.htm), Macromedia® Shockwave (e.g., http://kartoweb.itc.nl/webcartography/webmaps/dynamic/explov.htm), or as a Java applet (e.g., Descartes system http://www.ais.fraunhofer.de/descartes/IcaVisApplet/). These types of client-side interactive maps might be slow to download (Kraak and Brown 2001).

49.3.2 Zooming

Web map zooming options depend on users' systems, including installed plug-ins (some tools like WebView do not require any browser plug-ins), and also upon the presence of enough map detail to allow considerable enlargement. There are three distinct zooming strategies or options (Kraak and Brown 2001):

- *Static linear zooming.* The relation between zoom factor and map content is static. When zooming into the map, the image is linearly enlarged but the content of the map does not actually change. It can be done on the client side using an appropriate browser plug-in or applet. In this case the map is stored simply as an image. Vector-based images, such as those in SWF (Macromedia Shockwave File) and SVG formats, will keep their sharp character when enlarged, while raster-based images, such as those saved in GIF or JPEG formats, will show jagged edges. There is, however, an ideal scale (or scale range) to display any particular map, depending on the density and accuracy of map detail. If a map is enlarged too much very few details may be visible in the image window and the positional accuracy of the symbols may be much less than what the users expect at such scale.

- *Static stepped zooming.* In this case a series of maps of the same area is available, each one designed for a different scale or scale range. When the user requests to zoom in or out, the software automatically selects the most suitable map for the desired scale. This system is widely used on route planning sites and by companies such as MapQuest (http://mapquest.com/). It offers better results compared to static linear zooming.
- *Dynamic zooming (animated scaling).* In this zooming strategy there is a continuous direct relation between scale and map content. The larger the scale the more detail is shown in the image. A direct link between the image and some kind of database is necessary. Although not always required, the cartographic symbolization may change with scale. For instance a town represented by a point symbol at a small scale may turn into an area symbol upon zooming into the map. This approach is more expensive, requires more data, processing and bandwidth, and is not always needed.

A fourth zooming strategy, semantic zooming, has also been described (Spence 2001). With a conventional geometric zoom all objects change only their size; with semantic zoom they can additionally change shape, details (not merely size of existing details) or, indeed, their very presence in the display, with objects appearing/disappearing according to the context of the map at hand.

49.3.3 Advanced Solutions for Publishing Desktop GIS Maps and Data on the Web

Two main options exist for sharing desktop GIS maps and projects on the Web as sensitive clickable maps:

- *Maps generated in real time.* Dynamic publishing to the Web using a dedicated Internet map server that maintains a live connection with the underlying GIS project/database (see Section 49.3.3.1 below); or
- *Serving pre-generated maps.* Publishing a static snapshot of the project (representing the project's maps and underlying data at time of publishing) as clickable client-side imagemaps using tools like HTML ImageMapper extension for ESRI ArcView/ArcGIS from alta4 Geoinformatik AG, Germany (http://www.alta4.com/eng/products_e/im/index.php—also see HTML ImageMapper Map Gallery http://www.alta4.com/eng/products_e/im/mapgallery.php3), and WebView extension for ESRI ArcView/ArcGIS from ZEBRIS, Germany (http://www.zebris.com/english/produkte/wv_beschreibung.htm).

An online interactive map example produced by this author using alta4 HTML ImageMapper is described in Boulos (2004c) and is freely available to browse at http://healthcybermap.org/PCT/ratings/. It is noteworthy that HTML ImageMapper does not require any server-side software installation, and as such is much simpler to use than some other Internet GIS solutions like the client/server version of ALOV Map/TimeMap (http://alov.org/docs/setup.html).

The standalone versions of ALOV Map/TimeMap (http://alov.org/docs/quickstart.html) and JShape (http://skyscraper.fortunecity.com/redmond/829/jshape2.htm) Java applets, which don't require any server- side setup, are limited by the fact that they need to download the whole map shapefile from the Web server before they can start on the client side, and are thus not suitable for large data sets (the PCT boundary data set used in Boulos (2004c), for example, is about 50 MB in size in ESRI shape file format).

Boulos et al. (2002) describes a simple, low-cost method using ZEBRIS WebView to serve pregenerated hypermaps with dynamic database drill-down functionality (dynamic database links), thus providing some (but not all) of the functionalities of maps generated in real time, without the need for a dedicated Internet Map Server (Figure 49-8).

Since ZEBRIS WebView does not allow the dynamic generation (i.e., in real time—at the time the map is requested by and served to the client browser) of Web maps from ArcView/ArcGIS, some of the project's hypermaps will ultimately need to be manually regenerated using WebView when the underlying data change, if this change has implications on the

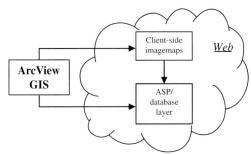

Figure 49-8 A partial workaround for pregenerated imagemap limitations. Tools like ZEBRIS WebView can be used to convert views in ArcView/ArcGIS to client-side imagemaps for the Web. The solution illustrated in this figure makes use of WebView HotLink functionality to implement a dynamic database drill-down that will always reflect the latest updates to the project's database. Clicking the different hotspots on the pregenerated client-side image-maps will trigger server-side preformulated SQL (Structured Query Language) queries against an underlying database on the same server where the client-side imagemaps are stored. This database should be the same updateable project database that ArcView/ArcGIS is connecting to. SQL queries can be coded in ASP or PHP pages for execution on the server (Boulos et al. 2002).

maps' appearance (e.g., changes in the data values of a choropleth map- shading variable). In cases when only map attribute data change without a corresponding effect on map appearance, nothing needs to be done; the same dynamic ASP or PHP query pages (unmodified) will always retrieve the latest updates (Boulos et al. 2002).

49.3.3.1 Dedicated Internet Map Server Solutions for Serving Maps with Dynamic Map Generation and Database Drill-down Functionalities

Advanced mapping applications running on the server side can be linked to the server software, e.g., using CGI. These applications can be used to provide live database access for browsing/querying a map database on the server. Using a dedicated map server, users could get a map depicting the latest figures from a database, which can come from another remote server, visualized with the colors and classification the user has requested (Frappier and Williams 1999; Kraak and Brown 2001).

ESRI ArcIMS (Internet Map Server— http://www.esri.com/software/arcgis/arcims/about/overview.html) solutions allow for an existing ArcView/ArcGIS project to be transparently ported to the Web with minimal effort (ESRI 2005a). Updates carried on the original project in ArcView/ArcGIS will also show automatically in real time in the Web front-end (see customer site examples at http://www.esri.com/software/internetmaps/—Figure 49-9). Almost all other major GIS vendors have already done this (Table 49-2), and although their approaches differ in detail, most use a combination of server-side and client-side components (e.g., a Java applet, an ActiveX control, or another client application type running in the client Web browser and connecting to the map server).

Table 49-2 Examples of commercial Internet map servers.

Product	Internet address
AutoDesk MapGuide	http://www.mapguide.com/
Bentley's Web Publisher	http://www.bentley.com/en-us/products/bentley+geo+web+publisher/
ER Mapper's Image Web Server	http://www.ermapper.com/ProductView.aspx?t=26
GE SmallWorld Internet Application Server	http://www.gepower.com/prod_serv/products/gis_software/en/sias.htm
Intergraph Geomedia WebMap Server	http://imgs.intergraph.com/gmwm/
MapInfo MapXtreme	http://extranet.mapinfo.com/products/Overview.cfm?productid=1849

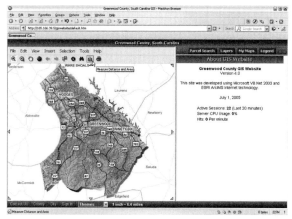

Figure 49-9 Greenwood County, South Carolina GIS (http://gis.greenwoodsc.gov/GISWeb-site/Default.htm). Written in Microsoft® Visual Basic .Net 2003 using the ActiveX connector, this ESRI ArcIMS client application is an example of a lightweight GIS Web site that loads quickly, even on a dial-up Internet connection. The site has an ArcReader look-and-feel (see http://www.esri.com/software/arcgis/arcreader/), and allows users to search, view, and print parcel maps and assess database information. Other layers include orthophotos, contours, and facilities. See included DVD for color version.

In one *earlier* typical ESRI ArcIMS setup, the ArcView GIS program takes on a role "similar to conventional CGI applications" running on the server. An ArcView extension, called Internet Map Server (IMS), is installed to receive commands from the Web browser via the Web server. A command can be, for example, a map query. It will be passed to and processed by ArcView GIS and the result (a map view) will be converted to a GIF or JPEG file and sent to the browser. An ArcIMS client application running in the client's Web browser (a Java applet called MapCafe in earlier ArcIMS implementations) is used to implement in the Web browser an interface similar to the standard ArcView GIS interface. Users can, for example, click the zoom button and drag a rectangle on the map displayed by the client application. This would result in the client application building a command to implement the required zoom action (IMS will receive this command and hand it to ArcView). The last item in this setup is a plug-in to the Web server software (called esrimap. dll in some implementations) that enables the server to find the appropriate ArcView GIS application to handle the request. ArcView GIS can be run on another computer to decrease server load and the server plug-in can even distribute requests among a number of computers running the same ArcView application. The client application can be customized and the IMS can handle all functionality within ArcView, including its built-in scripting language. This makes the system very flexible but also expensive and more difficult to set up and run (Kraak and Brown 2001; ESRI 2002).

The latest ESRI ArcIMS architecture (ESRI 2004a) is a bit different than the earlier ArcIMS setup described above, but the gist remains the same. ESRI (2005a) also provides video demos (available at http://www.esri.com/software/arcgis/arcims/about/demos.html) to get started with their latest version of ArcIMS and to learn how to serve ArcGIS maps in ArcIMS in three easy steps: (1) author a map in ArcGIS or ArcIMS Author, (2) publish it as an ArcIMS Service using ArcIMS Administrator, and (3) use the published Service (map) online.

Unfortunately, all these excellent features of dedicated Internet map server solutions do come at a cost (Boulos et al. 2002):

- Map server software licensing costs can be several thousands of US dollars; however, some Open Source options also exist. For example, MapServer, an Open Source development environment for constructing spatially-enabled Internet-Web applications (see http://mapserver.gis.umn.edu/ and MapServer application gallery http://mapserver.gis.umn.edu/gallery). MapServer, through the use of special libraries, can access various raster and vector data formats without data conversion (Mitchell 2005a; Boulos and Honda 2006);

- Expertise is required to install, customize and manage the Internet map server, whether it is a commercial product or Open Source. However, some easy-to-implement Open Source Web GIS server solutions have started to appear that target users with no prior technical experience in Web GIS or Internet map servers (Boulos and Honda 2006);
- Full access to the hosting Web server is required to install and manage software components, which is not always possible with mainstream (cheap) shared virtual hosting packages offered by most Web hosting providers. However, dedicated Internet map server (e.g., ArcIMS) Application Service and Hosting Providers are also available that can help solve this issue and the previous one too (expertise), though not without additional cost—see, for example, http://www.geocortex.net/hosting.html and http://www.geocortex.net/imf/; and
- Speed: client applications running in the client's Web browser (e.g., ESRI MapCafe Java applet in earlier ArcIMS implementations) might be slow to download, depending on nature/size of these client applications and the client's Internet connection speed.

49.3.3.2 Standards (GML, WFS, SVG), Integration and Interoperability

As stated above, many commercial Internet GIS programs have been developed, such as ESRI ArcIMS and the other examples listed in Table 49-2. There is no doubt these commercial Internet GIS programs have greatly increased the accessibility of GIS data and tools, but there remains two more problems associated with many of the current implementations of these products (Peng 2004).

The first problem is lack of true interoperability between them. This non-interoperability problem in turn has two aspects. The first aspect is a data interoperability issue; data created by different programs cannot be shared by others without data conversion. This causes problems for real-time data access, especially at the feature level. The second aspect is an access interoperability issue; data on the server can only be accessed by their own client. Other clients cannot access data due to proprietary access methods implemented by different Internet GIS programs. For example, a GeoMedia WebMap client cannot access data on an ESRI ArcIMS server. Enabling interoperability among heterogeneous systems and geospatial data is a challenging task for the development of Internet GIS, both technically and institutionally (Peng 2004).

The second problem is that the graphic output of most commercial Internet GIS programs is usually in raster image formats such as GIF and JPEG. This is, for example, the case with many ESRI ArcIMS implementations like the example shown in Figure 49-9. The graphic quality of raster images is limited and becomes blurred when zooming in.

To solve these two problems, a standards-based framework is proposed below that uses GML (Geography Markup Language) as a coding and data transporting mechanism to achieve data interoperability, WFS (Web Feature Service) as a data query mechanism to access and retrieve data from the heterogeneous systems at the feature level to achieve access interoperability, and SVG (a vector graphic format—see Section 49.2.7.1) to display GML data on the Web to improve the display quality of map graphics (Boulos 2004a; Peng 2004).

Aggregating disparate data sources to a common geography has always been a strength of GISs. The challenge of nationwide, regional and global coordinated efforts in case of natural or man-made disasters (for example), however, calls for aggregating the aggregates on short notice. For instance, if a disaster hits at the border of two cities, US states, or two EU (European Union) countries, will their two information silos be able to work together, sharing and combining data instantaneously? Today, many systems are based on closed or proprietary interfaces and formats and are difficult to integrate with brands and platforms in use by other organizations. Embracing open standards is the key to interoperability (Lowe 2002).

Interoperability allows spatial data silos distributed anywhere on the Web to be searched, located, retrieved and compiled, either by a Web GIS service provider or at an individual's desktop. The Open Geospatial Consortium (OGC—http://www.opengeospatial.org/) develops specifications to accommodate any operational differences and allow disparate Web GIS clients and desktop users to fully integrate Web-accessible spatial data resources (Croner 2003). OGC's ultimate goal is to enable the "spatial Web" with products that plug-and-play across different processing platforms, vendor brands, networks, and programming languages (Lowe 2002).

Founded in 1994, OGC is an international industry consortium of 284 companies, government agencies and universities participating in a consensus process to develop publicly available geoprocessing specifications. GML is the base language developed by OGC. GML is becoming the world standard for XML encoding of geographic features and geoprocessing service requests. The relevance of XML Web Services (see http://www.w3.org/2002/ws/) to spatial integration of disparate data sources is also obvious (OGC 2004a; Shi 2005). XML encoding of geodata, using GML and Web Services specifications and recommendations, makes it possible to display, overlay, and analyze geodata on any Web browser, even if the browser obtains views of different map layers from different remote map servers. For example, layering Web Services from two politically/administratively separate, but geographically contiguous, cities or regions would allow the integration of their independent data silos to answer questions about an emergency involving both—provided that issues of common semantics, data models and case definitions have been resolved (Croner 2003; Lowe 2002; Peng and Ming-Hsiang Tsou 2003). It is noteworthy that XML is also used for encoding spatial metadata, which are essential to aid the discovery of spatial data in a distributed environment (Croner 2003). Standards also exist for metadata (Boulos 2004a).

One of the keys to GML deployment is a companion specification, the OGC WFS. To get GML data, users query a Web server with an OGC Web Service Interface, collectively known as a Web Feature Server (WFS). The OGC interface enables standardized access to a feature store and enables users to add, update or retrieve GML data locally or across the Internet. Any data store can be used—users no longer need to care whether the underlying store is from ESRI, Oracle or IBM (Lake 2002).

By separating presentation from content, powerful maps can be made that offer enhanced functionality for users. GML contains map "content" only (e.g., where features are, and their geometry, type and attributes), but it does not provide any information about how that map data should be displayed. This is actually a benefit because different "stylesheets" can be applied to the geographic data to make it appear however the user wishes (Galdos Systems Inc. 2000; Galdos Systems Inc. 2002). By combining a selected map stylesheet with a WFS query, users are presented with a fully interactive and editable vector map that can be viewed in any Web browser (Lake 2002) (Figure 49-10).

Another key feature of GML is its ability to be "self describing" through the use of XML schema. Thanks to this feature, tools have been developed to model and load proprietary databases, e.g., Oracle Spatial databases, with geographic data supplied in GML formats (Lake 2002).

GML 2 lacked some important features like metadata support and several other geographic information prerequisites (Lake 2002). The latest GML 3 addresses the limitations of GML 2, while being backwards compatible with GML 2. New additions in GML 3 include support for metadata, units of measure, complex geometries, spatial and temporal reference systems (time information is essential in tracking applications like monitoring ambulance locations and in exploring the movement and growth of natural disasters), topology (the relationships between features, e.g., for use by routing applications popular in location-based services), gridded data, and default styles for feature and coverage visualization. The new release is modular, allowing users to pick out only the schemas or schema components that apply to their work, which simplifies and minimizes the size of implementations (Reichardt 2003; Cox et al. 2003).

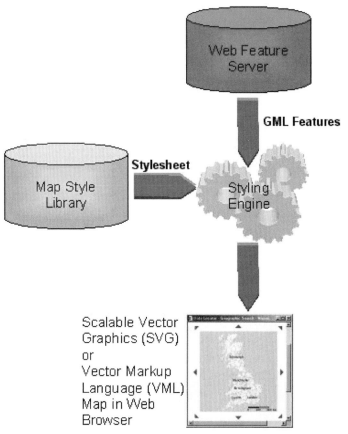

Figure 49-10 Diagram showing the main steps involved in GML (Geography Markup Language) map making. GML contains map "content" only (e.g., where features are, and their geometry, type and attributes), but does not provide any information about how that map data should be displayed. This allows different "stylesheets" to be applied to the geographic data to make it appear however the user wishes. By combining a selected map stylesheet with a Web Feature Service (WFS) query, users are presented with a fully interactive and editable vector map that can be viewed in any Web browser. Modified and adapted from Galdos Systems Inc. (2000). See included DVD for color version.

It is noteworthy that leading GIS companies like ESRI and MapInfo have started supporting OGC standards in their products (see, for example, http://www.esri.com/software/ standards/ and http://www.mapinfo.com/ogc). The Open Source GIS community is similarly, if not more actively, embracing OGC standards; for example, MapServer (described above) supports several Open Geospatial Consortium Web specifications, including WMS (Web Map Service—client/server), non-transactional WFS (client/server) and GML (see also other Open Source OGC-compliant Web GIS tools/products at: http://www.refractions.net/ and http://maptools.org/, and Mitchell (2005b) and Erle et al. (2005)). Boulos and Honda (2006) discuss the latest Open Source Web GIS solutions that can be used to publish users' maps on the Web and add to those maps additional layers retrieved from remote WMS servers.

Along the same vein, Ordnance Survey (OS), the UK's national mapping agency, is now also using OGC GML as the only geospatial data format for its MasterMap of Great Britain (http://www.ordnancesurvey.co.uk/oswebsite/products/osmastermap/). OS MasterMap boasts about 400 million geographic features in GML format. Each feature within OS MasterMap is assigned a unique 16-digit "topographic identifier" (TOID) that can be used by OS or its customers to reference any given feature in the database. This makes it much easier for users to associate other information to the spatial feature, to refer unambiguously to a particular feature, and, therefore, to share spatial information with other users (Lake 2002).

However, it should be noted that OGC standards and XML Web Services are only part of the solution to integration and interoperability. Other discipline-specific standards are also likely to be needed. For example, in the health sector, standards like HL7 (Health Level 7—http://www.hl7.org/) and clinical coding schemes like SNOMED CT (Systematised Nomenclature of Medicine–Clinical Terms—http://www.ihtsdo.org/), and ICD (International Classification of Diseases—http://www.cdc.gov/nchs/icd9.htm) are also equally important in some health GIS applications (Boulos 2004a).

Lowe (2002) also stresses the fact that technologies like XML and SOAP (Simple Object Access Protocol—involved in Web Services) are only part of the integration issue, and points to integrating geoprocessing and databases at other levels, and the related issues of optimizers and federated databases. Industry professionals now manage very large spatial databases. Often, client programs will pull a copy of the database spatial data into their own environment to process it instead of asking the database to do the processing. If the client program request happens to involve a very large database table, the copy-and-exchange process may drag on endlessly or even fail because of overload. This same potential problem awaits users of multiple feature-streaming map services (Lowe 2002).

Alternatively, if the spatial processing remains within the database environment, an optimizer program common to all professional databases will internally organize a response to the query that returns results in the fastest possible time. A query from the larger integrated system goes into the database and only the results come out, taking advantage of the database optimizer, reducing processing loads on the client that generated the question, and also reducing transmission loads (Lowe 2002).

Each database vendor's optimizer works best within its own specific database environment. A potential problem arises in case one wants to optimize the use of multiple databases when a query joins data from several different databases (from different vendors) at the same time. In the same spirit as the Web Services model, agencies can keep their existing heterogeneous database technology and use a federated database technology to unite the mix. IBM, for example, offers a federated database technology that simulates views of any other database tables in IBM DB2 database, offering a master view of all data holdings. Furthermore, the federated technology's optimizer is aware of the available processing resources in other databases and organizes query responses appropriately (Haas and Lin 2002; Lowe 2002).

49.4 Emerging and Future Trends in Interactive Web Cartography and Internet GIS

49.4.1 Online Geospatial Data Discovery and Metadata Repositories, Distributed GIS, On-demand Custom Geospatial Applications and Telegeoprocessing, and Mobile GIS and Location-based Services

With the rapid global growth of the Internet during recent years, the discovery of Internet-distributed geospatial data and the associated online geospatial metadata repositories/geospatial data clearinghouses have become among the most important and active research themes and service development areas in Internet GIS.

Among the best examples of geospatial metadata repositories for distributed geospatial data discovery and viewing is Geospatial One-Stop (http://www.whitehouse.gov/omb/egov/c-2-1-geo.html), a recent US NSDI (National Spatial Data Infrastructure)-related e-government initiative. Geospatial One-Stop is intended to revolutionize electronic government in the US by providing a geographic component for use in all Internet-based government activities across all government levels. The goals are to enable immediate discovery and "one-stop" access to spatial metadata and data via a single Internet location/interface for different kinds of analyses and improved decision-making, and to eliminate the redundancies of costs associated with duplicate efforts of spatial data collection, conversion between formats, production and dissemination (Croner 2003; US Federal Geographic Data Committee 2002).

To achieve its vision, the Geospatial One-Stop initiative has launched Geodata.gov (http://www.geodata.gov/—Figure 49-11), a Web-based portal for one-stop access to US maps, data and other spatial services that will simplify the ability of all levels of government, private sector, academia and citizens in the US to find and use spatial data, and learn more about spatial projects underway (Boulos 2004a).

Other online examples of geospatial data discovery and related services include EarthExplorer, a service that can be used to query and order satellite images, aerial photographs, and cartographic products through the US Geological Survey (http://edcsns17.cr.usgs.gov/EarthExplorer/), ESRI's Geography Network (http://www.geographynetwork.com/), Gigateway, a free Web service aimed at increasing awareness of and access to geospatial information in the

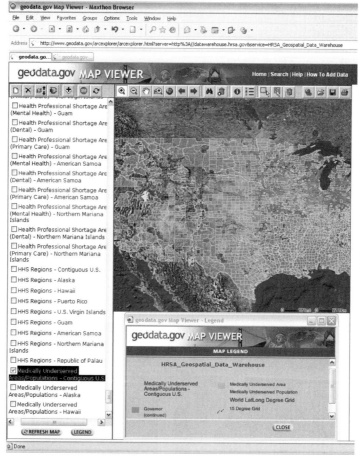

Figure 49-11 Screenshot of the "Health Resources and Services Administration (HRSA) Geospatial Data Warehouse - Feature Service" accessible via Geodata.gov (under the "Human Health and Disease" data category; other top-level Geodata.gov categories include Administrative and Political Boundaries; Agriculture and Farming; Atmosphere and Climate; Biology and Ecology; Business and Economic; Cadastral; Cultural, Society, and Demographic; Elevation and Derived Products; Environment and Conservation; Geological and Geophysical; Imagery and Base Maps; Inland Water Resources; Locations and Geodetic Networks; Oceans and Estuaries; Transportation Networks; and Utilities and Communication). The HRSA Geospatial Data Warehouse and its associated applications provide access to a broad range of information and HRSA-specific spatial layers about HRSA programs, related health resources, and demographic data useful for planning and policy purposes. The layer featured in this screenshot is the "Medically Underserved Areas/Populations–Contiguous US" data set. See the included DVD for the color version.

UK (http://www.gigateway.org.uk/), and Go-Geo!, a tool designed to help users find details about geospatial data sets and related resources within Great Britain tertiary education and beyond (http://www.gogeo.ac.uk/). (See also Kelly 2005.)

True distributed GIS applications and geoprocessing services that run and are accessed over the Web are becoming a reality, building on, and going beyond current data discovery, access and sharing points/services. Some online services like Google Maps expose an API (Application Programming Interface) allowing the development of custom third-party applications (see Section 49.4.2 and Figure 49-12 below). Other services like ESRI ArcWeb Services give users access to both GIS content and GIS capabilities—on demand when needed—and eliminate the overhead of purchasing, storing and maintaining large data sets (ESRI 2005b).

With ArcWeb Services, to use the same example, data storage, maintenance, and updates are handled by ESRI (the service provider). Users can directly access terabytes of dynamic, up-to-date content and capabilities using ArcGIS, or use ArcWeb Services to build their own unique Web-based applications (on-demand custom applications). Users can also batch geocode addresses using ArcWeb Services (ESRI 2005b). It is noteworthy that high-volume batch geocoding can also be done these days using services like EZ-Locate (http://www.geocode.com/), a dedicated online US geocoding service.

Readers interested in ArcWeb Services can freely browse some sample ArcWeb applications at http://gis.esri.com/showcase/showcase.cfm. The featured ArcWeb showcases include a real estate application, a geofence and tracking application that enables users to watch real-time as vehicles travel along their designated routes, a routing application, a weather reporting application, and a world photo album application that allows users to discover digital photographs published by others or to publish links to their own photographs based on geographic location (cf. Smugmug Maps described in Section 49.4.2).

Back in 2002, Xue et al. defined telegeoprocessing as a new discipline revolving around real-time spatial databases that are updated regularly by means of telecommunications systems in order to support problem solving and decision making at any time and any place. Telegeoprocessing involves the integration of remote sensing, GIS, GPS and telecommunications.

Grid-based, real-time distributed collaborative geoprocessing could also form the basis of a next-generation solution to data and computationally intensive geoprocessing applications that are extremely difficult to execute on conventional systems and networks (Shi et al. 2002). Grid computing allows non-collocated computers to work on and process data together, not just communicate and exchange data between each other. It is already a reality with many ongoing projects (see, for example, http://www.urec.cnrs.fr/rubrique207.html). It remains to be seen how telegeoprocessing and grid-based geoprocessing will further evolve and become incorporated into mainstream GIS over the next years.

Mobile phones and other digital devices are also rapidly gaining location awareness and Web connectivity, thus promising new spatial technology applications that will yield vast amounts of spatial information and open many new possibilities (Löytönen and Sabel 2004; Reichardt 2002), but not without their challenging security and privacy concerns (see also Section 49.4.2.2). Examples of such applications include in-the-field data entry and access, and many useful location-based information services (Boulos 2003b).

According to Dobson (2005), human-tracking applications are also going mainstream and becoming affordable using GPS (Global Positioning System) and RFID (Radio Frequency IDentification) technologies, but again not without associated privacy concerns. ObjectFX Corporation (http://www.objectfx.com/) is among the emerging market players of relevance to this discussion. ObjectFX provides real-time online mapping software that turns dynamic data into interactive views. This could prove useful for companies that use GPS services in their business and need to monitor the location and status of people or mobile assets. In fact, according to ObjectFX Web site, FedEx Custom Critical (http://customcritical.fedex.com/) is one of their customers. ObjectFX can provide such companies with near real-time and historical visualization of GPS data. Also of interest is the article by Shuping (2005) on the geo-temporal visualization of RFID.

Manual of Geographic Information Systems

49.4.2 Geographic Search Engines and Related Online Consumer Geoinformatics Services

The visual language of geography is a powerful, universal language. News providers are regularly using maps to help them tell their stories, e.g., this interactive map by USATODAY. com on the demographic shifts changing the Roman Catholic Church in the USA http:// www.usatoday.com/news/graphics/diocese_data_2004/flash.htm, the BBC News interactive UK Election 2005 map http://news.bbc.co.uk/1/shared/vote2005/flash_map/html/map05. stm and also their Born Abroad (An immigration map of Britain) http://news.bbc.co.uk/1/ shared/spl/hi/uk/05/born_abroad/html/overview.stm. Reuters news agency has even launched its own dedicated interactive map service http://www.alertnet.org/map/index.htm.

The use of maps as an information filter and search platform is rapidly gaining momentum on the Web. MetaCarta Geographic Text Search appliance (GTS—http://www.metacarta. com/) is a good example of such a use. Founded by a team of Massachusetts Institute of Technology (MIT) researchers in 1999 with funding from the US Central Intelligence Agency (CIA), MetaCarta bridges the gap between GIS and text search, allowing users to link information to geography and to discover geographical themes within their documents. MetaCarta GTS is able to scan unstructured documents from a variety of sources, including Internet and intranet sites, and to extract geographical references buried in them, intelligently handling any geographical name ambiguities or inconsistencies it encounters. Users can search a geo-parsed document collection for the occurrence of any keywords (geographical or otherwise). GTS presents search results as icons on a map of the region of interest. The results are linked to the full documents. Users can further narrow searches by zooming in to specific geographical locations (the map acting as a filter), and by modifying their query string (Boulos 2005a).

Mainstream Web search engines like Google (http://www.google.com/) and MSN/Live Search (http://www.live.com/?searchonly=true&mkt=en-us) have more recently joined the geographic search bandwagon by releasing their own dedicated geographic interfaces, which run in standard Web browsers and also provide the general public with detailed satellite imagery/aerial photography map layers that were once only available to experts and select user communities.

Google released Google Maps (http://maps.google.com/—a localized UK version is also available; see, for example, Paddington Station, London, W2 1RH (Satellite): http://maps. google.co.uk/maps?q=paddington+station&spn=0.018368,0.039669&t=k&hl=en). Google also released Google Earth (http://earth.google.com/), a fat client, standalone 3D (three-dimensional) desktop application that offers anyone with Internet access a planet's worth of imagery and other geographic information, allowing users to virtually sightsee exotic locales like Paris, France, and Maui (Hawaii), Grand Canyon and Niagara Falls in the US, as well as viewing points of interest such as local restaurants, hospitals, schools, and more. Google Earth uses KML (Keyhole Markup Language) to store data (Schutzberg 2005). It is noteworthy that NASA is also offering its own World Wind 3D application (http:// worldwind.arc.nasa.gov/) that lets users zoom from satellite altitude into any place on Earth.

Thanks to Google Maps API (Application Programming Interface—http://www.google. com/apis/maps/), many third-party applications and custom-annotated maps have begun to appear (Blankenhorn 2005). Two good UK examples of such applications/custom-annotated maps are the Health QOF (Quality and Outcomes Framework) Database map (http://www. gpcontract.co.uk/map.html) and London July 2005 Terrorist Attacks maps (Figure 49-12). Because Google Maps has its roots in XML, users were also able to produce their own custom-annotated Google maps, e.g., based on their own GPS (Global Positioning System) locational data, and to even tie-in images and video to create interactive multimedia maps, well before the API was publicly documented (Dybwad 2005).

Smugmug Maps (http://maps.smugmug.com/) is another good example of a third-party Google Maps application in action. Smugmug, a photo hosting Web site, plots geocoded photographs to their actual locations on Google Maps (or Google Earth via a KML Google Earth feed: http://www.smugmug.com/hack/feed.mg?Type=geoAll&format=kml20&Size=Tiny—Figure 49-13), and allows location-based searching of photographs all over the world (O'Reilly Radar 2005). All smugmug RSS (Really Simple Syndication) feeds are now geo-enabled. If a photo has latitude, longitude and altitude information (geographic metadata), it will show up in all feeds (see http://www.smugmug.com/help/rss-atom-feeds).

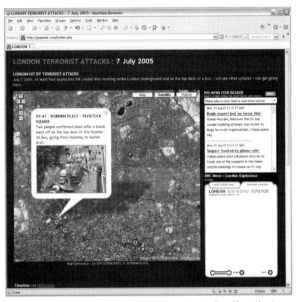

Figure 49-12 Screen shot of a third-party Google Maps application that provided a quick map and RSS (Really Simple Syndication) news feed about the London terrorist attacks, within a few hours of the blasts on 7 July 2005 (http://web.archive.org/web/20070308125907/http://geepster.com/london.php). The resulting page was far more attractive and informative than most news services, including the BBC, from which it took its news feed. The application also uses Google's satellite view of London. Users can right-click on the pins on the map to learn more about specific incidents, e.g., the Tavistock Square incident with a picture of the ruined bus (shown in this screenshot). See included DVD for color version.

Figure 49-13 A Smugmug KML photo feed in Google Earth. The KML feed is available at http://www.smugmug.com/hack/feed.mg?Type=geoAll&format=kml20&Size=Tiny and is intended to be opened by Google Earth desktop application, which can be downloaded at http://earth.google.com/. See included DVD for color version.

Manual of Geographic Information Systems

Sports Map (http://www.backstage.min-data.co.uk/sport/) is a further example of an interesting Google Maps application that uses BBC Sport RSS feeds (http://news.bbc.co.uk/sport1/hi/help/rss/3397215.stm). It allows users to zoom into UK areas of interest and then displays the nearest football club, along with all the latest BBC Sport news headlines and stories for that region.

Readers wanting to further explore Google Maps API might be interested in Google Mapki Knowledge Base and list of developer tools (http://mapki.com/index.php?title=Knowledge_Base). Google Mapki is a forum for sharing ideas, implementations, and help for the Google Maps API.

Microsoft's response to Google Maps and Google Earth comes in the form of MSN Virtual Earth (http://virtualearth.msn.com/). A distinguishing feature of MSN Virtual Earth is its "Locate Me" tool. Wired users can be located via their IP (Internet Protocol) address (this has been done for some time—see, for example, Boulos (2003b) and http://healthcybermap.semanticweb.org/ip.htm). Wireless users can download a small application that does locating based on connection to a Wi-Fi access point. MSN Virtual Earth also features aerial oblique imagery (45 degree angle views or "Eagle Eye Views") of major US metropolitan areas, provided by Pictometry International Corp. (http://www.pictometry.com/).

An MSN Virtual Earth Map Control/API (see http://www.viavirtualearth.com/vve/Dashboard/Default.ashx and http://www.viavirtualearth.com/vve/Resources/Default.notitia.ashx) allows users to create their own custom online maps and add their own data to MSN Virtual Earth. Boulos (2005b) presents three online interactive map examples created using Google Maps API, Google Earth KML, and MSN Virtual Earth Map Control (see http://www.healthcybermap.org/MSNVirtualEarth/). Along the same vein, Amazon.com was once also providing A9 Block View, an online Yellow Pages/map service that offers US maps with street-level photos (Figure 49-14).

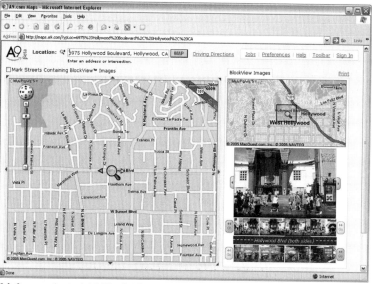

Figure 49-14 Screenshot of A9 Block View from Amazon.com (http://www.searchengine-journal.com/wp-trackback.php?p=3858) showing street-level photos from 6975 Hollywood Boulevard, Hollywood, CA, US. (A9 was using maps from MapQuest.com, Inc.) See included DVD for color version.

It is noteworthy that AJAX (Asynchronous JavaScript and XML) is adopted by Google's maps.google.com, Amazon's maps.a9.com, and MSN's virtualearth.msn.com (Pandey and Shukla 2005). Also some "standards" are also currently under development, collectively referred to as GeoRSS (see http://www.georss.org/), which enable the encoding of location in RSS (Really Simple Syndication) feeds. As RSS is becoming more and more prevalent as a way to publish and share information online, it has become increasingly important to extend it to allow location to be described in an interoperable manner, so that applications can request, aggregate, share and map geographically tagged feeds (Schutzberg 2006). Daden's Avian flu feed for Google Maps/Google Earth is a good example of a health-related geofeed in action (http://www.daden.co.uk/pages/000208.html).

Corresponding offerings from Yahoo! search engine have been modest by comparison, and include Yahoo! Maps (USA and Canada—http://maps.yahoo.com/) and an associated API (http://developer.yahoo.net/maps/). At the time of writing (September 2005), the Yahoo! Maps service does not offer any satellite imagery/aerial photography, but this might change in the near future. The latest traffic status/incidents, as well as Wi-Fi hotspots can be visualized on Yahoo! Maps. The ten-step Yahoo! map zoom control features handy captions about the expected level of details at key zoom steps: 8-state, 4-city, and 2-street levels (GlobeXplorer ImageAtlas http://www.globexplorer.com/ adopts a similar user-friendly zoom- level labeling approach—from nation to house level). Other countries are covered at: http://maps.yahoo.net/.

Dedicated Web browser toolbars and extensions have also started to appear, e.g., MutantMaps (http://www.mutantdesign.co.uk/maps/help.htm), a Mozilla Firefox toolbar that allows navigation between five popular mapping sites (Google Maps, MSN Virtual Earth, MultiMap.co.uk, TerraServer.com and 192.com), while preserving user's longitude, latitude and zoom levels, and gMapIt, another Mozilla Firefox extension that allows users to find directions from Google Maps based on publicly listed US phone numbers (http://www.gmacker.com/web/content/mapit/mapit.htm).

Some commentators have recently wondered if users will soon eschew ArcIMS (see Section 3.3.1) in favor of using Google Maps API to quickly create Web map applications. ESRI's response to all of the recent online consumer geoinformatics services we have described in this section was to announce (in 2005) its new partnership with National Geographic (http://www.nationalgeographic.com/), GlobeXplorer (http://www.globexplorer.com/), and TeleAtlas (http://www.teleatlas.com/), plus Geospatial One-Stop (GOS—http://www.geodata.gov/) and a few other partners like MDA (MacDonald, Dettwiler and Associates Ltd.—http://www.mda.ca/), to upgrade the National Geographic MapMachine (http://plasma.nationalgeographic.com/mapmachine/), a map service/online atlas that provides global map coverage for an extensive set of Earth science themes. MapMachine was first launched in November 1999 and is powered by ESRI's ArcWeb Services (for a description of the latter, please see Section 49.4.1). The planned upgrade aims at bringing satellite imagery, aerial photos, and street-level data to MapMachine users. Users will be able to access the service through a new viewer that is aimed at a mass audience, and appears to be ESRI's direct response to Google Earth and Microsoft Virtual Earth. However, one important difference from those services is that the ArcGIS back end will also allow users of the new service to accomplish much more sophisticated tasks, such as service area analysis. The next generation of MapMachine will also provide a link to GOS data and metadata to help users discover information about their area of interest or study. MapMachine will include capabilities for 3D globe services, allowing GIS users to "pull in" their own map services to overlay onto a globe. Also planned is the addition of ESRI's MapStudio (http://www.mapstudio.com/), an ArcWeb Services application used by many daily newspapers to create maps for printing, to enable users to create customized maps (Luccio 2005a).

49.4.2.1 The Geodata-rich Society and the Wikification of GIS, Maps and Satellite Imagery/Aerial Photography: Imaging and Geospatial Information for the Wide Masses

ESRI president Jack Dangermond recently predicted that the supply of satellite and aerial imagery will increase by two fold in the next few years. Availability will also increase greatly, via Web portals and online GIS services. This is all part of what Dangermond describes as a "geodata-rich society" that will have access to more geospatial information of all kinds, including, in addition to imagery, GPS/location data, geo-demographic data, and data from real-time monitoring (Luccio 2005b). The Internet is already the "foundation medium" to access, link and use all these data.

Satellite imagery and remote sensing are quickly entering the mainstream. Today, satellite imagery data are abundantly available from multiple sources, including companies such as Space Imaging (http://store.geoeye.com/), DigitalGlobe (http://digitalglobe.com/), GlobeXplorer (http://www.globexplorer.com/), Spot Image (http://www.spot.com/), ImageSat International (http://www.imagesatintl.com/), and EarthSat (http://www.earthsat.com/— an MDA company), and are used in hundreds of applications. But thanks to online consumer services like Terraserver (http://www.terraserver.com/), Google Earth, and Microsoft Virtual Earth (see Section 49.4.2), satellite imagery has also been made familiar and accessible to millions of people.

There is no doubt the different online consumer geoinformatics services that have been presented in Section 49.4.2, including the different geographic search interfaces from major Web search engine providers, have significantly contributed (in record time) to raising the general public interest in geography and satellite imagery. As millions of people start "playing" with these new online "gadgets" or "toys" from Google and Microsoft, many of them will soon start thinking about becoming active participants, sharing information and collaborating online (notions that have been rightly associated with the Web for quite a long time), rather than just being satisfied with a passive information consumer/viewer role. The reader should note that it has been estimated that about 800 million persons are online today worldwide (Global Reach 2004).

However, although Google Maps API (and similar API offerings from other providers) enables users to deeply customize the standard provider's interface (Google Maps), and to create their own custom-annotated maps (custom applications based on Google Maps), such APIs remain difficult for the non-expert average user to exploit. This author expects the technology to further evolve to enable the average Web user to share geospatial information, to customize, annotate and publish his/her own online maps and related Web applications, and to collaborate with other users/online communities within an online customizable and collaborative mapping environment, all without the need for any prior programming knowledge or expertise (cf. Toucan Navigate collaborative GIS http://www.infopatterns.com/products/toucannavigate2007.aspx).

The current "wiki" concept is not far from this vision. A wiki (from Hawaiian *wiki,* to hurry, swift) is a collaborative Web site whose content can be edited by anyone who has access to it (American Heritage Dictionaries/Answers.com 2004). Perhaps the best example of a wiki in action today is "Wikipedia–The Free Encyclopedia" (see the "wiki" entry in Wikipedia at http://en.wikipedia.org/wiki/Wiki). A related Web information sharing technology is the "blog." A blog (WeBLOG) is a Web site that contains dated entries in reverse chronological order (most recent first) about a particular topic. Functioning as an online journal, blogs can be written by one person or a group of contributors. Entries contain commentary, images, and links to other Web sites; a search facility may also be included (Freedman/Answers.com 2005—see the "blog" entry in Wikipedia at http://en.wikipedia.org/wiki/Blog).

Wikis, and in particular Wikipedia, have grown very popular in recent months and years (Connor 2005). Wikis represent a promising principle that can significantly transform the Internet information age. Special conferences have been and are being organized to discuss this interesting Web phenomenon of wikis; for example, Wikimania 2005, the First International Wikimedia Conference, 4–8 August 2005, Frankfurt am Main, Germany (http://wikimediafoundation.org/wiki/Wikimania), and the ACM (Association for Computing Machinery)-sponsored WikiSym 2005, the 2005 International Symposium on Wikis, 17–18 October 2005, San Diego, California, US (http://www.wikisym.org/).

Along the same lines, it is not difficult to imagine the development in the very near future of "geowikis," "mapwikis," geo-enabled blogs, "mapblogs" (imagine, for example, people with an Internet-connected, GPS-enabled mobile device wanting to blog their movements, and share their activity spaces and geo-referenced news with other online users for various purposes), and even geo-enabled, mappable Web/RSS feeds and map feeds (see the Smugmug KML photo feed example in Section 49.4.2/Figure 49-13 above). In fact, some early geowiki examples have already found their way on the Web; see, for example, http://www.wikyblog. com/Map/Guest/Home, http://www.geowiki.com/, and also worldKit GeoWiki, a publicly editable map application http://worldkit.org/doc/geowiki.php (a simple online demo of worldKit GeoWiki to which anyone can add their own data is available at http://worldkit. org/geowiki/— also of interest from the same provider is the worldKit geocoder, a free online worldwide city geocoder http://brainoff.com/geocoder/).

Another example is the Katrina Information Map (http://scipionus.com/), which was built using Google Maps (Thompson 2005). Katrina Information Map was conceived for use by people affected by Hurricane Katrina (August 2005) and their relatives who have, or are trying to find, information about the status of specific locations affected by the storm and its aftermath. Users having information about the status of an area that is not yet on the map can easily contribute to the map by adding/appending their information to it. (Readers interested in Hurricane Katrina's online maps and imagery in general might also find the following two sites useful: http://www.esri.com/disaster_response/katrina-rita.html, http:// ngs.woc.noaa.gov/katrina/ and http://msnbc.msn.com/apps/ve/katrina.htm.)

But 3D interactive and real-time mapping in 3D virtual worlds like Second Life (http:// secondlife.com/) is perhaps the most cutting-edge Web mapping trend around (Boulos and Burden 2007). Andrew Hudson-Smith and his team at the Centre for Advanced Spatial Analysis (CASA), University College London, also have an extensive portfolio of GIS-related projects in Second Life (http://digitalurban.blogspot.com/2007/12/geographic-data-in-second-life.html). In their popular "Digital Urban" blog (http://digitalurban.blogspot.com/), they frequently refer to Google Earth and Second Life as "Three Dimensional Collaborative (Multi-User) Geographic Information Systems." Second Life and Google Earth (and the related platforms that will definitely follow in the near future) are indeed promising environments for public participation and collaboration-type outreach activities, providing a good basis for what can be referred to as "The People's Atlas" or neogeography (McFedries 2007). CASA's Second Life GIS projects include (i) Virtual London (http://digitalurban.blogspot. com/2007/11/virtual-london-removed-from-second-life.html), (ii) a new approach to importing geographic terrains into Second Life as tabletops (http://digitalurban.blogspot. com/2007/11/importing-geographic-terrains-into.html), and (iii) an Arc (ESRI) to Second Life project (http://digitalurban.blogspot.com/2007/11/arc-to-second-life-geographic-data.html).

The possibilities and potentials are endless. This is what this author calls the ultimate "wikification" of GIS, maps and satellite imagery/aerial photography. If the majestic Tate Museum in London is currently posting captions from its visitors next to its greatest works of art (Wreden 2005), why shouldn't online maps (even those from very reputable sources like the National Geographic Society) allow a similar approach?!

49.4.2.2 Associated Individual Privacy, National Security, Data Confidentiality and Copyrights/Digital Rights Management Issues

As geospatial technology progresses and becomes more readily available to the wide masses around the world who are connected to the Internet, the interrelated issues of GIS and map data confidentiality/individual privacy, and even national security start to surface, calling for further examination of, and research into these delicate aspects of Internet GIS and Web maps (Boulos 2004a; Boulos et al. 2005b; Entchev 2005; Francica 2005b; Yahoo News/ Associated Press 2005; Barlow 2005).

For example, in public health worldwide, any public identification of an individual's health status and residence, regardless of level of contagion or risk, is usually prohibited with very few exceptions, e.g., Megan's Law in the US, which allows the release of residential information on registered child sex offenders to the public by local government (Croner 2003; Boulos 2004a). In fact, thanks to the latter law, we have a service like the Georgia Sex Offender Maps http://www.georgia-sex-offenders.com/maps/, which was built using Google Maps. SARS (Severe Acute Respiratory Syndrome) mapping in Hong Kong in 2003 using disaggregate case data at individual-building level in near real-time was another noticeable exception to this well-established public health confidentiality rule, and also a unique and rare GIS opportunity that resulted in some very comprehensive public Internet mapping services (Boulos 2004b).

On another level, following the September 2001 terrorist events in the US, many federal and local spatial databases, e.g., "critical infrastructure" spatial data, were assessed by their holding agencies as a potential liability to national security and withdrawn from the Internet or public dissemination. The current concern is to find an appropriate balance between public access to spatial information and protection of information considered a priority for national security (Croner 2003; Boulos 2004a).

But despite all these undeniable, legitimate and real concerns about Internet GIS and map data privacy and confidentiality, many of the doubts and misgivings that are raised concerning these aspects of Internet GIS seem to be ill founded, or at least exaggerated. Entchev (2005) has wisely stated, "Let us not cripple the GIS system to meet some vague privacy perceptions."

Another thorny Internet GIS issue that needs to be addressed is that of data and map copyrights. Conner (2005) has rightly described online maps as a copyright minefield. Copyrighted geo-data and maps are usually more difficult and expensive to acquire and use.

But as geo-data become more important in everything from blogs through mobile phones to finding lost people, free maps could make more and more of a difference (Conner 2005). However, someone needs to pay the bill for such "free" maps, and so finding sustainable commercial models for adoption by online geo-data and Web map providers is becoming of prime importance these days (Francica 2005a). Examples of such commercial models include ad-sponsored map services, and low-cost, added-value paid services supporting the free service like Google Earth plus http://earth.google.com/earth_plus.html and Google Earth Pro http://earth.google.com/earth_pro.html.

The Open Geospatial Consortium's work on Geospatial Digital Rights Management (GeoDRM) is also poised to become an important enabler in the context of geo-data and map copyrights (OGC 2004b). A great deal of work has already been done in the area of data ownership and rights management for the online e-book, video and music industries, with some mature working solutions already in existence from companies like Macrovision (http://www.macrovision.com/), Microsoft (http://www.microsoft.com/windows/windowsmedia/ forpros/drm/default.mspx and http://www.microsoft.com/windows/ie/downloads/addon/ rm.mspx) and Adobe (https://aractivate.adobe.com/). Such developments are of interest to the geospatial community in that many geospatial data providers need to control or track who has access to their data and how the data are used. The lack of a GeoDRM capability has been identified as a major barrier to the broader adoption of Web-based geospatial technologies. The mission of OGC GeoDRM Working Group is to coordinate and mature the development and validation of work being done on digital rights management for the geospatial community (OGC 2004b).

49.5 Conclusion

With the advent of Internet GISs and mapping, GIS information products (GIS output) have become accessible to a much wider audience of decision makers who have no GIS software skills. Many Internet GIS applications have been developed, spanning many disciplines.

Web-based maps offer the following extra features not found in conventional paper-based maps: (1) real-time or near-real-time map updates based on the latest data sets; (2) interactivity—desktop GIS-like functionality, e.g., drill-down and zooming, map querying, measuring distances on the map, and switching map layers on and off; and (3) availability to larger audiences/wider and more rapid dissemination of information compared to other publishing media.

Non-animated (static) interactive Web maps, also called clickable maps, imagemaps or hypermaps, can be served either as client-side imagemaps or as server-side imagemaps, depending on where mouse-click coordinates are resolved.

Map comprehension can be enhanced with appropriate use of visual and typographic variables, and the application of sound cartographic design principles regarding cartographic symbols, map contrast and visual hierarchy, and map scale and abstraction. An important visual variable is value (differences in gray values or lighter/darker shades of the same color), which conveys differences in order or relative quantity and forms the basis of choropleth maps. Pictorial symbols are much easier to understand than geometric or abstract symbols, especially when a clear metaphorical relation exists between the symbol and the user's real world knowledge.

The ideal Web map should not be too large in both file size and image size to download quickly and cope with the limited size of display screens. The Web offers many techniques for adding interactivity to maps and responding to mouse events (e.g., zooming, panning, and MapTips on mouse-over for map feature labeling), which can be used to make smaller, "smarter" maps that are faster to download and provide additional functionality. Other important online map usability and accessibility issues were also discussed.

Raster image formats used in Web maps include GIF (CompuServe Graphical Interchange Format), which is more suitable for line-art images and images with solid colors, and JPEG (Joint Photographic Experts Group), the best compression algorithm for maps making considerable use of color blends. Web map designers should test both formats on the maps they intend to produce, varying the various parameters available, including JPEG compression/quality settings and GIF palette sizes to find which setting gives good results (usually assuming a 256-color configuration), while maintaining a reasonably small file size. A third raster image format in use today is PNG (Portable Network Graphic).

Vector graphic formats offer scalable, editable, interactive and more bandwidth-efficient Web maps. Three formats were briefly discussed, namely the Vector Markup Language (VML), the Scalable Vector Graphics format (SVG) and Macromedia® Flash.

There are three main online map zooming strategies: static linear zooming, static stepped zooming and dynamic zooming. A fourth zooming strategy, semantic zooming, has also been described.

Currently two main options exist for sharing desktop GIS maps and projects on the Web as sensitive clickable maps. The cheaper and simpler option is to publish a static snapshot of the project as clickable client-side imagemaps representing the project's maps and underlying data at time of publishing. The other option is dynamic publishing to the Web using a dedicated

Internet map server that maintains a live connection with the underlying GIS project/database, but this could be an expensive and complex solution to acquire, run and maintain.

The issues of standards, integration and interoperability in relation to the latter publishing option were also covered, and a standards-based implementation framework has been proposed that uses GML (Geography Markup Language) as a coding and data transporting mechanism to achieve data interoperability, WFS (Web Feature Service) as a data query mechanism to access and retrieve data from heterogeneous systems at the feature level to achieve access interoperability, and SVG to display GML data on the Web to improve the display quality of map graphics.

Finally, emerging and future trends in interactive Web cartography and Internet GIS were presented, including online geospatial data discovery and metadata repositories, distributed GIS, on-demand custom geospatial applications and telegeoprocessing, mobile GIS and location-based services, geographic search engines and related online consumer geoinformatics services (including the Google Maps/Google Earth phenomenon).

We also discussed the notions of the geodata-rich society and the wikification of GIS, and maps and satellite imagery/aerial photography, which promise to bring imaging and geospatial information and collaboration to the wide masses. We also briefly touched on the associated individual privacy, national security, data confidentiality, and copyrights/digital rights management issues.

Throughout the text, we provided illustrative examples of freely available online maps covering various applications and developed using a wide range of technologies, as well as pointers to real-world solutions and tools (including Open Source products) that interested readers can use to develop and optimize such maps.

References

Note: All URLs were accessed 11 January 2008.

American Heritage Dictionaries/Answers.com. 2004. Definition of 'wiki.' In *The American Heritage® Dictionary of the English Language,* 4th ed. Boston, Mass.: Houghton Mifflin Company. <http://www.answers.com/topic/wiki>

Barlow, K. (for The World Today). 2005. Google Earth prompts security fears. ABC (Australian Broadcasting Corporation) News Online. 8 Aug 05. <http://www.abc.net.au/news/indepth/featureitems/s1432602.htm>

Blankenhorn, D. 2005. Google map API transforms the Web. ZDNet.com Open Source Blog. 12 Jul 05. <http://blogs.zdnet.com/open-source/?p=374>

Boulos, M. N. 2003a. The use of interactive graphical maps for browsing medical/health Internet information resources. *Int. J. Health Geogr.* 2:1. <http://www.ncbi.nlm.nih.gov/entrez/query.fcgi?cmd=Retrieve&db=pubmed&dopt=Abstract&list_uids=12556244>

Boulos, M. N. 2003b. Location-based health information services: A new paradigm in personalised information delivery. *Int. J. Health Geogr.* 2:2. <http://www.ncbi.nlm.nih.gov/entrez/query.fcgi?cmd=Retrieve&db=pubmed&dopt=Abstract&list_uids=12556243>

Boulos, M. N. 2004a. Towards evidence-based, GIS-driven national spatial health information infrastructure and surveillance services in the United Kingdom. *Int. J. Health Geogr.* 3:1. <http://www.ncbi.nlm.nih.gov/entrez/query.fcgi?cmd=Retrieve&db=pubmed&dopt=Abstract&list_uids=14748927>

Boulos, M. N. 2004b. Descriptive review of geographic mapping of severe acute respiratory syndrome (SARS) on the Internet. *Int. J. Health Geogr.* 3:2. <http://www.ncbi.nlm.nih.gov/entrez/query.fcgi?cmd=Retrieve&db=pubmed&dopt=Abstract&list_uids=14748926>

Boulos, M. N. 2004c. Web GIS in practice: An interactive geographical interface to English Primary Care Trust performance ratings for 2003 and 2004. *Int. J. Health Geogr.* 3:16. <http://www.ncbi.nlm.nih.gov/entrez/query.fcgi?cmd=Retrieve&db=pubmed&dopt=Abstract&list_uids=15282027>

Boulos, M. N. 2005a. On geography and medical journalology: a study of the geographical distribution of articles published in a leading medical informatics journal between 1999 and 2004. *Int. J. Health Geogr.* 4:7. <http://www.ncbi.nlm.nih.gov/entrez/query.fcgi?cmd=Retrieve&db=pubmed&dopt=Abstract&list_uids=15788097>

Boulos, M. N. 2005b. Web GIS in practice III: Creating a simple interactive map of England's Strategic Health Authorities using Google Maps API, Google Earth KML, and MSN Virtual Earth Map Control. *Int. J. Health Geogr.* 4:22. <http://www.ij-healthgeographics.com/content/pdf/1476-072X-4-22.pdf>

Boulos, M. N., A. V. Roudsari and E. R. Carson. 2002. A simple method for serving Web hypermaps with dynamic database drill-down. *Int. J. Health Geogr.* 1:1. <http://www.ncbi.nlm.nih.gov/entrez/query.fcgi?cmd=Retrieve&db=pubmed&dopt=Abstract&list_uids=12437788>

Boulos, M. N. and G. P. Phillipps. 2004. Is NHS dentistry in crisis? 'Traffic light' maps of dentists distribution in England and Wales. *Int. J. Health Geogr.* 3:10. <http://www.ncbi.nlm.nih.gov/entrez/query.fcgi?cmd=Retrieve&db=pubmed&dopt=Abstract&list_uids=15134580>

Boulos, M. N., C. Russell and M. Smith. 2005a. Web GIS in practice II: Interactive SVG maps of diagnoses of sexually transmitted diseases by Primary Care Trust in London, 1997–2003. *Int. J. Health Geogr.* 4:4. <http://www.ncbi.nlm.nih.gov/entrez/query.fcgi?cmd=Retrieve&db=pubmed&dopt=Abstract&list_uids=15655078>

Boulos, M. N., Q. Cai, J. A. Padget and G. Rushton. 2005b. Using software agents to preserve individual health data confidentiality in micro-scale geographical analyses. *Journal of Biomedical Informatics* 39(2):160–170. <http://dx.doi.org/10.1016/j.jbi.2005.06.003>

Boulos, M. N. and K. Honda. 2006. Web GIS in practice IV: Publishing your health maps and connecting to remote WMS sources using the Open Source UMN MapServer and DM Solutions MapLab. *Int. J. Health Geogr.* 5:6. <http://www.ij-healthgeographics.com/content/pdf/1476-072X-5-6.pdf>

Boulos, M. N. and D. Burden. 2007. Web GIS in practice V: 3-D interactive and real-time mapping in Second Life. *Int. J. Health Geogr.* 6:51. <http://www.ij-healthgeographics.com/content/pdf/1476-072x-6-51.pdf>

Brewer, C. A. 2005. *Designing Better Maps: A Guide for GIS Users*. Oakland, Calif.: ESRI Press.

Connor, A. 2005. Rewriting the rule books. BBC News WEBLOG WATCH - The Magazine's review of weblogs. 15 Aug 05. <http://news.bbc.co.uk/2/hi/uk_news/magazine/4152860.stm>

Cox, S., P. Daisey, R. Lake, C. Portele and A. Whiteside, eds. 2003. *OpenGIS® Geography Markup Language (GML) Implementation Specification Version 3.00 (OGC 02-023r4)*. Wayland, Mass.: Open GIS Consortium, Inc. <http://www.opengeospatial.org/standards/gml>

Croner, C. M. 2003. Public health, GIS, and the Internet. *Annu. Rev. Public Health* 24:57–82. <http://web.archive.org/web/20050512193125/http://www.fgdc.gov/whatsnew/gis_internet.pdf>

Crossley, D. and T. Boston. 1995. A generic map interface to query geographic information using the World Wide Web. In *Proceedings of the Fourth International World Wide Web Conference*, Boston, Mass.,11–14 December. <http://www.w3.org/Conferences/WWW4/Papers/australia/>

Dobson, J. 2005. Human-tracking goes mainstream. *Directions Magazine* 27 Jun 05. <http://www.directionsmag.com/article.php?article_id=883>

Dodge, M. 2000. Accessibility to information within the Internet: How can it be measured and mapped? In *Information, Place, and Cyberspace,* edited by D. G. Janelle and D. C. Hodge, 187–204. Berlin: Springer Verlag. <http://www.casa.ucl.ac.uk/martin/varenius_accessibility.pdf>

Dybwad, B. 2005. HOW-TO: Make your own annotated multimedia Google map. Engadget.com. 8 Mar 05. <http://www.engadget.com/entry/1234000917034960/>

Entchev, A. 2005. GIS and privacy. *Directions Magazine* 24 Mar 05. <http://www.directionsmag.com/article.php?article_id=810>

Erle, S., R. Gibson and J. Walsh. 2005. *Mapping Hacks (Tips & Tools for Electronic Cartography).* 1st ed. Sebastopol, Calif.: O'Reilly. Free sample content available at <http://www.oreilly.com/catalog/mappinghks/index.html>

ESRI (Environmental Systems Research Institute, Inc.). 2002. ArcView Internet Map Server Frequently Asked Questions - Archived Copy. <http://web.archive.org/web/20020613081159/http://www.esri.com/software/arcview/extensions/imsfaq.html>

———. 2004a. ArcIMS® 9 Architecture and Functionality - An ESRI® White Paper. May 04. <http://www.esri.com/library/whitepapers/pdfs/arcims9-architecture.pdf>

———. 2004b. Section 508—Accessibility. 1 Dec 04. <http://www.esri.com/software/section508/index.html>

———. 2005a. ArcIMS Internet Map Server. <http://www.esri.com/software/arcgis/arcims/index.html>

———. 2005b. ArcWeb Services. <http://www.esri.com/software/arcwebservices/>

Federal IT Accessibility Initiative. 2005. Home Page (Section 508: The Road to Accessibility). <http://www.section508.gov/>

Francica, J. 2005a. The End of Free Maps, Directions Media All Points Blog. 9 Jun 05. <http://www.allpointsblog.com/archives/304-The-End-of-Free-Maps.html>

———. 2005b. The Flow of Free, Location-Aware News in the Age of Terrorism, Directions Media All Points Blog. 18 Jul 05. <http://www.allpointsblog.com/archives/391-The-Flow-of-Free,-Location-Aware-News-in-the-Age-of-Terrorism.html>

Frappier, J. and D. Williams. 1999. An overview of the National Atlas of Canada. Pages 261–267 in *Proceedings of the 19th International Cartographic Conference ICC99,* Ottawa, Canada, 16–20 August, Canadian Institute of Geomatics. Edited by C. P. Keller.

Freedman, A./Answers.com. 2005. Definition of blog, from Computer Desktop Encyclopedia, Computer Language Company Inc. 2005. <http://www.answers.com/topic/blog>

Galdos Systems Inc. 2000. Making Maps With Geography Markup Language (GML). <http://www.galdosinc.com/files/MakingMapsInGML2.pdf>

———. 2002. Why GML: Top 10 Benefits of Using GML. <http://web.archive.org/web/20060714010536/http://www.galdosinc.com/technology-whygml.html>

Global Reach. 2004. Global Internet Statistics (by Language). 30 Mar 04. <http://www.glreach.com/globstats/index.php3>

Graham, I. 1998. Introduction to HTML: ISMAP server-side programs. Jan 1998. <http://www.utoronto.ca/webdocs/HTMLdocs/NewHTML/serv-ismap.html>

Haas, L. and E. Lin. 2002. IBM Federated Database Technology—An IBM DB2 Developer Domain article. Mar 02. San Jose, Calif.: International Business Machines Corporation. <http://www.ibm.com/developerworks/db2/library/techarticle/0203haas/0203haas.html>

Harrower, M. and C. A. Brewer. 2003. Colorbrewer.org: An online tool for selecting colour schemes for maps. *The Cartographic Journal* 40(1):27–37. <http://dx.doi.org/10.1179/000870403235002042>

Harwell, R. 2004. Web mapping with SVG. *Directions Magazine* 5 Nov 04. <http://www.directionsmag.com/article.php?article_id=693>

Held, G., T. Ullrich, A. Neumann and A. M. Winter. 2004. Comparing SWF (ShockWave Flash) and SVG (Scalable Vector Graphics) file format specifications. 29 Mar 04. <http://www.carto.net/papers/svg/comparison_flash_svg/>

Jacobson, R. D. 1998. Navigating maps with little or no sight: an audio-tactile approach. Pages 95–102 in *Proceedings of the Workshop on Content Visualization and Intermedia Representations (CVIR'98)*, 15 August, University of Montreal, Montreal, Quebec, Canada. <http://acl.ldc.upenn.edu/W/W98/W98-0214.pdf>

Kahn, P. and K. Lenk. 2001. *Mapping Web Sites*. Hove, U.K.: RotoVision.

Kelly, M. C. 2005. The road taken: the evolution of GIS data clearinghouses from FTP to map services. *Directions Magazine* 15 Aug 05. <http://www.directionsmag.com/article.php?article_id=932>

Kraak, M.-J. 2004. The role of the map in a Web-GIS environment. *J. Geograph Syst.* 6:83–93. <http://dx.doi.org/10.1007/s10109-004-0127-2>

Kraak, M.-J. and A. Brown. 2001. *Web Cartography: Developments and Prospects*. London: Taylor & Francis. <http://kartoweb.itc.nl/webcartography/webbook/>

Lake, A. 2002. Will GML enable an accessible geo-Web? *GeoWorld* 15(7):42–44. <http://web.archive.org/web/20061016224415/http://www.geoplace.com/gw/2002/0207/0207gml.asp>

Lie, H. W. 1991. The electronic broadsheet: All the news that fits the display. Section 4: Navigation. Master's thesis, The Media Arts and Sciences Section, School of Architecture and Planning, Massachusetts Institute of Technology, Cambridge, Mass. <http://www.w3.org/People/howcome/TEB/www/hwl_th_6.html>

Lowe, J. W. 2002. Homeland homework: Reconfiguring for wider spatial integration. *Geospatial Solutions* 12:42–45. <http://www.geospatial-online.com/geospatialsolutions/article/articleDetail.jsp?id=22143>

Löytönen, M. and C. Sabel. 2004. Mobile phone positioning systems and the accessibility of health services. Pages 277–286 in *GIS in Public Health Practice,* edited by R. Maheswaran and M. Craglia. Boca Raton, Fla: CRC Press.

Luccio, M. 2005a. Partnership to expand MapMachine. *GIS Monitor* 28 Jul 05. <http://www.gismonitor.com/news/newsletter/archive/072805.php>

———. 2005b. Editor's Introduction. *GIS Monitor* 18 Aug 05. <http://www.gismonitor.com/news/newsletter/archive/081805.php>

McCauley, J. D., K.C.S. Navulur, B. A. Engel and R. Srinivasan. 1996. Serving GIS data through the World Wide Web. In *Proceedings of the Third International Conference/Workshop on Integrating GIS and Environmental Modeling*, 21–25 January, Santa Fe, New Mexico, USA. <http://abe.www.ecn.purdue.edu/~engelb/ncgia96/engel.html>

McFedries, P. 2007. The new geographers. *IEEE Spectrum* 44(12-INT):64. <http://www.spectrum.ieee.org/dec07/5738>

Mitchell, T. 2005a. *Web Mapping Illustrated (Using Open Source GIS Toolkits)*. Sebastopol, Calif.: O'Reilly. Free chapter available at <http://www.oreilly.com/catalog/webmapping/chapter/ch03.pdf>

———. 2005b. Chameleon Web mapping framework. *Directions Magazine* 15 Jul 05 <http://www.directionsmag.com/article.php?article_id=914>

Monmonier, M. 1996. *How to Lie with Maps*. Chicago, Ill.: University of Chicago Press.

Natural Resources Canada. 2003. Using Audio-Tactile Maps in SVG Format. 25 Jun 03. <http://web.archive.org/web/20060711235629/http://www.tactile.nrcan.gc.ca/page.cgi?url=recherche_research/svg2_e.html>

OGC (Open Geospatial Consortium, Inc.). 2004a. OGC Web Services, Phase 2 (OWS-2). <http://www.opengeospatial.org/initiatives/?iid=7>

———. 2004b. Geo Digital Rights Management (GeoDRM) WG (GeoDRM WG). <http://web.archive.org/web/20061210184323rn_1/www.opengeospatial.org/projects/groups/geodrmwg>

Olson, J. M. and C. A. Brewer. 1997. An evaluation of color selections to accommodate map users with color-vision impairments. *Annals of the Association of American Geographers* 87:103–134.

Ordnance Survey. 2003. GIS Files: Expert GIS Concepts - Web GIS - v2.0. Mar 03. <http://www.ordnancesurvey.co.uk/oswebsite/gisfiles/section6/page5.html>

O'Reilly Radar. 2005. SmugMug Maps goes live. 18 Aug 05. <http://radar.oreilly.com/archives/2005/08/smugmug_maps_go.html>

Pandey, P. K. and R. Shukla. 2005. The big changes in internet GIS. *GIS@development* 9(10). <http://www.gisdevelopment.net/magazine/years/2005/oct/webgis_tsou44_1.htm>

Peng, Z.-R. 2004. GML, WFS, SVG and the future of internet GIS. *GIS@development* 8(7). <http://www.gisdevelopment.net/magazine/years/2004/july/38.shtml>

Peng, Z.-R. and M.-H. Tsou. 2003. *Internet GIS: Distributed Geographic Information Services for the Internet and Wireless Networks*. Hoboken, NJ: John Wiley & Sons. <http://map.sdsu.edu/gisbook/>

Reichardt, M. E. 2002. XML's Role in the Geospatial Information Revolution. <http://www.gsa.gov/gsa/cm_attachments/GSA_DOCUMENT/11-MReichardt-OGC_R2GXI1_0Z5RDZ-i34K-pR.htm>

———. 2003. Press Announcement: OGC Approves GML 3 – Feb 03. Wayland, Mass.: Open GIS Consortium, Inc. <http://web.archive.org/web/20031105010658/http://www.opengis.org/docs/2003/20030205_GML3_PR.pdf>

Schutzberg, A. 2005. KML gets two thumbs up from file format experts. *Directions Magazine* 23 Jul 05. <http://www.directionsmag.com/article.php?article_id=919>

———. 2006. Fun with GeoRSS *Directions Magazine* 09 Jun 06. <http://www.directionsmag.com/article.php?article_id=2197&trv=1>

Shi, X. 2005. WSDL Web services for GIS applications with SVG viewer. *Directions Magazine* 29 Apr 05. <http://www.directionsmag.com/article.php?article_id=849>

Shi, Y., A. Shortridge and J. Bartholic. 2002. Grid computing for real time distributed collaborative geoprocessing. In *Proceedings of ISPRS (International Society for Photogrammetry and Remote Sensing) Commission IV Symposium on Geospatial Theory, Processing and Applications,* 9–12 July, Ottawa, Canada. <http://web.archive.org/web/20050518113415/http://www.isprs.org/commission4/proceedings/pdfpapers/128.pdf>

Shuping, D. 2005. Geo-temporal visualization of RFID. *Directions Magazine/Location Intelligence Magazine* 31 Aug 05. <http://www.directionsmag.com/article.php?article_id=951>

Siekierska, E. and A. Müller. 2003. Tactile and audio-tactile maps within the Canadian 'government on-line' program. *The Cartographic Journal* 40(3):299–304. <http://dx.doi.org/10.1179/000870403225013050>

Skupin, A. 2000. From metaphor to method: Cartographic perspectives on information visualization. In *Proceedings of IEEE Symposium on Information Vizualization (InfoVis 2000)*, 9-10 October, Salt Lake City, Utah. <http://web.archive.org/web/20050425022016/http://www.geog.uno.edu/~askupin/research/infovis2000/figures/>

Spence, R. 2001. *Information Visualization*. Essex, U.K.: ACM Press.

Strømberg, P. M. 2005. Richness or reach? *Directions Magazine* 30 Mar 05. <http://www.directionsmag.com/article.php?article_id=812>

Theseira, M. 2002. Using internet GIS technology for sharing health and health related data for the West Midlands Region. *Health Place* 8(1):37–46.

Thompson, B. 2005. Net offers map help after the flood. BBC News (Technology). 2 Sep 05. <http://news.bbc.co.uk/1/hi/technology/4208070.stm>

Tiits, K. 2003. Usability of geographic information systems in Internet - A case study of journey planners. Master's thesis, Institute of Geography, Tartu University, Tartu, Estonia. <http://www.hot.ee/kyllitiits/Kylli_Tiits_MSc_thesis.pdf>

———. 2004. GIS software usability – is it more than a marketing slogan? *Directions Magazine* 10 May 04. <http://directionsmag.com/article.php?article_id=557>

United States Access Board. 2001. Web-based Intranet and Internet Information and Applications (1194.22) - Why do client-side imagemaps provide better accessibility? 21 Jun 01. <http://www.access-board.gov/sec508/guide/1194.22.htm#(f)>

US Federal Geographic Data Committee. 2002. E-Government Geospatial One-Stop (US). <http://web.archive.org/web/20050307193537/http://www.fgdc.gov/geo-one-stop/>

W3C (World Wide Web Consortium). 2005. Policies Relating to Web Accessibility. 14 Feb 05. <http://www.w3.org/WAI/Policy/>

Wreden, N. 2005. Case Studies: Wikification in Action - blog entry. 13 Jul 05. <http://fusionbrand.blogs.com/fusionbrand/2005/07/case_studies_wi.html>

Xue, Y., A. P. Cracknell and H. D. Guo. 2002. Telegeoprocessing: The integration of remote sensing, Geographic Information System (GIS), Global Positioning System (GPS) and telecommunication. *Int. J. of Remote Sensing* 23(9):1851–1893.

Yahoo News/Associated Press. 2005. South Korea discusses security concerns with U.S. over Google Earth. 31 Aug 05. <http://www.asiancanadian.net/2005/08/south-korea-discusses-security.html>

CHAPTER 50
Advancement of Web Standards and Techniques for Developing Hypermedia GIS on the Internet

Shunfu Hu

50.1 Introduction

In the past two decades, there has been rapid growth of the use of multimedia technology in computer-assisted mapping systems. In the 1980s, interactive maps and electronic atlases were characterized by an intuitive graphical user interface that allowed the user to manipulate the map features (points, lines or polygons) through a computer mouse (Peterson 1995). The link to multimedia information was achieved through superimposing "hotspots" on the cartographic features of the map or on scanned aerial photographs. Interactivity became a key feature of the interactive maps, which allowed the user to explore more detailed information in the area predefined by the map developer. Examples of early electronic atlases included the Domesday Project and Goode's *World Atlas* (Openshaw and Mounsey 1987; Rhind et al. 1988; Espenshade 1990).

The early 1990s saw the development of hypermaps. Coined by Laurini and Milleret-Raffort (1990), the term "hypermap" was described as multimedia hypertext documents with geographical access. In other words, the hypermap was an interactive, digital multimedia map that allowed users to zoom and find locations using a hyperlinked gazetteer (Cotton and Oliver 1994). The underlying principle of the hypermap was the concept of hypertext. By activating predefined hyperlinks, it is possible for the user to connect a hypertext to other non-linear text information (Nielson 1990). If the hypertext is linked to multimedia information, the term "multimedia hypertext" or "hypermedia" is used (Stefanakis et al. 2006). Therefore, hypermap is also called "cartographic application of hypertext" or "hypermedia mapping" (Cartwright 1999). The development of hypermaps was made possible with Apple's Hypercard software developed for the Macintosh computer and released in 1987 (Raveneau et al. 1991). Examples of hypermaps include the Glasgow Online digital atlas (Raper 1991) and HYPERSNIGE (Camara and Gomes 1991).

Most recently, integration of the hypermedia system, which features hypertext, hyperlinks and multimedia, and geographic information systems (GIS) has resulted in hypermedia GIS (Hu 1999, 2003, 2004a; Hu et al. 2000, 2003). In general, GIS is used to capture, retrieve, manipulate, and display geographic information tied to a common coordinate system and is featured by linking cartographic features with their alphanumeric attributes to perform spatial analysis (Burrough 1986; Star and Estes 1989; Clarke 1995). A standalone GIS software package such as Environmental Systems Research Institute (ESRI) ArcView 3.3 or ArcGIS 9.2, known as discrete GIS, is typically installed on a personal computer (PC). This traditional GIS is a closed, centralized system that incorporates a user graphical interface, programs, and data. Each system is platform dependent (e.g., Unix, Windows, or Macintosh) and application dependent (e.g., ArcView, ArcGIS, or MapInfo). Data to be used in the discrete GIS application can be stored on a computer hard drive or an external drive. Access to the data set by the GIS program is usually fast.

In recent years, GIS applications have seen advancement on the Internet. Known as distributed GIS, web-based GIS is defined as GIS applications and the related data that are distributed via the Internet. Web-based GIS applies the dynamic client/server concept in performing GIS analysis tasks through standard interfaces of the World Wide Web (WWW).

Web-based GIS applications provide GIS database query or interactive map exploration on the Internet (Peng and Nebert 1997; Abel et al. 1998; Peng 1999; Dragicevic et al. 2000; Meyer et al. 2001; Zhang and Wang 2001; Chang and Park 2004).

In this chapter, discussion is focused on two issues. The first is the advancement of web standards, including Hypertext Markup Language (HTML), Extensible Markup Language (XML), Scalable Vector Graphics (SVG), and Asynchronous JavaScript and XML (Ajax). The second is a variety of the techniques for developing hypermedia maps and hypermedia GIS applications on the Internet, including scripting in HTML, using Web development tools, scripting in SVG, using custom development software, and scripting in Virtual Reality Macro Language (VRML). Finally, a case study of the development of hypermedia GIS application for the Florida Everglades is presented.

50.2 Advancement of Web Standards

The techniques for developing hypermedia GIS applications on the Internet closely follow the advancement of the web standards on the Internet, from Hypertext Markup Language (HTML), to Extensible Markup Language (XML), to Scalable Vector Graphics (SVG), and now to Ajax. The use of a better web standard provides faster map delivery with higher map quality on the Internet. The web standards are discussed below.

50.2.1 HTML

In the early stage of web development, HTML was the popular programming language used for the Internet across a variety of platforms and browsers. HTML operates through a series of codes placed within an ASCII (or text) document. These codes are translated by a web browser into specific labels of formats to be displayed on the user's computer screen, and on which the user can take action by clicking hotspots or hypertext to explore more information on the Internet.

50.2.2 XML

XML is a recent web standard developed by the World Wide Web Consortium (W3C). XML transports structured text between client and server. Like HTML, XML is an easy language to understand and use. It is platform independent and supports internationalization and localization. While it looks similar in coding style, XML differs from HTML in that programmers define the rules. Unlike HTML, which has a fixed element set and defines how to display content in browsers, XML has a custom element set that defines the content to be displayed. That is why the end-user sees different flavors of XML. Within GIS, Geography Markup Language (GML) and ArcXML are the two most predominant variations. GML is an XML-based encoding standard for geographic information developed by the OpenGIS Consortium (OGC). The objective is to allow Internet browsers the ability to view web-based mapping without additional components or viewers.

50.2.3 SVG

SVG, the latest web standard vector graphics developed by the W3C, is becoming a popular choice for rendering maps on the Internet (Harwell 2004). SVG, using XML encoding, is a text-based graphics language that describes images with vector shapes, text, and embedded raster graphics.

SVG can be used in conjunction with OGC standards, specifically OGC's Web Feature Service (WFS), a standard vector mapping web service, and GML, a standard XML encoding for communicating the resulting maps across the web. Because both GML and SVG use XML encoding, it is very straightforward to convert between the two using an XML Style

Language Transformation (XSLT). SVG files provide high-quality graphics on the web, in print, and on handheld devices. SVG is designed to work effectively across platforms, output resolutions, color spaces, and a range of available bandwidths.

The SVG Viewer is required to view web content that contains SVG files. The Viewer integrates with the web browser as a plug-in. The Viewer allows the user to interact with SVG graphics. An example of the SVG Viewer includes the Adobe SVG Viewer 3.03 (Adobe Systems Incorporated 2008).

50.2.4 Ajax

Ajax stands for Asynchronous JavaScript and XML. Ajax is a group of interrelated web development techniques used for creating interactive web applications or rich Internet applications. With Ajax, web applications can retrieve data from the server asynchronously in the background without interfering with the display and behavior of the existing page (Moore 2008). Examples of such techniques used in Ajax include: 1) standards-based presentation using Extensible HTML (or XHTML) and Cascading Style Sheets (CSS); 2) dynamic display and interaction using the Document Object Model; 3) data interchange and manipulation using XML and XSLT; 4) asynchronous data retrieval using XMLHttpRequest; and 5) use of JavaScript (Garrett 2005). The benefit of using Ajax for web mapping applications is its fast response and enhanced interaction.

50.3 Techniques for Developing Hypermedia Maps and Hypermedia GIS Applications on the Internet

50.3.1 Scripting in HTML

The original HTML allowed only text, while later pictures, graphics and various types of lists and link types were added. More recently, elements such as fill-in forms, clickable (or interactive) maps, sound, animation, and video are possible and may be linked via HTML. The documents one sees on a WWW browser are usually HTML documents. What makes an HTML file worthwhile is the browser's interpretation of its formatting codes, so that a link appears as a highlighted item, a list appears with associated bullets or numbers, and a graphic (usually in raster format) appears as the picture it represents.

Most cartographic animations that are available through the Internet were in the "animated GIF" format, an extension to the popular picture file format, or one of the two major movie formats—MPEG or QuickTime. Embedding a QuickTime movie into a hypertext document can be accomplished through the <EMBED> tag. The syntax looks similar to <IMG> in a standard HTML. They both have an SRC, WIDTH, and HEIGHT parameter. These parameters are required and tell the Internet browser which media to display, and specify its width and height.

The most common examples of animated maps on the Internet are those of weather patterns that depict the movement of clouds as seen on television weather forecasts. The movement of cloud patterns associated with hurricanes is especially suited for viewing as an animation (Peterson 2001; Peterson 2003).

50.3.2 Using Web Development Tools

With Macromedia Director developed by the Adobe Corporation (www.adobe.com), one can develop interactive maps, or map animation for use on the Internet. For interactive maps, hotspots can be superimposed to provide links to access to multimedia information. The final product can be saved in Shockwave movie format and delivered on the Internet. Internet users must have a Shockwave player (a freeware from Adobe) as a plug-in for the web browser to play the Shockwave movie. Other web development tools include Adobe's Flash and Portable Document Format (PDF). Flash is capable of displaying animated maps in vector format. Flash format is designed for presenting vector-based interactive and animated graphics with sound for the web. With Adobe's Indesign, one can place a map animation or virtual reality (VR) within a PDF file, which is a unique type of cross platform file format.

50.3.3 Scripting in SVG

SVG supports scripting, making it ideal for interactive, data-driven, and personalized graphics. Event handlers such as "onmouseover" and "onclick" can be assigned to any SVG graphical object, making the format suitable for interactive applications (Harwell 2004). Neumann (2005) used SVG scripting to develop an interactive choropleth map for the demonstration of social patterns such as population older than 60 years of age, and population with secondary education in Vienna, Austria.

50.3.4 Scripting in VRML

Virtual Reality Modeling Language (VRML) has been used recently to develop three-dimensional (3D) visualization of cartographic features on the Internet (Ottoson 2003; Hu 2004b). Hu (2004b) demonstrated an approach to incorporating ArcView GIS and VR techniques for the visualization of 3D landscapes of a watershed on the Internet. Both 3D and VR visualization of a spatial environment provide the user interaction with the environment and immersion into the environment so the user can have better understanding about the spatial environment under investigation. Furthermore, VRML browsers and plug-ins are usually inexpensive and widely available; the 3D virtual worlds that GIS developers make from existing geographic data can be accessible to a wide audience. Geographical VR products can be created in commercial GIS software packages such as ArcGIS 3D Analyst and ArcScene from ESRI.

50.3.5 Scripting in Ajax

Ajax scripting makes web-based applications more responsive and interactive. Therefore, Ajax has been quickly adopted by the GIS community for developing GIS applications on the Internet. Sayar et al. (2006) discuss integrating AJAX models into the browser-based GIS Visualization Web Services systems.

50.4 Web-based Hypermedia GIS Application: A Case Study

The development of a web-based hypermedia GIS is demonstrated here using a portion of vegetation database derived from an Everglades research project (Welch et al. 1999; Hu 1999). The focus area corresponds to the US Geographical Survey (USGS) Long Pine Key and Pa-Hay-Okee Lookout Tower 1:24,000-scale topographic quadrangles in Everglades National Park. The data sets include a GIS database and a multimedia database. The GIS database contains digital maps of vegetation patterns in ArcView shape file format, and is tied to the Universal Transverse Mercator (UTM) coordinate system and North American Datum of 1983 (NAD83). The multimedia database contains descriptive text, ground photographs, digital video clips, and audio segments highlighting the characteristics of Everglades vegetation plant communities, individual species, and invasive exotics, as well as plant-animal interactions, hurricane damage, and post-fire vegetation succession provided by Seavey Field Guides, Inc. (www.seaveyfieldguides.com/index.htm).

The web-based hypermedia GIS was based upon interactions between three components: 1) a web-based GIS application developed using ESRI ArcIMS software to manipulate spatial data sets in ArcView shape files of vegetation patterns; 2) a standard web homepage created using a hypertext markup language (HTML) and designed to manipulate multimedia information such as hypertext, hyperlinks, ground photographs, digital video, and sound; and 3) a mechanism linking the web-based GIS application and web-based interactive multimedia application. Microsoft Internet Information Server (MIIS) was employed as the web server and ESRI ArcIMS as the map server.

Figure 50-1 is an illustration of the web-based hypermedia GIS application for visualizing the Florida Everglades vegetation database. GIS functions such as Zoom, Pan, and Identify were developed to manipulate the vector GIS database. For instance, the user can select the "Zoom in" option and zoom into an area on the digital map, then select the "Identify" option, click on any polygon, and the alphabetic letters representing the dominant vegetation category in that polygon will be displayed on the screen (Figure 50-2). Further, the user can select the "Hyperlink" option, click on one polygon, and be directed to the web-based interactive multimedia site containing a ground panoramic view of that plant community and a descriptive text, including hypertext and hyperlinks, about the plant community (Figure 50-3). Video clips can be either linked directly to a map feature or to a hypertext.

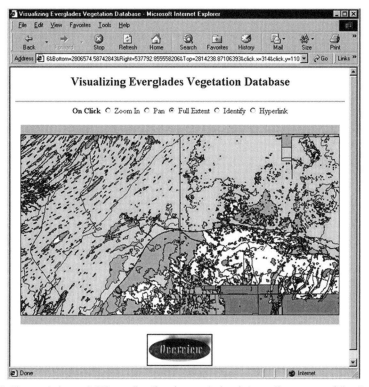

Figure 50-1 The web-based GIS application for exploring interactive maps of the Everglades vegetation database: Pa-Hay-Okee Lookout Tower (left) and Long Pine Key (right) quadrangles in Everglades National Park.

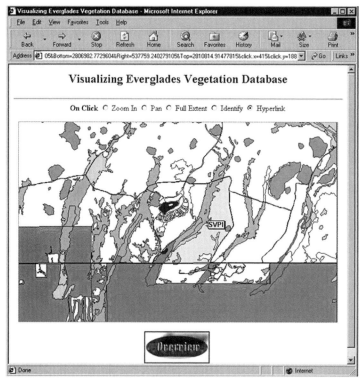

Figure 50-2 The web-based GIS functions allow the user to zoom in and pan the digital maps, and to retrieve the dominant vegetation category for a selected polygon from the GIS database.

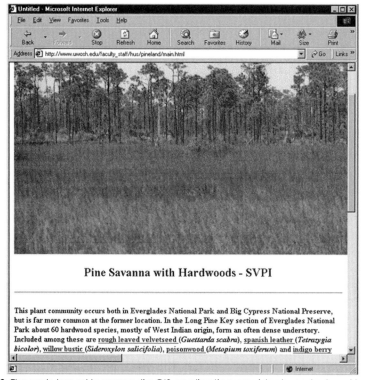

Figure 50-3 The web-based hypermedia GIS application provides hypertext and hyperlinks to explore multimedia information about the Everglades natural environment.

50.5 Summary and Discussion

The development of web-based GIS applications has experienced a great deal of technological advancement in recent years due to several factors. The first was the improvement of the web standards on the Internet, from HTML, to XML, SVG, and Ajax. Being the latest web standard, Ajax, with its ability to display richer maps faster and more interactively, is becoming a popular format used to distribute maps by web-based GIS applications, and to facilitate mapping interoperability.

The case study of web-based hypermedia GIS application that links web-based GIS applications and interactive multimedia provides the Internet user with user-friendly tools to interactively explore both digital maps and associated multimedia information through standard web browsers. The web-based hypermedia GIS approach has the potential to generate many interests in the GIS community. First, web-based GIS applications have been limited in the past in dealing with only cartographic features (i.e., points, lines, and polygons), and alphanumeric attribute data. With the integration of multimedia information in GIS, the web-based hypermedia GIS is able to handle geographic information in any format (e.g., spatial data in image/vector format, attribute data in alphanumeric format, and multimedia data in the form of text, graphics, photographs, and digital video). It provides the essential concepts and techniques for many new GIS applications such as visualization and spatial decision support systems.

Second, from the user's perspective, the integrated multimedia-GIS approach provides a multi-sensory learning environment. This chapter has demonstrated the feasibility of integrating multimedia and GIS technologies on the Internet to popularize geographic information by delivering it to more people and in more understandable ways. For instance, agencies and organizations with the task of developing GIS databases, especially applications in natural resource management, can employ the web-based multimedia GIS technique to disseminate the various types of collected information through the Internet to end users (i.e, resource managers, policy maker, researchers, planners, and the general public) with different levels of computer skills, thus enabling truly interactive collaboration among all interested parties towards the common goal of resource preservation.

In all, with the advancement of web standards such as Ajax and the availability of performance-improving map servers, the Internet user is able to explore a high-quality digital map interactively on the Internet as well as multimedia information associated with the features on the map.

References

Abel, D., T. Kerry, R. Ackland and S. Hungerford. 1998. An exploration of GIS architectures for internet environments. *Computers, Environment and Urban Systems* 22: 1.

Adobe Systems Incorporated. 2008. Scalable Vector Graphics, Adobe SVG Viewer. <http://www.adobe.com/svg/viewer/install/main.html> Accessed 15 October 2008.

Burrough, P. A. 1986. Principles of Geographical Information Systems for Land Resources Assessment. New York: Oxford Science Publications.

Camara, A. and A. L. Gomes. 1991. HYPERSNIGE: A navigation system for geographic information. Pages 175–179 in *Proceedings of EGIS'91,* Second European Conference on GIS, Brussels.

Cartwright, W. 1999. Development of multimedia. Pages 11–30 in *Multimedia Cartography.* Edited by W. Cartwright, M. P. Peterson and G. Gartner. New York: Springer.

Chang, Y. and H. Park. 2004. Development of a web-based geographic information system for the management of borehole and geological data. *Computers and Geosciences* 30: 8.

Clarke, K. C. 1995. *Analytical and Computer Cartography.* 2nd ed. Englewood Cliffs, N.J.: Prentice Hall.

Cotton, B. and R. Oliver. 1994. *The Cyberspace Lexicon – An Illustrated Dictionary of Terms from Multimedia to Virtual Reality.* London: Phaidon Press Ltd.

Dragicevic, S., S. Balram and J. Lewis. 2000. The role of web GIS tools in the environmental modeling and decision-making process. In *4th International Conference on Integrating GIS and Environmental Modeling (GIS/EM4): Problems, Prospects and Research Needs.* Banff, Alberta, Canada, 2–8 September 2000.

Espenshade, E.B.J. 1990. *Goode's World Atlas.* 19th ed. Chicago: Rand McNally.

Garrett, J. J. 2005. Ajax: a new approach to web applications. <http://www.adaptivepath.com/ideas/essays/archives/000385.php> Accessed on 21 October 2008.

Harwell, R. 2004. Web mapping with SVG. <http://www.directionsmag.com/article.php?article_id=693&trv=1> Accessed on 15 October 2008.

Hu, S. 1999. An integrated multimedia approach to the utilization of an Everglades vegetation database. *Photogrammetric Engineering and Remote Sensing* 65:2.

———. 2003. Multimedia GIS: Analysis and visualization of spatio-temporal and multimedia geographic information. *Geographic Information Sciences* 9 (2):90–96.

———. 2004a. Design issues associated with discrete and distributed hypermedia GIS. *GIScience & Remote Sensing* 41 (4):330–342.

———. 2004b. Use of GIS, remote sensing and virtual reality in flooding hazardous modeling, assessment and visualization. *Papers of Applied Geography Conference* 27:203–211.

Hu, S., A. O. Gabriel and C. Lancaster. 2000. An integrated multimedia approach for wetland management and planning of Terrell's Island, Winnebago Pool Lakes, Wisconsin. *The Wisconsin Geographer* 15–16:34–44.

Hu, S., A. O. Gabriel and L. R. Bodensteiner. 2003. Inventory and characteristics of wetland habitat on the Winnebago Upper Pool Lakes, Wisconsin, USA: An integrated multimedia-GIS approach. *Wetland* 23 (1):82–94.

Laurini, R. and R. Milleret-Raffort. 1990. Principles of geomatic hypermaps. Pages 642–655 in *Proceedings of the 4th International Symposium on Spatial Data Handling,* Vol. 2, Zurich, Switzerland.

Meyer, J., R. Sugumaran, J. Davis and C. Fulcher. 2001. Development of a web-based watershed level environmental sensitivity screen tool for local planning using multi-criteria evaluation. In *Proceedings of ASPRS Annual Conference,* 23–27 April 2001, St. Louis, Missouri.

Moore, J. 2008. What is Ajax? <http://www.riaspot.com/articles/entry/What-is-Ajax-> Accessed on 15 October 2008.

Nielson, J. 1990. *Hypertext and Hypermedia.* Boston: Academic Press Professional.

Neumann, A. 2005. Choroplethe map with interchangeable statistic variables. <http://www.carto.net/papers/svg/samples/wien.shtml> Accessed on 15 October 2008.

Openshaw, S. and H. Mounsey. 1987. Geographic information systems and the BBC's Domesday interactive videodisk. *International Journal of Geographical Information Systems* 1:2.

Ottoson, P. 2003. Three-dimensional visualization on the Internet. Pages 247–270 in *Maps and the Internet.* Edited by M. P. Peterson. Amsterdam: Elsevier.

Peng, Z. and D. Nebert. 1997. An internet-based GIS data access system. *Journal of the Urban and Regional Information Systems Association* 9:1.

Peng, Z. R. 1999. An assessment framework for the development of internet GIS. *Environment and Planning B: Planning and Design* 26:1.

Peterson, M. P. 1995. *Interactive and Animated Cartography.* Upper Saddle River, N.J.: Prentice Hall.

———. 2001. The development of map distribution through the World Wide Web. In *Proceedings of the 19th International Cartographic Conference.* Beijing, China.

————. 2003. *Maps on the Internet.* Amsterdam: Elsevier.

Raper, J. 1991. Spatial data exploration using hypertext techniques. Pages 920–928 in *Proceedings of EGIS'91, Second European Conference on GIS,* Brussels, Belgium.

Raveneau, J. L., M. Miller, Y. Brousseau and C. Dufour. 1991. Micro-atlases and the diffusion of geographic information: An experiment with Hypercard. Pages 263–268 in *GIS: The Microcomputer and Modern Cartography.* Edited by D. R. Fraser Taylor. Oxford: Pergamon Press.

Rhind, D. P., P. Armstrong and S. Openshaw. 1988. The Domesday machine: A nationwide geographical information system. *Geographical Journal* 154 (1):56–58.

Sayar, A., M. Pierce and G. Fox. 2006. Integrating AJAX approach into GIS visualization web services. Page 169 in *Proceedings of International Conference on Internet and Web Applications and Services/Advanced International Conference on Telecommunications.* IEEE Computer Society, Washington, D.C.

Star, J. and J. Estes. 1989. *Geographic Information Systems: An Introduction.* Englewood Cliffs, N.J.: Prentice Hall.

Stefanakis, E., M. P. Peterson, C. Armenakis and V. Delis, eds. 2006. *Geographic Hypermedia: Concepts and Systems.* New York: Springer-Verlag Berlin Heidelberg.

Welch, R., M. Madden and R. F. Doren. 1999. Mapping the Everglades. *Photogrammetric Engineering and Remote Sensing* 65: 2.

Zhang, X. and Y. Q. Wang. 2001. Web based spatial decision support for ecosystem management. In *Proceedings of ASPRS Annual Conference,* 23–27 April 2001, St. Louis, Missouri.

CHAPTER 51

Introduction to the Spatial Sensor Web

Vincent Tao and *Steve H.L. Liang*

"The most profound technologies are those that disappear. They weave themselves into the fabric of everyday life until they are indistinguishable from it." (Weiser 1991) The Web is an excellent example of such a technology – it is no longer exciting because it has become part of our life. However, Web-enabled technologies are continuously advancing, challenging our vision and even our dreams. The Sensor Web is one of these emerging web-enabled technologies.

51.1 Introduction

Neil Gross' (1999) article, "The Earth Will Don an Electronic Skin," provides a compelling explanation of the Sensor Web concept: *"In the next century, planet Earth will don an electronic skin. It will use the Internet as a scaffold to support and transmit sensations. This skin is already being stitched together. It consists of millions of embedded electronic measuring devices: thermostats, pressure gauges, pollution detectors, cameras, microphones, glucose sensors, EKGs, electroencephalographs. These will probe and monitor cities; endangered species; the atmosphere; our ships, highway traffic; fleets of trucks; our conversations; our bodies—even our dreams."* Figure 51-1 illustrates the concept of the Sensor Web as an electronic skin of planet Earth.

Figure 51-1 The concept of the Sensor Web as an electronic skin of planet Earth. See included DVD for color version.

With the ongoing development of cheaper miniature and smart sensors; abundant fast and ubiquitous computing devices; wireless and mobile communication networks; and autonomous and intelligent software agents, the Sensor Web is rapidly emerging as an extremely powerful technological framework for geospatial data collection, fusion and distribution. The Sensor Web is a Web-centric, open, interconnected, intelligent and dynamic network of sensors; it presents a new vision for how we deploy sensors, collect data, and fuse and distribute information. In short, the Sensor Web is a revolutionary concept toward achieving collaborative, coherent, consistent and consolidated sensor data collection, fusion and distribution.

The Sensor Web has attracted enormous attention from both the research and business communities. In a 2006 article in the journal *Nature* Butler (2006) describes the Sensor Web as a computing platform that will be fully established by the year 2020, predicting that it will include *"tiny computers that constantly monitor ecosystems, buildings and even human bodies."* A *BusinessWeek* article entitled "Sensor Revolution" (Green 2003) lists the Sensor Web as one of the four technological waves that will change the way we live in the next decade. In a special report from the National Science Foundation (NSF 2005) the Sensor Web is described as *"the world's first electronic nervous system."* The report predicts that the Sensor Web will be the next biggest computing revolution since the PC (Personal Computer) revolution of the '80s and the Internet revolution of the '90s.

We have seen the technological revolution of the world-wide web (WWW) into what is now considered to be a "global computer," connecting enormous computing resources around the world. We now anticipate the 'next wave' in sensing, where the Sensor Web represents a "global sensor" that links to massive numbers of distributed sensors and sensor databases. The Sensor Web is an information infrastructure, a backbone to support sensor information publishing, distribution and discovery.

In the present chapter, first the concept of the Spatial Sensor Web (SSW) is introduced, wherein we describe an architectural framework that serves as a basic reference for the SSW system implementation. We then describe a real-world SSW implementation as a use case, in order to illustrate SSW architecture and applications. Finally, future SSW research trajectories and challenges are outlined and discussed.

51.2 What Is the Sensor Web?

51.2.1 Sensor

A sensor is a device capable of detecting and responding to physical stimuli such as movement, light, heat, etc. Today, over a hundred physical, chemical and biological properties can be measured and monitored by sensors. Within the context of geospatial data collection, given differences in sensing range, sensors are typically classified as either *in situ* sensors (i.e., ground-based) or remote sensors (i.e., typically carried by airplane or satellite). Sensors are everywhere, ranging from commodity sensors designed for everyday use, such as webcams, thermometers, microphones, smoke detectors, cell phones, etc., to specialized sensors for science and engineering applications, such as meteorological stations, seismic monitors, and radar detectors. Technical advances are allowing sensors to be smaller, lighter, and more energy-efficient. New technologies, such as Microelectromechanical Systems (MEMS), allow sensors to be constructed at the micro- or even nano-scale (Warneke et al. 2001), resulting not only in great portability but also allowing exciting new and powerful applications.

51.2.2 Sensor Network

A sensor network (SN) is a computerized network consisting of spatially distributed sensors with the purpose of cooperatively monitoring physical and environmental conditions, such as temperature, humidity, precipitation, wind, streamwater levels, or pollutants (Akyildiz et al. 2002a). In addition to being comprised of one or more sensors, each node in a sensor network is typically equipped with a communication device which incorporates a variety of means of transmitting data, from wires to cellular phones and to microwave radios. Currently sensors are often connected via wireless networks, allowing for much greater flexibility than is otherwise possible (Hill et al. 2000). Although the development of sensor networks was originally motivated by military applications such as battlefield surveillance, sensor networks are now used in many civilian applications, including home automation, health monitoring, weather forecasting, habitat monitoring, flood alerting, navigation, etc.

51.2.3 Sensor Web

Although sensor networks have, to date, been deployed for a variety of applications, the communication links among these have typically been lacking. Consequently there is a strong and immediate need to connect these heterogeneous sensor networks and sensors. For instance, coastal zone emergency management draws on data that has been obtained from a variety of sources and collected through different sensor networks, such as the seismic monitoring network, weather network, traffic network, flood monitoring stations, etc. This virtual sensor system consists of interconnected sensor networks or individual networked sensors.

Within this context, the Sensor Web (SW) is defined as a world-wide information infrastructure of sensor networks that are typically connected via the WWW. In other words, the internet protocol and web services are used for communication among sensor networks and sensors. This allows heterogeneous sensing resources, such as sensor data and sensor processing, to be connected and accessible from anywhere at any time. Some references refer to this system as the World Wide Sensor Web (WWSW), because connection among these sensor networks is typically achieved via the World Wide Web. It is worth noting that the concept of the Sensor Web is evolving, as evidenced by the wide range of Sensor Web definitions (Delin and Jackson 2001; Teillet et al. 2002; Estrin et al. 2003; Gibbons et al. 2003; Tao et al. 2004; Liang et al. 2005). Some references do not differentiate between the sensor network and the Sensor Web.

By combining heterogeneous sensor networks or individual networked sensors, the Sensor Web can provide an enormous amount of timely, comprehensive, continuous and multiresolution data and information for a variety of applications, including precision agriculture, environmental monitoring, habitat monitoring, transportation, homeland security, defense, and even planetary exploration.

Given that the network protocols, information models and data formats in use by various sensors and sensor networks are often proprietary in nature, establishing a single unified virtual sensor system is very challenging. This stems primarily from the fact that intelligent sensor planning, control, dispatch, and alerting within the sensor web requires a high level of interoperability. It is interesting that the current status of Sensor Web research and development is similar to that of the WWW about a decade ago, where there were many *ad hoc* computing networks that were not connected together. The emergence of the WWW has brought with it great interoperability, allowing for the vision of a single unified virtual global computing system to become a reality.

51.2.4 Spatial Sensor Web

Sensing is essentially a spatial sampling process whereby each sensor observation can generally be associated with location information. Without information on their spatial context, sensed results are much less meaningful and useful. Therefore, the term "spatial is special" is particularly relevant to Sensor Web development. The following example effectively illustrates the importance of the spatial characteristics of the Sensor Web:

> *Mike is driving from suburban to downtown Toronto for a meeting. Before his arrival he uses his PDA phone to send out a query to a parking/routing sensor web agent. Mike specifies the destination of his meeting in the query. His PDA phone's built-in GPS also provides Mike's current location in the query message. Through the sensor web service a network of parking sensors in the specified location is connected in order to find the availability of the parking space. The traffic monitoring network is also connected to provide real-time traffic information. The agent constantly sends Mike a dynamic navigation map with driving directions, while constantly taking into consideration real-time traffic conditions and the closest location of the parking space.*

51.3 Architecture of the Spatial Sensor Web (SSW)

This section of the paper describes an architectural framework for constructing an SSW. It also introduces a new way of thinking about architectural issues related to the SSW (i.e., both in terms of SSW applications and constraints).

The SSW architecture is comprised of three major components: (1) Service component, (2) Communication component, and (3) Information models component.

51.3.1 First Component: Spatial Sensor Web Services

The SSW is a global virtual sensing system comprised of SSW services. The SSW services are sensor-focused spatial web services. These are network-enabled entities that provide the sensor networks' resources through the exchange of messages via the WWW.

The web service approach represents the convergence of Web protocols and the service-oriented model (W3C 2004; Huhns and Singh 2005). The use of the Web protocols minimizes the amount of infrastructure that needs to be deployed, thereby helping to achieve pervasive deployment, and assuring Internet-wide interoperability at the level of message exchange. The adoption of the service-oriented model, which commonly treats all sensor network data and processing as services, encapsulates diverse sensor network implementations behind common interfaces and enables consistent resource access and interchange across heterogeneous sensor networks.

In terms of the roles of the services in the SSW architecture, the SSW services can be grouped into four categories: (1) sensor services, (2) processing services, (3) storage services, and (4) discovery services.

51.3.1.1 Sensor Services

Sensor services are the virtual representation of the sensor networks. The purpose of the sensor services is to collect scientific data. Within the context of the Spatial Sensor Web, an SSW sensor service offers sensor network resources according to customized spatial-temporal requests.

At the implementation level, an SSW sensor service is in fact a web-based computing system that is comprised of a system of sensor hardware/software, sensor gateways and a sensor web server. The right side of Figure 51-2 shows a general tiered architecture of an SSW sensor service. The computing power and storage capabilities increase from the bottom tier to the top tier. The lowest level consists of the sensors that take physical energy of the environment as inputs and output estimated values of the observed phenomenon. The sensors transmit their collected data through a local sensor network (LSN) to at least one sensor gateway and this gateway then transmits the data collected from the local sensors via a local transmit network (LTN) to the sensor web server. The sensor web server then offers the data through one or more standardized SSW service interfaces. Users or other SSW services can access the data by invoking the standardized SSW service interfaces regardless of the underlying implementation details of the sensor networks.

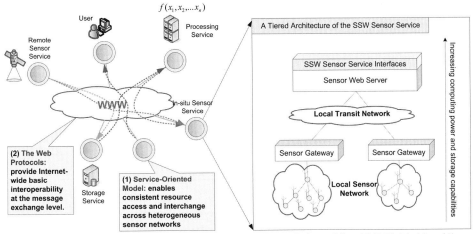

Figure 51-2 Left side of the figure shows a service-oriented view of the SSW. Right side of the figure shows a tiered architecture of the SSW sensor service using wireless sensor network as an example. See included DVD for color version.

Some of the more advanced sensors may be equipped with computing power and tasking capabilities. These sensors support programming languages and can be reprogrammed on the fly. The advances of MEMS and semiconductor technologies enable sensors to be integrated with control and signal processing electronics in a single, compact package (Hill et al. 2000). While the traditional sensor outputs only raw data, this type of "smart" sensor uses the embedded computing power to process the collected data and output a higher level of information derived from the raw data. Furthermore, some of the more advanced sensors may dynamically change their behaviors according to the environment, user's commands, or information from other SSW services (Delin and Jackson 2001).

In terms of the degree of connectivity and the sensors' embedded processing power, the SSW services can be further classified into three different groups: (1) data collection sensor services, (2) collaborative sensor services, and (3) intelligent sensor services.

51.3.1.1.1 Data Collection Sensor Services

Data collection sensor services are simple data collectors. These services take inputs only from the physical environment, and output the estimated value of the observed phenomenon. These nodes don't take inputs from other services. The parameters and methodologies of the data collection sensor services cannot be changed and modified according to user commands or other SSW services. Therefore, there is only one method of connection between the data collection services and other SSW services. A sensor service using factory preprogrammed data loggers is an example of a data collection sensor service.

51.3.1.1.2 Collaborative Sensor Services

Collaborative sensor services collect and react. They can accept inputs not only from the physical environment but also from other SSW services via the WWW. They react to different input values and then change their parameters accordingly (e.g., increasing sampling rate, changing field-of-view) and methodologies (e.g., accept new processing algorithms). In the collaborative sensor service architecture, the three tiers communicate and coordinate with one another in order to collaborate. Each tier needs to have bi-directional communication and computing capabilities.

We use the following scenario to illustrate the concept of the collaborative sensor service. A water-level sensor service connects to a precipitation sensor service. In the normal sensing mode, the water-level sensor performs hourly measurements, saves the data locally, and transmits aggregated data back to the sensor server daily in order to save battery power.

However, if, during the past 24 hour period, more than 10 cm of rain has fallen in the neighboring region, the water-level sensor service wakes up the sensors, increases the sampling rate, and transmits the collected data back to the sensor server in real-time.

51.3.1.1.3 Intelligent Sensor Service

Intelligent sensor services collect, react, and perhaps most importantly, make decisions (e.g., event prediction). These intelligent sensor services take inputs from the environment and other services, change their sensors' behaviors according to the inputs, and use computing models to assimilate and fuse the inputs, and then output not only processed data but also a higher level of knowledge (e.g., event prediction, object tracking).

51.3.1.2 Processing Services

Processing services are science data transformers or processors. A processing service provides functions to assimilate and fuse sensor data with auxiliary data (e.g., GIS data sets) in a manner determined by user-specified parameters. A processing service takes inputs from other services (e.g., real-time observations from sensor services or historical data from storage services), based on the underlying methodologies and parameters (e.g., weather forecast models), and then outputs higher level information (e.g., weather forecast). A processing service is a re-useable computing component, and is not associated with specific sensor service instances or specific applications.

51.3.1.3 Storage Services

Storage services are data archives for the SSW. A storage service takes two types of inputs: (1) the data from other sensor services or processing services, and (2) the spatial-temporal query requests from other services. The core of an SSW's storage service is a spatial-temporal database with interfaces linking to the SSW. A storage node outputs historical data according to spatial-temporal queries from other services.

51.3.1.4 Discovery Services

Discovery services allow users and other services to classify, register, describe, search, maintain, and access information about the SSW services. The discovery service is to the SSW as the search engine is to the WWW. Discovery services store information on sensor types, observation types, online sensor instances, SSW service types, and online SSW service instances.

51.3.1.5 Standard Service Framework

A standardized service framework is critical in order to virtualize the heterogeneous sensor networks with services and allow the sensor networks to access and interchange each other's resources. Such a standardized service framework consists of two elements: (1) a set of standard interface definitions, and (2) standard semantics for service interactions. Currently several research communities are focusing on defining such an interoperable service framework for the Sensor Web; examples of this include OpenDAP (Cornillon et al. 2003) and Open Geospatial Consortium (OGC) Sensor Web Services (Botts et al. 2006).

51.3.2 Second Component: Communication

The SSW's communication component facilitates the exchange of information throughout the SSW. Ultimately, all data need to be propagated to the WWW and this is why some references refer to the Sensor Web (SW) as the World-Wide Sensor Web (WWSW).

However, unlike normal computer networks, the design of the SSW's communication component has some unique considerations, which stem from the fact that sensor systems are systems that are deployed in the field (Akyildiz et al. 2002b; Polastre 2003). Different field environments impose different challenges for sensors. Key challenges include power and communication constraints resulting from remote locations, extreme weather conditions, etc. Therefore, the implementations of the communication component will vary considerably depending upon the environment and the unique functional and performance requirements of the SSW (e.g., Quantity of Service and Quality of Service). The underlying implementation of the SSW's communication must take into account media (e.g., wireless or cable), topologies (e.g., fully connected, mesh, star, right, tree or bus), network protocols (e.g., TCP/IP), and routing (e.g., fixed or adaptive routing), etc.

There are three tiers of networks through which a sensor in the field can propagate its data to the Web: (1) local sensor network, (2) local transit network and (3) wide area network. Figure 51-3 shows the three tiers of the SSW's communication components.

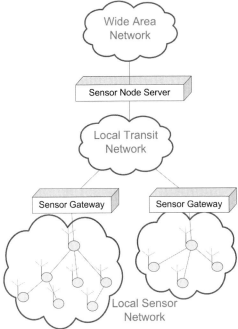

Figure 51-3 Three tiers of SSW's communication component: (from bottom to top) (1) Local Sensor Network, (2) Local Transit Network, and (3) Wide Area Network. See included DVD for color version.

51.3.2.1 Local Sensor Network (LSN)

The lowest level consists of the local sensor network (LSN) that links sensors within a confined geographic area. The LSN permits the sensors of the SSW to transmit and receive information between one another. In most cases, the LSN for *in situ* sensors uses a wireless network because it is inexpensive and easy to deploy in comparison to a wired network (Hill et al. 2000). The sensors transmit their data through the LSN to at least one sensor gateway. The sensor gateway is responsible for transmitting sensor data from the sensors to a local transit network (LTN).

The LSN is a network deployed in the field. Therefore, the settings and configurations of this network are often dependent on the constraints imposed by the environment. For example, low power and low data rate communication protocols, such as ZigBee (IEEE 2006), are often used for geographically dense deployment of tiny and power-sensitive sensors (e.g., deploy thousands of nodes in a plant for industrial control). For sparse deployment in a rural area it is possible to link a series of sensors through *ad hoc* long-range radio networks where inter-node hopping extends the range.

In many cases, power consumption is the dominant design constraint when deploying the LSN in the field. When designing and deploying an LSN, in-network aggregation is an important technique used to overcome these energy constraints. The in-network aggregation technique aggregates data drawn from different sources, thereby eliminating redundancy, minimizing the number of transmissions and thus saving energy (Krishnamachari et al. 2002).

51.3.2.2 Local Transit Network (LTN)

The LTN connects the sensor gateway to the sensor web server of the SSW. Similar to the LSN, each LTN's design has different characteristics with respect to expected robustness, bandwidth, energy efficiency, cost, and manageability (Polastre 2003). The LTN may also use the above-mentioned low-power and low-data-rate communication protocols, such as ZigBee. However, the LTN often requires a larger bandwidth and as a result, needs to use a higher data rate and power-intensive communication protocols. Common communication protocols for the LTN are satellite phone and mobile phone network for rural area deployment, and the 802.11 family of protocols for urban-area deployment. For rural deployments, solar panels and rechargeable batteries are often required in order to supply the power needed for the LTN.

51.3.2.3 Wide Area Network (WAN)

The WAN (e.g., the Internet) is a global communication infrastructure for data transmission and exchange. The WAN permits the SSW's sensor services, storage services and computing services to transmit and receive information between one another at a global scale. The Internet and Web protocols will be used for most SSW communications between the SSW services. Other protocols, for example the GRID (Foster et al. 2002), will be used for some special applications with particular requirements for Quality of Service (QoS).

51.3.3 Third Component: Information Models

The purpose of the SSW's information models is to allow the SSW's services to **understand** and **use** the exchanged data from other services. We call this capacity for being able to understand and use data from, and exchange data with, other SSW services 'information interoperability.'

The unsuccessful NASA mission to Mars with Mars Climate Orbiter in 1999 is a real-world example that proved the importance of SSW information interoperability (CNN.com 1999). According to NASA's official explanation of the crash: *"The 'root cause' of the loss of the spacecraft was the failed translation of English units into metric units in a segment of ground-based, navigation-related mission software."* In the above example, the sensors' collected data values were misunderstood and misused, due to the lack of proper information models that describe the data and relationships among different data types.

51.3.3.1 SSW Information Models

Widely accepted and used information model standards have proven successful for achieving information interoperability (Libicki 1995). A standard information model framework for the SSW shall define at least three basic types of information: (1) it shall identify what the basic entities involved in the SSW are; 2) it shall define the conceptual information models for the entities including their properties, their relationships, and the operations that can be performed on them; 3) it shall specify the physical data encodings (e.g., binary format specifications or XML schemas) according to the previously developed information models.

Several SSW related standard information frameworks have been developed and used by different communities; examples include OGC Sensor Web information models and encodings (Cox 2005), Network Common Data Form (NetCDF) (Unidata Undated), Earth

Science Markup Language (ESML) (Ramachandran et al. 2004), Semantic Web for Earth and Environmental Terminology (SWEET) (Robert and Michael 2005), etc.

51.4 Use Case: Intelligent Sensor Web for Integrated Earth Sensing (ISIES)

51.4.1 Introduction

Here we use the Intelligent Sensor Web for Integrated Earth Sensing System (ISIES) as a real-world example to illustrate the SSW architecture we introduced above.

ISIES system is an integrated Earth sensing sensor web system for improved crop and rangeland yield predictions. ISIES was deployed at two test sites (i.e., annual cropping and rangeland) in southern Alberta. The ISIES system was developed to integrate data from heterogeneous *in situ* and remote sensors automatically to provide maps of leaf area index, soil moisture and biomass, as well as improved predictions of crop and rangeland yield. Figure 51-4 shows the ISIES sensors deployed at the rangeland test site.

Figure 51-4 Two views of the rangeland test site of the ISIES (Smith et al. 2005). See included DVD for color version.

The ISIES project is funded in part by Precarn Incorporated and by its participants: MDA, Natural Resources Canada (NRCan), Agriculture and Agri-food Canada (AAFC), York University, and Radarsat International (RSI). Detailed information regarding the project and project participants can be found at the ISIES web site (http://isies.mda.ca/) and in ISIES publications (Liang and Tao 2005; Smith et al. 2005; Teillet et al. 2005).

51.4.2 ISIES Architecture

51.4.2.1 Service-oriented View of the ISIES System

From a high-level service-oriented view, the ISIES is constructed of several SSW services, the functions of which can be controlled by the chaining of individual services. By using standard web service interfaces to virtualize the ISIES sensors, the ISIES sensing resources become independent, re-usable SSW services. These can be invoked by external users or SSW services for different applications.

Figure 51-5 shows a service-oriented view of the ISIES system. ISIES has three different types of the SSW services: (1) *In situ* Sensor Service; (2) Remote Sensor Service; and (3) Processing Service. Logically, ISIES contains seven SSW service instances. It is worth noting that, although the implementation of the service partly shares the same communication networks and some of the service instances run on the same physical machines, logically they are independent SSW services.

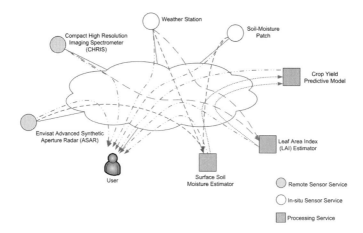

Figure 51-5 A service-oriented view of the ISIES system. See included DVD for color version.

The following describes the service flow of the ISIES service architecture. The ISIES final products, namely biomass and crop yield prediction, are the result of the chaining of the ISIES services. The ISIES system consists of three processing services (surface soil moisture estimator, Leaf Area Index (LAI) estimator, and crop yield predictive model), two remote sensor services (Envisat Advanced Synthetic Aperture Radar (ASAR) and Compact High Resolution Imaging Spectrometer (CHRIS)), and two *in situ* sensor services (weather/precipitation station and soil-moisture patch). The surface soil moisture estimator takes inputs from weather stations, soil-moisture patch, and ASAR, and uses a Bayesian estimator to estimate surface soil moisture. The LAI estimator takes data from the weather station and CHRIS as inputs, and outputs LAI maps. Finally, the crop yield predictive model uses the estimated surface soil moisture and LAI from the two processing services as inputs, and outputs biomass and yield prediction maps as the final ISIES product.

The users of the ISIES system (i.e., farmers) use a GIS client to access the ISIES products by invoking the ISIES services via the WWW. Figure 51-6 shows a screen capture of the Web GIS client of the ISIES.

Figure 51-6 Screen capture of the Web GIS client for the ISIES. The map window shows the crop yield prediction map from the crop yield predictive model service. The icon ANT1 indicates the location of the weather station. User can query the weather station by clicking on the ANT1 icon, and specify the time parameters for the query. The ISIES client will then invoke the weather station sensor service via the Web for the sensor's data. See included DVD for color version.

51.4.3 Physical Architecture of the ISIES *in situ* Sensor Services

We now describe the ISIES *in situ* sensor service's system architecture, the functionality of each individual component, and how these operate together. The ISIES *in situ* sensor service consists of the following three tiers: (1) *In situ* sensors, (2) SmartCore as the sensor gateway, and (3) ISIES sensor web server. Figure 51-7 shows the system architecture of the ISIES *in situ* sensor service.

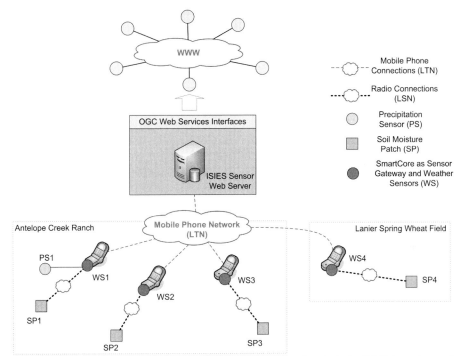

Figure 51-7 System architecture of the ISIES *in situ* sensor service. See included DVD for color version.

51.4.3.1 Sensors and the Local Sensor Network (LSN)

The ISIES *in situ* sensor service uses three types of sensors: (1) weather sensors, (2) precipitation sensors, and (3) soil moisture patches. Temporally, as a rule, the sensors obtain measurements every hour throughout the growing season. Spatially, the sensors are deployed according to the various spatial sampling requirements for accurate weather (10-25 km), precipitation (0.1-1 km), and soil moisture measurements (1-10 m).

The weather station and precipitation sensor only need to be sparsely deployed whereas the soil moisture sensors need to be densely deployed in order to meet the spatial sampling requirements. As a result, the weather station and precipitation are hard-wired to the sensor gateway (i.e., SmartCore), and the soil moisture patch uses short-range radio frequency as the LSN to communicate to the SmartCore.

51.4.3.2 Sensor Gateway and Local Transit Network (LTN)

ISIES uses the SmartCore (designed by Canada Centre for Remote Sensing, CCRS) as the sensor gateway that links the sensors in the field and the ISIES sensor web server in Vancouver, British Columbia. The SmartCore is a compact sensor gateway that controls sensor data traffic autonomously and communicates with the ISIES sensor web server in Vancouver wirelessly in two-way mode. In the ISIES configuration, the SmartCore uses the digital cellular network as the LTN to communicate with the ISIES sensor web server. The SmartCore device also supports the use of satellite modems for rural deployments.

The SmartCore devices have local computing and storage capabilities. They can be programmed locally or remotely to carry out aggregate calculations in the field and also to send special event alerts to the ISIES sensor web server.

51.4.4 The SSW Information Models

ISIES uses the OGC sensor web information framework (Cox et al. 2003; Cox 2005) as its information interoperability framework. The information framework consists of three major information models: (1) Observation, (2) Sensor and (3) Phenomenon. It looks obvious that they are three different types of entity, but people often tend to mix the three different entities together. Figure 51-8 shows the three entities and their relationships.

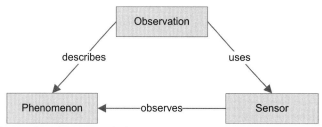

Figure 51-8 Observation, Sensor, and Phenomenon are three basic types of entity in the ISIES sensor web information framework. The arrows between the three boxes depict the relationship among the three types of entity: Observations uses Sensors to observe Phenomena.

At the implementation level, the ISIES information component is in fact a software module that takes the raw/aggregated data values from the sensors or processing models as inputs, combines them with associated metadata, and transforms them into structured and meaningful OGC sensor web information models and XML encodings ready for exchange via the Web. Figure 51-9 shows an example of using OGC's Observation and Measurement model (O&M) (Cox 2005) to describe a data instance collected by an ISIES weather station.

```
- <om:Observation>
  - <om:eventTime>
    - <gml:TimeInstant>
        <gml:timePosition>2005-10-09T10:32:00Z</gml:timePosition>
      </gml:TimeInstant>
  </om:eventTime>
  <om:eventLocation xlink:href="#location1457656"/>
  <om:procedure xlink:href="urn:isies:sensor:ANT1HB05.xml"/>
  <om:observedProperty xlink:href="urn:isies:phenomenon:airtemperature"/>
  - <om:featureOfInterest>
    - <om:Station>
      - <om:position>
        - <gml:Point gml:id="location1457656">
            <gml:pos gml:srsName="EPSG:4326">50.5971 -112.1794</gml:pos>
          </gml:Point>
      </om:position>
    </om:Station>
  </om:featureOfInterest>
  <om:result uom="urn:isies:unit:temperature:celsius">9.25</om:result>
</om:Observation>
```

Figure 51-9 An example of using OGC's observation information model to describe a data instance collected by an ISIES weather station. The above example shows the six basic properties of an observation model: (1) time—10:32:00Z on 9 October 2005, (2) location—coordinate reference system EPSG 4326, latitude 50.5971 north, longitude 112.1794 west, (3) feature of interest—Station, (4) phenomenon—air temperature in Celsius, (5) procedure—query sensor ANT1HB05, and (6) results—9.25° Celsius. See included DVD for color version.

51.5 Challenges

The Spatial Sensor Web is emerging as a promising information system for monitoring, and possibly actuating, the physical world. The vision of the SSW is realized by mixing and matching the heterogeneous sensing resources, despite underlying implementations, in order to create a higher level of knowledge (e.g., decision making or event prediction). However, the SSW also poses new challenges to the current GIS research community. Below we introduce and outline the three key challenges posed by the SSW: (1) scalability, (2) interoperability, and (3) dynamicity.

51.5.1 Scalability

The source of the scalability challenge stems from the enormous number of SSW sensors and users. Recent advances in MEMS technologies make it feasible and economically viable to deploy enormous amounts of low-cost, low-power, *in situ* wireless sensors in order to continuously monitor our environment. We also see that more and more advanced remote sensing satellites have been, or are scheduled to be, launched. It is envisioned that there will soon be millions to billions of sensors deployed all over the world by different organizations for a wide range of different applications. We also envision that there will be enormous numbers of users simultaneously accessing the SSW for a variety of applications. Scalability is an unavoidable yet crucial challenge to the SSW system.

51.5.2 Interoperability

The source of the interoperability challenge stems from the heterogeneity of sensors. The ability to access, interchange, understand, and use heterogeneous sensing is the main objective of the SSW. This complete interoperability is what differentiates the SSW from standard sensor networks. In fact, the above-mentioned ability is built upon interoperability at each level of the SSW architecture. The web service approach and standardized information model approach described in this chapter only deal with the technical level of the SSW interoperability issues. There are other levels of issues, such as the semantic level, which still need to be solved.

51.5.3 Dynamicity

The third key challenge for the SSW is related to 'dynamics'; this stems from four related realities: (1) Sensors can be mobile (i.e., mobile sensors change their locations dynamically); (2) Sensors can be dynamic and transient (i.e., sensors may join and depart the system frequently); for example, when an earthquake occurs, many existing sensors may cease to function, and many new sensors may immediately be deployed; (3) Sensors typically deliver data in streams (i.e., queries over data streams need to be processed dynamically because sensor streams represent real-world events that need responses); (4) Users' request patterns are highly dynamic; for example, when a disaster occurs near to a sensor, large numbers of users may suddenly take interest in this specific sensor. The associated high surge of user requests can bring down the service node associated with the sensor. Based on the above four aspects of SSW dynamism, a new dynamic sensor-focused GIS architecture is required.

51.6 Summary

The fundamental revolution in the Spatial Sensor Web vision lies in its interoperable, dynamic, and scalable "connectivity." This is an evolving framework enabled by many emerging technologies. Similar to our understanding of the WWW 10 year ago, our understanding of the SSW is in its infancy. However, despite this, it is only a matter of time before the SSW weaves itself into the fabric of our lives.

Acknowledgements

The authors would like to acknowledge and express appreciation for funds received from GEOIDE, CRESTech, and OGC. Additional acknowledgement is extended to the ISIES team, Mr. Haiyang Han, Dr. Feng Wang at GeoTango Inc., and Mr. Björn Prenzel and Mr. Jing Lu at York University.

References

Akyildiz, I. F., W. Su, Y. Sankarasubramaniam and E. Cayirci. 2002a. A Survey on Sensor Networks. *IEEE Communications Magazine* 40:102–114.

———. 2002b. Wireless Sensor Networks: A Survey. *Computer Networks* 38:393–422.

Botts, M., A. Robin, J. Davidson and I. Simonis. 2006. OpenGIS Sensor Web Enablement Architecture Document. Technical Report, Open Geospatial Consortium, Inc. 69pp.

Butler, D. 2006. 2020 Computing: Everything, Everywhere. *Nature* 440:402–405.

CNN.com. 1999. NASA: Human Error Caused Loss of Mars Orbiter. <http://www.cnn.com/TECH/space/9911/10/orbiter.02/> Accessed 20 January 2008.

Cornillon P., J. Gallagher and T. Sgouros. 2003. OpenDAP: Accessing Data in a Distributed, Heterogeneous Environment. *Data Science Journal* 2:164–174.

Cox, S. 2005. Observations and Measurements. OGC-05-087. Wayland, Mass.: Open Geospatial Consortium, Inc.

Cox, S., P. Daisey, R. Lake, C. Portele and A. Whiteside. 2003. OpenGIS Geography Markup Language (GML) Implementation Specification v.3.0, OGC-02-023r4. Wayland, Massachusetts: Open Geospatial Consortium, Inc.

Delin, K. A. and S. P. Jackson. 2001. *The Sensor Web: A New Instrument Concept*. In *SPIE's Symposium on Integrated Optics*. San Jose, California, USA.

Estrin, D., W. Michener and G. Bonito. 2003. *Environmental Cyberinfrastructure Needs for Distributed Sensor Networks. A Report from a National Science Foundation Sponsored Worshop,* 12-14 August, Scripps Institute of Oceanography.

Foster, I., C. Kesselman, J. M. Nick and S. Tuecke. 2002. Grid Services for Distributed System Integration. *Computer* 35:37–46.

Gibbons, P. B., B. Karp, Y. Ke, S. Nath and S. Srinivasan. 2003. IrisNet: an architecture for a worldwide sensor web. *IEEE Pervasive Computing* 2:22–33.

Green, H. 2003. The future of technologies - tech wave 2: the sensor revolution. <http://www.businessweek.com/magazine/content/03_34/b3846622.htm> Accessed 16 December 2008.

Gross, N. 1999. The earth will don an electronic skin. *BusinessWeek* 30 Aug. 30 1999. <http://www.businessweek.com/1999/99_35/b3644024.htm> Accessed 18 December 2008.

Hill J., R. Szewczyk, A. Woo, S. Hollar, D. Culler and K. Pister. 2000. System architecture directions for networked sensors. *ACM SIGPLAN Notices* 35:93–104.

Huhns, M. N. and M. P. Singh. 2005. Service-oriented computing: key concepts and principles. *IEEE Internet Computing* 9:75–81.

IEEE. 2006. IEEE 802.15.4 Wireless Personal Area Networks Task Group 4 Standards. <http://standards.ieee.org:80/getieee802/download/802.15.4-2006.pdf > Accessed 16 December 2008.

Krishnamachari, L., D. Estrin and S. Wicker. 2002. The impact of data aggregation in wireless sensor networks. Pages 575–578 in *ADSN – 1st International Workshop on Assurance in Distributed Systems and Networks, Workshop of the 22nd International Conference on Distributed Computing Systems, 2002 Proceedings.* Held in Vienna, Austria, 2-5 July 2002.

Liang, S.H.L., A. Croitoru and V. Tao. 2005. A distributed geospatial infrastructure for sensor web. *Computers and Geosciences* 31:221–231.

Liang, S.H.L. and V. Tao. 2005. Design of an integrated OGC spatial sensor web client. *Proceedings of the 13th International Conference on Geoinformatics,* held in Toronto, Canada, 17-19 August 2005.

Libicki, M. C. 1995. *Standards: The Rough Road to the Common Byte.* Technical Report, Institute for National Strategic Studies, National Defense University. Washington, DC: Center for Advanced Concepts and Technology. 46 pp.

NSF (National Science Foundation). 2005. The Sensor Revolution. <http://www.nsf.gov/news/special_reports/sensor/index.jsp> Accessed 18 December 2008.

Polastre, J. R. 2003. *Design and Implementation of Wireless Sensor Networks for Habitat Monitoring.* Ms.c. Thesis, Department of Electrical Engineering and Computer Sciences, University of California at Berkeley. 67 pp.

Ramachandran, R., S. J. Graves, H. Conover and K. Moe. 2004. Science Markup Language (ESML): a solution for scientific data-application interoperability problem. *Computers & Geosciences* 30:117–124.

Robert G. R. and J. P. Michael. 2005. Knowledge Representation in the Semantic Web for Earth and Environmental Terminology (SWEET). *Computers & Geosciences* 31:1119–1125.

Smith, A. M., C. Nadeau, J. Freemantle, H. Wehn, P. M. Teillet, I. Kehler, N. Daub, G. Bourgeois and R. D. Jong. 2005. *Leaf Area Index from CHRIS Satellite Data and Applications in Plant Yield Estimation. Proceedings of the 26th Canadian Symposium on Remote Sensing.* Held in Wolfville, Nova Scotia, Canada, 14–16 June 2005. Ontario, Canada: Canadian Aeronautics and Space Institute.

Tao, V., S.H.L. Liang, A. Croitoru, Z. M. Haider and C. Wang. 2004. GeoSWIFT: Open geospatial sensing services for sensor web. In *GeoSensor Networks,* edited by A. Stefanidis and S. Nittel, 267–274. Boca Raton, CRC Press.

Teillet, P. M., A. Chichagov, G. Fedosejevs, R. P. Gauthier, G. Ainsley, M. Maloley, M. Guimond, C. Nadeau, H. Wehn, A. Shankaie, J. Yang, M. Cheung, A. Smith, G. Bourgeois, R. de Jong, V. C. Tao, S.H.L. Liang, J. Freemantle and M. Salopek. 2005. Overview of an intelligent sensorweb for integrated earth sensing project. *Proceedings of the 26th Canadian Symposium on Remote Sensing.* Held in Wolfville, Nova Scotia, Canada, 14–16 June 2005. Ontario, Canada: Canadian Aeronautics and Space Institute. <http://earth.esa.int/pub/ESA_DOC/PROBA/Teillet_Sensorweb_2005.pdf> Accessed 18 December 2008.

Teillet, P. M., R. P. Gauthier, A. Chichagov and G. Fedosejevs. 2002. Towards integrated Earth sensing: advanced technologies for in situ sensing in the context of Earth observation. *The Canadian Journal of Remote Sensing* 28 (6):713–718.

Unidata. Undated. netCDF (network Common Data Form). <http://www.unidata.ucar.edu/software/netcdf/> Accessed 18 December 2008.

W3C. 2004. Web Service Architecture. <http://www.w3.org/TR/ws-arch/> Accessed 18 December 2008.

Warneke, B., M. Last, B. Liebowitz and K.S.J. Pister. 2001. Smart dust: communicating with a cubic-millimeter computer. *IEEE Computer Magazine* 34 (1):44–51.

Weiser, M. 1991. The computer for the 21st century. *Scientific American* 265 (3):94–104.

SECTION 8
GIS Reaches Out: Applications

CHAPTER 52
Using GIS to Track Lyme Disease

Tonny J. Oyana and *Jessica Gaffney Clark*

52.1 Lyme Disease: Background and Transmission

Lyme disease has the dubious distinction of being "The Great Imitator" within the medical community, since its perplexing symptoms can run the gamut from headaches to facial paralysis. This disease can be acquired by participating in activities that most people would not consider risky: taking a walk in the park, playing outdoors with a pet, or hiking in a scenic state park. In fact, many people do not recall seeing the culprits, *Ixodes scapularis*, or deer ticks, biting their skin. And why would they? Even the *Ixodes scapularis* adults are minuscule, a mere three mm long. Nymphal ticks, the juveniles that are responsible for 90% of human Lyme disease infection, are even smaller; the size of the period at the end of this sentence (Pesticide Education Office 2007).

Ticks cannot fly, and their little legs are fairly useless when it comes to personal locomotion. Instead, the opportunistic buggers *quest,* which is their equivalent of hailing a cab; they climb to the top of nearby vegetation and wait for a large animal to come along, which can be an ambling dog, the deer that got the dog's attention, or the person chasing after the dog chasing after the deer (Clark 2006).

Nymphs live in habitats shared by white-footed mice (*Peromyscus leucopus*) where larvae fed the previous late summer. This habitat, which also happens to be *Odocoileus virginianus* (white-tailed deer) habitat, is best described as woodlands: bushy, low shrub woodland edge regions and grassy areas that border woodlands. The mice travel in trails and nest almost anywhere they can find a sheltered depression. Nymphal tick activity coincides with human outdoor activity, and peak human infection symptoms occur in early July (Pesticide Education Office 2007).

A nymph will begin sucking the blood of the host, often without its knowledge. The highest probability of transmission occurs when nymphs and adults are abundant and a tick feeds on a person for more than 12 hours (Fischer 1999). If it is not detected and completely removed within 48 hours, the spirochete *Borrelia burgdorferi* in the tick's gut will be transmitted to the host.

First, we describe how GIS is useful in the study of Lyme disease, as well as the scale and limitations involved. Then, we present a review of the literature concerning historical, current, and potential future tick presence in Illinois as well as a rationale for the variable choices for our model. Then we outline our model construction methodology and the heuristics involved with creating the perceptual map from point locations. Finally we present the results of our model and analyses of our findings, as well as implications, conclusions, and future directions and recommendations.

52.1.1 Lyme Disease Symptoms and Prevention

A key component of early diagnosis is recognition of the characteristic Lyme disease rash, known as *Erythema migrans* or EM. It often looks like a bull's-eye and is observed in about 80% of Lyme disease patients (Lyme Disease 2005), although this statistic varies somewhat. Besides the EM, the other symptoms of the early localized stage of Lyme disease include swelling of lymph glands, fatigue, headache, achiness, muscle and joint pain, and chills (Eppes 2003). Most cases of Lyme disease can be cured with antibiotics, especially if treatment is begun early in the course of illness (Lyme Disease 2005; Eppes 2003). However, a small percentage of patients with Lyme disease have symptoms that last months to years, even after treatment with antibiotics (Lyme Disease 2005).

If the symptoms are not recognized (or the rash is not present) and antibiotics are not administered in time, the person can develop the next stages of Lyme disease, early disseminated stage and late stage. Early disseminated stage can appear two weeks to three months after the tick bite and can include multiple rashes not at the site of the bite, severe headache, severe fatigue, enlarged lymph nodes, stiffness in the joints and neck, facial paralysis (Bell's palsy) and meningitis (Eppes 2003). Several months to years later, approximately 60% of patients with untreated infection will get late stage or tertiary Lyme disease; they will have intermittent bouts of arthritis with severe joint pain and swelling (Lyme Disease 2005; Eppes 2003). While Lyme disease is not life-threatening, it does threaten one's quality of life.

Approximately 5% of untreated patients may develop chronic neurological symptoms months to years after the infection, which include shooting pains, numbness or tingling in the hands and feet, and problems with concentration and short-term memory (Lyme Disease 2005). The neurological symptoms can be so severe that many cases of Lyme disease have been previously misdiagnosed as multiple sclerosis (MS), chronic fatigue immune deficiency (CFIDS), amyotropic lateral sclerosis (ALS), lupus, Alzheimer's disease, Parkinson's disease, and other neurodegenerative syndromes.

52.1.2 Requirements for Lyme Disease Establishment

There are four factors necessary for the successful establishment of Lyme disease in humans. As the deer tick is the main vector for Lyme disease, the first two are the presence of the deer tick and the availability and suitability of tick *hosts,* or the organisms from which a parasite obtains its nutrition and/or shelter. Next, there needs to be adequate opportunities for human or domestic animal exposure, and lastly, the host reservoirs for the bacteria need to be competent, which means that if the tick bites the host, there is a high likelihood that the tick will get the disease in its bloodstream.

White-tailed deer in particular are critical for the survival and propagation of deer ticks, since they are the principal reproductive vectors, or carriers, of Lyme disease. Also crucial are white-footed mice, since they are the ticks' main means of local mobility, and the main reservoir for the Lyme disease spirochete, *Borrelia burgdorferi.* Unless these four factors are present, *Borrelia* will not become established in the environment.

52.2 GIS and Epidemiological Data

Under the broad umbrella of epidemiology is the study of the distribution and determinants of diseases. The spread of disease has both spatial and temporal components, and using GIS to map patterns of space and time can help predict future outbreaks and target areas with high risk. GIS has the ability to store, retrieve, manipulate, and analyze spatial data, which makes it a useful mechanism to study disease patterns over large areas with high resolution (Glass et al. 1994). GIS can be used to pinpoint the exact locations of infection, identify exploratory variables and combine multiple layers of information, which can be used to discern critical disease information. GIS can also track and monitor disease surveillance systems. It has been widely adapted as a major research tool that supports epidemiological research.

Furthermore, data like soil type, vegetation cover, and demographic conditions can be combined with distributions of disease and disease vectors into a database that can then be used to quantify spatial relationships between risk factors and vector or disease distributions, as well as for the prediction of their spread and establishment (Kitron et al. 1992; Kitron et al. 1991a). As a vector-borne disease, Lyme disease incidence is closely related to the distribution and abundance of the Lyme disease vectors, which are in turn limited by environmental factors. By mapping these environmental factors, we can see what areas provide suitable habitat for the tick vectors, mouse reservoirs, and deer hosts. In doing so, we will see what areas have the right conditions needed for a Lyme disease outbreak. This knowledge is useful as it can focus educational efforts to reduce tick exposure and increase motivation to use appropriate preventative measures when tick exposure is unavoidable (Frank et al. 2002).

Many researchers have used GIS to create risk maps for Lyme disease. In a study on canine seroprevalence, Guerra et al. (2001) found that the spatial pattern of seropositive dogs is correlated to human incidence of Lyme disease and abundance of the tick vector, *Ixodes scapularis*. They used GIS to compare the environmental data with the location of the dog to determine environmental risk factors.

Millstein (1999) created a program with which one can analyze data for the detection of areas at high risk for becoming foci for Lyme disease. He applied fuzzy logic inferences to a public health problem for which the relationships between vectors and their habitats were not clearly understood. Schulze et al. (1991) created a way to assess specific geographic areas for potential and actual risk of Lyme disease transmission with characterization of vegetative types and tick surveys, respectively.

Glass et al. (1994) used a neighborhood approach to characterize the habitat in regions where deer were collected, since sampling data as points suggested an inaccurately high level of sampling accuracy (Monmonier 1991 in Glass et al. 1994). Glass et al. (1992) combined selected environmental data with cases of Lyme disease to identify significant risk factors.

52.3 Scale of Study

In GIS model creation, it is important to keep in mind the scale of the study. As Nicholson and Mather (1996, p. 712) so eloquently state, "a critical element in studying transmission dynamics of Lyme disease and other tick-transmitted infections will be first to discern the appropriate scale at which to assess human risk." We have chosen to carry out our model at the county level, because at this level the land cover is sufficiently diverse and there remains a useful unit for reporting disease, since Health Departments generally cover one county.

Even though the study undertaken by Glass et al. (1994) was at the county level, the results of their model were consistent with previously identified factors affecting tick abundance, which indicates that little information was lost, even though the scale was fairly coarse (Glass et al. 1994). Dennis et al.'s comprehensive reported distribution of *Ixodes scapularis* in the United States used county-level data as well (1998). Many studies have been conducted at the county level, such as Allan et al. (2003) and LoGiudice et al. (2003) in Dutchess County, New York; Kitron et al. (1991a) in Rock Island County, Illinois, and many others. In the available literature, the county was by far the most common geographic unit, as well as the fact that most tick distribution data was available by county and could be correlated with human case rate data at the county level (Dennis et al. 1998).

However, the current national distribution map is an admittedly crude indicator of transmission risk, and they used county-level data since that was the least common geographic unit that could be linked with human case data (Dennis et al. 1998). Furthermore, Dennis et al. (1998) asserted that a sufficiently detailed ecological risk map is now needed for targeting various Lyme disease prevention strategies.

52.3.1 Limitations of Scale with GIS

The major concern of Glass et al. (1992) was to determine whether or not the level of geographic resolution in an existing GIS would be sufficient for detecting factors associated with disease. This is a valid concern. The factors that we have chosen that affect ticks are widely available in GIS databases, yet only indirectly predict abundance. The most important factors of localized tick habitat suitability are things like a thick litter layer and a high herbaceous vegetation layer (Ginsberg 2006). However, we did not have this high-resolution remotely sensed data available. Therefore, although using GIS in this way has its disadvantages and does not provide a complete picture of Lyme disease, we are still able to discern some useful insights regarding Lyme disease.

52.4 Lyme Disease in Illinois

There is a clear, focal pattern of Lyme disease risk in the United States with the greatest risk occurring in the northeast and upper midwest regions, which include the states of Wisconsin, Michigan, Minnesota, and Illinois (Fish and Howard 1999). In 1998, Wisconsin had the highest percent increase in Lyme disease cases nationally (Jobe et al. 2007; Illinois Department of Health 2007). The Illinois Department of Public Health believes that Lyme disease is a serious problem in Illinois. Studies such as Dennis et al. (1998) have shown that deer ticks are present in every county in Illinois and cases have been reported in at least two-thirds of these counties, according to the Centers for Disease Control and Prevention (Kitron and Kazmierczak 1997).

52.4.1 Historical Tick Presence in Illinois

Hunter and collection data is a valuable resource that allows us to trace the expansion of ticks from the highly endemic area of Wisconsin to northern Illinois and then southward. In the central northwest, deer ticks were first verified in Wisconsin in 1968. Shortly thereafter, in 1970, the first human case of Lyme disease was documented. The ticks then moved to northwestern Illinois in 1987, and by 1988, they had firmly established a tick foci in northwestern Illinois, signified by the abundance of ticks collected in Castle Rock State Park (Cortinas et al. 2002). It is clear that the geographical locations where ticks are found are expanding in a southerly direction.

In 1987, deer ticks were detected in Castle Rock State Park in northwestern Illinois, and in 1988, *Borrelia burgdorferi* spirochetes were isolated from ticks collected in this location (Kitron et al. 1991b; Nelson et al. 1991). The ticks' ranges were bound by the Wisconsin border on the north, the Rock River on the east and south, and the Mississippi on the west (Kitron et al. 1991a). But they were not bound for long.

Deer ticks were found in northern Kankakee and Putnam counties in 1988, followed shortly after by Brown and Peoria counties in 1989 (Bouseman and Nelson 2000). Bureau, Carroll, LaSalle, Scott, and Schuyler counties followed in 1990, Grundy and Will counties in 1991, and then Marshall and Tazewell counties in 1992. This geographic expansion has continued its steady pursuit southward.

In the spring of 1997, *Ixodes scapularis* were found in Carroll, Grundy, and Will counties in northern Illinois that were infected with *Borrelia*, verified by successful lab cultures of the spirochete. Furthermore, during the 1997 deer hunt, ticks were found on deer in Grundy and LaSalle counties (Cortinas et al. 2002). Previously, known infected ticks were only from Ogle and Rock Island counties. The trend indicates the geographical expansion of ticks, since the researchers noted larger populations of the ticks in the three counties (Bouseman and Nelson 1998). Cortinas et al. (2002) also found an increase in the expansion of ticks, as evidenced by how many captured deer were infected with *Borrelia burgdorferi*. Counties with reported populations are generally clustered around concentrated areas of established populations (Dennis et al. 1998).

In the spring of 2000, deer ticks were found at sites in Fulton and Peoria counties (Bouseman and Nelson 2000), which lends credence to the opinion of Guerra et al. (2002) that they are following the Illinois River south to the Mississippi (Kaye 2002).

52.4.2 Current Tick Presence in Illinois

Tick collections from that 1987 hunting season indicated that blacklegged ticks were already infiltrating northern Illinois but had not yet been found in central or southern Illinois (Bouseman et al. 1990). It is important to keep in mind that these data from 1987 are close to twenty years old now, in 2006. According to more recent data, blacklegged ticks have been found in central Illinois, specifically Clark County and Monroe County, the latter of which is only a county away from Jackson County (IDPH 2005). The crux of the problem is that current assessment aims are incomplete and untrustworthy.

Most presence/absence assessment is done on the county level. Since ticks are not spread uniformly throughout a county, one researcher can find them some places in the county and another researcher can look and not find any at all, even within the same year. When DuPage County was cited as a probable exposure locale, the IDPH (Illinois Department of Public Health) and local tick surveillance experts found the blacklegged tick in two small areas of the east branch of the DuPage River (Bestudik 2006). However, many other areas of DuPage County were surveyed and no blacklegged ticks were found (Bestudik 2006). This shows that even if a researcher is thorough in the areas that they sample, they still may not find ticks in counties that actually have them, and the assessment of "absence" could be false.

The definitive distribution of blacklegged ticks in the United States used "indirect means" and "supplemental sources" to indicate tick presence (reported or established) (Dennis et al. 1998). This includes published sources where the county data is made explicit, as well as supplemental sources such as local bulletins and Personal Communications (Dennis et al. 1998). As they explicitly state: "tick collection methods used have been nonstandardized, dissimilar, and rarely applied systematically in time and space (Dennis et al. 1998, p. 633)." Frank et al. (2002) agree; tick data are often unavailable, out of date, costly, and difficult to collect. Thus, tick sampling efforts, while worthwhile, do not show a complete indication of tick presence.

Bouseman et al. (1990) stated that little stands in the way of dispersal of Lyme disease in Illinois as far as the availability of suitable hosts is concerned, as white-tailed deer and *Peromyscus* mice are widely distributed and locally abundant in Illinois. As recently as 2005, the number of cases of Lyme disease increased by an amazing 45% to a total of 126, setting a new record for the state (Bergman 2005; Newbart 2006). Not only has the number gone up in each of the last five years, but "we are starting to see a steady increase," says Joan Bestudik, education coordinator in communicable diseases for the state agency (Newbart 2006, p. 1). Figure 52-1 is a 2005 map of locations where deer ticks are known to be distributed in Illinois. However, this is a very conservative estimate.

Figure 52-1 Map of tick locations in Illinois. Known geographic distribution of *Ixodes scapularis* by county in Illinois in 2005. Counties where ticks are established are those where the deer tick has been found repeatedly in the environment: the CDC criteria for "established" ticks are at least six ticks or two life stages (larvae, nymphs, adults) identified. "Present" refers to the counties where additional reports suggest that the deer tick is present and may be established. Courtesy: Illinois Department of Public Health Prevention and Control 2005. See included DVD for color version.

52.4.3 Future Tick Presence in Illinois

Wisconsin is a state that is already highly endemic to Lyme disease, and over the past twenty years, this disease has slowly been making its way south into northern Illinois. Because all the necessary components for the transmission of the Lyme disease spirochete and the fulfillment of the tick life cycle are available throughout much of the state of Illinois, *Ixodes scapularis* and Lyme disease can spread and become established in large portions of Illinois (Bouseman et al. 1990). It is also clear that the range is increasing, and the expansion of blacklegged ticks is characterized by extension from well-established areas to contiguous regions and may be limited by availability of suitable habitat (Dennis et al. 1998).

Southern Illinois is a unique area, shaped and scraped by history, first by glacial movement and then afterwards by significant land use changes. Guerra et al. (2001) suggest that glacial outwash and tick endemicity are linked. As far as land use is concerned, it has changed significantly in the past hundred years.

Land use has changed historically to unknowingly provide circumstances that favor the spread of Lyme disease. In the early 1800s, Illinois had nearly 14 million acres of forest habitat, but today less than 20% (4 million acres) of this original pre-settlement forest habitat remains. According to the US Geological Survey report, subsequent land conversion for agriculture, urbanization, mining, and transportation corridors have eliminated forest habitat in Illinois (Herkert et al. 1993).

These farms that were formerly cleared forests were subsequently abandoned in the late 1800s and 1900s, causing succession of the fields to second-growth forests (Patnaude and Mather 2000). When farmland reverted to woodland, deer proliferated, white-footed mice were plentiful, and the deer ticks thrived (Steere et al. 2004). Soil moisture and land cover, as found near rivers and along the coast, were favorable for tick survival. Next, these areas became heavily populated with both humans and deer, as more rural wooded areas became wooded suburbs in which deer were without predators and hunting was prohibited (Steere et al. 2004; Guerra et al. 2002).

Ticks in southern Illinois have been increasing over time. Subsequent deer examinations found that tick prevalence and intensity on white-tailed deer in Crab Orchard National Wildlife Refuge in a study conducted from 1980–1983 was higher than a study done 14 years prior (Montgomery and Hawkins 1967 as reviewed by Nelson et al. 1984). However, this should be treated as speculative, since the differences could be attributed to potential differences in the season and climactic conditions in which deer are captured (Nelson et al. 1984).

52.5 Model Rationale

For our model, we first searched the literature to find common deterministic threads that would predict tick habitat well without duplicating previously done work in this area, both figuratively and literally. One of our strongest sources, "Predicting the Risk of Lyme Disease: Habitat Suitability for *Ixodes scapularis* in the North Central United States" (Guerra et al. 2002), utilized a number of abiotic factors to predict tick habitat suitability, including elevation, soil order and type, forest moisture, bedrock geology, and others. It created a risk model for Illinois, Wisconsin, and a small portion of Michigan. However, it looked at tick habitat suitability, and we noticed that in order for Lyme disease to really be a problem, you need ticks, white-footed mice, and white-tailed deer. White-tailed deer are the means by which they move throughout an area—their main means of transportation.

White-footed mice are the disease reservoirs; since ticks do not pass on the *Borrelia burgdorferi* to their larvae, each generation of ticks must acquire the disease on their own, and mice carry the disease and pass it on to ticks that feed on them. Their role cannot be overstated; without white-footed mice to act as a competent reservoir, even if there are bountiful ticks and deer, Lyme disease will not be a threat to the human population.

Reservoir competency is the ability for an animal to transmit the disease that it carries to other animals. For comparison purposes, white-footed mouse reservoir competency is around 95%. The main reservoir for *Borrelia burgdorferi* in the South is the lizard, and it has a very low reservoir competency; lower than 5%. This means that if an infected tick bites a lizard and infects it, there is only a 5% chance that subsequent non-diseased ticks that attach to that lizard will pick up the spirochete from this lizard. As a result, very few ticks carry the spirochete, and even if people are bitten often, they will not frequently get Lyme disease from these bites.

Deer, however, are prime movers of the ticks so they can bring in new infected ticks often, and if there becomes a large competent reservoir base, then Lyme disease can spread throughout a region. We found it encouraging that the paper that we referenced above did find that southern Illinois had an 80% habitat suitability index, even though this model was not tested by looking for ticks there. However, in areas where they did look for ticks (northern Illinois and Wisconsin), they found the model to be almost 90% accurate. We will use our perceptual map to see how accurate our model is. That model, however, was created on a larger scale, and ours will be much more specific. We believe that this will help local people know their risk better, since it is easier for most people to see where they live and recreate on a county map as opposed to a state or regional map. In this way, our risk map will greatly help the residents of Jackson County, Illinois.

Although others have successfully trapped live animals and tested ticks that they found for the Lyme disease agent, *Borrelia burgdorferi*, we have neither the time nor the resources to undertake this level of intensive testing. Also, time is against us in another sense; the tick population peaks in July, and by the time that this chapter is completed, ticks will still not be at their most aggressive or abundant, so testing this model via specimen collection may not accurately indicate the true level of ticks. Fortunately, it can be tested by looking at locations where our sources have seen ticks and seeing how they coincide with or complement our model, as we describe in Section 52.8.1.

52.5.1 Tick Habitat Suitability Model: Variable Choices

Numerous studies of habitat suitability of *Ixodes scapularis* have used a GIS to detect significant associations between tick presence and environmental variables. They were in two categories. For our intents and purposes, the first category is factors that can be measured with GIS and the second category is factors that cannot be measured with a GIS. Variables measured using a GIS include: patch size (Allan et al. 2003; Duffy et al. 1994; Glass et al. 1995), river/stream proximity (Nicholson and Mather 1996; Bouseman et al. 1990; Cortinas et al. 2002), land cover (Nicholson and Mather 1996; Glass et al. 1992, 1995), slope and aspect (Glass et al. 1995, 1995; Ginsberg et al. 2004), and elevation (Nicholson and Mather 1996; Guerra et al. 2002; Glass et al. 1992, 1995).

There are also those that cannot be measured with a GIS at the necessary scale but can be validated with the narrative accounts: wooded edges (Guerra et al. 2002; Glass et al. 1995; Cortinas et al. 2002) and forest cover (Nicholson and Mather 1996; Fischer 1999; Furlanello et al. 1997).

It takes more than deer-tick presence to spread Lyme disease, however. Host presence (Buskirk et al. 1998; Cortinas et al. 2002; Furlanello et al. 1997) is also important since Lyme disease requires adequate and suitable reservoirs as well as means of transportation for the tick vector.

52.5.1.1 Patch Size

A study of fragmentation, mice, and Lyme disease found that the five smallest forest fragments in the study had seven times as many infected nymphs per square meter as the larger fragments and that the nymphal infection prevalence in the smallest fragments was much higher than in nymphal populations inhabiting more continuous forest at a nearby site (Allan et al. 2003). When patch size was less than two acres, there is an increased risk for Lyme disease (Allan et al. 2003). Furthermore, nymph density, nymphal infection prevalence, and the density of infected nymphs were all negative functions of patch area (Allan et al. 2003).

To reduce the risk of Lyme disease, Allan et al. (2003) suggested that decreasing fragmentation of deciduous forests, especially where there is already high incidence of Lyme disease, would be an important first step. The risk of Lyme disease decreased with increasing distance from the forest edge in a study conducted in Maryland (Glass et al. 1995; Dennis et al. 1998).

Allan et al. (2003) used a GIS to take data from aerial photos and classify them into land-cover types, and then used simple regression analyses to see if nymph density, nymphal infection prevalence and the density of infected nymphs were a significant function of patch area. LoGiudice et al. (2003) found that both nymphal infection prevalence and density of infected nymphs are inversely related to fragment size, with nymphal infection prevalence of over 80% observed in extremely small forest fragments, under one hectare.

Forest fragmentation and loss is even more pronounced in southern Illinois. Instead of one large contiguous forest area, the landscape now consists of small isolated forested blocks. Forest has survived best in the Shawnee Hills along the southern part of the counties and in the lowlands along the Wabash and Ohio rivers (Herkert et al. 1993).

Forest fragmentation has a profound effect on its ubiquitous resident, the white-footed mouse. Since populations of habitat generalists such as white-footed mice are regulated by predators and competitors, small patches with little competition cause white-footed mice to thrive, especially when diversity is low (Rosenblatt et al. 1999 in Mahan and O'Connell 2005; Allan et al. 2003). When mice are more plentiful, the fraction of tick meals taken from these competent vectors is very high, and thus nymphal infection prevalence and the density of infected nymphs increase as a result (Allan et al. 2003). In other words, what would be considered poor habitat to most other animals, such as that near urban centers with lots of pavement, is not only not bad for white-footed mice, but it even helps them. Furthermore, the more urban the environment is, the higher the likelihood for human-rodent interaction. Large bottomland forest patches in southern Illinois have low white-footed mouse abundance, so the trend towards patchiness also increases the risk (Barko et al. 2003).

The study undertaken by Wilder and Meikle (2004) concerning ticks, mice, and fragmentation found an odd occurrence; there is a lower rate of infection on mice in small forest fragments. Wilder and Meikle (2004) postulate that perhaps if smaller fragments have higher food abundance, then mice could be much more able to fight off parasites. Landscape could also affect prevalence, since habitat use by deer affects juvenile tick densities in subsequent years, and white-tailed deer use small patches differently (Wilder and Meikle 2004).

52.5.1.2 River/Stream Proximity

River proximity is consistently noted as an environmental factor with a strong correlation to tick density (Cortinas et al. 2002; Kitron et al. 1991a). Spatial analysis showed that tick presence on deer was associated with proximity to rivers (Kitron 1991a, 1992; Cortinas et al. 2002). Cortinas et al. (2002) also believe that *Ixodes scapularis* and Lyme disease will continue their southern expansion to the next major riparian corridor, the Illinois River. The river corridor is forested and provides suitable habitat for *Peromyscus leucopus* mice and white-tailed deer (Cortinas et al. 2002). Bouseman et al. (1990) indicated that it is likely that deer migration along rivers, such as the Rock River and perhaps the Mississippi River, could explain a great deal about the known distribution of *Ixodes scapularis* in Illinois.

In the hardwood forests of the Northeast and along damp river corridors across the Midwest, ecological factors appear to favor the prevalence of Lyme disease and the ticks that carry it (Pleasant 2004). In Baltimore County, Maryland, for example, a county in the northeast that is endemic to Lyme disease, the area with the highest incidence of disease extends along vegetational corridors bordering the Gunpowder Falls river system and associated reservoirs (Frank et al. 2002). In an adjacent county, a similar pattern follows the runs of Broad Creek and Deer Creek (Frank et al. 2002). These forests are aligned along larger rivers and creeks in an environment that is ideal for the transmission of Lyme disease (Frank et al. 2002).

Rivers are a likely route for the spread of Lyme disease (Bouseman et al. 1990). While the geographic distribution of *Ixodes scapularis* is limited by agriculture, its spread does favor river corridors—most recently, the Illinois River (Guerra et al. 2002). Dr. Uriel Kitron mused about this connection, "Tick densities have been increasing in counties along the southern portion of the Illinois River, where the habitat is favorable" (Perea 2002, p. 1).

Cortinas et al. (2002) have been tracking tick's travels for the past decade in Illinois, and predict further movement south along the Illinois River to the Mississippi River (Kaye 2002). This fact would favor southern Illinois for habitat suitability, as it is bounded by the Mississippi and Ohio Rivers. Glass et al. (1992, 1995), Cortinas et al. (2002) and other studies have utilized distance from streams, rivers, or other bodies of water.

52.5.1.3 Land Cover

Tick populations may be limited to river corridors because a great deal of land in Illinois is used for agriculture (Anonymous 2001). Tick abundance was negatively associated with the percentage of land used for urban/residential purposes, the percentage of the land that was classified as wetlands, and the percentage of privately owned non-recreational land (Glass et al. 1994).

Urban development is negatively associated with tick density, affecting the tick populations directly and the host populations indirectly (Steere et al. 2004; Glavanakov et al. 2001). Glass et al. (1992) found that residence in highly developed areas decreased risk threefold. In his GIS model, Glass et al. (1995) used a land cover/land use database with 20 categories, including urban property, agricultural land, forested land, and undeveloped residential land. Canines in forested and urban areas are twice as likely to be seropositive as those in agricultural areas (Guerra et al. 2001).

Guerra et al. (2002) found a negative association of tick presence with grasslands and conifer forests and a positive association with deciduous and dry/mesic and dry forests. This could be due to the fact that Glass et al. (1995) found this to be true in his East Coast study, and Guerra et al.'s (2002) study took place in Illinois and Wisconsin. Guerra et al. (2002) grouped land cover data in five ordinal categories: agriculture, grasslands, coniferous forest (≥ 75% of trees maintain leaves all year), deciduous forest (>75% of trees shed foliage in response to seasonal change), and mixed forest (neither deciduous nor coniferous make up > 75% of land cover).

52.5.1.4 Slope and Aspect

Kitron et al. (1992) found that from 0 km to 10 km, the slope for infested deer was greater than the slopes for uninfested deer using the Poisson process. Having derived slope and aspect information from the digital elevation model via IDRISI's surface analysis module, Glass et al. (1992) found that risk of disease was higher on steep slopes. Furlanello et al. (2002) found that northerly exposures are preferred. When Glass et al. (1995) evaluated linear trends of continuous variables such as slope and aspect, they found that the risk of disease increased with increasing steepness of slope.

52.5.1.5 Elevation

In the landmark study of canines as Lyme disease sentinels—or individuals that are potentially susceptible to an infection or infestation that are being monitored for the appearance or recurrence of the causative pathogen according to *Merriam-Webster's Medical Dictionary*—

they found that canine seropositivity is positively correlated with increased elevation (Guerra et al. 2001). Elevation was also a statistically significant environmental factor in numerous studies (Schulze et al. 1984 in Glass et al. 1994) and some (Furlanello et al. 1997, 2002) even go so far as to say that it is the primary variable.

The two Furlanello references found that the presence of ticks rapidly changed at 1,100 m (328 feet). Furlanello et al. (1997) said that tick presence drastically decreased below 1,100 m (328 feet) while Furlanello et al. (2002) found that ticks were likely to be present above 1,124 m (406 feet) only in grass highlands with deer, and also likely to be present below 1,124 m everywhere except in the medium-high mountain vegetation and where deer are absent. These conflicting reports are interesting in that they were reported by the same group of researchers. The latter study is more detailed, and so it shows that elevation is a factor but other variables, such as deer, must be present in order for ticks to be present. Glass et al. (1995) found that risk increased with altitude, but decreased at the highest altitude. However, Guerra et al. (2002) did not find elevation to be an important discriminator in the model.

Elevation also seems to play a particular role in the north-central United States, especially Wisconsin and Illinois, where higher elevations tend to be found in places that were once covered by glaciers, and associations between glacial outwash and tick endemicity have been suggested (Guerra et al. 2001). Not only is there a significant association between seropositivity and increasing elevation, but these observations would also lend credence to the hypothesis that southern Illinois is more than suitable for tick populations (Guerra et al. 2001; Cortinas et al. 2002).

52.5.2 Lyme Disease Hosts

While Lyme disease risk is most closely associated with deer tick habitat suitability, it also requires white-footed mice and white-tailed deer to flourish. Lyme disease risk factors include vertebrate host populations, the levels of pathogens in those populations, and the characteristics of the habitat (Riehle and Paskewitz 1996). The availability of the host in time and space is an important determinant of tick bionomics, or the relationship between the ticks and their environment (Cortinas et al. 2002). The geographic range of the pathogen *Borrelia burgdorferi* is largely determined by host-related factors (Cortinas et al. 2002). Synchronized host-vector encounters are also necessary to complete the Lyme disease life cycle, although this may be more crucial in Europe than in the United States (Cortinas et al. 2002).

Cortinas et al. (2002) suggest that in southern Illinois, host-related factors are dissimilar to northern Illinois, and the inclusion of this parameter to the model may result in different risk probabilities, even in areas with similar habitat profiles. There are more hosts in southern Illinois than northern Illinois, so this could mean that this region is more suitable than previously believed.

52.5.2.1 White-footed Mouse (Reservoir) Presence

In a study of small mammals in Illinois parks with varying degrees of human modification, Mahan and O'Connell (2005) found that the most abundant and most widely distributed species was the white-footed mouse. While other mammals (or birds) may play a role as host for the larval stage of tick (Kitron et al. 1991b), mice are the most competent and prevalent. Mice have wide habitat tolerances, and can occur in pristine forests as well as degraded woodlots (Krohne and Hoch 1999). Nymphs are carried by mice (with their high infection rates) and transferred to vegetation where humans traverse, making them the most important stage for human infection (Ginsberg et al. 2004).

52.5.2.2 White-footed Mouse (Reservoir) Competence

Reservoir competence is characterized by three components: susceptibility of the host to infection when bitten by an infected vector; the ability of the pathogen to magnify and persist in the host; and the efficiency of the host at transmitting the spirochete to feeding vectors (Richter et al. 2000 as referenced in LoGiudice et al. 2003). Even if vector densities are high, disease risk to humans can still be low if vector infection prevalence is low (LoGiudice et al. 2003). Because larval ticks that feed to repletion from white-footed mice, *Peromyscus leucopus*, have a high (up to 90%) probability of becoming infected with Borrelia burgdorferi, the distribution of ticks on this host is an important determinant of Lyme disease risk to humans (Shaw et al. 2003; LoGiudice et al. 2003).

Mice tend to use larvae-infested habitats much more thoroughly than other less-competent rodent species, and therefore attract more questing larvae and nymphs, which would lead to a dramatically higher initial infection rate. However, mice are also effective groomers, so many ticks are bitten off (Shaw et al. 2003). Kitron et al. (1991b) found that rather then encountering clumps of larvae, mice encountered single larvae in a nonrandom way; only single larvae successfully attach to mice following these encounters.

Lyme prevalence is related to the proportion of nymphs infected and the density of infected nymphs (Ostfeld and Keesing 2000 as reviewed by Ostfeld et. al 2002). Nymphal infection prevalence is expected to increase with the increasing density of white-footed mice, since decreasing species diversity in the host community reduces the availability of incompetent reservoir hosts, as the ticks that feed on them have a low likelihood of getting *Borreliosis* from incompetent hosts (Ostfeld and Keesing 2000 in Ostfeld et. al 2002).

As far as range expansion is concerned, as hosts with small home ranges, mice can limit range expansion if they divert a sufficient number of ticks from feeding on more mobile hosts (Madhav et al. 2004). Obviously, mice are not the primary modes by which ticks expand their geographic range; this role is fulfilled by the white-tailed deer.

52.5.2.3 White-tailed Deer Presence

Deer were almost completely eradicated from Illinois in the early part of this century, and their numbers have been growing ever since (Bouseman et al. 1990). In an interview about blacklegged tick presence, Roberto Cortinas stated, "Deer used to be a limiting factor when they were rather sparse in Illinois, but now deer are abundant statewide" (Perea 2002, p. 1). Overpopulation of deer and expansion of their habitat into suburban areas have resulted in increased interactions between deer and humans (Bouseman et al. 1990).

Deer are crucial for the existence of *Ixodes scapularis*. They are their prime reproductive host, as adults attach to deer in the fall before they are ready to lay eggs (Steere et al. 2004; Allan et al. 2003). Since deer ticks are dependent on white-tailed deer as reproductive hosts, the density of the larvae is correlated to deer abundance (Allan et al. 2003). Kitron et al. (1991a) found that infested deer were clustered around an endemic focus of *Ixodes scapularis*, unlike uninfested deer, which had no such clustering.

Deer abundance alone does not determine nymphal distribution but certainly contributes to it (Ginsberg et al. 2004). Others say that the spread of deer ticks in Illinois *may* be associated with deer dispersal (Bouseman et al. 1990; Kitron et al. 1992). As hosts with high tick burdens and large ranges, deer play a critical role in the range expansion of ticks, and not surprisingly, the geographical distribution of the deer tick is strongly associated with deer presence (Kitron et al. 1992; Wilson et al. 1985; Glass et al. 1992; Madhav et al. 2004).

52.6 Model Testing

The next step beyond habitat modeling with environmental factors is testing the robustness of their models. Some researchers, like Guerra et al. (2002), will create a model and then test it by sampling in areas that are predicted to be highly suitable and areas that are highly unsuitable to see if the model was able to accurately predict tick presence. Most researchers, however, collect ticks first and then see what the common characteristics of areas with high tick densities are. Ticks are sampled a number of ways.

52.6.1 Historical and Medical

Dennis et al. (1998), among others, used historical data found in the literature to determine tick presence. Cortinas et al. (2002) and Fish and Howard (1999) used human case data to see if ticks were present in a certain locale. The drawbacks to these methods are the fact that tick surveys are done very unevenly, with no consistency whatsoever over time or over space. An area that was not tested for ticks might have ticks and it wouldn't be noted in the literature. Literature is helpful to see if ticks have been found at all in the past in an area, but can not be used to definitively determine tick presence.

Human case data can be helpful if the patient history is also included, especially travel to other parts of the state or country where they could have been exposed to ticks (Cortinas et al. 2002; Fish and Howard 1999). However, the locations, if provided, are home addresses of the patients, which may or may not be where the patient was exposed to ticks. Also, the patient locations are considered confidential, so that data is not available anyway. Furthermore, if people acquire Lyme disease in an area where ticks are not known to be present, doctors may have a difficult time diagnosing it as Lyme disease, and the number of cases would probably be artificially low.

52.6.2 Community Methods

Dennis et al. (1998) received tick submissions from concerned citizens and used those as part of his methodology to see where ticks are found. However, mapping of ticks submitted for identification is subject to certain biases, which limits its utility for predicting human risk (Stone et al. 2005). Submission rates can vary depending on population, education, and local concern, and results show little about disease transmission, particularly in disease-emergent areas where infection rates may lag behind tick distribution (Stone et al. 2005).

Questionnaires are another way to determine a community's exposure to ticks and Lyme disease. Daniels et al. (2000) sent questionnaires to 1,000 random residences. His 32% response rate provided data on tick bite incidence and history of tick-borne illness for 405 properties. In our study, we are doing a more focused, albeit nonrandom, sampling effort to target people who may be at risk for Lyme disease. From knowledge gained from members of the community, we will verify our tick locations.

52.6.3 Tick Dragging

Drag sampling, or dragging strips of cloth through vegetation to gather questing ticks, is used in many studies to expand geographical knowledge of tick locations (Bouseman et al. 1990; Dennis et al. 1998; Cortinas et al. 2002; Riehle and Paskewitz 1996; Frank et al. 1998; Jones and Kitron 1999; Lindgren et al. 2000; Falco and Fish 1989; Allan et al. 2003; Kitron et al. 1991b; Nicholson and Mather 1996; Furlanello et al. 2002; Ginsberg et al. 2004; Guerra et al. 2002). Dragging and flagging are very similar; the only difference being that in dragging, the cloth remains on the ground, while with flagging, the cloth is secured to a long stick and waved like a flag over vegetation. The differences are subtle and often the terms are used interchangeably. June and July are prime months to drag because nymphal *Ixodes scapularis*, which are responsible for most human infections, are the most abundant, active, and aggressive during these months (Cortinas et al. 2002).

Drag sampling is a very effective means of gathering ticks and assessing population abundance. According to Falco and Fish (1989), drag sampling is better than animal trapping or CO_2 -baited traps, since one can reach the greatest area per unit of effort with drag sampling. Furthermore, since ticks are seeking hosts, and a piece of flannel cloth nicely simulates a human that would brush up against a questing tick, this is a good method for finding tick populations that represent particular risk to humans (Falco and Fish 1989).

A number of studies identify tick populations in a geographic region by tick dragging. In Sweden, Lindgren et al. (2000) used a standardized cloth-dragging method to corroborate indications of a boundary in Sweden where tick densities were said to decrease dramatically. Jones and Kitron (1999) directly captured host-seeking ticks using a modified tick drag (Siegel et al. 1990 in Jones and Kitron 1999) consisting of 12 weighted flannel strips attached to a canvas body. Both sides of the 400-m section of trail were dragged for a total distance of 800 m. Drags were examined at 100-m intervals; ticks were removed and placed in 70% ethanol for later examination (Jones and Kitron 1999). J. K. Bouseman invented a specific method for tick-dragging that increased the capture of questing ticks, which involved doubled strips of cotton flannel and weighted sinkers, to help the researcher comb through the vegetation and thus increase thoroughness (1990).

Drag sampling can be a way to look for ticks in places that are not yet infested. Susan Paskewitz, Wisconsin entomologist at the College of Agricultural and Life Sciences, notes: "we know there are areas where human cases have been reported but we don't know anything about the tick populations—we've sampled but we haven't found ticks. The ticks may be very clustered, and we need a more intensive sampling to pick them up" (Riehle and Paskewitz 1996, pp. 934–935). Dragging can be an effective means to evaluate tick presence and abundance.

In a tick drag, tick populations are generally classified as "reported" if there are less than six ticks found and only one life stage identified and "established" if there are more than six ticks found or more than one life stage identified (Dennis et al. 1998). We will use the same definitions.

Tick dragging should be done when the ticks are most abundant and active. Nymphs, the ticks that most commonly bite humans, are most active in late May and early June (Fischer 1999). They begin to appear in the spring and increase in abundance until a peak is reached in early June, after which the numbers begin to decline (Frank et al. 1998). We are currently exploring this approach as a means to further validate the Lyme disease model.

52.7 Model Construction Methodology

Like Guerra et al. (2002) and Cortinas et al. (2002), we used pre-existing GIS data layers obtained from outside sources to build our model. The trustworthy sources of our layers were governmental agencies like the Illinois Department of Agriculture (IDA), the Illinois Natural History Survey (INHS), and the United States Geological Survey (USGS). The model was accomplished in three steps: (1) identification of variables to include in the model, (2) weighting, scoring/ranking, and standardizing all of the variables, and (3) processing and prepping the variables to create the final model. Most of the variables were in raster format, except for the Jackson County shape file and streams layer, which consisted of vectors. All of the GIS processing was done with ERDAS Imagine 8.7 (Leica Geosystems GIS and Mapping, LLC, Atlanta, Ga.), ArcView GIS 3.3 (ESRI, Redlands, Calif.), and ArcMap 9.0 (ESRI, Redlands, Calif.).

The pixel size used for analysis was 30 m × 30 m. Guerra et al. (2002), one of the existing models of Lyme disease risk via tick habitat suitability in the Midwest that we have heavily cited, used a resolution of 2.5 km^2. Therefore, our finer resolution is an attractive feature of this model. Most of our data was at 30 m × 30 m resolution as well, so our data allowed us this high degree of specificity.

The locations identified by subjects on our *Atlas* are at a very high resolution: each grid is 1 in^2, which covers 2,000 ft on each side, or 609.6 m (LCCW 2001). Each square is the equivalent of 371,612.15 m^2, which is approximately 20 times the size of each pixel.

52.7.1 Step 1: Variable Identification

Primarily, we identified numerous variables that were shown by the literature to affect tick habitat suitability and Lyme disease risk, as described in the model rationale. After seeing what information was available and determining what layers we could use at the right scale and in the correct coordinate systems and projections as well, we shortened our list considerably. For example, the determinant of soil type and orders may have been very helpful, and even Dr. Uriel Kitron told us that "soil may be the most important" (Kitron 2006). Unfortunately, STATSGO (where others obtained their soil information) warned against using it on small-scale projects, which ours is, so we decided against using it as a factor. We decided to use forest patch area, proximity to water, land cover, slope, aspect, and elevation for tick habitat.

52.7.2 Step 2: Weighting, Ranking, and Standardizing Variables

We chose our weights by expert opinion, which is meant to eliminate subjectivity. Keeping this in mind, we carefully selected our experts accordingly. We scrutinized our literature review many times to find the strongest papers written by the most well-regarded experts in entomology and disease biology, zoology, and epidemiology and ranked them by how many times we cited a source that they co-authored and by how relevant their work was geographically to our study area. For example, Dr. Uriel Kitron is a Co-Director at the Center for Zoonoses Research as well as an esteemed professor of Epidemiology at University of Illinois Champaign-Urbana. He was also instrumental in creating the model with Guerra et al. (2002). He also contributed to no fewer than 12 of our references. Also, we had contacted him over a year ago to ask for guidance and advice on our topic and he was very friendly and helpful. These factors made him an ideal person whose opinion we could solicit. Further uncertainty was reduced by utilizing fuzzy logic to calculate our weights (Table 52-1).

We sent out two phases of queries; one in the beginning of May and another reminder query in June, each time giving them a two-week deadline. Of our ten tick experts, six agreed, three sent their apologies but said that they were either going on vacation or were otherwise pressed for time, and one never replied. However, only three people completed the questionnaire. One problem with our tick weights is the fact that only three of our ten tick experts responded, so it may not be a representative sample. However, our top choice was one of those who responded, even though he pointed out some of the difficulty inherent in quantifying tick risk in such a dichotomous fashion.

We calculated our weights by recording pairwise comparisons in an Excel spreadsheet. We then found weights using two different methods. First, we constructed individual importance tables for each expert and normalized the values and then averaged all of the experts' normalized numbers. Summing rows and dividing by the minimum gave us the following weights (in the second column of Table 52-1). For the second method, we just averaged all of the people's findings and plugged those values into an already created table in IDRISI 32 Release 2 (Clark Labs, Worcester, Mass.). This gave us the weights using fuzzy logic inferences so they are more accurate, and we use them in the model.

Table 52-1 Deriving weights.

Factor	Using the Analytical Hierarchy Process	Using Fuzzy Logic in IDRISI 32
Forest patch area	3.580	0.2567
River/stream proximity	1.768	0.1134
Land cover	4.784	0.3045
Slope	2.501	0.0865
Aspect	2.309	0.1422
Elevation	1.000	0.0967

Notice that the weights using Analytical Hierarchy Process (AHP) were significantly larger than those using fuzzy logic. The weights from IDRISI also summed to 1.00, so each was a percentage. There are some inconsistencies between the AHP weights and the fuzzy logic weights because AHP has "Elevation" as the lowest value (thus indicating the least significant variable), yet with fuzzy logic, slope is the lowest value. The slope varies so gradually in Jackson County it was the weakest factor going into the model. Using the weights derived via fuzzy logic was the best choice.

52.7.3 Step 3: Layer Creation

First of all, we made sure that all of our layers were projected in the same geographic coordinate system (North American Datum 1983) and all had the same projection (UTM Zone 16). We made our layers beautiful with ColorBrewer (Brewer et al. 2002). Our data sets are described in detail in Table 52-2 and the specific techniques for each layer are described in more detail below.

Table 52-2 Data sets used in construction of the model's data layers.

NAME	DESCRIPTION
Jackson County Shape	This shape file came from the ESRWE data set, which included all counties in the United States.
River/stream proximity	The water features layer came from the Illinois Department of Natural Resources Landcover data from 1996. We obtained it from the Digital data set of Illinois CD-ROM, Volume 1; Illinois Geographic Information System, Springfield, Illinois, USA.
Elevation Slope Aspect	This National Elevation Dataset (NED) was developed by the US Geological Survey, EROS Data Center, in 1999. The primary initial source data are 7.5 minute elevation data for the conterminous United States.
Land Cover Forest type Forest patch area Land use around forest patch	In 1999, the US Department of Agriculture National Agricultural Statistics Service (NASS), the Illinois Department of Agriculture (IDA), and the Illinois Department of Natural Resources (IDNR) formed a cooperative interagency initiative to produce statewide land cover information on a recurring basis. All of the TM/ETM+ satellite imagery were geometrically corrected and co-registered to a transverse Mercator projection with a UTM zone 16 grid and NAD83 datum. The TM/ETM+ multispectral imagery possesses a 30 m × 30 m (98.4 ft × 98.4 ft) ground spatial resolution, which means that the resulting Land Cover of Illinois 1999–2000 Classification data is suitable for GIS and mapping applications at a scale of approximately 1:100,000 (1" = 8,333') or smaller.
Mouse predicted distribution	This came from the Illinois Natural History Survey Vertebrate Distribution Model. This habitat model was created from literature and data on the known ranges for vertebrate species from survey data records and field experiences of biologists. They took habitat descriptions and broke them down into vegetative communities and other environmental factors (e.g., elevation, riparian, edges used, wetland type, etc.), and then identified those areas within a known range to create a predicted distribution map. It is a 30 m × 30 m model that was created in 2004.

52.7.3.1 Jackson County Shape

First, we clipped this county shape file from the ESRI data set that came with ArcMap. It had a sufficiently large extent since it was within the layer of counties for the entire United States. We then set each subsequent layer to the same extent.

52.7.3.2 River/Stream Proximity

We used this county boundary layer to clip the statewide stream layer for only Jackson County. Then we created a multiple-ring buffer around the water features with distances of 15.24, 30.48, 60.96, 121.42, and 243.84 m (after converting from the units in ft given in the expert questionnaire). After that we converted these features to a raster, changed the cell size to 30 m, and reclassified it into the above five categories.

52.7.3.3 Elevation

After downloading this DEM, we clipped it to our county shape file and projected it into UTM Zone 16. Then we reclassified the categories based on the range of elevations found in Jackson County, all of which are separated by over 300 ft. We converted our ranges to meters: 94.977 m–112.776 m, 112.776 m–121.92 m, 121.92 m–152.4 m, 152.4 m–182.88 m, and 182.88 m–259.775 m. Originally the last category was "greater than 600 ft" (182.88 m) but since the highest one was 259.774 we made that our highest value.

52.7.3.4 Slope

From the elevation grid, we created the slope grid (and the aspect grid) by using Spatial Analyst surface tools in the ArcMap Toolbox and specified our output in degrees. We also manually reclassified these into five categories as specified in the literature: 0–0.5, 0.5–1, 1–1.5, 1.5–2, and >2 (Table 52-3). Unfortunately, these classes were not expressive of the variety of data. Even Dr. Kitron remarked on the survey, "Your range is again off. Slope helps drainage, but we are probably talking about the barely noticeable 2 degrees" (Kitron 2006). As with the patch sizes in Table 52-4, we had to use Jenks' Natural Breaks method. The new values have a wider range of variability but are much more representative of the slope variation within Jackson County. Fortunately, slope is one of the least important categories, so even these minor changes are not going to have a significant effect on our model.

Table 52-3 New slope ranges.

Category	Experts Ranking	Our Range (degrees)	Natural Breaks
1	1	0–0.5	0–1
2	2	0.5–1	1–5
3	3	1–1.5	5–10
4	4	1.5–2	10–16
5	4	> 2	16–49

52.7.3.5 Aspect

From the elevation grid, we created another grid by using Spatial Analyst surface tools in the ArcMap Toolbox and specified our output in degrees. We also manually reclassified these into five categories (flat, north, east, south, and west) by degrees: flat = -1–0, north = 0–90, east = 90–180, south = 180–270, and west = 270–360.

52.7.3.6 Land Cover

We downloaded the Imagine file from the Illinois Department of Agriculture. We used ERDAS Imagine 8.7 to classify land cover into categories specified by Barko et al. (2003): urban/other, cropland, grassland, coniferous, and deciduous. For the most part, the categories remained grouped with the categories that they were created under, except grassland (previously under agriculture) and coniferous (previously under forest) became their own categories. After the AOI (Area of Interest) was reclassified, we clipped it to the Jackson County shape file.

52.7.3.7 Forest Types

We used ERDAS Imagine 8.7 once again to reclassify forest into distinct types: upland and dry upland were one category, and dry/mesic upland, mesic upland, partial canopy/savannah upland, and coniferous rounded out the other four categories. Then we reclassified the AOI so we just had one category of forest (including all of the forest land cover categories: upland/dry upland, dry/mesic upland, mesic upland, partial canopy/savannah upland, and coniferous) from which we would determine patch size. We also clipped it to the Jackson County shape file.

52.7.3.8 Patch Size

We went through the same steps to reclassify all forest land cover into one type, so we would have a forest-only layer. Then we extracted by attribute for value = 1 so that the grid only contained forest and did not count the areas of other land cover found within forest as "patches." Then we used the Region Group tool to group contiguous patches, making the regions more inclusive by using eight neighbors instead of four. We looked at the attribute table and sorted in descending order to see that the largest patch was 92,307 m² and the smallest one was 1 m² since each pixel is 30 m × 30 m or 900 m². We used ArcView 3.3 to reclassify them by count. Two of our categories (20,000 m–45,000 m and 45,000 m–80,000 m) did not have any values so we could not translate our categories from the expert questionnaire directly. Instead, we used Jenks' Natural Breaks method. This gave values that were very close to the ones that we specified initially, although they are not the exact values for the ranges that we had our experts rank, which introduces some error. Dr. Uriel Kitron noted, "There is a minimum size patch to maintain transmission—it may be byond [sic] 8 hectare" (Kitron 2006) and this category remains.

Table 52-4 New patch size ranges.

Category	Experts Ranking	Our Range (in ha)	Count = m²	Natural Breaks
1	1	0 – 0.5	0 – 5,000	1 – 2,189
2	2	0.5 – 2	5,000 – 20,000	2,190 – 7,219
3	3	2 – 4.5	20,000 – 45,000	7,220 – 15,044
4	5	4.5 – 8	45,000 – 80,000	15, 045 – 19,700
5	4	> 8	80,000 – 100,000	19,701 – 92,307

52.7.3.9 Mouse Habitat

This layer was not used in the model but was used to add the reservoir dimension to the tick model. If there are no mice, there will be no Lyme disease. This layer was also obtained from the Illinois Gap Analysis Project in their section on Vertebrate Modeling. We only clipped this layer with the Jackson County shape file and increased its transparency to see how it validated or complemented the model.

52.7.4 Step 4: Model Creation

First of all, we reclassed the variables with the numerical rankings of the experts averaged and then ordered. In this final reclassification, 1 = high risk, 5 = low risk, and 6 = NoData. We then wrote the following expression with the raster calculator and created Final_Model:

```
Final_Model=[patch - patch] * 0.2567 + [stream - stream] * 0.1134
+ [landcover - landcover] * 0.3045 + [slope2 - slope2] * 0.0865 +
[aspect - aspect] * 0.1422 + [elevation - elevation] * 0.0967
```

Since NoData values were included, we had to extract this layer by the mask of Jackson County to get our model in the right shape once more. Then we changed the symbology from continuous to classified and once again used five categories with Jenks' Natural Breaks method. In order to analyze this model, we re-added the layers as they were before the final reclassification, so that we could see how much each variable affected the model geographically.

52.8 Study Design

A study protocol was developed to assess whether or not people were exposed to deer ticks, and if they were, whether or not they have developed Lyme disease. After it was submitted and approved by the Human Subjects Committee on 10 March 2006, we proceeded to recruit our sample through a series of steps. We decided to enlist subjects in four departments at Southern Illinois University (SIU) that we determined would have a tendency to work and/or spend time outdoors: Forestry, Zoology, Outdoor Recreation, and Geography. Living and working outdoors increases tick exposure and risk of Lyme disease, and we believe that those who have spent a great deal of time outdoors as part of their education might be especially knowledgeable about ticks in this region.

Although SIU is not entirely representative of Jackson County, we are accepting the bias that this sample selection entails and believe that choosing SIU as our starting point for our sample selection does not alter the general message from these findings.

Since we are merely looking for any indication of the presence of deer ticks, a completely representative sample is not required. While we cannot say that the ticks that our subjects tell us exist in specific locations within Jackson County must also exist in other parts of Jackson County, we can suggest that they exist in the aforementioned places and may, or may not, exist elsewhere. Using the locations given to us, we mapped locations determined by subjects to have deer tick presence, as well as lone star and dog tick presence (since they have similar habitat requirements). We used these gleaned locations to test the habitat suitability model.

52.8.1 Creating a Perceptual Map

When our subjects gave us their locations, and indicated their certainty about the location where the bite occurred, and certainty that it was in Jackson County (Figure 52-2), we had them show us the locations in our *Atlas* (LCCW 2001). This *Atlas* contains 37 map pages and each page has a grid with lettered (A–J) and numbered (1–13) coordinates; each grid is 1 in² and covers 2,000 ft on each side or 0.38 mi. This served a useful purpose; in many cases, people could tell us exactly where they were bitten, down to the nearest square on the grid. We recorded each page number and alphanumeric coordinate, for example 35-G7.

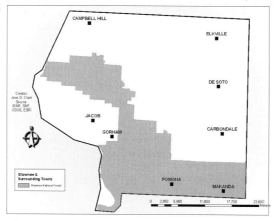

Figure 52-2 Study region, Jackson County, Illinois.

Three of the 32 people (#10, #23, and #26) who described deer tick encounters in Jackson County said that the encounter could have happened in more than one specific location, so these encounters are not mapped. If people have encounter possibilities on the same page, although not directly adjacent, we will still allow it. However, for the people who indicated multiple adjacent cells (the most was six); we included all of the possible cells. Fortunately, most could specify their location to the nearest alphanumeric coordinate; some even told us exactly where they were within the coordinate.

Another benefit of asking the participants to identify where they were with the *Atlas* is that they could also see if their locations were outside of Jackson County. A few locations were barely outside of county limits and others were still in the *Atlas* even though technically over the border, so we allowed them. We asked them to describe those encounters after we had asked all of our questions, so that we could still learn about the vegetative characteristics of the areas where they had definitely seen ticks, but still make sure that the data we gained for our quantitative analysis was based solely on experiences in Jackson County.

Since the locations that we received were both precise and accurate, as determined by the level of certainty indicated by our subjects, we decided to create layers from locations of tick presence and use them as a means to validate the GIS model. Precision was also greatly increased by the use of addresses as point locations. When we asked subjects for their addresses, we assured them that they would only be used as point locations and in no way would their address or identity be compromised by allowing us to use this information. All agreed and provided us with addresses if they were available. Our first step was to take the coordinates and draw in pencil the approximate location where they found their ticks. This location was chosen by using the following heuristics:

For coordinates that corresponded to single square grids, we placed the point in the center, unless otherwise specified (i.e., "#37: NE quadrant of 9-I1).

If multiple adjacent squares were specified, we chose the center of the group by placing our point in the center of center square.

We removed the locations with non-adjacent squares (#26, #10, and #23 all had multiple places where they could have encountered their tick), since if someone had three places where they may have picked up their tick, then two of them might not have ticks.

If an address was given in addition to the coordinate, we looked it up with Google Maps to see exactly where the location was within the specified square.

If the same address was given for "other" ticks and deer ticks, we inputted both.

For a large unspecified area (i.e., #34: Little Grassy Lake), we placed a point in the center of the square nearest the road, closest to the most recreational symbols (i.e., campgrounds; places people are likely to go).

We placed points in the center of the squares as described above, unless this point would have been underwater.

One person gave us a business name in lieu of coordinates in the *Atlas*, so we found the address of the business and plotted it directly on the *Atlas*, as well as other locations that only provided addresses and not grid coordinates, using Google Maps. In this way we could validate our subjects' accuracy concerning the location of the address on the *Atlas* grid. In only two cases were we able to determine that the subject had made a mistake and plotted it in the wrong box. In the first case, #7 gave us the grid number 29-F11, when her location was really in 29-G10. The same thing happened with #31; he said his house was in 35-E3 when in fact it was in E2. We placed them in the correct grid rather than the mistaken one given by the subject. We double checked all addresses with Google Maps, even those that had coordinates, to eliminate errors and self-reporting mistakes such as these. Even though there is uncertainty introduced by this method, the results are still useful for the purpose of validating and comparing this perceptual map to the model.

After the locations were ascertained, we used an Excel spreadsheet to compute decimal degrees from latitude and longitude. After converting this into a database file, we were able to plot these locations, as well as added attribute columns of type of tick and type of location, so we could compare these graphically.

52.9 Model Results

After we created the model with ArcMap 9.0, we looked at individual layers to see how they contributed to the final habitat suitability score (Figure 52-3). Not surprisingly, the model was strongly influenced by patch size and land cover, which were two of the layers with the heaviest weights. However, it did yield some surprises.

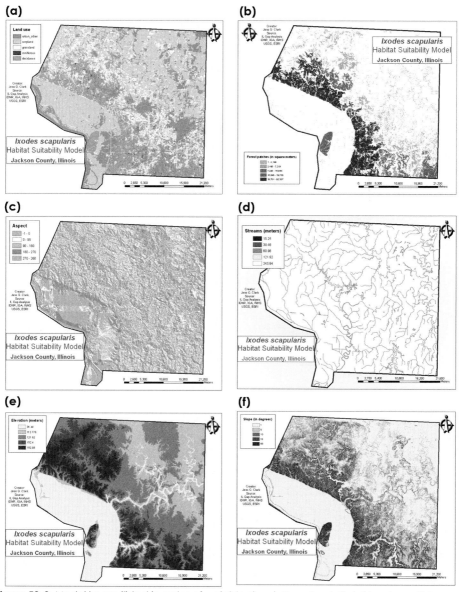

Figure 52-3 Model layers (listed in order of weights given). From top left: (a) land use, (b) patch size, (c) aspect, (d) stream proximity, (e) elevation, and (f) slope. See included DVD for color version.

We then took a closer look at some layers to see what sort of geographically pertinent information would become clear during a visual inspection of the model (Figure 52-4). Even though land cover has the heaviest weight and deciduous forest was seen to be ideal for ticks, the Shawnee Forest is only mildly suitable, with the second and third (of five) most suitable categories. The areas that are the most suitable are in a diagonal region from the southeast part of the county to the northwest part of the county. There are also red bits in the tiny patches of forest in the upper east side of the map, which also happens to be nearest the areas of lowest risk. This means that ticks must traverse from patch to patch if they are going to survive and spread. The largest contiguous areas of blue east of the Shawnee National Forest are in urban centers, and Carbondale and Murphysboro are clearly visible. However, there are still category 1 risk patches directly adjacent to category 5 risk patches, which could mean that humans and ticks can interact heavily in those small areas.

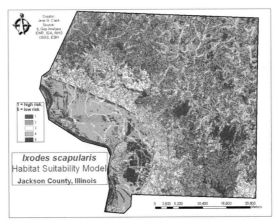

Figure 52-4 *Ixodes scapularis* Habitat Suitability Model. See included DVD for color version.

We noticed that most of the areas with higher habitat suitability are in the northwest and southeast quadrants of Jackson County. Since many of our layers were influenced by elevation (land cover, streams, slope, and aspect) we made the elevation layer more transparent and laid it over a gray-scale model, so areas with higher suitability were darker (Figure 52-5). There appear to be a few shadowy sections in the northeast quadrant of this map, but besides that, this corner appears uniformly flat and unsuitable. Of course, there are tiny suitable areas within this region but none that are clearly visible at this map size.

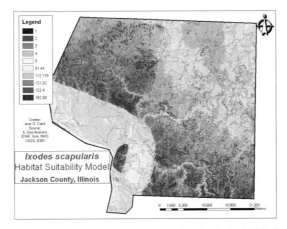

Figure 52-5 Transparent elevation overlaid on model. See included DVD for color version.

Manual of Geographic Information Systems

Figure 52-6 shows how well mouse habitat suitability matches up with the Shawnee National Forest, especially the small patches. The southeast area is much more clearly visible as being risky. We believe that this model complements the tick model well, since both mice and ticks are necessary in the cycle of Lyme disease transmission, although the mouse ranges are much higher.

There is a noticeable line of demarcation that runs parallel to the western boundary of the county—the floodplain. This area has low tick habitat suitability for numerous reasons, including flat aspect, low elevations, some urban settlements, longer distances from streams, and cropland, which is considered the worst land cover for ticks. Even though Guerra et al. (2002), Bouseman et al. (1990) and others have hypothesized that ticks are following river corridors in a southerly direction down the state, this seems like difficult terrain to traverse since the habitat is so unsuitable except for slivers right next to the river on the far side.

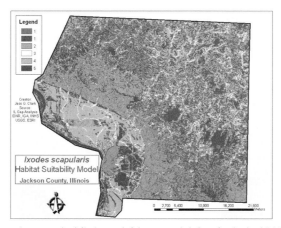

Figure 52-6 Transparent mouse habitat overlaid on model. See included DVD for color version.

52.9.1 Where Ticks Were Found

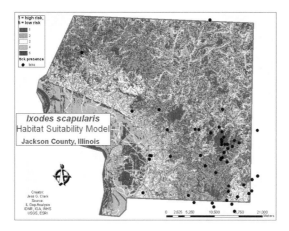

Figure 52-7 Tick locations from perceptual sightings map. See included DVD for color version.

This model appears to be robust; the perceptual sightings map coincides well with our model for predicted habitat (Figure 52-7). The southeast corner definitely has the bulk of the ticks, but Carbondale is also in that corner, so using SIU as our sample site may have caused an overrepresentation of this area. However, we are not using the perceptual sightings map to see where ticks are not; only for places where they have been spotted. From our narratives we learned not only locations of deer ticks, but also locations of dog ticks and lone star ticks as well. There is not a clear pattern for dog ticks or lone star ticks, but deer ticks seem to have a few distinct clusters (see Figure 52-8).

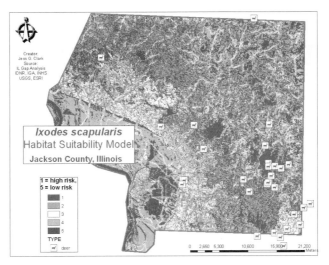

Figure 52-8 Deer tick locations. See included DVD for color version.

There is a cluster of ticks in this blue area (the city of Carbondale), which would initially appear as though our model showed ticks in areas where habitat is supposed to be poor. However, when we zoom in further (Figure 52-9), we can see that for the most part, the ticks are in fact found in locations that are highly suitable. The reason that the blue portion is so unsuitable is because urban land cover is not particularly tick-friendly. However, some patches are extremely suitable for deer ticks, even within this urban matrix. This poses a greater risk to humans because they do not have to go far from their front doors to encounter a deer tick.

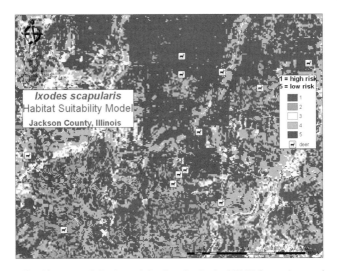

Figure 52-9 Magnified image of Carbondale. See included DVD for color version.

52.9.2 Characteristics of Locations

Locations were coded into three distinct categories (Figure 52-10). Category 3 is "recreation," which includes Lakes (Cedar Lake, Campus Lake, etc.), state parks (Giant City State Park), and other known recreational sites (such as Little Grand Canyon), determined by the frequency of ticks found there. Category 2 is "home," which is categorized as a private residence belonging to the person, a relative, friend, or even acquaintance. These were very convenient to map because most of the people who gave us these locations also gave us their addresses. The final category, "other" (coded as 1), was truly miscellaneous, and included places such as outside Morris Library, near the rugby fields, woods near Southern Hills, animal clinics, as well as secret mushroom hunting and fossil hunting spots.

In regards to the latter two categories, we repeatedly assured the subject who was telling us about his morel mushroom spots that we greatly appreciated his trusting us with valuable information. We told him that the location would not be identifiable in any way and that the only person seeing it would be our advisor and me. He said he appreciated that. He also said he was leaving Carbondale in May so it was not too big of a deal if someone found out, since he would not be able to go there anymore. We believe he told us the truth about his locations because we assured him of his privacy. Kitron et al. (1992) noted that inaccuracies may result in data gleaned from hunter reports, since they could either inadvertently make a reporting mistake or could be reluctant to disclose a favorite hunting spot. However, we believe that we were able to minimize this uncertainty by double checking addresses and by handling subjects' concerns about private locations.

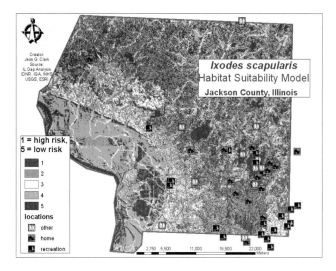

Figure 52-10 Locations of tick encounters (all tick types). See included DVD for color version.

It is clear that recreation locations are highly aggregated in this aforementioned lower limit of the *Atlas* and the home locations are in almost a circular formation around Carbondale. Even though the subjects were made to limit their encounters to those found within Jackson County borders, we realized that two locations that were still found on coordinates in the *Atlas* were technically on the other side of the county boundary line. However, the map makes it look as though many more are over the border.

Considering that this county shape is taken from ESRI's map of counties in the United States, it has an extremely large extent, but is limited in that the borders might be a little bit off. However, the decimal degrees for each location were painstakingly calculated from the *Atlas*, which is of a much larger scale than the US counties map, so we are more inclined to believe that the locations from the *Atlas* map are both more accurate and more precise. The furthest one from the border, according to the map, is a bit over 2,000 m off. That location is actually only 2,000 ft (or 600 m) from the border. Furthermore, this location was given to us by grid number and so we chose a central point, so the actual location could have been even closer. The model is more fully validated in the following section.

52.10 Analysis of Findings

Our model demonstrates that the habitat suitability variables listed in the literature review were accurate predictors of deer tick habitat. The model reinforces what is already discussed in the literature, that patch size (Allan et al. 2003; Duffy et al. 1994; Glass et al. 1995), river/stream proximity (Nicholson and Mather 1996; Bouseman et al. 1990; Cortinas et al. 2002), land cover (Nicholson and Mather 1996; Glass et al. 1992, 1995), slope and aspect (Glass et al. 1995; Ginsberg et al. 2004), and elevation (Nicholson and Mather 1996; Guerra et al. 2002; Glass et al. 1992, 1995) are all necessary components of tick habitat. Common vegetative characteristics cited by our subjects included wooded landscapes, tall grasses, off-trail locations, areas near water, and places where deer also frequented. Even though they could not be measured with GIS, they help designate good areas within those indicated by the model where it would be beneficial to return to drag for ticks at a later date.

The most suitable habitat is on the fringe of the Shawnee National Forest and in tiny patches within the urban matrix. Aside from that, urban areas are not highly suitable. People who had the rash from ticks on their properties lived further from town in locations bordered by woods and grassy fields. The perception sightings map of deer ticks (Figure 52-8) shows that there is a large cluster of positively identified sites in the southeastern portion of Jackson County, and numerous people reported finding ticks in Giant City State Park, Little Grassy Lake, and the Touch of Nature Center near Crab Orchard Refuge.

It would appear as though ticks are coming to Jackson County from the southeast, due to abundant suitable habitat and recreational opportunities that abound in this region. However, the locations given were no doubt affected by the fact that all of the people we spoke with were around the university area, so in order to really test this, we would need to do a random sampling effort for a much larger number of people. Another limitation of this model is the fact that weighting is subjective.

52.11 Findings from Perceptual Map

Because 37.10% of our locations plotted were residences, our perceptual map was more precise and accurate than one that was based solely on gridded coordinates. It showed that the areas determined by our model were in fact areas where ticks have been found. Not only were ticks found in these places, but the ticks were also intimately involved with the humans that they interacted with (i.e., biting them). Since an EM is sufficient criteria for an official case of Lyme disease, we do not believe that it is overstating our results to assert that areas where people got bull's-eye rashes are areas that have an increased Lyme disease risk.

Furthermore, by mapping locations where people have had their deer-tick bite and subsequent bull's-eye rash, we can track Lyme disease and assess local risk. Of the six, we have chosen two to illustrate further with maps created of their locations. Only one person, #22, of the six people with EM had his location not coincide with the risky red portion of the model; the other five fit perfectly. #22 indicated that he had had an EM at least twice before, although his rashes did not resemble classic bull's-eyes, instead they were less well-known variants that are also considered EM. He never had any symptoms and never went to the doctor, so it is possible that his rashes were something else, even though he reported heavy tick densities around his property, as well as numerous tick bites.

This is the story of #24, who works here at SIU and has lived in Jackson County with his wife for many years. One weekend, he was exploring some newly purchased property by walking around an unmowed portion of the perimeter. Two days later, he found a tick on his back and scratched it off himself. While he was outside the time limit, albeit barely, he did not remove it successfully, which could have sealed his fate more than the delay did. Another two days passed and he (and his wife) noticed a bull's-eye on his back the size of a pie plate that was such a perfect circle, it "looked like you painted it on there." Shortly thereafter he went to the doctor, who gave him antibiotics, but did not recall any blood work ordered. When we asked him if the doctor performed any Lyme disease tests, he replied, "If he did, he didn't get back to us with the results."

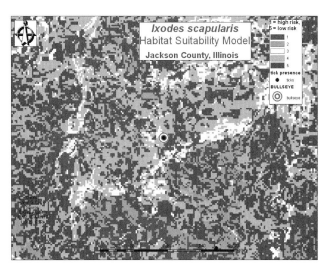

Figure 52-11 Location of #24's deer-tick bite and residence. See included DVD for color version.

Figure 52-11 is zoomed in to the scale of 1:24,000, which is the same as the scale used in the *Atlas*. Since this was a residence, we were able to map this location with a great deal of precision.

#16 is an outdoor enthusiast who has lived here all her life. A few years ago she recalled reading an article on Lyme disease in the local newspaper, where she learned what it was and that the bull's-eye rash was a symptom. She told us how fortuitous this was, since the next day she was at Green Earth II and was bitten by a lone star tick. When we asked her if the bite hurt, she replied, "Yeah, that one did. That's why we knew something was wrong." Sure enough, the next day she found a huge bull's-eye rash on her arm. Her 6 in (15.2 cm) bull's-eye was large even by conservative EM standards, which says that a rash must be at least 5 cm in order to be considered as such.

"It's one thing to have a tick; it's another to see a big bull's-eye," she told me. She went to the doctor that same day. Her assessment of EM was confirmed by the doctor who called it "textbook," and sent her off with a round of antibiotics. The weird part? She was "certain" that it was a lone star tick that bit her, not a deer tick. The location that she gave us fit the model perfectly (Figure 52-12).

Wait, what? This doesn't make sense. Lone stars are not known to be able to transmit the Lyme disease spirochete *Borrelia burgdorferi* to people, although some have speculated that it could carry the spirochete. Perhaps it is not *Borrelia burgdorferi* at all; this could potentially be a case of *Borrelia lonestari*, or STARI (Southern Tick-Associated Rash Illness), a newly discovered and little-known relative of the *Borrelia burgdorferi* spirochete that is carried and transmitted by lone star ticks. Since no serological tests were performed, which can distinguish STARI from Lyme disease, there is no way to know what she had. STARI also causes

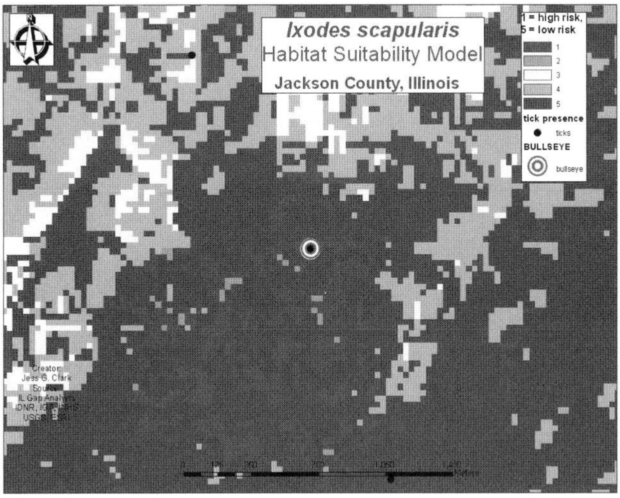

Figure 52-12 Location of #16's lone star tick bite. See included DVD for color version.

the EM, but does not cause the other symptoms of Lyme disease. However, only 1–3% of lone star ticks are infected with this spirochete, although a thorough assessment of risk of infection has not been conducted (Overstreet 2007). Since this is such a newly discovered illness, there is very little that is currently known about this disease. In the future, more knowledge about this strain of *Borrelia* could shed some light on Lyme disease risk in this region, since they are so closely related.

52.12 Implications of Findings from Perceptual Map

To the east of the aforementioned group of recreational areas (Giant City State Park, Little Grassy Lake, etc.) are the borders for Union County and Williamson County, both counties where people had many more deer-tick stories to tell, but chose to tell us about other encounters when we said that we were only quantitatively recording those that we found in Jackson County. They could still recall local (Jackson County) tick bites but said that deer ticks were much more of a problem in these neighboring counties. Since ticks do not respect political boundaries, it is not surprising that ticks would spill over into Jackson County from these other counties and find havens in suitable habitats close by, as validated by our habitat suitability map as well (see Figure 52-8).

However, we posted flyers around SIU and most of the people that we spoke with were somehow affiliated with SIU, so it is not surprising that so many are clustered around SIU's location—in the southeast quadrant of the county. It is also not surprising that people recreate close by, and why Giant City State Park, Little Grand Canyon, and Touch of Nature, to name a few, are popular places to visit. For this reason, we are not making any assumptions about where ticks are not, only where they are. We can only infer about the area near SIU campus, and at the very least, we can say with confidence that they are present in Jackson County, especially the lower southeast portion.

It is difficult to test the hypothesis that Kitron (1991a) and Cortinas (2002) espouse—that ticks are moving slowly southward down the state of Illinois. We have very little data about tick presence or absence in that area, since only two of the points are even in the upper half of the county. However, our model shows that agricultural land is the least suitable tick habitat, and we believe that the great expanse of agricultural land in central Illinois, as well as in the Mississippi River floodplain, would be a highly limiting deterrent.

Manual of Geographic Information Systems

However, there is nothing to stop *Ixodes* from coming through the south or east. *Ixodes* is common in the southeast as well (see Figure 52-13). The only reason why the southern United States is not a high-risk region is because the main disease vectors (lizards) are only marginally competent. If the ticks move north to southern Illinois, where the mice and deer are abundant, we believe that this region, specifically Jackson and the other southern counties, could be the site for the establishment of the next large Lyme disease endemic.

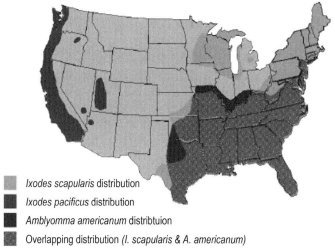

Ixodes scapularis distribution

Ixodes pacificus distribution

Amblyomma americanum distribtuion

Overlapping distribution (I. scapularis & A. americanum)

Figure 52-13 Tick distribution in the United States. Courtesy: Centers for Disease Control and Prevention. <http://www.cdc.gov/ncidod/dvrd/ehrlichia/Epidemiology/ehrldist.gif> See included DVD for color version.

52.13 Conclusions

The main strengths of this work are combining the approaches of GIS for habitat modeling and qualitative and quantitative analysis. In the process of writing this report, the *Chicago Sun-Times* published a report saying that Lyme disease cases in Illinois have increased 45% over the past year, a monumental jump (Newbart 2006). We are not surprised. It is our belief that Lyme disease is alive and well (so to speak) in Jackson County. Deer ticks may not be fully established, but they are definitely reported by our subjects. Furthermore, the places where they were found by subjects coincide with the places that the model predicted as good habitat, thus validating the model. It is crucial that we make people aware of this problem/ issue so that they can take preventative and mediating action and establish countermeasures to protect themselves against ticks. The quantitative model and interviews complement each other nicely, and create a more complete picture of Lyme disease risk in this region.

52.14 Future Directions and Recommendations

We would like to continue this research by adding factors (such as soil) to the model and testing it via tick dragging. By using similar methods to create and overlay mouse and deer models with the deer-tick model, we could learn about Lyme disease risk on an even finer scale. We would also like to test similar approaches by replicating this study in other geographical areas. In this study, the model has been successful and we anticipate finding similar success elsewhere.

Acknowledgments

This research was funded by an SIU Graduate School Fellowship. We would like to thank the experts who contributed their knowledge and expertise to the model as well as Charles Clark for his support and guidance. All research was approved by the Southern Illinois University Carbondale Human Investigation Review Board in accordance with national and institutional guidelines for the protection of human subjects.

References

Allan, B. F., F. Keesing and R. S. Ostfeld. 2003. Effect of forest fragmentation on Lyme disease risk. *Conservation Biology* 17 (1):267–272.

Anonymous. 2001. The Epidemiology of Infectious Diseases in Illinois. Illinois Department of Public Health. <http://www.idph.state.il.us/health/infect/infectdiseaserpt01.pdf> Accessed 29 September 2008.

Barko, V. A., G. A. Feldhamer, M. C. Nicholson and D. K. Davie. 2003. Urban habitat: A determinant of white-footed mouse (*Peromyscus leucopus*) abundance in Southern Illinois. *Southeastern Naturalist* 2 (3):369–376.

Bergman, M. 2005. Lyme Disease: From Ticks to Aches. <http://suntimes.healthology.com/infectious-diseases/lyme-disease/article932.htm> Accessed 9 October 2008.

Bestudik, J. 2006. Surveillance is information for action: Lyme disease. Illinois Department of Public Health. *Illinois Infectious Disease Report* 3 (2):3–5.

Bouseman, J. K. and J. A. Nelson. 1998. Lyme disease alert. *Illinois Natural History Survey Reports* 350(5).

———. 2000. Illinois River Valley ticks. *Illinois Natural History Survey Reports* 365(6).

Bouseman, J. K., U. Kitron, C. E. Kirkpatrick, J. Siegel and K. S. Todd Jr. 1990. Status of *Ixodes dammini* (Acari: Ixodidae) in Illinois. *Journal of Medical Entomology* 27 (4):556–560.

Brewer, C. A., M. Harrower and Pennsylvania State University. ColorBrewer. <http://www.personal.psu.edu/cab38/ColorBrewer/ColorBrewer_intro.html> Accessed 29 September 2008.

Buskirk, J. V. and R. S. Ostfeld. 1998. Habitat heterogeneity, dispersal, and local risk to Lyme disease. *Ecological Applications* 8 (2):365–378.

Clark, J. G. 2006. Tracking Lyme disease: A hybrid approach to assess potential locations of risk in Jackson County. Masters of Science thesis, Southern Illinois University-Carbondale, Carbondale, Illinois.

Cortinas, M. R., M. A. Guerra, C. Jones and U. Kitron. 2002. Detection, characterization, and prediction of tick-borne disease foci. *International Journal of Medical Microbiology* 291 (33):11–20.

Daniels, T. J., R. C. Falco and D. Fish. 2000. Estimating population size and drag sampling efficiency for the black-legged tick (Acari:Ixodidae). *Journal of Medical Entomology* 37:357–363.

Dennis, D. T., T. S. Nekomoto, J. C. Victor, W. S. Paul and J. Piesman. 1998. Reported distribution of *Ixodes scapularis* and *Ixodes pacificus* (Acari: Ixodidae) in the United States. *Journal of Medical Entomology* 35:629–638.

Duffy, D. C., S. R. Campbell, D. Clark, C. DiMotta and S. Gurney. 1994. *Ixodes scapularis* (Acari: Ixodidae) deer tick mesoscale populations in natural areas: Effects of deer, area, and location. *Journal of Medical Entomology* 31 (1):152–158.

Eppes, S. 2003. KidsHealth for parents. Infections, Lyme Disease. 1995–2008 The Nemours Foundation. <http://kidshealth.org/parent/infections/bacterial_viral/lyme.html> Accessed 29 September 2008.

Falco, R. C. and D. Fish. 1989. Potential for exposure to tick bites in recreational parks in a Lyme disease endemic area. *American Journal of Public Health* 79 (1):12–15.

Fischer, J. 1999. Investigating the Role of White-footed Mice in the Transmission of Lyme Disease on Fire Island, New York. The Roosevelt Wildlife Station: Conservation and Education Research. <http://www.esf.edu/ResOrg/RooseveltWildlife/Research/LymeDisease/Lymedisease.htm> Accessed 29 September 2008.

Fish, D. and C. A. Howard. 1999. Appendix methods used for creating a national Lyme disease risk map. *CDC Morbidity and Mortality Weekly* Report 48 (7):21–24.

Frank, D. H., D. Fish and F. H. Moy. 1998. Landscape features associated with Lyme disease risk in a suburban residential environment. *Landscape Ecology* 13 (1):27–36.

Frank, C., A. D. Fix, C. A. Pena and G. T. Strickland. 2002. Mapping Lyme disease incidence for diagnostic and preventive decisions, Maryland. *Emerging Infectious Diseases* 8 (4):427–429.

Furlanello, C., S. Merler and C. Chemini. 1997. Tree-based classifiers and GIS for biological risk forecasting. Pages 316–323 in *Advances in Intelligent Systems*. Edited by F. C. Morabito. Amsterdam: IOS Press.

Furlanello, C., S. Merler, S. Menegon, S. Mancuso and G. Bertiato. 2002. New WEBGIS technologies for geo-location of epidemiological data: An application for the surveillance of the risk of Lyme borreliosis disease. *Giornale Italiano di Aritmologia e Cardiostimolazione* 5 (1):241–245.

Ginsberg, H. S. 2006. Email communication, 10 May.

Ginsberg, H. S., E. Zhioua, S. Mitra, J. Fischer, P. A. Buckley, F. Verret, B. H. Underwood and F. B. Buckley. 2004. Woodland type and spatial distribution of nymphal *Ixodes scapularis* (Acari: Ixodidae). *Environmental Entomology* 33 (5):1266–1273.

Glass, G. E., J. M. Morgan III, ¬D. T. Johnson, P. M. ¬Noy, E. Israel ¬and B. S. Schwartz. 1992. Infectious disease epidemiology and GIS: A case study of Lyme disease. *Geo Info Systems* 3:65–69.

Glass, G. E., F. P. Amerasinghe, J. M. Morgan III, and T. W. Scott. 1994. Predicting *Ixodes scapularis* abundance on white-tailed deer using geographic information systems. *American Journal of Tropical Medicine and Hygiene* 51 (5):538–544.

Glass, G. E., B. S. Schwartz, J. M. Morgan III, D. T. Johnson, P. M. Noy and E. Israel. 1995. Environmental risk factors for Lyme disease identified with geographic information systems. *American Journal of Public Health* 85 (7):944–948.

Glavanakov, S. D., J. White, T. Caraco, A. Lapenis, G. R. Robinson, B. K. Szymanski and W. A. Maniatty. 2001. Lyme disease in New York State: Spatial pattern at a regional scale. *American Journal of Tropical Medicine and Hygiene* 65:538–545.

Guerra, M. A., E. D. Walker and U. Kitron. 2001. Canine surveillance system for Lyme *Borreliosis* in Wisconsin and Northern Illinois: Geographic distribution and risk factor analysis. *American Journal of Tropical Medicine and Hygiene* 65 (5):546–552.

Guerra, M. A., E. D. Walker, C. Jones, S. Paskewitz, R. M. Cortinas, A. Stancil, L. Beck, M. Bobo and U. Kitron. 2002. Predicting the risk of Lyme disease: Habitat suitability for *Ixodes scapularis* in the North Central United States. Center of Disease Control and Prevention. *Emerging Infectious Diseases* 8 (3):289–297.

Herkert, J. R., R. E. Szafoni, V. M. Kleen and J. E. Schwegman. 1993. Habitat establishment, enhancement, and management for forest and grassland birds in Illinois. Division of Natural Heritage, Illinois Department of Conservation. *Natural Heritage Technical Publication #1.* Springfield, Ill.

Illinois Department of Public Health (IDPH). 2005. Known Geographic Distribution of *Ixodes scapularis* by county in Illinois 2005, Health Beat, Lyme disease. Illinois Department of Public Health. <http://www.idph.state.il.us/envhealth/tick_dist.htm> Accessed 29 September 2008.

————. 2007. Lyme Disease in Illinois. Springfield, Ill. <http://www.idph.state.il.us/health/infect/LymeDiseaseHlthProviderInfo.pdf> Accessed 14 October 2008.

Jobe, D. A., J. A. Nelson, M. D. Adam and S. A. Martin Jr. 2007. Lyme Disease in Urban Areas, Chicago. Volume 13, Number 11–November 2007 Letter. <http://www.cdc.gov/eid/content/13/11/1799.htm> Accessed 14 October 2008.

Jones, C. J. and U. Kitron. 1999. Populations of *Ixodes Scapularis* (Acari: Ixodidae) are modulated by drought at a Lyme disease focus in Illinois. *Journal of Medical Entomology* 37 (3):408–415.

Kaye, D. 2002. Lyme disease moves south along rivers. *Clinical Infectious Diseases 34* (1):1–2.

Kitron, U. 2006. Email communication, 2 June.

Kitron, U. and J. J. Kazmierczak. 1997. Spatial analysis of the distribution of Lyme disease in Wisconsin. *American Journal of Epidemiology* 145 (6):558–566.

Kitron, U., J. K. Bouseman and C. J. Jones. 1991a. Use of the ARC/INFO GIS to study the distribution of Lyme disease ticks in an Illinois county. *Preventative Veterinary Medicine* 11:243–248.

Kitron U., C. J. Jones and J. K. Bouseman. 1991b. Spatial and temporal dispersion of immature *Ixodes dammini* on *Peromyscus leucopus* in northwestern Illinois. *Journal of Parasitology* 77 (6):945–949.

Kitron, U., C. J. Jones, J. K. Bouseman, J. A. Nelson and D. L. Baumgartner. 1992. Spatial analysis of the distribution of *Ixodes dammini* (Acari: Ixodidae) on white-tailed deer in Ogle County, Illinois. *Journal of Medical Entomology* 29 (2):259–266.

Krohne, D. T. and G. A. Hoch. 1999. Demography of *Peromyscus leucopus* populations on habitat patches: The role of dispersal. *Canadian Journal of Zoology* 77:1247–1253.

LCCW (Lick Creek Cartographic Works). 2001. *Jackson County, Illinois Road and Recreation Atlas.* Carbondale, Ill.: Lick Creek Cartographic Works.

Lindgren, E., L. Talleklint and T. Polfeldt. 2000. Impact of climatic change on the northern latitude limit and population density of the disease-transmitting European tick *Ixodes ricinus*. *Environmental Health Perspectives* 108 (2):119–123.

LoGiudice, K., R. S. Ostfeld, K. A. Schmidt and F. Keesing. 2003. The ecology of infectious disease: Effects of host diversity and community composition on Lyme disease risk. *Proceedings of the National Academy of the Sciences of the United States of America* 100 (2):567–571.

Lyme Disease. 2005. Division of Vector-Borne Infectious Diseases (DVBID), Lyme Disease. Centers for Disease Control and Prevention. <http://www.cdc.gov/ncidod/dvbid/lyme/ld_humandisease_symptoms.htm> Accessed 29 September 2008.

Madhav, N. K., J. S. Brownstein, J. I. Tsao and D. Fish. 2004. A dispersal model for the range expansion of blacklegged tick (Acari: Ixodidae). *Journal of Medical Entomology* 41 (5):842–852.

Mahan, C. G. and T. J. O'Connell. 2005. Small mammal use of suburban and urban parks in central Pennsylvania. *Northeastern Naturalist* 12 (3):307–314.

Millstein, J. 1999. Detecting sites at risk of becoming foci for Lyme disease. Funded by National Institute for Health (NIH). Applied Biomathematics. <http://www.ramas.com/lyme.htm> Accessed 29 September 2008.

Monmonier, M. 1991. *How to Lie with Maps.* Chicago: University of Chicago Press.

Nelson, T. A., K. Y. Grubb and A. Woolf. 1984. Ticks on white-tailed deer fawns from Southern Illinois. *Journal of Wildlife Diseases* 20:300–302.

Nelson, J. A., J. K. Bouseman, U. Kitron, S. M. Callister, B. Harrison, M. J. Bankowski, M. E. Peeples, B. J. Newton and J. F. Anderson. 1991. Isolation and characterization of *Borellia burgdorferi* from Illinois *Ixodes dammini*. *Journal of Clinical Microbiology* 29 (8):1732–1734.

Newbart, D. 2006. Lyme disease ticks found here: Infected insects turn up in DuPage, Cook forest preserves. *The Chicago Sun-Times.* <http://www.highbeam.com/doc/1P2-1619912. html> Accessed 9 October 2008.

Nicholson, M. C. and T. N. Mather. 1996. Methods for evaluating Lyme disease risks using geographic information systems and geospatial analysis. *Journal of Medical Entomology* 33 (5):711–720.

Ostfeld, R. S., F. Keesing, E. M. Schauber and K. A. Schmidt. 2002. The ecological context of infectious disease: Diversity, habitat fragmentation, and Lyme disease risk in North America. Pages 207–219 in *Conservation Medicine: Ecological Health in Practice.* Edited by A. Aguirre, R. S. Ostfeld, C. A. House, G. Tabor and M. Pearl. New York: Oxford University Press. <http://www.cals.wisc.edu/media/news/11_96/1196tick_survey.html> Accessed 29 September 2008.

Overstreet, M. 2007. Spirochete infections: Lyme disease and southern tick-associated rash illness. *Critical Care Nursing Clinics of North America* 19 (1):39–42.

Patnaude, M. and T. N. Mather. 2000. *Ixodes scapularis* Say (Arachnida: Acari: Ixodidae). Featured Creatures: Department of Entomology and Nematology, Publication Number: EENY–143. Copyright 2000–2008 University of Florida. <http://creatures.ifas.ufl. edu/urban/medical/deer_tick.htm> Accessed 29 September 2008.

Perea, P. J. 2002. Ticked off? Deer hunters harvest more than just bucks and does. *Outdoor Illinois* 10 (11):17. <http://www.lib.niu.edu/2002/oi021117.html> Accessed 9 October 2008.

Pesticide Education Office. 2007. Ticks, Mites, Bedbugs & Lice. Chapter 14 in Structural/ Health Related Pesticide Applicator Training Manual. University of Nebraska Lincoln. <http://pested.unl.edu/pesticide/UserFiles/File/strucchapt_14.pdf> Accessed 14 October 2008.

Pleasant, B. 2004. The lowdown on Lyme disease. *Mother Earth News* Issue 203. <http:// www.motherearthnews.com/DIY/2004_April_May/The_Lowdown_on_Lyme_Disease> Accessed 29 September 2008.

Riehle, M. and S. M. Paskewitz. 1996. *Ixodes scapularis* (Acari: Ixodidae): Status and changes in prevalence and distribution in Wisconsin between 1981 and 1994 measured by deer surveillance. *Journal of Medical Entomology* 33:933–938.

Schulze, T. L., R. C. Taylor, G. C. Taylor and E. M. Bosler. 1991. Lyme disease: A proposed ecological index to assess areas of risk in northeastern United States. *American Journal of Public Health* 81 (6):714–718.

Shaw, M. T., F. Keesing, R. McGrail and R. S. Ostfeld. 2003. Factors influencing the distribution of larval blacklegged ticks on rodent hosts. *American Journal of Tropical Medicine and Hygiene* 68 (4):447–452.

Steere, A. C., J. Coburn and L. Glickstein. 2004. The emergence of Lyme disease. *Journal of Clinical Investigations* 113 (8):1093–1101. The American Society for Clinical Investigation. <http://www.jci.org/cgi/content/full/113/8/1093?ck=nck> Accessed 9 October 2008.

Stone, E. G., E. H. Lacombe and P. W. Rand. 2005. Antibody testing and Lyme disease risk. *Emerging Infectious Diseases.* <http://www.cdc.gov/ncidod/EID/vol11no05/04-0381. htm> Accessed 29 September 2008.

Wilder, S. M. and D. B. Meikle. 2004. Prevalence of deer ticks (*Ixodes Scapularis*) on white-footed mice (*Peromyscus leucopus*) in forest fragments. *Journal of Mammalogy* 85 (5):1015–1018.

Wilson, M. L., G. H. Adler and A. Spielman. 1985. Correlation between deer abundance and that of the deer tick *Ixodes dammini* (Acari: Ixodidae). *Annals of the Entomological Society of America* 78:172–176.

CHAPTER 53

Transportation Applications of Geographic Information Systems

Jean-Claude Thill

53.1 Introduction

It has been argued very effectively by some of the most prominent geographers (e.g., Ullman 1954; Gould 1991) that transportation and the various modalities of spatial interactions constitute one of the pillars of the geographic sciences because they are instrumental in creating many of the structures that are represented as information layers in mainstream geographic information systems (GISs). In fact, pioneering GIS research and development of the 1980s rests on foundations laid by researchers who revolutionized the field of transportation with new perspectives (Goodchild 2000; Thill 2000a), such as the concepts of network, topology, and connectivity (e.g., Garrison and Marble 1962), a systems approach to transportation planning (Wilson 1974; Stopher and Meyburg 1975), analytical modeling and optimization (e.g., Rushton et al. 1973), and information systems and data models (e.g., NCHRP 1974; Schneider 1983).

The relationship between transportation and GIS has now come full circle as the community of transportation researchers and practitioners has largely embraced GIS technologies and a new paradigm that reasserts the significance of the geographic space in the operation of transportation systems, in the delivery of mobility and accessibility benefits to individuals, businesses, and public institutions, and in the production of economic, social, and environmental externalities. Today, applications of GIS to the domain of transportation and mobility systems (GIS-T) are legion as evidenced by an abundance of published literature (e.g., Zhao 1997; Lang 1999; Thill 2000b; Miller and Shaw 2001). In short, "GIS-T have 'arrived'" (Miller and Shaw 2001, p. 3).

GIS-T is more than just another domain of application of core GIS principles. Transportation and mobility systems have a number of data modeling, visualization, and computational requirements that set GIS-T apart from others. This chapter discusses a number of principles that GIS-T share and have been successfully implemented in commercial off-the-shelf computer applications. It starts with an overview of real-world problems of transportation and mobility that are handled efficiently by GIS-T approaches. A selection of fundamental concepts, functionalities, and models that support GIS-T is then investigated. Finally, some of these issues are examined more closely through a discussion of GIS-based analysis of accessibility.

53.2 GIS-T Applications

GIS-T has become one of the most active domains of socioeconomic applications of GIS. GIS-T pertains to transportation and mobility systems. The first and foremost subject matter is the layers of the built environment that enable the movement of passengers and commodities between geographically-referenced places. Fixed assets (infrastructure), including roadways, rail lines, and auxiliary components such as toll plazas, rest stops, traffic signs and markings, intermodal facilities, and others, very early on were the center of attention for several reasons. First, early land information systems and population census information systems relied on transportation lines to delineate polygons and to reference data records against the national postal addressing scheme (O'Neill and Harper 2000). More importantly for the specific development of GIS-T, though, is the fact that these facilities are an integral part of the humanized landscape of the planet that GIS aimed at cataloging and inventorying; transportation infrastructure was finally analyzed spatially in and of itself. In addition, state agencies vested with responsibilities in transportation operations and private transportation operators have had a long tradition of maintaining inventories of their infrastructure (Petzold and Freund 1990). This was traditionally accomplished through transportation information systems, but the migration to GIS occurred once specific GIS-T models and functionality had been developed (most notably dynamic segmentation) above and beyond core spatial analysis functions of GIS (overlay, buffering, geocoding, etc.).

Mobile transportation assets (vehicles) and travel or shipping activities taking place on the transportation system have become part of the growing portfolio of GIS-T applications. Today, GIS-T is a core integrative element of the metropolitan transportation planning process (Nyerges 1995), from data collection (Wolf et al. 2001) to data analysis and modeling (Choi and Kim 1996), and of various operational facets of traffic management such as congestion management systems (Quiroga 2000) and traffic signal coordination and traffic simulation (Sarasua 1994). Transit planning and operations have been one of the most active areas for GIS-T implementations (Chapleau et al. 1996; Horner and Grubesic 2001), including interactive internet-based and wireless applications (Peng and Huang 2000). Vehicle routing applications are commonplace (Sutton and Visser 2004), and deployments of intelligent transportation systems (ITS) rely increasingly on geospatial information systems (Ralston 2000; Golledge 2002). Finally, accident analysis has recently undergone significant transformation by embracing geospatial technologies (Levine et al. 1995; Steenberghen et al. 2004), especially spatial statistical methods that recognize that vehicular crashes are constrained to happen on a network space (Yamada and Thill 2007).

Finally, GIS-T has been applied to the study of impacts of transportation systems on other systems, singularly natural environment systems and land use systems. Complex GIS-based simulation models integrating the economic and behavioral processes of land-use markets and processes of travel demand and distribution across a metropolitan area are now widely available (Johnston and de la Barra 2000; Waddell 2002; Vorraa 2004). The study of the environmental impacts of transportation is primarily motivated by the desire to mitigate the severity of the effects. Thanks to GIS technologies, analysts and decision makers have on hand impact estimates that are not only more precise but also more disaggregated on the basis of neighborhoods and local communities (Alexander and Waters 2000; Bachman et al. 2000).

53.3 Essentials of GIS-T

Transportation and mobility systems have traditionally been conceived within the framework of network-based models (Haggett and Chorley 1969; Lowe and Moryadas 1975). The network is a set of points (nodes) and a set of lines (arcs) that represents connections between the points. This approach stems from the fact that many types of movements can be seen as constrained by some physical infrastructure: cars ride on streets, pedestrians walk on sidewalks, train ride of rail lines, and so on. Hence the infrastructure supporting movements forms a network endowed by some connectivity property and movement can take place on routes spanning portions of the network or its entirety.

53.3.1 The Vector Model

Neither the field view of space (raster model) nor the discrete view (vector model), which have framed GIS since their inception, can adequately support a broad class of network models (Goodchild 2000). The node-arc-area (NAA) model of vector GIS databases assumes planarity (two-dimensional embedding), whereby nodes mark the intersection of all lines in the network, while real-world transportation networks are commonly non-planar. For instance, overpasses, underpasses, and tunnels are topological features of modern highway networks, just as air travel routes evolve at different elevations. The US Bureau of the Census has made use of the NAA model for its Topologically Integrated Geographic Encoding and Referencing (TIGER) files (Marx 1986). Instead of being a liability, planarity of the TIGER files was essential to maintain topological consistency across layers of geographic features, as well as to provide block-based address attribution, a basic functional requirement of the TIGER system.

53.3.2 The Network Model

Given the shortcomings of the NAA model, real-world connectivity situations encountered in transportation are handled by GIS network models in different ways. Some models, such as Geographic Data Files (GDF) and UNETRANS, extend the basic NAA model by relaxing the planar requirement. The European GDF format (CEN 1995) uses a feature-based model with non-planar network components built from primitive elements collected in a planar graph (reference network). The recent UNETRANS object-oriented data model (Curtin et al. 2003) incorporates similar considerations through a relational data structure.

A simpler, yet rather inefficient, alternative consists in having a z-attribute in the reference network to represent the presence or absence of network connectivity where links visually intersect. Another "fix" could entail the coding of allowed movements about nodes of the reference network so that lack of connectivity between two edges would be represented by a forbidden move. Relational turn tables are simple devices to implement this solution (Goodchild 1998). Full-blown three-dimensional (3D) network models have also been proposed. A notable effort in this direction is Lee's (2001) topological Node-Relation Structure (NRS), which uses Poincaré duality to simplify the complex spatial relationships between entities in a 3D space. The NRS is also effective in modeling the interior space of buildings.

When it is applied to transportation and mobility analysis, the NAA model suffers from other crippling limitations inherited from representing the network space by a set of discrete entities (nodes and arcs) referenced to the surface of the earth through a coordinate system. Arcs and nodes are assumed to be homogeneous and the spatial resolution of information or pertinent events pinned to the network space cannot be changed to anything other than the native resolution of network entities. Often times this places excessive restrictions on the functionality of a GIS-T to model real-world transportation conditions. In traffic analysis, traffic speed is bound to vary markedly along the length of a particular highway segment. Car crashes, speed zones, and traffic signs are only a few examples of events that need to be located down to a granularity finer than the node or the arc. Through dynamic segmentation of network elements, zero- and one-dimensional events can be directly linearly referenced to the elements of the one-dimensional network space such as routes, streets, or alignments (Nyerges 1990). UNETRANS, GDF, and other transportation data models support this functionality.

With origins that can be traced back to the early years of state highway systems, linear referencing is sometimes regarded as an out-of-date information technology (Bespalko et al. 1996). Quite the contrary, however, new enterprise GIS-T data models (e.g., Vonderohe et al. 1997; Scarponcini 1999; Dueker and Butler 2000) that support multiple one- and two-dimensional location referencing systems have given a second life to linear referencing by offering a framework to handle large legacy databases and data sharing between agencies and GIS data users operating under different standards. The benefits of linear referencing have also been recognized in recent technological leaps in intelligent transportation systems (ITS). Coupled with an ITS datum, the linear referencing of vehicle trajectories is instrumental to information consistency and exchange in on-board navigation systems. Several standards for the transmission of location references of static or dynamic objects among different ITS components (motorists and traffic management center, emergency management services and traffic management center, for instance) also incorporate linear referencing expressions (Noronha and Goodchild 2000).

Transportation network models are compatible with multiple representations of transportation assets and activities. Touring routes and bus routes are logical entities that can be represented as route systems of different feature classes related to the topological network through dynamic segmentation (Miller et al. 1995; Curtin et al. 2003). The same principles can be put to use in representing activity-based travel survey data in a fashion that preserves the tri-fold indexing of paths with respect to time, location, and the person (Wang and Cheng 2001). Many GIS-T applications also incorporate carriageway- and lane-based representations (Goodchild 1998) as substitutes for, or in addition to, the common representation of transportation assets by their centerline. For instance, facility management operations need a detailed inventory of infrastructures that need to be maintained, along with their design characteristics and current condition. A requirement of on-board navigational systems is to provide to motorists precise and reliable turn-by-turn instructions on maneuvers to be operated (for instance, "bear to the left into a turning lane to prepare for a left turn") and information on the layout of the roadway to be traversed. Finally, multiple representations may be justified to dissociate or duplicate data layers on the basis of the different functionalities required for the application. In the transportation domain, it is not uncommon to dissociate the data layers used for graphical visualization at one or multiple scales from the layers used in computationally intensive procedures such as dynamic routing and traffic assignment (Frank et al. 2000).

53.3.3 The Raster Model

Raster models are fundamentally a regular or irregular tessellation of the plane. The fundamental unit of regular raster models is a cell of given size, usually square by convenience, but possibly also triangular or hexagonal. Space can also be partitioned into non-uniform areas: the triangulated irregular network (TIN) is the most common irregular tessellation. In the context of transportation applications, the value assigned to each cell is some measure of the "cost" (or impedance) incurred while traversing the cell. If the impedance is a measure dependent on direction, such as the slope of the topography, there is no simple way to assign an impedance value to a cell. Bell et al. (2002) proposed an algorithm that uses slope and aspect data in conjunction with the direction of the feature to model to dynamically assign impedance values to raster cells.

Like for the network model, the impedance variable can reflect a very wide range of considerations, from the approximate mileage involved in traversing a cell, to the monetary cost, the effort, or the time associated to this movement. Distance-based analyses proceed by cumulating the impedance as one moves from one cell to adjacent cells within the tessellation. Algorithms have been developed for this purpose in two-dimensional (Eastman 1989; de Smith 2004) and three-dimensional raster spaces (Scott 1994).

Owing to its extreme simplicity, the raster model has the advantage of being easier and faster to set up and process than the network model. This model can be used to represent movements and trajectories with a couple rather simple spatial operations: overlaying a grid over the infrastructure or track layer and turning "on" the cells that intersect with the infrastructure or track features. In any case, the result depends on the granularity of the model (size of the cells). The raster model has been used to represent the movements of various classes of agents, including pedestrians (Jiang 1999; Batty 2001), wildlife, unmanned aerial vehicles and other robots (Casas et al. 2007).

In a raster model, precision can be increased by reducing the cell size. Scalability is accomplished by adapting the cell size to the size of the geographic frame of reference and to the granularity to be captured by the model. Pedestrian raster models typically cover a small area, but aim at tracking movements down to a couple feet. Hence, cells should be on the order of a one-to-two feet resolution. On the other hand, in case grid cells are rather coarse, no further information on the placement of the trajectory is gained if the movement that is modeled is constrained to occur on some infrastructure (cars on the highway). In fact, the rasterization process may be superfluous as the tessellation incorporates a great deal of data that is unnecessary for analyzing movement.

53.3.4 Hybrid Models

Data models that combine elements of the discrete, continuous, and network models can be very handy for handling complex transportation realities, such as the interfacing between multiple modal systems and working complementarily at different scales. Figure 53-1 illustrates three different types of hybrid data models. In Figure 53-1a, a network model serves to represent transportation assets at a rather coarse scale. In this instance, the assets are facilities used by pedestrians, sidewalks, square, or indoor pedestrian mall. Within the walkable space, a fine-resolution raster model captures the micro-scale trajectory of pedestrians within a certain time period via a square mesh (Schelhorn et al. 1999). In their recent model of pedestrian travel behavior, Hoogendoorn and Bovy (2005) use the limit case of continuous-space model, where cells are infinitesimally small and continuous trajectories are treated.

Figure 53-1b depicts a network model used to represent transportation assets, while a raster model serves as an extension of the mobility systems that covers the portion of the geographic space that is not served by the fixed transportation assets. Connectors are shown to link the "off-road" portion of the system to the fixed assets. This hybrid model would be well suited for modeling intermodal urban movements involving one or more pedestrian segments and/or one or more vehicular segments. Outside urbanized areas, the featured model will effectively represent the movements of all-terrain vehicles on and off the road.

The hybrid model in Figure 53-1c is partly a network and partly continuous. The network model may be for a fixed transportation asset, such as a light rail line and assorted stations (Hsiao et al. 1997), or a helipad (Flanigan et al. 2005). The discrete model may consist of points, lines, or polygons related to the transportation features. In the featured case, circular buffers of predefined radius are drawn around stations to model each station's pedestrian catchment area. In transportation analysis, discrete models typically involve generalized features (e.g., a regular catchment area instead of a list of land parcels within an acceptable walk time from a station) well matched to the various techniques of spatial analysis that GIS is best known for.

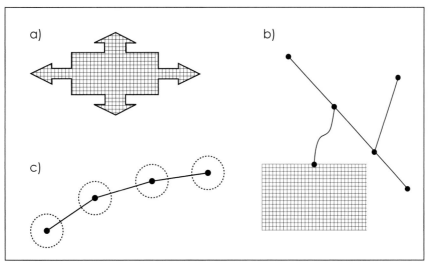

Figure 53-1 Three cases of hybrid transportation data models.

Manual of Geographic Information Systems

53.3.5 GIS-T Operations and Capabilities

The application of GIS to transportation problems relies on all the conventional functionality of GIS in terms of data management, visualization, query, and analysis. It is clear in a case where the data model that is adopted uses, for a part or the whole, a raster or discrete representation, conventional tools of spatial analysis (overlay, geocoding, buffering, etc.) are indispensable. In addition, GIS-T applications increasingly call for added functionality to handle new data sources (e.g., real-time streaming of GPS tracks, intermittent data logging from wireless distributed networks of mobile sensors), new visualization requirements (e.g., time-dependent, multi-resolution views), the proliferation of dynamic data supplied by ITS services (e.g., spatial data fusion and data mining), and new real-time operational imperatives (e.g., dynamic traffic assignment, real-time navigation assistance).

The network modeling perspective brings along a host of requirements that have made GIS-T the powerful integrative problem-solver it is today. Techniques of network analysis (e.g., connectivity measure, shortest path analysis, routing, network flow modeling, and spatial interaction modeling) and the quartet of techniques of the four-step urban transportation modeling system (trip generation, trip distribution, modal split, and traffic assignment) form the core capability of any GIS-T (Waters 1999). In short, to paraphrase Thill (2000a), GIS-T "is more than just one more domain of application of generic GIS functionality" (p. 7).

53.4 Accessibility Analysis

Accessibility is a core concept of transportation, as it captures the situational advantage garnered by a place in relation to its surroundings, given the transportation systems that are assumed to serve this location. Accessibility analysis has become a mainstay of regional economic development analysis, logistics, and transit planning, among others. Accessibility analysis can be conducted on network or raster data, or data layers in a hybrid model. This section uses this concept to illustrate issues of GIS-T implementation.

53.4.1 The Concept of Accessibility

Accessibility has been defined by Dalvi (1978) as the ease with which some land-use activity can be reached from a reference location, using a particular transportation system or a combination thereof. Accessibility is related to various concepts such as nearness, proximity, and actual or potential interactions between places or people in different locations (Ingram 1971). The term "integral accessibility" refers to the one location's accessibility with respect to a set of potential destinations (a one-to-many relationship), while "relative accessibility" is the nearness between two locations only (a one-to-one relationship). The concept of integral accessibility is used here.

53.4.2 Measuring Accessibility

Settling on a measure of accessibility is not an easy task. A vast array of indicators of accessibility has been proposed over the decades. Even though several different views on measuring accessibility have been presented, two basic elements are consistently found in definitions and measurement methods. Those two elements are the cost involved in overcoming spatial separation between sites (impedance) and the anticipated satisfaction of conducting some activity at a destination (attractiveness). Ultimately, accessibility is about securing a desired level of attractiveness by overcoming the impedance to reach the destination.

Following the established tradition, accessibility is here considered as a strict spatial construct. Some researchers also incorporate space-time travel constraints to operationalize the concept of accessibility (Miller 1991). The simplest kind of indicator of the accessibility of a place is a series of interpolated lines on a map that connects locations of equal travel time or distance away from the reference location. Such time contour lines are known as isochrones. An example of isochrones is given in Figure 53-2. O'Sullivan et al. (2000) used an isochrone approach to depict accessibility patterns on Glasgow, Scotland's, multimodal public transit system.

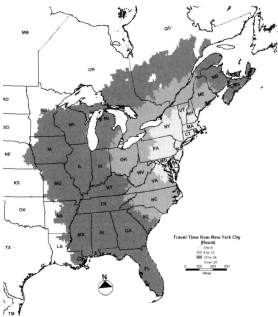

Figure 53-2 6-, 12-, and 24-hour isochrones of commercial vehicle operations out of New York City (NYC) on the North American Highway System. See included DVD for color version.

There are four main conventional approaches to quantitatively measuring the accessibility of a reference location. They are the distance approach, the topological approach, the gravity approach, and the cumulative-opportunity approach. Comparisons of different measures of accessibility (Kwan 1998; Thill and Kim 2005) have concluded that, while they are partly redundant, each indicator brings its specific perspective to the notion of accessibility that is not captured by the others.

Gravity-type accessibility measures are the most popular of all the measures. They cumulate the attractiveness of each potential travel destination after applying a discounting factor given by some function of their spatial separation from the reference location. More formally, the gravity accessibility of reference location i is given by $GR_i = \sum_j A_j f(d_{ij})$, where A_j denotes the attractiveness of location j, d_{ij} is a measure of spatial separation (distance or time) between i and j, and $f(.)$ is some travel impedance function. Several impedance functions are commonly used, including the inverse power $f(d_{ij}) = d_{ij}^{-\alpha}$, the negative exponential $\exp(-\beta d_{ij})$, and the Gaussian function $\exp(d_{ij}^{-2}/\gamma)$, where α, β, and γ are non-negative parameters. Parameterization of gravity-type accessibility measures is necessary before implementing any of these measures.

The cumulative type of accessibility measure tallies the number or degree of choices that one can reach for a certain time or distance. According to this principle, accessibility is defined by a maximum travel separation band, which is operationalized by a time or distance contour line. Potential destinations within the same isoline from the reference location are treated in one of two different ways. With a rectangular impedance function, all potential destinations within the same contour line are weighted equally. The accessibility of location i is calculated as $CUM_i = \sum_j A_j f(d_{ij})$, where $f(d_{ij}) = 1$ if $d_{ij} \leq T$, and $f(d_{ij}) = 0$ otherwise; T is a non-negative spatial separation threshold (time or distance), and A_j and d_{ij} are defined as earlier. Alternatively, a negative linear impedance function can be used to weight potential destinations in inverse relation to the spatial separation from the reference location. In this case, the accessibility measure is $CUM_i = \sum_j A_j f(d_{ij})$, where $f(d_{ij}) = 1 - d_{ij}/T$ if $d_{ij} \leq T$, and $f(d_{ij}) = 0$ otherwise.

53.4.3 Accessibility Analysis in GIS

Accessibility analysis on network data follows a fairly standard sequence of steps as presented in Figure 53-3. This schema is adapted loosely from the process articulated by Liu and Zhu (2004). Given that accessibility modeling at a large scale can be computationally taxing, the steps of this process need to be carefully thought through so as to maximize the efficiency of the approach.

The first step of the process consists in setting up the research question, getting familiar with its circumstances, selecting an accessibility indicator consistent with the objectives of the study, and specifying this indicator. A number of elements are necessary to fully specify an accessibility measure. The spatial unit for which accessibility is measured, as well as potential destinations, is a point feature (e.g., housing unit or place of business) or an area (e.g., a census tract or grid cell). In the later case, the polygon may be portrayed by a representative point, such as its centroid. The degree of spatial disaggregation of spatial units depends on the scale of the analysis, as well as on data availability. What spatial units are treated as origins (reference locations) and potential destinations is determined by the purpose of the analysis. Other specification considerations include the attractiveness indicator, the mode of travel allowed between origins and destinations (one or multiple modes), the travel impedance measure, and other numerical parameters that are either imputed or estimated. A list of data requirements is derived from the accessibility model specification process.

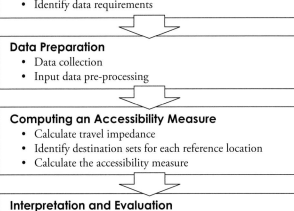

Figure 53-3 Process of accessibility analysis (adapted from Liu and Zhu 2004).

In the second step, data identified as inputs of the accessibility model are collected from primary or secondary sources. A multi-layered GIS database is designed and populated with these inputs. Often times, multiple transportation networks must be handled to reflect the multimodal nature of some (or all) of the trips involved. A commuting trip may start at home with a two-block walk to a nearby bus stop, continue with a two-mile bus ride to the city center, and end with a two-and-a-half block walk from the downtown bus station to the place of work. While the pedestrian legs of this trip could be disregarded altogether, their inclusion in the analysis increases the precision of the accessibility model, while also enabling us to look at the role of walking in framing accessibility and at impediments to gaining access to vehicle-based modes of transportation. The task of modeling an inter-modal network can be challenging and calls for special attention. In Lim and Thill (2008), the transfer of containerized freight from a rail line to a truck at a terminal is modeled via a rail-access link and a highway-access link (both of which are real), and a terminal-transfer link (which is virtual) (Figure 53 4). In this study, shipment origins and destinations are zip code centroids that are connected to the transportation networks through rail and highway connectors (Figure 53-5). All links have an impedance attribute imputed by the associated mode of shipment allowed. Various strategies involving intelligent combinations of network modeling techniques exist to model intermodal networks in GIS-T. In Lim and Thill (2008), a single network is created from the single-mode networks and assorted access, transfer, and connecting links. Miller et al. (1995) proposed another approach based on a single reference network using dynamic segmentation. New object-oriented models, such as UNETRANS, have the flexibility of not requiring the actual merger of networks, but instead rely on feature classes that control transfers between individual networks.

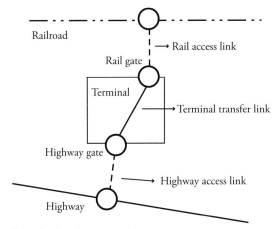

Figure 53-4 Intermodal freight network modeling.

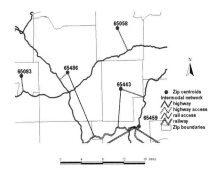

Figure 53-5 Rail and highway connectors in an intermodal freight transportation system. See included DVD for color version.

Manual of Geographic Information Systems

The third step of the accessibility analysis process (Figure 53-3) involves calculating travel impedances for all the origin-destination pairs identified in the first step. These travel impedances are typically for the "shortest path" between a given pair of locations and are computed using a shortest path algorithm, such as the famed Dijkstra algorithm. Travel impedances are usually stored in a data structure enabling ready retrieval in the form of a matrix and linking with attribute tables. The accessibility measure can now be evaluated by performing the algebraic manipulations required by the measure, as described in Section 53.4.2.

The final step focuses on visualization, interpretation, and evaluation of the results of the accessibility model. It is common practice to use interpolated choropleth maps as a means to visualize the spatial variation of accessibility over the study region. For instance, an instance of gravity-type accessibility modeling of freight distribution researched by Lim and Thill (2008) is presented in Figure 53-6. Linear interpolation was used in this case to smooth out the detail of the accessibility surface that was not of interest. Accessibility is shown to be high throughout large sections of the Midwest and the Southeast. It is rather low in the northeastern states and gradually falls from the midsection of the country to the Pacific Coast.

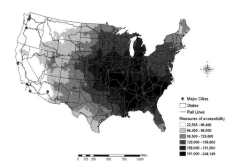

Figure 53-6 Map of containerized freight shipping accessibility (Lim and Thill 2008).

Other visualization techniques that are also very effective include three-dimensional map forms. Figure 53-7 shows examples of SUNY-Buffalo campus maps depicting the gravity accessibility of campus buildings from the vantage point of the Creek Side Village A, a student apartment complex. One map assumes that travel would be on foot, while the other uses impedances consistent with riding the campus shuttle system. The height of each destination building represents the accessibility from Creek Side Village A; it is color-coded by interval. The juxtaposition of the two maps highlights the accessibility gained by remote campus locations resulting from the operation of a shuttle service.

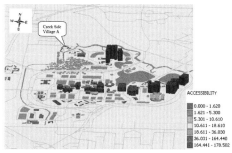 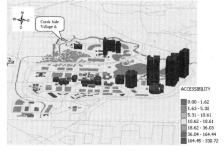

(a) Travel on Foot   (b) Travel on Shuttle Bus.

Figure 53-7 Three-dimensional maps of accessibility on foot (a) and by shuttle bus (b) on SUNY-Buffalo's North Campus from Creek Side Village A (Oh 2004). See included DVD for color version.

Figure 53-8 is another example of integral accessibility maps at SUNY-Buffalo. This time a three-dimensional floating mesh represents the surface of pedestrian accessibility to academic buildings. Great discrepancies in accessibility are easily noticeable on this map.

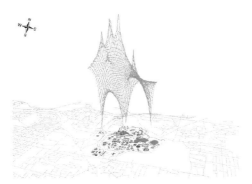

Figure 53-8 Three-dimensional mesh map of accessibility to academic buildings on SUNY-Buffalo's North Campus (Oh 2004). See included DVD for color version.

53.5 Conclusions

Transportation is one of the cornerstones of modern societies. The advent of GIS presented an opportunity for transportation researchers and practitioners to return to the roots of the transportation profession and adopt a more spatially-aware perspective in their study and operation of transportation and mobility systems. Today GIS-T is alive and well. To complement core GIS capabilities and functionalities, data models, analytical methods, and visualization techniques addressing the specific requirements of transportation applications have been developed.

GIS-T has flourished to become one of the most active domains of socioeconomic applications of GIS. Principles of GIS-T have now also found wide use in associated domains of application, particularly in logistics, business geographics, urban land-use planning, disaster management and emergency response, and location-based services. A convergence of technologies is under way.

The current trend towards the availability of more spatially-disaggregated data that is also dynamic and multispectral in nature suggests that GIS will increasingly be able to support research that aims at linking spatial-temporal processes, such as mobility of people, freight, and ideas, to structures of spatial landscapes that are newly created or amended from previous epochs. The systems approach is also noticeable in the everyday uses of geospatial information technologies. Location-based services and advanced traffic information systems are enabled by GIS-T principles to deliver information to subscribers. In turn, subscribers serve as probes gathering information that is fed back to the system for processing and dissemination. Endowed with spatial information, users may amend their behaviors in a number of different ways, hence contributing bit-by-bit to the evolution of structures of human society imprinted in the Earth's landscapes.

References

Alexander, S. M. and N. M. Waters. 2000. The effects of highway transportation corridors on wildlife: A case study of Banff National Park. *Transportation Research Part C* 8:307–320.

Bachman, W., W. Sarasua, S. Hallmark and R. Guensler. 2000. Modeling regional mobile emissions in a geographic information system framework. *Transportation Research Part C* 8:205–229.

Batty, M. 2001. Agent-based pedestrian modeling. *Environment and Planning B* 28:321–326.

Bell, T. and G. Lock. 2000. Topographic and cultural influences on walking the Ridgeway in later prehistoric times. Pages 85–100 in *Beyond the Map: Archaeology and Spatial Technologies,* edited by G. Lock. Amsterdam, The Netherlands: IOS Press.

Bell, T., A. Wilson and A. Wickham. 2002. Tracking the Samnites: Landscape and communications in the Sangro Valley, Italy. *American Journal of Archaeology* 106:169–186.

Bespalko, S., J. H. Ganter and M. D. Van Meter. 1996. Geospatial data for ITS. Pages 209–226 in *Converging Infrastructures: Intelligent Transportation and the National Information Infrastructure,* edited by L. M. Branscomb and J. Keller. Cambridge, Mass.: MIT Press.

Casas, I., A. Malik, E. M. Delmelle, M. H. Karwan and R. Batta. 2007. An automated network generation procedure for routing of unmanned aerial vehicles (UAVs) in a GIS environment. *Networks and Spatial Economics* 7 (2):153-176.

Chapleau, R., B. Allard and M. Trepanier. 1996. Transit path calculation supported by a special GIS-transit information system. *Transportation Research Record* 1521:98–110.

Choi, K. and T. J. Kim. 1996. A hybrid travel demand model with GIS and expert systems. *Computers, Environment and Urban Systems* 20 (4):247–259.

CEN (Comité Européen de Normalisation). 1995. *Geographic Data Files (GDF)*. European Standard, First Draft, Version 3.0, October 12, 1995, CEN Technical Committee 278 Road Transport and Traffic Telematics, Working Group 7.2.

Curtin, K., V. Noronha, M. Goodchild and S. Grisé. 2003. *ArcGIS Transportation Data Model (UNETRANS)*. ESRI Inc. and Regents of the University of California, Redlands, Calif. <http://www.geog.ucsb.edu/~curtin/unetrans/TransAttributes03.zip> Accessed 7 October 2006.

Dalvi, M. Q. 1978. Behavioral modeling, accessibility, mobility and need: Concepts and measurement. Pages 639–653 in *Behavioural Travel Modelling,* edited by D. A. Hensher and P. R. Stopher. London: Croom Helm.

de Smith, M. J. 2004. Distance transforms as a new tool in spatial analysis, urban planning, and GIS. *Environment and Planning B* 31:85–104.

Dueker, K. J. and J. A. Butler. 2000. A geographic information system framework for transportation data sharing. *Transportation Research Part C* 8:13–36.

Eastman, J. R. 1989. Pushbroom algorithms for calculating distances in raster grids. Pages 288–297 in *Proceedings of AutoCarto 9,* Baltimore, Md., ACSM-ASPRS.

Flanigan, M., A. Blatt, L. Lombardo, D. Mancuso, M. Miller, D. Wiles, H. Pirson, J. Hwang, J.-C. Thill and K. Majka. 2005. Air medical coverage and the correlation with reduced highway fatality rates: Use of ADAMS as a research tool. *Air Medical Journal* 24 (4):151–163.

Frank, W. C., J.-C. Thill and R. Batta. 2000. Spatial decision support system for hazardous material routing. *Transportation Research Part C: Emerging Technologies* 8:337–359.

Garrison, W. L. and D. F. Marble. 1962. The structure of transportation networks. *U. S. Army Transportation Command, Technical Report* 62-11:73–88.

Golledge, R. G. 2002. Dynamics and ITS: Behavioural responses to information available from ATIS. Pages 81–126 in *In Perpetual Motion: Travel Behavior Research Opportunities and Application Challenges,* edited by H. S. Mahmassani. New York: Pergamon.

Goodchild, M. F. 1998. Geographic information systems and disaggregate transportation modeling. *Geographical Systems* 5:19–44.

———. 2000. GIS and transportation: Status and challenges. *Geoinformatica* 4:127–139.

Gould, P. 1991. Dynamic structures of geographic space. Pages 3–30 in *Collapsing Space and Time: Geographic Aspects of Communications and Information,* edited by S. D. Brunn and T. R. Leinbach. New York: HarperCollins.

Haggett, P. and R. J. Chorley. 1969. *Network Analysis in Geography.* London: Edward Arnold.

Hoogendoorn, S. P. and P.H.L. Bovy. 2005. Pedestrian travel behavior modeling. *Networks and Spatial Economics* 5 (2):193–216.

Horner, M. W. and T. H. Grubesic. 2001. A GIS-based planning approach to locating urban rail terminals. *Transportation* 28 (10):55–77.

Hsiao, S., J. Lu, J. Sterling and M. Weatherford. 1997. Use of geographic information system for analysis of transit pedestrian access. *Transportation Research Record* 1604:50–59.

Ingram, D. R. 1971. The concept of accessibility: A search for an operational form. *Regional Studies* 5:101–107.

Jiang, B. 1999. SimPed: Simulating pedestrian flows in a virtual urban environment. *Journal of Geographic Information and Decision Analysis* 3 (1):21–30.

Johnston, R. A. and T. de la Barra. 2000. Comprehensive regional modeling for long-range planning: Linking integrated urban models and geographic information systems. *Transportation Research A* 34:124–136.

Kwan, M-P. 1998. Space-time and integral measures of individual accessibility: A comparative analysis using a point-based framework. *Geographical Analysis* 30:191–216.

Lang, L. 1999. *Transportation GIS.* Redlands, Calif.: ESRI Press.

Lee, J. 2001. 3D data model for representing topological relations of urban features. *Proceedings of the ESRI User Conference,* San Diego, Calif. ESRI Inc., Redlands, Calif. <http://gis.esri.com/library/userconf/proc01/professional/papers/pap565/p565.htm> Accessed 6 October 2006.

Lee, K. and H.-Y. Lee. 1998. A new algorithm for graph-theoretic nodal accessibility measurement. *Geographical Analysis* 30 (1):1–14.

Levine, N., K. E. Kim and L. H. Nitz. 1995. Spatial analysis of Honolulu motor vehicle crashes: I. Spatial patterns. *Accident Analysis and Prevention* 27 (5):663–674.

Lim, H. and J.-C. Thill. 2008. Intermodal freight transportation and regional accessibility in the United States. *Environment and Planning A,* 40:2006–2025.

Liu S. and X. Zhu. 2004. An integrated GIS approach to accessibility analysis. *Transactions in GIS* 8 (1):45–62.

Lowe, J. C. and S. Moryadas. 1975. *The Geography of Movement.* Prospect Heights, Ill.: Waverland Press.

Marx, R. W. 1986. The TIGER system: Automating the geographic structure of the United States census. *Government Publications Review* 13:181–201.

Miller, H. J. 1991. Modeling accessibility using space-time prism concepts within geographical information systems. *Int J of Geographical Information Systems* 5 (2):287–301.

Miller, H. J. and S.-L. Shaw. 2001. *Geographic Information Systems for Transportation. Principles and Applications.* New York: Oxford University Press.

Miller, H. J., J. D. Storm and M. Bowen. 1995. GIS design for multimodal network analysis. Pages 750–759 in *GIS/LIS'95 Proceedings,* Nashville, Tenn., 14–16 November 1995.

NCHRP. 1974. *Highway Location Reference Methods.* Synthesis of Highway Practice 21, Transportation Research Board, Washington, D. C.

Noronha, V. and M. F. Goodchild. 2000. Map accuracy and location expression in transportation–reality and prospects. *Transportation Research Part C* 8:53–69.

Nyerges, T. 1990. Locational referencing and highway segmentation in a geographic information system. *ITE Journal* 60 (3):27–31.

———. 1995. Geographic information system support for urban/regional transportation analysis. Pages 240–265 in *The Geography of Urban Transportation,* edited by S. Hanson. New York: Guilford Press.

Oh, G. 2004. *Accessibility on the University at Buffalo, SUNY Campus: A Comparison between Pedestrian and Shuttle Modes.* M.A. Project Report, University at Buffalo, Buffalo, New York.

O'Neill, W. and E. A. Harper. 2000. Implementation of linear referencing systems in GIS. Pages 79–98 in *Urban Planning and Development Applications of GIS,* edited by S. Easa and Y. Chan. Reston, Va.: American Society of Civil Engineers.

O'Sullivan, D., A. Morrison and J. Shearer. 2000. Using desktop GIS for the investigation of accessibility by public transport: An isochrone approach. *Int J of Geographical Information Science* 14:85–104.

Peng, Z.-R. and R. Huang. 2000. Design and development of interactive trip planning for web-based transit information systems. *Transportation Research Part C* 8:409–425.

Petzold, R. G. and D. M. Freund. 1990. Potential for geographic information systems in transportation planning and highway infrastructure management. *Transportation Research Record* 1261:1–9.

Quiroga, C. 2000. Performance measures and data requirements for congestion management systems. *Transportation Research Part C* 8:287–306.

Ralston, B. 2000. GIS and ITS traffic assignment: Issues in dynamic user-optimal assignments. *GeoInformatica* 4 (2):231–243.

Rushton, G., M. F. Goodchild and L. Ostresh. 1973. *Computer Programs for Location-Allocation Problems.* Monograph No. 6, Department of Geography, University of Iowa, Iowa City, Iowa.

Sarasua, W. A. 1994. A GIS-based traffic signal coordination and information management system. *Microcomputers in Civil Engineering* 9 (4):235–250.

Scarponcini, P. 1999. Generalized model for linear referencing. Pages 53–59 in *Advances in Geographic Information Systems, Proceedings of the 7th International Symposium, ACM GIS'99,* ACM, 2–6 November 1999, Kansas City, Mo.

Schelhorn, T., D. O'Sullivan and M. Thurstain-Goodwin. 1999. *STREETS: An Agent-based Pedestrian Model.* Working Paper 9, Centre for Advanced Spatial Analysis, University College London, London, UK.

Schneider, J. B. 1983. A review of a decade of applications of computer graphics software in the transportation field. *Computers, Environment and Urban Systems* 9:1–20.

Scott, M. S. 1994. The development of an optimal path algorithm in three-dimensional raster space. Pages 687–696 in *Proceedings of GIS/LIS,* ASPRS, Phoenix, Arizona, 25–27 October 1994.

Steenberghen, T., T. Dufays, I. Thomas and B. Flahaut. 2004. Intra-urban location and clustering of road accidents using GIS: A Belgian example. *Int J of Geographical Information Science* 18 (2):169–181.

Stopher, P. R. and A. H. Meyburg. 1975. *Urban Transportation Modelling and Planning.* Lexington, Ky.: D.C. Heath and Co.

Sutton, J. C. and J. Visser. 2000. The role of GIS in routing and logistics. Pages 357–374 in *Handbook of Transport Geography and Spatial Systems,* edited by D. A. Hensher, K. J. Button, K. E. Haynes and P. R. Stopher. New York: Elsevier.

Thill, J.-C. 2000a. Geographic information systems for transportation in perspective. *Transportation Research Part C* 8:3–12.

———. 2000b. *Geographic Information Systems in Transportation Research.* New York: Pergamon.

Thill, J.-C. and M. Kim. 2005. Trip making, induced travel demand, and accessibility. *Journal of Geographical Systems* 7:229–248.

Ullman, E. L. 1954. Geography as spatial interaction. *Annals of the Association of American Geographers* 44:283–294.

Vonderohe, A., C. Chou, F. Sun and T. Adams. 1997. *A Generic Data Model for Linear Referencing Systems.* Research Results Digest 218, National Cooperative Highway Research Program, Transportation Research Board, Washington, D.C.

Vorraa, T. 2004. Cube Land–integration between land use and transportation. Pages 111–117 in *Urban Transport X: Urban Transport and the Environment in the 21st Century,* edited by C. A. Brebbia and L. C. Wadhwa. Southampton, UK: WIT Press.

Waddell, P. 2002. UrbanSim: Modeling urban development for land use, transportation and environmental planning. *Journal of the American Planning Association* 68 (3):297–314.

Wang, D. and T. Cheng. 2001. A spatio-temporal data model for activity-based transport demand modeling. *Int J of Geographic Information Science* 15 (6):561–585.

Waters, N. M. 1999. Transportation GIS: GIS-T. Pages 827–844 in *Geographic Information Systems: Principles, Techniques, Management and Applications,* edited by P. Longley, M. Goodchild, D. Maguire and D. Rhind. New York: John Wiley & Sons.

Wilson, A. G. 1974. *Urban and Regional Models in Geography and Planning.* New York: John Wiley & Sons.

Wolf, J., R. Guensler and W. Bachman. 2001. Elimination of the travel diary: Experiment to derive trip purpose from global positioning system travel data. *Transportation Research Record* 1768:125–134.

Yamada, I. and J.-C. Thill. 2007. Local indicators of network-constrained clusters in spatial point patterns. *Geographical Analysis* 39 (3):268–292.

Zhao, Y. 1997. *Vehicle Location and Navigation Systems.* Boston: Artech House.

CHAPTER 54

GIS and Decision Making: The Gap Analysis Program (GAP)

Liz Kramer, Alexa J. McKerrow, Leonard G. Pearlstine, Frank J. Mazzotti, David M. Stoms and *Jill Maxwell*

54.1 An Overview of the Gap Analysis Program (GAP)

Gap analysis is a scientific method for identifying the degree to which native animal species and natural communities are represented in our present-day mix of conservation lands. Those species and communities not adequately represented in the existing network of conservation lands constitute conservation "gaps." The purpose of the Gap Analysis Program (GAP) is to provide broad geographic information on the status of all species and their habitats in order to provide land managers, planners, scientists and policy makers with the information they need to make better-informed decisions.

A gap project has four objectives:
- Map the natural land cover of a region;
- Predict the potential occurrence of the terrestrial vertebrate species;
- Produce a database of conservation lands within a defined region;
- Document the occurrence of natural communities and vertebrate species in lands managed for the long-term conservation of biodiversity.

54.1.1 Land Cover Classification and Mapping

Gap analysis relies on maps of dominant natural land cover types as the fundamental spatial component of the analysis for terrestrial environments (Scott et al. 1993). A major assumption for GAP is that vegetation patterns are integrators of the physical and chemical factors that control the environment of a landscape and thus determine the overall pattern of biodiversity of that landscape, and they can be used as a surrogate for habitat types in conservation evaluations (Specht 1975; Austin 1991). These vegetation patterns cannot be acceptably mapped from any single source of remotely sensed imagery; therefore, ancillary data, previous maps and field surveys are used. And often topographic modeling plays a key role in identifying vegetation patterns.

There is no standard methodology or mapping protocol established for gap analysis. What is standardized is the classification scheme that all programs use. The classification system used by many GAP projects is referred to as the National Vegetation Classification System (NVCS). The basic assumptions and definitions for this system have been described by Jennings (1993).

While GAP projects in the western states have had some success creating thematic maps of vegetation distributions using this classification system, its use in the east is more problematic. This is due to several factors, including the spatial complexity and tendency of communities in the east to grade into one another rather than adhere to strict boundaries. In addition, the coarse spectral and spatial resolution of the source data, Landsat TM (Thematic Mapper), as well as the paucity of adequate ancillary data such as detailed soil maps, makes classification of many alliances difficult at best. While advances in vegetation community sampling using videography, classification procedures, ancillary data sources, and image stratification may ultimately ameliorate this situation, at present accurate statewide mapping of alliances from Landsat TM data in the eastern US is probably impossible. However, adherence to a classification system compatible with more detailed classes that might be mapped in the future has many benefits. More recently, broader mapping units have been drawn that are more suitable to efforts on the scale of GAP. These mapping units are known as ecological systems (Comer et al. 2003; NatureServe 2009).

Landsat TM imagery is used as the base data layer for all GAP projects and each project develops independent methodologies for creating its vegetation classifications. Most methodologies are based upon pixel-level modeling. Often the methods are limited by the type, age and extent of coverage of ancillary data available for a region. The methodologies are described as part of each state and regional report and can be found at http://gapanalysis.nbii.gov.

54.1.2 Predicted Animal Species Distributions and Species Richness

The purpose of the GAP vertebrate species maps is to provide precise information about the current predicted distribution of individual native species within their general ranges. With this information, better estimates can be made about the actual amounts of habitat area and the nature of its configuration.

GAP maps are produced at a nominal scale of 1:100,000 or better. Gap analysis uses the predicted distributions of animal species to evaluate their conservation status relative to existing land management (Scott et al. 1993). However, the maps of species distributions may be used to answer a variety of management, planning and research questions relating to individual species or groups of species. In addition to the maps, great utility may be found in the consolidated specimen collection records and literature that are assembled into databases used to produce the maps.

Previous to this effort there were no maps available, digital or otherwise, showing the likely present-day distribution of species by habitat type across their ranges. Because of this, ordinary species (i.e., those not threatened with extinction or not managed as game animals) are generally not given sufficient consideration in land-use decisions in the context of large geographic regions or in relation to their actual habitats. Their decline because of incremental habitat loss can, and does, result in one threatened or endangered species "surprise" after another. Frequently, the records that do exist for an ordinary species are truncated by state boundaries. Simply creating a consistent spatial framework for storing, retrieving, manipulating, analyzing and updating the totality of our knowledge about the status of each animal species is one of the most necessary and basic elements for preventing further erosion of biological resources.

54.1.3 Land Stewardship

Often the first product developed for a gap analysis is a map of the conservation lands throughout the region. Assembling the conservation lands database before putting together a land cover and vertebrate distribution map offers an important advantage. Since many of the lands included in the database support native plant communities, the conservation lands database provides a sampling frame of lands which allow public access for ground-truthing both vegetation and vertebrate occurrences.

To fulfill the analytical mission of GAP, it is necessary to compare the mapped distribution of elements of biodiversity with their representation in different categories of land ownership and management. As will be explained in the Analysis section, these comparisons do not measure viability, but are a start to assessing the likelihood of future threat to a biotic element through habitat conversion—the primary cause of biodiversity decline. We use the term "stewardship" in place of "ownership" in recognition that legal ownership does not necessarily equate to the entity charged with management of the resource, and that the mix of ownership and managing entities is a complex and rapidly changing condition not suitably mapped by GAP. At the same time, it is necessary to distinguish between stewardship and protection status in that a single category of land stewardship such as a national forest may contain several degrees of management for biodiversity.

The purpose of comparing biotic distribution with stewardship is to provide a method by which land stewards can assess their relative amount of responsibility for the management of a species or plant community, and identify other stewards sharing that responsibility. This information can reveal opportunities for cooperative management of that resource, which directly supports the primary mission of GAP to provide objective, scientific information to decision makers and managers to make informed decisions regarding biodiversity. Another possibility is that a steward that has previously borne the major responsibility for managing a species may, through such analyses, identify a more equitable distribution of that responsibility. We emphasize, however, that GAP only identifies private land as a homogenous category and does not differentiate individual tracts or owners, unless the information was provided voluntarily to recognize a long-term commitment to biodiversity maintenance.

After comparison to stewardship, the next necessary step is to compare biotic occurrence to categories of management status. The purpose of this comparison is to identify the need for change in management status for the distribution of individual elements or areas containing high degrees of diversity. Such changes can be accomplished in many ways that do not affect the stewardship status. While it will eventually be desirable to identify specific management practices for each tract, and whether they are beneficial or harmful to each element, GAP currently uses a scale of 1 to 4 to denote relative degree of maintenance of biodiversity for each tract. A status of "1" denotes the highest, most permanent level of maintenance, and "4" represents the lowest level of biodiversity management, or unknown status. If a land steward institutes a program backed by legal and institutional arrangements that are intended for permanent biodiversity maintenance, we use that as the guide for assigning protection status.

The characteristics used to determine protection status are as follows:

- Permanence of protection from conversion of natural land cover to unnatural (human-induced barren, exotic-dominated, arrested succession).
- Relative amount of the tract managed for natural cover.
- Inclusiveness of the management, i.e., single feature or species versus all biota.
- Type of management and degree that it is mandated through legal and institutional arrangements.

The four status categories can generally be defined as follows (after Scott et al. 1993; Edwards et al. 1995; Crist et al. 1995):

Status 1: An area having permanent protection from conversion of natural land cover and a mandated management plan in operation to maintain a natural state within which disturbance events (of natural type, frequency, and intensity) are allowed to proceed without interference or are mimicked through management.

Status 2: An area having permanent protection from conversion of natural land cover and a mandated management plan in operation to maintain a primarily natural state, but which may receive use or management practices that degrade the quality of existing natural communities.

Status 3: An area having permanent protection from conversion of natural land cover for the majority of the area, but subject to extractive uses of either a broad, low-intensity type or localized intense type. It also confers protection to federally listed endangered and threatened species throughout the area.

Status 4: Lack of irrevocable easement or mandate to prevent conversion of natural habitat types to anthropogenic habitat types. Allows for intensive use throughout the tract. Also includes those tracts for which the existence of such restrictions or sufficient information to establish a higher status is unknown.

54.1.4 Analysis Based on Stewardship and Management Status

As described earlier the primary objective of GAP is to provide information on the distribution and status of several elements of biological diversity. This is accomplished by first producing maps of land cover, predicted distributions for selected animal species, and land stewardship and management status. Intersecting the land stewardship and management map with the distribution of the elements results in tables that summarize the area and percent of total mapped distribution of each element in different land stewardship and management categories.

Although GAP "seeks to identify habitat types and species not adequately represented in the current network of biodiversity management areas," it is unrealistic to create a standard definition of "adequate representation" for either land cover types or individual species (Noss et al. 1995). A practical solution to this problem is to report both percentages and absolute area of each vegetation type in biodiversity management areas and allow the user to determine which types are adequately represented in natural areas. There are many other factors that should be considered in such determinations such as (a) historic loss or gain in distribution, (b) nature of the spatial distribution, (c) immediate versus long term risk, and (d) degree of local adaptation among populations of the biotic elements that are worthy of individual conservation consideration. In a state such as Georgia, with a long history of human manipulation, loss of habitat from historic levels may be a particularly important issue. Such analyses are beyond the scope of this project, but we encourage their application coupled with field confirmation of the mapped distributions. As a coarse indicator of the status of the elements, we do provide a breakdown along three levels of representation (10%, 20% and 50%) that have been recommended in the literature as necessary amounts of conservation (Noss and Cooperrider 1994; Noss 1991; Odum and Odum 1972; Specht et al. 1974; Ride 1975).

Currently, land cover types and terrestrial vertebrates are the primary focus of GAP's mapping efforts. However, other components of biodiversity, such as aquatic organisms or selected groups of invertebrates may be incorporated into GAP distributional data sets. Where appropriate, GAP data may also be analyzed to identify the location of a set of areas in which most or all land cover types or species are predicted to be represented. The use of "complementarity" analysis, that is, an approach that additively identifies a selection of locations that may represent biodiversity rather than "hot spots of species richness" may prove most effective for guiding biodiversity maintenance efforts. Several quantitative techniques have been developed that facilitate this process (see Pressey et al. 1993; Williams et al. 1996; Csuti et al. 1997; Soule and Sanjayan 1998). These areas become candidates for field validation and may be incorporated into a system of areas managed for the long-term maintenance of biological diversity.

54.1.5 Conclusion

The value of GAP data sets goes well beyond the final GAP analysis. These data have been used by federal, state, and local government agencies, as well as a number of NGO's (nongovernmental organizations) for multiple projects including setting targets for land acquisition, planning green space, developing research questions etc. The following sections represent just a few of the many projects that have used and are using GAP data.

54.2 Gap Analysis Data in Decision Making: Case Studies from North Carolina

54.2.1 Introduction

In many states, the Gap Analysis Program (GAP) provided the first statewide database linking land cover and vertebrate species with a comprehensive view of conservation lands. That previously rare perspective is becoming less so as access to the data sets improves through continued work and standardization. As described in Section 54.4, conservation planning tools and the Gap Analysis Program have both evolved making tremendous strides in data specificity and the use of new technologies. Together they can serve as a framework for decision making. In the following discussion, specific case studies based on the use of the Gap Analysis Program data in North Carolina are described.

54.2.2 Southeastern Biodiversity Project

A site and landscape-oriented approach to conservation planning was taken in the Southeastern Biodiversity Project (Hall and Schafale 1999). In this project, sites with concentrations of rare plants and animals and core areas connected through corridors and buffers (landscape-oriented) were delineated to create a landscape scale network of priority conservation lands in the Southeastern Coastal Plain. Three key partners in the project included the North Carolina Natural Heritage Program, North Carolina Museum of Natural Sciences and North Carolina Gap Analysis Project (NC-GAP).

A two-stage approach was used to identify both species goals and plant community specific goals, as well as the landscape level objectives. First, sites were ranked using traditional Natural Heritage and The Nature Conservancy protocols. At that time, there were 351 documented records for the priority conservation targets in the 1.7 million hectare study area. The known sites (element occurrence records) used in the analysis included occurrences of plants (192), vertebrates (62), invertebrates (44) and plant communities or specialized habitats (53). Sixty-two of those elements were considered critically imperiled (G1) or imperiled (G2). Twenty-eight of the sites were globally significant for the elements they supported.

In addition to ranking of sites, analyses of habitat distributions throughout were completed. For this landscape-level analysis, a series of habitat guilds (a group of animals that depend on similar habitats) were defined and the element occurrence records for the species in those guilds were mapped. For species guilds with a tight linkage between the defined habitat and a specific land cover type the NC-GAP land cover map was used to map the distribution of that habitat. Concentrations of element occurrence records, mapped habitat, and existing conservation lands were then used to identify corridors and core areas that would best serve each guild. Twenty-one habitat indicator guilds were identified for the project (e.g. maritime forests and shrub, sand hills, acidic shrub lands, Atlantic white cedar forests, etc.). For each guild, a complete description was developed which included threats, distribution, and maps of the important core areas and connectors.

Finally, priorities set through this analysis were incorporated directly into the North Carolina Natural Heritage Program's conservation plan and similar assessments were initiated in other parts of the state using the same methods.

54.2.3 Gap Ecosystem Data Explorer Tool

In order to make GAP data sets more accessible, the GAP Ecosystem Data Explorer (GEDE) Tool was developed in collaboration with the Roanoke-Tar-Neuse-Cape Fear (RTNCF) Ecosystem Team of the US Fish and Wildlife Service. The goal of the project was to provide biologists access to the GAP data without a need for extensive GIS training. The GEDE

Tool was therefore built to allow for a variety of skill levels. Within the tool, the user selects an area of interest; the tool then identifies the species with ranges that intersect the study area and presents them in a list of species that can be "explored". The user can view and map species in the area of interest in either the single species mode (range, occurrence records, and predicted distribution) or multiple species mode (species richness). In the RTNCF edition of the tool, detailed soils and National Wetlands Inventory data were included at the request of the refuge biologists. Those layers, along with roads, streams, land cover, land ownership and management can all be viewed and mapped along with the GAP vertebrate distribution data.

While the tool does allow novice GIS users to quickly interact with the GAP data set, advanced users can use the data within the tool or "step out" to conduct more sophisticated analyses. Some of the applications specific to the refuges in the RTNCF Ecosystem include developing moist soils maps, mapping habitats for the comprehensive conservation plans, and using land cover and stewardship data for fire and management planning. At the ecosystem planning level, the GEDE Tool and GAP data have been used in habitat suitability analyses. At the Wildlife Habitat Management Office, the tool was used in an analysis to identify priority carbon sequestration areas.

In that analysis, the biologist was interested in mapping areas within the RTNCF Ecosystem where concentrations of potentially restorable sites were adjacent to lands that currently supported Partners in Flight (PIF) priority bird species. The GAP land cover was used to identify agricultural fields and shrub lands occurring on high organic matter soils. In parallel, a species-richness map of seven PIF species was created and concentrations of those priority birds were mapped. Finally, adjacency was used to identify the areas that, if restored, could optimize both carbon sequestration and core habitat patch size for priority birds (Figure 54-1).

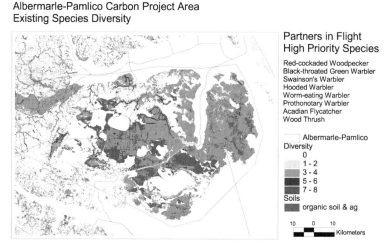

Figure 54-1 Potential carbon sequestration areas adjacent to areas with high species richness for Partners in Flight priority species. See included DVD for color version.

The first edition of the GEDE tool was developed as a desktop application, with ArcView 3.2 and the Spatial Analyst Extension. The decision to develop it for the desktop was based on the current standards within the agency, as well as the remoteness of the refuges and the general lack of high-speed connections. With the increase in high-speed access, the functionality of the GEDE Tool is being updated to a web-based platform. The Online GAP Data Explorer Tool (OGDET) will serve data from a wide variety of GAP efforts including state and regional projects.

54.2.4 North Carolina State Wildlife Action Plan

In order to continue to be eligible for funding through the State Wildlife Grants Program, each of the state wildlife agencies needed to complete a Comprehensive Wildlife Action Plan in 2005. That plan is intended to serve as a guide for the management of wildlife and research activities in the state. In many cases, GAP data were used at some point in the development of the action plans (Maxwell 2005; see also Section 54.1). In North Carolina, the land cover and species-predicted distribution data were used to help address two of the key elements required for the plans: information on the distribution and abundance of wildlife species and the location and relative condition of key habitats and community types (NCWRC 2005; IAFWA 2002).

Specifically, the land cover types mapped by NC-GAP were relabeled (cross-walked) to match the key habitat types identified by Wildlife Resource Commission biologists and used to describe and map the distribution of those habitats across the state. Those habitat types were then used as the basis for the agency-specific habitat-association matrix. In addition, species richness maps for amphibians, birds, mammals and reptiles based on the NC GAP's predicted distributions were created using the priority vertebrates named in the plan (34 amphibians, 81 birds, 34 mammals and 36 reptiles, Figure 54-2). Finally, the land management status data provided a backdrop for the discussion of conservation strategies with respect to private vs. public land.

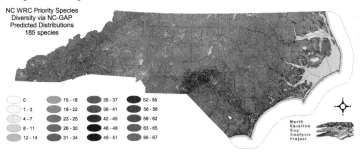

Figure 54-2 Statewide species richness for 185 species identified as priorities for management by the North Carolina Wildlife Resources Commission. See included DVD for color version.

54.2.5 Onslow Bight LANDFIRE Partnership

The North Carolina Field office of The Nature Conservancy has taken the lead on a LANDFIRE application in the Onslow Bight Region of North Carolina. That team works at the local level to identify methods for applying GIS and remotely sensed data in fire management. One of the main objectives of the project is to work with a wide array of partners to make fire management more effective with respect to restoration of the fire-dependent ecosystems and for wildfire prevention. Some of the partners involved in the project include the US Forest Service, North Carolina Division of Forestry, Camp Lejeune (US Marine Corps), and the US Fish and Wildlife Service. Within the Onslow Bight, land ownership is complex, with federal, state, and private ownership intermingled. The interaction of ownership with the complexity of the landscape makes managing fire on the landscape truly challenging.

Historically, the Coastal Plain supported vast acreages of fire-dependent ecosystems (Frost 1998; Noss 1988). Several of those ecosystems are critically endangered with a small fraction of the former acreage still intact (Noss et al. 1995; Schafale 1999). Four of those, including dry longleaf pine (*Pinus palustris*) woodlands, mesic longleaf woodlands, pond pine (*Pinus serotina*) woodlands and non-alluvial wetlands were considered in an analysis of habitat change in the bight area. To map the changes in habitat availability, a pre-settlement vegetation map (developed for the project) and the NC-GAP land cover map were used as the basis for modeling predicted species occurrences for two time periods. For each habitat, the mapped distribution in the pre-settlement map was considered the full extent for that ecosystem and used as the baseline. For example, for the areas mapped as dry longleaf pine woodlands in the pre-settlement map, the predicted distribution for the dry longleaf pine species guild was compared to the predicted distributions within the same areas based on the 1992 NC-GAP map. In the final analysis, areas predicted to be supporting less than half the species in that habitat guild were identified. Finally, the habitat change information was mapped along with the existing conservation priorities, for use by resource managers in identifying sites where fire management and restoration activities coincide.

54.3 Application of FL GAP to Florida Everglades's Restoration

FL GAP was completed in 2000 with land cover classification of 1993/1994 Landsat TM imagery (Pearlstine et al. 2002) and habitat modeling of Florida's terrestrial vertebrate species. Updates to land cover and vertebrate modeling for the Southeastern US are now in progress at the Biodiversity and Spatial Information Center (BaSIC) at North Carolina State University in partnership with the University of Georgia and Auburn University.

The Everglades are an approximately 100-km wide by 160-km long wetland prairie created by the overflow of Lake Okeechobee spreading water in a slow-moving sheet flow across a very shallow elevation change from the Lake to the mangrove-lined margins of Florida Bay and the Gulf of Mexico (Figure 54-3). Several factors combined to subject the Everglades to massive engineering projects continuing through the mid seventies to drain the marshes with 1,700 miles of levees and canals, 150 control structures and 16 pump stations: the desire to transform "wasteland" into rich agricultural land, devastating floods in 1903, 1946 and 1947, and improved water supplies for growing urban and agricultural areas (Figure 54-3). Losses of the Everglades' ecological productivity have been attributed to changes in water quantity, hydroperiods, spatial patterns of water distribution, and water quality (Davis and Ogden 1994). Development and farming of mesic and hydric forest communities adjacent to the Everglades proper may also have had a profound impact on ecological stability and wildlife habitat value by reducing nesting habitat (Robertson and Frederick 1994) and changing connectivity of wetland and upland systems (Pearlstine et al. 1997).

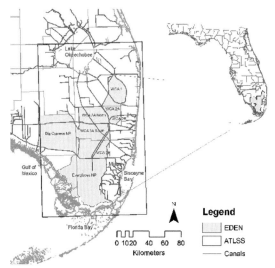

Figure 54-3 The Greater Everglades and the ATLSS and EDEN study area. See included DVD for color version.

Recognition in the eighties of the failure to maintain water quality and hydrological conditions that sustain public lands and their habitats led to settlements including the Everglades Forever Act (State of Florida 1995), the US Army Corps of Engineers Restudy (US Army Corps of Engineers 1998) and the Comprehensive Everglades Restoration Program (CERP). CERP was funded in 2000 and continues as the central multi-agency decision-making program for evaluation, implementation and monitoring of scientific research and engineering projects to restore Everglades ecological productivity while maintaining south Florida urban and agricultural water supplies and flood protection.

Two modeling efforts in south Florida have used FL GAP land cover mapping as a mechanism to produce finer scale spatial data of hydrological features across the Everglades than would have otherwise been possible. A decision tool has also been developed with FL GAP wildlife models for US Fish and Wildlife Service (USFWS) examination of multi-species project impacts.

Across Trophic Level System Simulation (ATLSS) models have been developed through a US Geological Survey (USGS)/University of Tennessee project cooperatively with many individuals. ATLSS models are being used for assessment of wildlife response to alternative hydrologic scenarios of Everglades restoration (DeAngelis et al. 1998; Curnutt et al. 2000). The models include spatially-explicit individual-based and species index models that include fishes, alligator, wading birds, Cape Sable sea side sparrow, snail kite, Florida panther and deer. Hydrological modeling of Everglades restoration alternatives is provided by the South Florida Water Management District (SFWMD) and output resolution is a 2-mile × 2-mile grid cell simulating continuous sheet flow and canalized flow of water from above Lake Okeechobee to Florida Bay (Figure 54-4). Resolution of the SFWMD model is too coarse to adequately provide input for responses of fauna to local variation in the landscape. To address this problem, ATLSS generates a high-resolution topography (HRT) and high-resolution hydroperiod by combining the 2×2-mile SFWMD hydrologic model output with FL GAP 30-m × 30-m land cover and estimated land cover community–flood durations associations.

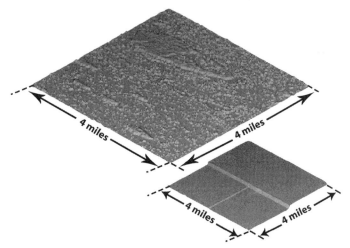

Figure 54-4 Example high-resolution hydrology (upper left) illustrating the finer scale of resolution achieved by combining FL GAP land cover with modeled 2×2-mile spatial resolution SFWMD hydrology model output (lower right). Redrawn from Duke-Sylvester (1999), used with permission.

A range of suitable hydroperiods for each land cover type in FL GAP was estimated based on literature values. Using 1986-1995 averaged stage height data from SFWMD hydrology model runs, the HRT model adjusts local ground surface elevation for each grid cell in the FL GAP map to provide that location with a hydroperiod which is appropriate for the habitat type given for the cell. Using the HRT as the new base map, high resolution hydrology was created from daily runs of alternative hydrological scenarios. Total water volume of each 2×2-mile hydrologic cell is preserved as it is redistributed across the 30×30-m HRT cells within the 2×2-mile cell. The result is an interpolation of heterogeneous water depths at a finer scale appropriate for modeling biotic response to changing hydrology (Figure 54-4).

From 1995 through 2007, the US Geological Survey has been completing high accuracy (± 15 cm vertical accuracy) elevation data (HAED) surveys positioned approximately every 400 m throughout the Everglades (Desmond 2003). As this survey work was completed, HAED has been utilized to supplement the modeling of ground elevation, reducing the areas for which the HRT method is utilized. The FL GAP continues to provide a critically needed source of information which supplements that from the HAED surveys and is still used in areas where HAED is not available. The HRT model has been expanded through the development of a High Resolution MultiData Source Topography (HMDT) which utilizes HAED, the HRT and other available data sources (Duke-Sylvester 2003). The FL GAP is also a critical input to an array of other ATLSS species models and the Vegetation Succession and Fire Models (Duke-Sylvester 2006).

The **Everglades Depth Estimation Network** (EDEN; USGS/University of Florida) provides real-time telemetered water-level data and derives other hydrologic characteristics, such as water depth, recession rates, time since last dry period, water-surface slope, and hydroperiod as continuous modeled surfaces across the Everglades landscape in daily time steps at a spatial resolution of 400 m × 400 m. A network of approximately 250 water-stage recorders is polled daily to provide data used to create a statistical surface of water stage (Palaseanu and Pearlstine 2008). Water depth is computed as the difference between modeled water stage and modeled ground elevations. A surface of ground elevations has been created by kriging of ground-surface elevation data collected by the USGS at over 50,000 points (Desmond 2003) with 400-meter resolution (Jones and Price 2007).

Data sets for validation of EDEN-modeled water depths are the depth measurements frequently taken by other investigators as part of their biological monitoring and research activities. Nearly 60,000 observations of water depth have been compiled from 35 investigators in the period 1996–2004. At issue in attempting to validate depths with these data is the scale discrepancy between point observations by field investigators and 400-m-resolution averaged depths. Depths in sloughs (deeper, water lily and open water) as well as emergent vegetation (shallower, primarily sawgrass) often occur within modeled grid cells.

An effort is now underway to adjust depths reported in each grid cell based on proportions of vegetation communities found in the cell. Being able to account for variation within grid cells enables use of field point data for validations, but more importantly, it also enhances use of EDEN-modeled depths for biological simulations (Jones 2001). FL GAP and SFWMD (Rutchey et al. 2005) vegetation classifications are being used to estimate patterns of land cover that are associated with variation in local elevation. For these analyses, land cover has been aggregated into 6 major communities: 1) slough or open water, 2) wet prairie, 3) ridge or saw grass and emergent marsh, 4) exotics and cattail, 5) upland and 6) other (mostly wetland shrub and wetland forested). Relationships between community types, USGS elevation data points and field observed water depths are being compiled and the expected output will be attributed grid cells with expected depths at each community type and proportion of the cell that is each community type. Because relationships between depth and community type for slough in particular varies in different areas of the Everglades as a result of changing geology, part of this effort will be to stratify relationships based on a cluster analysis of landscape features. The exact nature of the clustering and functional associations is still in planning.

The USFWS **South Florida Multi-Species Recovery Plan** (MSRP) was completed to help achieve the objectives of threatened and endangered species recovery and restoration and maintenance of biological diversity (US Fish and Wildlife Service 1999). Multi-species plans allow planners to consider similar threats and conflicting demands for species recovery within the same ecosystems. The MSRP identifies recovery needs of 68 threatened and endangered species and 23 natural communities in the 19 southernmost counties in Florida.

In South Florida, changes in species distribution and health are predominantly the result of hydrologic and fire-regime changes and loss and fragmentation of habitat. A landscape or ecosystem-based evaluation addresses the needs of the majority of these species simultaneously. Many of questions being asked are familiar to users of GAP products nationally: How well do existing conservation lands protect habitat for endangered species? How well do individual terrestrial vertebrate threatened and endangered species serve as "umbrella species" protecting habitats for other threatened and endangered species and all other terrestrial vertebrates? How will Everglades' restoration projects affect habitat for threatened and endangered species?

To respond to those questions, the University of Florida obtained potential habitat models for south Florida's 56 mammals, 123 reptiles and amphibians, and 194 breeding birds from the Florida Gap Project (Pearlstine et al. 2002). The Florida Gap models for 22 of 24 of the listed terrestrial vertebrates identified in the MSRP were enhanced whenever possible by updating information used in the Gap models and adding information on home range size, dispersal distance and use of adjacent habitats.

To assess the potential contribution of protection of habitat for each threatened and endangered species for overall protection of vertebrate biological diversity, species richness information for terrestrial vertebrates from the Florida GAP project (Pearlstine et al. 2002) was overlaid with maps of potential habitat for each of the threatened and endangered species. The amount of habitat outside of areas designated as conservation lands ranges from 0 for the Cape Sable Seaside Sparrow to over 90% for the Atlantic Salt Marsh Snake and Audubon's Crested Caracara. On average, about 45% of areas identified as potential habitat for threatened and endangered species is on conservation lands.

The number of threatened and endangered species that overlap some part of a particular threatened-and-endangered-species potential habitat ranges from 2 to 15. Examining the proportional area of overlap between threatened and endangered species helps maximize recovery efforts by better quantifying where protection of habitat for one species benefits other species. Combining this quantitative analysis with a map of where species overlaps occur can provide information that can be used to target specific land areas for protection. For example there is approximately 11,590 ha of potential habitat not on conservation lands where Indigo Snake and Grasshopper Sparrow overlap, primarily around Kissimmee Prairie, Lake Istokpoga, and Kissimmee River area of south-central Florida.

When areas of importance for a target threatened and endangered species also provide habitat for a number of other species, they should be priority areas for protection. With species such as the Bald Eagle and Eastern Indigo Snake where there is a wide range of species-richness values under the umbrella of their potential habitats, this information can be useful by identifying potential habitat for threatened and endangered species that also has high vertebrate richness such as areas to the west of Lake Okeechobee and areas in northern Osceola County. When analysis looks for areas with both high numbers of threatened-and-endangered-species potential habitat and very high overall vertebrate-species richness, only a very small amount (1,679 ha) of area is identified in South Florida.

A graphical decision interface was created to make these and other related habitat analyses readily available to natural resource planners for flexible queries and evaluations of alternative restoration scenarios. With the interface, users can create their own landscape maps of South Florida or an appropriate subset of South Florida, based on past, present or future scenarios and run species models on these scenarios. An extensive set of analytical tools allows users to compare and contrast different landscape scenarios and species models and analyses. A multi-criteria evaluation tool allows users to incorporate multiple factors into a planning strategy. Since there are hundreds of species models and certain species models (particularly the suite of 22 MSRP species models) can take time to run, what is particularly useful is the ability to batch scenario runs, leave the application running many models on many scenarios at once.

54.4 Gap Analysis Data in Decision Making: Conservation Planning Tools to Fill the Gaps

54.4.1 The Co-Evolution of Gap Analysis and Conservation Planning Tools

The purpose of gap analysis is to find the gaps, that is, to identify either species or vegetation types (a "set of features") that are not protected to the desired percentage of their range (Scott et al. 1993). Despite the complexity of generating spatial data about the patterns of biodiversity and conservation management (described in other sections of this chapter), finding the gaps involves a relatively simple GIS overlay analysis. Conservation planners had been exploring systematic methods to fill the gaps since the early 1980s (Kirkpatrick 1983). The completion of the first gap analysis projects in Idaho, Oregon and California, which found the gaps in species and vegetation representation in those states, gave planners renewed incentive to develop and apply new methods for filling the gaps. Planners quickly discovered that designing a reserve or conservation area network that would efficiently fill the identified gaps was not a trivial spatial problem, however, and that it required tools beyond those available in commercial GIS packages. Simply ranking sites by a biodiversity index, such as the number of species they contained, while feasible for GIS analysis (Stoms 1992), would not generally create efficient reserve networks (Pressey and Nicholls 1989). Eventually geographers had the insight that selecting a set of sites to protect species populations to some target level was operationally analogous to selecting sites for facilities for a socially-desired level of safety or other services for human populations (Church et al. 1996; ReVelle et al. 2002), which already had a rich body of methods in the field of location science. GAP needed the location science community to help solve previously intractable conservation problems, and the algorithm developers building conservation planning software tools needed GAP (and similar) spatial data to develop and test their products. Thus began a productive collaboration of the conservation community with the location science community and software developers to implement the techniques. This section provides a brief history of that collaboration. We begin with the co-evolution of planning tools and algorithms that have used GAP data. We next summarize some of the issues in conservation biology that have been addressed with these tools and GAP data. Finally the section will suggest some directions for future collaboration.

The central goal of conservation planning is to fill the gaps, that is, to represent all the species and habitats to some desired level that would ensure their persistence (Margules and Pressey 2000). Conservation planners have always known, however, that minimizing the cost or area to fill the gaps would be essential to win political, economic and social acceptance (Kirkpatrick 1983). Kirkpatrick's insight was that the best way to achieve that objective was to select sites that were as distinct from each other as possible. This led to the concept of "complementarity," which refers to a site's marginal contribution to the set of conservation representation goals relative to the existing set of conservation areas. With each new conservation area added to the set, the complementarity of remaining sites is altered. This explains why standard GIS functions cannot solve the problem. Although the spatial database containing GAP data includes the biodiversity content and conservation status of each site, the method also necessitates tracking how much of each species or habitat is currently protected in the entire planning region and recalculating the complementarity value of each site. Initial efforts to solve the problem required planners, usually ecologists, to develop custom software to read text-based data tables (perhaps extracted from a GIS database) and perform simple greedy heuristic algorithms to select a good set of candidate conservation areas. Finally in the mid-1990s, researchers fully recognized the analogies of the conservation area selection problem with the location set covering problem (Church et al. 1996; ReVelle et al. 2002), although this was presaged by the Australians, Cocks and Baird (1989). At that point the collaboration took off as developers offered up new models to solve the problem faster and faster or to a higher level of optimality (i.e., minimal cost or area) than was possible with the original heuristics (Csuti et al. 1997).

Initial applications from location science followed the standard practice of the time by setting goals to simply protect one or more occurrences of each species. Kiester et al. (1996) used data on the presence of species and vegetation types from the Idaho Gap Analysis Project to identify potential conservation areas. Church et al. (1996) then introduced an important modification to the problem by asking instead how close one could come to filling the gaps with a limited budget or available land area, a form of the maximal covering location problem, and demonstrated its application using presence data from the California Gap Analysis Project. This was followed shortly thereafter in a milestone paper that compared algorithms for solving the maximal covering location problem for conservation areas using presence data from the Oregon GAP (Csuti et al. 1997). However, GAP distribution data is polygonal so that areal extent can be calculated. Targets can be set as areal or percentage amounts. Partly through the influence of the GAP spatial data model, algorithms were later revised to select conservation areas that met targets in terms of percentage of each distribution rather than by occurrences based on presence-only data. Davis et al. (1996) demonstrated one of these new models with GAP data from California's Sierra Nevada ecoregion and later from the Columbia Plateau ecoregion spanning parts of several states (Davis et al. 1999).

Other refinements soon followed as conservation practitioners demanded that models increasingly represent the real-world conservation problem and found opportunities in the GAP databases. For one thing, planners recognized that the size of a site was not a particularly good proxy for its conservation cost. Although relative land value data is rarely available at state or regional scales (Polasky et al. 2001), planners began using GIS modeling with GAP and related data to estimate the relative difficulty of conserving sites using factors such as land ownership and ecological condition (Davis et al. 1996, 1999). Not only do sites vary in their financial or management cost, but they also vary in the quality or suitability of the habitat for each species. Church et al. (2000) used California GAP data on vertebrate species habitat suitability to examine the tradeoffs between overall representation and high quality representation for a fixed number of sites selected. Other researchers used Oregon GAP data that documented the confidence of occurrence records as probabilities (in contrast to simple presence) to maximize the expected number of species represented (Polasky et al. 2000). Planners may also prefer spending scarce resources on the most rare or imperiled species. Gerrard et al. (1997) demonstrated the differential weighting of species in the California GAP database, followed later by Arthur et al. (2004) for Oregon.

The spatial configuration of conservation area networks is also important for effectiveness. Therefore algorithm developers began to generate models that would allow sites to be clustered into large, compact conservation areas with minimal perimeter. Two of the early implementations were first demonstrated with the California GAP data for the Sierra Nevada ecoregion (Andelman et al. 1999; Fischer and Church 2003). This refinement has been particularly challenging because it requires not only tracking the complementarity of each site but also the spatial topology of adjacency of polygons, which is beyond the typical spatial relationships in other applications of location science. Current solutions are still relatively coarse in that they attempt to minimize total perimeter but cannot direct clustering to small conservation areas that need expanding the most. Thus there is only a tenuous link between a conservation planner's objective and the mechanism in the model to achieve it.

By the late 1990s, conservation planning tools had reached the point that they could be used not just by researchers but on personal computers by conservation organizations. The Nature Conservancy contracted for a prototype application to assist in developing an ecoregional plan for the Columbia Plateau using a research model (Davis et al. 1999). This led to the development of a selection algorithm coupled with ArcView GIS to allow the organization to do its own planning with limited map query and viewing capability (Andelman et al. 1999). Kelley et al. (2002) describe another public-domain planning tool embedded in ArcView with similar GIS functionality, which they demonstrated with the Texas GAP database.

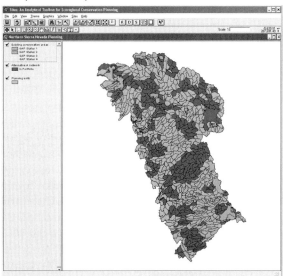

Figure 54-5 User interface for the Sites conservation planning software that selects a portfolio of conservation areas to meet representation targets. The software, like similar tools, loosely couples a selection model with GIS tools for preselecting sites and displaying results. See included DVD for color version.

54.4.2 GAP and Conservation Biology Issues

In addition to providing empirical data for testing and evaluating conservation area selection models by developers, GAP data have also been used in these models by researchers to analyze fundamental scientific questions in conservation biology. What species, plant communities, or landscape types should be used to design conservation area networks for all biodiversity? How much area is enough to ensure persistence? What spatial units should be used as candidate sites, and how large should they be? And how should threats be accommodated?

One perennial debate is over the question of which subset of known biodiversity will simultaneously fill the gaps for the remaining species. This surrogacy issue as it is known springs from the problem that we know more about the distribution of some species than we do others. For some species we know nothing at all. GAP addresses this issue pragmatically by using vertebrates, which are relatively well-known in the United States, and vegetation types as surrogates for plants, invertebrates and genetic diversity (Scott et al. 1993). In their analysis of Idaho GAP data, Kiester et al. (1996) coined the term "sweep analysis" to describe the process of measuring the effectiveness of one group of taxa or vegetation types in protecting the others. In other words, if you select areas based on one set of surrogates, how many other species or vegetation types do you sweep along into the conservation area network? They found that endangered species, terrestrial vertebrates, and vegetation types were not good surrogates for sweeping each other, a general finding that has been reinforced in other studies. A similar study was conducted with GAP data from California's Sierra Nevada region for terrestrial vertebrates and plant communities (Davis et al. 1996).

Setting scientifically-defensible targets is another recurring challenge. In conservation planning, it may be more important to understand the effects of different target levels on the size of the conservation area network and the opportunity costs of foregone resource uses. Conducting sensitivity analyses with systematically varying target levels for GAP data is one approach that has been used to address this question (Davis et al. 1996, Justus et al. 2008).

All conservation area selection models choose from a set of sites or planning units. The size and configuration of planning units has been a recurring debate among conservation scientists. GAP data has been used in some of the studies that tested the effect of different planning unit systems on the efficiency of alternative conservation area networks. For instance, Davis et al. (1996) applied a model with identical surrogates and targets on GAP data in the northern Sierra Nevada ecoregion but with planning units based on two different hierarchical levels of watershed classification. Justus et al. (2008) did a more systematic analysis with West Virginia GAP data.

Finally, conservation practitioners have confronted planners and algorithm developers with a critical real-world issue with selecting conservation area networks—namely that networks are almost never implemented instantaneously. They are implemented gradually at the rate that funding is available and that desired sites become available for acquisition. Candidate sites in a plan may be lost to other uses before they can be purchased. Thus the reality is that conservation occurs in dynamic environments that may be ill-served by static plans. GAP does not produce data on threats, but such data can be integrated with GAP data to prioritize conservation action. For instance, Costello and Polasky (2004) ranked sites by threat based on an urban growth model to represent species from the California GAP database. Davis et al. (2006) created another ranking method to represent plant communities and other objectives that formally accounted for land use threats using California GAP data.

54.4.3 Future Directions and Conclusions

As GAP and conservation planning tools have both matured, the expectations for better data and tools have continued to keep pace. Directions for improvements can be grouped into three classes: advances in conservation science, advances in planning practice, and GIS data modeling to support conservation planning. Gap analysis is fundamentally based on measuring biodiversity patterns. Planning tools to date have largely adopted that paradigm to represent those patterns within conservation areas. This has in general been a conscious choice because pattern is amenable to mapping and GIS analysis. Conservation scientists recognize that ecological and evolutionary processes must also be maintained (summarized in Sarkar et al. 2006). Some processes can be represented in GIS as spatial features (e.g., river gorges, interfaces between soil types, elevation gradients) and thus treated by the pattern representation approach. More challenging are the processes characterized by interactions between sites (Stoms et al. 2005), such as export of freshwater from headwaters into wetlands further down the watershed. Protecting the wetland will not ensure the persistence of its biodiversity if upstream land use activities taint the quality of water flowing into it. These types of processes are challenging in terms of our scientific knowledge of the impacts of land uses on flows of materials and the myriad biological responses to those changes, and GAP needs to move beyond the concept of representation to encompass persistence (Cowling et al. 1999). It is also incumbent upon GIScience to design new data structures that can store data on dynamic ecological relationships between sites.

Plan implementation is typically another dynamic process, taking years or decades to complete. As noted above, many aspects of the landscape are apt to change during the implementation period. This leads to a modification of the conservation planning problem to accommodate scheduling such that the maximum biodiversity is preserved at the end of the planning period (Costello and Polasky 2004; Davis et al. 2006; Sarkar et al. 2006).

Most of the conservation area selection models are only loosely coupled with GIS software. They use a variety of algorithms for finding optimal or good solutions. Yet all use virtually the same data, although the required inputs may have different data structures. Stoms (2003) has suggested that formalizing a core geospatial data model would benefit algorithm developers, researchers, and practitioners through standardized data management and transformation that could be made compatible with any planning tool (Figure 54-5).

This section has focused on the co-evolution of GAP and conservation planning tools. We would be remiss to fail to emphasize that much of the pioneering development of such tools has occurred outside the USA, particularly in Australia, South Africa, Europe, and more recently in marine systems. Nevertheless, GAP has been instrumental in the development and deployment of such tools. GAP was an eager consumer of these tools, while tool developers in turn were consumers of the data, and conservation principles, produced by GAP.

54.5 Using GAP Data to Address Conservation Issues: Some Examples

54.5.1 Introduction

One of the Gap Analysis Program's (GAP) primary objectives has always been to provide objective scientific data about land cover and species habitat to the on-the-ground natural resources professionals and planners who can use the data to make decisions about how to best plan for and manage those resources. Since its inception in 1989, GAP has worked with nearly 500 cooperating state and federal agencies, academic and nonprofit institutions, and businesses. Nationwide, the program has cooperated with multiple Department of the Interior bureaus, the Department of Defense, the Environmental Protection Agency, and NatureServe (USGS 2003) to complete state and regional-level projects. GAP products—including land cover maps and data, predicted species distributions maps and models, land stewardship maps, species richness information, and even the GAP methodology itself—have been key tools in more than 230 conservation and planning projects, not including the state and regional projects that are the program's focus (Maxwell 2005). As of January 2009 all GAP state projects have been completed, and a regional GAP project for the Southwestern US states of Arizona, New Mexico, Utah, Wyoming and Colorado had also been completed. The intent of this section of the chapter is to provide a brief overview of how GAP data has been used to develop wildlife corridors, plan open space, and identify high priority targets for conservation. The following brief case studies are included to highlight some applications of GAP data.

54.5.2 Spokane County Land Use Planning in Washington State

Spokane County, Washington, is the most populous county in mainly rural eastern Washington. Rapid population growth in the early '90s made planning under the Washington State Growth Management Act mandatory for Spokane County and its cities. By the end of the decade, the county's population was 417,939. (Spokane County 2006). Spokane County planners worked with Washington State GAP researchers to incorporate GAP land cover and species distribution data into their county comprehensive plan. The plan called for identification, improvement, and protection of fish and wildlife habitat. It also stressed the importance of minimizing habitat fragmentation by protecting open space and connecting corridors to create an open-space network for wildlife habitat (Stevenson 1998).

Planners chose species representation and richness as the criteria for evaluating biodiversity across the county. They used a five-step process to identify potential lands.

- Representation was used to ensure that each species was included at least once in the final network of conservation lands. Stevenson (1998) produced a map showing the minimum amount of Spokane County that could be protected while including at least one occurrence of each GAP species that was predicted to occur in the county.
- Stevenson (1998) established increments of richness for each taxonomic group (mammals, birds, reptiles and amphibians) in the county to identify areas of high species richness for potential reserves. A decision-rule was created, which specified that individual polygons should be selected based on the number of species predicted to be present. Planners set selection threshold levels at 75% for mammals and birds and at 50% for reptiles and amphibians. All polygons selected based on these thresholds were initially considered for protection.
- A map depicting the connections between potential preserves was created by selecting those polygons between potential reserves and representative areas with the highest species richness and natural land cover values.
- These three maps were combined to create an initial map of lands to be considered for the open-space network.
- The resulting selection was refined using a map of housing parcels, roads, topography, wetlands, land use and zoning, growth area boundaries for neighboring cities, and utility and railroad rights-of-way. The final map showed an open-space network that comprised 30% of the county.

County planners used the map to update their Comprehensive Plan and Critical Areas Ordinance. The Washington Department of Fish and Game evaluated proposed properties for acquisition under the state's Conservation Futures program; and the Inland Northwest Land Trust also used the map to inform decisions about their land protection work (Stevenson 1998).

54.5.3 Analyzing Wildlife Movement Corridors

As development pressures increase, habitat fragmentation has become one of the biggest threats to wildlife species (Ehrlich 1986; Harris 1984). Development or preservation of corridors of native habitat that can facilitate animal movement from one protected area to another (Noss 1987; Hobbs 1992) could minimize this threat. The next two examples will discuss how GAP data have been used to identify potential corridors.

54.5.3.1 Identifying Wildlife Corridors in Montana

Researchers with the Craighead Environmental Research Institute developed a methodology to identify wildlife corridors. They used Montana GAP vertebrate data to determine potential corridors for grizzly bear, elk and cougar moving across the core protected areas of the Northern Rockies (i.e., Salmon-Selway, Northern Continental Divide, and Greater Yellow-stone Ecosystems; Walker and Craighead 1997).

To model regional scale corridor routes, Walker and Craighead (1997) used the following assumptions:

- Suitable corridors for a species are comprised primarily of preferred habitat types for those species.
- Humans pose problems for successful transit--their residences attract animals that become habituated to easy access to food, and residential areas also displace animals from important habitat.
- Current human developments are permanent. Walker and Craighead did not analyze routes that would involve closing roads or altering land management practices.

- The least-cost path, based upon the previous assumptions, offers an animal the greatest probability of survival in traversing the entire distance.

Three coverages were developed to model the potential corridors:
- Walker and Craighead (1997) used GAP's 30-m raster land cover data to create a coverage of habitat quality. Each type was rated from 0 (unsuitable) to 3 (highly preferred) according to its perceived suitability for grizzly bear, cougar and elk. Each 1-km^2 cell was assigned a habitat quality value for each of the three species.
- Because all three species (i.e., grizzly bear, elk and cougar) are known to prefer a mixture of cover and open areas, Walker and Craighead (1997) devised a second coverage based on the length of forest/grassland and forest/shrub-land boundaries for the region. The total length of forest/shrub-land edge and forest/grassland interface was derived for each cell
- Finally, a coverage of road density was used as a measure of human development and use. Landscape and habitat variables were integrated at 1 km^2 because Walker and Craighead (1997) believed that this scale offered a reasonable balance between the original fine scale (30 m) habitat data, the continuous roads data, and other broader-scale data. (Each cell was also assigned a weighting based on road density.)

These coverages were combined to derive a measure of overall habitat suitability. Walker and Craighead (1997) then used ArcInfo GRID functions "costdistance" and "corridors" with the habitat suitability data as inputs, to calculate possible corridors between core protected areas, along with critical barriers, bottlenecks, and other high risk habitat. Least-cost travel grids were generated for each combination of the three species and the three core protected areas. These consisted of cumulative distances calculated for every cell from the nearest point of a protected area weighted by the habitat suitability of the grid. The analysis helped to identify high priority conservation areas for wildlife that could improve the connectivity between protected areas.

Walker and Craighead (1997) posit that although this approach focused on a small number of umbrella species, the corridors it delineated include important habitat for the majority of wildlife in the area in question.

54.5.3.2 Modeling Wildlife Habitat Corridors in the Greater Grand Staircase-Escalante Ecosystem

Hartley and Aplet (2001) expanded Walker and Craighead's (1997) methodology and used GAP data to identify potential animal movement corridors for black bear, bald eagle, desert bighorn sheep, mountain lion, and peregrine falcon in the Greater Grand Staircase-Escalante Ecosystem (GGSEE), Utah. This ecosystem was established as a National Monument in September 1996 in part because it acts as a landscape connector between Glen Canyon National Recreation Area, and Canyonlands, the Grand Canyon, Bryce Canyon and Capitol Reef National Parks. However, the location and size of key connectors were unresolved in the initial Management Plan for the monument (Hartley and Aplet 2001).

Hartley and Aplet (2001) identified potential travel corridors for the chosen species in GGSEE by modeling the spatial relationship between roads and the species' preferred habitat. Species habitat suitability data (90-m resolution) and vegetation coverages were obtained from the Utah Gap Analysis Project. Road data were obtained from two digital sources—US Census Bureau TIGER files and US Forest Service Cartographic Feature files.

Using ARC/INFO's GRID module's "corridor" function and GAP habitat suitability data, a cost surface grid was derived for each species. Hartley and Aplet (2001) defined sources as the federally protected areas in the vicinity of GGSEE. These included Zion National Park, Bryce Canyon National Park, Capitol Reef National Park, and the Glen Canyon National Recreation Area. Movement between these areas was said to be across a "cost surface" which represented species habitat. The concept of cost surfaces rests on the assumption that low quality habitat "costs more" to cross than high quality habitat does, because low quality habitat provides less food and cover than high quality habitat.

GAP habitat data were divided into five classes (critical, high value, significant value, low value, and no value). To develop the cost surface, each of these classes was assigned a numerical value between one and five. Once developed, each cost surface was modified to increase costs according to the influence of roads. The road data were first divided into high volume and low volume road classes. High volume roads were buffered 1600 m, while low volume roads were buffered 400 m. The final cost surface for each species was created by multiplying the habitat grid cell weights by the road grid cell weights.

The next step involved creating cost distance grids to quantify the cost of a species moving away from a protected source area. For example, to identify a corridor between Capitol Reef and Bryce National Parks, a separate grid was created to represent the cost to a species of moving away from each park. These two grids were combined using ARC/INFO GRID module's "corridor" function. Within the resulting grid, the corridor was comprised of the lowest value cells. To isolate this corridor, a mask was applied to eliminate all but the lowest 1% cell values.

Twenty corridors in total were identified, one for each species traveling between the four protected areas. These corridors were combined by adding the values for individual grid cells to create a composite corridor.

Using GAP data in this way can help identify and prioritize lands that could be prioritized for conservation because of their potential value as corridors.

54.5.4 Establishing Acreage Goals for Farmland Preservation

Cropland acreage in the United States declined from 420 million acres in 1982 to 368 million acres in 2003, a decrease of about 12%. The net decline between 1997 and 2003 was 8 million acres, or about 2 percent (USDA 2007). This conversion of farmland to urban development resulted in a loss of habitat for plant and animal species. Habitat loss is one of the biggest threats to biodiversity across the US (Wilcove et al. 1998). As pressure from urban development escalates, cropland preservation has become a key issue in Michigan. One of the key threats to preservation is the fragmentation of cropland. Adelaja et al. (2006) are developing a framework for prioritizing extant Michigan farmland which is at risk of conversion to urban development. Adelaja et al. (2006) used GAP land cover data as a key input in their analysis to help Michigan policy-makers identify potential targets for a statewide farmland preservation program.

The approach adopted by Adelaja et al. (2006) included three major components. First, farmland was ranked in each county, according to a score it received in an in-depth analysis based on a combination of 22 agricultural, economic, socio-economic and land-use characteristics, which analysts felt could reveal the resiliency of farmlands. Second, the results of these rankings were visually displayed with a series of maps created using GIS. And finally, four different preservation scenarios were developed based on this analysis.

The 22 resiliency measures included in the analysis were: prime farmland, unique farmland, biodiversity, farm viability, economic support, value added potential, product diversity, proximity to farmers' markets, proximity to food processors, proximity to grain elevators, commodity viability, proximity to customers, livestock local demand, income demographics, ethnic diversity, tourism, amenity value, population pressure, farm size diversity, farm contiguity, competition of land use, and current preservation.

GAP land cover data were used in developing five of these resiliency measures. GAP land cover classes of non-vegetated farmland, row crops, forage crops and orchards/vineyards/nursery were combined with prime farmland information from the State Soil Geographic (STATSGO) database to create a base farmland layer. Each county was then given a score representing the percent of its land that consisted of prime farmland. Farms were also given a score representing the percentage of their farmland that was dedicated to growing unique crops (i.e., crops not commonly grown elsewhere in the state).

GAP land cover data were also used to create values indicating the proximity of farmland to farmer's markets, grain elevators, and food processors, which Adelaja et al. (2006) deemed to be beneficial to farmers because they would tend to enhance a farm's profitability.

Finally, GAP data were used to assess contiguity, another factor that Adelaja et al. (2006) argued was beneficial to farms because it facilitates labor and equipment sharing. To assess contiguity, land use clusters of 0.25 mi2 (65.56 hectares) were created from the GAP Michigan Land Cover. Then adjacent clusters that contained >10% of agricultural land were merged. Counties were given a score indicating the level of contiguity of farms in the county with a high score indicating high contiguity of farms.

Adelaja et al. (2006) scored each county from 1 to 10 based on each of the resiliency indicators. The scoring indicated how each county compared with respect to other counties for that indicator. A high score, 7 and above, indicated that the county was strong based on a particular resiliency measure; while a score of 3 or below indicated weakness. They displayed these rankings in maps. Planners than developed four different preservation scenarios based on the values placed on different priorities

54.5.5 Conclusion

These examples are a small representation of the issues that GAP data can be used to address. In addition to corridor development, site prioritization and open-space planning, GAP has been used for threat assessment, fire modeling, assessing changes in land cover, identifying critical habitat, watershed planning, refuge management, and predicting species dispersal patterns (Maxwell 2005). GAP data and information about other applications of GAP data are freely available over the internet at http://gapanalysis.nbii.gov/applications.

References

Adelaja S., M. B. Lake, M. Colunga-Garcia, M. Hamm, J. Bingen, S. Gage and M. Heller. 2006. Acreage and Funding Goals for Farmland Preservation in Michigan: Targeting Resiliency, Diversity and Flexibility. A Viable Agriculture Report. Michigan State University Land Policy Institute, Report # 2006-1. <http://www.landpolicy.msu.edu/modules.php?name=Documents&op=viewlive&sp_id=142> Accessed 25 November 2008.

Andelman, S., I. Ball, F. Davis and D. Stoms. 1999. *Sites V 1.0: An Analytical Toolbox for Designing Ecoregional Conservation Portfolios.* Santa Barbara: University of California.

Arthur, J. L., J. D. Camm, R. G. Haight, C. A. Montgomery and S. Polasky. 2004. Weighing conservation objectives: Maximum expected coverage versus endangered species protection. *Ecological Applications* 14:1936–1945.

Austin, M. P. 1991. Vegetation: data collection and analysis. In *Nature Conservation: Cost Effective Biological Surveys and Data Analysis,* edited by C. R. Margules and M. P. Austin, 37–41. East Melbourne, Australia: CSIRO.

Church, R., R. Gerrard, A. Hollander and D. Stoms. 2000. Understanding the tradeoffs between site quality and species presence in reserve site selection. *Forest Science* 46:157–167.

Church, R. L., D. M. Stoms and F. W. Davis. 1996. Reserve selection as a maximal covering location problem. *Biological Conservation* 76:105–112.

Cocks, K. D. and I. A. Baird. 1989. Using mathematical programming to address the multiple reserve selection problem: An example from the Eyre Peninsula, South Australia. *Biological Conservation* 49:113–130.

Comer P., D. Faber, R. Evans, S. Gawler, C. Josse, G. Kittel. S. Menard, M. Pyne, M. Reid, K. Schulz, K. Snow and J. Teague. 2003. Ecological Systems of the United States: A Working Classification of US Terrestrial Systems. Arlington, Va.: NatureServe.

Costello, C. and S. Polasky. 2004. Dynamic reserve site selection. *Resource and Energy Economics* 26: 157–174.

Cowling, R. M., R. L. Pressey, A. T. Lombard, P. G. Desmet and A. G. Ellis. 1999. From representation to persistence: Requirements for a sustainable system of conservation areas in the species-rich Mediterranean-climate desert of southern Africa. *Diversity and Distributions* 5:51–71.

Crist, P., B. Thompson and J. Prior-Magee. 1995. A dichotomous key of land management categorization, unpublished. Las Cruces, N. M.: New Mexico Cooperative Fish and Wildlife Research Unit.

Csuti, B., S. Polasky, P. H. Williams, R. L. Pressey, J. D. Camm, M. Kershaw, A. R. Kiester, B. Downs, R. Hamilton, M. Huso and K. Sahr. 1997. A comparison of reserve selection algorithms using data on terrestrial vertebrates in Oregon. *Biological Conservation* 80:83–97.

Curnutt, J. L., J. Comiskey, M. P. Nott and L. J. Gross. 2000. Landscape-based spatially explicit species index models for Everglades restoration. *Ecological Applications* 10(6):1849–1860.

Davis, F. W., C. J. Costello and D. M. Stoms. 2006. Efficient conservation in a utility-maximization framework. *Ecology and Society* 11:33. <http://www.ecologyandsociety.org/vol11/iss1/art33/> Accessed 25 November 2008.

Davis, F. W., D. M. Stoms and S. Andelman. 1999. Systematic reserve selection in the USA: An example from the Columbia Plateau Ecoregion. *Parks* 9:31–41.

Davis, F. W., D. M. Stoms, R. L. Church, W. J. Okin and K. N. Johnson. 1996. Selecting biodiversity management areas. In *Sierra Nevada Ecosystem Project: Final Report to Congress, vol. II, Assessments and scientific basis for management options,* 1503–1528. Davis, Calif.: University of California, Centers for Water and Wildlands Resources.

Davis, S. M. and J. C. Ogden, editors. 1994. *Everglades: The Ecosystem and its Restoration.* Delray Beach, Fla.: St. Lucie Press.

DeAngelis, D. L., L. J. Gross, M. A. Huston, W. F. Wolff, D. M. Fleming, E. J. Comiskey and S. M. Sylvester. 1998. Landscape modeling for Everglades ecosystem restoration. *Ecosystems* 1:64–75.

Desmond, G. D. 2003. Measuring and Mapping the Topography of the Florida Everglades for Ecosystem Restoration. US Geological Survey Fact Sheet, 021-03.

Duke-Sylvester, S. 1999. The ATLSS High Resolution Topography (HRT) and High Resolution Hydrology (HRH) models. <http://www.tiem.utk.edu/~sylv/HTML/Work/Everglades/Presentation/GEER/presentation.ppt#20> Accessed 25 November 2008.

Duke-Sylvester, S. 2003. The ATLSS High Resolution Multi-Data Source Topography (HMDT). <http://atlss.org/~sylv/HTML/Everglades/HMDT-ShortReport/index.html> Accessed 25 November 2008.

Duke-Sylvester, S. 2006. *Applying Landscape-scale Modeling to Everglades Restoration.* Doctor of Philosophy Dissertation, University of Tennessee, Knoxville, Tenn. 187 pp.

Edwards, T. C., Jr., C. H. Homer, S. D. Bassett, A. Falconer, R. D. Ramsey and D. W. Wight. 1995. Utah Gap Analysis: An Environmental Information System. Technical Report 95-1. Utah Cooperative Fish and Wildlife Research Unit. Logan, Utah: Utah State University.

Ehrlich, P. R. 1986. The loss of diversity. In *Biodiversity,* edited by E. O. Wilson, 21–27. Washington, D.C.: National Academy Press.

Fischer, D. T. and R. L. Church. 2003. Clustering and compactness in reserve site selection: An extension of the Biodiversity Management Area Selection model. *Forest Science* 49:555–565.

Frost, C. C. 1998. Presettlement fire frequency regimes of the United States: A first approximation. *Tall Timbers Fire Ecology Conference Proceedings* (20):70–81.

Gerrard, R. A., R. L. Church, D. M. Stoms and F. W. Davis. 1997. Selecting conservation reserves using species covering models: Adapting the ARC/INFO GIS. *Transactions in GIS* 2:45–60.

Hall, S. P. and M. P. Schafale. 1999. *Conservation Assessment of the Southeast Coastal Plain of North Carolina, Using Site-oriented and Landscape Oriented Analyses.* Report to the USGS - Biological Resources Division. 250 pp. <http://www.ncnhp.org/Images/SEcoastalplain.pdf> Accessed 25 November 2008.

Harris, L. 1984. The Fragmented Forest: Island Biogeography Theory and the Preservation of Biotic Diversity. Chicago: University of Chicago Press.

Hartley, D. and G. H. Aplet. 2001. Modeling wildlife habitat corridors in the Greater Grand Staircase-Escalante Ecosystem. In *Proceedings of the Fifth Biennial Conference of Research on the Colorado Plateau,* edited by van Riper III, C., K. A. Thomas and M. A. Stuart. US Geological Survey/FRESC Report Series. USGSFRESC/COPL/2001/21. Flagstaff, Ariz.

Hobbs, R. J. 1992. The role of corridors in conservation: Solution or bandwagon? *Trends in Ecology and Evolution* 7:389–392.

International Association of Fish and Wildlife Agencies (IAFWA). 2002. IAFWA GUIDING PRINCIPLES: Guiding Principles for States to Consider in Developing Comprehensive Wildlife Conservation Plans/Wildlife Action Plans and Wildlife Conservation Strategies (Plans-Strategies) for the State Wildlife Grant and Wildlife Conservation and Restoration Programs. Washington, D.C. <http://www.ncwildlife.org/pg07_wildlifespeciescon/WAP_AppendixA.pdf> Accessed 6 February 2009.

Jennings, M. D. 1993. Natural terrestrial cover classification: assumptions and definitions. Gap Analysis Technical Bulletin 2. Moscow. Idaho: Idaho Cooperative Fish and Wildlife Research Unit. 28pp.

Jones, J. W. 2001. Image and in situ data integration to derive sawgrass density information for surface-flow modeling in the Everglades. In *Remote Sensing and Hydrology 2000 Proceedings*, edited by M. Owe, K. Brubaker, J Ritchie and A. Rango, 507–512. International Association of Hydrological Sciences Symposium held in Santa Fe, New Mexico, April 2000. Publication no. 267. Wallingford, Oxfordshire, UK: IAHS Press.

Jones, J. W. and S. D. Price. 2007. *Everglades Depth Estimation Network (EDEN) Digital Elevation Model Research and Development.* US Geological Survey Open-File Report 2007–1034. 29 pp.

Justus, J., T. Fuller and S. Sarkar. 2008. Influence of representation targets on the total area of conservation-area networks. *Conservation Biology* 22:673–682.

Kelley, C., J. Garson, A. Aggarwal and S. Sarkar. 2002. Place prioritization for biodiversity reserve network design: A comparison of the SITES and ResNet software packages for coverage and efficiency. *Diversity and Distributions* 8:297–306.

Kiester, A. R., J. M. Scott, B. Csuti, R. F. Noss, B. Butterfield, K. Sahr and D. White. 1996. Conservation prioritization using GAP data.. *Conservation Biology* 10:1332–1342.

Kirkpatrick, J. B. 1983. An iterative method for establishing priorities for selection of nature reserves: an example from Tasmania. *Biological Conservation* 25:127–134.

Margules, C. R. and R. L. Pressey. 2000. Systematic conservation planning. *Nature* 405:243–253.

Maxwell, J. 2005. Using GAP data as a tool for natural resource management. Poster presented at 2005 Gap Analysis Conference and Interagency Symposium, Reno, Nev.

NatureServe. 2009. NatureServe Explorer: An online encyclopedia of life. <www.natureserve. org/explorer/> Accessed 22 February 2009.

North Carolina Wildlife Resources Commission. 2005. North Carolina Wildlife Action Plan. Raleigh, N.C. 577 pp. <http://www.ncwildlife.org/Plan/index.htm> Accessed 25 November 2008.

Noss, R. F. 1987. Protecting natural areas in fragmented landscapes. *Natural Areas Journal* 7:2–13.

Noss, R. F. 1988. "The longleaf pine landscape of the southeast: almost gone and almost forgotten." *Endangered Species Update* 1 5(5):1–5 (17077).

Noss, R. F. 1991. Unpublished report to the Fund for Animals in Washington, D.C.

Noss, R. F., and A. Y. Cooperrider. 1994. Saving Nature's Legacy. Washington, D.C.: Island Press. 416 pp.

Noss, R. F., E. T. LaRoe III and J. M. Scott. 1995. *Endangered Ecosystems of the United States: A Preliminary Assessment of Loss and Degradation.* Biological Report 28. Washington, D.C.: USDI National Biological Service. 58 pp.

Odum, E. P. and H. T. Odum. 1972. Natural areas as necessary components of. man's total environment. In *Transactions of the 37th North American Wildlife and Natural Resources Conference,* 178-189. Washington, D.C.: Wildlife Management Institute.

Palaseanu, M. and L. Pearlstine. 2008. Estimation of water surface elevations for the Everglades, Florida. *Computers and Geosciences* 34:815–826.

Pearlstine, L. G., L. A. Brandt, F. Mazzotti and W. M. Kitchens. 1997. Fragmentation of pine flatwood communities converted for ranching and citrus. *Landscape and Urban Planning* 38:159–169.

Pearlstine, L., S. Smith, L. Brandt, C. Allen, W. Kitchens and J. Stenberg. 2002. Assessing state-wide biodiversity in the Florida Gap Analysis Project. *Journal of Environmental Management* 66(2):127–144.

Polasky, S., J. D. Camm and B. Garber-Yonts. 2001. Selecting biological reserves cost-effectively: An application to terrestrial vertebrate conservation in Oregon. *Land Economics* 77:68–78.

Polasky, S., J. D. Camm, A. R. Solow, B. Csuti, D. White and R. Ding. 2000. Choosing reserve networks with incomplete species information. *Biological Conservation* 94:1–10.

Pressey, R. L., C. J. Humphries, C. R. Margules, R. I. Vane-Wright and P. H. Williams. 1993. Beyond opportunism: Key principles for systematic reserve selection. *Trends in Ecology and Evolution* 8:124–128

Pressey, R. L. and A. O. Nicholls. 1989. Efficiency in conservation planning--scoring versus iterative approaches. *Biological Conservation* 50:199–218.

ReVelle, C. S., J. C. Williams and J. J. Boland. 2002. Counterpart models in facility location science and reserve selection science. *Environmental Modeling and Assessment* 7:71–80.

Ride, W.L.D. 1975. Towards an integrated system: a study of selection and acquisition of national parks and nature reserves in Western Australia. In *A national system of ecological reserves in Australia* edited by F. Fenner, 64–85. Canberra: Australian Academy of Science.

Robertson, W. B. and P. C. Frederick. 1994. The faunal chapters: Contexts, synthesis, and departures. In *Everglades: The Ecosystem and Its Restoration,* edited by S. M. Davis and J. C. Ogden, 709–37. Delray Beach, Fla.: St. Lucie Press.

Rutchey, K., L. Vilchek and M. Love. 2005. *Development of a Vegetation Map for Water Conservation Area 3.* Technical Publication ERA #421. West Palm Beach, Fla.: South Florida Water Management District.

Sarkar, S., R. L. Pressey, D. P. Faith, C. R. Margules, T. Fuller, D. M. Stoms, A. Moffett, K. Wilson, K. J. Williams, P. H. Williams and S. Andelman. 2006. Biodiversity conservation planning tools: Present status and problems for the future. *Annual Review of Environment and Resources* 31:123–159.

Schafale, M. P. 1999. Nonriverine Wet Hardwood Forests in North Carolina – Status and Trends. North Carolina Natural Heritage Program. Department of Environment, Health and Natural Resources, Raleigh, NC. 14 pp.

Scott, J. M., F. Davis, B. Csuti, R. Noss, B. Butterfield, C. Groves, H. Anderson, S. Caicco, F. D'Erchia, J.T.C. Edwards, J. Ulliman and R. G. Wright. 1993. Gap analysis: A geographic approach to protection of biological diversity. *Wildlife Monographs* 123:1–41.

Soule, M. and M. Sanjayan. 1998. Conservation targets: Do they help? *Science* 279:2060–2061.

Specht, R. L. 1975. The Report and its recommendations. In *A National System of Ecological Reserves in Australia.*, edited by F. Fenner, 11–16. Report No. 19. Canberra, Australia: Australian Academy of Sciences.

Specht, R. L., E. M. Roe and V. H. Boughlon. 1974. Conservation of major plant communities in Australia and Papua New Guinea. Australian Journal of Botany Supplement Series. Supplement No. 7.

Spokane County Department of Building and Planning. 2006. Comprehensive Plan Summary & 5 Year Update. Spokane, Washington.

State of Florida. 1995. Everglades Forever Act. §373.4592. Florida Statutes, Tallahassee, FL.

Stevenson, M. 1998. Applying gap analysis to county land use planning in Washington State. *Gap Analysis Bulletin* No. 7.

Stoms, D. M. 1992. Effects of habitat map generalization in biodiversity assessment. *Photogrammetric Engineering and Remote Sensing* 58:1587–1591.

Stoms, D. M. 2003. Linking GIS and reserve selection algorithms: Towards a geospatial data model. Biogeography Lab, Bren School of Environmental Science and Management, University of California Santa Barbara. <http://www.biogeog.ucsb.edu/pubs/ Technical%20Reports/Reserve_Selection_Data_Model.pdf> Accessed 2 March 2009.

Stoms, D. M., F. W. Davis, S. J. Andelman, M. H. Carr, S. D. Gaines, B. S. Halpern, R. Hoenicke, S. G. Leibowitz, A. Leydecker, E.M.P. Madin, H. Tallis and R. R. Warner. 2005. Integrated coastal reserve planning: making the land-sea connection. *Frontiers in Ecology and the Environment* 3:429–436.

US Army Corps of Engineers. 1998. Central and Southern Florida project (C&SF) Comprehensive Review Study. US Army Corps of Engineers, Jacksonville District, Jacksonville, Fla.

US Department of Agriculture (USDA) Natural Resources Conservation Service. 2007. National Resources Inventory 2003 Annual NRI. <http://www.nrcs.usda.gov/technical/ NRI/2003/Landuse-mrb.pdf> Accessed 22 January 2009.

US Fish and Wildlife Service. 1999. South Florida multi-species recovery plan. Atlanta, Georgia.

US Geographical Survey, National Biological Information Infrastructure. 2003. Gap Analysis Program Overview and History.

Walker, R., and L. Craighead. 1997. Analyzing Wildlife Movement Corridors in Montana Using GIS. Environmental Sciences Research Institute. Proceedings of ESRI User's Conference. <http://gis.esri.com/library/userconf/proc97/proc97/to150/pap116/p116. htm> Accessed 5 November 2008.

Wilcove, D., D. Rothstein, J. Dubow, A. Phillips and E. Losos. 1998. Quantifying threats to imperiled species in the United States. *BioScience* 48:607–615.

Williams, P., D. Gibbons, C. Margules, A. Rebelo, C. Humphries and R. Pressey. 1996. A comparison of richness hotspots, rarity hotspots, and complementary areas for conserving diversity of British Birds. *Conservation Biology*. 10 (1):155–174.

CHAPTER 55

Application of Remote Sensing and GIS for Coastal Zone Management

Shailesh Nayak

55.1 Introduction

The coastal zone is an area of interaction between terrestrial and marine/tidal processes. Coastal zones worldwide are under increasing pressure due to high rates of human population growth, development of various industries (tourism, chemical, petrochemical, fishing, aquaculture, shipping, mining, etc.), discharge of municipal sewage and industrial waste effluents, and offshore petroleum exploration activities. This industrial development on coastlines has resulted in degradation of coastal ecosystems and diminished the living resources of Exclusive Economic Zones in the form of coastal and marine biodiversity and productivity. Apart from these, the coastal processes of erosion, deposition, sediment transport, as well as sea-level rise, continuously modify the shoreline and affect coastal ecosystems, while local events, such as cyclones and floods occur very rapidly and pose serious threats to human life and property. Human activities also induce certain changes or accelerate the process of change. The loss of habitats, severe coastal erosion, sedimentation in ports and harbors, and municipal and industrial pollution are of major concern to coastal zone managers. Scientific data on coastal wetlands, land use, landforms, shorelines and water quality are required periodically to ensure that environmentally effective coastal zone management is practiced. There is an urgent need to conserve the coastal ecosystems and habitats, including individual plant species and communities, so that their current and potential usefulness to people is not impaired. By promoting wise use of coastal resources, annual yields can be assured in perpetuity.

Environmentally effective coastal zone management requires accurate, up-to-date and comprehensive scientific data on which policy decisions can be based. A basic problem is limited availability of geographic data on the coastal zone. Although conventional maps are quite useful, they do not provide up-to-date information. Since the coastal zone is very dynamic, periodic mapping is vital for planning effective strategies. Satellite data have been used for studying various components of the coastal environment. During the last 30 years, availability of remote sensing data has ensured synoptic and repetitive coverage for the entire Earth. This information has been extremely useful in the generation of spatial information of various scales and with reasonable classification and control accuracy. In India, coastal wetlands, land use and landform and shoreline-change maps have been produced at 1:250,000, 1:50,000 and 1:25,000 scale using IRS LISS I, II and III, Landsat MSS/TM and SPOT data. These maps have been used by many central and state-level agencies for a variety of purposes. The acceptability of such coastal maps by the Government of India for delineating high and low tide line to identify coastal regulation zones has spurred large-scale use by state governments and private entrepreneurs. During the 1990s, using high-resolution IRS LISS III and PAN merged data, detailed maps of coral reefs and mangroves were prepared for the first time in India.

The availability of 1–5 m very high resolution and stereo data from *Ikonos, Quickbird, Resourcesat* and *Cartosat* greatly facilitate the preparation of local level maps at 1:5,000 scale and larger. The easy access to high spatial resolution data, along with multispectral characteristics, repetitive coverage and development of geographic information system databases, has provided new impetus to coastal zone management models.

55.2 Coastal Issues and Classification Systems

Classification systems for various themes of the coastal zone have been designed based on coastal zone issues to be addressed, definitions of coastal zones, kinds of changes encountered and characteristics of remote sensing data.

55.2.1 Critical Issues

The following issues are critical in the context of coastal zone management (Nayak 2000).

55.2.1.1 Coastal Habitats

- Availability of benchmark or reference data (base line data)
- Preservation, conservation and monitoring of vital and critical habitats, e.g., coral reefs, mangroves, etc.
- Appropriate site selection for industries, landfall points, aquaculture, recreational activities, etc.
- Assessment of conditions in regulation zones, areas-under-construction setback lines, mega-cities, etc.
- Reclamation of wetland for agricultural and industrial purposes

55.2.1.2 Coastal Processes

- Planning and implementation of coastal protection measures (erosion, flood protection, salt water intrusion, etc.)
- Interactions between developmental activities and modification of coastal processes
- Impact of dam construction on shoreline equilibrium
- Suspended sediment dynamics
- Changes in bottom topography

55.2.1.3 Coastal Hazards

- Cyclones, storm surges, tsunamis
- Coastal erosion
- Sea-level rise and possible effects
- Non-point and point pollution
- Phytoplankton blooms

55.2.1.4 Availability of Resources and Their Utilization

- Sand-mining
- Fishery exploration
- Fish stock assessment
- Seaweeds resources

55.2.2 Definition of Coastal Zone

The coastal zone has been variously defined based on administrative, physical, and environmental units or simply based on fixed distances from shore. Each definition has its merits and demerits. We define this zone as the area that includes coastal waters, wetlands and adjacent shore lands influenced by marine waters or vice versa (Nayak et al. 1989). Wetlands have been defined as transitional land between terrestrial and aquatic (marine) systems where the water table is at or near the surface (Cowardin et al. 1979). The high and low tide lines define boundaries of coastal wetlands. Shore land comprises coastal watersheds, flood-prone areas, deltaic features and sand features. The seaward boundary of coastal water has been chosen as the 50 m isobath line. Based on these basic definitions, classification systems for various coastal themes have been developed.

55.2.3 Types of Environmental Change

There are four different types of changes that can take place in the coastal zone.

- Seasonal change: this change covers a time frame of one year and follows a known, logical and ordered sequence. Although they are expected, it is useful to know the varying rate and pattern of seasonal changes, e.g., changes in the growth of seaweeds.
- Long-term change: this change occurs slowly and/or infrequently and may or may not be localized. It will cover a span of one to many years, e.g., changes in mangroves and shorelines.
- Short-term change: in this type of change, the coastal landscape generally returns to its earlier condition. It may cover a period of a few days to months, e.g., annual tides and floods.
- Constant change: this change occurs when water is moving along with material, e.g., suspended sediments and currents.

55.2.4 Characteristics of Remote Sensing Data

The accuracy of the mapping of coastal zones and the assessment of changes depends on the type of data used. It is necessary to take into account the following characteristics before mapping is undertaken.

- Area: the study area size should be in accordance with the type of sensor data to be used. Low spatial resolution data, for example, should not be used for very small regions and vice versa.
- Spatial resolution of data: the level of information should be selected in such a way that low-resolution data (70–100 m, IRS LISS I, IRS WiFS, Landsat MSS), medium resolution (20–40 m, IRS LISS II and III, Landsat TM, SPOT) and high-resolution data (5 m or larger, IRS Pan or LISS IV), and very high resolution data (~1–5 m, Ikonos, Cartosat) can provide appropriate detail of features of interest.
- Spectral change: the classes or changes to be mapped should have distinct spectral signatures.
- Spatial change: in principle, any class or change of the order of one pixel can be detected. However, it is observed that a class should be on the order of 12–15 pixels and should have a distinct spectral signature to be detected and mapped. High-contrast linear features like road, shoreline, etc., however, can be identified and mapped at smaller than a single pixel size.
- Time frame: the optimum period to derive information is very important. Many times, more than two seasons of data are required to derive information. The present repeat cycles of remote sensing satellites of 1-24 days are quite adequate for capturing all kinds of environmental changes, as well as understanding coastal geomorphic and ecological processes. In areas of estuaries, gulfs and sheltered areas, tidal condition at the time of data acquisition also needs to be checked.
- Length of record: Satellite data are available since 1972. This is very useful for assessing the conditions during the past 35 years. Earlier maps and aerial photos also are quite useful.

55.2.5 Classification System

Various classification systems for coastal wetland, land use, landform and coral reefs have been designed primarily keeping the ultimate user in mind. Attention has been given to the first two levels of categorization because information at these levels is required for nationwide, state-level or regional analysis. Coastal classification systems have been designed based on the following criteria (Nayak et al. 1991).

- Accuracy: the minimum level of interpretation accuracy required for various coastal themes should be 85% at 90% confidence level. The accuracy of all classes should be more or less the same.

- Repeatability: a classification system should provide repetitive results from one interpreter to another, as well as from one time to another.
- Applicability: a classification system should be applicable to a large area, such as an entire country or a region.
- Suitability: a classification system should be suitable for use with remote-sensing data obtained during different times of the year and by different sensors.
- Flexibility: a classification system should allow use of a higher level of information collected by ground survey, aerial data or high-resolution satellite data. Aggregation of categories and comparison with future data should be possible.

A classification system designed for mapping of the coastal regulation zone is given in Table 55-1.

55.3 Satellite Data Interpretation and Generation of Maps

55.3.1 Analysis Procedure

A variety of satellite data such as IRS LISS I, II, III and PAN, Landsat MSS and TM, and SPOT HRV are available for coastal zone mapping and assessment. IRS LISS II/III and SPOT are ideally suited, as the spatial resolution is around 0.04–0.1 ha (Carter 1982). Coastal land use and wetland classification up to Level II can be carried out using such data. False-color composites (FCCs) made using green (0.52–0.59 μm), red (0.62–0.69 μm) and infrared (0.77–0.86 μm) are often used to enhance land-water boundary delineation and vegetation characteristics.

The period of time for which satellite data is required depends on the final use. For coastal regulation zone (CRZ) mapping in India, satellite data collected during 1990–1991 was selected to provide a condition assessment of wetlands just before CRZ notification became operational. Satellite data collected in December–February are generally used to show the reproductive cycle of vegetation in wetland areas. Satellite data collected during low-tide time is ideal, because both low tide line and high tide can be drawn. FCC transparencies or geocoded paper prints on 1:50,000/1:25,000-scale can be used. FCC transparencies should be enlarged using optical enlargers. The minimum area to be mapped is 2 mm × 2 mm or 9 pixels.

The high tide line is delineated based on tonal discontinuity, the result of water leaving its mark wherever it travels, observed on satellite images. Apart from the presence of mangroves, certain landforms should be taken into account. In our studies of coastal India, the extent of tidal influence in estuaries and creeks was taken up from the Survey of India topographical maps. Various wetland and land use categories were delineated based on an image interpretation key (Nayak et al. 1991, 1992). Field checks were carried out to verify doubtful areas. Based on field information, necessary modifications were carried out to prepare final maps.

Each map covering 1° × 1°, 15′ × 15′ or 7.5′ × 7.5′, depending on scale and corresponding to the Survey of India (SOI) index, was prepared. Edge-map matching with adjoining sheets was completed before final map preparation.

55.3.2 Map Representation Scheme

The systematic presentation of coastal wetland and land use features was accomplished by the following methods (Nayak et al. 1991).

- Color coding: color coding is the best presentation method when the number of classes is small.

Table 55-1 Classification system for coastal land use mapping.

Level	Level II	Level III
Agricultural land		
Forest	Natural	
	Man-made	
Wetland	Estuary	
	Lagoon	
	Creek	
	Bay	
	Tidal flat/mudflat	
	Sand/beach/spit/bar	
	Coral reef	
	Rocky coast	
	Mangroves	Dense
		Sparse
	Salt marsh/marsh vegetation	
	Other vegetation (Scrub/grass/algae/sea weeds)	
Barren land	Sandy area/dunes	
	Mining area/dumps	
	Rock outcrops/gullied/ eroded/ badlands	
Built-up land	Habitation	
	Habitation with vegetation	
	Open/vacant land	
	Transportation	Roads
		Railways
		Harbor/jetty
		Airport
		Waterways
Other features	Reclaimed area	
	Salt pans	
	Aquaculture ponds	
	Ponds/lakes	
	Rivers/streams	
	Drains/outfalls/effluents	
	Sea wall/embankments	
High tide line		
Low tide line		
District/state boundary		
CRZ boundary		

- Numerical system: in this system, the number of digits equals the level of categorization. For example, classes 12 and 13 are sublevels of class 1. This system helps in rapid integration of subcategories and permits continuation of the next level of categorization for adding finer details. However, this system lacks visual orientation.
- Graphic symbols: symbols, such as dots, stipples, cross-hatching, etc., allow visual orientation to each discrete level. However, this system is only useful when the number of classes is small.
- Alphabetical code: this code has the advantage of suggesting a logical name for each category, but also impedes interpretation at the detailed level. Its length can cause placement problems for small-sized categories.

In all coastal mapping projects in this analysis of India's coastal zones, a hybrid system was adopted.

55.3.3 Estimation of Accuracy of Maps

The ultimate quality of maps depends upon the type of satellite data, scale of mapping, size of minimum area to be mapped, analysis procedure, amount of ground information collected, skill of interpreter, method of digitization, etc. The specification for input satellite data (bands, season, etc.), scale, classification system, interpretation key and minimum mapping unit was established for all analyses of India's coastal zones. At each stage, quality checks were carried out during the analysis of satellite data. Generally quality checks are carried out as follows:

- Preliminary interpretation checks
- Final interpretation checks
- Digitization checks
- Estimation of classification and control accuracy

55.3.3.1 Estimation of Classification Accuracy

The classification accuracy of coastal zone mapping was tested on a sample basis assuming a binomial distribution for the probability of success/failure of sample tests. Sample size was decided using look up tables, prepared using a binomial probability distribution model (Aronoff 1982). Each segment was chosen as a size of 2×2 mm (50×50 m on 1:25,000 scale). This procedure ensures that there is only one predominant class in each segment to satisfy binomial condition. This segment should be large enough to be easily locatable on the ground.

Initially 300 points were selected by pseudo-random sampling and plotted on the final thematic map (the points falling in the sea as well as far inland were rejected). The selected points were transferred on 1:25,000/1:50,000 scale topographical maps to evaluate approachability of these sites/locations. To reach these areas, approach by the nearest village, cultural features, water wells, and other landmarks needed to be used to confirm the position. Wherever no such cultural features were available, the topographic/geomorphic information, like creeks, rock outcrop, islands, etc., was used to reach the sample sites. The use of GPS is recommended. Before visiting the area the tidal conditions were checked in coastal regions, as the points falling in the inter-tidal area could be approached only during low tide. The observations were noted in the specified Performa. A confusion matrix was then drawn and the accuracy estimated. The confusion matrix scored each observation according to the class in which it was mapped and the true class as observed on the ground. Such a matrix is used to provide an indication of the reliability of the maps.

55.3.3.2 Control Accuracy of Maps

Control accuracy essentially gives an idea of overall geometric quality of maps. It is specified in terms of the following:

- Scale variation: represents overall deviation in map scale (expressed as percentage).
- Non linearity in scale: represents variation in scale across the map and expressed in percentage.
- Planimetric error: residual error after taking care of simple rotation and shift in origin and expressed in mm. All residual errors after a shift in origin and appropriate rotation are combined into planimetric error. This essentially represents probable error in measuring distances or areas in the map. Root Mean Square (RMS) deviation at control points after accounting for a simple rotation and translation is computed and presented as overall planimetric error in mm.

Discussions with various agencies led to the identification of control accuracy requirements as follows.

- Scale variation   < 5%
- Non linearity in scale < 5%
- Planimetric error   < 1 mm at specified scale

55.4 Coastal Habitats

Coastal habitats, especially wetlands, coral reefs, mangroves, salt marshes, and sea grass beds, are rapidly being cleared for urban, industrial, and recreational growth, as well as for aquaculture ponds. Estimates of coastal habitat loss are not available. In India, for example in Kochi, many wetlands have been drained for development and for prevention of malaria. The construction of canals for flood control, especially on the east coast of India, has resulted in loss of wetlands.

Information on loss of tidal wetlands is important. Such wetlands provide a vital link in the marine energy flow through transfer of solar energy into forms readily usable by a wide variety of estuarine organisms. Wetlands are responsible for maintaining reproductive fisheries not only by way of catch but as feeding, spawning and nursery grounds as well. Approximately 90% of the world's marine fish catch (measured by weight) is produced in these areas. Thus, degradation of coastal habitats can have long term consequences for fish populations. Apart from this, they also serve as a buffer for the mainland against ocean storms and protect the coast from erosion. Knowledge about areal extent, condition and destructive uses of wetlands is vital for coastal management programs.

55.4.1 Nationwide Status of Coastal Habitats

Baseline information on coastal habitat is both critical and vital to India's ecology, economy and safety for millions of coastal inhabitants. Accordingly, mapping of associated shore land features along the entire Indian coast at 1:250,000/1:50,000 scale has been carried out for macro level planning. The major findings of the inventory of the Indian coast carried out through visual interpretation of multispectral IRS LISS II and Landsat TM data include degradation of mangroves, worsening condition of coral reefs, and reclamation of lagoons, mudflats and mangroves for agriculture and aquaculture (Desai et al. 1991; Nayak et al. 1991, 1992; Nayak 1996). Classification accuracy of wetland/landform maps at 1:250,000/1:50,000 scale is 85–90% at a 90% confidence level.

55.4.2 Vital/Critical Habitats

55.4.2.1 Coral Reefs

Knowledge about the extent and condition of coral reefs is useful in planning conservation and preventive measures to protect this fragile system. Coral reef features such as type (fringing, atoll, platform, patch, coral heads, sand cays, etc.), reef-flat, reef vegetation, degraded reef, lagoons, live corals and coralline shelf have been mapped using IRS LISS II and III data at 1:50,000 scale for the Indian reefs. These maps can be used as a basic input for identifying the boundaries of protected areas and biosphere reserves. Degraded condition of coral reef is indicated by mud deposition (Bahuguna and Nayak 1998). The felling of mangroves and clearing of forests have increased sedimentation and affected live coral and species diversity. It was also possible to map uncharted extensive coralline shelf, atolls, and coral heads, live coral platforms, coral pinnacles and new coral growth.

Coral reefs exhibit distinctive patterns of morphological and ecological zones, which are determined by the morphology; spatial and temporal variations; interaction between the hydrodynamic processes (waves and tides); geomorphic processes (sediment generation, sorting and transport); and ecological processes (abundance, composition, growth form, cover and productivity in biological communities). IRS-1C LISS III and PAN merged data have been found to be extremely useful for coral reef zonation studies (Nayak et al. 1996). Communities of species and/or substrata often exhibit considerable variability and several distinct communities may inhabit each morphological zone. Geomorphological zones tend to have more distinct boundaries than ecological habitats that tend to exhibit change along gradients (e.g., progressive changes in species composition with changing depths). The classification system for zoning the reefs based on morphological and ecological components using satellite data has evolved for the entire Indian coral reef with region-specific modifications (Navalgund and Bahuguna 1999; Nayak and Bahuguna 2001). After appropriate image corrections and image-enhancement techniques, each IRS LISS III (of period 1998–2001) data set was classified using a Maximum Likelihood Classifier. Sufficient ground information was used to generate signatures for coral reef classification. While IRS LISS III and PAN merged data permitted boundary delineation and the generation of signatures, contextual editing was performed to improve the classification accuracy. The classification accuracy of these maps is 85–95% at a 90% confidence level.

55.4.2.2 Mangroves

Mangroves are very important as they help in the production of detritus and organic matter, and the recycling of nutrients, thus enriching the coastal waters and supporting the benthic population of the sea. They support the most fundamental need of the coastal people—food, fuel, shelter and monetary earnings. The earlier estimate of mangroves was 6,740 sq km, but this was not accurate. The estimate by the Space Applications Centre based on IRS data is 4,460 sq km. Forest Survey of India has been providing estimates of mangroves since 1987. At many places, mangroves are degraded and destroyed due to conversion of these areas for agriculture—aquaculture on the east coast and industrial purposes on the west coast. At most places, mangroves are used as fuel and fodder; however, this impact is not very significant. Dense mangroves were mapped in delta regions of the east coast, fringing the coast on the Andaman and Nicobar Islands, estuarine regions of Maharashtra and Goa, Gulf of Kachchh, delta regions of Kori creek on Gujarat coast, and Pichavaram and Vedaranyam mangrove forest in Tamil Nadu coast. Mangroves are shrub type and degraded in Karnataka. The mangroves in Kerala occur in small patches and are mostly degraded.

Mangrove vegetation shows distinct zonation characterized by the presence of particular species within specific physico-chemical environments. The dominant genus of a particular mangrove zone is dependent upon the extent and frequency of inundation under tidal waves, salinity, and soil characteristics. Information regarding different mangrove community zonation is a vital remote sensing-based input to a GIS for a biodiversity assessment and for preparing management plans for conservation.

Major community zonation of mangroves has been carried out using IRS-1C LISS III data (a combination of red, near-infrared and middle-infrared bands) for selected mangrove habitats. Spectral signatures of the major homogeneous, as well as heterogeneous, communities of mangroves have been established, e.g., *Avicennia, Rhizophora, Ceriops, Heritiera, Excoecaria, Sonneratia, Xylocarpus,* etc. This distinction was possible because of the different spectral properties of canopies of two types of mangroves produced by a combination of individual vegetative components, effects of plant growth, density and height. This new information is extremely useful for biodiversity studies. The delineation of small sand bodies (the smallest is 0.25 ha), which demarcated the past position of shorelines, provided vital clues in understanding the growth pattern of this coast. It was observed that anthropogenic activities such as development of new ports, suspended particulate matter (SPM), housing, aquaculture and attack by pests all degrade and destroy mangroves.

55.4.2.3 Change Detection

Marine national parks and wildlife sanctuaries have been established in India for protecting vital/critical habitats such as mangroves, coral reefs and wetlands. Seasonal and long-term changes are normally encountered in the condition of habitats. The repetitive coverage of the IRS satellites is quite adequate for monitoring such changes. It is necessary, however, to also use GIS techniques to monitor these areas to assess the impact of conservation measures as well as anthropogenic activities. In one such study, in the Marine National Park, Jamnagar, on the Gujarat coast, significant changes in the mangrove vegetation and coral reef area were observed during the period 1975 to 1998 (Nayak et al. 1989). Degradation of both ecosystems continued until 1985 due to mining of coralline sand and use of mangroves as fuel and fodder. In 1983, this area was declared a marine park (protected area). Extensive measures were initiated for conservation of mangrove and coral reef areas by marine park authorities. This resulted in reversing the trend of degradation after 1985 and has certainly helped towards restoring the environment. However, industrialization, development of ports, and other activities have again put these ecosystems under stress, as is evident from recent satellite data.

55.4.3 Coastal Regulation Zone

The increasing pressure on the coastal zone due to global concentrations of populations in coastal areas, development of industries, discharge of waste effluents and municipal sewage, and increases in coastal recreational activities, has adversely affected the coastal environment. In India, coastal stretches of bays, estuaries, backwaters, seas, and creeks that are influenced by tidal action up to 500 m from High Tide Line (HTL) and the land between the Low Tide Line (LTL) and the HTL have been declared as the Coastal Regulation Zone (CRZ). Maps showing wetland features between HTL and LTL and coastal land use features up to 500 m from HTL at 1:25,000 scale for the entire Indian coast using IRS LISS II and SPOT data have been prepared. The accurate demarcation of HTL and LTL is important, as they control boundaries of regulation zones. The HTL and LTL have been delineated based on tonal discontinuity. These maps provide conditions of land use and wetlands during 1990–1991, just before the announcement of CRZ by the Government of India. These maps have about 85% classification accuracy at 90% confidence level. The planimetric accuracy is about 60–75 m. The use of IRS-1C/1D data (the spatial resolution of the merged product is 5.8 m) has improved the planimetric accuracy to 15–20 m. CRZ maps are being used by the state governments to prepare coastal zone management plans. In 1992, the Government of India issued a notification to use satellite data for the preparation of 1:25,000-scale maps for regional planning. Efforts are continuing to use remote sensing data and GIS on the cadastral level.

The recent development along coasts has resulted in environmental deterioration. Broad-scale migration to Mumbai, Chennai, Calcutta and many other cities is expected in the next 15 years. The human population in coastal regions is likely to increase from 15% of the total Indian population in 2001 to almost 40% in the next decade. The population pressure and increased industrialization and urbanization will further deteriorate the coastal and marine environment. The broad-scale withdrawal of ground water for domestic and industrial use may lead to ingress of sea water. Municipal sewage, industrial wastewater, shipping, fishing industry, ocean dumping, tourism, oil spills, and non-point sediments are likely to be main sources of pollution. Broad-scale reclamation of wetlands is expected for residential, industrial, commercial and agricultural purposes. Multi-date IRS data and GIS have been used to prove legally that mangrove areas have been reclaimed in Mumbai. All these activities will lead to degradation/destruction of vital and critical habitats such as mangroves, coral reefs, sand dunes, and wetlands, and diminish the living resources of the Exclusive Economic Zone in the form of coastal and marine biodiversity and productivity. It is necessary to plan future development in an integrated manner.

55.4.4 Brackish Water Aquaculture Site Selection

Brackish water aquaculture has tremendous potential due to ever-increasing demand for prawns. In India, the aquaculture development started essentially to provide employment in rural coastal areas, as well as to increase exports to developed countries. Aquaculture development and planning require comprehensive data on land use and water resources. Remote sensing data provide information on these aspects due to repeat coverage, multi-spectral characteristics and synoptic view. Intensive commercial aquaculture practiced both in developed and developing countries is growing in popularity among many Asian and Latin American countries as an export industry. It could have harmful consequences as it replaces coastal mangrove habitats, thereby reducing breeding grounds of wild stocks. Broad-scale encroachment has been noticed in many mangrove areas.

IRS LISS II data has been utilized to prepare coastal land use maps at 1:50,000 scale along the Indian coast. These maps show wetland features between high and low water lines and land use features of the adjoining shore (up to 1.5 km from high waterline). The land use/wetland information has been used for evaluating the quality of the surrounding coastal waters, as this information was not available in most of the cases. Presence of saltpans and aquaculture ponds indicates availability of brackish water. The areas under mangroves/marsh are of vital/critical concern and were avoided for site selection. The spatial distribution of mud/tidal flat areas, which are most suitable from a substrate condition point of view, indicate potential areas available for brackish water aquaculture. These maps have been used by the Central and State Fisheries departments for evaluating proposed sites as well as for selecting new sites and reassessing potential for brackish water aquaculture. These maps, along with other engineering, biological, meteorological, socioeconomic and infrastructure-related parameters, were integrated using GIS for evaluating site suitability (Gupta et al. 1995). Site evaluation was carried out to determine whether a site was suitable or not and to appreciate requirements that might be needed to make the site suitable. This procedure has ensured that development for brackish water aquaculture will result in minimal damage to the ecology of the area.

55.4.5 Island Ecosystem

An island ecosystem information system has been developed for zoning the entire Andaman and Nicobar Islands into environmental-sensitive management zones for the purpose of preservation, conservation, development and utilization. A database was created using various thematic layers, viz., 1) 116 coastal wetland/land use maps at 1:25,000 scale using master grids, 2) 50 maps of coral reefs at 1:50,000 scale, 3) transport, and 4) habitation. The Survey of India standards were used for generating master grids, map projections and map registration.

A query shell (or interface) for the island ecosystem was developed to assign preservation, conservation, development and utilization zoning using an Environmental Appraisal Model, which assesses the environment, based on the conditions of coastal habitats. This shell has the capability to: 1) retrieve all the database information individually of one theme or perform integration of different themes, 2) generate theme-wise or zone-wise information with area for the entire Andaman or Nicobar Islands, and 3) generate output maps for land use, coral reef and coral reef zonation and land use zonation maps by giving the Survey of India topographical map number as the input or by selection of Area of Interest (AOI) interactively. It generates map-specific legend, scale, and map-index with a standard map format.

55.5 Coastal Processes

Worldwide, many coastal areas are being eroded, threatening the life and property of local populations. One of the major requirements of planning coastal protection work is to understand coastal processes of erosion, deposition, sediment transport, flooding and sea-level changes, which continuously modify the shoreline. Multi-date satellite data have been used to study shoreline change and coastal landforms to understand coastal processes. Satellite data available before 2000 had limited use for shoreline change studies; however, currently available high-resolution images are extremely useful for large-area sediment-transport studies and for detecting long term change in an entire coastline.

55.5.1 Shoreline Changes and Landforms

Shoreline is one of the most rapidly changing landforms. The accurate demarcation and monitoring of shoreline (long term, seasonal and short term changes) are necessary for understanding coastal processes. The historical and functional approaches to study shoreline changes along with various landforms help in deciphering the coastal processes operating in an area. The rate of shoreline change varies depending on factors such as the intensity of causative forces, warming of oceanic waters and melting of continental ice.

Shoreline change mapping for the periods 1967–1968, 1985–1989 and 1990–1992 for the entire Indian coast has been carried out using Landsat MSS/TM and IRS LISS II data at 1:250,000 and 1:50,000 scale. The silting of lagoons, progradation and degradation of spits, shoreline erosion, shifting of river mouths, formation of shoals and changes in estuary configuration have been noted (Nayak et al. 1992, 1997). The mere mapping of shoreline and its rate of change, however, is not sufficient information for understanding the coastal processes operating in an area. Information on near-shore water flow is also required. Data from IRS-P4 (Oceansat-1) (OCM) have been used to provide information on near-shore flow.

Landforms are best studied on aerial photographs in conjunction with topographical maps. In the absence of stereo aerial photography, considerable information on dynamics of topography may be inferred from sequential satellite data. IRS-1C/1D and Cartosat 1 panchromatic stereo data having 5.6 m and 2.5 m spatial resolution, respectively, are extremely useful for such studies.

The study of landforms provides clues to the processes operating in an area. This information, when combined with shoreline change, points towards possible causes for change.

Coastal landform studies of the Gulf of Khambhat at 1:250,000 scale using IRS LISS II FCC (Shaikh et al. 1989) indicated that it is possible to evaluate the role of sea level changes, neo-tectonics and sediment transport in shaping the present day landforms. Dunes, relict alluvium, terraces and paleo-mudflats indicate changes in sea level that had occurred in the past. The vast and extensive mud flats suggest that net transport of sediments in the Gulf is towards the land. The high earthen cliffs of Mahi and Narmada estuarine areas indicate the presence of neo-tectonic activities. Similar maps at 1:250,000 and 1:50,000 scales for the entire country have been prepared.

55.5.2 Impact of Developmental Activities

Rapid industrialization has led to development of new ports or expansion of existing facilities. The groins, sea walls, breakwaters and other protective structures have secondary effects resulting in downstream erosion. Erosion has been observed north of Visakhapatnam, Paradip, and Ennore, north of Madras, near Nagapattiam and Kanyakumari ports on the east coast of India, while deposition has been observed south of these ports. These changes are attributed to construction of artificial barriers like breakwaters, jetties, etc. (Nayak et al. 1992).

Knowledge about suspended sediment movement helps in understanding near-shore water flow. In one such study, a sediment plume of circular shape emerging from the Kochi Harbor was identified. This plume indicated a sharp contact with the sediments along the coast indicating two different water masses. It was also observed that sediment concentration was more on the northern side of the plume compared to the southern side. Such behavior was observed in all seasons. This clearly indicated that the plume was acting as an obstruction to the sediment movement, resulting in erosion on the southern coast and deposition on the northern coast. This behavior is also seen in the shoreline-change maps. The effect of break-waters on suspended sediment has been recorded near Tuticorin (Chauhan et al. 1996). The IRS-P4 Ocean Colour Monitor (OCM) data are extremely useful to study sediment dispersal and sediment transport studies due to their two-day repeat cycle. OCM data have been used for computing advective velocity of surface currents.

55.5.3 Impact of Dam Construction on Shoreline Equilibrium

It has been realized that construction of dams on rivers significantly alters coastal environments, at least for some time. The cooling pond of the Dhuvaran Thermal Power Station on the northern bank of the Mahi estuary in the Gulf of Khambhat had experienced severe erosion during 1979–1981, with an average rate of erosion of 0.5 m per day. This area also experiences very high tidal range. The analysis of multi-date satellite imagery indicated significant shoreline changes in the Mahi estuary between 1972 and 1988 (Nayak and Sahai 1985). These changes were attributed to construction of dams on the Mahi and Panam rivers in upstream regions during 1975. The repetitive nature of satellite data had helped to understand the estuarine behavior. Remedial measures in the form of a diaphragm wall and spurs have certainly helped to check the erosion. The effect of protective structures on the shoreline was also visible on the images. New beach had developed near Dhuwaran after 1986. New creeks have developed in the beach indicating the possibility of widespread erosion. The construction of dams along with the neo-tectonic activity are the main causes of unstable behavior of the estuary.

55.5.4 Coastal Salinity

The agricultural land around coastal regions is being affected by ingress of seawater. Using pre and post monsoon images from Landsat TM (1986) and employing condition of vegetation, color, association, and location as criteria, salt encrustation, saline, slightly saline, and non saline areas were delineated at 1:250,000 scale. The salt encrustation and saline areas, slightly saline areas, and non saline areas correlated well with soils having electrical conductivity (EC) of more than 2 mmhos, 1–2 mmhos and less than 1 mmhos (1:2 Soil:Water extract). Using this technique, the coastal belt around the Gulf of Khambhat (northwest Indian coast) was mapped. This region is severely affected by salinity/alkalinity that has resulted in low productivity throughout the area. It was observed that a 0.9 M ha area has been affected by salinity. The classification accuracy of these maps is 91%. Thus remote-sensing data were effectively used for mapping and monitoring of salt-affected areas.

55.5.5 Suspended Sediment Dynamics

Suspended sediments are easily observed on satellite imagery. They help in studying dynamic relationships between sediment input, transport and deposition. Tides play an important role in the movement of suspended sediments and fronts (Nayak and Sahai 1985). In the Gulf of Khambhat, a large tidal range gives rise to strong tidal currents and provides the mechanism for transport of suspended sediments. The net transport of sediments is towards land, as evidenced by extensive mudflats. The observation of suspended sediments suggests that during the monsoon, sediments brought in by various river systems remain in suspension and start settling down with the onset of winter season. A similar study was carried out in the Hooghly estuary area (Nayak et al. 1996). Successive images from IRS-P4 OCM were used for understanding impacts of tides on sediments in tide- and wave-dominated regions. Many interesting features, such as fronts, eddies, gyres, plumes, etc., were found. As the suspended sediments carry absorbed chemicals and fronts are associated with pollutants, knowledge about their movement will help in predicting the waste effluent transportation path. The wastewater discharge from a titanium plant has been traced near Thiruvananthapuram and efforts are being made to link this information to human activities upstream.

55.5.6 Bathymetry

Knowledge about depth values is important for coastal zone managers and navigators, for exploration and exploitation of non-living and living resources, operations on engineering structures and ocean circulation studies. Tides and currents constantly modify the submerged land, which proves hazardous for navigators. Updating the bathymetric charts by conventional methods is time-consuming and expensive. Remote sensing is a relatively cheap and fast method for periodically updating navigational routes. It is also a useful technique for detecting new reefs and shoals. The principal advantage of satellite data is the repetitive coverage. IRS data can be used for updating medium- and small-scale nautical charts. Techniques have been developed to retrieve depth values using high-resolution satellite data in shallow parts of the sea (Vyas and Andharia 1988) and Synthetic Aperture Radar (SAR) image data in coastal regions (Kumar et al. 1999). It was observed that inferred depths vary about 10–15% as compared to published charts. However, these methods are not routinely used for updating navigational charts.

55.6 Coastal Hazards

The coastal zone is subject to various cyclic and random processes, both natural and man-made, which continuously modify the region. Protection of human life, property and natural ecosystems from various hazards is a major concern. The major hazards are cyclones and associated tidal floods, tsunamis, coastal erosion, pollution and sea-level rise and its impact.

55.6.1 Cyclones and Storm Surges

Tropical cyclones constitute one of the most destructive natural disasters that affect India, especially along its east coast. The impact is greatest over coastal regions, which bear the brunt of strong winds, heavy rainfall and flooding. Remote sensing data and GIS have been utilized for tracking, monitoring and forecasting cyclones, and assessing damage and preventive measures. INSAT data have been regularly utilized to monitor the track of cyclones and forecast their crossing point on land. IRS-1C/1D data provide base imagery to assess the damage caused by cyclones and IRS WiFS data is used to delineate inundated areas (Nayak et al. 2001). LISS III data were found to be quite useful to assess damage caused to agriculture and horticulture areas, and PAN data provided input on structural damage caused to large buildings. NDVI images generated using OCM/WiFS data also provided information related to damage caused to vegetation. It was observed that mangroves are affected marginally. During the inundation associated with surge, the roots of mangroves get covered by mud and, thus, they cannot breathe. Hence defoliation occurs, but as the mud gets washed off by subsequent tides, mangroves recover in three to four months. This also suggests that mangrove afforestation should be taken up in cyclone-prone areas to protect life and property in coastal regions. Remote sensing data can be used to identify sites for such afforestation. The important aspect of damage assessment is to provide input within two to three days of the event, so that necessary relief measures can be activated. Many times, optical data are not useful due to cloud cover during the event and it is necessary to instead use radar data.

55.6.2 Coastal Erosion

Coastal erosion is a continuous and predictable process that causes damage on a relatively finer scale compared to cyclones. Large amounts of funds are spent in India to protect the shoreline by constructing sea walls. Details about shoreline changes have been discussed in the Section 55.5.1.

55.6.3 Sea-level Rise and Possible Effects

The present-day coastal landforms are essentially manifestations of geomorphic processes, neo-tectonics and relative sea-level fluctuations occurring in the last two million years. The sea-level rise is highly variable both in terms of time and space. Estimates for the Indian coast vary from 0.5 to 2.2 mm per year. It is expected that the rise in sea level will lead to erosion, tidal shift, seawater ingression and degradation of the coastal ecosystem. The Indian coast is a "trailing edge" coast and hence it is low-lying, with extensive sedimentary plains and wide continental shelves. Thus the projected relative sea-level rise would likely affect this type of coast. The response of different ecosystems to sea-level rise has been evaluated based on their characteristics for the Gujarat coast (Nayak 1994). Lowlands comprising barren mudflats may be inundated and eroded, as they comprise loose or less compacted sediments. In estuarine areas, the tidal amplitude will increase and bring changes in the current pattern. Human activities have significantly altered the sediment transport from land to estuaries. In flat and low-lying areas, the high tide line will move landward and inundate intertidal slopes. In such areas, the mangroves will colonize the adjoining non-vegetated tracts. Large areas in delta regions have degraded mangroves. The sparse mangrove will not be able to trap sufficient sediments to make new land. In such areas, the mangroves will be further degraded and possibly destroyed. Ultimately, the destruction of mangroves will lead to erosion of the coast. The growing deltas will not be affected. Most of the spits on the east coast of India are growing, which indicates pronounced depositional activity, especially the Krishna and the Gautami-Godavari deltas (central-east coast of India) and coast near the Vedaranyam part of the Cauveri delta. The estuarine areas will recede and tidal influence will be felt further landward. Seawater will enter groundwater further inland, especially in the area having cavernous limestone rocks and areas affected by past sea-level changes. It seems that conservation of mangrove areas will be one of the important steps to mitigate impact of the sea-level rise.

55.6.4 Non-point and Point Pollution

55.6.4.1 Non-point Pollution

Turbidity/suspended sediments and color/chlorophyll are indicators of water quality. Chlorophyll indicates trophic status, nutrient load and the possibility of pollutants in coastal waters. Suspended sediments affect navigation, fisheries, aquatic life, and the recreation potential of sea resorts. They also carry absorbed chemicals and create an environmental problem. Suspended sediments are easily observed on satellite images, which help in studying dynamic relationships between sediment input, transport and deposition. High sediment load is hazardous to many developmental activities. Knowledge about suspended sediment movement helps in predicting the waste effluent transportation path. Methods need to be developed to link concentrations of suspended sediment to upstream activities. In this regard, IRS-P4 OCM images are extremely useful.

Chlorophyll *a* has been recognized as an important environmental parameter for monitoring water quality. Satellite data show color variations and correlate well with chlorophyll. IRS-P3 MOS-B data have been extensively used to measure spatial variations in chlorophyll in the Arabian Sea. However, it has not been linked to any pollutants. The IRS-P4 OCM has provided excellent data to measure chlorophyll over large areas every two days. Attempts are being made to correlate eutrophic waters with high discharge of nutrient-rich waters.

55.6.4.2 Point Pollution

Municipal sewage and industrial waste are major types of pollution observed on the coast. Such waste out-falls are difficult to detect as near-shore waters are turbid, although some of the effluents have color and can be detected. One such waste out-fall from a titanium factory near Thiruvananthapuram was traced using high-resolution satellite data. Indian coastal waters are relatively free from pollution except for a few pockets around industrialized zones and large cities.

55.6.4.3 Oil Pollution

Oil rises to the surface and spreads across the water body, thus making oil spills amenable to remote detection. Few case studies have been carried out in India. In one such study, IRS-P4 OCM data was used for monitoring an oil slick that occurred in the Gulf of Kachchh. However, even the two-day repeat cycle of OCM data is not adequate for monitoring the slick in tide-dominated areas. Satellite surveillance is possible if high-resolution remote-sensing geo-synchronous satellite data are available.

55.6.5 Phytoplankton Blooms

Phytoplankton blooms are known to occur under various conditions in the near-shore regions. They may occur with cyclic regularity in certain regions where certain optimum environmental conditions prevail in marine waters. These plankton blooms produce certain toxins, which adversely affect fish and other organisms. The bloom usually takes place rather suddenly and may spread with amazing speed, changing the color of surface water to red, green or hay color. IRS-P4 OCM data have been used to monitor such blooms in the Arabian Sea.

55.6.6 Tsunami Warning System

GIS technology has been effectively used to develop a tsunami warning system in India since the devastating tsunami of December 2004 (Nayak and Srinivasa Kumar 2008). A state-of-the-art warning center has been established at INCOIS with all the necessary computational and communication infrastructure that enables reception of real-time data from the network of national and international seismic stations, tide gauges and bottom pressure recorders (BPRs). Earthquake parameters are computed during the 15 minutes or less time of occurrence. A database of pre-run scenarios for travel times and run-up height has been created using the Tunami N2 model. At the time of an event, the closest scenario is picked from the database for generating advisories. Water-level data enables confirmation or cancellation of a tsunami. Tsunami bulletins are then generated based on decision-support rules and disseminated to the concerned authorities for action, following a standard operating procedure. The criteria for generation of advisories (warning/alert/watch) are based on the tsunamigenic potential of an earthquake, travel time (i.e., time taken by the tsunami wave to reach the particular coast) and likely inundation. The performance of the system was tested on 12 September 2007 during an earthquake of magnitude 8.4 off the Java coast. The system performed as designed. It was possible to generate advisories in time for the administration to react and possible evacuation was avoided.

55.7 Living Resources

55.7.1 Ocean Color

Phytoplankton forms the first link in the ocean food chain and gives an indication about the standing stock of green biomass, which helps in predicting the third level of productivity. The varying levels of phytoplankton pigment (Chlorophyll *a*) and other constituents impart color varying from bluish to greenish to brownish. Hence, a satellite-based observing system having narrow spectral bands in the visible region provides better insight into our understanding of ocean productivity. It also provides better understanding of the role played by ocean productivity in the uptake of carbon dioxide from the atmosphere. IRS-P4 OCM has been providing ocean color data every two days for the Indian regions. Various models are under development to estimate primary productivity.

55.7.2 Identification of Potential Fishing Zones

India has high potential for marine fisheries development. The present fish production in the country is mainly from the coastal waters (up to depths of 50 m). The remote sensing satellites with their capability to monitor large spatial areas over oceans have proven to be of substantial economic benefit, particularly for a nation like India that has a long coastline and extensive EEZ. NOAA AVHRR data were being used in India to determine sea surface temperature (SST) at 1.1 km^2 pixel level over Indian oceanic regions on a daily and weekly (composite) basis by the National Remote Sensing Agency, Hyderabad. The Potential Fishing Zone (PFZ) maps are generated based on oceanographic features such as thermal boundaries, fronts, eddies, rings, gyres, meanders and upwelling regions visible on three to four day composite maps of SST (Narain et al. 1992). The extensive validation of these SST estimates with data collected by research vessels/drifting buoys has shown an accuracy of measurement of ±0.7° C. The major inadequacy of thermal infrared sensors is that they will measure ocean surface temperature only through a cloud-free atmosphere and strictly the ocean skin temperature.

Currently, GIS-integrated forecasts using ocean color, SST, bathymetry, and surface currents are routinely provided to fishermen (Nayak et al. 2007). The relationship between chlorophyll and SST has been studied (Solanki et al. 1998). The data from the OCM sensor, launched in May 1999 on-board IRS-P4 (Oceansat-1), has been used to develop GIS models that integrate chlorophyll information along with SST as a first step towards providing a fishery forecast to more accurately predict likely availability of fishes. It was observed that this technique is quite useful for pelagic fishery. The information on surface wave, wind, topography, and coastal circulation using microwave data that will become operationally available from planned series by different space-faring nations will also assist in developing an integrated model for fishery forecasting. Apart from this, new resources through sea ranching and mariculture in enclosed and semi-enclosed bodies will have to be tapped.

55.8 Conceptual Model for Coastal Zone Management

The management of coastal zones requires data on varied aspects as discussed earlier. Information exists in the form of thematic maps, as well as data in non-spatial formats. It is difficult to integrate these data conventionally, so it is necessary to develop a computer based information system composed of a comprehensive and integrated set of data designed for decision making. In this remote sensing based management plan, basic input about coastal areas is derived from remote sensing data. Integration of this thematic data with other secondary data leads to initial zoning.

The first step of data analysis is to divide the coastal zone depending on its environmental sensitivity. The index of environmental sensitivity can be decided based on criteria such as presence or absence of mangroves, coral reefs and erosional/depositional status of coast. It is proposed to divide the coast into four zones, i.e., preservation, conservation, utilization and development zones (Nayak et al. 1997). This information is available on wetland/land form and shoreline change maps at 1:250,000/1:50,000 scales. The second step includes validation of these zones and is carried out using similar maps at 1:50,000/1:25,000 scale along with other ground information such as flora, fauna, archaeological sites, areas under erosion, substrate conditions, biomass available, living and non-living resources, relief and pollution. The third step is to assess the causes of change in habitat conditions, make recommendations for conserving critical areas and creating a buffer around vital areas, prepare dredging schedules, and designate sites for industrial and engineering activities. The interaction between various activities conducted in the coastal zones is also assessed. This ensures judicious development of India's coastal zone without endangering the environment and ecology.

Acknowledgements

I am extremely grateful to Dr. A.K.S. Gopalan, former Director, Space Applications Centre, Ahmedabad, for his constant encouragement and support. Thanks are due to my colleagues Dr. Anjali Bahuguna, Dr. M.C. Gupta, Shri H.B. Chauhan, Dr. H.U. Solanki, and Dr. T. Srinivaskumar for helping me with the preparation of this chapter.

References

Aronoff, S. 1982. Classification accuracy: A user approach. *Photogrammetric Engineering and Remote Sensing* 48 (8):1299–1307.

Bahuguna, A. and S. Nayak. 1998. Coral reefs of Indian Coast. *Scientific Note.* SAC/RSA/ RSAG/DOD COS/SN/16/98. Space Applications Centre, Ahmedabad, India.

Carter, V. 1982. Application of remote sensing to wetlands. Pages 284–300 in *Remote Sensing for Resource Management.* Edited by C. Johannssen and J. L. Sanders. Soil Conservation Society of America, Iowa.

Chauhan, P., S. Nayak, R. Ramesh, R. Krishnamoorthy and S. Ramachandran. 1996. Remote sensing of suspended sediments along the Tamil Nadu Coastal waters. *J. Ind. Soc. Remote Sensing* 24 (3):105–114.

Cowardin, L. M., V. Carter, F. Golet and E. LaRoe. 1979. *Classification of Wetlands and Deep Water Habitats of the United States.* FWS/OBS-79/31. Office of Biological Services, US Fish and Wildlife Services, Washington, D.C.

Desai, P. S., A. Narain, S. R. Nayak, B. Manikiam, S. Adiga and A. N. Nath. 1991. IRS 1A applications for coastal and marine resources. *Current Science* 61 (3 & 4):204–208.

Gupta, M. C., S. Nayak, A. Bahuguna, M. Shaikh, A. Patel, A. Sinha and K. M. Parmar. 1995. Brackish water aquaculture site selection using techniques of Geographical Information System (GIS). *Scientific Note.* RSAM/SAC/CMASS/SN/08.95. Space Applications Centre, Ahmedabad, India.

Kumar, R., A. Sarkar and P. C. Pandey. 1999. Estimation of ocean depths off Goa coast using ERS-1 synthetic aperture radar data. *Continental Shelf Research* 19:171–181.

Narain, A., S. Beenakumari and M. Raman. 1992. Observation of a persistent coastal upwelling off Gujarat by NOAA AVHRR and its implication on fisheries. Pages 337–341 in *Remote Sensing Applications and Geographic Information Systems: Recent Trends.* New Delhi: Tata-McGraw Hill.

Navalgund, R. R. and A. Bahuguna. 1999. Applications of remote sensing and GIS in coastal zone management and conservation: Indian experience. Pages 121–146 in *Proceedings of UN-ESCAP/ISRO Science Symp. on Space Technology for Improving Quality of Life in Developing Countries: A Perspective for the Next Millennium.* N. Delhi, India, 15–17 Nov. 1999.

Nayak, S. 1994. Application of remote sensing data for estimation of impact of sea level rise along the Gujarat Coast. Pages 337–348 in *Global Change Studies. Scientific Results from ISRO Geosphere Biosphere Programme.* ISRO-GBP-SR-42-94. ISRO, Bangalore, India.

———. 1996. Monitoring the coastal environment of India using satellite data. *Science, Technology & Development* 14 (2):100–120.

———. 2000. Critical issues in coastal zone management and role of remote sensing. Pages 77–98 in *Subtle Issues in Coastal Management.* Indian Institute of Remote Sensing, Dehradun, India.

Nayak S. and A. Bahuguna. 2001. Application of RS data to monitor mangroves and other coastal vegetation of India. *Indian Journal of Marine Sciences* 30 (4) Dec:195–213.

Nayak, S. and B. Sahai. 1985. Coastal morphology: A case study in the Gulf of Khambhat (Cambay). *Inter. J. Remote Sens.* 6 (3 & 4):559–568.

Nayak, S. and T. Srinivasa Kumar. 2008. Addressing the risk of tsunami in the Indian Ocean. *J. of South Asia Disaster Studies* 1 (1):45–57.

Nayak, S., A. Pandeya, M. C. Gupta, C. R. Trivedi, K. N. Prasad and S. A. Kadri. 1989. Application of satellite data for monitoring degradation of tidal wetlands of the Gulf of Kachchh, Western India. *Acta Astronautica* 20:171–178.

Nayak, S., A. Bahuguna, M. Shaikh, R. S. Rao, C. R. Trivedi, K. N. Prasad, S. A. Kadri, P. H. Vaidya, V. B. Patel, S. H. Oza, S. S. Patel, T. Ananda Rao, A. N. Sheriff and P.V. Suresh. 1991. Manual for mapping of coastal wetlands/land forms and shoreline changes using satellite data. *Technical Note.* IRS UP/SAC/MCE/SN/32/91. Space Applications Centre, Ahmedabad, India.

Nayak, S., A. Bahuguna, M. Shaikh, H. B. Chauhan, R. S. Rao, A. Arya, J. P. Aggarwal, B. N. Srivastava, A. Patel, P. H. Vaidya, P. S. Dwivedi, A. G. Untawale, T. G. Jagtap, R. Chinna, N. Sankha, R. M. Dayal, Devendranath, T. Y. More, D. B. Dhera, A. V. Patil, H. S. Mahadar, T. Ananda Rao, A. N. Sheriff, P. V. Suresh, N.J.K. Nair, G. Shankar, S. Nalinakumar, V. Guruswamy, T. Fraklin, V. Kandaswamy, M. Ramalingam, S. Sanjeevi, R. Vaidyanadhan, R. Ramesh, N. K. Das, R. C. Samal and P. Kumar. 1992. Coastal Environment. *Scientific Note.* RSAM/SAC/COM/SN/11/92. Space Applications Centre, Ahmedabad, India.

Nayak, S., P. Chauhan, H. B. Chauhan, A. Bahuguna and A. Narendra Nath. 1996. IRS 1C applications for coastal zone management. *Current Science* 70 (7):614–618.

Nayak, S., A. Bahuguna, P. Chauhan, H. B. Chauhan and R. S. Rao. 1997. Remote sensing applications for coastal environmental management in India. *MAEER'S MIT PUNE JOURNAL, Special Issue on Coastal Environmental Management* 4 (15 & 16):113–125.

Nayak, S., R. K. Sarangi and A. S. Rajawat. 2001. Application of IRS P4 OCM data to study impact of cyclone on coastal environment of Orissa. *Current Science* 80 (9):1208–1213.

Nayak, S., T. Srinivaskumar and M. Nagarajakumar. 2007. Satellite-based fishery service in India. Pages 256–257 in *The Full Picture*. Published on behalf of Group on Earth Observations (GEO). Geneva, Switzerland: Tudor Rose.

Shaikh, M. G., S. R. Nayak, P. N. Shah and B. B. Jambusaria. 1989. Coastal land form mapping around the Gulf of Khambhat using Landsat TM data. *Jour. Ind. Soc. Remote Sens.* 17 (1):41–48.

Solanki, H. U., R. M. Dwivedi and S. Nayak. 1998. Relationship between IRS MOS-B derived chlorophyll and NOAA AVHRR SST: A case study in the NW Arabian Sea, India. Pages 438–442 in *Proceedings 2nd Inter. Workshop on MOS-IRS and Ocean Color*. Institute of Space Sensor Technology, Berlin, Germany, 10–12 June 1998.

Vyas, N. K. and H. I. Andharia. 1988. Coastal bathymetric studies from space imagery. *Marine Geodesy* 12:177.

CHAPTER 56
GIS in Support of Ecological Indicator Development

Carol A. Johnston, Terry N. Brown, Tom Hollenhorst, Peter T. Wolter, Nicholas P. Danz and *Gerald J. Niemi*

56.1 Introduction

Concern over environmental degradation has prompted establishment of ecological monitoring programs, but measuring and interpreting the myriad factors that influence or respond to environmental quality is impossible. Ecological indicators, defined as "a measure, an index of measures, or a model that characterizes an ecosystem or one of its critical components," have increasingly been used by monitoring programs to optimize their ability to detect environmental degradation (Jackson et al. 2000; Kurtz et al. 2001). Ecological indicators can be used to assess the condition of the environment, to provide an early warning signal of changes in the environment, or to diagnose the cause of an environmental problem (Dale and Beyeler 2001; Niemi and McDonald 2004). Indicators may characterize anthropogenic stress or ecological responses to stress, and may reflect biological, chemical, or physical attributes of ecological condition.

Geographic information system (GIS) technology can be instrumental in various aspects of the development and testing of ecological indicators (Johnston 1998). This chapter provides a case study of how GIS was used in the Great Lakes Environmental Indicators (GLEI) project, which had the overall goal of developing indicators of ecological condition for the US coastal region of the Laurentian Great Lakes. This chapter primarily illustrates the use of GIS in support of a massive field observational study designed to develop indicators from field measurements of a variety of many biota: plants, diatoms, fish, aquatic macroinvertebrates, birds, and amphibians. Particular emphasis was placed on the role of GIS in sampling design (deciding what sites to sample), response design (how to sample at a site), and calculating independent variables for statistical analyses. A secondary focus of this chapter is to demonstrate the use of GIS for landscape scale indicators.

Given that the Great Lakes basin is huge and spans four UTM zones (15, 16, 17, and 18) and 767,000 km², special consideration had to be given to a map projection that could be consistently applied across the region. An Albers Equal-Area Projection was adopted, which is a conic projection that uses two true-scale standard parallels, and is well suited for land masses that extend in an east to west orientation. All GLEI Project GIS databases used the following geographic format combination:

- Albers Equal-Area Projection
- North American Datum 1983 (NAD83)
- Geodetic Reference System of 1980 (GRS80), and
- Map units in meters

Custom parameter specifications used by the GLEI Project were:
- Central Meridian: − 96
- Reference Latitude: 23
- Standard Parallel 1:   29.5
- Standard Parallel 2:   45.5
- False Easting: 0
- False Northing: 0

All GIS analyses for the GLEI Project were done with ArcView™ 3.3 and ArcInfo™ (ESRI, Redlands, Calif.), with the aid of various extensions, Avenue™ scripts, and macros written with the Arc Macro Language (AML).

56.2 Dividing Coast and Basin into Units for Study Site Selection

An initial challenge to developing ecological indicators was to subdivide the US Great Lakes shoreline into meaningful units for field sampling that could be used to develop indicators. This was a daunting task, given that the US Great Lakes shoreline spans a distance of 17,017 km (Botts and Krushelnicki 1987). An important starting point was to conceptualize the geographic extent of the coastal ecosystems being studied, as well as the geographic extent of areas that could influence coastal systems but were not necessarily part of the coast. The following terminology was developed:

- Sampling domain – the maximum extent of the area within which field sampling will occur.
- Field site – the target ecosystem actually sampled, usually an individual wetland.
- Sampling locations – points, quadrats or transects within the sampling unit where actual field measurements were made.
- Stressor domain – area that is a source of anthropogenic stress to the sampling domain, which usually extends beyond the sampling domain (such as a watershed or an airshed). Stresses are physical, chemical, or biological entities that can bring about adverse ecological changes in ecosystem properties (US EPA 1998).

For coastal wetlands, the sampling domain has a subaqueous component in shallow waters lakeward of the shoreline, and a shoreland component landward of the shoreline (Figure 56-1). In the GLEI Project, the limits of the sampling domain varied by the response organism being sampled. For example, the lakeward limit for the wetland vegetation team was the limit of emergent hydrophytic vegetation (Reed 1988), but the lakeward limit for the fish team was determined by water depth rather than the presence of vegetation. GIS databases depicting subaqueous features were less common than those depicting terrestrial features, so the availability of GIS data differed across the sampling domain.

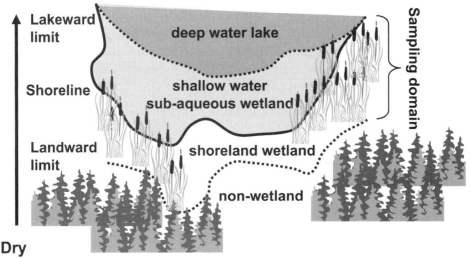

Figure 56-1 The sampling domain for coastal wetlands includes areas that are lakeward and landward of the shoreline.

56.2.1 Shoreline Segments

The shoreline between lake and land was a key feature of our sampling domain, and one which is depicted on multiple GIS data sources. The GLEI project utilized a shoreline depiction that was part of the US EPA River Reach File 3 (RF3, http://www.epa.gov/waters/doc/rfindex.html), modified to classify Strahler (1957) stream orders, obtained from the US Environmental Protection Agency National Health and Environmental Effects Research Laboratory in Corvallis, Oregon (Olsen 2001). This modified database allowed us to identify all streams of second order or larger that drained into the Great Lakes, which was why it was chosen over other possibilities (e.g., the National Hydrography Dataset). The nodes of intersection between stream mouths and the shoreline were used to subdivide the shoreline into reaches, and each reach was bisected using a custom computer program written in Avenue™ (ESRI, Redlands, Calif.). The two reach halves on either side of a stream mouth were then combined into a shoreline "segment" (Figure 56-2). This process resulted in 762 segments of the Great Lakes shoreline from the Minnesota–Canada border to Cape Vincent, New York, on Lake Ontario.

56.2.2 Segment-Sheds

The stressor domain was considered to be all land draining to the shoreline, which was defined topographically using the National Elevation Dataset (http://ned.usgs.gov/) (Gesch et al. 2002). Stressor domain subdivisions, called "segment-sheds," were also defined topographically as the area of land draining to each shoreline segment (Danz et al. 2005). The segment-sheds ranged in area from a 0.3 km^2 segment-shed surrounding an unnamed stream on Lake Huron to the 16,938 km^2 Maumee River segment-shed in Ohio, Indiana, and southern Michigan, draining to Lake Erie.

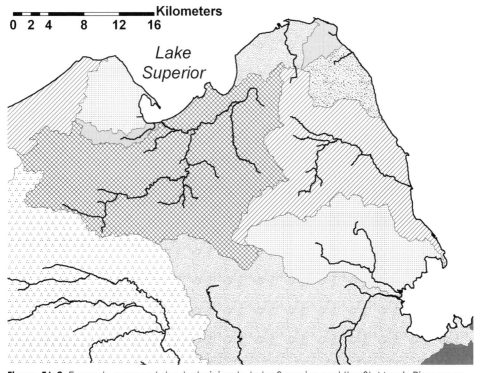

Figure 56-2 Example segment sheds draining to Lake Superior and the St. Mary's River near Sault Ste. Marie, Michigan. Each segment shed consists of the drainage area surrounding a second order-or-higher stream.

Manual of Geographic Information Systems

An online "segment browser" was developed as a graphical interface to provide the field biologists with a geographical context for each segment. Each shoreline segment and segment-shed was displayed in a GIS on a choice of two backgrounds, a digital raster graphic or a digital orthophotoquad, and converted to a jpeg snapshot of the composite image. The resultant four images were posted to a web interface that could be browsed online. Although the segment browser was not as versatile as an interactive map server, it provided a GIS-derived product that could be displayed quickly over the internet and required no GIS expertise.

56.2.3 Coastal Wetlands

Wetlands were the main coastal ecosystem of interest, so wetland databases were obtained for the eight Great Lakes states (Table 56-1). Digital products from the National Wetland Inventory (NWI) were deemed most useful because of their detail and relative consistency across the region. Digital versions of Michigan's National Wetland Inventory maps were not available through NWI, but a preliminary digital version was available from the Michigan Geographic Data Library, in which quadrangle coverages had been mapjoined into county coverages. The state of Wisconsin had conducted its own wetland inventory, which was similar to NWI (Johnston and Meysembourg 2002), and digital copies of the Wisconsin Wetlands Inventory for the Great Lakes drainage basin were obtained through a cooperative agreement with the US EPA. The NWI maps for most of Ohio had not been digitized, so the raster Ohio Wetland Inventory (OWI) had to be used. The OWI was derived from 1987 Landsat Thematic Mapper satellite imagery, and utilized only six wetland classes: wet forest, open water, shallow marsh, shrub/scrub wetland, wet meadow, and farmed wetland. The digital wetland maps were used as a source of data in the selection of field sites because wetlands were the sampling domain for most of the teams studying the different biota groups.

Table 56-1 Wetland GIS databases used by the GLEI project.

State	Source	URL
Minnesota, Illinois, Indiana, New York, Pennsylvania, small portion of Ohio	National Wetlands Inventory	http://www.nwi.fws.gov/
Wisconsin	Wisconsin Wetlands Inventory	http://dnr.wi.gov/wetlands/mapping.html
Michigan	Michigan Geographic Data Library	http://www.mcgi.state.mi.us/mgdl/
Ohio	Ohio Wetlands Inventory	http://www.dnr.state.oh.us/gims

56.3 Characterizing Anthropogenic Stress Gradients for Study Site Selection

Understanding the relationship between human activity and ecological response is essential to the process of indicator development; an indicator is not useful unless it varies predictably across a gradient of stress (Dale and Beyeler 2001). Although potential indicators can be shown to be responsive to stress in laboratory or field experiments, for large observational studies the best way to demonstrate responsiveness is by evaluating the potential indicator at sites along a gradient from relatively pristine to highly disturbed (US EPA 1998). Thus, different levels of stress must be present in the sample.

The goal of site selection was to obtain an unbiased group of sampling units that were suitable for developing indicators of ecological condition and represented the full range of anthropogenic stress across the Great Lakes (Danz et al. 2005). The segment-shed was the geographic unit used as the basis for characterizing stress in the site selection process.

Environmental profiles were created for each segment-shed to ensure that field sampling sites were distributed across the stressor and environmental gradients in the Great Lakes basin. Using primarily public sources, we collected GIS data for seven categories of human disturbance and environmental variation (Table 56-2), summarizing values by segment-shed for a total of 207 variables. For each variable, a single value was computed for each of the 762 segment-sheds (Figure 56-3).

GIS methods for computing segment-shed values for the individual stressor variables varied according to the data aggregation of the original data source. Only one raster GIS database was used—the National Land Cover Dataset (NLCD), which is a continuous grid of 30 m × 30 m pixels. The original data were subdivided by segment-shed boundaries, and land cover classes were summarized as a proportion of segment-shed area (e.g., proportion of evergreen forest, proportion of high intensity residential).

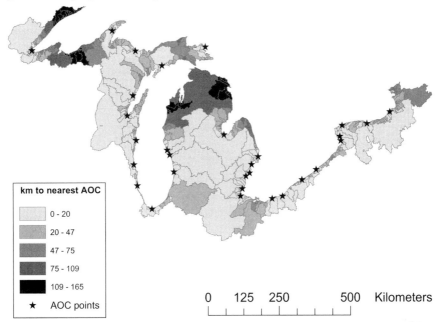

km to nearest AOC

	0 - 20
	20 - 47
	47 - 75
	75 - 109
	109 - 165
★	AOC points

0   125   250   500   Kilometers

Figure 56-3 The 762 GLEI segment-sheds, categorized by distance to nearest Area of Concern (AOC).

Point data sets were summarized by segment-shed in several different ways, depending on point density and data type. Point data obtained from the US EPA Toxic Release Inventory (TRI) consisted of 5,681 facility locations (points) throughout the US Great Lakes basin, for which there were 1,789,063 records for discharges of 330 chemicals (Figure 56-4). The TRI and the EPA's National Pollutant Discharge Elimination System (NPDES) data set was summarized as a count of the number of facilities per unit area of segment-shed. Point locations of mines, mine processing plants, and electric power plants were summarized as a count of the number of facilities in the segment-shed divided by the unit length of shoreline segment. Point measurements of atmospheric deposition collected at monitoring stations of the National Atmospheric Deposition Program (NADP) (Figure 56-5) were obtained for the year 2000. Because these data were widely scattered, they were interpolated and averaged for each segment-shed (Kg ha^{-1} yr^{-1}). Point locations of Great Lakes "Areas of Concern" (AOC) were obtained from the US EPA Region 5 office in Chicago (Bolka 2001), and the Euclidean distance from each segment-shed to the nearest AOC point was calculated. AOCs were identified by the 1978 Great Lakes Water Quality Agreement (http://www.epa.gov/glnpo/aoc/index.html), and are areas where serious impairment of beneficial uses of water or biota (swimming, fishing, drinking, navigation, etc.) is known to exist, or where environmental quality criteria are exceeded to the point that such impairment is likely.

Manual of Geographic Information Systems

Table 56-2 Sources of spatial data sets used for the seven categories of environmental variation.

Category	Database	Aggregation of Source Data	Description or Source
Agriculture and agricultural chemicals	Spatial Data in Geographic Information System Format on Agricultural Chemical Use, Land Use, and Cropping Practices in the United States	county	http://pubs.usgs.gov/wri/wri944176/ (Battaglin and Goolsby 1995)
	Potential Priority Watersheds for Protection of Water Quality from Nonpoint Sources Related to Agriculture	8-digit hydrologic units	http://www.nrcs.usda.gov/technical/ NRI/pubs/wqpost2.html (Kellogg et al. 1997)
	USDA NRCS 1997 National Resources Inventory	8-digit hydrologic units	http://www.nrcs.usda.gov/technical/ NRI/1997/summary_report/ (USDA 2001a, b)
	USDA NRCS 2001 Nutrient Management	8-digit hydrologic units	http://ias.sc.egov.usda.gov/parmsreport/ nutrient.asp (USDA 2001c)
	USGS SPARROW Total N, Total P	8-digit hydrologic units	http://water.usgs.gov/nawqa/sparrow/ wrr97/results.html (Smith et al. 1997)
Atmospheric deposition	National Atmospheric Deposition Program (NADP)	points	http://nadp.sws.uiuc.edu/ (Dossett and Bowersox 1999)
Soils	State Soil Geographic (STATSGO) Database	STATSGO map units	http://www.ncgc.nrcs.usda.gov/ products/datasets/statsgo/ (USDA 1994)
Land use and land cover	National Land Cover Dataset 1992 (NLCD 1992)	30 m × 30 m pixels	http://landcover.usgs.gov/ natllandcover.php (Vogelmann et al. 2001)
Shoreline modification	Great Lakes Environmental Research Laboratory (GLERL) Medium Resolution Vector Shoreline Data	lines	http://www.glerl.noaa.gov/data/char/ glshoreline.html (Stewart and Pope 1993)
Point and non-point pollution	USEPA Toxic Release Inventory Facilities in the United States	points	http://www.epa.gov/waterscience/ basins/metadata/tri.htm
	EPA/OW Permit Compliance System for CONUS (NPDES permits)	points	http://www.epa.gov/waterscience/ basins/metadata/pcs.htm
	Electric Power Plants of the United States (non-nuclear)	points	http://dss1.er.usgs.gov/ftp/appbasin/ ap_Ppall.met (USGS 1997)
	USGS Mineral and Metal Operations	points	http://tin.er.usgs.gov/mineplant/ (USGS 1998)
Human population density and development	Distance to Nearest Great Lakes Area of Concern (AOC)	points	http://www.epa.gov/glnpo/aoc/ index.html
	Census 2000 Census Block Data	census block	http://www.census.gov/geo/www/ census2k.html
	Census 2000 TIGER/Line Files	lines, census block polygons	http://www.census.gov/geo/www/tiger/ tiger2k/tgr2000.html

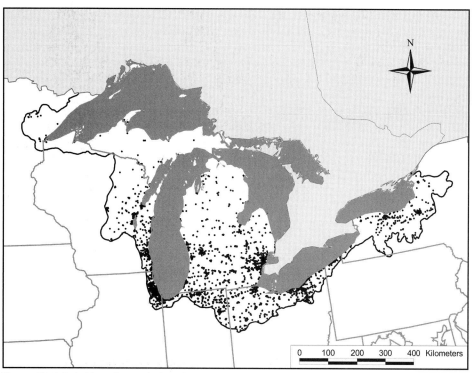

Figure 56-4 Point locations of 5,681 facilities listed in the Toxic Release Inventory data set for the US Great Lakes basin.

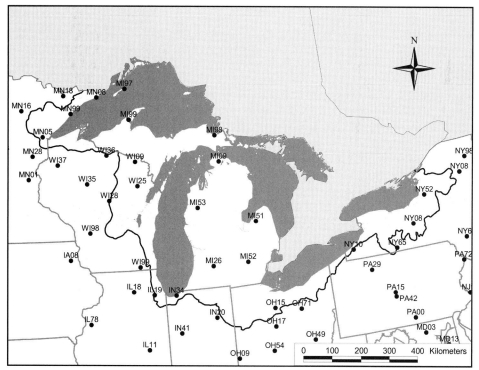

Figure 56-5 National Atmospheric Deposition Program monitoring sites near the Great Lakes.

Manual of Geographic Information Systems

Two types of linear data sets were used: roads and shoreline protection (Table 56-2). Linear road data were obtained from Census 2000 TIGER/line files, clipped with segment-shed boundaries, and summarized as length of roads per unit area of segment-shed. Shoreline protection had been mapped as linear segments of the Great Lakes shoreline (Stewart and Pope 1993), but those segments did not match the 762 segments developed by the GLEI project. Therefore, the different shoreline protection classes (no protection, minor protection, moderate protection, high protection, artificial shoreline, and non-structural protection) were summarized as length of protection class per unit length of GLEI segment.

Several data sources were aggregated by various types of polygons: counties, USGS hydrologic units, census blocks, and STATSGO map units. The US Census and STATSGO provide polygon boundaries for their map units as part of the data dissemination, but polygon boundaries had to be obtained from independent sources for counties and hydrologic units (Table 56-3). To calculate segment-shed values from these data sets, the polygons were intersected with the segment-shed boundaries, and area weighted averages were calculated for each variable of interest. The census block polygon data were rasterized prior to data analysis to simplify area weighting.

Table 56-3 Databases for geographic aggregation of stress data.

Database	Source	Description
Ecological Units of the Eastern United States	http://www.srs.fs.usda.gov/econ/data/keys/	Hierarchical ecoregion system developed by the US Forest Service (Keys et al. 1995).
County Boundaries	http://water.usgs.gov/lookup/getspatial?county100	1:100,000-scale base map of US county boundaries.
USGS Hydrologic Cataloging Units	http://water.usgs.gov/GIS/huc.html	Hierarchical regionalization system for major drainage basins developed by the USGS. Each hydrologic unit is numbered by a unique hydrologic unit code (HUC) consisting of two to eight digits based on the level of classification.

Because there was a large amount of redundancy in the full set of 207 environmental variables, principal component analysis (PCA) was used to remove redundancy and to reduce dimensionality within each category of environmental variables (Danz et al. 2005). Preliminary analysis of the environmental data had revealed major differences in primary environmental gradients between the two ecoprovinces (Keys et al. 1995) within the US Great Lakes basin, the Laurentian Mixed Forest and the Eastern Broadleaf Forest. Hence, the PCA analysis was done separately for those two ecoprovinces. A polygon file of the ecoprovince boundaries (Table 56-3) was used to assign segment-sheds to the two ecoprovince groups. Following this procedure, a cluster analysis was performed to group the segment sheds within each province into clusters with similar stress profiles. Segment sheds for sampling within each cluster were then randomly selected from available and accessible segment sheds within each cluster (Danz et al. 2005).

56.4 Establishing Sampling Locations within Field Sites and Providing Custom Maps for Field Use

Two fundamental sources of background information for field crews were Digital Raster Graphics (DRG) and Digital Orthophotoquads (DOQ). There are 563 1:24,000 quadrangles that intersect the Great Lakes coast from Minnesota to New York State, so the task of just compiling DRGs and DOQs for the Great Lakes coast required substantial effort.

After wetland field sites were randomly chosen within each segment-shed, GIS techniques were used to pre-select sampling locations within the wetlands. Randomization of sampling is required to obtain a statistically valid ecological sample, but the initial starting point used by field biologists is usually *not* random—it is often determined by the easiest access point to the field site (e.g., a road to a wetland, a boat ramp). For wetland vegetation sampling, we removed such bias by pre-selecting sampling transects and providing field scientists with the GPS coordinates to locate those transects within the wetland. An ArcView extension called Sample (http://www.quantdec.com/sample) was used to randomize transect placement within areas mapped by national and state wetland inventories as emergent wetland vegetation. Each transect intersected a randomly selected point generated by the Sample program (Figure 56-6a), and was oriented to be perpendicular to the perceived water depth gradient, extending from open water to the upland boundary or to a shrub-dominated wetland zone, if present (Figure 56-6b). Total transect length and target number of sample quadrats was determined in proportion to the size of the wetland to be sampled (20 quadrats/60 ha, with a minimum of 10–15 and 20 m of transect length/quadrat). Transect endpoint coordinates were uploaded into a handheld global positioning system (GPS) for use by wetland vegetation field crews (further described in section 56.5.2). Field crews were also provided with custom maps that depicted a regional context for the field site (e.g. roads, boat ramps, nearby towns) in the form of a clipped DRG (Figure 56-6c), as well as field site maps depicting transect locations on a DRG and a DOQ background. These were made available to widely distributed field teams via the website.

56.5 Georeferencing and Displaying Field Results

Documentation of sample location was provided by Global Positioning System (GPS) readings made by the field teams (Figure 56-7). This information was needed for future resampling of field sites, and to show which portion of a wetland field site was sampled by each of the field teams (Figure 56-8). The GPS readings also automatically provided the time and date of sampling. Consumer-grade handheld GPS units were used, which provided locational accuracy within a precision of a few meters; documenting sample locations more precisely would have required more expensive and bulkier GPS equipment.

A protocol was adopted that standardized collection and simplified processing of GPS data from the field teams. This protocol was designed primarily for Garmin GPS units, which were most commonly used in this project, but applied to other brands as well. This provided a consistent means to:

- Collect GPS waypoints representing sampling locations for each GLEI field team.
- Collect GPS tracking points for the entire time each team was at a particular study site.
- Link GPS waypoint IDs with field sample IDs.
- Maintain accurate date and time information for both waypoints and tracking data.
- Provide an interface to the GLEI database for uploading GPS tracking and waypoint data to the project website, thereby improving efficiency at processing GPS data.
- Supply GPS information rapidly to project investigators and other field teams who may be working in the same area.

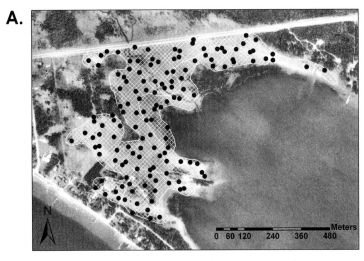

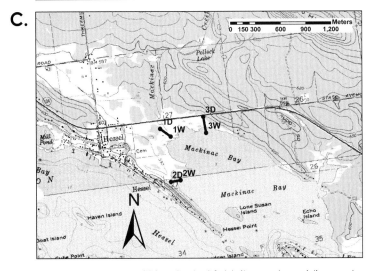

Figure 56-6 Establishing transects within selected field sites and providing custom maps for field use. A. Initial set of random sample points within areas mapped as emergent wetland. B. Transect that intersects a selected sample point and is perpendicular to the perceived water depth gradient. C. Clipped DRG that provides a regional context for the field site.

Figure 56-7 Field GPS unit attached to a tripod collects location data while field crew makes vegetation measurements.

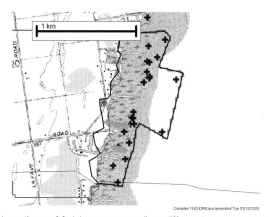

Figure 56-8 Sample locations of field crews sampling different response organisms. Asterisk indicates the single bird sampling location, scattered dark crosses indicate fish and macroinvertebrate sampling locations, and overlapping light crosses indicate vegetation sampling transects.

GPS Utility software, version 4.04.0, was used for uploading and downloading GPS data (current version as of 30 July 2008 is 4.94). This Windows-based program was compatible with the Garmin GPS MAP76 and Garmin 12 CX GPS units that were used in the field. The software is available for download at http://www.gpsu.co.uk (*GPS Utility* Limited, Hampshire, UK). A freeware version of *GPS Utility* is available with restricted storage and transfer capability (100 waypoints, 500 trackpoints), but registering the software for a nominal fee provides additional functionality. Although the following instructions were written for older GPS units that have a serial port interface between GPS and computer, the steps would be comparable for contemporary GPS units.

Manual of Geographic Information Systems

56.5.1 Settings for *GPS Utility*

To assure the accuracy and compatibility of collected GPS points, it was essential that they be downloaded using common settings in the *GPS Utility* software.

56.5.1.1 Set Datum to NAD 83

The NAD 83 datum was used during GPS data collection, as well as for final display. Thus, it was essential to set the datum to NAD 83 in *GPS Utility*. If NAD 83 was not the datum displayed on the tab in the upper right tab in the file box, it was set using the following steps (NOTE: the datum tab will only be present if a file, new or containing data, is open in *GPS Utility*):

    a. With *GPS Utility* open, select NEW under FILE.
    b. Under VIEW, select DATUM (otherwise click on datum display tab in upper right of box).
    c. If NAD 83 is present in the left box, select it and click OK.
    d. If NAD 83 is not present on the left, click on ADD, and scroll through the list of datums to NAD 83. Click on OK.
    e. Once set, NAD 83 will remain on the *GPS Utility* run on a given computer. If changes to program have been made, however, the datum will need to be verified.

56.5.1.2 Export of Database Fields

The *GPS Utility* default settings for export of database fields did not include all of the fields needed for the GLEI data and had to be added manually.

    a. With new file or file containing recently downloaded GPS points open in *GPS Utility*, select DATABASE FIELDS under OPTIONS.
    b. There should be adequate width allocated to each of the 9 fields except "symbol." If any of the other fields have a "0" in the WIDTH column, highlight it and manually enter an adequate number (if unsure, enter 15).
    c. When finished, click on OK and exit.

56.5.1.3 Set Coordinate Format to Decimal Degrees

    a. Check to see that the Coordinate Format is set to Decimal Degrees in *GPS Utility* (denoted by D.dddddd in the drop down box in the upper right of an open file box).
    b. If this is not the setting, click on the drop down arrow and select decimal degrees.

56.5.2 Loading Waypoints into the GPS Unit

Field teams requiring approximate sampling locations extracted from the GIS map data (e.g., transect endpoint coordinates) were given a text file (xxxx.txt) of target waypoints prior to field work. These waypoints were uploaded to their GPS unit(s) using *GPS Utility* as follows:

    a. Plug GPS unit into serial port of computer.
    b. Open *GPS Utility*.
    c. In *GPS Utility* open text file of waypoints (xxxx.txt).
    d. In *GPS Utility* select GPS, then UPLOAD ALL to load waypoints to GPS unit (it shouldn't matter what datum *GPS Utility* is set to, but for consistency set it to NAD 83). Waypoints are then loaded onto the GPS unit.
    e. "Incompatible Symbol Sets" may appear; if it does, click YES to continue.

56.5.3 GPS Field Use

Two types of locational data were collected in the field: waypoints (e.g., points sampled, entry point into the field site) and trackpoints. Trackpoints provide a trace of the path followed within the field site and were useful for backtracking data, analysis of effort, and efficiency investigations. After setting the GPS to collect trackpoints, the field team started the GPS unit to collect locations for the duration of work at the study site.

56.5.3.1 Locate Target Waypoints

a. Turn GPS unit on, press page button to proceed past opening screens.
b. Press MENU, and MENU again for main menu.
c. Check to make sure that GPS unit is set to Garmin (default) mode (this is found off the Main Menu SETUP INTERFACE tab).
d. Scroll down to desired point and press ENTER.
e. If the point is not visible, switch from "search for nearest" to "search by name." Press MENU to select waypoints by name or nearest. Press ENTER.
f. Select desired waypoint and press ENTER.
g. Scroll to desired waypoint, and select by pressing ENTER.
h. Scroll over to "Goto" button and select by pressing ENTER.
i. Press PAGE button to select desired Goto interface, and follow bearing or compass heading.

56.5.3.2 Record Waypoints Identifying Specific Locations of Sampling Activities

a. At a desired sampling location, with GPS on, press and hold ENTER.
b. Mark Waypoint interface will appear. Press MENU and choose AVERAGE LOCATION, press ENTER.
c. Average Location interface will appear, counting measurements, wait for an acceptable estimated accuracy (less than 6 meters), and press ENTER to save.
d. Mark Waypoint interface re appears with the waypoint ID on the top. Record this ID with the date and time on the data sheet. It is possible to change this ID, but we recommend simply recording the ID number and allowing the GPS unit to increment the ID for each waypoint.
e. Move to next point.
f. To distinguish waypoints associated with each study site, record the ID, date and time for the first and last waypoint collected at each site. This information is needed when the points are uploaded to the GLEI database.

56.5.4 Export Data from GPS Unit to Portable Computer

Waypoints were downloaded from GPS units to a portable (i.e., laptop) computer each night as follows:

a. Plug GPS unit into serial port of computer.
b. Open *GPS Utility* (set *GPS Utility* settings as specified in Section 56.5.1).
c. Under GPS select CONNECT to connect to GPS unit.
d. In *GPS Utility* click GPS, then DOWNLOAD ALL to load waypoints and trackpoints from the GPS unit to the computer.
e. A record for each waypoint will appear in *GPS Utility*, if the waypoint view is selected. Trackpoints can also be viewed by selecting the trackpoint view.
f. Choose SAVE AS with the type set to "text (.txt)" and navigate to an appropriate directory and save downloaded GPS points with appropriate file name. For reference, file names should indicate the segment number, subcomponent team, the GPS unit and the date the data were collected.
g. Close this file before downloading new points from GPS unit (select CLEAR ALL under RECORD separately with Waypoints and Tracks selected under VIEW).
h. It is a good idea to clear all downloaded and saved data from *GPS Utility* before shutting down the program for the day.
i. Once points have been downloaded from GPS unit and saved through *GPS Utility*, clear all Waypoints and Trackpoints from GPS Unit (see instructions on following pages under appropriate unit).

56.5.5 Upload GPS Data to the GLEI Database

After GPS points from the field were downloaded and saved to a computer using *GPS Utility*, they were entered into the GLEI database as soon as an internet connection was available. A custom web interface was developed by the GLEI project for this purpose, but this function could be provided by conventional spreadsheet or database software.

56.6 Calculating Stressors at Specific Site Locations

Segment-sheds were subdivisions of the stressor domain used in the sample selection process, but indicator development required that stressors be quantified for the specific field sites sampled. Two new sets of boundaries were developed: boundaries circumscribing the actual sampling locations (Figure 56-9a), and boundaries circumscribing the land draining to each field site (Figure 56-9b). The field site boundaries were digitized interactively while displaying sampling locations over DOQs and/or wetland maps, and the land draining to each field site (i.e., field site watersheds) was computed using DEM data, using the same methods that had previously been used for the segment-sheds. These field site watersheds were smaller than segment-sheds, thereby tightening the coupling between stressor data and downslope wetlands, and hence the relationships between environmental stress and ecological response variables.

The same stressor variables computed for segment-sheds were also calculated for field site watersheds. A smaller set of variables was computed for the area within the field site boundaries, including new variables that would have been too time-consuming to gather for all of the segment-sheds. For example, an estimate of local hydrologic modification was developed by measuring the length of features that likely disrupt the natural flow and fluctuation of water (e.g., road beds, dikes, ditches) within wetland field site boundaries, and dividing by field site area (Bourdaghs 2004). These hydrologic modifications were identified from DRGs and DOQs, because there was no existing digital data set that provided such information.

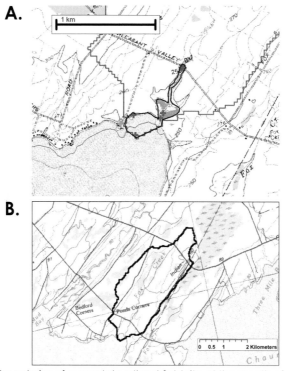

Figure 56-9 A. Boundaries of a coastal wetland field site at the mouth of Fox Creek on northern Lake Ontario. B. Boundaries of the watershed draining to the Fox Creek field site.

56.7 Developing Maps of Stressor Change (Land Use)

The US Great Lakes basin is experiencing land use changes that could greatly affect the condition of coastal and nearshore environments (Cummings 1978; Johnston 1992; Thorp et al. 1997). The pace of land use change, particularly in urban and suburban areas, far exceeds that predicted by population growth alone. Urbanization from 1970 to 1990 in the Chicago metropolitan area, for example, increased its developed area by 19.0%, with only a 2.2% increase in population (Auch et al. 2004).

Land use change over time is an important landscape scale indicator of development pressure. Land change results from changing human demographics, natural resource uses, agricultural technologies, economic priorities, and land tenure systems. Different land uses impose different environmental stresses on natural plant and animal communities, with consequent implications to water quality, climate, ecosystem goods and services, economic welfare, and human health (Gutman et al. 2004).

GIS provides the computational tools for quantifying land use change, but caution must be used with the use of existing data sets for land change calculations because differences are often due to map and registration errors rather than real change (Congalton and Green 1999). Differences in land use classification systems over time prevent quantitative comparisons of change, because differences between databases may merely represent different interpretations of land use.

We sought to measure decadal land use/land cover (LULC) change by comparing the 1992 National Land Cover Data (NLCD) with the 2001 Coastal Change Analysis Program (C CAP) data set for the Great Lakes basin (Vogelman et al. 2001; US NOAA 2003). The two data sets had 21 and 22 land cover classes, respectively, and were both derived from Landsat Thematic Mapper satellite imagery with a 30-m spatial resolution. Although this would seem like a routine process given the apparent similarities between the two data sets, a number of modifications were necessary to minimize spurious differences that were not true changes over time.

Although the number of classes used by the two LULC data sets was comparable, there was not a one-to-one correspondence between all classes in the two data sets. For example, the NLCD data set mapped more classes of agricultural land than did the C-CAP data set, and the C-CAP data set mapped more classes of wetland type than did the NLCD data (Figure 56-10). Rules had to be developed to cross-walk the two sets of classes.

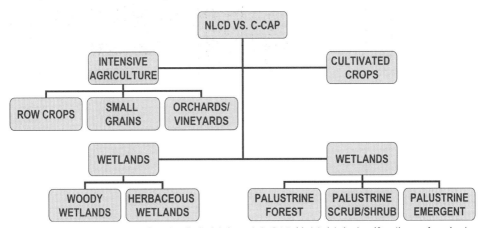

Figure 56-10 Comparison of NLCD (left side) and C-CAP (right side) classifications of agricultural and wetland cover types.

Additionally, it was apparent from inspection of the two data layers that ancillary data had been used to map roads on the C-CAP map but not the NLCD maps. To ensure that both maps were consistent in their depiction of roads, all major paved road vectors in the 1992 and 2001 TIGER databases were selected and converted to 30 m raster data, which were overlaid on the corresponding NLCD and C-CAP data sets. A final product was an enhanced US Great Lakes Land Cover 2001 data layer (Figure 56-11) that could be compared with the 1992 NLCD data layer to detect land cover change during the 1990s.

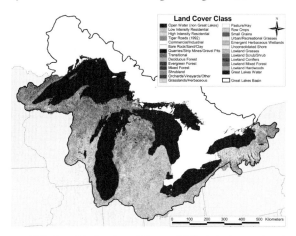

Figure 56-11 Enhanced US Great Lakes Land Cover 2001 data layer developed by GLEI for land cover change detection analysis. See included DVD for color version.

After the change detection matrix was computed for the two data layers, transition cases were examined for logical consistency and edited where needed. Vegetation transitions were checked to ensure that inferred succession rates were reasonable. For example, a time 1 to time 2 change from "transitional" to "hardwood forest" was deemed feasible, but a change from "grassland" to "hardwood forest" was not, because establishment of a mature forest is impossible over such a short time span within the climatic region of the Great Lakes. Details of the process used to develop the change map are detailed by Wolter and colleagues (Wolter et al. 2006).

56.8 Summary

GIS and GPS technologies were an integral part of the GLEI project, and were used in many different ways:
- Subdividing the sampling domain into manageable and ecologically relevant units for study site selection.
- Characterizing anthropogenic stress gradients for study site selection.
- Randomization of sampling locations prior to field work.
- Georeferencing and display of final sampling locations after field work.
- Providing custom maps for use by field crews.
- Defining geographic extents of areas sampled and the watersheds draining to those areas.
- Calculating anthropogenic stressors within areas sampled and the watersheds draining to them.
- Developing landscape scale indicators from mapped data.

Existing GIS data were fundamental to the success of the project, ranging from base layers such as DRGs and DOQs to stressor data sources such as atmospheric deposition and pesticide runoff. Although the increasing availability of GIS data layers was very beneficial, there was still considerable effort required to use them for ecological indicator development:

data discovery, database acquisition, database manipulation (e.g., clipping, calculation, projection, edge matching, cross walking classifications, generalization), and quality assurance and quality control (QA/QC). This effort resulted in extensive GIS holdings that provided benefits to a number of subsequent ecological GIS applications, providing benefits beyond the GLEI project.

Most importantly, the GLEI GIS data set and its user interfaces provided a common framework to unify the work of dozens of GLEI investigators spread across the Great Lakes basin into a cohesive whole. Without this common framework, the development of integrated indicators of ecological condition would not have been possible.

56.9 Acknowledgements

We are grateful to Bruce Pengra for assistance with figure preparation, and to Connie Host, Paul Meysembourg, Jim Salés, and Gerald Sjerven for assistance in compiling GIS data. This research was supported by a grant from the National Aeronautics and Space Administration (NAG5 11262 Sup 5) and through a cooperative agreement with US Environmental Protection Agency's Science to Achieve Results (STAR) Estuarine and Great Lakes (EaGLe) program through funding to Great Lakes Environmental Indicators (GLEI), US EPA Agreement R828675 00. Although the research described in this article was funded in part by the US Environmental Protection Agency, it has not been subjected to the Agency's required peer and policy review and therefore does not necessarily reflect the views of the Agency and no official endorsement should be inferred.

References

Auch, R., J. Taylor and W. Acevado. 2004. *Urban Growth in American Cities: Glimpses of U.S. Urbanization.* US Geological Survey Circular 1252. US Geological Survey, Sioux Falls, S.D.

Battaglin, W. A. and D. A. Goolsby. 1995. *Spatial Data in Geographic Information System Format on Agricultural Chemical Use, Land Use, and Cropping Practices in the United States.* US Geological Survey, Water Resources Investigations Report 94-4176.

Bolka, Barry. 2001. Meeting at US EPA Region 5 Office, Chicago, Ill., 16 February.

Botts, L. and B. Krushelnicki. 1987. *The Great Lakes: An Environmental Atlas and Resource Book.* Environment Canada, Toronto, Ontario and United States Environmental Protection Agency, Chicago, Illinois.

Bourdaghs, M. 2004. Properties and performance of the floristic quality index in Great Lakes coastal wetlands. Master's thesis, University of Minnesota, Duluth, Minn.

Congalton, R. and K. Green. 1999. *Assessing the Accuracy of Remotely Sensed Data: Principles and Practices.* Boca Raton, Fla.: CRC/Lewis Press.

Cummings, T. R. 1978. *Agricultural Land Use and Water Quality in the Upper St. Joseph River Basin, Michigan.* US Geological Survey Open-File Report 78-950.

Dale, V. H. and S. C. Beyeler. 2001. Challenges in the development and use of ecological indicators. *Ecological Indicators* 1:3–10.

Danz, N. P., R. R. Regal, G. J. Niemi, V. Brady, T. Hollenhorst, L. B. Johnson, G. E. Host, J. M. Hanowski, C. A. Johnston, T. Brown, J. Kingston and J. R. Kelly. 2005. Environmentally stratified sampling design for the development of Great Lakes environmental indicators. *Environmental Monitoring and Assessment* 102:41–65.

Dossett, S. R. and V. C. Bowersox. 1999. *National Trends Network Site Operation Manual. NADP Manual 1999–01.* National Atmospheric Deposition Program Office at the Illinois State Water Survey, Champaign, Ill. <http://nadp.sws.uiuc.edu/lib/manuals/opman.pdf> Accessed 1 August 2008.

Gesch, D., M. Oimoen, S. Greenlee, C. Nelson, M. Steuck and D. Tyler. 2002. The national elevation dataset. *Photogrammetric Engineering and Remote Sensing* 68:5–12.

Gutman, G., A. Janetos, C. Justice, E. Moran, J. Mustard, R. Rindfuss, D. Skole and B. J. Turner II, eds. 2004. *Land Change Science: Observing, Monitoring, and Understanding Trajectories of Change on the Earth's Surface.* New York: Kluwer Academic Publishers.

Jackson, L. E., J. C. Kurtz and W. S. Fisher, eds. 2000. *Evaluation Guidelines for Ecological Indicators.* EPA/620/R-99/005. US Environmental Protection Agency, Office of Research and Development, Research Triangle Park, N.C., USA.

Johnston, C. A. 1992. Land use activities and western Lake Superior water quality. Pages 145–162 in *Making a Great Lake Superior. Proc. 1990 Int. Conf. on Remedial Action Plans in the Lake Superior Basin.* Edited by J. Vander Wal and P. D. Watts. Lakehead University Centre for Northern Studies Occasional Paper #9, Thunder Bay, Ontario.

————. 1998. *Geographic Information Systems in Ecology.* Oxford, England: Blackwell Science.

Johnston, C. A. and P. Meysembourg. 2002. Comparison of the Wisconsin and National Wetlands Inventories. *Wetlands* 22:386–405.

Kellogg, R. L., S. Wallace, K. Alt and D. W. Goss. 1997. Potential Priority Watersheds for Protection of Water Quality from Nonpoint Sources Related to Agriculture. *52nd Annual Soil and Water Conservation Service Conference,* Toronto, Ontario, Canada, 22–25 1997. <http://www.nrcs.usda.gov/technical/NRI/pubs/wqpost2.html> Accessed 28 July 2008.

Keys, J. E., Jr., C. A. Carpenter, S. L. Hooks, F. G. Koeneg, W. H. McNab, W. Russell and M. L. Smith. 1995. *Ecological Units of the Eastern United States: First Approximation.* Map (scale 1:3,500,000). Technical Publication R8-TP 21. US Dept. of Agriculture, Forest Service, Atlanta, Georgia.

Kurtz, J. C., L. E. Jackson and W. S. Fisher. 2001. Strategies for evaluating indicators based on guidelines from the Environmental Protection Agency's Office of Research and Development. *Ecological Indicators* 1:49–60.

Niemi, G. J. and M. E. McDonald. 2004. Application of ecological indicators. *Annual Review of Ecology, Evolution, and Systematics* 35:89–111.

Olsen, Tony. 2001. Unpublished material presented at GLEI Advisory Committee meeting at the University of Minnesota, Duluth, Minn., 20 April.

Reed, P. B. 1988. *National List of Vascular Plant Species That Occur in Wetlands.* US Fish & Wildlife Service Biological Report 88(24). Superintendent of Documents, Washington, DC.

Smith, R. A., G. E. Schwarz and R. B. Alexander. 1997. Regional interpretation of water quality monitoring data. *Water Resources Research* 33 (12):2781–2798.

Stewart, C. J. and J. Pope. 1993. *Erosion Processes Task Group Report.* International Joint Commission Great Lakes Water Level Reference Study, Working Committee 2, Land Use and Management.

Strahler, A. N. 1957. Quantitative analysis of watershed geomorphology. *Transactions of the American Geophysical Union* 8:913–920.

Thorp, S., R. Rivers and V. Pebbles. 1997. Impacts of changing land use. *State of the Lakes Ecosystem Conference (SOLEC) 1996.* EPA 905-R-97-015d. US EPA Great Lakes National Program Office, Chicago, Ill. <http://www.epa.gov/glnpo/solec/solec_1996/The_Impacts_of_Changing_Land_Use.PDF> Accessed 1 August 2008.

USDA (US Department of Agriculture). 1994. *State Soil Geographic Data Base (STATSGO), Data Users Guide.* Natural Resources Conservation Service, Fort Worth, Texas. US Department of Agriculture Miscellaneous Publication Number 1492.

————. 2001a. *1997 National Resources Inventory (revised December 2000).* CD-ROM, Version 1. Natural Resources Conservation Service, Washington, DC, and Statistical Laboratory, Iowa State University, Ames, Iowa.

———. 2001b. *A Guide for Users of 1997 NRI Data Files.* CD-ROM, Version 1. Natural Resources Conservation Service, Washington, DC, and Statistical Laboratory, Iowa State University, Ames, Iowa.

———. 2001c. *FY 2001 Nutrient Management, Key Conservation Treatments.* Performance and Results Measurement System, Natural Resources Conservation Service, US Department of Agriculture. <http://ias.sc.egov.usda.gov/parmsreport/nutrient.asp> Accessed 1 August 2008.

US EPA (US Environmental Protection Agency). 1998. *Guidelines for Ecological Risk Assessment.* EPA/630/R 95/002Fa, Federal Register 63(93), 26846–26924.

USGS (US Geological Survey). 1997. *Electric Power Plants of the United States (non nuclear).* USGS Open File Report 97-172. John Tully, compiler. USGS, Eastern Energy Team, Reston, Va.

———. 1998. *Mineral and Metal Operations.* Reston, Va.: US Geological Survey.

US NOAA (US National Oceanic and Atmospheric Administration). 2003. *Great Lakes Accuracy Assessment Report.* Internal report prepared by Perot Systems Government Services. Project 12, Event 9.01, Coastal Services Center, Charleston, S.C.

Vogelmann, J. E., S. M. Howard, L. Yang, C. R. Larson, B. K. Wylie and N. Van Driel. 2001. Completion of the 1990s national land cover data set for the conterminous United States from Landsat Thematic Mapper data and ancillary data sources. *Photogrammetric Engineering and Remote Sensing* 67 (6):650–662.

Wolter, P. T., C. A. Johnston and G. J. Niemi. 2006. Land use change in the U.S. Great Lakes basin 1992 to 2001. *Journal of Great Lakes Research* 32: 607–628.

CHAPTER 57

The Role of GIS in Military Strategy, Operations and Tactics

Steven D. Fleming, Michael D. Hendricks and *John A. Brockhaus*

57.1 Introduction

The United States military has used geospatial information in every conflict throughout its history of warfare. Until the last quarter century, geospatial information used by commanders on the battlefield was in the form of paper maps. Of note, these maps played pivotal roles on the littoral battlegrounds of Normandy, Tarawa and Iwo Jima (Greiss 1984; Ballendorf 2003). Digital geospatial data were employed *extensively* for the first time during military actions on Grenada in 1983 (Cole 1998). Since then, our military has conducted numerous operations while preparing for many like contingencies (Cole 1998; Krulak 1999). US forces have and will continue to depend on maps—both analog and digital—as baseline planning tools for military operations that employ both Legacy and Objective Forces (Murray and O'Leary 2002).

Important catalysts involved in transitioning the US military from dependency on analog to digital products include: (1) the Global Positioning System (GPS); (2) unmanned aerial vehicles (UAVs); (3) high-resolution satellite imagery; and (4) geographic information systems (GISs) (NIMA 2003). In addressing these four important catalysts, this review is first structured to include a summary of geospatial data collection technologies, traditional and state-of-the-art, relevant to military operations and, second, to examine GIS integration of these data for use in military applications. The application that will be addressed is the development and analysis of littoral warfare (LW) databases used to assess maneuvers in coastal zones (Fleming et al. 2008).

57.2 Geospatial Data Collection Technologies

Three major data collection categories used in populating GIS databases include: (1) field data collection and GPS; (2) aerial reconnaissance; and (3) satellite reconnaissance. Discussed here, these collection methods provide a complementary mix of platforms and technologies for gathering information about operational areas.

57.2.1 Field Data Collection and GPS

There are numerous methods of collecting raw data in the field for direct input into geospatial databases. These methods are most often used when the required data do not exist in any other readily available format, such as maps, photographs or satellite images. Field data also are frequently collected when "ground truthing" of remotely sensed data is required. Traditional manual surveying techniques make use of levels and theodolites for directly collecting field measurements. Modern digital equivalents of these manual techniques have been developed so that data collected are stored in digital format ready for direct input into a GIS. Examples here include total stations (high-precision theodolites with electronic distance measurement [EDM] and data logger capabilities), hand-held laser range finders and digital compasses. A universal military locating system, GPS, was designed and fully introduced to the military by the late 1980s. During this time, global missions for US forces expanded dramatically, often requiring immediate information about "place" anywhere on Earth. Joint operations between services became the norm for how America's military planned and executed tasks. A common system for providing key location data for friendly units, enemy targets and critical terrain was required.

Joint US combat operations in Grenada (1983) demonstrated the need for improved positioning technology. Although US forces prevailed as a result of large amounts of non-standard geospatial data between services, the conflict was not an efficient, well-coordinated effort by any measure of warfighting (Cole 1998). Since then, GPS integration and employment has accelerated, becoming the answer to many location-based challenges brought about by mission and interoperability changes.

The GPS, including satellites and monitoring equipment, undergoes constant improvement cycles to increase accuracy, reliability and capability. Currently, military GPS receivers reliably provide position accuracies to within one meter (GPS JPO 2000). These receivers have been made smaller, more accurate and easier to use. Microelectronics have made them very affordable so that every individual, weapon system and command post can share the technology, making available the benefits of a reliable, accurate worldwide navigation and positioning system (Huybrechts 2004).

The user-equipment segment of GPS consists of the military receivers, antennae and other GPS-related equipment. Global positioning system receivers are used on aircraft, ships at sea, ground vehicles or hand-carried by individuals. They convert satellite signals into position, velocity and time estimates for navigation, positioning and time dissemination. Most of the user equipment is employed by more than one branch of military service with very few (if any) having utility for just one.

System devices and GPS-aided weapons have been employed in numerous warfighting applications including navigation and positioning, weapon guidance, targeting and fire control, intelligence and imagery, attack coordination, search and rescue, force location, communication network timing and force deployment/logistics (NAVSTAR 2001). Major benefits of GPS realized in these applications include: (1) improved position accuracy; (2) more accurate weapon placement; (3) enhanced systems performance; and (4) time synchronization (GPS JPO 2000). Table 57-1 provides a detailed listing of benefits derived from GPS employment.

The GPS has a bright future; it is being improved to preserve the advantages it brings to the battlefield and to prevent its vulnerability to attack (GISDevelopment 2004). The vulnerability of the GPS includes terrorist use as demonstrated by the tragic events of 11 September 2001 where al Qaeda loyalists exploited GPS technology in guiding airliners into their targets on the US mainland.

Changes designed to better support the warfighter in an evolving threat environment are planned. They will provide more flexibility through more portable systems as well as military anti-jam capability, meaning that GPS accuracy will be maintained closer to the target in a high-jamming environment. In this, the GPS has recently been linked to commercial laptop computers and personal data assistants. Overall, GPS will provide a more secure, robust military signal service, assuring acquisition of the GPS signal when needed in a hostile electronic environment (Kimble and Veit 2000). Ongoing changes will deny an enemy the military advantage of GPS, thereby protecting friendly force operations and preserving peaceful GPS use outside areas of operations (SPAWAR 2001).

Table 57-1 Military benefits resulting from employing GPS.

Table 57-1 Military benefits resulting from employing GPS.

Improved Position Accuracy	Accurate Weapon Placement
Mine Countermeasures	Saved Ordnance
Search and Rescue	Improved "Kill Ratios"
Special and Night Operations	Increased Efficiency
Intelligence Assessments	Demoralized Enemy
Logistics Support & Tanker Ops	Reduced Exposure to Hostile Fires
Enhanced Systems Performance	**Time Synchronization**
Standoff Land Attack Missile	Command and Control
Patriot	Secure Communications
Artillery and Armored Vehicles	Coordinated Operations
Sensors	Joint Operations
Attack Aircraft	Special Operations

57.2.2 Aerial Reconnaissance

There are numerous methods of collecting data via aerial reconnaissance for use in military operations. Some methods have been used for many years, while others make use of relatively new technologies. Included here is a discussion of two primary methods of employing airborne reconnaissance platforms to populate military geospatial databases: (1) air photos and digital images; and (2) sensor data obtained with UAVs.

57.2.2.1 Air Photographs and Digital Images

Aerial photographs have been traditionally used for over 75 years in mapping littoral regions (NOAA 1997). Taken from specially designed aerial camera systems, several different types of aerial photographs have been used routinely by military intelligence sources. These include simple black-and-white (panchromatic), color and color infrared. Color-infrared systems assist military analysts in camouflage detection mandates.

Current aerial photographs show changes that have taken place since the making of a map. For this reason, in military operations, maps and aerial photographs complement each other. More information can be gained by using the two together than by using either alone. Detailed in Table 57-2, aerial photographs (or digital images) provide many advantages over an analog map for military applications.

Over the past decade, digital images have been used increasingly in populating military databases. Scanning analog photographs or collecting scenes with digital cameras mounted on aircraft are the two primary means of generating digital images. In the latter use, digital cameras for collecting panchromatic, color and color-infrared images are designed around a matrix (array) of charge-coupled device (CCD) imaging elements. Camera features such as completely electronic forward motion compensation (FMC) and 12-bit per pixel radiometric resolution ensure image quality (Intergraph Corporation 2006). Significant advances in sensor technology have stemmed from subdividing spectral ranges of radiation into bands (intervals of continuous wavelengths), allowing digital camera sensors in several bands to form multispectral (MS) images. For multispectral data, the total spectral range is normally between 0.4 and 0.9 μm for visual and near infrared (IR). An advantage over aerial photos, digital images enable rapid image enhancement, zoom viewing and classification via supervised or unsupervised methods.

Table 57-2 Advantages of aerial photographs over analog maps
(Department of the Army 2001).

Photos provide a current pictorial view of the ground that no map can equal.
Photos are more readily obtained; they may be in the hands of the user within a few hours after they are taken. A map may take months to prepare.
Photos may be taken of places that are inaccessible to ground soldiers.
Photos show military features that do not appear on maps.
Photos provide a day-to-day comparison of selected areas, permitting evaluations to be made of enemy activity.
Photos provide a permanent and objective record of the day-to-day changes with the area.
Photos are often used to obtain data not available from other secondary sources, such as location and the extent of certain areas of interest.

Another popular technology, imaging spectroscopy (also known as hyperspectral remote sensing) allows a sensor on a moving platform to gather reflected radiation from ground targets where a special detector system records up to 200+ narrow spectral channels simultaneously over a range from 0.38 to 2.50 μm (JPL 2004). With such detail, the ability to detect and identify individual materials or classes greatly improves. Airborne Visible/Infrared Imaging Spectrometer (AVIRIS), one such hyperspectral sensor operated since 1987, consists of four spectrometers with a total of 224 individual bands, each with a spectral bandwidth of 10 nm and a spatial resolution of 20 m (Lillesand et al. 2007).

A new form of digital imagery, light detection and ranging (lidar) is a very powerful and versatile remote sensing tool. It has a broad range of applications and is extremely well suited for monitoring combat zones. One noteworthy application of lidar technology is the Scanning Hydrographic Operational Airborne Lidar Survey (SHOALS) system (Guenther et al. 1998). This bathymetric mapping application uses a technique known as airborne lidar bathymetry (ALB) or airborne lidar hydrography (ALH) where lidar is employed to rapidly and accurately measure seabed depths and topographic elevations, surveying large areas and far exceeding the capabilities and efficiency of traditional survey methods (Guenther et al. 1998). Other uses of lidar include terrestrial data collections, which further detail the context and elevation of terrain. When aerial lidar data and electro-optical imagery are merged, accurate and current terrain modeling products are possible.

In addition to these digital technologies, thermal remote sensing, operating primarily in the 8–14 μm but also in the 3–5 μm wavelength region of the spectrum, produces data that aid in identifying materials by their thermal properties. Finally, radio detection and ranging (radar), an active microwave system, has been flown on both military and civilian platforms because of its ability (for certain wavelengths) to penetrate clouds. Aircraft-mounted synthetic aperture radar (SAR) is the most popular radar device used in military mapping operations.

57.2.2.2 Sensor Data Obtained with UAVs

Although the use of aerial photographs and digital images for military applications has seen modest increase over the past few years, UAV exploitation has grown tremendously. The ability to provide real-time or near real-time data about the terrain they fight on and the enemy they face has always been a goal of the military intelligence community (Mahnken 1995). Unmanned aerial vehicles have made that goal a reality at many levels of war, becoming a valuable tool for ground commanders in preparation and execution of missions. With increasingly more UAVs populating the littoral battlespace, coupled with robust communications systems for distribution of the information they gather, these data may soon be available to every soldier and marine.

Unmanned aerial vehicles are remotely piloted or self-piloted aircraft that carry cameras, sensors, communications equipment or other payloads (Reinhardt et al. 1999). Not a new idea, the UAV has been employed by military units since the late 1950s (Pike 2003). Until the last 15 years, however, their usefulness was viewed as limited because the analog data they collected were not accessible (in most all cases) until after they returned from their missions. Digital technology changed this paradigm. As a result, since the early 1990s, the Department of Defense (DoD) has employed UAVs to satisfy surveillance requirements in close range, short range and endurance categories. Initially, close range was defined to be within 50 km, short range was defined as within 200 km, and endurance range was set as anything beyond. By the late 1990s, the close and short range categories were combined. The current classes of these vehicles are the tactical UAV and the endurance category.

Numerous digital multispectral, hyperspectral and radar sensor platforms are used on-board both tactical and endurance UAVs for military applications in a variety of regions. As the ability to move data more quickly and in greater volume improves, military commanders now receive current details of battlefield events like never before. Commanders are trained warfighters; they have a basic understanding of aerial photos/video, but typically are *not* trained in the interpretation of IR and radar data. For simple utility purposes, much of the tactical data gathered for military use by these systems are high-resolution multispectral images, predominantly from the visual portion of the electromagnetic spectrum. Average spatial ground resolutions now routinely achieved by these systems are on the order of 1 m. Systems collecting IR, thermal and radar data are quickly approaching similar resolutions (FAS 1996).

In all cases of UAV employment, tactical control stations (TCS) are used to control the vehicles and their on-board systems. The TCS is the hub where all software and communications links reside as well as connectivity links to other battlefield command, control, communication, computers and intelligence (C4I) systems (FAS 1999b).

Tactical commanders routinely control UAVs from within their command posts. Three tactical UAVs (TUAVs) are discussed here. The Pioneer was procured beginning in 1985 as an initial UAV capability to provide imagery intelligence for tactical commanders on land and sea at ranges out to 185 km. Used temporarily by the Army, it is currently only used by the US Navy (FAS 2000a). The Outrider was designed to provide follow-on, interim support to Army tactical commanders with near real-time imagery intelligence at ranges up to 200 km. This system, still in limited use, helped developers create the systems' capabilities requirement for future TUAV design (FAS 2000b). The resulting product, now in extensive use, was the Joint Tactical UAV or Hunter. This system was developed to provide ground and maritime forces with real-time and near real-time imagery intelligence at ranges up to 200 km and extensible to 300+ km by using another Hunter as an airborne relay (FAS 2001a).

Complementing TUAVs, Endurance UAVs have seen tremendous application and experienced great success over the past five years for military commanders, particularly in Afghanistan and Iraq. The medium altitude endurance UAV is called the Predator. This vehicle provides imagery intelligence to satisfy Joint Task Force and Theater commanders at ranges out to 830 km (FAS 2001b). Global Hawk and Darkstar are high-altitude endurance UAVs. These latter two vehicles are used for missions requiring long-range deployment, wide-area surveillance or prolonged acquisition over the target area. They are both directly deployable from the continental United States to any theater of operations (FAS 1999a; FAS 2001c).

Micro unmanned aerial vehicles (MAV) are currently under development. Experiments are being conducted to explore the military relevance of MAVs for future operations and to develop and demonstrate flight-enabling technologies for very small aircraft (less than 15 cm in any dimension) (FAS 2000c). As portable systems capable of receiving and utilizing image data proliferate the littoral battlefield, data volume will continue to be a challenge. Communication systems designed to monitor, control and filter bandwidth at different levels of warfighting (strategic, operational or tactical) will play critical roles in "moving" the data. When combined, the aerial reconnaissance data collection methods provide an important resource for populating military databases. These technological benefits offered by the various systems are a tremendous improvement to the intelligence assets available to military forces only a few years ago.

57.2.3 Satellite Reconnaissance

There are a growing number of satellites orbiting the earth, collecting valuable military data and returning these data to ground stations all over the world. Satellite remote sensing has the ability to provide complete, cost-effective, repetitive spatial and temporal data coverage. Tasks such as the assessment and monitoring of various conditions can be carried out over large regions. Classified and, increasingly, unclassified systems have and continue to be successfully used by intelligence organizations to provide critical information to military units.

57.2.3.1 Classified Systems

Satellite imaging systems have long been the workhorse of the military intelligence community. Classified satellite systems are primarily used for the collection of intelligence information about military activities of foreign countries. These satellites can detect missile launches or nuclear explosions in space and acquire/record radio and radar transmissions while passing over other nations. There are four basic types of reconnaissance satellites: (1) optical-imaging satellites that have light sensors designed to detect enemy weapons on the ground; (2) radar-imaging satellites that are able to observe the Earth through cloud cover; (3) signals-intelligence or ferret satellites that are sophisticated radio receivers capturing the radio and microwave transmissions emitted from any country on Earth; and (4) relay satellites that make military satellite communications around the globe much faster by transmitting data from spy satellites to stations on Earth (Galactics 1997). The first two will be discussed in detail as part of this review.

Starting in the 1960s, the US began launching reconnaissance satellites with the first series called Discoverer. As these satellites circled the Earth in polar orbits, on-board cameras recorded photographs (Pike 2000). The next series of US spy satellites was given the code name Keyhole, or KH for short. They mostly performed routine surveillance or weapons targeting. Traveling in elliptical orbits at low altitudes of 140 km at perigee, they either took wide-area photographs of large land masses or close-up photos of special interest objects (McDonald 1995; Pike 2000). The early KH satellites—*Corona, Argon,* and *Lanyard*—were used through the early 1970s to assess the former Soviet Union's long-range bombers and ballistic missile production and deployment (McDonald 1995; Pike 2000). The resulting photographs were used to produce maps and charts for DoD and other US government mapping programs.

In June 1971, the KH-9 satellite deployed. Weighing 30,000 pounds and placed in an orbit that at times came within 150 km of the Earth, it was nicknamed Big Bird because of its extraordinarily large size. Big Bird employed two cameras to obtain both area-surveillance images and close-up photos. On the latter photos, it was reported that objects as small as 20 cm could be distinguished (McDonald 1995; Pike 2000). The Big Bird satellites were launched at the rate of about two a year from 1971 to 1984; 19 successful launches were followed by one failure, on 18 April 1986, in which the booster exploded after takeoff. The Big Bird's major limitation was its relatively short life span, which started out at some 52 days. By 1978, it was extended to 179 days and the average orbital life was 138 days with a maximum of 275 days achieved in 1983 (McDonald 1995; Pike 2000).

In the early 1970s, another major US classified initiative, the Defense Satellite Program (DSP), was established. The satellites from this program, a key part of North America's early warning system, detect missile launches, space launches and nuclear detonations. Operated by Air Force Space Command, the satellites feed warning data to North American Aerospace Defense Command (NORAD) and US Space Command early warning centers at Cheyenne Mountain Air Force Base, Colorado. The first launch of a DSP satellite took place in the early 1970s, and, since that time, they have provided an uninterrupted early warning capability to the US. The system's capability was demonstrated during Desert Shield/Storm when the satellites detected the launch of Iraqi SCUD missiles, providing warning to civilian populations and coalition forces in Israel and Saudi Arabia (USAF 2004).

In December of 1988, the National Aeronautical and Space Administration (NASA) launched the $500 million *Lacrosse* satellite. *Lacrosse's* main attribute, like most spy satellites, is its image sensor. *Lacrosse* uses SAR technology to detect objects only 1 m across, the level of detail necessary to identify military hardware. Instead of providing a constant stream of images like most radars, *Lacrosse* records a series of snapshots as it arcs over the Earth (Pike 2000). *Lacrosse* also actively beams microwave energy to the ground and reads the weak return signals reflected into space. This allows the satellite to "see" objects on Earth that would otherwise be obscured by cloud cover and darkness. In order to send out these signals, however, *Lacrosse* has very substantial power needs that are met with solar panels larger than would be found on most satellites its size. *Lacrosse* uses a rectangular antenna, 15 m long and 3 m wide, which is very different from the standard mechanical antenna (Pike 2000). This antenna is covered by rows and columns of small transmitting and receiving elements that help *Lacrosse* pick up the faint return signals bouncing back from the Earth. Today, the National Reconnaissance Office continues to design, build, launch and operate classified satellites. Its future looks promising with over $25 billion planned for the next two decades (USAF 2004).

57.2.3.2 Unclassified Satellite Systems Producing High-resolution Images

Although the military has had and continues to have its share of classified satellite programs, commercial systems are now producing data with comparably high spatial resolution (Behling and McGruther 1998). Historically, remote sensor data with spatial resolutions corresponding to 0.5–10 m are required to adequately define the high-frequency detail that characterizes the urban scene (Welch 1982). Military databases demand similar detail, as many of the features found in the urban scene are common to LW data sets. Because of their ability to provide high-resolution spatial data, these systems are useful in most military mapping applications at large scale (examples provided in Table 57-3).

Table 57-3 Commercial high-resolution satellites and their sensor systems (Wilson and Davis 1998; DigitalGlobe 2004 and 2008; Orbital Sciences 2006; and GeoEye 2008).

SYSTEM	Ikonos	QuickBird	OrbView-3	WorldView-1
Date of Launch	September 1999	October 2001	June 2003	September 2007
Orbital Parameters	Altitude: 681 km Orbit type: sun-sync. Orbit time: 98 min	Altitude: 450 km Orbit type: sun-sync. Orbit time: 93.4 min	Altitude: 470 km Orbit type: sun-sync. Orbit time: 98 min	Altitude: 496 km Orbit type: sun-sync. Orbit time: 94.6 min
Sensor Parameters	Spatial Resolution 1m (pan) 4 m (XS) Spectral Resolution Panchromatic 0.45 - 0.90 μm Multispectral #1: Blue 0.45 - 0.52 #2: Green 0.52 - 0.60 #3: Red 0.63 - 0.69 #4: Near IR 0.76 - 0.90 Radiometric Resolution: 11 - bit Swath Width: 11 km at nadir	Spatial Resolution 0.61 m (pan) 2.5 m (XS) Spectral Resolution Panchromatic 0.445 - 0.90 μm Multispectral #1: Blue 0.45 - 0.52 #2: Green 0.52 - 0.60 #3: Red 0.63 - 0.69 #4: Near IR 0.76 - 0.89 Radiometric Resolution: 11 - bit Swath Width: 2.12 degrees (nominal 16.5 km at nadir – can be 14 – 34 km; altitude dependent)	Spatial Resolution 1m (pan) 4 m (XS) Spectral Resolution Panchromatic 0.45 - 0.90 μm Multispectral #1: Blue 0.45 - 0.52 #2: Green 0.52 - 0.60 #3: Red 0.625 - 0.695 #4: Near IR 0.76 - 0.90 Radiometric Resolution: 11 - bit Swath Width: 8 km at nadir	Spatial Resolution 0.55 m (pan) Spectral Resolution Panchromatic 0.45 - 0.90 μm Radiometric Resolution: 11 - bit Swath Width: 17.6 km at nadir
Data Parameters	Scene Size: 13 km by 13 km	Scene Size: 16.5 km by 16.5 km in-orbit stereo pairs	Scene Size: User defined	Scene Size: 17.6 km by 14 km at nadir

Commercial satellite images are primarily characterized by significant *spatial* resolution improvements over the well-known Landsat and SPOT satellite images and are useful for mapping applications at large scale. Three noteworthy high-resolution systems—*Ikonos, QuickBird* and *OrbView-3*—have some unique qualities (GeoEye 2008; DigitalGlobe 2004; Orbital Sciences 2006). In September 1999, with the successful launch and deployment of *Ikonos* by Space Imaging (now GeoEye, Inc.), high-resolution satellite images exploded onto the commercial market scene (GeoEye 2008). Just over two years later (October 2001), DigitalGlobe launched the *QuickBird* satellite (DigitalGlobe 2004). *Ikonos* provides panchromatic and 4-band multispectral images of 1- and 4-m resolutions, respectively, whereas QuickBird generates panchromatic images of 0.61-m and multispectral images of 2.44-m pixel resolutions. *OrbView-3,* launched in June 2003, has very similar technical capabilities as the *Ikonos* and *QuickBird* satellites. The greatest advantage is its repeat cycle, re-visiting (through sensor "pointability") ground tracks every one to three days to provide extraordinary temporal resolution required for assessing rapidly occurring changes on the Earth's surface (such as flooding or volcanic activity). All of these systems provide high-resolution multi-spectral data that are suitable for military mapping, change detection and the assessment of threats. Stereo images suitable for generating digital elevation models (DEMs) and large-scale mapping also can be obtained by these systems (Dial and Grodecki 2003; Haverkamp and Poulsen 2003). Other commercial satellite systems that provide high-resolution images (e.g., WorldView-1, GeoEye-1) have been and are continuing to be launched, providing additional resources to military organizations.

57.3 Data Integration and GIS Applications in Military Environments

GIS technology allows for the use of digital data in developing and employing tailored, current battlefield information to military commanders. In the 1990s, DoD performed work in GIS that focused primarily on database design/population and software development (Satyanarayana and Yogendran 2001). Numerous digital data formats are available for incorporation into large-scale military mapping projects. Previously discussed, many of these are the result of various data collection methods currently in use and they facilitate military and civilian organizations supporting DoD.

GIS analysis has been effectively demonstrated for military base (also known as garrison) operations. This affords installation personnel from multiple organizations with the impressive capability to successfully answer questions related to geographic inventory, analysis and modeling (GISO 2001). Although garrison operations are important, this example does not demonstrate the possible applications of GIS for military commanders. The remainder of this review will focus on the relevant GIS functions for use in combat operations followed by a discussion of current and planned developments of GIS technology for our armed services.

57.3.1 GIS and Its Role in Military Applications

Two major components of a GIS include a geographic database and software that includes different types of analysis functions. These spatial analysis functions distinguish a GIS from other information systems (Peuquet and Marble 1990; Mcguire et al. 1991). The use of spatial and non-spatial attributes in the database to answer questions about the changing world facilitates the study of real-world processes by developing and applying models (Burrough and McDonnell 1998). Such models often illuminate underlying trends in geographic data, making new information available and accessible through digital maps. The organization of databases into map layers provides rapid access to data elements required for geographic analysis.

There are four major groups of analytical functions: (1) data query; (2) overlay operations; (3) neighborhood analysis; and (4) connectivity operations (Aronoff 1991; Maguire et al. 1991; Lo and Yeung 2002). Critical to military operations, the rapid and selective retrieval, display, measurement and reclassification of information from a database (data query) are fundamental to every GIS. Overlay operations are important as well to military decision makers. Just as plastic acetate attached to a map has been historically used to show different components of the battlefield, overlay functions efficiently integrate layers of geospatial data and result in the creation of new spatial elements.

Neighborhood analysis involves the search and assessment of geospatial data surrounding a target location followed by calculation and/or assignment of a value. The generation of DEMs—the interpolation of a continuous surface from discrete points of elevation for terrain analysis—is an example of a neighborhood analysis that is important in military applications. The DEMs provide realistic terrain shape necessary for accurate visualization of the battlefield. Finally, connectivity operations are based on interconnecting logical components of a process or model. Those important to military operations include intervisibility (line-of-sight), seek (or stream) functions, buffering and spread analysis.

Buffers, calculated circular or square areas from a given point or series of points, are frequently required in combat planning/execution to establish radii or zones around critical locations and key terrain (e.g., weapon impact areas or search and rescue zones) (ESRI 1998; ESRI 2002). Spread functions evaluate phenomena that accumulate with distance (Aronoff 1991). One final military application of this type of analysis is terrain trafficability—predicting the time needed to traverse terrain with variable conditions. The trafficability, or ease and speed of movement, varies with the type of ground cover, topography, mode of transport and season of travel (Aronoff 1991; Fleming et al. 2008).

GIS technology is rapidly moving from its historic niche usage of installation inventory and monitoring within defense organizations to becoming a critical defense-wide infrastructure. The importance of GIS is based on the fact that defense operations depend on battlespace awareness—and the battlespace is geographic. This involves more than an understanding of location—geography is a science that creates a framework for understanding the relationships between all battlespace entities. This, in turn, develops knowledge from the flood of data. Defense-wide spatial infrastructures break down the divisional "stovepipes" of the separate military services—Army, Navy, Air Force and Marines—to provide a common framework for handling mapping, charting, geodesy and imagery across all defense systems. This is important because it avoids having the government pay time and time again for the same core functions to be developed for each system. Defense-wide spatial infrastructures also ensure that the warfighter receives the latest capabilities from the commercial off-the-shelf (COTS) community where information technology (IT) innovation occurs.

The idea of delivering interoperability is only part of the rationale of creating a defense-wide infrastructure. Far more important is the contribution that this spatial information infrastructure contributes to network-centric operations that represent a revolution in military affairs that is affecting every nation in the world. Such an infrastructure fundamentally reengineers defense organizations, doctrines, and systems to take full advantage of the capabilities of modern information technology. A GIS is the critical infrastructure that connects the three concepts of network-centric operations: 1) situational awareness (intelligence, surveillance and reconnaissance, or ISR); 2) command, control, computers, communications and intelligence assessment, or C4I; and 3) precision engagement. Sensors in the ISR domain are being directly coupled to geodatabases that are then distributed and replicated into the C4I domain to support decision making. They are then, in turn, distributed and replicated into the precision engagement domain to coordinate and target weapon systems. GIS is the COTS technology that makes all of this possible and affordable. ArcGIS, for example, supports the very scalable and rich geodatabases that are populated from a wide range of sensors, distributed and replicated across low to high-bandwidth networks. The advanced analysis and dissemination of information supports precision engagement, thus permitting the more effective employment of existing weapon systems. Numerous products are now being created for military use, such as assessments of cross country mobility, mobility corridors, zones of entry, aerial concealment, line-of-sight and fields-of-fire, as well as perspective views and fly-throughs of 3D terrain (Fleming et al. 2008).

The Cross Country Mobility (CCM) product demonstrates the off-road speed for a vehicle as determined by the terrain scenario and vehicle type (Figure 57-1). The makeup of the CCM includes surface traction and resistance, slope, vehicle dynamics, obstacles and vegetation. The CCM is used to develop the best axis of advance for a particular course of action or development of an engagement area.

The aerial concealment overlays describe the most suitable areas to conceal a force from overhead detection. This overlay is important to judge where enemies may be located, especially in areas where guerilla forces may be operating. This overlay also may be used by friendly forces to develop concealed movement routes and staging areas. Concealment may be provided by woods, underbrush, tall grass or cultivated vegetation. This product is predicated on canopy closure information within the vegetation layer. Line-of-Sight Profiles show an area of direct observation possible from one location to another based on digital elevation data or a digital surface model (DSM), if available. This line-of-sight analysis is used to anticipate enemy positions, plan locations for communications platforms and develop engagement areas (Figure 57-2); obviously, the higher the resolution of the DEM (or DSM), the more accurate the results.

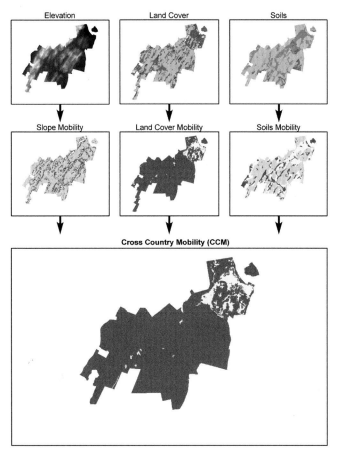

Figure 57-1 Example of a cross-country mobility product. See included DVD for color version.

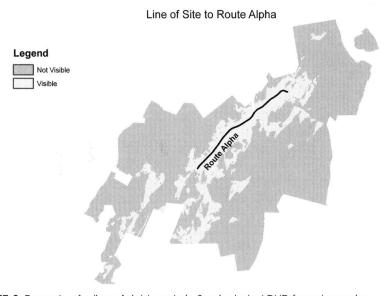

Figure 57-2 Example of a line-of-sight analysis. See included DVD for color version.

Manual of Geographic Information Systems

A virtual representation of a view of an area or target at a specified altitude, azimuth or angle of attack from a point position is called a perspective view. This product replicates a photograph of an area of battlespace (Figure 57-3). A perspective view can be used in many aspects of the military. Typically it is used to visualize an objective area or important battlefield terrain. A visual preview of an area of interest along a specified flight line at a specified altitude and angle viewed from inside the aircraft is called a fly-through. This product is used by commanders to visualize maneuver areas and plan operations. Operators can display roads, rivers, operational graphics and a text to enhance the visualization of the terrain.

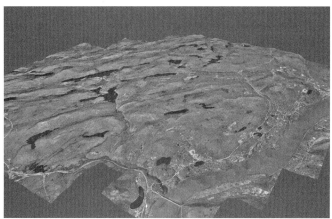

Figure 57-3 Example of a perspective view for battlefield assessment. See included DVD for color version.

57.3.2 Current and Future Military Applications in the Armed Services

Substantial ongoing research efforts by each service employ digitization and GIS analysis to aid in combat decision making by commanders and their staffs. Full digitization of the battlefield, however, will demand an extensive technological leap—the complete embracing of digital geospatial data and the means of exploiting it with GIS at all levels of war. This condition is, arguably, still some time from now. For the foreseeable future, paper maps and GIS will be complementary. The defense community has been using digital data in training and combats for about a decade, primarily confined to strategic and air systems (JCS 1997). Its use on the battlefield, long predicted, has also been leveraged at the tactical level of war by deployable systems (PEO-C3S 1997). Tremendous growth is now being realized as the importance of GIS technology on the battlefield is recognized. GIS allows for efficient representation of the ever-changing battlespace and provides for rapid transmission of that information over their robust communications infrastructure.

As discussed earlier, paper maps have two major limitations. First, they often do not adequately provide relevant information to individual commanders leading diverse organizations on complex missions, and, second, they quickly become out of date and therefore inaccurate. In addition, every paper map represents a compromise between the needs of differing users, none of whom receive the ideal product. Employing GIS, users are able to create (or have created) timely custom products that depict information that they need (Evans et al. 2000). The modern battlefield changes rapidly; the analog map product cannot. This is a critical limitation on today's fast-moving battlefield where weapon systems are capable of significant alteration of the real world. A GIS can help solve this problem, but only if the problem is clearly acknowledged and effectively addressed. To accomplish this, three things must happen. First, proper GIS models of the real world must be developed, validated and implemented. Second, data must be properly maintained. Finally, human intervention must apply a "sanity check" after each step in the decision process; where problems are determined, inspections of the models and/or data are required.

At the direction of the US National Geospatial-Intelligence Agency (NGA), an effort to leverage and consolidate GIS technology for military commanders (in all services) is now being developed. Northrop Grumman is the prime contractor for NGA's Commercial/Joint Mapping Tool Kit (C/JMTK) Program. The C/JMTK will be a standardized, commercial, comprehensive tool kit of software components for the management, analysis and visualization of map and map-related information (Northrop Grumman 2002). The commercial software companies involved in this plan include the Environmental Systems Research Institute, Inc. (ESRI), ERDAS, Inc., Analytical Graphics, Inc. (AGI), and Great-Circle Technologies. The developing foundation of the C/JMTK is ESRI's ArcGIS framework, which includes Spatial Analyst, 3D Analyst, and Military Overlay Editor (MOLE), extended by the ArcSDE database engine and distributed by the ArcIMS Internet server. These products provide a seamless package that give unprecedented capabilities in viewing map and map-related information along with tools to support the analysis and storage of map data (Birdwell et. al 2004). The program integrates the best of government and industry into a common, long-term solution that will advance operational mission application development into the next generation of interoperable systems for the warfighter (ESRI 2003).

Taking full advantage of such inventions as the C/JMTK, it is envisioned that the Objective Force—the planned future combat systems for 2020 and beyond—will operate on four war-fighting tenets: (1) see first; (2) understand first; (3) act first; and (4) finish decisively (JCS 1997). Unprecedented intelligence, surveillance and reconnaissance capabilities, coupled with other ground, air and space sensors networked into a common integrated operational picture, will enable forces to accurately see individual components of enemy units, friendly units and the terrain. Data integration systems will enable decision makers to have a synthesized Common Operational Picture (COP) (JCS 1997). Using the COP, Objective Force commanders will be able to leverage the intellect, experience and tactical intuition of leaders at multiple levels in order to identify enemy strengths and conceptualize future plans. As commanders decide on a course of action, they will be able to instantaneously disseminate their intent to all appropriate levels, affording maximum time for subordinate levels to conduct requisite troop-leading procedures. The time gained through effective use of these information technologies should permit Objective Force units to seize and retain the initiative, building momentum quickly for decisive outcomes.

Seeing and understanding first gives commanders and their units the situational awareness to engage at times and places with methods of their own choosing. Objective Force units will be able to move, shoot and reengage faster than the enemy. It is planned that target acquisition systems will see farther than the enemy in all conditions and environments. The intent, here, is to deny the enemy any respite or opportunity to regain the initiative. Objective Force units will be able to understand the impact of events and synchronize their own actions. Finally, Objective Force units should finish decisively by quickly destroying the enemy's ability to continue the fight. Units will be able to maneuver by both ground and air to assume tactical and operational positions of advantage through which they will continue to fight the enemy and pursue subsequent military objectives.

Although these advances will not eliminate battlefield confusion, the resulting battlespace awareness should improve situational knowledge, decrease response time, and make the battlefield considerably more transparent to those who achieve it. The integration of geospatial technologies and GIS will likely provide an improvement in warfighting success. Commanders will be able to attack targets successfully with fewer platforms and less ordnance while achieving objectives more rapidly and with reduced risk. Strategically, this improvement will enable more rapid power projection. Operationally, within the theater, these capabilities will mean a more rapid transition from deployment to full operational capability. Tactically, individual warfighters will be empowered as never before, with an array of detection, targeting and communications equipment that will greatly magnify the power of small units. As a result, US forces will improve their capability for rapid worldwide deployment.

57.4 Conclusions

There are numerous critical and advanced image data collection technologies that now define unprecedented military intelligence, surveillance and reconnaissance capabilities. These advances enhance the detectability of features and targets across the littoral battlespace, improving distance ranging, "turning" night into day for some classes of operations, reducing the risk of friendly fire incidents (fratricide) and further accelerating operational tempo (JCS 1997). On the horizon, improvements in information and systems integration technologies will significantly impact future military operations by providing decision makers with accurate information in a timely manner. The fusion of information with the integration of sensors, platforms and command organizations will allow operational tasks to be accomplished rapidly, efficiently and effectively.

References

Aronoff, S. 1991. *Geographic Information Systems: A Management Perspective.* Ottawa, Canada: WDL Publications.

Ballendorf, D. A. 2003. The Battle for Tarawa: A Validation of the U.S. Marines. The Patriot Files. <http://www.tiggertags.com/patriotfiles/forum/showthread.php?t=24504> Accessed 17 November 2008.

Behling, T. and K. McGruther. 1998. Satellite reconnaissance of the future. *Joint Forces Quarterly* Spring Edition 23–30.

Birdwell, T., J. Klemunes and D. Oimoen. 2004. Tracking the dirty battlefield during Operation Iraqi Freedom with the tactical minefield database. *Mil Intel Muster* Winter 2003/2004 Edition. Redlands, Calif.: Environmental Systems Research Institute (ESRI).

Burrough, P. A. and R. A. McDonnell. 1998. *Principles of Geographical Information Systems.* Oxford: Oxford University Press.

Cole, R. H. 1998. Grenada, Panama and Haiti: Joint operational reform. *Joint Forces Quarterly* Autumn/Winter Edition 57–64.

Dial, G. and J. Grodecki. 2003. Applications of Ikonos imagery. *Proceedings of the ASPRS 2003 Annual Convention,* 5–9 May, Anchorage, Alaska. Bethesda, Md.: American Society for Photogrammetry and Remote Sensing, unpaginated CD-ROM.

DigitalGlobe. 2004. QuickBird Imagery Products. DigitalGlobe, Inc., Longmont, Colo. <http://www.digitalglobe.com/index.php/6> Accessed 17 November 2008.

———. 2008. WorldView-1 Product Quick Reference Guide. DigitalGlobe, Inc., Longmont, Colo. <http://www.digitalglobe.com/file.php/545/WV-1_Product_QR_Guide.pdf> Accessed 17 November 2008.

ESRI. 1998. *The Role of Geographic Information Systems on the Electronic Battlefield.* Redlands, Calif.: Environmental Systems Research Institute (ESRI).

———. 2002. *ArcGIS in Defense.* Redlands, Calif.: Environmental Systems Research Institute (ESRI).

———. 2003. *C/JMTK Technical Overview: An Introduction to the Architecture, Technologies and Capabilities.* Defense Technology Center. Vienna, Va.: Environmental Systems Research Institute (ESRI).

Evans, S., J. Murday and R. Lawrence. 2000. *Moving GIS into the Ocean Realm: Meeting the Need for Intelligent Data.* Redlands, Calif.: Environmental Systems Research Institute (ESRI).

FAS (Federation of American Scientists). 1996. UAV Annual Report FY 96. Washington, D.C. <http://www.fas.org/irp/agency/daro/uav96/content.html> Accessed 17 November 2008.

———. 1999a. BQM-145A Medium Range UAV. Washington, D.C. <http://www.fas.org/irp/program/collect/mr-uav.htm> Accessed 17 November 2008.

———. 1999b. UAV Tactical Control System (TCS). Washington, D.C. <http://www.fas.org/irp/program/collect/uav_tcs.htm> Accessed 17 November 2008.

———. 2000a. Pioneer Short Range (SR) UAV. Washington, D.C. <http://www.fas.org/irp/program/collect/pioneer.htm> Accessed 17 November 2008.

———. 2000b. Tactical Unmanned Aerial Vehicle (TUAV). Washington, D.C. <http://www.fas.org/irp/program/collect/cr-tuav.htm> Accessed 17 November 2008.

———. 2000c. Tactical Unmanned Aerial Vehicle (TUAV) Close Range - Tactical Unmanned Aerial Vehicle (CR-TUAV). Washington, D.C. <http://www.fas.org/irp/program/collect/cr-tuav.htm> Accessed 17 November 2008.

———. 2001a. Hunter Short Range (SR) UAV. Washington, D.C. <http://www.fas.org/irp/program/collect/hunter.htm> Accessed 17 November 2008.

———. 2001b. RQ-1 Predator MAE UAV. Washington, D.C. <http://www.fas.org/irp/program/collect/predator.htm> Accessed 17 November 2008.

———. 2001c. Unmanned Aerial Vehicle Battlelab (UAVB). Washington, D.C. <http://www.fas.org/irp/agency/aia/cyberspokesman/97jul/pg13.htm, Accessed 17 November 2008.

Fleming, S., T. Jordan, M. Madden, E. L. Usery and R. Welch. 2008. GIS applications for military operations in coastal zones. *International Journal for Photogrammetry and Remote Sensing.* In press.

Galactics. 1997. Types and Uses. Reconnaissance Satellites. SchoolNet Digital Collections, Canada. <http://epe.lac-bac.gc.ca/100/205/301/ic/cdc/satellites/english/index.html> Accessed 17 November 2008.

GeoEye. 2008. Imagery Sources. Ikonos. GeoEye, Inc. Dulles, Va. <http://www.geoeye.com/CorpSite/products/imagery-sources/Default.aspx#ikonos> Accessed 17 November 2008.

GISO (Geographic Information Systems Office). 2001. *Integrated Geographic Information Repository (IGIR) 2001.* Camp Lejeune, N.C.

GISDevelopment. 2004. GPS: A Military Perspective. Uttar Pradesh, India. <http://www.gisdevelopment.net/technology/gps/techgp0048a.htm> Accessed 17 November 2008.

GPS JPO (GPS Joint Program Office). 2000. NAVSTAR GPS, NOVELLA on User Equipment (UE) Acquisition, Second edition, 4 July 2000. El Segundo, Calif. <http://www.kilkaari.com/kfiles/UENPublic.pdf> Accessed 17 November 2008.

Greiss, T. E. 1984. *The Second World War.* Department of History, United States Military Academy. New York: West Point.

Guenther, G. C., M. W. Brooks and P. E. LaRocque. 1998. New capabilities of the SHOALS airborne lidar bathymeter. *Proceedings of the 5th International Conference on Remote Sensing for Marine and Coastal Environments*, ERIM International, 5–7 October, San Diego, Calif., Vol. I, 47–55 [reprinted in 2000, *Remote Sensing of the Environment* 73: 247–255.]

Haverkamp, D. and R. Poulsen. 2003. Change detection using Ikonos imagery. *Proceedings of the ASPRS 2003 Annual Convention,* 5–9 May, Anchorage, Alaska. Bethesda, Md.: American Society for Photogrammetry and Remote Sensing, unpaginated CD-ROM.

Huybrechts, S. 2004. *The High Ground.* National Defense University, National War College, Washington, D.C.

Intergraph Corporation. 2006. Intergraph Z/I Imaging DMC (Digital Mapping Camera) System. Intergraph Corporation, Huntsville, Ala. <http://www.intergraph.com/assets/documents/florida_DOT.pdf> Accessed 17 November 2008.

JCS (Office of the Chairman of the Joint Chiefs of Staff). 1997. *JV2010.* The Pentagon, Washington, D.C.

JPL (Jet Propulsion Laboratory). 2004. AVIRIS. General Overview. California Institute of Technology, Pasadena, Calif. <http://aviris.jpl.nasa.gov/html/aviris.overview.html> Accessed 17 November 2008.

Kimble, K and R. Veit. 2000. SPACE – the next area of responsibility. Joint Forces Quarterly Autumn/Winter Edition 20–23.

Krulak, C. 1999. Operational maneuver from the sea. *Joint Forces Quarterly* Spring Edition 78–86.

Lillesand, T., R. Kiefer and J. Chipman. 2007. *Remote Sensing and Image Interpretation,* 6th ed. New York: Wiley and Sons.

Lo, C. P. and A. Yeung. 2002. *Concepts and Techniques of Geographic Information Systems.* Upper Saddle River, N.J.: Prentice Hall.

McDonald, R. A. 1995. Opening the Cold War sky to the public: Declassifying satellite reconnaissance imagery. *Photogrammetric Engineering and Remote Sensing* 61 (4):385–390.

Maguire, D. J., M. F. Goodchild and D. W. Rhind. 1991. *Geographical Information Systems: Principles and Applications.* 2 volumes. London: Longman.

Mahnken, T. 1995. War in the information age. *Joint Forces Quarterly* Winter Edition 39–43.

Murray, T. and T. O'Leary. 2002. Military transformation and legacy forces. *Joint Forces Quarterly* Spring Edition 20–27.

NAVSTAR (GPS Program Office). 2001. Global Positioning Systems Wings. Los Angeles, Calif. <http://gps.losangeles.af.mil/> Accessed 17 November 2008.

NIMA. 2003. *Geospatial Intelligence Capstone Document.* National Imagery and Mapping Agency (NIMA), Washington, D.C.

NOAA. 1997. Airline photography and shoreline mapping. Washington, D.C. <http://www.oceanservice.noaa.gov/topics/navops/mapping/welcome.html> Accessed 17 November 2008.

Northrop Grumman. 2002. *Commercial/Joint Mapping Toolkit.* TASC-NG, Northrop Grumman, Chantilly, Va.

Orbital Sciences. 2006. OrbView-3 Fact Sheet. Orbital Sciences Corporation. Dulles, Va. <http://www.orbital.com/NewsInfo/Publications/OV3_Fact.pdf> Accessed 17 November 2008.

PEO-C3S. 1997. *The Warfighters Digital Information Resource Guide.* Ft. Hood, Tex. Program Executive Office–Command, Control, Communication and Security, Army Materiel Command.

Peuquet, D. J. and D. F. Marble, eds. 1990. *Introductory Readings in Geographic Information Systems.* London: Taylor & Francis.

Pike, J. 2000. Imagery Intelligence – Military Space Programs. Federation of American Scientists, Washington, D.C. <http://www.fas.org/spp/military/program/imint/> Accessed 17 November 2008.

———. 2003. Unmanned Aerial Vehicles (UAVs). Federation of American Scientists, Alexandria, Va. <http://www.fas.org/irp/program/collect/uav.htm> Accessed 17 November 2008.

Reinhardt, J., J. James and E. Flanagen. 1999. Future employment of UAVs – Issues of jointness. *Joint Forces Quarterly* Summer Edition 36–41.

Satyanarayana, P. and S. Yogendran. 2001. Military Applications of GIS. <http://www.gisdevelopment.net/application/military/overview/militaryf0002.htm> Accessed 17 November 2008.

SPAWAR (System Center). 2001. US Navy Global Positioning System (GPS) and Navigation Systems. United States Navy, San Diego, Calif. <http://www.spawar.navy.mil/depts/d30/d31/> Accessed 17 November 2008.

USAF (United States Air Force). 2004. U.S. Air Force Online Encyclopedia. Commandant, Air Command and Staff College, Air University, Air Education and Training Command, Maxwell Air Force Base, Ala. <http://www.au.af.mil/au/database/projects/ay1996/acsc/96-004/index.htm> Accessed 17 November 2008.

Welch, R. 1982. Spatial resolution requirements for urban studies. *International Journal of Remote Sensing* 3 (2):139–146.

Wilson, T. and C. Davis. 1998. *Naval EarthMap Observer (NEMO) Satellite.* New Research Laboratory, Washington, D.C.

Zhou, G. and R. Li. 2000. Accuracy evaluation of ground control points from Ikonos high-resolution satellite imagery. *Photogrammetric Engineering and Remote Sensing* 66 (9):1103–1112.

Zimmer, L. S. 2002. Testing the spatial accuracy of GIS data. *Professional Surveyor* 22 (1):21–28.

Other Sources

Bolstad, P. 2004. GIS Fundamentals. A First Text on GIS Fundamentals. University of Minnesota, St. Paul, Minn. <http://bolstad.gis.umn.edu/gisbook.html> Accessed 17 November 2008.

Braud, D. H. and W. Feng. 1998. Semi-automated construction of the Louisiana coastline digital land/water boundary using Landsat TM satellite imagery. Department of Geography & Anthropology, Louisiana State University, Louisiana Applied Oil Spill Research and Development Program, *OSRAPD Technical Report Series 97-002.*

Caton, J. 1995. Joint warfare and military dependence on space. *Joint Forces Quarterly* Winter Edition 48–53.

Chan, K. 1999. *DIGEST – A Primer for the International GIS Standard.* Boca Raton, Fla.: CRC Press LLC.

Comer, R., G. Kinn, D. Light and C. Mondello. 1998. Talking digital. *Photogrammetric Engineering and Remote Sensing* December 1998, 64 (12): 1139–1142.

DMS (Defense Mapping School). 1997. *NIMA Standard Hardcopy Imagery and Mapping Products.* Fort Belvoir, Va.: Defense Mapping Agency (DMA).

Department of the Army. 1994. FM 5-4300-00-1: Planning and Design of Roads, Airfields, and Heliports in the Theater of Operations – Road Design. Department of the Army, Washington, D.C. <http://www.globalsecurity.org/military/library/policy/army/fm/5-430-00-2/index.html> Accessed 17 September 2008.

Department of the Army. 2001. FM 3-25.26: Map Reading and Land Navigation. Department of the Army, Washington, D.C. <http://www.globalsecurity.org/military/library/policy/army/fm/3-25-26/index.html> Accessed 17 November 2008.

———. 2003. *TC 5-230: Army Geospatial Guide for Commanders and Planners.* Headquarters, Department of the Army, Washington, D.C.

Di, K., R. Ma and R. Li. 2001. Deriving 3-D shorelines from high-resolution Ikonos satellite images with rational functions. *Proceedings of the ASPRS 2001 Annual Convention,* 25–27 April, St. Louis, Mo. Bethesda, Md.: American Society for Photogrammetry and Remote Sensing, unpaginated CD-ROM.

Di, K., J. Wang, R. Ma and R. Li. 2003. Automatic shoreline extraction from high-resolution Ikonos satellite imagery. *Proceedings of the ASPRS 2003 Annual Convention,* 5–9 May, Anchorage, Alaska. Bethesda, Md.: American Society for Photogrammetry and Remote Sensing, unpaginated CD-ROM.

Emap International. 2002. *QuickBird – Aerial Photography Comparison Report.* Reddick, Fla.: Emap International.

ERDAS. 2000. *Imagine User's Guide.* Atlanta, Ga.: ERDAS, Inc.

ESRI. 2002. *COTS GIS: The Value of a Commercial Geographic Information System.* Redlands, Calif.: Environmental Systems Research Institute (ESRI).

Gibeaut, J. C. 2000. Texas shoreline change project: Gulf of Mexico shoreline change from the Brazos River to Pass Cavallo. *Report of the Texas Coastal Coordination Council pursuant to National Oceanic and Atmospheric Administration Award No. NA870Z0251.* The University of Texas at Austin, October, 2000.

Grodecki, J. and G. Dial. 2002. Ikonos geometric accuracy validation. *Proceedings of the Mid-Term Symposium in Conjunction with Pecora 15/Land Satellite Information IV Conference,* 10–15 November, Denver, Colo., International Society for Photogrammetry and Remote Sensing.

Ingham, A. E. 1992. *Hydrography for Surveyors and Engineers.* London: Blackwell Scientific Publications.

JCS. 1999. *Joint Pub 2-03: Joint Tactics, Techniques, and Procedures for Geospatial Information and Services Support to Joint Operations.* The Pentagon, Washington, D.C.

Li, R. 1997. Mobile mapping: An emerging technology for spatial data acquisition. *Photogrammetric Engineering and Remote Sensing* 63 (9): 1165–1169.

Li, R., G. Zhou, N. J. Schmidt, C. Fowler and G. Tuell. 2002. Photogrammetric processing of high-resolution airborne and satellite linear array stereo images for mapping applications. *International Journal of Remote Sensing* 23 (20):4451–4473.

Lillesand, T. and R. Kiefer. 1999. *Remote Sensing and Image Interpretation,* 4th ed. New York: Wiley and Sons.

McCaffrey, B. 2000. Lessons of Desert Storm. *Joint Forces Quarterly* Winter Edition 12–17.

Millett, N. and S. Evans. 2002. *Hydrographic Data Management Using GIS Technologies.* Redlands, Calif.: Environmental Systems Research Institute (ESRI).

Moore, L. 2003. *Viewscales and Their Effect on Data Display – The National Map Catalog Technical Discussion Paper.* Reston, Va.: USGS.

Niedermeier, A., E. Romaneessen and S. Lehner. 2000. Detection of coastlines in SAR images using wavelet methods. *IEEE Transactions on Geoscience and Remote Sensing* 38 (5):2270–2281.

NIMA. 1998a. LWD Prototype 2 Littoral Warfare Data. National Imagery and Mapping Agency (NIMA). Washington, D.C., digital file.

———. 1998b. Camp Lejeune Military Installation Map, 1:50,000 scale, Reprinted 3-1998. National Imagery and Mapping Agency (NIMA), Washington, D.C.

———. 2000. Digital Geographic Information Exchange Standard (DIGEST), Version 2.1. Relational database in Microsoft Access format. National Imagery and Mapping Agency (NIMA), Washington, D.C.

———. 2001. Digital Nautical Chart (DNC) Eastern United States, Series DNCD, Item 017, Edition 15. National Imagery and Mapping Agency (NIMA), Washington, D.C.

———. 2002. *Assessing the Ability of Commercial Sensors to Satisfy Littoral Warfare Data Requirements, Agreement # NMA 201-00-1-1006* (18 January 2002), Cooperative Agreement between NIMA and the UGA Foundation, National Imagery and Mapping Agency (NIMA), Washington, D.C.

NOAA. 2003. Our Restless Tides. National Oceanic and Atmospheric Administration (NOAA), Washington, D.C. <http://co-ops.nos.noaa.gov/restles1.html> Accessed 17 November 2008.

Pike, J. 1998. National Image Interpretability Rating Scales, Image Intelligence Resource Program. Federation of American Scientists, Washington, D.C. <http://www.fas.org/irp/imint/niirs.htm> Accessed 17 November 2008.

———. 2003. Marine Corps Base Camp Lejeune. GlobalSecurity.org, Alexandria, Va. <http://www.globalsecurity.org/military/facility/camp-lejeune.htm> Accessed 17 November 2008.

USMA (United States Military Academy). 2001. Soil Textural Triangle. *Academic Study Guide EV203 (Terrain Analysis)*. West Point, N.Y.: United States Military Academy (USMA).

USGS (US Geological Survey). 1952. New River Inlet, SC and Browns Inlet, SC, 1:24,000-scale Topographic Quadrangles, compiled 1952 with planimetric photorevisions 1972 and 1988. US Geological Survey, St. Louis, Mo.

———. 2003. The National Elevation Data Set Fact Sheet. US Geological Survey, St. Louis, Mo. <http://egsc.usgs.gov/isb/pubs/factsheets/fs10602.html > Accessed 17 November 2008.

Welch, R. 1972. Quality and applications of aerospace imagery. *Photogrammetric Engineering* April Edition 379–398.

Welch, R., S. Fleming, T. Jordan and M. Madden. 2003. *Assessing the Ability of Commercial Sensors to Satisfy Littoral Warfare Data Requirements*. Center for Remote Sensing and Mapping Science, University of Georgia, Athens, Ga.

Zeiler, M. 1999. *Modeling our World – The ESRI Guide to Geodatabase Design*. Redlands, Calif.: Environmental Systems Research Institute (ESRI).

CHAPTER 58

Adirondack GIS: Resources, Wilderness and Management

*Eileen B. Allen, Raymond P. Curran, Sunita S. Halasz,
Stacy McNulty, John W. Barge, Andy Keal,
and Michale J. Glennon*

58.1 Introduction

At approximately 6 million acres (2.4 million ha) the Adirondack Park in northeastern New York State (Figure 58-1) is the largest park in the conterminous United States, larger than Yellowstone, Yosemite, Everglades, Olympic, Arches, and Great Smoky Mountains National Parks combined. The Park contains the largest wilderness area east of the Mississippi River and is part of the UNESCO Champlain-Adirondack Biosphere Reserve. The Adirondacks are primarily forested with numerous settlements and limited agriculture. The Park is a unique mixture of 56% private land and 44% public land. Approximately 132,000 people are permanent residents in the Park, yet it is within a day's drive of nearly 90 million people.

The original Forest Preserve, the nucleus of the Park, was defined in 1892 by the New York State (NYS) Legislature to help protect water and timber resources. An Adirondack Park was created in 1892 and over the years expanded to include nearly all of the Adirondack Ecological Zone, approximately defined by the 1,000-foot (305-m) topographic contour. In 1894, an amendment to the New York State Constitution (Article XIV) proclaimed that "The lands of the State, now owned or hereafter acquired, constituting the forest preserve as now fixed by law, shall be forever kept as wild forest lands. They shall not be leased, sold or exchanged, or be taken by any corporation, public or private, nor shall the timber thereon be sold, removed or destroyed." This unique constitutional amendment provided protection to the state lands and bequeathed the natural beauty of the Adirondacks to the public in perpetuity. However, development pressures continued to increase on private lands. Finally, the NYS Legislature formed the Adirondack Park Agency in 1971 to address land use issues on both private and state lands within the Park. Today, the NYS Department of Environmental Conservation has the care, custody, and control of public lands, while the NYS Adirondack Park Agency (APA) provides management guidance for state lands and oversees private lands, wetlands, and designated rivers within the Park.

Figure 58-1 The Adirondack Park in New York State.

The Adirondacks are a dome of ancient rocks about 160 miles wide and one mile high whose surface was reshaped by Pleistocene glaciers. The result is a complex topography that embraces a wide diversity of habitats and wildlife (Figure 58-2). Broadly described as temperate deciduous forest, the region has over 11,111 lakes and ponds; thousands of miles of rivers; a variety of wetland types including freshwater marshes, swamps, rich fens and bogs; forest types that include temperate broadleaf deciduous and boreal coniferous; small areas of alpine meadows; and rich wildlife resources. There are ongoing efforts by numerous groups in New York State to identify large contiguous parcels and special habitats that merit protection.

Not only does the Park have rich resources, it has a rich history of mapping those resources. Innovations starting in the late 1800s continue today, as GIS has become an essential tool for planning, management, and scientific inquiry. From historical maps to present day digital files, there has been a consistent eye toward mapping quality and data sharing. Consequently, databases developed for resource assessment have become indispensable for regional resource planning and management and serve as an important research and educational tool. In this chapter, Ray Curran (Adirondack Sustainable Communities and formerly APA) and Sunita Halasz (APA) explore the mapping history and developments at the APA. The Lookup System allows non-map and non-computer staff at the APA to access a wealth of spatial data and is described by its ingenious creator, John Barge (APA). Eileen Allen (State University of NY at Plattsburgh) has helped dozens of students learn about GIS by supervising the development of digital watershed and wetland files. The collaborative Unit Management Planning-GIS project is spearheaded and described by Stacy McNulty (Adirondack Ecological Center). Michale Glennon (Wildlife Conservation Society) has used GIS for innovative work developing a biotic integrity index and in conservation planning. Finally, Andy Keal (Wildlife Conservation Society and co-author of *The Adirondack Atlas*) describes the development of *The Adirondack Atlas,* the first effort to bring together a broad array of spatial resources and explanatory text to help us understand this intriguing region.

Figure 58-2 Relief map of the Adirondack Park. See included DVD for color version.

58.2 A History of Mapping at the Adirondack Park Agency

58.2.1 Introduction

The Adirondacks were once "lands unknown." Explorers/land surveyors/cartographers like Verplanck Colvin and Seneca Ray Stoddard were some of the first to chart the northern New York State wilderness and provide a reference for others to follow. In fact, wilderness surveyor Colvin's descriptions of the Adirondacks' natural assets, including his maps, drawings and accompanying narrative over several decades in the latter half of the nineteenth century, were instrumental in awakening the conscience of New Yorkers to the destruction of mountain forests and watersheds. Eventually, his work indirectly led to protection of this spectacular natural area. Colvin's first Adirondack Survey, begun in 1872, started a trend in mapping excellence for the Adirondacks (Figure 58-3) that continues today even as current technologies are embraced (Figure 58-4). The remarkable growth in knowledge about landscape relationships in the Adirondacks essentially depends upon the mapping quality of the Park resources.

New York's current exemplary land use and regulatory program governing the Adirondacks is founded upon maps and mapping techniques. Land management issues such as forest trends, water quality, endangered species, land use capacity, recreational management of state lands, and regulated wetlands are all enhanced by maps. Most map information has benefited from or migrated to a Geographic Information System (GIS). Hence, a chronology of mapping by the NYS APA from the early 1970s to the present is useful to our discussion of how maps form the basis for Adirondack Park land management.

Figure 58-3 Mapping the Adirondacks in the 1880s. Clockwise from upper left: carrying surveying equipment down Hurricane Mountain; signal tower on Mount Marcy used for triangulation; measuring elevation on a steep slope; packing in survey party provisions; tent over transit and workers help prevent wind from vibrating the instrument; the Grand Theodolite. Courtesy: Adirondack Museum.

Figure 58-4 Mapping Adirondack wetlands in the 1990s. Clockwise from the upper left: wetland delineation on color infrared aerial photos with IIS SIS 95 Zoom Stereoscope; QA/QC digital file plot against original manuscript; photo-to-base map transfer using the IIS Stereo Zoom TransferScope; Plattsburgh State Remote Sensing Laboratory assistants visiting the APA; field checking in an Adirondack bog; digitizing in PC ArcInfo. See included DVD for color version.

58.2.2 Birth of an Agency and a Plan

In the late 1960s it was apparent to New York State as well as national conservation leaders that the Adirondacks were threatened by change from within. Although about 40% of the Adirondacks were constitutionally protected as "forever wild" Forest Preserve lands, trends in both the deterimental impacts on state land from recreational uses and incompatible uses on the private lands threatened key natural resources of the Park. In part, geography exacerbated the influence of private lands over adjoining state lands because of the patchwork and intertwined nature of land holdings. It is no wonder, then, that a solution using maps was found.

In the late 1960s and early 1970s, the staff of the Temporary Study Commission on the Future of the Adirondacks, later tranforming into the Adirondack Park Agency (APA), assembled a collection of maps covering the entire Park. The group of individuals that assembled these maps set a standard for high-quality map information that continues today.

Resource and human-use information was systematically collected and mapped onto base maps, in this case 58 15-minute series USGS paper maps covering the 6-million-acre (2.4-million-ha) Park area. Much of the information contained in the maps was based on first-hand knowledge of the land, or it was transferred/rectified from soils and other maps. These maps provided the APA with a detailed, region-wide perspective of the natural resources of the Park and their ability to support varying levels of development.

Using Ian McHarg's 1969 land capability approach from *Design with Nature,* a system of tissue tracings converted to Mylar overlays was employed to compile, rate, and analyze the interrelationships of resources and landscape characteristics on each quadrangle. This method built a land suitability and land sensitivity analysis that was ultimately used to produce the Official Adirondack Park Land Use and Development Plan (Figures 58-5 and 58-6). Significantly, the Plan was enacted as law by the NYS Legislature, thereby instituting a series of regulations based upon a map and the actual legislation of a map. The legislature even conceived of the need to maintain the map and adjust it based upon the discovery of new information. The fundamental importance of maps was thus reinforced. In view of the importance of maps to its program, it is not surprising that the APA has continued to support the development of quality mapping information.

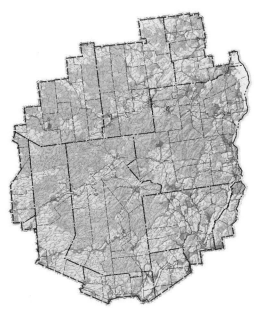

Figure 58-5 Official Adirondack Park Land Use and Development Plan. See included DVD for color version.

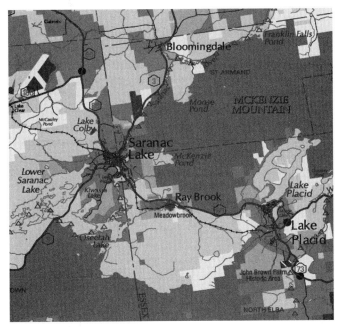

Figure 58-6 Official Adirondack Park Land Use and Development Plan, detail showing the Saranac Lake/Lake Placid area. See included DVD for color version.

58.2.3 From Pigeon Holes to Silicon Wafers – Map Storage at the Agency

The first maps used to support the planning and regulatory function of the APA were rolled up and stored in "pigeon hole" racks, which made organization and map preservation difficult. It was a very great improvement when the APA developed a system of mapping based on the 7.5-minute USGS map series and installed file drawers for each of the approximately 200 quadrangle sheets covering the Adirondack Park (Figure 58-7). That organization system, what some term a manual GIS, allowed easy transition to storage and organization of the information in digital form on computer servers.

Figure 58-7 Map flat files at the NYS Adirondack Park Agency. Today, using GIS, map information is stored more efficiently on the computer. It can be accessed readily by the user and precisely overlaid with other map information to aid decision making.

58.2.4 Geographic Information Systems

In the 1980s and 1990s, to address, in part, the growing need to manage the vast map information resources, the APA began using digital map information, bringing the Adirondacks to the forefront of modern mapping. A digital GIS has the advantages of allowing for information updates, map reproduction at a variety of scales, overlay and analyses of combined spatial information sources, and allowing easy access and recall of historic information (a "digital institutional memory").

The APA was the first workplace installation of ERDAS GIS software in the world. Previously GIS software was found primarily in academic environments. Over the period 1979–1981, the APA used ERDAS' image analysis capabilities to prepare a forest cover map and detect changes in the forest over time (Curran and Banta 1982). It was a good test of very large-area resource mapping and temporal analysis using satellite data. Eventually the APA migrated to the ESRI ArcInfo software environment as the technology matured and the cost of such a system was reduced.

As any new GIS user knows, the learning curve is steep. Digital map information is so valuable, however, that it had to be made available somehow to all staff to improve regulatory timeliness. APA staff solved the problem by developing an easy-to-use GIS "Lookup System" that allows access to natural resource and ownership information through a simple, customized menu system available on each staff member's desktop computer. It is used daily at the APA and is directly responsible for markedly improved response quality and timeliness to client requests for information.

The APA developed a web GIS application in ArcIMS (Internet Map Server) allowing individuals all over the world to access the Park-wide APA Land Classification data layer. A variety of Adirondack stakeholders, including realtors, lawyers, and surveyors, found the site useful and innovative. The website was one of four sites prominently featured on New York State's Web Banner until the events of September 11, 2001, forced all New York State internet map services and data access to be temporarily terminated. Data availability and access to online map data have been restored and continue to improve with changes in technology and data availability. Widespread data access either through file downloads or internet map services is expected by the public. New York State offers map-based data from many state agencies at http://www.nysegov.com/map-NY.cfm. The NYS Adirondack Park Agency land classification data layer may be viewed at http://www.apa.state.ny.us/gis/FacsimileMap.html.

58.2.5 Innovative Mapping Projects

The APA was involved in a number of innovative projects in partnership with local governments and other state and federal agencies. One of these was the Essex County Forest Feasibility Study, which was a Resource, Conservation and Development project with Essex County, the Soil Conservation Service, and SUNY College of Environmental Science and Forestry (SUNY ESF) (Craul et al. 1987). These groups worked together to use available map information to identify lands with the highest potential for timber harvest. In the 1990s, the APA was very involved in the Northern Forest Lands Inventory project, which developed map information on natural and cultural resources for the Adirondacks and New England (SUNY ESF 1995). These projects and the partnerships that were developed helped to bring mapping in the Adirondacks to the place it is today.

58.2.5.1 Wetlands Mapping

New York's wetlands inventory was originally conceived of as a statewide wildlife habitat management tool. One of the earliest applications of the inventory was in the Adirondack Park to enhance the description of wetlands to protect them as critical environmental areas under the APA's regulatory program. When New York passed its statewide wetlands law in 1975, the preparation of wetland maps as a regulatory tool was mandated. Initially in the Park, to generate regulatory maps, black-and-white 1:24,000-scale photos were used for interpretation, then eventually high-resolution quad-centered color infrared photography became available. The US Fish and Wildlife Service National Wetlands Inventory (NWI) techniques (Cowardin et al. 1979) were introduced into the Park wetlands inventory and inevitably led the way to digital mapping and GIS analyses.

In the Park, the regulatory wetlands maps served in part to alleviate landowner concern about the nebulous location of wetlands and potential "over-regulation" of upland areas. The maps were to be a concrete and reliable statement about the relationship of wetlands to land ownerships. However, as cartographers and wetland managers recognize, a map is only a representation of the real world; a thin wetland boundary line on a map may represent 100 feet on the ground. Therefore the representation of a wetland on the map is imprecise at best. In addition wetlands are dynamic and subject to landscape changes (from shifting climatic conditions and beaver activity) while a map, particularly a paper map, remains the same. Therefore, the APA, like other regulatory agencies, does not rely solely on maps to determine its wetland jurisdiction but employs full-time wetland field biologists to provide on-the-ground verification, a process called delineation. As the regulatory wetland mapping process proceeded and was completed for almost half the Park, the realization of the regulatory map inadequacies became apparent. This led to a new phase of detailed wetland mapping through

US Environmental Protection Agency (EPA)-funded grant projects aimed at producing the best quality digital wetlands maps for use in landowner notification, landscape assessment, and ongoing planning activities. The process continues to be one of the notable qualities of the data availability in the Adirondacks and a primary objective of state agency efforts in the Adirondacks.

58.2.5.2 EPA-funded Projects

The Adirondack landscape averages approximately 15% wetlands by area, with numerous polygons (over 2,100 polygons in one 7.5-minute quadrangle) and a great diversity of wetland cover types. Since 1993, the APA has received grants from the US EPA's State Wetlands Protection Program to address watershed-scale environmental issues (http://apa. state.ny.us/Research/epa_projects.htm). In preparation for each of these projects, digital maps of wetlands and watersheds at a scale of 1:24,000 were prepared. As of December 2005, over 70% of the Park was mapped and digitized, with projects in the unmapped areas underway. The data have been used to aid in regulatory review, to help landowners understand their property better, and for numerous scientific studies at local universities. Digitizing the data from these maps allows better quality assurance controls because the wetland boundaries interpreted from aerial photos are revisited an additional time. Digital data have also allowed the APA to compare temporally differing data sets to determine wetland trends and to characterize landscape-level properties of wetlands, such as the composition and size of geographically complex wetland systems. These data sets are of high quality and provide an extremely valuable insight to understanding the role of natural processes in the Adirondacks.

58.2.5.3 Shared Adirondack Park Geographic Information CD-ROM

In 2001, APA staff realized that the volume of digital data available from the APA had reached a point where it could be compiled into one large digital library for ease of distribution. The *Shared Adirondack Park Geographic Information CD-ROM* (*Shared GIS CD*) (http://apa.state.ny.us/gis/shared/index.html) set contains over 50 geographic natural and cultural resource data layers from eleven state and federal agencies. Since its release in July 2001, the APA has distributed over 500 copies. Much of the high start-up cost of a GIS originates with data development. The *Shared GIS CD* gives new users, especially local governments with financial challenges, the advantage of an instant, comprehensive database for the Park. Feedback from users has been extremely positive and enthusiastic. APA staff uses the *Shared GIS CD* as a tool to encourage local governments, Forest Preserve foresters, and others to begin to use GIS as a decision-making tool.

58.2.5.4 Adirondack GIS Users Group

The creation of the *Shared GIS CD* had many benefits. In addition to serving as a focal point of GIS training sessions, it has led to the creation of the Adirondack GIS Users Group (www.adkgis.org), a partnership of state, local, not-for-profit, for-profit, and educational organizations. The Adirondack GIS Users Group is a forum for sharing data, providing assistance/training, sharing new research projects and ideas, and keeping the Park's GIS community updated. In 2002, the group held training workshops for NYS Department of Environmental Conservation Forest Preserve planners, and local Code Enforcement Officers at Local Government Day and the Adirondack Research Consortium Meetings. The Adirondack GIS Users Group also hosted the 2004 Northeastern Arc User's Group in Lake Placid, New York, bringing local mapping expertise to the meeting. Since its inception, the Users Group has held meetings to discuss GIS in natural resource planning, emergency management, tourism promotion, and K-12 education. The Users Group has programs twice a year, directing their focus towards the collective GIS needs and questions gleaned from contacts with local governments and state agencies, not-for-profit groups, non-governmental organizations, educational institutions, and the public. This collective thinking and effort is a new chapter in Adirondack mapping that promises further innovation and focus on protecting a truly unique region.

Mapping and mapmakers in the Adirondacks have changed dramatically since 1872. Each decade seems to have brought new ideas, awareness, and technologies that impact how information gathering is approached, the ability to analyze and share that information, and increases in the public's thirst for understanding where they live. Verplanck Colvin would be both awed and pleased to witness how his efforts spawned a mapping legacy.

58.3 GIS in the Hands of Public Servants: The Adirondack Park Agency "Lookup System"

58.3.1 A Tool for Program Staff

Many organizations have implemented GIS intending to integrate it into their business process. Unfortunately, GIS is often either a mysterious "black box" process that is underutilized for daily tasks or a bottleneck in times of critical need. For over ten years, the New York State Adirondack Park Agency has put GIS on the desktops of public servants. Every day, the APA GIS "Lookup System" enables staff to provide information quickly to the public, make decisions, spot problems, and document business transactions. The Lookup System has become a critical tool in a time when public service demand and pressures on the Park have increased and greater government efficiency is required.

58.3.2 Extent of Need

The Adirondack Park covers an area of public and private lands in northern New York State roughly the size of Vermont. Approximately 132,000 year-round residents live in nearly 100 separate municipalities within the Park.

For over 30 years, the Adirondack Park Agency has regulated land use and development on all lands in the Park and administered laws protecting wetlands and designated rivers. Fifty-eight APA staff members provide services in Regulatory, Legal, Planning, Interpretive, and Administrative divisions.

Every year the APA handles an average of 7,000 phone calls from the public requesting advice on whether a permit is needed for an intended project. The location of a project helps determine the jurisdictional requirements for the three laws the APA administers: the Adirondack Park Agency Act; the New York State Freshwater Wetlands Act; and the New York State Wild, Scenic, and Recreational Rivers System Act. Staff also must determine if an APA permit was ever issued for a property or if it has a pending enforcement case. Along with phone calls, the APA writes approximately 1,000 legally binding letters each year informing a land owner or developer whether or not they need a permit. In addition, the APA processes over 300 permits a year, more than 400 permit applications, about 300 enforcement cases, provides advice to local governments, and reviews numerous state land management plans. Map amendments to the Official Adirondack Park Land Use and Development Plan (Figures 58-5 and 58-6) are processed, environmental interpretive programs developed, and scientific research on subjects such as acid precipitation, invasive plant species, and wetland function are conducted

Mapped information plays a key role in almost all of the APA responsibilities. Locating, interpreting, and extracting spatial information from maps on file at the APA has been a time-consuming job. APA staff, including attorneys, accumulated a great deal of experience using engineer scales to transfer boundaries between zoning, wetlands, and tax parcel maps (Figure 58-8).

Figure 58-8 Boundaries were transferred between zoning, wetlands, and tax parcel maps using engineer scales before the APA GIS Lookup System. See included DVD for color version.

The results of labor invested in rescaling, transferring, and documenting spatial information for business transactions were filed away in project folders. Future retrieval of this information added to the list of tasks to be done for each new project. Index maps and master summary tables for cross-referencing past transactions were difficult to keep up to date and cumbersome to access.

Besides hardcopy maps, staff would routinely refer to air photos, permit files on microfiche, oversized engineering drawings, and other printed documents. These documents were stored in many locations throughout the APA headquarters including the attic and basement (Figure 58-9).

Figure 58-9 Hardcopy maps in the APA Headquarters attic. This was part of the map filing system used before GIS. See included DVD for color version.

58.3.3 Plan and Implementation

In 1988, the Governor's Executive Budget included a plan for improved computing capability at the Adirondack Park Agency. Three of the plan's objectives approved by the Legislature included:

- Integrate existing Planning, Legal, Natural Resource, and Operations computer use into one unified APA-wide system;
- Provide immediate computer access to land ownership and APA jurisdictional information indexed to tax parcel number and owners' names to improve response time to public jurisdictional inquiries; and
- Initiate cross-referencing of permits and computerized resource maps (such as soils, landcover, and hydrography).

The APA acquired a Prime minicomputer with ESRI ArcInfo software and eight graphics terminals. After initial training and data development a text-based parcel ownership search and reporting system was created. In the early 1990s, the APA purchased a UNIX Data General workstation, which became the platform for the map-based Lookup System. The system in use today is largely the same as that made available to staff in 1994.

58.3.4 A Simple Interface

The Lookup System interface consists of several menus along with the map view window. Menus employ buttons with descriptive labels, which open up additional menus and forms. The system was designed prior to the common graphic user interface seen in many desktop applications today and does not include the use of icons to open tasks.

On start up, the main menu (Figure 58-10) provides the most common tasks including searching parcel data by number or owner name, finding a named place, and searching APA transactions. When searching parcel data, users select a municipality and then can either enter numbers or names or pick from pop-up lists. In the case of entering numbers, the system will right and left justify entries and add leading or trailing zeros as needed.

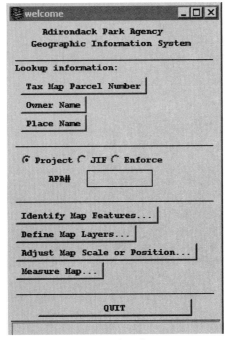

Figure 58-10 Main menu of the Lookup System interface.

Manual of Geographic Information Systems

When parcel query results yield a match, a form displays property information and options to map the property and print information. Some of the results include the owner name and address, the parcel number, the deed reference, acreage, use classification, and whether an APA transaction occurred on the parcel at a prior time. The user can also view additional ecological information with the parcel overlay such as proximity to aquifers, New York Natural Heritage Program rare species occurrences, ecozones, and significant biological sites. Users are also given the opportunity to record the parcel as a new APA transaction. In this case, the user enters the unique transaction number, which is then stored with key parcel result items.

The Lookup System main menu also provides access to additional tools. With "Identify Map Features," users can point at items on the map and retrieve attribute information about them. When selecting APA transaction locations for instance, users can retrieve and open permit text documents. Users can also select soil association areas and retrieve septic suitability ratings.

Menus are also provided to "Define Map Layers" used to turn map layers on and off and to "Adjust Map Scale or Position" to resize and reposition the map view area. The "Measure Map" menu provides tools to measure distance, area, direction, projected coordinates, and elevation of any location on the map. The Lookup System also provides a linear referencing tool to locate mile marker increments along state highway routes.

58.3.5 Data – Ready or Not, Here It Is

One of the Lookup System objectives is to provide staff with access to the best available GIS data (Table 58-1). Ideally, all layers would be geographically and temporally complete and mapped at a scale with sufficient detail for reliable decision making. This is not always the case. Instead, the staff is provided data when and where available along with knowledge to carefully interpret it. The majority of spatial layers are mapped at a regional scale of 1:24,000 or one inch equals 2,000 feet. Parcel data, available from county tax mapping offices, is larger scale while some of the ecological layers are from smaller scale, less detailed map sources.

The aquifer layer is one of the small-scale data layers. In New York State, aquifers are mapped at 1:250,000 scale or one inch equals about four miles. APA Resource Analysis staff considers this an important layer in assessing cumulative impacts of development. Having no other source of aquifer data, GIS staff created a "fuzzy-edged" polygon aquifer layer for use in parcel overlay analysis. One quarter mile buffers were created inside and outside of each aquifer polygon. Using this layer, the Lookup System reports if a parcel is wholly outside mapped aquifers; outside mapped aquifers but within a quarter mile of an aquifer; inside an aquifer but within a quarter mile of its edge; or wholly inside a mapped aquifer. In this way users are given the spatial relationship between different scaled data, which they can use as a preliminary estimation.

Table 58-1 Lookup System map layers.

Planimetric Base Features	Significant Biological Sites
Tax Map Parcels	Natural Heritage Sites
20 foot Topographic Contours	Ecozones
APA Land Classification Zoning	Archeological Sites
Designated Rivers System	Airphoto Inventory Index
Project Permit Review Sites	Percent Slope
Wetland Field Visits	Shaded Relief
Regulatory and NWI Wetlands	Conservation Easement

58.3.6 Process and Products

Users of the APA Lookup System have a desktop icon on their PC, which enables them to automatically log in to a server, start ArcInfo and open the map and menu interface. The system is currently delivered to staff over a local area network using Citrix Metaframe to connect desktop PCs to a dual processor PC application server. In this configuration, GIS software does not have to be loaded on desktop PCs. A small number of server user accounts are shared among desktop users. These accounts accommodate default printing to the nearest color and black-and-white laser printers. Inactive users are logged out after 10 minutes to make the ArcInfo licenses available.

The Lookup System runs within ESRI's Arc Macro Language environment. The system was programmed and is maintained by an APA GIS staff member. There are 84 program and menu files used to run the system. Server, network and printer support are provided by an APA IT staff person.

Connectivity from the Lookup System to other office documents is accomplished by using unique transaction numbers and a standardized file-naming scheme. Mapped APA transactions store a unique identifier in their spatial attributes. When the identifier matches an operating system folder or files, the results are listed for the user. In this way, links to specific permit documents or scanned engineering drawings do not have to be hard coded in the attribute tables. The transaction code is used as a wild card to search for available documents.

Products from the Lookup System include print, image, and tabular information. Along with viewing map and descriptive information on-screen, users can create and send custom map prints to a local printer. Maps can include custom titles, a legend, scale bar and locator map along with highlighted query results. Maps can also be saved as images to the user's network hard drive for inclusion in email or text documents. Descriptive information for query results includes parcel ownership, address, size, and location as well as where the parcel is located relative to various ecological and jurisdictional map layers. Users can also generate address lists from the Lookup System to be included in mail-merge word processing applications.

58.3.7 Directions and Recommendations

Reliable data has contributed most to the Lookup System's success. APA staff believes the replacement of parcel center points with parcel polygon boundaries is the greatest improvement to the Lookup System. Parcel boundaries better enable staff to determine and visualize project jurisdiction and impacts on natural resources. They also help staff generate reliable adjacent land owner lists for project notification. Most of the 12 counties in the Adirondack Park now produce parcel GIS data and provide it for free. The APA recognizes how important local government GIS programs are and seeks to provide leadership in sharing data, applications, and experience.

The APA has found having unique, standard transaction identification numbers valuable for cataloging APA business. An example would be J2005-0346 for the Jurisdiction Department's 346th project in 2005. Items such as spatial data, file system folders, microfiche, and all other transaction records are identified with these numbers. The Lookup System is able to locate and label transactions and retrieve text and image documents with these numbers. It is essential to ensure that these numbers are used in their standard format and applied to all types of records. APA staff feels that complete historic coverage of all past APA transactions is important. Elimination of gaps in the historic data record provides confidence in APA determinations and reduces additional required research time.

Associating parcel identification numbers with all APA transactions has also proved valuable in locating and cataloging parcel-related APA business. Unique town codes and tax map numbers are keyed for each transaction in carefully formatted database fields. Transaction data with these tax map numbers can be cross-referenced with parcel point or polygon data and shown on a map. The Lookup System is thus able to flag queried parcels for past

APA actions. Difficulties in using tax map numbers arise when projects encompass multiple parcels and when tax map numbers change due to subdivision or merging of parcels. Therefore, it is useful to record a tax roll year with the tax map number.

Lastly, staff point to the system's ease of use. Simple mapping and query tools, tailored to specific needs are packaged in an intuitive interface. After a few minutes of orientation to tools and options, staff is able to use the Lookup System for their work responsibilities. The system provides desktop access in a "one-stop-shop" environment where mapped information can be used alongside other office applications. Jurisdictional Inquiry staff points out that telephone questions that used to take one week to assemble answers for can now be answered during the original phone call.

The future of the Lookup System will include several changes. Improvements in delineation and data entry of incoming APA transactions within the Lookup System are needed. Printing tools to create routine sets of thematic maps and tools to create large-format maps have been requested. Tools to import GPS data and digital photo locations may be designed to assist field staff. Orthophoto imagery is now available and in demand by staff. This imagery is not readily supported in ArcPlot, which may lead to the entire system being ported to ArcMap. ArcMap has additional advantages such as a common programming language, the ability to re-project data on-the-fly, and the powerful Geodatabase storage format.

The Lookup System is a tool that does not replace a site visit, but provides APA staff with an initial view of existing site characteristics when assessing jurisdictional inquiries, reviewing permit projects, evaluating violations, or delineating wetlands. The Lookup System is also a *digital institutional memory*. As more and more staff retire who spent the majority of their careers at the APA, their replacements can rely on the Lookup System to provide a spatial history of APA jurisdiction around the Park. Many permit review officers that utilize the Lookup System are not experts in using spatial data or computers. Their workload does not allow them time to explore the steep learning curve of GIS nor does it let them use data that are disorganized or not readily available. GIS has provided both a spatial and database resource essential to the work of the APA and the Lookup System has made the GIS useable.

58.4 NYS Adirondack Park Watersheds and Wetlands: GIS Database Development and Education

58.4.1 Introduction

The 1975 NYS Freshwater Wetlands Act charged the NYS Adirondack Park Agency with protecting wetlands within the Adirondack Park. It has long been recognized that to manage wetlands effectively, a highly accurate and credible inventory was necessary to understand wetland types, relationships, and distributions. However, available technology and resources have dictated how an inventory would be created rather than being dictated by the needs. In the early 1970s, wetlands were interpreted and delineated on 1968 1:24,000 black-and-white aerial photography for use as a general planning and wildlife management tool. Another inventory was conducted on 1:80,000 black-and-white quad-centered transparencies in the 1980s to address regulatory delineation needs. However, both of these inventories were attempted before GIS was widely available.

By the late 1980s GIS software was maturing into a tool that could be utilized for practical data development. Software routines were prepackaged so that data developers could use out-of-the-box tools with minimal additional programming. Hardware was becoming more affordable, although it was still expensive, slow, and with minimal memory and storage capabilities by today's standards.

Flat maps at the Adirondack Park Agency were organized by 7.5-minute topographic map sheets and the digital data were organized in the same manner. This was fortuitous because the early versions of PC ArcInfo had limited file sizes and some of the digitized wetland maps nearly exceeded software limits. As software and hardware increased in capabilities and size, final digital wetland files were appended into GIS layers encompassing the study area. However, the old 7.5-minute map file organization has continued to be helpful for data development, project tracking, and quality assurance.

ESRI products (PC ArcInfo, ArcInfo, ArcView, and ArcGIS) comprised most of the software used for Adirondack Park database creation. The development of ESRI capabilities dovetailed nicely with project needs. For example, just as researchers were struggling with how to analyze complex watersheds (i.e., where one body of water flows into another), ArcInfo regionalization was developed. When ArcView first became available, it was quickly adopted as a critical tool for quality assurance, which would have been extremely difficult without it.

Current watershed and wetland digital data production is a combination of map analysis; optical remote sensing; analog data transfer; and digital scanning, editing, and attributing. As of this writing, air photo transparencies are still widely available and the APA has a complete library of orthophoto transparencies at 1:24,000 scale. Analog methods of wetland delineations, cover typing, and transfer to base maps have been reliable, yet it is likely that a new phase of wetlands mapping and GIS data layer development will be explored in the near future. Digital orthophotos are readily available now and digital air photos will be flown within the next few years to support the continued orthophoto development in New York State. As digital mapping cameras become widely used and affordable software becomes more robust, it is expected that the next phase of wetlands database creation will be fully digital.

Watershed and wetlands digital data development depended heavily on student help. These projects were important training grounds for students interested in GIS, remote sensing, and wetlands management. Since students have been included in all phases of the work, the projects provided unusual insight and experience for the undergraduate college students into an ongoing regional research project involving numerous agencies.

58.4.2 Regional Watershed Projects

Since 1993, the APA has received many grants from the US EPA State Wetlands Protection Program. As of late 2005, over 70% of the 6-million-acre (2.4-million-ha) park has been mapped and funding has been secured to map the remainder of the Park. These projects have used a watershed approach (Figure 58-11) and each has allowed exploration of critical management issues particular to the watershed in question and resulting in a detailed inventory of wetland resources. The wetlands inventories were necessary to answer management questions, but were each conducted using consistent techniques so that results could be compared across watershed boundaries.

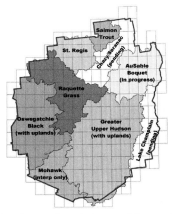

Figure 58-11 Regional watershed mapping status as of December 2005. Gray lines indicate 7.5' USGS topographic map quadrangles. See included DVD for color version.

The Oswegatchie/Black Watershed (Roy et al. 1996, 1997) was mapped to evaluate acid deposition impacts to ponded waters and wetlands. Watersheds and wetlands were delineated to build a database to assess the impacts of the 1991 Federal Clean Air Act amendments and the 1985 NYS Acid Deposition Control Act standards.

In the Upper Hudson Watershed, the goal was to develop a methodology to assess the cumulative impact to wetlands, which result from numerous, often small, wetland changes due to human activities (Primack et al. 2000). The focus of mapping the St. Regis (Halasz et al. 2000) and the Salmon/Trout, Raquette, and Grasse Watersheds (Karasin et al. 2002) was on locating large peatlands and identifying large single-ownership parcels. These projects included a literature search of large wetland complexes in the Park, which was keyed to the GIS database. These latter projects included a significant public outreach component that allowed the APA to understand issues of importance to both the large landowners and land managers in these regions of the Park.

The Mohawk Watershed (LaPoint et al. 2003) has been mapped on aerial photograph overlays but not transformed into digital data. The AuSable/Boquet Watershed is being mapped with the goal of identifying areas suitable for wetland restoration. Funding has been secured for the Saranac River and Lake Champlain Watersheds; these last two projects will complete the first detailed wetlands mapping for the entire Adirondack Park.

The regional watershed projects all included wetlands mapping from color infrared aerial photography. Regional watershed boundaries were delineated from USGS topographic maps for all of the projects to date. Ponded water and major riverine watersheds were identified in all but the Mohawk and AuSable/Boquet Watersheds. Upland land cover classifications from 1990–1997 Landsat TM imagery were completed for the Oswegatchie/Black and Upper Hudson projects.

58.4.3 Watersheds

The Adirondack Park was divided into nine watershed regions for the purposes of the EPA watershed projects (Figure 58-11). The study area "pour points" were defined by the outflow of the named river or lake system intersection with the Adirondack Park Boundary. Both the study area and smaller ponded and riverine watersheds were delineated on 7.5-minute 1:24,000 and 7.5x15-minute 1:25,000 USGS topographic maps. Surficial topography is complicated in the Adirondacks: some watershed boundaries can be difficult to ascertain and occasionally the boundaries actually have wetlands straddling them. In addition, beaver activity can create small ponds and change the flow direction of existing small ponds and wetlands (Figure 58-12). Since wetlands and ponds can change in the Adirondacks, the most recent topographic maps available were used as base maps for watershed delineation. Topographic maps were used as much as possible to determine watershed boundaries, but in areas where the watershed could not be ascertained, aerial photographs were used and, where necessary, fieldwork was conducted.

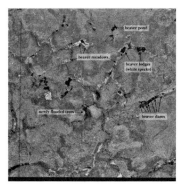

Figure 58-12 Aerial photograph showing beaver activity. Original imagery is 1:40,000 NAPP color infrared photo number 8036-11 flown 2 May 1997. See included DVD for color version.

Watersheds were defined as the pour points for ponded waters listed in the NYS Department of Environmental Conservation Adirondack Lakes Survey Corporation (ALSC) Database. This database listed all ponds that had been sampled at any time by the NYS Department of Environmental Conservation or the ALSC, some of which had been sampled in the 1940s. However, the database included some sampled ponds that did not appear as ponded water on recent topographic maps. These were eliminated from the delineated watersheds. In some watersheds, such as the Upper Hudson and Raquette, large areas remained unsubdivided by ponded watersheds. In these, major riverine areas formed the basis for additional watershed divisions.

Two individuals delineated watersheds separately. A third person compared the delineations and resolved the differences. Watersheds were then digitized into an existing tic database using ArcInfo and tablet digitizers. As each quadrangle was finished, a hard copy plot was made and compared against the original delineation. Digital files were edgematched and the files appended to create a single project watershed file of both the study area watershed and the subregional watersheds.

58.4.4 Wetlands

Wetlands were delineated on acetate overlays attached to 1985–1986 1:58,000 National High Altitude Photography (NHAP) half quad centered color infrared (CIR) transparencies (Oswegatchie/Black Watershed) and 1994–1999 1:40,000 National Aerial Photography Program (NAPP) quarter quad centered CIR transparencies (all other regional watershed projects). Delineations were done using NWI protocols (Cowardin et al. 1979) with some label convention modifications as needed in the Adirondacks. An Image Interpretations System (IIS, formerly Bausch & Lomb) SIS 95 Zoom Stereoscope was used for delineations. Fieldwork consisted of traveling accessible roads and marking wetland types on USGS topographic maps. In some inaccessible areas, canoes or motorized boats were used for field checking.

The 1:40,000 CIR quarter quad transparencies were well suited for regional wetland mapping. The number of photographs per quadrangle was more manageable than at a larger scale and an appropriate amount of detail could be observed for wetlands mapping. The vertical exaggeration was excellent for interpreting the wetland landscape position, texture for vegetation types, and height for vegetation structure. Color infrared leaf-off imagery was ideal for identifying water bodies and discriminating vegetation types. Some types of vegetation, such as submerged aquatics, were not identifiable with this imagery. The minimum mapping unit at this scale is about 0.25 acre (0.1 ha).

Several types of displacement are inherent in aerial photography. Most notable in this imagery were radial and topographic displacement. Consequently, photo enlargements cannot be used as maps. Photo delineations of wetlands were transferred to hard copy orthophoto transparencies using an IIS Stereo Zoom TransferScope. Orthophotos were preferred to topographic maps as base maps because in many areas of the Adirondacks there is so little ground control that forest edges and even individual trees may become control points. Separate overlays were made for wetland delineations and for wetland labels. Each of these overlays had numerous quality control steps.

Wetlands delineation overlays were scanned and the raster-to-vector conversion was conducted with Able Software R2V. From this point, students would be responsible for much of the digital editing and file attribution. Rigorous Quality Assurance/Quality Control (QA/QC) standards were written and students had to develop an understanding of Adirondack wetland cover types, labeling conventions, remote sensing techniques, GIS software editing options, and common sense about how wetlands work in the environment (for example, a large pond would likely have a permanent riverine outflow). Students were responsible for individual quadrangle wetland files from initial scanning to final inclusion in the database. Wetland quadrangle files were edgematched and appended together and regional wetland file compilations were subjected to an additional set of QA/QC protocols before the final files were accepted by the APA.

58.4.5 Upland Land Cover

Upland land cover was developed using supervised classification techniques and ERDAS Imagine image processing software for the Oswegatchie/Black and Upper Hudson Watersheds. Since the wetlands data were delineated in detail from aerial photographs and were in a GIS format, they were masked from the study area. Both leaf-on and leaf-off Landsat TM scenes were used to identify land cover classes in the two watersheds within the following categories: deciduous forest, mixed forest, coniferous forest, open/barren land, open with vegetation, and open without vegetation. Wetlands and open water classes were based upon the wetland GIS files.

58.4.6 Educational Benefits

The development of a digital watershed/wetlands database has had many positive impacts. While exploring questions of interest to the EPA and the APA, the APA has been able to develop a database to help evaluate the wetland resources they are legally mandated to manage. This large consistent database has sparked other projects, such as research to identify Bicknell thrush habitat and to support Forest Preserve Unit Management Planning. Public outreach, a goal of the project, has helped educate Park residents about the importance of wetlands, natural resource management issues, and both GIS and remote sensing tools.

One of the most important impacts of the wetland delineation work has been the educational opportunity provided. Over 30 undergraduate students have been hired or have conducted independent studies to help create the database. Demonstrations of the project have exposed students to remote sensing and GIS in introductory and advanced remote sensing classes, introductory GIS classes, and wetlands ecology classes. Students have had the opportunity to participate in a project where both state government (APA and ALSC) and researchers from several disciplines (geomorphology, ecology, remote sensing, and GIS) work together to create the database and analyze the underlying problem addressed by each grant.

Fortunately, the original database design was flexible, the techniques were transferable to newer software versions, and there was detailed project documentation to ensure a consistent workflow. Initially there was no way of knowing that the first digital wetlands mapping project in 1993 would evolve into a Park-wide database, yet the product consistency and detailed metadata enabled this to occur. Advances in GIS seemed to happen just when needed. PC ArcInfo and ArcInfo have been ideal for database creation, while ArcView was essential in QA/QC of the detailed wetlands maps.

For undergraduate college students the projects have been invaluable and the regional watershed projects could not have been done without their help. The components of good project design such as consistent data development protocols, clear workflow, comprehensive QA/QC procedures, and meticulous metadata were also critical for good education. Most students have not been exposed to data development from multiple resources and the problems associated with it, error checking, developing QA/QC procedures, problem resolution, and keeping detailed notes of their work. For educational purposes and project consistency, several items became critical:

- Parsing the project into discrete tasks. This is important to keep students from being overwhelmed, especially with detailed long-term projects.
- Giving students data ownership. Each student was responsible for bringing individual quadrangles into the digital environment from the quadrangle wetland overlay scan through all QA/QC procedures to the final product.
- Training students in every step of the entire database creation project with a series of small projects even if it was not their major task. This allowed them to see where

their work fit into the project. It also helped them anticipate the sources of errors and mistakes from different types of data and helped them understand the workflow.

- Using students to train other students. This ensured project consistency and increased understanding, and student-to-student descriptions often resulted in better communication of complex ideas.
- Developing an arena where students could gain self-reliance and confidence. Students were encouraged to develop answers to problems and then validate the answers with project faculty.
- Recognizing that much of what the students did was tedious work and helping them develop ways to minimize error.
- Giving students responsibility. Students reported their work in a variety of venues, including project meetings.

Creating the digital watershed/wetland database for the Adirondacks has taken many years. Fortunately, the same research team at the Remote Sensing Laboratory, State University of NY at Plattsburgh, has conducted the photo interpretation in all of the regional projects, ensuring wetland delineation and labeling consistency. However, the process of developing digital files from analog sources has involved many people from both educational and state agencies. As a consequence, establishing a consistent workflow and QA/QC procedures has been essential. The primary process necessary to build both of these tasks was the anticipation of the sources of potential errors and mistakes. An in-house laboratory manual was developed to teach students the basic steps in digitizing and quality control for both the watershed and wetland data files. A digitizer menu for NWI wetland cover type labels minimized attribute entry errors, as did a QA/QC checklist used for each quadrangle. A project index map was created and used for each step of the project to indicate stages of completeness and which analyst was working on each file. Such extensive projects rely on student help and these projects have been an excellent training ground for students interested in map interpretation, remote sensing, GIS, and wetlands management.

58.5 UMP-GIS: Enabling Better Public Land Management in the Adirondack Park

Through a collaborative effort called the Unit Management Planning-GIS project, a group of partners is developing innovative data layers, tools, and approaches to natural resource inventory, with the goal of improved management, protection, and enjoyment of public lands in the Adirondack Park.

58.5.1 Background

Full appreciation for the progress of the UMP-GIS project requires some background in the Adirondack Park history and administration. Roughly half of the 6-million-acres (2.4-million-ha) in the Adirondack Park are state-owned. Collectively called the Forest Preserve and administered by the New York State Department of Environmental Conservation (DEC), recreation is allowed but timber harvest is restricted on these public lands. Similarly, the Adirondack Park Agency regulates many activities on privately owned lands, from private waterfront homes to town centers to industrial timberlands. The spatial distribution of Forest Preserve and private land in the Park is a matrix. This combination of divided responsibility for land management and the complex arrangement of land ownership means that unlike other parks, land stewardship and planning human usage can conflict.

The Forest Preserve is divided into smaller units (Figure 58-13), each guided by a Unit Management Plan (UMP). A UMP contains an inventory of natural resources such as vegetative communities and wetlands, as well as facilities such as lean-tos, snowmobile trails or boat launches. Once inventory for a unit is completed, alternative management scenarios are compared and a plan and timeline are developed. The UMP outlines steps needed to protect the lands and waters from overuse, degradation and loss of biodiversity, and designates appropriate forms of recreation for the unit (http://www.dec.ny.gov/lands/4979.html).

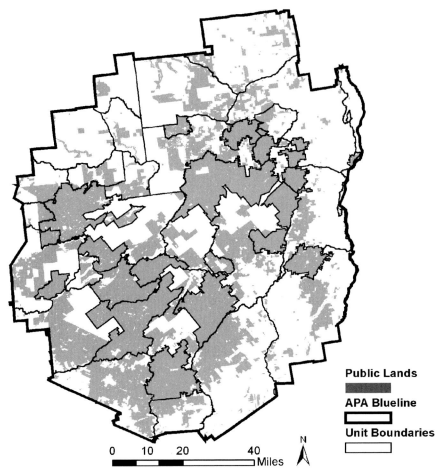

Public Lands

APA Blueline

Unit Boundaries

0   10   20   40
Miles

N

Figure 58-13 Forest Preserve lands and Unit Management Planning areas in the Adirondack Park.

A UMP must be completed before any facilities can be constructed or altered. Yet since the inception of the Adirondack Park Agency Act and adoption of UMPs in 1971, few UMPs have been completed or revised, despite a planned schedule of revision every five years. In 1999, the New York State Governor, George Pataki, announced that all unfinished UMPs would be completed in 5 years. Not long after the governor's announcement, a group of individuals met through the Adirondack Research Consortium to discuss ways to assist in finding, collecting and interpreting data on the resources in each unit. The outcome of this meeting was the formation of the UMP-GIS project, whose partners include the DEC, APA, New York Natural Heritage Program (NHP), the Association for the Protection of the Adirondacks, Audubon Society of New York, the Adirondack Nature Conservancy, Wildlife Conservation Society, and many others. The UMP-GIS project is led by researchers at the Adirondack Ecological Center, a biological research station of the State University of New York College of Environmental Science and Forestry (http://www.esf.edu/aec). The goal of the project is to provide DEC planners with an efficient method to access inventory data using GIS to result in improved land management in the Adirondack Park.

58.5.2 The UMP-GIS Project

The project objectives are fourfold:

- *Assemble the GIS database.* Establish a collection of data layers from diverse sources.
- *Provide interpretation and analysis.* Offer GIS, statistical and ecological expertise. The GIS is used to explore questions such as "Should a campsite be relocated away from a sensitive wetland?"
- *Provide technical support to DEC planners.* Enable the planners to focus on planning by maximizing efficient use of GIS software.
- *Maintain a data library for future users.* Ensure high-quality, well-documented, consistent data that is compatible with existing DEC databases and flexible for inclusion of data in the future. Provide metadata (data documentation).

Project partners recognized that GIS should be the means of organizing data and focusing land management on the entire Adirondack region for several reasons. The ecological data and expertise in spatial analysis needed were distributed across many state agencies, universities, and non-profit organizations, some of which have had a history of conflict regarding public land management decisions. Some Forest Preserve units are fragmented, consisting of several separate parcels or including privately owned inholdings (Figure 58-14). Furthermore, activities on adjacent lands impact public land, and natural communities extend across political boundaries.

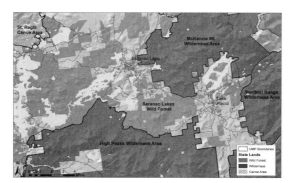

Figure 58-14 Saranac Lakes Wild Forest unit boundary, showing adjacent public and private lands. Courtesy: Steve Signell, SUNY-ESF. See included DVD for color version.

Prior to the implementation of the UMP-GIS project, DEC planners had to seek out and compile data, often evaluating each data layer independently. UMPs contained different data sources, differed widely in their inclusion of information, and standards for map presentation were lacking. Consider a proposal to establish a new hiking trail. This decision should be based on soil erosion potential, topography, existing trails, trail access points, and distance to sensitive natural resources. The UMP-GIS project helps planners integrate information for the inventory in a GIS, analyze the information, compare alternative trail locations, and create map output for draft plan documents and public meetings.

Setting priorities for public land involves interest groups with different goals and can be a contentious process. GIS provides a way to identify management alternatives using information rather than agendas. If members of an interest group or the public perceive a UMP-GIS partner as having undue influence, the integrity of the entire UMP process might be called into question. Therefore, successful cooperation relies on the restriction of decision making to the sole purview of the DEC. Towards this end, UMP-GIS partners have agreed to provide information without advancing a position advocated by their organizations. Trust is a critical component of the project.

Initially, sharing data was a significant issue to UMP-GIS partners, due to concerns about improper data usage and sensitive information. A Memorandum of Agreement was signed by UMP-GIS partners specifying that only the creator or original owner of the data retains the right to share data with a third party. Data are not shared with other partners, nor are data sets made public, without the owner's consent. A separate Memorandum of Understanding details the data protection specifically for NHP data, due to the sensitive, non-public nature of the information (for example, bald eagle nest locations). Security of data sets is assured and there is no question as to the quality or accuracy of data.

Much of the effort expended under the auspices of the UMP-GIS collaboration is directed toward individual training of the DEC planners. Input from GIS and natural resource experts and feedback from planners is used to tailor training sessions and data delivery to the needs of each decision maker. A combination of informal one-on-one training of DEC planners, group training, and technical support was provided, resulting in improved GIS skills, enhanced usage of natural resource data, and strengthened communication between agency personnel.

58.5.3 GIS Analyses and Products

New inventory layers for UMP were developed by collecting, reprojecting, and reformatting data and synthesizing existing spatial and ecological data. One example is the potential deer yard model (a yard is an area with coniferous cover where white-tailed deer in northern regions take shelter during winter). The model includes suitable elevation, slope, and land cover types, using criteria based on published deer research and studies conducted in the central Adirondacks (Figure 58-15). The model indicates where further field surveys should be conducted to identify critical winter habitat that warrants protection.

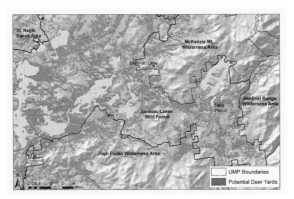

Figure 58-15 Potential white-tailed deer winter yard model for Saranac Lakes Wild Forest. Courtesy: Steve Signell, SUNY-ESF. See included DVD for color version.

Another set of GIS layers was created using the *New York State Breeding Bird Atlas* (BBA) (Andrle and Carroll 1988). The BBA is a statewide field survey for presence of breeding birds conducted twice since 1980 (http://www.dec.ny.gov/animals/7312.html). The new GIS layers contain locations of rare species, species richness, and other information summarized spatially. The BBA summary also has data on bird guilds, groups of ecologically similar species (for example, cavity-nesting birds). A data layer was included of species that increased or decreased in distribution over time; coupled with additional data this could help identify targets for conservation action. The Park-wide BBA data layers may also delineate areas with recreational potential (for instance, opportunities for wetland bird watching).

Several GIS tools have been developed to aid DEC planners as well. Some are for map display and include standardized colors, markers, fonts, legends, and scale guidelines. This helps planners create consistent maps and facilitates comparison between Forest Preserve units. Another tool is a data dictionary for GPS data collection to make facilities inventory faster. A free GIS viewer (ArcReader) was used to distribute data to those who were not yet able to access the full GIS software capabilities.

Other tools are for analyzing alternative management scenarios. The Cost Path Analysis (CPA) tool measures the least expensive path from one point to another (Figure 58-16). Cost in the CPA refers to the potential environmental and economic cost of a new road or trail. CPA is particularly useful for siting new trails (e.g., snowmobile trails, hiking trails). Data sources used include existing trail and road data, New York Natural Heritage Program data on sensitive areas and rare species, wetlands, terrain, and APA land classification data as base data layers. The DEC planners then assign to each layer the cost of traveling over different features on a relative scale of 1 to 10 (lowest to highest cost). DEC planners generate alternatives by running the CPA tool with different cost values. CPA can help make the decision-making process more transparent, because the inputs for each alternative can be saved and the results shared and compared.

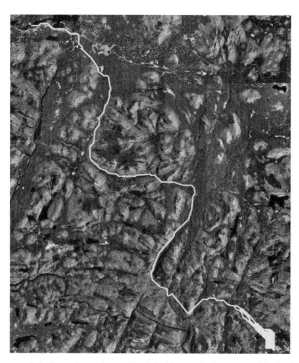

Figure 58-16 One possible iteration of the Cost Path Analysis applied to trail siting overlaid onto digital orthoimagery. This analysis included sensitive ecological communities, wetlands, rivers, trails, roads, and slope to determine the lowest cost trail from A to B. Map courtesy: Jennifer Gagnon, SUNY-ESF. Digital orthoimagery courtesy: NYS Office of Cyber Security and Critical Infrastructure Coordination. See included DVD for color version.

Another tool identifies the spatial arrangement and extent of habitat for species or guilds using a land cover map and a habitat association matrix (a table containing a list of species and each habitat type in which a species is found). With this tool the planner can compare a Forest Preserve unit with the area outside the unit to evaluate the proportion of a habitat type in public ownership. Resulting management strategies may differ depending on whether species are found in one area or dispersed throughout the Park. Indeed, Forest Preserve boundaries are not "hard," and protected lands are undoubtedly influenced by activities on adjacent lands. Spatial context, for example distance from roads (Figure 58-17), impacts land usage and is included in appropriate analyses.

Of course, the UMP-GIS data layers and tools have uses and limitations. The GIS is only as good as the data included within it and the ability of the user to manipulate that data to answer questions. Creation of GIS data and maps is the first step in managing land. GIS cannot and should not replace the need for field data collection; rather, UMP-GIS is designed to both incorporate and help direct on-the-ground assessments.

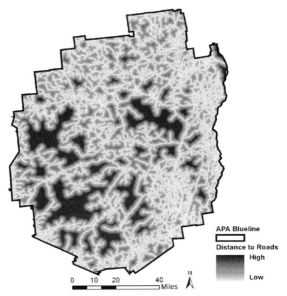

Figure 58-17 Distance from roads in the Adirondack Park (darker areas are farther from roads). See included DVD for color version.

58.5.4 Next Steps

The UMP-GIS collaboration is a multi-year project that continues to evolve. As new data sets become available and tools are developed, they will be incorporated into the GIS. Initially, data and tools were integrated into the DEC's existing GIS database. In the future, UMP-GIS data sets may be made public via a server or the NYS GIS Clearinghouse website (www.nysgis.state.ny.us). Sharing can improve planning for varied land uses while focusing attention on targeted data collection in critical areas.

This complementary approach to planning will be even more helpful as land ownership changes in the Adirondack Park. In recent years, thousands of acres have been added to the Forest Preserve through purchases or conservation easements; those lands will need to be inventoried and have UMPs completed. The next round of planning should benefit tremendously from the achievements of the UMP-GIS project collaboration.

58.5.5 Summary

Planning for recreational use and protection of public lands is greatly enhanced when natural resource information is brought together into a complete picture. That picture is becoming clearer in the Adirondack Park because of an extensive and centralized GIS database, innovative analyses and tools, solid training and collaboration of experts, and consistent maps and products—the outcomes of the UMP-GIS project. Much progress has been made, and we will continue to pursue the development of innovative data layers and approaches to issues of natural resource protection and enjoyment on state lands in the Adirondack Park.

58.6 Use of GIS in Adirondack Conservation Projects

58.6.1 Biotic Integrity

GIS was an invaluable tool for the research conducted as part of Dr. Michale Glennon's dissertation work at the State University of New York, College of Environmental Science and Forestry, in Syracuse, New York. Having grown up in the Adirondack Park, Dr. Glennon was keenly aware of the Adirondack Park Land Use and Development Plan and was very interested in what role this system might play in structuring wildlife communities on this landscape. The Shared Adirondack Park Geographic Information CD-ROM Ver. 1.0 (Adirondack Park Agency 2001), and with it a number of spatial data sets from the NYS Adirondack Park Agency (APA) were essential for this. One of those data sets was the official Adirondack Park Land Use and Development Plan Map (Figures 58-5 and 58-6) or the "fruit salad map" as it is affectionately referred to by some APA staff members because of the color scheme. This map, in conjunction with information from the original *Breeding Bird Atlas* (Andrle and Carroll 1988), was used to investigate the role of land use management in explaining patterns of biotic integrity across the Adirondack Park. The *Breeding Bird Atlas* was rendered much more user friendly by an ArcView extension produced by NYS Department of Environmental Conservation staff member Jim Daley, which enabled users to be able to select polygons and obtain a unique species list for any area of interest within the state. The combination of these two data sets, together with additional GIS layers provided by the APA *Shared GIS CD* (2001) and other GIS resources (e.g., roads, land cover, parcel centroids) allowed the investigation of integrity patterns of breeding bird communities in the Park.

Biotic integrity refers to "the capability of supporting and maintaining a balanced, integrated, adaptive community of organisms having a species composition, diversity, and functional organization comparable to that of the natural habitat of the region" (Karr and Dudley 1981). An Index of Biotic Integrity (IBI) provides a multiparameter tool for assessing the impacts of human activities on natural systems and incorporates aspects of community structure and function that show reliable responses to human disturbance factors (Glennon and Porter 2005). An Index of Biotic Integrity was developed using the bird atlas data and evaluated several questions including: (1) does biotic integrity differ among the different land use designations that characterize the Park, and (2) what factors control the patterns in biotic integrity across this landscape? Biotic integrity was lowest in the hamlets within the Adirondacks, and showed an increasing pattern along a gradient from hamlets to rural use, resource management, wild forest, and wilderness areas (Glennon and Porter 2005). The largest driver explaining the patterns in biotic integrity was the distance to roads. How far a particular atlas block was located from a road alone could explain 50% of the variability in biotic integrity in that block (Figure 58-18). The number of residential structures located within the block was also an important variable in explaining biotic integrity. All of this indicated that large, roadless forest regions in the Adirondacks are critical to maintaining biotic integrity on this landscape and pointed out the importance of private as well as public lands in long-term biodiversity conservation for the Park. This research is fully described in Glennon and Porter (2005).

58.6.2 Conservation Planning

The Wildlife Conservation Society (WCS) is engaged in conservation planning exercises for the Adirondack Park and other regions through the Living Landscapes Program (www.wcslivinglandscapes.org). The Living Landscapes Program uses a suite of focal, or landscape, species to identify and prioritize conservation planning across regions. Landscape species are defined as species that use large, ecologically diverse areas and often have significant impacts on the structure and function of natural ecosystems. Their requirements in time and space make landscape species particularly susceptible to human alteration and use of natural

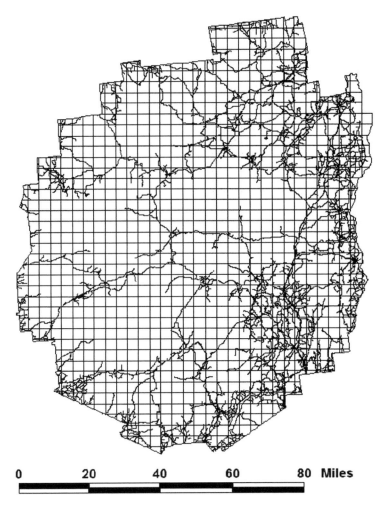

Figure 58-18 Map depicting the *Breeding Bird Atlas* (Andrle and Carroll 1988) blocks and roads located within the Adirondack Park, New York.

landscapes (Sanderson et al. 2002). The moose, black bear, common loon, marten, wood turtle, and three-toed woodpecker were chosen to capture the major habitat types and major threats to the Adirondack region. These species were chosen according to five criteria: (1) area requirements, (2) heterogeneity of habitat use, (3) vulnerability, (4) ecosystem functionality, and (5) cultural and/or economic significance. WCS believes that planning conservation actions to meet the needs of a suite of landscape species functions to identify the necessary area, condition, and configuration of habitats to meet the long-term ecological requirements for most species occurring in a wild landscape (Sanderson et al. 2002). Through mapping of the biological needs of each of these species, as well as the human-caused threats that have the potential to negatively impact their populations, areas where conservation action is most needed and most likely to be effective can be prioritized. For example, wood turtles are vulnerable to human-caused threats such as roads and water pollution. By mapping areas of high-quality wood turtle habitat that may be in close proximity to roads and/or pollution sources, regions of the Park in which conservation intervention may be most beneficial can be identified.

This work would not be possible without GIS and spatial data sets. Much use has been made of available land cover and land use information from the *Shared Adirondack Park*

Geographic Information CD-ROM (2001) as well as other sources to map biological needs of our landscape species. In addition, a number of spatial data sets have been used to map threats to these species. The capability of GIS technology to combine information across a number of data layers allows us to examine patterns in habitat quality and vulnerability and to set spatial priorities for conservation in the Adirondacks.

58.7 *The Adirondack Atlas:* A Geographic Portrait of the Adirondack Park

58.7.1 Why an Atlas?

The Adirondack Park has the largest wilderness area east of the Mississippi. It is important to understand why this land has had so much focused attention for conservation and preservation. The Adirondack Park is an area roughly the size of Vermont, and approximately 132,000 people live within its 6-million-acre boundary. There are also estimates that 3–12 million people visit the Park every year and that 90 million people live within a day's drive of the Adirondack Park. No matter what the real numbers are, the Park, which is owned half publicly and half privately, has a complex set of public policy issues. It also has a rich history rooted in geology, resource extraction, wilderness, recreation, wealth, poverty, and conservation. No matter which side of the issues you fall on or which part of the history interests you most, there are hundreds of thousands of people who care deeply about the Adirondack Park.

The Wildlife Conservation Society's Adirondack Communities & Conservation Program (ACCP) identified a number of projects to connect communities to conservation with the help of a multi-year grant from the Ford Foundation. In the summer of 1998, ACCP received the *Adirondack Park Geographic Information CD-ROM* from the NYS Adirondack Park Agency (Adirondack Park Agency 1998). In a short period of time an assortment of maps was produced using APA data with GIS software. These maps sparked great interest and conversation. For example, people who live in the Park and are involved in Adirondack Park issues had never seen a map showing all of the towns in the Park. A thematic map as simple as municipal boundaries gave the viewer a new sense of what towns existed and how towns are situated. One of the most important data layers on the CD is the Adirondack Park Land Use and Development Plan Map, also known as the "fruit salad map" (Figures 58-5 and 58-6). The "fruit salad map" became an essential ingredient for the initial maps because the land classification attributes could be extracted, combined or reclassified to show various land issues. Without really knowing it at the time, GIS data and software was becoming a tool that could help us tell stories and share information through the language of geography.

After seeing a number of positive reactions, it occurred to ACCP that few people had ever seen maps from GIS data sets. Dr. William Weber, North America Program Director for the Wildlife Conservation Society, thought that an atlas should be developed to show people a mirror of their community. Since then, ACCP has elected to use geographic data with a wide variety of topics and bring this information to the community level so that it can be viewed and digested, and individuals can make sense of the information on their own. The *Adirondack Park Geographic Information CD* was the catalyst that propelled ACCP into developing the atlas project. For the next five years, ACCP and Jerry Jenkins pulled together *The Adirondack Atlas: A Geographic Portrait of the Adirondack Park* (Jenkins and Keal 2004).

58.7.2 History of Data Assembly

During 1998 and 1999 spatial data were not readily accessible on the internet. The Adirondack Park Agency's first *Adirondack Park Geographic Information CD-ROM* (1998) was actually the best way at the time to share large GIS data sets. Between slow internet connections and lack of available electronic data, the shared CD was revolutionary and most helpful. Similar in content to the APA CD, another significant GIS data set covering this region was the Northern Forest Lands Inventory (NFLI) compiled during the early to mid 1990s (State University of New York College of Environmental Science and Forestry 1995). The NFLI data set was compiled by a team of GIS faculty and graduate students at SUNY ESF. The region of coverage for the data set is essentially all land north of Syracuse, the Mohawk River and Saratoga County in New York, which includes the Adirondack Park. The NFLI data set is a compilation of GIS data from many organizations such as the NYS Department of Transportation, NYS Division of Equalization and Assessment, United States Census Bureau, Niagara Mohawk Power Corporation and New York State Electric and Gas. Some data were digitized by the LA Group who created the region's first and only recreation and trails data sets for the complete NFLI region. Most of the NFLI data sets were never made available on the internet as they were compiled. In order for ACCP to obtain the data, it had to be physically picked up at SUNY ESF in Syracuse. In the fast changing world of technology, it is interesting to note that the NFLI data was distributed to ACCP on four 100 MB Zip-Disks.

Since the late 1990s many issues surrounding data availability have changed and made collecting data for the *Atlas* much easier. Within the first few years of the *Atlas* compilation, most Internet connections increased speed significantly, while a tremendous amount of GIS data became available on the Internet through the New York GIS Clearinghouse (www.nysgis.state.ny.us), the Cornell University Geospatial Information Repository (CUGIR) (http://cugir.mannlib.cornell.edu), and others. In addition, during this time many governmental agencies started sending out data on CDs and Zip-Disks upon request. In 2001, the APA updated the GIS data CD (Adirondack Park Agency 2001). The revised Shared Adirondack Geographic Information CD contained newly released and updated data sets. All of these advancements facilitated accessing data for *The Adirondack Atlas*.

By the year 2000, still early in the process, internal brainstorming on content and chapters began to shape the beginning of *The Adirondack Atlas*. It was becoming evident that not all of the data desired was in a GIS format or readily available to the team. The team also recognized that others in the community may have useful ideas to contribute to the *Atlas*. Heidi Kretser and Cali Brooks of ACCP set up a series of *Atlas* meetings with the public. During the meetings, groups of maps were shown to the attendees who commented on existing maps and gave recommendations on their cartographic appearance, content, and accuracy. In addition, attendees commented on potential topics and gave their opinions on adding other maps of interest to the *Atlas*. People in the community were happy to participate in this exercise, while simultaneously it fueled excitement about GIS data and the *Atlas*.

Between brainstorming among ACCP staff and public meetings the "to do" lists began to grow. For about three years, there was always a list of people to call from businesses, villages, towns, counties, and state agencies for all kinds of information. For example, hundreds of phone calls went into creating two interesting maps: the Arts map showing galleries, theaters, and event places (Figure 58-19); and the History map showing museums, visitor centers, and historic places (Figure 58-20). As can be imagined, there were always people and a few government agencies that did not want to share their information for a variety of reasons. However, almost everyone contacted during the research process shared an excitement to bring this information to life in the *Atlas*.

Many data sets had to be digitized into a GIS format or enhanced with further information. Two such examples are the historic railroads and the growth of the Adirondack Park's Forest Preserve lands (Figure 58-21). This required research and utilizing old maps and records from libraries and state agencies, such as the Adirondack Museum's library and the NYS Department of Environmental Conservation. Most data collected for the *Atlas* was recorded in an Excel spreadsheet. GIS provided the interface to then spatially manipulate

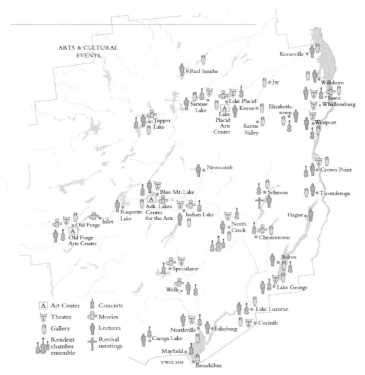

Figure 58-19 Arts and cultural events map from *The Adirondack Atlas*. Courtesy: Wildlife Conservation Society and Jerry Jenkins. See included DVD for color version.

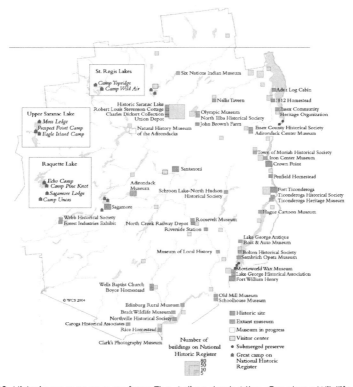

Figure 58-20 Historic resources map from *The Adirondack Atlas*. Courtesy: Wildlife Conservation Society and Jerry Jenkins. See included DVD for color version.

Manual of Geographic Information Systems

and assign the data to points, lines, or polygons. Preliminary maps were created in GIS to display all data collected. After that step, data were transferred into the final map format for production in Adobe Illustrator. To tell the complete story of the data, lead author Jenkins used the GIS data to make combined thematic maps such as the aforementioned Arts map, to display some data such as census data in graphs or charts, to create complex illustrations such as the Adirondack Waste Stream (Figure 58-22), or to simply convert existing spatial information into a final map in Adobe Illustrator. Figure 58-23 is an example of an *Atlas* page illustrating how combining spatial information with an explanatory narrative can benefit both depictions.

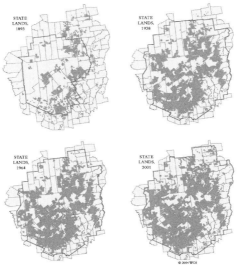

Figure 58-21 Growth of state-owned Forest Preserve lands in the Adirondack Park. Map from *The Adirondack Atlas*. Courtesy: Wildlife Conservation Society and Jerry Jenkins. See included DVD for color version.

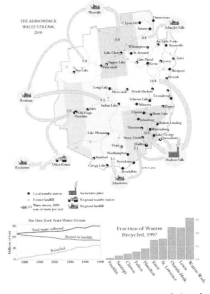

Figure 58-22 The Adirondack Waste Stream map uses a variety of symbols, pie charts, a bar graph, and a line graph to demonstrate the complex Adirondack waste stream. Map from *The Adirondack Atlas*. Courtesy: Wildlife Conservation Society and Jerry Jenkins. See included DVD for color version.

13-4 CAMPS, CAMPGROUNDS, COTTAGES

MANY people come to the Adirondacks not for exercise or adventure but simply to be near woods and water and pretty places. For every visitor that camps somewhere in the backcountry, many others stay at camps, campgrounds, cabins, and summer houses.

The Adirondacks have a large supply of these. The map shows 75 summer camps, 147 campgrounds with a total of 11,700 campsites, and 20,028 houses classified as seasonal residences. There are in addition some 28,800 other "unoccupied" houses that are used seasonally and 200 landlords who rent at least a thousand summer cottages (Map 13-5).

Many of the users of these houses, camps, and cottages are a group of somewhere between 100,000 and 200,000 regular visitors whom we call the other Adirondack population. They are neither residents nor tourists. Like residents they are tied to particular communities, and often make substantial economic contributions to those communities. Like tourists they live and vote elsewhere. What they are, in many cases, are people who live in one place but whose real home, the Adirondacks, is someplace else. In this they are not particularly unusual; the exigencies of life and work separate many of us from our real homes.

Because the other Adirondack population does not vote here, it is often ignored politically. This may be unfortunate because it disenfranchises many people who matter to the Adirondacks and to whom the condition and future of the park are very important. If this was not the case—if the Adirondack constituency was expanded to include everyone who supports the park and cares about it—the terms and outcomes of many Adirondack debates might be very different.

Figure 58-23 Places to Stay map from *The Adirondack Atlas*. Note how the text and map enhance each other. Courtesy: Wildlife Conservation Society and Jerry Jenkins. See included DVD for color version.

58.7.3 *Atlas* Summary and Facts

The project to create *The Adirondack Atlas* yielded 18 chapters (Table 58-2), 288 pages, 450 maps, 300 graphs, and illustrations with full notes and index. *The Adirondack Atlas* was released in June of 2004. The Adirondack Museum at Blue Mountain Lake, New York, provided many primary historical resources for the book from its comprehensive library collection of Adirondack books and manuscripts. The Museum was a research partner for the *Atlas* with the Wildlife Conservation Society and a publishing partner with Syracuse University Press.

Table 58-2 *The Adirondack Atlas* chapter contents.

• About the Adirondacks and the *Atlas*	• Schools & Colleges
• Environments	• Town Budgets & Local Taxes
• The Adirondack Park	• Vital Services
• Animals & Plants	• Business & Industry
• War, Settlement & Industry	• Media & Culture
• Forest Change	• Outdoor Recreation
• Vital Statistics	• Changing Towns
• Employers, Jobs & Income	• Pollution & Wastes
• Death, Injury, Disease & Crime	• Seven Questions about Change

References

Adirondack Park Agency. 2001. *Shared Adirondack Park Geographic Information CD-ROM* (2-disks). Version 1.0. Adirondack Park Agency, Ray Brook, NY.

————. 1998. *Adirondack Park Geographic Information Including GIS Data Layers from: Influences on Wetlands and Lakes in the Adirondack Park of New York State: A Catalog of Existing and New GIS Data Layers for the 400,000 Hectare Oswegatchie/Black River Watershed CD-ROM* (1 disk). Volume 1.4. Adirondack Park Agency, Ray Brook, NY.

Andrle, R. F. and J. R. Carroll, eds. 1988. *The Atlas of Breeding Birds in New York State.* Ithaca, NY: Cornell University Press.

Cowardin, L. M., V. Carter, F. C. Golet and E. T. LaRoe. 1979. *Classification of Wetlands and Deepwater Habitats of the United States.* FWS/OBS-79/31. Washington, D.C.: US Fish and Wildlife Service.

Craul, P., J. W. Barge and R. P. Curran. 1987. Essex County Forestland Soil-Site Feasibility Survey. Unpublished manuscript. State University of New York College of Environmental Science and Forestry, Syracuse, NY.

Curran, R. P. and J. S. Banta. 1982. Development of LANDSAT derived forest cover information for integration into Adirondack Park GIS. *National Conference on Energy Resource Management Proceedings,* 9–12 September 1982. NASA Conference Publication 2261, Vol. 2.

Glennon, M. J. and W. F. Porter. 2005. Effects of land use management on biotic integrity: An investigation of bird communities. *Biological Conservation* 126:499–511.

Halasz, S., R. P. Curran, D. M. Spada, K. M. Roy, J. W. Barge, E. B. Allen, G. K. Gruendling, W. A. Kretser and C. C. Cheeseman. 2000. *Watershed Protection of the St. Lawrence-Raquette Watershed with Special Consideration to Large Wetlands and Large Landownership. Part One: The St. Regis River Basin. Draft Final Report.* Adirondack Park Agency, Ray Brook, NY. <http://www.apa.state.ny.us/Research/stregis/cdintro.html> Accessed 24 Aug 2008.

Jenkins, J. and A. Keal. 2004. *The Adirondack Atlas: A Geographic Portrait of the Adirondack Park.* Syracuse, NY: Syracuse University Press.

Karasin, L, R. P. Curran, S. Halasz, D. M. Spada, J. W. Barge, E. B. Allen, D. J. Bogucki, K. M. Roy, C. Burkett and C. C. Cheeseman. 2002. *Final Report, November 2002; Watershed Protection of the St. Lawrence River Watershed with Special Consideration to Large Tracts of Land; Part Two: The Salmon/Trout, Raquette, and Grasse Watersheds.* State Wetlands Protection Program, US Environmental Protection Agency Grant No. #CD992644. Adirondack Park Agency, Ray Brook, NY. <http://www.apa.state.ny.us/Research/stlawr2/stlawrence2welcome.html> Accessed 24 Aug 2008.

Karr, J. R. and D. R. Dudley. 1981. Ecological perspective on water quality goals. *Environmental Management* 5:55–68.

LaPoint, S. D., R. P. Curran, S. S. Halasz, J. W. Barge, D. M. Spada, E. B. Allen and D. J. Bogucki. 2003. *Final Report, September 2003; Watershed Protection of the Mohawk River Watershed, Phase I.* State Wetlands Protection Program, US Environmental Protection Agency Grant No. #CD982246-01. Adirondack Park Agency, Ray Brook, NY. <http://www.apa.state.ny.us/Research/Mohawk_Final_Report.pdf> Accessed 24 Aug 2008.

McHarg, I. L. 1969. *Design with Nature.* Garden City, NY: Doubleday & Company.

Primack, A.G.B., D. M. Spada, R. P. Curran, K. M. Roy, J. Barge, B. Grisi, D. J. Bogucki, E. B. Allen, W. A. Kretser and C. C. Cheeseman. 2000. *Watershed Scale Protection for Adirondack Wetlands: Implementing a Procedure to Assess Cumulative Effects and Predict Cumulative Impacts from Development Activities to Wetlands and Watersheds in the Oswegatchie, Black and Greater Upper Hudson River Watersheds of the Adirondack Park, New York State, USA. Draft Final Report.* Adirondack Park Agency, Ray Brook, NY. <http://www.apa.state.ny.us/Research/uh/uhreporttitle.html> Accessed 24 Aug 2008.

Roy, K. M., R. P. Curran, J. W. Barge, D. M. Spada, D. J. Bogucki, E. B. Allen and W. A. Kretser. 1996. *Watershed Protection for Adirondack Wetlands: A Demonstration-Level GIS Characterization of Subcatchments of the Oswegatchie/Black River Watershed.* Adirondack Park Agency, Ray Brook, NY.

Roy, K. M., E. B. Allen, J. W. Barge, J. A. Ross, R. P. Curran, D. J. Bogucki, D. A. Franzi, W. A. Kretser, M. M. Frank, D. M. Spada and J. S. Banta. 1997. *Influences on Wetlands and Lakes in the Adirondack Park of New York State: A Catalog of Existing and New GIS Data Layers for the 400,000-Ha Oswegatchie/Black River Watershed - Final Report.* Adirondack Park Agency, Ray Brook, NY. <http://www.apa.state.ny.us/Research/OB2/HtmlDocs/siteindx.htm> Accessed 24 Aug 2008.

Sanderson, E. W., K. H. Redford, A. Vedder, P. B. Copolillo and S. E. Ward. 2002. A conceptual model for conservation planning based on landscape species requirements. *Landscape and Urban Planning* 58:41–56.

State University of New York College of Environmental Science and Forestry. 1995. *Northern Forest Lands Inventory GIS Database.* SUNY-ESF, Syracuse, NY.

CHAPTER 59

GIS in Precision Agriculture and Watershed Management

E. Lynn Usery, David D. Bosch, Michael P. Finn, Tasha Wells, Stuart Pocknee and *Craig Kvien*

59.1 Introduction

The technologies used to support the agriculture industry have changed significantly in the last 20 years. While genetic plant and animal research have improved varieties and yields, the introduction of information systems and precision management techniques have allowed reduced inputs, including nutrients, herbicides, and pesticides, increased yields, and reduced environmental pollution, particularly at the field and watershed scales. The technologies that have provided a base in these improvements in agricultural production and watershed protection include the Global Positioning System (GPS), remote sensing, geographic information systems (GIS), and variable rate technology (VRT) equipment. It is the purpose of this chapter to document the basis of these technologies and their use in precision agriculture and watershed management. The chapter is organized to present the basic technologies and methods used in precision agriculture and watershed management followed by the actual description and application of these technologies to these areas. We present the cartographic basis, followed by a description of remote sensing in agriculture, and provide some basic GIS operations critical to the application areas. We then describe precision agriculture, soil mapping, and watershed management and modeling.

59.2 Cartographic Basis

The rectification of geospatial agricultural data, including images, sample locations, interpolated surfaces, yield, and other data, require specification and transformations of spheroids, datums, map projection, and coordinate systems. A spheroid is the reference ellipsoid for the figure of the Earth. A particular coordinate system is constructed based on a specific reference spheroid, datum, and map projection.

59.2.1 Datums and Coordinate Systems

A datum is the basis of a coordinate system, including a spheroid of reference, an initial point, and a reference angle (not required for geocentric systems). Common datums used in precision agriculture and watershed modeling include the North American Datum of 1927 (NAD 27). This datum is the basis of most US maps constructed prior to 1980. The North American Datum 1983 (NAD 83) is also used extensively. It was developed with satellite positioning technology. For agricultural applications the World Geodetic System 1984 (WGS 84) datum using the WGS 84 spheroid is essentially identical to NAD 83 and is the basis of coordinates obtained with GPS. Between NAD 27 coordinates and NAD 83 coordinates, positional differences can be 200 m in plane-projected space in the continental US (Welch and Homsey 1997).

A map projection is a systematic transformation of spherical coordinates, latitude (φ) and longitude (λ), to a plane coordinate representation. For a detailed treatment of map projections see Snyder (1987) and Usery et al. (2008, Chapter 8, this volume). Common projections used in precision agriculture and watershed management are the transverse

Mercator and the Lambert conformal conic, which are the basis of the Universal Transverse Mercator (UTM) and state plane coordinate (SPC) systems, the preferred coordinate systems for large-scale (small areal extent) applications.

In the UTM system, the x coordinates are referred to as Eastings and the y coordinates as Northings. The system is projected from the spheroid in 6 degree zones on the transverse Mercator projection for areas 80 degrees south to 84 degrees north. The 6-degree wide projection zones achieve an accuracy of 1 part in 2,500 or a scale factor (SF) = 0.9996. The UTM coordinates are a worldwide system with units in meters. The central meridian of each zone, or normally the x = 0 m coordinate value, is offset to a value of 500,000 m to insure positive coordinate values throughout the zone.

The SPC system also uses Eastings and Northings. It is constructed by state with zones that are 158 miles or less in width. Similar to UTM zones, each SPC zone is independently projected, but the projection in use depends on the orientation of the long axis of the state (and zone). The transverse Mercator projection is used for north-south trending states such as Georgia or Illinois; the Lambert conformal conic projection is used for east-west trending states and zones such as Tennessee and North Carolina. States including Florida, New York and Alaska use both projections. The projection of a zone of 158 miles or less allows achievement of an accuracy of 1 part in 10,000 or an SF = 0.9999. The SPC system exists only in the US and uses official units of the US foot if developed from the NAD 27 datum and meters for NAD 83.

59.2.2 Global Positioning System (GPS) and Differential GPS

Accurate positioning is very important to agricultural managers, watershed modelers, and farmers employing precision techniques. Collectively, they need answers to the fundamental question of where on the Earth we are. From the launch of Sputnik in 1957, scientists and engineers have been using transmitted radio frequency signals from satellites to make range, or distance, observations. Today, satellites employing ranging procedures like those in the Global Positioning System (GPS) can routinely provide us with an accurate position on the Earth's surface to better than 10 m in absolute measures and 10 cm when relative techniques are used (Wells et al. 1986). These accomplishments have their roots dating to the early eighteenth century when Edmund Haley calculated the orbits of comets using Sir Isaac Newton's techniques. Haley predicted the return of a comet observed twice before in history (Armitage 1966).

The Swiss mathematician Leonard Euler followed Haley's accomplishment when he derived the first completely analytical method for solving a parabolic orbit (Bate et al. 1971). From Euler's era forward, we have used time and orbit determination as fundamental parameters of position. Employing these positioning methods, we now use techniques developed by the great mathematician Karl Friedrich Gauss. Namely, we minimize the errors inherent in observations of orbits to reach the greatest precision attainable in determining a position (or orbit) by accumulating the greatest number of perfect observations and adjusting them to agree with all observations in the best possible manner (Gauss 1857; O'Neil 1987). Further detail on point positioning can be found in Torge (1980) and in Vanicek and Krakiwsky (1986).

The Navigation Satellite Time and Ranging (NAVSTAR) GPS consists of satellites, the control system, and users. The US Department of Defense began work on NAVSTAR/GPS in 1973 and designed the array of satellites in six orbital planes inclined 55 degrees to the Earth's equator (Figure 59-1) (Wells et al. 1986). They orbit the Earth at 20,000 km and continuously transmit radio pulses at known times.

The radio ranging designed into the GPS uses a primary signal at 1575.42 MHz and a secondary signal at 1227.6 MHz. The primary, or $L1$, frequency has two modulations: the C/A-code and the P-code (Parkinson 1996). The C/A or Clear Acquisition code is a short pseudorandom noise code broadcast at a bit rate of 1.023 MHz. The P or Precision code is a

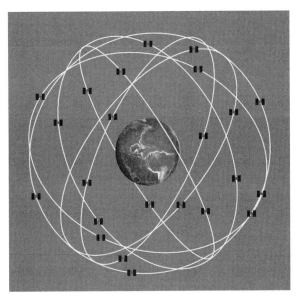

Figure 59-1 GPS satellite array (from NASA 2008).

very long code that is broadcast at 10.23 MHz (Wells et al. 1986; Van Dierendonck 1996). A third modulation of the carrier signal is the broadcast message or data modulation. This is used as a communication link through which the satellite transmits its location and corrections to the on-board atomic clock (Parkinson 1996). The control component consists of five stations around the planet, which track the satellites. A further four stations monitor the orbits, and the master control in Colorado Springs, Colorado, USA, corrects the navigational messages received from the satellites.

Receiving equipment measures the instant when pulses arrive to determine distances to the satellites. This distance determination is called radio ranging. It is a subcategory of general ranging techniques and is similar to triangulation in surveying. By measuring the precise time a signal is received from a known satellite, we have a value for the range vector, $\mathbf{r}_{ij}$ (Figure 59-2). Because we know precisely the position of the satellite that emitted the signal from its ephemeris, we know the satellite position vector, $\mathbf{r}_j$ (from the center of mass of the Earth). Thus, we can calculate the position vector of the receiver (antenna), $\mathbf{r}_i$ from the following equation:

$$\mathbf{r}_{ij} = \mathbf{r}_j - \mathbf{r}_i \tag{59-1}$$

Substituting the vector components, we have:

$$(X, Y, Z)_{ij} = (X, Y, Z)_j - (X, Y, Z)_i \tag{59-2}$$

This leads to the basic math model for ranging as follows:

$$F_{ij} = [(X_j - X_i)^2 + (Y_j - Y_i)^2 + (Z_j - Z_i)^2]^{0.5} - \mathbf{r}_{ij} = 0 \tag{59-3}$$

If $\mathbf{r}_i$ is stationary, then at least three non-coplanar ranges must be measured. This allows us to solve a series of equations for the three unknown components of $\mathbf{r}_i$ (Wells et al. 1986). Using algebraic equations, clock errors from both the satellites and the receiver are determined and adjusted to provide an exact location for the receiver (Herring 1996).

The system was originally designed for military and navigation uses and was quickly adopted for use with large survey control projects. Since the 1980s, when the GPS' most accurate timing was made available to civilian uses in addition to military uses, numerous practical applications of GPS have grown to fruition in addition to precision farming and

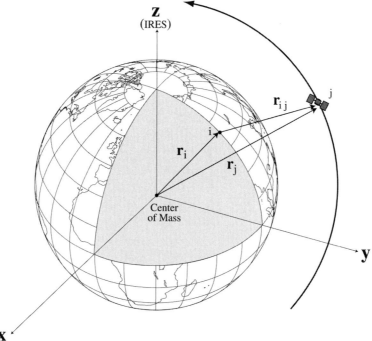

Figure 59-2 Schematic of basic terrestrial point positioning using satellite radio ranging.

watershed modeling, such as automobile navigation, geo-caching, and recreational fishing (Garmin 2000). The current GPS is composed of second-generation satellites, called Block II, which became fully operational in 1995, and is currently undergoing modernization through the GPS IIR-M series. There are currently 29 GPS satellites in orbit and each takes about 12 hours to orbit the Earth once (Grisso et. al. 2003; SpaceDaily 2006).

The GPS is a crucial technology in precision agriculture that allows farmers to obtain data of within-field unevenness of a variety of limiting factors such as soil acidity and crop yields (Goering 1993; Usery et al. 1995). Precision farming techniques can enhance production and diminish environmental pollution by reducing the amount of chemicals spread over a field. For example, recording the yield variability permits a farmer to explore, at a specific location within the field, the extremely low-yielding or extremely high-yielding areas (Auern-hammer et al. 1994; Usery et al. 1995).

Differential GPS (DGPS) works by having a reference station with a known location and a mobile GPS receiver. The reference station determines the errors in the satellite signals by measuring the ranges to each satellite using the received signals and then compares these ranges to the calculated ranges from its known (or true) location. The difference between these two ranges becomes the differential correction. For instance, to improve the accuracy of location of a GPS receiver on a tractor or harvester, the unit on the tractor applies these differential corrections to calculate the tractor's location more exactly (Figure 59-3). In the case of a roving receiver, carried by a person or mounted on a harvester, the receiver unit applies the correction by decoding the error code transmitted by the reference station. The reference station's signal includes the timing errors for each satellite and an additional "rate of change error" for them. Then the roving receiver interpolates its position precisely (Garmin 2000; Trimble 2006). Computer hardware and GIS software, used for incorporating geospatial data analyses and management, are critical components to couple with DGPS technology in order to maximize the results desired in modeling and farming.

However, not everything is perfect in this world of DGPS. Users should be aware that degraded positioning accuracies might occur when magnetic storms cause loss of signal lock

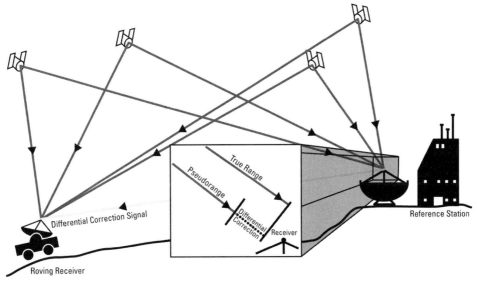

Figure 59-3 Example of applying a "differential correction" to a GPS receiver. See included DVD for color version.

(Skone 2001). In addition, because the reference station compensated errors vary with space, users can suffer accuracy degradation as the distance between their mobile receiver and the reference station increases. This distance factor is the most important factor determining DGPS accuracy and it is further exacerbated by a lack of inter-visibility, the inability of the reference and user to see the same satellites (Monteiro et al. 2005). Fortunately, in the US, on a daily basis, positional coordinates are computed for a network of ground-based GPS stations that comprise the National Continuously Operating Reference Stations (CORS). The US National Geodetic Survey provides this monitoring service to update coordinates and velocities of CORS sites. This update accounts for changes of antennae positions caused by human effects or natural effects, such as tectonic plate rotations, and is critical to rigorous DGPS positioning (Soler et al. 2003)

Although degraded positioning accuracies can occur, it is important to realize that the improved positional accuracy of a GPS receiver mounted on a tractor using DGPS can provide a location within centimeters (Wells et al. 1986). Thus, there are many uses of GPS point positioning that ultimately allow environmental scientists and agricultural managers, who rely greatly on exact positions, to answer to the fundamental questions of where am I now, or where do I need to apply pesticide. Knowing the specific location of variables, such as crop types, temperature, precipitation, slope, and soils, among others, and using this information in conjunction with DGPS, remote sensing, GIS, VRT and related technologies, farmers can increase profits through selective application of fertilizers, and land stewards can reduce risks by minimizing exposure of pesticides, herbicides, plant nutrients, and other chemicals to the ecosystem.

59.2.3 Scale

Scale can be defined in many ways. For example, *The Merriam-Webster Online Dictionary* (http://www.m-w.com/) contains seven primary definitions of scale; four as nouns and three as verbs. The noun definitions form into two classes: one dealing with weights and measures, and the other dealing with coverings on animals and plants (e.g., fish scales). For GIS and GIScience the definition of interest is the one dealing primarily with measurement. From engineering, surveying, and cartography we refer to scale as being a representative fraction (RF) of something portrayed schematically in ratio to its true size, as follows:

$$RF= \text{(portrayed size/actual size).} \qquad (59\text{-}4)$$

Scale is expressed in mapping and cartography as the ratio of a map distance to the distance on the Earth (with the distance on the map being expressed in unity). According to Robinson et al. (1995), scale can be stated verbally, graphically, or most commonly, via the RF, such as 1/500,000 or 1:500,000. They note that scale can be elusive because map scale varies as an inherent effect of the spherical or ellipsoidal dimensional transformation associated with map projections (see Usery et al. 2008, Chapter 8, this volume). Despite this elusiveness, cartographic scale is generally considered an absolute scale.

An RF of 1:500,000 indicates that one unit of linear horizontal distance on a map corresponds to 500,000 of the same units on the ground. Star and Estes (1990) give a nice mnemonic on the terms small scale and large scale, stating that an object on a small-scale map (e.g., 1:2,500,000) appears smaller than the same object on a large-scale map (e.g., 1:5,000).

In ecological and landscape studies, including agricultural and watershed management applications, alternative definitions are often used—such as field scale, basin scale, or watershed scale to continental and global scales. Spatial scale is essential to the understanding of natural and anthropogenic-influenced ecosystems. Often, terms of scale used in these disciplines become ones of relative scale or geographical extent. Lam et al. (2004) use scale with four meanings or types, measurement (resolution, grain), operational, observational (geographic/extent) and cartographic (representation). All of these meanings are used in agriculture and watershed management.

The problems with relative scale definitions are twofold. First, when they are used with terms such as small scale (small geographic area) and large scale (large geographic area), they are usually used in opposition to how they are used with the absolute scale meaning. For example, often the term field scale is used interchangeably to mean small scale and basin scale is used to mean large scale. This relative term for small scale (field scale) is opposite to what the absolute term means for small scale. In terms of absolute scale, a field would be considered large scale (large detail—as depicted on a 1:4,800-scale topographic map). Likewise, a large-scale (basin scale—say the lower Mississippi) relative term would be considered a small (or intermediate) scale in absolute terms (at a 1:500,000-scale topographic map, for example).

The second problem with the relative scale terms used this way is that there are no standard definitions. What does field scale mean quantitatively? How about basin scale? We know that a watershed or drainage basin is a finite area. The basin holds a network of channels (often with a unique pattern) that drain it, and it is separated from other basins by a divide (Ritter 1986). That which differentiates a basin or a sub-basin in the spatial domain is arbitrary. Due to oversimplifications of complex physical processes in the merging of digital spatial data for representations of phenomena such as soils, precipitation, and temperature—and the models that use them as input—issues of scale become of paramount importance to precision farming and watershed modeling (Goodrich and Woolhiser 1991; Song and James 1992). Singh (1995) defines the importance of scale to modeling as the size of a subwatershed within which the hydrologic response can be treated as homogeneous. Therefore, the optimal scale in a watershed is determined by the collective workings of multiple processes that generate hydrologic response and the availability of hydrologic data.

For the United States, the Hydrologic Units Codes (HUCs) used by the USGS allows us to impose some hierarchy into the relative scale problem in watershed management. Hydrologic Units are nested, from the smallest cataloging units, to accounting units, to subregions, to the largest regions. Nationally, there are 21 regions that contain either the drainage area of a major river or the combined drainage areas of a series of rivers (Seaber et al. 1987). Each of the four levels of hierarchy contains a 2-digit code. So a 2-digit HUC represents a region and an 8-digit HUC represents a cataloging unit. The cataloging units are often referred to as a watershed. The HUC coding system still leaves a relative scale problem but the hierarchy allows us some control and common understanding of scale. Currently, for the 48 contiguous states, there are 18 regions, 204 subregions, 324 accounting units, and 2,111 cataloging units (USGS 2006).

In watershed management and precision farming applications, moving between these relative scales on which a GIS practitioner can operate changes the discernible detail available for analyses. For example, Meentemeyer and Box (1987) note that scale changes may dramatically change watershed model structures because some variables change more significantly than others with a change in spatial scale. Further, they note that an increase in the size of a study area tends to increase the range of values for a landscape variable.

Scale issues are directly related to resolution issues, with which GIS practitioners and scientists deal routinely. Raster data sets are used repeatedly in precision farming and watershed modeling applications. These data sets are often derived from satellite observations or some gridded natural phenomena such as temperature or land cover. There are inherent scale-related errors in the creation of these types of data and any subsequent aggregations of them, collectively known as raster or grid resolution, or simply resolution (Usery et al. 2004). Resolution issues and errors are extremely important to all environmental modelers, not just watershed modelers, because, as Willmott and Johnson (2005) point out, many physical phenomena are highly variable over space and much of the phenomenon's spatial variability can be lost when the resolution is too low.

Ultimately, we can define scale in many ways. GIS practitioners concerning themselves with watershed management or precision farming need to be aware of the variety of definitions and uses of scale-related terms in this arena. If possible, an absolute measure of scale such as the RF is preferred. Perhaps a better way to deal with the various uses of terms like small scale and large scale among an interdisciplinary group of scientists and technicians is to use the terms broad scale or coarse scale on one end of the spectrum and narrow scale or fine scale at the other end.

59.3 Remote Sensing for Agriculture

Remote sensing is the science and art of obtaining information about an object, area, or phenomenon through the analysis of data acquired by a device that is not in contact with the object, area, or phenomenon under investigation (Lillesand et al. 2008). As you read these words, you are employing remote sensing. Your eyes are sensors and they respond to light reflected from this paper. Through the use of sensors, we collect data and analyze the data to produce information. With remote sensing, the data are signals from the electromagnetic spectrum. While the electromagnetic spectrum spans cosmic rays to radio waves, remote sensing commonly uses only specific portions of this spectrum. The visible spectrum from approximately 400 to 700 nm in wavelength, and the wavelengths to which our eyes are sensitive, is commonly used in remote sensing, but we extend the spectrum to include the infrared wavelengths in the range of 700 to 1,100 nm and refer to it as the photographic spectrum. Also, active sensors including longer wavelength radar and, recently, visible wavelength lidar are used for vegetation and canopy structure. We usually sense a range of wavelengths in remote sensing with each range referred to as a band. For the photographic spectrum, common bands are approximately 400 to 500 nm (blue), 500 to 600 nm (green), and 600 to 700 nm (red), and 700 to 1,100 nm (infrared) bands. While these wavelength ranges have been traditionally used for remote sensing of Earth phenomena, the introduction of hyperspectral sensors, i.e., sensors that detect narrow wavelength bands of 10 nm or so with hundreds of bands, has extended the use of remote sensing in agriculture and in many other areas. Also in recent years radar in the C, X, and L bands have been used for vegetation applications (Jensen et al. 2004).

A full literature of the theory and practice of remote sensing has been developed (Colwell 1983; Lillesand et al. 2004; Jensen 2005) and thus will not be replicated here. This section will focus specifically on the use of remote sensing, and particularly remotely sensed images, for agricultural applications. Agricultural phenomena commonly detected and recognized in remotely sensed images include vegetation types, vegetation health, biomass, soil types, soil moisture, and crop yield. Additional vegetation characteristics detectable and measurable include stress, moisture content, landscape ecology metrics, surface roughness, and canopy structure (Carter 1993; Carter et al. 1996; Frohn 1998; Jensen et al. 2004).

59.3.1 Basic Concept of Remote Sensing

The basic concept on which all remote sensing relies is that Earth objects reflect various wavelengths of light in differing amounts. We can identify reflectances of specific wavelength ranges for specific objects to develop a ***spectral signature.*** The spectral signature can be used to identify the Earth object in composite remotely sensed images or images with multiple wavelength bands. For agriculture, we commonly use this concept to identify vegetation types, such as crops versus forest or individual species of crops, grasses, and rangeland. The processing approaches to identify these crops include unsupervised and supervised classification (Lillesand et al. 2004; Jensen 2005), neural networks, and recently knowledge-based and learning classifiers using computer approaches, such as Holland classifiers, and genetic algorithms (Bandyopadhyay and Pal 2001; Liu et al. 2004; Yang and Yang 2004). The results of these classifications include not only species types, but also moisture content, stress levels, organic matter content, clay content, and many other characteristics of plants and soils that are useful for agricultural management.

A variety of processing techniques of remotely sensed images is required to make them useful for agricultural or watershed management. Some of these methods include geometric and radiometric correction, image enhancement, feature extraction, and mapping. Specific techniques with particular applications in agriculture include indices computed by image band combinations, band ratios, and image transformations resulting in measures of plant characteristics such as biomass, chlorophyll and moisture content. Some of these methods include the Vegetation Index (VI), the Normalized Difference Vegetation Index (NDVI), and the Kauth-Thomas or Tasseled Cap transformation, which yields images of brightness, greenness, and wetness (Figure 59-4). These and other methods are documented in Jensen (2007) and available in standard remote sensing software packages such as ERDAS Imagine (Leica Geosystems 2006) and ENVI (RSI 2006).

For remote sensing of agricultural information, spatial resolution determines the application and accuracy. Low resolution, for example 1 km as with Advanced Very High Resolution Radiometer (AVHRR) images, implies broad area analysis. Applications include determining land cover, crop inventory, and weather patterns. A good example of this is the use of NDVI generated from AVHRR to show the greening of America over the early spring and summer (KARS 2005). Field-specific applications require high resolution data, such as the width of a harvester—1 to 10 m. For weed and insect detection, the spacing determines the resolution requirements, commonly in the range of 1 to 3 m.

Other agricultural applications include detection and measurement of soil properties and soil inventory (Johannsen et al. 2006). Soil properties affect the reflectance in an image based on the influence of soil conditions on response. Such reflectance changes allow mapping soil patterns, moisture, organic content, clay and other mineral components and other characteristics.

The spectral reflectance in remotely sensed images is a product of vegetation response, which can be used to define soil management zones. Once management zones are defined they can be sampled as separate units and nutrient applications can be adjusted as needed (SSMG-22).

Figure 59-4a Brightness image of Augusta, Georgia, area, extracted from an Enhanced Thematic Mapper Plus (ETM+) image.

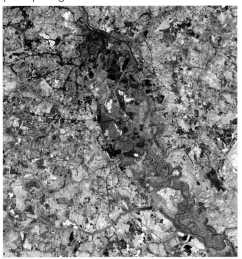

Figure 59-4b Greenness image of Augusta, Georgia, area, extracted from ETM+ image.

Figure 59-4c Wetness image of Augusta, Georgia, area, extracted from an ETM+ image.

Manual of Geographic Information Systems

For crop inventory and yield prediction, remotely sensed images can be used for crop identification, inventory of areas planted, estimation of potential harvest, and yield prediction based on NDVI. Commercial programs and government authorities are using remote sensing for monitoring production and compliance with regulations (CROPINS 2006; USDA 2006).

Soil nutrient detection from images is based on pixel reflectances that result from soil characteristics (e.g., color) that are related to organic content. From organic content one can predict nitrogen (N) release to plants. Leaf greenness related to chlorophyll content, which directly relates to N, can also be detected. Discoloration may be a result of potassium (K) deficiency. While possible, particularly with newer hyperspectral images and methods, other chemicals are usually difficult to detect from image data.

Detection of vegetation change is a major application of remotely sensed images. For example, green and infrared bands highlight volume of vegetation (biomass) and indicate plant vigor. These bands can be used to create crop vigor maps and change detection in the vitality of plants. Images can also be used to determine crop injury, for example, from hail and wind damage. An example is a project to determine damage to corn in Indiana from a tornado. The project used before and after images, and the corn downed from winds was detectable.

Images can be used for crop residue evaluation, such as determining the Farm Service Administration's (FSA) erodibility values that require minimum levels of crop residue to reduce wind and water erosion. Remote sensing is a good method for managing the supervision of standards. For example, Landsat Thematic Mapper (TM) images can be used to distinguish different types of crop residue.

Detecting crop stress from drought, weed, insects, erosion, and nutrient deficiency is a significant application of remote sensing to agriculture. These types of problems all result in stress that is detectable through vegetation indices, such as the brightness, greenness, and wetness images from a Tasseled Cap transformation. Change detection, based on before and after stress conditions, can be used to help manage crop vigor and growth.

A standard result from remote sensing is determination of land cover. While land use usually cannot be determined, the land cover can, with categories dependent on spectral and spatial resolution. Land cover maps from remotely sensed images are used by the USDA to assess regulation compliance, among other applications.

59.4 GIS Databases, Manipulation, and Processing

59.4.1 Building a GIS Database for Precision Farming— A Generalized Approach

Building a database for precision agriculture and watershed management requires specific data for the type of management application to be implemented. However, there is a set of common data and procedures that establish the basic GIS framework and allow manipulation and processing of the data to support the management application. These common data sets and a procedural framework for building a GIS database follow.

59.4.1.1 Data Acquisition

1) Determine the coordinate system to be used for the GIS database. Account for spheroid, datum, projection, and coordinates (see Chapter 8, Section 3, this volume). Good choices are the State Plane and UTM coordinates, for which the listed parameters are standard and accuracies are defined.

2) Establish high-accuracy base information.

A) Acquire high-accuracy (0.1 m root-mean-square error [RMSE] or better) ground control points (GCPs) over the field for which the database is to be constructed. Ideally, GPS equipment will be used on targeted points, which will appear in high-resolution photos acquired over the field. A minimum of 15 to 20 points are usually needed depending on field size.

B) Acquire high-resolution, high-quality base image data. At a minimum, one of the following is necessary:

   i) High-resolution photograph (nominal scale of 1:5,000 or better) or digital image (0.1 to 1 m pixel digital equivalent resolution) over the field under bare soil conditions with targeted GCPs. This is the preferred base data.

   ii) Existing US Geological Survey digital orthophotographs of 1 m resolution.

C) Digital elevation data with 1 m or better post spacing of elevation values and vertical accuracy of 0.1 to 0.5 RMSE.

3) Acquire field data.

A) Acquire any existing soils maps from USDA, etc.

B) Acquire soil samples on a grid spacing of one acre or better with GPS coordinates in the chosen system. A fine grid sample or field stratified samples are the best choices.

C) Acquire field history information on a variable within field basis or summaries for entire field if no variable data exist. Assimilate all information known to be available. Examples include:

   i) Crop production histories including types of crops, yields, rotation cycles, etc.

   ii) Yield data.

   iii) Tillage history.

   iv) Field maps of compacted areas, terracing, and others.

   v) Field input maps or histories (fertilizers, lime, herbicides, pesticides, and others).

   vi) All other data available.

4) Acquire additional data to suit applications of database. These may include:

A) Electromagnetic induction images.

B) Vegetation index maps.

C) Others.

5) Acquire data throughout the growing season and at harvest time as variable maps, if applied that way:

A) Periodic aerial images (color infrared).

B) Input applications (nutrient, herbicide, pesticide, and others).

C) Scouting reports.

D) Yield data.

59.4.1.2 Data Processing

1) Convert all map and image data to the selected coordinate system.

A) For base photo (bare soil image), rectify using GCPs. If the field is reasonably flat, this simple process will be sufficient. However, for a field with considerable relief (20 m or more) use of a high-resolution DEM in a differential rectification process may be needed. This may require contracting a GIS or photogrammetric consulting service to use the DEM to create the orthophotos since most GIS only support simple rectification with GCPs as we performed in the laboratory exercises.

B) Transform all other data layers to match the base image using the GCPs with standard rectification methods (see Chapter 8, Section 3, this volume).

2) Interpolate point data sets (e.g., soil samples) to a raster grid and match the cell size to base data or desired resolution.

3) Convert scouting data or other samples to polygon or raster map formats.

4) Convert vector data to raster form as needed to match processing and modeling needs.

5) Build models from data as required to support applications.
   A) Models should be built conceptually in graphic or diagrammatic form prior to any attempts to use data layers in models.
6) Implement models to determine desired results.

59.4.1.3 Building a GIS Database for Watershed Modeling

Follow similar procedures for precision farming beginning with the control and base information. Layers will vary but will typically include:
1) Base image (orthophoto) and map (Digital Raster Graphic of topographic map).
2) Digital elevation data.
3) Land cover at appropriate resolution and classification system for size and application of study area.
4) Soils data from NRCS Soil Surveys.
5) Precipitation data.
6) Other data as required for specific application. For example, for watershed analysis, point samples of water flow and water quality are usually needed.

59.4.2 Data Manipulation and Processing

While there are many GIS processing operations that are necessary for precision agriculture and watershed management, several specific operations including spatial interpolation and modeling are particularly important and are discussed briefly below. The description in this section is to demonstrate the application of these operations to specific agriculture and watershed data sets.

59.4.2.1 Spatial Interpolation

Sampling for soil chemicals, nutrient content, and organic matter is invariably performed with point locations, often in a gridded pattern, but also along soil management zones or as stratified random samples (Pocknee 2000). Elevation data are commonly generated as samples from images or other sources and yield data results in a dense set of posting points with a yield value at each location. To be useful in a GIS analysis context, these samples of all these types and others are usually interpolated to create a surface representation. While there is a variety of interpolation algorithms (see Lam 1983; Burrough and McDonnell 1998), a few algorithms dominate the use in interpolation of agricultural data and will be discussed here. These commonly used algorithms, available in GIS software, include inverse distance weighting, spline methods, and kriging.

Inverse distance weighting is a linear interpolation method. It is implemented through the ideas of Thiessen polygons (Boots 1986) with a gradual change in a trend surface and uses weighted moving averages for computation of values at unknown locations. The basic concept is that data points are weighted by the inverse of their distance to the estimation point. This weighting has the effect of giving more influence to nearby data points than those farther away. Figure 59-5 shows the location of sample points within a field. Figure 59-6 shows the resulting interpolation for phosphorus (P) concentration using inverse distance weighting.

Splines are curves represented by piecewise polynomials that have continuous first and second derivatives. For interpolation purposes, the unknown points are fitted to the spline equation, but for geospatial data, this often requires fitting points to surfaces where the surfaces are represented with bicubic splines (Burrough and MacDonnell 1998). Figure 59-7 shows the same field from Figure 59-5 with P interpolated using bicubic splines.

A regionalized variable is one in which spatial variation can be expressed as the sum of three components: 1) a structural component having constant mean or trend; 2) a random, but spatially correlated component, which is the variation of the regionalized variable; and

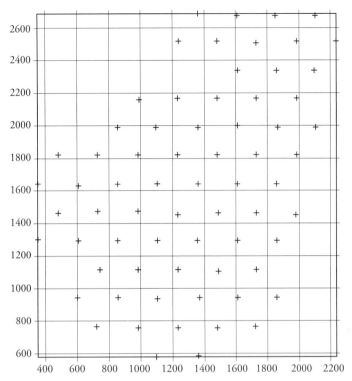

Figure 59-5 Soil sample locations to be used to interpolate a surface for phosphorus.

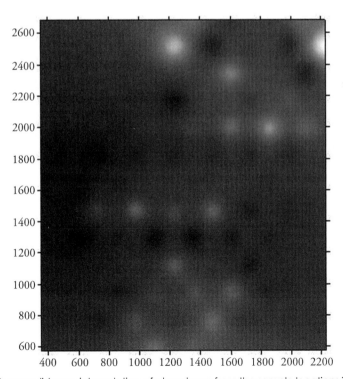

Figure 59-6 Inverse distance interpolation of phosphorus from the sample locations in Figure 59-5.

Manual of Geographic Information Systems

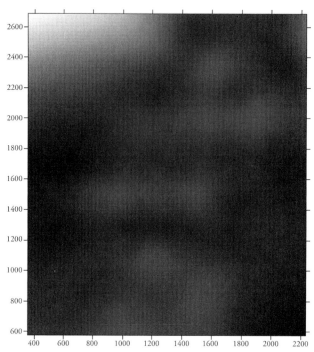

Figure 59-7 Spline interpolation of phosphorus from the sample locations in Figure 59-5.

3) random noise, which is a spatially uncorrelated residual error term. These variables are represented mathematically as:

$$Z(x) = m(x) + \varepsilon'(x) + \varepsilon'' \qquad (59\text{-}5)$$

where,

$m(x)$ is deterministic describing the structural component of Z,

$\varepsilon'(x)$ denotes the stochastic, locally varying but spatially dependent residuals from $m(x)$, and

ε'' is residual spatially independent noise with zero mean and variance.

Using an assumption that the variances of differences in the attribute or Z value for a set of sample points depends only on the distance (lag or h) and direction between the points, i.e., intrinsic stationarity, an equation for the spatially dependent component called semivariance, $\gamma(x)$, can be developed. For details of the derivation of $\gamma(x)$, see Bailey and Gatrell (1995) and Burrough and MacDonnell (1998).

A plot of semivariance, $\gamma(x)$, from the sample data against the lag distance (h) called an experimental variogram provides a basis for quantitative description of regionalized variation. The variogram provides useful information for interpolation, sampling design, and determining spatial patterns. The key components of the variogram that provide information are the sill, range, and nugget, which provide parameter weights to control the kriging interpolation process. The sill is where the graph levels and implies that beyond these values of lag (h) there is no spatial dependence. The range is the rise from the lowest value to the sill and determines spatial dependence of the sample points. The range helps determine the window size to use in weighted moving averages. The nugget is the y-intercept and provides an estimate of ε'', representing spatially uncorrelated random noise. It is difficult or impossible to determine an appropriate set of weights for interpolating unknown points from the experimental variogram from the sample data, thus, a fit of one of several standard models is used. These standard models include the spherical, exponential, linear, and the Gaussian. The linear model is used when the graph never levels and the Gaussian is used for smooth variation and small nugget variance.

For use of the variogram in spatial analysis and interpolation, one should not interpolate but use the mean value if the nugget dominates local variation. A noisy variogram indicates too few samples; usually 50 to 100 data points are required to get a stable variogram. The range determines optimum search window sizes. If a hole effect exists, i.e., a dip at distances greater than the range, this may indicate a periodicity effect or something similar. A large range shows long-range variation and if the Gaussian model fits, then smooth variation exists among the sample points and on the surface (Burrough and MacDonnell 1998).

As an example, Figure 59-8 shows the experimental variogram computed for P from the sample points for the field in Figure 59-5. Figure 59-9 shows the P values interpolated to a surface using kriging with weights determined from the variogram.

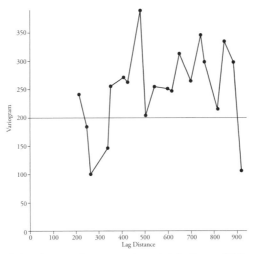

Figure 59-8 Experimental variogram from sample points in Figure 59-5.

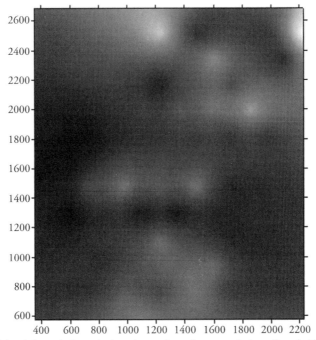

Figure 59-9 Kriging interpolation of phosphorus from the sample locations in Figure 59-5.

Manual of Geographic Information Systems

59.4.2.2 Spatial Modeling

One common approach for analyzing geospatial data is to create layers of single themes (thematic maps) of geographic phenomena and treat these layers as variables to which a set of operations can be applied. The approach is similar to mathematical algebra and is referred to as map algebra. Commonly, this approach uses a raster data model and each raster layer or map becomes a variable. A sequence of mathematical and spatial operators applied to the raster layers can be used to generate the desired output geospatial data set. The approach was developed by Tomlin (1984). The approach is flexible and allows a variety of operations to be completed to generate desired results. As with image processing operators, the map algebra operators are classified into three types: (1) point operators that work with single pixels, (2) neighborhood operators that work with a small group of pixels around the pixel of interest to determine the value of a single output pixel, and (3) global operators that use all pixels in the data set to determine the output.

The concept of map algebra can be used to implement a spatial modeling procedure referred to as *cartographic modeling*. Effectively a *cartographic model* is a sequence of map variables (layers) connected by map algebra operators that result in a spatial distribution of a particular geographic phenomenon. For example, a simple cartographic model can be implemented to use an elevation matrix to generate a slope map. The following equation shows the form of the operation:

$$slope = differentiate\ (elevation) \tag{59-6}$$

where,
slope is the resulting map of slope,
differentiate is the function that determines the first derivative of the elevation surface at each point, and
elevation is the elevation map.

Note that the *differentiate* operator only requires a single operand and is commonly implemented as a finite difference approximation on a pixel neighborhood basis (often a 3 x 3 filter) on a grid-based elevation matrix.

The Universal Soil Loss Equation (USLE) can be implemented as a cartographic model. Six parameters are combined to determine an estimate of soil loss as shown in the following equation.

$$A = R * K * L * S * C * P \tag{59-7}$$

where,
A is the annual soil loss in tons/ha,
R is the erosivity of rainfall,
K is the erodibility of the soil,
L is the slope length in meters,
S is the slope in percent,
C is the cultivation parameter, and
P is the protection parameter

If we develop a raster data layer for each of the parameters for a particular watershed area and place them in a multiplicative model (Figure 59-10), we can compute an estimate of the soil loss for the watershed.

We can also use a cartographic model to determine the profit potential for growing a particular crop. An example for profit from growing coffee is provided in Burrough (1986). In the example, the price of coffee is determined from the suitability of the soil while the price is modified by transportation costs, which are determined from distance to a road and the terrain.

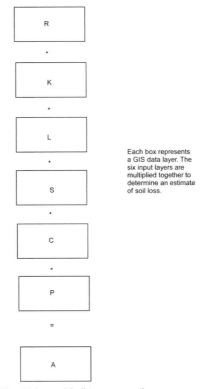

Figure 59-10 Cartographic model of the Universal Soil Loss equation.

59.5 Precision Agriculture

In theory, precision agriculture is a management philosophy that espouses matching inputs to exact needs everywhere. It is commonly referred to by other names such as precision farming, farming by the foot, farming by satellite, site specific management and target farming. In reality, many practices often are grouped under the banner of precision agriculture whose only common link is a reliance on advanced geospatial technologies.

Precision agriculture allows measuring and characterizing spatial variability, thereby aiding our ability to manage that variability. Management decisions may aim to overcome the variability through variable rate application of inputs, tolerate the variability, if for instance the cost of remedial action is too high, or enhance the variability through management practices that optimize the yield potential of different areas within a field. Variation is a feature of all farming systems and may be both spatial and temporal. Spatial variability results from changes in physical properties, such as soil type, texture, moisture and fertility. In addition to spatial variability, variability exists from season to season. Seasonal changes, temporal variations, are usually driven by unforeseen factors such as the weather, which in non-irrigated environments can be extremely challenging to manage.

In the United States precision agriculture began to emerge in the early 1990s (Usery et al. 1995). Although research into field spatial variability dates back considerably further, it was not until the maturation of several enabling technologies that precision agriculture began to take shape. For examples of precision agriculture projects see NESPAL (2008b).

59.5.1 Yield Monitoring

A yield monitor is a device that is mounted on crop-harvesting machines that measures yield as the crop is harvested. A yield monitor consists of several components, including a variety of sensors that measure harvester parameters, a yield sensor and a GPS that provides positional information. The GPS readings are coupled to the sensory measurements and logged to an on-board computer. A yield monitor quantifies the variability in yield and enables the generation of a thematic map, which highlights spatial variability. Yield monitors are commercially available for most major row crops (Figure 59-11) and have wide acceptance in the agricultural community, particularly for grain production.

Figure 59-11 Yield monitor for cotton. Note the GPS antenna in the upper center of the cotton picker. Courtesy: NESPAL. See included DVD for color version.

Yield monitors function by measuring either the volumetric or mass flow of material. Crop mass is measured by weighing the material in a basket or bin, or measuring the mass flow rate of the material as it passes a specific point on the harvester. Weighing techniques are straightforward and utilize load cells to measure the change in weight as the crop is harvested. Load cell designs are used to monitor peanut, citrus and vegetable crops.

There are several techniques employed to measure the mass of material as it is being transported through a section of the harvester. Grain harvesters have employed three primary methods. One utilizes an impact plate that measures the force exerted by material as it exits the grain elevator. Another is a technique that bombards radiation onto the cross-section of flowing grain. A receiver on the other side receives radiation after interacting with the flow of material. Changes in the transmitted radiation is measured and correlated to yield. Radiometric sensors are commonly used in European harvesters but are not available in the United States. A third technique involves a capacitive sensor that detects the dielectric changes in the mass of material as it passes capacitive plates. The dielectric properties are sensitive to moisture and foreign materials, which must be taken into account if the sensor is to be accurate.

Light-based techniques have become popular in measuring yield on mechanical cotton pickers. In this method, light-emitters are placed on one side of the air duct and photodiodes (light detectors) on the opposite side. As cotton flows through the duct, the beam of light to the detectors is broken. The mass of material is then correlated to the amount of time the light beam is broken between the emitter and detector.

The principle behind volumetric measurement is based on filling a compartment of a known volume and then releasing that volume into a bin or basket. There are two principle systems used. The first relies on a paddle wheel mounted below the discharge spout on a clean grain elevator. A level sensor above the paddle wheel indicates when the grain has filled to a specific depth. The wheel then rotates, drops one load of grain, and begins accumulating another load. The rate at which the paddle wheel turns is correlated to the yield. The second type of volumetric measurement technique employs a light detector that indicates the volume of grain on a grain elevator rotation. The yield is calculated from the speed of the grain elevator and the volume of each rotation.

59.5.2 Variable Rate Technology and Variable Rate Irrigation

59.5.2.1 Variable Rate Technology

Variable rate technology (VRT) refers to any equipment designed to allow the rate of farm inputs to be precisely controlled and varied while the machine is in operation. There are two types of variable rate technology: map-based and real-time sense and treat. Map-based VRT is most widely used and requires a GPS receiver, computer controller and a regulated drive mechanism mounted on the applicator. Application rates are preprogrammed into computer memory using information collected prior to the application. Maps are based on soil tests, aerial photographs, or other information collected prior to application. The GPS provides positional information that is used to determine the appropriate application rate. Equipment such as planters, fertilizer spreaders and liquid sprayers can be equipped to vary single or multiple inputs simultaneously.

Real-time sense and treat systems are dynamic systems that use a sensor to measure a physical or crop property and apply an input in response to this measurement. These systems are often used in weed control using an optical sensor to measure the presence or absence of weeds within a crop. Other optical systems use the reflectance properties of plants to measure crop reflectance and create a vegetative index that is used to vary the application of crop inputs such as nitrogen and growth regulators.

Precision farming recognizes the concept that agricultural fields are not uniform. Many fields will have variable topographic and soil conditions that will result in variability in biophysical parameters such as crop growth, weed and pest infestations and yield. Yield mapping, remote sensing, and intensive soil sampling are frequently used to characterize the variability present. This information can be extremely informative in gaining an understanding of the underlying agronomic processes within a field. The value of this insight hinges on how we can utilize the information.

This is where variable rate application technology (VRA or VRT) offers value. The use of VRT enables the rate of farm inputs to be precisely controlled and varied while a machine is in operation. This often involves the use of a combination of electronic controllers and variable rate pumps, motors, or valves, however this need not be the case. However, VRT may rely on a low technology solution where the operator manually adjusts the rate or application of inputs based on cues he/she detects with his/her own senses and expert knowledge of the field.

There is a large volume of research examining variable rate fertilizer applications, particularly in grain cropping systems. Issues that need to be addressed when developing a method for varying fertilizer application within a field include the cost, time, and complexity of variable rate application. Studies have long supported the common sense notion that soil fertility can vary tremendously between regions within a field. Site-specific nutrient management can offer economic rewards as long as manageable nutrient variability exists in the field.

The concept of spatially matching soil fertility inputs to field needs requires a knowledge of how the fertility varies, and hence a soil sampling strategy. Unlike the general concept, the choice of sampling pattern has probably been the most controversial issue in precision agriculture. In the early 1990s two lines of thought arose around the opposing theories of grid sampling and directed sampling. In the beginning the proponents of grid sampling achieved a strong following and soon this methodology was widely practiced. Grid sampling became synonymous with soil sampling for precision agriculture. Over the past several years, however, grid sampling has fallen from favor. The far more rational, flexible, and economically viable strategy, known as directed sampling (along with the associated concept of "zone management"), currently prevails as the method of choice (Pocknee 2000).

To implement a directed sampling scheme requires a field to be divided into zones of similar fertility. The delineation of these zones can be achieved in many ways but often involves the use of yield maps, aerial photos, soil maps, farmer knowledge, and field topography. The resultant map is used to direct the location of individual soil samples. A composite sample from each zone is analyzed and an application map can be generated using these results along with yield goals that are assigned to each zone. If soil test results do not indicate significant differences in nutrient levels between zones, variable rate application may not be warranted.

Management zones must be analyzed, evaluated and adjusted over time. They are not static and will change as the management style and capabilities of the farmer change. It may be prudent to combine zones that consistently perform similarly over time and to split zones that show more variability than initially thought. Changes in equipment must also be considered and may require an adjustment to zones.

Crucial to any variable rate application is the need to assess the response or outcome to the treatment. Measuring a response to different treatments requires on-farm experimentation. To optimize production for economic and environmental benefits, it is necessary to develop response functions to fertilizer. Variable rate technology coupled with yield mapping enables farmers complete flexibility in on-farm testing. Farm experimentation may be as simple as a strip plot design where test strips run the length of the field, or may be a sophisticated randomized block design where treatments are applied to the management zones of the field.

59.5.2.2 Variable Rate Irrigation

Variable-rate irrigation (VRI), also called site-specific or precision irrigation, is a relatively new concept in agriculture. VRI is a tool that involves the delivery of irrigation water in amounts that match the needs of individual areas within fields.

Most center-pivot (CP) irrigation systems currently in use apply a constant rate of water. Some systems can provide variable application rates in wedge-shaped sections of a field by varying the travel speed of the system through those areas. However, the ability to vary application rates over randomly shaped zones has not been possible with normal systems. Recent advances in technology and the availability of a commercial system has changed this. VRI can now deliver the appropriate amount of irrigation water to each area in a field. This is achieved by a combination of pivot speed control and cycling sprinklers on and off.

Variable soil properties and topography will cause corresponding variations in soil water properties such as water holding capacity, drainage rates, and infiltration rates. Many CP systems do not make complete circles nor overlap other pivots. Similarly, many CP systems have areas, such as waterways, ditches, ponds, or roadways, that are not cropped and do not need to be watered. Other fields may be irregularly shaped or contain multiple crops. All of these scenarios can benefit from a system with the ability to apply varying amounts of irrigation water to specific areas in a field.

One should not immediately assume that they must purchase new hardware to implement VRI. Some things that can be done inexpensively include:
- dividing a pivot into pie sections with different water needs and manually changing the pivot travel speed through the sections (this can also be done with newer control panels)
- using end-gun controls more effectively

- using manual valves to turn off individual sprinklers
- installing VRI controls on only part of a pivot, rather than on the entire pivot

High-resolution true-color images are a valuable tool for assessing a field's suitability for VRI. Images are an excellent low-cost tool that can aid the characterization of within-field variability, and identify and quantify features such as non-cropped intrusions or inclusions in a field. Non-cropped areas are parts of a field that are not planted such as ponds, depressed areas, roads, areas of trees and drainage lines. These areas are under the coverage of the irrigation system but do not require water. Using rectified images, the non-cropped areas of a field can be quantified and estimates of water savings using a VRI system calculated. Although this information needs to be ground-truthed by a site visit, this approach enables a quick assessment and screening tool for determining a field's suitability for VRI and the potential water savings. Developing a water application map for the VRI controller requires the assimilation of available spatial data such as yield maps, aerial photos, soil survey maps, and first-hand knowledge of the field to develop watering management zones.

In most agricultural systems around the world, the primary input driving crop production is water. Irrigation is a necessary part of many cropping systems due to insufficient or uneven rainfall. The potential for an irrigation system that delivers optimum amounts of irrigation water over an entire field is significant.

In the US there are approximately 150,000 center pivots watering over 21 million acres of cropland. In all parts of the US, agriculture is putting increased demands on limited water resources. Droughts and lawsuits have prompted a renewed interest in water conservation methods by the general public, which is becoming increasingly insistent that agriculture participate in conserving water. Studies on over 25 VRI pivots installed in Georgia have shown a 12–16% reduction in water use while increasing yield (NESPAL 2008a). The main savings were achieved through improved control of the system, allowing sprinklers to be turned off over non-cropped regions (ponds, drainage areas, field roads, etc.) within the field and bordering areas.

59.6 Soil Sampling, Mapping and Imaging

Geographic representations of soils have existed since 1899 when the US Department of Agriculture developed the soil survey program to help farmers evaluate suitable cropping and management practices for the soils on their farms. As more was learned about soils, the soil survey information was applied to other land uses. Modern soil surveys are used for many activities, such as highway construction, farm planning, tax assessment, forest management and ecological research, in addition to agriculture. Accurate soil descriptions are critical for many investigations related to land responses to natural and man-induced inputs.

Aerial photography has been used as an aid for soil mapping and a presentation base for the final maps since the 1920s. The use of these maps has greatly increased the precision of soil surveys and permitted extensive mapping at detailed scales (1:24,000 or greater). These map-based soil surveys have naturally evolved into geographic representations that can be integrated into GIS to make comparisons between soils and other geographic characteristics.

Soil surveys for most counties throughout the US were compiled by the National Cooperative Soil Survey (NCSS). The NCSS is a joint effort of the US Department of Agriculture (USDA) and other federal agencies, along with many state agencies and local agencies. County-level soil surveys have been available since the 1900s and refined surveys are still being developed. Currently, two levels of soil survey are available in GIS. These include the State Soil Geographic (STATSGO) data and the Soil Survey Geographic (SSURGO) data. The two data sets vary considerably in their level of detail and their availability. Both data sets can be downloaded from the USDA Natural Resources Conservation Service (NRCS) geospatial data gateway (USDA 2008a).

Manual of Geographic Information Systems

The STATSGO data are produced by generalizing the detailed county-level soil survey data. The mapping scale for STATSGO is 1:250,000 (with the exception of Alaska, which is 1:1,000,000). The minimum area mapped is approximately 625 ha (1,544 acres). The level of mapping provided in the STATSGO database is designed to be used for broad planning and management uses covering state, regional, and multi-state areas. The STATSGO data are available for the conterminous US, Alaska, Hawaii, and Puerto Rico. Digitizing of the STATSGO data is done in line segment (vector) format in accordance with NRCS digitizing standards. Map unit delineations match at state boundaries. Composition of soil map units was coordinated across state boundaries, so that component identities and relative extents would match. The STATSGO data are available in the USGS Digital Line Graph (DLG-3) optional distribution format. They are also available in ArcInfo 7.0 coverage and GRASS 4.13 vector formats (http://datagateway.nrcs.usda.gov/).

Each STATSGO map is linked to the Soil Interpretations Record (SIR) attribute database. The attribute database gives the proportional extent of the component soils and their properties for each map unit. The STATSGO map units each consist of 1 to 21 components. The SIR database includes over 25 physical and chemical soil properties, interpretations, and productivity. Examples of information that can be queried from the database are available water capacity, soil reaction, electrical conductivity, and flooding; building site development and engineering uses; cropland, woodland, rangeland, pastureland, and wildlife; and recreational development.

The SSURGO data are substantially more detailed than the STATSGO data. Mapping scales generally range from 1:12,000 to 1:63,360. The SSURGO data are the most detailed level of soil mapping done by the NRCS. Digitized SSURGO maps duplicate county-level soil survey maps and provide the data in GIS format. This level of mapping is designed for use by landowners, townships, and county natural resource planning and management. The SSURGO data are currently only available for selected counties and areas throughout the US (USDA 2008b). Completion of the SSURGO database is expected in 2008.

As with STATSGO, digitizing of the SSURGO data is done by line segment (vector) format in accordance with NRCS digitizing standards. The mapping bases meet national map accuracy standards. SSURGO data are distributed as a complete coverage for a soil survey area. The SSURGO data are linked to a National Soil Information System (NASIS) attribute database. The attribute database gives the proportionate extent of the component soils and their properties for each map unit. The SSURGO map units consist of 1 to 3 components each. Examples of information that can be queried from the database are available water capacity, soil reaction, electrical conductivity, and flooding; building site development and engineering uses; cropland, woodland, rangeland, pastureland, and wildlife; and recreational development.

The map extent for a SSURGO data set is a single soil survey area, which may consist of a county, multiple counties, or parts of multiple counties. A SSURGO data set consists of map data, attribute data, and metadata. SSURGO map data are available in ArcView shape files, ArcInfo coverages, and ArcInfo interchange file formats. The coordinate systems are geographic, UTM, and SPC. Attribute data are distributed in ASCII format and can be imported into a Microsoft Access database. Metadata are in ASCII and XML format.

59.7 Watershed Management, Analysis and Modeling

59.7.1 Watersheds and Watershed Management

A watershed is a geographic area that drains into a single stream or river because of topography. John Wesley Powell defined it as "that area of land, a bounded hydrologic system, within which all living things are inextricably linked by their common water course and where, as humans settled, simple logic demanded that they become part of a community."

Watersheds are separated by topographic ridges dividing the drainage systems. Because watersheds often contain different geologic features and differences in climate, they can often be used to divide regions for analysis and management purposes. Watersheds are important for drought and flooding investigations, as legal points of reference when dealing with water rights, and for managing the quality of the water contained therein. Since large watersheds are made up of many smaller watersheds, it is necessary to define the watershed in terms of a point of reference in the stream. This point is referred to as the watershed outlet. With respect to the outlet, the watershed consists of all the land area that sheds water to the outlet during a rainstorm. A watershed is defined by all points enclosed within an area from which rain falling at these points will contribute water to the outlet. Surface and groundwater flow patterns are not always defined by the same boundaries. Thus, watersheds defined by their surface drainage patterns and those defined by their subsurface drainage patterns are not always the same.

Human modifications of lands and waters directly alter delivery of water, sediments, and nutrients. Any activity that changes soil permeability, vegetation type or cover, water quality, quantity, or rate of flow at a location can change the characteristics of a stream or even the watershed at downstream locations. Land use practices such as clearing land for timber or agriculture, developing and maintaining roads, housing developments, and water diversions may have environmental consequences that greatly affect stream conditions even when the land use is not directly associated with a stream. Proper planning and adequate care in implementing projects can help ensure that one activity within a watershed does not detrimentally impact the downstream environment.

Because of similar climatic, geographic, and land management conditions within watersheds and their natural ties to communities, they often represent the most logical basis for managing water resources. The water resources within the watershed become the focal point for management considerations. Managers are able to gain a clearer understanding of the overall conditions in an area and the stressors that affect those conditions by examining the watershed as a whole. Watershed management can offer a stronger foundation for uncovering the many stressors affecting a watershed. The result is a better assessment of the actions required to protect or restore the resource. Because of these factors, watersheds, and hydrologic models that describe surface and subsurface flow within these watersheds, are commonly used for land management decisions.

Watersheds often contain similar vegetation, crop production, soils, topography, and climatic patterns. Increasingly state agencies are turning toward watershed management as a means of achieving water quantity and quality goals. Land owners can be organized because they often share similar production methods and goals. Because watersheds are defined by natural features, they represent a logical basis for managing natural resources. Conditions that affect the overall state of the watershed can often be recognized and addressed in a consistent manner. Watershed management provides a framework for integrated decision making, where we strive to: (1) assess the nature and status of the watershed; (2) define short-term and long-term goals; (3) determine objectives and actions needed to achieve selected goals; (4) assess both benefits and costs of each action; (5) implement desired actions; (6)

evaluate the effects of actions and progress toward goals; and (7) re-evaluate goals and objectives as part of an iterative process.

Besides the environmental pay-offs, watershed approaches can have the added benefits of saving time and money. Whether the task is monitoring, modeling, issuing permits, or reporting, a watershed framework offers many opportunities to simplify and streamline the workload. Efficiency is increased by providing a common focus and resources can be pooled to focus on a single geographic area. By coordinating their efforts, agencies can complement and reinforce each others' activities, avoid duplication, and leverage resources to achieve greater results.

Watershed protection can lead to greater awareness and support from the public. Once individuals become aware of and interested in their watershed, they often become more involved in decision making as well as hands-on protection and restoration efforts. Through such involvement, watershed approaches build a sense of community, help reduce conflicts, increase commitment to the actions necessary to meet environmental goals, and ultimately, improve the likelihood of success for environmental programs.

However, land traditionally has often been managed at a very local scale. Changes in land management historically were, and still largely are, implemented by private land owners with little regard to the location of the managed land in relation to other parcels of land. While very effective at the local field scale, these types of management implementation policies often have a very localized impact and make it difficult to obtain larger, regional impacts. People have varying goals and values relative to uses of local land and water resources. Overcoming the obstacles created by private land management requires time and an understanding of the motivation of the land owners. Watershed management requires use of the social, ecological, and economic sciences. Common goals for land and water resources must be developed among people of diverse social backgrounds and values. The decision process must weigh the economic benefits and costs of alternative actions, and blend current market dynamics with considerations of long-term sustainability of the ecosystem.

Watershed management is an iterative process of integrated decision making regarding uses and modifications of lands and waters within a watershed. This process provides a chance for stakeholders to balance diverse goals and uses for environmental resources, and to consider how their cumulative actions may affect long-term sustainability. As a form of ecosystem management, watershed management encompasses the entire watershed system, from uplands and headwaters, to floodplain wetlands and river channels. It focuses on the processing of water, sediments, nutrients, and toxics downslope through this system. Of principle concern is management of the watersheds' water budget, which is the routing of precipitation through the pathways of evaporation, infiltration, overland flow, groundwater recharge, and groundwater discharge (Woolhiser 1982). This routing of groundwater and overland flow defines the delivery patterns to particular streams, lakes, and wetlands and largely shapes the nature of these aquatic systems.

59.7.2 Watershed Modeling

Because of the complex nature of watersheds and the many factors related to watershed management, it is often useful to incorporate a watershed model into the process. A watershed model is an abstraction of the natural processes. Representation of the natural system obviously requires simplification. This simplification allows managers to place the system into a manageable context, which can be used to examine the impact of different conditions on watershed processes.

Models can take many shapes and forms. They can be mathematical representations of the natural processes or physical prototypes of the watershed. For more complex examinations, mathematical models are typically used. Mathematical models can be empirical, based largely upon observed data, or theoretical, based upon mathematical representations of the physical processes occurring within the watershed. Empirical models can be very useful if an extensive data base of observed data is available. An empirical model is a mathematical representation of the observed data. In reality, most physically based watershed models contain empirical components because of the complexity of the natural system.

Management of watersheds with the goals of balancing the needs of the general public with environmental concerns in those regions often requires complex descriptions of the processes occurring therein. Significant advances have been made with physically based watershed models with the advent of modern computer systems and GIS. These watershed models can incorporate many of the natural geologic and climatic features into the modeling system by making full use of the GIS packages. These models allow a very thorough examination of the impact of watershed scale management on the quantity and quality of water within the watershed. Several reviews of watershed models have been previously published (Singh and Frevert 2005, 2002a, b; Singh 1995; Parsons et al. 2001). The accuracy of the models has improved with the introduction of computers and the capacity that they brought to increase the level of complexity of these models.

Many currently available models are physically based, attempting to describe the geophysical processes occurring on the landscape with mathematical equations. These models require considerable input data, most of which is spatially related. Physical characteristics of watersheds, such as the soil, topography, and land use, vary spatially. GIS allows spatial representation of the features, improving the accuracy of the watershed models. Many watershed models have quickly adapted to using the geographical data handling capabilities of GIS, both to ease the development of input data sets and for the analysis of model output. The GIS allows the compiling and processing of these input data sets with relative ease and significantly reduces the rather cumbersome task of assembling input data for large-scale watershed models.

Several watershed models have been developed and modified to take advantage of GIS databases. The models that best utilize the full capabilities of the GIS are classified as distributed models. Distributed models attempt to explicitly account for the variability of processes by varying the input, boundary conditions, and watershed characteristics through space. Because it is impossible in practice to capture the full variability of natural processes, all distributed models perform some degree of lumping, or simplification, with the input characteristics.

The assembly of the distributed climatic, geologic, and land use data necessary for accurately representing characteristics in the watershed can become an insurmountable task. However, the ability of the GIS to extract, overlay, and delineate watershed characteristics greatly simplifies the development of the input information. Integration of GIS with watershed modeling accomplishes several functions, namely, design, calibration, modification, and comparison of watershed models. These systems also facilitate watershed division so that specific sections of the watershed can be modeled. GIS also facilitates closer examination of relationships between model outputs and the spatial characteristics of the watershed.

A few of the commonly used Watershed Models that utilize GIS are briefly discussed here. This review is provided as illustration of the tools and not as a complete description of this topic. For additional discussions on the topic the reader is guided to texts that focus solely on Watershed Models (Singh and Frevert 2005, 2002a, b; Singh 1995).

59.7.2.1 AGNPS and AnnAGNPS

The single event Agricultural Non-Point Source (AGNPS) model was developed in the early 1980s by the Agricultural Research Service (ARS) in cooperation with the Minnesota Pollution Control Agency, and the Natural Resource Conservation Service (NRCS) (Young et al. 1989, 1995). The model was developed to analyze and provide estimates of runoff water quality resulting from single storm events from agricultural watersheds ranging in size from a few hectares to 20,000 ha. The AGNPS model is grid or raster based. The grid-based cells were used to track surface drainage patterns throughout the watershed. The use of grid cells led to adaptation of the model into several GIS platforms (Mitchell et al. 1993; Srinivasan et al. 1994; Rode et al. 1995; Tim and Jolly 1994).

In the early 1990s, a cooperative team of ARS and NRCS scientists was formed to develop an annualized continuous-simulation version of the model, AnnAGNPS (Cronshey and Theurer 1998). AnnAGNPS is the pollutant-loading component for a suite of models referred to as AGNPS 2001. AGNPS 2001 includes GIS routines for developing model input and analysis of model output. The GIS tool automates many of the input data preparation steps needed for use with large watershed systems. Interfaces have been developed to facilitate the use of GIS and the AnnAGNPS model (Xiao 2005).

59.7.2.2 SWAT

SWAT is a river basin model developed to quantify the impact of land management practices in large watersheds (Arnold et al. 1998). SWAT was a modification of the Simulator for Water Resources in Rural Basins (SWRRB) model (Williams et al. 1985). SWAT was developed to predict the impact of land management practices on water, sediment, and agricultural chemical yields in large watersheds with varying soils, land use, and management conditions over long periods of time. The major components of the model include hydrology, weather, erosion and sediment transport, soil temperature, crop growth, agrichemical transport, and agricultural management. The model has been successfully applied on watersheds up to 598,538 km^2 in area (Srinivasan et al. 1997).

To facilitate data input for these large areas, GIS tools have been developed for SWAT (Di Luzio et al. 2004). Currently SWAT 2000 is incorporated into AVSWAT-2000, an ArcView extension and a graphical user interface for the model (Di Luzio et al. 2004). The GIS framework simplifies input development and output analysis. The interface can be used to: 1) generate specific parameters from user-specified GIS coverages; 2) create SWAT input data files; 3) establish agricultural management scenarios; 4) control and calibrate SWAT simulations; and 5) extract and organize model output data for charting and display (Di Luzio et al. 2002). The GIS tool also includes features to delineate the watershed drainage.

59.7.2.3 BASINS

The US EPA and its counterparts in states and pollution control agencies are increasingly emphasizing watershed assessments. To facilitate this, the EPA developed the Better Assessment Science Integrating point and Nonpoint Sources (BASINS) system (US EPA 2007). BASINS integrates a geographic information system (GIS), national watershed and meteorological data, and state-of-the-art environmental assessment and modeling tools into one software package. BASINS contains three models for estimating watershed loading. These models include a simplified GIS-based nonpoint-source loading model (PLOAD) and two physically based watershed loading and transport models, Hydrologic Simulation Program-Fortran (HSPF) and SWAT.

59.8 Conclusions

Application of advanced information technologies to agriculture and watershed management has become standard procedure for many farmers, agronomists, and hydrologists. The improvements brought by these technologies include precise positioning allowing determination of nutrient, herbicide, and pesticide needs on a variable infield basis. The same positioning allows determination of yields for exact portions in a field so that correct amounts of inputs can be applied to improve yields and minimize environmental degradation. Watershed modeling has been improved since inputs to models can now be automatically generated and finer resolution data can be used to produce better models of watershed activity. The technologies of GPS, remote sensing, GIS and VRT provide a significant advance to agriculture and watershed managers and a return to the public through improved agriculture with reduced environmental damage.

References

Armitage, A. 1966. *Edmond Halley.* London: Thomas Nelson and Sons Ltd.

Arnold, J. G., R. Srinivasan, R. S. Muttiah and J. R. Williams. 1998. Large area hydrologic modeling and assessment-Part I: Model development. *Journal of the American Water Resources Association* 34:1.

Auernhammer, H., M. Demmel, T. Muhr, J. Rottmeier and K. Wild. 1994. GPS for yield mapping on combines. *Computers and Electronics in Agriculture* 11:53–68.

Bate, R. R., D. D. Mueller and J. E. White. 1971. *Fundamentals of Astrodynamics.* New York: Dover Publications, Inc.

Bailey, T. C. and A. C. Gatrell. 1995. *Interactive Spatial Data Analysis.* Essex, England: Longman Scientific and Technical.

Bandyopadhyay, S. and S. K. Pal. 2001. Pixel classification using variable string genetic algorithms with chromosome differentiation. *IEEE Transactions on Geoscience and Remote Sensing* 39 (2):303–308.

Boots, B. N. 1986. *Voronoi (Thiessen) Polygons.* CATMOG 45. Norwich, UK: Geo Books.

Burrough, P. A. and R. A. McDonnell. 1998. *Principles of Geographical Information Systems.* New York: Oxford University Press.

Carter, G. A. 1993. Responses of leaf spectral reflectance to plant stress. *American Journal of Botany* 80 (3):239–243.

Carter, G. A., W. G. Cibula and R. L. Mikker. 1996. Narrow-band reflectance imagery compared with thermal imagery for early detection of plant stress. *Journal of Plant Physiology* 148:515–522.

Colwell, R. N., ed. 1983. *Manual of Remote Sensing.* Volume 1, *Theory, Instruments, and Techniques,* Volume 2, *Interpretation and Applications.* Falls Church, Va.: American Society of Photogrammetry.

Cronshey, R. G. and F. D. Theurer. 1998. AnnAGNPS - Non-point pollutant loading model. Pages 1–9 to 1–16 in *Proceedings of the First Federal Interagency Hydrologic Modeling Conference,* 19–23 April 1998, Las Vegas, Nevada.

CROPINS. 2006. CROPINS: The Remote Sensing Crop Information System. <http://directory.eoportal.org/d_ann.php?an_id=4523> Accessed 1 August 2008.

Di Luzio, M., R. Srinivasan and J. G. Arnold. 2004. A GIS-coupled hydrological model system for the watershed assessment of agricultural nonpoint and point sources of pollution. *Transactions in GIS* 8 (1):113–136.

Di Luzio, M., R. Srinivasan, J. G. Arnold and S. L. Neitsch. 2002. *ArcView Interface for SWAT2000: User's Guide.* Blackland Research and Extension Center. Temple, Texas. BRC Report 02-07.

Frohn, R. C. 1998. *Remote Sensing for Landscape Ecology.* Boca Raton, Fla.: Lewis.

Garmin. 2000. *GPS Guide for Beginners.* Olathe, Kan.: Garmin International, Inc.

Gauss, C. F. 1857. Theoria Motus. Translated by Charles H. Davis. 1963. *Theory of the Motion of the Heavenly Bodies Moving About the Sun in Conic Sections.* New York: Dover Publications, Inc.

Goering, C. E. 1993. Recycling a concept. *Agricultural Engineering.* November: 25.

Goodrich, D. C. and D. A. Woolhiser. 1991. Catchment hydrology. *Reviews of Geophysics* 29:202–209.

Grisso, R., R. Oderwald, M. Alley and C. Heatwole. 2003. Precision Farming Tools: Global Positioning System (GPS). *Virginia Cooperative Extension Publication* 442–503.

Herring, T. A. 1996. The Global Positioning System. *Scientific American* February:44–50.

Jensen, J. R. 2007. *Remote Sensing of the Environment: An Earth Resource Perspective.* 2d ed. Upper Saddle River, N.J.: Prentice Hall.

———. 2005. *Introductory Digital Image Processing.* 3d ed. Upper Saddle River, N.J.: Prentice Hall.

Jensen, J., A. Saalfeld, F. Broome, D. Cowen, K. Price, L. Lapine and E. L. Usery. 2004. Spatial data acquisition and integration. Pages 17–60 in *A Research Agenda for Geographic Information Science,* edited by R. B. McMaster and E. L. Usery. Boca Raton, Fla.: CRC Press.

Johannsen, C. J., P. G. Carter, D. K. Morrison, B. Erickson and K. Ross. 2006. Potential applications of remote sensing. *Site Specific Management Guidelines, SSMG-22.* <http://www.ppi-far.org/ssmg> Accessed 26 August 2008.

KARS. 2005. *The GreenReport.* Kansas Remote Sensing Center. <http://www.kars.ku.edu/products/greenreport.shtml> Accessed 26 August 2008.

Lam, N. 1983. Spatial interpolation methods: A review. *The American Cartographer* 10 (2):129–149.

Lam, N., D. Catts, D. Quattrochi, D. Brown and R. B. McMaster. 2004. Scale. Pages 93–128 in *A Research Agenda for Geographic Information Science,* edited by R. B. McMaster and E. L. Usery. Boca Raton, Fla.: CRC Press.

Leica Geosystems. 2006. ERDAS Imagine. <http://gi.leica-geosystems.com/LGISub1x33x0.aspx> Accessed 26 August 2008.

Lillesand, T. M., R. W. Kiefer and J. W. Chipman. 2008. Remote Sensing and Image Interpretation. 6th ed. New York: John Wiley & Sons.

———. 2004. *Remote Sensing and Image Interpretation.* 5th ed. New York: John Wiley & Sons.

Liu, Z., A. Liu, C. Wang and Z. Niu. 2004. Evolving neural network using real coded genetic algorithm (GA) for multispectral image classification. *Future Generation Computer Systems* 20 (2):1119–1129.

Meentemeyer, V. and E. O. Box. 1987. Scale effects in landscape studies. In *Landscape Heterogeneity and Disturbance, Ecological Studies.* Vol. 64. Edited by M. G. Turner. New York: Springer-Verlag.

Mitchell, J. K., B. A. Engel, R. Srinivasan and S.S.Y. Wang. 1993. Validation of AGNPS for small watersheds using an integrated AGNPS/GIS system. *Water Resources Bulletin* 29 (5):833–842.

Monteiro, L. S., T. Moore and C. Hill. 2005. What is the accuracy of DGPS? *Journal of Navigation* 58:207–225.

NASA (National Aeronautic and Space Administration). 2008. Orbiter and Radio Metric Systems Group: GIPSY-OASIS Software. <http://gipsy.jpl.nasa.gov/orms/goa/> Accessed 26 August 2008.

NESPAL. 2008a. Enhancing Irrigation Efficiencies. <http://nespal.cpes.peachnet.edu/Water/vri.asp> Accessed 26 August 2008.

NESPAL. 2008b. National Environmentally Sound Production Agricultural Laboratory. <http://nespal.cpes.peachnet.edu/precisionag.html> Accessed 26 August 2008.

O'Neil, P. V. 1987. *Advanced Engineering Mathematics.* 2d ed. Belmont, Calif.: Wadsworth Publishing Company.

Parkinson, B. W. 1996. Introduction and heritage of NAVSTAR, the Global Positioning System. In *Global Positioning System: Theory and Applications,* edited by B. W. Parkinson and J. J. Spilker Jr. Washington, D.C.: American Institute of Aeronautics and Astronautics, Inc.

Parsons, J. E., D. L. Thomas and R. L. Huffman, eds. 2001. Agricultural non-point source water quality models: Their use and application. *Southern Coop. Ser. Bulletin #398.* ISBN:1-58161-398-9. <http://www3.bae.ncsu.edu/Regional-Bulletins/Modeling-Bulletin/modeling-bulletin.pdf> Accessed 26 August 2008.

Pocknee, S. 2000. The management of within-field soil spatial variability. Ph.D. diss., University of Georgia, Athens, Georgia.

Ritter, D. F. 1986. *Process Geomorphology.* 2d ed. Dubuque, Iowa: W. C. Brown.

Robinson, A. H., J. L. Morrison, P. C. Muehrcke, A. J. Kimmerling and S. C. Guptill. 1995. *Elements of Cartography.* 6th ed. New York: John Wiley & Sons.

Rode, M., S. Grunwald and H.-G. Frede. 1995. Modeling of water quality using AGNPS and GIS. *J. of Rural Engineering and Development* 36 (2):63–68.

RSI (Research Systems, Inc.). 2006. ENVI – Get the Information You Need from Imagery. <http://www.ittvis.com/envi/index.asp> Accessed 26 August 2008.

Seaber, P. R., F. P. Kapinos and G. L. Knapp. 1987. Hydrologic unit maps. *U.S. Geological Survey Water-Supply Paper 2294.* United States Government Printing Office. <http://pubs.usgs.gov/wsp/wsp2294/#pdf> Accessed 26 August 2008.

Singh, V. P., ed. 1995. *Computer Models of Watershed Hydrology.* Highlands Ranch, Colo.: Water Resources Publications.

Singh, V. P. and D. K. Frevert, eds. 2005. *Watershed Models.* Boca Raton, Fla.: CRC Press.

———, eds. 2002a. *Mathematical Modeling of Large Watershed Hydrology.* Highlands Ranch, Colo.: Water Resources Publications.

———, eds. 2002b. *Mathematical Modeling of Small Watershed Hydrology and Applications.* Highlands Ranch, Colo.: Water Resources Publications.

Skone, S. H. 2001. The impact of magnetic storms on GPS receiver performance. *Journal of Geodesy* 75 (9-10):457–468.

Snyder, J. P. 1987. *Map Projections: A Working Manual.* US Geological Survey Professional Paper 1395, US Government Printing Office, Washington, D.C.

Soler, T., R. A. Snay, R. H. Foote and M. W. Cline. 2003. Maintaining accurate coordinates for the National CORS Network. *Proceedings of the FIG (International Federation of Surveyors) Working Week,* Paris, France, 13–17 April. <http://www.ngs.noaa.gov/CORS/Articles/FIGParis.pdf> Accessed 26 August 2008.

Song, Z. and L. D. James. 1992. An objective test for hydrologic scale. *Water Resources Bulletin* 28 (5):833–844.

SpaceDaily. 2006. Latest GPS Bird Ready For Launch From Cape Canaveral. <http://www.spacedaily.com/reports/Latest_GPS_Bird_Ready_For_Launch_From_Cape_Canaveral_999.html> Accessed 26 August 2008.

Srinivasan, R., B. A. Engel, J. R. Wright and L. G. Lee. 1994. The impact of GIS-derived topographic attributes on the simulation of erosion using AGNPS. *Applied Engineering in Agriculture* 10 (4):561–566.

Srinivasan, R., T. S. Ramanarayanan, R. Jayakrishnan and H. Wang. 1997. Hydrologic modeling of Rio Grande/Rio Bravo basin. *ASAE Paper No. 972236.* St. Joseph, Mich.: ASAE.

Star, J. and J. Estes. 1990. *Geographic Information Systems: An Introduction.* Englewood Cliffs, N.J.: Prentice-Hall, Inc.

Tim, U. S. and R. Jolly. 1994. Evaluating agricultural nonpoint-source pollution using integrated geographic information systems and hydrologic/water quality model. *J. Environ. Qual.* 23 (1):25–35.

Tomlin, C. D. 1984. Digital cartographic modeling techniques in environmental planning. Ph.D. thesis, School of Forestry and Environmental Studies, Yale University, New Haven, Conn.

Torge, W. 1980. *Geodesy, an Introduction.* Berlin: Walter de Gruyter.

Trimble. 2006. *GPS Tutorial.* <http://www.trimble.com/gps/index.shtml> Accessed 26 August 2008.

USDA. 2006. U.S. Department of Agriculture, Remote Sensing Crop Residue Cover, <http://www.ars.usda.gov/research/projects/projects.htm?accn_no=410784> 26 August 2008.

———. 2008a. US Department of Agriculture, Geospatial Data Gateway. <http://datagateway.nrcs.usda.gov/> Accessed 2 October 2008.

———. 2008b. US Department of Agriculture, Soil Survey Geographic (SSURGO) Database.<http://soils.usda.gov/survey/geography/ssurgo/> Accessed 2 October 2008.

US EPA (Environmental Protection Agency). 2007. Better Assessment Science Integrating Point & Nonpoint Sources. <http://www.epa.gov/waterscience/basins/> Accessed 26 August 2008.

USGS. 2006. 1:250,000-scale Hydrologic Units of the United States. <http://water.usgs.gov/GIS/metadata/usgswrd/XML/huc250k.xml> Accessed 26 August 2008.

Usery, E. L., S. Pocknee and B. Boydell. 1995. Precision farming data management using geographic information systems. *Photogrammetric Engineering & Remote Sensing* 61 (11):1383–1391.

Usery, E. L., M. P. Finn, D. J. Scheidt, S. Ruhl, T. Beard and M. Bearden. 2004. Geospatial data resampling and resolution effects on watershed modeling: A case study using the agricultural non-point source pollution model. *Journal of Geographical Systems* 6 (3):286–309.

Usery, E. L., M. P. Finn and C. J. Mugnier. 2008. Coordinate Systems and Map Projections. In *Manual of Geographic Information Systems*, edited by M. Madden, Chapter 8 in this volume. Bethesda, Md.: ASPRS.

Van Dierendonck, A. J. 1996. GPS receivers. In *Global Positioning System: Theory and Applications*. Edited by B. W. Parkinson and J. J. Spilker Jr. Washington, D.C.: American Institute of Aeronautics and Astronautics, Inc.

Vanicek, P. and E. J. Krakiwsky. 1986. *Geodesy: The Concepts*. 2d revised ed. Amsterdam, The Netherlands: North Holland.

Welch, R. and A. Homsey. 1997. Datum shifts for UTM coordinates. *Photogrammetric Engineering and Remote Sensing* 63 (4):371–375.

Wells, D. E., N. Beck, D. Delikaraoglou, A. Kleusberg, E. J. Krakiwsky, G. Lacharpelle, R. B. Langley, M. Nakiboglu, K. P. Schwarz, J. M. Tranquilla and P. Vanicek. 1986. *Guide to GPS Positioning*. Fredericton, NB, Canada: Canadian GPS Associates.

Williams, J. R., A. D. Nicks and J. G. Arnold. 1985. Simulator for water resources in rural basins. *Am. Soc. Civil Eng. J. Hydr. Engr.* 111 (6):970–986.

Willmott, C. J. and M. L. Johnson. 2005. Resolution errors associated with gridded precipitation fields. *International Journal of Climatology* 25:1957–1963.

Woolhiser, D. A. 1982. Hydrologic system synthesis. Pages 3–16 in Hydrologic Modeling of Small Watersheds. Edited by C. T. Haan, H. P. Johnson and D. L. Brankensiek. *American Society of Agricultural Engineers. Monograph Number 5*. St. Joseph, Mich.

Yang, M. D. and Y. F. Yang. 2004. Genetic algorithm for unsupervised classification of remote sensing imagery. Pages 395–402 in *Image Processing: Algorithms and Systems III, Proceedings of the SPIE*, Volume 5298, edited by E. R. Dougherty, J. T. Astola and K. O. Egiazarian.

Young, R. A., C. A. Onstad, D. D. Bosch and W. P. Anderson. 1995. AGNPS: An agricultural nonpoint source model. Pages 1011–1020 in *Computer Models of Watershed Hydrology*. Edited by V. P. Singh. Highlands Ranch, Colo: Water Resources Publications.

Young, R. A., C. A. Onstad, D. D. Bosch and W. P. Anderson. 1989. AGNPS: A nonpoint-source pollution model for evaluating agricultural watersheds. *J. of Soil and Water Conservation* 44 (2):168–173.

Xiao, H. 2005. An integrated GIS-AnnAGNPS modeling interface for non-point source pollution assessment. <http://gis.esri.com/library/userconf/proc03/p1059.pdf> Accessed 26 August 2008.

CHAPTER 60

Extraterrestrial GIS

Trent M. Hare, Randy L. Kirk,
James A. Skinner Jr. and *Kenneth L. Tanaka*

60.1 Introduction

Extraterrestrial GIS is simply the application of GIS technologies to study planetary bodies other than the Earth. Because GIS expands upon digital cartography, the use of this technology fits naturally into planetary research. The most obvious difference in using extraterrestrial data sets is the shape definition of the planetary body, such as a planet or an orbiting moon, or an interplanetary body like a comet or asteroid. Fortunately, nearly all larger bodies in our solar system have defined and documented geodetic parameters allowing most GIS applications to work with extraterrestrial data sets.

Planetary researchers generally use GIS technologies for characterizing and analyzing a planet's geologic and structural history, and their studies rely upon the accurate rendering of surface features. However, in recent years, the utility of GIS for planetary research has expanded to include online data dissemination, mission planning, mission support, and modeling of planetary processes.

60.2 History

Planetary data acquisition truly exemplifies the term "remote sensing," commonly defined as collecting imagery at a distance (Greeley and Batson 1990). Because of the great distances traveled by spacecraft, nearly all planetary data sets are returned digitally. As a result, software processing tools have been required since 1965 when the *Mariner 4* spacecraft transmitted its first digital pictures of Mars (Figure 60-1). Despite this early introduction of digital data sets, it was not until the late 1980s that digital cartographic processing of these transmissions was technically practical. Imaging systems have evolved as well. While the original systems were simple television cameras with radio frequency (RF) transmitters, the robotic satellites that are currently orbiting Mars, the Moon, and Saturn use the same remote sensing technologies that are applied to terrestrial research. These technologies include CCD framing cameras and line scanners, synthetic aperture radar, laser altimeters, visible and infrared spectrometers, and many other styles of detectors.

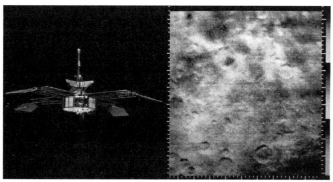

Figure 60-1 The image on the left shows the *Mariner 4* satellite. *Mariner 4* was the first spacecraft to successfully transmit digital images of another planet. The image on the right shows the seventh picture (frame 07B) of Mars taken by the *Mariner 4* spacecraft on 15 July 1965. This frame is considered to be the first image to clearly show impact craters. The image is centered at 14 S, 186 E and is approximately 262 km across. Courtesy: NASA.

Integrating these diverse data sets within a cartographic framework can be difficult even though GIS applications have been used for planetary projects for a number of years (Figure 60-2) (Mardon 1992; Carr 1995; Hare et al. 1998; Roddy et al. 1998; Dohm et al. 2001). For GIS applications to be primarily useful for planetary research, the data must be radiometrically calibrated, cosmetically enhanced, and spatially registered to the body. This process for planetary data sets can be extremely challenging as explained in more detail in a later section. Furthermore, for GIS as for other mapping technologies such as remote sensing (RS), photogrammetry or image processing, off-the-shelf software packages are generally written for terrestrial applications so using planetary data sets can be difficult or impossible because of assumptions that the coordinate reference system is Earth-based. Therefore, planetary researchers have become accustomed to writing their own applications that address their particular needs. But with the advent of powerful, inexpensive computers and the advances in extremely robust and easy-to-use software suites, the cost of creating and maintaining in-house applications is increasingly difficult to justify. Finally, during early planetary mapping projects there was typically only one digital base available, reducing the need to overlay and analyze multiple layers. Over the last decade, as many more missions have been launched, multiple data layers are now common. For example, the Moon and Mars now have dozens of diverse data sets including multiple global image mosaics, topography, geology, and mineral maps. The many missions to these and more distant bodies planned by the United States and other nations for the coming years will greatly increase the available data and the need for a GIS approach.

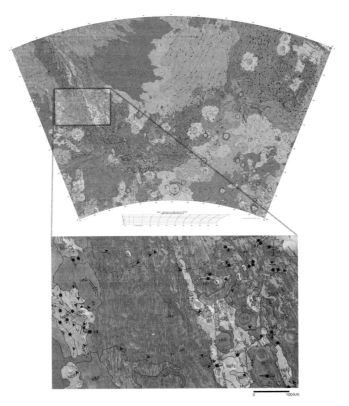

Figure 60-2 Geologic and paleotectonic layers for the Thaumasia region of Mars from the USGS Geologic Investigations Series Map I-2650. This was one of the first planetary geologic maps published by the USGS that was created and analyzed by using a GIS (Dohm et al. 2001). The area zoomed in shows the density of features one can capture in the digital domain. See included DVD for color version.

60.3 Planetary Coordinate Reference Systems

As stated before, the most significant difference between extraterrestrial and earth-based GIS is the definition of the body to be mapped. Most planetary bodies are defined as simple spheres because of the lack of information regarding the true shape. Interestingly, for Mars, there is enough information available via detailed altimetric profiling to determine that the best-fit ellipsoid is triaxial. However, because of the kilometer-sized departures from this ellipsoid (and the theoretical complications of mapping a triaxial body), an oblate spheroid has always been used as a reference surface. Even for highly irregular or nearly triaxial bodies like comets and asteroids, a simple sphere is generally fit so that cartographic packages can more easily utilize the data. Thus any cartographic application that allows custom coordinate reference systems (CRS) should be able to utilize planetary data.

60.3.1 IAU and IAG

Authority over the coordinate systems and cartographic constants of planetary bodies resides with the International Astronomical Union (IAU). In 1976, the IAU established a Working Group on the Cartographic Coordinates and Rotational Elements of Planets and Satellites. This group reports triennially on the preferred rotation rate, spin axis, prime meridian, and reference surface for planets and satellites (Seidelmann et al. 2002). Since 1985, the group has also been co-sponsored by the International Association of Geodesy (IAG).

One of the challenges in using off-the-shelf GIS applications for planetary data sets originates from the fact that the IAU has approved two coordinate systems for each body. In 1970, the IAU general assembly approved the use of "planetographic" coordinates for extra-terrestrial mapping, consisting of geographic latitudes (i.e., measured normal to the ellipsoid, equivalent to geographic or geodetic latitude on the Earth) and the direction of positive longitude chosen such that the longitudes seen by an outside observer increase with time. This longitude convention follows traditional astronomical usage, and results in positive West longitude for Mars and many other bodies, which is unfortunately opposite of the direction used for Earth. At the same time, the IAU also approved "planetocentric" coordinates, which use geocentric latitudes and positive East longitudes, giving a right-handed spherical coordinate system. The two types of latitude differ slightly at mid-latitudes, but are identical at the equator and poles.

An acceptable method to resolve this issue is to define a spherical reference surface. This technique forces the use of planetocentric latitude system but has the potential to cause errors in the map scale. Thus it may be still necessary for planetary scientists to modify their existing application or use custom-written software if they want to work with planeto-centric latitudes on elliptical defined bodies. Also, nearly all off-the-shelf applications do not natively support a positive West longitude system. Most users are forced to use a positive East longitude system for all cases (planetocentric and planetographic) even though it may not comply with IAU/IAG guidelines.

60.3.2 Standardizing CRSs for GIS

Sharing extraterrestrial data across multiple GIS applications has proven problematic because defining planetary CRSs in standardized GIS/RS raster formats, like Geotiff (Ritter and Ruth 2000), is not well supported even though it is possible to do so. By interfacing with geospatial standards bodies such as the Open Geospatial Consortium (OGC), we may help facilitate better planetary support both in existing and in the next generation of GIS/RS file formats.

The OGC was created to help establish and promote a series of formats and Internet protocols for sharing GIS resources (i.e., geospatial data). The OGC is comprised of over two hundred hardware and software companies, universities, and research facilities. Some of the goals are (1) to make geospatial information easy to find, (2) to allow easy access and acquisition of data sets, and (3) to permit data from different sites to be integrated, registered and analyzed (OGC 2005).

A key aspect of shared data sets and/or online streamed data sets is the ability to synchronize the different data sets to a single map projection and defined CRS so that data register correctly. Currently, the OGC, and its various web mapping standards and various file formats primarily use the coded CRS definitions and coordinate transformation descriptions as defined by the European Petroleum Survey Group in 2005 (EPSG 2005). This group is now called the Oil and Gas Producers (OGP) however the standard will keep the EPSG namespace for the CRS definitions. If a needed CRS is not part of the EPSG database, and no extraterrestrial definitions are, it can be defined as a custom setting. However, using these custom settings is generally suitable only for private use, and generic applications would not be able to access the details of the CRS since they are not advertised. Although including planetary CRS definitions within the EPSG standard is an option, it is not ideal because the original intent for the EPGS database is for the oil and gas industry. The alternative approaches to solving this problem include (1) adding planetary CRS definitions into the OGC, (2) creating an online-accessible CRS registry catalog, and (3) modifying the current standards. Each of these items is discussed below.

The first method would be to generate a new set of numeric IDs specifically for planetary CRSs and reference it within the OGC. Currently, web mapping servers generally default to using the numeric EPSG codes. For example "4326" is the EPSG identifier for Earth's "WGS 84" geographic CRS. Projection definitions are also separately coded within the EPSG standard (e.g., EPSG:9801 = Lambert Conic Conformal with one standard parallel). A drawback is that the numeric codes simultaneously try to define popular combinations of CRSs, projections, and the projection's parameters. The IDs were generated by trying to catalog all the widely used cartographic mapping series from all countries (e.g., EPSG:32612 = WGS 84 / UTM zone 12N, EPSG:21413 = Beijing 1954 / Gauss-Kruger zone 13). The number of potential combinations is limitless and would be even greater with planetary definitions included. The OGC has previously reserved numeric codes for 41000 to 41999 to allow for new static definitions. But the number of planetary definitions needed could easily surpass the available 999 definitions given all the combinations of planetary CRSs and map projections. Thus the generation of an independent set of codes gives the planetary community the most flexibility. The most likely name space for these coded values would be "IAU" appended with the year it was last published by the IAU/IAG. Thus for the last published reference for the IAU/IAG in the year 2000, the name space would be "IAU2000." For example, the Mars CRS definition might be "IAU2000:49900." The number 499 has been previously used by the Navigation and Ancillary Information Facility (NAIF) for Mars. To be better aligned with NAIF we will use their planetary codes as the base for this new standard.

The second alternative would be to support an online CRS registry catalog. This method is currently part of the web mapping standards but not widely supported in client applications. The web mapping server would reference a Uniform Resource Locator (URL) address and have the registry return the CRS using a Well-Known Text (WKT) string. Because the WKT is not exceedingly long, some have argued that the entire string could be simply passed inline with the standard map requests. This would offer the most flexibility, as the CRS or map projection parameters would never have to be locked into a specific code. This method would be hard to utilize in a file format as the application would need Internet access. The OGP introduced a CRS registry service in 2007 and clients have already begun to use it. Hopefully, the planetary community can host a registry using the same engine such that the registries will function using the same methods. The Mars CRS would look similar to the following example:

```
GEOGCS ("GCS_Mars_2000",DATUM("D_Mars_2000",SPHEROID("Mars_2000_IAU_IAG",
3396190.0,169.8944472236118)), PRIMEM("Reference_Meridian",0.0),
UNIT("Degree",0.0174532925199433))
```

The third alternative mentioned above would be to modify the current web mapping standard to allow a little more flexibility. One such method would be to separate the CRS code, the map projection code, and the projection's parameters. Such a definition would allow web mapping services to accept requests like "CRS, Projection, Unit, and parameter list." However, this would still require placing the planetary CRS into a coded system. The largest shortcoming to this method is that particulars of the map projection may not be explicitly spelled out. For example, does the projection use a simple sphere or ellipse equation or does the transformation require a datum translation?

There are benefits and drawbacks to all three proposed methods. With help from the planetary community and OGC members, a common solution can soon be implemented. Supporting planetary CRSs is a critical aspect that needs to be solved in order for the planetary community to make better use of the latest GIS and remote sensing technologies.

60.4 Collecting Cartographic Data Sets

As stated earlier, the instruments flown to extraterrestrial bodies include similar technologies that are currently used for terrestrial research. This includes the standard CCD framing image systems and hyperspectral line scanners, but current missions are also collecting data from instruments like thermal emission and gamma-ray spectrometers, topographic laser and subsurface sounding radar altimeters, which allow researchers to derive, for example, mineral and density maps or even gravity and climate models.

60.4.1 Lunar Missions

In 1964, the *Ranger VII* spacecraft, flown by the United States, transmitted 4,316 high-resolution TV photos during its last 13 minutes of flight before impacting the Moon (Heacock et al. 1965; Hall 2006). These images represent the first large catalog of planetary remotely sensed images as taken by a robotic satellite. In the following years, the United States and the Soviet Union completed many successful robotic missions including orbiters, landers, and several sample-return missions to the Moon. We must also mention the six successful Apollo manned-missions in the late 1960s and early 1970s. During these early missions the best cartographic quality imagery was taken by the Lunar Orbiter missions. These five missions returned near-global imagery at less than 100-m resolution. However, after the mid 1970s, the Moon was not as feverishly studied, as most of the attention was turned toward more distant planetary bodies such as Venus and Mars.

Not until 1990, when the Japan Aerospace Exploration Agency (JAXA) sent the test mission Hiten back to the Moon was interest in lunar geodesy reinvigorated. In 1994, the United States, in a joint test mission between the Strategic Defense Initiative Organization and NASA, launched the *Clementine Orbiter* to acquire multi-band imagery of the lunar surface. This mission returned a wealth of data sets, including most notably those taken by its ultraviolet/visible, near-infrared, and high-resolution cameras as well as a laser altimeter instrument, which returned a modest amount of topographic data. In 1998, the US sent the *Lunar Prospector* satellite to globally map lunar crustal composition, gravity, and magnetic fields. Since early 2000, there is once again widespread interest in missions to the Moon by Japan, China, India, the United States, and the European Space Agency (ESA). In 2007, the Japanese *Selene* and the Chinese *Chang'e* satellites were successfully launched and are currently returning wonderful images and even high-resolution video. While the previous lunar catalogs, starting in 1960s, were groundbreaking, they will easily be surpassed in quality, accuracy, resolution and quantity by these new missions.

60.4.2 Mars Missions

Beginning in the early 1960s, the United States and Soviet Union launched robotic probes toward Mars. A variety of technical failures doomed most of the Soviet missions and several of the American ones, but the successes gradually revealed a new world. The US *Mariner 4* imaged about 1% of Mars at 1-km resolution in 1965, revealing a heavily-cratered, moon-like surface. In 1967, *Mariners 6* and *7* imaged the entire planet at resolutions of a few km and selected areas at 100 m to 1 km. In 1971, *Mariner 9* became the first probe to orbit another planet successfully (Soviet missions also reached orbit but, unlike Mariner, were unable to delay their observations until a global dust storm, which obscured most of the surface, had subsided). *Mariner 9* revealed enormous volcanoes and canyon systems, as well as smaller channels resembling rivers that raised hopes that Mars had a more Earth-like climate in the past. In 1976, the US Viking mission dispatched two identical orbiters and landers to Mars. These provided over 50,000 images covering the globe at ~300-m resolution, global color and stereo at a few km, and extensive regional coverage at higher resolutions down to 8 m. The *Viking Orbiter* images formed the basis of Mars cartography (and geoscience) for the following two decades (Batson et al. 1995; Kieffer et al. 1992).

The exploration and mapping of Mars resumed in 1997 with the arrival of the *Mars Global Surveyor (MGS)* orbiter, carrying among other instruments, the Mars Orbiter Camera (MOC) with 1.5-m resolution narrow angle and 240-m resolution wide-angle pushbroom scanners, and the Mars Orbiter Laser Altimeter (MOLA) (Malin et al. 1992). Both the cameras and the altimeter have produced data of tremendous value for both cartography and science, but it is the MOLA topography that has truly revolutionized the mapping of Mars. Before its oscillator failed in 2001, the instrument made more than 600 million measurements of the Martian surface with a vertical precision of better than 1 m (Smith et al. 1999). A comparison of the altimetric profiles at the millions of locations where they cross one another allowed errors in the spacecraft orbit to be corrected, and effectively turned each MOLA measurement into a ground control point with an absolute accuracy better than 100 m horizontally and approaching 1 m vertically. The MOLA data set thus provides the ultimate source of geodetic control for mapping with other data.

Additional robotic spacecraft have been dispatched to explore Mars at every opportunity since 1997, with both US and non-US missions planned for the future. The NASA 2001 *Mars Odyssey Orbiter* carries a Thermal Imaging System (THEMIS) that obtains visible images at 18 m/pixel and thermal infrared images at 100 m/pixel (Christensen et al. 2004). Global coverage of infrared images may be obtained by the end of the Odyssey extended mission, with ~50% of the planet imaged in the visible wave lengths, and preliminary planning has already taken place for the production of a new generation of high-resolution global image mosaics from the data. *MGS* and *Odyssey* were joined at the beginning of 2004 by the European Space Agency's *Mars Express* orbiter, which carries a High Resolution Stereo Camera capable of 12 m/pixel imaging with simultaneous three-line stereo and four-filter color at slightly lower resolution (Neukum et al. 2004). In late 2005, the *Mars Reconnaissance Orbiter* (MRO) was successfully launched equipped with a multi-line scanner called HiRISE that is returning images at an amazing 25 cm/pixel resolution of the Martian surface (McEwen et al. 2002).

60.4.3 Other Missions

The Moon and Mars have been emphasized above because of the number of missions and the amount of data returned; however, most other planets, moons, and several minor bodies have also been studied. Missions to Venus, Mercury, and the outer planets have been successfully flown. For example, with its launch in 1989, the *Magellan* satellite mapped 98% of the Venusian surface (Saunders et al. 1992). Because of Venus' thick atmosphere, a synthetic aperture radar instrument was flown to capture the surface. Once the veil was lifted, the images revealed a geologically young surface scarred with large volcanoes and fissure systems. Conversely, as seen during three passes of the *Mariner 10* satellite in the early 1970s, Mercury was determined to be rocky with extremely old cratered surface. Unfortunately, only about 50% of Mercury's surface was captured, and the images failed to reveal a high level of detail concerning how the surface formed. Many of the unanswered issues concerning Mercury will be addressed; in 2004, NASA launched the *Messenger* satellite, which will begin global imaging of the planet in 2011 (Figure 60-3) (Solomon et al. 2001).

Figure 60-3 Artist's concept of the *Messenger* satellite at Mercury. Courtesy: NASA/Johns Hopkins University. See included DVD for color version.

Even though the gas giants, Jupiter, Saturn, Uranus, and Neptune, have no solid surfaces that can be mapped using a GIS, they each have a number of solid (rock or rock-ice) moons. The geologic features of these bodies are enormously diverse and the mapping and study of their surfaces continues to be a very active field of planetary research. One of these bodies, Titan, has an atmosphere denser than Earth's that in January of 2005 was penetrated by the ESA's *Huygens* probe before landing on the surface. Results from this probe and the radar instrument on the *Cassini* satellite show what appear to be an undetermined type of erosional "shoreline" and associated drainage channels near the landing site, as well as channels, dunes, impact craters, and probable volcanoes.

Lastly, several missions have returned data from asteroids and comets. The asteroid Eros was extensively imaged and then soft-landed on by the *Near Earth Asteroid Rendezvous (NEAR)* satellite. In September of 2005, the comet Tempel 1 was also purposely impacted by the Deep Impact mission at a speed of 23,000 mph. During the impact, the main body of the satellite imaged the ejected debris. Other asteroids have been imaged by the *Galileo* spacecraft, and additional comet nuclei have been studied by the Giotto, Vega, Deep Space 1, and Stardust missions.

60.5 Working with Extraterrestrial Data

NASA's Planetary Data System (PDS), since the 1980s, has been chartered to archive and distribute the original raw scientific data from the National Aeronautics and Space Administration (NASA) (McMahon 1996). The archived data sets use an Object Description Language (ODL) to describe document contents. ODL is a simple language made by "keyword = value" pairs and has also been called PVL (Parameter Value Line). Usually, the ASCII encoded ODL label is at the beginning of the simple binary data file or resides in a detached file. Because of the increase in image size and catalog volumes, the PDS has approved the use of the JPEG2000 format. This format meets the minimum requirements of PDS because it is able to use lossless compression methods and the format has an open/nonproprietary specification so software is freely available for decoding.

The PDS archives also contain a limited number of map-projected and derived data sets. Transforming PDS data into a form that is spatially located can be a complex process and usually requires specialized image-processing software like ISIS (Integrated Software for Imagers and Spectrometers) or VICAR (Video Image Communication and Retrieval, Figure 60-4), which are discussed further below. This is important to point out because a GIS system generally requires base images to be spatially registered. Unfortunately, the PDS map-projected data sets are not compatible with most GIS or RS applications. The Planetary Interactive GIS-on-the-Web Analyzable Database (PIGWAD) project, hosted at the United States Geological Survey (USGS), has generated many tutorials and tools to help make this data conversion easier (Hare et al. 2003). The freely available tools can parse the PDS labels and generate GIS-compatible headers for direct reading or conversion to another format. This site also provides many of the PDS map-projected data sets as well as data sets like vector maps (e.g., geology, structural, channels, craters) that are not supported by the PDS for live access and download in a GIS-compatible format. The Map-A-Planet website, also hosted at USGS, was created to also help distribute these PDS map-projected holdings (Garcia et al. 2001). The web interface allows the user to extract a desired subsection of a pre-existing global map and reformat it with the desired output resolution and projection, and offers a minimal GIS referencing via an image worldfile. The future adoption by the PDS of GIS-compatible formats like Geotiff or a geospatial version of JPEG2000 would enhance the utility of Map-A-Planet significantly and the PDS catalogs in general. In the summer of 2008, the HiRISE team was the first to release their map-projected PDS archive using the geoJPEG2000 standard, which essentially uses the same method of the Geotiff standard to incorporate a geospatial header.

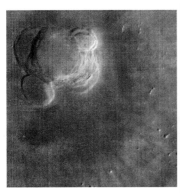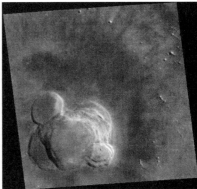

Figure 60-4 The raw MOC frame m0703041 on the left shows Ascraeus Mons caldera. The image on the right has been radiometrically and cosmetically corrected and map projected to a sinusoidal projection. Courtesy: NASA/JPL/MSSS.

60.5.1 Cartographic and Geodetic Software

As stated above, the bulk of the PDS data holdings are cataloged in the original raw instrument space. Thus to use these data sets in cartographic applications they must first be spatially referenced (the equivalent, for bodies other than Earth, of being "georeferenced"). The ISIS and VICAR image processing systems were created to do just that and excel at radiometrically, cosmetically, and geometrically correcting extraterrestrial images and combining them into mosaics (Figure 60-4).

In the early 1990s, ISIS grew directly from the Planetary Image Cartography System (PICS) (Torson and Becker 1997; Gaddis et al. 1997; Anderson et al. 2004). Both ISIS and PICS were developed by the Branch of Astrogeology at the USGS. To list a few examples, ISIS is currently used to process images from the Mars Odyssey THEMIS, Mars Global Surveyor MOC imagery, to revive the old Lunar Orbiter images, and it will be used for upcoming MRO and Messenger missions.

The VICAR image processing system began in 1966 at the Jet Propulsion Laboratories (JPL) (Stanfill and Girard 1986). This system processed data from missions like Ranger, Surveyor, Mariner, Viking and Voyager. It has now been upgraded and ported to nearly all modern operating systems and is still used by many facilities to process data for Earth, Mars and other bodies. VICAR is currently being used for the Cassini mission and the Mars Express HRSC instrument.

It should be noted that the use of ISIS, VICAR, or other software to transfer image data to a map-projected space does not guarantee that the resulting map coordinates of features are strictly accurate. Positional accuracy depends on the ability to measure or reconstruct the position of the spacecraft and the pointing of the instrument as each observation was taken, as well as on knowledge of the shape and rotation of the target body. The predicted (as opposed to reconstructed) position and pointing data that are sometimes used to make uncontrolled mosaics can be wildly erroneous, especially for pre-1990s missions, leading to maps with visible mismatches at image boundaries and systematic position errors of many kilometers. Both ISIS and VICAR are, therefore, used with associated photogrammetric software that estimates improved position, pointing, and planetary parameters based on measurements of matching ground features where images overlap. This process of geodetic control reduces positional errors but cannot, of course, eliminate them entirely. In fact, the lack of ground control for most planets (laser altimetry data now provides an effective substitute for Mars) leads to generally greater errors than are encountered in mapping the Earth. An initial lack of topographic data for each planet or satellite introduces further positional errors in the early maps, because parallax distortions cannot be corrected for. This is no longer an issue for Mars, where images are now routinely orthorectified by using the MOLA global altimetry data set, but it is still a problem for other bodies.

Both ISIS and VICAR can be used for scientific analysis and geologic mapping in addition to their primary functions of producing digital map bases, but in recent times such analysis functions have mainly been handed over to more capable GIS and RS applications. We note, however, that in 1975 a VICAR-derived GIS system called the Image-Based Information System (IBIS) was developed at JPL (Zobrist and Bryant 1980). It was designed to be a comprehensive GIS that worked with raster imagery, and tabular and vector data (originally called graphics). The IBIS system proved to be very useful for many diverse applications, such as image rubbersheeting and mosaicking, multispectral classification, stereo-matching and other cartographic applications. Unfortunately, IBIS and the successor IBIS-2 are no longer supported.

60.5.2 Streamed Data Distribution

Since the explosion of Internet usage in the 1990s, the planetary community has embraced it as a method of data delivery. Before wide-spread Internet availability, data sets were archived on volumes of CDs or DVDs and distributed to libraries and researchers. Now PDS data sets are almost exclusively available by online methods. This largely includes the use of File Transfer Protocol (FTP) sites and other download protocols. Unfortunately, like the Earth mapping community, planetary researchers are quickly facing huge data volumes that may be extremely hard to even offer as downloads. Thus many organizations have been researching other alternatives and there are currently many excellent planetary online mapping sites. These resources allow the client to make use of the spatially co-registered data without the need to purchase commercial software or download large data sets. Most sites also have a set of specialized tools used to query the data. Nearly all planetary facilities, US and non-US alike, are producing these specialized applications. However, the GIS capabilities provided by most of these sites are limited, and the data must still be downloaded and converted into a compatible format in order to perform more advanced analyses. An ultimate goal for all these facilities is to ensure their websites are interoperable for cartographic applications. This will allow the end user to load different co-registered layers from different facilities into a single application. A term used by the OGC describes this process as "geo-enabling" the web.

Some of the planetary facilities like Arizona State University (ASU), JPL, Ohio State University (OSU) and United States Geological Survey (USGS) publicly support extraterrestrial online mapping via the OGC's Web Mapping Service (WMS) online mapping protocol or Environmental Systems Research Incorporated's (ESRI) ArcIMS software (Gorelick et al. 2003; Hare and Tanaka 2004). For example, currently the ASU, JPL and USGS sites may be viewed independently or co-registered in a single application (Figure 60-5). There are still some hurdles to overcome to better support planetary data sets, as described in the OGC section above, but the technology is quickly maturing. Other facilities that are associated with ESA, JAXA, and NASA have also expressed interest in supporting the technology and most likely already have servers running. Additional OGC technologies that allow more robust data analysis such as Web Feature Services (WFS) or Web Coverage Servers (WCS) will certainly also be supported in the near future.

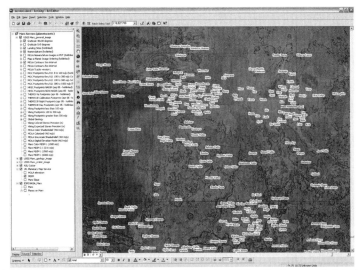

Figure 60-5 An example of the application ArcMap, by ESRI, with streamed data layers from USGS, ASU, JPL, and ESRI Mars servers. The layers displayed are a Mars image base from JPL's onMars server, thermal inertia data set from ASU's JMars server, and geology, MOLA 1-km contours, and nomenclature from the USGS server. See included DVD for color version.

60.6 Example Use Cases

60.6.1 Mapping

60.6.1.1 Geologic

GIS is increasingly used in geologic mapping and topical studies of planetary surfaces, as data sets become spatially registered in GIS format and planetary scientists become more familiar with GIS tools (Figure 60-2, Figure 60-6). Completed and ongoing formal geologic studies incorporating GIS thus far encompass Mars, the Moon, Venus, and Jupiter's satellites Europa, Ganymede, and Io (Dohm et al. 2001; Gregg et al. 2004, 2005). GIS provides analytical tools that can assist in interpretations based on spatially-based measurements and associations of geologic materials and structures (e.g., Byrne and Murray 2002; Dohm and Tanaka 1999; Kolb and Tanaka 2001; Tanaka et al. 1998). Planetary photogeologic mapping techniques are being tested and optimized using GIS on terrestrial sites; such development may in turn benefit photogeologic mapping studies on Earth using spatial data sets acquired from airborne and spacecraft platforms (Tanaka et al. 2004).

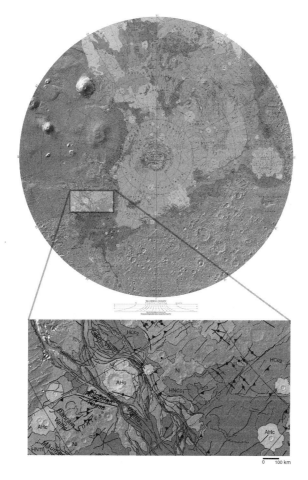

Figure 60-6 Geologic map of the northern plains of Mars from the USGS Geologic Investigations Series Map I-2888. This is the first published map that covers a significant part of Mars using topography and image data from both the Mars Global Surveyor and Mars Odyssey missions. The entire map was compiled using GIS (Tanaka et al. 2005). The area zoomed in shows the full detail of the digital map. See included DVD for color version.

60.6.1.2 Impact Crater Studies

The record left by impact craters on planetary surfaces, Earth included, provides important constraints concerning the early evolution of the Solar System as well as a systemic catalog for examining the age and geologic character of the impacted surface. Crater densities are fundamental for determining the surface age of geologic units on the Moon, Mars, and other planetary bodies, based on an assumed population of crater-forming bodies as a function of time since the formation of the Solar System. As such, size-frequency distributions provide a proxy for the age of a given planet's surface. GIS-based studies allow rapid statistical delineation of crater populations based on spatial associations and groupings. For example, by identifying a population of degraded craters and a spatially-correlative population of "pristine" craters, researchers are able to assess the age, intensity, and in some cases, the duration of surface processes. Moreover, GIS-based analysis provides a means to determine the geologic character of the impacted surface. For example, the size-frequency distribution of fluidized crater ejecta compared to ballistic (or dry) crater ejecta provides a proxy for assessing the distribution of subsurface ground-volatiles. Multi-data set statistics performed within a GIS allow repeatable, documented geologic and temporal associations that are critical to assessing the history of planetary surfaces (Barlow et al. 2003; Hartmann 1977; Tanaka 1986).

60.6.1.3 Topography

Accurate topographic information is of interest at all phases of planetary exploration and scientific investigation, from landing site selection to the quantitative analysis of the morphologic record of surface processes. The availability of extraterrestrial topographic data has unfortunately been limited in the past, but the situation is rapidly changing.

To gather digital elevation models (DEMs), the planetary community has had a long history of using photogrammetric techniques (Wu 2005). Stereotopographic mapping of the Moon, Mars, Mercury, Venus, several of the large satellites of the outer planets, the asteroid Eros, and the comets Borelly and Wild 2 has been accomplished with photogrammetric software. However, novel procedures must frequently be developed to deal with problems of planetary data sets, such as the need to use large numbers of small images, nonuniform image coverage, poor image overlap, and lack of true ground control (Kirk, Rosiek et al. 2001; Rosiek et al. 2000). Some sensors, such as the Magellan Synthetic Aperture Radar (SAR) and Mars Global Surveyor Mars Orbiter Camera (MOC), also require the development of specialized sensor model software (Howington-Kraus et al. 2001; Kirk, Howington-Kraus et al. 2001). When stereo images are not available, photoclinometry or shape-from-shading techniques have also been used to derive topographic data sets (Kirk et al. 2003). Laser (and to a much lesser extent, radar) altimetry systems are increasingly important as sources of global topographic data. As discussed previously, the Mars Orbiter Laser Altimeter (MOLA) data set provides absolute horizontal as well as vertical coordinates for Mars with unprecedented precision. The Lunar Orbiter Laser Altimeter that will be flown in 2008 is expected to provide an equally revolutionary advance in mapping of the Moon. Image-based topomapping techniques nonetheless continue to be valuable because of their ability to fill in topographic details at higher spatial resolution than can be provided by altimetry. The Mars Express HRSC camera, which is the first instrument designed specifically for stereoscopic imaging of Mars, has eliminated many of the difficulties of topomapping with non-dedicated camera systems and is likely to collect global stereo image coverage at a resolution substantially better than 50 m.

Where topographic data are available, standard GIS tools can be applied to generate slope and aspect maps, contours, hillshades, watersheds, drainage delineations, viewshed (line-of-sight analysis), 3D visualization, and volume calculations (e.g., Grant et al. 2003; Hynek et al. 2003; Fenton 2003). The MOLA Mars topography and these derived data sets have even been used as the main mapping base for a recent geologic map of the northern plains of Mars (Figure 60-6) (Tanaka et al. 2005).

60.6.1.4 3D Visualization

3D visualization is an important aspect of planetary studies when surface visual reference is unavailable. With the exception of the few lander and manned missions to the Moon and the Martian lander and rover missions, there is no other way to visualize features from the surface. GIS applications now easily provide the power to visualize planetary surfaces whether it is a simple extruded image (Figure 60-7), a 3D Globe (e.g., NASA's World Wind application; Figure 60-8), or even an immersive 3D cave-like environment. One such cave-based planetary visualization system, called Advanced Visualization in Solar System Exploration and Research (ADVISER), currently being developed at Brown University, promises an immersive scientific visualization environment for Mars research (Figure 60-9) (Head et al. 2005).

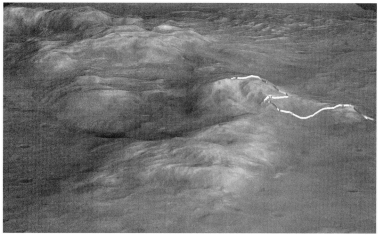

Figure 60-7 A perspective view of the (informally named) Columbia Hills at the MER Spirit landing site. The sinuous line displays the path of the Spirit rover, as defined by OSU, to the top of the approximately 100-m-tall hill in the foreground. The draped image is MOC frame e0300012. The DEM was photogrammetrically derived by the USGS using MOC frames e0300012 and r0200357. A 1.5 Z exaggeration is set for the scene.

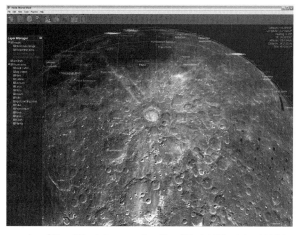

Figure 60-8 An example of the 3D globe application World Wind, by NASA, with the streamed Lunar Clementine base map and nomenclature layers turned on. The globe is centered on the crater Tycho. This application also demonstrates the usefulness of WMS and WMS-like technologies for streaming large planetary data sets. See included DVD for color version.

Manual of Geographic Information Systems

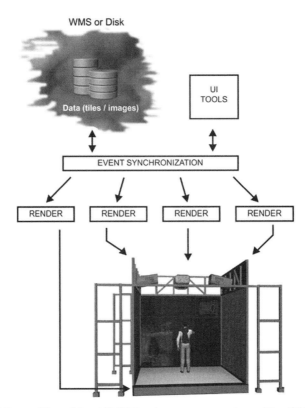

Figure 60-9 Artist's rendition of the ADVISER system components and the CAVE immersive virtual reality system. Courtesy: Brown University.

60.6.1.5 Automatic Feature Extraction

Automatic feature extraction is ideally suited for planetary data sets. Planetary surfaces generally change very little except over long geologic time scales, and they lack vegetation. There are many interesting techniques currently being developed for use on planetary data (Brumby et al. 2003). One such application involves refining techniques to automatically recognize impact craters (Bue and Stepinski 2007; Kim and Muller 2003). Another application combines slope and aspect functions to help identify wrinkle ridges to study potential subsurface ice deposits (Frigeri et al. 2004). Automatic identification of other feature types such as volcanoes has been suggested as a research goal, but can be very challenging because of the diversity of original forms and degradation states of such complex features.

60.6.2 Mission Support

60.6.2.1 Landing Site Selection

In 2001, the *Mars Exploration Rover (MER) GIS CD-ROM* was created at the USGS (Hare and Tanaka 2001). The included application was used to analyze potential landing sites for the rovers. It included topography, base imagery, thermal inertia, and geologic data sets, among other layers. It allowed the user to graphically choose candidate landing sites and would return information for all the layers that were necessary to meet the engineering constraints for safely landing rovers. For example, the MOLA altimetry would be used to determine if the elevation was low enough to allow the descent vehicle to slow down. Sites might also be excluded if the thermal inertia, which is a proxy for rock abundance, was too low (indicating very dusty conditions) or too high (indicating an area too rocky for a safe landing and/or for rover trafficability). Figure 60-10 shows a simple GIS flow-diagram or landing site suitability model that encapsulates this process. Another GIS-like application, Mars-O-Web, hosted at NASA AMES, gives scientists the ability to query and analyze the Mars data sets via an online interface in much the same way. It includes several online querying, profiling, and 3D visualization tools (Gulick and Deardorff 2003).

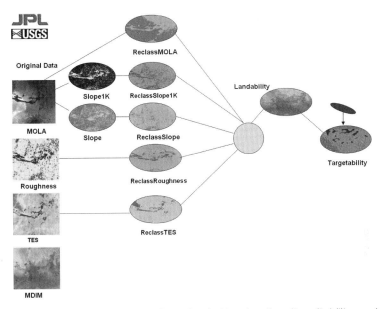

Figure 60-10 An example flow diagram for a simple Mars landing site suitability model. This type of modeling process will be used in selecting sites for future extraterrestrial landers and rovers. Courtesy: JPL/USGS. See included DVD for color version.

60.6.2.2 Real-time Robotic Mission Support

Because of the ability to quickly add and change data layers, GIS has begun to play an important role in supporting real-time mission operations. Ohio State University (OSU) has implemented an online GIS interface to track the two MER rovers as they move across the surface (Figure 60-11) (Li et al. 2004). Another such application, called JMars, by Arizona State University, was originally conceived to help researchers plan Mars observations for the THEMIS instrument aboard the *Mars Odyssey* satellite (Gorelick et al. 2003). However, the application was modified to run in two modes, targeting and map projected. Both modes have multiple data layers with which the user can interact. When in the map projection mode, JMars basically functions like a GIS application but uses streamed data sets. Another streaming application, created at the USGS, is helping to coordinate the Cassini Orbiter instrument teams for the limited number of available *Titan* fly-bys. It allows the teams to visualize the planned image footprints for each instrument and coordinate targeting changes to optimize science returns. The company Applied Coherent Technology Corporation (ACT) has also supported many planetary missions with targeting and cartographic tools. Their Rapid Environmental Assessment Composition Tools (REACT) is a GIS at its core and has abilities to overlay, process, and analyze multiple data sets.

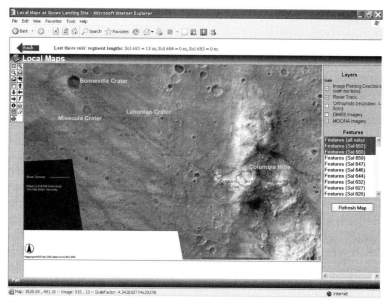

Figure 60-11 The online interface for the MER Spirit rover, implemented by the OSU Mapping and GIS Laboratory. The data sets are rapidly updated during the mission so that the researchers have access to the latest traverse information and linked images. Courtesy: Ohio State University. See included DVD for color version.

60.6.2.3 Future Manned Mission Support

The push to return humans to the Moon and eventually Mars is already underway. NASA's Constellation Program (CxP) was created to meet these goals and consists of a subtask called the Lunar Mapping and Modeling Project (LMMP). Under LMMP, GIS tools and GIS-compatible data sets will be used and created to support surface operation needs including CxP planning and real-time operations. The GIS infrastructure developed by LMMP will also support the planetary science community, educational facilities and the general public. There is a strong requirement that all products utilize standard formats, which will be made available for download, but also made available via streamable GIS web protocols, as defined by the OGC, such that the data can be easily consumed by a host of GIS applications. Some of the data sets that LMMP will be responsible for include controlled global albedo (visible image), topographic, temperature, gravity, geochemical, and mineralogical bases to name a few (Cohen et al. 2008).

60.7 Conclusion

The future will see growing uses of GIS technologies and spatial analysis in extraterrestrial research. As discussed above, these tools are being used in a wide variety of research, mission planning, and mission support tasks. Increasingly, specialists outside the realm of planetary geologic mapping are also expressing interest in using GIS technology, for example, astrobiologists, planetary protection experts, and planetary hardware designers, who are striving to develop more autonomous robots that can catalog and analyze multiple types of information without human interaction. The volume of data now in hand is already difficult for planetary researchers to keep up with, and new missions to the Moon, Mars, and beyond are planned by multiple countries or are already on their way. The volume and complexity of the data sets that will come from these missions guarantee that the exploitation of GIS technologies to assist extraterrestrial research has barely begun.

Acknowledgments

We are grateful to the following people for discussions, edits or figures prior to publication: S. Byrne, D. Galuszka, J. Richie, J. Blue, L. Hare, J. Head, R. Li, L. Plesea, E. Dobinson, D. Curkendall, M. Rosiek, B. Redding, B. Archinal, A. Howington-Kraus, N. Gorelick, V. Gulick, E. Kolb, C. Fortezzo, F. Chuang, A. Frigeri, W. Fink, and J. Dohm, who always recognized the significance of GIS for planetary research. We must also thank those in the planetary mapping community who continue to support advances in the field and keep the planetary GIS discussion site active with new topics and issues.

References

Anderson, J. A., S. C. Sides, D. L. Soltesz, T. L. Sucharski and K. J. Becker. 2004. Modernization of the integrated software for imagers and spectrometers. *Lunar Planet Science Conference XXXI.* Lunar and Planetary Institute, Houston, Tex., Abstract 2039 (CD-ROM). <http://isis.astrogeology.usgs.gov/> Accessed 11 September 2008.

Barlow, N. G., C. W. Barnes, O. S. Barnouin-Jha, J. M. Boyce, C. R. Chapman, F. M. Costard, R. A. Craddock, J. B. Garvin, R. Greeley, T. M. Hare, R. O. Kuzmin, P. J. Mouginis-Mark, H. E. Newsom, S.E.H. Sakimoto, S. T. Stewart and L. A. Soderblom. 2003. Utilizing GIS in Martian impact crater studies. *International Society for Photogrammetry and Remote Sensing Working Group IV/9: Extraterrestrial Mapping Workshop, "Advances in Planetary Mapping 2003."* 22 March 2003, Lunar and Planetary Institute, Houston, Tex.

Batson, R. M. and E. M. Eliason. 1995. Digital maps of Mars. *Photogrammetric Engineering & Remote Sensing* 6: 1499–1507.

Brumby, S. P., C. S. Plesko and E. Asphaug. 2003. Evolving automated feature extraction algorithms for planetary science. *International Society for Photogrammetry and Remote Sensing Working Group IV/9: Extraterrestrial Mapping Workshop, "Advances in Planetary Mapping 2003."* 22 March 2003, Lunar and Planetary Institute, Houston, Tex.

Bue, B. D. and T. F. Stepinski. 2007. Machine detection of Martian impact craters from digital topography data. Pages 265–274 in *IEEE Transactions on Geoscience and Remote Sensing,* Vol. 45, Issue 1.

Byrne, S. and B. C. Murray. 2002. North polar stratigraphy and the paleo-erg of Mars. *Journal of Geophysical Research (Planets)* 107 (E6).

Carr, M. H. 1995. The Martian drainage system and the origin of networks and fretted channels. *Journal of Geophysical Research* 100:7479–7507.

Christensen, P. R., B. M. Jakosky, H. H. Kieffer, M. C. Malin, H. Y. McSween Jr., K. Nealson, G. L. Mehall, S. H. Silverman, S. Ferry, M. Caplinger and M. Ravine. 2004. The Thermal Emission Imaging System (THEMIS) for the Mars 2001 Odyssey Mission. *Space Science Review* 110:85–130.

Cohen, B. A., M. E. Nall, R. A. French, K. G. Muery and A. R. Lavoie. 2008. The lunar mapping and modeling project (LMMP). *Lunar Planet Science Conference XXIX.* Lunar and Planetary Institute, Houston, Tex., Abstract 1640.

Dohm, J. M. and K. L. Tanaka. 1999. Geology of the Thaumasia region, Mars: Plateau development, valley origins, and magmatic evolution. *Planetary & Space Science* 47:411–431.

Dohm, J. M., K. L. Tanaka and T. M. Hare. 2001. *Geologic, Paleotectonic, and Paleoerosional Maps of the Thaumasia Region, Mars.* USGS Geologic Investigations Series Map I-2650.

EPSG. Coordinate Reference System Definition, European Petroleum Survey Group. <http://www.epsg.org/> Accessed 11 September 2008.

Fenton, L. K. 2003. Aeolian processes on Mars: Atmospheric modeling and GIS analysis. Ph.D. diss., California Institute of Technology, Pasadena, California. <http://resolver.caltech.edu/CaltechETD:etd-03052003-124751> Accessed 11 September 2008.

Frigeri, A., C. Federico, C. Pauselli and G. Minelli. 2004. Identifying wrinkle ridges from Mars MGS and Viking mission data: Using GRASS GIS in planetary geology. *Transactions in GIS* 8 (2).

Gaddis, L. R., J. Anderson, K. Becker, T. Becker, D. Cook, K. Edwards, E. Eliason, T. Hare, H. Kieffer, E. M. Lee, J. Mathews, L. Soderblom, T. Sucharski and J. Torson. 1997. An overview of the integrated software for imaging spectrometers. *Lunar Planetary Science Conference XXVIII.* Lunar and Planetary Institute, Houston, Tex., Abstract 1226 (CD-ROM). <http://isis.astrogeology.usgs.gov/> Accessed 11 September 2008.

Garcia, P. A., E. M. Eliason and J. M. Barrett. 2001. Creating cartographic image maps on the web using PDS MAP-A-PLANET. *Lunar and Planetary Science Conference XXXII.* Lunar and Planetary Institute, Houston, Tex., Abstract 2046 (CD-ROM).

Grant, J. A. and C. Fortezzo. 2003. Hypsometric analyses of martian basins. *Lunar Planetary Science Conference XXXIV.* Lunar and Planetary Institute, Houston, Tex., Abstract 1123 (CD-ROM).

Greeley, R. and R. M. Batson, eds. 1990. *Planetary Mapping.* Cambridge, UK: Cambridge University Press.

Gregg, T.K.P., K. L. Tanaka and R. S. Saunders, eds. 2004. *Abstracts of the Annual Meeting of Planetary Geologic Mappers.* Flagstaff, Ariz., US Geological Survey Open-File Report 2004-1289. <http://pubs.usgs.gov/of/2004/1289/> Accessed 11 September 2008.

————. 2005. *Abstracts of the Annual Meeting of Planetary Geologic Mappers.* Washington, D.C., US Geological Survey Open-File Report 2005-1271. <http://pubs.usgs.gov/of/2005/1271/> Accessed 11 September 2008.

Gorelick, N. S., M. Weiss-Malik, B. Steinberg and S. Anwar. 2003. JMARS: A multimission data fusion application. *Lunar Planetary Science Conference XXXIV.* Lunar and Planetary Institute, Houston, Tex., Abstract 2057 (CD-ROM).

Gulick, V. C. and D. G. Deardorff. 2003. Mars Data Visualization and E/PO with Marsoweb. *Lunar and Planetary Science Conference XXXIV.* Lunar and Planetary Institute, Houston, Tex., Abstract 2081 (CD-ROM).

Hall, R. C. 2006. Lunar Impact: A History of Project Ranger. NASA SP-4210. <http://history.nasa.gov/SP-4210/pages/Cover.htm> Accessed 11 September 2008.

Hare, T. M. and K. L. Tanaka. 2001. Planetary Interactive GIS-on-the-Web Analyzable Database – PIGWAD. *International Cartographic Conference,* Beijing, China, abstract (CD-ROM). <http://webgis.wr.usgs.gov> Accessed 11 September 2008.

————. 2004. Expansion in geographic information services for PIGWAD. *Lunar Planet Science Conference XXXV.* Lunar and Planetary Institute, Houston, Tex., Abstract 1765 (CD-ROM).

Hare, T. M., J. M. Dohm and K. L. Tanaka. 1998. GIS and its application to planetary research. *Lunar Planet Science Conference XXXI.* Lunar and Planetary Institute, Houston, Tex., Abstract 515 (CD-ROM). <http://webgis.wr.usgs.gov> Accessed 11 September 2008.

Hare, T. M., K. L. Tanaka and J. A. Skinner Jr. 2003. GIS 101 for planetary research. *International Society for Photogrammetry and Remote Sensing Working Group IV/9: Extraterrestrial Mapping Workshop, "Advances in Planetary Mapping 2003."* 22 March 2003, Lunar and Planetary Institute, Houston, Tex. <http://astrogeology.usgs.gov/Projects/ISPRS/MEETINGS/Houston2003/abstracts/Hare_isprs_mar03.pdf> Accessed 11 September 2008.

Hartmann, W. K. 1977. Cratering in the solar system. *Scientific American* 236:84–99.

Heacock, R. L., G. P. Kuiper, E. M. Shoemaker, H. C. Urey and E. A. Whitaker. 1965. *Ranger VII, Part II, Experimenters' Analyses and Interpretations.* Technical Report No. 32-700. Pasadena, Calif.: Jet Propulsion Laboratory.

Head III, J. W., A. van Dam, S. G. Fulcomer, A. Forsbeg, Prabhat, G. Rosser and S. Milkovich. 2005. ADVISER: Immersive scientific visualization applied to Mars research and exploration. *Photogrammetric Engineering & Remote Sensing* October 2005 71 (10):1219–1225.

Howington-Kraus, E., R. L. Kirk, D. Galuszka, T. M. Hare and B. Redding. 2001. Validation of the USGS sensor model for topographic mapping of Venus using Magellan radar stereoimagery. *International Society for Photogrammetry and Remote Sensing Working Group IV/9: Extraterrestrial Mapping Workshop "Planetary Mapping 2001."* November 2001, virtual workshop.

Hynek, B. M. and R. J. Philips. 2003. New data reveal mature, integrated drainage systems on Mars indicative of past precipitation. *Geology* 31 (9):757–760.

Kieffer, H. H., B. M. Jakosky, C. W. Snyder and M. S. Matthews, eds. 1992. *Mars.* Tucson, Ariz.: University of Arizona Press.

Kim, J. and J. Muller. 2003. Automated impact crater detection on images and DEMs. *International Society for Photogrammetry and Remote Sensing Working Group IV/9: Extraterrestrial Mapping Workshop, "Advances in Planetary Mapping 2003."* 22 March 2003, Lunar and Planetary Institute, Houston, Tex.

Kirk, R. L., E. Howington-Kraus and B. A. Archinal. 2001. High resolution digital elevation models of Mars from MOC narrow angle stereoimages. *International Society for Photogrammetry and Remote Sensing Working Group IV/9: Extraterrestrial Mapping Workshop "Planetary Mapping 2001."* November 2001, virtual workshop.

Kirk, R. L., M. Rosiek, E. Howington-Kraus, E. Eliason, B. Archinal and E. Lee. 2001. Planetary geodesy and cartography at the USGS, Flagstaff: Moon, Mars, Venus, and beyond. *International Cartographic Conference,* Beijing, China, abstract (CD-ROM).

Kirk, R. L., J. M. Barrett and L. A. Soderblom. 2003. Photoclinometry made simple…? *International Society for Photogrammetry and Remote Sensing Working Group IV/9: Extraterrestrial Mapping Workshop, "Advances in Planetary Mapping 2003."* 22 March 2003, Lunar and Planetary Institute, Houston, Tex. <http://astrogeology.usgs.gov/Projects/ISPRS/MEETINGS/Houston2003/abstracts/Kirk_isprs_mar03.pdf> Accessed 11 September 2008.

Kolb, E. J. and K. L. Tanaka. 2001. Geologic history of the polar regions of Mars based on Mars Global Surveyor data: II. Amazonian Period. *Icarus* 154 (1):22–39.

Li, R., K. Di, F. Xu and J. Wang. 2004. Landing site mapping and rover localization for the 2003 Mars Exploration Rover mission: Technology and experimental results. *International Archives of Photogrammetry & Remote Sensing,* XXXV, B, Geo-Imagery Bridging Continents, Istanbul (DVD-ROM).

Malin, M. C., G. E. Danielson, A. P. Ingersoll, H. Masursky, J. Veverka, M. A. Ravine and T. A. Soulanille. 1992. The Mars Observer camera. *Journal of Geophysical Research* 97(E5):7699–7718.

Mardon, A. A. 1992. Utilization of geographic information system in lunar mapping. Pages 37–39 *Lunar and Planetary Institute.* Joint Workshop on New Technologies for Lunar Resource Assessment. 6 April 1992, Santa Fe, N.M.

McEwen, A. S., W. A. Delamere, E. M. Eliason, J. A. Grant, V. C. Gulick, C. J. Hansen, K. E. Herkenhoff, L. Keszthlyi, R. L. Kirk, M. T. Mellon, S. W. Squyres, N. Thomas and C. Weitz. 2002. HiRISE: The High Resolution Imaging Science Experiment for Mars Reconnaissance Orbiter. *Lunar Planet Science Conference XXXI.* Lunar and Planetary Institute, Houston, Tex.

McMahon, S. K. 1996. Overview of the Planetary Data System. *Planetary Space Science* 44 (1):3–12.

Neukum, G., R. Jaumann and the HRSC Co-Investigator and Experiment Team. 2004. HRSC-The High Resolution Stereo Camera of Mars Express. *European Space Agency Special Publication,* ESA SP-1240, 17–35.

OGC. FAQ's - OGC's Purpose and Structure. <http://www.opengeospatial.org/resources/?page=faq> Accessed 11 September 2008.

Ritter, N. and M. Ruth. 2000. GeoTIFF Format Specification, Revision 1.0, Version 1.8.2, December 2000. <http://www.remotesensing.org/geotiff/spec/geotiffhome.html> Accessed 11 September 2008.

Roddy, D. J., N. R. Isbell, C. L. Mardock, T. M. Hare, M. B. Wyatt, L. M. Soderblom and J. M. Boyce. 1998. I. Martian impact craters, ejecta blankets, and related morphologic features: Computer digital inventory in Arc/Info and Arcview format. *Lunar Planet Science Conference XXXI.* Lunar and Planetary Institute, Houston, Tex., Abstract 1874 (CD-ROM).

Rosiek, M. R., R. L. Kirk and E. Howington-Kraus. 2000. Digital elevation models derived from small format lunar images. *American Society for Photogrammetry and Remote Sensing Conference,* held 22–26 May 2000, Washington, D.C. (CD-ROM).

Saunders, R. S., A. J. Spear, P. C. Allin, R. S. Austin, A. L. Berman, R. C. Chandlee, J. Clark, A. V. DeCharon, E. M. DeJong, D. G. Griffith, J. M. Gunn, S. Hensley, W.T.K. Johnson, C. E. Kirby, K. S. Leung, D. T. Lyons, G. A. Michaels, J. Miller, R. B. Morris, A. D. Morrison, R. G. Piereson, J. F. Scott, S. J. Shaffer, J. P. Slonski, E. R. Stofan, T. W. Thompson and S. D. Wall. 1992. Magellan mission overview. *Journal Geophysical Research* 97 (E8):13067–13090.

Seidelmann, P. K., V. K. Abalakin, M. Bursa, M. E. Davies, C. De Bergh, J. H. Lieske, J. Oberst, J. L. Simon, E. M. Standish, P. Stooke and P. C. Thomas. 2002. Report of the IAU/IAG Working Group on cartographic coordinates and rotational elements of the planets and satellites: 2000. *Celestial Mechanics and Dynamical Astronomy* 82:83–110.

Smith, D., G. Neumann, P. Ford, R. E. Arvidson, E. A. Guinness and S. Slavney. 1999. Mars global surveyor laser altimeter precision experiment data record. *NASA Planetary Data System,* MGS-M-MOLA-3-PEDR-L1A-V1.0.

Solomon, S. C., R. L. McNutt Jr., R. E. Gold, M. H. Acuña, D. N. Baker, W. V. Boynton, C. R. Chapman, A. F. Cheng, G. Gloeckler, J. W. Head III, S. M. Krimigis, W. E. McClintock, S. L. Murchie, S. J. Peale, R. J. Phillips, M. S. Robinson, J. A. Slavin, D. E. Smith, R. G. Strom, J. I. Trombka and M. T. Zuber. 2001. The MESSENGER mission to Mercury: Scientific objectives and implementation. *Planetary Space Science* 49:1445–1465.

Stanfill, D. and M. Girard. 1986. An Introduction to the VICAR Image Processing Executive, JPL Document D-4309. <http://www-mipl.jpl.nasa.gov/external/vicar.html> Accessed 11 September 2008.

Tanaka, K. L. 1986. The stratigraphy of Mars. *Proceedings of the 17th Lunar and Planetary Science Conference,* Part 1. *Journal of Geophysical Research* 91, supplement, E139-E158.

Tanaka, K. L., J. M. Dohm, J. H. Lias and T. M. Hare. 1998. Erosional valleys in the Thaumasia Region of Mars: Hydrothermal and seismic origins. *Journal of Geophysical Research* 103:31,407–31,419.

Tanaka, K. L., L. S. Crumpler, J. M. Dohm, T. M. Hare and J. A. Skinner Jr. 2004. Assessing photogeologic mapping techniques in reconstructing the geologic history of Mars. *Lunar Planetary Science Conference XXXV.* March 2004, Lunar and Planetary Institute, Houston, Tex., Abstract 2109 (CD-ROM).

Tanaka, K. L., J. A. Skinner Jr. and T. M. Hare. 2005. *Geologic Map of the Northern Plains of Mars.* US Geologic Survey, Science Investigation Map SIM-2888.

Torson, J. M. and K. J. Becker. 1997. ISIS - A software architecture for processing planetary images. Pages 1443–1444 in *Lunar Planetary Science Conference XXVIII*. Lunar and Planetary Institute, Houston, Tex. <http://isis.astrogeology.usgs.gov/> Accessed 11 September 2008.

Wu, S.S.C. 2005. Extraterrestrial mapping. Pages 1063–1090 in *Manual of Photogrammetry, 5th ed.* Edited by J. C. McGlone. Bethesda, Md.: American Society for Photogrammetry and Remote Sensing.

Zobrist, A. L. and N. A. Bryant. 1980. Designing an Image Based Information System. Pages 177–197 in *Pictorial Information Systems: Lecture Notes in Computer Science*. Vol. 80. Edited by Shi-Kuo Chang and King-sun Fu. Berlin/Heidelberg: Springer.

CHAPTER 61
GIScience in Archeology: Ancient Human Traces in Automated Space

Minna A. Lönnqvist and *Emmanuel Stefanakis*

61.1 Introduction

61.1.1 Dimensions of Archeology: Humans, Space and Time

Archeology, *archaios logos,* is the "discipline of antiquity" that involves studying past materials and traces left by humans. All archeological *finds* are connected with humans, space and time. In the US and in a number of other countries, archeology is a subfield of anthropology. Any object that has been constructed or modified by a human being is an *artifact,* while *ecofacts* are natural remains connected with humans. Besides artifacts and ecofacts, archeologists search and study *structures* and *features,* such as stains, footprints, trails and tracks left by humans on the ground (see, e.g., Clarke 1978; Renfrew and Bahn 2000).

Archeology is a mixture of art and science. It is a relatively young discipline that emerged from the collection of curiosities along with the Renaissance. Pompeii, an ancient Roman town buried by a volcanic eruption of Vesuvius in A.D. 79 in Italy, is considered to be the site of the world's oldest archeological dig; since 1748 it has continued over two hundred years without a break (Slayman 1997). After the first excavations, Napoleon's invasion and expedition to Egypt in 1798 produced a monumental work known as *Description of Egypt* (*Description de l'Égypte*). It is a series of books in which observations of ancient monuments are connected with data collected by natural scientists (Figure 61-1). The work of the expedition also reflects the development of archeology as a discipline that actually acts on the border of human socio-cultural studies and natural sciences. In archeology this border and the emphasis of either is flexible and constantly changing, depending on the perspectives of scientific schools and approaches.

Figure 61-1 An illustration of the Giza pyramids and the sphinx from Napoleon's expedition work *Description of Egypt,* Second Edition, Paris 1829, Vol. V, Table 8: Giza. The Giza pyramids and the sphinx are archeological monuments that have always been known and have stood the test of time for thousands of years. Note that the famous sphinx was partly buried under the sand before its body was exposed by excavation. The sphinx has been recorded, documented, studied and conserved by the American Research Centre (ARCE) in Egypt.

Methods and experiments of field archeology have been largely adapted and modified from geology, geography, anthropology, biology, chemistry and physics. For example, archeology uses cartographic coordinate systems and three-dimensional geometric recordings of site and "find" locations. Maps provide a major tool to visually display find/site locations, their surroundings and topology in archeology. Archeologists make conclusions from the dispersion, distribution and patterns of recovered remains. Distribution maps, now often in digital or electronic form, are simple and traditional ways in archeology to display finds/sites and their locations. In addition, however, spatial analyses help to answer particular questions concerning the locations and relations of finds in space. Through computerization of archeology, GIS (Geographic Information Systems) have come to play a key role in collecting, storing, visualizing and analyzing the spatial dimensions of materials and traces left by humans in the past. GIS, archeology and digital technology in the third millennium have especially been discussed by Zubrow (2006).

The application of GIS, along with the computerization of archeology, has evolved from the needs of archeologists to explicitly and efficiently handle spatial data. Data layers in GIS have provided the means to connect archeological findings to different contexts—whether spatial or periodical—and to those that are not necessarily visible on the ground. However, there is a difficulty to exactly define the temporal coevality and comparability of remains by GIS and distribution maps. One is able to report and reconstruct the event and place of finding artifacts, *but not necessarily define exactly where and when the remains were originally produced, used and deposited.* Relativity is always found present to some extent, when evaluating the comparability and compatibility of archeological remains in relation to each other. (This has initiated critical views and even questions of information manipulation with maps.) Archeologist have developed different approaches in order to comprehend the past worlds and possible situations concerning the formation processes of archeological finds and different methods, both relative and absolute, in dating the remains.

However, like in archeology, unanswered spatial questions still remain pertaining to the temporal dimensions of GIS. Understanding time and its structure is and will be one of the major challenges of archeology as well as GIS in the future.

61.1.2 Computerization of Archeology, New Schools and Application of GIS

61.1.2.1 Computers, Arizona and the "New Archeology"

Archeologists have used computers for data recording since the 1950s. (Richards and Ryan 1985). In the US, incidentally or not, Arizona—which is known as the "nation's field laboratory for anthropology"—has been the place where experiments in the field of computer use were first efficiently executed in archeology (Gaines 1984), and where a fresh way to look at the archeological record started emerging in the 1960s. Then a scientific school of archeology with the fresh vision known as the "New Archeology" was initiated by Lewis Binford from the University of Arizona. Modern—often environmentally oriented—research designs, questions, clear empirically testable hypotheses, and methods that required collecting large databases and use of statistical approaches and methods were employed. Cultural Resource Management (CRM) in the US became a field in which new requirements and needs expanded, and in that field computers provided a way to control and analyze large site inventories. Universities—in the state of Arizona especially—applied automated archeological information systems to meet the needs of CRM. (Gaines 1984). The 1970s was the decade of breakthroughs in several frontiers of archeology, both in terms of tools and theories. It was the decade of the real implementation of computer technology into archeology.

61.1.2.2 Arizona (US) - Cambridge (UK), Spatiality and GIS

While the University of Arizona became the center for new approaches on the American continent, the University of Cambridge in Britain appeared to be its counterpart in fresh developments taking place in the old world. While Lewis Binford launched new agendas and strategies, David L. Clarke at Cambridge clearly defined the methods of the New Archeology in his *Analytical Archaeology* (London 1968). He also edited such profound works as *Models in Archaeology* (1972) and *Spatial Archaeology* (1977). Apart from David L. Clarke, Ian Hodder and Clive Orton were the researchers to further execute spatial approaches and cartographic methods to demonstrate topological, contingent and contextual features in analyzing the distribution of the archeological finds, e.g., in *Spatial Analysis in Archaeology* (1976). Besides spatial analyses and modeling Hodder also employed simulation to study and analyze archeological data (Hodder 1978).

The Granite Reef archeological project in the American Southwest during 1979–82 is considered to represent the first real application of GIS into archeology using data layers of different environmental information in raster form. Digital elevation models (DEMs) were introduced in those years to archeological research as well (Kvamme 1995). At the University of Arizona, K. L. Kvamme was one of the first archeologists to apply GIS into regional analyses and archeological site prediction models for finding new sites in the US (see, e.g., Kvamme 1980). Predictive models were the field in which GIS initially was executed for the benefit of management of archeological remains. The All-American Pipeline project in the US became one of the major projects using computers in archeology. The project, which began in 1986, further developed automation of the capture, analysis and retrieval of archeological information. In 1990 the first monograph dealing with GIS and archeology was published in the US bearing the title *Interpreting Space: GIS and Archaeology* (Allen et al. 1990). It was followed in 1995 by the European counterpart *Archaeology and Geographical Information Systems: A European Perspective* (Lock and Stančič 1995). These works have been steadily followed by GIS and archeology publications such as *Anthropology, Space, and Geographic Information Systems* (Aldenderfer and Maschner 1996) and *Geographic Information Systems in Archaeology* (Conolly and Lake 2006).

61.2 GIS as a Tool for Archeological Heritage Management

61.2.1 Archeological Heritage Management

The primary needs for archeological recording and documentation are considered to consist of 1) management and 2) research. Management aims to save the cultural and archeological heritage while the purpose of research is to answer specific questions of human past. National archeological inventories and databases can form a foundation for an information system that helps both management and research. GIS provides an efficient platform and tool for constructing such an information system for predicting, storing and analyzing large amounts of archeological information.

By the archeological heritage we mean the material culture of past societies that has survived to the present day and the process through which it is re-evaluated and re-used in the present. The property, ownership and patrimony of the cultural and archeological heritage are defined in national and federal laws. In addition, among archeologists and anthropologists there is a view that besides national and federal interests, universal ownership by mankind related to specific sites exists. Indigenous people may also have ethnic or/and religious interests in archeological sites, and tourism has its own interest in experiencing the common past and heritage. However, questions of property, ownership and patrimony raise continuing debates and problems around the world. Aboriginals and natives may have different views on the cultural and archeological heritages from state policy managers; and after having received their independence the earlier colonized states or subjects often try to claim the antiquities originating in their areas back from their earlier imperial occupiers who may have transported them elsewhere, e.g., to their museums (Skeates 2000).

Management starts with searching, discovering and defining archeological finds/sites for protection, preservation, conservation and storage purposes (Figure 61-2). An ideal situation would be that UNESCO under the United Nations, for example, could receive and store all the national archeological inventories and databases of the member states for the sake of the common heritage protection and information preservation in the case of possible wars and natural catastrophes. A vivid recent example is the planning and outbreak of the war in Iraq (see, e.g., Farchakh 2003). On 11 March 2003, UNESCO actually notified the Secretary-General of the United Nations and the US Department of State about the hazards an intervention would create for culturally significant sites, and it made available a detailed map of the locations of Iraqi archeological sites and museums. UNESCO also provided information on its mechanisms of protection for objects in museums and archives (Mairitsch 2004).

Digital data concerning the cultural heritage sector is the resource and outcome of computerized processes and information systems. It can store and present the data of cultural heritage, but can also be art and cultural heritage in its own right—cf. UNESCO Charter on the Preservation of Digital Heritage (UNESCO 2008) and see, e.g., discussion in Cameron and Kenderline (2007).

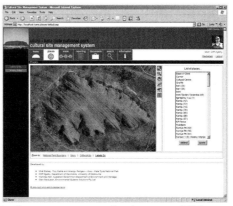

Figure 61-2 The Uluru – Kata Tjuta National Park heritage management system uses GIS for documentation, preservation and display of the locations of important features at Uluru in Australia. Uluru is the world's largest monolith, a rock formation that is a native site encompassing considerable rock art. It also belongs to the World Heritage Sites defined by UNESCO. Courtesy: Cliff Ogleby, See included DVD for color versions.

61.2.2 From Archeological Inventories and Databases to Archeological Information Systems and GIS

The ideal situation is when the national archeological databases form an information system that serves both specialists and a larger public. In such a system, archeological information can be connected to serve a number of other city/communal, regional, county or state regulated needs for the benefit of land-use planning and management. Layers of information concerning, for instance, cadastre, traffic and mining can be analyzed and monitored in association with the archeological information. Queries concerning particular areas and sites can be made in different layers, associating archeological information with land use and locating hazards.

In the layered data based on GIS, a hierarchical organization or a data model can be built with databases starting from sites in the World Heritage List in a given country (the US has currently 20 sites in the World Heritage List). This can be extended to nationally important sites and sites on local levels offering layers with cartographic information, imagery, textual sources and other archival data connected with the site. Efficient and manifold systems enable one to carry out queries of all the archeological objects with their register numbers evolving from the site under study and the information of the stacks where they are situated today— see, e.g., the Louvre database (Musée du Louvre 2009) on the World Wide Web. Attached information can be extended to ethnographic interviews for the benefit of ethnoarcheology, even with audiovisual data such as ethnic music or different dialects for linguistic studies.

Through GIS programs, special buffer zones can be added for presenting the needed zones or spaces for the site and structure protection. This is an efficient way to illustrate the boundaries of protected areas for land-use planning and field monitoring. Different colors and/or raster covers of the zones may distinctly inform the levels of global, national and local importance of the buffered site. Visible buffer zones in online services on the World Wide Web—accessed through mobile phones, communicators or PDAs—can also provide the boundaries of the protected areas and restrictions to visitors while they are moving in national parks and archeological sites.

The US National Park Service (NPS) comprising national Archeology and Ethnography services has planned and organized the National Archeological Database (NADB). In 1992 it became an online system, free of charge, via the World Wide Web. It has helped to solve a number of data access and information exchange problems, but needs for developing the system for different demands still exist, e.g., in the way the data are organized and retrieved. The NPS launched its own website in 1995 offering navigation possibilities with locations, texts, images and audio-visual data. Location information, however, is limited to county and state levels in order to protect the archeological sites. Sensitive information is not available and specific coordination is avoided in the public online system (Canouts 1999). Later on major difficulties have been found in managing the accumulating stocks of relics in the NPS regulated Archeology and Ethnography Programs. This has led to the Curation Crisis (Childs 1995). The NPS National Center for Preservation Technology & Training (NCPTT) offers a means for technical assistance (NCPTT undated), especially urgent in the case of natural catastrophes such as caused by hurricanes.

As far as certain cultures and periods are concerned, particular projects have been launched for creating databases and electronic atlases. For example, the electronic GIS-based atlas of ancient Maya sites has been under construction by W.R.T. Witschey and C. T. Brown supported by the Middle American Research Institute and the Science Museum of Virginia (see more on http//mayagis.smv.org).

In Europe Greece has had several GIS-based minor projects and a larger one called POLEMON comprising information technology services for recording, documenting, managing and presenting the Greek cultural heritage. In POLEMON, the Global Access System in the National Archives of Monuments information system offers the basis on which the federal database can be connected. The Global Access System is built on an object-oriented Semantic Index System (SIS) (Bekiari et al. 1999). In 1997, the Istituto Centrale del Restauro of the Italian Ministry for Cultural Goods and Environment finished an information system project which interacts with risk information. It has resulted in a Risk Map which integrates GIS with different types of danger to the cultural heritage in every region of Italy. The database includes over 51,000 monuments of which there are churches, villas, palaces and other buildings. A total of 5,200 buildings have wall paintings to be protected. (Bodo and Cicerchia 2000). In addition, in Spain a complete archeological inventory was initiated for the Community of Madrid in 1985 and has since been developed for an information system using GIS (Bosqued et al. 1996).

In developing countries problems often exist in planning, organizing and implementing an archeological information system. However, Jordan has developed the national archeological database of sites and monuments known as JADIS (Jordan Antiquities Database and Information System). It was started as early as 1990 as a co-operative program between the Department of Antiquities of Jordan and the American Center of Oriental Research (ACOR). JADIS is a modular system which, with the site location information, provides possibilities for future additions with images and maps and integration with GIS (Palumbo 1994). Similarly, the Egyptian Antiquities Information System (EAIS) pilot project was initiated in 2000 by the Finnish developing aid and had a goal to become an umbrella for all the antiquities pertaining to Egypt (Figure 61-3). It employs a Microsoft Access® database system.

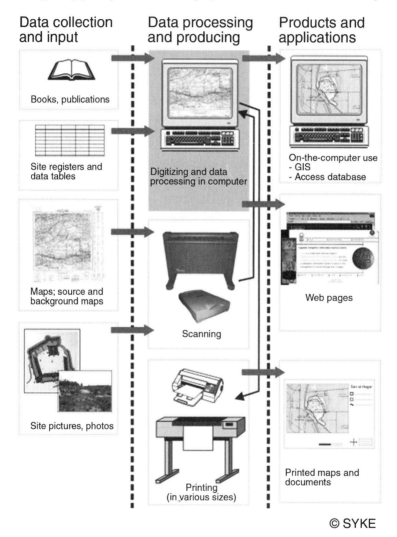

© SYKE

Figure 61-3 The Egyptian Antiquities Information System (EAIS) scheme showing the input, storage and output of data processing. Courtesy: SYKE, Mikko Hynninen, The Finnish Centre of Environment. See included DVD for color version.

61.2.3 Information Standards—the Data Standard

The classification of archeological material to different types and periods has been one of the central methods used in archeology from the invention of typology influenced by the

Darwinian evolutionary thinking. The classification, besides being a scientific approach to the cultural material, offers possibilities for organizing large amounts of artifacts for their identification and retrieval in stacks and museums. The classification and standardization of the archeological information to make it comparable and compatible has also been one of the key issues in recording and documentation in recent years. The data standards define what information is to be recorded, how it should be recorded and the means for supporting the information systems (Quine 1999). Such data standards have been developed for cultural heritage management and research, but the development should be such that the comparability and compatibility with other sources of land-use information are taken into account in the planning process as well. It has become clear that *the standardized information, if available, should be applied already during the field work when the first observations and recording take place.* For that purpose, special standardized field forms would be of use. It would be ideal if the commonly accepted principles were to be included in the antiquities legislation in different countries, meeting the requirement followed in all fieldwork projects, so that the data are harmonized also for scientific use.

The CIDOC (Le Comité international pour la documentation) standard is an international data standard for recording archeological sites and monuments. It has been developed by the International Council of Museums (ICOM) under UNESCO (Quine 1999; see also Core Data Standard for archeological sites and monuments http://www.object-id.com/heritage/int2.html). The US NPS site information conforms to the CIDOC standard (Canouts 1999). The RecorDIM under CIPA—the ICOMOS (International Council of Monuments and Sites under UNESCO)/ISPRS (International Society for Photogrammetry and Remote Sensing) Committee for Documentation of Cultural Heritage—with its working groups and task groups is a five year initiative (2002–2007) improving the data recording systems and standards. It cooperated, for example, with ICOMOS, English Heritage, the Getty Conservation Institute, Public Works and Government Services of Canada, and World Monument Fund. Members of the Open Geospatial Consortium are successfully developing standards so that different geospatial software packages and geospatial data sources can all interoperate. In building up a GIS, one should take into account that the produced cartographic and spatial information must be uniform with the national standards. This will ensure the usability of the data also in the future.

In data standards, the required minimum information should be defined. The US *Managing Archeological Collections* website (NPS 2007) provides information on how artifact collections can be processed, cataloged and archived; and it discusses the standards and support from the local stacks. The glossaries or thesauri to define archeological sites are also important and helpful. English Heritage (1999) provides thesauri on national monuments; English Heritage and the Royal Commission on the Historic Monuments of England (EH/RCHME 1997) has published *Archaeological Objects Thesaurus*. These standards are to be used in archeological and architectural records. The European Council has also provided archeological Bronze Age terminology in English and French (Barber and Regteren Altena 1999, Barger 1999). These vocabularies, glossaries and thesauri are aimed at using standardized terminology in recording, documenting, data feeding and accessing for the need of computerized heritage management. The use of these is recommended for building databases and applying GIS for heritage management and for scientific research.

Three-dimensional representations provide a tool to visualize archeological sites and finds, especially in museum exhibitions and collections. The 3D reconstructions can be connected to GIS-based archeological information systems. Virtual museums offer a new way of visiting the collections through the internet. See, e.g., Oriental Institute Virtual Museum (Oriental Institute 2009). Possibilities of multimedia applications to display archeological finds in 3D are various in GIS.

61.3 Searching for Archeological Sites with GIS

61.3.1 Background Data Sources

61.3.1.1 Archival Information

Archeological research procedure in action concerns 1) collecting finds and site data through prospecting, surveying and excavating, 2) classifying and analyzing finds/sites through different methods and 3) building models, hypotheses and theories (Clarke 1978).

The search for archeological sites has traditionally started from preliminary studies at an office and has been based on archival information: old registers, maps, photographs and other historical data (accounts, correspondence, newspapers, interviews, etc.), even ethnographic information, folklore and myths. Archeological fieldwork is a costly operation when archeologists, geodesists and natural scientists are sent with whole laboratories to distant places. All the preparative measures—including prospecting with computerized information, archival and library work—save money, reduce working hours and offer possibilities for better defining the research questions and goals for the fieldwork. It is even possible to construct the working plan as a digital or virtual model for the benefit of a field survey or an excavation. In rescue operations and low budget operations belonging to the responsibilities of the heritage management, preparative measures unfortunately do not always take place, e.g., in urgent *ad hoc* rescue works or when budgeting of an operation has not primarily been influenced by archeologists. The easily and quickly accessible online databases with context information as in GIS are therefore of particular importance.

Religious traditions, mythical information, epics as well as sagas have also been used in tracing archeological sites. Site names are part of geographic information. Linguistic studies of site names and their etymology have been of particular value. Site names can be discerned from maps, and ethnographic interviews for acquiring local names and traditions may be carried out among native people. One of the famous examples of successful use of ancient epics and site names is the search for Troy and Mycenae by Heinrich Schliemann in the 19ᵗʰ century using Homer's *The Iliad* and *The Odyssey*. Ancient Ilion at Hissarlik in the Troad situated on the western coast of modern Turkey, for example, came to be identified with the ancient city of Troy by Heinrich Schliemann. In 1869, Schliemann published a short study *Ithaque, Le Péloponnès, Troie: Recherchers archéologiques* using philological methods in order to support his theory. Now a similar human interest in modern times is continuing in the search for Odysseus' home island, Ithaca. The search is now available through NASA's World Wind 3D visualization software with virtual close-ups (NASA 2004).

All the available information including old data such as maps, photographs, graphics and written sources, can be digitized, applied to GIS and connected to the site registers as attribute information and as visual layers for the purpose of prospecting new sites and managing the already known ones. This preliminary work concerns both survey and excavation, as well as archeological underwater studies. At its best the preparation for a field survey and/or an excavation is based not only on old archival information, whether in paper or digital form, but also on acquiring remotely sensed data such as satellite images, aerial photographs, digital elevation models (DEMs) and employing geophysical studies for prospecting the area. The predictive models for site identification, such as developed in GIS, can be constructed using different types of data sources. (See, e.g., the use of photogrammetry and image archives for preparing an archeological excavation of a Crusader castle in Grussenmeyer and Yasmine 2004l) A new ontology is needed for storing digital data for the Cultural Heritage Information management sector. CIDOC-CRM is a recommended standard for encoding digital information that enables data storage, retrieval, sharing and dissemination (Généraux and Niccolucci 2006).

61.3.1.2 Remote Sensing from Space to Air and under the Ground

Remote sensing has taken a central role in archeological prospecting methods during the past decades (see e.g., Scollar 1990). Besides aerial photographs, satellite technologies have provided a new asset in remotely sensing ancient remains and an important source to be used in GIS. Satellite images offer a means to study large areas, and their study and analyses save working hours on the field. Images are important information carriers not only for the naked eye, but also as numeric mosaics studied through GIS. The information they provide comes in the *raster form*, and is therefore usually used as background and/or attribute data for archeological site information in GIS. The satellite images and photographs have to be rectified for the needed coordinate system before they can serve prospecting, mapping, surveying and as background images for mapping find/site locations. Photogrammetric methods can be applied in mapping with aerial photographs or satellite images and analyzing the data. Control points are used in rectifying and georeferencing photographs and images. Maps and GPS data captured from the field are employed in creating control points for remote sensing, e.g., for rectifying images into a needed coordinate system.

Aerial photography was used in archeology already in the early 20th century, first employing camera platforms in kites and hot air balloons. However, the real breakthrough of archeological remote sensing took place with the introduction of airplanes and in the aerial surveys in the course of the First World War. Father A. Poidebard even used the advantages of different seasons, vegetation and daylights in his aerial prospecting work in the Near East in the 1920s (Figure 61-4). Aerial archeology has developed into a branch of its own with a specific journal (Brooks and Johannes 1990; Bewley and Rączkowski 2002). A useful sensitive method for tracing shallow and buried archeological remains is analyzing vegetation cover through aerial photography, because the plants and soil humidity respond to the structures. The approach has been especially fruitful in crop sites. Spectral properties of different vegetation covers can also be analyzed in order to trace structures. The choice of films and filters affects how much can be detected, and how much information the photograph is carrying of the area under study (Scollar 1990). John Rowe (1953) is mentioned to be the first American anthropologist applying the use of aerial photography to ethnographic research in order to retrieve settlement and agricultural information.

Figure 61-4 Poidebard's pioneering work in aerial archeology was published in 1934. The aerial photograph in the background was taken during a flight over Palmyra and shows the ancient tower tombs in the Syrian desert. Source: Poidebard 1934.

Manual of Geographic Information Systems

The developments in the "space archeology" was coeval with the application of computers and GIS into archeology. W. J. Stringer and J. P. Cook (1974) utilized satellite remote sensing in locating archeological village sites in Alaska in the 1970s. Remote sensing with aerial photographs and satellite images has been especially important in studying sites that are difficult to access on the ground. The first conference on the application of satellite images in archeological research took place in Boston in 1980. Archeologists J. I. Ebert, Thomas R. Lyons (see Lyons and Ebert 1978; Ebert and Lyons 1983; Ebert et al. 1997) and Tom Sever, researcher from NASA, had started to use satellite images in archeological research in the 1970s and 1980s. The Piedras Negras, a Maya Indian habitation area in Central America, became a focus of research from 1988 onwards. With the aid of satellite images and electro-magnetic devices, the project identified ancient forts and roads. (Wiseman 1998a, 1998b). The detection of the lost town of Ubar in Oman, the site appearing in *The Thousand and One Nights,* is to be mentioned, among other early projects in archeological satellite image prospecting, as well as the delineation of *The Silk Road* in the Taklaman desert in China in 1992 (Evans et al. 1994).

The earliest archeological mapping projects using NASA's remotely sensed radar imaging system starting in 1996 focused on the ancient Khmer capital of *Angkor,* one of the UNESCO's world heritage sites in Cambodia in the middle of a jungle. NASA's DC-8 equipped with Airborne Synthetic Aperture Radar (AIRSAR) was used in searching for new underground features such as ancient river channels of the site as well as tracing new structures on the nearby Kapilapura mound. The knowledge of the ancient hydrology of the site especially helps in its conservation by preventing further degradation and in restoring some of the original canals and reservoirs (Holloway 1995; SPMH 1998). Radar-based remotely sensed data can be utilized in landscape modeling, as offered, e.g., by NASA Shuttle Radar Mission (SRTM) 2000, providing the readily available DEM data. (See Figure 61-5, e.g., from Lönnqvist and Törmä 2004).

Figure 61-5 3D landscape model displaying the mountain of Jebel Bishri in Syria and the Euphrates Valley from the east. The model is produced by fusing NASA X-SAR mission radar data in DEM tiles and Landsat-7 ETM data. Data copyrights: DLR (German Aerospace Centre) and Eurimage. The model was constructed by Markus Törmä 2004. Courtesy: SYGIS - the Jebel Bishri Project, University of Helsinki. See included DVD for color version

Earlier difficulties in the application of remotely sensed data in archeology consisted of the high costs of satellite images with adequate spatial resolution, and lack of know-how to extract information from them. In the 1990s, the opening and availability of the US Central Intelligence Agency's (CIA) old CORONA satellite photograph archives deriving from the 1960s and the 1970s improved the situation. An easy access to purchase the low-cost photographs (a few dozen dollars per photograph) from the archives via the World Wide

Web has been a benefit (EROS data center: http://edc.usgs.gov/products/satellite/declass1. html). Today, several low-budget archeological projects (provided that the photographs cover the area under the study) may also use satellite data with adequate spatial resolution. The CORONA satellite photographs offer, in black and white negatives, even 1.8-m spatial resolution. (Kennedy 1998; Ur 2003; Lönnqvist and Törmä 2003, 2004). The data should be used in digitized form. It is also advisable to obtain stereo-coverage and use GIS software (Figure 61-6) in order to recognize site and structure information in relief. Before ordering the photographs it is also important to carefully check from the preview photos available online that clouds do not prevent visibility over the targeted study or mapping area.

ARC/INFO® by ESRI is commonly used by archeologists for mapping, GIS analysis and predictive modeling—see, for example, the creation of probability models in the Madaba Plains Project (Christopherson et al. 1996). Although ArcGIS® is the current version of the ESRI GIS software, many archeologists have limited resources for upgrades and, therefore, turn to available versions of GIS software and freeware. A freely available geographic information system, GRASS, can be an answer for low-budget projects to address spatial archeological data collecting, mining and analysis. Successfully used by a survey project in western Texas and southern New Mexico, GRASS even has the capability to present data in 3D (Brandon et al. 1999).

ArcView® GIS software by ESRI can be used on personal desktop computers for mapping and building GIS databases without requiring advanced geospatial or analysis expertise. It can be used in mapping from remotely sensed data, as well as for producing field maps, with the advantage of compatibility and comparability with the data produced in ARC/INFO. Therefore, ArcView has often been chosen by the practitioners of heritage management for integration with data produced by ARC/INFO in different fields of land-use management. IDRISI is another GIS software program that can be used on personal computers, and it offers a handy tool for building predictive models. The georeferencing capability of MapInfo® mapping software contributes to the analysis of satellite images for archeology applications and its add-on tool, MapInfo® Vertical Mapper™, can be used for conducting spatial analyses. ERDAS® ViewFinder (formerly MapSheets and MapSheets Express) is a software package that can be downloaded free of charge from the internet for archeological mapping from satellite data. CAD/CAM graphic software also enable 3D data display and virtual landscape reconstruction.

Figure 61-6 Mapping and GIS software commonly used in Archeology.

Landsat images, which can be downloaded free of charge from the US Geological Survey (USGS) EROS Data Center via the World Wide Web, are useful for environmental studies providing attribute information for the areas under study and allowing the construction of predictive models for finding new archeological sites. However, the spatial resolution of Landsat data only reveals sites, roads and monumental remains in dozens of meters in scale, and structures such as the sites occupied by mobile hunter-gatherer and nomads are generally unrecognizable. In addition to this, a number of other satellite images offer data for environmental studies. For example, SPOT images have largely been used for environmental analyses, modeling and tracing sites in several archeological projects (Madry and Crumley

1990). The best results for detecting sites and structures, however, can currently be attained with Ikonos and QuickBird images available in the market and offering as good as 0.6-m spatial resolution, but the price of the images is high and the minimum area of coverage from the supplier may be limited, e.g., 8 km × 8 km.

Remote sensing by non-destructive *geophysical methods* includes techniques using different equipment ranging from, e.g., a *Ground Penetrating Radar (GPR)* and a magnetometer to metal detectors. The purpose of these is to trace underground structures, sites and materials without excavating areas (see Barba 2003; Eppelbaum and Itkis 2003). Soil samples can also be used for chemical analyses, such as phosphate and phosphorus analyses for tracing ancient human activities and determining sites. While geophysical experiments and soil sample collecting are carried out, the locations of the data capture have to be defined and recorded. Recorded and coordinated information can be inserted into a GIS and the significance of the results analyzed.

61.4 The Data Capture and Documentation in Action

61.4.1 From Landscapes and Sites to Locations and Boundaries

All the archeological landscapes, sites and artifacts derive from some spatial location. The definition of a location cartographically requires, however, a preliminary decision on whether an archeological discovery has been made. This decision-making process concerns as much artifacts, structures and features as sites. A special problem in archeology is: *How to define a site*? In remotely sensed data, certain observed patterns may be defined as structures or sites. On the ground one faces the question of whether a site consists of two/three discovered objects in "a square meter or two" or only of immovable remains such as structures, stains, tracks, carvings and paintings. With a few movable artifacts lacking connection to visible immovable remains on the ground there is always a possibility that the objects have moved from their original or previous site/sites by natural forces or having been dropped by passers-by. What is important is that every archeological site is a relic of human activities, and only the human dimension and action make it a site with cultural implications. Archeological sites are often defined loosely as "any locus of past human behaviour" (Banning 2002). There are natural sites that have become "archeological sites" only through the application of information about ancient traditions. They are unaltered sites, i.e., they are somehow impressive natural formations used for different, often cultic or ritual, purposes, but this does not necessarily imply they are structures or objects produced or modified by humans (Bradley 2000).

When *the definition of an archeological site is applied to a location*, its position and its dimensions are to be measured and defined (Figure 61-7). Earlier the locations were only relatively defined and described by distances related to known sites, such as cities, towns or villages, roads and geographic features. Nowadays, however, the coordinates (x, y) are the scientific parameters to define an archeological discovery in space, and they are the parameters that are used in GIS. They can be defined by degrees, minutes, seconds in a spherical coordinate system or expressed in the planar coordinate systems such as the Lambert and Universal Transversal Mercator (UTM) coordinate systems. Several countries also use their national grid systems, but the UTM is a universal system often used in Global Positioning Systems (GPS) and provides the location information with compatibility and comparability to a number of other available maps and data layers. An archeological find spot or a site can be inserted through GPS or digitized for the use in GIS as a cartographic layer with coordinate information by *vectorized* contours or/and *raster* fillings for the use in the GIS. The previously mentioned mapping software programs are used (see Figure 61-6), and *ArcView* is often preferred in producing maps from the digital data that have been captured.

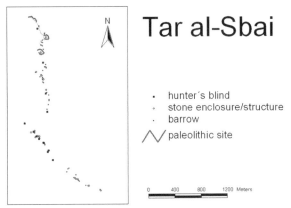

Figure 61-7 Recording and two-dimensional mapping of archeological sites, structures and artifacts at Tar al-Sbai, the western edge of Jebel Bishri, in Syria. Distribution of find types are mapped and displayed on a panchromatic Landsat-7 ETM+ satellite image. Mapping has been carried out by Markus Törmä in 2005, Jari Okkonen in 2005, Kirsi Lorentz in 2000, Donald Lillqvist in 2000 and Mervi Saario in 2000. Courtesy: SYGIS - the Jebel Bishri Project, University of Helsinki. See included DVD for color version

The two-dimensional (x, y) coordinates provide a horizontal view, i.e., a plan, surface or map, while the vertical view provides the third dimension: heights (z) or depths and thus reliefs. The three-dimensional recording system has been used in archeology since the 19th century, from General Pitt Rivers' times (Harris 1989). Its use applies not only to sites and structures, but also to single finds, the positions of which in the ground bear a special meaning for archeologists. The position, even the angle in which, for example, stone tools are recovered carries information on working techniques and activity sites for a specialist to analyze. *Three-dimensional coordinate recording* offers the possibility to record, document and examine the finds and sites from various angles and perspectives and to display shapes with multilateral effects (Figure 61-8). Three-dimensional recording can be carried out with electro-optical distance measuring (EDM) equipment, laser scanners and by using photo-grammetric methods in GIS. Recording 3D coordinates has made it possible to visualize artifacts, structures and sites in the context of physical environments, landscapes and underwater worlds.

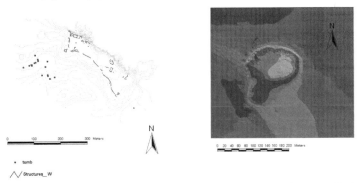

Figure 61-8 Mapping in three dimensions: a Late Roman–Byzantine fort at Tell Tabus on the Northeastern edge of Jebel Bishri in Syria and an elevation model of Tell Tibne on the Euphrates. Mapping has been carried out by Jari Okkonen in 2004-2005, Kenneth Lönnqvist in 2004-2005, Nils Anfinset in 2004 and Tuula Okkonen in 2004. Courtesy: SYGIS – the Jebel Bishri Project, University of Helsinki. See included DVD for color version

A field survey, besides its archeological goals serving management and/or research purposes, can provide empirical data, e.g., for the mentioned satellite mapping and image interpretation. Soil and other material samples (collected with special field data record forms) accompanied by coordinate information provide empirical information for identifying the material signatures in remote sensing.

As far as the dimensions of a site are concerned, in archeology it may be a hard task to determine the actual boundaries, especially if parts or the majority of the remains are buried beneath the ground surface and remain invisible. However, one needs to make a preliminary definition of the site boundaries for preserving and protecting the site and for the needs of its further research. The UNESCO recommends that the coordinates are to be taken from the center of a site/structure to define its location, and therefore for such central recording the site boundaries should be known. GIS provides new possibilities for modifying the boundaries in the course of finding or redefining the exact limits of a site (Figure 61-9). It also offers the means to monitor the development and condition of structures and sites by different methods. Remote sensing is one method to monitor the state of preservation, e.g., through aerial photos or videos and satellite images to update the site databases with the latest information.

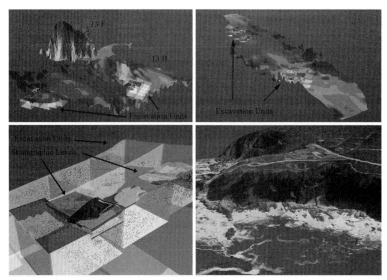

Figure 61-9 Three-dimensional GIS modeling at Pinnacle Point in South Africa. The upper left image shows caves 13B and 13F with excavations and the upper right image shows the floor surface of cave 13B looking north with excavation and geologic stratigraphy units. The lower left image is a close-up of the eastern group of 13B excavation units showing multicolored geologic stratigraphy, georeferenced photographs of unique features, plotted artifacts and excavation units. The lower right image is a high-oblique view of Pinnacle Point, Mossel Bay showing the location of caves 13B and 13F in the cliff face. Courtesy: Erich Fisher, University of Florida. See included DVD for color version

In defining the site boundaries, it is also advisable to change the perspectives. The use of an airborne or spaceborne platform enables one to detect traces that are not necessarily visible on the ground; some structures and features can only be clearly delineated from the aerial perspective or from stereo-pair imagery. Some surveys may comprise the digging of ditches, sondages (test pits) or trenches. Broad-scale excavations open possibilities to better define the limits and the nature of a site, and study its cultural implications. An excavation offers means to study deeper, to protect and to preserve the site, but at the same time it is a destructive act as layers of evidence from the past are removed from one over another, never to be restored to their original state.

61.4.2 Equipped for Digital Data Capture, Storing and Mining in the Field

61.4.2.1 Lighter Portables

A spade and a trowel are an archeologist's traditional companions, and despite technical developments from satellite images to ground penetrating radar, the traditional ground tools have not lost their importance. However, archeologists are not keen on carrying heavy tools or equipment while surveying in different terrains and over long distances. Therefore, portable technical devices are recommended. (See an example of Mobile GIS survey by Tripcevich 2004). The data capture procedure that begins *in situ* in the field keeps the data updated. The use of GPS is now a necessity in a field survey and an excavation. If the special survey instruments such as small pocket PCs with local field software including GIS are available, they are handy to use. Some of them can collect and store considerable amounts of data. Archeologists usually use GPS with separate laptops or hand-held PCs, but today there are all-in-one devices in which GPS is included in the hand-held computer that has data capture software and navigating possibilities. According to some experts, the all-in-one solutions still have handicaps in flexibility especially in using GIS (Bedford 2004).

Climate and weather constraints, such as sandstorms and rains, can cause problems for using computers during a field survey. For example, it is possible to use hand-held water-proof PCs which vary from withstanding a drop of water to a shower or being immersed to a meter in depth. The solution can also be computerized field forms that are printed and manually filled *in situ* for the information to be later fed into computer databases. Digital papers and pens are also available; these can be connected to data feeding systems. Manual recording and documentation are always a security measure and they form "a back-up file." If a technical problem occurs with the equipment, the unique data is not lost forever, but can be retrieved. This is essential in archeology because of changing circumstances, for example, in digging up layers or collecting finds and samples in loci. On excavation, an entire computer laboratory can be established when an expedition is working continuously in a specific location.

Mobile phones and/or *communicators* may help in various ways in the field as well; the services and possibilities they offer are constantly increasing from GPS to cameras and beyond. In distant places without Global System for Mobile Communications (GSM) stations, satellite telephones can be used. Some *binoculars* now offer different geographic devices, such as displaying the geographic directions or infrared capabilities for seeing in darkness. For a surveyor, these are valuable instruments for looking at a terrain from afar and planning a survey procedure. Easily portable objects such as *compasses* also are available in digital form by army suppliers for determining the geographic directions. *Digital hand-held altimeters* included in barometers for weather forecasts are available from army suppliers as well. The altimeters, however, need to be regularly calibrated.

Digital cameras are well-suited for documentation in surveys and excavations, although several antiquities authorities still require traditional black and white negatives as well as color slides. *Cameras,* films and lenses of different types are used for *photogrammetric recording and documenting* sites and structures, especially architectural and underwater remains for the sake of their conservation (Figure 61-10). Digital images also record, store and provide information on the sizes of the documented remains. The benefits of seeing the immediate results in digital camera screens is valuable in the field, especially if finds cannot be transported to museums to be photographed and if secure photograph developing conditions are not at hand. However, it is clear that digital information storing and preservation are major problems to be solved in antiquities information systems. The experiences of archival preservation of black and white negatives and prints extend period of over a hundred years, and the information they store are continuously attainable even today. They can always be digitized for the purpose of GIS.

Figure 61-10 The image above is a panoramic stitched image of an art shelter at Uluru in Australia. Below is photogrammetric recording and documenting at Uluru using a modified Hasselblad film camera on the stereomount and photogrammetric targets inside a shelter. Courtesy: Cliff Ogleby, University of Melbourne. See included DVD for color version

Electronic devices, such as *digital calipers and scales* for measuring and weighing small objects can now be acquired and connected to computers for measuring and documenting finds already in the field. *Laser scanners* are the most recent devices for measuring artifacts and producing reconstructed images of objects.

61.4.2.2 Heavier Portables

Desktops or laptops storing remotely sensed data, maps and other site data, are important to be at hand, at least in survey/excavation headquarters of a base camp. Compared to the hand-held lighter portable computers, their screens are larger and data can visually be better detected and analyzed. In developing countries where there is often a lack of proper maps, archeologists may create their own field maps with the aid of computers and satellite images. In remote places, the limited availability of electricity can cause a problem. This can, however, be solved with solar panels if the climate conditions are suitable. For example, arctic places do not always offer enough sun light for such panels.

Apart from electro-optical distance measuring (EDM) equipment, a tacheometer or simply a "total station" (Figure 61-11), a laser scanner is a piece of equipment used in surveying and conducting site excavations today. EDMs and laser scanners have largely replaced surveying levels and theodolites. They are used to capture and store digital location information in 3D format. Laser scanners even record colors. However, an EDM and a laser scanner with a tripod are heavier portable equipment (e.g., 4–5 kg without tripods) that also require regular recharging of batteries, up to as often as once or twice a day. An EDM usually needs a reflector, but today there also are reflectorless devices available. A piece of equipment, such as an EDM, may be difficult to carry into remote places, and sometimes even impossible to use in deep ravines or steep rock slopes. In these situations there is no possibility to build a platform on which the EDM can be set in an upright position or where the reflector prism can be used. However, for difficult terrains, handheld PC solutions with a laser scanner are available.

Figure 61-11 Recording and mapping archeological sites and artifacts with Electro-optical Distance Measuring (EDM) equipment. See included DVD for color version

To begin, an excavation area is typically laid out in a grid, either on a visible plan on the ground and/or invisibly stored in an EDM or a laser scanner, usually in 1 m × 1 m squares. An EDM or a laser scanner is used for recording structures, as well as features and the location of small finds *in situ*. The captured data downloaded into a computer can be used for creating maps with different mapping software, e.g., ArcView and ArcGIS and CAD/CAM software. In an excavation, each layer or *stratum* is dug away either in natural or arbitrary layers (see Harris 1989). A stratigraphic excavation proceeds layer by layer, whether in small sondages or trenches, exposing large open areas. Stratigraphic excavation especially benefits from the three-dimensional recording (see, e.g., digital recording of stratigraphic excavations by Doneus et al. 2003). Baulks, the faces of which provide standing sections, are needed for correlating and studying the succession of the layers. For the succession and study of the relation of the phases, computer models based on schematic diagrams such as the *Harris Matrix* (see Harris 1989) can be built at an excavation. Such a model helps in planning the procedure of an excavation. The Harris Matrix type data model also can be connected to the visual documentation, such as plans and photographs, which can then be analyzed in a GIS.

As previously mentioned, geophysical methods are used in prospecting; they also are applied to surveys and excavations. A defined area or a plot can be chosen for particular studies to be carried out with a *ground penetrating radar* (GPR) or a magnetometer in order to detect structures or other phenomena, such as hollow tombs under the ground surface (Figure 61-12). In the case of ground penetrating radars, the equipment is quite large and heavy when associated with laptops to follow the signals. Therefore, it is often impossible to apply their use to complete broad-scale survey operations. The depth of the radar penetration is dependent on the frequency (MHz) of the antennas used and the soil conditions in the area. The captured data with site location information can be connected to GIS.

Tracing underground anomalies
such as ancient graves with
a ground penetrating radar.
Reflections analysed on
a computer screen below.

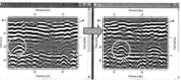

Figure 61-12 Ground penetrating radar recording, documentation and study along the Euphrates was carried out by Joseph Pedrez Rodés and Charlotte Børlit. Courtesy: SYGIS- the Jebel Bishri Project, University of Helsinki. See included DVD for color version

Photogrammetry is well-suited for documentation of rock drawings, inscriptions, paintings, mosaics, plaster covers of buildings and for reconstruction of finds. An admirable effort for building platforms for photogrammetric documentation at considerable heights has been carried out in Iran at the King Dareios famous Behistun inscriptions that were first documented by Dane Carsten Niebuhr and later interpreted by British officer Henry Rawlinson in the 19th century (Zolfaghari and Malian 2004). In the case of the large Buddha statues hewn in a rock in the Bamyan Valley in Afghanistan and destroyed by the Taliban militia in 2001, a new study using photogrammetric methods on three metric photographs taken in 1970 has enabled the reconstruction of the statues to begin (see Gruen et al. 2003, 2004).

After the data capture with various pieces of equipment, such as an EDM or a laser scanner, topographic site models can be produced with different software to visualize and study the shapes and contours of the sites. The CAM/CAD software especially enables one to graphically display digitally recorded site or structure lines in three dimensions. They have been successfully used in, e.g., the GIZA Plateau Mapping Project by the Oriental Institute of Chicago mapping laboratory (Oriental Institute 2009). Mapping in 3D can be aimed to create virtual images of the site under study. Virtual reconstructions are here to stay (Forte and Siliotti 1997). An example of virtual site reconstruction can be visited, for example, on the web site of the Great Kiva in Chaco Canyon in northwestern New Mexico (Kantner 2005). Planning virtual reconstructions with guided tours may also be the only solution to display to tourists the sites that are in hazard like Queen Nefertari's tomb in the Valley of Kings at Thebes in Egypt. The famous paintings of the tomb do not bear constant exposition to changing atmospheric condition while the tomb is kept open for tourists (Holloway 1995).

61.5   Past Environments and Landscapes to GISs

61.5.1 Environments as Dynamic Factors

Environments have become a modern focus of archeology, and they are no longer simply providing background illustrations for archeological works. Environmental archeology evolved decades ago, and now in archeology, environments form dynamic and influential backgrounds that can be extended from local to regional, continental and even global levels. They serve as attribute data for archeological information, as referred in separate and specific layers in GIS. Geo-archeology has become a special branch that can be defined as a subclass of environmental archeology using Earth sciences in its research methods and concepts. In addition to geo-archeology, different scientific fields have been applied to environmental archeology including paleobotany or palynology, zoo-archeology and bio-archeology. (Butzer 1982; Reitz et al. 1996). Paleoclimatological reconstructions also help us to understand changes in the environment. Whether humans have contributed to climate changes is more a subject of modern debates concerning global warming and not so much of an archeological dilemma. Human impact on the environment, however, is currently studied by archeologists. Salinization, overstocking and disappearance of woodlands are issues that archeologists need to deal with in studying ancient situations.

All the mentioned fields related to environmental studies have enjoyed benefits from GIS, and they in turn have offered data for database building for GIS after the applications of the systems and methods into archeology. Bio-archeology is one of the latest fields that can reveal and reflect different aspects of socio-economic features and interaction between humans and the environment. Different layers of information open vistas to the formations of the sites in their environmental context and may answer various questions on the interaction between environmental processes, materials and outcomes of the sites. As previously mentioned, environmental information gained through remote sensing is also used for predictive models to locate archeological sites.

Reconstructing paleoenvironments is one of the challenges of archeology today, and information related to the particular type of environment is used as the basis of arguments in interpreting cultural development and change. Studies of paleoenvironments are usually carried out by natural scientists who use proxy data collected from different fields of environmental studies ranging from geology, palynology, osteology and bio-anthropology to oceanology. The environmental data can be layered into different time periods in GIS signifying the conditions in past archeological periods and analyzed as contexts of the studied finds/sites (Figure 61-13). Archeology has, in turn, helped in the process of recognizing that a region such as the Sahara in northern Africa has not been desert throughout prehistory. GIS also offers a means to analyze environmental changes and conditions through collected data. The rate and distribution of erosion can be calculated. The spread and rate of desertification can be traced. Ancient rivers and their channel changes can be traced and the rate of water flow analyzed. Remotely sensed data can be used as a source of information for analyzing a number of these phenomena.

Figure 61-13 Sea-level modeling at Pinnacle Point in South Africa. The top image shows modeled sea levels during Marine Isotope Stage 5e (beginning ~130,000 years ago). The bottom image is a plan view perspective of the same model showing the transparent MIS 5e overlaid upon a recent georectified image of Pinnacle Point. Cave 13B is represented in red. Courtesy: Erich Fisher, Lydia Pyne and Curtis Marean. See included DVD for color version

61.5.2 Modeling Landscapes and Their Dimensions

Landscapes consist of spatial entities formed by human action in the environment including archaeological sites and cultural heritage in general. They are archeological entities that are especially nowadays studied through GIS. Positive results have been gained in using GIS in cultural landscapes in Norway (Boaz and Uleberg 1995). With DEM data and CAM/CAD software, environment and landscape studies have further opened new possibilities at hand for topographical, topological and contextual associations in archeology (Figure 61-14). However, critical views to the differences between reality and virtuality are needed. An augmented reality is a good way to express differences with the reality and the researcher's interpretation, as well as to show developments and changes of a given site through the past.

Figure 61-14 Geologic modeling in cave 13F at Pinnacle Point in South Africa. The upper left image is a photograph of Cave 13F as it appears in real life. The upper right image is Cave 13F as it is modeled in the mDGIS including polygons representing individual cave features. The lower left image shows Cave 13F in relation to sea levels during Marine Isotope Stage (MIS) 11. The lower right image shows the modeled infilling of Cave 13F by sand during MIS 6 and MIS 5d-4. Courtesy: Erich Fisher, Lydia Pyne, and Curtis Marean. See included DVD for color version

In archeology, the study of landscapes was brought into focus in the course of increasing spatial studies in the 1990s (see, e.g., Smith 1995; Knapp and Ashmore 2000). In 1992, UNESCO defined the value of cultural landscapes, which allowed the inclusion of sites without monumental or urban nature under the interest of protection and preservation (see Skeates 2000). The 3D viewing of photographs with stereoscopic equipment and methods is nearly as old as photography, and it is continuously used in photogrammetric study of ancient landscapes, as well as prospecting archeological sites and structures. Special software packages for stereoscopic analyses are also currently available.

Archeologists have an important role in understanding the long-term development and change in landscapes, and, therefore, they should be involved in landscape planning (Fairclough and Rippon 2002). Tradition and memory are closely associated with landscapes and every generation may have its own interpretation for a landscape. Views, vistas and panoramas can be studied through different GIS analyses. The concept of a "sacred landscape" has been applied for stages of cherished folklore traditions, cults and rituals. The sites attached to the traditions such as those of native people, e.g., aboriginals, are often associated with these kinds of concepts. In the cultural landscapes of Canada, for instance, sacred mountains appear to be associated with stories and traditions. The locations of archeological sites in the landscape have a meaning which archeologists try to analyze and interpret. GIS with attribute information has provided an efficient tool to store and analyze environmental–landscape connections of archeological sites and objects.

An example of a data model for a GIS-based landscape analysis has been constructed for the famous Nazca lines or geoglyphs in their landscape at Palpa in Peru (produced by the Nazca culture 100/0 b.c.–a.d. 600/700) by an inter-disciplinary project. The purpose was to develop a method that could cover the large, and in some cases nearly inaccessible, area of geoglyphs by recording and analyzing them. In the study aerial images were used for mapping and modeling the landscape in 3D. A GIS has been under construction to analyze and interpret the significance of the geoglyphs to the Nazca culture. The first step in building up the GIS has been taken in designing a conceptual data model that is working as a basis and will be inserted into the information system. (Lambers and Sauerbier 2003).

61.5.3 Underwater Worlds Towards GIS

Gradually digital documentation, application of different data sources and GIS have also been applied to the management and studies of underwater archeology, also referred to as maritime, marine or nautical archeology. (See, e.g., Breman 2003). The Institute of Nautical Archaeology (INA 2007) offers information and data including attainable imagery concerning different projects in the field.

The concept of underwater archeology has extended the field beyond the scope of maritime or nautical archeology. Apart from shipwreck sites and harbors, sunken settlements and remains in inland lakes or rivers have been included in this research field. There also are several sites, from prehistoric caves to towns, which have submerged and are currently the target of underwater archeology. For example, outside the Isle of Wight a submerged prehistoric landscape has been found with flint objects dating to ca. 5000 B.C. The Act 2002 in United Kingdom amended sites under the seabed to be included in the definition of ancient monuments. England's Historic Seascape project now emphasizes the introduction of GIS mapping to the historical sites of marine character. Fifty-eight designated wreck sites are under the development plan (Oxley 2005).

A coordinated grid layout of the excavation area under water also is essential in order to locate, record and document the finds *in situ*. Underwater differential GPS is used for location data capture. Photogrammetry with special underwater cameras is widely used in documentation of underwater archeological remains from wrecks to small finds. An example of the search for pottery fragments and their ultimate reconstruction with photogrammetric studies is the survey of ancient dolia in a wreck on the coast of Italy (Canciani et al. 2003).

A famous Uluburun Late Bronze Age shipwreck near Kaç on the southern coast of Turkey has inspired a number of studies. An internet website to study the wreck is provided by a project of The Ellis School in Pittsburgh, Pennsylvania

(http://sara.theellisschool.org/~shipwreck/ulusplash.html). One is able to study the finds in the shipwreck through an interactive GIS-based diving.

61.6 GIS in Archeological Research as an Analytical Tool

61.6.1 Original Sites, Their Inventories or "Empty Spaces"

The dispersion or distribution of archeological finds can be studied and modeled by different spatial analyses in GIS. Distances between finds and clustering of finds, for example, bear a significance which is left for an archeologist to interpret in each case. But the ways in which the finds appear and are preserved are the outcome of different processes that do not necessarily represent the situation of the original production process, use or time of burial of the cultural material. Critical thinking is needed in assessing patterns of the past remains left by humans. Understanding the site formation process is essential in order not to draw simplistic conclusions of the observed spatial patterns.

There is a concept in archeology known as *the "Pompeii premise"* that emerged in the academic debates between Lewis Binford (1981) and Michael Schiffer (1985). The debates and behavioral studies by Schiffer have changed our understanding of archeological formation processes and the outcome of site inventories. Earlier it was thought that the artifacts at sites represent original and undisturbed inventories much like in the case of Pompeii. This famous city was buried by the eruption of Vesuvius in Italy in A.D. 79 and found preserved in a mummified state after ca. 1700 years. Seldom, however, is the past fossilized, mummified or frozen for the benefit of the present archeological research to discover, like the town of Pompeii with its seemingly original and intact site inventory. Site inventories can differ and are divided into sites of production, sites of use and sites of discard. (See, e.g., Schiffer 1987).

Beside human behavior and impact, nature is an influential factor in changing the contents of archeological inventories. Depositional processes are linked with erosion. Two major mechanisms have been defined to distinctively affect site formation: 1) natural formation processes (N-transforms); and 2) cultural transformation processes (C-transforms) (Renfrew and Bahn 2000). As a spatial discipline, archeology tends to derive conclusions from what is seen and what is not seen on different surfaces. Distribution maps, now produced by automated geographic information, are used as bases of arguments. An oscillation in the amounts and types of finds in different periods has often been explained with different environmental or social reasons. Models have been constructed for displaying, predicting and explaining such oscillations. GIS also can be used in approaching studies of depositional processes in archeology. Different stratigraphic layers can be displayed and the development of structures and sites better analyzed and understood.

Seemingly empty spaces can, in the course of archeological studies, become meaningful areas with human traces. For example, deserts and steppes were long seen as "empty spaces" devoid of human habitation. Interestingly, earlier nomads were thought to leave little, if nothing, behind for archeologists to trace their past activities. Nomads tend to use perishable materials and as mobile people, like hunter-gatherers, they do not leave much remains of permanent nature. However, new methods benefiting from anthropology, and especially ethnoarcheology, have enabled archeologists to detect nomadic remains and see deserts and steppes as areas providing clear traces of human activities. Despite the lessons learned from "empty spaces," arguments *ex silentio* still continue to be used and are supporting different theories and models without any awareness of the impact of possible post-burial processes. For example, it is precarious to argue that an ancient social group of hunter-gatherers living on a seashore did not go fishing because archeologists have not found any fishing hooks in the ancient habitation sites of the region. In addition, erosion is a dynamic factor in the Mediterranean world, and the simplistic conclusions concerning the lack of evidence may be drawn if geo-archeological processes are not fully taken into account in studying the sites.

61.6.2 Spatial Analyses and Modeling Using GIS

Site catchment analysis is a method for studying site location strategies. It is based on finding the resources which individual sites exploited and relied on for their existence (Clarke 1978). This analysis has been especially used in hunter-gatherer studies in studying the surrounding areas within certain distances of finds. GIS offers association of layered environmental background information from different data sources and capabilities for spatially tracing the natural resources that evidently have been exploited within a certain radius of a site. However, information concerning the development of environment and landscape during the period under scrutiny has to be at hand for one to be able to draw relevant conclusions. See, for example, a site catchment study of the human activity and paleoenvironment in the Jomon Period in Japan by Watanabe (2004).

Special activity locations are studied in *spatial intra-site analyses.* At the micro-scale, patterns of soot in soil may reflect activities such as cooking. Distribution studies of debitage (i.e., waste material left from the construction of stone tools) may lead to the identification of flint workshops or quarries. Studies of the relationship between domestic spaces and finds has been carried out using GIS (see, e.g., Meffert 1995 and Csáki et al. 1995). On the other hand, *spatial inter-site analyses* open views to inter-connections of the sites over larger geographic areas. This can be started at the micro scale by studying components of pottery and origins of clays and then proceeding to larger networks of production. Macro-scale studies may, for example, result in maps of the dispersion of hill-forts in a studied country or caravan routes in a desert reflecting long-distance trade.

Geographic theories and models have been integrated into archeology in order to trace larger territorial and political units. Christaller's central place theory (Christaller 1966) has been applied to archeology and used in its GIS analyses. Peerpolities of early states and empires have been defined as spatial socio-political entities with distinct cultural implications (Renfrew 1986). Their remains are believed to have been dispersed in certain regularity as a reflection of governmental purposes and power. GIS analyses such as the creation of *Thiessen* or *Voronoi* polygons have been especially useful to assess territorial or political units through archeological evidence. An example of their application in studying territories in arche-ology is the study of Danebury, an English Iron Age hill fort, by Lock and Harris (1996). Difficulties in executing large inter-site analyses between sites and cultures is faced when one gathers information of sites that have been surveyed and recorded during the past two hundred years without information of exact locations. In this case, the researcher has to "dig up" the old information that once was recorded or left unrecorded and standardize it for the use of modern GIS.

GIS-based *viewshed analyses* have been applied for studying territorial, political, ethnic or religious questions. They have employed the benefits of topographic maps and digital elevation models. However, possible ancient environments have to be taken into account, e.g., presently bare areas may have been forests that prevented visibility in ancient times. The intervisibility of the sites, such as between forts, barrows—see, e.g., a cumulative viewshed analysis by Wheatley (1995), megalithic structures or rock carvings has hypothetically borne a particular significance for ancient people. This assumption has worked as a basis for several archeological viewshed analyses. A military organization with watch towers and signaling systems has traditionally been thought to be dependent on intervisibility. A viewshed analysis was, for example, carried out between three Late Roman forts on the Euphrates in order to study their time of construction and military organization (Lönnqvist et al. 2005) (Figure 61-15). Furthermore, territoriality has been assumed to have gained benefits from intervisibility. The intervisibility of ritual sites is also based on similar assumptions, when no written records are available. It needs to be noted that the concept of visibility itself has changed during past decades. In the 1970s, it was largely understood as an environmental variable. Through the 1980s and 1990s, however, it has come to represent more of a human perceptual act differing from sole static dependence on environment to include its abstract impact on humans. Mobility has come to play a new part in visibility analysis since it is understood that views are changing while the viewer moves in the landscape. Higuchi's insights (Wheatley and Gillings 2000) have affected viewshed analysis by leveling the stages depending on distance in which a view is revealed. Different parameters, such as haziness and object background also are taken into account. A horizon is acknowledged as one of the most important factors affecting the visibility (Wheatley and Gillings 2000).

Figure 61-15 A viewshed analysis of Late Roman forts at Qreiye, Tabus and Tibne along the Euphrates. Elevation data and ground control points were used in the analysis carried out by Markus Törmä in 2005. The analysis is displayed on a Landsat -7 ETM image. Courtesy: SYGIS - the Jebel Bishri Project, University of Helsinki. See included DVD for color version

Natural phenomena such as shore displacement can be studied and analyzed in GIS. Shore displacement serves as a relative dating method in isostatically rising parts of the world (see, e.g., Nunez et al. 1995). Possibilities of calculations with GIS also apply to the natural phenomena such as river migrations, salinization, erosion, spread of deforestation and desertification. The rate of population movements such as migration, transportation of goods and trade and the diffusion of ideas can be studied and displayed through GIS. Even datings, such as crossdatings through Egyptian calendaric years, and the rate of inception of particular innovations can be displayed in GIS.

Astroarcheology has become a recognized field of scientific archeology, and it already has a journal of its own. Spatial orientations of ancient structures including tombs belong to this field. The azimuths are measured and their astronomical significance is analyzed. Empirical testings of solar and lunar sightings can also elucidate the orientations of ancient buildings and even date their construction times. For the visibility of celestial objects, GIS programs can be used to reconstruct ancient landscapes and test the visibilities. Software such as Sky Map can be utilized to reconstruct the ancient skies at the studied sites in given years and hours. Ancient texts also are used to evaluate the meaning of astronomy to ancient societies. (see e.g., Lönnqvist and Lönnqvist 2002, 2004) (Figure 61-16).

In cultural resource management (CRM), *predictive models* are often based on the general rule that sites cluster around certain environmental features (Altschul 1990; Warren 1990) such as waterways and springs. *Practical Applications of GIS for Archaeologists,* is a book that deals with predictive modeling in archeology (Westcott and Brandon 2000). From predictive models, prehistoric site distribution has been traced, for example, in Montana, in the US (Carmichael 1990). The web-based US *Minnesota Archaeological Predictive Model* (Minnesota DoT 2005) is also an example of tools used successfully to predict archeological sites in a landscape. The *Fort Drum Project* is a case study in historical predictive modeling in the Fort Drum military reservation area (see Hasenstab and Resnick 1990). Besides the environmental indications referred to above, there are approaches that offer possibilities to quantitatively trace the find/site locations of archeological remains in specified spaces (Marozas and Zack 1990). Simulation models are used for quantifying possible dispersion of archeological remains and of ancient natural sources, and for building predictive models (Van West and Kohler 1996).

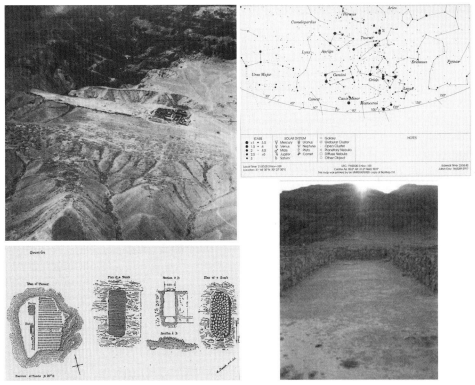

Figure 61-16 Spatial and archeoastronomical studies have been carried out on the orientations of the architectural structures and cemetery at Qumran on the shore of the Dead Sea in Israel. The upper left image shows the site in a 1950s aerial photograph acquired by the Jordanian Air Force. In the center below is a old "map" by C. Clermont-Ganeau from 1867 showing the consistent 20° East of North orientations of the settlement and the graves of the cemetery. Courtesy: Photo, École Biblique et Archéologique Française, Jerusalem, and map from Palestine Exploration Fund, London. The upper right map displays the Qumran sky at night in 160 B.C. reconstructed by astrophycisist Reino Anttila in 1996, University of Helsinki with SkyMap software. Lower right image captures a summer solstice sunset as seen at Qumran showing the lining of the main communal structure Locus 77 in June 1996. Courtesy: Minna Lönnqvist, Kenneth Lönnqvist and Reino Anttila in 2002.

61.7 GIS as an Analytical Tool: Contribution to the Archeological Research

In this Section we provide a simplified example of GIS analysis for an excavation site selection which is the result of a geographic analysis and may be supported by GIS commercial packages (Section 61.7.1). Finally, we conclude by mentioning the research and industrial directions towards an enhanced geographic analysis (Section 61.7.2).

61.7.1 Excavation Site Selection Based on a Sequence of GIS Operations

A simplified example for the task of site selection for a future excavation using GIS analysis functions is given next. The basic approach is to create a set of constraints which restrict the activity, and a set of opportunities which are conducive to the activity. The procedure of site selection is based on these sets of constraints and opportunities, and consists of a sequence of operations which extract the best locations for the activity.

In the simplified situation that follows, the set of constraints and opportunities consists of the following.

- vacant area (i.e., non-built area)
- dry land
- level and smooth site (i.e., slope < 10%)
- south-facing slope

A wider set could be taken into account, but this subset is enough to illustrate some basic data-interpreting operations available in GIS.

In addition, all candidate sites should have an adequate size to satisfy the needs of the activity (e.g., between 1 and 1.5 sq km).

The whole task requires as input three layers of the region under examination:

- hypsography layer: the three-dimensional surface of the region (altitude values),
- development layer: it depicts the existing infrastructure of the region (e.g., roads, buildings, etc.), and
- moisture layer: it depicts the soil moisture of the region (e.g., lakes, wet-lands, dry-lands, etc.).

For the purposes of processing, an artificial layer of the area under study called *dummy* is used, which consists of all individual locations of the area with no attribute values assigned to them (i.e., no zones are defined). This layer is commonly used in Zonal (search) operations as the first operand (notice that the second operand is the mask layer).

The procedure of site selection, based on the sets of constraints and opportunities determined above, may consist of the sequence of operations shown in Figure 61-17. The general syntax adopted for the operations is this: *new-layer* = Operation-class (operation-subclass) of *existing-layer.*

1. Vacant areas: A new layer of vacant areas is produced from the layer of development by classifying, generalizing and finally performing a selective search on the result (Figure 61-17a).
   - *development-classes* = Local (classification) of *development*
   - *vacant-developed* = Local (generalization) of *development-classes*
   - *vacant* = Zonal (search) of *dummy* and *vacant-developed*

2. Dry lands: A new layer of dry lands is produced from the layer of moisture by classifying, reducing detail and performing a selective search on the result (Figure 61-17b).
   - *moisture-classes* = Local (classification) of *moisture*
   - *dry-wet* = Local (generalization) of *moisture-classes*
   - *dry* = Zonal (search) of *dummy* and *dry-wet*

3. Level sites: A new layer of level and smooth sites is produced from the layer of hypsography by computing, classifying, generalizing and finally performing a selective search on the result (Figure 61-17c).
   - *slope* = Focal (surfacial) of *hypsography*
   - *slope-classes* = Local (classification) of *slope*
   - *level-steep* = Local (generalization) of *slope-classes*
   - *level* = Zonal (search) of *dummy* and *level-steep*

4. South-facing areas: A new layer of south-facing areas is produced from the layer of hypsography by computing, classifying, generalizing and finally performing a selective search on the aspects (Figure 61-17d).
   - *aspect* = Focal (surfacial) of *hypsography*
   - *aspect-classes* = Local (classification) of *aspect*
   - *south-north* = Local (generalization) of *aspect-classes*
   - *south* = Zonal (search) of *dummy* and *south-north*

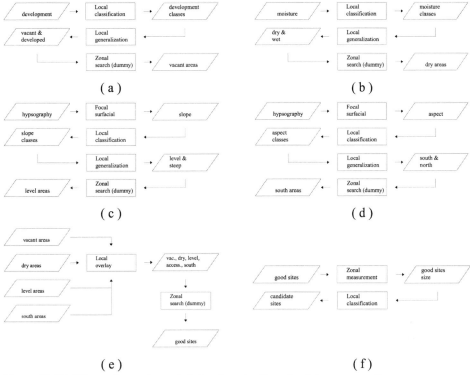

Figure 61-17 GIS zonal search analysis for site selection of a future excavation.

5. Good-sites: A new layer of sites that satisfy the set of constraints and opportunities is produced by the successive overlay of layers produced in the previous steps. Finally, good sites are highlighted by performing a selective search on the result (Figure 61-17e).
   - *vacant-dry* = Local (overlay) of *vacant* and *dry*
   - *vacant-dry-level* = Local (overlay) of *level* and *vacant-dry*
   - *vacant-dry-level-south* = Local (overlay) of *south* and *vacant-dry-level*
   - *good-sites* = Zonal (search) of *dummy* and *vacant-dry-level-south*

6. Candidate sites: A new layer of sites that satisfy the set of constraints and opportunities and have adequate size is produced from the layer of good sites by measuring the sizes of zones and highlighting those that are within the predefined size interval (Figure 61-21f).
   - *good-sites-size* = Zonal (measurement) of *good-sites*
   - *candidate-sites* = Zonal (search) of *dummy* and *good-sites-size*

61.7.2 Advanced Analytical Capabilities

The discussion above focuses on the basic analytical functionality offered by current GIS software packages. Researchers and professionals have spent considerable effort enhancing this functionality. The enhanced functionality may definitely be used to support archeological activities (Stefanakis and Sellis 1998). In the next paragraphs we briefly mention the directions of developments in GIS technology.

On the one hand, a more advanced decision-making process can be achieved by incorporating fuzzy logic methodologies into data analysis operations. Fuzzy set theory (Zadeh 1965, 1988) provides useful concepts and tools for representing geographic information. Additionally, fuzzy analysis procedures are clearly relevant to many areas of GIS and archeology because they provide a means of dealing with uncertain data (Wang et al. 1990; Leung and

Leung 1993; Altman 1994; Stefanakis et al.. 1999). Several commercial GIS software packages (such as IDRISI, ArcGIS, etc.) already provide fuzzy analytical functionality to their users.

On the other hand, during the last decade geographic-information scientists are more and more focused on the dimension of time and how it can be modeled and handled (Peuquet 1984, 2002). Spatio-temporal database systems are already present in the market (e.g., Oracle®) and current technology may support their management efficiently. In combination with uncertainty modeling and reasoning, an advanced tool for fuzzy spatio-temporal analysis is more and more available to scientists, professionals and researchers.

Finally, we should mention current developments in computer science in the area of data mining and knowledge discovery. Several application domains such as geography, archeology, molecular biology, astronomy, etc., produce a tremendous amount of data. All these data require automated analytical tools to be managed (Han and Kamber 2000). Hence, there is an increasing need for efficient data mining methods to extract useful information contained implicitly in GIS data. Very effective methods and tools have been developed to extract hidden knowledge and useful information from large geographic databases (Han et al. 2001, Miller and Han 2001). All of the above are novel tools and open new horizons in archeological research.

61.8 Case Studies—Applications of GIS in Archeology

The scope of this Section is to briefly present a couple of applications of GIS in archeology to offer the reader an idea of the capabilities of current technology in cultural management.

61.8.1 The ARCHAEOTOOL Prototype

In this Section we present the *ARCHAEOTOOL* prototype (EPET-II Project, funded by the General Secretariat of Research and Technology of Greece). ARCHAEOTOOL is a computer-based tool to assist archeological excavations and recordings; it has been implemented by coupling a commercial PC-GIS package with a commercial PC-DBMS, in order to meet the requirements of end-users by providing them a system with high usability.

Archeological excavations constitute the way to approach societies and civilizations of the past, especially when written evidences are poor or non-existent. However, the destructive nature of the excavation procedure makes necessary the adoption of documentation methods for all accumulated information, so that a systematic and objective recording of archeological findings and stratigraphic relations is achieved. The contemporary needs regarding both the excavation methodology and systematic documentation of information lead to the accumulation of a large set of data of various types, such as texts, catalogs, pictures, drawings, measurements, etc. The recording, management, and scientific exploitation of information derived from excavations constitute a hard task, which can be significantly assisted by the use of computer technology.

The research project ARCHAEOTOOL (1997) resulted in the implementation of a prototype system, which adopts the capabilities offered by computer technology, and is able to assist the recording and exploitation of the accumulated information from the excavation procedure. This prototype system has been adopted to assist the prehistoric excavation of Akrotiri in Thira Island and the classical excavation of Dion in Olympus Mountain.

The prototype consists of a set of tools which offer the following: a) simplification of the work performed by the archeologists on a daily basis and improvement of the results quality; b) an integrated and accurate recording of both thematic and geometric information characterizing the excavation findings; c) generation of thematic maps of excavation areas based on the information attached to them; and d) support of both geometric and thematic queries posed by archeologists as part of their research activities.

The general architecture of the prototype system appears in Figure 61-18. A traditional DBMS is used to handle the thematic information assigned to excavation findings, while a commercial GIS package has been adopted to describe their geometry. Common ID values assigned to the entities of both systems serve as links of thematic with geometric information. The two systems provide user-friendly interfaces and may be called and work independently. However, special functions are implemented to control and assure the consistency between the two databases (thematic and geometric).

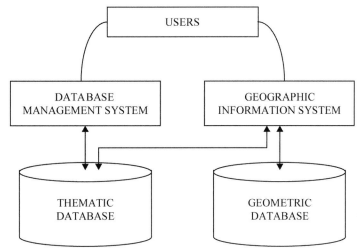

Figure 61-18 ARCHAEOTOOL system architecture.

Figure 61-19 illustrates the workflow for recording the information accumulated by the excavation procedure. All products from excavation (i.e., diary pages, filled forms, photographs, plans, etc.) are inserted into the databases daily. Appropriate forms are designed and implemented to support this task. In addition, the system provides the possibility to generate all plans and blank forms required for the excavation of the next day.

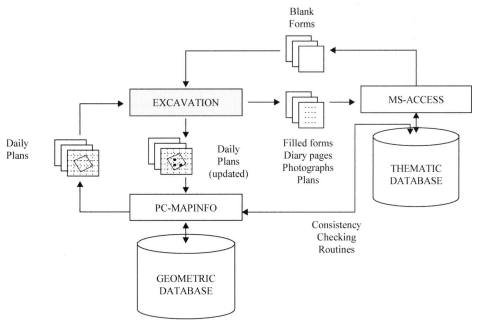

Figure 61-19 The general workflow for recording the information accumulated from the excavation.

Manual of Geographic Information Systems

The commercial systems used for the implementation of the prototype for the Akrotiri excavation are the MS-Access and the PC-MapInfo in the roles of the DBMS and GIS, respectively. A set of tools has been implemented in both systems to meet the users' requirements. Specifically, those tools assist in the following tasks: a) the generation of appropriate paper forms and diary pages to record information accumulated by the excavation procedure; b) the updating of the thematic database by inserting data of the filled paper forms into the system; c) the querying of the thematic database; d) the generation of findings catalogs and diary pages, so that publications may be supported; e) the statistical analysis and evaluation of excavation data; f) the generation of the excavation area plans; g) the updating of the geometric database; h) the searching of both the thematic and geometric databases; and i) the generation of thematic maps based on results derived from the statistical analysis.

A more advanced architecture, as the one shown in Figure 61-20, also has been built for the needs of the ARCHAEOTOOL project. This architecture provides a common interface to both subsystems, which has been developed using a graphic and object-oriented programming tool for MS-Windows® 95 (i.e., Visual Basic). Figure 61-21 illustrates a typical screen of the ARCHAEOTOOL prototype.

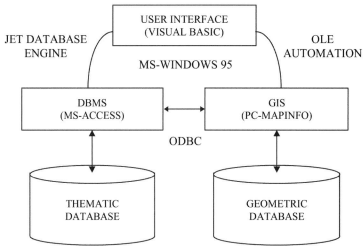

Figure 61-20 An advanced architecture for ARCHAEOTOOL system.

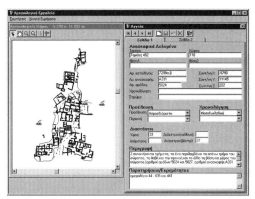

Figure 61-21 A typical screen of the ARCHAEOTOOL system.

61.8.2 The ARCHEOGUIDE Project

ARCHEOGUIDE (Augmented Reality-based Cultural Heritage On-site Guide) was a project funded by the EU IST (Information Society Technologies) Framework (Archeoguide 2002), with industrial partners from Greece, Germany, Italy and Portugal, and with the Hellenic Ministry of Culture in the role of the user. The project has been developed for the archeological site of ancient Olympia, Greece.

Project objectives are twofold. The first objective is to develop a virtual Guide to cultural heritage sites that will provide personalized tours to the visitors according to their individual preferences and provide virtual reconstruction of selected monuments through augmented reality techniques (Guile 2004). The second objective is to develop a database of scientific information about such sites that is readily accessible to archeologists and other scientists from remote locations through intranet/Internet.

The core idea is as follows. The visitor arrives at a cultural heritage site and is provided with a high-tech "wearable computer," which includes a lightweight portable computer and a head mounted display (glasses, earphone, speaker, camera, etc.). Then the visitor enters personal preferences and proceeds to walk through a tour customized by the system in response to the personal preferences specified. The visitor is then able to "see" a reconstructed ancient monument placed where the ruins lie (Figure 61-22). While listening to the information the system provides, the visitor may interact with the system, e.g., request more information from the system.

Figure 61-22 Augmented reality of reconstructed ancient monument placed where the ruins lie as seen by a visitor with a wearable computer and special glasses. See included DVD for color version.

The overall system architecture of the augmented reality system is shown in Figure 61-23. The user is equipped with a head-mounted display. Its position and look direction are calculated by positioning and photogrammetric techniques. The server provides the user with personalized information through the wireless network and based on the authored scenario.

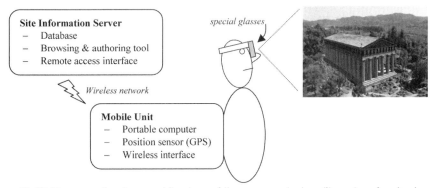

Figure 61-23 The overall system architecture of the augmented reality system for viewing reconstructed ancient monuments. See included DVD for color version.

Manual of Geographic Information Systems

The site information server comprises three separate modules: (a) the data manager; (b) the GIS authoring tool; and (c) a set of tour editors (edit schedules, scripts, new media objects in the DBMS). The GIS authoring tool is where all multimedia information (images, video and sound) is assigned its spatial dimension.

Finally, several other tools have been developed during the project, such as the navigation tool (Figure 61-24a), the internet tool (Figure 61-24b), etc., which support the visitor in an actual or virtual visit to the archeological site of Olympia.

Figure 61-24 The Internet tool (a) and the navigation tool (b) for actual or virtual visits to archeological sites. See included DVD for color version.

References

Aldenderfer, M. and H.D.G. Maschner. 1996. *Anthropology, Space, and Geographic Information Systems.* Spatial Information Systems. New York: Oxford University Press. 294 pp.

Allen, K.M.S., S. W. Green and E.B.W. Zubrow, editors. 1990. *Interpreting Space: GIS and Archaeology.* Series: Applications of Geographic Information Systems. London: Taylor & Francis. 398 pp.

Altman, D. 1994. Fuzzy set theoretic approaches for handling imprecision in spatial analysis. *International Journal of Geographical Information Systems* 8:271–289.

Altschul, J. H. 1990. Red flag models: The use of modeling in management contexts. In *Interpreting Space: GIS and Archaeology*, edited by K.M.S. Allen, S. W. Green and E.B.W. Zubrow, 226–238. London: Taylor & Francis.

Archaeotool. 1997. *User's Guide.* Unpublished manuscript, project deliverable submitted to General Secretariat of Research and Technology of Greece.

Archeoguide. 2002. Archeoguide: Augmented Reality-based Cultural Heritage On-site Guide. <http://archeoguide.intranet.gr/> Accessed 20 March 2009.

Banning, E. B. 2002. *Archaeological Survey.* Manuals in Archaeological Method, Theory and Technique. New York: Kluwer Academic/Plenum Publishing. 273 pp.

Barba, L. 2003. The geophysical study of buried archaeological remains and the preservation of the architectural patrimony of Mexico City. Pages 401–405 in *Proceedings of the XIXth International Symposium CIPA 2003, New Perspectives to Save the Cultural Heritage,* Antalya, Turkey, 30 September – 04 October 2003. *The International Archives of Photogrammetry, Remote Sensing and Spatial Information Sciences,* Vol. XXXIV, Part 5/C15, edited by O. Altan. Turkey: CIPA – The ICOMOS/ISPRS Committee for Documentation of Cultural Heritage.

Barber, M. 1999. Archaeological monuments: standardization and explanation. In *Our Fragile Heritage, Documenting the Past for the Future,* edited by H. J. Hansen and G. Quine, 113–122. Copenhagen: The National Museum of Denmark.

Barber, M. and van Regteren Altena, J. F. , eds. 1999. *European Bronze Age Monuments: A Multilingual Glossary of Archaeological Terminology.* Strasbourg: Council of Europe Publishing.

Bedford, M. 2004. Field data capture solutions. *GEO:connexion* 3 (July/Aug):56–60.

Bekiari, C., D. Calomirakis, P. Constantopoulos and P. Pantos. 1999. POLEMON: a project to computerize the monuments records of the Greek Ministry Of Culture. In *Our Fragile Heritage, Documenting the Past for the Future*, edited by H. J. Hansen and G. Quine, 139–146. Copenhagen: The National Museum of Denmark.

Bewley, R. H. and W. Rączkowski, eds. 2002. *Aerial Archaeology, Developing Future Practice.* Amsterdam: IOS Press.

Binford, L. R. 1981. Behavioural Archaeology and the "Pompeii Premise". *Journal of Anthropological Research* 37:195–208.

Boaz, J. S. and E. Uleberg. 1995. The potential of GIS-based studies of Iron Age cultural landscapes in eastern Norway. In *Archaeology and Geographic Information Systems*, edited by G. Lock and Z. Stančič, 249–259. London: Taylor & Francis.

Bodo, C. and A. Cicerchia. 2000. A risk map of the cultural heritage in Italy. *World Culture Report 2000, Cultural Diversity, Conflict and Pluralism.* Paris, UNESCO publishing, Paris. 150 pp.

Bosqued, C. B., J. B. Presyler and J. Espiago. 1996. The role of GIS in the management of archaeological data: an example of application for the Spanish administration. In *Anthropology, Space and Geographic Information Systems*, edited by M. Aldenderfer and H.D.G. Maschner, 190–201. New York: Oxford University Press.

Bradley, R. 2000. *An Archaeology of Natural Places.* London: Routledge. 160 pp.

Brandon, J., T. Kludt and M. Neteler. 1999. Archaeology and GIS – The Linux Way. *Linux Journal* <http://www.linuxjournal.com/article/2983> Unpaginated. Accessed 27 March 2009.

Breman, J. 2003. Marine archaeology goes underwater with GIS. *Journal of GIS in Archaeology.* Vol. I, April:23–31.

Brooks, R. R. and D. Johannes. 1990. *Phytoarchaeology.* Volume 3 of Historical, Ethno- and Economy Botany Series, edited by T. R. Dudley. Portland, Ore.: Dioscorides Press. 224 pp.

Butzer, K. 1982. *Archaeology as Human Ecology: Method and Theory for a Contextual Approach.* Cambridge: Cambridge University Press. 364 pp.

Cameron, F. and S. Kenderline, eds. 2007. *Theorizing Digital Cultural Heritage, A Critical Discourse.* Boston: MIT Press.

Canciani, M., P. Gambogi, F. G. Romano, G. Cannata and P. Drap. 2003. Low cost digital photogrammetry for underwater archaeological sites survey and artifact isertion, the case study of the Dolia wreck in Secche della Meloria – Livorno – Italia. *The International Archives of the Photogrammetry, Remote Sensing and Spatial Information Sciences.* Vol. XXXIV, Part 5/W12, pp. 95–100. <http://www.commission5.isprs.org/wg4/workshop_ancona/proceedings/20.pdf> Accessed 27 March 2009.

Canouts, V. 1999. The U.S. National Archaeological Database: points of access. In *Our Fragile Heritage, Documenting the Past for the Future*, edited by H. J. Hansen and G. Quine, 165–174. Copenhagen: The National Museum of Denmark.

Carmichael, D. L. 1990. GIS predictive modeling of prehistoric site distribution in central Montana. In *Interpreting Space: GIS and Archaeology*, edited by K.M.S. Allen, S. W. Green and E.B.W. Zubrow, 216–225. London: Taylor & Francis.

Childs, S. T. 1995. The curation crisis. Common Ground: Archeology and Ethnography in the Public Interest. 7(4): unpaginated. <http://www.nps.gov/archeology/cg/fd_vol7_num4/crisis.htm> Accessed 26 March 2009.

Christaller, W. 1966. *Central Places in Southern Germany*. Englewood Cliffs, N.J.: Prentice Hall. 230 pp.

Christopherson, G. L., D. P. Guertin and K. A. Borstad. 1996. GIS and Archaeology: Using ARC/INFO to Increase our Understanding of Ancient Jordan. ESRI User Conference 1996. <http://gis.esri.com/library/userconf/proc96/TO150/PAP119/P119.HTM> and <http://www.casa.arizona.edu/MPP/p119/p119.html> Accessed 26 March 2009.

Clarke, D. L., ed. 1972. *Models in Archaeology*. London: Methuen.

Clarke, D. L., ed. 1977. *Spatial Archaeology*. Boston: Academic Press.

Clarke, D. L. 1978. *Analytical Archaeology*, 2nd edition. London: Methuen.

Conolly, J. and M. Lake. 2006. *Geographic Information Systems in Archaeology*. Cambridge Manuals in Archaeology. Cambridge: Cambridge University Press.

Csáki, G., E. Jerem and F. Redö. 1995. Data recording and GIS applications in landscape and intra-site analysis: case studies in progress at the Archaeological Institute of the Hungarian Academy of Sciences. In *Archaeology and Geographic Information Systems*, edited by G. Lock and Z. Stančič, 85–99. London: Taylor & Francis .

Doneus, M., W. Neubauer and N. Studnicka. 2003. Digital recording of stratigraphic excavations. In *Proceedings of the XIXth International Symposium CIPA 2003, New Perspectives to Save the Cultural Heritage*, held in Antalya (Turkey) 30 September – 04 October 2003. *The International Archives of Photogrammetry, Remote Sensing and Spatial Information Sciences*, Vol. XXXIV, Part 5/C15, edited by O. Altan, 451–456. Turkey: CIPA – The ICOMOS/ISPRS Committee for Documentation of Cultural Heritage.

Ebert, J. I., author-editor, and J.-M. Dubois, M. Pinsonneault, B. A. Marozas, J. W. Walker, A. Lind, J. T. Parry, L. Wandsnider and E. Camilli, contributing authors. 1997. Archaeology and cultural resource management. In *Manual of Photographic Interpretation*, 2nd edition, edited by W. R. Philipson, 555–590. Bethesda, Md.: American Society for Photogrammetry and Remote Sensing.

Ebert, J. I. and T. R. Lyons, author-editors; B. W. Bevan, E. L. Camilli, S. Dennett, D. L. Drager, R. Fanale, N. Hartmann, H. Muessig and I. Scollar, contributing authors. 1983. Archaeology, anthropology, and cultural resources management. In *Manual of Remote Sensing*, 2nd edition, edited by R. N. Colwell, Vol II: Interpretation and Applications, edited by J. E. Estes, 1233–1304. Falls Church, Va.: American Society of Photogrammetry.

EH/RCHME (English Heritage & Royal Commission on the Historical Monuments of England). 1997. Archaeological Objects Thesaurus. <http://www.mda.org.uk/archobj/archcon.htm> Accessed 28 March 2009.

English Heritage. 1999. National Monuments Record Thesauri. <http://thesaurus.english-heritage.org.uk/> Accessed 27 March 2009.

Eppelbaum, L. V. and S. E. Itkis. 2003. Geophysical Examination of the Christian Archaeological Site Emmaus-Nicopolis (Central Israel). In *Proceedings of the XIXth International Symposium CIPA 2003, New Perspectives to Save the Cultural Heritage*, held in Antalya (Turkey) 30 September – 04 October 2003. *The International Archives of Photogrammetry, Remote Sensing and Spatial Information Sciences*, Vol. XXXIV, Part 5/C15, edited by O. Altan, 395–400. Turkey: CIPA – The ICOMOS/ISPRS Committee for Documentation of Cultural Heritage.

Evans, D. L., E. R. Stofan, T. D. Jones and L. M. Godwin. 1994. Earth from sky: radar systems carried aloft by the space shuttle Endeavour provide a new perspective of the Earth´s environment. *Scientific American* 271 (6/Dec.):70–75.

Fairclough, G. and S. Rippon. 2002. Conclusion: archaeological management of Europe´s cultural landscape. In *Europe´s Cultural Landscape: Archaeologists and the Management of Changes*. EAC Occasional paper 2, edited by G. Fairclough and S. Rippon, 201–206. Brussels, Belgium.

Farchakh, J. 2003. The scepter of war, protecting Iraq's museum collections and archaeological sites in the event of an invasion. *Archaeology* May/June:14–15.

Forte, M. and A. Siliotti. 1997. *Virtual Archaeology, Re-Creating Ancient Worlds*. London: Thames & Hudson. 288 pp.

Gaines, S. 1984. The impact of computerized information systems on American archaeology: an overview of the past decade. In *Information Systems in Archaeology*, edited by R. Martlew, 63–76. Gloucester: Alan Sutton Publishing Limited.

Généraux, M. and F. Niccolucci. 2006. Extraction and mapping of CIDOC-CRM encoding from texts and other digital formats. In *The e-volution of Information Communication Technology in Cultural Heritage, Where Hi-Tech Touches the Past: Risks and Challenges for the 21st Century*, short papers from the joint event CIPA/VAST/EG/EuroMed 2006, held in Nicosia, Cyprus 30 October – 4 November 2006, edited By M. Ioannides, D. Arnold, F. Niccolucci and K. Mania. Budapest: EPOCH Publications.

Gruen, A., F. Remondino and L. Zhang. 2003. Computer reconstruction and modeling of the Great Buddha Statue in Bamiyan, Afghanistan. In *Proceedings of the XIXth International Symposium CIPA 2003, New Perspectives to Save the Cultural Heritage*, held in Antalya (Turkey) 30 September – 04 October 2003. *The International Archives of Photogrammetry, Remote Sensing and Spatial Information Sciences*, Vol. XXXIV, Part 5/C15, edited by O. Altan, 440–445. Turkey: CIPA – The ICOMOS/ISPRS Committee for Documentation of Cultural Heritage. <http://www.photogrammetry.ethz.ch/general/persons/fabio/cipa_buddha.pdf> Accessed 31 March 2009.

Gruen, A., F. Remondino and L. Zhang. 2004. 3D modeling and visualization of large cultural heritage sites at very high resolution: the Bamiyan Valley and its standing Buddhas. In *Proceedings of the International Society for Photogrammetry and Remote Sensing XXth Congress, The International Archives of the Photogrammetry, Remote Sensing and Spatial Information Sciences*, Vol. XXXV, Part B, edited by O. Altan, 603–608. Turkey: Organizing Committee of the XXth International Congress for Photogrammetry and Remote Sensing. <http://www.photogrammetry.ethz.ch/general/persons/fabio/buddha_istanbul.pdf> Accessed 31 March 2009.

Grussenmeyer, P. and J. Yasmine. 2004. Photogrammetry for the preparation of archaeological excavation, a 3d restitution according to modern and archive images of Beaufort Castle landscape (Lebanon). In *Proceedings of the International Society for Photogrammetry and Remote Sensing XXth Congress, The International Archives of the Photogrammetry, Remote Sensing and Spatial Information Sciences*, Vol. XXXV, Part B, edited by O. Altan, 809–814. Turkey: Organizing Committee of the XXth International Congress for Photogrammetry and Remote Sensing.

Guile, J. 2004. *Augmented Reality*. New York: AuthorHouse.

Han, J. and M. Kamber. 2000. *Data Mining: Concepts and Techniques*. The Morgan Kaufmann Series in Data Management Systems. San Francisco: Morgan Kaufmann.

Han, J., M. Kamber and A.K.H. Tung. 2001. Spatial clustering methods in data mining: a survey. In *Geographic Data Mining and Knowledge Discovery*, edited by H. J. Miller and J. Han, 1–28. New York: Taylor & Francis.

Harris, E. C. 1989. *Principles of Archaeological Stratigraphy*. London: Academic Press. 170 pp.

Hasenstab, R. J. and B. Resnick. 1990. GIS in historical predictive modeling: the Fort Drum Project. In *Interpreting Space: GIS and Archaeology*, edited by K.M.S. Allen, S. W. Green and E.B.W. Zubrow, 284–306. London: Taylor & Francis.

Hodder, I. 1978. *Simulation Studies in Archaeology*. New Directions in Archaeology. Cambridge: Cambridge University Press. 139 pp.

Hodder, I. and C. Orton. 1976. *Spatial Analysis in Archaeology*. New Studies in Archaeology 1. (ed. Renfrew, Colin) Cambridge, New York: Cambridge University Press.

Holloway, M. 1995. The preservation of past. *Scientific American* May:78–81.

INA (Institute of Nautical Archaeology). 2007. Welcome to the Institute of Nautical Archaeology! <http://ina.tamu.edu/inamain.htm> Accessed 27 March 2009.

Kantner, J. 2005. Sipapu—Chetrl Ketl Great Kiva. <http://sipapu.ucsb.edu/html/kiva.html> Accessed 27 March 2009.

Kennedy, D. 1998. Declassified satellite photographs and archaeology in the Middle East: case studies from Turkey. *Antiquity* 72 (277):553–561.

Knapp, B. A. and W. Ashmore. 2000. Archaeological landscapes: constructed, conceptualized, ideational. In *Archaeologies of Landscapes, Contemporary Perspectives*, edited by W. Ashmore and B. A. Knapp, 1–30. Oxford: Oxford: Blackwell.

Kvamme, K. L. 1980. Predictive model of site location in the Glenwood Springs Resource Area. *A Class II Cultural Resource Inventory of the Bureau of Land Managements's Glenwood Springs Resource Area.* Report submitted to US Bureau of Land management, Grand Junction District, Colorado, Montrose, Nickers and Associates, Colorado.

Kvamme, K. L. 1989. Geographic information systems in regional archaeological research and data management. In *Method and Theory in Archaeology*, Vol. 1, edited by M. B. Schiffer, 13–203. Tucson: University of Arizona Press.

Kvamme, K. L. 1995. A view from across the water: the North American experience in archaeological GIS. In *Archaeology and Geographic Information Systems,* edited by G. Lock and Z. Stančič, 1–14. London: Taylor & Francis.

Lambers, K. and M. Sauerbier. 2003. A data model for a GIS-based analysis of the Nasca Lines at Palpa (Peru). In *Proceedings of the XIXth International Symposium CIPA 2003, New Perspectives to Save the Cultural Heritage,* held in Antalya (Turkey) 30 September – 04 October 2003, 713–718. *The International Archives of Photogrammetry, Remote Sensing and Spatial Information Sciences*, Vol. XXXIV, Part 5/C15. Turkey: CIPA – The ICOMOS/ISPRS Committee for Documentation of Cultural Heritage.

Leung, Y. and K. S. Leung. 1993. An intelligent expert system shell for knowledge-based GIS: 1. The tools, 2. Some applications. *International Journal of Geographical Information Systems* 7:189–213.

Lock, G. R. and T. M. Harris. 1996. Danebury Revisited: An English Iron Age Hillfort in a Digital Landscape. In *Anthropology, Space and Geographic Information Systems*, edited by M. Aldenderfer and H.D.G. Maschner, 214–240. New York: Oxford University Press.

Lock, G. and Z. Stančič. 1995. *Archaeology and Geographic Information Systems.* London: Taylor & Francis.

Lönnqvist, M. and K. Lönnqvist. 2002. *Archaeology of the Hidden Qumran, The New Paradigm.* Helsinki: Helsinki University Press. 377 pp.

Lönnqvist, K. and M. Lönnqvist. 2004. A spatial approach to the ruins of Qumran at the Dead Sea. In *Proceedings of the International Society for Photogrammetry and Remote Sensing XXth Congress,* The International Archives of the Photogrammetry, Remote Sensing and Spatial Information Sciences, Vol. XXXV, Part B, edited by O. Altan, 616–621. Turkey: Organizing Committee of the XXth International Congress for Photogrammetry and Remote Sensing.

Lönnqvist, M. and M. Törmä. 2003. SYGIS - The Fnnish archaeological project in Syria. In *Proceedings of the XIXth International Symposium CIPA 2003, New Perspectives to Save the Cultural Heritage,* held in Antalya (Turkey) 30 September – 04 October 2003, 609–614. *The International Archives of Photogrammetry, Remote Sensing and Spatial Information Sciences*, Vol. XXXIV, Part 5/C15. Turkey: CIPA – The ICOMOS/ISPRS Committee for Documentation of Cultural Heritage.

Lönnqvist, M. and M. Törmä. 2004. Different implications of a spatial boundary, Jebel Bishri between the Desert and the Sown in Syria. In *The International Archives of the Photogrammetry, Remote Sensing and Spatial Information Sciences*, Vol. XXXV, Part B, edited by O. Altan, 897–902. Turkey: Organizing Committee of the XXth International Congress for Photogrammetry and Remote Sensing.

Lönnqvist, M., K. Lönnqvist, M. S. Whiting, M. Törmä, M. Nunez and J. Okkonen. 2005. Documenting, Identifying and Protecting a Late Roman – Byzantine Fort at Tabus on the Euphrates, International Cooperation to Save the World's Cultural Heritage. In *Proceedings of the XX International Symposium CIPA 2005,* held in Turin (Italy) 26 September -01 October 2005. Vol. 1, edited by S. Dequal, pp. 427–432. *The International Archives of Photogrammetry, Remote Sensing and Spatial Information Sciences*, Vol. XXXVI-5/C34. Italy: CIPA – The ICOMOS/ISPRS Committee for Documentation of Cultural Heritage.

Lyons, T. R. and J. Ebert, eds. 1978. *Remote Sensing and Nondestructive Archaeology.* Washington, DC: National Park Service and University of Mexico. 72 pp.

Madry, S.L.H. and C. L. Crumley. 1990. An application of remote sensing and GIS in a regional archaeological settlement pattern analysis: the Arroux River valley, Burgundy, France. In *Interpreting Space: GIS and Archaeology*, edited by K.M.S. Allen, S. W. Green and E.B.W. Zubrow, 364–380. London: Taylor & Francis.

Mairitsch, M. 2004. Cultural Heritage in Iraq, UNESCO-Activities from 1976 to 2003. In *Looted, destroyed, forgotten: protecting cultural heritage in war and world cultural heritage in Iraq*, summary documentation of the symposium Austrian Commission for UNESCO, held in Graz and Vienna 11-13 June 2003, 2nd enlarged edition, 15–19. Vienna: Austrian Commission for UNESCO.

Marozas, B. A. and J. A. Zack. 1990. GIS and archaeological site location. In *Interpreting Space: GIS and Archaeology*, edited by K.M.S. Allen, S. W. Green and E.B.W. Zubrow, 165–172. London: Taylor & Francis.

Meffert, M. 1995. Spatial relations in Roman Iron Age settlements in the Assendelver Polders, The Netherlands. In *Archaeology and Geographic Information Systems*, edited by G. Lock and Z. Stančič, 287–299. London: Taylor & Francis.

Miller, H. J. and J. Han, eds. 2001. *Geographic Data Mining and Knowledge Discovery*. New York: Taylor & Francis. 338 pp.

Minnesota DoT (Department of Transportation). 2005. Mn/Model Statewide Archaeological Predictive Model. <http://www.mnmodel.dot.state.mn.us/> Accessed 27 March 2009.

Musée du Louvre. 2009. Louvre Museum Official Website: Atlas database of exhibits. <http://cartelen.louvre.fr/cartelen/visite?srv=crt_frm_rs&langue=fr&initCritere=true> Accessed 27 March 2009.

NASA (National Aeronautics and Space Administration). 2004. NASA World Wind helps solve 3,000 year old mystery of ancient Ithaca, the island home of Homer's Odysseus. <http://worldwind.arc.nasa.gov/odysseus.html> Accessed 27 March 2009.

NCPTT (National Center for Preservation Technology and Training). Undated. NCPTT. <http://www.ncptt.nps.gov/> Accessed 26 March 2009.

NPS (National Park Service). 2007. Archeology Program. <http://www.nps.gov/archeology/collections/> Accessed 26 March 2009.

Nuñez, M., A. Vikkula and T. Kirkinen. 1995. Perceiving time and space in an isostatically rising region. In *Archaeology and Geographic Information Systems*, edited by G. Lock and Z. Stančič, 141-151. London: Taylor & Francis.

Oriental Institute. 2009. The Giza Plateau Mapping Project (GPMP). <http://oi.uchicago.edu/OI/PROJ/GIZ/Giza.html> Accessed 27 March 2009.

Oxley, I. 2005. English Heritage and Maritime Archaeology, the first three years. *Conservation Bulletin* 48 (Spring):4–7.

Palumbo, G., ed. 1994. *JADIS, The Jordan Antiquities Database and Information System, A Summary of the Data*. Amman: American Center of Oriental Research (ACOR). 334 pp.

Peuquet, D. J. 1984. A conceptual framework and comparison of spatial data models. *Cartographica* 21 (4):66–113.

Peuquet, D. J. 2002. *Representations of Space and Time.* New York: Guilford Press. 380 pp.

Poidebard, A. 1934. *La Trace de Rome dans le Désert de Syrie, Le Limes de Trajan à la Conquète Arabe, Recherches Aériennes (1925-1932). Bibliothèque archéologique et historique, tome XVIII.* Texte, Atlas. Paris: Paul Geuthner.

Quine, G. 1999. The CIDOC standard: an international data standard for recording archaeological sites and monuments. In *Our Fragile Heritage, Documenting the Past for the Future,* edited by H. J. Hansen and G. Quine, 105–121. Copenhagen: The National Museum of Denmark.

Reitz, E, J, L. A. Newsom and S. J. Scudder. 1996. Issues in environmental archaeology. In *Case Studies in Environmental Archaeology,* Interdisciplinary Contributions to Archaeology, edited by E. Reitz, L. A. Newsom and S. Scudder, 3–16. New York: Plenum Press.

Renfrew, C. 1986. Introduction: peer polity interaction and socio-political change. In *Peer Polity Interaction and Socio-political Change,* edited by C. Renfrew and J. F. Cherry, 1–18. New Directions in Archaeology. Cambridge: Cambridge University Press.

Renfrew, C. and P. Bahn. 2000. *Archaeology: Theories and Methods.* London: Thames & Hudson. 640 pp.

Richards, J. D. and N. S. Ryan. 1985. *Data processing in archaeology.* Cambridge Manuals in Archaeology. Cambridge: Cambridge University Press. 232 pp.

Rowe, J. 1953. Technical aids in anthropology: a historical survey. In *Anthropology Today,* edited by A. Kroeber, 895–940. Chicago: University of Chicago Press.

Schiffer, M. 1985. Is there a "Pompeii Premise" in archaeology? *Journal of Anthropological Research* 41:18–41.

Schiffer, M. 1987. *Formation Processes of the Archaeological Record.* Albuquerque: University of New Mexico Press. 428 pp.

Scollar, I. 1990. *Archaeological Prospecting and Remote Sensing.* Topics in Remote Sensing 2. Cambridge: Cambridge University Press. 674 pp.

Skeates, R. 2000. *Debating the Archaeological Heritage.* Duckworth Debates in Archaeology. London: Duckworth. 160 pp.

Slayman, A. L. 1997. The new Pompeii, excavations beneath the A.D. 79 level illuminate the history of the famous Roman resort. *Archaeology* 50 (6):26–34.

Smith, N. 1995. Towards a study of ancient Greek landscapes: the Perseus GIS. In *Archaeology and Geographic Information Systems,* edited by G. Lock and Z. Stančič, 239–248. London: Taylor & Francis.

SPMH. 1998. Remapping Angkor, remote sensing undermines old beliefs. *Archaeology* 51 (3 May/June):24.

Stefanakis, E. and T. Sellis. 1998. Enhancing operations with spatial access methods in a database management system for GIS. *Cartography and Geographic Information Systems* 25 (1):16–32.

Stefanakis, E., M. Vazirgiannis and T. Sellis. 1999. Incorporating fuzzy set methodologies in a DBMS repository for the application domain of GIS. *International Journal of Geographical Information Science* 13 (7):657–675.

Stringer, W. J. and J. P. Cook. 1974. *Feasibility Study for Locating Archaeological Village Sites by Satellite Remote Sensing Techniques.* Tech. Rep. Geophysics Institute, University of Alaska, Fairbanks.

Tripcevich, N. 2004. Mobile GIS in archaeological survey. *The SAA Archaeological Record* May:17–22.

UNESCO. 2008. UNESCO Charter on the Preservation of the Digital Heritage. <http://portal.unesco.org/ci/en/ev.php-URL_ID=13367&URL_DO=DO_TOPIC&URL_SECTION=201.html> Accessed 27 March 2009.

Ur, J. 2003. CORONA satellite photography and ancient road networks: a northern Mesopotamian case study. *Antiquity* 77 (295):102–115.

Van West, C. and T. A. Kohler. 1996. A time to rend, a time to sew, new perspectives on northern Anasagi sociological development in later prehistory. In *Anthropology, Space, and Geographic Information Systems*, edited by M. Aldenderfer and H.D.G. Maschner, 107–131. Spatial Information Systems. New York: Oxford University Press.

Wang, F., G. B. Hall and Subaryono. 1990. Fuzzy information representation and processing in conventional GIS software: database design and application. *International Journal of Geographical Information Systems* 4:261–283.

Warren, R. E. 1990. Predictive modeling of archaeological site location: a case study in the Midwest. In *Interpreting Space: GIS and Archaeology*, edited by K.M.S. Allen, S. W. Green and E.B.W. Zubrow, 201–215. London: Taylor & Francis.

Watanabe, N. 2004. A Study of Tempo-Spatial Change of Interaction between the Human Activity and Paleoenvironment in Jomon Period, Japan. In *Proceedings of the International Society for Photogrammetry and Remote Sensing XXth Congress, The International Archives of the Photogrammetry, Remote Sensing and Spatial Information Sciences*, Vol. XXXV, Part B, edited by O. Altan, 520–525. Turkey: Organizing Committee of the XXth International Congress for Photogrammetry and Remote Sensing.

Westcott, K. and J. Brandon, eds. 2000. *Practical Applications of GIS for Archaeologists*. London: Taylor & Francis. 160 pp.

Wheatley, D. 1995. Cumulative viewshed analysis: a GIS-based method for investigating intervisibility, and its archaeological application. In *Archaeology and Geographic Information Systems*, edited by G. Lock and Z. Stančič, 171–185. London: Taylor & Francis.

Wheatley, D. and M. Gillings. 2000. Vision, perception and GIS: developing enriched approaches to the study of archaeological visibility. In *Beyond the Map,* edited by G. Lock, 1–27. *Archaeology* and Spatial Technologies Series A: Life Sciences – Vol. 32. Amsterdam: IOS Press.

Wiseman, J. 1998a. Eagle eye at NASA. *Archaeology* 51 (4):12–17.

Wiseman, J. 1998b. Reforming academia. *Archaeology* 51 (5):27–30.

Zadeh, L. A. 1965. Fuzzy sets. *Information and Control* 8:338–353.

Zadeh, L. A. 1988. Fuzzy logic. *IEEE Computer* 21:83–93.

Zolfaghari, M. and A. Malian. 2004. Documentation of the Documentations of the King of the Kings. In *Proceedings of the International Society for Photogrammetry and Remote Sensing XXth Congress, The International Archives of the Photogrammetry, Remote Sensing and Spatial Information Sciences, Vol. XXXV, Part B,* edited by O. Altan, 530–535. Turkey: Organizing Committee of the XXth International Congress for Photogrammetry and Remote Sensing. <http://www.isprs.org/congresses/istanbul2004/comm5/papers/611.pdf> Accessed 20 March 2009.

Zubrow, E.B.W. 2006. Digital archaeology, a historical context. In *Digital Archaeology, Bridging Method and Theory,* edited by T. L. Evans and P. Daly, 10–31. New York: Routledge.

CHAPTER 62

Supporting Curriculum Development in Geographic Information Science and Technology: The GIS&T Body of Knowledge

David DiBiase, Michael DeMers, Ann Johnson, Karen Kemp,
Ann Taylor Luck, Brandon Plewe and *Elizabeth Wentz*

As applications of GIS and related geospatial technologies have proliferated in government, industry, and academia, the demand for educated workers has increased. In response to this demand, a loosely organized geospatial "education infrastructure" has taken shape in the US. The geospatial education infrastructure includes the educational institutions, government agencies, and private businesses that provide formal and informal education and training opportunities at all levels to learners of all ages (DiBiase et al. 2006). At present, however, it appears that demand still outpaces the supply of adequately educated graduates.

Not long ago, one informed observer estimated that "the shortfall in producing individuals with an advanced level of GIS education is around 3,000 to 4,000 [annually] in the U.S. alone" (Phoenix 2000, p. 13). Predicting a "serious shortfall of professionals and trained specialists who can utilize geospatial technologies in their jobs," the National Aeronautics and Space Administration (NASA) sponsored an ambitious workforce development initiative in 1997 (Gaudet et al. 2003, p. 21). The Assistant Secretary for Labor and Training of the US Department of Labor (DoL) has pointed to survey data indicating that "87 percent of geospatial product and service providers … had difficulty filling positions requiring geospatial technology skills" (DeRocco 2004, p. 2). In 2003, DoL identified geospatial technologies as a "high-growth" industry (Department of Labor n.d.). Respondents to an ASPRS industry survey complained not only about the "shortage of trained workers emerging from educational programs," but also about "the lack of the required skill sets among many of the graduates" (Mondello et al. 2004, p. 13).

Some educators share the view that the geospatial education infrastructure is turning out too few qualified graduates. In 1998, Duane Marble published an influential critique of the "low-level, non-technical" character of GIS education in undergraduate degree programs (Marble 1998, p. 28). Unlike the early days of GIS education, when the primitive state of the technology required students to master computer programming skills, Marble pointed out that latter-day students, and some instructors, believe that all one has to do to become a GIS professional is to master the standard functions of commercial off-the-shelf (COTS) software. Thus, graduates are no longer prepared "to make substantial contributions to the ongoing development of GIS technology" (Marble 1998, p. 1).

Marble identified a "pyramid" of six competency levels that undergraduate degree programs should prepare students to achieve (Figure 62-1). Public awareness of geospatial technologies constitutes the base of the pyramid. One level above the base is the relatively large number of workers who need to be prepared for careers involving "routine use" of COTS software and related geospatial technologies. A somewhat smaller number of graduates needed to work with "higher level modeling applications" within COTS software must possess knowledge and skills in spatial analysis, computer programming, and database management systems. More demanding and fewer still are "application design and development" roles that require workers to create software applications rather than to simply use them. Specialists responsible for "system design" require advanced analytical as well as technical skills, including system analysis, database design and development, user interface design, and programming. Finally, the peak of the pyramid represents the relatively small number (perhaps 10,000 or more worldwide) of individuals whose sophisticated understanding of geography, spatial analysis, computer science and information technology prepares them to lead "research and software development" teams within software companies, government agencies, and in universities. Marble (1998) argued that the base of the pyramid is expanding "at explosive rate while the upper levels have been permitted to crumble" (p. 29).

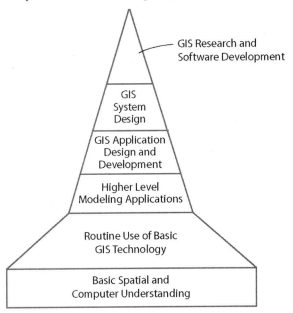

Figure 62-1 Six competency levels required for workers in geographic information science and technology (after Marble 1998). © 2006 Association of American Geographers and University Consortium for Geographic Information Science. Used by permission. All rights reserved.

62.1 GIS Curricula

Notwithstanding Marble's critique, and others, "GIS instructors in higher education have shown an almost exemplary concern for teaching" (Unwin 1997, p. 2). Since the late 1970s, GIS, cartography, and remote sensing educators have proposed frameworks to guide curriculum planning (e.g., Marble 1979, 1981; Dahlberg and Jensen 1986), published numerous textbooks (e.g., Burrough 1982; DeMers 1996; Longley et al. 2000), developed educational software products (e.g., GISTutor, OSU Map-for-the-PC, Map II), convened panel discussions, workshops, and entire international conferences devoted to teaching and learning, such as the 1997 GIS in Higher Education Conference (e.g., Goodchild 1985; Poiker 1985; Gilmartin and Cowen 1991), investigated professional job titles, salaries, qualifications, and recommended coursework (e.g., Huxhold 1991, 2000; Wikle 1994), compiled lists of core topics (e.g., Macey 1997), published local, national, and international GIS course syllabi and curricula (e.g., Tobler 1977; Nyerges and Chrisman 1989; Unwin 1990; Goodchild and Kemp 1992; Kemp and Frank 1996; Foote 1996), and demonstrated the propriety of including GIScience within general education curricula (DiBiase 1996). Concerns about educational effectiveness have also given rise to several national-scale geospatial curriculum development efforts, including those described below.

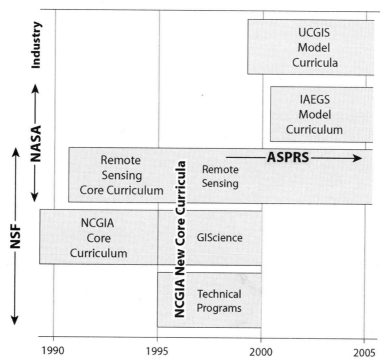

Figure 62-2 National-scale geospatial education curriculum initiatives in the US, 1988–2006. © 2006 Association of American Geographers and University Consortium for Geographic Information Science. Used by permission. All rights reserved.

62.1.1 NCGIA GIS Core Curriculum

The National Science Foundation's 1987 solicitation for a National Center for Geographic Information Analysis (NCGIA) included as one of its four goals "to augment the nation's supply of experts in GIS and geographic analysis in participating disciplines" (National Science Foundation 1987). In 1988, shortly after winning the NSF award, the NCGIA consortium of the University of California at Santa Barbara, State University of New York at Buffalo, and the University of Maine developed and distributed for comment "a detailed outline for a three-course sequence of 75 one-hour units" (Goodchild and Kemp 1992, p. 310). Fifty leading scholars and practitioners were recruited to prepare draft units. Over 100 institutions worldwide agreed to implement the resulting three-course sequence (Introduction to GIS, Technical Issues in GIS, and Application Issues in GIS) and to share assessment data with NCGIA. Lecture notes and laboratory exercises were revised extensively in response to user comments (e.g., Coulson and Waters 1991), and were subsequently published as the *NCGIA Core Curriculum* in July 1990. The *Core Curriculum* has had a significant impact—over 1,500 institutions requested the original print version.

62.1.2 NCGIA GIScience Core Curriculum

In 1995, NCGIA announced plans to develop a completely revised and expanded *Core Curriculum in GIScience* intended to account for developments in the field since the original 1990 *Core Curriculum.* The new curriculum was to include lecture notes corresponding to at least 176 hour-long units organized in a "tree" structure with fundamental geographic concepts as root nodes from which four branches—Fundamental Geographic Concepts for GIS, Implementing Geographic Concepts in GIS, Geographic Information Technology in Society, and Application Areas and Case Studies—were to spring. According to senior editor Karen Kemp (Kemp 2005), "Over a period of about four years, several new units were commissioned and a few units from the original core curriculum were updated bringing the total of units available in the new curriculum up to about one-third of the units envisioned in the new tree. While it was intended that the new curriculum become a living document with revisions to the tree structure and additions to the available units continuing indefinitely, by August 2000 the momentum for a document of this type slowed." The partial *NCGIA Core Curriculum in GIScience* (last updated in 2000) remains available for review at http://www.ncgia.ucsb.edu/giscc/.

62.1.3 Remote Sensing Core Curriculum

One of NCGIA's research initiatives (I-12) concerned the integration of remote sensing and geographic information systems. The need for educational materials that promoted integration was recognized early on. In 1992, NCGIA empanelled a steering committee responsible for guiding development of a remote sensing core curriculum. Initially the committee focused on four courses: Introduction to Air Photo Interpretation and Photogrammetry, Overview of Remote Sensing of the Environment, Introductory Digital Image Processing, and Applications in Remote Sensing. The four original courses, along with four subsequent ones, now appear as "volumes" at the project Web site (http://www.r-s-c-c.org/). The materials resemble online textbooks rather than lecture notes. Course authors, who include well-known educators and researchers, contributed voluntarily and without compensation. In 1995 the project secured funding from NASA to support student assistants to format and test the volumes. In 1997, ASPRS agreed to support the Remote Sensing Core Curriculum project (RSCC) through 2012. The RSCC is listed on the NCGIA Core Curricula Web site (http://www.ncgia.ucsb.edu/pubs/core.html) as an unofficial complement to its *Core Curricula for Technical Programs and GIScience*.

62.1.4 IAEGS Model Curriculum

In 2001, the University of Mississippi secured a $9 million contract from NASA to create resources that would increase the capacity of higher education institutions to prepare students for careers in the remote sensing industry. By 2005, the University's Institute for Advanced Education in Geospatial Sciences (IAEGS) had developed its own course management system and materials comprising thirty non-instructor-led, online undergraduate courses in remote sensing and geospatial technology (http://geoworkforce.olemiss.edu/). With ASPRS's help and NASA's funding, IAEGS commissioned sixteen leading educators and researchers to outline their model curriculum. Prospective course authors were invited to submit proposals; selected authors earned $80,000 each. Courses consist of text, graphics, animations, interactive quizzes, and other content delivered through IAEGS's delivery system. The initiative is based upon a novel business model that targets institutions that wish to offer remote sensing education but lack the necessary faculty resources. Adopters pay per-student fees and are expected to provide local faculty points-of-contact. Although they were intended for licensing by higher education institutions, most early adopters of IAEGS courses have been government agencies and private firms (Luccio 2005).

62.1.5 UCGIS Model Curricula

This most ambitious initiative arose from a set of eight education challenges identified at the 1997 UCGIS Summer Assembly in Bar Harbor, Maine. One challenge concluded that "improving GIScience education requires the specification and assessment of curricula for a wide range of student constituencies" (Kemp and Wright 1997, p. 4). A Task Force, chaired by Marble, was formed in 1998. With backing from leading GIS software vendors, the Task Force set out to create a new undergraduate curriculum in geographic information science and technology. In time, the plural "curricula" was adopted to emphasize the project's goal of supporting multiple curricular pathways tailored to the requirements of the diverse occupations and application areas that rely upon geospatial technologies. In 2003 the Task Force issued a "Strawman Report" that presented an ambitious vision of how higher education should prepare students for success in the variety of professions that rely upon geospatial technologies (Marble et al. 2003). Three characteristics of the UCGIS Model Curricula vision distinguish it from other curriculum planning projects in the geospatial realm. These include: (1) its broad and integrative conception of the "Geographic Information Science and Technology" (GIS&T) knowledge domain; (2) its adaptability to particular institutions; and (3) its adaptability to individual educational goals. Each of these distinguishing characteristics is considered below.

1. **Broad conception of the geospatial field:** As illustrated in Figure 62-3 below, the GIS&T knowledge domain envisioned by the Task Force and its successors encompasses three subdomains, including:

   • **Geographic Information Science,** the multidisciplinary research enterprise that addresses the nature of geographic information and the application of geospatial technologies to basic scientific questions;

   • **Geospatial Technology,** the specialized set of information technologies that support data acquisition, data storage and manipulation, data analysis, and visualization of georeferenced data; and

   • **Applications of GIS&T,** the increasingly diverse uses of geospatial technology in government, industry, and academia. The number and variety of fields that apply geospatial technologies is suggested in Figure 62-3 by the stack of "various application domains."

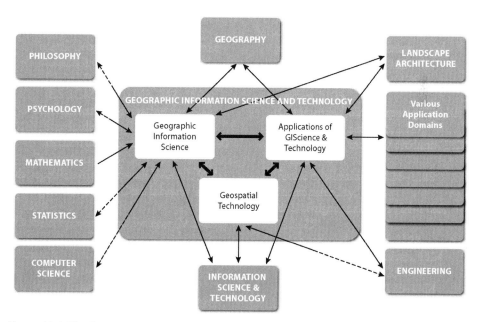

Figure 62-3 The three sub-domains comprising the GIS&T domain, in relation to allied fields. Two-way relations that are half-dashed represent asymmetrical contributions between allied fields. © 2006 Association of American Geographers and University Consortium for Geographic Information Science. Used by permission. All rights reserved.

2. **Adaptable to varied institutions:** From the outset the Task Force envisioned a curriculum that would be adaptable to the special circumstances of academic institutions and departments, as well as to learners and employers. The Task Force's successors describe a diverse "GIS&T education infrastructure" (Figure 62-4) that cultivates a range of competency levels—from basic awareness to research and development—through a lifetime of learning—from primary and secondary schools through postbaccalaureate and professional education (DiBiase et al. 2006).

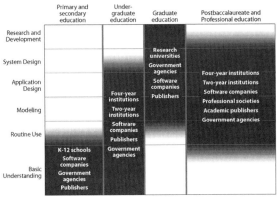

Figure 62-4 The GIS&T education infrastructure. Columns represent sectors of formal education that span a lifetime of learning. Rows correspond to levels of competency as described in Marble (1998). Shaded bands represent the competency levels cultivated within each sector. Informal education spans the learner's lifetime in parallel with formal education. © 2006 Association of American Geographers and University Consortium for Geographic Information Science. Used by permission. All rights reserved.

3. **Adaptable to individual educational goals:** Recognizing the multidisciplinary nature of the field, the Model Curricula Task Force envisioned an *adaptive* curriculum that students and advisors could tailor to suit individual aims. The GIS&T Model Curricula was to specify multiple curricular "pathways" by which learners could traverse the "bodies of knowledge" of GIS&T and allied fields, leading to a set of educational outcomes that reflect the diverse requirements of the various employers that comprise the geospatial industry. Figure 62-5 illustrates one such pathway.

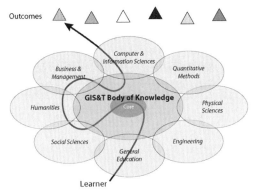

Figure 62-5 A prototypical curricular pathway through the bodies of knowledge of GIS&T and allied domains, leading to one of several educational outcomes. The pyramidal shape of symbols that represent outcomes imply that learners may achieve a range of mastery levels, from basic understanding to the advanced expertise needed for GIS&T research and development. © 2006 Association of American Geographers and University Consortium for Geographic Information Science. Used by permission. All rights reserved.

62.1.5.1 The GIS&T Body of Knowledge (BoK 1/e)

Determined to strengthen the role of computer science and information technology in GIS education, the Task Force consciously emulated the form and process of the *Computing Curricula* developed jointly by the Association of Computing Machinery (ACM) and Institute for Electrical and Electronic Engineers (IEEE). Accordingly, the Task Force began by attempting to specify the comprehensive set of knowledge areas and their constituent units and topics, which comprise a body of knowledge for the GIS&T domain. Following over seven years of deliberations involving more than 70 contributors and reviewers, the Association of American Geographers published the 1ˢᵗ edition of the *GIS&T Body of Knowledge* (*BoK 1/e*) in 2006. Like the bodies of knowledge included in recent computing curricula, *BoK 1/e* represents the GIS&T knowledge domain as a hierarchical list of knowledge areas, units, topics, and educational objectives. The ten knowledge areas and 73 units that comprise *BoK 1/e* are shown in Table 62-1. Twenty-six "core" units (those in which all graduates of a degree or certificate program should be able to demonstrate some level of mastery) are shown in bold type. Not shown are the 329 topics that make up the units, or the 1,660 education objectives by which topics are defined.

Table 62-1 Knowledge areas and units comprising BoK 1/e. Core units are indicated with bold type. © 2006 Association of American Geographers and University Consortium for Geographic Information Science. Used by permission. All rights reserved.

Knowledge Area AM. Analytical Methods
Unit AM1 Academic and analytical origins
Unit AM2 Query operations and query languages
Unit AM3 Geometric measures
Unit AM4 Basic analytical operations
Unit AM5 Basic analytical methods
Unit AM6 Analysis of surfaces
Unit AM7 Spatial statistics
Unit AM8 Geostatistics
Unit AM9 Spatial regression and econometrics
Unit AM10 Data mining
Unit AM11 Network analysis
Unit AM12 Optimization and location-allocation modeling

Knowledge Area CF. Conceptual Foundations
Unit CF1 Philosophical foundations
Unit CF2 Cognitive and social foundations
Unit CF3 Domains of geographic information
Unit CF4 Elements of geographic information
Unit CF5 Relationships
Unit CF6 Imperfections in geographic information

Knowledge Area CV. Cartography and Visualization
Unit CV1 History and trends
Unit CV2 Data considerations
Unit CV3 Principles of map design
Unit CV4 Graphic representation techniques
Unit CV5 Map production
Unit CV6 Map use and evaluation

Knowledge Area DA. Design Aspects
Unit DA1 The scope of GIS&T system design
Unit DA2 Project definition
Unit DA3 Resource planning
Unit DA4 Database design
Unit DA5 Analysis design
Unit DA6 Application design
Unit DA7 System implementation

Knowledge Area DM. Data Modeling
Unit DM1 Basic storage and retrieval structures
Unit DM2 Database management systems
Unit DM3 Tessellation data models
Unit DM4 Vector and object data models
Unit DM5 Modeling 3D, temporal, and uncertain phenomena

Knowledge Area DN. Data Manipulation
Unit DN1 Representation transformation
Unit DN2 Generalization and aggregation
Unit DN3 Transaction management of geospatial data

Knowledge Area GC. Geocomputation
Unit GC1 Emergence of geocomputation
Unit GC2 Computational aspects and neurocomputing
Unit GC3 Cellular Automata (CA) models
Unit GC4 Heuristics
Unit GC5 Genetic algorithms (GA)
Unit GC6 Agent-based models
Unit GC7 Simulation modeling
Unit GC8 Uncertainty
Unit GC9 Fuzzy sets

Knowledge Area GD. Geospatial Data
Unit GD1 Earth geometry
Unit GD2 Land partitioning systems
Unit GD3 Georeferencing systems
Unit GD4 Datums
Unit GD5 Map projections
Unit GD6 Data quality
Unit GD7 Land surveying and GPS
Unit GD8 Digitizing
Unit GD9 Field data collection
Unit GD10 Aerial imaging and photogrammetry
Unit GD11 Satellite and shipboard remote sensing
Unit GD12 Metadata, standards, and infrastructures

Knowledge Area GS. GIS&T and Society
Unit GS1 Legal aspects
Unit GS2 Economic aspects
Unit GS3 Use of geospatial information in the public sector
Unit GS4 Geospatial information as property
Unit GS5 Dissemination of geospatial information
Unit GS6 Ethical aspects of geospatial information and technology
Unit GS7 Critical GIS

Knowledge Area OI. Organizational and Institutional Aspects
Unit OI1 Origins of GIS&T
Unit OI2 Managing the GI system operations and infrastructure
Unit OI3 Organizational structures and procedures
Unit OI4 GIS&T workforce themes
Unit OI5 Institutional and inter-institutional aspects
Unit OI6 Coordinating organizations (national and international)

One of the most extensive of the ten knowledge areas in the *BoK 1/e* is "Analytical Methods." Twelve units, three of which are core units, comprise knowledge area AM (see Table 62-1). Fifty-nine topics, defined in terms of 281 educational objectives, comprise the twelve units. In many cases, objectives span the six "cognitive levels" and first three "knowledge types" identified in the *Taxonomy of Education Objectives* (Anderson and Krathwohl 2001). Also provided at the end of the knowledge area are references to 34 "key readings." An example core unit—AM4—appears in Table 62-2.

Table 62-2 Topics and educational objectives comprising core unit AM4: Basic analytical operations, from the Analytical Methods knowledge area of BoK 1/e. © 2006 Association of American Geographers and University Consortium for Geographic Information Science. Used by permission. All rights reserved.

Unit AM4 Basic analytical operations (*core unit*)
This small set of analytical operations is so commonly applied to a broad range of problems that their inclusion in software products is often used to determine if that product is a "true" GIS. Concepts on which these operations are based are addressed in Unit CF3 Domains of geographic information and Unit CF5 Relationships.

Topic AM4-1 Buffers
• Compare and contrast raster and vector definitions of buffers
• Explain why a buffer is a contour on a distance surface
• Outline circumstances in which buffering around an object is useful in analysis

Topic AM4-2 Overlay
• Explain why the process "dissolve and merge" often follows vector overlay operations
• Explain what is meant by the term "planar enforcement"
• Outline the possible sources of error in overlay operations
• Exemplify applications in which overlay is useful, such as site suitability analysis
• Compare and contrast the concept of overlay as it is implemented in raster and vector domains
• Demonstrate how the geometric operations of intersection and overlay can be implemented in GIS
• Demonstrate why the georegistration of data sets is critical to the success of any map overlay operation
• Formalize the operation called map overlay using Boolean logic

Topic AM4-3 Neighborhoods
• Discuss the role of Voronoi polygons as the dual graph of the Delaunay triangulation
• Explain how the range of map algebra operations (local, focal, zonal, and global) relate to the concept of neighborhoods
• Explain how Voronoi polygons can be used to define neighborhoods around a set of points
• Outline methods that can be used to establish non-overlapping neighborhoods of similarity in raster data sets
• Create proximity polygons (Thiessen/Voronoi polygons) in point data sets
• Write algorithms to calculate neighborhood statistics (minimum, maximum, focal flow) using a moving window in raster data sets

Topic AM4-4 Map algebra
• Describe how map algebra performs mathematical functions on raster grids
• Describe a real modeling situation in which map algebra would be used (e.g., site selection, climate classification, least-cost path)
• Explain the categories of map algebra operations (i.e., local, focal, zonal, and global functions)
• Explain why georegistration is a precondition to map algebra
• Differentiate between map algebra and matrix algebra using real examples
• Perform a map algebra calculation using command line, form-based, and flow charting user interfaces

---

62.1.5.2 Current Status and Next Steps

The notion of a community-developed and maintained body of knowledge is central to the Model Curricula vision. Publication of the *BoK 1/e* was a milestone in the ongoing effort to fulfill that vision. Still, as of this writing, much remains to be done.

A vital next step is to collect descriptions of various curricular pathways by which learners should traverse the *BoK 1/e*, and the bodies of knowledge of allied fields, as they pursue their particular individual goals. A collection of pathway descriptions will help the diverse institutions that make up the GIS&T education infrastructure to approach curriculum planning and revision strategically. By referencing their curricula to a common body of knowledge, accredited colleges and universities should find it easier to execute articulation agreements that facilitate the needs of a mobile workforce to transfer credits among institutions. Articulation agreements should also facilitate the actual and virtual exchanges of students and faculty members that enrich education for all concerned. Promoting synergies among the components of the education infrastructure is one of the most important potential benefits of the *BoK 1/e*.

A second next step will be to produce a revised second edition of the *BoK*. The 1st edition will likely prove not to be "representative of the views of a majority of the broad GIS&T community" (Marble et al. 2003, p. 27). Educators and practitioners will justifiably criticize *BoK 1/e* on various grounds, including the extent to which the three subdomains (Geographic Information Science, Geospatial Technology, and Applications of GIS&T) are adequately represented; which topics and objectives are included or left out; how topics are parsed into knowledge areas; the extent to which the necessary range of competency levels is supported; how the topical material should be sequenced; and how cross-cutting themes are distributed among knowledge areas. Rather than a single document like *BoK 1/e*, a second edition may be split into a family of specialized bodies of knowledge for GIS and Cartography, Remote Sensing, and perhaps Positioning. (A precedent is the current *Computing Curricula*, which includes bodies of knowledge and curricula for computer science, computer engineering, information systems, software engineering, and information technology.) In any case, given that the GIS&T field will continue to evolve in response to innovative science, disruptive new technologies, and imaginative applications, it is important that the GIS&T community remain committed to revising and improving its body (or bodies) of knowledge periodically, as has been done in allied fields. The cooperation of professional societies like ASPRS, AAG, GITA, and others, is needed to answer this challenge.

As geospatial technologies become more and more critical to the economy and security of the US, higher expectations for the accountability of the GIS&T education infrastructure seem likely to follow. Voluntary certification programs are available for practitioners who seek credentials that attest to their qualifications (i.e., ASPRS' Certified Mapping Scientist and Technologist programs in GIS/LIS, and the GIS Certification Institute's GIS Professional certification program, which is referenced to *BoK 1/e*). But unlike the allied fields of computer science and engineering, most of the academic departments that offer certificate and degree programs in GIS and GIScience are not subject to accreditation—the process by which organizations demonstrate their educational effectiveness. Few educators seem eager to embrace accreditation voluntarily, both because conventional accreditation mechanisms seem ill-suited to multidisciplinary fields like GIS&T, and because the associated paperwork is onerous (Obermeyer and Onsrud 1997). However, alternative accreditation processes that emphasize self-assessment, and that are adaptable to unique institutional contexts, have been proposed (DiBiase 2003). How effective is the GIS&T education infrastructure in producing qualified graduates? Community-based initiatives like the *GIS&T Body of Knowledge* are one way to make sure that the GIS&T field is able to answer this legitimate question on its own terms.

62.2 Acknowledgements

This chapter was adapted from the *GIS&T Body of Knowledge,* with kind permission of the publisher, the Association of American Geographers (AAG). The *GIS&T Body of Knowledge* is available at http://www.aag.org/bok © 2006 by the AAG and the University Consortium for Geographic Information Science (UCGIS); all rights reserved. Illustrations realized by Barbara Trapido-Lurie. The authors wish to thank our sponsors, including Intergraph, GE Smallworld, and especially ESRI, which provided financial support that helped to defray volunteers' travel expenses. Thanks also to the National Academies of Science, which commissioned some of the research reported in Sections 62.1.2 and 62.1.3, and graciously permitted its publication as part of the *GIS&T Body of Knowledge.*

References

Anderson, L. W. and D. R. Krathwohl, eds., with contributions by P. W. Airasian, K. A. Cruikshank, R. E. Mayer, P. R. Pintrich, J. Raths and M. C. Wittrock. 2001. *A Taxonomy for Learning, Teaching, and Assessing.* New York: Longman.

Burrough, P. 1982. *Principles of Geographic Information Systems for Land Resources Assessment.* Clarendon, Oxford: Oxford Science Publications.

Coulson, M.R.C. and N. M. Waters. 1991. Teaching the NCGIA curriculum in practice: Assessment and evaluation. *Cartographica* 28 (3):94–102.

Dahlberg, R. E. and J. R. Jensen. 1986. Education for cartography and remote sensing in the service of an information society: The U.S. case. *The American Cartographer* 13 (1):51–71.

DeMers, M. 1996. *Fundamentals of Geographic Information Systems.* New York: John Wiley & Sons.

Department of Labor (no date.) The President's High Growth Job Training Initiative. <http://www.doleta.gov/BRG/JobTrainInitiative> Accessed 21 August 2008.

DeRocco, E. S. 2004. Speech at AACC & ACCT National Legislative Summit. 10 February 2004, Washington, D.C. <http://www.doleta.gov/whatsnew/Derocco_speeches/AACC%20-%20Legislative.cfm> Accessed 21 August 2008.

DiBiase, D. 1996. Rethinking laboratory education for an introductory course on geographic information. *Cartographica* 33 (4):61–72.

———. 2003. On accreditation and the peer review of geographic information science education. *Journal of the Urban and Regional Information Systems Association* 15 (1):7–14. <http://www.urisa.org/files/Dibiasevol15no1.pdf> Accessed 21 August 2008.

DiBiase, D., M. DeMers, A. Johnson, K. Kemp, A. Taylor-Luck, B. Plewe and E. Wentz, eds. 2006. *Geographic Information Science and Technology Body of Knoweldge.* UCGIS Education Committee, 1st ed. Washington, D.C.: Association of American Geographers.

Foote, K. 1996. The Geographer's Craft Project. <http://www.colorado.edu/geography/gcraft/contents.html> Accessed 21 August 2008.

Gaudet, C., H. Annulis and J. Carr. 2003. Building the geospatial workforce. *URISA Journal* 15 (1):21–30. <http://www.urisa.org/files/Gaudetvol15no1.pdf> Accessed 21 August 2008.

Gilmartin, P. and D. Cowen. 1991. Educational essentials for today's and tomorrow's jobs in cartography and geographic information systems. *Cartography and Geographic Information Systems* 18 (4):262–267.

Goodchild, M. F. 1985. Geographic information systems in undergraduate geography: A contemporary dilemma. *The Operational Geographer* 8:34–38.

Goodchild, M. F. and K. K. Kemp. 1992. NCGIA education activities: The core curriculum and beyond. *International Journal of Geographical Information Systems* 6 (4):309–320.

Huxhold, W. 1991. The GIS profession: Titles, pay, qualifications. *Geo Info Systems* March:12–22.

————, ed. 2000. *Model Job Descriptions for GIS Professionals.* Chicago: Urban and Regional Information Systems Association.

Kemp, K. 2005. Email to first author. 17 May 2005.

Kemp, K. K. and A. U. Frank. 1996. Toward consensus on a European GIS curriculum: The international post-graduate course on GIS. *International Journal of Geographical Information Systems* 10 (4):477–497.

Kemp, K. and R. Wright. 1997. UCGIS identifies GIScience education priorities. *Geo Info Systems* 7 (9):16–18, 20.

Longley, P., M. F. Goodchild, D. Maguire and D. Rhind. 2000. *Geographical Information Systems and Science.* Chichester, England: John Wiley & Sons.

Luccio, M. 2005. Institute for Advanced Education in Geospatial Sciences. *GIS Monitor* June 2005. <http://www.gismonitor.com/news/newsletter/archive/060205.php> Accessed 21 August 2008.

Macey, S. 1997. Identifying key GIS concepts—What the texts tell us. *International Conference on GIS in Higher Education,* held in Chantilly, Va., Oct 30–Nov 2, 1997.

Marble, D. F. 1979. Integrating cartographic and geographic information systems education. Pages 493–499 in *Technical Papers, 39th Annual Meeting of the American Congress on Surveying and Mapping.* Washington, D.C.: ACSM.

————. 1981. Toward a conceptual model for education in digital cartography. Pages 302–310 in *Technical Papers, 41st Annual Meeting of the American Congress on Surveying and Mapping.* Washington, D.C.: ACSM.

————. 1998. Rebuilding the top of the pyramid. *ArcNews* 20 (1):1, 28–29.

Marble, D. F., and members of the Model Curricula Task Force. 2003. *Strawman Report: Model Curricula.* Alexandria, Va.: University Consortium for Geographic Information Science.

Mondello, C., G. F. Hepner and R. A. Williamson. 2004. 10-Year industry forecast, Phases I-III, Study documentation. *Photogrammetric Engineering and Remote Sensing* January:7–58.

National Science Foundation. 1987. *Solicitation: National Center for Geographic Information and Analysis.* Washington D.C.: National Science Foundation.

Nyerges, T. and N. R. Chrisman. 1989. A framework for model curricula development in cartography and geographic information systems. *Professional Geographer* 41 (3):283–293.

Obermeyer, N. O. and H. Onsrud. 1997. Educational policy and GIS: Accreditation and certification. Alexandria, Va.: University Consortium for Geographic Information Science. <http://www.ucgis.org/priorities/education/priorities/a&c.htm> Accessed 21 August 2008.

Phoenix, M. 2000. Geography and the demand for GIS education. *Association of American Geographers Newsletter.* June:13.

Poiker, T. K. 1985. Geographic information systems in the geographic curriculum. *The Operational Geographer* 8: 38–41.

Tobler, W. R. 1977. Analytical cartography. *The American Cartographer* 3 (1):21–31.

Unwin, D. J. 1990. A syllabus for teaching geographical information systems. *International Journal of Geographical Information Systems* 4 (4):457–465.

————. 1997. Curriculum design for GIS. NCGIA core curriculum in GIScience. <http://www.ncgia.ucsb.edu/giscc/units/u159/u159.html> Accessed 21 August 2008.

Wikle, T. A. 1994. Survey defines background coursework for GIS education. *GIS World* 7 (6):53–55.

INDEX

Page numbers in *italic type* refer to figures or tables.

A

A9 Block View, 962, *962*
Abbreviations, 290, 353–354
Able Software R2V, 1151
Absolute time, 312
Abstract data-type (ADT) model, 322
Accessibility
    analysis in GIS, 1042–1045, *1042*
    cumulative-type measures, 1041–1042
    freight shipping, *1043*
    gravity-type measures, 1041
    isochrones, 1041, *1041*
    measuring, 1040–1042
    transportation and, 1040
    See also Transportation GIS
Accuracy, 204, 204–207, 210–212, 244, *244*
    of data, 254, *254*
    positional, 206, *206*
    temporal, *206*, 207
    See also Precision
Accuracy assessment
    metadata, 226
    methods of, 225, 226–232
    need for, 225
    qualitative methods, 226–228, *227, 228*
    quantitative methods, 228–232, *229, 230*
    spatial data sets, 225–233
Acronyms. *See* Abbreviations
Across Trophic Level System Simulation (ATLSS) models, 1059–1060
Activity-travel patterns. *See* Human movement
Ad-hoc wireless systems, 927
Adams, P. C., 437
Adams, T. M., 321, *323*
Adelaja, S., 1069–1070
Adironack Park, biotic integrity, 1159
The Adirondack Atlas, 1136, 1161–1165, *1163, 1164, 1165*
Adirondack Park, 1135–1136, *1135, 1136*
    Adirondack Park Geographic Information CD, 1161

beaver activity, 1150
mapping history, 1137, *1137*
Saranac Lakes Wild Forest, *1155, 1156*
Shared Adirondack Park Geographic Information CD-ROM, 1142, 1160
watershed regions, 1150–1151
See also New York State
Adirondack Park Agency (APA), 1148
    beginnings, 1138
    conservation planning, 1159061
    EPA-funded projects, 1142
    GIS educational benefits, 1152–1153
    GIS quality control, 1152, 1153
    GIS use, 1140–1141
    GIS Users Group, 1142–1143
    land cover classification, 1152
    legal responsibilities, 1143
    map storage, 1140, *1140,* 1144, *1144,* 1148–1149
    mapping by, 1138, *1139*
    mapping history, 1137–1143, *1137, 1138*
    mapping projects, 1141
    Official Adirondack Park Land Use and Development Plan, 1138, *1139,* 1159
    map amendments, 1143, *1144*
    wetlands delineation, 1151–1152
    wetlands mapping, 1141–1142
Adirondack Park Agency (APA) Lookup System, 1143–1148
    data reliability, 1147
    future changes, 1148
    interface, 1145–1146, *1145*
    map layers, 1146, *1146*
    need for, 1143–1144
    parcel identification numbers, 1147–1148
    planning and implementation, 1145
    process, 1147
    products, 1147
    watershed data, 1148–1149
    watershed mapping, 1149–1150, *1149*
    watershed projects, 1149–1150
    wetlands data, 1148–1149

Direct linear transformation (DLT) of image geometry, 735, 739
Direct reconstruction of image geometry, 734, *734*
Dirnböck, T., 462
Discord, Expert opinion, *243*
Discord measure, for ambiguity, 201, *243*
Discordant classifications, 252, *252*
Discrepancies generated by different image orientation methods and numbers of control points
　　for IKONOS image, 737–738, *738*
　　for QuickBird image, 736–737, *737*
Discrete time, 313
Diva-GIS, 297
DLR HRSC-AX, *732*
DMC+4 (BLMIT-1) satellite, *729*
Dobson, J., 959
Dobson, J. E., 784
Domesday Project, 975
Dragicevic, S., 322, *323*
Drechsler, M., 466
Duan, N., 317, *323*
Dublin Core metadata standard, 150
Dunham, M. H., 496
Dynamic modeling, GIS and, 13
Dyreson, C. E., 313

E

EADS Astrium, 729
Earth Explorer, 958
Earth Resources Technology Satellite (ERTS-1) *See* Landsat-1 satellite
Earthquakes
　　Sumatra-Andaman Islands (2004), 866–867, *867, 868*
　　in Turkey (1976-1999), *849*
　　See also Natural hazards assessment
EarthSat, 964
Ebert, J. I., 1230
Ecological conceptual framework, 447–449, *449*, 482–483
　　quantifying patterns of, 461–466
　　scale, 451
　　spatially explicit models in, 460–466
Ecological models, 445, 483
　　Agent-based, 461, 474
　　applying GIS to, 453–456, *453–456*
　　area description, 458

case studies of, 466–479
climate change effect on wildlife, 467–468, *469, 470*
cougar movement causality, 474–476, *476*
to define species distribution, 457–459
for designing field experiments, 460
error analysis, 477–478, *478*
fire impacts on landscapes, 459
forest growth, 480
forest succession, 480
habitat fitness assessment, 470–473, *473*
Habitat Suitability Index, 461–462
home range analysis, 458
individual-based, 461
land management, 459
landscape pattern change, 459
metapopulation analysis, 458–459
natural disturbances, 459
patterns derived from location attributes, 462
plant distribution, 460
population viability analysis, 480
reserve design, 458–459
sensitivity analysis, 479, *479*
software for, 480
spatial modeling, 445–447, 452, *481–482*
spatially explicit, 456–460
for spatially-oriented theoretical studies, 459
species sensitivity to atmospheric carbon dioxide, 467–470, *469, 470*
stochastic events, 477, *477*
suitability models, 461–462
uncertainty in, 452, 477, 477–479, *478, 479*
Ecological monitoring systems, 1095
Ecological processes
　　action/reaction, 450
　　boundaries, 449
　　decision-making, 450, 452
　　ecosystem change, 451–452
　　herding process, 450
　　individuals within, 449, 450
　　mental requirements, 450, 460
　　patterns, 450–451
　　physical requirements, 450, 460
　　space, importance of, 446–447

G

Page numbers in *italic type* refer to figures or tables.

P

Y

Z